Pressure Conversion Factors

	Pa	bar	atm	torr
1 Pa =	1	10^{-5}	$9.869\,23 \times 10^{-6}$	$7.500\,62 \times 10^{-3}$
1 bar =	10^5	1	0.986 923	750.062
1 atm =	$1.013\,25 \times 10^5$	1.013 25	1	760
1 torr =	133.322	$1.333\,22 \times 10^{-3}$	$1.315\,79 \times 10^{-3}$	1

Some Commonly Used Non-SI Units

Unit	Quantity	Symbol	SI value
Angstrom	length	Å	10^{-10} m $= 100$ pm
Micron	length	μ	10^{-6} m
Calorie	energy	cal	4.184 J (defined)
Debye	dipole moment	D	3.3356×10^{-30} C $\cdot$ m
Gauss	magnetic field strength	G	10^{-4} T

Greek Alphabet

Alpha	A	α	Iota	I	ι	Rho	P	ρ
Beta	B	β	Kappa	K	κ	Sigma	Σ	σ
Gamma	Γ	γ	Lambda	Λ	λ	Tau	T	τ
Delta	Δ	δ	Mu	M	μ	Upsilon	Υ	υ
Epsilon	E	ϵ	Nu	N	ν	Phi	Φ	ϕ
Zeta	Z	ζ	Xi	Ξ	ξ	Chi	X	χ
Eta	H	η	Omicron	O	o	Psi	Ψ	ψ
Theta	Θ	θ	Pi	Π	π	Omega	Ω	ω

E_h	cm^{-1}	Hz
$2.293\,710 \times 10^{17}$	$5.034\,11 \times 10^{22}$	$1.509\,189 \times 10^{33}$
$3.808\,798 \times 10^{-4}$	83.5935	$2.506\,069 \times 10^{12}$
$3.674\,931 \times 10^{-2}$	8065.54	$2.417\,988 \times 10^{14}$
1	$2.194\,7463 \times 10^5$	$6.579\,684 \times 10^{15}$
$4.556\,335 \times 10^{-6}$	1	$2.997\,925 \times 10^{10}$
$1.519\,830 \times 10^{-16}$	$3.335\,64 \times 10^{-11}$	1

Problems and Solutions

to accompany

McQuarrie and Simon

PHYSICAL CHEMISTRY: A MOLECULAR APPROACH

Heather Cox

University Science Books
Mill Valley, California

University Science Books
www.uscibooks.com

Production Manager: *Susanna Tadlock*
Designer: *Robert Ishi*
Illustrator: *John Choi*
Compositor: *Eigentype*
Printer & Binder: *The P. A. Hutchison Company*

This book is printed on acid-free paper.

Printed in the United States of America
10 9 8 7

Contents

vi   Contents

Preface

This manual contains complete solutions to every one of the more than 1400 problems in *Physical Chemistry: A Molecular Approach* by Donald A. McQuarrie and John D. Simon.

Many of the problems in this text involve the manipulation of experimental data or empirically derived equations. In most cases, a graphing program (such as *Kaleidagraph*) is the most appropriate tool to use, and most students of physical chemistry already have experience with these programs.

For some problems, however, that are unnecessarily tedious or time-consuming using a simple graphing program, I recommend using a program like *Mathematica* or *MathCad* for related calculations and graphs. The plots of standard molar quantities against temperature using empirical equations[1], for which a function must be defined over varying temperature ranges, are simpler and pedagogically more useful with a program that allows such definitions. Also, because the program is saved, the work is easily checked and easily corrected, and the calculations of standard molar quantities using the same empirical equations[2] follow naturally. Although molecular statistical thermodynamic problems[3] can be done using paper and pencil, they are also more useful to the student when performed with a program such as *Mathematica*, thus allowing the student to spend more time working with the equations and less time writing out constants to be multiplied together.

Unfortunately, the learning curve on a program like *Mathematica* is rather steep, and instead of being a time-saving tool, it can easily become a source of frustration while one struggles to learn how to define functions or how to form a data set. I have posted several sample *Mathematica* files on the web site for this book at http://www.uscibooks.com in order to provide a practical and useful demonstration of its applications with regard to some problems in physical chemistry. This will provide the student or professor with some exposure to the available problem-solving methods, and enable those with only a basic knowledge of *Mathematica* to use it in ways that may have been difficult without a guide.

I have attempted to make this manual as accurate as possible, and would appreciate being informed of any errors that are present.

—Heather Cox

---

[1] Problems 19-45, 21-15, 21-17, 21-19, 21-26, 22-25 and 22-26.

[2] Problems 21-14, 21-16, 21-18 and 21-20 through 21-25.

[3] Problems 21-30 through 21-39, 23-37 through 23-41, and 24-34 through 24-52, to name those that are the most time consuming.

Acknowledgments

John Simon and Donald McQuarrie devoted much time and energy to making sure that these solutions were both clear and correct, and I am indebted to them both for their substantial help. I would also like to thank Bruce and Kathy Armbruster for giving me the opportunity to work on this manual and providing their support, and Don, Carole, and Bo McQuarrie, who encouraged me every step of the way, and without whose generous assistance, hospitality, and general warmth, I would not have been able to embark on—let alone complete—this project.

Problems and Solutions

to accompany

McQuarrie and Simon

PHYSICAL CHEMISTRY: A MOLECULAR APPROACH

CHAPTER 1

The Dawn of the Quantum Theory

PROBLEMS AND SOLUTIONS

1–1. Radiation in the ultraviolet region of the electromagnetic spectrum is usually described in terms of wavelength, λ, and is given in nanometers (10^{-9} m). Calculate the values of ν, $\tilde{\nu}$, and E for ultraviolet radiation with $\lambda = 200$ nm and compare your results with those in Figure 1.11.

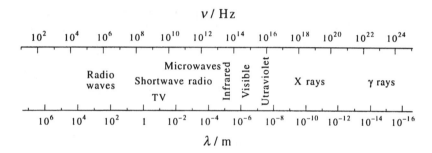

$\nu\,/\,\mathrm{Hz}$

FIGURE 1.11
The regions of electromagnetic radiation.

$$\nu = \frac{c}{\lambda} = \frac{2.998 \times 10^8 \text{ m·s}^{-1}}{200 \times 10^{-9} \text{ m}} = 1.50 \times 10^{15} \text{ Hz}$$

$$\tilde{\nu} = \frac{1}{\lambda} = \frac{1}{200 \times 10^{-7} \text{ cm}} = 5.00 \times 10^4 \text{ cm}^{-1}$$

$$E = \frac{hc}{\lambda} = \frac{(6.626 \times 10^{-34} \text{ J·s})(2.998 \times 10^8 \text{ m·s}^{-1})}{200 \times 10^{-9} \text{ m}} = 9.93 \times 10^{-19} \text{ J}$$

These results correspond to those expected when considering Figure 1.11.

1–2. Radiation in the infrared region is often expressed in terms of wave numbers, $\tilde{\nu} = 1/\lambda$. A typical value of $\tilde{\nu}$ in this region is 10^3 cm^{-1}. Calculate the values of ν, λ, and E for radiation with $\tilde{\nu} = 10^3$ cm^{-1} and compare your results with those in Figure 1.11.

$$\nu = c\tilde{\nu} = (2.998 \times 10^8 \text{ m·s}^{-1})(1 \times 10^5 \text{ m}^{-1}) = 3 \times 10^{13} \text{ Hz}$$

$$\lambda = \frac{1}{\tilde{\nu}} = \frac{1}{1 \times 10^5 \text{ m}^{-1}} = 1 \times 10^{-5} \text{ m}$$

$$E = h\nu = (6.626 \times 10^{-34} \text{ J·s})(3 \times 10^{13} \text{ Hz}) = 2 \times 10^{-20} \text{ J}$$

Again, we can predict these results from Figure 1.11.

1–3. Past the infrared region, in the direction of lower energies, is the microwave region. In this region, radiation is usually characterized by its frequency, ν, expressed in units of megahertz (MHz), where the unit, hertz (Hz), is a cycle per second. A typical microwave frequency is 2.0×10^4 MHz. Calculate the values of ν, λ, and E for this radiation and compare your results with those in Figure 1.11.

$$\nu = 2.0 \times 10^4 \text{ MHz} \left(\frac{1 \times 10^6 \text{ Hz}}{1 \text{ MHz}} \right) = 2.0 \times 10^{10} \text{ s}^{-1}$$

$$\lambda = \frac{c}{\nu} = \frac{2.998 \times 10^8 \text{ m·s}^{-1}}{2.0 \times 10^{10} \text{ s}^{-1}} = 1.5 \times 10^{-2} \text{ m}$$

$$E = h\nu = (6.626 \times 10^{-34} \text{ J·s})(2.0 \times 10^{10} \text{ s}^{-1}) = 1.3 \times 10^{-23} \text{ J}$$

This is illustrated in Figure 1.11.

1–4. Planck's principal assumption was that the energies of the electronic oscillators can have only the values $E = nh\nu$ and that $\Delta E = h\nu$. As $\nu \to 0$, then $\Delta E \to 0$ and E is essentially continuous. Thus, we should expect the nonclassical Planck distribution to go over to the classical Rayleigh-Jeans distribution at low frequencies, where $\Delta E \to 0$. Show that Equation 1.2 reduces to Equation 1.1 as $\nu \to 0$. (Recall that $e^x = 1 + x + (x^2/2!) + \cdots$, or, in other words, that $e^x \approx 1 + x$ when x is small.)

$$d\rho(\nu, T) = \frac{8\pi h}{c^3} \frac{\nu^3 d\nu}{e^{h\nu/k_B T} - 1} \tag{1.2}$$

For small x, $e^x \approx 1 + x$. As $\nu \to 0$, $h\nu/k_B T \to 0$, and

$$d\rho(\nu, T) = \frac{8\pi h}{c^3} \frac{\nu^3 d\nu}{1 + \frac{h\nu}{k_B T} - 1} = \frac{8\pi h \nu^3 k_B T d\nu}{c^3 h\nu}$$

$$= \frac{8\pi \nu^2 k_B T d\nu}{c^3}$$

which is the classical Rayleigh-Jeans distribution (Equation 1.1).

1–5. Before Planck's theoretical work on blackbody radiation, Wien showed empirically that (Equation 1.4)

$$\lambda_{max} T = 2.90 \times 10^{-3} \text{ m·K}$$

where λ_{max} is the wavelength at which the blackbody spectrum has its maximum value at a temperature T. This expression is called the Wien displacement law; derive it from Planck's theoretical expression for the blackbody distribution by differentiating Equation 1.3 with respect to λ. *Hint*: Set $hc/\lambda_{max} k_B T = x$ and derive the intermediate result $e^{-x} + (x/5) = 1$. This problem cannot be solved for x analytically but must be solved numerically. Solve it by iteration on a hand calculator, and show that $x = 4.965$ is the solution.

The Planck distribution law for blackbody radiation is

$$\rho_\lambda(T) d\lambda = \frac{8\pi hc}{\lambda^5} \frac{d\lambda}{e^{hc/\lambda k_B T} - 1} \tag{1.3}$$

As suggested, we let $x = hc/(\lambda_{max} k_B T)$, and then

$$\frac{d}{d\lambda}\left[\rho_\lambda(T)\right] = \frac{d}{d\lambda}\left(\frac{8\pi hc}{\lambda^5}\frac{1}{e^{hc/\lambda k_B T} - 1}\right)$$

$$0 = 8\pi hc\left[\frac{-5}{\lambda_{max}^6 (e^x - 1)} - \frac{e^x\left(\frac{-x}{\lambda_{max}}\right)}{\lambda_{max}^5 (e^x - 1)^2}\right]$$

$$xe^x = 5(e^x - 1)$$

To solve, we iterate and find $x = hc/\lambda_{max} k_B T = 4.965$, or $\lambda_{max} T = 2.90 \times 10^{-3}$ m·K.

1–6. At what wavelength does the maximum in the radiant energy density distribution function for a blackbody occur if (a) $T = 300$ K? (b) $T = 3000$ K? (c) $T = 10\,000$ K?

From Equation 1.4,

a.
$$\lambda_{max} = \frac{2.90 \times 10^{-3} \text{ m·K}}{300 \text{ K}} = 9.67 \times 10^{-6} \text{ m}$$

b.
$$\lambda_{max} = \frac{2.90 \times 10^{-3} \text{ m·K}}{3000 \text{ K}} = 9.67 \times 10^{-7} \text{ m}$$

c.
$$\lambda_{max} = \frac{2.90 \times 10^{-3} \text{ m·K}}{10\,000 \text{ K}} = 2.90 \times 10^{-7} \text{ m}$$

1–7. Sirius, one of the hottest known stars, has approximately a blackbody spectrum with $\lambda_{max} = 260$ nm. Estimate the surface temperature of Sirius.

From Equation 1.4,

$$T = \frac{2.90 \times 10^{-3} \text{ m·K}}{\lambda_{max}} = \frac{2.90 \times 10^{-3} \text{ m·K}}{260 \times 10^{-9} \text{ m}} = 1.12 \times 10^4 \text{ K}$$

1–8. The fireball in a thermonuclear explosion can reach temperatures of approximately 10^7 K. What value of λ_{max} does this correspond to? In what region of the spectrum is this wavelength found (cf. Figure 1.11)?

From Equation 1.4,

$$\lambda_{max} = \frac{2.90 \times 10^{-3} \text{ m·K}}{1 \times 10^7 \text{ K}} = 3 \times 10^{-10} \text{ m}$$

This corresponds to the X-ray region in the electromagnetic spectrum.

1–9. Calculate the energy of a photon for a wavelength of 100 pm (about one atomic diameter).

$$E = \frac{hc}{\lambda} = \frac{(6.626 \times 10^{-34}\ \text{J·s})\,(2.998 \times 10^8\ \text{m·s}^{-1})}{1 \times 10^{-10}\ \text{m}} = 2 \times 10^{-15}\ \text{J}$$

1–10. Express the Planck distribution law in terms of λ (and $d\lambda$) by using the relationship $\lambda\nu = c$.

$$\rho_\nu(T)d\nu = \frac{8\pi h}{c^3}\,\frac{\nu^3 d\nu}{e^{h\nu/k_\text{B}T} - 1} \tag{1.2}$$

We know that $\nu\lambda = c$, so $d\nu = -cd\lambda/\lambda^2$. Substituting, we obtain

$$\rho_\lambda(T)d\lambda = \frac{8\pi h}{c^3}\,\frac{-c^4 d\lambda}{\lambda^5(e^{hc/\lambda k_\text{B}T} - 1)}$$

$$= \frac{8\pi hc}{\lambda^5}\,\frac{d\lambda}{e^{hc/\lambda k_\text{B}T} - 1}$$

where we have dropped the negative sign for convenience. It occurs because $d\nu$ and $d\lambda$ have opposite signs.

1–11. Calculate the number of photons in a 2.00 mJ light pulse at (a) 1.06 μm, (b) 537 nm, and (c) 266 nm.

a.
$$E_\text{photon} = h\nu = \frac{hc}{\lambda}$$

$$= \frac{(6.626 \times 10^{-34}\ \text{J·s})(2.998 \times 10^8\ \text{m·s}^{-1})}{1.06 \times 10^{-6}\ \text{m}}$$

$$= 1.87 \times 10^{-19}\ \text{J · photon}^{-1}$$

Since 2.00 mJ of energy are contained in the light pulse,

$$\text{Number of photons} = \frac{2.00 \times 10^{-3}\ \text{J}}{1.87 \times 10^{-19}\ \text{J · photon}^{-1}} = 1.07 \times 10^{16}\ \text{photons}$$

Parts (b) and (c) are done in the same manner to find

b. 5.41×10^{15} photons   **c.** 2.68×10^{15} photons

1–12. The mean temperature of the Earth's surface is 288 K. Calculate the wavelength at the maximum of the Earth's blackbody radiation. What part of the spectrum does this wavelength correspond to?

From Equation 1.4,

$$\lambda_\text{max} = \frac{2.90 \times 10^{-3}\ \text{m·K}}{288\ \text{K}} = 1.01 \times 10^{-5}\ \text{m}$$

This is in the infrared region of the spectrum.

1–13. A helium-neon laser (used in supermarket scanners) emits light at 632.8 nm. Calculate the frequency of this light. What is the energy of a photon generated by this laser?

$$E = \frac{hc}{\lambda} = \frac{(6.626 \times 10^{-34} \text{ J·s})(2.998 \times 10^8 \text{ m·s}^{-1})}{632.8 \times 10^{-9} \text{ m}}$$

$$= 3.139 \times 10^{-19} \text{ J}$$

1–14. The power output of a laser is measured in units of watts (W), where one watt is equal to one joule per second. ($1 \text{ W} = 1 \text{ J·s}^{-1}$) What is the number of photons emitted per second by a 1.00 mW nitrogen laser? The wavelength emitted by a nitrogen laser is 337 nm.

$$E_{\text{photon}} = h\nu = \frac{hc}{\lambda}$$

$$= \frac{(6.626 \times 10^{-34} \text{ J·s})(2.998 \times 10^8 \text{ m·s}^{-1})}{337 \times 10^{-9} \text{ m}}$$

$$= 5.89 \times 10^{-19} \text{ J · photon}^{-1}$$

$$1.00 \times 10^{-3} \text{ J·s}^{-1} \frac{1}{5.89 \times 10^{-19} \text{ J · photon}^{-1}} = 1.70 \times 10^{15} \text{ photon·s}^{-1}$$

1–15. A household lightbulb is a blackbody radiator. Many light bulbs use tungsten filaments that are heated by an electric current. What temperature is needed so that $\lambda_{\text{max}} = 550$ nm?

From Equation 1.4,

$$T = \frac{2.90 \times 10^{-3} \text{ m·K}}{\lambda_{\text{max}}} = \frac{2.90 \times 10^{-3} \text{ m·K}}{550 \times 10^{-9} \text{ m}} = 5300 \text{ K}$$

1–16. The threshold wavelength for potassium metal is 564 nm. What is its work function? What is the kinetic energy of electrons ejected if radiation of wavelength 410 nm is used?

We will use Equation 1.7 to find ϕ and then use Equation 1.6 to find the kinetic energy.

$$\nu_0 = \frac{c}{\lambda_0} = \frac{2.998 \times 10^8 \text{ m·s}^{-1}}{564 \times 10^{-9} \text{ m}} = 5.32 \times 10^{14} \text{ Hz}$$

$$\phi = h\nu_0$$

$$= (6.626 \times 10^{-34} \text{ J·s})(5.32 \times 10^{14} \text{ Hz}) = 3.52 \times 10^{-19} \text{ J}$$

$$\text{KE} = h\nu - \phi = \frac{hc}{\lambda} - \phi$$

$$= \frac{(6.626 \times 10^{-34} \text{ J·s})(2.998 \times 10^8 \text{ m·s}^{-1})}{410 \times 10^{-9} \text{ m}} - 3.52 \times 10^{-19} \text{ J}$$

$$= 1.32 \times 10^{-19} \text{ J}$$

1–17. Given that the work function of chromium is 4.40 eV, calculate the kinetic energy of electrons emitted from a chromium surface that is irradiated with ultraviolet radiation of wavelength 200 nm.

$$\phi = 4.40 \text{ eV} \left(\frac{1.602 \times 10^{-19} \text{ J}}{1 \text{ eV}} \right) = 7.05 \times 10^{-19} \text{ J}$$

$$h\nu = \frac{hc}{\lambda} = \frac{(6.626 \times 10^{-34} \text{ J·s})(2.998 \times 10^{8} \text{ m·s}^{-1})}{200 \times 10^{-9} \text{ m}}$$

$$= 9.93 \times 10^{-19} \text{ J}$$

$$\text{KE} = h\nu - \phi$$

$$= 9.93 \times 10^{-19} \text{ J} - 7.05 \times 10^{-19} \text{ J} = 2.88 \times 10^{-19} \text{ J}$$

1–18. When a clean surface of silver is irradiated with light of wavelength 230 nm, the kinetic energy of the ejected electrons is found to be 0.805 eV. Calculate the work function and the threshold frequency of silver.

From Equation 1.6,

$$\phi = h\nu - \text{KE} = \frac{hc}{\lambda} - \text{KE}$$

$$= \frac{(6.626 \times 10^{-34} \text{ J·s})(2.998 \times 10^{8} \text{ m·s}^{-1})}{230 \times 10^{-9} \text{ m}} - 0.805 \text{ eV} \left(\frac{1.602 \times 10^{-19} \text{ J}}{1 \text{ eV}} \right)$$

$$= 8.64 \times 10^{-19} \text{ J} - 1.29 \times 10^{-19} \text{ J} = 7.35 \times 10^{-19} \text{ J}$$

From Equation 1.7,

$$\nu_0 = \frac{\phi}{h} = \frac{7.35 \times 10^{-19} \text{ J}}{6.626 \times 10^{-34} \text{ J·s}} = 1.11 \times 10^{15} \text{ Hz}$$

1–19. Some data for the kinetic energy of ejected electrons as a function of the wavelength of the incident radiation for the photoelectron effect for sodium metal are

λ/nm	100	200	300	400	500
KE/eV	10.1	3.94	1.88	0.842	0.222

Plot these data to obtain a straight line, and calculate h from the slope of the line and the work function ϕ from its intercept with the horizontal axis.

First, we use the equation $\nu\lambda = c$ to convert wavelength to frequency and the conversion factor $1\ eV = 1.6022 \times 10^{-19}$ J. Then we can plot kinetic energy (in joules) as a function of frequency (in Hz):

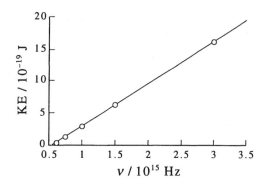

The best fit of a line to the data is $y = -3.5891 \times 10^{-19}$ J $+ \left(6.5918 \times 10^{-34}\right)$ J·s, which gives a slope of $h = 6.60 \times 10^{-34}$ J·s. The threshhold freqency is the frequency when the kinetic energy is equal to zero, and can be obtained from the x-intercept:

$$y = 0 = -3.5891 \times 10^{-19}\ \text{J} + \left(6.5918 \times 10^{-34}\ \text{J·s}\right) x$$

Solving for x,

$$x = \nu_0 = \frac{3.5891 \times 10^{-19}\ \text{J}}{6.5918 \times 10^{-34}\ \text{J·s}} = 5.44 \times 10^{14}\ \text{Hz}$$

Equation 1.7 then gives

$$\phi = h\nu_0 = \left(6.60 \times 10^{-34}\ \text{J·s}\right)\left(5.44 \times 10^{14}\ \text{Hz}\right) = 3.59 \times 10^{-19}\ \text{J}$$

1–20. Use the Rydberg formula (Equation 1.10) to calculate the wavelengths of the first three lines of the Lyman series.

$$\tilde{\nu} = \frac{1}{\lambda} = 109\ 680 \left(\frac{1}{n_1^2} - \frac{1}{n_2^2}\right)\ \text{cm}^{-1} \qquad (n_2 > n_1)$$

To find the first line in the Lyman series, substitute 1 for n_1 and 2 for n_2:

$$\tilde{\nu}_2 = 109\ 680 \left(\frac{1}{1^2} - \frac{1}{2^2}\right)\ \text{cm}^{-1} = 82\ 260\ \text{cm}^{-1}$$

$$\lambda_2 = \frac{1}{\tilde{\nu}_2} = 121.566\ \text{nm}$$

The second line occurs when $n_2 = 3$, at 102.571 nm, and the third line occurs when $n_2 = 4$, at 97.2526 nm.

1–21. A line in the Lyman series of hydrogen has a wavelength of 1.03×10^{-7} m. Find the original energy level of the electron.

In the Lyman series, $n_1 = 1$ in Equation 1.10. Thus

$$\frac{1}{\lambda} = 109\,680 \left(1 - \frac{1}{n_2^2}\right) \text{cm}^{-1}$$

$$\frac{109\,680 \text{ cm}^{-1}}{n_2^2} = -\frac{1}{1.03 \times 10^{-5} \text{ cm}} + 109\,680 \text{ cm}^{-1}$$

$$n_2 = 2.98 \approx 3$$

because n must be an integer.

1–22. A ground-state hydrogen atom absorbs a photon of light that has a wavelength of 97.2 nm. It then gives off a photon that has a wavelength of 486 nm. What is the final state of the hydrogen atom?

First, we find the value of n_2, the state of the hydrogen atom that is obtained upon absorption, by using Equation 1.10 with $n_1 = 1$.

$$\frac{1}{97.2 \times 10^{-7} \text{ cm}} = 109\,680 \left(1 - \frac{1}{n_2^2}\right) \text{cm}^{-1}$$

$$n_2 = 4$$

We can now use Equation 1.10 with $n_1 = 4$ to find the final state of the hydrogen atom:

$$\frac{1}{97.2 \times 10^{-7} \text{ cm}} = 109\,680 \left(\frac{1}{4^2} - \frac{1}{n_2^2}\right) \text{cm}^{-1}$$

$$n_2 = 2$$

The final state of the hydrogen atom is $n = 2$.

1–23. Show that the Lyman series occurs between 91.2 nm and 121.6 nm, that the Balmer series occurs between 364.7 nm and 656.5 nm, and that the Paschen series occurs between 820.6 nm and 1876 nm. Identify the spectral regions to which these wavelengths correspond.

We can use the Rydberg formula (Equation 1.10) to determine the ranges of the specified series. The maximum wavelength can be found by taking the smallest n_2 allowed ($n_2 = n_1 + 1$) and the minimum wavelength can be found by using the largest n_2 allowed ($n_2 = \infty$).

Lyman series

maximum wavelength
$$\frac{1}{\lambda} = 109\,680 \left(1 - \frac{1}{2^2}\right) \text{cm}^{-1}$$
$$\lambda = 121.6 \text{ nm}$$

minimum wavelength
$$\frac{1}{\lambda} = 109\,680 \left(1 - \frac{1}{\infty}\right) \text{cm}^{-1}$$
$$\lambda = 91.2 \text{ nm}$$

This corresponds to the ultraviolet region of the spectrum.

Balmer series

maximum wavelength

$$\frac{1}{\lambda} = 109\,680 \left(\frac{1}{2^2} - \frac{1}{3^2} \right) \text{ cm}^{-1}$$

$$\lambda = 656.5 \text{ nm}$$

minimum wavelength

$$\frac{1}{\lambda} = 109\,680 \left(\frac{1}{2^2} - \frac{1}{\infty} \right) \text{ cm}^{-1}$$

$$\lambda = 364.7 \text{ nm}$$

This corresponds to the near ultraviolet region of the spectrum.

Paschen series

maximum wavelength

$$\frac{1}{\lambda} = 109\,680 \left(\frac{1}{3^2} - \frac{1}{4^2} \right) \text{ cm}^{-1}$$

$$\lambda = 1875.6 \text{ nm}$$

minimum wavelength

$$\frac{1}{\lambda} = 109\,680 \left(\frac{1}{3^2} - \frac{1}{\infty} \right) \text{ cm}^{-1}$$

$$\lambda = 820.6 \text{ nm}$$

This corresponds to the near infrared region of the spectrum.

1–24. Calculate the wavelength and the energy of a photon associated with the series limit of the Lyman series.

We have found that the minimum wavelength of the Lyman series is 91.2 nm (Problem 1–23). So

$$E = \frac{hc}{\lambda} = \frac{\left(6.626 \times 10^{-34} \text{ J·s}\right) \left(2.998 \times 10^8 \text{ m·s}^{-1}\right)}{91.2 \times 10^{-9} \text{ m}} = 2.18 \times 10^{-18} \text{ J}$$

1–25. Calculate the de Broglie wavelength for (a) an electron with a kinetic energy of 100 eV, (b) a proton with a kinetic energy of 100 eV, and (c) an electron in the first Bohr orbit of a hydrogen atom.

We use Equation 1.12 ($\lambda = h/p$) in all cases to find λ.

a.

$$\text{KE} = \frac{mv^2}{2}$$

$$100 \text{ eV} \left(\frac{1.602 \times 10^{-19} \text{ J}}{1 \text{ eV}} \right) = \frac{(v^2)\left(9.109 \times 10^{-31} \text{ kg}\right)}{2}$$

$$v = 5.93 \times 10^6 \text{ m·s}^{-1}$$

So

$$\lambda = \frac{h}{p} = \frac{h}{mv}$$

$$= \frac{6.626 \times 10^{-34} \text{ J·s}}{\left(9.109 \times 10^{-31} \text{ kg}\right)\left(5.93 \times 10^6 \text{ m·s}^{-1}\right)}$$

$$= 1.23 \times 10^{-10} \text{ m} = 0.123 \text{ nm}$$

b. Replace m_e with m_p in (a) to find $\lambda = 2.86 \times 10^{-3}$ nm.

c. We must first determine the velocity of an electron in the first Bohr orbit of a hydrogen atom. From Equation 1.16, we see that

$$v = \frac{nh}{2\pi m_e r} \qquad (1.16)$$

and we know

$$r = \frac{\varepsilon_0 h^2 n^2}{\pi m_e e^2} \qquad (1.17)$$

Substituting Equation 1.17 for r into Equation 1.16 gives

$$v = \frac{e^2}{2nh\varepsilon_0}$$

For $n = 1$,

$$v = \frac{(1.602 \times 10^{-19}\ \text{C})^2}{2(1)(6.626 \times 10^{-34}\ \text{J·s})(8.854 \times 10^{-12}\ \text{C}^2\ \text{J}^{-1}\ \text{m}^{-1})}$$

$$= 2.188 \times 10^6\ \text{m·s}^{-1}$$

So

$$\lambda = \frac{h}{p} = \frac{h}{mv}$$

$$= \frac{6.626 \times 10^{-34}\ \text{J·s}}{(9.109 \times 10^{-31}\ \text{kg})(2.188 \times 10^6\ \text{m·s}^{-1})}$$

$$= 3.32 \times 10^{-10}\ \text{m} = 0.332\ \text{nm}$$

1–26. Calculate (a) the wavelength and kinetic energy of an electron in a beam of electrons accelerated by a voltage increment of 100 V and (b) the kinetic energy of an electron that has a de Broglie wavelength of 200 pm (1 picometer $= 10^{-12}$ m).

---

a.

$$\text{Electron charge} \times \text{Potential} = \text{KE/electron}$$

$$(1.602 \times 10^{-19}\ \text{C})(100\ \text{V}) = 1.602 \times 10^{-17}\ \text{J/electron}$$

$$v = \sqrt{\frac{2\,(\text{KE})}{m}} = \sqrt{\frac{2\,(1.602 \times 10^{-17}\ \text{J})}{9.109 \times 10^{-31}\ \text{kg}}} = 5.93 \times 10^6\ \text{m·s}^{-1}$$

$$\lambda = \frac{h}{mv} = \frac{6.626 \times 10^{-34}\ \text{J·s}}{(9.109 \times 10^{-31}\ \text{kg})(5.93 \times 10^6\,\text{m·s}^{-1})} = 1.23 \times 10^{-10}\ \text{m}$$

b. Use Equation 1.12 to find the velocity of the electron:

$$v = \frac{h}{m\lambda} = \frac{6.626 \times 10^{-34}\ \text{J·s}}{(9.109 \times 10^{-31}\ \text{kg})(200 \times 10^{-12}\,\text{m})} = 3.64 \times 10^6\ \text{m·s}^{-1}$$

$$\text{KE} = \frac{mv^2}{2} = \frac{(9.109 \times 10^{-31}\ \text{kg})(3.46 \times 10^6\ \text{m·s}^{-1})^2}{2} = 6.02 \times 10^{-18}\ \text{J}$$

1–27. Through what potential must a proton initially at rest fall so that its de Broglie wavelength is 1.0×10^{-10} m?

We will use Equation 1.12 to find the velocity of the proton. Then we will calculate the kinetic energy and the potential needed to supply the kinetic energy.

$$v = \frac{h}{m\lambda} = \frac{6.626 \times 10^{-34} \text{ J·s}}{\left(1.672 \times 10^{-27} \text{ kg}\right)\left(1 \times 10^{-10} \text{ m}\right)} = 4.0 \times 10^3 \text{ m·s}^{-1}$$

Thus

$$KE = \frac{mv^2}{2} = \frac{\left(1.672 \times 10^{-27} \text{ kg}\right)\left(4.0 \times 10^3 \text{ m·s}^{-1}\right)^2}{2}$$
$$= 1.3 \times 10^{-20} \text{ J}$$

and

$$\text{Potential} = \frac{KE/\text{proton}}{\text{proton charge}} = \frac{1.3 \times 10^{-20} \text{ J}}{1.602 \times 10^{-19} \text{ C}} = 0.082 \text{ V}$$

1–28. Calculate the energy and wavelength associated with an α particle that has fallen through a potential difference of 4.0 V. Take the mass of an α particle to be 6.64×10^{-27} kg.

An alpha particle is a helium nucleus, so it has a +2 charge.

$$\alpha \text{ particle charge} \times \text{Potential} = KE/\alpha \text{ particle}$$
$$\left[2\left(1.602 \times 10^{-19} \text{ C}\right)\right](4.0 \text{ V}) = 1.28 \times 10^{-18} \text{ J}/\alpha \text{ particle}$$

$$KE = \frac{mv^2}{2} = \frac{p^2}{2m}$$
$$p = \sqrt{2(KE)m} = \sqrt{2\left(6.64 \times 10^{-27} \text{ kg}\right)\left(1.28 \times 10^{-18} \text{ J}\right)}$$
$$= 1.30 \times 10^{-22} \text{ kg·m·s}^{-1}$$
$$\lambda = \frac{h}{p} = \frac{6.626 \times 10^{-34} \text{ J·s}}{1.30 \times 10^{-22} \text{ kg·m·s}^{-1}} = 5.08 \times 10^{-12} \text{ m}$$
$$= 5.08 \text{ pm}$$

1–29. One of the most powerful modern techniques for studying structure is neutron diffraction. This technique involves generating a collimated beam of neutrons at a particular temperature from a high-energy neutron source and is accomplished at several accelerator facilities around the world. If the speed of a neutron is given by $v_n = (3k_B T/m)^{1/2}$, where m is the mass of a neutron, then what temperature is needed so that the neutrons have a de Broglie wavelength of 50 pm? Take the mass of a neutron to be 1.67×10^{-27} kg.

We are given

$$v_n = \left(\frac{3k_B T}{m}\right)^{1/2}$$

and so (Equation 1.12)

$$\lambda = \frac{h}{mv_n} = \frac{h}{\left(3mk_B T\right)^{1/2}}$$

Solving for T gives

$$T = \frac{h^2}{3mk_B\lambda^2}$$

$$= \frac{\left(6.626 \times 10^{-34} \text{ J·s}\right)^2}{3\left(1.67 \times 10^{-27} \text{ kg}\right)\left(1.381 \times 10^{-23} \text{ J·K}^{-1}\right)\left(50 \times 10^{-12} \text{ m}\right)^2}$$

$$= 2500 \text{ K}$$

1–30. Show that a small change in the speed of a particle, Δv, causes a change in its de Broglie wavelength, $\Delta \lambda$, of

$$|\Delta\lambda| = \frac{|\Delta v|\lambda_0}{v_0}$$

where v_0 and λ_0 are its initial speed and de Broglie wavelength, respectively.

For a small change, $\Delta v = dv$ and $\Delta\lambda = d\lambda$. Then

$$\lambda = \frac{h}{mv}$$

$$d\lambda = \frac{-h}{mv_0^2}dv = \frac{-h}{mv_0}\frac{dv}{v_0}$$

$$|d\lambda| = \frac{h}{mv_0}\frac{|dv|}{v_0}$$

$$|\Delta\lambda| = \lambda_0\frac{|\Delta v|}{v_0}$$

1–31. Derive the Bohr formula for $\tilde{\nu}$ for a nucleus of atomic number Z.

For a nucleus of charge Z, the attractive force of the nucleus is Ze and Equation 1.14 becomes

$$\frac{(Ze)(e)}{4\pi\varepsilon_0 r^2} = \frac{m_e v^2}{r}$$

In the subsequent formulas, e^2 is replaced by Ze^2, and Equation 1.22 becomes

$$E_n = \frac{-m_e Z^2 e^4}{8\varepsilon_0^2 h^2 n^2}$$

Likewise, Equation 1.24 becomes

$$\tilde{\nu} = \frac{m_e Z^2 e^4}{8\varepsilon_0^2 ch^3}\left(\frac{1}{n_1^2} - \frac{1}{n_2^2}\right) = Z^2 R_H \left(\frac{1}{n_1^2} - \frac{1}{n_2^2}\right)$$

which is the Bohr formula for $\tilde{\nu}$ for a nucleus of atomic number Z.

1–32. The series in the He$^+$ spectrum that corresponds to the set of transitions where the electron falls from a higher level into the $n = 4$ state is called the Pickering series, an important series in solar astronomy. Derive the formula for the wavelengths of the observed lines in this series. In what region of the spectrum does it occur? (See Problem 1–31.)

From Problem 1–31, we have

$$\tilde{\nu} = Z^2 R_{\mathrm{H}} \left(\frac{1}{n_1^2} - \frac{1}{n_2^2} \right)$$

For the Pickering series in the helium spectrum, $Z = 2$ and $n_2 = 4$. Thus

$$\tilde{\nu} = 4(109\,680 \text{ cm}^{-1}) \left(\frac{1}{4^2} - \frac{1}{n_1^2} \right) \qquad n_1 = 5, 6, 7, ...$$

For $n_1 = 5$, $\tilde{\nu} = 9871$ cm^{-1}, or $\lambda = 1.0131 \times 10^{-6}$ m. This is in the visible region of the electomagnetic spectrum.

1–33. Using the Bohr theory, calculate the ionization energy (in electron volts and in kJ·mol^{-1}) of singly ionized helium.

To find the ionization energy of singly ionized helium, we can consider the case where we move an electron from the ground state ($n = 1$) to an infinite distance from the nucleus. Then, from Problem 1–31,

$$\tilde{\nu} = Z^2 R_{\mathrm{H}} \left(\frac{1}{n_1^2} - \frac{1}{n_2^2} \right)$$

$$= 2^2 (109\,680 \text{ cm}^{-1}) \left(\frac{1}{1^2} - \frac{1}{\infty^2} \right)$$

$$= 4.3872 \times 10^5 \text{ cm}^{-1}$$

or

$$E = hc\tilde{\nu} = 8.72 \times 10^{-18} \text{ J} = 5250 \text{ kJ·mol}^{-1} = 54.4 \text{ eV}$$

1–34. Show that the speed of an electron in the nth Bohr orbit is $v = e^2/2\varepsilon_0 nh$. Calculate the values of v for the first few Bohr orbits.

We derived this relationship in Problem 1.25(c), where we found that $v = 2.188 \times 10^6$ m·s^{-1} for $n = 1$. Likewise, for $n = 2$, $v_2 = 1.094 \times 10^6$ m·s^{-1}, and for $n = 3$, $v_3 = 7.292 \times 10^5$ m·s^{-1}.

1–35. If we locate an electron to within 20 pm, then what is the uncertainty in its speed?

By definition, $\Delta p = m\Delta v$, and Heisenberg's Uncertainty Principle (Equation 1.26) states that $\Delta x \Delta p \geq h$. Then

$$\Delta x(m\Delta v) \geq h$$

$$\Delta v \geq \frac{h}{m\Delta x}$$

$$\geq \frac{6.626 \times 10^{-34} \text{ J·s}}{(9.109 \times 10^{-31} \text{ kg})(20 \times 10^{-12} \text{ m})}$$

$$\geq 3.6 \times 10^7 \text{ m·s}^{-1}$$

The minimum uncertainty in the velocity of the electron is 3.6×10^7 m·s^{-1}.

1–36. What is the uncertainty of the momentum of an electron if we know its position is somewhere in a 10 pm interval? How does the value compare to momentum of an electron in the first Bohr orbit?

Using Equation 1.26:

$$\Delta x \Delta p \geq h$$

$$\Delta p \geq \frac{h}{\Delta x} = \frac{6.626 \times 10^{-34} \text{ J·s}}{10 \times 10^{-12} \text{ m}}$$

$$\geq 6.6 \times 10^{-23} \text{ kg·m·s}^{-1}$$

We can calculate the momentum of an electron in the first Bohr radius by using v from Problem 1–25(c):

$$p = m_e v = (9.109 \times 10^{-31} \text{ kg})(2.188 \times 10^6 \text{ m·s}^{-1}) = 1.993 \times 10^{-24} \text{ kg·m·s}^{-1}$$

The uncertainty of the momentum of an electron somewhere in a 10 pm interval is about thirty times greater than the momentum of an electron in the first Bohr radius.

1–37. There is also an uncertainty principle for energy and time:

$$\Delta E \Delta t \geq h$$

Show that both sides of this expression have the same units.

The units of ΔE are J and those of Δt are seconds, and Planck's constant has units of J·s.

1–38. The relationship introduced in Problem 1–37 has been interpreted to mean that a particle of mass m ($E = mc^2$) can materialize from nothing provided that it returns to nothing within a time $\Delta t \leq h/mc^2$. Particles that last for time Δt or more are called *real particles*; particles that last less than time Δt are called *virtual particles*. The mass of the charged pion, a subatomic particle, is 2.5×10^{-28} kg. What is the minimum lifetime if the pion is to be considered a real particle?

Use the condition for real particles defined above:

$$\Delta t \geq \frac{h}{mc^2} = \frac{6.626 \times 10^{-34} \text{ J·s}}{(2.5 \times 10^{-28} \text{ kg})(2.998 \times 10^8 \text{ m·s}^{-1})^2}$$

$$\geq 2.9 \times 10^{-23} \text{ s}$$

1–39. Another application of the relationship given in Problem 1–37 has to do with the excited state energies and lifetimes of atoms and molecules. If we know that the lifetime of an excited state is 10^{-9} s, then what is the uncertainty in the energy of this state?

Using the relationship in Problem 1–37 gives

$$\Delta E \geq \frac{h}{\Delta t} = \frac{6.626 \times 10^{-34} \text{ J} \cdot \text{s}}{1 \times 10^{-9} \text{ s}} = 7 \times 10^{-25} \text{ J}$$

1–40. When an excited nucleus decays, it emits a γ-ray. The lifetime of an excited state of a nucleus is of the order of 10^{-12} s. What is the uncertainty in the energy of the γ-ray produced? (See Problem 1–37.)

$$\Delta E \geq \frac{h}{\Delta t} = \frac{6.626 \times 10^{-34} \text{ J} \cdot \text{s}}{1 \times 10^{-12} \text{ s}} = 7 \times 10^{-22} \text{ J}$$

1–41. In this problem, we will prove that the inward force required to keep a mass revolving around a fixed center is $f = mv^2/r$. To prove this, let us look at the velocity and the acceleration of a revolving mass. Referring to Figure 1.12, we see that

$$|\Delta \mathbf{r}| \approx \Delta s = r\Delta\theta \tag{1.27}$$

if $\Delta\theta$ is small enough that the arc length Δs and the vector difference $|\Delta \mathbf{r}| = |\mathbf{r}_1 - \mathbf{r}_2|$ are essentially the same. In this case, then

$$v = \lim_{\Delta t \to 0} \frac{\Delta s}{\Delta t} = r \lim_{\Delta t \to 0} \frac{\Delta\theta}{\Delta t} = r\omega \tag{1.28}$$

where $\omega = d\theta/dt = v/r$.

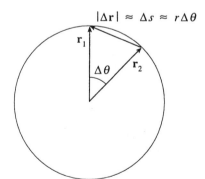

FIGURE 1.12
Diagram for defining angular speed.

If ω and r are constant, then $v = r\omega$ is constant, and because acceleration is $\lim_{t \to 0}(\Delta v/\Delta t)$, we might wonder if there is any acceleration. The answer is most definitely *yes* because velocity is a vector quantity and the direction of $\mathbf{v}$, which is the same as $\Delta \mathbf{r}$, is constantly changing even though its magnitude is not. To calculate this acceleration, draw a figure like Figure 1.12 but expressed in terms of v instead of r. From your figure, show that

$$\Delta v = |\Delta \mathbf{v}| = v\Delta\theta \tag{1.29}$$

is in direct analogy with Equation 1.27, and show that the particle experiences an acceleration given by

$$a = \lim_{\Delta t \to 0} \frac{\Delta v}{\Delta t} = v \lim_{\Delta t \to 0} \frac{\Delta \theta}{\Delta t} = v\omega \qquad (1.30)$$

Thus we see that the particle experiences an acceleration and requires an inward force equal to $ma = mv\omega = mv^2/r$ to keep it moving in its circular orbit.

In analogy to Figure 1.12, we can draw a figure expressed in terms of velocity:

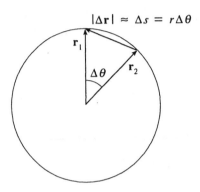

From this figure, we see that $\Delta v = |\Delta \mathbf{v}| = v\Delta \theta$. Then

$$a = \lim_{\Delta t \to 0} \frac{\Delta v}{\Delta t} = v \lim_{\Delta t \to 0} \frac{\Delta \theta}{\Delta t} = v\omega$$

and the particle accelerates with acceleration $v\omega$.

1–42. Planck's distribution (Equation 1.2) law gives the radiant energy density of electromagnetic radiation emitted between v and $v + dv$. Integrate the Planck distribution over all frequencies to obtain the total energy emitted. What is its temperature dependence? Do you know whose law this is? You will need to use the integral

$$\int_0^\infty \frac{x^3 dx}{e^x - 1} = \frac{\pi^4}{15}$$

Equation 1.2 gives

$$\rho_v(T)dv = \frac{8\pi h}{c^3} \frac{v^3 dv}{e^{hv/k_B T} - 1}$$

Now we will integrate this expression over all values of v.

$$\int_0^\infty \rho_v(T)dv = \frac{8\pi h}{c^3} \int_0^\infty \frac{v^3 dv}{e^{hv/k_B T} - 1}$$

Let $x = hv/k_B T$, so $dx = hdv/k_B T$. Then

$$\int_0^\infty \rho_v(T)dv = \frac{8\pi h}{c^3} \left(\frac{k_B T}{h}\right)^4 \int_0^\infty \frac{x^3 dx}{e^x - 1}$$

$$= \frac{8\pi^5 (k_B T)^4}{15h^3 c^3}$$

The temperature dependence is T^4. This relation is called the Stefan-Boltzmann law.

1-43. Can you derive the temperature dependence of the result in Problem 1-42 without evaluating the integral?

Yes. The quantity x is unitless, so the integral $\int_0^\infty \frac{x^3 dx}{e^x-1}$ in Problem 1-42 is also unitless. Therefore, the temperature dependence of the total energy emitted is T^4.

1-44. Ionizing a hydrogen atom in its electronic ground state requires 2.179×10^{-18} J of energy. The sun's surface has a temperature of ≈ 6000 K and is composed, in part, of atomic hydrogen. Is the hydrogen present as H(g) or H$^+$(g)? What is the temperature required so that the maximum wavelength of the emission of a blackbody ionizes atomic hydrogen? In what region of the electromagnetic spectrum is this wavelength found?

We can use Equation 1.4 to find λ_{max} at 6000 K. Then, using $E = h\nu = hc/\lambda$, we can calculate the energy available on the surface of the sun.

$$\lambda_{max} = \frac{2.90 \times 10^{-3}\ \text{m}\cdot\text{K}}{6000\ \text{K}} = 4.833 \times 10^{-7}\ \text{m}$$

$$E = \frac{hc}{\lambda} = 4.11 \times 10^{-19}\ \text{J}$$

This is less energy than that required to ionize a hydrogen atom; therefore, most hydrogen is present in the form of H(g) in the sun. To ionize hydrogen, blackbody emissions must provide at least 2.179×10^{-18} J of energy. Thus

$$\lambda_{max} = \frac{(6.626 \times 10^{-34}\ \text{J}\cdot\text{s})(2.998 \times 10^8\ \text{m}\cdot\text{s}^{-1})}{2.179 \times 10^{-18}\ \text{J}} = 9.12 \times 10^{-8}\ \text{m}$$

$$T = \frac{2.90 \times 10^{-3}\ \text{m}\cdot\text{K}}{9.12 \times 10^{-8}\ \text{m}} = 3.18 \times 10^4\ \text{K}$$

This wavelength is in the UV region of the electromagnetic spectrum.

Complex Numbers

PROBLEMS AND SOLUTIONS

A–1. Find the real and imaginary parts of the following quantities:

   a. $(2-i)^3$ **b.** $e^{\pi i/2}$ **c.** $e^{-2+i\pi/2}$ **d.** $(\sqrt{2}+2i)e^{-i\pi/2}$

---

   a. $(2-i)^3 = 8 - 12i + 6i^2 - i^3 = 2 - 11i$
      $\text{Re}(z) = 2;\ \text{Im}(z) = -11$

   b. $e^{\pi i/2} = \cos(\pi/2) + i\sin(\pi/2) = i$
      $\text{Re}(z) = 0;\ \text{Im}(z) = 1$

   c. $e^{-2+i\pi/2} = e^{-2}e^{i\pi/2} = e^{-2}i$
      $\text{Re}(z) = 0;\ \text{Im}(z) = e^{-2}$

   d. $(\sqrt{2}+2i)e^{-i\pi/2} = i(\sqrt{2}+2i)\sin(-\pi/2) = 2 - i\sqrt{2}$
      $\text{Re}(z) = 2;\ \text{Im}(z) = -\sqrt{2}$

A–2. If $z = x + 2iy$, then find

   a. $\text{Re}(z^*)$ **b.** $\text{Re}(z^2)$ **c.** $\text{Im}(z^2)$ **d.** $\text{Re}(zz^*)$ **e.** $\text{Im}(zz^*)$

---

   a. $\text{Re}(z^*) = \text{Re}(x - 2iy) = x$

   b. $\text{Re}(z^2) = \text{Re}[(x+2iy)^2] = \text{Re}(x^2 + 4ixy - 4y^2) = x^2 - 4y^2$

   c. $\text{Im}(z^2) = \text{Im}[(x+2iy)^2] = 4xy$

   d. $\text{Re}(zz^*) = \text{Re}\left[(x+2iy)(x-2iy)\right] = \text{Re}(x^2 + 4y^2) = x^2 + 4y^2$

   e. $\text{Im}(zz^*) = \text{Im}(x^2 - 4y^2) = 0 = 0$

A–3. Express the following complex numbers in the form $re^{i\theta}$:

   a. $6i$ **b.** $4 - \sqrt{2}i$ **c.** $-1 - 2i$ **d.** $\pi + ei$

Use Equations A.8 and A.9 to determine the values of r and θ in the expression $re^{i\theta}$.

a.
$$6i = r\cos\theta + ir\sin\theta$$
$$r = 6$$
$$\theta = \tan^{-1}\left(\frac{6}{0}\right)$$
$$\theta = \frac{\pi}{2}$$

so

$$6i = 6e^{i\pi/2}$$

b.
$$4 - \sqrt{2}i = r\cos\theta + ir\sin\theta$$
$$r = \sqrt{16+2} = 3\sqrt{2}$$
$$\theta = \tan^{-1}\left(\frac{-\sqrt{2}}{4}\right) = -0.340$$

so

$$4 - \sqrt{2}i = 3\sqrt{2}\,e^{-0.340i}$$

c.
$$-1 - 2i = r\cos\theta + ir\sin\theta$$
$$r = \sqrt{1+4} = \sqrt{5}$$
$$\theta = \tan^{-1}\left(\frac{-2}{-1}\right) = 1.11$$

so

$$-1 - 2i = \sqrt{5}\,e^{1.11i}$$

d.
$$\pi + ei = r\cos\theta + ir\sin\theta$$
$$r = \sqrt{\pi^2 + e^2}$$
$$\theta = \tan^{-1}\left(\frac{e}{\pi}\right) = 0.7130$$

so

$$\pi + ei = \sqrt{\pi^2 + e^2}\,e^{0.713i}$$

A–4. Express the following complex numbers in the form $x + iy$:

a. $e^{\pi/4i}$ **b.** $6e^{2\pi i/3}$ **c.** $e^{-(\pi/4)i + \ln 2}$ **d.** $e^{-2\pi i} + e^{4\pi i}$

---

a.
$$e^{\pi/4i} = \cos\left(\frac{\pi}{4}\right) + i\sin\left(\frac{\pi}{4}\right)$$
$$= \frac{1}{\sqrt{2}} + i\frac{1}{\sqrt{2}}$$

b.
$$6e^{2\pi i/3} = 6\cos\left(\frac{2\pi}{3}\right) + 6i\sin\left(\frac{2\pi}{3}\right)$$
$$= -3 + 3\sqrt{3}i$$

c.
$$e^{-(\pi/4)i + \ln 2} = 2e^{-\pi i/4} = 2\left[\cos\left(\frac{-\pi}{4}\right) + i\sin\left(\frac{-\pi}{4}\right)\right] = 2\left(\frac{1}{\sqrt{2}} - i\frac{1}{\sqrt{2}}\right)$$
$$= \sqrt{2} - \sqrt{2}i$$

d.
$$e^{-2\pi i} + e^{4\pi i} = \cos(-2\pi) + i\sin(-2\pi) + \cos(4\pi) + i\sin(4\pi)$$
$$= 2$$

A–5. Prove that $e^{i\pi} = -1$. Comment on the nature of the numbers in this relation.

$$e^{i\pi} = \cos(\pi) + i\sin(\pi) = -1$$

This is an amazing equation. It shows that a transcendental number (e), raised to the product of an imaginary number (i) and another transcendental number (π), is equivalent to an integer.

A–6. Show that

$$\cos\theta = \frac{e^{i\theta} + e^{-i\theta}}{2}$$

and that

$$\sin\theta = \frac{e^{i\theta} - e^{-i\theta}}{2i}$$

Using Equation A.6,

$$e^{i\theta} = \cos\theta + i\sin\theta$$
$$e^{-i\theta} = \cos\theta - i\sin\theta$$

Adding these two expressions gives

$$e^{i\theta} + e^{-i\theta} = 2\cos\theta$$
$$\frac{e^{i\theta} + e^{-i\theta}}{2} = \cos\theta$$

and subtracting the first two expressions gives

$$e^{i\theta} - e^{-i\theta} = 2i\sin\theta$$
$$\frac{e^{i\theta} - e^{-i\theta}}{2i} = \sin\theta$$

A–7. Use Equation A.7 to derive

$$z^n = r^n(\cos\theta + i\sin\theta)^n = r^n(\cos n\theta + i\sin n\theta)$$

and from this, the formula of De Moivre:

$$(\cos\theta + i\sin\theta)^n = \cos n\theta + i\sin n\theta$$

Beginning with Equation A.7,

$$z = r(\cos\theta + i\sin\theta)$$
$$z^n = r^n(\cos\theta + i\sin\theta)^n \qquad (1)$$

We also know that

$$z = re^{i\theta}$$

$$z^n = r^n e^{ni\theta} \qquad (2)$$

Equating Equations 1 and 2 gives

$$z^n = r^n(\cos\theta + i\sin\theta)^n = r^n(\cos n\theta + i\sin n\theta)$$

Let $r = 1$ to obtain De Moivre's formula:

$$(\cos\theta + i\sin\theta)^n = \cos(n\theta) + i\sin(n\theta)$$

A–8. Use the formula of De Moivre, which is given in Problem A–7, to derive the trigonometric identities

$$\cos 2\theta = \cos^2\theta - \sin^2\theta$$

$$\sin 2\theta = 2\sin\theta\cos\theta$$

$$\cos 3\theta = \cos^3\theta - 3\cos\theta\sin^2\theta$$

$$= 4\cos^3\theta - 3\cos\theta$$

$$\sin 3\theta = 3\cos^2\theta\sin\theta - \sin^3\theta$$

$$= 3\sin\theta - 4\sin^3\theta$$

Use De Moivre's formula with $n = 2$:

$$(\cos\theta + i\sin\theta)^2 = \cos 2\theta + i\sin 2\theta$$

$$\cos^2\theta + 2i\sin\theta\cos\theta - \sin^2\theta = \cos 2\theta + i\sin 2\theta$$

Equating the real and imaginary parts of this last equation gives

$$\cos 2\theta = \cos^2\theta - \sin^2\theta$$

$$\sin 2\theta = 2\sin\theta\cos\theta$$

Likewise, for $n = 3$

$$\left(\cos^2\theta + 2i\sin\theta\cos\theta - \sin^2\theta\right)(\cos\theta + i\sin\theta) = \cos 3\theta + i\sin 3\theta$$

$$\cos^3\theta + 3i\sin\theta\cos^2\theta - 3\cos\theta\sin^2\theta - i\sin^3\theta = \cos 3\theta + i\sin 3\theta$$

Equationg the real and imaginary parts of this last equation gives

$$\cos 3\theta = \cos^3\theta - 3\cos\theta\sin^2\theta$$

$$= 4\cos^3\theta - 3\cos\theta$$

$$\sin 3\theta = 3\cos^2\theta\sin\theta - \sin^3\theta$$

$$= 3\sin\theta - 4\sin^3\theta$$

A–9. Consider the set of functions

$$\Phi_m(\phi) = \frac{1}{\sqrt{2\pi}}e^{im\phi} \quad m = 0, \pm 1, \pm 2, \ldots$$

$$0 \leq \phi \leq 2\pi$$

First show that

$$\int_0^{2\pi} d\phi\, \Phi_m(\phi) = 0 \qquad \text{for all value of } m \neq 0$$

$$= \sqrt{2\pi} \quad m = 0$$

Now show that

$$\int_0^{2\pi} d\phi\, \Phi_m^*(\phi)\Phi_n(\phi) = 0 \quad m \neq n$$

$$= 1 \quad m = n$$

$$\Phi_m(\phi) = \frac{1}{\sqrt{2\pi}}e^{im\phi} \qquad m = 0, \pm 1, \pm 2, \ldots \qquad 0 \leq \phi \leq 2\pi$$

Let $m = 0$. Then

$$\int_0^{2\pi} \frac{1}{\sqrt{2\pi}}d\phi = \frac{2\pi}{\sqrt{2\pi}} = \sqrt{2\pi}$$

If $m \neq 0$,

$$\int_0^{2\pi} \frac{1}{\sqrt{2\pi}}e^{im\phi}d\phi = \int_0^{2\pi} \frac{1}{\sqrt{2\pi}}\cos m\phi\, d\phi + \int_0^{2\pi} \frac{i}{\sqrt{2\pi}}\sin m\phi\, d\phi \tag{1}$$

Each integral in Equation 1 is equal to zero, because they are evaluated over one full cycle of the function. Now consider

$$\int_0^{2\pi} d\phi\, \Phi_m^*(\phi)\Phi_n(\phi) = \frac{1}{2\pi}\int_0^{2\pi} d\phi\, e^{-im\phi}e^{in\phi}$$

$$= \frac{1}{2\pi}\int_0^{2\pi} d\phi\, e^{i(n-m)\phi}$$

If $n \neq m$, we can define $n - m = k$ and the resulting integral is identical to that in Equation 1, and so has a value of 0. If $n = m$,

$$\int_0^{2\pi} d\phi\, \Phi_m^*(\phi)\Phi_m(\phi) = \frac{1}{2\pi}\int_0^{2\pi} d\phi = 1$$

A–10. This problem offers a derivation of Euler's formula. Start with

$$f(\theta) = \ln(\cos\theta + i\sin\theta) \tag{1}$$

Show that

$$\frac{df}{d\theta} = i \tag{2}$$

Now integrate both sides of Equation 2 to obtain

$$f(\theta) = \ln(\cos\theta + i\sin\theta) = i\theta + c \tag{3}$$

where c is a constant of integration. Show that $c = 0$ and then exponentiate Equation 3 to obtain Euler's formula.

$$f(\theta) = \ln(\cos\theta + i\sin\theta)$$

$$\frac{df}{d\theta} = \frac{-\sin\theta + i\cos\theta}{\cos\theta + i\sin\theta}$$

$$= \frac{i(\cos\theta + i\sin\theta)}{\cos\theta + i\sin\theta} = i$$

Integrating gives

$$\int df = \int i\,d\theta$$

$$f = i\theta + c$$

Because $f(\theta) = \ln 1 = 0$ when $\theta = 0$, c must equal 0. We can exponentiate this last result to obtain

$$\cos\theta + i\sin\theta = e^{i\theta}$$

A–11. We have seen that both the exponential and the natural logarithm functions (Problem A–10) can be extended to include complex arguments. This expression is generally true of most functions. Using Euler's formula and assuming that x represents a real number, show that $\cos ix$ and $-i\sin ix$ are equivalent to real functions of the real variable x. These functions are defined as the hyperbolic cosine and hyperbolic sine functions, $\cosh x$ and $\sinh x$, respectively. Sketch these functions. Do they oscillate like $\sin x$ and $\cos x$? Now show that $\sinh ix = i\sin x$ and that $\cosh ix = \cos x$.

Recall from Problem A-6

$$\sin u = \frac{e^{iu} - e^{-iu}}{2i} \qquad \cos u = \frac{e^{iu} + e^{-iu}}{2}$$

Let $u = ix$. Then

$$\sin ix = \frac{e^{-x} - e^{x}}{2i} \qquad \cos ix = \frac{e^{-x} + e^{x}}{2}$$

$$-i\sin ix = -\frac{e^{-x} - e^{x}}{2} \qquad \cos ix = \frac{e^{-x} + e^{x}}{2}$$

$$\sinh x = -\frac{e^{-x} - e^{x}}{2} \qquad \cosh x = \frac{e^{-x} + e^{x}}{2}$$

so $-i\sin ix = \sinh x$ and $\cos ix = \cosh x$.

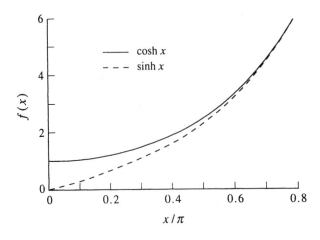

In the figure cosh x is represented by the solid line and sinh x is represented by the dashed line. Clearly these functions do not oscillate like cos x and sin x. Finally,

$$\sinh ix = -\frac{e^{-ix} - e^{-ix}}{2} = i \sin x$$

$$\cosh ix = \frac{e^{-ix} + e^{ix}}{2} = \cos x$$

A–12. Evaluate i^i.

$$i = e^{i\pi/2}$$

$$i^i = e^{i^2\pi/2} = e^{-\pi/2}$$

$$\approx 0.2079$$

A–13. The equation $x^2 = 1$ has two distinct roots, $x = \pm 1$. The equation $x^N = 1$ has N distinct roots, called the N roots of unity. This problem shows how to find the N roots of unity. We shall see that some of the roots turn out to be complex, so let's write the equation as $z^N = 1$. Now let $z = re^{i\theta}$ and obtain $r^N e^{iN\theta} = 1$. Show that this must be equivalent to $e^{iN\theta} = 1$, or

$$\cos N\theta + i \sin N\theta = 1$$

Now argue that $N\theta = 2\pi n$, where n has the N distinct values $0, 1, 2, \ldots, N - 1$ or that the N roots of units are given by

$$z = e^{2\pi in/N} \qquad n = 0, 1, 2, \ldots, N - 1$$

Show that we obtain $z = 1$ and $z = \pm 1$, for $N = 1$ and $N = 2$, respectively. Now show that

$$z = 1, -\frac{1}{2} + i\frac{\sqrt{3}}{2}, \quad \text{and} \quad -\frac{1}{2} - i\frac{\sqrt{3}}{2}$$

for $N = 3$. Show that each of these roots is of unit magnitude. Plot these three roots in the complex plane. Now show that $z = 1, i, -1,$ and $-i$ for $N = 4$ and that

$$z = 1, -1, \frac{1}{2} \pm i\frac{\sqrt{3}}{2}, \quad \text{and} \quad -\frac{1}{2} \pm i\frac{\sqrt{3}}{2}$$

for $N = 6$. Plot the four roots for $N = 4$ and the six roots for $N = 6$ in the complex plane. Compare the plots for $N = 3$, $N = 4$, and $N = 6$. Do you see a pattern?

Start with $z^N = 1$. Let $z = re^{i\theta}$ to get $r^N e^{iN\theta} = 1$. The magnitude of z must be unity, so $r = 1$. Therefore, we have $e^{iN\theta} = 1$, or $\cos N\theta + i \sin N\theta = 1$. The solutions to this equation are $N\theta = 0, 2\pi, 4\pi, \ldots$, or $N\theta = 2\pi n$, where $n = 0, 1, 2, \ldots$ The integer n has an upper limit of $N - 1$, because for $n = N$ $\theta = 2\pi n/N$, which is equivalent to the value obtained when $n = 0$. (At this point, we have gone around a complete circle.) Therefore,

$$z = e^{2\pi in/N} \qquad n = 0, 1, 2, \ldots, N - 1$$

If $N = 1$, then $\theta = 0$, and so $z = 1$.

If $N = 2$, then $\theta = 0$ or π, and $z = \pm 1$.

If $N = 3$, then $\theta = 0, 2\pi/3$, and $4\pi/3$, and so $z = 1, -\frac{1}{2} \pm \frac{\sqrt{3}}{2}i$.

If $N = 4$, then $\theta = 0$, $\pi/2$, π, and $3\pi/2$, and so $z = \pm 1$ or $\pm i$.

And finally, if $N = 6$, then $\theta = 0$, $\pi/6$, $\pi/3$, $\pi/2$, $2\pi/3$, and $5\pi/6$, and so

$$z = \pm 1, -\frac{1}{2} \pm \frac{\sqrt{3}}{2}i, \frac{1}{2} \pm \frac{\sqrt{3}}{2}i.$$

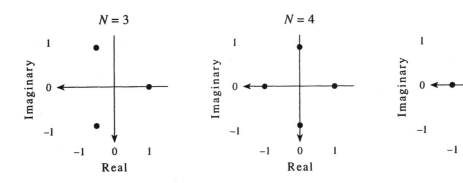

All the roots for a given value of N lie on the circumference of a unit circle. One root lies along the x-axis, at $x = 1$, and the others are distributed symmetrically on the unit circle, so that the angles between all the roots are equal. (This means that the locations of the roots in the complex plane form regular polygons.)

A–14. Using the results of Problem A–13, find the three distinct roots of $x^3 = 8$.

In the previous problems, we found that for $z^3 = 1$ we would obtain the roots

$$z = 1, -\frac{1}{2} + i\frac{\sqrt{3}}{2}, \text{ and } -\frac{1}{2} - i\frac{\sqrt{3}}{2}$$

Because $x^3 = 8$, we must multiply these roots by two (if $r^3 = 8$, then $r = 2$) to find the roots of $x^3 = 8$. Then

$$x = 2, -1 \pm i\sqrt{3}$$

Notice that each root has a length of two.

CHAPTER 2

The Classical Wave Equation

PROBLEMS AND SOLUTIONS

2–1. Find the general solutions to the following differential equations.

a. $\dfrac{d^2y}{dx^2} - 4\dfrac{dy}{dx} + 3y = 0$ **b.** $\dfrac{d^2y}{dx^2} + 6\dfrac{dy}{dx} = 0$ **c.** $\dfrac{dy}{dx} + 3y = 0$

d. $\dfrac{d^2y}{dx^2} + 2\dfrac{dy}{dx} - y = 0$ **e.** $\dfrac{d^2y}{dx^2} - 3\dfrac{dy}{dx} + 2y = 0$

For all these equations, let $y = e^{\alpha x}$ as in Example 2–1, so $dy/dx = \alpha e^{\alpha x}$ and $d^2y/dx^2 = \alpha^2 e^{\alpha x}$.

a.
$$\frac{d^2y}{dx^2} - 4\frac{dy}{dx} + 3y = 0$$
$$\alpha^2 - 4\alpha + 3 = 0$$
$$(\alpha - 1)(\alpha - 3) = 0$$
$$\alpha = 1,\ 3$$

The general solution is therefore $y(x) = c_1 e^{3x} + c_2 e^{x}$.

b.
$$\frac{d^2y}{dx^2} + 6\frac{dy}{dx} = 0$$
$$\alpha^2 + 6\alpha = 0$$
$$\alpha(\alpha + 6) = 0$$
$$\alpha = 0,\ -6$$

The general solution is $y(x) = c_1 + c_2 e^{-6x}$.

c.
$$\frac{dy}{dx} + 3y = 0$$
$$\alpha + 3 = 0$$
$$\alpha = -3$$

The general solution is $y(x) = c_1 e^{-3x}$.

d.

$$\frac{d^2y}{dx^2} + 2\frac{dy}{dx} - y = 0$$

$$\alpha^2 + 2\alpha - 1 = 0$$

$$\alpha = \frac{-2 \pm \sqrt{4+4}}{2} = -1 \pm \sqrt{2}$$

The general solution is $y(x) = c_1 e^{(-1+\sqrt{2})x} + c_2 e^{(-1-\sqrt{2})x}$.

e.

$$\frac{d^2y}{dx^2} - 3\frac{dy}{dx} + 2y = 0$$

$$\alpha^2 - 3\alpha + 2 = 0$$

$$(\alpha - 2)(\alpha - 1) = 0$$

$$\alpha = 1,\ 2$$

The general solution is $y(x) = c_1 e^{2x} + c_2 e^x$.

2–2. Solve the following differential equations:

a. $\dfrac{d^2y}{dx^2} - 4y = 0 \qquad y(0) = 2;\ \dfrac{dy}{dx}(\text{at } x = 0) = 4$

b. $\dfrac{d^2y}{dx^2} - 5\dfrac{dy}{dx} + 6y = 0 \qquad y(0) = -1;\ \dfrac{dy}{dx}(\text{at } x = 0) = 0$

c. $\dfrac{dy}{dx} - 2y = 0 \qquad y(0) = 2$

First, we need to find the general solution. Then we will impose the boundary conditions to find c_1 and c_2.

a.

$$\frac{d^2y}{dx^2} - 4y = 0$$

$$\alpha^2 - 4 = 0$$

$$\alpha = \pm 2$$

The general solution is therefore $y(x) = c_1 e^{2x} + c_2 e^{-2x}$. The first derivative of $y(x)$ is

$$\frac{dy}{dx} = 2c_1 e^{2x} - 2c_2 e^{-2x}$$

Using these boundary conditions on $y(x)$ and dy/dx at $x = 0$ gives

$$2 = c_1 e^0 + c_2 e^0 \qquad 4 = 2c_1 - 2c_2$$
$$2 = c_1 + c_2 \qquad\qquad 2 = c_1 - c_2$$

Adding these two equations gives

$$4 = 2c_1$$

from which we find

$$2 = c_1 \quad \text{and} \quad 0 = c_2$$

The solution is $y(x) = 2e^{2x}$.

b.
$$\frac{d^2 y}{dx^2} - 5\frac{dy}{dx} + 6y = 0$$

$$\alpha^2 - 5\alpha + 6 = 0$$

$$(\alpha - 2)(\alpha - 3) = 0$$

$$\alpha = 2,\ 3$$

The general solution is $y(x) = c_1 e^{2x} + c_2 e^{3x}$. Using the boundary conditions gives

$$y = c_1 e^{2x} + c_2 e^{3x} \qquad \frac{dy}{dx} = 2c_1 e^{2x} + 3c_2 e^{3x}$$

$$-1 = c_1 e^0 + c_2 e^0 \qquad 0 = 2c_1 + 3c_2$$

$$-1 = c_1 + c_2 \qquad c_1 = -\frac{3c_2}{2}$$

Combining these results gives

$$-1 = -\frac{c_2}{2}$$

or

$$-3 = c_1 \qquad 2 = c_2$$

The solution is $y(x) = -3e^{2x} + 2e^{3x}$.

c.
$$\frac{dy}{dx} - 2y = 0$$

$$\alpha - 2 = 0$$

$$\alpha = 2$$

The general solution is $y(x) = c_1 e^{2x}$. The condition $y(0) = 2$ gives

$$2 = c_1 e^0$$

The solution is $y(x) = 2e^{2x}$.

2–3. Prove that $x(t) = \cos \omega t$ oscillates with a frequency $\nu = \omega/2\pi$. Prove that $x(t) = A \cos \omega t + B \sin \omega t$ oscillates with the same frequency, $\omega/2\pi$.

The functions $\cos t$ and $\sin t$ oscillate with a frequency of $\nu = 1/2\pi$, since they go through one complete cycle every 2π radians. The functions $\cos(\omega t)$ and $\sin(\omega t)$ go through ω complete cycles every 2π radians, so they oscillate with a frequency of $\nu = \omega/2\pi$. A linear combination of these functions (for example, $A \cos \omega t + B \sin \omega t$) will oscillate with the same frequency, $\omega/2\pi$.

2–4. Solve the following differential equations:

a. $\dfrac{d^2 x}{dt^2} + \omega^2 x(t) = 0 \qquad x(0) = 0;\ \dfrac{dx}{dt}(\text{at } t = 0) = v_0$

b. $\dfrac{d^2 x}{dt^2} + \omega^2 x(t) = 0 \qquad x(0) = A;\ \dfrac{dx}{dt}(\text{at } t = 0) = v_0$

Prove in both cases that $x(t)$ oscillates with frequency $\omega/2\pi$.

We solved the general case of the differential equation given in Example 2–4, so we know that $x(t) = c_3 \cos \omega t + c_4 \sin \omega t$. Now we apply the given initial conditions.

a.
$$x(t) = c_3 \cos \omega t + c_4 \sin \omega t \qquad \frac{dx}{dt} = -\omega c_3 \sin \omega t + \omega c_4 \cos \omega t$$

$$0 = c_3(1) + c_4(0) \qquad v_0 = 0 + \omega c_4$$

$$0 = c_3 \qquad \frac{v_0}{\omega} = c_4$$

The solution to the differential equation under these conditions is $x(t) = \dfrac{v_0}{\omega} \sin \omega t$.

b.
$$x(t) = c_3 \cos \omega t + c_4 \sin \omega t \qquad \frac{dx}{dt} = -\omega c_3 \sin \omega t + \omega c_4 \cos \omega t$$

$$A = c_3(1) + c_4(0) \qquad v_0 = 0 + \omega c_4$$

$$A = c_3 \qquad \frac{v_0}{\omega} = c_4$$

The solution to the differential equation under these conditions is $x(t) = A \cos \omega t + \dfrac{v_0}{\omega} \sin \omega t$.

Both of these solutions can be written in the form $x(t) = A \cos \omega t + B \sin \omega t$, which (as shown in Problem 2–3) oscillates with frequency $\omega/2\pi$.

2–5. The general solution to the differential equation

$$\frac{d^2 x}{dt^2} + \omega^2 x(t) = 0$$

is

$$x(t) = c_1 \cos \omega t + c_2 \sin \omega t$$

For convenience, we often write this solution in the equivalent forms

$$x(t) = A \sin(\omega t + \phi)$$

or

$$x(t) = B \cos(\omega t + \psi)$$

Show that all three of these expressions for $x(t)$ are equivalent. Derive equations for A and ϕ in terms of c_1 and c_2, and for B and ψ in terms of c_1 and c_2. Show that all three forms of $x(t)$ oscillate with frequency $\omega/2\pi$. *Hint*: Use the trigonometric identities

$$\sin(\alpha + \beta) = \sin \alpha \cos \beta + \cos \alpha \sin \beta$$

and

$$\cos(\alpha + \beta) = \cos \alpha \cos \beta - \sin \alpha \sin \beta$$

$$x(t) = A \sin(\omega t + \phi)$$
$$= A \sin \omega t \cos \phi + A \cos \omega t \sin \phi$$
$$= c_1 \cos \omega t + c_2 \sin \omega t$$

where we define $c_1 = A \sin\phi$ and $c_2 = A\cos\phi$. These equations can be solved for A and ϕ in terms of c_1 and c_2:

$$c_1^2 + c_2^2 = A^2(\sin^2\phi + \cos^2\phi)$$
$$A = \left(c_1^2 + c_2^2\right)^{1/2}$$
$$\phi = \sin^{-1}\frac{c_2}{\left(c_1^2 + c_2^2\right)^{1/2}} = \tan^{-1}\frac{c_2}{c_1}$$

Likewise,

$$x(t) = B\cos(\omega t + \phi)$$
$$= B\cos\omega t\cos\psi - B\sin\omega t\sin\psi$$
$$= c_1\cos\omega t + c_2\sin\omega t$$

where we define $c_1 = B\cos\psi$ and $c_2 = -B\sin\psi$. Solving for B and ϕ in terms of c_1 and c_2 gives

$$c_1^2 + c_2^2 = B^2\left(\cos^2\psi + \sin^2\psi\right)$$
$$B = \left(c_1^2 + c_2^2\right)^{1/2}$$
$$\psi = \cos^{-1}\frac{c_1}{\left(c_1^2 + c_2^2\right)^{1/2}} = \tan^{-1}\frac{c_2}{c_1}$$

Because ϕ and ψ are constants, $x(t) = A\sin(\omega t + \phi)$ and $x(t) = B\cos(\omega t + \phi)$ oscillate with a frequency of $\nu = \omega/2\pi$. We showed in Problem 2–3 that $x(t) = A\cos\omega t + B\sin\omega t$ oscillates with the frequency $\omega/2\pi$.

2–6. In all the differential equations we have discussed so far, the values of the exponents α that we have found have been either real or purely imaginary. Let us consider a case in which α turns out to be complex. Consider the equation

$$\frac{d^2y}{dx^2} + 2\frac{dy}{dx} + 10y = 0$$

If we substitute $y(x) = e^{\alpha x}$ into this equation, we find that $\alpha^2 + 2\alpha + 10 = 0$ or that $\alpha = -1 \pm 3i$. The general solution is

$$y(x) = c_1 e^{(-1+3i)x} + c_2 e^{(-1-3i)x}$$
$$= c_1 e^{-x}e^{3ix} + c_2 e^{-x}e^{-3ix}$$

Show that $y(x)$ can be written in the equivalent form

$$y(x) = e^{-x}(c_3\cos 3x + c_4\sin 3x)$$

Thus we see that complex values of the α's lead to trigonometric solutions modulated by an exponential factor. Solve the following equations.

a. $\dfrac{d^2y}{dx^2} + 2\dfrac{dy}{dx} + 2y = 0$

b. $\dfrac{d^2y}{dx^2} - 6\dfrac{dy}{dx} + 25y = 0$

c. $\dfrac{d^2y}{dx^2} + 2\beta\dfrac{dy}{dx} + (\beta^2 + \omega^2)y = 0$

d. $\dfrac{d^2y}{dx^2} + 4\dfrac{dy}{dx} + 5y = 0 \qquad y(0) = 1;\ \dfrac{dy}{dx}(\text{at } x = 0) = -3$

To show that the above expressions for $y(x)$ are equivalent, use the expression $e^{i\theta} = \cos\theta + i\sin\theta$ (Equation A.6). [Recall that $\cos(-\theta) = \cos(\theta)$ and that $\sin(-\theta) = -\sin(\theta)$.]

$$
\begin{aligned}
y(x) &= c_1 e^{-x} e^{3ix} + c_2 e^{-x} e^{-3ix} \\
&= e^{-x}\left[c_1\cos(3x) + c_1 i\sin(3x) + c_2\cos(-3x) + c_2 i\sin(-3x)\right] \\
&= e^{-x}\left[(c_1 + c_2)\cos(3x) + i(c_1 - c_2)\sin(3x)\right] \\
&= e^{-x}\left(c_3\cos 3x + c_4\sin 3x\right)
\end{aligned}
$$

where $c_3 = c_1 + c_2$ and $c_4 = (c_1 - c_2)i$. Now we can solve the differential equations as we did in Problem 2–1.

a.

$$\frac{d^2 y}{dx^2} + 2\frac{dy}{dx} + 2y = 0$$

$$\alpha^2 + 2\alpha + 2 = 0$$

$$\alpha = \frac{-2 \pm \sqrt{4-8}}{2} = -1 \pm i$$

The general solution is $y(x) = c_1 e^{(-1+i)x} + c_2 e^{(-1-i)x}$. Using the expression derived in the first part of this problem, we can write the solution as $y(x) = e^{-x}\left[c_3\cos(x) + c_4\sin(x)\right]$.

b.

$$\frac{d^2 y}{dx^2} - 6\frac{dy}{dx} + 25y = 0$$

$$\alpha^2 - 6\alpha + 25 = 0$$

$$\alpha = \frac{6 \pm \sqrt{36-100}}{2} = 3 \pm 4i$$

The general solution is $y(x) = c_1 e^{(3+4i)x} + c_2 e^{(3-4i)x}$ or $y(x) = e^{3x}\left[c_3\cos(4x) + c_4\sin(4x)\right]$.

c.

$$\frac{d^2 y}{dx^2} + 2\beta\frac{dy}{dx} + (\beta^2 + \omega^2)y = 0$$

$$\alpha^2 + 2\beta\alpha + (\beta^2 + \omega^2) = 0$$

$$\alpha = \frac{-2\beta \pm \sqrt{4\beta^2 - 4\beta^2 - 4\omega^2}}{2} = -\beta \pm i\omega$$

The general solution is $y(x) = c_1 e^{(-\beta+\omega i)x} + c_2 e^{(-\beta-\omega i)x}$ or $y(x) = e^{-\beta x}\left[c_3\cos(\omega x) + c_4\sin(\omega x)\right]$.

d.

$$\frac{d^2 y}{dx^2} + 4\frac{dy}{dx} + 5y = 0$$

$$\alpha^2 + 4\alpha + 5 = 0$$

$$\alpha = \frac{-4 \pm \sqrt{16-20}}{2} = -2 \pm i$$

The general solution is $y(x) = c_1 e^{(-2+i)x} + c_2 e^{(-2-i)x}$ or $y(x) = e^{-2x}\left[c_3\cos(x) + c_4\sin(x)\right]$. We now use the conditions on $y(x)$ and dy/dx at $x = 0$ to find c_3 and c_4:

$$y(0) = 1 = e^0\left[c_3\cos(0) + c_4\sin(0)\right] = c_3 = 1$$

$$\frac{dy}{dx}(0) = -3 = -2e^0\left[\cos(0) + c_4\sin(0)\right] + e^0\left[-\sin(0) + c_4\cos(0)\right] = -2 + c_4$$

and so

$$c_4 = -1$$

The specific solution is therefore $y(x) = e^{-2x}(\cos x - \sin x)$.

2–7. This problem develops the idea of a classical harmonic oscillator. Consider a mass m attached to a spring as shown in Figure 2.8. Suppose there is no gravitational force acting on m so that the only force is from the spring. Let the relaxed or undistorted length of the spring be x_0. Hooke's law says that the force acting on the mass m is $f = -k(x - x_0)$, where k is a constant characteristic

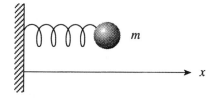

x **FIGURE 2.8**
A body of mass m connected to a wall by a sping.

of the spring and is called the force constant of the spring. Note that the minus sign indicates the direction of the force: to the left if $x > x_0$ (extended) and to the right if $x < x_0$ (compressed). The momentum of the mass is

$$p = m\frac{dx}{dt} = m\frac{d(x - x_0)}{dt}$$

Newton's second law says that the rate of change of momentum is equal to a force

$$\frac{dp}{dt} = f$$

Replacing $f(x)$ by Hooke's law, show that

$$m\frac{d^2(x - x_0)}{dt^2} = -k(x - x_0)$$

Upon letting $\xi = x - x_0$ be the displacement of the spring from its undistorted length, then

$$m\frac{d^2\xi}{dt^2} + k\xi = 0$$

Given that the mass starts at $\xi = 0$ with an intial velocity v_0, show that the displacement is given by

$$\xi(t) = v_0 \left(\frac{m}{k}\right)^{1/2} \sin\left[\left(\frac{k}{m}\right)^{1/2} t\right]$$

Interpret and discuss this solution. What does the motion look like? What is the frequency? What is the amplitude?

Substituting for f from Hooke's law into Newton's second law,

$$\frac{dp}{dt} = -k(x - x_0)$$

$$\frac{d}{dt}\left(m\frac{dx}{dt}\right) = -k(x - x_0)$$

$$m\frac{d^2x}{dt^2} = -k(x - x_0)$$

Now let $\xi(t) = x - x_0$, so $md^2\xi/dt^2 + k\xi = 0$, and use the initial conditions of $\xi(0) = 0$ and $d\xi/dt = v_0$ at $t = 0$. Solving this differential equation by the method used in Problem 2–6 gives a general solution of $\xi(t) = c_3 \cos \omega t + c_4 \sin \omega t$. Using the initial conditions, we find that

$$\xi(0) = 0 = c_3$$

$$\frac{d\xi}{dt} = -\omega c_3 \sin \omega t + \omega c_4 \cos \omega t$$

$$v_0 = \omega c_4$$

$$c_4 = \frac{v_0}{\omega}$$

so the solution that satisfies the initial conditions is

$$\xi(t) = \frac{v_0}{\omega} \sin \omega t = v_0 \left(\frac{m}{k}\right)^{1/2} \sin\left[\left(\frac{m}{k}\right)^{1/2} t\right]$$

The function $\xi(t)$, which describes the time evolution of the oscillator, shows that the motion is sinusoidal with frequency $(1/2\pi)(k/m)^{1/2}$ and amplitude $v_0(m/k)^{1/2}$.

2–8. Consider the linear second-order differential equation

$$\frac{d^2y}{dx^2} + a_1(x)\frac{dy}{dx} + a_0(x)y(x) = 0$$

Note that this equation is linear because $y(x)$ and its derivatives appear only to the first power and there are no cross terms. It does not have constant coefficients, however, and there is no general, simple method for solving it like there is if the coefficients were constants. In fact, each equation of this type must be treated more or less individually. Nevertheless, because it is linear, we must have that if $y_1(x)$ and $y_2(x)$ are any two solutions, then a linear combination,

$$y(x) = c_1 y_1(x) + c_2 y_2(x)$$

where c_1 and c_2 are constants, is also a solution. Prove that $y(x)$ is a solution.

Let $y(x) = c_1 y_1(x) + c_2 y_2(x)$. Then

$$\frac{dy}{dx} = c_1 \frac{dy_1}{dx} + c_2 \frac{dy_2}{dx}$$

$$\frac{d^2y}{dx^2} = c_1 \frac{d^2y_1}{dx^2} + c_2 \frac{d^2y_2}{dx^2}$$

$$\frac{d^2y}{dx^2} + a_1(x)\frac{dy}{dx} + a_0(x)y(x) = c_1 \frac{d^2y_1}{dx^2} + c_2 \frac{d^2y_2}{dx^2} + a_1 \left[c_1 \frac{dy_1}{dx} + c_2 \frac{dy_2}{dx}\right] + a_0 \left(c_1 y_1 + c_2 y_2\right)$$

$$= c_1 \left[\frac{d^2y_1}{dx^2} + a_1 \frac{dy_1}{dx} + a_0 y_1\right] + c_2 \left[\frac{d^2y_2}{dx^2} + a_1 \frac{dy_2}{dx} + a_0 y_2\right]$$

$$= c_1(0) + c_2(0) = 0$$

If $y_1(x)$ and $y_2(x)$ are solutions, $y(x) = c_1 y_1(x) + c_2 y_2(x)$ is also a solution.

2–9. We will see in Chapter 3 that the Schrödinger equation for a particle of mass m that is constrained to move freely along a line between 0 and a is

$$\frac{d^2\psi}{dx^2} + \left(\frac{8\pi^2 mE}{h^2}\right)\psi(x) = 0$$

with the boundary condition

$$\psi(0) = \psi(a) = 0$$

In this equation, E is the energy of the particle and $\psi(x)$ is its wave function. Solve this differential equation for $\psi(x)$, apply the boundary conditions, and show that the energy can have only the values

$$E_n = \frac{n^2 h^2}{8ma^2} \qquad n = 1, 2, 3, \ldots$$

or that the energy is quantized.

Using the solution found in Example 2–4, with $\omega^2 = 8\pi^2 mE/h^2$, we find

$$\psi = C \sin kx + D \cos kx$$

where $k = (8\pi^2 mE/h^2)^{1/2}$. Applying the boundary condition $\psi(0) = 0$ gives $\psi = C \sin kx$. The boundary condition $\psi(a) = C \sin ka = 0$ requires that $ka = n\pi$ with $n = 1, 2, \ldots$. Therefore

$$\psi(x) = C \sin \frac{n\pi x}{a} \qquad n = 1, 2, 3, \ldots$$

and

$$ka = n\pi$$

or

$$\sqrt{\frac{8\pi^2 mE}{h^2}} = \frac{n\pi}{a}$$

Solving for E gives

$$E_n = \frac{n^2 h^2}{8ma^2} \qquad n = 1, 2, 3, \ldots$$

where we have added the subscript "n" to E to explicitly show the dependence of E on the integer n.

2–10. Prove that the number of nodes for a vibrating string clamped at both ends is $n - 1$ for the nth harmonic.

Equation 2.25 gives us a formula for u_n where u_n represents harmonic motion in a normal mode:

$$u_n(x, t) = A_n \cos\left(\omega_n t + \phi_n\right) \sin \frac{n\pi x}{l} \qquad (2.25)$$

We want to find the number of nodes, or the number of solutions to $u_n(x)$ where $u_n = 0$. (We can ignore the time-dependent portion of $u_n(x)$ because the nodes are not time-dependent.) Then $u_n(x, t) = 0$ if

$$\sin \frac{n\pi x}{l} = 0$$

or

$$\frac{n\pi x}{l} = 0, \pi, 2\pi, \ldots$$

Solving for x gives

$$x = 0, \frac{l}{n}, \frac{2l}{n}, \ldots$$

We know that, physically, $0 \le x \le l$. Furthermore, by definition, $x = 0$ and $x = l$ are not nodes (they are the ends of the string). Therefore nodes occur at

$$x = \frac{l}{n}, \frac{2l}{n}, \ldots, \frac{(n-1)l}{n}$$

and so the nth normal mode has $n - 1$ nodes.

2–11. Prove that

$$y(x, t) = A \sin\left[\frac{2\pi}{\lambda}(x - vt)\right]$$

is a wave of wavelength λ and frequency $\nu = v/\lambda$ traveling to the right with a velocity v.

All sine and cosine functions oscillate in a wave-like manner, so $y(x, t)$ is a wave. We can write $y(x, t)$ as

$$y(x, t) = A \sin[Bx - Ct], \qquad \text{where} \qquad B = \frac{2\pi}{\lambda} \qquad \text{and} \qquad C = \frac{2\pi v}{\lambda}$$

From this expression, we see that

$$\text{wavelength} = \frac{2\pi}{B} = \lambda$$

$$\text{frequency} = \frac{C}{2\pi} = \frac{v}{\lambda} = \nu$$

A standing wave has the equation $y_s = A \sin k x_s$. The wave equation presented in this problem is related to the equation of a standing wave by $x = x_s + vt$. The point x_s is arbitrary, and so we set it equal to 0. This gives $x = vt$, so the wave is traveling right with velocity v.

2–12. Sketch the normal modes of a vibrating rectangular membrane and convince yourself that they look like those shown in Figure 2.6.

The sketch will look (as the problem states) just like Figure 2.6.

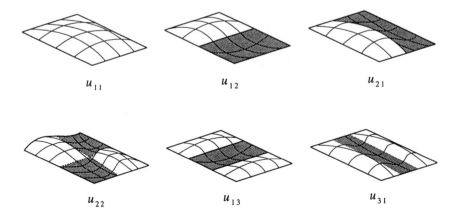

u_{11} u_{12} u_{21}

u_{22} u_{13} u_{31}

2–13. This problem is the extension of Problem 2–9 to two dimensions. In this case, the particle is constrained to move freely over the surface of a rectangle of sides a and b. The Schrödinger equation for this problem is

$$\frac{\partial^2 \psi}{\partial x^2} + \frac{\partial^2 \psi}{\partial y^2} + \left(\frac{8\pi^2 m E}{h^2}\right)\psi(x, y) = 0$$

with the boundary conditions

$$\psi(0, y) = \psi(a, y) = 0 \quad \text{for all } y, \quad 0 \le y \le b$$

$$\psi(x, 0) = \psi(x, b) = 0 \quad \text{for all } x, \quad 0 \le x \le a$$

Solve this equation for $\psi(x, y)$, apply the boundary conditions, and show that the energy is quantized according to

$$E_{n_x n_y} = \frac{n_x^2 h^2}{8ma^2} + \frac{n_y^2 h^2}{8mb^2} \qquad \begin{array}{l} n_x = 1, 2, 3, \ldots \\ n_y = 1, 2, 3, \ldots \end{array}$$

We will solve this partial differential equation using the technique of separation of variables. By letting $\psi(x, y) = X(x)Y(y)$, we can write

$$\frac{\partial^2 \psi}{\partial x^2} + \frac{\partial^2 \psi}{\partial y^2} + \left(\frac{8\pi^2 m E}{h^2}\right)\psi(x, y) = 0$$

$$Y\frac{\partial^2 X}{\partial y^2} + X\frac{\partial^2 Y}{\partial x^2} + \left(\frac{8\pi^2 m E}{h^2}\right)XY = 0$$

$$\frac{1}{X}\frac{\partial^2 X}{\partial x^2} + \frac{1}{Y}\frac{\partial^2 Y}{\partial y^2} = -\frac{8\pi^2 m E}{h^2}$$

We can now separate the above equation into two differential equations:

$$\frac{1}{X}\frac{\partial^2 X}{\partial x^2} = -p^2 \quad \text{and} \quad \frac{1}{Y}\frac{\partial^2 Y}{\partial y^2} = -q^2$$

where $p^2 + q^2 = 8\pi^2 m E/h^2$ (cf. Equations 2.34 through 2.37). These two equations have identical forms and were solved in Problem 2–9. Using the result of Problem 2–9 and applying the boundary conditions gives

$$X(x) = B \sin px = B \sin \frac{n_x \pi x}{a} \qquad n_x = 1, 2, 3, \ldots$$

$$Y(y) = D \sin qy = D \sin \frac{n_y \pi y}{b} \qquad n_y = 1, 2, 3, \ldots$$

Because $\psi(x, y) = X(x)Y(y)$, $\psi(x, y) = BD \sin \frac{n_x \pi x}{a} \sin \frac{n_y \pi y}{b}$. Also

$$p^2 + q^2 = \frac{8\pi^2 m E}{h^2}$$

and so

$$\left(\frac{n_x^2 \pi^2}{a^2} + \frac{n_y^2 \pi^2}{b^2}\right)\frac{h^2}{8\pi^2 m} = E$$

or

$$E_{n_x n_y} = \frac{n_x^2 h^2}{8ma^2} + \frac{n_y^2 h^2}{8mb^2} \qquad \begin{array}{l} n_x = 1, 2, 3, \ldots \\ n_y = 1, 2, 3, \ldots \end{array}$$

where we have used the subscripts on E to denote explicitly the dependence of the energy on the values of n_x and n_y.

2–14. Extend Problems 2–9 and 2–13 to three dimensions, where a particle is constrained to move freely throughout a rectangular box of sides a, b, and c. The Schrödinger equation for this system is

$$\frac{\partial^2 \psi}{\partial x^2} + \frac{\partial^2 \psi}{\partial y^2} + \frac{\partial^2 \psi}{\partial z^2} + \left(\frac{8\pi^2 mE}{h^2}\right)\psi(x, y, z) = 0$$

and the boundary conditions are that $\psi(x, y, z)$ vanishes over all the surfaces of the box.

As in Problem 2–13, we can separate the variables to produce three differential equations, one for each dimension:

$$\frac{\partial^2 X}{\partial x^2} + \frac{8\pi^2 mE}{h^2}X = p^2$$

$$\frac{\partial^2 Y}{\partial y^2} + \frac{8\pi^2 mE}{h^2}Y = q^2$$

$$\frac{\partial^2 Z}{\partial Z^2} + \frac{8\pi^2 mE}{h^2}Z = r^2$$

where $p^2 + q^2 + r^2 = 8\pi^2 mE/h^2$. Following the method described in Problem 2–13, we find

$$\psi(x, y, z) = A \sin\frac{n_x \pi x}{a} \sin\frac{n_y \pi y}{b} \sin\frac{n_z \pi z}{c} \qquad \begin{array}{l} n_x = 1,\ 2,\ 3,\ \ldots \\ n_y = 1,\ 2,\ 3,\ \ldots \\ n_z = 1,\ 2,\ 3,\ \ldots \end{array}$$

and

$$E_{n_x n_y n_z} = \frac{n_x^2 h^2}{8ma^2} + \frac{n_y^2 h^2}{8mb^2} + \frac{n_z^2 h^2}{8mc^2} \qquad \begin{array}{l} n_x = 1,\ 2,\ 3,\ \ldots \\ n_y = 1,\ 2,\ 3,\ \ldots \\ n_z = 1,\ 2,\ 3,\ \ldots \end{array}$$

2–15. Show that Equations 2.46 and 2.48 are equivalent. How are G_{nm} and ϕ_{nm} in Equation 2.48 related to the quantities in Equation 2.46?

$$T_{nm}(t) = G_{nm}\cos(\omega_{nm}t + \phi_{nm}) \tag{2.48}$$

Using the trigonometric identities from Problem 2–5, we write this as

$$T_{nm}(t) = G_{nm}\left[\cos(\omega_{nm}t)\cos\phi_{nm} - \sin(\omega_{nm}t)\sin\phi_{nm}\right]$$

$$= G_{nm}\cos(\omega_{nm}t)\cos\phi_{nm} - G_{nm}\sin(\omega_{nm}t)\sin\phi_{nm}$$

$$= E_{nm}\cos\omega_{nm}t + F_{nm}\sin\omega_{nm}t \tag{2.46}$$

where $G\cos\phi_{nm} = E_{nm}$ and $-G\sin\phi_{nm} = F_{nm}$.

Many problems in classical mechanics can be reduced to the problem of solving a differential equation with constant coefficients (cf. Problem 2–7). The basic starting point is Newton's second law, which says that the rate of change of momentum is equal to the force acting on a body. Momentum p equals mv, and so if the mass is constant, then in one dimension we have

$$\frac{dp}{dt} = m\frac{dv}{dt} = m\frac{d^2x}{dt^2} = f$$

If we are given the force as a function of x, then this equation is a differential equation for x(t), which is called the trajectory of the particle. Going back to the simple harmonic oscillator discussed in Problem 2–7, if we let x be the displacement of the mass from its equilibrium position, then Hooke's law says that f(x) = −kx, and the differential equation corresponding to Newton's second law is

$$\frac{d^2x}{dt^2} + kx(t) = 0$$

a differential equation that we have seen several times.

2–16. Consider a body falling freely from a height x_0 according to Figure 2.9a. If we neglect air resistance or viscous drag, the only force acting upon the body is the gravitational force mg. Using the coordinates in Figure 2.9a, mg acts in the same direction as x and so the differential equation corresponding to Newton's second law is

$$m\frac{d^2x}{dt^2} = mg$$

Show that

$$x(t) = \frac{1}{2}gt^2 + v_0 t + x_0$$

where x_0 and v_0 are the initial values of x and v. According to Figure 2.9a, $x_0 = 0$ and so

$$x(t) = \frac{1}{2}gt^2 + v_0 t$$

If the particle is just dropped, then $v_0 = 0$ and so

$$x(t) = \frac{1}{2}gt^2$$

Discuss this solution. Now do the same problem using Figure 2.9b as the definition of the various quantities involved, and show that although the equations may look different from those above, they say exactly the same thing because the picture we draw to define the direction of x, v_0, and mg does not affect the falling body.

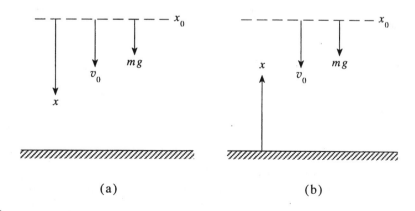

(a)   (b)

FIGURE 2.9
(a) A coordinate system for a body falling from a height x_0, and (b) a different coordinate system for a body falling from a height x_0.

We solve the equation (written using the coordinates in Figure 2.9a)

$$m\frac{d^2x}{dt^2} = mg$$

by integrating twice to obtain

$$x = \frac{gt^2}{2} + ct + k$$

At $t = 0$, $x = 0$, and so $k = 0$. Likewise, at $t = 0$, $dx/dt = v_0$, and so $c = v_0$. Thus, we obtain the result

$$x(t) = \frac{1}{2}gt^2 + v_0 t \tag{1}$$

When $v_0 = 0$, we have $x(t) = \frac{1}{2}gt^2$, which is a formula for the acceleration of a falling body from rest; the distance $x(t)$ increases quadratically with time. Newton's equation in the coordinate system of Figure 2.9b is

$$m\frac{d^2x}{dt^2} = -mg$$

whose general solution is

$$x(t) = at^2 + bt + c$$

The particle is falling from an initial height of x_0, so $x(0) = x_0$. Also $dx/dt = -v_0$ initially, and so

$$x(t) = -\frac{gt^2}{2} - v_0 t + x_0 \tag{2}$$

Both Equations 1 and 2 say the very same thing. For example, to find the time that it takes for the mass to strike the ground, let $x(t) = x_0$ in Equation 1 and $x(t) = 0$ in Equation 2 to obtain

$$\frac{1}{2}gt^2 + v_0 t = x_0$$

in each case.

2–17. Derive an equation for the maximum height a body will reach if it is shot straight upward with a velocity v_0. Refer to Figure 2.9b but realize that in this case v_0 points upward. How long will it take for the body to return to its initial position, $x = 0$?

Using the coordinate system of Figure 2.9b and Equation 2 derived in Problem 2–16 (with $v_0 = -v_0$ and $x(0) = 0$), we find that

$$x(t) = -gt^2 + v_0 t$$

and

$$\frac{dx}{dt} = -gt + v_0$$

To determine how long it will take for the body to return to earth, we first calculate how long it will take the body to reach its maximum height. At its maximum height, the velocity is zero ($dx/dt = 0$). Therefore the time needed to reach the maximum height, t_{max}, is given by

$$0 = -gt_{max} + v_0$$

or

$$t_{max} = \frac{v_0}{g}$$

The maximum height that the mass will attain is $x(t_{max}) = v_0^2/2g$. From the instant the body is shot, it will take $2t_{max}$ (or $2v_0/g$) to return to earth because it takes the same amount of time for the body to return from its maximum height as it takes the body to reach that height.

2–18. Consider a simple pendulum as shown in Figure 2.10. We let the length of the pendulum be l and assume that all the mass of the pendulum is concentrated at its end as shown in Figure 2.10. A physical example of this case might be a mass suspended by a string. We assume that the motion of the pendulum is set up such that it oscillates within a plane so that we have a problem in plane polar coordinates. Let the distance along the arc in the figure describe the motion of the pendulum, so that its momentum is $mds/dt = mld\theta/dt$ and its rate of change of momentum is $mld^2\theta/dt^2$. Show that the component of force in the direction of motion is $-mg\sin\theta$, where the minus sign occurs because the direction of this force is opposite to that of the angle θ. Show that the equation of motion is

$$ml\frac{d^2\theta}{dt^2} = -mg\sin\theta$$

Now assume that the motion takes place only through very small angles and show that the motion becomes that of a simple harmonic oscillator. What is the natural frequency of this harmonic oscillator? *Hint:* Use the fact that $\sin\theta \approx \theta$ for small values of θ.

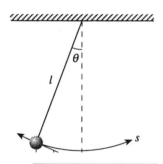

FIGURE 2.10
The coordinate system describing an oscillating pendulum.

The component of the force mg along the arc in Figure 2–10 is $mg\sin\theta$, but in a direction opposite to the motion. Newton's law states that the change in momentum is equal to the forces acting on the body. Therefore

$$m\frac{d^2s}{dt^2} = -mg\sin\theta$$

Since $s = l\theta$,

$$ml\frac{d^2\theta}{dt^2} = -mg\sin\theta \qquad (a)$$

For small angles, $\sin\theta \approx \theta$ and (a) becomes

$$\frac{d^2\theta}{dt^2} + \frac{g}{l}\theta = 0$$

The general solution to this equation is (Example 2–4)

$$x(t) = c_1 \cos t\sqrt{\frac{g}{l}} + c_2 \sin t\sqrt{\frac{g}{l}}$$

The natural frequency of the pendulum is $(g/l)^{1/2}$.

2–19. Consider the motion of a pendulum like that in Problem 2–18 but swinging in a viscous medium. Suppose that the viscous force is proportional to but oppositely directed to its velocity; that is,

$$f_{viscous} = -\lambda\frac{ds}{dt} = -\lambda l\frac{d\theta}{dt}$$

where λ is a viscous drag coefficient. Show that for small angles, Newton's equation is

$$ml\frac{d^2\theta}{dt^2} + \lambda l\frac{d\theta}{dt} + mg\theta = 0$$

Show that there is no harmonic motion if

$$\lambda^2 > \frac{4m^2g}{l}$$

Does it make physical sense that the medium can be so viscous that the pendulum undergoes no harmonic motion?

Now we have both the force of gravity and the viscous force acting on the system, so, again by Newton's Law,

$$ml\frac{d^2\theta}{dt^2} = -mg\sin\theta - \lambda l\frac{d\theta}{dt}$$

$$ml\frac{d^2\theta}{dt^2} + mg\sin\theta + \lambda l\frac{d\theta}{dt} = 0$$

For small angles $\sin\theta \approx \theta$, so

$$\frac{d^2\theta}{dt^2} + \frac{\lambda}{m}\frac{d\theta}{dt} + \frac{g}{l}\theta = 0 \tag{a}$$

Substituting $\theta(t) = e^{\alpha t}$ into (a) and dividing through by $\theta(t)$ gives

$$\alpha^2 + \frac{\lambda}{m}\alpha + \frac{g}{l} = 0$$

Solving for α gives

$$\alpha = -\frac{\lambda}{2m} \pm \frac{\left(\frac{\lambda^2}{m^2} - \frac{4g}{l}\right)^{1/2}}{2} = -\frac{\lambda}{2m} \pm \left(\frac{\lambda^2}{4m^2} - \frac{g}{l}\right)^{1/2}$$

and so the solution to the differential equation is (Problem 2–6)

$$\theta(t) = e^{-\lambda t/2m}\left(c_1 e^{\beta t} + c_2 e^{\beta t}\right)$$

where $\beta = \left(\dfrac{\lambda^2}{4m^2} - \dfrac{g}{l} \right)^{1/2}$. If $\lambda^2 < 4m^2 g/l$, then β is imaginary and the motion is harmonic. However, if $\lambda^2 \geq 4m^2 g/l$, then β is real and there is no harmonic motion. The viscosity is so large that the pendulum simply approaches its vertical position without oscillating.

2–20. Consider two pendulums of equal lengths and masses that are connected by a spring that obeys Hooke's law (Problem 2–7). This system is shown in Figure 2.11. Assuming that the motion takes place in a plane and that the angular displacement of each pendulum from the vertical is small, show that the equations of motion for this system are

$$m \frac{d^2 x}{dt^2} = -m\omega_0^2 x - k(x - y)$$

$$m \frac{d^2 y}{dt^2} = -m\omega_0^2 y - k(y - x)$$

where ω_0 is the natural vibrational frequency of each isolated pendulum, [i.e., $\omega_0 = (g/l)^{1/2}$] and k is the force constant of the connecting spring. In order to solve these two simultaneous differential equations, assume that the two pendulums swing harmonically and so try

$$x(t) = A e^{i\omega t} \qquad y(t) = B e^{i\omega t}$$

Substitute these expressions into the two differential equations and obtain

$$\left(\omega^2 - \omega_0^2 - \frac{k}{m} \right) A = -\frac{k}{m} B$$

$$\left(\omega^2 - \omega_0^2 - \frac{k}{m} \right) B = -\frac{k}{m} A$$

Now we have two simultaneous linear homogeneous algebraic equations for the two amplitudes A and B. We shall learn in MathChapter E that the determinant of the coefficients must vanish in order for there to be a nontrivial solution. Show that this condition gives

$$\left(\omega^2 - \omega_0^2 - \frac{k}{m} \right)^2 = \left(\frac{k}{m} \right)^2$$

Now show that there are two natural frequencies for this system, namely,

$$\omega_1^2 = \omega_0^2 \quad \text{and} \quad \omega_2^2 = \omega_0^2 + \frac{2k}{m}$$

Interpret the motion associated with these frequencies by substituting ω_1^2 and ω_2^2 back into the two equations for A and B. The motion associated with these values of A and B are called *normal modes* and any complicated, general motion of this system can be written as a linear combination of these normal modes. Notice that there are two coordinates (x and y) in this problem and two

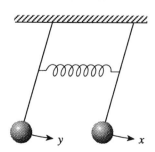

FIGURE 2.11
Two pendulums coupled by a spring that obeys Hooke's law.

normal modes. We shall see in Chapter 13 that the complicated vibrational motion of molecules can be resolved into a linear combination of natural, or normal, modes.

The quantities $x(t)$ and $y(t)$ are the distances along the arcs in Figure 2.11. In terms of angles (see Problem 2–18), we can write

$$m\frac{d^2x}{dt^2} = -mg\theta_x - k(x-y)$$

where $x = l\theta_x$, $y = l\theta_y$, and we assume small displacements from the vertical. In terms of x and y we then have

$$m\frac{d^2x}{dt^2} = -\frac{mg}{l}x - k(x-y) = -m\omega_0^2 x - k(x-y)$$

$$m\frac{d^2y}{dt^2} = -\frac{mg}{l}y - k(y-x) = -m\omega_0^2 y - k(y-x)$$

which are the equations of motion in the problem [we have taken $\omega_0 = (g/l)^{1/2}$]. Notice that the force due to the connecting spring acts in opposite directions in the equations for x and y. Substituting $x(t) = Ae^{i\omega t}$ and $y(t) = Be^{i\omega t}$ into these equations, we find

$$-m\omega^2 Ae^{i\omega t} = -Am\omega_0^2 e^{i\omega t} - k(A-B)e^{i\omega t}$$

$$\left(\omega^2 - \omega_0^2 - \frac{k}{m}\right)A = -\frac{k}{m}B$$

$$-m\omega^2 Be^{i\omega t} = -Bm\omega_0^2 e^{i\omega t} - k(B-A)e^{i\omega t}$$

$$\left(\omega^2 - \omega_0^2 - \frac{k}{m}\right)B = -\frac{k}{m}A$$

One method of solving these coupled equations involves using determinants (MathChapter E). Specifically, the determinant of the coefficients of A and B must vanish or

$$\begin{vmatrix} \omega^2 - \omega_0^2 - \dfrac{k}{m} & \dfrac{k}{m} \\ \dfrac{k}{m} & \omega^2 - \omega_0^2 - \dfrac{k}{m} \end{vmatrix} = \left(\omega^2 - \omega_0^2 - \frac{k}{m}\right)^2 - \left(\frac{k}{m}\right)^2 = 0$$

This is true if

$$\omega_1^2 = \omega_0^2 \quad \text{or} \quad \omega_2^2 = \omega_0^2 + 2\frac{k}{m}$$

In the first case $\omega_1 = \omega_0$, and the two pendulums swing in phase. In the second case, they swing against each other, 180° out of phase.

2–21. Problem 2–20 can be solved by introducing center-of-mass and relative coordinates (cf. Section 5–2). Add and subtract the differential equations for $x(t)$ and $y(t)$ and then introduce the new variables

$$\eta = x + y \quad \text{and} \quad \xi = x - y$$

Show that the differential equations for η and ξ are independent. Solve each one and compare your results to those of Problems 2–20.

The two coupled differential equations from Problem 2–20 are

$$m\frac{d^2x}{dt^2} = -m\omega_0^2 x - k(x - y)$$

$$m\frac{d^2y}{dt^2} = -m\omega_0^2 y - k(y - x)$$

Adding these two equations gives

$$\frac{d^2}{dt^2}(x + y) + \omega_0^2(x + y) = 0$$

Subtracting these two equations gives

$$\frac{d^2}{dt^2}(x - y) + \omega_0^2(x - y) + \frac{2k}{m}(x - y) = 0$$

Using $\eta = x + y$ and $\xi = x - y$, we obtain independent equations for η and ξ,

$$\frac{d^2\eta}{dt^2} + \omega_0^2\eta = 0$$

$$\frac{d^2\xi}{dt^2} + \left(\omega_0^2 + \frac{2k}{m}\right)\xi = 0$$

Note that these equations are equations for harmonic motion of frequency ω_0 and $(\omega_0^2 + \frac{2k}{m})^{1/2}$, in agreement with the results of Problem 2–20.

Probability and Statistics

PROBLEMS AND SOLUTIONS

B–1. Consider a particle to be constrained to lie along a one-dimensional segment 0 to a. We will learn in the next chapter that the probability that the particle is found to lie between x and $x + dx$ is given by

$$p(x)dx = \frac{2}{a} \sin^2 \frac{n\pi x}{a} dx$$

where $n = 1, 2, 3, \ldots$. First show that $p(x)$ is normalized. Now calculate the average position of the particle along the line segment. The integrals that you need are (*The CRC Handbook of Chemistry and Physics* or *The CRC Standard Mathematical Tables*, CRC Press)

$$\int \sin^2 \alpha x dx = \frac{x}{2} - \frac{\sin 2\alpha x}{4\alpha}$$

and

$$\int x \sin^2 \alpha x dx = \frac{x^2}{4} - \frac{x \sin 2\alpha x}{4\alpha} - \frac{\cos 2\alpha x}{8\alpha^2}$$

If $p(x)$ is normalized, then $\int_0^a p(x)dx = 1$.

$$\int_0^a p(x)dx = \int_0^a \frac{2}{a} \sin^2 \frac{n\pi x}{a} dx$$

$$= \left[\frac{2}{a} \left(\frac{x}{2} - \frac{\sin 2n\pi a^{-1}x}{4n\pi a^{-1}} \right) \right]_0^a$$

$$= \frac{2}{a} \left[\frac{a}{2} - \frac{\sin 2n\pi}{4n\pi a^{-1}} - 0 + \frac{\sin 0}{4n\pi a^{-1}} \right]$$

$$= \frac{2}{a} \left(\frac{a}{2} \right) = 1$$

Thus, $p(x)$ is normalized. To find the average position of the particle along the line segment, use Equation B.12:

$$\langle x \rangle = \int_0^a x p(x)dx = \int_0^a x \frac{2}{a} \sin^2 \frac{n\pi x}{a} dx$$

$$= \frac{2}{a} \left[\frac{x^2}{4} - \frac{x \sin 2n\pi a^{-1}x}{4n\pi a^{-1}} - \frac{\cos 2n\pi a^{-1}x}{8n^2\pi^2 a^{-2}} \right]_0^a$$

$$= \frac{2}{a} \left[\frac{a^2}{4} - \frac{a \sin 2n\pi}{4n\pi a^{-1}} - \frac{\cos 2n\pi}{8n^2\pi^2 a^{-2}} + \frac{\cos 0}{8n^2\pi^2 a^{-2}} \right]$$

$$= \frac{2}{a} \left[\frac{a^2}{4} - \frac{1}{8n^2\pi^2 a^{-2}} + \frac{1}{8n^2\pi^2 a^{-2}} \right] = \frac{2}{a} \left(\frac{a^2}{4} \right)$$

$$= \frac{a}{2}$$

B-2. Calculate the variance associated with the probability distribution given in Problem B–1. The necessary integral is (*CRC tables*)

$$\int x^2 \sin^2 \alpha x\, dx = \frac{x^3}{6} - \left(\frac{x^2}{4\alpha} - \frac{1}{8\alpha^3}\right)\sin 2\alpha x - \frac{x\cos 2\alpha x}{4\alpha^2}$$

Use Equation B.13:

$$\langle x^2\rangle = \int_0^a x^2 p(x)dx = \frac{2}{a}\int_0^a x^2 \sin^2\frac{n\pi x}{a}dx$$

$$= \frac{2}{a}\left[\frac{x^3}{6} - \left(\frac{x^2}{4n\pi a^{-1}} - \frac{1}{8n^3\pi^3 a^{-3}}\right)\sin 2n\pi a^{-1}x - \frac{x\cos 2n\pi a^{-1}x}{4n^2\pi^2 a^{-2}}\right]_0^a$$

$$= \frac{2}{a}\left[\frac{a^3}{6} - \left(\frac{a^2}{4n\pi a^{-1}} - \frac{1}{8n^3\pi^3 a^{-3}}\right)\sin 2n\pi - \frac{a\cos 2n\pi}{4n^2\pi^2 a^{-2}} - 0\right]$$

$$= \frac{2}{a}\left(\frac{a^3}{6} - \frac{a^3}{4n^2\pi^2}\right)$$

$$= \frac{a^2}{3} - \frac{a^2}{2n^2\pi^2}$$

The variance σ^2 is given by

$$\sigma^2 = \langle x^2\rangle - \langle x\rangle^2 \tag{B.8}$$

Using the result of Problem B–1 and the above result for $\langle x^2\rangle$ gives

$$\sigma^2 = \frac{a^2}{3} - \frac{a^2}{2n^2\pi^2} - \frac{a^2}{4}$$

$$= \frac{a^2}{12} - \frac{a^2}{2n^2\pi^2}$$

B-3. Using the probability distribution given in Problem B–1, calculate the probability that the particle will be found between 0 and $a/2$. The necessary integral is given in Problem B–1.

The probability that the particle will lie within the region 0 to $a/2$ is given by $\int_0^{a/2} p(x)dx$ (Equation B.10).

$$\int_0^{a/2} p(x)dx = \int_0^{a/2} \frac{2}{a}\sin^2\frac{n\pi x}{a}dx$$

$$= \frac{2}{a}\int_0^{a/2}\sin^2\frac{n\pi x}{a}dx$$

$$= \frac{2}{a}\left[\frac{x}{2} - \frac{\sin 2n\pi a^{-1}x}{4n\pi a^{-1}}\right]_0^{a/2}$$

$$= \frac{2}{a}\left[\frac{a}{4} - \frac{\sin 2n\pi}{8n\pi a^{-1}} + \frac{\sin 0}{4n\pi a^{-1}}\right]$$

$$= \frac{2}{a}\left(\frac{a}{4}\right) = \frac{1}{2}$$

The probability of the particle being found in exactly half the box is 0.5.

B–4. Prove explicitly that

$$\int_{-\infty}^{\infty} e^{-\alpha x^2}\,dx = 2\int_{0}^{\infty} e^{-\alpha x^2}\,dx$$

by breaking the integral from $-\infty$ to ∞ into one from $-\infty$ to 0 and another from 0 to ∞. Now let $z = -x$ in the first integral and $z = x$ in the second to prove the above relation.

$$\int_{-\infty}^{\infty} e^{-\alpha x^2}\,dx = \int_{-\infty}^{0} e^{-\alpha x^2}\,dx + \int_{0}^{\infty} e^{-\alpha x^2}\,dx$$

We can let $z = -x$ in the first integral and $z = x$ in the second and write

$$\int_{-\infty}^{\infty} e^{-\alpha x^2}\,dx = -\int_{\infty}^{0} e^{-\alpha z^2}\,dz + \int_{0}^{\infty} e^{-\alpha x^2}\,dx$$

$$= \int_{0}^{\infty} e^{-\alpha z^2}\,dz + \int_{0}^{\infty} e^{-\alpha z^2}\,dz$$

$$= 2\int_{0}^{\infty} e^{-\alpha z^2}\,dz = 2\int_{0}^{\infty} e^{-\alpha x^2}\,dx$$

B–5. By using the procedure in Problem B–4, show explicitly that

$$\int_{-\infty}^{\infty} xe^{-\alpha x^2}\,dx = 0$$

$$\int_{-\infty}^{\infty} xe^{-\alpha x^2}\,dx = \int_{-\infty}^{0} xe^{-\alpha x^2}\,dx + \int_{0}^{\infty} xe^{-\alpha x^2}\,dx$$

We can let $x = -z$ in the first integral to get

$$\int_{-\infty}^{\infty} xe^{-\alpha x^2}\,dx = \int_{\infty}^{0} ze^{-\alpha z^2}\,dz + \int_{0}^{\infty} xe^{-\alpha x^2}\,dx$$

$$= -\int_{0}^{\infty} ze^{-\alpha z^2}\,dz + \int_{0}^{\infty} xe^{-\alpha x^2}\,dx$$

$$= -\int_{0}^{\infty} xe^{-\alpha x^2}\,dx + \int_{0}^{\infty} xe^{-\alpha x^2}\,dx$$

$$= 0$$

B–6. We will learn in Chapter 25 that the molecules in a gas travel at various speeds, and that the probability that a molecule has a speed between v and $v + dv$ is given by

$$p(v)\,dv = 4\pi \left(\frac{m}{2\pi k_{\mathrm{B}} T}\right)^{3/2} v^2 e^{-mv^2/2k_{\mathrm{B}} T}\,dv \qquad 0 \le v < \infty$$

where m is the mass of the particle, k_{B} is the Boltzmann constant (the molar gas constant R divided by the Avogadro constant), and T is the Kelvin temperature. The probability distribution of molecular speeds is called the Maxwell-Boltzmann distribution. First show that $p(v)$ is normalized, and then determine the average speed as a function of temperature. The necessary integrals are (*CRC tables*)

$$\int_{0}^{\infty} x^{2n} e^{-\alpha x^2}\,dx = \frac{1\cdot 3\cdot 5\cdots(2n-1)}{2^{n+1}\alpha^n}\left(\frac{\pi}{\alpha}\right)^{1/2} \qquad n \ge 1$$

and

$$\int_0^\infty x^{2n+1} e^{-\alpha x^2} dx = \frac{n!}{2\alpha^{n+1}}$$

where $n!$ is n factorial, or $n! = n(n-1)(n-2)\cdots(1)$.

First, we demonstrate that $p(v)$ is normalized by showing that $\int_0^\infty p(v)dv = 1$:

$$\int_0^\infty p(v)dv = 4\pi \left(\frac{m}{2\pi k_B T}\right)^{3/2} \int_0^\infty v^2 e^{-mv^2/2k_B T} dv$$

$$= 4\pi \left(\frac{m}{2\pi k_B T}\right)^{3/2} \frac{2k_B T}{4m} \left(\frac{2\pi k_B T}{m}\right)^{1/2}$$

$$= \pi^{3/2} \left(\frac{m}{2\pi k_B T}\right)^{3/2} \left(\frac{2k_B T}{m}\right)^{3/2}$$

$$= 1$$

Using Equation B.12, we write

$$\langle v \rangle = 4\pi \left(\frac{m}{2\pi k_B T}\right)^{3/2} \int_0^\infty v^3 e^{-mv^2/2k_B T} dv$$

$$= 4\pi \left(\frac{m}{2\pi k_B T}\right)^{3/2} \left[2\left(\frac{m}{2k_B T}\right)^2\right]^{-1}$$

$$= 2\pi \left(\frac{m}{2\pi k_B T}\right)^{3/2} \left(\frac{2k_B T}{m}\right)^2$$

$$= \left(\frac{8k_B T}{\pi m}\right)^{1/2}$$

B–7. Use the Maxwell-Boltzmann distribution in Problem B–6 to determine the average kinetic energy of a gas-phase molecule as a function of temperature. The necessary integral is given in Problem B–6.

Kinetic energy, KE, is defined as $KE = \frac{1}{2}mv^2$, so $\langle KE \rangle = \frac{1}{2}m\langle v^2 \rangle$. Using Equation B.13, we write

$$\langle v^2 \rangle = 4\pi \left(\frac{m}{2\pi k_B T}\right)^{3/2} \int_0^\infty v^4 e^{-mv^2/2k_B T} dv$$

$$= 4\pi \left(\frac{m}{2\pi k_B T}\right)^{3/2} \frac{3}{8} \left(\frac{2k_B T}{m}\right)^2 \left(\frac{2\pi k_B T}{m}\right)^{1/2}$$

$$= \frac{3k_B T}{m}$$

And so $E = \frac{1}{2}m\langle v^2 \rangle = \frac{3}{2}k_B T$.

CHAPTER 3

The Schrödinger Equation and a Particle in a Box

PROBLEMS AND SOLUTIONS

3–1. Evaluate $g = \hat{A} f$, where $\hat{A}$ and f are given below:

$\hat{A}$	f
(a) SQRT	x^4
(b) $\dfrac{d^3}{dx^3} + x^3$	e^{-ax}
(c) $\displaystyle\int_0^1 dx$	$x^3 - 2x + 3$
(d) $\dfrac{\partial^2}{\partial x^2} + \dfrac{\partial^2}{\partial y^2} + \dfrac{\partial^2}{\partial z^2}$	$x^3 y^2 z^4$

a. $\text{SQRT}(x^4) = \pm x^2$

b. $\dfrac{d^3 e^{-ax}}{dx^3} + x^3 e^{-ax} = -a^3 e^{-ax} + x^3 e^{-ax} = e^{-ax}\left(x^3 - a^3\right)$

c. $\displaystyle\int_0^1 \left(x^3 - 2x + 3\right) dx = \left.\dfrac{x^4}{4} - x^2 + 3x\right|_0^1 = \dfrac{9}{4}$

d. $\dfrac{\partial^2(x^3 y^2 z^4)}{\partial x^2} + \dfrac{\partial^2(x^3 y^2 z^4)}{\partial y^2} + \dfrac{\partial^2(x^3 y^2 z^4)}{\partial z^2} = 6xy^2 z^4 + 2x^3 z^4 + 12x^3 y^2 z^2$

3–2. Determine whether the following operators are linear or nonlinear:

a. $\hat{A} f(x) = \text{SQR} f(x)$ [square $f(x)$]

b. $\hat{A} f(x) = f^*(x)$ [form the complex conjugate of $f(x)$]

c. $\hat{A} f(x) = 0$ [multiply $f(x)$ by zero]

d. $\hat{A} f(x) = [f(x)]^{-1}$ [take the reciprocal of $f(x)$]

e. $\hat{A} f(x) = f(0)$ [evaluate $f(x)$ at $x = 0$]

f. $\hat{A} f(x) = \ln f(x)$ [take the logarithm of $f(x)$]

An operator $\hat{A}$ is linear if $\hat{A}\left[c_1 f_1(x) + c_2 f_2(x)\right] = c_1 \hat{A} f_1(x) + c_2 \hat{A} f_2(x)$ (Equation 3.9).

a.
$$\hat{A}\left[c_1 f_1(x) + c_2 f_2(x)\right] = \left[c_1 f_1(x) + c_2 f_2(x)\right]^2$$
$$= c_1^2 f_1(x)^2 + 2c_1 f_1(x) c_2 f_2(x) + c_2^2 f_2(x)^2$$
$$c_1 \hat{A} f_1(x) + c_2 \hat{A} f_2(x) = c_1 \left[f_1(x)\right]^2 + c_2 \left[f_2(x)\right]^2$$
$$\neq \hat{A}\left[c_1 f_1(x) + c_2 f_2(x)\right]$$

Nonlinear.

b.
$$\hat{A}\left[c_1 f_1(x) + c_2 f_2(x)\right] = c_1^* f_1^*(x) + c_2^* f_2^*(x)$$
$$c_1 \hat{A} f_1(x) + c_2 \hat{A} f_2(x) = c_1 f_1^*(x) + c_2 f_2^*(x)$$
$$\neq \hat{A}\left[c_1 f_1(x) + c_2 f_2(x)\right]$$

Nonlinear.

c.
$$\hat{A}\left[c_1 f_1(x) + c_2 f_2(x)\right] = 0$$
$$c_1 \hat{A} f_1(x) + c_2 \hat{A} f_2(x) = c_1 f_1(x)0 + c_2 f_2(x) = 0$$
$$= \hat{A}\left[c_1 f_1(x) + c_2 f_2(x)\right]$$

Linear.

d.
$$\hat{A}\left[c_1 f_1(x) + c_2 f_2(x)\right] = \left[c_1 f_1(x) + c_2 f_2(x)\right]^{-1}$$
$$c_1 \hat{A} f_1(x) + c_2 \hat{A} f_2(x) = \frac{c_1}{f_1(x)} + \frac{c_2}{f_2(x)}$$
$$\neq \hat{A}\left[c_1 f_1(x) + c_2 f_2(x)\right]$$

Nonlinear.

e.
$$\hat{A}\left[c_1 f_1(x) + c_2 f_2(x)\right] = c_1 f_1(0) + c_2 f_2(0)$$
$$= c_1 \hat{A} f_1(x) + c_2 \hat{A} f_2(x)$$

Linear.

f.
$$\hat{A}\left[c_1 f_1(x) + c_2 f_2(x)\right] = \ln\left[c_1 f_1(x) + c_2 f_2(x)\right]$$
$$c_1 \hat{A} f_1(x) + c_2 \hat{A} f_2(x) = c_1 \ln f_1(x) + c_2 \ln f_2(x)$$
$$\neq \hat{A}\left[c_1 f_1(x) + c_2 f_2(x)\right]$$

Nonlinear.

3–3. In each case, show that $f(x)$ is an eigenfunction of the operator given. Find the eigenvalue.

$\hat{A}$	$f(x)$
(a) $\dfrac{d^2}{dx^2}$	$\cos \omega x$
(b) $\dfrac{d}{dt}$	$e^{i\omega t}$
(c) $\dfrac{d^2}{dx^2} + 2\dfrac{d}{dx} + 3$	$e^{\alpha x}$
(d) $\dfrac{\partial}{\partial y}$	$x^2 e^{6y}$

a. $\hat{A}f(x) = \dfrac{d^2(\cos \omega x)}{dx^2} = -\omega^2 \cos \omega x; \quad \text{eigenvalue} = -\omega^2$

b. $\hat{A}f(x) = \dfrac{d(e^{i\omega t})}{dt} = i\omega e^{i\omega t}; \quad \text{eigenvalue} = i\omega$

c. $\hat{A}f(x) = \dfrac{d^2(e^{\alpha x})}{dx^2} + 2\dfrac{d(e^{\alpha x})}{dx} + 3e^{\alpha x} = \left(\alpha^2 + 2\alpha + 3\right)e^{\alpha x}; \quad \text{eigenvalue} = \alpha^2 + 2\alpha + 3$

d. $\hat{A}f(x) = \dfrac{\partial x^2 e^{6y}}{\partial y} = 6x^2 e^{6y}; \quad \text{eigenvalue} = 6$

3–4. Show that $(\cos ax)(\cos by)(\cos cz)$ is an eigenfunction of the operator,

$$\nabla^2 = \frac{\partial^2}{\partial x^2} + \frac{\partial^2}{\partial y^2} + \frac{\partial^2}{\partial z^2}$$

which is called the Lapacian operator.

$$\nabla^2(\cos ax)(\cos by)(\cos cz) = \frac{\partial^2(\cos ax)(\cos by)(\cos cz)}{\partial x^2} + \frac{\partial^2(\cos ax)(\cos by)(\cos cz)}{\partial y^2}$$

$$+ \frac{\partial^2(\cos ax)(\cos by)(\cos cz)}{\partial z^2}$$

$$= -a^2(\cos ax)(\cos by)(\cos cz) - b^2(\cos ax)(\cos by)(\cos cz)$$

$$-c^2(\cos ax)(\cos by)(\cos cz)$$

$$= -(a^2 + b^2 + c^2)(\cos ax)(\cos by)(\cos cz)$$

The eigenvalue of the eigenfunction $(\cos ax)(\cos by)(\cos cz)$ is $-(a^2 + b^2 + c^2)$.

3–5. Write out the operator $\hat{A}^2$ for $\hat{A} =$

a. $\dfrac{d^2}{dx^2}$
 b. $\dfrac{d}{dx} + x$
 c. $\dfrac{d^2}{dx^2} - 2x\dfrac{d}{dx} + 1$

Hint: Be sure to include $f(x)$ before carrying out the operations.

For all cases, we need to determine an expression for $\hat{A}^2$ where

$$\hat{A}^2 f(x) = \hat{A}\left[\hat{A} f(x)\right]$$

a. $\hat{A}[\hat{A} f(x)] = \dfrac{d^2}{dx^2}\left[\dfrac{d^2 f(x)}{dx^2}\right] = \dfrac{d^4 f(x)}{dx^4}$

The operator $\hat{A}^2$ is then

$$\hat{A}^2 = \frac{d^4}{dx^4}$$

b.
$$\hat{A}\left[\hat{A} f(x)\right] = \left(\frac{d}{dx} + x\right)\left[\frac{df(x)}{dx} + xf(x)\right]$$

$$= \frac{d^2 f(x)}{dx^2} + x\frac{df(x)}{dx} + f(x)\frac{dx}{dx} + x\frac{df(x)}{dx} + x^2 f(x)$$

$$= \frac{d^2 f(x)}{dx^2} + 2x\frac{df(x)}{dx} + f(x) + x^2 f(x)$$

So $\hat{A}^2$ is written as

$$\hat{A}^2 = \frac{d^2}{dx^2} + 2x\frac{d}{dx} + 1 + x^2$$

c.
$$\hat{A}\left[\hat{A} f(x)\right] = \left(\frac{d^2}{dx^2} - 2x\frac{d}{dx} + 1\right)\left[\frac{d^2 f(x)}{dx^2} - 2x\frac{df(x)}{dx} + f(x)\right]$$

$$= \frac{d^4 f(x)}{dx^4} - 4x\frac{d^3 f(x)}{dx^3} + (4x^2 - 2)\frac{d^2 f(x)}{dx^2} + f(x)$$

So $\hat{A}^2$ is written as

$$\hat{A}^2 = \frac{d^4}{dx^4} - 4x\frac{d^3}{dx^3} + (4x^2 - 2)\frac{d^2}{dx^2} + 1$$

3–6. In Section 3–5, we applied the equations for a particle in a box to the π electrons in butadiene. This simple model is called the free-electron model. Using the same argument, show that the length of hexatriene can be estimated to be 867 pm. Show that the first electronic transition is predicted to occur at 2.8×10^4 cm^{-1}. (Remember that hexatriene has six π electrons.)

We assume that the π electrons move along a straight line consisting of three C=C bond lengths (3×135 pm), two C–C bond lengths (2×154 pm), and the distance of a carbon atom radius at each end (2×77.0 pm) or a total distance of 867 pm. Because there are six π electrons in one molecule, the first electronic transition occurs between the $n = 3$ and the $n = 4$ electronic states. Using Equation 3.21, the energy of this transition is

$$\Delta E = \frac{h^2}{8m_e a^2}(4^2 - 3^2)$$

$$= \frac{(6.626 \times 10^{-34} \text{ J·s})^2}{8(9.11 \times 10^{-31} \text{ kg})(867 \times 10^{-12} \text{ m})^2}(16 - 9)$$

$$= 5.61 \times 10^{-19} \text{ J} = 2.82 \times 10^4 \text{ cm}^{-1}$$

3–7. Prove that if $\psi(x)$ is a solution to the Schrödinger equation, then any constant times $\psi(x)$ is also a solution.

Because $\hat{H}$ is a linear operator,

$$\hat{H}[c\psi(x)] = c\hat{H}\psi(x) = cE\psi(x) = E[c\psi(x)]$$

and so $c\psi(x)$, where c is any constant, is a solution.

3–8. Show that the probability associated with the state ψ_n for a particle in a one-dimensional box of length a obeys the following relationships:

$$\text{Prob}(0 \leq x \leq a/4) = \text{Prob}(3a/4 \leq x \leq a) = \begin{cases} \dfrac{1}{4} & n \text{ even} \\[2mm] \dfrac{1}{4} - \dfrac{(-1)^{\frac{n-1}{2}}}{2\pi n} & n \text{ odd} \end{cases}$$

and

$$\text{Prob}(a/4 \leq x \leq a/2) = \text{Prob}(a/2 \leq x \leq 3a/4) = \begin{cases} \dfrac{1}{4} & n \text{ even} \\[2mm] \dfrac{1}{4} + \dfrac{(-1)^{\frac{n-1}{2}}}{2\pi n} & n \text{ odd} \end{cases}$$

For a particle in a one-dimensional box of length a, we know that (Equation 3.27) $\psi_n = (2/a)^{1/2}\sin(n\pi x/a)$. Now

$$\text{Prob}(c \leq x \leq d) = \int_c^d \psi^*(x)\psi(x)dx$$

$$= \frac{2}{a}\int_c^d \sin^2 \frac{n\pi x}{a}dx$$

$$= \frac{2}{a}\left[\frac{x}{2} - \frac{\sin 2n\pi x/a}{4(n\pi/a)}\right]_{x=c}^{x=d}$$

$$= \left[\frac{x}{a} - \frac{\sin 2n\pi x/a}{2n\pi}\right]_{x=c}^{x=d}$$

$$= \frac{d}{a} - \frac{\sin 2n\pi d/a}{2n\pi} - \frac{c}{a} + \frac{\sin 2n\pi c/a}{2n\pi}$$

$$= \frac{d-c}{a} - \left(\frac{1}{2n\pi}\right)\left[\sin\left(\frac{2n\pi d}{a}\right) - \sin\left(\frac{2n\pi c}{a}\right)\right]$$

For all regions under consideration, $d - c = a/4$. We can now calculate the probability associated with ψ_n for each of the four regions. (Recall that $\sin n\pi = \sin 2n\pi = 0$ for integer n.)

c	d	Prob($c \leq x \leq d$)
0	$\dfrac{a}{4}$	$\dfrac{1}{4} - \dfrac{1}{2n\pi}\sin\left(\dfrac{n\pi}{2}\right)$
$\dfrac{a}{4}$	$\dfrac{a}{2}$	$\dfrac{1}{4} - 0 + \dfrac{1}{2n\pi}\sin\left(\dfrac{n\pi}{2}\right)$
$\dfrac{a}{2}$	$\dfrac{3a}{4}$	$\dfrac{1}{4} - \dfrac{1}{2n\pi}\sin\left(\dfrac{3n\pi}{2}\right) + 0$
$\dfrac{3a}{4}$	a	$\dfrac{1}{4} - 0 + \dfrac{1}{2n\pi}\sin\left(\dfrac{3n\pi}{2}\right)$

For n even, $\sin\dfrac{n\pi}{2} = 0$; for n odd, $\sin\dfrac{n\pi}{2} = (-1)^{(n-1)/2}$ and $\sin\dfrac{3n\pi}{2} = -(-1)^{(n-1)/2}$. Therefore

$$\text{Prob}(0 \leq x \leq a/4) = \text{Prob}(3a/4 \leq x \leq a) = \begin{cases} \dfrac{1}{4} & n \text{ even} \\[2mm] \dfrac{1}{4} - \dfrac{(-1)^{(n-1)/2}}{2\pi n} & n \text{ odd} \end{cases}$$

and

$$\text{Prob}(a/4 \leq x \leq a/2) = \text{Prob}(a/2 \leq x \leq 3a/4) = \begin{cases} \dfrac{1}{4} & n \text{ even} \\[2mm] \dfrac{1}{4} + \dfrac{(-1)^{(n-1)/2}}{2\pi n} & n \text{ odd} \end{cases}$$

3–9. What are the units, if any, for the wave function of a particle in a one-dimensional box?

The normalized wavefunction $\psi(x)$ must be unitless. The normalization constant for $\psi(x)$ is $\sqrt{2/a}$ (from Equation 3.27); therefore, the wavefunction must have units of $m^{-1/2}$.

3–10. Using a table of integrals, show that

$$\int_0^a \sin^2\frac{n\pi x}{a}\,dx = \frac{a}{2}$$

$$\int_0^a x\sin^2\frac{n\pi x}{a}\,dx = \frac{a^2}{4}$$

and

$$\int_0^a x^2\sin^2\frac{n\pi x}{a}\,dx = \left(\frac{a}{2\pi n}\right)^3\left(\frac{4\pi^3 n^3}{3} - 2n\pi\right)$$

All these integrals can be evaluated from

$$I(\beta) = \int_0^a e^{\beta x}\sin^2\frac{n\pi x}{a}\,dx$$

Show that the above integrals are given by $I(0)$, $I'(0)$, and $I''(0)$, respectively, where the primes denote differentiation with respect to β. Using a table of integrals, evaluate $I(\beta)$ and then the above three integrals by differentiation.

In MathChapter B, Problems B–1 and B–2, we evaluated the integrals $\int_0^a p(x)dx$ and $\int_0^a xp(x)dx$ where $p(x)$ was given by

$$p(x)dx = \frac{2}{a}\sin^2\frac{n\pi x}{a}dx$$

Using the results from Problems B–1 and B–2 gives the numerical results for the integrals considered in the first part of the problem. Now consider

$$I(\beta) = \int_0^a e^{\beta x}\sin^2\frac{n\pi x}{a}dx$$

Taking the first and second derivatives of $I(\beta)$ with respect to β gives

$$I'(\beta) = \int_0^a xe^{\beta x}\sin^2\frac{n\pi x}{a}dx$$

$$I''(\beta) = \int_0^a x^2e^{\beta x}\sin^2\frac{n\pi x}{a}dx$$

The corresponding expressions for $I(0)$, $I'(0)$, and $I''(0)$ are

$$I(0) = \int_0^a \sin^2\frac{n\pi x}{a}dx$$

$$I'(0) = \int_0^a x\sin^2\frac{n\pi x}{a}dx$$

$$I''(0) = \int_0^a x^2\sin^2\frac{n\pi x}{a}dx$$

Generally, from a table of integrals, we can write

$$\int e^{\beta x}\sin^2 bxdx = \frac{e^{\beta x}}{2\beta} - \frac{e^{\beta x}}{\beta^2 + 4b^2}\left(\frac{\beta\cos 2bx}{2} + b\sin 2bx\right)$$

and so

$$I(\beta) = \frac{e^{\beta a} - 1}{2\beta} - \frac{\beta}{2}\left(\frac{e^{\beta a} - 1}{\beta^2 + 4n^2\pi^2a^{-2}}\right)$$

Now we use the Maclaurin series and the series expansion of $e^{\beta a}$ (MathChapter J):

$$I(\beta) = I(0) + \beta I'(0) + \frac{\beta^2}{2}I''(0) + O(\beta^3)$$

$$I(\beta) = \frac{a}{2} + \frac{a^2}{4}\beta + \frac{a^3}{12}\beta^2 - \frac{1}{4n^2\pi^2a^{-2}}\left(\frac{\beta^2 a}{2} + \frac{\beta^3 a^2}{4}\right) + O(\beta^3)$$

$$= \frac{a}{2} + \frac{a^2}{4}\beta + \left(\frac{a^3}{6} - \frac{a^3}{4\pi^2n^2}\right)\frac{\beta^2}{2} + O(\beta^3)$$

Therefore,

$$I(0) = \frac{a}{2} \qquad I'(0) = \frac{a^2}{4} \qquad I''(0) = \frac{a^3}{6} - \frac{a^3}{4\pi^2n^2}$$

3–11. Show that

$$\langle x \rangle = \frac{a}{2}$$

for all the states of a particle in a box. Is this result physically reasonable?

$$\langle x \rangle = \int_0^a x \psi^*(x) \psi(x) dx$$

$$= \frac{2}{a} \int_0^a x \sin^2 \frac{n\pi x}{a} dx$$

$$= \frac{2}{a} \cdot \frac{a^2}{4} = \frac{a}{2}$$

For any n, $\langle x \rangle = a/2$. This result is physically reasonable and is discussed in detail in the text.

3–12. Show that $\langle p \rangle = 0$ for all states of a one-dimensional box of length a.

From Equation 3.37,

$$\langle p \rangle = \frac{2}{a} \int_0^a \sin \frac{n\pi x}{a} \left(-i\hbar \frac{d}{dx} \right) \sin \frac{n\pi x}{a} dx$$

$$= -\frac{2i\hbar n\pi}{a^2} \int_0^a \sin \frac{n\pi x}{a} \cos \frac{n\pi x}{a} dx$$

$$= 0$$

This result holds for any integer value of n.

3–13. Show that

$$\sigma_x = (\langle x^2 \rangle - \langle x \rangle^2)^{1/2}$$

for a particle in a box is less than a, the width of the box, for any value of n. If σ_x is the uncertainty in the position of the particle, could σ_x ever be larger than a?

Use Equations 3.31 and 3.32 for $\langle x \rangle$ and $\langle x^2 \rangle$:

$$\langle x \rangle = \frac{a}{2} \qquad \langle x^2 \rangle = \frac{a^2}{3} - \frac{a^2}{2n^2\pi^2}$$

$$\sigma_x = (\langle x^2 \rangle - \langle x \rangle^2)^{1/2} = \left(\frac{a^2}{3} - \frac{a^2}{2n^2\pi^2} - \frac{a^2}{4} \right)^{1/2}$$

$$= a \left(\frac{1}{12} - \frac{1}{2\pi^2 n^2} \right)^{1/2}$$

For $\sigma_x \geq a$,

$$\left(\frac{1}{12} - \frac{1}{2\pi^2 n^2} \right)^{1/2} \geq 1$$

$$\frac{\pi^2 n^2 - 6}{12\pi^2 n^2} \geq 1$$

$$\pi^2 n^2 - 6 \geq 12\pi^2 n^2$$

$$\pi^2 n^2 \geq 12\pi^2 n^2 + 6$$

This inequality cannot be satisfied for any value of n, so $\sigma_x < a$ for all n.

3–14. Using the trigonometric identity

$$\sin 2\theta = 2 \sin\theta \cos\theta$$

show that

$$\int_0^a \sin\frac{n\pi x}{a} \cos\frac{n\pi x}{a} dx = 0$$

$$\int_0^a \sin\frac{n\pi x}{a} \cos\frac{n\pi x}{a} dx = \frac{1}{2}\int_0^a \sin\frac{2\pi nx}{a} dx$$

$$= \frac{1}{2}\left(\frac{-a}{2\pi n}\cos\frac{2\pi nx}{a}\right)\Big|_0^a = 0$$

3–15. Prove that

$$\int_0^a e^{\pm i 2\pi nx/a} dx = 0 \qquad n \neq 0$$

$$\int_0^a e^{\pm i 2\pi nx/a} dx = \int_0^a \cos\frac{2\pi nx}{a} dx \pm i \int_0^a \sin\frac{2\pi nx}{a} dx$$

The integrals on the right side of this equation go over complete cycles of the cosine and sine functions, and so both are equal to zero if $n \neq 0$. If $n = 0$, the integral on the left side of this equation is a and the first integral on the right side of this equation is also a.

3–16. Using the trigonometric identity

$$\sin\alpha \sin\beta = \frac{1}{2}\cos(\alpha - \beta) - \frac{1}{2}\cos(\alpha + \beta)$$

show that the particle-in-a-box wave functions (Equations 3.27) satisfy the relation

$$\int_0^a \psi_n^*(x)\psi_m dx = 0 \qquad m \neq n$$

(The asterisk in this case is superfluous because the functions are real.) If a set of functions satisfies the above integral condition, we say that the set is *orthogonal* and, in particular, that $\psi_m(x)$ is orthogonal to $\psi_n(x)$. If, in addition, the functions are normalized, then we say that the set is *orthonormal*.

If $n \neq m$,

$$\int_0^a \psi_n^*(x)\psi_m dx = \frac{2}{a}\int_0^a \sin\frac{n\pi x}{a} \sin\frac{m\pi x}{a} dx$$

$$= \frac{1}{a}\int_0^a \cos\frac{(n-m)\pi x}{a} dx - \frac{1}{a}\int_0^a \cos\frac{(n+m)\pi x}{a} dx$$

$$= \frac{1}{(n-m)\pi} \sin \frac{(n-m)\pi x}{a} \bigg|_0^a - \frac{1}{(n+m)\pi} \sin \frac{(n+m)\pi x}{a} \bigg|_0^a$$

$$= \frac{1}{(n-m)\pi} \sin(n-m)\pi - \frac{1}{(n+m)\pi} \sin(n+m)\pi$$

$$= 0$$

because $\sin N\pi = 0$ for integer values of N. Note that if $n = m$,

$$\int_0^a \psi_n^*(x)\psi_n dx = \frac{1}{a} \int_0^a dx - \frac{1}{a} \int_0^a \cos \frac{2n\pi x}{a} dx$$

$$= 1$$

and so the particle-in-a-box wave functions are orthonormal.

3–17. Prove that the set of functions

$$\psi_n(x) = (2a)^{-1/2} e^{i\pi n x/a} \qquad n = 0, \pm 1, \pm 2, \ldots$$

is orthonormal (cf. Problem 3–16) over the interval $-a \leq x \leq a$. A compact way to express orthonormality in the ψ_n is to write

$$\int_{-a}^a \psi_m^*(x)\psi_n dx = \delta_{mn}$$

The symbol δ_{mn} is called a Kroenecker delta and is defined by

$$\delta_{mn} = 1 \qquad \text{if } m = n$$
$$= 0 \qquad \text{if } m \neq n$$

For $n = m$, we have

$$\int_{-a}^a \psi_m^*(x)\psi_n dx = \frac{1}{2a} \int_{-a}^a e^{-in\pi x/a} e^{in\pi x/a} = \frac{1}{2a} \int_{-a}^a dx = 1$$

For $n \neq m$,

$$\int_{-a}^a \psi_m^*(x)\psi_n dx = \frac{1}{2a} \int_{-a}^a e^{i\pi(n-m)x/a} dx$$

$$= \frac{1}{2a} \int_{-a}^a \left[\cos \frac{\pi(n-m)x}{a} + i \sin \frac{\pi(n-m)x}{a} \right] dx$$

Both integrals here are taken over complete cycles of sine and cosine, and so vanish. This set of functions is therefore orthonormal over the stated interval.

3–18. Show that the set of functions

$$\phi_n(\theta) = (2\pi)^{-1/2} e^{in\theta} \qquad 0 \leq \theta \leq 2\pi$$

is orthonormal (Problem 3–16).

When $n = m$, we have

$$\int_0^{2\pi} \phi_n^*(\theta)\phi_n(\theta) d\theta = \frac{1}{2\pi} \int_0^{2\pi} e^{i(n-n)\theta} d\theta = \frac{1}{2\pi} \int_0^{2\pi} d\theta = 1$$

When $n \neq m$, we have

$$\int_0^{2\pi} \phi_m^*(\theta)\phi_n(\theta)d\theta = \frac{1}{2\pi}\int_0^{2\pi} e^{i(n-m)\theta}d\theta$$

We encountered the above integral in Problem 3–17 and showed that it is equal to zero, so the set of functions ϕ_n is orthonormal.

3–19. In going from Equation 3.34 to 3.35, we multiplied Equation 3.34 from the left by $\psi^*(x)$ and then integrated over all values of x to obtain Equation 3.35. Does it make any difference whether we multiplied from the left or the right?

It does not make a difference whether we multiply from the left or the right. Realize that $\hat{H}\psi_n(x)$ is just a function; $\hat{H}$ has already operated on $\psi_n(x)$.

3–20. Calculate $\langle x \rangle$ and $\langle x^2 \rangle$ for the $n = 2$ state of a particle in a one-dimensional box of length a. Show that

$$\sigma_x = \frac{a}{4\pi}\left(\frac{4\pi^2}{3} - 2\right)^{1/2}$$

We will use the equations for a particle in a one-dimensional box

$$\langle x \rangle = \frac{a}{2} \tag{3.31}$$

$$\langle x^2 \rangle = \frac{a^2}{3} - \frac{a^2}{2n^2\pi^2} \tag{3.32}$$

$$\sigma_x = \frac{a}{2\pi n}\left(\frac{\pi^2 n^2}{3} - 2\right)^{1/2} \tag{3.33}$$

For $n = 2$,

$$\langle x \rangle = \frac{a}{2}$$

$$\langle x^2 \rangle = \frac{a^2}{3} - \frac{a^2}{8\pi^2}$$

$$\sigma_x = \frac{a}{4\pi}\left(\frac{4\pi^2}{3} - 2\right)^{1/2}$$

3–21. Calculate $\langle p \rangle$ and $\langle p^2 \rangle$ for the $n = 2$ state of a particle in a one-dimensional box of length a. Show that

$$\sigma_p = \frac{h}{a}$$

We will use the equations from the chapter

$$\langle p \rangle = 0 \qquad (3.38)$$

$$\langle p^2 \rangle = \frac{n^2 \pi^2 \hbar^2}{a^2} = \sigma_p^2 \qquad (3.40)$$

$$\sigma_p = \frac{n\pi\hbar}{a} = \frac{nh}{2a}$$

For $n = 2$, $\sigma_p = \dfrac{h}{a}$ and $\langle p^2 \rangle = \dfrac{h^2}{a^2}$.

3–22. Consider a particle of mass m in a one-dimensional box of length a. Its average energy is given by

$$\langle E \rangle = \frac{1}{2m} \langle p^2 \rangle$$

Because $\langle p \rangle = 0$, $\langle p^2 \rangle = \sigma_p^2$, where σ_p can be called the uncertainty in p. Using the Uncertainty Principle, show that the energy must be at least as large as $\hbar^2/8ma^2$ because σ_x, the uncertainty in x, cannot be larger than a.

From Equation 3.43 and the condition $\sigma_x \le a$,

$$\frac{\hbar}{2\sigma_p} < \sigma_x \le a$$

Then

$$\frac{\hbar}{2a} \le \sigma_p$$

and so

$$\frac{\hbar^2}{4a^2} \le \sigma_p^2 \qquad (1)$$

We are given that $\langle p^2 \rangle = \sigma_p^2$, so we write

$$\frac{\sigma_p^2}{2m} = \frac{\langle p^2 \rangle}{2m} = \langle E \rangle \qquad (2)$$

Substituting Equation 2 into Equation 1 gives

$$\frac{\hbar^2}{8ma^2} \le \langle E \rangle$$

3–23. Discuss the degeneracies of the first few energy levels of a particle in a three-dimensional box when all three sides have a different length.

$$E = \frac{h^2}{8m} \left(\frac{n_x^2}{a^2} + \frac{n_y^2}{b^2} + \frac{n_z^2}{c^2} \right) \qquad (3.57)$$

Even if $a \neq b \neq c$, the energy levels will not necessarily be degenerate.

3–24. Show that the normalized wave function for a particle in a three-dimensional box with sides of length a, b, and c is

$$\psi(x, y, z) = \left(\frac{8}{abc}\right)^{1/2} \sin\frac{n_x \pi x}{a} \sin\frac{n_y \pi y}{b} \sin\frac{n_z \pi z}{c}$$

We can separate $\psi(x, y, z)$ into three one-dimensional wavefunctions $X(x)$, $Y(y)$, and $Z(z)$ such that $\psi(x, y, z) = X(x)Y(y)Z(z)$. These three one-dimensional wavefunctions have the same form and boundary conditions as the one we treated in Section 3–5, and so (as in Section 3–9)

$$X(x) = A_x \sin\frac{n_x \pi x}{a} \qquad n_x = 1, 2, 3, \ldots$$

$$Y(y) = A_y \sin\frac{n_y \pi y}{b} \qquad n_y = 1, 2, 3, \ldots$$

$$Z(z) = A_z \sin\frac{n_z \pi z}{c} \qquad n_z = 1, 2, 3, \ldots$$

To normalize ψ, we require that

$$\psi(x, y, z)\psi^*(x, y, z) = 1 = (A_x A_y A_z)^2 \int_0^a \sin^2\frac{n_x \pi x}{a}dx \int_0^b \sin^2\frac{n_y \pi y}{b}dy \int_0^c \sin^2\frac{n_z \pi z}{c}dz$$

In Problem 3–10, we learned that $\int_0^a \sin^2\frac{n\pi x}{a}dx = \frac{a}{2}$. Thus

$$\psi(x, y, z)\psi^*(x, y, z) = 1 = (A_x A_y A_z)^2 \left(\frac{a}{2}\right)\left(\frac{b}{2}\right)\left(\frac{c}{2}\right)$$

or

$$A_x A_y A_z = \left(\frac{8}{abc}\right)^{1/2}$$

3–25. Show that $\langle \mathbf{p} \rangle = 0$ for the ground state of a particle in a three-dimensional box with sides of length a, b, and c.

$$\hat{\mathbf{P}} = -i\hbar \left(\mathbf{i}\frac{\partial}{\partial x} + \mathbf{j}\frac{\partial}{\partial y} + \mathbf{k}\frac{\partial}{\partial z}\right) \tag{3.58}$$

$$\langle \mathbf{p} \rangle = \int_0^a dx \int_0^b dy \int_0^c dz\,\psi^*(x, y, z)\hat{\mathbf{P}}\psi(x, y, z)$$

$$= -i\hbar\mathbf{i} \int_0^a \sin\frac{n_x \pi x}{a}\frac{\partial}{\partial x}\left(\sin\frac{n_x \pi x}{a}\right)dx \int_0^b \sin^2\frac{n_y \pi y}{b}dy \int_0^c \sin^2\frac{n_z \pi z}{c}dz$$

$$-i\hbar\mathbf{j} \int_0^a \sin^2\frac{n_x \pi x}{a}dx \int_0^b \sin\frac{n_y \pi y}{b}\frac{\partial}{\partial y}\left(\sin\frac{n_y \pi y}{b}\right)dy \int_0^c \sin^2\frac{n_z \pi z}{c}dz$$

$$-i\hbar\mathbf{k} \int_0^a \sin^2\frac{n_x \pi x}{a}dx \int_0^b \sin^2\frac{n_y \pi y}{b}dy \int_0^c \sin\frac{n_z \pi z}{c}\frac{\partial}{\partial z}\left(\sin\frac{n_z \pi z}{a}\right)dz$$

Each of these three sets of integrals has a multiplicative factor like $\int_0^a \sin \dfrac{n_x \pi x}{a} \cos \dfrac{n_x \pi x}{a} dx$ that results from taking a derivative. We have previously shown (Problem 3–14) that such an integral is equal to zero, and so we have $\langle \mathbf{p} \rangle = 0$ for a particle in a three-dimensional box.

3–26. What are the degeneracies of the first four energy levels for a particle in a three-dimensional box with $a = b = 1.5c$?

$$E = \frac{h^2}{8m} \left(\frac{n_x^2 + n_y^2 + 2.25 n_z^2}{a^2} \right) \tag{3.57}$$

Energy level	(n_x, n_y, n_z)	Degeneracy	$E/(h^2/8ma^2)$
E_{111}	(1, 1, 1)	1	4.25
E_{211}	(2, 1, 1)(1, 2, 1)	2	7.25
E_{221}	(2, 2, 1)	1	10.25
E_{112}	(1, 1, 2)	1	11
E_{311}	(3, 1, 1)(1, 3, 1)	2	12.25
E_{212}	(2, 1, 2)(1, 2, 2)	2	14
E_{133}	(1, 3, 3)	1	30.25

3–27. Many proteins contain metal porphyrin molecules. The general structure of the porphyrin molecule is

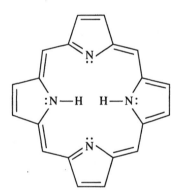

This molecule is planar and so we can approximate the π electrons as being confined inside a square. What are the energy levels and degeneracies of a particle in a square of side a? The porphyrin molecule has 18 π electrons. If we approximate the length of the molecule by 1000 pm, then what is the predicted lowest energy absorption of the porphyrin molecule? (The experimental value is $\approx 17\,000$ cm^{-1}.)

$$E_{n_x n_y} = \frac{h^2}{8ma^2} \left(n_x^2 + n_y^2 \right) \tag{3.57}$$

$E_{n_x n_y}$ is singly degenerate for $n_x = n_y$ and doubly degenerate for $n_x \neq n_y$.

Energy level	(n_x, n_y)	Degeneracy
E_{11}	(1, 1)	1
E_{12}	(1, 2) (2, 1)	2
E_{22}	(2, 2)	1
E_{31}	(3, 1) (1, 3)	2
E_{23}	(3, 2) (2, 3)	2
E_{41}	(4, 1) (1, 4)	2
E_{33}	(3, 3)	1
E_{42}	(4, 2) (2, 4)	2
E_{43}	(4, 3) (3, 4)	2

Because each energy level can hold 2 π-electrons, the lowest energy absorption of the porphyrin molecule will be that which excites an electron from the E_{42} state to the E_{43} state. Then

$$\Delta E = \frac{h^2}{8m_e a^2}\left(n_{x,2}^2 + n_{y,2}^2 - n_{x,1}^2 - n_{y,1}^2\right)$$

$$= \frac{(6.626 \times 10^{-34}\ \text{J·s})^2}{8\,(9.11 \times 10^{-31}\ \text{kg})\,(1000 \times 10^{-12}\ \text{m})^2}\,(4^2 + 3^2 - 4^2 - 2^2)$$

$$= 3.012 \times 10^{-19}\ \text{J}$$

Because $E = hc\tilde{\nu}$, $\tilde{\nu} = 1.52 \times 10^4\ \text{cm}^{-1}$.

3–28. The Schrödinger equation for a particle of mass m constrained to move on a circle of radius a is

$$-\frac{\hbar^2}{2I}\frac{d^2\psi}{d\theta^2} = E\psi(\theta) \qquad 0 \le \theta \le 2\pi$$

where $I = ma^2$ is the moment of inertia and θ is the angle that describes the position of the particle around the ring. Show by direct substitution that the solutions to this equation are

$$\psi(\theta) = Ae^{in\theta}$$

where $n = \pm(2IE)^{1/2}/\hbar$. Argue that the appropriate boundary condition is $\psi(\theta) = \psi(\theta + 2\pi)$ and use this condition to show that

$$E = \frac{n^2\hbar^2}{2I} \qquad n = 0,\ \pm 1,\ \pm 2,\ \ldots$$

Show that the normalization constant A is $(2\pi)^{-1/2}$. Discuss how you might use these results for a free-electron model of benzene.

We can write the differential equation as

$$\frac{d^2\psi}{d\theta^2} + \frac{2EI}{\hbar^2}\psi(\theta) = 0$$

Substituting $\psi(\theta) = Ae^{in\theta}$ gives

$$-n^2 Ae^{in\theta} + \frac{2EI}{\hbar^2}Ae^{in\theta} = 0$$

or $n = \pm(2IE)^{1/2}/\hbar$. Because the particle is moving in a circle, it must return to any designated point after traveling 2π radians: $\psi(\theta) = \psi(\theta + 2\pi)$. Then

$$e^{in\theta} = e^{in(\theta + 2\pi)}$$

$$1 = e^{i2\pi n}$$

$$1 = \cos 2\pi n + i \sin 2\pi n$$

This is true only if n is an integer. In that case,

$$E = \frac{n^2\hbar^2}{2I} \qquad n = 0, \pm1, \pm2, \ldots$$

To normalize $\psi(x)$, we require that

$$A^2 \int_0^{2\pi} e^{in\theta} e^{-in\theta} d\theta = 1$$

$$A^2 \int_0^{2\pi} d\theta = 1$$

so

$$A = \frac{1}{\sqrt{2\pi}}$$

There are six π electrons in benzene. If we model the electrons of a benzene molecule as described above, there will be two electrons in each of the three energy levels $n = 0$ and ±1. The first electronic transition would be a $n = 1 \rightarrow n = 2$ transition, and the frequency associated with this transition would be given by the expression

$$\tilde{\nu} = \frac{\hbar}{4\pi c I}(2^2 - 1^2)$$

3–29. Set up the problem of a particle in a box with its walls located at $-a$ and $+a$. Show that the energies are equal to those of a box with walls located at 0 and $2a$. (These energies may be obtained from the results that we derived in the chapter simply by replacing a by $2a$.) Show, however, that the wave functions are not the same and in this case are given by

$$\psi_n(x) = \frac{1}{a^{1/2}} \sin \frac{n\pi x}{2a} \qquad n \text{ even}$$

$$= \frac{1}{a^{1/2}} \cos \frac{n\pi x}{2a} \qquad n \text{ odd}$$

Does it bother you that the wave functions seem to depend upon whether the walls are located at $\pm a$ or 0 and $2a$? Surely the particle "knows" only that it has a region of length $2a$ in which to move and cannot be affected by where you place the origin for the two sets of wave functions. What does this tell you? Do you think that any experimentally observable properties depend upon where you choose to place the origin of the x-axis? Show that $\sigma_x \sigma_p > \hbar/2$, exactly as we obtained in Section 3–8.

The general solution of the Schrödinger equation for a particle in a one-dimensional box is (Section 3–5)

$$\psi(x) = A \cos kx + B \sin kx \qquad k = \frac{(2mE)^{1/2}}{\hbar}$$

We have the boundary conditions $\psi(-a) = \psi(a) = 0$, so

$$\psi(-a) = A\cos(-ka) + B\sin(-ka) = A\cos ka - B\sin ka = 0$$

and

$$\psi(a) = A\cos(ka) + B\sin(ka) \quad = A\cos ka + B\sin ka = 0$$

Adding and subtracting these two equations gives

$$A\cos ka = 0 \quad \text{and} \quad B\sin ka = 0$$

The general solution to these equations is to set

$$k = \frac{n\pi}{2a}$$

where $n = 1, 2, \ldots$ and to satisfy the boundary conditions by setting $B = 0$ when n is odd and $A = 0$ when n is even. Thus

$$\psi_n(x) = B\sin\frac{n\pi x}{2a} \qquad n \text{ even}$$

$$= A\cos\frac{n\pi x}{2a} \qquad n \text{ odd}$$

The normalization constants A and B are both equal to $a^{-1/2}$. We find E through the defined variable k:

$$\frac{(2mE)^{1/2}}{\hbar} = \frac{n\pi}{2a}$$

$$E = \frac{h^2 n^2}{32ma^2}$$

When we solved the Schrödinger equation for the boundary conditions $\psi(0) = \psi(2a) = 0$ (Section 3–5), we found

$$E_n = \frac{h^2 n^2}{8m(2a)^2} \qquad n = 1, 2, \cdots$$

which is the same result as that for a box with walls located at $\pm a$. Realize that the wave functions are independent of where the walls are located; however, how we define our coordinate system will change the way we express the wave function mathematically. No experimentally observable properties depend upon how we define our coordinate system - the coordinate system is a purely hypothetical construct which does not impact any observable system. Since σ_x and σ_p are observable properties, $\sigma_x \sigma_p \geq \hbar/2$ as in Section 3.8.

3–30. For a particle moving in a one-dimensional box, the mean value of x is $a/2$, and the mean square deviation is $\sigma_x^2 = (a^2/12)[1 - (6/\pi^2 n^2)]$. Show that as n becomes very large, this value agrees with the classical value. The classical probability distribution is uniform,

$$p(x)dx = \frac{1}{a}dx \qquad 0 \leq x \leq a$$

$$= 0 \qquad \text{otherwise}$$

Classically,

$$\langle x \rangle = \frac{1}{a}\int_0^a x\,dx = \frac{a}{2} \quad \text{and} \quad \langle x^2 \rangle = \frac{1}{a}\int_0^a x^2\,dx = \frac{a^2}{3}$$

so

$$\sigma^2 = \langle x^2 \rangle - \langle x \rangle = \frac{a^2}{12}$$

For the particle in a box,

$$\lim_{n \to \infty} \sigma^2 = \lim_{n \to \infty} \frac{a^2}{12}\left[1 - \frac{6}{\pi^2 n^2}\right] = \frac{a^2}{12}$$

3–31. This problem shows that the intensity of a wave is proportional to the square of its amplitude. Figure 3.7 illustrates the geometry of a vibrating string. Because the velocity at any point of the string is $\partial u / \partial t$, the kinetic energy of the entire string is

$$K = \int_0^l \frac{1}{2}\rho\left(\frac{\partial u}{\partial t}\right)^2 dx$$

where ρ is the linear mass density of the string. The potential energy is found by considering the increase of length of the small arc PQ of length ds in Figure 3.7. The segment of the string along that arc has increased its length from dx to ds. Therefore, the potential energy associated with this increase is

$$V = \int_0^l T(ds - dx)$$

where T is the tension in the string. Using the fact that $(ds)^2 = (dx)^2 + (du)^2$, show that

$$V = \int_0^l T\left\{\left[1 + \left(\frac{\partial u}{\partial x}\right)^2\right]^{1/2} - 1\right\} dx$$

Using the fact that $(1 + x)^{1/2} \approx 1 + (x/2)$ for small x, show that

$$V = \frac{1}{2}T \int_0^l \left(\frac{\partial u}{\partial x}\right)^2 dx$$

for small displacements. The total energy of the vibrating string is the sum of K and V and so

$$E = \frac{\rho}{2}\int_0^l \left(\frac{\partial u}{\partial t}\right)^2 dx + \frac{T}{2}\int_0^l \left(\frac{\partial u}{\partial x}\right)^2 dx$$

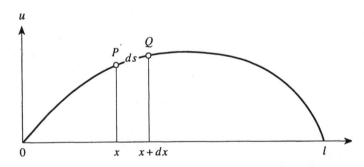

FIGURE 3.7
The geometry of a vibrating string.

We showed in Chapter 2 (Equations 2.23 through 2.25) that the nth normal mode can be written in the form

$$u_n(x, l) = D_n \cos(\omega_n t + \phi_n) \sin \frac{n\pi x}{l}$$

where $\omega_n = v n \pi / l$. Using this equation, show that

$$K_n = \frac{\pi^2 v^2 n^2 \rho}{4l} D_n^2 \sin^2(\omega_n t + \phi_n)$$

and

$$V_n = \frac{\pi^2 n^2 T}{4l} D_n^2 \cos^2(\omega_n t + \phi_n)$$

Using the fact that $v = (T/\rho)^{1/2}$, show that

$$E_n = \frac{\pi^2 v^2 n^2 \rho}{4l} D_n^2$$

Note that the total energy, or intensity, is proportional to the square of the amplitude. Although we have shown this proportionality only for the case of a vibrating string, it is a general result and shows that the intensity of a wave is proportional to the square of the amplitude. If we had carried everything through in complex notation instead of sines and cosines, then we would have found that E_n is proportional to $|D_n|^2$ instead of just D_n^2. Generally, there are many normal modes present at the same time, and the complete solution is (Equation 2.25)

$$u(x, t) = \sum_{n=1}^{\infty} D_n \cos(\omega_n t + \phi_n) \sin \frac{n\pi x}{l}$$

Using the fact that (see Problem 3–16)

$$\int_0^l \sin \frac{n\pi x}{l} \sin \frac{m\pi x}{l} dx = 0 \qquad \text{if } m \neq n$$

show that

$$E_n = \frac{\pi^2 v^2 \rho}{4l} \sum_{n=1}^{\infty} n^2 D_n^2$$

We begin with the potential energy associated with the vibrating string,

$$V = \int_0^l T(ds - dx)$$

$$= \int_0^l T\left[(dx^2 + du^2)^{1/2} - dx\right]$$

$$= \int_0^l T\left\{\left[1 + \left(\frac{\partial u}{\partial x}\right)^2\right]^{1/2} - 1\right\} dx$$

Using the fact that $(1 + x)^{1/2} \approx 1 + x/2$ for small x, we obtain

$$V \approx \int_0^l T\left[1 + \frac{1}{2}\left(\frac{\partial u}{\partial x}\right)^2 - 1\right] dx \approx \frac{T}{2} \int_0^l \left(\frac{\partial u}{\partial x}\right)^2 dx$$

Starting with the equation for the nth normal mode

$$u_n(x, l) = D_n \cos(\omega_n t + \phi_n) \sin \frac{n\pi x}{l}$$

we obtain the partial derivatives

$$\frac{\partial u_n}{\partial t} = -\omega_n D_n \sin(\omega_n t + \phi_n) \sin \frac{n\pi x}{l}$$

and

$$\frac{\partial u_n}{\partial x} = \frac{n\pi}{l} D_n \cos(\omega_n t + \phi_n) \cos \frac{n\pi x}{l}$$

where $\omega_n = v n\pi / l$. Thus,

$$\begin{aligned}
K_n &= \frac{\rho}{2} \int_0^l \left(\frac{\partial u_n}{\partial t}\right)^2 dx \\
&= \frac{\rho}{2} \int_0^l \omega_n^2 D_n^2 \sin^2(\omega_n t + \phi_n) \sin^2 \frac{n\pi x}{l} dx \\
&= \frac{\omega_n^2 D_n^2 \rho}{2} \int_0^l \sin^2(\omega_n t + \phi_n) \sin^2 \frac{n\pi x}{l} dx \\
&= \frac{n^2 \pi^2 v^2 \rho}{2l^2} D_n^2 \sin^2(\omega_n t + \phi_n) \int_0^l \sin^2 \frac{n\pi x}{l} dx \\
&= \frac{n^2 \pi^2 v^2 \rho}{4l} D_n^2 \sin^2(\omega_n t + \phi_n)
\end{aligned}$$

and

$$\begin{aligned}
V_n &= \frac{T}{2} \int_0^l \left(\frac{\partial u_n}{\partial x}\right)^2 dx \\
&= \frac{T}{2} \int_0^l \frac{n^2 \pi^2}{l^2} D_n^2 \cos^2(\omega_n t + \phi_n) \cos^2 \frac{n\pi x}{l} dx \\
&= \frac{T n^2 \pi^2}{2l^2} D_n^2 \cos^2(\omega_n t + \phi_n) \int_0^l \cos^2 \frac{n\pi x}{l} dx \\
&= \frac{T n^2 \pi^2}{4l} D_n^2 \cos^2(\omega_n t + \phi_n)
\end{aligned}$$

where $T = \rho v^2$. The total energy is given by

$$\begin{aligned}
E_n &= V_n + K_n \\
&= \frac{n^2 \pi^2 v^2 \rho}{4l} D_n^2 \sin^2(\omega_n t + \phi_n) + \frac{T n^2 \pi^2}{4l} D_n^2 \cos^2(\omega_n t + \phi_n) \\
&= \frac{n^2 \pi^2 v^2 \rho}{4l} D_n^2 \sin^2(\omega_n t + \phi_n) + \frac{v^2 \rho n^2 \pi^2}{4l} D_n^2 \cos^2(\omega_n t + \phi_n) \\
&= \frac{n^2 \pi^2 v^2 \rho}{4l} D_n^2 \left[\sin^2(\omega_n t + \phi_n) + \cos^2(\omega_n t + \phi_n)\right] \\
&= \frac{n^2 \pi^2 v^2 \rho}{4l} D_n^2
\end{aligned}$$

Now, using the complete solution, we find

$$\sum_{n=1}^{\infty} u_n(x, t) = \sum_{n=1}^{\infty} D_n \cos(\omega_n t + \phi_n) \sin \frac{n\pi x}{l}$$

$$\frac{\partial u_n}{\partial t} = -\sum_{n=1}^{\infty} \omega_n D_n \sin(\omega_n t + \phi_n) \sin \frac{n\pi x}{l}$$

$$\frac{\partial u_n}{\partial x} = \sum_{n=1}^{\infty} \frac{n\pi}{l} D_n \cos(\omega_n t + \phi_n) \cos \frac{n\pi x}{l}$$

To find V and K, use the identity from Problem 3–16

$$\int_0^l \sin\frac{n\pi x}{l}\sin\frac{m\pi x}{l}dx = \int_0^l \cos\frac{n\pi x}{l}\cos\frac{m\pi x}{l}dx = \frac{l}{2}\delta_{nm}$$

We then have

$$\sum_{n=1}^{\infty} K_n = \frac{\rho}{2}\int_0^l \left(\frac{\partial u_n}{\partial t}\right)^2 dx$$

$$= \frac{\rho}{2}\int_0^l \left(\sum_{n=1}^{\infty} -\omega_n D_n \sin(\omega_n t + \phi_n)\sin\frac{n\pi x}{l}\right)\left(\sum_{m=1}^{\infty} -\omega_m D_m \sin(\omega_m t + \phi_m)\sin\frac{m\pi x}{l}\right)dx$$

$$= \frac{\rho}{2}\sum_{n=1}^{\infty}\sum_{m=1}^{\infty}\omega_n\omega_m D_n D_m \sin(\omega_n t + \phi_n)\sin(\omega_m t + \phi_m)\int_0^l \sin\frac{n\pi x}{l}\sin\frac{m\pi x}{l}dx$$

$$= \frac{\rho}{2}\sum_{n=1}^{\infty}\omega_n^2 D_n^2 \sin^2(\omega_n t + \phi_n)\left(\frac{l}{2}\right)$$

$$= \frac{\rho\pi^2 v^2}{4l}\sum_{n=1}^{\infty} n^2 D_n^2 \sin^2(\omega_n t + \phi_n)$$

and

$$\sum_{n=1}^{\infty} V_n = \frac{T}{2}\int_0^l \left(\frac{\partial u_n}{\partial x}\right)^2 dx$$

$$= \frac{T}{2}\int_0^l \left(\sum_{n=1}^{\infty}\frac{n\pi}{l} D_n \cos(\omega_n t + \phi_n)\cos\frac{n\pi x}{l}\right)\left(\sum_{m=1}^{\infty}\frac{m\pi}{l} D_m \cos(\omega_m t + \phi_m)\cos\frac{m\pi x}{l}\right)dx$$

$$= \frac{T\pi^2}{2l^2}\sum_{n=1}^{\infty}\sum_{m=1}^{\infty} nm D_n D_m \cos(\omega_n t + \phi_n)\cos(\omega_m t + \phi_m)\int_0^l \cos\frac{n\pi x}{l}\cos\frac{m\pi x}{l}dx$$

$$= \frac{T\pi^2}{2l^2}\sum_{n=1}^{\infty} n^2 D_n^2 \cos^2(\omega_n t + \phi_n)\left(\frac{l}{2}\right)$$

$$= \frac{\pi^2 v^2 \rho}{4l}\sum_{n=1}^{\infty} n^2 D_n^2 \cos^2(\omega_n t + \phi_n)$$

Finally,

$$\sum_{n=1}^{\infty} E_n = \sum_{n=1}^{\infty} V_n + \sum_{n=1}^{\infty} K_n$$

$$= \frac{\pi^2 v^2 \rho}{4l}\sum_{n=1}^{\infty} n^2 D_n^2 \cos^2(\omega_n t + \phi_n) + \frac{\rho\pi^2 v^2}{4l}\sum_{n=1}^{\infty} n^2 D_n^2 \sin^2(\omega_n t + \phi_n)$$

$$= \frac{\pi^2 v^2 \rho}{4l}\sum_{n=1}^{\infty} n^2 D_n^2 \left[\cos^2(\omega_n t + \phi_n) + \sin^2(\omega_n t + \phi_n)\right]$$

$$= \frac{\pi^2 v^2 \rho}{4l}\sum_{n=1}^{\infty} n^2 D_n^2$$

3–32. The quantized energies of a particle in a box result from the boundary conditions, or from the fact that the particle is restricted to a finite region. In this problem, we investigate the quantum-mechanical problem of a free particle, one that is not restricted to a finite region. The potential energy $V(x)$ is equal to zero and the Schrödinger equation is

$$\frac{d^2\psi}{dx^2} + \frac{2mE}{\hbar^2}\psi(x) = 0 \qquad -\infty < x < \infty$$

Note that the particle can lie anywhere along the x-axis in this problem. Show that the two solutions of this Schrödinger equation are

$$\psi_1(x) = A_1 e^{i(2mE)^{1/2}x/\hbar} = A_1 e^{ikx}$$

and

$$\psi_2(x) = A_2 e^{-i(2mE)^{1/2}x/\hbar} = A_2 e^{-ikx}$$

where

$$k = \frac{(2mE)^{1/2}}{\hbar}$$

Show that if E is allowed to take on negative values, then the wave functions become unbounded for large x. Therefore, we will require that the energy, E, be a positive quantity. We saw in our discussion of the Bohr atom that negative energies correspond to bound states and positive energies correspond to unbound states, and so our requirement that E be positive is consistent with the picture of a free particle. To get a physical interpretation of the states that $\psi_1(x)$ and $\psi_2(x)$ describe, operate on $\psi_1(x)$ and $\psi_2(x)$ with the momentum operator $\hat{P}$ (Equation 3.11), and show that

$$\hat{P}\psi_1 = -i\hbar\frac{d\psi_1}{dx} = \hbar k\psi_1$$

and

$$\hat{P}\psi_2 = -i\hbar\frac{d\psi_2}{dx} = -\hbar k\psi_2$$

Notice that these are eigenvalue equations. Our interpretation of these two equations is that ψ_1 describes a free particle with fixed momentum $\hbar k$ and that ψ_2 describes a particle with fixed momentum $-\hbar k$. Thus, ψ_1 describes a particle moving to the right and ψ_2 describes a particle moving to the left, both with a fixed momentum. Notice also that there are no restrictions on k, and so the particle can have any value of momentum. Now show that

$$E = \frac{\hbar^2 k^2}{2m}$$

Notice that the energy is not quantized; the energy of the particle can have any positive value in this case because no boundaries are associated with this problem. Last, show that $\psi_1^*(x)\psi_1(x) = A_1^* A_1 = |A_1|^2 = $ constant and that $\psi_2^*(x)\psi_2(x) = A_2^* A_2 = |A_2|^2 = $ constant. Discuss this result in terms of the probabilistic interpretation of $\psi^*\psi$. Also discuss the application of the Uncertainty Principle to this problem. What are σ_p and σ_x?

From Example 2–4, we know that the solutions to this Schrödinger equation are

$$\psi_1(x) = A_1 e^{ikx} \qquad\qquad \psi_2(x) = A_2 e^{-ikx}$$

where

$$k = \frac{\sqrt{2mE}}{\hbar}$$

Suppose E can be less than zero. Then $k \rightarrow i(-2mE)^{1/2}/\hbar$ and

$$\lim_{x \to \infty} \psi_2(x) = \lim_{x \to \infty} A_2 e^{x\sqrt{-2mE}/\hbar}$$

diverges. Therefore, E must be positive. Using Equation 3.11 for the momentum operator, we find

$$\hat{P}\psi_1 = -i\hbar\frac{d\psi_1}{dx} = -i\hbar\frac{d}{dx}\left(A_1 e^{ikx}\right)$$
$$= -i^2\hbar k A_1 e^{ikx} = \hbar k A_1 e^{ikx} = \hbar k \psi_1(x)$$
$$\hat{P}\psi_2 = -i\hbar\frac{d\psi_2}{dx} = -i\hbar\frac{d}{dx}\left(A_2 e^{-ikx}\right)$$
$$= i^2\hbar k A_2 e^{-ikx} = -\hbar k A_2 e^{-ikx} = -\hbar k \psi_1(x)$$

For a free particle, all energy is kinetic energy. The possible values of the momentum of a free particle are $\hbar k$ and $-\hbar k$. Using $E = p^2/2m$, we find that

$$E = \frac{p^2}{2m} = \frac{(\pm\hbar k)^2}{2m} = \frac{\hbar^2 k^2}{2m}$$

Finally,

$$\psi_1^*(x)\psi_1(x) = \left(A_1^* e^{-ikx}\right)\left(A_1 e^{ikx}\right)$$
$$= A_1^* A_1 = |A_1|^2 = \text{constant}$$
$$\psi_2^*(x)\psi_2(x) = \left(A_2^* e^{ikx}\right)\left(A_2 e^{-ikx}\right)$$
$$= A_2^* A_2 = |A_2|^2 = \text{constant}$$

Since $\psi^*(x)\psi(x)$ is a constant, the particle is equally likely to be found anywhere along the x axis. Thus, there is an infinite uncertainty in the location of the particle. This is consistent with the Uncertainty Principle because the momentum of the particle is known exactly ($\sigma_p = 0$).

3–33. Derive the equation for the allowed energies of a particle in a one-dimensional box by assuming that the particle is described by standing de Broglie waves within the box.

The de Broglie relationship is

$$\lambda = \frac{h}{p} \tag{1.12}$$

Because the waves are standing waves, an integral number of half-wavelengths will fit into the box, or

$$\frac{n\lambda}{2} = a \qquad \text{and} \qquad \frac{nh}{2p} = a$$

Solving for p gives

$$p = \frac{nh}{2a}$$

and the corresponding energy is

$$E = \frac{mv^2}{2} = \frac{p^2}{2m} = \frac{1}{2m}\frac{n^2 h^2}{4a^2} = \frac{n^2 h^2}{8ma^2}$$

3–34. We can use the Uncertainty Principle for a particle in a box to argue that free electrons cannot exist in a nucleus. Before the discovery of the neutron, one might have thought that a nucleus of atomic number Z and mass number A is made up of A protons and $A - Z$ electrons, that is, just enough electrons such that the net nuclear charge is $+Z$. Such a nucleus would have an atomic number Z and mass number A. In this problem, we will use Equation 3.41 to *estimate* the energy of an electron confined to a region of nuclear size. The diameter of a typical nucleus is approximately 10^{-14} m. Substitute $a = 10^{-14}$ m into Equation 3.41 and show that σ_p is

$$\sigma_p \gtrsim 3 \times 10^{-20} \text{ kg} \cdot \text{m} \cdot \text{s}^{-1}$$

Show that

$$E = \frac{\sigma_p^2}{2m} = 5 \times 10^{-10} \text{ J}$$
$$\approx 3000 \text{ MeV}$$

where millions of electron volts (MeV) is the common nuclear physics unit of energy. It is observed experimentally that electrons emitted from nuclei as β radiation have energies of only a few MeV, which is far less than the energy we have calculated above. Argue, then, that there can be no free electrons in nuclei because they should be ejected with much higher energies than are found experimentally.

$$\sigma_p = \frac{n\pi\hbar}{a} = \frac{nh}{2a} \tag{3.41}$$

For $n = 1$ and $a = 10^{-14}$ m,

$$\sigma_p = \frac{6.626 \times 10^{-34} \text{ J} \cdot \text{s}}{2(1 \times 10^{-14} \text{ m})} \approx 3 \times 10^{-20} \text{ kg} \cdot \text{m} \cdot \text{s}^{-1}$$

$$E = \frac{mv^2}{2} = \frac{p^2}{2m} = \frac{\sigma_p^2}{2m} \tag{Problem 3–23}$$

and in this case

$$E \approx \frac{(3 \times 10^{-20} \text{ kg} \cdot \text{m} \cdot \text{s}^{-1})^2}{2(9.11 \times 10^{-31} \text{ kg})} \approx 5 \times 10^{-10} \text{ J} \approx 3000 \text{ MeV}$$

Because no nuclei emit electrons with energies on the order of 1000 MeV, we assume such electrons do not exist in nuclei; therefore, there can be no free electrons in nuclei. (Electrons with energies of only a few MeV are the result of a neutron decaying into a proton and an electron in the nucleus.)

3–35. We can use the wave functions of Problem 3–29 to illustrate some fundamental symmetry properties of wave functions. Show that the wave functions are alternately symmetric and antisymmetric or even and odd with respect to the operation $x \to -x$, which is a reflection through the $x = 0$ line. This symmetry property of the wave function is a consequence of the symmetry of the Hamiltonian operator, as we now show. The Schrödinger equation may be written as

$$\hat{H}(x)\psi_n(x) = E_n\psi_n(x)$$

Reflection through the $x = 0$ line gives $x \to -x$ and so

$$\hat{H}(-x)\psi_n(-x) = E_n\psi_n(-x)$$

Now show that $\hat{H}(x) = \hat{H}(-x)$ (i.e., that $\hat{H}$ is symmetric) for a particle in a box, and so show that

$$\hat{H}(x)\psi_n(-x) = E_n\psi_n(-x)$$

Thus, we see that $\psi_n(-x)$ is also an eigenfunction of $\hat{H}$ belonging to the same eigenvalue E_n. Now, if only one eigenfunction is associated with each eigenvalue (the state is nondegenerate), then argue that $\psi_n(x)$ and $\psi_n(-x)$ must differ only by a multiplicative constant [i.e., that $\psi_n(x) = c\psi_n(-x)$]. By applying the inversion operation again to this equation, show that $c = \pm 1$ and that all the wave functions must be either even or odd with respect to reflection through the $x = 0$ line because the Hamiltonian operator is symmetric. Thus, we see that the symmetry of the Hamiltonian operator influences the symmetry of the wave functions. A general study of symmetry uses group theory, and this example is actually an elementary application of group theory to quantum-mechanical problems. We will study group theory in Chapter 12.

Consider the wavefunctions found in Problem 3–29. For odd n

$$\psi_n(-x) = \frac{1}{a^{1/2}} \cos\left(-\frac{n\pi x}{2a}\right) = \frac{1}{a^{1/2}} \cos\left(\frac{n\pi x}{2a}\right) = \psi_n(x)$$

and the wavefunctions for odd n are symmetric. For even n

$$\psi_n(-x) = \frac{1}{a^{1/2}} \sin\left(-\frac{n\pi x}{2a}\right) = -\frac{1}{a^{1/2}} \sin\left(\frac{n\pi x}{2a}\right) = -\psi_n(x)$$

and the wavefunctions for even n are antisymmetric. We will now show that $\hat{H}(x)$ is an even function of x:

$$\hat{H}(x) = -\frac{\hbar^2}{2m}\frac{d^2}{dx^2}$$

$$\hat{H}(-x) = -\frac{\hbar^2}{2m}\frac{d^2}{d(-x)^2} = -\frac{\hbar^2}{2m}\frac{d^2}{dx^2} = \hat{H}(x)$$

Now we have shown that $E_n\psi_n(-x) = \hat{H}(x)\psi_n(-x) = \hat{H}(x)\psi_n(x)$. If the state is nondegenerate, $\psi_n(-x) = c\psi_n(x)$. Repeating the operation, we find

$$\psi_n(-x) = c\psi_n(x) = c^2\psi_n(-x)$$

which leads to the conclusion that $c = \pm 1$ and consequently to the conclusion that all $\psi_n(x)$ are either even or odd with respect to reflection through the $x = 0$ line. Thus the symmetry of the Hamiltonian influences the symmetry of its eigenfunctions (assuming a nondegenerate system).

Vectors

PROBLEMS AND SOLUTIONS

C–1. Find the length of the vector $\mathbf{v} = 2\mathbf{i} - \mathbf{j} + 3\mathbf{k}$.

Use Equation C.5:

$$v = |\mathbf{v}| = (v_x^2 + v_y^2 + v_z^2)^{1/2}$$
$$= \left[2^2 + (-1)^2 + 3^2\right]^{1/2}$$
$$= \sqrt{14}$$

C–2. Find the length of the vector $\mathbf{r} = x\mathbf{i} + y\mathbf{j}$ and of the vector $\mathbf{r} = x\mathbf{i} + y\mathbf{j} + z\mathbf{k}$.

Use Equation C.5 for the length of a vector:

$$r = |\mathbf{r}| = (r_x^2 + r_y^2 + r_z^2)^{1/2}$$
$$= \left(x^2 + y^2 + 0^2\right)^{1/2}$$
$$= \left(x^2 + y^2\right)^{1/2}$$

$$r = |\mathbf{r}| = (r_x^2 + r_y^2 + r_z^2)^{1/2}$$
$$= \left(x^2 + y^2 + z^2\right)^{1/2}$$

C–3. Prove that $\mathbf{A} \cdot \mathbf{B} = 0$ if $\mathbf{A}$ and $\mathbf{B}$ are perpendicular to each other. Two vectors that are perpendicular to each other are said to be orthogonal.

If $\mathbf{A}$ and $\mathbf{B}$ are perpendicular to each other, then the angle θ between them is 90°. From the definition of the scalar product (Equation C.6),

$$\mathbf{A} \cdot \mathbf{B} = |\mathbf{A}||\mathbf{B}| \cos \theta = |\mathbf{A}||\mathbf{B}| \cos 90° = 0$$

C–4. Show that the vectors $\mathbf{A} = 2\mathbf{i} - 4\mathbf{j} - 2\mathbf{k}$ and $\mathbf{B} = 3\mathbf{i} + 4\mathbf{j} - 5\mathbf{k}$ are orthogonal.

Using Equation C.9,

$$\mathbf{A} \cdot \mathbf{B} = A_x B_x + A_y B_y + A_z B_z$$
$$= (2)(3) + (-4)(4) + (-2)(-5)$$
$$= 0$$

C–5. Show that the vector $\mathbf{r} = 2\mathbf{i} - 3\mathbf{k}$ lies entirely in a plane perpendicular to the y axis.

The y axis can be represented by the vector $\mathbf{A} = y\mathbf{j}$, where y can take on any real value. If $\mathbf{r}$ is perpendicular to the y axis, the scalar product of this vector and the vector $\mathbf{r}$ should be 0:

$$\mathbf{r} \cdot \mathbf{A} = r_x A_x + r_y A_y + r_z A_z$$
$$= (2)(0) + (0)(y) + (-3)(0)$$
$$= 0$$

C–6. Find the angle between the two vectors $\mathbf{A} = -\mathbf{i} + 2\mathbf{j} + \mathbf{k}$ and $\mathbf{B} = 3\mathbf{i} - \mathbf{j} + 2\mathbf{k}$.

Using Equation C.9,

$$\mathbf{A} \cdot \mathbf{B} = A_x B_x + A_y B_y + A_z B_z$$
$$= -3 - 2 + 2 = -3$$

Then, from Equation C.6,

$$|\mathbf{A}||\mathbf{B}| \cos \theta = -3$$
$$\cos \theta = \frac{-3}{(1 + 4 + 1)^{1/2} (9 + 1 + 4)^{1/2}}$$
$$\theta = 109°$$

C–7. Determine $\mathbf{C} = \mathbf{A} \times \mathbf{B}$ given that $\mathbf{A} = -\mathbf{i} + 2\mathbf{j} + \mathbf{k}$ and $\mathbf{B} = 3\mathbf{i} - \mathbf{j} + 2\mathbf{k}$. What is $\mathbf{B} \times \mathbf{A}$ equal to?

Using Equation C.15 for the cross product of two vectors gives

$$\mathbf{A} \times \mathbf{B} = (A_y B_z - A_z B_y)\mathbf{i} + (A_z B_x - A_x B_z)\mathbf{j} + (A_x B_y - A_y B_x)\mathbf{k}$$
$$= (4 + 1)\mathbf{i} + (3 + 2)\mathbf{j} + (1 - 6)\mathbf{k}$$
$$= 5\mathbf{i} + 5\mathbf{j} - 5\mathbf{k}$$

From Equation C.13, we know that $\mathbf{A} \times \mathbf{B} = -\mathbf{B} \times \mathbf{A}$, so

$$\mathbf{B} \times \mathbf{A} = -5\mathbf{i} - 5\mathbf{j} + 5\mathbf{k}$$

C–8. Show that $\mathbf{A} \times \mathbf{A} = 0$.

Again, use Equation C.15 for the cross product of two vectors:

$$\mathbf{A} \times \mathbf{A} = (A_y A_z - A_z A_y)\mathbf{i} + (A_z A_x - A_x A_z)\mathbf{j} + (A_x A_y - A_y A_x)\mathbf{k}$$
$$= 0$$

C–9. Using Equation C.14, prove that $\mathbf{A} \times \mathbf{B}$ is given by Equation C.15.

$$\mathbf{A} = A_x \mathbf{i} + A_y \mathbf{j} + A_z \mathbf{k}$$
$$\mathbf{B} = B_x \mathbf{i} + B_y \mathbf{j} + B_z \mathbf{k}$$
$$\mathbf{A} \times \mathbf{B} = \left(A_x \mathbf{i} + A_y \mathbf{j} + A_z \mathbf{k}\right) \times \left(B_x \mathbf{i} + B_y \mathbf{j} + B_z \mathbf{k}\right) \tag{1}$$

Because A_x, A_y, A_z, B_x, B_y, and B_z are scalars, we can write Equation 1 as

$$\mathbf{A} \times \mathbf{B} = A_x B_x (\mathbf{i} \times \mathbf{i}) + A_x B_y (\mathbf{i} \times \mathbf{j}) + A_x B_z (\mathbf{i} \times \mathbf{k})$$
$$+ A_y B_x (\mathbf{j} \times \mathbf{i}) + A_y B_y (\mathbf{j} \times \mathbf{j}) + A_y B_z (\mathbf{j} \times \mathbf{k})$$
$$+ A_z B_x (\mathbf{k} \times \mathbf{i}) + A_z B_y (\mathbf{k} \times \mathbf{j}) + A_z B_z (\mathbf{k} \times \mathbf{k})$$

Now we substitute Equations C.14 into this expression to obtain

$$\mathbf{A} \times \mathbf{B} = 0 + A_x B_y \mathbf{k} - A_x B_z \mathbf{j} - A_y B_x \mathbf{k} + 0 + A_y B_z \mathbf{i} + A_z B_x \mathbf{j} - A_z B_y \mathbf{i} + 0$$
$$= \left(A_y B_z - A_z B_y\right)\mathbf{i} + \left(A_z B_x - A_x B_z\right)\mathbf{j} + \left(A_x B_y - A_y B_x\right)\mathbf{k}$$

which is Equation C.15.

C–10. Show that $|\mathbf{L}| = mvr$ for circular motion.

By definition, $\mathbf{L} = \mathbf{r} \times \mathbf{p}$ (Equation C.17). Because $\mathbf{p} = m\mathbf{v}$, we can write $\mathbf{L} = m\mathbf{r} \times \mathbf{v}$. For circular motion, the angle between the direction of travel $\mathbf{r}$ and the velocity vector $\mathbf{v}$ is always $\theta = 90°$. Using Equation C.12 gives

$$m\mathbf{r} \times \mathbf{v} = m|\mathbf{r}||\mathbf{v}|\mathbf{c}\sin\theta$$
$$= mrv\mathbf{c}\sin 90°$$
$$= mvr\mathbf{c}$$

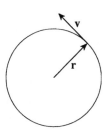

The vector $\mathbf{c}$ is a unit vector of length one, and so $|\mathbf{L}| = mvr$.

C–11. Show that

$$\frac{d}{dt}(\mathbf{A} \cdot \mathbf{B}) = \frac{d\mathbf{A}}{dt} \cdot \mathbf{B} + \mathbf{A} \cdot \frac{d\mathbf{B}}{dt}$$

and

$$\frac{d}{dt}(\mathbf{A} \times \mathbf{B}) = \frac{d\mathbf{A}}{dt} \times \mathbf{B} + \mathbf{A} \times \frac{d\mathbf{B}}{dt}$$

Take

$$\mathbf{A} = A_x(t)\mathbf{i} + A_y(t)\mathbf{j} + A_z(t)\mathbf{k}$$
$$\mathbf{B} = B_x(t)\mathbf{i} + B_y(t)\mathbf{j} + B_z(t)\mathbf{k}$$

The first derivatives of these vectors are

$$\frac{d\mathbf{A}}{dt} = \frac{dA_x(t)}{dt}\mathbf{i} + \frac{dA_y(t)}{dt}\mathbf{j} + \frac{dA_z(t)}{dt}\mathbf{k}$$
$$\frac{d\mathbf{B}}{dt} = \frac{dB_x(t)}{dt}\mathbf{i} + \frac{dB_y(t)}{dt}\mathbf{j} + \frac{dB_z(t)}{dt}\mathbf{k}$$

Equation C.9 gives

$$\mathbf{A} \cdot \mathbf{B} = A_x(t)B_x(t) + A_y(t)B_y(t) + A_z(t)B_z(t)$$

So

$$\frac{d}{dt}(\mathbf{A} \cdot \mathbf{B}) = \frac{d}{dt}\left[A_x(t)B_x(t)\right] + \frac{d}{dt}\left[A_y(t)B_y(t)\right] + \frac{d}{dt}\left[A_z(t)B_z(t)\right]$$
$$= \frac{dA_x}{dt}B_x + \frac{dA_y}{dt}B_y + \frac{dA_z}{dt}B_z + \frac{dB_x}{dt}A_x + \frac{dB_y}{dt}A_y + \frac{dB_z}{dt}A_z$$
$$= \frac{d\mathbf{A}}{dt} \cdot \mathbf{B} + \mathbf{A} \cdot \frac{d\mathbf{B}}{dt}$$

Equation C.15 gives

$$\mathbf{A} \times \mathbf{B} = (A_yB_z - A_zB_y)\mathbf{i} + (A_zB_x - A_xB_z)\mathbf{j} + (A_xB_y - A_yB_x)\mathbf{k}$$

So

$$\frac{d}{dt}(\mathbf{A} \times \mathbf{B}) = \frac{d}{dt}(A_yB_z - A_zB_y)\mathbf{i} + \frac{d}{dt}(A_zB_x - A_xB_z)\mathbf{j} + \frac{d}{dt}(A_xB_y - A_yB_x)\mathbf{k}$$
$$= \left(\frac{dA_y}{dt}B_z + \frac{dB_z}{dt}A_y - \frac{dA_z}{dt}B_y - \frac{dB_y}{dt}A_z\right)\mathbf{i}$$
$$+ \left(\frac{dA_z}{dt}B_x + \frac{dB_x}{dt}A_z - \frac{dA_x}{dt}B_z - \frac{dB_z}{dt}A_x\right)\mathbf{j}$$
$$+ \left(\frac{dA_x}{dt}B_y + \frac{dB_y}{dt}A_x - \frac{dA_y}{dt}B_x - \frac{dB_x}{dt}A_y\right)\mathbf{k}$$
$$= \left(\frac{dA_y}{dt}B_z - \frac{dA_z}{dt}B_y\right)\mathbf{i} + \left(\frac{dA_z}{dt}B_x - \frac{dA_x}{dt}B_z\right)\mathbf{j}$$
$$+ \left(\frac{dA_x}{dt}B_y - \frac{dA_y}{dt}B_x\right)\mathbf{k} + \left(\frac{dB_z}{dt}A_y - \frac{dB_y}{dt}A_z\right)\mathbf{i}$$
$$+ \left(\frac{dB_x}{dt}A_z - \frac{dB_z}{dt}A_x\right)\mathbf{j} + \left(\frac{dB_y}{dt}A_x - \frac{dB_x}{dt}A_y\right)\mathbf{k}$$
$$= \frac{d\mathbf{A}}{dt} \times \mathbf{B} + \mathbf{A} \times \frac{d\mathbf{B}}{dt}$$

C–12. Using the results of Problem C–11, prove that

$$\mathbf{A} \times \frac{d^2\mathbf{A}}{dt^2} = \frac{d}{dt}\left(\mathbf{A} \times \frac{d\mathbf{A}}{dt}\right)$$

From Problem C-11,

$$\frac{d}{dt}\left(\mathbf{A} \times \frac{d\mathbf{A}}{dt}\right) = \frac{d\mathbf{A}}{dt} \times \frac{d\mathbf{A}}{dt} + \mathbf{A} \times \frac{d}{dt}\left(\frac{d\mathbf{A}}{dt}\right)$$

The first term on the right side of expression is zero (Problem C–8). Thus,

$$\frac{d}{dt}\left(\mathbf{A} \times \frac{d\mathbf{A}}{dt}\right) = \mathbf{A} \times \frac{d}{dt}\left(\frac{d\mathbf{A}}{dt}\right)$$

$$= \mathbf{A} \times \frac{d^2\mathbf{A}}{dt^2}$$

C–13. In vector notation, Newton's equations for a single particle are

$$m\frac{d^2\mathbf{r}}{dt^2} = \mathbf{F}(x, y, z)$$

By operating on this equation from the left by $\mathbf{r}\times$ and using the result of Problem C–12, show that

$$\frac{d\mathbf{L}}{dt} = \mathbf{r} \times \mathbf{F}$$

where $\mathbf{L} = m\mathbf{r} \times d\mathbf{r}/dt = \mathbf{r} \times md\mathbf{r}/dt = \mathbf{r} \times m\mathbf{v} = \mathbf{r} \times \mathbf{p}$. This is the form of Newton's equations for a rotating system. Notice that $d\mathbf{L}/dt = 0$, or that angular momentum is conserved if $\mathbf{r} \times \mathbf{F} = 0$. Can you identify $\mathbf{r} \times \mathbf{F}$?

We start with Newton's equations,

$$m\frac{d^2\mathbf{r}}{dt^2} = \mathbf{F}(x, y, z)$$

Operating from the left by $\mathbf{r}\times$ gives

$$\mathbf{r} \times m\frac{d^2\mathbf{r}}{dt^2} = \mathbf{r} \times \mathbf{F}$$

We can use the results of Problem C–12 to rewrite this last equation as

$$m\frac{d}{dt}\left(\mathbf{r} \times \frac{d\mathbf{r}}{dt}\right) = \mathbf{r} \times \mathbf{F}$$

$$\frac{d}{dt}(\mathbf{r} \times \mathbf{p}) = \mathbf{r} \times \mathbf{F}$$

$$\frac{d\mathbf{L}}{dt} = \mathbf{r} \times \mathbf{F}$$

The quantity $\mathbf{r} \times \mathbf{F}$ is called torque.

CHAPTER 4

Some Postulates and General Principles
of Quantum Mechanics

PROBLEMS AND SOLUTIONS

4–1. Which of the following candidates for wave functions are normalizable over the indicated intervals?

a. $e^{-x^2/2}$ $(-\infty, \infty)$ **b.** e^x $(0, \infty)$ **c.** $e^{i\theta}$ $(0, 2\pi)$

d. $\sinh x$ $(0, \infty)$ **e.** xe^{-x} $(0, \infty)$

Normalize those that can be normalized. Are the others suitable wave functions?

Only the functions given by (a), (c), and (e) can be normalized. The functions given by (b) and (d) diverge as $x \to \infty$. If a function cannot be normalized, it is not a suitable wave function. Therefore (b) and (d) are not suitable wave functions. We now normalize the functions given by (a), (c), and (e).

a.
$$A^2 \int_{-\infty}^{\infty} e^{-x^2} dx = 1$$

$$2A^2 \int_0^{\infty} e^{-x^2} dx = 2A^2 \left(\frac{\pi}{4}\right)^{1/2} = 1$$

$$A = \left(\frac{1}{\pi}\right)^{1/4}$$

recalling that e^{-x^2} is an even function.

c.
$$A^2 \int_0^{2\pi} e^{-i\theta} e^{i\theta} d\theta = 1$$

$$A^2 \int_0^{2\pi} d\theta = A^2(2\pi) = 1$$

$$A = \left(\frac{1}{2\pi}\right)^{1/2}$$

e.
$$A^2 \int_0^{\infty} x^2 e^{-2x} dx = 1$$

$$\frac{A^2}{4} = 1$$

$$A = 2$$

4–2. Which of the following wave functions are normalized over the indicated two-dimensional intervals?

a. $e^{-(x^2+y^2)/2}$ $\begin{array}{l} 0 \le x < \infty \\ 0 \le y < \infty \end{array}$ **b.** $e^{-(x+y)/2}$ $\begin{array}{l} 0 \le x < \infty \\ 0 \le y < \infty \end{array}$

c. $\left(\dfrac{4}{ab}\right)^{1/2} \sin\dfrac{\pi x}{a} \sin\dfrac{\pi y}{b}$ $\begin{array}{l} 0 \le x \le a \\ 0 \le y \le b \end{array}$

Normalize those that aren't.

---

a.

$$\int_0^\infty dx \int_0^\infty dy\, e^{-(x^2+y^2)} = \int_0^\infty dx\, e^{-x^2} \int_0^\infty dy\, e^{-y^2}$$

$$= \left(\frac{\pi}{4}\right)^{1/2} \left(\frac{\pi}{4}\right)^{1/2}$$

$$= \frac{\pi}{4}$$

Therefore, to normalize the function, it must be multiplied by $\dfrac{2}{\sqrt{\pi}}$.

b.

$$\int_0^\infty dx \int_0^\infty dy\, e^{-(x+y)} = \int_0^\infty dx\, e^{-x} \int_0^\infty dy\, e^{-y} = 1$$

This function is normalized.

c.

$$\left(\frac{4}{ab}\right) \int_0^a dx\, \sin^2\frac{\pi x}{a} \int_0^b dy\, \sin^2\frac{\pi y}{b} = \left(\frac{4}{ab}\right)\left(\frac{a}{2}\right)\left(\frac{b}{2}\right) = 1$$

This function is normalized.

4–3. Why does $\psi^*\psi$ have to be everywhere real, nonnegative, finite, and of definite value?

This is required if $\psi^*\psi$ is to be a measure of probability.

4–4. In this problem, we will prove that the form of the Schrödinger equation imposes the condition that the first derivative of a wave function be continuous. The Schrödinger equation is

$$\frac{d^2\psi}{dx^2} + \frac{2m}{\hbar^2}[E - V(x)]\psi(x) = 0$$

If we integrate both sides from $a - \epsilon$ to $a + \epsilon$, where a is an arbitrary value of x and ϵ is infinitesimally small, then we have

$$\left.\frac{d\psi}{dx}\right|_{x=a+\epsilon} - \left.\frac{d\psi}{dx}\right|_{x=a-\epsilon} = \frac{2m}{\hbar^2} \int_{a-\epsilon}^{a+\epsilon} [V(x) - E]\psi(x)dx$$

Now show that $d\psi/dx$ is continuous if $V(x)$ is continuous.

Suppose now that $V(x)$ is *not* continuous at $x = a$, as in

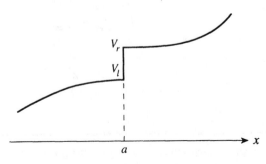

Show that

$$\frac{d\psi}{dx}\bigg|_{x=a+\epsilon} - \frac{d\psi}{dx}\bigg|_{x=a-\epsilon} = \frac{2m}{\hbar^2}[V_l + V_r - 2E]\psi(a)\epsilon$$

so that $d\psi/dx$ is continuous even if $V(x)$ has a *finite* discontinuity. What if $V(x)$ has an infinite discontinuity, as in the problem of a particle in a box? Are the first derivatives of the wave functions continuous at the boundaries of the box?

We start with

$$\frac{d\psi}{dx}\bigg|_{x=a+\epsilon} - \frac{d\psi}{dx}\bigg|_{x=a-\epsilon} = \frac{2m}{\hbar^2}\int_{a-\epsilon}^{a+\epsilon}[V(x) - E]\psi(x)dx \qquad (1)$$

Because $V(x)$ is continuous, $\lim_{\epsilon\to0} V(a \pm \epsilon) = V(a)$. We already know that $\psi(x)$ is continuous, so

$$\lim_{\epsilon\to0}\frac{2m}{\hbar^2}\int_{a-\epsilon}^{a+\epsilon}[V(x) - E]\psi(x)dx = \frac{2m}{\hbar^2}[V(a) - E]\,\psi(a)\lim_{\epsilon\to0}\int_{a-\epsilon}^{a+\epsilon}dx = 0 \qquad (2)$$

Combining Equations 1 and 2 shows that

$$\lim_{\epsilon\to0}\left(\frac{d\psi}{dx}\bigg|_{x=a+\epsilon} - \frac{d\psi}{dx}\bigg|_{x=a-\epsilon}\right) = 0$$

and therefore, $d\psi/dx$ is continuous. Now suppose that $V(x)$ has a finite discontinuity at $x = a$. We divide the integral into two parts, $a - \epsilon$ to a and a to $a + \epsilon$:

$$\frac{2m}{\hbar^2}\int_{a-\epsilon}^{a+\epsilon}[V(x) - E]\psi(x)dx = \frac{2m}{\hbar^2}\left[\int_{a-\epsilon}^{a}[V_l(x) - E]\psi(x)dx + \int_{a}^{a+\epsilon}[V_r(x) - E]\psi(x)dx\right]$$

$$= \frac{2m}{\hbar^2}\left\{[V_l(a) - E]\,\psi(a)\int_{a-\epsilon}^{a}dx + [V_r(a) - E]\,\psi(a)\int_{a}^{a+\epsilon}dx\right\}$$

$$= \frac{2m}{\hbar^2}[V_l(a) + V_r(a) - 2E]\,\psi(a)\epsilon$$

Because $\lim_{\epsilon\to0}[V_l(a) + V_r(a) - 2E]\,\psi(a)\epsilon = 0$, $d\psi/dx$ remains continuous even though $V(x)$ has a finite discontinuity. If $V(x)$ has an infinite discontinuity at an arbitrary point a, however, we cannot approach a from one side and therefore cannot integrate the expression $\frac{2m}{\hbar^2}\int_{a-\epsilon}^{a+\epsilon}[V(x) - E]\psi(x)dx$. This implies that $d\psi/dx$ is no longer a continuous function at a. For this reason, the first derivatives of the wave functions of a particle in a box are not continuous at (and only at) the boundaries of the box.

4–5. Determine whether the following functions are acceptable or not as state functions over the indicated intervals.

a. $\dfrac{1}{x}$ $(0, \infty)$

b. $e^{-2x}\sinh x$ $(0, \infty)$

c. $e^{-x}\cos x$ $(0, \infty)$

d. e^{x} $(-\infty, \infty)$

---

a. Unacceptable because it cannot be normalized.

b. Acceptable.

c. Acceptable.

d. Unacceptable because it cannot be normalized.

4-6. Calculate the values of $\sigma_E^2 = \langle E^2 \rangle - \langle E \rangle^2$ for a particle in a box in the state described by

$$\psi(x) = \left(\frac{630}{a^9}\right)^{1/2} x^2(a-x)^2 \qquad 0 \le x \le a$$

Using Postulate 4, $\langle E \rangle$ is given by

$$\langle E \rangle = \int_0^a \psi^*(x) \hat{H} \psi(x) dx \tag{4.11}$$

and $\langle E^2 \rangle$ is given by

$$\langle E^2 \rangle = \int_0^a \psi^*(x) \hat{H}^2 \psi(x) dx$$

We are interested in σ_E^2 for the state described by

$$\psi^*(x) = \psi(x) = \left(\frac{630}{a^9}\right)^{1/2} x^2(a-x)^2$$

To evaluate $\langle E \rangle$ and $\langle E^2 \rangle$, we will need to know the first four derivatives of ψ. These are given by

$$\frac{d\psi(x)}{dx} = \left(\frac{630}{a^9}\right)^{1/2} \left[2x(a-x)^2 - 2x^2(a-x)\right]$$

$$\frac{d^2\psi(x)}{dx^2} = \left(\frac{630}{a^9}\right)^{1/2} \left[2(a-x)^2 - 4x(a-x) - 4x(a-x) + 2x^2\right]$$

$$= \left(\frac{630}{a^9}\right)^{1/2} \left(2a^2 - 12ax + 12x^2\right)$$

$$\frac{d^3\psi(x)}{dx^3} = \left(\frac{630}{a^9}\right)^{1/2} (-12a + 24x)$$

$$\frac{d^4\psi(x)}{dx^4} = \left(\frac{630}{a^9}\right)^{1/2} (24)$$

Using these results,

$$\langle E \rangle = \int_0^a \psi^*(x) \left(-\frac{\hbar^2}{2m} \frac{d^2\psi(x)}{dx^2}\right) dx$$

$$= -\frac{\hbar^2}{2m} \int_0^a \left(\frac{630}{a^9}\right) x^2(a-x)^2 \left(2a^2 - 12ax + 12x^2\right) dx$$

$$= -\frac{630\hbar^2}{2ma^9} \int_0^a x^2(a-x)^2 \left(2a^2 - 12ax + 12x^2\right) dx$$

$$= -\frac{630\hbar^2}{2ma^9} \left[\int_0^a 2a^2x^2(a-x)^2 dx + \int_0^a -12ax^3(a-x)^2 dx + \int_0^a 12x^4(a-x)^2 dx\right]$$

or

$$\langle E \rangle = \frac{6\hbar^2}{ma^2} \tag{1}$$

We used the general integral

$$\int_0^1 x^m(1-x)^n dx = \frac{m!n!}{(m+n+1)!} \tag{2}$$

to evaluate the integrals in the last step to obtain Equation 1. We now evaluate $\langle E^2 \rangle$, which is given by

$$\langle E^2 \rangle = \int_0^a \psi^*(x) \hat{H}^2 \psi(x) dx$$

Substituting in the appropriate quantities gives

$$\langle E^2 \rangle = \int_0^a \psi^*(x) \left(\frac{\hbar^4}{4m^2} \frac{d^4 \psi(x)}{dx^4} \right) dx$$

$$= \frac{630\hbar^4}{4m^2 a^9} \int_0^a x^2(a-x)^2(24)dx$$

or

$$\langle E^2 \rangle = \frac{126\hbar^4}{m^2 a^4} \tag{3}$$

Using Equations 1 and 3 gives

$$\sigma_E^2 = \langle E^2 \rangle - \langle E \rangle^2 = \frac{126\hbar^4}{m^2 a^4} - \frac{36\hbar^4}{m^2 a^4} = \frac{90\hbar^4}{m^2 a^4}$$

4–7. Consider a free particle constrained to move over the rectangular region $0 \le x \le a$, $0 \le y \le b$. The energy eigenfunctions of this system are

$$\psi_{n_x, n_y}(x, y) = \left(\frac{4}{ab} \right)^{1/2} \sin \frac{n_x \pi x}{a} \sin \frac{n_y \pi y}{b} \quad \begin{array}{l} n_x = 1, 2, 3, \ldots \\ n_y = 1, 2, 3, \ldots \end{array}$$

The Hamiltonian operator for this system is

$$\hat{H} = -\frac{\hbar^2}{2m} \left(\frac{\partial^2}{\partial x^2} + \frac{\partial^2}{\partial y^2} \right)$$

Show that if the system is in one of its eigenstates, then

$$\sigma_E^2 = \langle E^2 \rangle - \langle E \rangle^2 = 0$$

If the system is in one of its eigenstates, then

$$\hat{H} \psi_{n_x n_y} = E_{n_x n_y} \psi_{n_x n_y}$$

$$\hat{H}^2 \psi_{n_x n_y} = E^2_{n_x n_y} \psi_{n_x n_y}$$

Therefore,

$$\langle E \rangle = \int\int \psi^*_{n_x n_y} \hat{H} \psi_{n_x n_y} = \int\int \psi^*_{n_x n_y} E_{n_x n_y} \psi_{n_x n_y}$$

$$= E_{n_x n_y} \int\int \psi^*_{n_x n_y} \psi_{n_x n_y} = E_{n_x n_y}$$

$$\langle E^2 \rangle = \int\int \psi^*_{n_x n_y} \hat{H}^2 \psi_{n_x n_y} = \int\int \psi^*_{n_x n_y} E^2_{n_x n_y} \psi_{n_x n_y}$$

$$= E^2_{n_x n_y} \int\int \psi^*_{n_x n_y} \psi_{n_x n_y} = E^2_{n_x n_y}$$

and $\sigma_E^2 = \langle E^2 \rangle - \langle E \rangle^2 = 0$.

4–8. The momentum operator in two dimensions is

$$\hat{\mathbf{P}} = -i\hbar \left(\mathbf{i}\frac{\partial}{\partial x} + \mathbf{j}\frac{\partial}{\partial y} \right)$$

Using the wave function given in Problem 4–7, calculate the value of $\langle p \rangle$ and then

$$\sigma_p^2 = \langle p^2 \rangle - \langle p \rangle^2$$

Compare your result with σ_p^2 in the one-dimensional case.

We find

$$\langle p \rangle = \int_0^a dx \int_0^b dy\, \psi_{n_x n_y}^*(x, y) \left[-i\hbar \left(\mathbf{i}\frac{\partial}{\partial x} + \mathbf{j}\frac{\partial}{\partial y} \right) \right] \psi_{n_x n_y}(x, y)$$

$$= -\frac{4i\hbar}{ab} \int_0^a dx \int_0^b dy \sin\frac{n_x\pi x}{a} \sin\frac{n_y\pi y}{b} \left(\mathbf{i}\frac{n_x\pi}{a} \cos\frac{n_x\pi x}{a} \sin\frac{n_y\pi y}{b} + \mathbf{j}\frac{n_y\pi}{b} \sin\frac{n_x\pi x}{a} \cos\frac{n_y\pi y}{b} \right)$$

$$= 0$$

and

$$\langle p^2 \rangle = \int_0^a dx \int_0^b dy\, \psi_{n_x n_y}^*(x, y) \left[-\hbar^2 \left(\frac{\partial^2}{\partial x^2} + \frac{\partial^2}{\partial y^2} \right) \right] \psi_{n_x n_y}(x, y)$$

$$= \frac{4\hbar^2}{ab} \int_0^a dx \int_0^b dy \sin\frac{n_x\pi x}{a} \sin\frac{n_y\pi y}{b} \left(\frac{n_x^2\pi^2}{a^2} + \frac{n_y^2\pi^2}{b^2} \right) \sin\frac{n_x\pi x}{a} \sin\frac{n_y\pi y}{b}$$

$$= \pi^2\hbar^2 \left(\frac{n_x^2}{a^2} + \frac{n_y^2}{b^2} \right) = \frac{h^2}{4} \left(\frac{n_x^2}{a^2} + \frac{n_y^2}{b^2} \right)$$

$$\sigma_p^2 = \langle p^2 \rangle - \langle p \rangle^2 = \langle p^2 \rangle = \frac{h^2}{4} \left(\frac{n_x^2}{a^2} + \frac{n_y^2}{b^2} \right)$$

This is an extension of the result in the one-dimensional case (Problem 3–25).

4–9. Suppose that a particle in a two-dimensional box (cf. Problem 4–7) is in the state

$$\psi(x, y) = \frac{30}{(a^5 b^5)^{1/2}} x(a - x)y(b - y)$$

Show that $\psi(x, y)$ is normalized, and then calculate the value of $\langle E \rangle$ associated with the state described by $\psi(x, y)$.

First, we show that $\psi(x, y)$ is normalized.

$$\int_0^a dx \int_0^b dy\, \psi^*\psi = \frac{900}{a^5 b^5} \int_0^a dx\, x^2(a - x)^2 \int_0^b dy\, y^2(b - y)^2$$

$$= \frac{900}{a^5 b^5} \left[\frac{(2)(2a^5)}{5!} \right]\left[\frac{(2)(2b^5)}{5!} \right]$$

$$= 1$$

We used Equation 2 of Problem 4–6 to evaluate the integrals over dx and dy. To find $\langle E \rangle$, we evaluate

$$\langle E \rangle = \frac{900}{a^5 b^5} \int_0^a dx \int_0^b dy \, x(a-x)y(b-y)$$

$$\left[-\frac{\hbar^2}{2m} \left(\frac{d^2}{dx^2} + \frac{d^2}{dy^2} \right) (ax - x^2)(by - y^2) \right]$$

$$= \frac{900\hbar^2}{ma^5 b^5} \int_0^a dx \int_0^b dy \, x(a-x)y(b-y)[y(b-y) + x(a-x)]$$

$$= \frac{900\hbar^2}{ma^5 b^5} \left[\int_0^a dx \int_0^b dy \, x(a-x)y^2(b-y)^2 + \int_0^a dx \int_0^b dy \, x^2(a-x)^2 y(b-y) \right]$$

$$= \frac{900\hbar^2}{ma^5 b^5} \left[\left(\frac{a^3}{3!} \right) \left(\frac{4b^5}{5!} \right) + \left(\frac{4a^5}{5!} \right) \left(\frac{b^3}{3!} \right) \right]$$

$$= \frac{5\hbar^2}{m} \left(\frac{1}{a^2} + \frac{1}{b^2} \right)$$

4–10. Show that

$$\psi_0(x) = \pi^{-1/4} e^{-x^2/2}$$

$$\psi_1(x) = (4/\pi)^{1/4} x e^{-x^2/2}$$

$$\psi_2(x) = (4\pi)^{-1/4}(2x^2 - 1)e^{-x^2/2}$$

are orthonormal over the interval $-\infty < x < \infty$.

All of the functions are real, so $\psi^*(x) = \psi(x)$. We want to show that

$$\int_{-\infty}^{\infty} \psi_n^*(x)\psi_m(x)dx = \delta_{nm}$$

$$\int_{-\infty}^{\infty} \psi_0^2(x)dx = \frac{2}{\pi^{1/2}} \int_0^{\infty} e^{-x^2}dx = 1$$

$$\int_{-\infty}^{\infty} \psi_1^2(x)dx = \frac{4}{\pi^{1/2}} \int_0^{\infty} x^2 e^{-x^2}dx = 1$$

$$\int_{-\infty}^{\infty} \psi_2^2(x)dx = \frac{1}{2\pi^{1/2}} \int_0^{\infty} (2x^2 - 1)^2 e^{-x^2}dx$$

$$= \frac{1}{\pi^{1/2}} \left[4 \left(\frac{3\pi^{1/2}}{8} \right) - 4 \left(\frac{\pi^{1/2}}{4} \right) + \frac{\pi^{1/2}}{2} \right] = 1$$

$$\int_{-\infty}^{\infty} \psi_0(x)\psi_1(x) = \left(\frac{2}{\pi} \right)^{1/2} \int_{-\infty}^{\infty} x e^{-x^2}dx = 0$$

$$\int_{-\infty}^{\infty} \psi_0(x)\psi_2(x) = \frac{1}{(2\pi)^{1/2}} \left[\int_{-\infty}^{\infty} 2x^2 e^{-x^2}dx - \int_{-\infty}^{\infty} e^{-x^2}dx \right]$$

$$= \frac{1}{(2\pi)^{1/2}} \left(\pi^{1/2} - \pi^{1/2} \right) = 0$$

$$\int_{-\infty}^{\infty} \psi_1(x)\psi_2 = \frac{1}{\pi^{1/2}} \int_{-\infty}^{\infty} x(2x^2 - 1)e^{-x^2} = 0$$

The last integral is easy to evaluate because the integrand is an odd function and the integral is over a symmetric interval.

4–11. Show that the polynomials

$$P_0(x) = 1, \qquad P_1(x) = x, \quad \text{and} \quad P_2(x) = \tfrac{1}{2}(3x^2 - 1)$$

satisfy the orthogonality relation

$$\int_{-1}^{1} P_l(x) P_n(x) dx = \frac{2\delta_{ln}}{2l + 1}$$

where δ_{ln} is the Kroenecker delta (Equation 4.30).

Again, all the functions are real, so $P_n^*(x) = P_n(x)$.

$$\int_{-1}^{1} P_0^2(x) dx = \int_{-1}^{1} dx = 2$$

$$\int_{-1}^{1} P_1^2(x) dx = \int_{-1}^{1} x^2 dx = \frac{2}{3}$$

$$\int_{-1}^{1} P_2^2(x) dx = \frac{1}{2} \int_{0}^{1} (9x^4 - 6x^2 + 1) dx = \frac{2}{5}$$

$$\int_{-1}^{1} P_0(x) P_1(x) dx = \int_{-1}^{1} x dx = 0$$

$$\int_{-1}^{1} P_0(x) P_2(x) dx = \int_{0}^{1} (3x^2 - 1) dx = 0$$

$$\int_{-1}^{1} P_1(x) P_2(x) dx = \frac{1}{2} \int_{-1}^{1} (3x^3 - x) dx = 0$$

4–12. Show that the set of functions $(2/a)^{1/2} \cos(n\pi x/a)$, $n = 0, 1, 2, \ldots$ is orthonormal over the interval $0 \le x \le a$.

Because the functions are real, $\psi_n^*(x) = \psi_n(x)$.

$$\int_{0}^{a} \psi_n(x) \psi_m(x) dx = \frac{2}{a} \int_{0}^{a} \cos \frac{n\pi x}{a} \cos \frac{m\pi x}{a} dx$$
$$= \delta_{nm}$$

This integral was solved explicitly in Problem 3–18.

4–13. Prove that if δ_{nm} is the Kroenecker delta

$$\delta_{nm} = \begin{cases} 1 & n = m \\ 0 & n \ne m \end{cases}$$

then

$$\sum_{n=1}^{\infty} c_n \delta_{nm} = c_m$$

and

$$\sum_{n} \sum_{m} a_n b_m \delta_{nm} = \sum_{n} a_n b_n$$

These results will be used later.

The sum $\sum_{n=1}^{\infty} c_n \delta_{nm} = c_m$ because every term is equal to zero except for those where $n = m$. Now

$$\sum_n \sum_m a_n b_m \delta_{nm} = \sum_n a_n \left(\sum_m b_m \delta_{nm} \right) = \sum_n a_n b_n$$

where we used the first result to evaluate $\sum_m b_m \delta_{nm}$.

4–14. Determine whether or not the following pairs of operators commute.

	$\hat{A}$	$\hat{B}$
(a)	$\dfrac{d}{dx}$	$\dfrac{d^2}{dx^2} + 2\dfrac{d}{dx}$
(b)	x	$\dfrac{d}{dx}$
(c)	SQR	SQRT
(d)	$x^2 \dfrac{d}{dx}$	$\dfrac{d^2}{dx^2}$

a.

$$\hat{A}\hat{B}f = \frac{d}{dx}\left(\frac{d^2 f}{dx^2} + 2\frac{df}{dx} \right) = \frac{d^3 f}{dx^3} + 2\frac{d^2 f}{dx^2}$$

$$\hat{B}\hat{A}f = \left(\frac{d^2}{dx^2} + 2\frac{d}{dx} \right)\frac{df}{dx} = \frac{d^3 f}{dx^3} + 2\frac{d^2 f}{dx^2}$$

$$\hat{A}\hat{B}f = \hat{B}\hat{A}f$$

This pair of operators commutes.

b.

$$\hat{A}\hat{B}f = x\frac{df}{dx}$$

$$\hat{B}\hat{A}f = \frac{d}{dx}(xf) = f + x\frac{df}{dx}$$

$$\hat{A}\hat{B}f \neq \hat{B}\hat{A}f$$

This pair of operators does not commute.

c.

$$\hat{A}\hat{B}f = [\text{SQRT}(f)]^2 = \left(\pm f^{1/2} \right)^2 = f$$

$$\hat{B}\hat{A}f = \text{SQRT}(f^2) = \pm f$$

$$\hat{A}\hat{B}f \neq \hat{B}\hat{A}f$$

This pair of operators does not commute.

d.

$$\hat{A}\hat{B}f = x^2 \frac{d}{dx}\left(\frac{d^2 f}{dx^2} \right) = x^2 \frac{d^3 f}{dx^3}$$

$$\hat{B}\hat{A}f = \frac{d^2}{dx^2}\left(x^2 \frac{df}{dx} \right) = x^2 \frac{d^3 f}{dx^3} + 4x\frac{d^2 f}{dx^2} + 2\frac{df}{dx}$$

$$\hat{A}\hat{B}f \neq \hat{B}\hat{A}f$$

This pair of operators does not commute.

4–15. In ordinary algebra, $(P + Q)(P - Q) = P^2 - Q^2$. Expand $(\hat{P} + \hat{Q})(\hat{P} - \hat{Q})$. Under what conditions do we find the same result as in the case of ordinary algebra?

$$(\hat{P} + \hat{Q})(\hat{P} - \hat{Q}) = \hat{P}^2 - \hat{Q}^2 + \hat{Q}\hat{P} - \hat{P}\hat{Q} = \hat{P}^2 - \hat{Q}^2 + \left[\hat{Q}, \hat{P}\right]$$

In order for this result to equal the result found in ordinary algebra, $\hat{P}$ and $\hat{Q}$ must commute.

4–16. Evaluate the commutator $[\hat{A}, \hat{B}]$, where $\hat{A}$ and $\hat{B}$ are given below.

	$\hat{A}$	$\hat{B}$
(a)	$\dfrac{d^2}{dx^2}$	x
(b)	$\dfrac{d}{dx} - x$	$\dfrac{d}{dx} + x$
(c)	$\displaystyle\int_0^x dx$	$\dfrac{d}{dx}$
(d)	$\dfrac{d^2}{dx^2} - x$	$\dfrac{d}{dx} + x^2$

---

a.

$$\hat{A}\hat{B}f = \frac{d^2}{dx^2}(xf) = 2\frac{df}{dx} + x\frac{d^2f}{dx^2}$$

$$\hat{B}\hat{A}f = x\frac{d^2f}{dx^2}$$

$$[\hat{A}, \hat{B}] = \hat{A}\hat{B}f - \hat{B}\hat{A}f = 2\frac{d}{dx}$$

b.

$$\hat{A}\hat{B}f = \left(\frac{d}{dx} - x\right)\left(\frac{df}{dx} + xf\right) = \frac{d^2f}{dx^2} + f - x^2f$$

$$\hat{B}\hat{A}f = \left(\frac{d}{dx} + x\right)\left(\frac{df}{dx} - xf\right) = \frac{d^2f}{dx^2} - f - x^2f$$

$$[\hat{A}, \hat{B}] = 2$$

c.

$$\hat{A}\hat{B}f = \int_0^x dx'\frac{df}{dx'} = f(x) - f(0)$$

$$\hat{B}\hat{A}f = \frac{d}{dx}\int_0^x dx' f(x') = f(x)$$

$$[\hat{A}, \hat{B}]f = -f(0)$$

d.

$$\hat{A}\hat{B}f = \left(\frac{d^2}{dx^2} - x\right)\left(\frac{df}{dx} + x^2f\right) = \frac{d^3f}{dx^3} + x^2\frac{d^2f}{dx^2} + 3x\frac{df}{dx} + (2 - x^3)f$$

$$\hat{B}\hat{A}f = \left(\frac{d}{dx} + x^2\right)\left(\frac{d^2f}{dx^2} - xf\right) = \frac{d^3f}{dx^3} + x^2\frac{d^2f}{dx^2} - x\frac{df}{dx} - (1 + x^3)f$$

$$[\hat{A}, \hat{B}] = 4x\frac{d}{dx} + 3$$

4–17. Referring to Table 4.1 for the operator expressions for angular momentum, show that

$$[\hat{L}_x, \hat{L}_y] = i\hbar \hat{L}_z$$

$$[\hat{L}_y, \hat{L}_z] = i\hbar \hat{L}_x$$

and

$$[\hat{L}_z, \hat{L}_x] = i\hbar \hat{L}_y$$

(Do you see a pattern here to help remember these commutation relations?) What do these expressions say about the ability to measure the components of angular momentum simultaneously?

$$\hat{L}_x \hat{L}_y f = \left[-i\hbar \left(y\frac{\partial}{\partial z} - z\frac{\partial}{\partial y} \right) \right] \left[-i\hbar \left(z\frac{\partial f}{\partial x} - x\frac{\partial f}{\partial z} \right) \right]$$

$$= -\hbar^2 \left(y\frac{\partial f}{\partial x} + yz\frac{\partial^2 f}{\partial z \partial x} - xy\frac{\partial^2 f}{\partial z^2} - z^2\frac{\partial^2 f}{\partial x \partial y} + xz\frac{\partial^2 f}{\partial y \partial z} \right)$$

$$\hat{L}_y \hat{L}_x f = \left[-i\hbar \left(z\frac{\partial}{\partial x} - x\frac{\partial}{\partial z} \right) \right] \left[-i\hbar \left(y\frac{\partial f}{\partial z} - z\frac{\partial f}{\partial y} \right) \right]$$

$$= -\hbar^2 \left(yz\frac{\partial^2 f}{\partial x \partial z} - z^2\frac{\partial^2 f}{\partial x \partial y} - xy\frac{\partial^2 f}{\partial z^2} + x\frac{\partial f}{\partial y} + xz\frac{\partial^2 f}{\partial z \partial y} \right)$$

$$\left[\hat{L}_x, \hat{L}_y \right] = -\hbar^2 \left(y\frac{\partial}{\partial x} - x\frac{\partial}{\partial y} \right) = i\hbar \left[-i\hbar \left(x\frac{\partial}{\partial y} - y\frac{\partial}{\partial x} \right) \right]$$

$$= i\hbar \hat{L}_z$$

In the same way, we can show $[\hat{L}_y, \hat{L}_z] = i\hbar \hat{L}_x$ and $[\hat{L}_z, \hat{L}_x] = i\hbar \hat{L}_y$. The pattern involves the cyclic permutation of x, y, and z. Since no combination of the operators $\hat{L}_x$, $\hat{L}_y$, and $\hat{L}_z$ commutes, it is not possible to simultaneously measure any two of the three components of angular momentum to arbitrary precision (as discussed in Section 4–6).

4–18. Defining

$$\hat{L}^2 = \hat{L}_x^2 + \hat{L}_y^2 + \hat{L}_z^2$$

show that $\hat{L}^2$ commutes with each component separately. What does this result tell you about the ability to measure the square of the total angular momentum and its components simultaneously?

$$\hat{L}_x^2 \hat{L}_x - \hat{L}_x \hat{L}_x^2 = 0$$

$$\hat{L}_y^2 \hat{L}_x - \hat{L}_x \hat{L}_y^2 = \hat{L}_y^2 \hat{L}_x - \left(\hat{L}_y \hat{L}_x + i\hbar \hat{L}_z \right) \hat{L}_y$$

$$\hat{L}_z^2 \hat{L}_x - \hat{L}_x \hat{L}_z^2 = \hat{L}_z^2 \hat{L}_x - \left(\hat{L}_z \hat{L}_x - i\hbar \hat{L}_y \right) \hat{L}_z$$

$$\left[\hat{L}^2, \hat{L}_x \right] = 0 + \hat{L}_y^2 \hat{L}_x - \hat{L}_y \hat{L}_x \hat{L}_y - i\hbar \hat{L}_z \hat{L}_y + \hat{L}_z^2 \hat{L}_x - \hat{L}_z \hat{L}_x \hat{L}_z + i\hbar \hat{L}_y \hat{L}_z$$

$$= \hat{L}_y \left(\hat{L}_y \hat{L}_x - \hat{L}_x \hat{L}_y \right) + \hat{L}_z \left(\hat{L}_z \hat{L}_x - \hat{L}_x \hat{L}_z \right) - i\hbar \left(\hat{L}_z \hat{L}_y - \hat{L}_y \hat{L}_z \right)$$

$$= \hat{L}_y \left(-i\hbar \hat{L}_z \right) + \hat{L}_z \left(i\hbar \hat{L}_y \right) - i\hbar \left(-i\hbar \hat{L}_x \right)$$

$$= i\hbar \left(\hat{L}_z \hat{L}_y - \hat{L}_y \hat{L}_z \right) - \hbar^2 \hat{L}_x$$

$$= i\hbar \left(-i\hbar \hat{L}_x \right) - \hbar^2 \hat{L}_x = 0$$

In a similar way, we can show that $\hat{L}^2$ commutes with $\hat{L}_y$ and $\hat{L}_z$. This result tells us that we can simultaneously measure the square of the total angular momentum and any of its components to arbitrary precision (as discussed in Section 4–6).

4–19. In Chapter 6 we will use the operators

$$\hat{L}_+ = \hat{L}_x + i\hat{L}_y$$

and

$$\hat{L}_- = \hat{L}_x - i\hat{L}_y$$

Show that

$$\hat{L}_+ \hat{L}_- = \hat{L}^2 - \hat{L}_z^2 + \hbar \hat{L}_z$$

$$[\hat{L}_z, \hat{L}_+] = \hbar \hat{L}_+$$

and that

$$[\hat{L}_z, \hat{L}_-] = -\hbar \hat{L}_-$$

$$\hat{L}_+ \hat{L}_- = \left(\hat{L}_x + i\hat{L}_y\right)\left(\hat{L}_x - i\hat{L}_y\right) = \hat{L}_x^2 + i\hat{L}_y\hat{L}_x - i\hat{L}_x\hat{L}_y + \hat{L}_y^2$$

$$= \hat{L}_x^2 + \hat{L}_y^2 + i[\hat{L}_y, \hat{L}_x] = \hat{L}_x^2 + \hat{L}_y^2 + \hbar\hat{L}_z = \hat{L}^2 - \hat{L}_z^2 + \hbar\hat{L}_z$$

where we used the fact that $\hat{L}_x^2 + \hat{L}_y^2 = \hat{L}^2 - \hat{L}_z^2$.

$$\left[\hat{L}_z, \hat{L}_+\right] = \hat{L}_z\hat{L}_x + i\hat{L}_z\hat{L}_y - \hat{L}_x\hat{L}_z - i\hat{L}_y\hat{L}_z$$

$$= [\hat{L}_z, \hat{L}_x] + i[\hat{L}_z, \hat{L}_y] = i\hbar\hat{L}_y + \hbar\hat{L}_x$$

$$= \hbar\hat{L}_+$$

$$\left[\hat{L}_z, \hat{L}_-\right] = [\hat{L}_z, \hat{L}_x] - i[\hat{L}_z, \hat{L}_y] = i\hbar\hat{L}_y - \hbar\hat{L}_x$$

$$= -\hbar\hat{L}_-$$

4–20. Consider a particle in a two-dimensional box. Determine $[\hat{X}, \hat{P}_y]$, $[\hat{X}, \hat{P}_x]$, $[\hat{Y}, \hat{P}_y]$, and $[\hat{Y}, \hat{P}_x]$.

From Equation 3.56, we have

$$\psi(x, y) = \left(\frac{4}{ab}\right)^{1/2} \sin\frac{n_x\pi x}{a} \sin\frac{n_y\pi y}{b}$$

a.

$$\hat{X}\hat{P}_y\psi = x\left(-i\hbar\frac{\partial\psi}{\partial y}\right)$$

$$= -i\hbar\left(\frac{4}{ab}\right)^{1/2}\frac{n_y\pi x}{b}\sin\frac{n_x\pi x}{a}\cos\frac{n_y\pi y}{b}$$

$$\hat{P}_y\hat{X}\psi = -i\hbar\frac{\partial}{\partial y}\left(x\left(\frac{4}{ab}\right)^{1/2}\sin\frac{n_x\pi x}{a}\sin\frac{n_y\pi y}{b}\right)$$

$$= -i\hbar\left(\frac{4}{ab}\right)^{1/2}\frac{n_y\pi x}{b}\sin\frac{n_x\pi x}{a}\cos\frac{n_y\pi y}{b}$$

$$\left[\hat{X}, \hat{P}_y\right] = \hat{X}\hat{P}_y - \hat{P}_y\hat{X} = 0$$

b.

$$\hat{X}\hat{P}_x\psi = x\left(-i\hbar\frac{\partial\psi}{\partial x}\right)$$

$$= -i\hbar\left(\frac{4}{ab}\right)^{1/2}\frac{n_x\pi x}{a}\cos\frac{n_x\pi x}{a}\sin\frac{n_y\pi y}{b}$$

$$\hat{P}_x\hat{X}\psi = -i\hbar\frac{\partial}{\partial x}\left(x\left(\frac{4}{ab}\right)^{1/2}\sin\frac{n_x\pi x}{a}\sin\frac{n_y\pi y}{b}\right)$$

$$= -i\hbar\left(\frac{4}{ab}\right)^{1/2}\left(\sin\frac{n_x\pi x}{a}\sin\frac{n_y\pi y}{b} + \frac{n_x\pi x}{a}\cos\frac{n_x\pi x}{a}\sin\frac{n_y\pi y}{b}\right)$$

$$= -i\hbar\psi - i\hbar\left(\frac{4}{ab}\right)^{1/2}\frac{n_x\pi x}{a}\cos\frac{n_x\pi x}{a}\sin\frac{n_y\pi y}{b}$$

$$\left[\hat{X},\hat{P}_x\right] = \hat{X}\hat{P}_x - \hat{P}_x\hat{X} = i\hbar$$

c.

$$\hat{Y}\hat{P}_y\psi = y\left(-i\hbar\frac{\partial\psi}{\partial y}\right)$$

$$= -i\hbar\left(\frac{4}{ab}\right)^{1/2}\frac{n_y\pi y}{b}\sin\frac{n_x\pi x}{a}\cos\frac{n_y\pi y}{b}$$

$$\hat{P}_y\hat{Y}\psi = -i\hbar\frac{\partial}{\partial y}\left(y\left(\frac{4}{ab}\right)^{1/2}\sin\frac{n_x\pi x}{a}\sin\frac{n_y\pi y}{b}\right)$$

$$= -i\hbar\left(\frac{4}{ab}\right)^{1/2}\left(\sin\frac{n_x\pi x}{a}\sin\frac{n_y\pi y}{b} + \frac{n_y\pi y}{b}\sin\frac{n_x\pi x}{a}\cos\frac{n_y\pi y}{b}\right)$$

$$= -i\hbar\psi - i\hbar\left(\frac{4}{ab}\right)^{1/2}\frac{n_y\pi y}{b}\sin\frac{n_x\pi x}{a}\cos\frac{n_y\pi y}{b}$$

$$\left[\hat{Y},\hat{P}_y\right] = \hat{Y}\hat{P}_y - \hat{P}_y\hat{Y} = i\hbar$$

d.

$$\hat{Y}\hat{P}_x\psi = y\left(-i\hbar\frac{\partial\psi}{\partial x}\right)$$

$$= -i\hbar\left(\frac{4}{ab}\right)^{1/2}\frac{n_x\pi y}{a}\cos\frac{n_x\pi x}{a}\sin\frac{n_y\pi y}{b}$$

$$\hat{P}_x\hat{Y}\psi = -i\hbar\frac{\partial}{\partial x}\left(y\left(\frac{4}{ab}\right)^{1/2}\sin\frac{n_x\pi x}{a}\sin\frac{n_y\pi y}{b}\right)$$

$$= -i\hbar\left(\frac{4}{ab}\right)^{1/2}\frac{n_x\pi y}{a}\cos\frac{n_x\pi x}{a}\sin\frac{n_y\pi y}{b}$$

$$\left[\hat{Y},\hat{P}_x\right] = \hat{Y}\hat{P}_x - \hat{P}_x\hat{Y} = 0$$

4–21. Can the position and total angular momentum of any electron be measured simultaneously to arbitrary precision?

Yes. The position and total angular momentum operators are vector operators given by $\hat{\mathbf{R}} = \mathbf{i}\hat{x} + \mathbf{j}\hat{y} + \mathbf{k}\hat{z}$ and $\hat{\mathbf{L}} = \mathbf{i}\hat{L}_x + \mathbf{j}\hat{L}_y + \mathbf{k}\hat{L}_z$. We are interested in whether $\hat{\mathbf{R}}$ and $\hat{\mathbf{L}}$ commute.

$$[\hat{\mathbf{R}}, \hat{\mathbf{L}}] = [\mathbf{i}\hat{x} + \mathbf{j}\hat{y} + \mathbf{k}\hat{z}, \mathbf{i}\hat{L}_x + \mathbf{j}\hat{L}_y + \mathbf{k}\hat{L}_z]$$
$$= [\hat{x}, \hat{L}_x] + [\hat{y}, \hat{L}_y] + [\hat{z}, \hat{L}_z]$$
$$= 0$$

where we have used the fact that $\mathbf{i} \cdot \mathbf{i} = \mathbf{j} \cdot \mathbf{j} = \mathbf{k} \cdot \mathbf{k} = 1$ and $\mathbf{i} \cdot \mathbf{j} = \mathbf{j} \cdot \mathbf{k} = \mathbf{k} \cdot \mathbf{i} = 0$ (MathChapter C). Therefore, the position and total angular momentum of any electron can be measured simultaneously to arbitrary precision.

4–22. Can the angular momentum and kinetic energy of a particle be measured simultaneously to arbitrary precision?

Yes. The proof is similar to that in Problem 4–21.

$$[\hat{\mathbf{K}}, \hat{\mathbf{L}}] = [\hat{K}_x, \hat{L}_x] + [\hat{K}_y, \hat{L}_y] + [\hat{K}_z, \hat{L}_z]$$
$$= 0$$

We see from this that the kinetic energy and angular momentum of the electron can be simultaneously measured to arbitrary precision.

4–23. Using the result of Problem 4–20, what are the "uncertainty relationships" $\Delta x \Delta p_y$ and $\Delta y \Delta p_x$ equal to?

Because $[\hat{X}, \hat{P}_y]$ and $[\hat{Y}, \hat{P}_x]$ are both equal to zero (Problem 4–20), the "uncertainty relationships" are $\Delta x \Delta p_y = \Delta y \Delta p_x = 0$. This means that the quantities x, p_y and y, p_x can be measured simultaneously to arbitrary precision.

4–24. We can define functions of operators through their Taylor series (MathChapter I). For example, we define the operator $\exp(\hat{S})$ by

$$e^{\hat{S}} = \sum_{n=0}^{\infty} \frac{(\hat{S})^n}{n!}$$

Under what conditions does the equality

$$e^{\hat{A}+\hat{B}} = e^{\hat{A}} e^{\hat{B}}$$

hold?

Let $\hat{I}$ be the identity operator. Then

$$e^{\hat{A}+\hat{B}} = \hat{I} + \hat{A} + \hat{B} + \frac{(\hat{A}+\hat{B})^2}{2!} + O[(\hat{A}+\hat{B})^3]$$
$$= \hat{I} + \hat{A} + \hat{B} + \frac{\hat{A}^2}{2} + \frac{\hat{B}^2}{2} + \frac{\hat{A}\hat{B}}{2} + \frac{\hat{B}\hat{A}}{2} + O[(\hat{A}+\hat{B})^3]$$

and

$$e^{\hat{A}}e^{\hat{B}} = \left[\hat{I} + \hat{A} + \frac{\hat{A}^2}{2} + O(\hat{A}^3)\right]\left[\hat{I} + \hat{B} + \frac{\hat{B}^2}{2} + O(\hat{B}^3)\right]$$

$$= \hat{I} + \hat{A} + \hat{B} + \frac{\hat{A}^2}{2} + \frac{\hat{B}^2}{2} + \hat{A}\hat{B} + O[(\hat{A} + \hat{B})^3]$$

These two expressions are equivalent only if $[\hat{A}, \hat{B}] = 0$; in other words, only if $\hat{A}$ and $\hat{B}$ commute.

4–25. In this chapter, we learned that if ψ_n is an eigenfunction of the time-independent Schrödinger equation, then

$$\Psi_n(x, t) = \psi_n(x)e^{-iE_n t/\hbar}$$

Show that if ψ_m and ψ_n are both stationary states of $\hat{H}$, then the state

$$\Psi(x, t) = c_m \psi_m(x)e^{-iE_m t/\hbar} + c_n \psi_n(x)e^{-iE_n t/\hbar}$$

satisfies the time-dependent Schrödinger equation.

Postulate 5 gives the time-dependent Schrödinger equation as

$$\hat{H}\Psi(x, t) = i\hbar \frac{\partial \Psi(x, t)}{\partial t}$$

We will substitute Ψ into each side of this equation separately to show that the equivalence holds. The left side becomes

$$\hat{H}\Psi = \hat{H}\left[c_m \psi_m e^{-iE_m t/\hbar} + c_n \psi_n e^{-iE_n t/\hbar}\right]$$

$$= E_m c_m \psi_m e^{-iE_m t/\hbar} + E_n c_n \psi_n e^{-iE_n t/\hbar}$$

and the right side is

$$i\hbar \frac{\partial \Psi}{\partial t} = i\hbar\left(\frac{-iE_m}{\hbar}c_m \psi_m e^{-iE_m t}\hbar - \frac{iE_n}{\hbar}c_n \psi_n e^{-iE_n t}\hbar\right)$$

$$= c_m E_m \psi_m e^{-iE_m t/\hbar} + c_n E_n \psi_n e^{-iE_n t/\hbar}$$

4–26. Starting with

$$\langle x \rangle = \int \Psi^*(x, t)x\Psi(x, t)dx$$

and the time-dependent Schrödinger equation, show that

$$\frac{d\langle x \rangle}{dt} = \int \Psi^* \frac{i}{\hbar}(\hat{H}x - x\hat{H})\Psi dx$$

Given that

$$\hat{H} = -\frac{\hbar^2}{2m}\frac{d^2}{dx^2} + V(x)$$

show that

$$\hat{H}x - x\hat{H} = -2\frac{\hbar^2}{2m}\frac{d}{dx} = -\frac{\hbar^2}{m}\frac{i}{\hbar}\hat{P}_x = -\frac{i\hbar}{m}\hat{P}_x$$

Finally, substitute this result into the equation for $d\langle x\rangle/dt$ to show that

$$m\frac{d\langle x\rangle}{dt} = \langle \hat{P}_x\rangle$$

Interpret this result.

$$\langle x\rangle = \int \Psi^* x\Psi\, dx$$

$$\frac{d\langle x\rangle}{dt} = \int \frac{\partial \Psi^*}{\partial t} x\Psi\, dx + \int \Psi^* x\frac{\partial \Psi}{\partial t}\, dx$$

Using Postulate 5 to express $\partial\Psi/\partial t$, we can write

$$\frac{d\langle x\rangle}{dt} = -\frac{1}{i\hbar}\int (\hat{H}\Psi)^* x\Psi\, dx + \frac{1}{i\hbar}\int \Psi^* x(\hat{H}\Psi)\, dx$$

Because $\hat{H}$ is Hermitian, this is equivalent to

$$\frac{d\langle x\rangle}{dt} = -\frac{1}{i\hbar}\int \Psi^* \hat{H}x\Psi\, dx + \frac{1}{i\hbar}\int \Psi^* x(\hat{H}\Psi)\, dx$$

$$= \frac{i}{\hbar}\int \Psi^*(\hat{H}x - x\hat{H})\Psi\, dx \qquad (1)$$

Now, using the given expression for $\hat{H}$, we find that

$$(\hat{H}x - x\hat{H})f = \left[\frac{-\hbar^2}{2m}\frac{d^2}{dx^2} + V(x)\right]xf - x\left[\frac{-\hbar^2}{2m}\frac{d^2f}{dx^2} + V(x)f\right]$$

$$= \frac{-\hbar^2 x}{2m}\frac{d^2f}{dx^2} = \frac{-2\hbar^2}{2m}\frac{df}{dx} + V(x)xf + \frac{+\hbar^2 x}{2m}\frac{d^2f}{dx^2} - xV(x)f$$

$$(\hat{H}x - x\hat{H}) = \frac{-\hbar^2}{m}\frac{d}{dx} = \frac{-\hbar^2}{m}\frac{i}{\hbar}\hat{P}_x$$

$$= \frac{-i\hbar}{m}\hat{P}_x \qquad (2)$$

Substituting Equation 2 into Equation 1 gives

$$m\frac{d\langle x\rangle}{dt} = \langle \hat{P}_x\rangle$$

which is the quantum mechanical equivalent of the classical definition of linear momentum, $p = mv$.

4–27. Generalize the result of Problem 4–26 and show that if F is any dynamical quantity, then

$$\frac{d\langle F\rangle}{dt} = \int \Psi^* \frac{i}{\hbar}(\hat{H}\hat{F} - \hat{F}\hat{H})\Psi\, dx$$

Use this equation to show that

$$\frac{d\langle \hat{P}_x\rangle}{dt} = \left\langle -\frac{dV}{dx}\right\rangle$$

Interpret this result. This last equation is known as *Ehrenfest's theorem*.

Replace x by $\hat{F}$ in the first part of the previous problem to show that

$$\frac{d\langle F \rangle}{dt} = \int \Psi^* \frac{i}{\hbar} (\hat{H}\hat{F} - \hat{F}\hat{H})\Psi dx \tag{1}$$

Now consider the case where $\hat{F} = \hat{P}_x$:

$$\hat{H}\hat{P}_x = -i\hbar \hat{H}\frac{d}{dx}$$

$$= -\frac{\hbar^2}{2m}(-i\hbar)\frac{d^3}{dx^3} - i\hbar V(x)\frac{d}{dx}$$

$$\hat{P}_x\hat{H} = -i\hbar\left(-\frac{\hbar^2}{2m}\right)\frac{d^3}{dx^3} - i\hbar V(x)\frac{d}{dx} - i\hbar\frac{dV}{dx}$$

$$\left[\hat{H}, \hat{P}_x\right] = \hat{H}\hat{P}_x - \hat{P}_x\hat{H} = i\hbar\frac{dV}{dx}$$

Substituting this result into Equation 1 gives

$$\frac{d\langle \hat{P}_x \rangle}{dt} = -\int \Psi^* \frac{dV}{dx}\Psi dx = \left\langle -\frac{dV}{dx}\right\rangle$$

Ehrenfest's theorem is the quantum mechanical equivalent of Newton's law, $F = ma$.

4–28. The fact that eigenvalues, which correspond to physically observable quantities, must be real imposes a certain condition on quantum-mechanical operators. To see what this condition is, start with

$$\hat{A}\psi = a\psi \tag{1}$$

where $\hat{A}$ and ψ may be complex, but a must be real. Multiply Equation 1 from the left by ψ^* and then integrate to obtain

$$\int \psi^*\hat{A}\psi d\tau = a\int \psi^*\psi d\tau = a \tag{2}$$

Now take the complex conjugate of Equation 1, multiply from the left by ψ, and then integrate to obtain

$$\int \psi \hat{A}^*\psi^* d\tau = a^* = a \tag{3}$$

Equate the left sides of Equations 2 and 3 to give

$$\int \psi^*\hat{A}\psi d\tau = \int \psi \hat{A}^*\psi^* d\tau \tag{4}$$

This is the condition that an operator must satisfy if its eigenvalues are to be real. Such operators are called Hermitian operators.

We start with

$$\hat{A}\psi = a\psi \tag{1}$$

Multiplying Equation 1 from the left by ψ^* and then integrating gives

$$\int \psi^*\hat{A}\psi d\tau = \int \psi^*a\psi d\tau = a\int \psi^*\psi d\tau = a \tag{2}$$

The complex conjugate of Equation 1 is

$$\hat{A}^*\psi^* = a^*\psi^*$$

Multiplying this expression from the left by ψ and then integrating gives

$$\int \psi \hat{A}^*\psi^*d\tau = \int \psi a^*\psi^*d\tau = a^*\int \psi\psi^*d\tau = a \qquad (2)$$

because we have imposed the restriction that a is real. Equating Equations 2 and 3 shows that if a is real, then

$$\int \psi^*\hat{A}\psi d\tau = \int \psi \hat{A}^*\psi^*d\tau$$

4–29. In this problem, we will prove that not only are the eigenvalues of Hermitian operators real but that their eigenfunctions are orthogonal. Consider the two eigenvalue equations

$$\hat{A}\psi_n = a_n\psi_n \quad \text{and} \quad \hat{A}\psi_m = a_m\psi_m$$

Multiply the first equation by ψ_m^* and integrate; then take the complex conjugate of the second, multiply by ψ_n, and integrate. Subtract the two resulting equations from each other to get

$$\int_{-\infty}^{\infty} \psi_m^*\hat{A}\psi_n dx - \int_{-\infty}^{\infty} \psi_n\hat{A}^*\psi_m^*dx = (a_n - a_m^*)\int_{-\infty}^{\infty} \psi_m^*\psi_n dx$$

Because $\hat{A}$ is Hermitian, the left side is zero, and so

$$(a_n - a_m^*)\int_{-\infty}^{\infty} \psi_m^*\psi_n dx = 0$$

Discuss the two possibilities $n = m$ and $n \neq m$. Show that $a_n = a_n^*$, which is just another proof that the eigenvalues are real. When $n \neq m$, show that

$$\int_{-\infty}^{\infty} \psi_m^*\psi_n dx = 0 \qquad m \neq n$$

if the system is nondegenerate. Are ψ_m and ψ_n necessarily orthogonal if they are degenerate?

Carrying out the stated steps gives

$$\int \psi_m^*\hat{A}\psi_n d\tau = a_n\int \psi_m^*\psi_n d\tau$$

$$\int \psi_n\hat{A}^*\psi_m^*d\tau = a_m^*\int \psi_n\psi_m^*d\tau = a_m^*\int \psi_m^*\psi_n d\tau$$

Subtracting these two expressions gives

$$\int \psi_m^*\hat{A}\psi_n d\tau - \int \psi_n\hat{A}^*\psi_m^*d\tau = (a_n - a_m^*)\int \psi_m^*\psi_n d\tau = 0$$

We set this last equation equal to zero because (as stated in the question) $\hat{A}$ is Hermitian. If $n = m$, then the integral $\int \psi_n^*\psi_n d\tau$ is equal to one and so the above equation tells us that $a_n = a_m^* (= a_n^*)$. In other words, a must be real. If $n \neq m$, for a non-degenerate system $(a_n - a_m^*)$ will be nonzero and so

$$\int \psi_m^*\psi_n d\tau = 0 \qquad n \neq m$$

Note that ψ_m and ψ_n are not necessarily orthogonal if they are degenerate, because a_n can equal a_m^*.

4–30. All the operators in Table 4.1 are Hermitian. In this problem, we show how to determine if an operator is Hermitian. Consider the operator $\hat{A} = d/dx$. If $\hat{A}$ is Hermitian, it will satisfy Equation 4 of Problem 4–28. Substitute $\hat{A} = d/dx$ into Equation 4 and integrate by parts to obtain

$$\int_{-\infty}^{\infty} \psi^* \frac{d\psi}{dx} dx = \left. \psi^* \psi \right|_{-\infty}^{\infty} - \int_{-\infty}^{\infty} \psi \frac{d\psi^*}{dx} dx$$

For a wave function to be normalizable, it must vanish at infinity, so the first term on the right side is zero. Therefore, we have

$$\int_{-\infty}^{\infty} \psi^* \frac{d}{dx} \psi \, dx = - \int_{-\infty}^{\infty} \psi \frac{d}{dx} \psi^* dx$$

For an arbitrary function $\psi(x)$, d/dx does *not* satisfy Equation 4 of Problem 4–28, so it is *not* Hermitian.

We will use the fact that $\int v \, du = uv - \int u \, dv$. Let ψ^* be v and $\frac{d\psi}{dx} dx$ be du. Then

$$\int_{-\infty}^{\infty} \psi^* \frac{d\psi}{dx} dx = \left. \psi^* \psi \right|_{-\infty}^{\infty} - \int_{-\infty}^{\infty} \psi \frac{d\psi^*}{dx} dx$$

$$= - \int_{-\infty}^{\infty} \psi \frac{d}{dx} \psi^* dx$$

4–31. Following the procedure in Problem 4–30, show that the momentum operator is Hermitian.

$$\int_{-\infty}^{\infty} \psi^* \hat{P}_x \psi \, dx = \int_{-\infty}^{\infty} \psi^* \left(-i\hbar \frac{d\psi}{dx} \right) dx$$

$$= -i\hbar \left. \psi^* \psi \right|_{-\infty}^{\infty} + \int_{-\infty}^{\infty} i\hbar \psi \frac{d\psi^*}{dx} dx$$

$$= \int_{-\infty}^{\infty} \left[-i\hbar \frac{d}{dx} \psi(x) \right]^* \psi \, dx$$

$$= \int_{-\infty}^{\infty} \left(\hat{P}_x^* \psi^* \right) \psi \, dx$$

The momentum operator is Hermitian.

4–32. Specify which of the following operators are Hermitian: id/dx, d^2/dx^2, and id^2/dx^2. Assume that $-\infty < x < \infty$ and that the functions on which these operators operate are appropriately well behaved at infinity.

We must determine whether the operator satisfies the condition

$$\int_{-\infty}^{\infty} f^*(x) \hat{A} f(x) dx = \int_{-\infty}^{\infty} f(x) \hat{A}^* f^*(x) dx$$

If the operator satisfies this equation, then it is Hermitian (Section 4–5).

a.

$$\int_{-\infty}^{\infty} f^*\left(i\frac{df}{dx}\right)dx = i\int_{-\infty}^{\infty} f^*\frac{df}{dx}dx = i\left[\left.\int_{-\infty}^{\infty} f^*f\right| - \int_{-\infty}^{\infty} f\frac{df^*}{dx}dx\right]$$

$$= -i\int_{-\infty}^{\infty} f\frac{df^*}{dx}dx = \int_{-\infty}^{\infty} f\left(-i\frac{d}{dx}\right)f^*dx$$

$$= \int_{-\infty}^{\infty} f\left(i\frac{d}{dx}\right)^* f^*dx$$

This operator is Hermitian.

b.

$$\int_{-\infty}^{\infty} f^*\left(\frac{d^2 f}{dx^2}\right)dx = \left.\int_{-\infty}^{\infty} f^*\frac{df}{dx}\right| - \int_{-\infty}^{\infty}\frac{df^*}{dx}\frac{df}{dx}dx$$

$$= -\left.\int_{-\infty}^{\infty}\frac{df^*}{dx}f\right| + \int_{-\infty}^{\infty}\frac{d^2 f^*}{dx^2}f\,dx$$

$$= \int_{-\infty}^{\infty} f\left(\frac{d^2}{dx^2}\right)f^*dx = \int_{-\infty}^{\infty} f\left(\frac{d^2}{dx^2}\right)^* f^*dx$$

This operator is Hermitian.

c.

$$\int_{-\infty}^{\infty} f^*\left(i\frac{d^2 f}{dx^2}\right)dx = \left.\int_{-\infty}^{\infty} f^*i\frac{df}{dx}\right| - i\int_{-\infty}^{\infty}\frac{df^*}{dx}\frac{df}{dx}dx$$

$$= -i\left.\int_{-\infty}^{\infty} f\frac{df^*}{dx}\right| + i\int_{-\infty}^{\infty} f\frac{d^2 f^*}{dx^2}dx$$

$$= -\int_{-\infty}^{\infty} f\left(i\frac{d^2}{dx^2}\right)^* f^*dx$$

This operator is not Hermitian.

Problems 4–33 through 4–38 deal with systems with piece-wise constant potentials.

4–33. Consider a particle moving in the potential energy

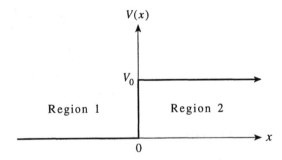

whose mathematical form is

$$V(x) = \begin{matrix} 0 & x < 0 \\ V_0 & x > 0 \end{matrix}$$

where V_0 is a constant. Show that if $E > V_0$, then the solutions to the Schrödinger equation in the two regions (1 and 2) are (see Problem 3–32)

$$\psi_1(x) = Ae^{ik_1 x} + Be^{-ik_1 x} \qquad x < 0 \tag{1}$$

and

$$\psi_2(x) = Ce^{ik_2 x} + De^{-ik_2 x} \qquad x > 0 \tag{2}$$

where

$$k_1 = \left(\frac{2mE}{\hbar^2}\right)^{1/2} \quad \text{and} \quad k_2 = \left(\frac{2m(E - V_0)}{\hbar^2}\right)^{1/2} \tag{3}$$

As we learned in Problem 3–32, e^{ikx} represents a particle traveling to the right and e^{-ikx} represents a particle traveling to the left. Let's consider a particle traveling to the right in region 1. If we wish to exclude the case of a particle traveling to the left in region 2, we set $D = 0$ in Equation 2. The physical problem we have set up is a particle of energy E incident on a potential barrier of height V_0. The squares of the coefficients in Equation 1 and 2 represent the probability that the particle is traveling in a certain direction in a given region. For example, $|A|^2$ is the probabillity that the particle is traveling with momentum $+\hbar k_1$ (Problem 3–32) in the region $x < 0$. If we consider many particles, N_0, instead of just one, then we can interpret $|A|^2 N_0$ to be the number of particles with momentum $\hbar k_1$ in the region $x < 0$. The number of these particles that pass a given point per unit time is given by $v|A|^2 N_0$, where the velocity v is given by $\hbar k_1/m$. Now apply the conditions that $\psi(x)$ and $d\psi/dx$ must be continuous at $x = 0$ (see Problem 4–4) to obtain

$$A + B = C$$

and

$$k_1(A - B) = k_2 C$$

Now define a quantity

$$R = \frac{v_1 |B|^2 N_0}{v_1 |A|^2 N_0} = \frac{\hbar k_1 |B|^2 N_0/m}{\hbar k_1 |A|^2 N_0/m} = \frac{|B|^2}{|A|^2}$$

and show that

$$R = \left(\frac{k_1 - k_2}{k_1 + k_2}\right)^2$$

Similarly, define

$$T = \frac{v_2 |C|^2 N_0}{v_1 |A|^2 N_0} = \frac{\hbar k_2 |C|^2 N_0/m}{\hbar k_1 |A|^2 N_0/m} = \frac{k_2 |C|^2}{k_1 |A|^2}$$

and show that

$$T = \frac{4k_1 k_2}{(k_1 + k_2)^2}$$

The symbols R and T stand for reflection coefficient and transmission coefficient, respectively. Give a physical interpretation of these designations. Show that $R + T = 1$. Would you have expected the particle to have been reflected even though its energy, E, is greater than the barrier height, V_0? Show that $R \to 0$ and $T \to 1$ as $V_0 \to 0$.

Region 1 ($x < 0$):

$$-\frac{\hbar^2}{2m}\frac{d^2\psi_1}{dx^2} = E\psi_1$$

The solution to this differential equation is

$$\psi_1(x) = Ae^{ik_1 x} + Be^{-ik_1 x} \qquad k_1 = \left(\frac{2mE}{\hbar^2}\right)^{1/2}$$

Region 2 ($x > 0$):

$$-\frac{\hbar^2}{2m}\frac{d^2\psi_2}{dx^2} + V_0\psi_2 = E\psi_2$$

giving

$$\psi_2(x) = Ce^{ik_2x} + De^{-ik_2x} \qquad k_2 = \left[\frac{2m(E - V_0)}{\hbar^2}\right]^{1/2}$$

Let $D = 0$ as stated in the problem. Now we impose the boundary conditions $\psi(0) = 0$ and $d\psi/dx = 0$ at $x = 0$, so

$$\psi_1(0) = \psi_2(0) \qquad \left.\frac{d\psi_1}{dx}\right|_{x=0} = \left.\frac{d\psi_2}{dx}\right|_{x=0}$$

This gives

$$A + B = C \quad (1) \qquad k_1A - k_1B = k_2C \quad (2)$$

If we multiply Equation 1 by k_2 and subtract the result from Equation 2, we can obtain

$$\frac{B}{A} = \frac{k_1 - k_2}{k_2 + k_1}$$

$$R = \frac{|B|^2}{|A|^2} = \left(\frac{k_1 - k_2}{k_1 + k_2}\right)^2 \tag{3}$$

If we multiply Equation 1 by k_1 and add the result to Equation 2, we can obtain

$$\frac{C}{A} = \frac{2k_1}{k_1 + k_2}$$

$$T = \frac{k_2|C|^2}{k_1|A|^2} = \frac{4k_1k_2}{(k_1 + k_2)^2} \tag{4}$$

The quantity R is the proportion of the wavefunction that is reflected back into region 1 and T is the proportion of the wavefunction that is transmitted into region 2. Therefore, $R + T = 1$, which is in agreement with the sum of Equations 3 and 4.

$$R + T = \frac{4k_1k_2 + k_1^2 - 2k_1k_2 + k_2^2}{(k_1 + k_2)^2} = \frac{(k_1 + k_2)^2}{(k_1 + k_2)^2} = 1$$

In a classical treatment of the problem, the particle would not be reflected when its energy was greater than the barrier height. This is a quantum mechanical effect. As $V_0 \to 0$, $k_1 \to k_2$, and so $R \to 0$ and $T \to 1$.

4–34. Show that $R = 1$ for the system described in Problem 4–33 but with $E < V_0$. Discuss the physical interpretation of this result.

Region 1 ($x < 0$):

$$-\frac{\hbar^2}{2m}\frac{d^2\psi_1}{dx^2} = E\psi_1$$

giving

$$\psi_1(x) = Ae^{ik_1x} + Be^{-ik_1x} \qquad k_1 = \left(\frac{2mE}{\hbar^2}\right)^{1/2}$$

Region 2 $(x > 0)$:

$$\frac{\hbar^2}{2m}\frac{d^2\psi_2}{dx^2} + (E - V_0)\psi_2 = 0$$

giving

$$\psi_2(x) = Ce^{\beta x} + De^{-\beta x} \qquad \beta = \left[\frac{2m(V_0 - E)}{\hbar^2}\right]^{1/2}$$

Now, β must be real because $E < V_0$. This means that for ψ to remain finite as $x \to \infty$, C must equal zero. The boundary conditions give

$$\psi_1(0) = \psi_2(0) \qquad \left.\frac{d\psi_1}{dx}\right|_{x=0} = \left.\frac{d\psi_2}{dx}\right|_{x=0}$$

$$A + B = D \qquad ik_1A - ik_1B = -\beta D$$

We can now use these relationships between the coefficients to solve for R as in Problem 4–33.

$$\frac{B}{A} = \frac{\beta + ik_1}{-\beta + ik_1}$$

$$R = \left|\frac{B}{A}\right|^2 = \frac{(\beta + ik_1)(\beta - ik_1)}{(-\beta + ik_1)(-\beta - ik_1)} = 1$$

This result tells us that all the particles will be reflected by the barrier.

4–35. In this problem, we introduce the idea of *quantum-mechanical tunneling*, which plays a central role in such diverse processes as the α-decay of nuclei, electron-transfer reactions, and hydrogen bonding. Consider a particle in the potential energy regions as shown below.

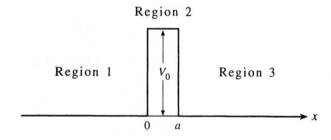

Mathematically, we have

$$V(x) = \begin{array}{ll} 0 & x < 0 \\ V_0 & 0 < x < a \\ 0 & x > a \end{array}$$

Show that if $E < V_0$, the solution to the Schrödinger equation in each region is given by

$$\psi_1(x) = Ae^{ik_1x} + Be^{-ik_1x} \qquad x < 0 \qquad (1)$$

$$\psi_2(x) = Ce^{k_2x} + De^{-k_2x} \qquad 0 < x < a \qquad (2)$$

and

$$\psi_3(x) = Ee^{ik_1 x} + Fe^{-ik_1 x} \qquad x > a \qquad (3)$$

where

$$k_1 = \left(\frac{2mE}{\hbar^2}\right)^{1/2} \quad \text{and} \quad k_2 = \left(\frac{2m(V_0 - E)}{\hbar^2}\right)^{1/2} \qquad (4)$$

If we exclude the situation of the particle coming from large positive values of x, then $F = 0$ in Equation 3. Following Problem 4–33, argue that the transmission coefficient, the probability the particle will get past the barrier, is given by

$$T = \frac{|E|^2}{|A|^2} \qquad (5)$$

Now use the fact that $\psi(x)$ and $d\psi/dx$ must be continuous at $x = 0$ and $x = a$ to obtain

$$A + B = C + D \qquad ik_1(A - B) = k_2(C - D) \qquad (6)$$

and

$$Ce^{k_2 a} + De^{-k_2 a} = Ee^{ik_1 a} \qquad k_2 Ce^{k_2 a} - k_2 De^{-k_2 a} = ik_1 Ee^{ik_1 a} \qquad (7)$$

Eliminate B from Equations 6 to get A in terms of C and D. Then solve Equations 7 for C and D in terms of E. Substitute these results into the equation for A in terms of C and D to get the intermediate result

$$2ik_1 A = [(k_2^2 - k_1^2 + 2ik_1 k_2)e^{k_2 a} + (k_1^2 - k_2^2 + 2ik_1 k_2)e^{-k_2 a}]\frac{Ee^{ik_1 a}}{2k_2}$$

Now use the relations $\sinh x = (e^x - e^{-x})/2$ and $\cosh x = (e^x + e^{-x})/2$ (Problem A–11) to get

$$\frac{E}{A} = \frac{4ik_1 k_2 e^{-ik_1 a}}{2(k_2^2 - k_1^2)\sinh k_2 a + 4ik_1 k_2 \cosh k_2 a}$$

Now multiply the right side by its complex conjugate and use the relation $\cosh^2 x = 1 + \sinh^2 x$ to get

$$T = \left|\frac{E}{A}\right|^2 = \frac{4}{4 + \dfrac{(k_1^2 + k_2^2)^2}{k_1^2 k_2^2}\sinh^2 k_2 a}$$

Finally, use the definition of k_1 and k_2 to show that the probability the particle gets through the barrier (even though it does not have enough energy!) is

$$T = \frac{1}{1 + \dfrac{v_0^2}{4\varepsilon(v_0 - \varepsilon)}\sinh^2(v_0 - \varepsilon)^{1/2}} \qquad (8)$$

or

$$T = \frac{1}{1 + \dfrac{\sinh^2[v_0^{1/2}(1 - r)^{1/2}]}{4r(1 - r)}} \qquad (9)$$

where $v_0 = 2ma^2 V_0/\hbar^2$, $\varepsilon = 2ma^2 E/\hbar$, and $r = E/V_0 = \varepsilon/v_0$. Figure 4.3 shows a plot of T versus r. To plot T versus r for values of $r > 1$, you need to use the relation $\sinh ix = i \sin x$ (Problem A–11). What would the classical result look like?

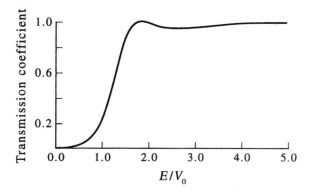

FIGURE 4.3
A plot of the probability that a particle of energy E will penetrate a barrier of height V_0 plotted against the ratio E/V_0 (Equation 9 of Problem 4–35).

Region 1: $-\dfrac{\hbar^2}{2m}\dfrac{d^2\psi}{dx^2} = E\psi_1$, giving

$$\psi_1(x) = Ae^{ik_1x} + Be^{-ik_1x} \qquad x < 0 \tag{1}$$

Region 2: $-\dfrac{\hbar^2}{2m}\dfrac{d^2\psi}{dx^2} + V_0\psi_2 = E\psi_2$, giving

$$\psi_2(x) = Ce^{k_2x} + De^{-k_2x} \qquad 0 < x < a \tag{2}$$

Region 3: $-\dfrac{\hbar^2}{2m}\dfrac{d^2\psi}{dx^2} = E\psi_3$, giving

$$\psi_3(x) = Ee^{ik_1x} + Fe^{-ik_1x} \qquad x > a \tag{3}$$

where

$$k_1 = \left(\frac{2mE}{\hbar^2}\right)^{1/2} \quad \text{and} \quad k_2 = \left(\frac{2m(V_0 - E)}{\hbar^2}\right)^{1/2} \tag{4}$$

Let $F = 0$ (given in the problem). Now we will use the boundary conditions

$$\psi_1(0) = \psi_2(0) \qquad \left.\frac{d\psi_1}{dx}\right|_{x=0} = \left.\frac{d\psi_2}{dx}\right|_{x=0}$$

to find

$$A + B = C + D \qquad ik_1(A - B) = k_2(C - D) \tag{6}$$

and the boundary conditions

$$\psi_2(a) = \psi_3(a) \qquad \left.\frac{d\psi_2}{dx}\right|_{x=a} = \left.\frac{d\psi_3}{dx}\right|_{x=a}$$

to find

$$Ce^{k_2a} + De^{-k_2a} = Ee^{ik_1a} \qquad k_2Ce^{k_2a} - k_2De^{-k_2a} = ik_1Ee^{ik_1a} \tag{7}$$

Following the steps suggested in the problem,

$$ik_1[A - (C + D - A)] = k_2(C - D)$$

$$2ik_1A = (ik_1 + k_2)C + (ik_1 - k_2)D \tag{8}$$

and

$$2k_2 C e^{k_2 a} = (k_2 + ik_1)E e^{ik_1 a}$$

$$C = \frac{(k_2 + ik_1)E e^{ik_1 a}}{2k_2 e^{k_2 a}} \tag{9}$$

$$2k_2 D e^{-k_2 a} = (k_2 - ik_1)E e^{ik_1 a}$$

$$D = \frac{(k_2 - ik_1)E e^{ik_1 a}}{2k_2 e^{-k_2 a}} \tag{10}$$

Substituting Equations 9 and 10 into Equation 8 gives

$$2ik_1 A = (ik_1 + k_2)\frac{(k_2 + ik_1)E e^{ik_1 a}}{2k_2 e^{k_2 a}} + (ik_1 - k_2)\frac{(k_2 - ik_1)E e^{ik_1 a}}{2k_2 e^{-k_2 a}}$$

$$= \frac{E e^{ik_1 a}}{2k_2}\left[\frac{(k_2 + ik_1)^2}{e^{k_2 a}} + \frac{-(ik_1 - k_2)^2}{e^{-k_2 a}}\right]$$

$$= \frac{E e^{ik_1 a}}{2k_2}\left[(k_2^2 - k_1^2 + 2ik_1 k_2)e^{-k_2 a} + (k_1^2 - k_2^2 + 2ik_1 k_2)e^{k_2 a}\right]$$

which can be written as

$$\frac{E}{A} = \frac{4ik_1 k_2 e^{-ik_1 a}}{k_1^2(e^{k_2 a} - e^{-k_2 a}) - k_2^2(e^{k_2 a} - e^{-k_2 a}) + 2ik_1 k_2(e^{k_2 a} + e^{-k_2 a})}$$

$$= \frac{4ik_1 k_2 e^{-ik_1 a}}{2(k_2^2 - k_1^2)\sinh k_2 a + 4ik_1 k_2 \cosh k_2 a} \tag{11}$$

To find T, we need to evaluate

$$T = \left|\frac{E}{A}\right|^2$$

Using Equation 11, we obtain

$$T = \frac{(4ik_1 k_2 e^{-ik_1 a})(-4ik_1 k_2 e^{-ik_1 a})}{(2(k_2^2 - k_1^2)\sinh k_2 a + 4ik_1 k_2 \cosh k_2 a)(2(k_2^2 - k_1^2)\sinh k_2 a - 4ik_1 k_2 \cosh k_2 a)}$$

$$= \frac{16k_1^2 k_2^2}{4(k_2^2 - k_1^2)^2 \sinh^2 k_2 a + 16k_1^2 k_2^2 \cosh^2 k_2 a}$$

Using the identity $\cosh^2 x = 1 + \sinh^2 x$, we have

$$T = \frac{16k_1^2 k_2^2}{(4k_2^4 - 8k_1^2 k_2^2 + 4k_1^4)\sinh^2 k_2 a + 16k_1^2 k_2^2 \sinh^2 k_2 a + 16k_1^2 k_2^2}$$

$$= \frac{16k_1^2 k_2^2}{4(k_2^2 + k_1^2)^2 \sinh^2 k_2 a + 16k_1^2 k_2^2} = \frac{4}{4 + \dfrac{(k_1^2 + k_2^2)^2}{k_1^2 k_2^2}\sinh^2 k_2 a}$$

Using the definitions of k_1 and k_2 in Equation 4 gives

$$T = \frac{4}{4 + \dfrac{\left[\dfrac{2mE}{\hbar^2} + \dfrac{2m(V_0 - E)}{\hbar^2}\right]^2}{\left(\dfrac{2mE}{\hbar^2}\right)\left[\dfrac{2m(V_0 - E)}{\hbar^2}\right]}\sinh^2\left[\left(\dfrac{2ma^2}{\hbar^2}\right)^{1/2}(V_0 - E)^{1/2}\right]}$$

Now let $v_0 = 2ma^2 V_0/\hbar^2$, $\varepsilon = 2ma^2 E/\hbar$, and $r = E/V_0 = \varepsilon/v_0$ to find

$$T = \frac{4}{4 + \dfrac{v_0^2}{\varepsilon(v_0 - \varepsilon)} \sinh^2(v_0 - \varepsilon)^{1/2}}$$

$$= \frac{1}{1 + \dfrac{v_0^2}{4\varepsilon(v_0 - \varepsilon)} \sinh^2(v_0 - \varepsilon)^{1/2}}$$

$$= \frac{1}{1 + \dfrac{\sinh^2[v_0^{1/2}(1 - r)^{1/2}]}{4r(1 - r)}}$$

The classical result would have a discontinuity at $E = V_0$; if $E < V_0$ the probability that the particle would penetrate the barrier would be zero, and if $E > V_0$ the probability that the particle would penetrate the barrier would be one.

4–36. Use the result of Problem 4–35 to determine the probability that an electron with a kinetic energy 8.0×10^{-21} J will tunnel through a 1.0 nm thick potential barrier with $V_0 = 12.0 \times 10^{-21}$ J.

The probability of the electron tunneling through the barrier is given by the expression for T in Problem 4–35,

$$T = \frac{1}{1 + \dfrac{\sinh^2[v_0^{1/2}(1 - r)^{1/2}]}{4r(1 - r)}}$$

First we calculate the values of the variables in this expression.

$$r = \frac{E}{V_0} = \frac{8}{12} = \frac{2}{3}$$

$$v_0 = \frac{2ma^2 V_0}{\hbar^2}$$

$$= \frac{2(9.1094 \times 10^{-31} \text{ kg})(1.0 \times 10^{-9} \text{ m})^2(12.0 \times 10^{-21} \text{ J})}{(1.0546 \times 10^{-34} \text{ J·s})^2}$$

$$= 2.0$$

Now substitute into Equation 7 from Problem 4–35:

$$T = \frac{1}{1 + \dfrac{\sinh^2[v_0^{1/2}(1 - r)^{1/2}]}{4r(1 - r)}}$$

$$= \frac{1}{1 + \dfrac{\sinh^2[(2.0)^{1/2}(\frac{1}{3})^{1/2}]}{4(\frac{2}{3})(\frac{1}{3})}}$$

$$= 0.52$$

4–37. Problem 4–35 gives that the probability of a particle of relative energy E/V_0 will penetrate a rectangular potential barrier of height V_0 and thickness a is

$$T = \frac{1}{1 + \dfrac{\sinh^2[v_0^{1/2}(1-r)^{1/2}]}{4r(1-r)}}$$

where $v_0 = 2mV_0a^2/\hbar^2$ and $r = E/V_0$. What is the limit of T as $r \to 1$? Plot T against r for $v_0 = 1/2$, 1, and 2. Interpret your results.

Use the fact that $\sinh x = x + x^3/6 + \cdots$ to write

$$T = \frac{1}{1 + \dfrac{v_0(1-r)}{4r(1-r)} + \cdots} = \frac{1}{1 + v_0/4} \qquad \text{as} \qquad r \to 1$$

In the graph below, the solid line is the graph of T versus r for $v_0 = 1/2$, the dashed line is the graph of T versus r for $v_0 = 1$, and the dotted line is the graph of T versus r for $v_0 = 2$.

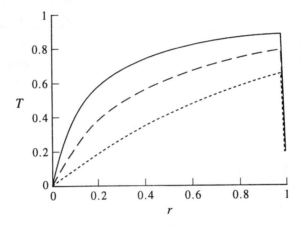

4–38. In this problem, we will consider a particle in a *finite* potential well

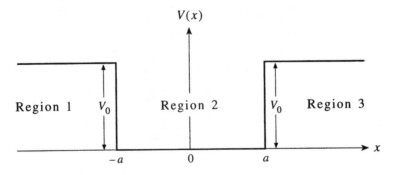

whose mathematical form is

$$V(x) = \begin{array}{ll} V_0 & x < -a \\ 0 & -a < x < a \\ V_0 & x > a \end{array} \qquad (1)$$

Note that this potential describes what we have called a "particle in a box" if $V_0 \to \infty$. Show that if $0 < E < V_0$, the solution to the Schrödinger equation in each region is

$$
\begin{aligned}
\psi_1(x) &= A e^{k_1 x} & x &< -a \\
\psi_2(x) &= B \sin \alpha x + C \cos \alpha x & -a &< x < a \\
\psi_3(x) &= D e^{-k_1 x} & x &> a
\end{aligned}
\tag{2}
$$

where

$$
k_1 = \left(\frac{2m(V_0 - E)}{\hbar^2} \right)^{1/2} \quad \text{and} \quad \alpha = \left(\frac{2mE}{\hbar^2} \right)^{1/2}
\tag{3}
$$

Now apply the conditions that $\psi(x)$ and $d\psi/dx$ must be continuous at $x = -a$ and $x = a$ to obtain

$$
A e^{-k_1 a} = -B \sin \alpha a + C \cos \alpha a
\tag{4}
$$

$$
D e^{-k_1 a} = B \sin \alpha a + C \cos \alpha a
\tag{5}
$$

$$
k_1 A e^{-k_1 a} = \alpha B \cos \alpha a + \alpha C \sin \alpha a
\tag{6}
$$

and

$$
-k_1 D e^{-k_1 a} = \alpha B \cos \alpha a - \alpha C \sin \alpha a
\tag{7}
$$

Add and subtract Equations 4 and 5 and add and subtract Equations 6 and 7 to obtain

$$
2C \cos \alpha a = (A + D) e^{-k_1 a}
\tag{8}
$$

$$
2B \sin \alpha a = (D - A) e^{-k_1 a}
\tag{9}
$$

$$
2\alpha C \sin \alpha a = k_1 (A + D) e^{-k_1 a}
\tag{10}
$$

and

$$
2\alpha B \cos \alpha a = -k_1 (D - A) e^{-k_1 a}
\tag{11}
$$

Now divide Equation 10 by Equation 8 to get

$$
\frac{\alpha \sin \alpha a}{\cos \alpha a} = \alpha \tan \alpha a = k_1 \qquad (D \neq -A \text{ and } C \neq 0)
\tag{12}
$$

and then divide Equation 11 by Equation 9 to get

$$
\frac{\alpha \cos \alpha a}{\sin \alpha a} = \alpha \cot \alpha a = -k_1 \quad \text{and} \quad (D \neq A \text{ and } B \neq 0)
\tag{13}
$$

Referring back to Equation 3, note that Equations 12 and 13 give the allowed values of E in terms of V_0. It turns out that these two equations cannot be solved simultaneously, so we have two sets of equations

$$
\alpha \tan \alpha a = k_1
\tag{14}
$$

and

$$
\alpha \cot \alpha a = -k_1
\tag{15}
$$

Let's consider Equation 14 first. Multiply both sides by a and use the definitions of α and k_1 to get

$$\left(\frac{2ma^2E}{\hbar^2}\right)^{1/2} \tan\left(\frac{2ma^2E}{\hbar^2}\right)^{1/2} = \left[\frac{2ma^2}{\hbar^2}(V_0 - E)\right]^{1/2} \tag{16}$$

Show that this equation simplifies to

$$\varepsilon^{1/2} \tan \varepsilon^{1/2} = (v_0 - \varepsilon)^{1/2} \tag{17}$$

where $\varepsilon = 2ma^2E/\hbar^2$ and $v_0 = 2ma^2V_0/\hbar^2$. Thus, if we fix v_0 (actually $2ma^2V_0/\hbar^2$), then we can use Equation 17 to solve for the allowed values of ε (actually $2ma^2E/\hbar^2$). Equation 17 cannot be solved analytically, but if we plot both $\varepsilon^{1/2} \tan \varepsilon^{1/2}$ and $(v_0 - \varepsilon)^{1/2}$ versus ε on the same graph, then the solutions are given by the intersections of the two curves. Show that the intersections occur at $\varepsilon = 2ma^2E/\hbar^2 = 1.47$ and 11.37 for $v_0 = 12$. The other value(s) of ε are given by the solutions to Equation 15, which are obtained by finding the intersection of $-\varepsilon^{1/2} \cot \varepsilon^{1/2}$ and $(v_0 - \varepsilon)^{1/2}$ plotted against ε. Show that $\varepsilon = 2ma^2E/\hbar^2 = 5.68$ for $v_0 = 12$. Thus, we see there are only three bound states for a well of depth $V_0 = 12\hbar^2/2ma^2$. The important point here is not the numerical values of E, but the fact that there is only a finite number of bound states. Show that there are only two bound states for $v_0 = 2ma^2V_0/\hbar^2 = 4$.

Region 1: $-\dfrac{\hbar^2}{2m}\dfrac{d^2\psi}{dx^2} + V_0\psi_1 = E\psi_1$, giving

$$\psi_1(x) = Ae^{k_1 x}$$

(We ignore the solution $c_2 e^{-k_1 x}$ because if the particle goes from region 2 into region 1, it must be traveling to the left.)

Region 2: $-\dfrac{\hbar^2}{2m}\dfrac{d^2\psi}{dx^2} = E\psi_2$, giving

$$\psi_2(x) = B\sin\alpha x + C\cos\alpha x$$

as in Example 2–4.

Region 3: $-\dfrac{\hbar^2}{2m}\dfrac{d^2\psi}{dx^2} + V_0\psi_3 = E\psi_3$, giving

$$\psi_3(x) = De^{-k_1 x} \qquad x > a$$

(We ignore the solution $c_1 e^{k_1 x}$ because if the particle goes from region 2 into region 3, it must be traveling to the right.) In the above expressions

$$k_1 = \left(\frac{2m(V_0 - E)}{\hbar^2}\right)^{1/2} \quad \text{and} \quad \alpha = \left(\frac{2mE}{\hbar^2}\right)^{1/2}$$

Now use the boundary condition equations $\psi_1(-a) = \psi_2(-a)$, $\psi_2(a) = \psi_3(a)$, $d\psi_1/dx = d\psi_2/dx$ at $x = -a$, and $d\psi_2/dx = d\psi_3/dx$ at $x = a$ to find

$$Ae^{-k_1 a} = -B\sin\alpha a + C\cos\alpha a \tag{4}$$

$$B\sin\alpha a + C\cos\alpha a = De^{-k_1 a} \tag{5}$$

$$k_1 Ae^{-k_1 a} = \alpha B\cos\alpha a + \alpha C\sin\alpha a \tag{6}$$

$$-k_1 De^{-k_1 a} = \alpha B\cos\alpha a - \alpha C\sin\alpha a \tag{7}$$

Adding and subtracting Equations 4 and 5 gives

$$2C \cos \alpha a = (A + D)e^{-k_1 a} \qquad (8)$$

$$2B \sin \alpha a = (D - A)e^{-k_1 a} \qquad (9)$$

Adding and subtracting Equations 6 and 7 gives

$$2\alpha C \sin \alpha a = k_1(A + D)e^{-k_1 a} \qquad (10)$$

$$2\alpha B \cos \alpha a = -k_1(D - A)e^{-k_1 a} \qquad (11)$$

Now divide Equation 10 by Equation 8 and Equation 11 by Equation 9 to obtain

$$\frac{\alpha \sin \alpha a}{\cos \alpha a} = \alpha \tan \alpha a = k_1 \qquad (D \neq -A \text{ and } C \neq 0) \qquad (12)$$

$$\frac{\alpha \cos \alpha a}{\sin \alpha a} = \alpha \cot \alpha a = -k_1 \qquad (D \neq A \text{ and } B \neq 0) \qquad (13)$$

We now have two sets of equations:

$$\alpha \tan \alpha a = k_1 \qquad (14)$$

$$\alpha \cot \alpha a = -k_1 \qquad (15)$$

The result that $A = \pm D$ implies that the chance of the particle leaving the finite well through one side of the well is equal to the chance that it will leave the finite well through its other side. Using the definitions of α and k_1 (Equation 3) and multiplying Equation 14 by a, we find

$$\left(\frac{2ma^2 E}{\hbar^2}\right)^{1/2} \tan \left(\frac{2ma^2 E}{\hbar^2}\right)^{1/2} = \left[\frac{2ma^2}{\hbar^2}(V_0 - E)\right]^{1/2} \qquad (16)$$

If we let $\varepsilon = 2ma^2 E/\hbar^2$ and $v_0 = 2ma^2 V_0/\hbar^2$, we obtain

$$\varepsilon^{1/2} \tan \varepsilon^{1/2} = (v_0 - \varepsilon)^{1/2} \qquad (17)$$

Likewise, we can obtain the expression

$$-\varepsilon^{1/2} \cot \varepsilon^{1/2} = (v_0 - \varepsilon)^{1/2} \qquad (18)$$

by going through the same procedure with Equation 15. The solutions to Equations 17 and 18 are shown graphically for $v_0 = 12$ in the captioned figure. The solutions for $v_0 = 4$ are shown in the figure below. Because these graphs show two intersections, there are two bound states for $v_0 = 4$.

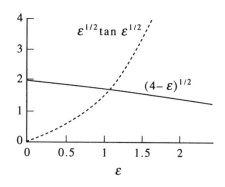
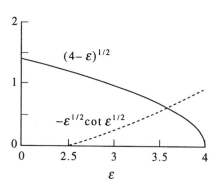

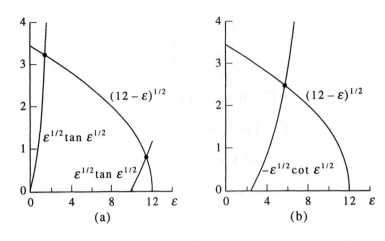

FIGURE 4.4

(a) Plots of both $\varepsilon^{1/2} \tan \varepsilon^{1/2}$ and $(12 - \varepsilon)^{1/2}$ versus ε. The intersections of the curves give the allowed values of ε for a one-dimensional potential well of depth $V_0 = 12\hbar^2/2ma^2$. (b) Plots of both $-\varepsilon^{1/2} \cot \varepsilon^{1/2}$ and $(12 - \varepsilon)^{1/2}$ plotted against ε. The intersection gives an allowed value of ε for a one-dimensional potential well of depth $V_0 = 12\hbar^2/2ma^2$.

Spherical Coordinates

PROBLEMS AND SOLUTIONS

D–1. Derive Equation D.2 from D.1.

Equations D.1 are

$$x = r \sin\theta \cos\phi \qquad y = r \sin\theta \sin\phi \qquad z = r \cos\theta \tag{D.1}$$

We use these equations to write $\tan\phi$ as

$$\tan\phi = \frac{\sin\phi}{\cos\phi} = \frac{r\sin\theta}{r\sin\theta}\left(\frac{\sin\phi}{\cos\phi}\right) = \frac{y}{x} \tag{1}$$

Likewise, we can write (using trigonometric identities)

$$
\begin{aligned}
r^2 &= r^2(\sin^2\theta + \cos^2\theta)(\sin^2\phi + \cos^2\phi) \\
&= r^2\sin^2\theta\sin^2\phi + r^2\cos^2\theta\sin^2\phi + r^2\sin^2\theta\cos^2\phi + r^2\cos^2\theta\cos^2\phi \\
&= (r\sin\theta\sin\phi)^2 + (r\sin\theta\sin\phi)^2 + (r\cos\theta)^2(\sin^2\phi + \cos^2\phi) \\
&= x^2 + y^2 + z^2 \\
r &= (x^2 + y^2 + z^2)^{1/2}
\end{aligned} \tag{2}
$$

and

$$
\begin{aligned}
z &= r\cos\theta \\
\cos\theta &= \frac{z}{(x^2 + y^2 + z^2)^{1/2}}
\end{aligned} \tag{3}
$$

Equations 1, 2, and 3 are Equations D.2.

D–2. Express the following points given in Cartesian coordinates in terms of spherical coordinates.

$$(x, y, z): \quad (1, 0, 0); \quad (0, 1, 0); \quad (0, 0, 1); \quad (0, 0, -1)$$

Use the equations derived in the previous problem (Equations D.2).

a. $(1, 0, 0)$

$$r = (x^2 + y^2 + z^2)^{1/2} = 1$$
$$\theta = \cos^{-1}\left(\frac{z}{r}\right) = \cos^{-1} 0 = \frac{\pi}{2}$$
$$\phi = \tan^{-1}\left(\frac{y}{x}\right) = \tan^{-1} 0 = 0$$

Spherical coordinates: $(1, \frac{\pi}{2}, 0)$

115

b. $(0, 1, 0)$

$$r = (x^2 + y^2 + z^2)^{1/2} = 1$$
$$\theta = \cos^{-1}\left(\frac{z}{r}\right) = \cos^{-1} 0 = \frac{\pi}{2}$$
$$\phi = \tan^{-1}\left(\frac{y}{x}\right) = \tan^{-1}\left(\frac{1}{0}\right) = \frac{\pi}{2}$$

Spherical coordinates: $(1, \frac{\pi}{2}, \frac{\pi}{2})$

c. $(0, 0, 1)$

$$r = (x^2 + y^2 + z^2)^{1/2} = 1$$
$$\theta = \cos^{-1}\left(\frac{z}{r}\right) = \cos^{-1}(1) = 0$$
$$\phi = \tan^{-1}\left(\frac{y}{x}\right) = \tan^{-1}\frac{0}{0}$$

(This means that ϕ can take on any value.) Spherical coordinates: $(1, 0, \phi)$

d. $(0, 0, -1)$

$$r = (x^2 + y^2 + z^2)^{1/2} = 1$$
$$\theta = \cos^{-1}\left(\frac{z}{r}\right) = \cos^{-1}(-1) = \pi$$
$$\phi = \tan^{-1}\left(\frac{y}{x}\right) = \tan^{-1}\frac{0}{0}$$

Spherical coordinates: $(1, \pi, \phi)$

D–3. Describe the graphs of the following equations:

a. $r = 5$,   **b.** $\theta = \pi/4$,   **c.** $\phi = \pi/2$

---

a. A sphere of radius 5 centered at the origin

b. A cone about the z-axis of internal angle $\frac{\pi}{4}$ **c.** The yz-plane

---

D–4. Use Equation D.3 to determine the volume of a hemisphere.

Let the radius of the hemisphere be a. A hemisphere corresponds to $0 \le \theta \le \pi$ and $0 \le \phi \le \pi$, so

$$dV = r^2 \sin\theta \, dr d\theta d\phi$$
$$V = \int_0^a r^2 dr \int_0^\pi \sin\theta d\theta \int_0^\pi d\phi = \frac{2\pi a^3}{3}$$

D–5. Use Equation D.5 to determine the surface area of a hemisphere.

Let the radius of the hemisphere be a. Then the surface area of a hemisphere is

$$dA = r^2 \sin\theta \, d\theta d\phi$$
$$A = a^2 \int_0^\pi \sin\theta d\theta \int_0^\pi d\phi = 2\pi a^2$$

D–6. Evaluate the integral

$$I = \int_0^\pi \cos^2\theta \sin^3\theta \, d\theta$$

by letting $x = \cos\theta$.

If $x = \cos\theta$, then $dx = -\sin\theta \, d\theta$, $x = 1$ for $\theta = 0$, and $x = -1$ for $\theta = \pi$. Then

$$I = \int_0^\pi \cos^2\theta \sin^3\theta \, d\theta = \int_0^\pi \cos^2\theta \sin\theta(1 - \cos^2\theta) \, d\theta$$

$$= -\int_1^{-1} dx \, x^2(1 - x^2) = \int_{-1}^1 x^2 \, dx - \int_{-1}^1 x^4 \, dx$$

$$= \frac{1}{3} + \frac{1}{3} + \frac{1}{5} + \frac{1}{5} = \frac{4}{15}$$

D–7. We will learn in Chapter 6 that a $2p_y$ hydrogen atom orbital is given by

$$\psi_{2p_y} = \frac{1}{4\sqrt{2\pi}} r e^{-r/2} \sin\theta \sin\phi$$

Show that ψ_{2p_y} is normalized. (Don't forget to square ψ_{2p_y} first.)

We want to show that

$$I = \int_0^\infty \int_0^\pi \int_0^{2\pi} \psi_{2p_y}^* \psi_{2p_y} r^2 \sin\theta \, dr \, d\theta \, d\phi = 1$$

First, we square ψ_{2p_y}:

$$\psi_{2p_y}^* \psi_{2p_y} = \frac{1}{32\pi} r^2 e^{-r} \sin^2\theta \sin^2\phi$$

Then

$$I = \frac{1}{32\pi} \int_0^\infty dr \, r^4 e^{-r} \int_0^\pi d\theta \sin^3\theta \int_0^{2\pi} \sin^2\phi \, d\phi$$

$$= \left(\frac{1}{32\pi}\right)(4!)\left(\frac{4}{3}\right)\left(\frac{2\pi}{2}\right) = 1$$

and we have shown that ψ_{2p_y} is normalized.

D–8. We will learn in Chapter 6 that a $2s$ hydrogen atomic orbital is given by

$$\psi_{2s} = \frac{1}{4\sqrt{2\pi}}(2 - r)e^{-r/2}$$

Show that ψ_{2s} is normalized. (Don't forget to square ψ_{2s} first.)

We want to show that

$$I = \int_0^\infty \int_0^\pi \int_0^{2\pi} \psi_{2s}^* \psi_{2s} r^2 \sin\theta \, dr \, d\theta \, d\phi = 1$$

First, we square ψ_{2s}:

$$\psi_{2s}^* \psi_{2s} = \frac{1}{32\pi}(2-r)^2 e^{-r}$$

Then

$$I = \frac{1}{32\pi} \int_0^\infty r^2(2-r)^2 e^{-r} dr \int_0^\pi d\theta \sin\theta \int_0^{2\pi} d\phi$$

$$= \frac{1}{32\pi}\left(4\int_0^\infty r^2 e^{-r} dr - 4\int_0^\infty r^3 e^{-r} dr + \int_0^\infty r^4 e^{-r} dr\right)(2)(2\pi)$$

$$= \frac{1}{8}[4(2) - 4(3!) + 4!] = \frac{1}{8}(8) = 1$$

and ψ_{2s} is normalized.

D–9. Show that

$$Y_1^0(\theta, \phi) = \left(\frac{3}{4\pi}\right)^{1/2} \cos\theta$$

$$Y_1^1(\theta, \phi) = \left(\frac{3}{8\pi}\right)^{1/2} e^{i\phi} \sin\theta$$

and

$$Y_1^{-1}(\theta, \phi) = \left(\frac{3}{8\pi}\right)^{1/2} e^{-i\phi} \sin\theta$$

are orthonormal over the surface of a sphere.

We can show that two functions f_i and f_j are orthonormal by demonstrating that

$$\int_0^\infty \int_0^\pi \int_0^{2\pi} f_1^* f_2 r^2 \sin\theta \, dr d\theta d\phi = \delta_{ij}$$

For the functions above,

$$\int Y_1^{0*} Y_1^0 = \frac{3}{4\pi} \int_0^\pi \sin\theta \cos^2\theta d\theta \int_0^{2\pi} d\phi$$

$$= \frac{3}{4\pi}\left(\frac{-\cos^3\theta}{3}\bigg|_0^\pi\right)(2\pi) = \frac{3}{2}\left(\frac{2}{3}\right) = 1$$

$$\int Y_1^{1*} Y_1^1 = \frac{3}{8\pi} \int_0^\pi d\theta \sin^3\theta \int_0^{2\pi} d\phi$$

$$= \frac{3}{8\pi}\left(\frac{4}{3}\right)(2\pi) = 1$$

$$\int Y_1^{-1*} Y_1^{-1} = \frac{3}{8\pi} \int_0^\pi d\theta \sin^3\theta \int_0^{2\pi} d\phi$$

$$= \frac{3}{8\pi}\left(\frac{4}{3}\right)(2\pi) = 1$$

$$\int Y_1^{0*} Y_1^1 = \int Y_1^{-1*} Y_1^0 = \frac{3}{4\sqrt{2}\pi} \int_0^\pi d\theta \cos\theta \sin^2\theta \int_0^{2\pi} e^{i\phi} d\phi = 0$$

$$\int Y_1^{0*} Y_1^{-1} = \frac{3}{4\sqrt{2\pi}} \int_0^\pi d\theta \cos\theta \sin^2\theta \int_0^{2\pi} e^{-i\phi} d\phi = 0$$

$$\int Y_1^{1*} Y_1^{-1} = \frac{3}{8\pi} \int_0^\pi d\theta \sin^3\theta \int_0^{2\pi} e^{2i\phi} d\phi = 0$$

(Recall from Problem 3–15 that $\int_0^{2\pi} e^{-i\phi} d\phi = \int_0^{2\pi} e^{i\phi} d\phi = 0$.)

D–10. Evaluate the average of $\cos\theta$ and $\cos^2\theta$ over the surface of a sphere.

To determine the average of $\cos\theta$ over the surface of a sphere, we must evaluate the integral

$$\int_0^\pi \sin\theta\cos\theta d\theta \int_0^{2\pi} d\phi = \frac{1}{2}(2\pi) \int_0^\pi \sin 2\theta d\theta = 0$$

Because this integral is equal to zero, the average of $\cos\theta$ is zero. For the average of $\cos^2\theta$ over the surface of a sphere, we must first evaluate the integral

$$\int_0^\pi \sin\theta\cos^2\theta d\theta \int_0^{2\pi} d\phi = 2\pi \int_{-1}^1 x^2 dx = \frac{4\pi}{3}$$

The average of $\cos^2\theta$ over the surface of a sphere is then given by

$$\frac{\int \cos^2\theta dA}{\int dA} = \frac{4\pi}{3(4\pi)} = \frac{1}{3}$$

D–11. We shall frequently use the notation $d\mathbf{r}$ to represent the volume element in spherical coordinates. Evaluate the integral

$$I = \int d\mathbf{r} e^{-r} \cos^2\theta$$

where the integral is over all space (in other words, over all posible values of r, θ and ϕ).

$$I = \int d\mathbf{r} e^{-r} \cos^2\theta$$

$$= \int_0^\infty dr r^2 e^{-r} \int_0^\pi d\theta \cos^2\theta \sin\theta \int_0^{2\pi} d\phi$$

$$= (2)\left(\frac{4\pi}{3}\right) = \frac{8\pi}{3}$$

D–12. Show that the two functions

$$f_1(r) = e^{-r} \cos\theta \qquad \text{and} \qquad f_2(r) = (2 - r)e^{-r/2} \cos\theta$$

are orthogonal over all space (in other words, over all possible values of r, θ and ϕ).

If $f_1(r)$ and $f_2(r)$ are orthogonal, then

$$I = \int d\mathbf{r} f_1^*(r) f_2(r) = 0$$

We now evaluate this integral as

$$
\begin{aligned}
I &= \int_0^\infty dr\, r^2 e^{-r} (2 - r) e^{-r/2} \int_0^\pi d\theta \cos^2\theta \sin\theta \int_0^{2\pi} d\phi \\
&= \int_0^\infty dr (2r^2 - r^3) e^{-3r/2} \left(\frac{4\pi}{3} \right) \\
&= \frac{8\pi}{3} \int_0^\infty dr\, r^2 e^{-3r/2} - \frac{4\pi}{3} \int_0^\infty dr\, r^3 e^{-3r/2} \\
&= \left(\frac{8\pi}{3} \right) \left(\frac{16}{27} \right) - \left(\frac{4\pi}{3} \right) \left(\frac{32}{27} \right) = 0
\end{aligned}
$$

CHAPTER 5

The Harmonic Oscillator and the Rigid Rotator: Two Spectroscopic Models

PROBLEMS AND SOLUTIONS

5–1. Verify that $x(t) = A \sin \omega t + B \cos \omega t$, where $\omega = (k/m)^{1/2}$ is a solution to Newton's equation for a harmonic oscillator.

Newton's equation for a harmonic oscillator is

$$\frac{d^2x}{dt^2} + \frac{k}{m}x = 0 \tag{5.3}$$

Substituting $x(t) = A \sin \omega t + B \cos \omega t$ into Newton's equation, we find

$$-A\omega^2 \sin \omega t - B\omega^2 \cos \omega t + \frac{k}{m}(A \sin \omega t + B \cos \omega t)$$

$$= -\frac{k}{m}(A \sin \omega t + B \cos \omega t) + \frac{k}{m}(A \sin \omega t + B \cos \omega t) = 0$$

where we have used the relationship $\omega = (k/m)^{1/2}$.

5–2. Verify that $x(t) = C \sin(\omega t + \phi)$ is a solution to Newton's equation for a harmonic oscillator.

Newton's equation for a harmonic oscillator is

$$\frac{d^2x}{dt^2} + \frac{k}{m}x = 0 \tag{5.3}$$

Substituting $x(t) = C \sin(\omega t + \phi)$, we find

$$-C\omega^2 \sin(\omega t + \phi) + \frac{k}{m}C \sin(\omega t + \phi) = -\frac{k}{m}C \sin(\omega t + \phi) + \frac{k}{m}C \sin(\omega t + \phi) = 0$$

where we have used the relationship $\omega = (k/m)^{1/2}$.

5–3. The general solution for the classical harmonic oscillator is $x(t) = C \sin(\omega t + \phi)$. Show that the displacement oscillates between $+C$ and $-C$ with a frequency ω radian·s^{-1} or $\nu = \omega/2\pi$ cycle·s^{-1}. What is the period of the oscillations; that is, how long does it take to undergo one cycle?

Consider the general solution $x(t) = C\sin(\omega t + \phi)$. Because the sine function varies from +1 to -1, the value of x varies from $+C$ to $-C$. To find the period of oscillation, we determine the smallest nonzero value of τ that satisfies the condition

$$\sin(\omega t + \phi) = \sin[\omega(t+\tau) + \phi]$$
$$= \sin(\omega t + \phi + \omega\tau)$$
$$= \sin(\omega t + \phi)\cos\omega\tau + \cos(\omega t + \phi)\sin\omega\tau$$

This condition is met when τ satisfies the two conditions

$$\cos\omega\tau = 1 \qquad \sin\omega\tau = 0$$

or, equivalently,

$$\omega\tau = 2\pi n \qquad n = 1, 2, \ldots$$

The smallest value of τ is then $2\pi/\omega$, which is the time it takes for the oscillator to undergo one cycle. The frequency of oscillation is

$$\nu = \frac{1}{\tau} = \frac{\omega}{2\pi}$$

5–4. From Problem 5–3, we see that the period of a harmonic vibration is $\tau = 1/\nu$. The average of the kinetic energy over one cycle is given by

$$\langle K \rangle = \frac{1}{\tau}\int_0^\tau \frac{m\omega^2 C^2}{2}\cos^2(\omega t + \phi)dt$$

Show that $\langle K \rangle = E/2$ where E is the total energy. Show also that $\langle V \rangle = E/2$, where the instantaneous potential energy is given by

$$V = \frac{kC^2}{2}\sin^2(\omega t + \phi)$$

Interpret the result $\langle K \rangle = \langle V \rangle$.

We start with the general equation for a harmonic oscillator $x = C\sin(\omega t + \phi)$ (Problem 5–2). The total energy of this oscillator is

$$E = K + U = \frac{m}{2}\left(\frac{dx}{dt}\right)^2 + \frac{kx^2}{2}$$

where we have used Equations 5.11 and 5.12 for K and U. Substituting $x(t)$ into this expression gives

$$E = \frac{m}{2}\omega^2 C^2 \cos^2(\omega t + \phi) + \frac{k}{2}C^2 \sin^2(\omega t + \phi)$$
$$= \frac{k}{2}C^2\left[\cos^2(\omega t + \phi) + \sin^2(\omega t + \phi)\right] = \frac{k}{2}C^2$$

Now

$$\langle K \rangle = \frac{1}{\tau}\int_0^\tau \frac{m\omega^2 C^2}{2}\cos^2(\omega t + \phi)dt$$
$$= \frac{m\omega C^2}{2\tau}\int_\phi^{\omega\tau+\phi}\cos^2 x\, dx$$

$$= \frac{m\omega C^2}{2\tau} \left[\frac{x}{2} + \sin 2x \right]_{\phi}^{\omega\tau+\phi}$$

$$= \frac{m\omega C^2}{2\tau} \left[\frac{\omega\tau}{2} + \frac{\sin 2(\omega\tau + \phi) - \sin 2\phi}{4} \right]$$

$$= \frac{m\omega C^2}{2\tau} \left[\frac{\omega\tau}{2} + \frac{\sin 2\omega\tau \cos 2\phi + \cos 2\omega\tau \sin 2\phi - \sin 2\phi}{4} \right]$$

$$= \frac{m\omega C^2}{2\tau} \left[\frac{\omega\tau}{2} + \frac{\sin 4\pi \cos 2\phi + \cos 4\pi \sin 2\phi - \sin 2\phi}{4} \right]$$

$$= \frac{m\omega^2 C^2}{4} = \frac{kC^2}{4} = \frac{E}{2}$$

where we have used the fact that $\tau = 2\pi/\omega$ (Problem 5–3). Likewise,

$$\langle V \rangle = \frac{kC^2}{2\tau} \int_0^\tau \sin^2(\omega t + \phi)dt$$

$$= \frac{kC^2}{2\tau} \int_0^\tau [1 - \cos^2(\omega t + \phi)]dt$$

$$= \frac{kC^2}{2} - \frac{kC^2}{2\tau\omega}\left(\frac{\omega\tau}{2}\right) = \frac{kC^2}{4} = \frac{E}{2}$$

The motion of a harmonic oscillator is such that $\langle K \rangle = \langle V \rangle$ over a cycle of motion.

5–5. Consider two masses m_1 and m_2 in one dimension, interacting through a potential that depends only upon their relative separation $(x_1 - x_2)$, so that $V(x_1, x_2) = V(x_1 - x_2)$. Given that the force acting upon the jth particle is $f_j = -(\partial V/\partial x_j)$, show that $f_1 = -f_2$. What law is this? Newton's equations for m_1 and m_2 are

$$m_1 \frac{d^2 x_1}{dt^2} = -\frac{\partial V}{\partial x_1} \quad \text{and} \quad m_2 \frac{d^2 x_2}{dt^2} = -\frac{\partial V}{\partial x_2}$$

Now introduce center-of-mass and relative coordinates by

$$X = \frac{m_1 x_1 + m_2 x_2}{M} \qquad x = x_1 - x_2$$

where $M = m_1 + m_2$, and solve for x_1 and x_2 to obtain

$$x_1 = X + \frac{m_2}{M}x \quad \text{and} \quad x_2 = X - \frac{m_1}{M}x$$

Show that Newton's equations in these coordinates are

$$m_1 \frac{d^2 X}{dt^2} + \frac{m_1 m_2}{M}\frac{d^2 x}{dt^2} = -\frac{\partial V}{\partial x}$$

and

$$m_2 \frac{d^2 X}{dt^2} - \frac{m_1 m_2}{M}\frac{d^2 x}{dt^2} = +\frac{\partial V}{\partial x}$$

Now add these two equations to find

$$M\frac{d^2 X}{dt^2} = 0$$

Interpret this result. Now divide the first equation by m_1 and the second by m_2 and subtract to obtain

$$\frac{d^2x}{dt^2} = -\left(\frac{1}{m_1} + \frac{1}{m_2}\right)\frac{\partial V}{\partial x}$$

or

$$\mu\frac{d^2x}{dt^2} = -\frac{\partial V}{\partial x}$$

where $\mu = m_1 m_2/(m_1 + m_2)$ is the reduced mass. Interpret this result, and discuss how the original two-body problem has been reduced to two one-body problems.

The forces acting on masses 1 and 2 are

$$f_1 = -\frac{\partial V}{\partial x_1} \qquad f_2 = -\frac{\partial V}{\partial x_2} = \frac{\partial V}{\partial x_1} = -f_1$$

where we obtain the second equality from the fact that $V(x_1, x_2) = V(x_1 - x_2)$. This is Newton's second law: for every action there is an equal and opposite reaction. Now use the definitions of center-of-mass, X, and relative coordinates, x,

$$X = \frac{m_1 x_1 + m_2 x_2}{M} \qquad (1) \qquad\qquad x = x_1 - x_2 \qquad (2)$$

Multiply Equation 1 by M and Equation 2 by m_2 and add to find

$$x_1 = X + \frac{m_2}{M}x \tag{3}$$

Now multiply Equation 1 by M and Equation 2 by m_1 and subtract to find

$$x_2 = X - \frac{m_1}{M}x \tag{4}$$

Substitute Equations 3 and 4 into Newton's equations to obtain

$$m_1\frac{d^2x_1}{dt^2} = m_1\frac{d^2\left(X + \frac{m_2}{M}x\right)}{dt^2} = m_1\frac{d^2X}{dt^2} + \frac{m_1 m_2}{M}\frac{d^2x}{dt^2} = -\frac{\partial V}{\partial x} \tag{5}$$

and

$$m_2\frac{d^2x_2}{dt^2} = m_1\frac{d^2\left(X - \frac{m_1}{M}x\right)}{dt^2} = m_2\frac{d^2X}{dt^2} - \frac{m_2 m_1}{M}\frac{d^2x}{dt^2} = \frac{\partial V}{\partial x} \tag{6}$$

where we have set $\partial V/\partial x_1 = \partial V/\partial x$. Add Equations 5 and 6 to find

$$M\frac{d^2X}{dt^2} = 0$$

The physical interpretation of this equation is that the center of mass moves at a constant velocity. Now divide Equation 5 by m_1 and Equation 6 by m_2 and subtract the results to obtain

$$\frac{d^2x}{dt^2} = -\left(\frac{1}{m_1} + \frac{1}{m_2}\right)\frac{\partial V}{\partial x} \tag{7}$$

If we define $\mu = m_1 m_2/(m_1 + m_2)$, then we can write Equation 7 as

$$\mu\frac{d^2x}{dt^2} = -\frac{\partial V}{\partial x}$$

This is the equation of mass for a body of mass μ moving under a force $-\partial V/\partial x$. We were able to reduce the two-body problem because we could express the forces acting on body 1 in terms of the forces acting on body 2.

5–6. Extend the results of Problem 5–5 to three dimensions. Realize that in three dimensions the relative separation is given by

$$r_{12} = [(x_1 - x_2)^2 + (y_1 - y_2)^2 + (z_1 - z_2)^2]^{1/2}$$

We can treat the x-, y-, and z-dimensions individually because these directions are orthogonal. The equations in the x-direction are the same as in Problem 5–5. To find the equations in the y and z-directions, we can substitute y or z in place of x (and Y or Z in place of X) in the equations of Problem 5–5. We then find that

$$\mu\frac{d^2x}{dt^2} = -\frac{\partial V}{\partial x} \qquad \mu\frac{d^2y}{dt^2} = -\frac{\partial V}{\partial y} \qquad \mu\frac{d^2z}{dt^2} = -\frac{\partial V}{\partial z}$$

or

$$\mu\frac{d^2\mathbf{r}}{dt^2} = -\nabla V$$

which is the three-dimensional extension of Problem 5–5.

5–7. Calculate the value of the reduced mass of a hydrogen atom. Take the masses of the electron and proton to be $9.109\,390 \times 10^{-31}$ kg and $1.672\,623 \times 10^{-27}$ kg, respectively. What is the percent difference between this result and the rest mass of an electron?

$$\mu = \frac{m_1 m_2}{m_1 + m_2} = \frac{(9.109\,390 \times 10^{-31}\ \text{kg})(1.672\,623 \times 10^{-27}\ \text{kg})}{1.673\,534 \times 10^{-27}\ \text{kg}}$$

$$= 9.104\,432 \times 10^{-31}\ \text{kg}$$

The percent difference between this result and the rest mass of an electron is

$$\frac{9.109\,390 \times 10^{-31}\ \text{kg} - 9.104\,432 \times 10^{-31}\ \text{kg}}{9.109\,390 \times 10^{-31}\ \text{kg}} = 0.05\%$$

5–8. Show that the reduced mass of two equal masses, m, is $m/2$.

Setting $m_1 = m_2 = m$, we find that

$$\mu = \frac{m_1 m_2}{m_1 + m_2} = \frac{m^2}{2m} = \frac{m}{2}$$

5–9. Example 5–2 shows that a Maclaurin expansion of a Morse potential leads to

$$V(x) = D\beta^2 x^2 + \cdots$$

Given that $D = 7.31 \times 10^{-19}$ J·molecule^{-1} and $\beta = 1.81 \times 10^{10}$ m^{-1} for HCl, calculate the force constant of HCl. Plot the Morse potential for HCl, and plot the corresponding harmonic oscillator potential on the same graph (cf. Figure 5.5).

Because $V(x) = kx^2/2$ for a harmonic oscillator,

$$k = 2D\beta^2 = 2(7.31 \times 10^{-19} \text{ J·molecule}^{-1})(1.81 \times 10^{10} \text{ m}^{-1})^2 = 479 \text{ N·m}^{-1}$$

The bond length of HCl (l_0) is 127.5 pm (Table 5.1). The graph below shows the Morse potential $V(x) = D[1 - e^{-\beta(l-l_0)}]^2$ (solid line) and the harmonic oscillator potential $V(x) = \frac{1}{2}k(l - l_0)^2$ (dashed line) for HCl.

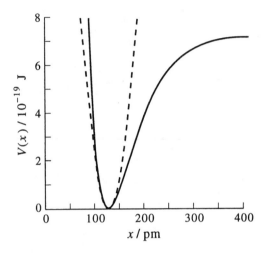

5–10. Use the result of Example 5–2 and Equation 5.34 to show that

$$\beta = 2\pi c \tilde{\nu} \left(\frac{\mu}{2D} \right)^{1/2}$$

Given that $\tilde{\nu} = 2886$ cm^{-1} and $D = 440.2$ kJ·mol^{-1} for H^{35}Cl, calculate β. Compare your result with that in Problem 5–9.

$$\tilde{\nu}_{\text{obs}} = \frac{1}{2\pi c} \left(\frac{k}{\mu} \right)^{1/2} \tag{5.34}$$

Example 5–2 shows that $k = 2D\beta^2$, which upon substituting into Equation 5.34 gives

$$\tilde{\nu}_{\text{obs}} = \frac{1}{2\pi c} \left(\frac{2D\beta^2}{\mu} \right)^{1/2}$$

Solving for β,

$$\beta = 2\pi c \tilde{\nu}_{\text{obs}} \left(\frac{\mu}{2D} \right)^{1/2}$$

For $H^{35}Cl$,

$$\beta = 2\pi c \tilde{\nu}_{obs} \left(\frac{\mu}{2D}\right)^{1/2}$$

$$= 2\pi (2.998 \times 10^8 \text{ m·s}^{-1}) \left(2886 \text{ cm}^{-1} \times \frac{100 \text{ cm}}{1 \text{ m}}\right)$$

$$\left[\frac{\left(\dfrac{1.008 \times 34.97}{1.008 + 34.97}\right)(1.661 \times 10^{-27} \text{ kg})}{2\left(\dfrac{440.2 \times 10^3 \text{ J·mol}^{-1}}{6.022 \times 10^{23} \text{ mol}^{-1}}\right)}\right]^{1/2}$$

$$= 1.81 \times 10^{10} \text{ m}^{-1}$$

which is the same as the value for β given in the previous problem.

5–11. Carry out the Maclaurin expansion of the Morse potential in Example 5–2 through terms in x^4. Express γ in Equation 5.24 in terms of D and β.

$$V(x) = D\left(1 - e^{-\beta x}\right)^2$$

$$= D\left\{1 - \left[1 - \beta x + \frac{1}{2}\beta^2 x^2 - \frac{1}{6}\beta^3 x^3 + O(x^4)\right]\right\}^2$$

$$= D\left[\beta x - \frac{1}{2}\beta^2 x^2 + \frac{1}{6}\beta^3 x^3 - O(x^4)\right]^2$$

$$= D\beta^2 x^2 \left[1 - \frac{1}{2}\beta x + \frac{1}{6}\beta^2 x^2 - O(x^3)\right]^2$$

$$= D\beta^2 x^2 \left[1 - \beta x + \frac{1}{3}\beta^2 x^2 + \frac{1}{4}\beta^2 x^2 + O(x^3)\right]$$

$$= D\left[\beta^2 x^2 - \beta^3 x^3 + \frac{7}{12}\beta^4 x^4 + O(x^5)\right]$$

By comparing this result to Equation 5.24, we see that

$$\gamma = -6D\beta^3$$

5–12. It turns out that the solution of the Schrödinger equation for the Morse potential can be expressed as

$$\tilde{E}_v = \tilde{\nu}\left(v + \frac{1}{2}\right) - \tilde{\nu}\tilde{x}\left(v + \frac{1}{2}\right)^2$$

where

$$\tilde{x} = \frac{hc\tilde{\nu}}{4D}$$

Given that $\tilde{\nu} = 2886 \text{ cm}^{-1}$ and $D = 440.2 \text{ kJ·mol}^{-1}$ for $H^{35}Cl$, calculate $\tilde{x}$ and $\tilde{\nu}\tilde{x}$.

$$\tilde{x} = \frac{(6.626 \times 10^{-34} \text{ J·s})(2.998 \times 10^8 \text{ m·s}^{-1})(2886 \text{ cm}^{-1})(100 \text{ cm·m}^{-1})}{4\left(\dfrac{440.2 \times 10^3 \text{ J·mol}^{-1}}{6.022 \times 10^{23} \text{ mol}^{-1}}\right)}$$

$$= 0.01961$$

$$\tilde{v}\tilde{x} = (2886 \text{ cm}^{-1})(0.01961) = 56.59 \text{ cm}^{-1}$$

5–13. In the infrared spectrum of $H^{79}Br$, there is an intense line at 2559 cm^{-1}. Calculate the force constant of $H^{79}Br$ and the period of vibration of $H^{79}Br$.

$$\tilde{v}_{obs} = \frac{1}{2\pi c}\left(\frac{k}{\mu}\right)^{1/2} \tag{5.34}$$

$$
\begin{aligned}
k &= (2\pi c \tilde{v}_{obs})^2 \mu \\
&= \left[2\pi(2.998 \times 10^{10} \text{ cm·s}^{-1})(2559 \text{ cm}^{-1})\right]^2 \\
&\quad \times \left[\frac{(1.008 \text{ amu})(78.92 \text{ amu})}{79.93 \text{ amu}}\right](1.661 \times 10^{-27} \text{ kg·amu}^{-1}) \\
&= 384 \text{ N·m}^{-1}
\end{aligned}
$$

The period of vibration is

$$T = \frac{1}{v} = \frac{1}{c\tilde{v}} = 1.30 \times 10^{-14} \text{ s}$$

5–14. The force constant of $^{79}Br^{79}Br$ is 240 N·m^{-1}. Calculate the fundamental vibrational frequency and the zero-point energy of $^{79}Br^{79}Br$.

$$\tilde{v}_{obs} = \frac{1}{2\pi c}\left(\frac{k}{\mu}\right)^{1/2} \tag{5.34}$$

$$\tilde{v}_{obs} = \frac{1}{2\pi(2.998 \times 10^{10} \text{ cm·s}^{-1})}\left\{\frac{240 \text{ N·m}^{-1}}{\left[\dfrac{(78.92 \text{ amu})^2}{78.92 \text{ amu} + 78.92 \text{ amu}}\right](1.66 \times 10^{-27} \text{ kg·amu}^{-1})}\right\}^{1/2}$$

$$= 321 \text{ cm}^{-1}$$

We use Equation 5.27 to find the zero point energy:

$$E_0 = \tfrac{1}{2}hv = \tfrac{1}{2}hc\tilde{v} = 3.19 \times 10^{-21} \text{ J}$$

5–15. Verify that $\psi_1(x)$ and $\psi_2(x)$ given in Table 5.3 satisfy the Schrödinger equation for a harmonic oscillator.

The Schrödinger equation for a harmonic oscillator is

$$\frac{d^2\psi_v}{dx^2} + \frac{2\mu}{\hbar^2}(E_v - \frac{1}{2}kx^2)\psi_v = 0 \tag{5.26}$$

where $E_v = h\nu(v + \frac{1}{2})$. From Table 5.3,

$$\psi_1(x) = \left(\frac{4\alpha^3}{\pi}\right)^{1/4} x e^{-\alpha x^2/2}$$

$$\psi_2(x) = \left(\frac{\alpha}{4\pi}\right)^{1/4} (2\alpha x^2 - 1)e^{-\alpha x^2/2}$$

where $\alpha = (k\mu)^{1/2}/\hbar$. Substituting ψ_1 into the Schrödinger equation with $v = 1$ gives

$$\frac{d^2\psi_1}{dx^2} + \frac{2\mu}{\hbar^2}\left(E_1 - \frac{1}{2}kx^2\right)\psi_1 = 0$$

$$\frac{d}{dx}\left[\left(\frac{4\alpha^3}{\pi}\right)^{1/4}\left(e^{-\alpha x^2/2} - \alpha x^2 e^{-\alpha x^2/2}\right)\right] + \frac{2\mu}{\hbar^2}\left(\frac{3h\nu}{2} - \frac{1}{2}kx^2\right)\left(\frac{4\alpha^3}{\pi}\right)^{1/4} x e^{-\alpha x^2/2} = 0$$

$$-3\alpha x e^{-\alpha x^2/2} + \alpha^2 x^3 e^{-\alpha x^2/2} + \left(\frac{3h\mu\nu}{\hbar^2} - \frac{k\mu x^2}{\hbar^2}\right)x e^{-\alpha x^2/2} = 0$$

$$-3\alpha x + \alpha^2 x^3 + \frac{3}{\hbar}(2\pi\nu)\mu x - \alpha^2 x^3 = 0$$

$$-3\alpha x + 3\alpha x = 0$$

where $k = (2\pi\nu)^2\mu$. Substituting ψ_2 into the Schrödinger equation with $v = 2$ gives

$$\frac{d^2\psi_2}{dx^2} + \frac{2\mu}{\hbar^2}\left(E_2 - \frac{1}{2}kx^2\right)\psi_2 = 0$$

$$\left(\frac{\alpha}{4\pi}\right)^{1/4}\frac{d}{dx}\left[4\alpha x e^{-\alpha x^2/2} - \alpha x(2\alpha x^2 - 1)e^{-\alpha x^2/2}\right]$$

$$+ \frac{2\mu}{\hbar^2}\left(\frac{5h\nu}{2} - \frac{1}{2}kx^2\right)\left(\frac{\alpha}{4\pi}\right)^{1/4}(2\alpha x^2 - 1)e^{-\alpha x^2/2} = 0$$

$$(5\alpha - 11\alpha^2 x^2 + 2\alpha^3 x^4)e^{-\alpha x^2/2} + \left(\frac{5h\mu\nu}{\hbar^2} - \frac{k\mu}{\hbar^2}x^2\right)(2\alpha x^2 - 1)e^{-\alpha x^2/2} = 0$$

$$5\alpha - 11\alpha^2 x^2 + 2\alpha^3 x^4 - \frac{5}{\hbar}(2\pi\nu)\mu + \alpha^2 x^2 + \frac{10}{\hbar}(2\pi\nu)\mu\alpha x^2 - 2\alpha^3 x^4 = 0$$

$$5\alpha - 11\alpha^2 x^2 - 5\alpha + 11\alpha^2 x^2 = 0$$

Both ψ_1 and ψ_2 are solutions to the Schrödinger equation.

5–16. Show explicitly for a harmonic oscillator that $\psi_0(x)$ is orthogonal to $\psi_1(x)$, $\psi_2(x)$, and $\psi_3(x)$ and that $\psi_1(x)$ is orthogonal to $\psi_2(x)$ and $\psi_3(x)$ (see Table 5.3).

From Table 5.3, we have

$$\psi_0(x) = \left(\frac{\alpha}{\pi}\right)^{1/4} e^{-\alpha x^2/2}$$

$$\psi_1(x) = \left(\frac{4\alpha^3}{\pi}\right)^{1/4} x e^{-\alpha x^2/2}$$

$$\psi_2(x) = \left(\frac{\alpha}{4\pi}\right)^{1/4} (2\alpha x^2 - 1)e^{-\alpha x^2/2}$$

$$\psi_3(x) = \left(\frac{\alpha^3}{9\pi}\right)^{1/4} (2\alpha x^3 - 3x)e^{-\alpha x^2/2}$$

There are five integrals we must evaluate to show orthogonality. Three have integrands that are odd functions of x, and so are zero.

$$\int_{-\infty}^{\infty} dx\psi_0(x)\psi_1(x) = \int_{-\infty}^{\infty} dx\psi_0(x)\psi_3(x) = \int_{-\infty}^{\infty} dx\psi_1(x)\psi_2(x) = 0$$

This leaves the integrals with even integrands to be evaluated explicitly.

$$\int_{-\infty}^{\infty} dx\psi_0(x)\psi_2(x) = 2\left(\frac{\alpha}{\pi}\right)^{1/4}\left(\frac{\alpha}{4\pi}\right)^{1/4}\int_0^{\infty} dx(2\alpha x^2 - 1)e^{-\alpha x^2}$$

$$= 2\left(\frac{\alpha}{\pi}\right)^{1/4}\left(\frac{\alpha}{4\pi}\right)^{1/4}\left[2\alpha\left(\frac{1}{4\alpha}\right)\left(\frac{\pi}{\alpha}\right)^{1/2} - \left(\frac{\pi}{4\alpha}\right)^{1/2}\right]$$

$$= 0$$

and

$$\int_{-\infty}^{\infty} dx\psi_1(x)\psi_3(x) = 2\left(\frac{4\alpha^3}{\pi}\right)^{1/4}\left(\frac{\alpha^3}{9\pi}\right)^{1/4}\int_0^{\infty} dx(2\alpha x^4 - 3x^2)e^{-\alpha x^2}$$

$$= 2\left(\frac{4\alpha^3}{\pi}\right)^{1/4}\left(\frac{\alpha^3}{9\pi}\right)^{1/4}\left[2\alpha\left(\frac{3}{8\alpha^2}\right)\left(\frac{\pi}{\alpha}\right)^{1/2} - 3\alpha^{1/2}\left(\frac{1}{4\alpha}\right)\left(\frac{\pi}{\alpha}\right)^{1/2}\right]$$

$$= 0$$

5–17. To normalize the harmonic-oscillator wave functions and calculate various expectation values, we must be able to evaluate integrals of the form

$$I_v(a) = \int_{-\infty}^{\infty} x^{2v}e^{-ax^2}dx \qquad v = 0, 1, 2, \ldots$$

We can simply either look them up in a table of integrals or continue this problem. First, show that

$$I_v(a) = 2\int_0^{\infty} x^{2v}e^{-ax^2}dx$$

The case $v = 0$ can be handled by the following trick. Show that the square of $I_0(a)$ can be written in the form

$$I_0^2(a) = 4\int_0^{\infty}\int_0^{\infty} dxdy e^{-a(x^2+y^2)}$$

Now convert to plane polar coordinates, letting

$$r^2 = x^2 + y^2 \quad \text{and} \quad dxdy = rdrd\theta$$

Show that the appropriate limits of integration are $0 \le r < \infty$ and $0 \le \theta \le \pi/2$ and that

$$I_0^2(a) = 4\int_0^{\pi/2} d\theta \int_0^{\infty} drr e^{-ar^2}$$

which is elementary and gives

$$I_0^2(a) = 4 \cdot \frac{\pi}{2} \cdot \frac{1}{2a} = \frac{\pi}{a}$$

or that

$$I_0(a) = \left(\frac{\pi}{a}\right)^{1/2}$$

Now prove that the $I_v(a)$ may be obtained by repeated differentiation of $I_0(a)$ with respect to a and, in particular, that

$$\frac{d^v I_0(a)}{da^v} = (-1)^v I_v(a)$$

Use this result and the fact that $I_0(a) = (\pi/a)^{1/2}$ to generate $I_1(a)$, $I_2(a)$, and so forth.

The function $I_v(a)$ is an even function, and so

$$I_v(a) = 2 \int_0^\infty x^{2v} e^{-\alpha x^2} dx$$

Because the function $I(a)$ depends only on a, we can write

$$I_0^2(a) = 4 \int_0^\infty dx e^{-ax^2} \int_0^\infty dy e^{-ay^2} = 4 \int_0^\infty \int_0^\infty dxdy e^{-a(x^2+y^2)}$$

We realize that we are integrating over the entire first quadrant, so our limits of integration in polar coordinates are $0 \le r < \infty, 0 \le \theta \le \pi/2$. In polar coordinates, $dx\,dy = r\,dr\,d\theta$ and $x^2 + y^2 = r^2$, and so we can write

$$I_0^2(a) = 4 \int_0^{\pi/2} d\theta \int_0^\infty dr r e^{-ar^2}$$

$$= 2\pi \left(\frac{1}{2a}\right) = \frac{\pi}{a}$$

Now differentiate $I_0(a)$ with respect to a:

$$\frac{dI_0}{da} = -\int_{-\infty}^\infty dx x^2 e^{-ax^2} = -I_1(a)$$

$$\frac{d^2 I_0}{da} = \int_{-\infty}^\infty dx x^4 e^{-ax^2} = I_2(a)$$

Extrapolating to the nth derivative gives the general solution

$$\frac{d^v I_0(a)}{da^v} = (-1)^v I_v(a)$$

Because $I_0(a) = (\pi/a)^{1/2}$, $I_1 = (\pi/a)^{1/2}/2a$, $I_2 = 3(\pi/a)^{1/2}/4a^2$, and so forth.

5–18. Prove that the product of two even functions is even, that the product of two odd functions is even, and that the product of an even and an odd function is odd.

Recall that an even function is one for which $f(x) = f(-x)$ and an odd function is one for which $f(x) = -f(-x)$. Let $P(x)$ be the product of two functions $f(x)$ and $g(x)$. For two even functions,

$$P(x) = f(x)g(x) = f(-x)g(-x) = P(-x)$$

so the product of two even functions is even. For two odd functions,

$$P(x) = f(x)g(x) = -f(-x)[-g(-x)] = f(-x)g(-x) = P(-x)$$

so the product of two odd functions is also even. For one odd and one even function,

$$P(x) = f(x)g(x) = [-f(-x)]g(-x) = -f(-x)g(-x) = -P(-x)$$

so the product of one odd and one even function is odd.

5–19. Prove that the derivative of an even (odd) function is odd (even).

If $f(x)$ is even, it can be represented by a power series of the form

$$f(x) = f_0 + f_2 x^2 + f_4 x^4 + O(x^6)$$

where the only allowed values of n in x^n are even. The derivative of this function is

$$f'(x) = 2f_2 x + 4f_4 x^3 + O(x^5)$$

which is an odd function expressed in a power series. Similarly, if $g(x)$ is odd, it can be represented by

$$g(x) = f_1 x + f_3 x^3 + f_5 x^5 + O(x^7)$$

where the only allowed values of n in x^n are odd, and its derivative is

$$g'(x) = f_1 + 3f_3 x^2 + 5f_5 x^4 + O(x^6)$$

which is an even function.

5–20. Show that

$$\langle x^2 \rangle = \int_{-\infty}^{\infty} \psi_2(x) x^2 \psi_2(x) dx = \frac{5}{2} \frac{\hbar}{(\mu k)^{1/2}}$$

for a harmonic oscillator. Note that $\langle x^2 \rangle^{1/2}$ is the square root of the mean of the square of the displacement (the *root-mean-square displacement*) of the oscillator.

From Table 5.3, $\psi_2(x) = \left(\dfrac{\alpha}{4\pi}\right)^{1/4} (2\alpha x^2 - 1)e^{-\alpha x^2/2}$. So

$$\langle x^2 \rangle = \int_{-\infty}^{\infty} \psi_2(x) x^2 \psi_2(x) dx$$

$$= 2\left(\frac{\alpha}{4\pi}\right)^{1/2} \int_0^{\infty} dx (2\alpha x^2 - 1)^2 x^2 e^{-\alpha x^2}$$

$$= \left(\frac{\alpha}{\pi}\right)^{1/2} \int_0^{\infty} dx (4\alpha^2 x^6 - 4\alpha x^4 + x^2)e^{-\alpha x^2}$$

$$= \left(\frac{\alpha}{\pi}\right)^{1/2} \left[4\alpha^2 \left(\frac{15}{16\alpha^3}\right) - 4\alpha \left(\frac{3}{8\alpha^2}\right) + \left(\frac{1}{4\alpha}\right) \right] \left(\frac{\pi}{\alpha}\right)^{1/2}$$

$$= \frac{5}{2\alpha} = \frac{5}{2} \frac{\hbar}{(\mu k)^{1/2}}$$

5–21. Show that

$$\langle p^2 \rangle = \int_{-\infty}^{\infty} \psi_2(x) \hat{P}^2 \psi_2(x) dx = \frac{5}{2} \hbar (\mu k)^{1/2}$$

for a harmonic oscillator.

From Table 5.3 and Table 4.1, $\psi_2(x) = \left(\dfrac{\alpha}{4\pi}\right)^{1/4}(2\alpha x^2 - 1)e^{-\alpha x^2/2}$ and $\hat{P} = -i\hbar d/dx$. So

$$\langle p^2 \rangle = \int_{-\infty}^{\infty} \psi_2(x)\hat{P}^2\psi_2(x)dx$$

$$= 2\left(\frac{\alpha}{4\pi}\right)^{1/2}\int_0^{\infty} dx(2\alpha x^2 - 1)e^{-\alpha x^2/2}\left\{-\hbar^2\frac{d^2}{dx^2}\left[(2\alpha x^2 - 1)e^{-\alpha x^2/2}\right]\right\}$$

$$= -\hbar^2\left(\frac{\alpha}{\pi}\right)^{1/2}\int_0^{\infty} dx(2\alpha x^2 - 1)e^{-\alpha x^2}\left(5\alpha - 11\alpha^2 x^2 + 2\alpha^3 x^4\right)$$

$$= -\hbar^2\left(\frac{\alpha}{\pi}\right)^{1/2}\int_0^{\infty} dx\left(4\alpha^4 x^6 - 24\alpha^3 x^4 + 21\alpha^2 x^2 - 5\alpha\right)e^{-\alpha x^2}$$

$$= -\hbar^2\left(\frac{\alpha}{\pi}\right)^{1/2}\left[4\alpha^4\left(\frac{15}{16\alpha^3}\right) - 24\alpha^3\left(\frac{3}{8\alpha^2}\right) + 21\alpha^2\left(\frac{1}{4\alpha}\right) - 5\alpha\left(\frac{1}{4}\right)\right]\left(\frac{\pi}{\alpha}\right)^{1/2}$$

$$= \hbar^2\frac{5}{2}\alpha = \frac{5}{2}\hbar(\mu k)^{1/2}$$

5–22. Using the fundamental vibrational frequencies of some diatomic molecules given below, calculate the root-mean-square displacement (see Problem 5–20) in the $v = 0$ state and compare it with the equilibrium bond length (also given below).

Molecule	$\tilde{\nu}/\text{cm}^{-1}$	l_0/pm
H_2	4401	74.1
$^{35}\text{Cl}^{35}\text{Cl}$	554	198.8
$^{14}\text{N}^{14}\text{N}$	2330	109.4

We will use Equation 5.34 to find k. Solving for k gives

$$k = (2\pi c\tilde{\nu})^2\mu$$

The root-mean-square displacement is given by $\langle x^2 \rangle^{1/2}$. For the ground state,

$$\langle x^2 \rangle = \int_{-\infty}^{\infty} \psi_0(x)x^2\psi_0(x)dx$$

$$= 2\left(\frac{\alpha}{\pi}\right)^{1/2}\int_0^{\infty} x^2 e^{-\alpha x^2}dx$$

$$= 2\left(\frac{\alpha}{\pi}\right)^{1/2}\left(\frac{\pi}{\alpha}\right)^{1/2}\left(\frac{1}{4\alpha}\right) = \frac{1}{2\alpha} = \frac{\hbar}{2(\mu k)^{1/2}}$$

$$= \frac{\hbar}{4\pi c\tilde{\nu}\mu}$$

and so the root-mean-square displacement of the molecule is

$$\langle x^2 \rangle^{1/2} = \left(\frac{\hbar}{4\pi c\tilde{\nu}\mu}\right)^{1/2}$$

For H_2,

$$\langle x^2 \rangle^{1/2} = \left\{\frac{1.055 \times 10^{-34}\ \text{J}\cdot\text{s}}{4\pi(2.998 \times 10^{10}\ \text{cm}\cdot\text{s}^{-1})(4401\ \text{cm}^{-1})\left[\left(\frac{1.008}{2}\right)(1.661 \times 10^{-27}\ \text{kg})\right]}\right\}^{1/2}$$

$$= 8.718 \times 10^{-12}\ \text{m} = 8.718\ \text{pm}$$

For $^{35}Cl^{35}Cl$,

$$\langle x^2 \rangle^{1/2} = \left\{ \frac{1.055 \times 10^{-34} \text{ J} \cdot \text{s}}{4\pi (2.998 \times 10^{10} \text{ cm} \cdot \text{s}^{-1})(554 \text{ cm}^{-1}) \left[\left(\frac{34.97}{2} \right) (1.661 \times 10^{-27} \text{ kg}) \right]} \right\}^{1/2}$$

$$= 4.172 \times 10^{-12} \text{ m} = 4.172 \text{ pm}$$

and finally, for $^{14}N^{14}N$,

$$\langle x^2 \rangle^{1/2} = \left\{ \frac{1.055 \times 10^{-34} \text{ J} \cdot \text{s}}{4\pi (2.998 \times 10^{10} \text{ cm} \cdot \text{s}^{-1})(2330 \text{ cm}^{-1}) \left[\frac{14.003}{2} (1.661 \times 10^{-27} \text{ kg}) \right]} \right\}^{1/2}$$

$$= 3.215 \times 10^{-12} \text{ m} = 3.215 \text{ pm}$$

These values are all much smaller than the equilibrium bond lengths.

5–23. Prove that

$$\langle K \rangle = \langle V(x) \rangle = \frac{E_v}{2}$$

for a one-dimensional harmonic oscillator for $v = 0$ and $v = 1$.

The operators for $V(x)$ and $K(x)$ given in Table 4.1, the expressions for $\psi_v(x)$ given in Table 5.3, and Equation 5.30 for the vibrational energy levels are

$$\hat{V}(x) = \frac{kx^2}{2}$$

$$\hat{K}(x) = -\frac{\hbar^2}{2\mu} \frac{d^2}{dx^2}$$

$$\psi_0(x) = \left(\frac{\alpha}{\pi} \right)^{1/4} e^{-\alpha x^2/2}$$

$$\psi_1(x) = \left(\frac{4\alpha^3}{\pi} \right)^{1/4} x e^{-\alpha x^2/2}$$

$$E_v = \hbar \left(\frac{k}{\mu} \right)^{1/2} \left(v + \frac{1}{2} \right)$$

For $v = 0$,

$$\langle K \rangle = \int_{-\infty}^{\infty} dx \, \psi_0(x) \left(\frac{-\hbar^2}{2\mu} \frac{d^2}{dx^2} \right) \psi_0(x)$$

$$= -\frac{\hbar^2}{\mu} \left(\frac{\alpha}{\pi} \right)^{1/2} \int_0^{\infty} dx \, e^{-\alpha x^2} \left[\alpha^2 x^2 - \alpha \right]$$

$$= -\frac{\hbar^2}{\mu} \left(\frac{\alpha}{\pi} \right)^{1/2} \left(\frac{\pi}{\alpha} \right)^{1/2} \left[\alpha^2 \left(\frac{1}{4\alpha} \right) - \alpha \left(\frac{1}{2} \right) \right] = \frac{\hbar^2 \alpha}{4\mu}$$

$$= \frac{\hbar}{4} \left(\frac{k}{\mu} \right)^{1/2} = \frac{E_0}{2}$$

and

$$\langle V \rangle = \int_{-\infty}^{\infty} dx \, \psi_0(x) \left(\frac{kx^2}{2} \right) \psi_0(x)$$

$$= k \left(\frac{\alpha}{\pi} \right)^{1/2} \int_0^{\infty} dx \, x^2 e^{-\alpha x^2}$$

$$= k \left(\frac{\alpha}{\pi} \right)^{1/2} \left(\frac{\pi}{\alpha} \right)^{1/2} \frac{1}{4\alpha} = \frac{k}{4\alpha} = \frac{\hbar}{4} \left(\frac{k}{\mu} \right)^{1/2} = \frac{E_0}{2}$$

For $v = 1$,

$$\langle K \rangle = \int_{-\infty}^{\infty} dx \, \psi_1(x) \left(\frac{-\hbar^2}{2\mu} \frac{d^2}{dx^2} \right) \psi_1(x)$$

$$= -\frac{\hbar^2}{\mu} \left(\frac{4\alpha^3}{\pi} \right)^{1/2} \int_0^{\infty} dx \, e^{-\alpha x^2} \left(\alpha^2 x^4 - 3\alpha x^2 \right)$$

$$= -\frac{\hbar^2}{\mu} \left(\frac{4\alpha^3}{\pi} \right)^{1/2} \left(\frac{\pi}{\alpha} \right)^{1/2} \left[\alpha^2 \left(\frac{3}{8\alpha^2} \right) - 3\alpha \left(\frac{1}{4\alpha} \right) \right]$$

$$= \frac{3\hbar^2 \alpha}{4\mu} = \frac{3\hbar}{4} \left(\frac{k}{\mu} \right)^{1/2} = \frac{E_1}{2}$$

and

$$\langle V \rangle = \int_{-\infty}^{\infty} dx \, \psi_1(x) \left(\frac{kx^2}{2} \right) \psi_1(x)$$

$$= k \left(\frac{4\alpha^3}{\pi} \right)^{1/2} \int_0^{\infty} dx \, x^4 e^{-\alpha x^2}$$

$$= k \left(\frac{4\alpha^3}{\pi} \right)^{1/2} \left(\frac{\pi}{\alpha} \right)^{1/2} \left(\frac{3}{8\alpha^2} \right) = \frac{3k}{4\alpha}$$

$$= \frac{3\hbar}{4} \left(\frac{k}{\mu} \right)^{1/2} = \frac{E_1}{2}$$

5–24. There are a number of general relations between the Hermite polynomials and their derivatives (which we will not derive). Some of these are

$$\frac{dH_v(\xi)}{d\xi} = 2\xi H_v(\xi) - H_{v+1}(\xi)$$

$$H_{v+1}(\xi) - 2\xi H_v(\xi) + 2v H_{v-1}(\xi) = 0$$

and

$$\frac{dH_v(\xi)}{d\xi} = 2v H_{v-1}(\xi)$$

Such connecting relations are called *recursion formulas*. Verify these formulas explicitly using the first few Hermite polynomials given in Table 5.2.

We will verify these formulas for $v = 0$, 1, and 2. The Hermite polynomials for these values of v are

$$H_0(\xi) = 1 \qquad H_1(\xi) = 2\xi \qquad H_2(\xi) = 4\xi^2 - 2$$

Using the first recursion formula, we have

$$\frac{d H_0(\xi)}{d\xi} = 2\xi\, H_0(\xi) - H_1(\xi)$$

or

$$0 = 2\xi(1) - 2\xi = 0$$

For $v = 1$,

$$\frac{d H_1(\xi)}{d\xi} = 2\xi\, H_1(\xi) - H_2(\xi)$$

$$2 = 2\xi(2\xi) - (4\xi^2 - 2) = 2$$

Using the second recursion fromula for $v = 1$, we have

$$H_2(\xi) - 2\xi\, H_1(\xi) + 2v\, H_0(\xi) = 0$$
$$4\xi^2 - 2 - 2\xi(2\xi) + 2 = 0$$

The third recursion formula is

$$\frac{d H_v(\xi)}{d\xi} = 2v\, H_{v-1}(\xi)$$

For $v = 1$, we have

$$2 = (2)(1)(1) = 2$$

and for $v = 2$, we have

$$8\xi = (2)(2)(2\xi) = 8\xi$$

5–25. Use the recursion formulas for the Hermite polynomials given in Problem 5–24 to show that $\langle p \rangle = 0$ and $\langle p^2 \rangle = \hbar(\mu k)^{1/2}(v + \frac{1}{2})$. Remember that the momentum operator involves a differentiation with respect to x, not ξ.

$$\langle p \rangle = \int_{-\infty}^{\infty} dx\, \psi_v(x)\, \hat{P}_x \psi_v(x)$$

$$= -i\hbar \int_{-\infty}^{\infty} dx\, \psi_v(x) \frac{d\psi_v(x)}{dx}$$

In Example 5–6 we learned that $H_v(x)$ is an even function of x if v is even and is an odd function of x if v is odd. Because $e^{-\alpha x^2/2}$ is an even function of x, ψ_v is an even function of x if v is even and an odd function of x if v is odd (Problem 5–18). Also, the quantity $d\psi_v/dx$ is an even function if v is odd and is an odd function if v is even (Problem 5–19). Therefore, the integrand of $\langle p \rangle$ will always be the product of one odd and one even function, and is therefore an odd function (Problem 5–18). Because the integrand is integrated over all space, $\langle p \rangle = 0$. Now consider

$$\langle p^2 \rangle = \int_{-\infty}^{\infty} dx\, \psi_v(x)\, \hat{P}^2 \psi_v(x)$$

$$= -\hbar^2 \int_{-\infty}^{\infty} dx\, \psi_v(x) \frac{d^2\psi_v(x)}{dx^2}$$

We use the Schrödinger equation for a harmonic oscillator

$$\frac{d^2\psi_v}{dx^2} + \frac{2\mu}{\hbar^2}\left(E_v - \frac{1}{2}kx^2\right)\psi_v(x) = 0$$

to write

$$\langle p^2 \rangle = 2\mu \int_{-\infty}^{\infty} dx\,\psi_v(x)\left(E_v - \frac{1}{2}kx^2\right)\psi_v(x)$$

$$= 2\mu E_v - k\mu \int_{-\infty}^{\infty} dx\,\psi_v(x)x^2\psi_v(x)$$

$$= 2\mu\hbar\omega\left(v + \frac{1}{2}\right)\frac{-k\mu}{\alpha^{3/2}}\int_{-\infty}^{\infty} d\xi\,N_v H_v(\xi)\xi^2 N_v H_v(\xi)e^{-\xi^2} \tag{1}$$

Using the second recursion relation in Problem 5–24, we have

$$\xi^2 H_v(\xi) = \xi\left[vH_{v-1}(\xi) + \frac{1}{2}H_{v+1}(\xi)\right]$$

$$= v\left[(v-1)H_{v-2}(\xi) + \frac{1}{2}H_v(\xi)\right] + \frac{1}{2}\left[(v+1)H_v(\xi) + \frac{1}{2}H_{v+2}(\xi)\right]$$

$$= v(v-1)H_{v-2}(\xi) + \left(v + \frac{1}{2}\right)H_v(\xi) + \frac{1}{4}H_{v+2}(\xi)$$

Substituting this result into the integral of Equation 1 above gives three integrals, only one of which is nonzero (Section 5–6), so

$$\langle p^2 \rangle = 2\mu\hbar\omega\left(v + \frac{1}{2}\right) - \frac{k\mu}{\alpha^{3/2}}\int_{-\infty}^{\infty} d\xi\,N_v H_v\xi N_v\left(v + \frac{1}{2}\right)H_v(\xi)e^{-\xi^2}$$

$$= 2\mu\hbar\omega\left(v + \frac{1}{2}\right) - \frac{k\mu}{\alpha}\left(v + \frac{1}{2}\right)\int_{-\infty}^{\infty}\psi_v(x)\psi_v(x)dx$$

$$= 2\hbar(\mu k)^{1/2}\left(v + \frac{1}{2}\right) - \hbar(\mu k)^{1/2}\left(v + \frac{1}{2}\right)$$

$$= \hbar(\mu k)^{1/2}\left(v + \frac{1}{2}\right)$$

5–26. It can be proved generally that

$$\langle x^2 \rangle = \frac{1}{\alpha}\left(v + \tfrac{1}{2}\right) = \frac{\hbar}{(\mu k)^{1/2}}\left(v + \tfrac{1}{2}\right)$$

and that

$$\langle x^4 \rangle = \frac{3}{4\alpha^2}(2v^2 + 2v + 1) = \frac{3\hbar^2}{4\mu k}(2v^2 + 2v + 1)$$

for a harmonic oscillator. Verify these formulas explicitly for the first two states of a harmonic oscillator.

$$\langle x^2 \rangle_{v=0} = \left(\frac{\alpha}{\pi}\right)^{1/2}\int_{-\infty}^{\infty} x^2 e^{-\alpha x^2}dx$$

$$= 2\left(\frac{\alpha}{\pi}\right)^{1/2}\left(\frac{\pi}{\alpha}\right)^{1/2}\left(\frac{1}{4\alpha}\right)$$

$$= \frac{1}{2\alpha} = \frac{\hbar}{2(\mu k)^{1/2}}$$

$$\langle x^4 \rangle_{v=0} = \left(\frac{\alpha}{\pi}\right)^{1/2} \int_{-\infty}^{\infty} x^4 e^{-\alpha x^2} dx$$

$$= 2\left(\frac{\alpha}{\pi}\right)^{1/2} \left(\frac{\pi}{\alpha}\right)^{1/2} \left(\frac{3}{8\alpha^2}\right)$$

$$= \frac{3}{4\alpha^2} = \frac{3\hbar^2}{4\mu k}$$

$$\langle x^2 \rangle_{v=1} = \left(\frac{4\alpha^3}{\pi}\right)^{1/2} \int_{-\infty}^{\infty} x^4 e^{-\alpha x^2} dx$$

$$= 2\left(\frac{4\alpha^3}{\pi}\right)^{1/2} \left(\frac{\pi}{\alpha}\right)^{1/2} \left(\frac{3}{8\alpha^2}\right)$$

$$= \frac{3}{2\alpha} = \frac{3\hbar}{2(\mu k)^{1/2}}$$

$$\langle x^4 \rangle_{v=1} = \left(\frac{4\alpha^3}{\pi}\right)^{1/2} \int_{-\infty}^{\infty} x^6 e^{-\alpha x^2} dx$$

$$= 2\left(\frac{4\alpha^3}{\pi}\right)^{1/2} \left(\frac{\pi}{\alpha}\right)^{1/2} \left(\frac{15}{16\alpha^3}\right)$$

$$= \frac{15}{4\alpha^2} = \frac{15\hbar^2}{4\mu k}$$

5-27. This problem is similar to Problem 3–35. Show that the harmonic-oscillator wave functions are alternately even and odd functions of x because the Hamiltonian operator obeys the relation $\hat{H}(x) = \hat{H}(-x)$. Define a reflection operator $\hat{R}$ by

$$\hat{R}u(x) = u(-x)$$

Show that $\hat{R}$ is linear and that it commutes with $\hat{H}$. Show also that the eigenvalues of $\hat{R}$ are ± 1. What are its eigenfunctions? Show that the harmonic-oscillator wave functions are eigenfunctions of $\hat{R}$. Note that they are eigenfunctions of both $\hat{H}$ and $\hat{R}$. What does this observation say about $\hat{H}$ and $\hat{R}$?

Consider the Schrödinger equation of a harmonic oscillator

$$\hat{H}(x)\psi_n(x) = E_n \psi_n(x)$$

Replace x by $-x$ and use the fact that $\hat{H}(x) = \hat{H}(-x)$ to obtain

$$\hat{H}\psi_n(-x) = E_n \psi_n(-x)$$

Both $\psi_n(-x)$ and $\psi_n(x)$ are eigenfunctions of $\hat{H}(x)$ corresponding to the eigenvalue E_n. Because the system is nondegenerate, these eigenfunctions can differ by only a multiplicative constant c. We can write this as $\psi_n(x) = c\psi_n(-x)$. But $\psi_n(-x) = c\psi_n(x)$, and so $c = \pm 1$ (as in Problem 3–35). Thus ψ_n is always either even or odd. Moreover, because $H_v(x)$ is even when v is even and odd when v is odd, and because

$$\psi_v(x) = N_v H_v(\alpha^{1/2}x)e^{-\alpha x^2/2} \tag{5.35}$$

$\psi_v(x)$ is even when v is even and odd when v is odd. Now define $\hat{R}$ as $\hat{R}u(x) = u(-x)$. $\hat{R}$ is linear because

$$\hat{R}\left[c_1 u_1(x) + c_2 u_2(x)\right] = c_1 u_1(-x) + c_2 u_2(-x)$$
$$= c_1 \hat{R}u_1(x) + c_2 \hat{R}u_2(x)$$

Because $\hat{R}\psi_n(x) = \psi_n(-x) = \pm\psi_n(x)$, we see that the eigenvalues of $\hat{R}$ are ± 1 and the eigenfunctions are $\psi_n(x)$. Because $\hat{H}$ and $\hat{R}$ have mutual eigenfunctions, they commute.

5–28. Use Ehrenfest's theorem (Problem 4–27) to show that $\langle p_x \rangle$ does not depend upon time for a one-dimensional harmonic oscillator.

$$\frac{d\langle p_x \rangle}{dt} = \left\langle -\frac{dV}{dx} \right\rangle = \langle -kx \rangle = 0$$

because $\langle x \rangle$ is the integral of an odd function. The fact that $d\langle p_x \rangle/dt = 0$ means that $\langle p_x \rangle$ does not depend upon time.

5–29. Show that the moment of inertia for a rigid rotator can be written as $I = \mu r^2$, where $r = r_1 + r_2$ (the fixed separation of the two masses) and μ is the reduced mass.

By definition, at the center of mass

$$m_1 r_1 = m_2 r_2 \qquad I = m_1 r_1^2 + m_2 r_2^2 \qquad \mu = \frac{m_1 m_2}{m_1 + m_2}$$

Now $r_1 = r - r_2$ and $r_2 = r - r_1$, so we can write the first equaiton as either $m_1(r - r_2) = m_2 r_2$ or $m_2(r - r_1) = m_1 r_1$. Solving these expressions for r_1 and r_2 gives

$$r_2 = \frac{m_1 r}{m_1 + m_2} \qquad r_1 = \frac{m_2 r}{m_1 + m_2}$$

Substituting these results into the expression of I gives

$$
\begin{aligned}
I &= m_1 r_1^2 + m_2 r_2^2 \\
&= \frac{m_1 m_2}{m_1 + m_2}\left[\frac{m_1 + m_2}{m_2}\left(\frac{m_2 r}{m_1 + m_2}\right)^2 + \frac{m_1 + m_2}{m_1}\left(\frac{m_1 r}{m_1 + m_2}\right)^2 \right] \\
&= \mu\left[\left(\frac{m_2}{m_1 + m_2} + \frac{m_1}{m_1 + m_2}\right) r^2 \right] \\
&= \mu r^2
\end{aligned}
$$

5–30. Consider the transformation from Cartesian coordinates to plane polar coordinates where

$$
\begin{aligned}
x &= r\cos\theta & r &= (x^2 + y^2)^{1/2} \\
y &= r\sin\theta & \theta &= \tan^{-1}\left(\frac{y}{x}\right)
\end{aligned}
\qquad (1)
$$

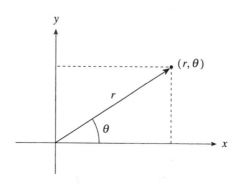

If a function $f(r, \theta)$ depends upon the polar coordinates r and θ, then the chain rule of partial differentiation says that

$$\left(\frac{\partial f}{\partial x}\right)_y = \left(\frac{\partial f}{\partial r}\right)_\theta \left(\frac{\partial r}{\partial x}\right)_y + \left(\frac{\partial f}{\partial \theta}\right)_r \left(\frac{\partial \theta}{\partial x}\right)_y \tag{2}$$

and that

$$\left(\frac{\partial f}{\partial y}\right)_x = \left(\frac{\partial f}{\partial r}\right)_\theta \left(\frac{\partial r}{\partial y}\right)_x + \left(\frac{\partial f}{\partial \theta}\right)_r \left(\frac{\partial \theta}{\partial y}\right)_x \tag{3}$$

For simplicity, we will assume r is constant so that we can ignore terms involving derivatives with respect to r. In other words, we will consider a particle that is constrained to move on the circumference of a circle. This system is sometimes called a *particle on a ring*. Using Equations 1 and 2, show that

$$\left(\frac{\partial f}{\partial x}\right)_y = -\frac{\sin\theta}{r}\left(\frac{\partial f}{\partial \theta}\right)_r \quad \text{and} \quad \left(\frac{\partial f}{\partial y}\right)_x = \frac{\cos\theta}{r}\left(\frac{\partial f}{\partial \theta}\right)_r \quad (r \text{ fixed}) \tag{4}$$

Now apply Equation 2 again to show that

$$\left(\frac{\partial^2 f}{\partial x^2}\right)_y = \left[\frac{\partial}{\partial x}\left(\frac{\partial f}{\partial x}\right)_y\right] = \left[\frac{\partial}{\partial \theta}\left(\frac{\partial f}{\partial x}\right)_y\right]_r \left(\frac{\partial \theta}{\partial x}\right)_y$$

$$= \left\{\frac{\partial}{\partial \theta}\left[-\frac{\sin\theta}{r}\left(\frac{\partial f}{\partial \theta}\right)_r\right]\right\}_r \left(-\frac{\sin\theta}{r}\right)$$

$$= \frac{\sin\theta\cos\theta}{r^2}\left(\frac{\partial f}{\partial \theta}\right)_r + \frac{\sin^2\theta}{r^2}\left(\frac{\partial^2 f}{\partial \theta^2}\right)_r \quad (r \text{ fixed})$$

Similarly, show that

$$\left(\frac{\partial^2 f}{\partial y^2}\right)_x = -\frac{\sin\theta\cos\theta}{r^2}\left(\frac{\partial f}{\partial \theta}\right)_r + \frac{\cos^2\theta}{r^2}\left(\frac{\partial^2 f}{\partial \theta^2}\right)_r \quad (r \text{ fixed})$$

and that

$$\nabla^2 f = \frac{\partial^2 f}{\partial x^2} + \frac{\partial^2 f}{\partial y^2} \longrightarrow \frac{1}{r^2}\left(\frac{\partial^2 f}{\partial \theta^2}\right)_r \quad (r \text{ fixed})$$

Now show that the Schrödinger equation for a particle of mass m constrained to move on a circle of radius r is (see Problem 3–28)

$$-\frac{\hbar^2}{2I}\frac{\partial^2 \psi(\theta)}{\partial \theta^2} = E\psi(\theta) \qquad 0 \leq \theta \leq 2\pi$$

where $I = mr^2$ is the moment of inertia.

First use Equations 1 to find the partial derivatives of r and θ with respect to x and y:

$$\left(\frac{\partial r}{\partial x}\right)_y = \left(\frac{\partial (x^2 + y^2)^{1/2}}{\partial x}\right)_y = \frac{x}{(x^2 + y^2)^{1/2}} = \cos\theta$$

$$\left(\frac{\partial r}{\partial y}\right)_x = \left(\frac{\partial (x^2 + y^2)^{1/2}}{\partial y}\right)_x = \frac{y}{(x^2 + y^2)^{1/2}} = \sin\theta$$

$$\left(\frac{\partial \theta}{\partial x}\right)_y = \left(\frac{\partial \tan^{-1}\left(\frac{y}{x}\right)}{\partial x}\right)_y = \frac{1}{1 + \left(\frac{y}{x}\right)^2}\left(-\frac{y}{x^2}\right)$$

$$= -\frac{y}{x^2 + y^2} = -\frac{r\sin\theta}{r^2} = -\frac{\sin\theta}{r}$$

$$\left(\frac{\partial \theta}{\partial y}\right)_x = \left(\frac{\partial \tan^{-1}\left(\frac{y}{x}\right)}{\partial y}\right)_x = \frac{1}{1+\left(\frac{y}{x}\right)^2}\left(\frac{1}{x}\right)$$

$$= \frac{x}{x^2 + y^2} = \frac{\cos\theta}{r}$$

Now substitute into Equations 2 and 3 to find Equations 4:

$$\left(\frac{\partial f}{\partial x}\right)_y = 0 + \left(\frac{\partial f}{\partial \theta}\right)_r \left(\frac{\partial \theta}{\partial x}\right)_y = -\frac{\sin\theta}{r}\left(\frac{\partial f}{\partial \theta}\right)_r$$

$$\left(\frac{\partial f}{\partial y}\right)_x = 0 + \left(\frac{\partial f}{\partial \theta}\right)_r \left(\frac{\partial \theta}{\partial y}\right)_x = \frac{\cos\theta}{r}\left(\frac{\partial f}{\partial \theta}\right)_r$$

where r is fixed. Now (keeping r fixed)

$$\left(\frac{\partial^2 f}{\partial x^2}\right)_y = \left[\frac{\partial}{\partial x}\left(\frac{\partial f}{\partial x}\right)_y\right] = \left[\frac{\partial}{\partial \theta}\left(\frac{\partial f}{\partial x}\right)_y\right]_r \left(\frac{\partial \theta}{\partial x}\right)_y$$

$$= \left\{\frac{\partial}{\partial \theta}\left[-\frac{\sin\theta}{r}\left(\frac{\partial f}{\partial \theta}\right)_r\right]\right\}_r \left(-\frac{\sin\theta}{r}\right)$$

$$= \frac{\sin\theta\cos\theta}{r^2}\left(\frac{\partial f}{\partial \theta}\right)_r + \frac{\sin^2\theta}{r^2}\left(\frac{\partial^2 f}{\partial \theta^2}\right)_r$$

and

$$\left(\frac{\partial^2 f}{\partial y^2}\right)_x = \left[\frac{\partial}{\partial y}\left(\frac{\partial f}{\partial y}\right)_x\right] = \left[\frac{\partial}{\partial \theta}\left(\frac{\partial f}{\partial y}\right)_x\right]_r \left(\frac{\partial \theta}{\partial y}\right)_x$$

$$= \left\{\frac{\partial}{\partial \theta}\left[\frac{\cos\theta}{r}\left(\frac{\partial f}{\partial \theta}\right)_r\right]\right\}_r \left(\frac{\cos\theta}{r}\right)$$

$$= -\frac{\sin\theta\cos\theta}{r^2}\left(\frac{\partial f}{\partial \theta}\right)_r + \frac{\cos^2\theta}{r^2}\left(\frac{\partial^2 f}{\partial \theta^2}\right)_r$$

giving

$$\nabla^2 f = \frac{\partial^2 f}{\partial x^2} + \frac{\partial^2 f}{\partial y^2}$$

$$= \frac{\sin^2\theta + \cos^2\theta}{r^2}\left(\frac{\partial^2 f}{\partial \theta^2}\right)_r$$

$$= \frac{1}{r^2}\left(\frac{\partial^2 f}{\partial \theta^2}\right)_r$$

The Schrödinger equation for the particle is

$$-\frac{\hbar^2}{2m}\nabla^2 \psi_n(\theta) = E\psi_n(\theta)$$

$$-\frac{\hbar^2}{2mr^2}\frac{\partial^2 \psi_n(\theta)}{\partial \theta^2} = E\psi_n(\theta)$$

$$-\frac{\hbar^2}{2I}\frac{\partial^2 \psi_n(\theta)}{\partial \theta^2} = E\psi_n(\theta)$$

5–31. Generalize Problem 5–30 to the case of a particle moving in a plane under the influence of a central force; in other words, convert

$$\nabla^2 = \frac{\partial^2}{\partial x^2} + \frac{\partial^2}{\partial y^2}$$

to plane polar coordinates, this time without assuming that r is a constant. Use the method of separation of variables to separate the equation for this problem. Solve the angular equation.

We can use the partial derivatives we found in the previous problem and Equations 2 and 3 to write

$$\left(\frac{\partial f}{\partial x}\right)_y = \cos\theta \left(\frac{\partial f}{\partial r}\right)_\theta - \frac{\sin\theta}{r}\left(\frac{\partial f}{\partial \theta}\right)_r,$$

$$\left(\frac{\partial f}{\partial y}\right)_x = \sin\theta \left(\frac{\partial f}{\partial r}\right)_\theta + \frac{\cos\theta}{r}\left(\frac{\partial f}{\partial \theta}\right)_r,$$

Now (as before)

$$\left(\frac{\partial^2 f}{\partial x^2}\right)_y = \left(\frac{\partial r}{\partial x}\right)_y \left[\frac{\partial}{\partial r}\left(\frac{\partial f}{\partial x}\right)_y\right] + \left(\frac{\partial \theta}{\partial x}\right)_y \left[\frac{\partial}{\partial \theta}\left(\frac{\partial f}{\partial x}\right)_y\right]$$

$$= \cos\theta \left[\cos\theta \left(\frac{\partial^2 f}{\partial r^2}\right)_\theta + \frac{\sin^2\theta}{r^2}\left(\frac{\partial f}{\partial \theta}\right)_r - \frac{\sin\theta}{r}\left(\frac{\partial^2 f}{\partial r \partial \theta}\right)\right]$$

$$\quad - \frac{\sin\theta}{r}\left[-\sin\theta \left(\frac{\partial f}{\partial r}\right)_\theta + \cos\theta \left(\frac{\partial^2 f}{\partial \theta \partial r}\right) - \frac{\cos\theta}{r}\left(\frac{\partial f}{\partial \theta}\right)_r - \frac{\sin\theta}{r}\left(\frac{\partial^2 f}{\partial \theta^2}\right)_r\right]$$

$$= \cos^2\theta \left(\frac{\partial^2 f}{\partial r^2}\right)_\theta - \frac{2\cos\theta\sin\theta}{r}\left(\frac{\partial^2 f}{\partial r \partial \theta}\right)$$

$$\quad + \frac{2\cos\theta\sin\theta}{r}\left(\frac{\partial f}{\partial \theta}\right)_r + \frac{\sin^2\theta}{r}\left(\frac{\partial f}{\partial r}\right)_\theta + \frac{\sin^2\theta}{r^2}\left(\frac{\partial^2 f}{\partial \theta^2}\right)_r$$

and

$$\left(\frac{\partial^2 f}{\partial y^2}\right)_x = \left(\frac{\partial r}{\partial y}\right)_x \left[\frac{\partial}{\partial r}\left(\frac{\partial f}{\partial y}\right)_x\right] + \left(\frac{\partial \theta}{\partial y}\right)_x \left[\frac{\partial}{\partial \theta}\left(\frac{\partial f}{\partial y}\right)_x\right]$$

$$= \sin\theta \left[\sin\theta \left(\frac{\partial^2 f}{\partial r^2}\right)_\theta - \frac{\cos\theta}{r^2}\left(\frac{\partial f}{\partial \theta}\right)_r + \frac{\cos\theta}{r}\left(\frac{\partial^2 f}{\partial r \partial \theta}\right)\right]$$

$$\quad + \frac{\cos\theta}{r}\left[\cos\theta \left(\frac{\partial f}{\partial r}\right)_\theta + \sin\theta \left(\frac{\partial^2 f}{\partial \theta \partial r}\right) - \frac{\sin\theta}{r}\left(\frac{\partial f}{\partial \theta}\right)_r + \frac{\cos\theta}{r}\left(\frac{\partial^2 f}{\partial \theta^2}\right)_r\right]$$

$$= \sin^2\theta \left(\frac{\partial^2 f}{\partial r^2}\right)_\theta + \frac{2\cos\theta\sin\theta}{r}\left(\frac{\partial^2 f}{\partial r \partial \theta}\right)$$

$$\quad - \frac{2\cos\theta\sin\theta}{r}\left(\frac{\partial f}{\partial \theta}\right)_r + \frac{\cos^2\theta}{r}\left(\frac{\partial f}{\partial r}\right)_\theta + \frac{\cos^2\theta}{r^2}\left(\frac{\partial^2 f}{\partial \theta^2}\right)_r$$

giving

$$\nabla^2 f = \left(\frac{\partial^2 f}{\partial x^2}\right)_y + \left(\frac{\partial^2 f}{\partial y^2}\right)_x = \left(\frac{\partial^2 f}{\partial r^2}\right)_\theta + \frac{1}{r}\left(\frac{\partial f}{\partial r}\right)_\theta + \frac{1}{r^2}\left(\frac{\partial^2 f}{\partial \theta^2}\right)_r$$

Now consider the Schrödinger equation of a particle moving in a plane under the influence of a central force:

$$-\frac{\hbar^2}{2\mu}\left[\frac{\partial^2 \psi}{\partial r^2} + \frac{1}{r}\frac{\partial \psi}{\partial r} + \frac{1}{r^2}\frac{\partial^2 \psi}{\partial \theta^2}\right] + V(r)\psi(r, \theta) = E\psi(r, \theta)$$

$$-\hbar^2\left[r^2\frac{\partial^2 \psi}{\partial r^2} + r\frac{\partial \psi}{\partial r}\right] - \hbar^2\frac{\partial^2 \psi}{\partial \theta^2} + 2\mu r^2[V(r) - E]\psi = 0$$

Let $\psi(r, \theta) = R(r)\Theta(\theta)$ to get

$$-\frac{\hbar^2}{R}\left[r^2\frac{d^2R}{dr^2} + r\frac{dR}{dr}\right] + 2\mu r^2[V - E] - \frac{\hbar^2}{\Theta}\frac{d^2\Theta}{d\theta^2} = 0$$

Separating the equation, we get the two equations

$$p^2 = -\frac{\hbar^2}{R}\left[r^2\frac{d^2R}{dr^2} + r\frac{dR}{dr}\right] + 2\mu r^2[V - E]$$

$$q^2 = -\frac{\hbar^2}{\Theta}\frac{d^2\Theta}{d\theta^2}$$

where $p^2 + q^2 = 0$. Using the second of these equations, we find

$$\frac{d^2\Theta}{d\theta^2} + \frac{q^2}{\hbar^2}\Theta = 0$$

The general solution to this equation (Example 2–4) is

$$\Theta(\theta) = c_1 e^{\pm ni\theta} = E\cos(n\theta + \phi)$$

where $n = q/\hbar$.

5–32. Using Problems 5–30 and 5–31 as a guide, convert ∇^2 from three-dimensional Cartesian coordinates to spherical coordinates.

This is an extremely long and tedious exercise in partial differentiation. We can avoid this tedium by approaching the problem another way. Let q_1, q_2, and q_3 be any suitable set of coordinates and let x, y, and z be given by

$$x = x(q_1, q_2, q_3) \qquad y = y(q_1, q_2, q_3) \qquad z = z(q_1, q_2, q_3)$$

For example, for spherical coordinates $q_1 = r$, $q_2 = \theta$ and $q_3 = \phi$. We give here without proof (the proof is actually straightforward, although lengthy) a general formula for ∇^2 in terms of q_1, q_2, and q_3:

$$\nabla^2 f = \frac{1}{h_1 h_2 h_3}\left[\frac{\partial}{\partial q_1}\left(\frac{h_2 h_3}{h_1}\frac{\partial f}{\partial q_1}\right) + \frac{\partial}{\partial q_2}\left(\frac{h_3 h_1}{h_2}\frac{\partial f}{\partial q_2}\right) + \frac{\partial}{\partial q_3}\left(\frac{h_1 h_2}{h_3}\frac{\partial f}{\partial q_3}\right)\right]$$

where

$$h_j^2 = \left(\frac{\partial x}{\partial q_j}\right)^2 + \left(\frac{\partial y}{\partial q_j}\right)^2 + \left(\frac{\partial z}{\partial q_j}\right)^2$$

We can now apply this formula to spherical coordinates, where $x = r\sin\theta\cos\phi$, $y = r\sin\theta\sin\phi$, and $z = r\cos\theta$. Using the above formula, we find

$$h_1^2 = \left(\frac{\partial x}{\partial r}\right)^2 + \left(\frac{\partial y}{\partial r}\right)^2 + \left(\frac{\partial z}{\partial r}\right)^2 = \sin^2\theta\cos^2\phi + \sin^2\theta\sin^2\phi + \cos^2\theta$$

$$h_1 = \left[\sin^2\theta(\cos^2\phi + \sin^2\phi) + \cos^2\theta\right]^{1/2} = 1$$

Likewise,

$$h_2^2 = \left(\frac{\partial x}{\partial \theta}\right)^2 + \left(\frac{\partial y}{\partial \theta}\right)^2 + \left(\frac{\partial z}{\partial \theta}\right)^2 = r^2 \cos^2 \theta \cos^2 \phi + r^2 \cos^2 \theta \sin^2 \phi + r^2 \sin^2 \theta$$

$$h_2 = r\left[\cos^2 \theta (\cos^2 \phi + \sin^2 \phi) + \sin^2 \theta\right]^{1/2} = r$$

and

$$h_3^2 = \left(\frac{\partial x}{\partial \phi}\right)^2 + \left(\frac{\partial y}{\partial \phi}\right)^2 + \left(\frac{\partial z}{\partial \phi}\right)^2 = r^2 \sin^2 \theta \sin^2 \phi + r^2 \sin^2 \theta \cos^2 \phi$$

$$h_3 = r\left[\sin^2 \theta (\sin^2 \phi + \cos^2 \phi)\right]^{1/2} = r \sin \theta$$

Then

$$\nabla^2 f = \frac{1}{r^2 \sin \theta}\left[\frac{\partial}{\partial r}\left(r^2 \sin \theta \frac{\partial f}{\partial r}\right) + \frac{\partial}{\partial \theta}\left(\sin \theta \frac{\partial f}{\partial \theta}\right) + \frac{\partial}{\partial \phi}\left(\frac{1}{\sin \theta}\frac{\partial f}{\partial \phi}\right)\right]$$

$$= \frac{1}{r^2}\frac{\partial}{\partial r}\left(r^2 \frac{\partial f}{\partial r}\right) + \frac{1}{r^2 \sin \theta}\frac{\partial}{\partial \theta}\left(\sin \theta \frac{\partial f}{\partial \theta}\right) + \frac{1}{r^2 \sin^2 \theta}\frac{\partial^2 f}{\partial \phi^2}$$

5–33. Show that rotational transitions of a diatomic molecule occur in the microwave region or the far infrared region of the spectrum.

Assuming that the diatomic molecule can be treated as a rigid rotator, the frequency of a rotational transition is

$$\nu = \frac{h}{4\pi^2 I}(J + 1) \qquad J = 0, 1, 2, \ldots \tag{5.60}$$

From Section 5–9, a typical moment of inertia for a diatomic molecule is 5×10^{-46} kg·m². The observed frequency is therefore an integral multiple of

$$\nu = \frac{6.626 \times 10^{-34} \text{ J·s}}{4\pi^2 (5 \times 10^{-46} \text{ kg·m}^2)} = 3.4 \times 10^{10} \text{ s}^{-1}$$

Frequencies in the vicinity of this value occur in the microwave or far infrared region of the spectrum.

5–34. In the far infrared spectrum of $H^{79}Br$, there is a series of lines separated by 16.72 cm^{-1}. Calculate the values of the moment of inertia and the internuclear separation in $H^{79}Br$.

Assuming that $H^{79}Br$ can be treated as a rigid rotator,

$$\tilde{\nu} = 2\tilde{B}(J + 1) \qquad J = 0, 1, 2, \ldots \tag{5.63}$$

$$\tilde{B} = \frac{h}{8\pi^2 c I} \tag{5.64}$$

The lines in the spectrum are separated by 16.72 cm^{-1}, so

$$\Delta\tilde{\nu} = 2\tilde{B} = \frac{2h}{8\pi^2 cI}$$

$$16.72 \text{ cm}^{-1} = \frac{2(6.626 \times 10^{-34} \text{ J·s})}{8\pi^2(2.998 \times 10^{10} \text{ cm·s}^{-1})I}$$

$$I = 3.35 \times 10^{-47} \text{ kg·m}^2$$

We can find μ for H^{79}Br:

$$\mu = \frac{(78.9)(1.01)}{79.91}(1.661 \times 10^{-27} \text{ kg}) = 1.653 \times 10^{-27} \text{ kg}$$

Now we can use the relationship $r = (I/\mu)^{1/2}$ to find r.

$$r = \left(\frac{3.35 \times 10^{-47} \text{ kg·m}^2}{1.653 \times 10^{-27} \text{ kg}}\right)^{1/2} = 1.42 \times 10^{-10} \text{ m} = 142 \text{ pm}$$

5–35. The $J = 0$ to $J = 1$ transition for carbon monoxide (^{12}C^{16}O) occurs at 1.153×10^5 MHz. Calculate the value of the bond length in carbon monoxide.

Assuming that ^{12}C^{16}O can be treated as a rigid rotator,

$$\nu = 2B(J + 1) \qquad J = 0, 1, 2, \ldots \qquad (5.61)$$

$$B = \frac{h}{8\pi^2 I} \qquad (5.62)$$

For the $J = 0$ to $J = 1$ transition,

$$\frac{1}{2}\nu = B = \frac{h}{8\pi^2 I}$$

$$\frac{1}{2}(1.153 \times 10^{11} \text{ s}^{-1}) = \frac{6.626 \times 10^{-34} \text{ J·s}}{8\pi^2 \mu r^2}$$

We can find μ and use the relationship $r = (I/\mu)^{1/2}$ to find r.

$$\mu = \frac{(12.00)(15.99)}{27.99}(1.661 \times 10^{-27} \text{ kg}) = 1.139 \times 10^{-26} \text{ kg}$$

$$r = \left[\frac{6.626 \times 10^{-34} \text{ J·s}}{4\pi^2(1.139 \times 10^{-26} \text{ kg})(1.153 \times 10^{11} \text{ s}^{-1})}\right]^{1/2}$$

$$= 1.13 \times 10^{-10} \text{ m} = 113 \text{ pm}$$

5–36. Figure 5.11 compares the probability distribution associated with the harmonic oscillator wave function $\psi_{10}(\xi)$ to the classical distribution. This problem illustrates what is meant by the classical distribution. Consider

$$x(t) = A\sin(\omega t + \phi)$$

which can be written as

$$\omega t = \sin^{-1}\left(\frac{x}{A}\right) - \phi$$

Now

$$dt = \frac{\omega^{-1}dx}{\sqrt{A^2 - x^2}} \tag{1}$$

This equation gives the time that the oscillator spends between x and $x + dx$. We can convert Equation 1 to a probability distribution in x by dividing by the time that it takes for the oscillator to go from $-A$ to A. Show that this time is π/ω and that the probability distribution in x is

$$p(x)dx = \frac{dx}{\pi\sqrt{A^2 - x^2}} \tag{2}$$

Show that $p(x)$ is normalized. Why does $p(x)$ achieve its maximum value at $x = \pm A$? Now use the fact that $\xi = \alpha^{1/2}x$, where $\alpha = (k\mu/\hbar^2)^{1/2}$, to show that

$$p(\xi)d\xi = \frac{d\xi}{\pi\sqrt{\alpha A^2 - \xi^2}} \tag{3}$$

Show that the limits of ξ are $\pm(\alpha A^2)^{1/2} = \pm(21)^{1/2}$, and compare this result to the vertical lines shown in Figure 5.11. [*Hint*: You need to use the fact that $kA^2/2 = E_{10}$ ($v = 10$).] Finally, plot Equation 3 and compare your result with the curve in Figure 5.11.

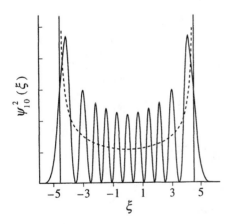

FIGURE 5.11
The probability distribution function of a harmonic oscillator in the $v = 10$ state. The dashed line is that for a classical harmonic oscillator with the same energy. The vertical lines at $\xi \approx \pm 4.6$ represents the extreme limits of the classical harmonic motion.

The variable ω is the angular velocity of the oscillator, defined as $\omega = 2\pi\nu$ where ν is in cycles per second. In going from $-A$ to A the function $x(t)$ goes through $\frac{1}{2}$ cycle, so

$$\omega = \frac{2\pi\left(\frac{1}{2}\text{ cycle}\right)}{t} \qquad t = \frac{\pi}{\omega}$$

Substitute $\omega^{-1} = t\pi^{-1}$ into Equation 1 and divide by t:

$$\frac{dt}{t} = \frac{dx}{\pi\sqrt{A^2 - x^2}}$$

Interpreting dt/t as a probability distribution in x, we find

$$p(x)dx = \frac{dx}{\pi\sqrt{A^2 - x^2}} \tag{2}$$

To show that this expression is normalized, we integrate over the time period we are observing:

$$\int_{-A}^{A} \frac{dx}{\pi\sqrt{A^2 - x^2}} = \frac{1}{\pi}\sin^{-1}\frac{x}{A}\Big|_{-A}^{A} = 1$$

The maximum values of $p(x)$ are at $x = \pm A$ because these are the points at which the classical harmonic oscillator has zero velocity. Substituting $\xi = \alpha^{1/2}x$ and $d\xi = \alpha^{1/2}dx$,

$$p(\xi)d\xi = \frac{d\xi}{\pi\sqrt{\alpha A^2 - \xi^2}}$$

Since the limits of x are $\pm A$, the limits of ξ are $\pm\alpha^{1/2}A = \pm\sqrt{\alpha A^2}$. But $kA^2/2 = E_{10} = \frac{21}{2}\hbar\omega$, so $A^2 = 21\hbar/(\mu k)^{1/2}$. Also, $\alpha = (\mu k)^{1/2}/\hbar$, so $(\alpha A^2)^{1/2} = (21)^{1/2} = 4.58$. The plot of Equation 3 is given by the dashed curve in Figure 5.11.

5–37. Compute the value of $\hat{L}^2 Y(\theta, \phi)$ for the following functions:

a. $1/(4\pi)^{1/2}$

b. $(3/4\pi)^{1/2}\cos\theta$

c. $(3/8\pi)^{1/2}\sin\theta e^{i\phi}$

d. $(3/8\pi)^{1/2}\sin\theta e^{-i\phi}$

Do you find anything interesting about the results?

Equation 5.52 is

$$\hat{L}^2 = -\hbar^2\left[\frac{1}{\sin\theta}\frac{\partial}{\partial\theta}\left(\sin\theta\frac{\partial}{\partial\theta}\right) + \frac{1}{\sin^2\theta}\frac{\partial^2}{\partial\phi^2}\right] \tag{5.52}$$

a. $\hat{L}^2\left[\dfrac{1}{(4\pi)^{1/2}}\right] = 0$ because $1/(4\pi)^{1/2}$ is independent of θ and ϕ.

b. $\hat{L}^2\left(\sqrt{\dfrac{3}{4\pi}}\cos\theta\right) = -\hbar^2\left(\dfrac{3}{4\pi}\right)^{1/2}[-2\cos\theta] = 2\hbar^2\left(\dfrac{3}{4\pi}\right)^{1/2}\cos\theta$

c. $\hat{L}^2\left(\sqrt{\dfrac{3}{8\pi}}\sin\theta e^{i\phi}\right) = -\hbar^2\left(\dfrac{3}{8\pi}\right)^{1/2}\left[\dfrac{\cos^2\theta - \sin^2\theta}{\sin\theta} - \dfrac{1}{\sin\theta}\right]e^{i\phi}$

$$= -\hbar^2\left(\dfrac{3}{8\pi}\right)^{1/2}[-2\sin\theta]e^{i\phi} = 2\hbar^2\left(\dfrac{3}{8\pi}\right)^{1/2}\sin\theta e^{i\phi}$$

d. $\hat{L}^2\left(\sqrt{\dfrac{3}{8\pi}}\sin\theta e^{-i\phi}\right) = \hat{L}^2\left(\sqrt{\dfrac{3}{8\pi}}\sin\theta e^{i\phi}\right)$, which we evaluated in c.

All of the spherical harmonics examined are eigenfunctions of $\hat{L}^2$ and the eigenvalues are multiples of $\hbar^2$. This is a general result.

Problems 5–38 through 5–43 develop an alternative method for determining the eigenvalues and eigenfunctions of a one-dimensional harmonic oscillator.

5–38. The Schrödinger equation for a one-dimensional harmonic oscillator is

$$\hat{H}\psi(x) = E\psi(x)$$

where the Hamiltonian operator is given by

$$\hat{H} = -\frac{\hbar^2}{2\mu}\frac{d^2}{dx^2} + \frac{1}{2}kx^2$$

where $k = \mu\omega^2$ is the force constant. Let $\hat{P}$ and $\hat{X}$ be the operators for momentum and position, respectively. If we define $\hat{p} = (\mu\hbar\omega)^{-1/2}\hat{P}$ and $\hat{x} = (\mu\omega/\hbar)^{1/2}\hat{X}$, show that

$$\hat{H} = \frac{\hat{P}^2}{2\mu} + \frac{k}{2}\hat{X}^2 = \frac{\hbar\omega}{2}(\hat{p}^2 + \hat{x}^2)$$

Use the definitions of $\hat{p}$ and $\hat{x}$ to show that

$$\hat{p} = -i\frac{d}{dx}$$

and

$$\hat{p}\hat{x} - \hat{x}\hat{p} = [\hat{p}, \hat{x}] = -i$$

Recall that $\hat{P} = -i\hbar(d/d\hat{X})$ and $\hat{X} = x$, so

$$\hat{H} = \frac{-\hbar^2}{2\mu}\frac{d^2}{dx^2} + \frac{kx^2}{2} = \frac{1}{2\mu}(-\hbar^2)\frac{d^2}{dx^2} + \frac{\mu\omega^2 x^2}{2}$$

$$= \frac{1}{2\mu}\hat{P}^2 + \frac{\mu\omega^2}{2}\hat{X}^2$$

$$= \frac{\mu\hbar\omega}{2\mu}\hat{p}^2 + \frac{\mu\omega^2\hbar}{2\mu\omega}\hat{x}^2$$

$$= \frac{\hbar\omega}{2}(\hat{p}^2 + \hat{x}^2)$$

Now

$$\hat{p} = (\mu\hbar\omega)^{-1/2}\hat{P} = \frac{1}{(\mu\omega\hbar)^{1/2}}\left(-i\hbar\frac{\partial}{\partial X}\right) = -i\frac{d}{d\hat{x}}$$

and

$$\hat{p}\hat{x}f - \hat{x}\hat{p}f = -i\left(f + x\frac{df}{dx}\right) + ix\frac{df}{dx} = -if$$

$$\hat{p}\hat{x} - \hat{x}\hat{p} = -i$$

5–39. We will define the operators $\hat{a}_-$ and $\hat{a}_+$ to be

$$\hat{a}_- = \frac{1}{\sqrt{2}}(\hat{x} + i\hat{p}) \quad \text{and} \quad \hat{a}_+ = \frac{1}{\sqrt{2}}(\hat{x} - i\hat{p}) \tag{1}$$

where $\hat{x}$ and $\hat{p}$ are given in Problem 5–38. Show that

$$\hat{a}_-\hat{a}_+ = \tfrac{1}{2}(\hat{x}^2 + i[\hat{p}, \hat{x}] + \hat{p}^2) = \tfrac{1}{2}(\hat{p}^2 + \hat{x}^2 + 1) \tag{2}$$

and that

$$\hat{a}_+\hat{a}_- = \tfrac{1}{2}(\hat{p}^2 + \hat{x}^2 - 1) \tag{3}$$

Now show that the Hamiltonian operator for the one-dimensional harmonic oscillator can be written as

$$\hat{H} = \frac{\hbar\omega}{2}(\hat{a}_-\hat{a}_+ + \hat{a}_+\hat{a}_-)$$

Now show that $\hat{a}_-\hat{a}_+ + \hat{a}_+\hat{a}_-$ is equal to $2\hat{a}_+\hat{a}_- + 1$ so that the Hamiltonian operator can be written as

$$\hat{H} = \hbar\omega\left(\hat{a}_+\hat{a}_- + \tfrac{1}{2}\right)$$

The operator $\hat{a}_+\hat{a}_-$ is called the number operator, which we will denote by $\hat{v}$, and using this definition we obtain

$$\hat{H} = \hbar\omega\left(\hat{v} + \tfrac{1}{2}\right) \tag{4}$$

Comment on the functional form of this result. What do you expect are the eigenvalues of the number operator? Without doing any calculus, explain why $\hat{v}$ must be a Hermitian operator.

$$\begin{aligned}
\hat{a}_-\hat{a}_+ &= \tfrac{1}{2}(\hat{x} + i\hat{p})(\hat{x} - i\hat{p}) \\
&= \tfrac{1}{2}\left[\hat{x}^2 + i(\hat{p}\hat{x} - \hat{x}\hat{p}) + \hat{p}^2\right] \\
&= \tfrac{1}{2}(\hat{p}^2 + \hat{x}^2 + 1) \tag{1}\\
\hat{a}_+\hat{a}_- &= \tfrac{1}{2}(\hat{x} - i\hat{p})(\hat{x} + i\hat{p}) \\
&= \tfrac{1}{2}\left[\hat{x}^2 + i(\hat{x}\hat{p} - \hat{p}\hat{x}) + \hat{p}^2\right] \\
&= \tfrac{1}{2}(\hat{p}^2 + \hat{x}^2 - 1) \tag{2}
\end{aligned}$$

Adding Equations 1 and 2 gives

$$\hat{p}^2 + \hat{x}^2 = \hat{a}_+\hat{a}_- + \hat{a}_-\hat{a}_+$$

and using this result and the result of Problem 5–38 gives

$$\hat{H} = \frac{\hbar\omega}{2}(\hat{p}^2 + \hat{x}^2) = \frac{\hbar\omega}{2}(\hat{a}_+\hat{a}_- + \hat{a}_-\hat{a}_+)$$

Now

$$2\hat{a}_+\hat{a}_- = \hat{p}^2 + \hat{x}^2 - 1$$

so

$$2\hat{a}_+\hat{a}_- + 1 = \hat{p}^2 + \hat{x}^2 = \hat{a}_+\hat{a}_- + \hat{a}_-\hat{a}_+$$

and we can write

$$\hat{H} = \frac{\hbar\omega}{2}(2\hat{a}_+\hat{a}_- + 1) = \hbar\omega\left(\hat{a}_+\hat{a}_- + \tfrac{1}{2}\right)$$

Letting $\hat{v} = \hat{a}_+ \hat{a}_-$, we find $\hat{H} = \hbar\omega(\hat{v} + \frac{1}{2})$. The eigenvalues of $\hat{v}$ must correspond to the v of Section 5–9. The operator $\hat{v}$ must be Hermitian because $\hat{H}$ is Hermitian.

5–40. In this problem, we will explore some of the properties of the operators introduced in Problem 5–39. Let ψ_v and E_v be the wave functions and energies of the one-dimensional harmonic oscillator. Start with

$$\hat{H}\psi_v = \hbar\omega \left(\hat{a}_+ \hat{a}_- + \frac{1}{2} \right) \psi_v = E_v \psi_v$$

Multiply from the left by $\hat{a}_-$ and use Equation 2 of Problem 5–39 to show that

$$\hat{H}(\hat{a}_- \psi_v) = (E_v - \hbar\omega)(\hat{a}_- \psi_v)$$

or that

$$\hat{a}_- \psi_v \propto \psi_{v-1}$$

Also show that

$$\hat{H}(\hat{a}_+ \psi_v) = (E_v + \hbar\omega)(\hat{a}_+ \psi_v)$$

or that

$$\hat{a}_+ \psi_v \propto \psi_{v+1}$$

Thus, we see that $\hat{a}_+$ operating on ψ_v gives ψ_{v+1} (to within a constant) and that $\hat{a}_- \psi_v$ gives ψ_{v-1} to within a constant. The operators $\hat{a}_+$ and $\hat{a}_-$ are called *raising* or *lowering operators*, or simply *ladder operators*. If we think of each rung of a ladder as a quantum state, then the operators $\hat{a}_+$ and $\hat{a}_-$ enable us to move up and down the ladder once we know the wave function of a single rung.

From the previous problem,

$$\hat{H}\psi_v = \hbar\omega \left(\hat{a}_+ \hat{a}_- + \frac{1}{2} \right) \psi_v = E_v \psi_v$$

Multiplying from the left by a_- gives

$$\hbar\omega \left(\hat{a}_- \hat{a}_+ \hat{a}_- + \frac{1}{2}\hat{a}_- \right) \psi_v = E_v \hat{a}_- \psi_v$$

Now use the relation $\hat{a}_- \hat{a}_+ = \hat{a}_+ \hat{a}_- + 1$ from Problem 5–39 to obtain

$$\hbar\omega \left(\hat{a}_+ \hat{a}_- + \frac{3}{2} \right) \hat{a}_- \psi_v = E_v \hat{a}_- \psi_v$$

$$\hbar\omega \left(\hat{a}_+ \hat{a}_- + \frac{1}{2} \right) \hat{a}_- \psi_v = (E_v - \hbar\omega) \hat{a}_- \psi_v$$

Because $\hat{H} = \hat{a}_+ \hat{a}_- + \frac{1}{2}$, we have

$$\hat{H}(\hat{a}_- \psi_v) = (E_v - \hbar\omega)(\hat{a}_- \psi_v)$$

$$\hat{a}_- \psi_v \propto \psi_{v-1}$$

Likewise, starting with

$$\hbar\omega\left(\hat{a}_+\hat{a}_- + \frac{1}{2}\right)\psi_v = E_v\psi_v$$

Multiplying from the left by a_+ gives

$$\hbar\omega\left(\hat{a}_+\hat{a}_+\hat{a}_- + \frac{1}{2}\hat{a}_+\right)\psi_v = E_v\hat{a}_+\psi_v$$

Now use the relation $\hat{a}_+\hat{a}_- = \hat{a}_-\hat{a}_+ - 1$ from Problem 5–39 to obtain

$$\hbar\omega\left(\hat{a}_+\hat{a}_- - \frac{1}{2}\right)\hat{a}_+\psi_v = E_v\hat{a}_+\psi_v$$

$$\hbar\omega\left(\hat{a}_+\hat{a}_- + \frac{1}{2} - 1\right)\hat{a}_+\psi_v = E_v\hat{a}_+\psi_v$$

$$\hbar\omega\left(\hat{a}_+\hat{a}_- + \frac{1}{2}\right)\hat{a}_+\psi_v = (E_v + \hbar\omega)\hat{a}_+\psi_v$$

Because $\hat{H} = \hat{a}_+\hat{a}_- + \frac{1}{2}$, we have

$$\hat{H}(\hat{a}_+\psi_v) = (E_v + \hbar\omega)(\hat{a}_+\psi_v)$$

$$\hat{a}_+\psi_v \propto \psi_{v+1}$$

5–41. Use the fact that $\hat{x}$ and $\hat{p}$ are Hermitian in the number operator defined in Problem 5–39 to show that

$$\int \psi_v^* \hat{v}\psi_v\, dx \geq 0$$

$$\int \psi_v^* \hat{v}\psi_v\, dx = \int \psi_v^* \hat{a}_+\hat{a}_-\psi_v\, dx$$

$$= \frac{1}{\sqrt{2}}\int \psi_v^* \hat{x}\hat{a}_-\psi_v\, dx - \frac{i}{\sqrt{2}}\int \psi_v^* \hat{p}\hat{a}_-\psi_v\, dx$$

$$= \frac{1}{\sqrt{2}}\int (\hat{x}\psi_v)^* \hat{a}_-\psi_v\, dx - \frac{i}{\sqrt{2}}\int (\hat{p}\psi_v)^* \hat{a}_-\psi_v\, dx$$

$$= \frac{1}{\sqrt{2}}\int (\hat{x}\psi_v)^* \hat{a}_-\psi_v\, dx + \frac{1}{\sqrt{2}}\int (i\hat{p}\psi_v)^* \hat{a}_-\psi_v\, dx$$

$$= \int (\hat{a}_-\psi_v)^*(\hat{a}_-\psi_v)\, dx$$

$$= \int |\hat{a}_-\psi_v|^2\, dx \geq 0$$

5–42. In Problem 5–41, we proved that $v \geq 0$. Because $\hat{a}_-\psi_v \propto \psi_{v-1}$ and $v \geq 0$, there must be some minimal value of v, v_{min}. Argue that $\hat{a}_-\psi_{v_{min}} = 0$. Now multiply $\hat{a}_-\psi_{v_{min}} = 0$ by $\hat{a}_+$ and use Equation 3 of Problem 5–39 to prove that $v_{min} = 0$, and that $v = 0, 1, 2, \ldots$.

The natural zero-point for $\hat{a}_-$ is when it acts on $\psi_{v_{min}}$, since the lowest quantum state has already been reached and ψ cannot be lowered any further. Therefore we define $\hat{a}_-\psi_{v_{min}} = 0$.

$$\hat{a}_-\psi_{v_{min}} = 0$$

$$\hat{a}_+\hat{a}_-\psi_{v_{min}} = 0$$

$$\hbar\omega\hat{a}_+\hat{a}_-\psi_{v_{min}} = 0$$

$$\frac{\hbar\omega}{2}(\hat{p}^2 + \hat{x}^2 - 1)\psi_{v_{min}} = 0$$

$$(\hat{H} - \frac{\hbar\omega}{2})\psi_{v_{min}} = 0$$

$$\hat{H}\psi_{v_{min}} = \frac{\hbar\omega}{2}\psi_{v_{min}}$$

Because Equation 4 of Problem 5–38 states that $\hat{H} = \hbar\omega(\hat{v} + \frac{1}{2})$, $\psi_{v_{min}}$ treated in this problem must be the eigenfunction of $\hat{v} = 0$. Therefore, $v_{min} = 0$.

5–43. Using the definition of $\hat{a}_-$ given in Problem 5–39 and the fact that $\hat{a}_-\psi_0 = 0$, determine the unnormalized wave function $\psi_0(x)$. Now determine the unnormalized wave function $\psi_1(x)$ using the operator $\hat{a}_+$.

$$\hat{a}_-\psi_0 = 0$$

$$\frac{1}{\sqrt{2}}(\hat{x} + i\hat{p})\psi_0 = 0$$

$$x\psi_0 + \frac{d\psi_0}{dx} = 0$$

$$\frac{d\psi_0}{\psi_0} = -xdx$$

$$\psi_0 = e^{-x^2/2}$$

Since $\psi_1 \sim \hat{a}_+\psi_0 \sim \hat{x}\psi_0 - i\hat{p}\psi_0$, and

$$\hat{x}\psi_0 - i\hat{p}\psi_0 = x\psi_0 - \frac{d\psi_0}{dx} = 2xe^{-x^2/2} = 2x\psi_0$$

we can write

$$\psi_1 \sim xe^{-x^2/2}$$

Problems 5–44 through 5–47 apply the idea of reduced mass to the hydrogen atom.

5–44. Given the development of the concept of reduced mass in Section 5–2, how do you think the energy of a hydrogen atom (Equation 1.22) will change if we do not assume that the proton is fixed at the origin?

$$E_n = -\frac{m_e e^4}{8\varepsilon_0^2 h^2 n^2} \tag{1.22}$$

Instead of using m_e, we will need to use μ, since the distance from the center of mass to the proton will not be zero.

5–45. In Example 1–8, we calculated the value of the Rydberg constant to be $109\,737$ cm^{-1}. What is the calculated value if we replace m_e in Equation 1.25 by the reduced mass? Compare your answer with the experimental result, $109\,677.6$ cm^{-1}.

From Problem 5–7, the reduced mass of hydrogen is $\mu = 9.104\,431 \times 10^{-31}$ kg $= 0.999\,455\,6m_e$.

$$R_{\mathrm{H}} = \frac{m_e e^4}{8\varepsilon_0^2 ch^3} \tag{1.25}$$

Replacing m_e with μ gives a new R_{H} value of

$$(109\,737.2 \text{ cm}^{-1})(0.999\,455\,6) = 109\,677.5 \text{ cm}^{-1}$$

which differs from the experimental result by about $1 \times 10^{-4}\%$.

5–46. Calculate the reduced mass of a deuterium atom. Take the mass of a deuteron to be $3.343\,586 \times 10^{-27}$ kg. What is the value of the Rydberg constant for a deuterium atom?

$$\mu = \frac{(9.109\,390 \times 10^{-31} \text{ kg})(3.343\,586 \times 10^{-27} \text{ kg})}{9.109\,390 \times 10^{-31} \text{ kg} + 3.343\,586 \times 10^{-27} \text{ kg}}$$
$$= 9.106\,909 \times 10^{-31} \text{ kg} = 0.999\,727\,7m_e$$

$$R_{\mathrm{H}} = (109\,737.2 \text{ cm}^{-1})(0.999\,727\,7) = 109\,707.3 \text{ cm}^{-1}$$

5–47. Calculate the ratio of the frequencies of the lines in the spectra of atomic deuterium and atomic hydrogen.

The ratio of the frequencies of the lines in these spectra is the same as the ratios of the Rydberg constants found in Problems 5–45 and 5–46:

$$\frac{109\,707.3 \text{ cm}^{-1}}{109\,677.5 \text{ cm}^{-1}} = 1.000\,272$$

CHAPTER 6

The Hydrogen Atom

6–1. Show that both $\hbar^2\nabla^2/2m_e$ and $e^2/4\pi\varepsilon_0 r$ have the units of energy (joules).

Recall that ∇^2 has units of m^{-2} (Section 5–8), so

$$\frac{\hbar^2\nabla^2}{2m_e} \sim J^2 \cdot s^2 \cdot m^{-2} \cdot kg^{-1} = J$$

$$\frac{e^2}{4\pi\varepsilon_0 r} \sim \frac{C^2}{C^2 \cdot s^2 \cdot kg^{-1} \cdot m^{-3} \cdot m} = kg \cdot m^2 \cdot s^{-2} = J$$

6–2. In terms of the variable θ, Legendre's equation is

$$\sin\theta \frac{d}{d\theta}\left(\sin\theta \frac{d\Theta(\theta)}{d\theta}\right) + (\beta^2 \sin^2\theta - m^2)\Theta(\theta) = 0$$

Let $x = \cos\theta$ and $P(x) = \Theta(\theta)$ and show that

$$(1-x^2)\frac{d^2 P(x)}{dx^2} - 2x\frac{dP(x)}{dx} + \left[\beta - \frac{m^2}{1-x^2}\right]P(x) = 0$$

Begin with Legendre's equation,

$$\sin\theta \frac{d}{d\theta}\left(\sin\theta \frac{d\Theta(\theta)}{d\theta}\right) + (\beta^2 \sin^2\theta - m^2)\Theta(\theta) = 0$$

Expanding the first term in θ gives

$$\sin^2\theta \frac{d^2\Theta}{d\theta^2} + \sin\theta\cos\theta \frac{d\Theta}{d\theta} + (\beta^2 \sin^2\theta - m^2)\Theta(\theta) = 0 \qquad (1)$$

Let $x = \cos\theta$ and $P(x) = \Theta(\theta)$. Then

$$\frac{d\Theta}{d\theta} = \frac{dP}{dx}\frac{dx}{d\theta} = -\sin\theta\frac{dP}{dx} = -(1-x^2)^{1/2}\frac{dP}{dx}$$

$$\frac{d^2\Theta}{d\theta^2} = \frac{d}{d\theta}\left(\frac{d\Theta}{d\theta}\right) = \frac{dx}{d\theta}\frac{d}{dx}\left[-(1-x^2)^{1/2}\frac{dP}{dx}\right]$$

$$= -\sin\theta\left[\frac{x}{(1-x^2)^{1/2}}\frac{dP}{dx} - (1-x^2)^{1/2}\frac{d^2 P}{dx^2}\right]$$

$$= -x\frac{dP}{dx} + (1-x^2)\frac{d^2 P}{dx^2}$$

155

Substituting these expressions into Equation 1 gives

$$(1-x^2)^2\frac{d^2P}{dx^2} - 2x(1-x^2)\frac{dP}{dx} + \left[\beta(1-x^2) - m^2\right]P = 0$$

$$(1-x^2)\frac{d^2P(x)}{dx^2} - 2x\frac{dP(x)}{dx} + \left(\beta - \frac{m^2}{1-x^2}\right)P(x) = 0$$

6–3. Show that the Legendre polynomials given in Table 6.1 satisfy Equation 6.23 with $m = 0$.

Letting $m = 0$, Equation 6.23 becomes

$$(1-x^2)\frac{d^2P}{dx^2} - 2x\frac{dP}{dx} + l(l+1)P = 0$$

For $l = 0$, $P_0(x) = 1$ and the equation is satisfied. For $l = 1$, $P_1(x) = x$ and Equation 6.23 becomes

$$-2x + 1(2)x = 0$$

For $l = 2$, $P_2(x) = \frac{1}{2}(3x^2 - 1)$ and Equation 6.23 becomes

$$(1-x^2)(3) - 2x(3x) + 2(3)\left[\frac{1}{2}(3x^2-1)\right] = 3 - 3x^2 - 6x^2 + 9x^2 - 3 = 0$$

For $l = 3$, $P_3(x) = \frac{1}{2}(5x^3 - 3x)$ and Equation 6.23 becomes

$$(1-x^2)15x - 2x\left[\frac{1}{2}(15x^2-3)\right] + 3(4)\left[\frac{1}{2}(5x^3-3x)\right] = 15x - 15x^3 - 15x^3 + 3x + 30x^3 - 18x = 0$$

For $l = 4$, $P_4(x) = \frac{1}{2}(35x^4 - 30x^2 + 3)$ and Equation 6.23 becomes

$$(1-x^2)\left[\frac{1}{2}(420x^2-60)\right] - 2x\left[\frac{1}{2}(140x^3-60x)\right] + 4(5)\left[\frac{1}{2}(35x^4-30x^2+3)\right]$$

$$= 210x^2 - 210x^4 - 30 + 30x^2 - 140x^4 + 60x^2 + 350x^4 - 300x^2 + 30 = 0$$

6–4. Show that the orthogonality integral for the Legendre polynomials, Equation 6.24, is equivalent to

$$\int_0^\pi P_l(\cos\theta)P_n(\cos\theta)\sin\theta d\theta = 0 \qquad l \neq n$$

Begin with

$$\int_{-1}^1 P_l(x)P_n(x)dx = 0 \qquad l \neq n \qquad (6.24)$$

Let $x = \cos\theta$ and $dx = -\sin\theta d\theta$ and write Equation 6.24 as an integral over θ, where θ ranges from π to 0:

$$\int_\pi^0 P_l(\cos\theta)P_n(\cos\theta)(-\sin\theta)d\theta = 0$$

Integrating from 0 to π and evaluating the result at the limits of integration yields

$$\int_0^\pi P_l(\cos\theta) P_n(\cos\theta) \sin\theta d\theta = 0$$

where $l \neq n$.

6–5. Show that the Legendre polynomials given in Table 6.1 satisfy the orthogonality and normalization conditions given by Equations 6.24 and 6.25.

We can write Equations 6.24 and 6.25 together as

$$\int_{-1}^1 P_l(x) P_n(x) dx = \frac{2\delta_{ln}}{2l+1}$$

Some examples of Legendre polynomials satisfying this condition are (for $l = 0$, 1 and 2)

$$\int_{-1}^1 P_0^2(x) dx = \int_{-1}^1 dx = 2$$

$$\int_{-1}^1 P_1^2(x) dx = \int_{-1}^1 x^2 dx = \frac{2}{3}$$

$$\int_{-1}^1 P_2^2(x) dx = \frac{1}{4} \int_{-1}^1 (3x^2 - 1)^2 dx = \frac{1}{4}\left(\frac{18}{5} - \frac{12}{3} + 2\right) = \frac{2}{5}$$

$$\int_{-1}^1 P_0(x) P_1(x) dx = \int_{-1}^1 x dx = 0$$

$$\int_{-1}^1 P_0(x) P_2(x) dx = \frac{1}{2} \int_{-1}^1 (3x^2 - 1) dx = 1 - 1 = 0$$

$$\int_{-1}^1 P_1(x) P_2(x) dx = \frac{1}{2} \int_{-1}^1 (3x^3 - x) dx = 0$$

6–6. Use Equation 6.26 to generate the associated Legendre functions in Table 6.2.

$$P_l^{|m|}(x) = (1 - x^2)^{|m|/2} \frac{d^m}{dx^m} P_l(x) \tag{6.26}$$

$$P_0^0(x) = (1 - x^2)^0 \frac{d^0}{dx^0} P_0(x) = P_0(x) = 1$$

$$P_1^0(x) = (1 - x^2)^0 \frac{d^0}{dx^0} P_1(x) = P_1(x) = x$$

$$P_1^1(x) = (1 - x^2)^{1/2} \frac{d}{dx} P_1(x) = (1 - x^2)^{1/2}$$

$$P_2^0(x) = (1 - x^2)^0 \frac{d^0}{dx^0} P_2(x) = P_2(x) = \frac{1}{2}(3x^2 - 1)$$

$$P_2^1(x) = (1 - x^2)^{1/2} \frac{d}{dx} P_2(x) = 3x(1 - x^2)^{1/2}$$

$$P_2^2(x) = (1 - x^2) \frac{d^2}{dx^2} P_2(x) = 3(1 - x^2)$$

$$P_3^0(x) = (1 - x^2)^0 \frac{d^0}{dx^0} P_3(x) = P_3(x) = \frac{1}{2}(5x^3 - 3x)$$

$$P_3^1(x) = (1 - x^2)^{1/2} \frac{d}{dx} P_3(x) = \frac{3}{2}(5x^2 - 1)(1 - x^2)^{1/2}$$

$$P_3^2(x) = (1 - x^2) \frac{d^2}{dx^2} P_3(x) = 15x(1 - x^2)$$

$$P_3^3(x) = (1 - x^2)^{3/2} \frac{d^3}{dx^3} P_3(x) = 15(1 - x^2)^{3/2}$$

6–7. Show that the first few associated Legendre functions given in Table 6.2 are solutions to Equation 6.23 and that they satisfy the orthonormality condition, Equation 6.28.

$$(1 - x^2)\frac{d^2 P}{dx^2} - 2x\frac{dP}{dx} + \left[l(l + 1) - \frac{m^2}{1 - x^2}\right]P = 0 \tag{6.23}$$

When $l = m = 0$, $P_0^0(x) = 1$ and Equation 6.23 is clearly satisfied. When $l = 1$ and $m = 0$, $P_1^0(x) = x$ and Equation 6.23 becomes

$$0 - 2x(1) + (2 - 0)x = 0$$

When $l = m = 1$, $P_1^1(x) = (1 - x^2)^{1/2}$ and Equation 6.23 becomes

$$0 = (1 - x^2)\left[-\frac{1}{(1 - x^2)^{1/2}} - \frac{x^2}{(1 - x^2)^{3/2}}\right] - 2x\left[-\frac{x}{(1 - x^2)^{1/2}}\right]$$

$$+ \left[2 - \frac{1}{(1 - x^2)}\right](1 - x^2)^{1/2}$$

$$= -(1 - x^2)^{1/2} - \frac{x^2}{(1 - x^2)^{1/2}} + \frac{2x^2}{(1 - x^2)^{1/2}}$$

$$+ 2(1 - x^2)^{1/2} - \frac{1}{(1 - x^2)^{1/2}}$$

$$= \frac{x^2 - 1}{(1 - x^2)^{1/2}} + (1 - x^2)^{1/2} = 0$$

The orthonormality condition is

$$\int_{-1}^{1} P_l^{|m|}(x) P_n^{|m|}(x)dx = \frac{2}{(2l + 1)}\frac{(l + |m|)!}{(l - |m|)!}\delta_{ln} \tag{6.28}$$

Examples of associated Legendre functions satisfying this condition are

$$\int_{-1}^{1} \left[P_1^1(x)\right]^2 dx = \int_{-1}^{1} dx(1 - x^2) = \frac{4}{3} = \frac{2}{3}\left(\frac{2!}{0!}\right)$$

$$\int_{-1}^{1} \left[P_2^1(x)\right]^2 dx = \int_{-1}^{1} dx(9x^2 - 9x^4) = \frac{12}{5} = \frac{2}{5}\left(\frac{3!}{1!}\right)$$

$$\int_{-1}^{1} P_1^1(x) P_2^1(x)dx = \int_{-1}^{1} dx\left[3x(1 - x^2)\right] = 0$$

6–8. There are a number of recursion formulas for the associated Legendre functions. One that we will have occasion to use in Section 13–12 is

$$(2l + 1)x P_l^{|m|}(x) = (l - |m| + 1) P_{l+1}^{|m|}(x) + (l + |m|) P_{l-1}^{|m|}(x)$$

Show that the first few associated Legendre functions in Table 6.2 satisfy this recursion formula.

Let $l = 1$ and $m = 0$:

$$3x P_1^0(x) \overset{?}{=} 2 P_2^0(x) + P_0^0(x)$$
$$3x^2 = (3x^2 - 1) + 1$$

Let $l = m = 1$:

$$3x P_1^1(x) \overset{?}{=} P_2^1(x) + 2 P_0^1(x)$$
$$3x(1 - x^2)^{1/2} = 3x(1 - x^2)^{1/2} + 0$$

$P_0^1(x) = 0$ because m cannot be greater than l. Let $l = 2$ and $m = 0$:

$$5x P_2^0(x) \overset{?}{=} 3 P_3^0(x) + 2 P_1^0(x)$$
$$\frac{5}{2}(3x^3 - x) \overset{?}{=} \frac{3}{2}(5x^3 - 3x) + 2x$$
$$\frac{15}{2}x^3 - \frac{5}{2}x = \frac{15}{2}x^3 - \frac{5}{2}x$$

Let $l = 2$ and $m = 1$:

$$5x P_2^1(x) \overset{?}{=} 2 P_3^1(x) + 3 P_1^1(x)$$
$$15x^2(1 - x^2)^{1/2} \overset{?}{=} (15x^2 - 3)(1 - x^2)^{1/2} + 3(1 - x^2)^{1/2}$$
$$15x^2(1 - x^2)^{1/2} = 15x^2(1 - x^2)^{1/2}$$

Let $l = 2$ and $m = 2$:

$$5x P_2^2(x) \overset{?}{=} P_3^2(x) + 4 P_1^2(x)$$
$$15x(1 - x^2) = 15x(1 - x^2) + 0$$

$P_1^2(x) = 0$, because m cannot be greater than l.

6–9. Show that the first few spherical harmonics in Table 6.3 satisfy the orthonormality condition, Equation 6.31.

$$\int_0^\pi d\theta \sin\theta \int_0^{2\pi} d\phi \, Y_l^m(\theta, \phi)^* Y_n^k(\theta, \phi) = \delta_{nl}\delta_{mk} \tag{6.31}$$

The normalization condition is

$$\int_0^{2\pi} d\phi \int_0^\pi d\theta \sin\theta \, |Y_0^0(\theta, \phi)|^2 = \int_0^{2\pi} d\phi \int_0^\pi d\theta \sin\theta (4\pi)^{-1}$$
$$= 2\pi \left(\frac{2}{4\pi}\right) = 1$$

$$\int_0^{2\pi} d\phi \int_0^\pi d\theta \sin\theta \, |Y_1^0(\theta, \phi)|^2 = \int_0^{2\pi} d\phi \int_0^\pi d\theta \sin\theta \left(\frac{3\cos^2\theta}{4\pi}\right)$$
$$= 2\pi \left(\frac{3}{4\pi}\right) \int_{-1}^1 dx \, x^2 = 1$$

$$\int_0^{2\pi} d\phi \int_0^\pi d\theta \sin\theta \, |Y_1^1(\theta, \phi)|^2 = \int_0^{2\pi} d\phi \int_0^\pi d\theta \sin\theta \left(\frac{3\sin^2\theta}{8\pi}\right)$$
$$= 2\pi \left(\frac{3}{8\pi}\right) \int_{-1}^1 dx(1 - x^2) = 1$$

The orthogonality condition is

$$
\int_0^{2\pi} d\phi \int_0^\pi d\theta \sin\theta \; Y_0^0(\theta, \phi)^* Y_1^0(\theta, \phi) = \int_0^{2\pi} d\phi \int_0^\pi d\theta \sin\theta \left(\frac{3^{1/2} \cos\theta}{4\pi} \right)
$$

$$
= 2\pi \left(\frac{3^{1/2}}{4\pi} \right) \int_{-1}^1 dx \, x = 0
$$

$$
\int_0^{2\pi} d\phi \int_0^\pi d\theta \sin\theta \; Y_0^0(\theta, \phi)^* Y_1^{\pm 1}(\theta, \phi) = \int_0^{2\pi} e^{\pm i\phi} d\phi \int_0^\pi d\theta \sin\theta \left(\frac{3^{1/2} \cos\theta}{2^{5/2}\pi} \right) = 0
$$

$$
\int_0^{2\pi} d\phi \int_0^\pi d\theta \sin\theta \; Y_1^0(\theta, \phi)^* Y_1^{\pm 1}(\theta, \phi) = \int_0^{2\pi} e^{\pm i\phi} d\phi \int_0^\pi d\theta \sin\theta \left(\frac{3^{1/2} \sin\theta \cos\theta}{2^{5/2}\pi} \right) = 0
$$

because we know that $\int_0^{2\pi} d\phi e^{\pm i\phi} = 0$.

6–10. Using explicit expressions for $Y_l^m(\theta, \phi)$, show that

$$
|Y_1^1(\theta, \phi)|^2 + |Y_1^0(\theta, \phi)|^2 + |Y_1^{-1}(\theta, \phi)|^2 = \text{constant}
$$

This is a special case of the general theorem

$$
\sum_{m=-l}^{+l} |Y_l^m(\theta, \phi)|^2 = \text{constant}
$$

known as Unsöld's theorem. What is the physical significance of this result?

$$
|Y_1^1(\theta, \phi)|^2 + |Y_1^0(\theta, \phi)|^2 + |Y_1^{-1}(\theta, \phi)|^2 = \frac{3 \sin^2\theta}{8\pi} + \frac{3 \cos^2\theta}{4\pi} + \frac{3 \sin^2\theta}{8\pi} = \frac{3}{4\pi}
$$

Unsöld's theorem states that the electron density in a filled subshell is spherically symmetric.

Converting Cartesian coordinates to spherical coordinates

In the following problems, we will often need to use the following equations relating Cartesian and spherical coordinates:

$$
x = r \sin\theta \cos\phi \qquad r = (x^2 + y^2 + z^2)^{1/2}
$$

$$
y = r \sin\theta \sin\phi \qquad \phi = \tan^{-1}\left(\frac{y}{x} \right)
$$

$$
z = r \cos\theta \qquad\qquad \theta = \cos^{-1}\left[\frac{z}{(x^2 + y^2 + z^2)^{1/2}} \right]
$$

From these expressions, we can derive the following relationships:

$$\left(\frac{\partial r}{\partial x}\right) = \sin\theta\cos\phi \qquad \left(\frac{\partial\theta}{\partial x}\right) = \frac{\cos\theta\cos\phi}{r} \qquad \left(\frac{\partial r}{\partial z}\right) = \cos\theta$$

$$\left(\frac{\partial\phi}{\partial x}\right) = -\frac{\sin\phi}{r\sin\theta} \qquad \left(\frac{\partial r}{\partial y}\right) = \sin\theta\sin\phi \qquad \left(\frac{\partial\phi}{\partial z}\right) = -\frac{\sin\theta}{r}$$

$$\left(\frac{\partial\theta}{\partial y}\right) = \frac{\cos\theta\sin\phi}{r} \qquad \left(\frac{\partial\phi}{\partial y}\right) = -\frac{\cos\phi}{r\sin\theta} \qquad \left(\frac{\partial\theta}{\partial z}\right) = 0$$

6–11. In Cartesian coordinates,

$$\hat{L}_z = -i\hbar\left(x\frac{\partial}{\partial y} - y\frac{\partial}{\partial x}\right)$$

Convert this equation to spherical coordinates, showing that

$$\hat{L}_z = -i\hbar\frac{\partial}{\partial\phi}$$

As in Problem 5–30, the chain rule of partial differentiation states that

$$\left(\frac{\partial f}{\partial x}\right) = \left(\frac{\partial f}{\partial r}\right)\left(\frac{\partial r}{\partial x}\right) + \left(\frac{\partial f}{\partial\phi}\right)\left(\frac{\partial\phi}{\partial x}\right) + \left(\frac{\partial f}{\partial\theta}\right)\left(\frac{\partial\theta}{\partial x}\right)$$

and that

$$\left(\frac{\partial f}{\partial y}\right) = \left(\frac{\partial f}{\partial r}\right)\left(\frac{\partial r}{\partial y}\right) + \left(\frac{\partial f}{\partial\phi}\right)\left(\frac{\partial\phi}{\partial y}\right) + \left(\frac{\partial f}{\partial\theta}\right)\left(\frac{\partial\theta}{\partial y}\right)$$

Using the relations between Cartesian and spherical coordinates, we find

$$\left(\frac{\partial f}{\partial x}\right) = \sin\theta\cos\phi\left(\frac{\partial f}{\partial r}\right) + \frac{\cos\theta\cos\phi}{r}\left(\frac{\partial f}{\partial\theta}\right) - \frac{\sin\phi}{r\sin\theta}\left(\frac{\partial f}{\partial\phi}\right)$$

$$\left(\frac{\partial f}{\partial y}\right) = \sin\theta\sin\phi\left(\frac{\partial f}{\partial r}\right) + \frac{\cos\theta\sin\phi}{r}\left(\frac{\partial f}{\partial\theta}\right) - \frac{\cos\phi}{r\sin\theta}\left(\frac{\partial f}{\partial\phi}\right)$$

Now,

$$\left(x\frac{\partial f}{\partial y} - y\frac{\partial f}{\partial x}\right) = r\sin\theta\cos\phi\left(\frac{\partial f}{\partial y}\right) - r\sin\theta\sin\phi\left(\frac{\partial f}{\partial x}\right)$$

$$= \left(r\sin^2\theta\cos\phi\sin\phi - r\sin^2\theta\cos\phi\sin\phi\right)\left(\frac{\partial f}{\partial r}\right)$$

$$+ \left(\sin\theta\cos\theta\cos\phi\sin\phi - \sin\theta\cos\theta\sin\phi\cos\phi\right)\left(\frac{\partial f}{\partial\theta}\right)$$

$$+ \left(\cos^2\phi + \sin^2\phi\right)\left(\frac{\partial f}{\partial\phi}\right)$$

$$= \frac{\partial f}{\partial\phi}$$

Therefore

$$\hat{L}_z = -i\hbar\left(x\frac{\partial}{\partial y} - y\frac{\partial}{\partial x}\right) = -i\hbar\frac{\partial}{\partial\phi}$$

6–12. Convert $\hat{L}_x$ and $\hat{L}_y$ from Cartesian coordinates to spherical coordinates.

We can use our expressions for $(\partial f/\partial x)$ and $(\partial f/\partial y)$ from the previous problem. The only new quantity needed is

$$\left(\frac{\partial f}{\partial z}\right) = \left(\frac{\partial f}{\partial r}\right)\left(\frac{\partial r}{\partial z}\right) + \left(\frac{\partial f}{\partial \phi}\right)\left(\frac{\partial \phi}{\partial z}\right) + \left(\frac{\partial f}{\partial \theta}\right)\left(\frac{\partial \theta}{\partial z}\right)$$

$$\left(\frac{\partial f}{\partial z}\right) = \cos\theta \left(\frac{\partial f}{\partial r}\right) - \frac{\sin\theta}{r}\left(\frac{\partial f}{\partial \phi}\right)$$

Then

$$\hat{L}_x = -i\hbar \left[y\frac{\partial}{\partial z} - z\frac{\partial}{\partial y} \right]$$

$$= -i\hbar \left[(r\sin\theta\cos\theta\sin\phi - r\sin\theta\cos\theta\sin\phi)\frac{\partial}{\partial r} + (-\sin^2\theta\sin\phi - \cos^2\theta\sin\phi)\frac{\partial}{\partial \theta} \right.$$

$$\left. - \cot\theta\cos\phi\frac{\partial}{\partial \phi} \right]$$

$$= -i\hbar \left(-\sin\phi\frac{\partial}{\partial \theta} - \cot\theta\cos\phi\frac{\partial}{\partial \phi} \right)$$

and

$$\hat{L}_y = -i\hbar \left[z\frac{\partial}{\partial x} - x\frac{\partial}{\partial z} \right]$$

$$= -i\hbar \left[(r\cos\theta\sin\theta\cos\phi - r\cos\theta\sin\theta\cos\phi)\frac{\partial}{\partial r} + (\cos^2\theta\cos\phi - \sin^2\theta\cos\phi)\frac{\partial}{\partial \theta} \right.$$

$$\left. - \cot\theta\sin\phi\frac{\partial}{\partial \phi} \right]$$

$$= -i\hbar \left(\cos\phi\frac{\partial}{\partial \theta} - \cot\theta\sin\phi\frac{\partial}{\partial \phi} \right)$$

6–13. Prove that $\hat{L}^2$ commutes with $\hat{L}_x$, $\hat{L}_y$, and $\hat{L}_z$ but that

$$[\hat{L}_x, \hat{L}_y] = i\hbar\hat{L}_z \qquad [\hat{L}_y, \hat{L}_z] = i\hbar\hat{L}_x \qquad [\hat{L}_z, \hat{L}_x] = i\hbar\hat{L}_y$$

(*Hint*: Use Cartesian coordinates.) Do you see a pattern in these formulas?

In Example 6–7, we showed that $\hat{L}^2$ commutes with $\hat{L}_z$. Because the labelling of x, y and z is arbitrary, $\hat{L}^2$ must also commute with $\hat{L}_y$ and $\hat{L}_x$. Recall that

$$\hat{L}_x = -i\hbar \left(y\frac{\partial}{\partial z} - z\frac{\partial}{\partial y} \right) \qquad \hat{L}_y = -i\hbar \left(z\frac{\partial}{\partial x} - x\frac{\partial}{\partial z} \right)$$

$$\hat{L}_z = -i\hbar \left(x\frac{\partial}{\partial y} - y\frac{\partial}{\partial x} \right)$$

and now find

$$[\hat{L}_x, \hat{L}_y] = \hat{L}_x\hat{L}_y - \hat{L}_y\hat{L}_x$$

$$= -\hbar^2\left[\left(y\frac{\partial}{\partial z} - z\frac{\partial}{\partial y}\right)\left(z\frac{\partial}{\partial x} - x\frac{\partial}{\partial z}\right) - \left(z\frac{\partial}{\partial x} - x\frac{\partial}{\partial z}\right)\left(y\frac{\partial}{\partial z} - z\frac{\partial}{\partial y}\right)\right]$$

$$= -\hbar^2\left[y\frac{\partial}{\partial x} + yz\frac{\partial^2}{\partial x\partial z} - xy\frac{\partial^2}{\partial z^2} - z^2\frac{\partial^2}{\partial x\partial y} + xz\frac{\partial^2}{\partial y\partial z}\right.$$

$$\left. - yz\frac{\partial^2}{\partial x\partial z} + z^2\frac{\partial^2}{\partial x\partial y} + xy\frac{\partial^2}{\partial z^2} - x\frac{\partial}{\partial y} - xz\frac{\partial^2}{\partial y\partial z}\right]$$

$$= -\hbar^2\left(y\frac{\partial}{\partial x} - x\frac{\partial}{\partial y}\right) = i\hbar\hat{L}_z$$

$$\left[\hat{L}_y, \hat{L}_z\right] = \hat{L}_y\hat{L}_z - \hat{L}_z\hat{L}_y$$

$$= -\hbar^2\left[\left(z\frac{\partial}{\partial x} - x\frac{\partial}{\partial z}\right)\left(x\frac{\partial}{\partial y} - y\frac{\partial}{\partial x}\right) - \left(x\frac{\partial}{\partial y} - y\frac{\partial}{\partial x}\right)\left(z\frac{\partial}{\partial x} - x\frac{\partial}{\partial z}\right)\right]$$

$$= -\hbar^2\left[z\frac{\partial}{\partial y} + xz\frac{\partial^2}{\partial x\partial y} - yz\frac{\partial^2}{\partial x^2} - x^2\frac{\partial^2}{\partial y\partial z} + xy\frac{\partial^2}{\partial x\partial z}\right.$$

$$\left. - xz\frac{\partial^2}{\partial x\partial y} + x^2\frac{\partial^2}{\partial y\partial z} + yz\frac{\partial^2}{\partial x^2} - y\frac{\partial}{\partial z} - xy\frac{\partial^2}{\partial x\partial z}\right]$$

$$= -\hbar^2\left(z\frac{\partial}{\partial y} - y\frac{\partial}{\partial z}\right) = i\hbar\hat{L}_x$$

$$\left[\hat{L}_z, \hat{L}_x\right] = \hat{L}_z\hat{L}_x - \hat{L}_x\hat{L}_z$$

$$= -\hbar^2\left[\left(x\frac{\partial}{\partial y} - y\frac{\partial}{\partial x}\right)\left(y\frac{\partial}{\partial z} - z\frac{\partial}{\partial y}\right) - \left(y\frac{\partial}{\partial z} - z\frac{\partial}{\partial y}\right)\left(x\frac{\partial}{\partial y} - y\frac{\partial}{\partial x}\right)\right]$$

$$= -\hbar^2\left[x\frac{\partial}{\partial z} + xy\frac{\partial^2}{\partial y\partial z} - xz\frac{\partial^2}{\partial y^2} - y^2\frac{\partial^2}{\partial x\partial z} + yz\frac{\partial^2}{\partial x\partial y}\right.$$

$$\left. - xy\frac{\partial^2}{\partial y\partial z} + y^2\frac{\partial^2}{\partial x\partial z} + xz\frac{\partial^2}{\partial y^2} - z\frac{\partial}{\partial x} - yz\frac{\partial^2}{\partial x\partial y}\right]$$

$$= -\hbar^2\left(x\frac{\partial}{\partial z} - z\frac{\partial}{\partial x}\right) = i\hbar\hat{L}_y$$

These formulas involve a cyclic permutation of x, y, and z.

6–14. It is a somewhat advanced exercise to prove generally that $\langle L_x\rangle = \langle L_y\rangle = 0$ (see, however, Problem 6–58), but prove that they are zero at least for the first few l, m states by using the spherical harmonics given in Table 6.3.

Because the labelling of x and y is arbitrary, if we can show that $\langle L_x\rangle = 0$ we will have also shown that $\langle L_y\rangle$ must equal zero.

$$\langle L_x\rangle = \int_0^{2\pi} d\phi \int_0^{\pi} d\theta\,\sin\theta\, Y_l^m(\theta, \phi)^*\hat{L}_x Y_l^m(\theta, \phi)$$

Recall from Problem 6–12 that in spherical coordinates

$$\hat{L}_x = -i\hbar \left(-\sin\phi \frac{\partial}{\partial\theta} - \cot\theta \cos\phi \frac{\partial}{\partial\phi} \right)$$

For $l = m = 0$, Y_0^0 is a constant and so $\langle L_x \rangle = 0$. For $l = 1$ and $m = 0$,

$$\hat{L}_x Y_1^0(\theta, \phi) = -i\hbar \left(\frac{3}{4\pi} \right)^{1/2} \sin\phi \sin\theta$$

$$\langle L_x \rangle = -i\hbar \left(\frac{3}{4\pi} \right) \int_0^{2\pi} d\phi \sin\phi \int_0^\pi d\theta \sin^2\theta \cos\theta$$

$$= -i\hbar \left(\frac{3}{4\pi} \right) (0) \int_0^\pi d\theta \sin^2\theta \cos\theta = 0$$

For $l = 1$ and $m = \pm 1$,

$$\hat{L}_x Y_1^{\pm 1}(\theta, \phi) = -i\hbar \left(\frac{3}{8\pi} \right)^{1/2} \left(-\sin\phi \frac{\partial}{\partial\theta} - \cot\theta \cos\phi \frac{\partial}{\partial\phi} \right) e^{\pm i\phi} \sin\theta$$

$$= i\hbar \left(\frac{3}{8\pi} \right)^{1/2} \left(e^{\pm i\phi} \sin\phi \cos\theta \pm i e^{\pm i\phi} \cos\theta \cos\phi \right)$$

$$\langle L_x \rangle = i\hbar \left(\frac{3}{8\pi} \right) \int_0^{2\pi} d\phi \int_0^\pi d\theta \sin\theta \left(\sin\phi \cos\theta \pm i \cos\theta \cos\phi \right)$$

$$= i\hbar \left(\frac{3}{8\pi} \right) \left[\int_0^{2\pi} \sin\phi\, d\phi \int_0^\pi d\theta \sin\theta \cos\theta \pm \int_0^{2\pi} d\phi \cos\phi \int_0^\pi i \sin\theta \cos\theta \right]$$

$$= i\hbar \left(\frac{3}{8\pi} \right) \left[(0) \int_0^\pi d\theta \sin\theta \cos\theta \pm (0) \int_0^\pi i \sin\theta \cos\theta \right] = 0$$

6–15. For an isolated hydrogen atom, why must the angular momentum vector **L** lie on a cone that is symmetric about the z-axis? Can the angular momentum operator ever point exactly along the z-axis?

The uncertainty principle prohibits **L** from lying along the z-axis. If we observe precise values of L^2 and L_z we cannot observe precise values of L_x and L_y; therefore (as discussed in Section 6–3) the angular momentum vector must lie on a cone that is symmetric about the z-axis. If it pointed exactly along the z-axis, we would know the precise values of L_x and L_y (both would be zero) as well as the exact values of L^2 and L_z, so this can never occur.

6–16. Referring to Table 6.5, show that the first few hydrogen atomic wave functions are orthonormal.

The orthonormality condition for hydrogen atomic wave functions is

$$\int_0^\infty dr\, r^2 \int_0^\pi d\theta \sin\theta \int_0^{2\pi} d\phi\, \psi_{n'l'm'}^*(r, \theta, \phi) \psi_{nlm}(r, \theta, \phi) = \delta_{nn'} \delta_{ll'} \delta_{mm'} \tag{6.51}$$

We first show that the first few hydrogen atomic wave functions are normalized. For ψ_{100},

$$\int d\tau\, \psi_{100}^* \psi_{100} = \int_0^\infty dr\, r^2 \int_0^\pi d\theta \sin\theta \int_0^{2\pi} d\phi\, \psi_{100}^* \psi_{100}$$

$$= \int_0^{2\pi} d\phi \int_0^\pi d\theta \sin\theta \int_0^\infty dr \, r^2 \left[\frac{1}{\pi} \left(\frac{Z}{a_0} \right)^3 e^{-2\sigma} \right]$$

$$= 4 \left(\frac{Z}{a_0} \right)^3 \int_0^\infty dr \, e^{-2\sigma} r^2$$

$$= 4 \int_0^\infty e^{-2\sigma} \sigma^2 d\sigma = 1$$

For ψ_{200},

$$\int d\tau \, \psi_{200}^* \psi_{200} = \int_0^\infty dr \, r^2 \int_0^\pi d\theta \sin\theta \int_0^{2\pi} d\phi \, \psi_{200}^* \psi_{200}$$

$$= \int_0^{2\pi} d\phi \int_0^\pi d\theta \sin\theta \int_0^\infty dr \, r^2 \left[\frac{1}{32\pi} \left(\frac{Z}{a_0} \right)^3 \left(4 - 4\sigma + \sigma^2 \right) e^{-\sigma} \right]$$

$$= \frac{1}{8} \left(\frac{Z}{a_0} \right)^3 \int_0^\infty dr \, r^2 \left(4 - 4\sigma + \sigma^2 \right) e^{-\sigma}$$

$$= \frac{1}{8} \int_0^\infty d\sigma \left(4\sigma^2 - 4\sigma^3 + \sigma^4 \right) e^{-\sigma}$$

$$= \frac{1}{8} \left(4 \cdot 2! - 4 \cdot 3! + 4! \right) = 1$$

For ψ_{210},

$$\int d\tau \, \psi_{210}^* \psi_{210} = \int_0^\infty dr \, r^2 \int_0^\pi d\theta \sin\theta \int_0^{2\pi} d\phi \, \psi_{210}^* \psi_{210}$$

$$= \int_0^{2\pi} d\phi \int_0^\pi d\theta \sin\theta \int_0^\infty dr \, r^2 \left[\frac{1}{32\pi} \left(\frac{Z}{a_0} \right)^3 \sigma^2 \cos^2\theta e^{-\sigma} \right]$$

$$= \frac{1}{16} \left(\frac{Z}{a_0} \right)^3 \int_0^\pi d\theta \sin\theta \cos^2\theta \int_0^\infty dr \, r^2 \sigma^2 e^{-\sigma}$$

$$= \frac{1}{16} \int_{-1}^1 dx \, x^2 \int_0^\infty d\sigma \sigma^4 e^{-\sigma} = \frac{1}{16} \left(\frac{2}{3} \right) (4!) = 1$$

For ψ_{211} or ψ_{21-1},

$$\int d\tau \, \psi_{21\pm1}^* \psi_{21\pm1} = \int_0^\infty dr \, r^2 \int_0^\pi d\theta \sin\theta \int_0^{2\pi} d\phi \, \psi_{21\pm1}^* \psi_{21\pm1}$$

$$= \int_0^{2\pi} d\phi \int_0^\pi d\theta \sin\theta \int_0^\infty dr \, r^2 \left[\frac{1}{64\pi} \left(\frac{Z}{a_0} \right)^3 \sigma^2 \sin^2\theta e^{-\sigma} \right]$$

$$= \frac{1}{32} \int_0^\pi d\theta \sin^3\theta \int_0^\infty dr \, r^2 \sigma^2 e^{-\sigma}$$

$$= \frac{1}{32} \int_{-1}^1 dx \, (1 - x^2) \int_0^\infty d\sigma \sigma^4 e^{-\sigma} = \frac{1}{32} \left(\frac{4}{3} \right) (4!) = 1$$

We now show that the first few hydrogen atomic wave functions are orthogonal:

$$\int d\tau \, \psi_{100}^* \psi_{200} = \int_0^\infty dr \, r^2 \int_0^\pi d\theta \sin\theta \int_0^{2\pi} d\phi \, \psi_{100}^* \psi_{200}$$

$$= \int_0^{2\pi} d\phi \int_0^\pi d\theta \sin\theta \int_0^\infty dr \, r^2 \left[\frac{1}{\pi^{1/2}} \left(\frac{Z}{a_0} \right)^{3/2} e^{-\sigma} \right] \left[\frac{1}{(32\pi)^{1/2}} \left(\frac{Z}{a_0} \right)^{3/2} (2 - \sigma) e^{-\sigma/2} \right]$$

$$= \frac{1}{\sqrt{2}} \left(\frac{Z}{a_0}\right)^3 \int_0^\infty dr \, r^2 (2-\sigma)e^{-3\sigma/2}$$

$$= \frac{1}{\sqrt{2}} \int_0^\infty d\sigma (2\sigma^2 - \sigma^3)e^{-3\sigma/2}$$

$$= \frac{1}{\sqrt{2}} \left[2 \cdot 2! \left(\frac{2}{3}\right)^3 - 3! \left(\frac{2}{3}\right)^4\right] = 0$$

In showing that either ψ_{100} or ψ_{200} is orthogonal to ψ_{210}, we find that the integral over θ is

$$\int_0^\pi d\theta \sin\theta \cos\theta = \int_{-1}^1 dx \, x = 0$$

and so these orbitals are orthogonal. Likewise, in showing that either ψ_{100}, ψ_{200} or ψ_{210} is orthogonal to ψ_{211} or ψ_{21-1}, the integral over ϕ is

$$\int_0^{2\pi} d\phi e^{\pm i\phi} = 0$$

and so these orbitals are orthogonal.

6–17. Show explicitly that

$$\hat{H}\psi = -\frac{m_e e^4}{8\varepsilon_0^2 h^2}\psi$$

for the ground state of a hydrogen atom.

The Hamiltonian operator for a hydrogen atom is (Equations 6.2 and 6.3)

$$-\frac{\hbar^2}{2m_e}\left[\frac{1}{r^2}\frac{\partial}{\partial r}\left(r^2\frac{\partial}{\partial r}\right) + \frac{1}{r^2 \sin\theta}\frac{\partial}{\partial\theta}\left(\sin\theta\frac{\partial}{\partial\theta}\right) + \frac{1}{r^2 \sin^2\theta}\frac{\partial^2}{\partial\phi^2}\right] + \frac{e^2}{4\pi\varepsilon_0 r}$$

and from Table 6.5 the ground state wave function of a hydrogen atom is

$$\psi_0 = \frac{1}{\pi^{1/2}a_0^{3/2}}e^{-r/a_0} \qquad a_0 = \frac{\varepsilon_0 h^2}{\pi m_e e^2}$$

We can therefore write

$$\hat{H}\psi_0 = -\frac{\hbar^2}{2m_e r^2}\frac{\partial}{\partial r}\left(r^2\frac{\partial\psi_0}{\partial r}\right) + \frac{e^2\psi_0}{4\pi\varepsilon_0 r}$$

$$= -\frac{h^2}{(4\pi^2)2m_e r^2}\frac{1}{\pi^{1/2}a_0^{3/2}}\left(-\frac{2r}{a_0}e^{-r/a_0} + \frac{r^2}{a_0^2}e^{-r/a_0}\right) - \frac{e^2}{4\pi\varepsilon_0 r}\frac{1}{\pi^{1/2}a_0^{3/2}}e^{-r/a_0}$$

$$= \left[\frac{e^2}{4\pi\varepsilon_0 r} - \frac{m_e e^4}{8\varepsilon_0^2 h^2} - \frac{e^2}{4\pi\varepsilon_0 r}\right]\psi_0$$

$$= -\frac{m_e e^4}{8\varepsilon_0^2 h^2}\psi_0$$

6–18. Show explicitly that

$$\hat{H}\psi = -\frac{m_e e^4}{32\varepsilon_0^2 h^2}\psi$$

for a $2p_0$ state of a hydrogen atom.

The Hamiltonian operator of a hydrogen atom is the same as in Problem 6–17, and the wave function of the $2p_0$ state is

$$\psi_{2p_0} = \frac{1}{(32\pi)^{1/2}a_0^{5/2}}re^{-r/2a_0}\cos\theta \qquad a_0 = \frac{\varepsilon_0 h^2}{\pi m_e e^2}$$

We can now write

$$
\begin{aligned}
\hat{H}\psi_{2p_0} &= -\frac{\hbar^2}{2m_e}\left[\frac{1}{r^2}\frac{\partial}{\partial r}\left(r^2\frac{\partial\psi_{2p_0}}{\partial r}\right) + \frac{1}{r^2\sin\theta}\frac{\partial}{\partial\theta}\left(\sin\theta\frac{\partial\psi_{2p_0}}{\partial\theta}\right)\right] + \frac{e^2}{4\pi\varepsilon_0 r}\psi_{2p_0} \\
&= -\frac{\hbar^2}{2m_e r^2}\frac{1}{(32\pi)^{1/2}a_0^{5/2}}\left[e^{-r/2a_0}\cos\theta\left(2r - \frac{r^2}{2a_0} - \frac{3r^2}{2a_0} + \frac{r^3}{4a_0^2}\right)\right] \\
&\quad + \frac{\hbar^2}{2m_e r^2}\frac{2r\cos\theta}{(32\pi)^{1/2}a_0^{5/2}}e^{-r/2a_0} + \frac{e^2}{4\pi\varepsilon_0 r}\frac{re^{-r/2a_0}}{(32\pi)^{1/2}a_0^{5/2}}\cos\theta \\
&= -\frac{\hbar^2}{2m_e}\frac{1}{(32\pi)^{1/2}a_0^{5/2}}\left(\frac{r}{4a_0^2}e^{-r/2a_0}\cos\theta\right) - \frac{1}{(32\pi)^{1/2}a_0^{5/2}}\left(\frac{\hbar^2}{2m_e a_0} - \frac{e^2}{4\pi\varepsilon_0}\right)e^{-r/2a_0}\cos\theta \\
&= -\frac{\hbar^2}{8m_e a_0^2}\psi_{2p_0} = -\frac{m_e e^4}{32\varepsilon_0^2 h^2}\psi_{2p_0}
\end{aligned}
$$

6–19. Given the first equality, show that the ground-state energy of a hydrogen atom can be written as

$$E_0 = -\frac{\hbar^2}{2m_e a_0^2} = -\frac{e^2}{8\pi\varepsilon_0 a_0} = -\frac{m_e e^4}{32\pi^2\varepsilon_0^2\hbar^2} = -\frac{m_e e^4}{8\varepsilon_0^2 h^2}$$

Recall that we can write a_0 as

$$a_0 = \frac{\varepsilon_0 h^2}{\pi m_e e^2} = \frac{4\pi\varepsilon_0\hbar^2}{m_e e^2}$$

We showed in Problem 6–17 that

$$\hat{H}\psi = -\frac{m_e e^4}{8\varepsilon_0^2 h^2}\psi$$

so we can write

$$
\begin{aligned}
E_0 &= -\frac{m_e e^4}{8\varepsilon_0^2 h^2} = -\frac{m_e e^4}{32\pi^2\varepsilon_0^2\hbar^2} = -\frac{e^2}{8\pi\varepsilon_0}\frac{m_e e^2}{4\pi\varepsilon_0\hbar^2} = -\frac{e^2}{8\pi\varepsilon_0 a_0} \\
&= -\frac{\hbar^2}{2m_e a_0}\frac{m_e e^2}{4\pi\varepsilon_0\hbar^2} = -\frac{\hbar^2}{2m_e a_0^2}
\end{aligned}
$$

6–20. Calculate the probability that a hydrogen $1s$ electron will be found within a distance $2a_0$ from the nucleus.

This problem is similar to Example 6–10. The wave function for the $1s$ orbital of hydrogen is

$$\psi_{100} = \frac{1}{\sqrt{\pi}} \left(\frac{1}{a_0}\right)^{3/2} e^{-\sigma}$$

where $\sigma = r/a_0$, and the probability that the electron will be found within a distance $2a_0$ from the nucleus is

$$\text{prob} = \int_0^{2\pi} d\phi \int_0^{\pi} d\theta \sin\theta \int_0^{2a_0} dr\, r^2 \frac{1}{\pi} \left(\frac{1}{a_0}\right)^3 e^{-2\sigma}$$

$$= 4 \left(\frac{1}{a_0}\right)^3 \int_0^{2a_0} dr\, r^2 e^{-2\sigma} = 4 \int_0^{2} d\sigma\, \sigma^2 e^{-2\sigma}$$

$$= 4 \left(\frac{1}{4} - \frac{13}{4} e^{-4}\right) = 1 - 13e^{-4} = 0.762$$

6–21. Calculate the radius of the sphere that encloses a 50% probability of finding a hydrogen $1s$ electron. Repeat the calculation for a 90% probability.

The probability that a $1s$ electron will be found within a distance Da_0 of the nucleus is given by

$$\text{prob}(D) = \int_0^{2\pi} d\phi \int_0^{\pi} d\theta \sin\theta \int_0^{Da_0} dr\, r^2 \frac{1}{\pi} \left(\frac{1}{a_0}\right)^3 e^{-2\sigma}$$

$$= 4 \int_0^{D} d\sigma\, \sigma^2 e^{-2\sigma} = 1 - e^{-2D}(2D^2 + 2D + 1)$$

We find that $D = 1.3$ for $\text{prob}(D) = 0.50$ and $D = 2.7$ for $\text{prob}(D) = 0.90$, so the 50% and 90% probability spheres have radii of $1.3a_0$ and $2.7a_0$, respectively.

6–22. Many problems involving the calculation of average values for the hydrogen atom require evaluating integrals of the form

$$I_n = \int_0^{\infty} r^n e^{-\beta r} dr$$

This integral can be evaluated readily by starting with the elementary integral

$$I_0(\beta) = \int_0^{\infty} e^{-\beta r} dr = \frac{1}{\beta}$$

Show that the derivatives of $I(\beta)$ are

$$\frac{dI_0}{d\beta} = -\int_0^{\infty} r e^{-\beta r} dr = -I_1$$

$$\frac{d^2 I_0}{d\beta^2} = \int_0^{\infty} r^2 e^{-\beta r} dr = I_2$$

and so on. Using the fact that $I_0(\beta) = 1/\beta$, show that the values of these two integrals are $-1/\beta^2$ and $2/\beta^3$, respectively. Show that, in general

$$\frac{d^n I_0}{d\beta^n} = (-1)^n \int_0^\infty r^n e^{-\beta r} dr = (-1)^n I_n$$

$$= (-1)^n \frac{n!}{\beta^{n+1}}$$

and that

$$I_n = \frac{n!}{\beta^{n+1}}$$

$$I_0(\beta) = \int_0^\infty e^{-\beta r} dr$$

$$\frac{d I_0}{d\beta} = \int_0^\infty \frac{\partial}{\partial \beta}\left(e^{-\beta r}\right) dr = -\int_0^\infty r e^{-\beta r} dr = -I_1$$

$$\frac{d^2 I_0}{d\beta^2} = \int_0^\infty \frac{\partial}{\partial \beta}\left(-r e^{-\beta r}\right) dr = \int_0^\infty r^2 e^{-\beta r} dr = I_2$$

Alternatively, since $I_0(\beta) = 1/\beta$,

$$\frac{d I_0}{d\beta} = \frac{\partial}{\partial \beta}\left(\frac{1}{\beta}\right) = -\frac{1}{\beta^2}$$

$$\frac{d^2 I_0}{d\beta^2} = \frac{\partial}{\partial \beta}\left(-\frac{1}{\beta^2}\right) = \frac{2}{\beta^3}$$

Generally,

$$\frac{d^n I_0}{d\beta^n} = (-1)^n \int_0^\infty r^n e^{-\beta r} dr$$

$$= (-1)^n \frac{n!}{\beta^{n+1}}$$

and so

$$I_n = \frac{n!}{\beta^{n+1}}$$

6–23. Prove that the average value of r in the $1s$ and $2s$ states is $3a_0/2Z$ and $6a_0/Z$, respectively.

The average value of r, $\langle r \rangle$, is given by

$$\langle r \rangle = \int_0^{2\pi} d\phi \int_0^\pi d\theta \sin\theta \int_0^\infty dr\, r^2 \psi_{nlm}(r, \theta, \phi)^* \psi_{nlm}(r, \theta, \phi)$$

The wave functions for the $1s$ and $2s$ states are (Table 6.5)

$$\psi_{100} = \frac{1}{\sqrt{\pi}}\left(\frac{Z}{a_0}\right)^{3/2} e^{-\sigma} \qquad \sigma = \frac{r}{a_0}$$

$$\psi_{200} = \frac{1}{\sqrt{32\pi}}\left(\frac{Z}{a_0}\right)^{3/2}(2 - \sigma)e^{-\sigma/2}$$

so

$$\langle r \rangle_{1s} = 4 \left(\frac{Z}{a_0} \right)^3 \int_0^\infty r^3 e^{-2\sigma}$$

$$= \frac{4a_0}{Z} \int_0^\infty d\sigma \sigma^3 e^{-2\sigma} = \frac{4a_0}{Z} \left(\frac{3!}{16} \right) = \frac{3}{2} \left(\frac{a_0}{Z} \right)$$

and

$$\langle r \rangle_{2s} = \frac{1}{8} \left(\frac{Z}{a_0} \right)^3 \int_0^\infty r^3 (4 - 4\sigma + \sigma^2) e^{-\sigma}$$

$$= \frac{a_0}{8Z} \int_0^\infty d\sigma (4\sigma^3 - 4\sigma^4 + \sigma^5) e^{-\sigma}$$

$$= \frac{a_0}{8Z} (4 \cdot 3! - 4 \cdot 4! + 5!) = \frac{6a_0}{Z}$$

6–24. Prove that $\langle V \rangle = 2\langle E \rangle$ and, consequently, that $\langle \hat{K} \rangle = -\langle E \rangle$, for a 2s electron.

The average potential energy of a hydrogen-like 2s electron is (Equation 6.1)

$$\langle V \rangle = \left\langle -\frac{e^2}{4\pi \varepsilon_0 r} \right\rangle$$

$$= \int_0^{2\pi} d\phi \int_0^\pi d\theta \sin\theta \int_0^\infty dr\, r^2 \psi_{2s}^* \left(-\frac{e^2}{4\pi \varepsilon_0 r} \right) \psi_{2s}$$

$$= -\frac{e^2}{32\pi \varepsilon_0} \left(\frac{Z}{a_0} \right)^3 \int_0^\infty dr\, r(2 - \sigma)^2 e^{-\sigma}$$

$$= -\frac{e^2}{32\pi \varepsilon_0} \left(\frac{Z}{a_0} \right) \int_0^\infty d\sigma (4\sigma - 4\sigma^2 + \sigma^3) e^{-\sigma}$$

$$= -\frac{Ze^2}{16\pi \varepsilon_0 a_0}$$

The total energies of a hydrogenlike atom are (Problem 6–34)

$$E_n = -\frac{Ze^2}{8\pi \varepsilon_0 a_0 n^2} = -\frac{Ze^2}{32\pi \varepsilon_0 a_0}$$

for $n = 2$, so $\langle V \rangle = 2\langle E \rangle$. Because $\langle \hat{K} \rangle + \langle V \rangle = E$, $\langle \hat{K} \rangle = -\langle E \rangle$.

6–25. By evaluating the appropriate integrals, compute $\langle r \rangle$ in the 2s, 2p, and 3s states of the hydrogen atom; compare your results with the general formula

$$\langle r_{nl} \rangle = \frac{a_0}{2} [3n^2 - l(l + 1)]$$

In Problem 6–23 we found the average value of r in a hydrogen 2s orbital to be $\langle r \rangle = 6a_0$. The wave functions for the 2p and 3s states are (Table 6.5)

$$\psi_{210} = \frac{1}{\sqrt{32\pi}} \left(\frac{Z}{a_0} \right)^{3/2} \sigma e^{-\sigma/2} \cos\theta$$

$$\psi_{300} = \frac{1}{81\sqrt{3\pi}} \left(\frac{Z}{a_0} \right)^{3/2} (27 - 18\sigma + 2\sigma^2) e^{-\sigma/3}$$

So for $Z = 1$, we have

$$\langle r_{2p} \rangle = \frac{1}{32\pi} \left(\frac{1}{a_0} \right)^3 \int_0^{2\pi} d\phi \int_0^\pi d\theta \sin\theta \cos^2\theta \int_0^\infty dr \, r^3 \sigma^2 e^{-\sigma}$$

$$= \frac{a_0}{16} \int_{-1}^1 dx x^2 \int_0^\infty d\sigma \sigma^5 e^{-\sigma} = \frac{a_0}{16} \left(\frac{2}{3} \right) (5!) = 5a_0$$

and

$$\langle r_{3s} \rangle = \frac{1}{3^9 \pi a_0^3} \int_0^{2\pi} d\phi \int_0^\pi d\theta \sin\theta \int_0^\infty dr \, r^3 (27 - 18\sigma + 2\sigma^2)^2 e^{-2\sigma/3}$$

$$= \frac{4a_0}{3^9} \int_0^\infty d\sigma \sigma^3 \left[3^6 + 4(3^4\sigma^2) + 4\sigma^4 - 4(3^5\sigma) + 4(3^3\sigma^2) - 8(3^2\sigma^3) \right] e^{-2\sigma/3}$$

$$= \frac{4a_0}{3^9} \left[3^6 \cdot 3! \left(\frac{3}{2} \right)^4 + 4 \cdot 3^4 \cdot 5! \left(\frac{5}{2} \right)^6 + 4 \cdot 7! \left(\frac{3}{2} \right)^8 \right.$$

$$\left. - 4 \cdot 3^5 \cdot 4! \left(\frac{3}{2} \right)^5 + 4 \cdot 3^3 \cdot 5! \left(\frac{3}{2} \right)^6 - 8 \cdot 3^2 \cdot 6! \left(\frac{3}{2} \right)^7 \right]$$

$$= \frac{27 a_0}{2}$$

Using the general formula

$$\langle r_{nl} \rangle = \frac{a_0}{2} [3n^2 - l(l+1)]$$

we obtain $6a_0$, $5a_0$ and $27a_0/2$ for the values of $\langle r_{nl} \rangle$ for the $2s$, $2p$ and $3s$ orbitals, in agreement with the above calculations.

6–26. Show that the first few hydrogen atomic orbitals in Table 6.6 are orthonormal.

See the solution to Problem 6–16.

6–27. Show that the two maxima in the plot of $r^2 \psi_{2s}^2(r)$ against r occur at $(3 \pm \sqrt{5})a_0$. (See Figure 6.3.)

We can write $r^2 \psi_{2s}^2(r)$ as

$$f(r) = r^2 \psi_{2s}^2(r)$$

$$= r^2 \frac{1}{32\pi} \left(\frac{1}{a_0} \right)^3 \left(2 - \frac{r}{a_0} \right)^2 e^{-r/a_0}$$

$$= \frac{r^2}{32\pi a_0^3} \left(2 - \frac{r}{a_0} \right)^2 e^{-r/a_0}$$

To determine the maxima of $f(r)$, we find the values of r for which $df(r)/dr = 0$:

$$\frac{df(r)}{dr} = \frac{1}{32\pi a_0^3} \left[2r \left(2 - \frac{r}{a_0} \right)^2 e^{-r/a_0} - \frac{2r^2}{a_0} \left(2 - \frac{r}{a_0} \right) e^{-r/a_0} - \frac{r^2}{a_0} \left(2 - \frac{r}{a_0} \right)^2 e^{-r/a_0} \right]$$

$$0 = 2 \left(2 - \frac{r_{max}}{a_0} \right) - \frac{2r_{max}}{a_0} - \frac{r_{max}}{a_0} \left(2 - \frac{r_{max}}{a_0} \right)$$

$$= 4a_0^2 - 2a_0 r_{max} - 2a_0 r_{max} - 2a_0 r_{max} + r_{max}^2$$

$$= r_{max}^2 - 6a_0 r_{max} + 4a_0^2$$

$$r_{max} = (3 \pm \sqrt{5})a_0$$

6–28. Calculate the value of $\langle r \rangle$ for the $n = 2$, $l = 1$ state and the $n = 2$, $l = 0$ state of the hydrogen atom. Are you surprised by the answers? Explain.

The average value of r, $\langle r \rangle$, is given by

$$\langle r \rangle = \int d\tau\, \psi_{nl}(r, \theta, \phi)^* \psi_{nl}(r, \theta, \phi)$$

We use the wave functions in Table 6.6 to find

$$\langle r \rangle_{20} = \frac{4\pi}{32\pi a_0^3} \int_0^\infty dr\, r^3 \left(2 - \frac{r}{a_0}\right)^2 e^{-r/a_0}$$

$$= \frac{a_0}{8} \int_0^\infty dx\, x^3 (2 - x)^2 e^{-x} = \frac{a_0}{8}(4 \cdot 3! - 4 \cdot 4! + 5!)$$

$$= 6a_0$$

and

$$\langle r \rangle_{21} = \frac{2\pi}{32\pi a_0^3} \int_0^\infty d\theta \sin\theta \cos^2\theta \int_0^\infty dr\, r^3 \left(\frac{r}{a_0}\right)^2 e^{-r/a_0}$$

$$= \frac{a_0}{16} \left(\frac{2}{3}\right) \int_0^\infty dx\, x^5 e^{-x}$$

$$= \frac{a_0}{16} \left(\frac{2}{3}\right)(5!) = 5a_0$$

These results show that an electron in the $2s$ orbital is farther from the nucleus (on average) than an electron in the $2p$ orbital. This is surprising, as we might expect the reverse to be true from our studies of multi-electron systems in general chemistry; note, however, that a one-electron hydrogen-like wave function differs from multi-electron wave functions (Chapter 8).

6–29. In Chapter 4, we learned that if ψ_1 and ψ_2 are solutions of the Schrödinger equation that have the same energy E_n, then $c_1\psi_1 + c_2\psi_2$ is also a solution. Let $\psi_1 = \psi_{210}$ and $\psi_2 = \psi_{211}$ (see Table 6.5). What is the energy corresponding to $\psi = c_1\psi_1 + c_2\psi_2$ where $c_1^2 + c_2^2 = 1$? What does this result tell you about the uniqueness of the three p orbitals, p_x, p_y, and p_z?

Recall that the energy of the hydrogen atom depends only on the value of n. Therefore, ψ_{211} and ψ_{210} have the same energy, E_2, and so (Chapter 4) the energy corresponding to $\psi = c_1\psi_1 + c_2\psi_2$ where $c_1^2 + c_2^2 = 1$ is also E_2. The three p orbitals (p_x, p_y, and p_z), therefore, are not a unique representation of the three degenerate orbitals for $n = 2$ and $l = 1$.

6–30. Show that the total probability density of the $2p$ orbitals is spherically symmetric by evaluating $\sum_{m=-1}^{1} \psi_{21m}^2$. (Use the wave functions in Table 6.6.)

$$\sum_{m=-1}^{1} \psi_{21m}^2 = \frac{1}{32\pi} \left(\frac{Z}{a_0}\right)^3 \sigma^2 e^{-\sigma} \left(\cos^2\theta + \sin^2\theta\cos^2\phi + \sin^2\theta\sin^2\phi\right)$$

$$= \frac{Z^3\sigma^2 e^{-\sigma}}{32\pi a_0^3} \left[\cos^2\theta + \sin^2\theta\left(\cos^2\phi + \sin^2\phi\right)\right]$$

$$= \frac{Z^3\sigma^2 e^{-\sigma}}{32\pi a_0^3} \left(\cos^2\theta + \sin^2\theta\right)$$

$$= \frac{Z^3\sigma^2 e^{-\sigma}}{32\pi a_0^3}$$

The sum depends only on the variable r (through σ), so the total probability density of the $2p$ orbitals is spherically symmetric.

6–31. Show that the total probability density of the $3d$ orbitals is spherically symmetric by evaluating $\sum_{m=-2}^{2} \psi_{32m}^2$. (Use the wave functions in Table 6.6.)

$$\sum_{m=-2}^{2} \psi_{32m}^2 = \frac{1}{81^2\pi} \left(\frac{Z}{a_0}\right)^3 \sigma^4 e^{-2\sigma/3} \left[\frac{(3\cos^2\theta - 1)^2}{6} + 2\sin^2\theta\cos^2\theta\cos^2\phi \right.$$

$$\left. + 2\sin^2\theta\cos^2\theta\sin^2\phi + \frac{\sin^4\theta\cos^2 2\phi}{2} + \frac{\sin^4\theta\sin^2 2\phi}{2}\right]$$

$$= \frac{Z^3\sigma^4 e^{-2\sigma/3}}{81^2\pi a_0^3} \left[\frac{(3\cos^2\theta - 1)^2}{6} + 2\sin^2\theta\cos^2\theta\left(\sin^2\phi + \cos^2\phi\right)\right.$$

$$\left. + \frac{\sin^4\theta\left(\cos^2 2\phi + \sin^2 2\phi\right)}{2}\right]$$

$$= \frac{Z^3\sigma^4 e^{-2\sigma/3}}{(81)^2 6\pi a_0^3} \left[(3\cos^2\theta - 1)^2 + 12\sin^2\theta\cos^2\theta + 3\sin^4\theta\right]$$

Now substitute $\sin^2\theta = 1 - \cos^2\theta$ into the above expression to get

$$\sum_{m=-2}^{2} \psi_{32m}^2 = \frac{Z^3\sigma^4 e^{-2\sigma/3}}{(81)^2 6\pi a_0^3} \left[9\cos^4\theta - 6\cos^2\theta + 1\right.$$

$$\left. + 12(1 - \cos^2\theta)\cos^2\theta + 3(1 - \cos^2\theta)^2\right]$$

$$= \frac{Z^3\sigma^4 e^{-2\sigma/3}}{(81)^2 6\pi a_0^3} \left[9\cos^4\theta - 6\cos^2\theta + 1 + 12\cos^2\theta - 12\cos^4\theta\right.$$

$$\left. + 3 - 6\cos^2\theta + 3\cos^4\theta\right]$$

$$= \frac{4Z^3\sigma^4 e^{-2\sigma/3}}{(81)^2 6\pi a_0^3} = \frac{2Z^3\sigma^4 e^{-2\sigma/3}}{(81)^2 3\pi a_0^3}$$

The sum depends only on the variable r (through σ), so the total probability density of the $3d$ orbitals is spherically symmetric.

6–32. Show that the sum of the probability densities for the $n = 3$ states of the hydrogen atom is spherically symmetric. Do you expect this to be true for all values of n? Explain.

In Problem 6–31 we showed that the sum of the probability densities of the $3d$ orbitals is spherically symmetric. The probability density of the $3s$ orbital is also spherically symmetric, and so we need only show that the sum of the probability densities of the $3p$ orbitals is spherically symmetric. The angular dependence of the $3p$ orbitals is the same as that of the $2p$ orbitals. In Problem 6–30 we showed that the sum of the squares of the $2p$ orbitals is spherically symmetric, and therefore the same must be true for the $3p$ orbitals. Thus, the sum of the probability densities for the $n = 3$ states of the hydrogen atom is spherically symmetric. We expect this to be the case for all values of n. Recall from Problem 6–10 that

$$\sum_{m=-l}^{+l} |Y_l^m(\theta, \phi)|^2 = \text{constant} \tag{1}$$

When we sum all of the probability densities corresponding to any given n, we evaluate a sum similar to that in Equation 1. Because the sum is equal to a constant, it cannot have any angular dependence and will depend only on r. Such a sum is spherically symmetric.

6–33. Determine the degeneracy of each of the hydrogen atomic energy levels.

The energy depends only on the quantum number n:

$$E_n = -\frac{e^2}{8\pi \varepsilon_0 a_0 n^2} \qquad n = 1, 2, \ldots \tag{6.45}$$

For a given value of n, there are $n - 1$ allowed values of l. For each l there are $2l + 1$ allowed values of m. All of these combinations of l and m are degenerate, and so the total number of energy sublevels for one value of n is

$$\sum_{l=0}^{n-1} (2l + 1) = \frac{2(n - 1)n}{2} + n = n^2$$

6–34. Set up the Hamiltonian operator for the system of an electron interacting with a fixed nucleus of atomic number Z. The simplest such system is singly ionized helium, where $Z = 2$. We will call this a hydrogenlike system. Observe that the only difference between this Hamiltonian operator and the hydrogen Hamiltonian operator is the correspondance that e^2 for the hydrogen atom becomes Ze^2 for the hydrogenlike ion. Consequently, show that the energy becomes (cf. Equation 6.44)

$$E_n = -\frac{m_e Z^2 e^4}{8\varepsilon_0^2 h^2 n^2} \qquad n = 1, 2, \ldots$$

Furthermore, now show that the solutions to the radial equation, Equation 6.47, are

$$R_{nl}(r) = -\left\{ \frac{(n - l - 1)!}{2n[(n + l)!]^3} \right\}^{1/2} \left(\frac{2Z}{na_0} \right)^{l+3/2} r^l e^{-Zr/na_0} L_{n+l}^{2l+1} \left(\frac{2Zr}{na_0} \right)$$

Show that the $1s$ orbital for this system is

$$\psi_{1s} = \frac{1}{\sqrt{\pi}} \left(\frac{Z}{a_0} \right)^{3/2} e^{-Zr/a_0}$$

and show that it is normalized. Show that

$$\langle r \rangle = \frac{3a_0}{2Z}$$

and that

$$r_{mp} = \frac{a_0}{Z}$$

Last, calculate the ionization energy of a hydrogen atom and a singly ionized helium atom. Express your answer in kilojoules per mole.

The Hamiltonian operator for a hydrogenlike system is

$$\hat{H} = -\frac{\hbar^2}{2m_e}\nabla^2 - \frac{Ze^2}{4\pi\varepsilon_0 r}$$

The only difference between this Hamiltonian and that of a hydrogen atom is that e^2 is replaced by Ze^2. Because e appears nowhere else, we can obtain the results of this problem by replacing e^2 by Ze^2 in the hydrogen atom results. For example, the expression for the energy of a hydrogen atom is

$$E_n = -\frac{m_e e^4}{8\varepsilon_0^2 h^2 n^2} \qquad n = 1, 2, \ldots \tag{6.44}$$

and so the energy of a hydrogenlike atom is given by

$$E_n = -\frac{m_e (Ze^2)^2}{8\varepsilon_0^2 h^2 n^2} = -\frac{m_e Z^2 e^4}{8\varepsilon_0^2 h^2 n^2} \qquad n = 1, 2, \ldots \tag{1}$$

The Bohr radius ($\varepsilon_0 h^2/\pi\mu e^2$ for hydrogen) for a hydrogenlike atom is

$$a_0 = \frac{\varepsilon_0 h^2}{\pi\mu Z e^2}$$

Consequently, we can replace a_0 by a_0/Z in Equation 6.47 to find

$$R_{nl}(r) = -\left\{\frac{(n-l-1)!}{2n[(n+l)!]^3}\right\}^{1/2} \left(\frac{2Z}{na_0}\right)^{l+3/2} r^l e^{-Zr/na_0} L_{n+l}^{2l+1}\left(\frac{2Zr}{na_0}\right)$$

The hydrogen atomic $1s$ orbital for a hydrogen atom is

$$\psi_{1s} = \frac{1}{\pi^{1/2} a_0^{3/2}} e^{-r/a_0}$$

and becomes, for a hydrogenlike atom of atomic number Z,

$$\psi_{1s} = \frac{1}{\sqrt{\pi}}\left(\frac{Z}{a_0}\right)^{3/2} e^{-Zr/a_0} \tag{2}$$

To see whether this function is normalized, we evaluate

$$\int d\tau\, \psi_{1s}^* \psi_{1s} = \int_0^{2\pi} d\phi \int_0^{\pi} d\theta \sin\theta \int_0^{\infty} dr\, r^2 \frac{Z^3}{\pi a_0^3} e^{-2Zr/a_0}$$

$$= \frac{4Z^3}{a_0^3}\int_0^{\infty} dr\, r^2 e^{-2Zr/a_0}$$

$$= \frac{4Z^3}{a_0^3}\frac{2! a_0^3}{8Z^3} = 1$$

We now evaluate $\langle r \rangle_{1s}$ using Equation 2.

$$\langle r \rangle_{1s} = \int_0^{2\pi} d\phi \int_0^{\pi} d\theta \sin\theta \int_0^{\infty} dr\, r^3 \left(\frac{Z^3}{\pi a_0^3}\right) e^{-2Zr/a_0}$$

$$= \frac{4Z^3}{a_0^3}\int_0^{\infty} dr\, r^3 e^{-2Zr/a_0} = \frac{4a_0}{Z}\int_0^{\infty} d\sigma\, \sigma^3 e^{-2\sigma}$$

$$= \frac{4a_0}{Z}\frac{3!}{16} = \frac{3a_0}{2Z}$$

The value of r_{mp} is given by

$$\frac{d}{dr}\psi_{1s}^{*}\psi_{1s}r_{mp}^{2} = 0$$

$$\frac{d}{dr}r_{mp}^{2}\frac{Z^{3}}{\pi a_{0}^{3}}e^{-2Zr_{mp}/a_{0}} = 0$$

$$2r_{mp}e^{-2Zr_{mp}/a_{0}} + \frac{-2Zr_{mp}^{2}}{a_{0}}e^{-2Zr_{mp}/a_{0}} = 0$$

$$a_{0} - Zr_{mp} = 0$$

$$r_{mp} = \frac{a_{0}}{Z}$$

The ionization energy is given by $-E_1$, and so we find from Equation 1

$$\text{IE} = \frac{m_e Z^2 e^4}{8\varepsilon_0^2 h^2} = (1312 \text{ kJ·mol}^{-1})Z^2$$

Therefore,

$$\text{IE}_{\text{H}} = 1312 \text{ kJ·mol}^{-1} = 13.60 \text{ eV} \quad \text{and} \quad \text{IE}_{\text{He}^+} = 5248 \text{ kJ·mol}^{-1} = 54.39 \text{ eV}$$

6–35. How does E_n for a hydrogen atom differ from Equation 6.44 if the nucleus is not considered to be fixed at the origin?

See the solution to Problem 5–44.

6–36. Determine the ratio of the ground-state energy of atomic hydrogen to that of atomic deuterium.

The energy of the ground state of a hydrogen atom is given by (Problem 6–35)

$$E = -\frac{\mu e^4}{8\varepsilon_0^2 h^2}$$

where μ is the reduced mass of the atom. The ratio of the ground-state energy of a hydrogen atom to a deuterium atom, $E_{\text{H}}/E_{\text{D}}$, is then

$$\frac{E_{\text{H}}}{E_{\text{D}}} = \frac{\mu_{\text{H}}}{\mu_{\text{D}}}$$

In Problems 5–7 and 5–46 we calculated that $\mu_{\text{H}} = 9.104\,431 \times 10^{-31}$ kg and that $\mu_{\text{D}} = 9.106\,909 \times 10^{-31}$ kg, so the ratio of the energies is

$$\frac{E_{\text{H}}}{E_{\text{D}}} = \frac{9.104\,431 \times 10^{-31} \text{ kg}}{9.106\,909 \times 10^{-31} \text{ kg}} = 0.999\,728$$

6–37. In this problem, we will prove the so-called *quantum-mechanical virial theorem*. Start with

$$\hat{H}\psi = E\psi$$

where

$$\hat{H} = -\frac{\hbar^2}{2m}\nabla^2 + V(x, y, z)$$

Using the fact that $\hat{H}$ is a Hermitian operator (Problem 4–28), show that

$$\int \psi^*[\hat{H}, \hat{A}]\psi \, d\tau = 0 \qquad (1)$$

where $\hat{A}$ is any linear operator. Choose $\hat{A}$ to be

$$\hat{A} = -i\hbar \left(x\frac{\partial}{\partial x} + y\frac{\partial}{\partial y} + z\frac{\partial}{\partial z} \right) \qquad (2)$$

and show that

$$[\hat{H}, \hat{A}] = i\hbar \left(x\frac{\partial V}{\partial x} + y\frac{\partial V}{\partial y} + z\frac{\partial V}{\partial z} \right) - \frac{i\hbar}{m}(\hat{P}_x^2 + \hat{P}_y^2 + \hat{P}_z^2)$$

$$= i\hbar \left(x\frac{\partial V}{\partial x} + y\frac{\partial V}{\partial y} + z\frac{\partial V}{\partial z} \right) - 2i\hbar\hat{K}$$

where $\hat{K}$ is the kinetic energy operator. Now use Equation 1 and show that

$$\left\langle x\frac{\partial V}{\partial x} + y\frac{\partial V}{\partial y} + z\frac{\partial V}{\partial z} \right\rangle = 2\langle \hat{K} \rangle \qquad (3)$$

Equation 3 is the quantum-mechanical virial theorem. Now show that if $V(x, y, z)$ is a Coulombic potential

$$V(x, y, z) = -\frac{Ze^2}{4\pi\varepsilon_0(x^2 + y^2 + z^2)^{1/2}}$$

then

$$\langle V \rangle = -2\langle \hat{K} \rangle = 2\langle E \rangle \qquad (4)$$

where

$$\langle E \rangle = \langle \hat{K} \rangle + \langle V \rangle$$

In Problem 6–24 we proved that this result is valid for a 2s electron. Although we proved Equation 4 only for the case of one electron in the field of one nucleus, Equation 4 is valid for many-electron atoms and molecules. The proof is a straightforward extension of the proof developed in this problem.

We first show that

$$\int \psi^*[\hat{H}, \hat{A}]\psi \, d\tau = 0 \qquad (1)$$

Writing out the commutator, we have

$$\int \psi^*\hat{H}\hat{A}\psi \, d\tau - \int \psi^*\hat{A}\hat{H}\psi \, d\tau \overset{?}{=} 0$$

Using the Hermitian property of $\hat{H}$, we can write this difference as

$$\int \left(\psi\hat{H} \right)^* \psi \, d\tau - \int \psi^*\hat{A}\left(\hat{H}\psi \right) d\tau \overset{?}{=} 0$$

We know that $\hat{H}\psi = E\psi$, so

$$\int \psi^*[\hat{H}, \hat{A}]\psi \, d\tau = E \int \psi^* \hat{A}\psi \, d\tau - E \int \psi^* \hat{A}\psi \, d\tau = 0$$

Now let

$$\hat{A} = -i\hbar\left(x\frac{\partial}{\partial x} + y\frac{\partial}{\partial y} + z\frac{\partial}{\partial z}\right) \tag{2}$$

We are given

$$\hat{H} = -\frac{\hbar^2}{2m}\nabla^2 + V(x, y, z)$$

Then

$$\hat{H}\hat{A}f = -i\hbar\left[-\frac{\hbar^2}{2m}\nabla^2 + V(x, y, z)\right]\left[x\frac{\partial f}{\partial x} + y\frac{\partial f}{\partial y} + z\frac{\partial f}{\partial z}\right]$$

$$= \frac{i\hbar^3}{2m}\left(2\frac{\partial^2 f}{\partial x^2} + x\frac{\partial^3 f}{\partial x^3} + y\frac{\partial^3 f}{\partial x^2 \partial y} + z\frac{\partial^3 f}{\partial x^2 \partial z} + x\frac{\partial^3 f}{\partial x \partial y^2}\right.$$

$$\left. + 2\frac{\partial^2 f}{\partial y^2} + y\frac{\partial^3 f}{\partial y^3} + z\frac{\partial^3 f}{\partial y^2 \partial z} + x\frac{\partial^3 f}{\partial x \partial z^2} + y\frac{\partial^3 f}{\partial y \partial z^2} + 2\frac{\partial^2 f}{\partial z^2} + z\frac{\partial^3 f}{\partial z^3}\right)$$

$$-i\hbar\left[V\left(x\frac{\partial f}{\partial x} + y\frac{\partial f}{\partial y} + z\frac{\partial f}{\partial z}\right)\right]$$

and

$$\hat{A}\hat{H}f = \frac{i\hbar^3}{2m}\left(x\frac{\partial}{\partial x} + y\frac{\partial}{\partial y} + z\frac{\partial}{\partial z}\right)\left(\frac{\partial^2 f}{\partial x^2} + \frac{\partial^2 f}{\partial y^2} + \frac{\partial^2 f}{\partial z^2}\right) - i\hbar\left(x\frac{\partial}{\partial x} + y\frac{\partial}{\partial y} + z\frac{\partial}{\partial z}\right)Vf$$

$$= \frac{i\hbar^3}{2m}\left(x\frac{\partial^3 f}{\partial x^3} + x\frac{\partial^3 f}{\partial x \partial y^2} + x\frac{\partial^3 f}{\partial x \partial z^2} + y\frac{\partial^3 f}{\partial x^2 \partial y} + y\frac{\partial^3 f}{\partial y^3} + y\frac{\partial^3 f}{\partial y \partial z^2} + z\frac{\partial^3 f}{\partial x^2 \partial z} + z\frac{\partial^3 f}{\partial z^3}\right)$$

$$-i\hbar\left[\left(x\frac{\partial V}{\partial x} + y\frac{\partial V}{\partial y} + z\frac{\partial V}{\partial z}\right)f + \left(x\frac{\partial f}{\partial x} + y\frac{\partial f}{\partial y} + z\frac{\partial f}{\partial z}\right)V\right]$$

Therefore,

$$\left[\hat{H}, \hat{A}\right] = \frac{i\hbar^3}{2m}\left(2\frac{\partial^2}{\partial x^2} + 2\frac{\partial^2}{\partial y^2} + 2\frac{\partial^2}{\partial z^2}\right) + i\hbar\left(x\frac{\partial V}{\partial x} + y\frac{\partial V}{\partial y} + z\frac{\partial V}{\partial z}\right)$$

$$= i\hbar\left(x\frac{\partial V}{\partial x} + y\frac{\partial V}{\partial y} + z\frac{\partial V}{\partial z}\right) - \frac{i\hbar}{m}\left(\hat{P}_x^2 + \hat{P}_y^2 + \hat{P}_z^2\right)$$

$$= i\hbar\left(x\frac{\partial V}{\partial x} + y\frac{\partial V}{\partial y} + z\frac{\partial V}{\partial z}\right) - 2i\hbar\hat{K}$$

where we know $\hat{P}$ and $\hat{K}$ from Table 4.1. Now, substituting this result into Equation 1 gives

$$\int \psi^*\left[i\hbar\left(x\frac{\partial V}{\partial x} + y\frac{\partial V}{\partial y} + z\frac{\partial V}{\partial z}\right) - 2i\hbar\hat{K}\right]\psi \, d\tau = 0$$

$$i\hbar\int \psi^*\left(x\frac{\partial V}{\partial x} + y\frac{\partial V}{\partial y} + z\frac{\partial V}{\partial z}\right)\psi \, d\tau = i\hbar\int \psi^* 2\hat{K}\psi \, d\tau$$

$$\left\langle x\frac{\partial V}{\partial x} + y\frac{\partial V}{\partial y} + z\frac{\partial V}{\partial z}\right\rangle = 2\langle\hat{K}\rangle \tag{3}$$

Now consider the Coulombic potential

$$V(x, y, z) = \frac{Ze^2}{4\pi\varepsilon_0(x^2 + y^2 + z^2)^{1/2}}$$

This gives

$$x\frac{\partial V}{\partial x} + y\frac{\partial V}{\partial y} + z\frac{\partial V}{\partial z} = -\frac{Ze^2}{4\pi\varepsilon_0}\left(\frac{x^2}{r^3} + \frac{y^2}{r^3} + \frac{z^2}{r^3}\right)$$

$$= -\frac{Ze^2}{4\pi\varepsilon_0 r} = -V$$

and so substituting into Equation 3 gives $\langle V \rangle = -2\langle \hat{K} \rangle$. Because $\langle \hat{K} \rangle + \langle V \rangle = E$, we can write

$$E - \langle \hat{K} \rangle = -2\langle \hat{K} \rangle$$

or $E = -\langle \hat{K} \rangle$ for a Coulombic potential.

6–38. Use the virial theorem (Problem 6–37) to prove that $\langle \hat{K} \rangle = \langle V \rangle = E/2$ for a harmonic oscillator (cf. Problem 5–23).

For a three-dimensional harmonic oscillator,

$$V(x, y, z) = \frac{k_x x^2}{2} + \frac{k_y y^2}{2} + \frac{k_z z^2}{2}$$

Therefore,

$$x\frac{\partial V}{\partial x} + y\frac{\partial V}{\partial y} + z\frac{\partial V}{\partial z} = k_x x^2 + k_y y^2 + k_z z^2 = 2V$$

and substituting into Equation 3 of Problem 6–37 gives $2\langle V \rangle = 2\langle \hat{K} \rangle$. Because $\langle \hat{K} \rangle + \langle V \rangle = E$, we can also write

$$\langle \hat{K} \rangle = \langle V \rangle = \tfrac{1}{2}E$$

6–39. The average value of r for a hydrogenlike atom can be evaluated in general and is given by

$$\langle r \rangle_{nl} = \frac{n^2 a_0}{Z}\left\{1 + \frac{1}{2}\left[1 - \frac{l(l+1)}{n^2}\right]\right\}$$

Verify this formula explicitly for the ψ_{211} orbital.

We first determine $\langle r \rangle_{21}$ directly:

$$\langle r \rangle_{21} = \int_0^{2\pi} d\phi \int_0^{\pi} \sin\theta d\theta \int_0^{\infty} r^2 dr \phi_{211}^* r \phi_{211}$$

$$= \frac{Z^3}{32\pi a_0^3} \int_0^{2\pi} d\phi \cos^2\phi \int_0^{\pi} d\theta \sin^3\theta \int_0^{\infty} dr r^3 \left(\frac{Zr}{a_0}\right)^2 e^{-Zr/a_0}$$

$$= \frac{a_0}{32\pi Z} \cdot \pi \cdot \frac{4}{3} \cdot \int_0^{\infty} dx x^5 e^{-x}$$

$$= \frac{a_0}{24Z}(5!) = \frac{5a_0}{Z}$$

Using the equation given in the problem,

$$\langle r \rangle_{21} = \frac{2^2 a_0}{Z} \left\{ 1 + \frac{1}{2} \left[1 - \frac{1(1+1)}{2^2} \right] \right\}$$

$$= \frac{4a_0}{Z} \frac{5}{4} = \frac{5a_0}{Z}$$

6–40. The average value of r^2 for a hydrogenlike atom can be evaluated in general and is given by

$$\langle r^2 \rangle_{nl} = \frac{n^4 a_0^2}{Z^2} \left\{ 1 + \frac{3}{2} \left[1 - \frac{l(l+1) - \frac{1}{3}}{n^2} \right] \right\}$$

Verify this formula explicitly for the ψ_{210} orbital.

We first determine $\langle r^2 \rangle_{21}$ directly:

$$\langle r^2 \rangle_{21} = \int_0^{2\pi} d\phi \int_0^\pi \sin\theta \, d\theta \int_0^\infty r^2 dr \, \phi_{210}^* r^2 \phi_{210}$$

$$= \frac{Z^3}{32\pi a_0^3} \int_0^{2\pi} d\phi \int_0^\phi d\theta \sin\theta \cos^2\theta \int_0^\infty \left(\frac{Zr}{a_0} \right)^2 e^{-Zr/a_0} r^4 dr$$

$$= \frac{2\pi Z^3}{32\pi a_0^3} \frac{2}{3} \int_0^\infty \left(\frac{Zr}{a_0} \right)^2 e^{-Zr/a_0} r^4 dr$$

$$= \frac{a_0^2}{24Z^2} \int_0^\infty dx \, x^6 e^{-x}$$

$$= \frac{a_0^2}{24Z^2} (6!) = \frac{30a_0^2}{Z^2}$$

Using the equation given in the problem,

$$\langle r^2 \rangle_{21} = \frac{2^4 a_0^2}{Z^2} \left\{ 1 + \frac{3}{2} \left[1 - \frac{1(1+1) - \frac{1}{3}}{2^2} \right] \right\}$$

$$= \frac{16a_0^2}{Z^2} \left[1 + \frac{3}{2} \left(\frac{7}{12} \right) \right]$$

$$= \frac{16a_0^2}{Z^2} \left(\frac{15}{8} \right) = \frac{30a_0^2}{Z^2}$$

6–41. The average values of $1/r$, $1/r^2$, and $1/r^3$ for a hydrogenlike atom can be evaluated in general and are given by

$$\left\langle \frac{1}{r} \right\rangle_{nl} = \frac{Z}{a_0 n^2}$$

$$\left\langle \frac{1}{r^2} \right\rangle_{nl} = \frac{Z^2}{a_0^2 n^3 (l + \frac{1}{2})}$$

and

$$\left\langle \frac{1}{r^3} \right\rangle_{nl} = \frac{Z^3}{a_0^3 n^3 l (l + \frac{1}{2})(l + 1)}$$

Verify these formulas explicitly for the ψ_{210} orbital.

We first evaluate these quantities directly:

$$\left\langle \frac{1}{r} \right\rangle_{21} = \int_0^{2\pi} d\phi \int_0^{\pi} \sin\theta d\theta \int_0^{\infty} r^2 dr \psi_{210}^* \frac{1}{r} \psi_{210}$$

$$= \frac{1}{32\pi} \left(\frac{Z}{a_0}\right)^3 \int_0^{2\pi} d\phi \int_0^{\pi} d\theta \sin\theta \cos^2\theta \int_0^{\infty} dr r^2 \frac{1}{r} \left(\frac{Zr}{a_0}\right)^2 e^{-Zr/a_0}$$

$$= \frac{Z^5}{16a_0^5} \cdot \frac{2}{3} \int_0^{\infty} dr r^3 e^{-Zr/a_0} = \frac{Z}{24a_0} \int_0^{\infty} dx x^3 e^{-x}$$

$$= \frac{Z}{24a_0}(3!) = \frac{Z}{4a_0}$$

$$\left\langle \frac{1}{r^2} \right\rangle_{21} = \int_0^{2\pi} d\phi \int_0^{\pi} \sin\theta d\theta \int_0^{\infty} r^2 dr \psi_{210}^* \frac{1}{r^2} \psi_{210}$$

$$= \frac{1}{32\pi} \left(\frac{Z}{a_0}\right)^3 \int_0^{2\pi} d\phi \int_0^{\pi} d\theta \sin\theta \cos^2\theta \int_0^{\infty} dr r^2 \frac{1}{r^2} \left(\frac{Zr}{a_0}\right)^2 e^{-Zr/a_0}$$

$$= \frac{Z^5}{16a_0^5} \cdot \frac{2}{3} \int_0^{\infty} dr r^2 e^{-Zr/a_0} = \frac{Z^2}{24a_0^2} \int_0^{\infty} dx x^2 e^{-x}$$

$$= \frac{Z^2}{24a_0^2}(2!) = \frac{Z^2}{12a_0^2}$$

$$\left\langle \frac{1}{r^3} \right\rangle_{21} = \int_0^{2\pi} d\phi \int_0^{\pi} \sin\theta d\theta \int_0^{\infty} r^2 dr \psi_{210}^* \frac{1}{r^3} \psi_{210}$$

$$= \frac{1}{32\pi} \left(\frac{Z}{a_0}\right)^3 \int_0^{2\pi} d\phi \int_0^{\pi} d\theta \sin\theta \cos^2\theta \int_0^{\infty} dr r^2 \frac{1}{r^3} \left(\frac{Zr}{a_0}\right)^2 e^{-Zr/a_0}$$

$$= \frac{Z^5}{16a_0^5} \cdot \frac{2}{3} \int_0^{\infty} dr r e^{-Zr/a_0} = \frac{Z^3}{24a_0^3} \int_0^{\infty} dx x e^{-x}$$

$$= \frac{Z^3}{24a_0^3}$$

Using the equation given in the problem,

$$\left\langle \frac{1}{r} \right\rangle_{21} = \frac{Z}{a_0(2)^2} = \frac{Z}{4a_0}$$

$$\left\langle \frac{1}{r^2} \right\rangle_{21} = \frac{Z^2}{a_0^2(2)^3(1+\frac{1}{2})} = \frac{Z^2}{12a_0^2}$$

$$\left\langle \frac{1}{r^3} \right\rangle_{21} = \frac{Z^3}{a_0^3(2)^3[1+\frac{1}{2}(2)]} = \frac{Z^3}{24a_0^3}$$

6–42. The designations of the d orbitals can be rationalized in the following way. Equation 6.63 shows that d_{xz} goes as $\sin\theta\cos\theta\cos\phi$. Using the relation between Cartesian and spherical coordinates, show that $\sin\theta\cos\theta\cos\phi$ is proportional to xz. Similarly, show that $\sin\theta\cos\theta\sin\phi$ (d_{yz}) is proportional to yz; that $\sin^2\theta\cos 2\phi$ ($d_{x^2-y^2}$) is proportional to $x^2 - y^2$; and that $\sin^2\theta\sin 2\phi$ (d_{xy}) is proportional to xy.

The relations between Cartesian and spherical coordinates are

$$x = r \sin\theta \cos\phi \qquad y = r \sin\theta \sin\phi \qquad z = r \cos\theta$$

Thus, for the d_{xz} and d_{yz} orbitals, we see that

$$r^2 \sin\theta \cos\theta \cos\phi = xz$$
$$\sin\theta \cos\theta \cos\phi \propto xz$$
$$r^2 \sin\theta \cos\theta \sin\phi = yz$$
$$\sin\theta \cos\theta \sin\phi \propto yz$$

Likewise, for the $d_{x^2-y^2}$ and d_{xy} orbitals,

$$r^2 \sin^2\theta \cos^2\phi - r^2 \sin^2\theta \sin^2\phi = x^2 - y^2$$
$$r^2 \sin^2\theta \cos 2\phi = x^2 - y^2$$
$$\sin^2\theta \cos 2\phi \propto x^2 - y^2$$
$$r^2 \sin^2\theta \cos\phi \sin\phi = xy$$
$$\tfrac{1}{2}r^2 \sin^2\theta \sin 2\phi = xy$$
$$\sin^2\theta \sin 2\phi \propto xy$$

Problems 6–43 through 6–47 examine the energy levels for a hydrogen atom in an external magnetic field.

6–43. Recall from your course in physics that the motion of an electric charge around a closed loop produces a magnetic dipole, μ, whose direction is perpendicular to the loop and whose magnitude is given by

$$\mu = iA$$

where i is the current in amperes ($\text{C} \cdot \text{s}^{-1}$) and A is the area of the loop (m^2). Notice that the units of a magnetic dipole are coulombs·meters²·seconds⁻¹ ($\text{C} \cdot \text{m}^2 \cdot \text{s}^{-1}$), or amperes·meters² ($\text{A} \cdot \text{m}^2$). Show that

$$i = \frac{qv}{2\pi r}$$

for a circular loop, where v is the velocity of the charge q and r is the radius of the loop. Show that

$$\mu = \frac{qrv}{2}$$

for a circular loop. If the loop is not circular, then we must use vector calculus and the magnetic dipole is given by

$$\mu = \frac{q(\mathbf{r} \times \mathbf{v})}{2}$$

Show that this formula reduces to the preceding one for a circular loop. Last, using the relationship $\mathbf{L} = \mathbf{r} \times \mathbf{p}$, show that

$$\mu = \frac{q}{2m}\mathbf{L}$$

Thus, the orbital motion of an electron in an atom imparts a magnetic moment to the atom. For an electron, $q = -|e|$ and so

$$\boldsymbol{\mu} = -\frac{|e|}{2m_e}\mathbf{L}$$

For a circular loop, the frequency with which a charge q will pass a given point is $v/2\pi r$ (the speed of the charge divided by the circumference of the circle). Then

$$i = qv = \frac{qv}{2\pi r}$$

The area A of the loop is πr^2, so

$$\mu = iA = \frac{qv\pi r^2}{2\pi r} = \frac{qrv}{2}$$

Using vector calculus, we find

$$\boldsymbol{\mu} = \frac{q(\mathbf{r} \times \mathbf{v})}{2} = \frac{qrv\sin\theta}{2} = \frac{qrv}{2}$$

because $\theta = \pi/2$ for a circular loop. Finally, to show the last relationship, recall that $\mathbf{p} = m\mathbf{v}$, so

$$\boldsymbol{\mu} = \frac{q(\mathbf{r} \times \mathbf{v})}{2} = \frac{qm(\mathbf{r} \times \mathbf{v})}{2m}$$

$$= \frac{q(\mathbf{r} \times \mathbf{p})}{2m} = \frac{q}{2m}\mathbf{L}$$

6–44. In Problem 6–43, we derived an expression for the magnetic moment of a hydrogen atom imparted by the orbital motion of its electron. Using the result that $L^2 = \hbar^2 l(l + 1)$, show that the magnitude of the magnetic moment is

$$\mu = -\beta_e[l(l + 1)]^{1/2}$$

where $\beta_e = \hbar|e|/2m_e$ is called the *Bohr magneton*. What are the units of β_e? What is its numerical value? A magnetic dipole in a magnetic field (**B**) has a potential energy

$$V = -\boldsymbol{\mu} \cdot \mathbf{B}$$

(We will discuss magnetic fields when we study nuclear magnetic resonance, NMR, in Chapter 14.) Show that the units of the intensity of a magnetic field are $J \cdot A^{-1} \cdot m^{-2}$. This set of units is called a *tesla* (T), so that we have $1\ T = 1\ J \cdot A^{-1} \cdot m^{-2}$. In terms of teslas, the units of the Bohr magneton, β_e, are $J \cdot T^{-1}$.

Taking the square root of both sides of the equation $L^2 = \hbar^2 l(l + 1)$ gives $|\mathbf{L}| = \hbar\,[l(l + 1)]^{1/2}$. Recall from Problem 6–43 that

$$\boldsymbol{\mu} = -\frac{|e|}{2m_e}\mathbf{L}$$

Substituting in $|\mathbf{L}| = \hbar\,[l(l + 1)]^{1/2}$ gives

$$|\boldsymbol{\mu}| = -\frac{|e|\hbar}{2m_e}[l(l + 1)]^{1/2} = -\beta_e[l(l + 1)]^{1/2}$$

Because the quantity $[l(l + 1)]^{1/2}$ is unitless, the units of β_e are the same as the units of μ ($C \cdot m^2 \cdot s^{-1}$). The numerical value of β_e is

$$\beta_e = -\frac{(1.602\ 177 \times 10^{-19}\ C)(1.054\ 572 \times 10^{-34}\ J \cdot s)}{2(9.109\ 390 \times 10^{-31}\ kg)}$$

$$= 9.274\ 007 \times 10^{-24}\ C \cdot m^2 \cdot s^{-1} = 9.274\ 007 \times 10^{-24}\ J \cdot T^{-1}$$

The units of **B** are the units of V divided by the units of μ, or

$$\frac{J}{C \cdot m^2 \cdot s^{-1}} = J \cdot A^{-1} \cdot m^{-2} = 1\ T$$

6–45. Using the results of Problems 6–43 and 6–44, show that the Hamiltonian operator for a hydrogen atom in an external magnetic field where the field is in the z direction is given by

$$\hat{H} = \hat{H}_0 + \frac{\beta_e B_z}{\hbar} \hat{L}_z$$

where $\hat{H}_0$ is the Hamiltonian operator of a hydrogen atom in the absence of the magnetic field. Show that the wave functions of the Schrödinger equation for a hydrogen atom in a magnetic field are the same as those for the hydrogen atom in the absence of the field. Finally, show that the energy associated with the wave function ψ_{nlm} is

$$E = E_n^{(0)} + \beta_e B_z m \tag{1}$$

where $E_n^{(0)}$ is the energy in the absence of the magnetic field and m is the magnetic quantum number.

The Hamiltonian operator for a hydrogen atom in an electromagnetic field can be expressed as

$$\hat{H} = \hat{H}_0 + V$$

where $\hat{H}_0$ is the Hamiltonian operator of an isolated hydrogen atom and $V = -\mu \cdot \mathbf{B}$ is the potential energy associated with the external magnetic field. If the external magnetic field is in the z direction, then only the z components of the vectors μ and **B** effect the potential energy of the atom, and we find

$$\hat{H} = \hat{H}_0 - \mu_z B_z = \hat{H}_0 + \frac{|e|}{2m_e} \hat{L}_z B_z$$

$$= \hat{H}_0 + \frac{\beta_e B_z}{\hbar} \hat{L}_z$$

The hydrogen atomic orbitals which satisfy the equation $\hat{H}_0 \psi_n = E_n^0 \psi_n$ are eigenfunctions of $\hat{L}_z$, and so they are also eigenfunctions of $\hat{H}$. Because $\hat{L}_z \psi_{nlm} = m\hbar \psi_{nlm}$,

$$\hat{H} \psi_{nlm} = \hat{H}_0 \psi_{nlm} + \frac{\beta_e B_z}{\hbar} \hat{L}_z \psi_{nlm}$$

$$E \psi_{nlm} = E_n^{(0)} \psi_{nlm} + \beta_e B_z m \psi_{nlm}$$

giving

$$E = E_n^{(0)} + \beta_e B_z m$$

6–46. Equation 1 of Problem 6–45 shows that a state with given values of n and l is split into $2l + 1$ levels by an external magnetic field. For example, Figure 6.8 shows the results for the $1s$ and $2p$ states of atomic hydrogen. The $1s$ state is not split ($2l + 1 = 1$), but the $2p$ state is split into three levels ($2l + 1 = 3$). Figure 6.8 also shows that the $2p$ to $1s$ transition in atomic hydrogen could (see Problem 6–47) be split into three distinct transitions instead of just one. Superconducting magnets have magnetic field strengths of the order of 15 T. Calculate the magnitude of the splitting shown in Figure 6.8 for a magnetic field of 15 T. Compare your result with the energy difference between the unperturbed $1s$ and $2p$ levels. Show that the three distinct transitions shown in Figure 6.8 lie very close together. We say that the $2p$ to $1s$ transition that occurs in the absence of a magnetic field becomes a *triplet* in the presence of the field. The occurrence of such multiplets when atoms are placed in magnetic fields is known as the *Zeeman effect*.

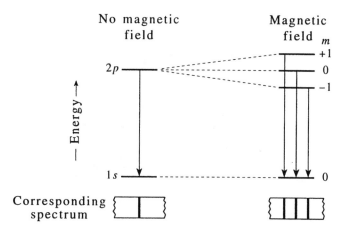

FIGURE 6.8
The splitting of the $2p$ state of the hydrogen atom in a magnetic field. The $2p$ state is split into three closely spaced levels. In a magnetic field, the $2p$ to $1s$ transition is split into three distinct transition frequencies.

In Problem 6–45 we derived the equation $E = E_n^{(0)} + \beta_e B_z m$. Thus

$$\Delta E = E_2 - E_1 = \beta_e B_z (m_2 - m_1)$$

For the $1s$ state $m = 0$ and for the $2p$ state $m = 0, \pm 1$. Thus $(m_2 - m_1) = 0, \pm 1$ and so the magnitude of the splitting shown for a magnetic field of 15 T is either 0 or

$$\Delta E = (9.274 \times 10^{-24} \text{ J·T}^{-1})(15 \text{ T})(1)$$
$$= 1.391 \times 10^{-22} \text{ J}$$

The energy difference between the unperturbed $2p$ and $1s$ energy levels is (Equation 1.11)

$$\Delta \tilde{\nu} = (109\ 737 \text{ cm}^{-1}) \left(\frac{1}{1^2} - \frac{1}{2^2} \right) = 82\ 303 \text{ cm}^{-1}$$
$$E_{2p} - E_{1s} = 1.635 \times 10^{-18} \text{ J}$$

The magnitude of the splitting caused by the magnetic field is on the order of 0.01% of the energy difference between the unperturbed energy levels.

6–47. Consider a transition between the $l = 2$ and the $l = 3$ states of atomic hydrogen. What is the total number of conceivable transitions between these two states in an external magnetic field? For light whose electric field vector is parallel to the direction of the external magnetic field, the selection rule is $\Delta m = 0$. For light whose electric field vector is perpendicular to the direction of the external magnetic field, the selection rule is $\Delta m = \pm 1$. In each case, how many of the possible transitions are allowed?

Recall from Problem 6–46 that a state with given values of n and l is split into $2l + 1$ levels by an external magnetic field. The $l = 2$ state will therefore be split into 5 states ($m = 0, \pm 1, \pm 2$) and the $l = 3$ state will be split into 7 states ($m = 0, \pm 1, \pm 2, \pm 3$), making a total of 35 possible transitions. Using the selection rule $\Delta m = 0$, five transitions are possible (when $m = 0$, $m = 1$, $m = -1$, $m = 2$, or $m = -2$ for both states). Using the selection rule $\Delta m = \pm 1$, the following ten transitions are allowed:

$l = 2$	$\rightarrow$	$l = 3$
$m = 0$		$m = 1$
$m = 0$		$m = -1$
$m = 1$		$m = 2$
$m = -1$		$m = -2$
$m = 1$		$m = 0$
$m = -1$		$m = 0$
$m = 2$		$m = 3$
$m = 2$		$m = 1$
$m = -2$		$m = -3$
$m = -2$		$m = -1$

Problems 6–48 through 6–57 develop the quantum-mechanical properties of angular momentum using operator notation, without solving the Schrödinger equation.

6–48. Define the two (not necessarily Hermitian) operators

$$\hat{L}_+ = \hat{L}_x + i\hat{L}_y \qquad \text{and} \qquad \hat{L}_- = \hat{L}_x - i\hat{L}_y$$

Using the results of Problem 6–13, show that

$$[\hat{L}_z, \hat{L}_+] = \hat{L}_z\hat{L}_+ - \hat{L}_+\hat{L}_z = \hbar\hat{L}_+$$
$$[\hat{L}_z, \hat{L}_-] = \hat{L}_z\hat{L}_- - \hat{L}_-\hat{L}_z = -\hbar\hat{L}_-$$

and

$$[\hat{L}^2, \hat{L}_+] = [\hat{L}^2, \hat{L}_-] = 0$$

In Problem 6–13, we showed that

$$[\hat{L}_x, \hat{L}_y] = ih\hat{L}_z \qquad [\hat{L}_y, \hat{L}_z] = ih\hat{L}_x \qquad [\hat{L}_z, \hat{L}_x] = ih\hat{L}_y$$

Now

$$\left[\hat{L}_z, \hat{L}_+\right] = \hat{L}_z\hat{L}_+ - \hat{L}_+\hat{L}_z$$

$$= \hat{L}_z\left(\hat{L}_x + i\hat{L}_y\right) - \hat{L}_x\hat{L}_z - i\hat{L}_y\hat{L}_z$$

$$= \hat{L}_z\hat{L}_x + i\hat{L}_z\hat{L}_y - \hat{L}_x\hat{L}_z - i\hat{L}_y\hat{L}_z$$

$$= (\hat{L}_z\hat{L}_x - \hat{L}_x\hat{L}_z) - i(\hat{L}_y\hat{L}_z - \hat{L}_z\hat{L}_y)$$

$$= \left[\hat{L}_z, \hat{L}_x\right] - i\left[\hat{L}_y, \hat{L}_z\right]$$

$$= i\hbar\hat{L}_y - i(i\hbar\hat{L}_x) = \hbar(\hat{L}_x + i\hat{L}_y) = \hbar\hat{L}_+$$

$$\left[\hat{L}_z, \hat{L}_-\right] = \hat{L}_z\hat{L}_- - \hat{L}_-\hat{L}_z$$

$$= \hat{L}_z\left(\hat{L}_x - i\hat{L}_y\right) - \hat{L}_x\hat{L}_z + i\hat{L}_y\hat{L}_z$$

$$= \hat{L}_z\hat{L}_x - i\hat{L}_z\hat{L}_y - \hat{L}_x\hat{L}_z + i\hat{L}_y\hat{L}_z$$

$$= (\hat{L}_z\hat{L}_x - \hat{L}_x\hat{L}_z) + i(\hat{L}_y\hat{L}_z - \hat{L}_z\hat{L}_y)$$

$$= \left[\hat{L}_z, \hat{L}_x\right] + i\left[\hat{L}_y, \hat{L}_z\right]$$

$$= i\hbar\hat{L}_y + i(i\hbar\hat{L}_x) = \hbar(-\hat{L}_x + i\hat{L}_y) = -\hbar\hat{L}_-$$

We also showed in Problem 6–13 that $\hat{L}^2$ commutes with $\hat{L}_x$, $\hat{L}_y$, and $\hat{L}_z$. Using this result, we find

$$\left[\hat{L}^2, \hat{L}_+\right] = \hat{L}^2\hat{L}_+ - \hat{L}_+\hat{L}^2$$

$$= \hat{L}^2\left(\hat{L}_x + i\hat{L}_y\right) - \hat{L}_x\hat{L}^2 - i\hat{L}_y\hat{L}^2$$

$$= \hat{L}^2\hat{L}_x + i\hat{L}^2\hat{L}_y - \hat{L}_x\hat{L}^2 - i\hat{L}_y\hat{L}^2$$

$$= (\hat{L}^2\hat{L}_x - \hat{L}_x\hat{L}^2) - i(\hat{L}_y\hat{L}^2 - \hat{L}^2\hat{L}_y)$$

$$= \left[\hat{L}^2, \hat{L}_x\right] - i\left[\hat{L}_y, \hat{L}^2\right] = 0$$

$$\left[\hat{L}^2, \hat{L}_-\right] = \hat{L}^2\hat{L}_- - \hat{L}_-\hat{L}^2$$

$$= \hat{L}^2\left(\hat{L}_x - i\hat{L}_y\right) - \hat{L}_x\hat{L}^2 + i\hat{L}_y\hat{L}^2$$

$$= \hat{L}^2\hat{L}_x - i\hat{L}^2\hat{L}_y - \hat{L}_x\hat{L}^2 + i\hat{L}_y\hat{L}^2$$

$$= (\hat{L}^2\hat{L}_x - \hat{L}_x\hat{L}^2) + i(\hat{L}_y\hat{L}^2 - \hat{L}^2\hat{L}_y)$$

$$= \left[\hat{L}^2, \hat{L}_x\right] + i\left[\hat{L}_y, \hat{L}^2\right] = 0$$

6–49. Show that

$$\hat{L}_-\hat{L}_+ = \hat{L}_x^2 + \hat{L}_y^2 + i\hat{L}_x\hat{L}_y - i\hat{L}_y\hat{L}_x$$

$$= \hat{L}^2 - \hat{L}_z^2 - \hbar\hat{L}_z$$

and

$$\hat{L}_+\hat{L}_- = \hat{L}^2 - \hat{L}_z^2 + \hbar\hat{L}_z$$

$$\hat{L}_-\hat{L}_+ = (\hat{L}_x - i\hat{L}_y)(\hat{L}_x + i\hat{L}y)$$
$$= \hat{L}_x^2 + \hat{L}_y^2 + i\hat{L}_x\hat{L}_y - i\hat{L}_y\hat{L}_x = \hat{L}^2 - \hat{L}_z^2 + i\left[\hat{L}_x, \hat{L}_y\right]$$
$$= \hat{L}^2 - \hat{L}_z^2 - \hbar\hat{L}_z$$
$$\hat{L}_+\hat{L}_- = (\hat{L}_x + i\hat{L}_y)(\hat{L}_x - i\hat{L}y)$$
$$= \hat{L}_x^2 + \hat{L}_y^2 + i\left[\hat{L}_y, \hat{L}_x\right] = \hat{L}^2 - \hat{L}_z^2 + \hbar\hat{L}_z$$

6–50. Because $\hat{L}^2$ and $\hat{L}_z$ commute, they have mutual eigenfunctions. We know from the chapter that these mutual eigenfunctions are the spherical harmonics, $Y_l^m(\theta, \phi)$, but we really don't need that information here. To emphasize this point, let $\psi_{\alpha\beta}$ be the mutual eigenfunctions of $\hat{L}^2$ and $\hat{L}_z$ such that

$$\hat{L}^2\psi_{\alpha\beta} = \beta^2\psi_{\alpha\beta}$$

and

$$\hat{L}_z\psi_{\alpha\beta} = \alpha\psi_{\alpha\beta}$$

Now let

$$\psi_{\alpha\beta}^{+1} = \hat{L}_+\psi_{\alpha\beta}$$

Show that

$$\hat{L}_z\psi_{\alpha\beta}^{+1} = (\alpha + \hbar)\psi_{\alpha\beta}^{+1}$$

and

$$\hat{L}^2\psi_{\alpha\beta}^{+1} = \beta^2\psi_{\alpha\beta}^{+1}$$

Therefore, if α is an eigenvalue of $\hat{L}_z$, then $\alpha + \hbar$ is also an eigenvalue (unless $\psi_{\alpha\beta}^{+1}$ happens to be zero). In the notation for the spherical harmonics that we use in the chapter, $\hat{L}_+Y_l^m(\theta, \phi) \propto Y_l^{m+1}(\theta, \phi)$.

We start with

$$\psi_{\alpha\beta}^{+1} = \hat{L}_+\psi_{\alpha\beta}$$

Operating by $\hat{L}_z$ from the left gives

$$\hat{L}_z\psi_{\alpha\beta}^{+1} = \hat{L}_z\hat{L}_+\psi_{\alpha\beta} = (\hat{L}_z\hat{L}_x + i\hat{L}_z\hat{L}_y)\psi_{\alpha\beta}$$
$$= \left(\left[\hat{L}_z, \hat{L}_x\right] + \hat{L}_x\hat{L}_z + i\left[\hat{L}_z, \hat{L}_y\right] + i\hat{L}_y\hat{L}_z\right)\psi_{\alpha\beta}$$
$$= \left(i\hbar\hat{L}_y + \hat{L}_x\hat{L}_z + \hbar\hat{L}_x + i\hat{L}_y\hat{L}_z\right)\psi_{\alpha\beta}$$
$$= \left(\hat{L}_+\hat{L}_z + \hbar\hat{L}_+\right)\psi_{\alpha\beta} = \hat{L}_+(\alpha + \hbar)\psi_{\alpha\beta}$$
$$= (\alpha + \hbar)\psi_{\alpha\beta}^{+1}$$

Finally,

$$\hat{L}^2 \psi_{\alpha\beta}^{+1} = \hat{L}^2 \hat{L}_+ \psi_{\alpha\beta}$$

$$= \left(\hat{L}^2 \hat{L}_x + i\hat{L}^2 \hat{L}_y\right) \psi_{\alpha\beta}$$

$$= \left(\left[\hat{L}^2, \hat{L}_x\right] + \hat{L}_x \hat{L}^2 + i\left[\hat{L}^2, \hat{L}_y\right] + i\hat{L}_y \hat{L}^2\right) \psi_{\alpha\beta}$$

$$= \left(\hat{L}_x \hat{L}^2 + i\hat{L}_y \hat{L}^2\right) \psi_{\alpha\beta} = \hat{L}_+ \beta^2 \psi_{\alpha\beta} = \beta^2 \psi_{\alpha\beta}^{+1}$$

6–51. Using $\hat{L}_-$ instead of $\hat{L}_+$ in Problem 6–50, show that if α is an eigenvalue of $\hat{L}_z$, then $\alpha - \hbar$ is also an eigenvalue (unless $\psi_{\alpha\beta}^{-1} = \hat{L}_- \psi_{\alpha\beta}$ happens to be zero). In the notation for the spherical harmonics that we use in the chapter, $\hat{L}_- Y_l^m(\theta, \phi) \propto Y_l^{m-1}(\theta, \phi)$.

We solve this problem by using the same approach as in Problem 6–50.

$$\hat{L}_z \psi_{\alpha\beta}^{-1} = \hat{L}_z \hat{L}_- \psi_{\alpha\beta} = \left(\hat{L}_z \hat{L}_x - i\hat{L}_z \hat{L}_y\right) \psi_{\alpha\beta}$$

$$= \left(\left[\hat{L}_z, \hat{L}_x\right] + \hat{L}_x \hat{L}_z - i\left[\hat{L}_z, \hat{L}_y\right] - i\hat{L}_y \hat{L}_z\right) \psi_{\alpha\beta}$$

$$= \left(i\hbar \hat{L}_y + \hat{L}_x \hat{L}_z - \hbar \hat{L}_x - i\hat{L}_y \hat{L}_z\right) \psi_{\alpha\beta}$$

$$= \left(\hat{L}_- \hat{L}_z - \hbar \hat{L}_-\right) \psi_{\alpha\beta} = \hat{L}_- (\alpha - \hbar) \psi_{\alpha\beta} = (\alpha - \hbar) \psi_{\alpha\beta}^{-1}$$

6–52. Show that each application of $\hat{L}_+$ to $\psi_{\alpha\beta}$ raises the eigenvalue by $\hbar$, so long as the result is nonzero.

In Problem 6–50, we showed that if

$$\hat{L}_z \phi_{\alpha\beta} = \alpha \phi_{\alpha\beta}$$

then

$$\hat{L}_z \hat{L}_+ \psi_{\alpha\beta} = (\alpha + \hbar) \psi_{\alpha\beta}^{+1}$$

Now applying $\hat{L}_+$ to $\psi_{\alpha\beta}^{+1}$ and operating with $\hat{L}_z$ gives

$$\hat{L}_z \hat{L}_+ \psi_{\alpha\beta}^{+1} = \hat{L}_+ \left(\hat{L}_z + \hbar\right) \psi_{\alpha\beta}^{+1}$$

where we have used the relation $\hat{L}_z \hat{L}_+ = \hat{L}_+ \hat{L}_z + \hbar \hat{L}_+$. Now $\hat{L}_z \psi_{\alpha\beta}^{+1} = (\alpha + \hbar)\psi_{\alpha\beta}^{+1}$, so

$$\hat{L}_z \hat{L}_+ \psi_{\alpha\beta}^{+1} = \hat{L}_+ (\alpha + 2\hbar) \psi_{\alpha\beta}^{+1} = (\alpha + 2\hbar) \psi_{\alpha\beta}^{+2}$$

which shows that the eigenvalue once again increases by $\hbar$. We can continue this process to show that $\hat{L}_z \psi_{\alpha\beta}^{+2} = (\alpha + 3\hbar)\psi_{\alpha\beta}^{+3}$, and so forth.

6–53. Show that each application of $\hat{L}_-$ to $\psi_{\alpha\beta}$ lowers the eigenvalue by $\hbar$, so long as the result is nonzero.

In Problem 6–51, we showed that if

$$\hat{L}_z \phi_{\alpha\beta} = \alpha \phi_{\alpha\beta}$$

then

$$\hat{L}_z \hat{L}_- \psi_{\alpha\beta} = (\alpha - \hbar)\psi_{\alpha\beta}^{-1}$$

Applying $\hat{L}_-$ to $\psi_{\alpha\beta}^{-1}$ and then operating by $\hat{L}_z$ gives

$$\hat{L}_z \hat{L}_- \psi_{\alpha\beta}^{-1} = \hat{L}_- \left(\hat{L}_z - \hbar\right)\psi_{\alpha\beta}^{-1}$$

where we have used the relation $\hat{L}_z \hat{L}_- = \hat{L}_- \hat{L}_z - \hbar \hat{L}_-$. Now $\hat{L}_z \psi_{\alpha\beta}^{-1} = (\alpha - \hbar)\psi_{\alpha\beta}^{-1}$, so

$$\hat{L}_z \hat{L}_- \psi_{\alpha\beta}^{-1} = \hat{L}_- (\alpha - 2\hbar)\psi_{\alpha\beta}^{-1} = (\alpha - 2\hbar)\psi_{\alpha\beta}^{-2}$$

Just as in Problem 6–52, we can continue this process to show that $\hat{L}_z \psi_{\alpha\beta}^{-2} = (\alpha - 3\hbar)\psi_{\alpha\beta}^{-3}$, and so forth.

6–54. According to Problem 6–48, $\hat{L}^2$ commutes with $\hat{L}_+$ and $\hat{L}_-$. Now prove that $\hat{L}^2$ commutes with $\hat{L}_+^2$ and $\hat{L}_-^2$. Now prove that

$$[\hat{L}^2, \hat{L}_\pm^m] = 0 \qquad m = 1, 2, 3, \ldots$$

$$[\hat{L}^2, \hat{L}_+^2] = \hat{L}^2 \hat{L}_+^2 - \hat{L}_+^2 \hat{L}^2 = \left(\hat{L}^2 \hat{L}_+\right)\left(\hat{L}_+\right) - \hat{L}_+\left(\hat{L}_+ \hat{L}^2\right)$$

Using the fact that $\hat{L}^2$ and $\hat{L}_+$ commute, we can rewrite this as

$$[\hat{L}^2, \hat{L}_+^2] = \hat{L}_+ \hat{L}^2 \hat{L}_+ - \hat{L}_+ \hat{L}^2 \hat{L}_+ = 0$$

We can show that $\hat{L}^2$ commutes with $\hat{L}_-^2$ by replacing $\hat{L}_+$ with $\hat{L}_-$ in the above proof. We can show that the general statement is true using a stepwise approach. For example, consider the case of $m = 3$. We have already shown that $\hat{L}^2$ commutes with $\hat{L}_+^2$ and $\hat{L}_-^2$. Then

$$[\hat{L}^2, \hat{L}_+^3] = \hat{L}^2 \hat{L}_+^3 - \hat{L}_+^3 \hat{L}^2 = \left(\hat{L}^2 \hat{L}_+\right)\left(\hat{L}_+^2\right) - \hat{L}_+^2\left(\hat{L}_+ \hat{L}^2\right)$$

Using the fact that $[\hat{L}^2, \hat{L}_+^2] = 0$, we can rewrite this as

$$[\hat{L}^2, \hat{L}_+^3] = \hat{L}_+\left(\hat{L}^2 \hat{L}_+\right)\hat{L}_+ - \hat{L}_+^2 \hat{L}^2 \hat{L}_+ = \hat{L}_+^2 \hat{L}^2 \hat{L}_+ - \hat{L}_+^2 \hat{L}^2 \hat{L}_+ = 0$$

We can show the general statement is true for any $m = n$ as long as we have proved it for $m = n - 1$. Again, the case with $\hat{L}_-$ is proved in the same way, substituting $\hat{L}_-$ for $\hat{L}_+$.

6–55. In Problems 6–50 through 6–53, we proved that if $\psi_{\alpha\beta}^{\pm m} = \hat{L}_\pm^m \psi_{\alpha\beta}$, then

$$\hat{L}_z \psi_{\alpha\beta}^{\pm m} = (\alpha \pm m\hbar)\psi_{\alpha\beta}^{\pm m} \qquad m = 0, 1, 2, \ldots$$

so long as the result is non-zero. The operators $\hat{L}_\pm$ are called raising ($\hat{L}_+$) or lowering ($\hat{L}_-$) operators because they raise or lower the eigenvalues of $\hat{L}_z$. They are also called ladder operators because the set of eigenvalues $\alpha \pm m\hbar$ form a ladder of eigenvalues. Use the result of Problem 6–54 to show that

$$\hat{L}^2 \psi_{\alpha\beta}^{\pm m} = \beta^2 \psi_{\alpha\beta}^{\pm m}$$

$$\hat{L}^2 \psi_{\alpha\beta}^{\pm m} = \hat{L}^2 \hat{L}_{\pm}^m \psi_{\alpha\beta}$$
$$= \hat{L}_{\pm}^m \hat{L}^2 \psi_{\alpha\beta}$$
$$= \hat{L}_{\pm}^m \beta^2 \psi_{\alpha\beta}$$
$$= \beta^2 \psi_{\alpha\beta}^{\pm m}$$

6–56. Start with

$$\hat{L}_z \psi_{\alpha\beta}^{\pm m} = (\alpha \pm m\hbar) \psi_{\alpha\beta}^{\pm m}$$

Operate on both sides with $\hat{L}_z$ and subtract the result from (Problem 6–55)

$$\hat{L}^2 \psi_{\alpha\beta}^{\pm m} = \beta^2 \psi_{\alpha\beta}^{\pm m}$$

to get

$$(\hat{L}^2 - \hat{L}_z^2) \psi_{\alpha\beta}^{\pm m} = (\hat{L}_x^2 + \hat{L}_y^2) \psi_{\alpha\beta}^{\pm m} = [\beta^2 - (\alpha \pm m\hbar)^2] \psi_{\alpha\beta}^{\pm m}$$

Because the operator $\hat{L}_x^2 + \hat{L}_y^2$ corresponds to a nonnegative physical quantity, show that

$$\beta^2 - (\alpha \pm m\hbar)^2 \geq 0$$

or that

$$-\beta \leq \alpha \pm m\hbar \leq \beta \qquad m = 0, \ 1, \ 2, \ \ldots$$

Because β is fixed, the possible values of m must be finite in number.

The steps outlined in the problem lead easily to the result

$$(\hat{L}^2 - \hat{L}_z^2) \psi_{\alpha\beta}^{\pm m} = (\hat{L}_x^2 + \hat{L}_y^2) \psi_{\alpha\beta}^{\pm m} = [\beta^2 - (\alpha \pm m\hbar)^2] \psi_{\alpha\beta}^{\pm m}$$

Because the operator $\hat{L}_x^2 + \hat{L}_y^2$ corresponds to a nonnegative physical quantity, we know that

$$\int d\tau \left(\psi_{\alpha\beta}^{\pm m} \right)^* \left(\hat{L}_x^2 + \hat{L}_y^2 \right) \psi_{\alpha\beta}^{\pm m} \geq 0$$

$$[\beta^2 - (\alpha \pm m\hbar)^2] \int d\tau \left(\psi_{\alpha\beta}^{\pm m} \right)^* \psi_{\alpha\beta}^{\pm m} \geq 0$$

$$\beta^2 - (\alpha \pm m\hbar)^2 \geq 0$$

Then

$$-\beta \leq \alpha \pm m\hbar \leq \beta \qquad m = 0, \ 1, \ 2, \ \ldots$$

6–57. Let α_{max} be the largest possible value of $\alpha \pm m\hbar$. By definition then, we have that

$$\hat{L}_z \psi_{\alpha_{max}\beta} = \alpha_{max} \psi_{\alpha_{max}\beta}$$

$$\hat{L}^2 \psi_{\alpha_{max}\beta} = \beta^2 \psi_{\alpha_{max}\beta}$$

and

$$\hat{L}_{+}\psi_{\alpha_{max}\beta} = 0$$

Operate on the last equation with $\hat{L}_{-}$ to obtain

$$\hat{L}_{-}\hat{L}_{+}\psi_{\alpha_{max}\beta} = 0$$
$$= (\hat{L}^2 - \hat{L}_z^2 - \hbar\hat{L}_z)\psi_{\alpha_{max}\beta}$$

and

$$\beta^2 = \alpha_{max}^2 + \hbar\alpha_{max}$$

Use a parallel procedure on $\psi_{\alpha_{min}\beta}$ to obtain

$$\beta^2 = \alpha_{min}^2 - \hbar\alpha_{min}$$

Now show that $\alpha_{max} = -\alpha_{min}$, and then argue that the possible values of the eigenvalues α of $\hat{L}_z$ extend from $+\alpha_{max}$ to $-\alpha_{max}$ in steps of magnitude $\hbar$. This is possible only if α_{max} is itself an integer (or perhaps a half-integer) times $\hbar$. Finally, show that this last result leads to

$$\beta^2 = l(l+1)\hbar^2 \qquad l = 0, \ 1, \ 2, \ \ldots$$

and

$$\alpha = m\hbar \qquad m = 0, \ \pm 1, \ \pm 2, \ \ldots, \ \pm l$$

Recall from Problem 6–49 that $\hat{L}_{-}\hat{L}_{+} = \hat{L}^2 - \hat{L}_z^2 - \hbar\hat{L}_z$. Then

$$\hat{L}_{-}\hat{L}_{+}\psi_{\alpha_{max}\beta} = \hat{L}_{-}(0) = 0 = (\hat{L}_{-}\hat{L}_{+})\psi_{\alpha_{max}\beta}$$
$$= \left(\hat{L}^2 - \hat{L}_z^2 - \hbar\hat{L}_z\right)\psi_{\alpha_{max}\beta}$$

Using our definitions in the beginning of this problem, we can evaluate the result for the operation of each operator on $\psi_{\alpha_{max}\beta}$ to get

$$0 = [\beta^2 - \alpha_{max}^2 - \hbar\alpha_{max}]\psi_{\alpha_{max}\beta}$$

This result implies that

$$\beta^2 - (\alpha_{max}^2 + \hbar\alpha_{max}) = 0$$

or, equivalently,

$$\beta^2 = \alpha_{max}^2 + \hbar\alpha_{max} \tag{1}$$

Similarly, for α_{min}, recall that $\hat{L}_{+}\hat{L}_{-} = \hat{L}^2 - \hat{L}_z^2 + \hbar\hat{L}_z$. Then

$$\hat{L}_{+}\hat{L}_{-}\psi_{\alpha_{min}\beta} = \hat{L}_{+}(0) = 0 = (\hat{L}_{+}\hat{L}_{-})\psi_{\alpha_{min}\beta}$$
$$= \left(\hat{L}^2 - \hat{L}_z^2 + \hbar\hat{L}_z\right)\psi_{\alpha_{min}\beta}$$

Using the definitions at the beginning of this problem (replacing α_{max} with α_{min}), we can substitute the result in for each operation and write

$$0 = [\beta^2 - \alpha_{min}^2 - \hbar\alpha_{min}]\psi_{\alpha_{min}\beta}$$

This result implies that

$$\beta^2 - (\alpha_{min}^2 - \hbar\alpha_{min}) = 0$$

or, equivalently,

$$\beta^2 = \alpha_{min}^2 - \hbar\alpha_{min}$$

Now we have two equations for β^2, one in terms of α_{max} and one in terms of α_{min}. Setting these equal to one another, we find

$$\alpha_{max}^2 + \hbar\alpha_{max} = \alpha_{min}^2 - \hbar\alpha_{min}$$

$$\alpha_{max} = \frac{-\hbar \pm \sqrt{\hbar^2 - 4\left(\hbar\alpha_{min} - \alpha_{min}^2\right)}}{2}$$

$$= -\frac{\hbar}{2} \pm \frac{2\alpha_{min} - \hbar}{2}$$

$$\alpha_{max} = -\alpha_{min} \quad \text{or} \quad \alpha_{min} - \hbar$$

However, α_{min} is the minimum value of α, so $\alpha_{max} \neq \alpha_{min} - \hbar$ and we can conclude that $\alpha_{max} = -\alpha_{min}$. According to Problem 6–55, α varies in integral steps of $\hbar$, so α must vary from α_{min} to α_{max} in steps of $\hbar$. This is only true if α_{max} is an integer or half-integer times $\hbar$. We call this multiplicative constant l and so

$$\alpha_{max} = l\hbar \qquad \alpha_{min} = -l\hbar$$

Because α varies in integral steps of $\hbar$, we can describe any value of α by

$$\alpha = m\hbar \qquad m = 0, \pm 1, \pm 2, \ldots, \pm l$$

(If $\alpha = 0$, then l must be an integer.) We can also write (from Equation 1)

$$\beta^2 = (l\hbar)^2 + \hbar(l\hbar) = \hbar^2(l^2 + l) = l(l+1)\hbar^2$$

where l is, of course, an integer.

6–58. According to Problems 6–50 and 6–51,

$$\hat{L}_+ Y_l^m(\theta, \phi) = \hbar c_{lm}^+ Y_l^{m+1}(\theta, \phi)$$

and

$$\hat{L}_- Y_l^m(\theta, \phi) = \hbar c_{lm}^- Y_l^{m-1}(\theta, \phi)$$

where we are using the notation $Y_l^m(\theta, \phi)$ instead of $\psi_{\alpha,\beta}$. Show that

$$\hat{L}_x Y_l^m(\theta, \phi) = \frac{\hbar c_{lm}^+}{2} Y_l^{m+1}(\theta, \phi) + \frac{\hbar c_{lm}^-}{2} Y_l^{m-1}(\theta, \phi)$$

and

$$\hat{L}_y Y_l^m(\theta, \phi) = \frac{\hbar c_{lm}^+}{2i} Y_l^{m+1}(\theta, \phi) - \frac{\hbar c_{lm}^-}{2i} Y_l^{m-1}(\theta, \phi)$$

Use this result to show that

$$\langle L_x \rangle = \langle L_y \rangle = 0$$

for any rotational state (see Problem 6–14).

Add the first two equations given in the problem to obtain

$$(\hat{L}_+ + \hat{L}_-)Y_l^m(\theta, \phi) = \hbar c_{lm}^+ Y_l^{m+1}(\theta, \phi) + \hbar c_{lm}^- Y_l^{m-1}(\theta, \phi)$$

$$2\hat{L}_x Y_l^m(\theta, \phi) = \hbar c_{lm}^+ Y_l^{m+1}(\theta, \phi) + \hbar c_{lm}^- Y_l^{m-1}(\theta, \phi)$$

$$\hat{L}_x Y_l^m(\theta, \phi) = \frac{\hbar c_{lm}^+}{2} Y_l^{m+1}(\theta, \phi) + \frac{\hbar c_{lm}^-}{2} Y_l^{m-1}(\theta, \phi) \qquad (1)$$

Subtract the first two equations given in the problem to obtain

$$(\hat{L}_+ - \hat{L}_-)Y_l^m(\theta, \phi) = \hbar c_{lm}^+ Y_l^{m+1}(\theta, \phi) - \hbar c_{lm}^- Y_l^{m-1}(\theta, \phi)$$

$$2i\hat{L}_y Y_l^m(\theta, \phi) = \hbar c_{lm}^+ Y_l^{m+1}(\theta, \phi) + \hbar c_{lm}^- Y_l^{m-1}(\theta, \phi)$$

$$\hat{L}_y Y_l^m(\theta, \phi) = \frac{\hbar c_{lm}^+}{2i} Y_l^{m+1}(\theta, \phi) - \frac{\hbar c_{lm}^-}{2i} Y_l^{m-1}(\theta, \phi) \qquad (2)$$

We will now use these results to show that $\langle L_x \rangle = \langle L_y \rangle = 0$ for any rotational state.

$$\langle L_x \rangle = \int Y_l^m(\theta, \phi)^* \hat{L}_x Y_l^m(\theta, \phi) \sin\theta d\theta d\phi$$

Substituting in from Equation 1 gives

$$\langle L_x \rangle = \frac{\hbar c_{lm}^+}{2} \int Y_l^m(\theta, \phi)^* Y_l^{m+1}(\theta, \phi) \sin\theta d\theta d\phi$$

$$+ \frac{\hbar c_{lm}^-}{2} \int Y_l^m(\theta, \phi)^* Y_l^{m-1}(\theta, \phi) \sin\theta d\theta d\phi$$

$$= 0$$

because the functions $Y_l^m(\theta, \phi)$ are orthogonal. Likewise,

$$\langle L_y \rangle = \int Y_l^m(\theta, \phi)^* \hat{L}_y Y_l^m(\theta, \phi) \sin\theta d\theta d\phi$$

Substituting in from Equation 2 gives

$$\langle L_y \rangle = \frac{\hbar c_{lm}^+}{2i} \int Y_l^m(\theta, \phi)^* Y_l^{m+1}(\theta, \phi) \sin\theta d\theta d\phi$$

$$- \frac{\hbar c_{lm}^-}{2i} \int Y_l^m(\theta, \phi)^* Y_l^{m-1}(\theta, \phi) \sin\theta d\theta d\phi$$

$$= 0$$

because the functions $Y_l^m(\theta, \phi)$ are orthogonal.

6–59. Show that

$$\hat{L}_+ = \hbar e^{i\phi} \left[\frac{\partial}{\partial \theta} + i \cot\theta \frac{\partial}{\partial \phi} \right]$$

and

$$\hat{L}_- = \hbar e^{-i\phi} \left[-\frac{\partial}{\partial \theta} + i \cot\theta \frac{\partial}{\partial \phi} \right]$$

From Equations 6.37 and the definitions of $\hat{L}_+$ and $\hat{L}_-$ (Problem 6–48),

$$\hat{L}_+ = \hat{L}_x + i\hat{L}_y$$

$$= -i\hbar\left(-\sin\phi\frac{\partial}{\partial\theta} - \cot\theta\cos\phi\frac{\partial}{\partial\phi}\right) + \hbar\left(\cos\phi\frac{\partial}{\partial\theta} - \cot\theta\sin\phi\frac{\partial}{\partial\phi}\right)$$

$$= \hbar\left[(\cos\phi + i\sin\phi)\frac{\partial}{\partial\theta} + i(\cos\phi + i\sin\phi)\cot\theta\frac{\partial}{\partial\phi}\right]$$

$$= \hbar e^{i\phi}\left[\frac{\partial}{\partial\theta} + i\cot\theta\frac{\partial}{\partial\phi}\right]$$

$$\hat{L}_- = \hat{L}_x - i\hat{L}_y$$

$$= -i\hbar\left(-\sin\phi\frac{\partial}{\partial\theta} - \cot\theta\cos\phi\frac{\partial}{\partial\phi}\right) - \hbar\left(\cos\phi\frac{\partial}{\partial\theta} - \cot\theta\sin\phi\frac{\partial}{\partial\phi}\right)$$

$$= \hbar\left[-(\cos\phi - i\sin\phi)\frac{\partial}{\partial\theta} + i(\cos\phi - i\sin\phi)\cot\theta\frac{\partial}{\partial\phi}\right]$$

$$= \hbar e^{-i\phi}\left[-\frac{\partial}{\partial\theta} + i\cot\theta\frac{\partial}{\partial\phi}\right]$$

Determinants

PROBLEMS AND SOLUTIONS

E–1. Evaluate the determinant

$$D = \begin{vmatrix} 2 & 1 & 1 \\ -1 & 3 & 2 \\ 2 & 0 & 1 \end{vmatrix}$$

Add column 2 to column 1 to get

$$\begin{vmatrix} 3 & 1 & 1 \\ 2 & 3 & 2 \\ 2 & 0 & 1 \end{vmatrix}$$

and evaluate it. Compare your result with the value of D. Now add row 2 of D to row 1 of D to get

$$\begin{vmatrix} 1 & 4 & 3 \\ -1 & 3 & 2 \\ 2 & 0 & 1 \end{vmatrix}$$

and evaluate it. Compare your result with the value of D above.

$$D = \begin{vmatrix} 2 & 1 & 1 \\ -1 & 3 & 2 \\ 2 & 0 & 1 \end{vmatrix} = 2 \begin{vmatrix} 3 & 2 \\ 0 & 1 \end{vmatrix} - \begin{vmatrix} -1 & 2 \\ 2 & 1 \end{vmatrix} + \begin{vmatrix} -1 & 3 \\ 2 & 0 \end{vmatrix} = 6 + 5 - 6 = 5$$

Adding column 1 to column 2, we find

$$D = \begin{vmatrix} 3 & 1 & 1 \\ 2 & 3 & 2 \\ 2 & 0 & 1 \end{vmatrix} = 3 \begin{vmatrix} 3 & 2 \\ 0 & 1 \end{vmatrix} - \begin{vmatrix} 2 & 2 \\ 2 & 1 \end{vmatrix} + \begin{vmatrix} 2 & 3 \\ 2 & 0 \end{vmatrix} = 9 + 2 - 6 = 5$$

which is the same result as the value of the original determinant. Adding row 1 to row 2 and evaluating the resulting determinant, we find

$$D = \begin{vmatrix} 1 & 4 & 3 \\ -1 & 3 & 2 \\ 2 & 0 & 1 \end{vmatrix} = \begin{vmatrix} 3 & 2 \\ 0 & 1 \end{vmatrix} - 4 \begin{vmatrix} -1 & 2 \\ 2 & 1 \end{vmatrix} + 3 \begin{vmatrix} -1 & 3 \\ 2 & 0 \end{vmatrix} = 3 + 20 - 18 = 5$$

Again, the value of this determinant is the same as that of the original determinant (in accordance with Rule 6).

E–2. Interchange columns 1 and 3 in D in Problem E–1 and evaluate the resulting determinant. Compare your result with the value of D. Interchange rows 1 and 2 of D and do the same.

Interchanging columns 1 and 3 gives

$$D = \begin{vmatrix} 1 & 1 & 2 \\ 2 & 3 & -1 \\ 1 & 0 & 2 \end{vmatrix} = \begin{vmatrix} 3 & -1 \\ 0 & 2 \end{vmatrix} - \begin{vmatrix} 2 & -1 \\ 1 & 2 \end{vmatrix} + 2 \begin{vmatrix} 2 & 3 \\ 1 & 0 \end{vmatrix} = 6 - 5 - 6 = -5$$

Likewise, interchanging rows 1 and 2 gives

$$D = \begin{vmatrix} -1 & 3 & 2 \\ 2 & 1 & 1 \\ 2 & 0 & 1 \end{vmatrix} = - \begin{vmatrix} 1 & 1 \\ 0 & 1 \end{vmatrix} - 3 \begin{vmatrix} 2 & 1 \\ 2 & 1 \end{vmatrix} + 2 \begin{vmatrix} 2 & 1 \\ 2 & 0 \end{vmatrix} = -1 - 4 = -5$$

The value of the determinant changes its sign when two rows or columns are interchanged, in accordance with Rule 3.

E–3. Evaluate the determinant

$$D = \begin{vmatrix} 1 & 6 & 1 \\ -2 & 4 & -2 \\ 1 & -3 & 1 \end{vmatrix}$$

Can you determine its value by inspection? What about

$$D = \begin{vmatrix} 2 & 6 & 1 \\ -4 & 4 & -2 \\ 2 & -3 & 1 \end{vmatrix}$$

For the first determinant, inspection shows that the first column is identical to the third column, so (by Rule 2) the value of this determinant is 0. For the second determinant, inspection reveals that the first column is twice the second column. This means (Rules 2 and 6) that the value of the second determinant is zero.

E–4. Find the values of x that satisfy the following determinantal equation

$$\begin{vmatrix} x & 1 & 1 & 1 \\ 1 & x & 0 & 0 \\ 1 & 0 & x & 0 \\ 1 & 0 & 0 & x \end{vmatrix} = 0$$

Expanding the determinant along its first row gives

$$x \begin{vmatrix} x & 0 & 0 \\ 0 & x & 0 \\ 0 & 0 & x \end{vmatrix} - \begin{vmatrix} 1 & 0 & 0 \\ 1 & x & 0 \\ 1 & 0 & x \end{vmatrix} + \begin{vmatrix} 1 & x & 0 \\ 1 & 0 & 0 \\ 1 & 0 & x \end{vmatrix} - \begin{vmatrix} 1 & x & 0 \\ 1 & 0 & x \\ 1 & 0 & 0 \end{vmatrix} = 0$$

Expanding each 3×3 determinant along its first row gives

$$x^2 \begin{vmatrix} x & 0 \\ 0 & x \end{vmatrix} - \begin{vmatrix} x & 0 \\ 0 & x \end{vmatrix} + \begin{vmatrix} 0 & 0 \\ 0 & x \end{vmatrix} - x \begin{vmatrix} 1 & 0 \\ 1 & x \end{vmatrix} - \begin{vmatrix} 0 & x \\ 0 & 0 \end{vmatrix} + x \begin{vmatrix} 1 & x \\ 1 & 0 \end{vmatrix} = 0$$

or

$$x^4 - x^2 + 0 - x^2 - 0 - x^2 = 0$$
$$x^4 - 3x^2 = 0$$
$$x^2(x^2 - 3) = 0$$
$$x = \pm\sqrt{3}, 0$$

The four solutions to the determinantal equation are $x = 0, 0$, and $\pm\sqrt{3}$.

E–5. Find the values of x that satisfy the following determinantal equation

$$\begin{vmatrix} x & 1 & 0 & 1 \\ 1 & x & 1 & 0 \\ 0 & 1 & x & 1 \\ 1 & 0 & 1 & x \end{vmatrix} = 0$$

Expanding the determinant along its first row gives

$$x\begin{vmatrix} x & 1 & 0 \\ 1 & x & 1 \\ 0 & 1 & x \end{vmatrix} - \begin{vmatrix} 1 & 1 & 0 \\ 0 & x & 1 \\ 1 & 1 & x \end{vmatrix} - \begin{vmatrix} 1 & x & 1 \\ 0 & 1 & x \\ 1 & 0 & 1 \end{vmatrix} = 0$$

Expanding each 3×3 determinant along its first row gives

$$x^2\begin{vmatrix} x & 1 \\ 1 & x \end{vmatrix} - x\begin{vmatrix} 1 & 1 \\ 0 & x \end{vmatrix} - \begin{vmatrix} x & 1 \\ 1 & x \end{vmatrix} + \begin{vmatrix} 0 & 1 \\ 1 & x \end{vmatrix} - \begin{vmatrix} 1 & x \\ 0 & 1 \end{vmatrix} + x\begin{vmatrix} 0 & x \\ 1 & 1 \end{vmatrix} - \begin{vmatrix} 0 & 1 \\ 1 & 0 \end{vmatrix} = 0$$

or

$$x^4 - x^2 - x^2 - x^2 + 1 - 1 - 1 - x^2 + 1 = 0$$
$$x^2(x + 2)(x - 2) = 0$$
$$x = \pm 2, 0$$

The four solutions to the determinantal equation are $x = 0, 0$, and ± 2.

E–6. Show that

$$\begin{vmatrix} \cos\theta & -\sin\theta & 0 \\ \sin\theta & \cos\theta & 0 \\ 0 & 0 & 1 \end{vmatrix} = 1$$

Expand this determinant about the third column:

$$\begin{vmatrix} \cos\theta & -\sin\theta & 0 \\ \sin\theta & \cos\theta & 0 \\ 0 & 0 & 1 \end{vmatrix} = \begin{vmatrix} \cos\theta & -\sin\theta \\ \sin\theta & \cos\theta \end{vmatrix} = \cos^2\theta + \sin^2\theta = 1$$

E–7. Solve the following set of equations using Cramer's rule

$$x + y = 2$$
$$3x - 2y = 5$$

Using Cramer's rule (Equations E.9 and E.10),

$$x = \frac{\begin{vmatrix} 2 & 1 \\ 5 & -2 \end{vmatrix}}{\begin{vmatrix} 1 & 1 \\ 3 & -2 \end{vmatrix}} = \frac{-9}{-5} = \frac{9}{5}$$

$$y = \frac{\begin{vmatrix} 1 & 2 \\ 3 & 5 \end{vmatrix}}{\begin{vmatrix} 1 & 1 \\ 3 & -2 \end{vmatrix}} = \frac{-1}{-5} = \frac{1}{5}$$

The solution to this set of the equations is $(\frac{9}{5}, \frac{1}{5})$.

E–8. Solve the following set of equations using Cramer's rule

$$x + 2y + 3z = -5$$
$$-x - 3y + z = -14$$
$$2x + y + z = 1$$

Following Example E–4,

$$x = \frac{\begin{vmatrix} -5 & 2 & 3 \\ -14 & -3 & 1 \\ 1 & 1 & 1 \end{vmatrix}}{\begin{vmatrix} 1 & 2 & 3 \\ -1 & -3 & 1 \\ 2 & 1 & 1 \end{vmatrix}} = \frac{-5\begin{vmatrix} -3 & 1 \\ 1 & 1 \end{vmatrix} - 2\begin{vmatrix} -14 & 1 \\ 1 & 1 \end{vmatrix} + 3\begin{vmatrix} -14 & -3 \\ 1 & 1 \end{vmatrix}}{\begin{vmatrix} -3 & 1 \\ 1 & 1 \end{vmatrix} - 2\begin{vmatrix} -1 & 1 \\ 2 & 1 \end{vmatrix} + 3\begin{vmatrix} -1 & -3 \\ 2 & 1 \end{vmatrix}}$$

$$x = \frac{(-5)(-4) - 2(-15) + 3(-11)}{(-4) - 2(-3) + 3(5)} = \frac{20 + 30 - 33}{-4 + 6 + 15} = \frac{17}{17} = 1$$

$$y = \frac{\begin{vmatrix} 1 & -5 & 3 \\ -1 & -14 & 1 \\ 2 & 1 & 1 \end{vmatrix}}{17} = \frac{\begin{vmatrix} -14 & 1 \\ 1 & 1 \end{vmatrix} + 5\begin{vmatrix} -1 & 1 \\ 2 & 1 \end{vmatrix} + 3\begin{vmatrix} -1 & -14 \\ 2 & 1 \end{vmatrix}}{17}$$

$$y = \frac{-15 - 15 + 3(27)}{17} = \frac{51}{17} = 3$$

$$z = \frac{\begin{vmatrix} 1 & 2 & -5 \\ -1 & -3 & -14 \\ 2 & 1 & 1 \end{vmatrix}}{17} = \frac{\begin{vmatrix} -3 & -14 \\ 1 & 1 \end{vmatrix} - 2\begin{vmatrix} -1 & -14 \\ 2 & 1 \end{vmatrix} + 5\begin{vmatrix} -1 & -3 \\ 2 & 1 \end{vmatrix}}{17}$$

$$z = \frac{11 - 2(27) - 25}{17} = \frac{-68}{17} = -4$$

The solution to this set of the equations is $(1, 3, -4)$.

CHAPTER 7

Approximation Methods

PROBLEMS AND SOLUTIONS

7–1. This problem involves the proof of the variational principle, Equation 7.4. Let $\hat{H}\psi_n = E_n\psi_n$ be the problem of interest, and let ϕ be our approximation to ψ_0. Even though we do not know the ψ_n, we can express ϕ formally as

$$\phi = \sum_n c_n \psi_n \tag{1}$$

where the c_n are constants. Using the fact that the ψ_n are orthonormal, show that

$$c_n = \int \psi_n^* \phi \, d\tau$$

We do not know the ψ_n, however, so Equation 1 is what we call a formal expansion. Now substitute Equation 1 into

$$E_\phi = \frac{\int \phi^* \hat{H} \phi \, d\tau}{\int \phi^* \phi \, d\tau}$$

to obtain

$$E_\phi = \frac{\sum_n c_n^* c_n E_n}{\sum_n c_n^* c_n}$$

Subtract E_0 from the left side of the above equation and $E_0 \sum_n c_n^* c_n / \sum_n c_n^* c_n$ from the right side to obtain

$$E_\phi - E_0 = \frac{\sum_n c_n^* c_n (E_n - E_0)}{\sum_n c_n^* c_n}$$

Now explain why every term on the right side is positive, proving that $E_\phi \geq E_0$.

By our formal definition,

$$\int \psi_m^* \phi \, d\tau = \int \psi_m^* \sum_n c_n \psi_n \, d\tau = \sum_n c_n \int \psi_m^* \psi_n \, d\tau = \sum_n c_n \delta_{nm} = c_m$$

Now find E_ϕ in terms of c_n:

$$E_\phi = \frac{\int \phi^* \hat{H} \phi \, d\tau}{\int \phi^* \phi \, d\tau}$$

$$\int \phi^* \hat{H} \phi \, d\tau = \sum_n \sum_m c_n^* c_m \int \psi_n^* \hat{H} \psi_m \, d\tau$$

$$= \sum_n \sum_m c_n^* c_m E_m \int \psi_n^* \psi_m \, d\tau$$

$$= \sum_n \sum_m c_n^* c_m E_m \delta_{nm} = \sum_n c_n^* c_n E_n$$

$$\int \phi^* \phi \, d\tau = \sum_n \sum_m c_n^* c_m \int \psi_n^* \psi_m \, d\tau$$

$$= \sum_n \sum_m c_n^* c_m \delta_{nm} = \sum_n c_n^* c_n$$

$$E_\phi = \frac{\sum_n c_n^* c_n E_n}{\sum_n c_n^* c_n}$$

Now subtract E_0 from each side:

$$E_\phi - E_0 = \frac{\sum_n c_n^* c_n (E_n - E_0)}{\sum_n c_n^* c_n}$$

By definition $E_n \geq E_0$, so the right-hand side of the equation is positive and $E_\phi \geq E_0$.

7–2. Using a Gaussian trial function $e^{-\alpha r^2}$ for the ground state of the hydrogen atom (see Equation 7.6 for $\hat{H}$), show that the ground-state energy is given by

$$E(\alpha) = \frac{3\hbar^2 \alpha}{2m_e} - \frac{e^2 \alpha^{1/2}}{2^{1/2} \varepsilon_0 \pi^{3/2}}$$

and that

$$E_{min} = -\frac{4}{3\pi} \frac{m_e e^4}{16\pi^2 \varepsilon_0^2 \hbar^2}$$

The Hamiltonian operator for the hydrogen atom is

$$\hat{H} = -\frac{\hbar^2}{2m_e r^2} \frac{d}{dr} \left(r^2 \frac{d}{dr} \right) - \frac{e^2}{4\pi \varepsilon_0 r} \tag{7.6}$$

The variational principle expresses E_ϕ as

$$E_\phi = \frac{\int \phi^* \hat{H} \phi \, d\tau}{\int \phi^* \phi \, d\tau} \tag{7.3}$$

Where ϕ is the trial wavefunction. In this problem $\phi = e^{-\alpha r^2}$, so the numerator of this equation is

$$\int \phi^* \hat{H} \phi \, d\tau = \int_0^\infty 4\pi r^2 \, dr \, e^{-\alpha r^2} \left[-\frac{\hbar^2}{2m_e r^2} \frac{d}{dr} \left(r^2 \frac{d}{dr} \right) - \frac{e^2}{4\pi \varepsilon_0 r} \right] e^{-\alpha r^2}$$

$$= -\frac{4\pi \hbar^2}{2m_e} \int_0^\infty dr \left(4\alpha^2 r^4 - 6\alpha r^2 \right) e^{-2\alpha r^2} - \frac{e^2}{\varepsilon_0} \int_0^\infty dr \, r e^{-2\alpha r^2}$$

$$= -\frac{4\pi \hbar^2}{2m_e} \left[-\frac{3}{8} \left(\frac{\pi}{2\alpha} \right)^{1/2} \right] - \frac{e^2}{\varepsilon_0} \left(\frac{1}{4\alpha} \right)$$

and the denominator is

$$\int \phi^* \phi d\tau = \int_0^\infty 4\pi r^2 dr e^{-2\alpha r^2}$$

$$= \frac{4\pi}{8\alpha} \left(\frac{\pi}{2\alpha} \right)^{1/2}$$

We can now write

$$E_\phi = \frac{-\dfrac{\hbar^2}{2m_e} \left[-\dfrac{3}{8} \left(\dfrac{\pi}{2\alpha} \right)^{1/2} \right] - \dfrac{e^2}{4\pi\varepsilon_0} \dfrac{1}{4\alpha}}{\dfrac{1}{8\alpha} \left(\dfrac{\pi}{2\alpha} \right)^{1/2}}$$

$$= \frac{3\hbar^2 \alpha}{2m_e} - \frac{e^2 \alpha^{1/2}}{2^{1/2} \varepsilon_0 \pi^{3/2}}$$

The minimum value of E_ϕ occurs when

$$\frac{dE_\phi}{d\alpha} = \frac{3\hbar^2}{2m_e} - \frac{e^2}{(2\pi)^{3/2}\varepsilon_0 \alpha_{\min}^{1/2}} = 0$$

or

$$\alpha_{\min} = \frac{m_e^2 e^4}{18\pi^3 \varepsilon_0^2 \hbar^4}$$

The value $E_{\min}$ is thus

$$E_{\min} = E(\alpha_{\min}) = -\frac{m_e e^4}{12\pi^3 \varepsilon_0^2 \hbar^2} = -\frac{4}{3\pi} \left(\frac{m_e e^4}{16\pi^2 \varepsilon_0^2 \hbar^2} \right)$$

7–3. Use a trial function $\phi(x) = 1/(1 + \beta x^2)^2$ to calculate the ground-state energy of a harmonic oscillator variationally. The necessary integrals are

$$\int_{-\infty}^\infty \frac{dx}{(1 + \beta x^2)^n} = \frac{(2n-3)(2n-5)(2n-7)\cdots(1)}{(2n-2)(2n-4)(2n-6)\cdots(2)} \frac{\pi}{\beta^{1/2}}$$

and

$$\int_{-\infty}^\infty \frac{x^2 dx}{(1 + \beta x^2)^n} = \frac{(2n-5)(2n-7)\cdots(1)}{(2n-2)(2n-4)\cdots(2)} \frac{\pi}{\beta^{3/2}}$$

The Hamiltonian operator for a harmonic oscillator is expressed as

$$\hat{H} = -\frac{\hbar^2}{2\mu} \frac{d^2}{dx^2} + \frac{kx^2}{2}$$

We now determine E_ϕ (Equation 7.3) using the trial function $\phi = 1/(1 + \beta x^2)^2$. The numerator is

$$\int_{-\infty}^\infty \phi^* \hat{H} \phi dx = \int_{-\infty}^\infty \frac{1}{(1 + \beta x^2)^2} \left[\frac{\hbar^2}{\mu} \frac{2\beta}{(1 + \beta x^2)^3} - \frac{\hbar^2}{\mu} \frac{12\beta^2 x^2}{(1 + \beta x^2)^4} + \frac{kx^2}{2(1 + \beta x^2)^2} \right] dx$$

$$= \frac{2\beta\hbar^2}{\mu} \left(\frac{7 \cdot 5 \cdot 3 \cdot 1 \cdot \pi}{8 \cdot 6 \cdot 4 \cdot 2 \cdot \beta^{1/2}} \right) - \frac{12\beta^2 \hbar^2}{\mu} \left(\frac{7 \cdot 5 \cdot 3 \cdot 1 \cdot \pi}{10 \cdot 8 \cdot 6 \cdot 4 \cdot 2 \cdot \beta^{3/2}} \right) + \frac{k}{2} \left(\frac{3 \cdot 1 \cdot \pi}{6 \cdot 4 \cdot 2 \cdot \beta^{3/2}} \right)$$

$$= \frac{7\pi\beta^{1/2}\hbar^2}{32\mu} + \frac{k\pi}{32\beta^{3/2}}$$

and the denominator is

$$\int_{-\infty}^{\infty} \phi^* \phi \, dx = \int_{-\infty}^{\infty} \frac{dx}{(1 + \beta x^2)^4} = \frac{5 \cdot 3 \cdot 1 \cdot \pi}{6 \cdot 4 \cdot 2 \cdot \beta^{1/2}} = \frac{5\pi}{16\beta^{1/2}}$$

We then find

$$\begin{aligned}
E_\phi &= \frac{\int_{-\infty}^{\infty} \phi^* \hat{H} \phi \, dx}{\int_{-\infty}^{\infty} \phi^* \phi \, dx} \\
&= \frac{7\pi \beta^{1/2} \hbar^2}{32\mu} \left(\frac{16\beta^{1/2}}{5\pi} \right) + \frac{k\pi}{32\beta^{3/2}} \left(\frac{16\beta^{1/2}}{5\pi} \right) = \frac{7}{10} \frac{\beta \hbar^2}{\mu} + \frac{1}{10} \frac{k}{\beta}
\end{aligned}$$

To find the minimum value of E_ϕ, we use

$$\frac{dE_\phi}{d\beta} = \frac{7\hbar^2}{10\mu} - \frac{k}{10\beta_{\min}^2} = 0$$

$$\beta_{\min} = \left(\frac{\mu k}{7\hbar^2} \right)^{1/2}$$

(We retain only the positive solution because $E_\phi \geq 0$.) Substituting this value back into the equation for E_ϕ gives

$$\begin{aligned}
E_{\min} &= \frac{7^{1/2}\hbar}{10} \left(\frac{k}{\mu} \right)^{1/2} + \frac{7^{1/2}\hbar}{10} \left(\frac{k}{\mu} \right)^{1/2} \\
&= \frac{7^{1/2}}{5}\hbar \left(\frac{k}{\mu} \right)^{1/2} = 0.529 \, \hbar \left(\frac{k}{\mu} \right)^{1/2}
\end{aligned}$$

compared to

$$E_{\text{exact}} = 0.500 \, \hbar \left(\frac{k}{\mu} \right)^{1/2}$$

The approximate value of E_ϕ differs from the exact value by about 6%.

7–4. If you were to use a trial function of the form $\phi(x) = (1 + c\alpha x^2)e^{-\alpha x^2/2}$, where $\alpha = (k\mu/\hbar^2)^{1/2}$ and c is a variational parameter to calculate the ground-state energy of a harmonic oscillator, what do you think the value of c will come out to be? Why?

The value of c will equal zero because $\phi(x) = e^{-\alpha x^2/2}$ is the exact form of the ground-state wave function.

7–5. Use a trial function of the form $\phi(r) = re^{-\alpha r}$ with α as a variational parameter to calculate the ground-state energy of a hydrogen atom.

The Hamiltonian operator for the hydrogen atom is given by Equation 7.6. We now find E_ϕ, using Equation 7.3. The numerator is

$$\begin{aligned}
\int \phi^* \hat{H} \phi \, d\tau &= 4\pi \int_0^\infty dr \, r^3 e^{-\alpha r} \left[-\frac{\hbar^2}{2m_e r^2} e^{-\alpha r} (2r - 4\alpha r^2 + \alpha^2 r^3) - \frac{e^2}{4\pi\varepsilon_0} e^{-\alpha r} \right] \\
&= -\frac{4\pi\hbar^2}{2m_e} \left[\frac{4}{(2\alpha)^3} - \frac{24\alpha}{(2\alpha)^4} + \frac{24\alpha^2}{(2\alpha)^5} \right] - \frac{e^2}{\varepsilon_0} \frac{6}{(2\alpha)^4} = \frac{\pi\hbar^2}{2m_e \alpha^3} - \frac{3e^2}{8\varepsilon_0 \alpha^4}
\end{aligned}$$

and the denominator is

$$\int \phi^* \phi d\tau = 4\pi \int_0^\infty dr r^4 e^{-2\alpha r} = 4\pi \left[\frac{24}{(2\alpha)^5} \right] = \frac{3\pi}{\alpha^5}$$

So

$$E_\phi = \frac{\int \phi^* \hat{H} \phi d\tau}{\int \phi^* \phi d\tau} = \frac{\alpha^2 \hbar^2}{6m_e} - \frac{\alpha e^2}{8\varepsilon_0 \pi}$$

To find the minimum value of α, we use

$$\frac{dE_\phi}{d\alpha} = \frac{2\alpha_{min} \hbar^2}{6m_e} - \frac{e^2}{8\varepsilon_0 \pi} = 0$$

$$\alpha_{min} = \frac{3m_e e^2}{8\varepsilon_0 \pi \hbar^2}$$

And finally

$$E_{min} = -\frac{3m_e e^4}{128\varepsilon_0^2 \pi^2 \hbar^2} = -\frac{3}{8} \frac{e^2}{4\pi \varepsilon_0 a_0}$$

compared to

$$E_{exact} = -\frac{1}{2} \frac{e^2}{4\pi \varepsilon_0 a_0}$$

The approximate value differs from the exact value by about 25%.

7–6. Suppose we were to use a trial function of the form $\phi = c_1 e^{-\alpha r} + c_2 e^{-\beta r^2}$ to carry out a variational calculation for the ground-state energy of the hydrogen atom. Can you guess without doing any calculations what c_1, c_2, α, and E_{min} will be? What about a trial function of the form $\phi = \sum_{k=1}^5 c_k e^{-\alpha_k r - \beta_k r^2}$?

The energy E_{min} will turn out to be the exact ground-state energy of atomic hydrogen because the exact ground-state wave function can be expressed by the trial function with the proper choice of c_1, c_2, and α. The parameter c_2 is zero and α is $1/a_0$. The parameter c_1 is the normalization constant, $1^{3/2}/(\pi^{1/2} a_0^{3/2})$. A similar result will be found for the more general trial function.

7–7. Use a trial function of the form $e^{-\beta x^2}$ with β as a variational parameter to calculate the ground-state energy of a harmonic oscillator. Compare your result with the exact energy $h\nu/2$. Why is the agreement so good?

The Hamiltonian operator for a harmonic oscillator is given in Problem 7–3. We now determine E_ϕ (Equation 7.3) using the trial function $\phi = e^{-\beta x^2}$. The numerator of E_ϕ is

$$\int_{-\infty}^\infty \phi^* \hat{H} \phi dx = \int_{-\infty}^\infty dx e^{-\beta x^2} \left(-\frac{\hbar^2}{2\mu} \frac{d^2}{dx^2} + \frac{kx^2}{2} \right) e^{-\beta x^2}$$

$$= -\frac{\hbar^2}{\mu} \int_0^\infty dx (4\beta^2 x^2 - 2\beta) e^{-2\beta x^2} + k \int_0^\infty dx x^2 e^{-2\beta x^2}$$

$$= \frac{\hbar^2 \beta}{2\mu} \left(\frac{\pi}{2\beta} \right)^{1/2} + \frac{k}{8\beta} \left(\frac{\pi}{2\beta} \right)^{1/2}$$

and the denominator is

$$\int_{-\infty}^{\infty} dx \phi^* \phi = \int_{-\infty}^{\infty} e^{-2\beta x^2} = 2 \int_0^{\infty} e^{-2\beta x^2} = \left(\frac{\pi}{2\beta}\right)^{1/2}$$

Therefore

$$E_\phi = \frac{\int_{-\infty}^{\infty} \phi^* \hat{H} \phi \, dx}{\int_{-\infty}^{\infty} dx \phi^* \phi} = \frac{\hbar^2 \beta}{2\mu} + \frac{k}{8\beta}$$

To find the minimum value of E_ϕ we first find β_{min} and then substitute into our equation for E_ϕ:

$$\frac{dE_\phi}{d\beta} = \frac{\hbar^2}{2\mu} - \frac{k}{8\beta_{min}^2} = 0$$

$$\beta_{min} = \left(\frac{\mu k}{4\hbar^2}\right)^{1/2}$$

$$E_{min} = \frac{\hbar}{2}\left(\frac{k}{\mu}\right)^{1/2}$$

This is the exact ground state energy. The exact energy is obtained because the form of the trial function is the same as that of the exact ground-state wave function

$$\psi_0(x) = \left(\frac{\alpha}{\pi}\right)^{1/2} e^{-\alpha x^2/2}$$

7–8. Consider a three-dimensional, spherically symmetric, isotropic harmonic oscillator with $V(r) = kr^2/2$. Using a trial function $e^{-\alpha r^2}$ with α as a variational parameter, estimate the ground-state energy. Do the same using $e^{-\alpha r}$. The Hamiltonian operator is

$$\hat{H} = -\frac{\hbar^2}{2\mu r^2}\frac{d}{dr}\left(r^2\frac{d}{dr}\right) + \frac{k}{2}r^2$$

Compare these results with the exact ground-state energy, $E = \frac{3}{2}h\nu$. Why is one of these so much better than the other?

For a three-dimensional, spherically symmetric, isotropic harmonic oscillator,

$$\hat{H} = -\frac{\hbar^2}{2\mu r^2}\frac{d}{dr}\left(r^2\frac{d}{dr}\right) + \frac{kr^2}{2}$$

Now we carry out a variational calculation to find the minimum value of E_ϕ when $e^{-\alpha r^2}$ is the trial function (using Equation 7.3). The numerator of E_ϕ is

$$\int \phi^* \hat{H} \phi \, d\tau = 4\pi \int_0^{\infty} dr \, r^2 e^{-\alpha r^2} \left[-\frac{\hbar^2}{2\mu r^2}\frac{d}{dr}\left(r^2\frac{d}{dr}\right) + \frac{kr^2}{2}\right] e^{-\alpha r^2}$$

$$= -\frac{4\pi\hbar^2}{2\mu}\int_0^{\infty} dr \left(4\alpha^2 r^4 - 6\alpha r^2\right) e^{-2\alpha r^2} + \frac{4\pi k}{2}\int_0^{\infty} dr \, r^4 e^{-2\alpha r^2}$$

$$= -\frac{4\pi\hbar^2}{2\mu}\left[\frac{3}{8}\left(\frac{\pi}{2\alpha}\right)^{1/2} - \frac{3}{4}\left(\frac{\pi}{2\alpha}\right)^{1/2}\right] + \frac{4\pi k}{2}\left[\frac{3}{32\alpha^2}\left(\frac{\pi}{2\alpha}\right)^{1/2}\right]$$

and the denominator is

$$\int \phi^* \phi \, d\tau = 4\pi \int_0^{\infty} dr \, r^2 e^{-2\alpha r^2} = 4\pi \frac{1}{8\alpha}\left(\frac{\pi}{2\alpha}\right)^{1/2}$$

Then

$$E_\phi = \frac{\int \phi^* \hat{H} \phi d\tau}{\int \phi^* \phi d\tau} = \frac{3\hbar^2 \alpha}{2\mu} + \frac{3k}{8\alpha}$$

We can now find the value of the minimum energy,

$$\frac{dE_\phi}{d\alpha} = \frac{3\hbar^2}{2\mu} - \frac{3k}{8\alpha_{min}^2} = 0$$

$$\alpha_{min} = \frac{(\mu k)^{1/2}}{2\hbar}$$

$$E_{min} = \frac{3}{2}\hbar \left(\frac{k}{\mu}\right)^{1/2}$$

This value of E_{min} is the exact result because the trial function has the same form as the exact wave function. Now repeat this procedure using $e^{-\alpha r}$ as the trial function. The numerator of E_ϕ is

$$\int \phi^* \hat{H} \phi d\tau = 4\pi \int_0^\infty r^2 e^{-\alpha r} \left[-\frac{\hbar^2}{2\mu r^2} \frac{d}{dr}\left(r^2 \frac{d}{dr}\right) + \frac{kr^2}{2} \right] e^{-\alpha r}$$

$$= -\frac{4\pi\hbar^2}{2\mu} \int_0^\infty dr \left(\alpha^2 r^2 e^{-2\alpha r} - 2\alpha r e^{-2\alpha r}\right) + \frac{4\pi k}{2} \int_0^\infty dr\, r^4 e^{-2\alpha r}$$

$$= -\frac{4\pi\hbar^2}{2\mu} \left(\frac{1}{4\alpha} - \frac{1}{2\alpha}\right) + \frac{4\pi k}{2} \left(\frac{3}{4\alpha^5}\right)$$

and the denominator is

$$\int \phi^* \phi d\tau = 4\pi \int_0^\infty dr\, r^2 e^{-2\alpha r} = 4\pi \frac{1}{4\alpha^3}$$

Then

$$E_\phi = \frac{\int \phi^* \hat{H} \phi d\tau}{\int \phi^* \phi d\tau} = \frac{\hbar^2 \alpha^2}{2\mu} + \frac{3k}{2\alpha^2}$$

and

$$\frac{dE_\phi}{d\alpha} = \frac{2\hbar^2 \alpha_{min}}{2\mu} - \frac{3k}{\alpha_{min}^3} = 0$$

giving

$$\alpha_{min} = \left(\frac{3\mu k}{\hbar^2}\right)^{1/4}$$

and

$$E_{min} = 3^{1/2}\hbar \left(\frac{k}{\mu}\right)^{1/2}$$

Notice that E_{min} is about 15% greater than the exact value.

7–9. Use a trial function of the form $e^{-\alpha x^2/2}$ to calculate the ground-state energy of a quartic oscillator, whose potential is $V(x) = cx^4$.

The Hamiltonian operator of a quartic oscillator is

$$\hat{H} = -\frac{\hbar^2}{2\mu}\frac{d^2}{dx^2} + cx^4$$

We will now carry out a standard calculation using Equation 7.3 to find the minimum value of E_ϕ given the trial function stated in the problem. The numerator of E_ϕ is

$$\int \phi^* \hat{H} \phi d\tau = \int_{-\infty}^{\infty} -\frac{\hbar^2}{2\mu}\left(-\alpha e^{-\alpha x^2} + \alpha^2 x^2 e^{-\alpha x^2}\right)dx + \int_{-\infty}^{\infty} dx c x^4 e^{-\alpha x^2}$$

$$= \frac{\hbar^2 \alpha}{\mu}\int_0^{\infty} dx e^{-\alpha x^2} - \frac{\hbar^2 \alpha^2}{\mu}\int_0^{\infty} dx\, x^2 e^{-\alpha x^2} + 2c\int_0^{\infty} dx\, x^4 e^{-\alpha x^2}$$

$$= \frac{\hbar^2 \alpha}{\mu}\left(\frac{\pi}{4\alpha}\right)^{1/2} - \frac{\hbar^2 \alpha}{4\mu}\left(\frac{\pi}{\alpha}\right)^{1/2} + \frac{6c}{8\alpha^2}\left(\frac{\pi}{\alpha}\right)^{1/2}$$

$$= \left(\frac{\hbar^2 \alpha}{4\mu} + \frac{3c}{4\alpha^2}\right)\left(\frac{\pi}{\alpha}\right)^{1/2}$$

and the denominator is

$$\int \phi^* \phi d\tau = 2\int_0^{\infty} e^{-\alpha x^2} = \left(\frac{\pi}{\alpha}\right)^{1/2}$$

Using these results, we find

$$E_\phi = \frac{\int \phi^* \hat{H} \phi d\tau}{\int \phi^* \phi d\tau} = \frac{\hbar^2 \alpha}{4\mu} + \frac{3c}{4\alpha^2}$$

The minimum value of α is then

$$\frac{dE}{d\alpha} = \frac{\hbar^2}{4\mu} - \frac{3c}{2\alpha_{min}^3} = 0$$

$$\alpha_{min} = \left(\frac{6\mu c}{\hbar^2}\right)^{1/3}$$

giving

$$E_{min} = \frac{3}{8}6^{1/3}\frac{c^{1/3}\hbar^{4/3}}{\mu^{2/3}}$$

7–10. Use a trial function of the form $\phi = \cos \lambda x$ with $-\pi/2\lambda < x < \pi/2\lambda$ and with λ as a variational parameter to calculate the ground-state energy of a harmonic oscillator.

For a harmonic oscillator,

$$\hat{H} = -\frac{\hbar^2}{2\mu}\frac{d^2}{dx^2} + \frac{kx^2}{2}$$

Note that in this problem we are given limits on x which will determine our limits of integration when we evaluate Equation 7.3. The numerator of E_ϕ is

$$\int \phi^* \hat{H} \phi d\tau = \int_{-\pi/2\lambda}^{\pi/2\lambda} dx \cos \lambda x\left[-\frac{\hbar^2}{2\mu}\left(-\lambda^2 \cos \lambda x\right) + \frac{kx^2}{2}\cos \lambda x\right]$$

$$= \frac{\hbar^2 \lambda^2}{2\mu}\int_{-\pi/2\lambda}^{\pi/2\lambda} dx \cos^2 \lambda x + \frac{k}{2}\int_{-\pi/2\lambda}^{\pi/2\lambda} dx\, x^2 \cos^2 \lambda x$$

$$= \frac{\hbar^2 \lambda}{2\mu} \int_{-\pi/2}^{\pi/2} dz \cos^2 z + \frac{k}{2\lambda^3} \int_{-\pi/2}^{\pi/2} dz \, z^2 \cos^2 z$$

$$= \frac{\hbar^2 \lambda}{\mu} \int_0^{\pi/2} dz \cos^2 z + \frac{k}{\lambda^3} \int_0^{\pi/2} dz \, z^2 \cos^2 z$$

$$= \frac{\hbar^2 \lambda}{\mu} \left(\frac{\pi}{4} \right) + \frac{k}{\lambda^3} \left(\frac{\pi^3}{48} - \frac{\pi}{8} \right)$$

and the denominator is

$$\int \phi^* \phi d\tau = \int_{-\pi/2\lambda}^{\pi/2\lambda} \cos^2 \lambda x \, dx = \frac{2}{\lambda} \int_0^{\pi/2} \cos^2 z \, dz = \frac{2}{\lambda} \left(\frac{\pi}{4} \right) = \frac{\pi}{2\lambda}$$

Then

$$E_\phi = \frac{\int \phi^* \hat{H} \phi d\tau}{\int \phi^* \phi d\tau} = \frac{\hbar^2 \lambda^2}{2\mu} + \frac{k}{\lambda^2} \left(\frac{\pi^2}{24} - \frac{1}{4} \right)$$

giving

$$\frac{dE_\phi}{d\lambda} = \frac{\hbar^2 \lambda_{min}}{\mu} - \frac{k}{\lambda_{min}^3} \left(\frac{\pi^2}{12} - \frac{1}{2} \right) = 0$$

$$\lambda_{min} = \left[\frac{k\mu}{2\hbar^2} \left(\frac{\pi^2}{6} - 1 \right) \right]^{1/4} \qquad = \frac{k_\mu \pi^2}{12\hbar^2} - \frac{k\mu}{2\hbar^2}$$

Therefore, we find that

$$E_{min} = \frac{\hbar^2}{2\mu} \left(\frac{k\mu}{2\hbar^2} \right)^{1/2} \left(\frac{\pi^2}{6} - 1 \right)^{1/2} + \frac{k}{4} \left(\frac{\pi^2}{6} - 1 \right)^{1/2} \left(\frac{2\hbar^2}{k\mu} \right)^{1/2}$$

$$= \left(\frac{\pi^2}{6} - 1 \right)^{1/2} \left(\frac{1}{2^{1/2}} \right) \hbar \left(\frac{k}{\mu} \right)^{1/2}$$

$$= 0.568 \, \hbar \left(\frac{k}{\mu} \right)^{1/2}$$

The value calculated here is about 13% greater than the exact value of $0.500 \, \hbar (k/\mu)^{1/2}$.

7–11. Use the variational method to calculate the ground-state energy of a particle constrained to move within the region $0 \leq x \leq a$ in a potential given by

$$V(x) = V_0 x \qquad \qquad 0 \leq x \leq \frac{a}{2}$$

$$= V_0(a - x) \qquad \frac{a}{2} \leq x \leq a$$

As a trial function, use a linear combination of the first two particle-in-a-box wave functions:

$$\phi(x) = c_1 \left(\frac{2}{a} \right)^{1/2} \sin \frac{\pi x}{a} + c_2 \left(\frac{2}{a} \right)^{1/2} \sin \frac{2\pi x}{a}$$

Recall that the particle-in-a-box wavefunctions are orthonormal. This means that $H_{12} = H_{21} = 0$, $S_{12} = S_{21} = 0$, and $S_{11} = S_{22} = 0$, so Equation 7.37 becomes

$$\begin{vmatrix} H_{11} - E & 0 \\ 0 & H_{22} - E \end{vmatrix} = (H_{11} - E)(H_{22} - E) = 0 \qquad (1)$$

Now we evaluate H_{11}, where $H_{11} = \int f_1 \hat{H} f_1 d\tau$.

$$H_{11} = \frac{2}{a} \int_0^{a/2} \sin \frac{\pi x}{a} \left(-\frac{\hbar^2}{2m} \frac{d^2}{dx^2} + V_0 x \right) \sin \frac{\pi x}{a} dx + \frac{2}{a} \int_{a/2}^{a} \sin \frac{\pi x}{a} \left[-\frac{\hbar^2}{2m} \frac{d^2}{dx^2} + V_0(a-x) \right] \sin \frac{\pi x}{a} dx$$

$$= \frac{\hbar^2 \pi^2}{a^3 m} \int_0^a \sin^2 \frac{\pi x}{a} dx + \frac{2V_0}{a} \int_0^{a/2} x \sin^2 \frac{\pi x}{a} dx + 2V_0 \int_{a/2}^a \sin^2 \frac{\pi x}{a} dx - \frac{2V_0}{a} \int_{a/2}^a x \sin^2 \frac{\pi x}{a} dx$$

Now we use the integrals discussed in Problem 3–10:

$$H_{11} = \frac{h^2 \pi^2}{4\pi^2 m a^3} \left(\frac{a}{2} \right) + \frac{2V_0}{a} \left(\frac{x^2}{4} - \frac{ax \sin \frac{2\pi x}{a}}{4\pi} - \frac{a^2 \cos \frac{2\pi x}{a}}{8\pi^2} \right) \Bigg|_0^{a/2}$$

$$+ 2V_0 \left(\frac{x}{2} - \frac{a}{4\pi} \sin \frac{2\pi x}{a} \right) \Bigg|_{a/2}^a - \frac{2V_0}{a} \left(\frac{x^2}{4} - \frac{ax \sin \frac{2\pi x}{a}}{4\pi} - \frac{a^2 \cos \frac{2\pi x}{a}}{8\pi^2} \right) \Bigg|_{a/2}^a$$

$$= \frac{h^2}{8ma^2} + \frac{2V_0}{a} \left(\frac{a^2}{16} + \frac{a^2 \pi}{8\pi^2} + \frac{a^2}{8\pi^2} \right) + 2V_0 \left(\frac{a}{2} - \frac{a}{4} \right) - \frac{2V_0}{a} \left(\frac{a^2}{4} - \frac{a^2}{8\pi^2} - \frac{a^2}{16} - \frac{a^2 \pi}{8\pi^2} \right)$$

$$= \frac{h^2}{8ma^2} + V_0 a \left(\frac{1}{4} + \frac{1}{\pi^2} \right)$$

In the same manner, we find that H_{22} is

$$H_{22} = \frac{2}{a} \int_0^{a/2} \sin \frac{2\pi x}{a} \left(-\frac{\hbar^2}{2m} \frac{d^2}{dx^2} + V_0 x \right) \sin \frac{2\pi x}{a} dx$$

$$= \frac{h^2}{2ma^2} + \frac{V_0 a}{4}$$

Substitute these values into Equation 1 and solve for E to find

$$E = \frac{h^2}{8ma^2} + V_0 a \left(\frac{1}{4} + \frac{1}{\pi^2} \right) \qquad \text{and} \qquad E = \frac{h^2}{2ma^2} + \frac{V_0 a}{4}$$

Given m, a, and V_0, we can determine which of these roots is E_{min}.

7–12. Consider a particle of mass m in the potential energy field described by

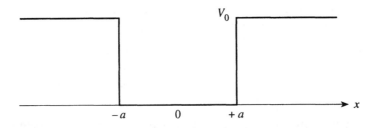

This problem describes a particle in a finite well. If $V_0 \to \infty$, then we have a particle in a box. Using $\phi(x) = l^2 - x^2$ for $-l < x < l$ and $\phi(x) = 0$ otherwise as a trial function with l as a variational parameter, calculate the ground-state energy of this system for $2mV_0 a^2/\hbar^2 = 4$ and 12. The exact ground-state energies are $0.530\hbar^2/ma^2$ and $0.736\hbar^2/ma^2$, respectively (see Problem 7–29).

We are given

$$\phi(x) = l^2 - x^2 \qquad -l < x < l$$

$$\hat{H} = -\frac{\hbar^2}{2m}\frac{d^2}{dx^2} \qquad -a < x < a$$

$$= -\frac{\hbar^2}{2m}\frac{d^2}{dx^2} + V_0 \qquad |x| \geq a$$

Note that $d^2\phi/dx^2 = -2$. We can then write

$$\int_{-\infty}^{\infty}\phi^*\hat{H}\phi\,dx = \int_{-l}^{-a}dx(l^2-x^2)\frac{\hbar^2}{m} + \int_{-l}^{-a}V_0(l^2-x^2)^2dx + \int_{-a}^{a}(l^2-x^2)\frac{\hbar^2}{m}dx$$

$$+ \int_{a}^{l}dx(l^2-x^2)\frac{\hbar^2}{m} + \int_{a}^{l}V_0(l^2-x^2)^2dx$$

$$= \int_{-l}^{l}\frac{\hbar^2}{m}(l^2-x^2)dx + 2V_0\int_{a}^{l}(l^2-x^2)^2dx$$

We are justified in writing this last equality because $(l^2-x^2)^2$ is clearly an even function (symmetric around the y-axis). Now

$$\int_{-\infty}^{\infty}\phi^*\hat{H}\phi\,dx = \frac{2\hbar^2}{m}\left(l^3 - \frac{l^3}{3}\right) + 2V_0\left[l^4(l-a) - \frac{2l^2}{3}(l^3-a^3) + \frac{l^5-a^5}{5}\right]$$

$$= \frac{4\hbar^2 l^3}{3m} + 2V_0\left(\frac{8l^5}{15} - al^4 + \frac{2}{3}a^3l^2 - \frac{a^5}{5}\right)$$

$$\int_{-\infty}^{\infty}\phi^*\phi\,dx = \int_{-l}^{l}(l^2-x^2)^2dx = \frac{16l^5}{15}$$

$$E_\phi = \frac{\int_{-\infty}^{\infty}\phi^*\hat{H}\phi\,dx}{\int_{-\infty}^{\infty}\phi^*\phi\,dx}$$

$$= \frac{15}{16}\left[\frac{4\hbar^2}{3ml^2} + \frac{2V_0}{15}\left(8 - 15\frac{a}{l} + \frac{10a^3}{l^3} - \frac{3a^5}{l^5}\right)\right]$$

$$= \frac{5}{16}\frac{\hbar^2}{ma^2}\left[\frac{4}{s^2} + \frac{\alpha}{5}\left(8 - \frac{15}{s} + \frac{10}{s^3} - \frac{3}{s^5}\right)\right]$$

where $\alpha = 2mV_0a^2/\hbar^2$ and $s = l/a$. Now minimize E_ϕ with respect to s:

$$\frac{\partial E_\phi}{\partial s} = -\frac{8}{s_{min}^3} + \frac{\alpha}{5}\left(\frac{15}{s_{min}^2} - \frac{30}{s_{min}^4} + \frac{15}{s_{min}^6}\right) = 0$$

$$0 = 3\alpha s_{min}^4 - 6\alpha s_{min}^2 - 8s_{min}^3 + 3\alpha$$

For $\alpha = 4$, this becomes

$$12s_{min}^4 - 8s_{min}^3 - 24s_{min}^2 + 12 = 0$$

$$3s_{min}^4 - 2s_{min}^3 - 6s_{min}^2 + 3 = 0$$

$$s_{min} = 1.6546$$

$$E_\phi = 0.6816\frac{\hbar^2}{ma^2}$$

which differs from the exact ground-state energy by about 30%. For $\alpha = 12$,

$$36s_{min}^4 - 72s_{min}^2 - 8s_{min}^3 + 36 = 0$$

$$9s_{min}^4 - 8s_{min}^3 - 18s_{min}^2 + 9 = 0$$

$$s_{min} = 0.68353$$

$$E_\phi = 0.6219\frac{\hbar^2}{ma^2}$$

which differs from the exact ground-state energy by about 40%.

7-13. Repeat the calculation in Problem 7–12 for a trial function $\phi(x) = \cos \lambda x$ for $-\pi/2\lambda < x < \pi/2\lambda$ and $\phi(x) = 0$ otherwise. Use λ as a variational parameter.

We are given

$$\phi(x) = \cos \lambda x \qquad -\frac{\pi}{2\lambda} < x < \frac{\pi}{2\lambda}$$

$$\hat{H} = -\frac{\hbar^2}{2m}\frac{d^2}{dx^2} \qquad -a < x < a$$

$$= -\frac{\hbar^2}{2m}\frac{d^2}{dx^2} + V_0 \qquad |x| \geq a$$

Note that $d^2\phi/dx^2 = -\lambda^2 \cos \lambda x$. We can then write

$$\int_{-\infty}^{\infty} \phi^* \hat{H} \phi \, dx = \int_{-\pi/2\lambda}^{-a} dx \frac{\hbar^2 \lambda^2}{2m} \cos^2 \lambda x + \int_{-\pi/2\lambda}^{-a} V_0 \cos \lambda x \, dx + \int_{-a}^{a} dx \frac{\hbar^2 \lambda^2}{2m} \cos^2 \lambda x$$

$$+ \int_{a}^{\pi/2\lambda} dx \frac{\hbar^2 \lambda^2}{2m} \cos^2 \lambda x + \int_{a}^{\pi/2\lambda} V_0 \cos^2 \lambda x \, dx$$

$$= \frac{\hbar^2 \lambda^2}{m} \int_{0}^{\pi/2\lambda} dx \cos^2 \lambda x + 2V_0 \int_{a}^{\pi/2\lambda} dx \cos^2 \lambda x$$

$$= \frac{\hbar^2 \lambda^2}{m}\left(\frac{\pi}{4\lambda}\right) + 2V_0\left(\frac{\pi}{4\lambda} - \frac{a}{2} - \frac{\sin 2\lambda a}{4\lambda}\right)$$

and

$$\int_{-\infty}^{\infty} \phi^* \phi \, dx = \int_{-\pi/2\lambda}^{\pi/2\lambda} dx \cos^2 \lambda x = \frac{\pi}{2\lambda}$$

Using Equation 7.3 then gives

$$E_\phi = \frac{\int_{-\infty}^{\infty} \phi^* \hat{H} \phi \, dx}{\int_{-\infty}^{\infty} \phi^* \phi \, dx}$$

$$= \frac{\hbar^2 \lambda^2}{2m} + 2V_0\left(\frac{1}{2} - \frac{a\lambda}{\pi} - \frac{\sin 2a\lambda}{2\pi}\right)$$

$$= \frac{\hbar^2}{2ma^2}s^2 + 2V_0\left(\frac{1}{2} - \frac{s}{\pi} - \frac{\sin 2s}{2\pi}\right)$$

$$= \frac{\hbar^2}{2ma^2}\left[s^2 + 2\alpha\left(\frac{1}{2} - \frac{s}{\pi} - \frac{\sin 2s}{2\pi}\right)\right]$$

where $\alpha = 2mV_0a^2/\hbar^2$ and $s = \lambda a$. Now minimize E_ϕ with respect to s:

$$\frac{dE_\phi}{ds} = 0$$

$$2s_{min} - \frac{2\alpha}{\pi} - \frac{2\alpha}{\pi}\cos 2s_{min} = 0$$

For $\alpha = 4$, we find $s_{min} = 0.92423$ and $E = 0.6381\hbar^2/ma^2$, which is about 20% greater than the exact value. For $\alpha = 12$, we find $s_{min} = 1.1689$ and $E = 0.8432\hbar^2/ma^2$, which is about 15% greater than the exact value.

7–14. Consider a particle in a spherical box of radius a. The Hamiltonian operator for this system is (see Equation 6.43)

$$\hat{H} = -\frac{\hbar^2}{2mr^2}\frac{d}{dr}\left(r^2\frac{d}{dr}\right) + \frac{\hbar^2 l(l+1)}{2mr^2} \qquad 0 < r \leq a$$

In the ground state, $l = 0$ and so

$$\hat{H} = -\frac{\hbar^2}{2mr^2}\frac{d}{dr}\left(r^2\frac{d}{dr}\right) \qquad 0 < r \leq a$$

As in the case of a particle in a rectangular box, $\phi(a) = 0$. Use $\phi(r) = a - r$ to calculate an upper bound to the ground-state energy of this system. There is no variational parameter in this case, but the calculated energy is still an upper bound to the ground-state energy. The exact ground-state energy is $\pi^2\hbar^2/2ma^2$ (see Problem 7–28).

To simplify the problem, first calculate the result of the Hamiltonian operator operating on ϕ:

$$\hat{H}\phi = \frac{\hbar^2}{m}\frac{1}{r}$$

Now we solve for the minimum value of E. The limits of integration come from the fact that the Hamiltonian is 0 everywhere except where $0 < r \leq a$.

$$\int \phi^* \hat{H}\phi\,d\tau = 4\pi \int_0^a r^2 dr\frac{\hbar^2}{mr}(a-r) = 4\pi \int_0^a \frac{\hbar^2}{m}r(a-r)dr$$

$$= \frac{4\pi\hbar^2}{m}\left(\frac{a^3}{2} - \frac{a^3}{3}\right) = \frac{4\pi\hbar^2}{m}\frac{a^3}{6}$$

$$\int \phi^*\phi\,d\tau = 4\pi \int_0^a r^2 dr(a-r)^2$$

$$= 4\pi\left(\frac{a^5}{3} - \frac{2a^5}{4} + \frac{a^5}{5}\right) = 4\pi\frac{a^5}{30}$$

$$E_\phi = \frac{\int \phi^* \hat{H}\phi\,d\tau}{\int \phi^*\phi\,d\tau} = \frac{5\hbar^2}{ma^2}$$

On comparison, we see that E_ϕ is about 1% greater than the exact ground-state energy of $\pi^2\hbar^2/2ma^2$.

7–15. Repeat the calculation in Problem 7–14 using $\phi(r) = (a-r)^2$ as a trial function.

Again, we first calculate the result of the Hamiltonian operator operating on ϕ:

$$\hat{H}\phi = -\frac{\hbar^2}{2m}\left(6 - \frac{4a}{r}\right)$$

Now

$$\int \phi^* \hat{H}\phi \, d\tau = 4\pi \left(-\frac{\hbar^2}{2m}\right)\int_0^a dr(6r^2 - 4ar)(a - r)^2$$

$$= -\frac{4\pi\hbar^2}{2m}\left(-\frac{2a^5}{15}\right) = \frac{4\pi\hbar^2 a^5}{15m}$$

$$\int \phi^* \phi \, d\tau = 4\pi \int_0^a r^2 dr(a - r)^4 = 4\pi\left(\frac{a^7}{105}\right)$$

$$E_\phi = \frac{\int \phi^* \hat{H}\phi \, d\tau}{\int \phi^* \phi \, d\tau} = \frac{7\hbar^2}{ma^2}$$

The approximate value of E_ϕ is about 40% greater than the exact ground-state energy of $\pi^2\hbar^2/2ma^2$. (In other words, the trial function we picked in the previous problem was a better approximation of the actual wave function.)

7–16. Consider a system subject to the potential

$$V(x) = \frac{k}{2}x^2 + \frac{\gamma}{6}x^3 + \frac{\delta}{24}x^4$$

Calculate the ground-state energy of this system using a trial function of the form

$$\phi = c_1\psi_0(x) + c_2\psi_2(x)$$

where $\psi_0(x)$ and $\psi_2(x)$ are the first two even wave functions of a harmonic oscillator. Why did we not include $\psi_1(x)$?

We do not include $\psi_1(x)$ in this problem because it is an odd function, so $\int_{-\infty}^{\infty} dx\psi_1(x)^2 = 0$. Since the Hamiltonian operator is not explicitly dependent on ψ, $\int_{-\infty}^{\infty} dx\psi_1^*(x)\hat{H}\psi_1(x) = 0$, so ψ_1 does not contribute to the wave function of a system with this Hamiltonian operator. We will use Equation 7.37 to determine E:

$$\begin{vmatrix} H_{11} - ES_{11} & H_{12} - ES_{12} \\ H_{21} - ES_{21} & H_{22} - ES_{22} \end{vmatrix} = 0$$

We will use the normalized wave functions of the harmonic oscillator, so $S_{nm} = \delta_{nm}$. Now we must find the H_{nm} components of the determinant:

$$H_{11} = \int dx\psi_0^*\hat{H}\psi_0 = E_0 + \frac{\gamma}{6}\langle x^3\rangle_{00} + \frac{\delta}{24}\langle x^4\rangle_{00}$$

$$H_{21} = \int dx\psi_2^*\hat{H}\psi_0 = 0 + \frac{\gamma}{6}\langle x^3\rangle_{20} + \frac{\delta}{24}\langle x^4\rangle_{20}$$

$$H_{12} = \int dx\psi_0^*\hat{H}\psi_2 = \int dx\psi_2^*\hat{H}\psi_0$$

$$H_{22} = \int dx\psi_2^*\hat{H}\psi_2 = E_2 + \frac{\gamma}{6}\langle x^3\rangle_{22} + \frac{\delta}{24}\langle x^4\rangle_{22}$$

where E_v is the energy level of a quantum mechanical harmonic oscillator and $\langle x^y \rangle_{nm} = \int_{-\infty}^{\infty} dx \psi_n^* x^y \psi_m$. Both ψ_0 and ψ_2 are even functions. Therefore, for any $\langle x^y \rangle_{nm}$ where y is odd, $\langle x^y \rangle_{nm} = 0$ and so

$$\langle x^3 \rangle_{00} = \langle x^3 \rangle_{20} = \langle x^3 \rangle_{22} = 0$$

Recall from Table 5.3 that

$$\psi_0(x) = \left(\frac{\alpha}{\pi}\right)^{1/4} e^{-\alpha x^2/2}$$

$$\psi_2(x) = \left(\frac{\alpha}{4\pi}\right)^{1/4} (2\alpha x^2 - 1)e^{-\alpha x^2/2}$$

Now calculate the $\langle x^4 \rangle_{nm}$:

$$\langle x^4 \rangle_{00} = \left(\frac{\alpha}{\pi}\right)^{1/2} \int_{-\infty}^{\infty} x^4 e^{-\alpha x^2} dx$$

$$= \left(\frac{\alpha}{\pi}\right)^{1/2} \frac{6}{8\alpha^2} \left(\frac{\pi}{\alpha}\right)^{1/2} = \frac{3}{4\alpha^2}$$

$$\langle x^4 \rangle_{20} = \left(\frac{\alpha}{2\pi}\right)^{1/2} \int_{-\infty}^{\infty} (2\alpha x^6 - x^4)e^{-\alpha x^2} dx$$

$$= \left(\frac{2\alpha}{\pi}\right)^{1/2} \left[2\alpha \frac{5 \cdot 3}{16\alpha^3} \left(\frac{\pi}{\alpha}\right)^{1/2} - \frac{3}{8\alpha^2} \left(\frac{\pi}{\alpha}\right)^{1/2} \right] = \frac{3}{2^{1/2}\alpha^2}$$

$$\langle x^4 \rangle_{22} = \left(\frac{\alpha}{\pi}\right)^{1/2} \int_0^{\infty} dx(4\alpha^2 x^4 - 4\alpha x^2 + 1)x^4 e^{-\alpha x^2/2}$$

$$= \left(\frac{\alpha}{\pi}\right)^{1/2} \left(4\alpha^2 \frac{7 \cdot 5 \cdot 3}{32\alpha^4} - 4\alpha \frac{5 \cdot 3}{16\alpha^3} + \frac{3}{8\alpha^2} \right) \left(\frac{\pi}{\alpha}\right)^{1/2} = \frac{39}{4\alpha^2}$$

Combining the above results and using Equation 5.27 for E_v, we find that

$$H_{11} = E_0 + \frac{\gamma}{6} \langle x^3 \rangle_{00} + \frac{\delta}{24} \langle x^4 \rangle_{00}$$

$$= \frac{\hbar\omega}{2} + \frac{\delta}{24} \left(\frac{3}{4\alpha^2}\right)$$

$$H_{21} = \frac{\gamma}{6} \langle x^3 \rangle_{20} + \frac{\delta}{24} \langle x^4 \rangle_{20}$$

$$= \frac{\delta}{24} \left(\frac{3}{2^{1/2}\alpha^2}\right) = H_{12}$$

$$H_{22} = E_2 + \frac{\gamma}{6} \langle x^3 \rangle_2 + \frac{\delta}{24} \langle x^4 \rangle_2$$

$$= \frac{5\hbar\omega}{2} + \frac{\delta}{24} (394\alpha^2)$$

where $\omega = (k/\mu)^{1/2}$. Rreturn to Equation 7.37:

$$0 = \begin{vmatrix} H_{11} - E & H_{12} \\ H_{21} & H_{22} - E \end{vmatrix}$$

$$0 = (H_{11} - E)(H_{22} - E) - H_{12}^2$$

$$0 = H_{11}H_{22} - (H_{11} + H_{22})E + E^2 - H_{12}^2$$

$$0 = E^2 - \left(3\hbar\omega + \frac{7\delta}{16\alpha^2}\right)E + \frac{5}{4}(\hbar\omega)^2 + \frac{27}{96}\frac{\hbar\omega\delta}{\alpha^2} + \frac{13}{(32)^2}\frac{\delta^2}{\alpha^4} - \frac{\delta^2}{128\alpha^4}$$

Solving this equation for E gives

$$E = \frac{3}{2}\hbar\omega + \frac{7\delta}{32\alpha^2} - \frac{1}{2}\left[(2\hbar\omega)^2 + \frac{3}{2}\frac{\hbar\omega\delta}{\alpha^2} + \frac{11}{64}\frac{\delta^2}{\alpha^4}\right]^{1/2}$$

7–17. It is quite common to assume a trial function of the form

$$\phi = c_1\phi_1 + c_2\phi_2 + \cdots + c_n\phi_n$$

where the variational parameters and the ϕ_n may be complex. Using the simple, special case

$$\phi = c_1\phi_1 + c_2\phi_2$$

show that the variational method leads to

$$E_\phi = \frac{c_1^*c_1 H_{11} + c_1^*c_2 H_{12} + c_1 c_2^* H_{21} + c_2^*c_2 H_{22}}{c_1^*c_1 S_{11} + c_1^*c_2 S_{12} + c_1 c_2^* S_{21} + c_2^*c_2 S_{22}}$$

where

$$H_{ij} = \int \phi_i^* \hat{H}\phi_j \, d\tau = H_{ji}^*$$

and

$$S_{ij} = \int \phi_i^* \phi_j \, d\tau = S_{ji}^*$$

because $\hat{H}$ is a Hermitian operator. Now write the above equation for E_ϕ as

$$c_1^*c_1 H_{11} + c_1^*c_2 H_{12} + c_1 c_2^* H_{21} + c_2^*c_2 H_{22}$$
$$= E_\phi(c_1^*c_1 S_{11} + c_1^*c_2 S_{12} + c_1 c_2^* S_{21} + c_2^*c_2 S_{22})$$

and show that if we set

$$\frac{\partial E_\phi}{\partial c_1^*} = 0 \quad \text{and} \quad \frac{\partial E_\phi}{\partial c_2^*} = 0$$

we obtain

$$(H_{11} - E_\phi S_{11})c_1 + (H_{12} - E_\phi S_{12})c_2 = 0$$

and

$$(H_{21} - E_\phi S_{21})c_1 + (H_{22} - E_\phi S_{22})c_2 = 0$$

There is a nontrivial solution to this pair of equations if and only if the determinant

$$\begin{vmatrix} H_{11} - E_\phi S_{11} & H_{12} - E_\phi S_{12} \\ H_{21} - E_\phi S_{21} & H_{22} - E_\phi S_{22} \end{vmatrix} = 0$$

which gives a quadratic equation for E_ϕ. We choose the smaller solution as an approximation to the ground-state energy.

Start with (Equation 7.3)

$$E_\phi = \frac{\int d\tau \phi^* \hat{H}\phi}{\int d\tau \phi^*\phi}$$

Now substitute $\phi = c_1\phi_1 + c_2\phi_2$ to obtain

$$\int d\tau \phi^* \hat{H} \phi = \int d\tau (c_1^* \phi_1^* + c_2^* \phi_2^*) \hat{H}(c_1\phi_1 + c_2\phi_2)$$

$$= c_1^* c_1 \int d\tau \phi_1^* \hat{H} \phi_1 + c_2^* c_1 \int d\tau \phi_2^* \hat{H} \phi_1 + c_1^* c_2 \int d\tau \phi_1^* \hat{H} \phi_2 + c_2^* c_2 \int d\tau \phi_2^* \hat{H} \phi_2$$

$$= c_1^* c_1 H_{11} + c_2^* c_1 H_{21} + c_1^* c_2 H_{12} + c_2^* c_2 H_{22}$$

$$\int d\tau \phi^* \phi = \int d\tau (c_1^* \phi_1^* + c_2^* \phi_2^*) \hat{H}(c_1\phi_1 + c_2\phi_2)$$

$$= c_1^* c_1 \int d\tau \phi_1^* \phi_1 + c_2^* c_1 \int d\tau \phi_2^* \phi_1 + c_1^* c_2 \int d\tau \phi_1^* \phi_2 + c_2^* c_2 \int d\tau \phi_2^* \phi_2$$

$$= c_1^* c_1 S_{11} + c_2^* c_1 S_{21} + c_1^* c_2 S_{12} + c_2^* c_2 S_{22}$$

and so

$$E_\phi = \frac{c_1^* c_1 H_{11} + c_2^* c_1 H_{21} + c_1^* c_2 H_{12} + c_2^* c_2 H_{22}}{c_1^* c_1 S_{11} + c_2^* c_1 S_{21} + c_1^* c_2 S_{12} + c_2^* c_2 S_{22}}$$

Because $\hat{H}$ is Hermitian, we can write

$$H_{ij} = \int \phi_i^* \hat{H}_j \phi_j d\tau = H_{ji}^*$$

and

$$S_{ij} = \int \phi_i^* \phi_j d\tau = S_{ji}^*$$

Now we multiply both sides of our equation for E_ϕ by $\int d\tau \phi^* \phi$ to obtain

$$E_\phi(c_1^* c_1 S_{11} + c_2^* c_1 S_{21} + c_1^* c_2 S_{12} + c_2^* c_2 S_{22}) = c_1^* c_1 H_{11} + c_2^* c_1 H_{21} + c_1^* c_2 H_{12} + c_2^* c_2 H_{22}$$

Now differentiate with respect to c_1^* and set the derivative equal to zero to find the minimum value of E_ϕ with respect to c_1^*:

$$\frac{\partial E_\phi}{\partial c_1^*}(c_1^* c_1 S_{11} + c_2^* c_1 S_{21} + c_1^* c_2 S_{12} + c_2^* c_2 S_{22}) + E_\phi(c_1 S_{11} + c_2 S_{12}) = c_1 H_{11} + c_2 H_{12}$$

$$\frac{\partial E_\phi}{\partial c_1^*} = \frac{c_1 H_{11} + c_2 H_{12} - E_\phi(c_1 S_{11} + c_2 S_{12})}{c_1^* c_1 S_{11} + c_2^* c_1 S_{21} + c_1^* c_2 S_{12} + c_2^* c_2 S_{22}}$$

$$0 = \frac{c_1 H_{11} + c_2 H_{12} - E_\phi(c_1 S_{11} + c_2 S_{12})}{c_1^* c_1 S_{11} + c_2^* c_1 S_{21} + c_1^* c_2 S_{12} + c_2^* c_2 S_{22}}$$

$$0 = c_1(H_{11} - E_\phi S_{11}) + c_2(H_{12} - E_\phi S_{12})$$

Repeat for c_2^* to find the minimum value of E_ϕ with respect to c_2^*:

$$\frac{\partial E_\phi}{\partial c_2^*}(c_1^* c_1 S_{11} + c_2^* c_1 S_{21} + c_1^* c_2 S_{12} + c_2^* c_2 S_{22}) + E_\phi(c_1 S_{21} + c_2 S_{22}) = c_1 H_{21} + c_2 H_{22}$$

$$\frac{\partial E_\phi}{\partial c_1^*} = \frac{c_1 H_{21} + c_2 H_{22} - E_\phi(c_1 S_{21} + c_2 S_{22})}{c_1^* c_1 S_{11} + c_2^* c_1 S_{21} + c_1^* c_2 S_{12} + c_2^* c_2 S_{22}}$$

$$0 = \frac{c_1 H_{21} + c_2 H_{22} - E_\phi(c_1 S_{21} + c_2 S_{22})}{c_1^* c_1 S_{11} + c_2^* c_1 S_{21} + c_1^* c_2 S_{12} + c_2^* c_2 S_{22}}$$

$$0 = c_1(H_{21} - E_\phi S_{21}) + c_2(H_{22} - E_\phi S_{22})$$

7–18. This problem shows that terms in a trial function that correspond to progressively higher energies contribute progressively less to the ground-state energy. For algebraic simplicity, assume that the Hamiltonian operator can be written in the form

$$\hat{H} = \hat{H}^{(0)} + \hat{H}^{(1)}$$

and choose a trial function

$$\phi = c_1 \psi_1 + c_2 \psi_2$$

where

$$\hat{H}^{(0)}\psi_j = E_j^{(0)}\psi_j \qquad j = 1, 2$$

Show that the secular equation associated with the trial function is

$$\begin{vmatrix} H_{11} - E & H_{12} \\ H_{12} & H_{22} - E \end{vmatrix} = \begin{vmatrix} E_1^{(0)} + E_1^{(1)} - E & H_{12} \\ H_{12} & E_2^{(0)} + E_2^{(1)} - E \end{vmatrix} = 0 \qquad (1)$$

where

$$E_j^{(1)} = \int \psi_j^* \hat{H}^{(1)} \psi_j \, d\tau \quad \text{and} \quad H_{12} = \int \psi_1^* \hat{H}^{(1)} \psi_2 \, d\tau$$

Solve Equation 1 for E to obtain

$$E = \frac{E_1^{(0)} + E_1^{(1)} + E_2^{(0)} + E_2^{(1)}}{2}$$
$$\pm \frac{1}{2}\left\{[E_1^{(0)} + E_1^{(1)} - E_2^{(0)} - E_2^{(1)}]^2 + 4H_{12}^2\right\}^{1/2} \qquad (2)$$

If we arbitrarily assume that $E_1^{(0)} + E_1^{(1)} < E_2^{(0)} + E_2^{(1)}$, then we take the positive sign in Equation 2 and write

$$E = \frac{E_1^{(0)} + E_1^{(1)} + E_2^{(0)} + E_2^{(1)}}{2} + \frac{E_1^{(0)} + E_1^{(1)} - E_2^{(0)} - E_2^{(1)}}{2}$$
$$\times \left\{1 + \frac{4H_{12}^2}{[E_1^{(0)} + E_1^{(1)} - E_2^{(0)} - E_2^{(1)}]^2}\right\}^{1/2}$$

Use the expansion $(1 + x)^{1/2} = 1 + x/2 + \cdots, \; x < 1$ to get

$$E = E_1^{(0)} + E_1^{(1)} + \frac{H_{12}^2}{E_1^{(0)} + E_1^{(1)} - E_2^{(0)} - E_2^{(1)}} + \cdots \qquad (3)$$

Note that if $E_1^{(0)} + E_1^{(1)}$ and $E_2^{(0)} + E_2^{(1)}$ are widely separated, the term involving H_{12}^2 in Equation 3 is small. Therefore, the energy is simply that calculated using ψ_1 alone; the ψ_2 part of the trial function contributes little to the overall energy. The general result is that terms in a trial function that correspond to higher and higher energies contribute less and less to the total ground-state energy.

Because we can assume that the ψ_j are normalized, we know that

$$S_{11} = S_{22} = 1 \qquad S_{12} = S_{21} = 0$$

and because $\hat{H}$ is Hermitian, we know that $H_{12} = H_{21}$. Now we find the values of the H_{ij}:

$$\hat{H}\psi_1 = \hat{H}^{(0)}\psi_1 + \hat{H}^{(1)}\psi_1 = E_1^0\psi_1 + E_1^{(1)}\psi_1$$

$$\hat{H}\psi_2 = \hat{H}^{(0)}\psi_2 + \hat{H}^{(1)}\psi_2 = E_2^0\psi_2 + E_2^{(1)}\psi_2$$

$$H_{11} = \int d\tau\, \psi_1^* \hat{H}\psi_1 = E_1^{(0)} + E_1^{(1)}$$

$$H_{22} = \int d\tau\, \psi_2^* \hat{H}\psi_2 = E_2^{(0)} + E_2^{(1)}$$

$$H_{12} = \int d\tau\, \psi_1^* \hat{H}\psi_2 = E_2^{(0)} \int d\tau\, \psi_1^* \psi_2 + \int d\tau\, \psi_1^* \hat{H}^{(1)}\psi_2$$

$$= \int d\tau\, \psi_1^* \hat{H}^{(1)}\psi_2$$

Substituting these results into Equation 7.37

$$\begin{vmatrix} H_{11} - ES_{11} & H_{12} - ES_{12} \\ H_{21} - ES_{21} & H_{22} - ES_{22} \end{vmatrix} = 0 \tag{7.37}$$

gives

$$\begin{vmatrix} E_1^{(0)} + E_1^{(1)} - E & H_{12} \\ H_{12} & E_2^{(0)} + E_2^{(1)} - E \end{vmatrix} = 0 \tag{1}$$

Expanding this determinant gives

$$\left[E_1^{(0)} + E_1^{(1)} - E \right]\left[E_2^{(0)} + E_2^{(1)} - E \right] - H_{12}^2 = 0$$

$$E^2 - E\left[E_1^{(0)} + E_1^{(1)} + E_2^{(0)} + E_2^{(1)} \right] + \left[E_1^{(0)} + E_1^{(1)} \right]\left[E_2^{(0)} + E_2^{(1)} \right] - H_{12}^2 = 0$$

Now we solve for E:

$$\begin{aligned}
E &= \frac{E_1^{(0)} + E_1^{(1)} + E_2^{(0)} + E_2^{(1)}}{2} \pm \frac{1}{2}\left\{ \left[E_1^{(0)} + E_1^{(1)} \right]^2 + \left[E_2^{(0)} + E_2^{(1)} \right]^2 \right. \\
&\qquad \left. + 2\left[E_1^{(0)} + E_1^{(1)} \right]\left[E_2^{(0)} + E_2^{(1)} \right] - 4\left[E_1^{(0)} + E_1^{(1)} \right]\left[E_2^{(0)} + E_2^{(1)} \right] + 4H_{12}^2 \right\}^{1/2} \\
&= \frac{E_1^{(0)} + E_1^{(1)} + E_2^{(0)} + E_2^{(1)}}{2} \pm \frac{1}{2}\left\{ \left[E_1^{(0)} + E_1^{(1)} - E_2^{(0)} - E_2^{(1)} \right]^2 + 4H_{12}^2 \right\}^{1/2} \\
&= \frac{E_1^{(0)} + E_1^{(1)} + E_2^{(0)} + E_2^{(1)}}{2} + \frac{\left[E_1^{(0)} + E_1^{(1)} - E_2^{(0)} - E_2^{(1)} \right]}{2} \\
&\qquad \times \left\{ 1 + \frac{4H_{12}^2}{\left[E_1^{(0)} + E_1^{(1)} - E_2^{(0)} - E_2^{(1)} \right]^2} \right\}^{1/2} \\
&= E_1^{(0)} + E_1^{(1)} + \frac{H_{12}^2}{E_1^{(0)} + E_1^{(1)} - E_2^{(0)} - E_2^{(1)}} + O(H_{12}^4)
\end{aligned}$$

where we have chosen the positive sign in front of the square root term because $E_1 < E_2$.

7–19. We will derive the equations for first-order perturbation theory in this problem. The problem we want to solve is

$$\hat{H}\psi = E\psi \tag{1}$$

where

$$\hat{H} = \hat{H}^{(0)} + \hat{H}^{(1)}$$

and where the problem

$$\hat{H}^{(0)}\psi^{(0)} = E^{(0)}\psi^{(0)} \tag{2}$$

has been solved exactly previously, so that $\psi^{(0)}$ and $E^{(0)}$ are known. Assuming now that the effect of $\hat{H}^{(1)}$ is small, write

$$\psi = \psi^{(0)} + \Delta\psi$$
$$E = E^{(0)} + \Delta E \tag{3}$$

where we assume that $\Delta\psi$ and ΔE are small. Substitute Equations 3 into Equation 1 to obtain

$$\hat{H}^{(0)}\psi^{(0)} + \hat{H}^{(1)}\psi^{(0)} + \hat{H}^{(0)}\Delta\psi + \hat{H}^{(1)}\Delta\psi = E^{(0)}\psi^{(0)} + \Delta E\psi^{(0)} + E^{(0)}\Delta\psi + \Delta E\Delta\psi \tag{4}$$

The first terms on each side of Equation 4 cancel because of Equation 2. In addition, we will neglect the last terms on each side because they represent the product of two small terms. Thus, Equation 4 becomes

$$\hat{H}^{(0)}\Delta\psi + \hat{H}^{(1)}\psi^{(0)} = E^{(0)}\Delta\psi + \Delta E\psi^{(0)} \tag{5}$$

Realize that $\Delta\psi$ and ΔE are the unknown quantities in this equation. Note that all the terms in Equation (5) are of the same order, in the sense that each is the product of an unperturbed term and a small term. We say that this equation is first order in the perturbation and that we are using first-order perturbation theory. The two terms we have neglected in Equation 4 are second-order terms and lead to second-order (and higher) corrections. Equation 5 can be simplified considerably. Multiply both sides from the left by $\psi^{(0)*}$ and integrate over all space to get

$$\int \psi^{(0)*}[\hat{H}^{(0)} - E^{(0)}]\Delta\psi\, d\tau + \int \psi^{(0)*}\hat{H}^{(1)}\psi^{(0)}\, d\tau = \Delta E \int \psi^{(0)*}\psi^{(0)}\, d\tau \tag{6}$$

The integral in the last term in Equation 6 is unity because $\psi^{(0)}$ is taken to be normalized. More important, however, is that the first term on the left side of Equation 6 is zero. Use the fact that $\hat{H}^{(0)} - E^{(0)}$ is Hermitian to show that

$$\int \psi^{(0)*}[\hat{H}^{(0)} - E^{(0)}]\Delta\psi\, d\tau = \int \{[\hat{H}^{(0)} - E^{(0)}]\psi^{(0)}\}^*\Delta\psi\, d\tau$$

But according to Equation 2, the integrand here vanishes. Thus, Equation 6 becomes

$$\Delta E = \int \psi^{(0)*}\hat{H}^{(1)}\psi^{(0)}\, d\tau \tag{7}$$

Equation 7 is called the *first-order correction* to $E^{(0)}$. To first order, the energy is

$$E = E^{(0)} + \int \psi^{(0)*}\hat{H}^{(1)}\psi^{(0)}\, d\tau + \text{higher order terms}$$

Substitute Equations 3 into Equation 1 to find Equation 4:

$$\left[\hat{H}^{(0)} + \hat{H}^{(1)}\right]\left[\psi^{(0)} + \Delta\psi\right] = \left[E^{(0)} + \Delta E\right]\left[\psi^{(0)} + \Delta\psi\right]$$

$$\hat{H}^{(0)}\psi^{(0)} + \hat{H}^{(1)}\psi^{(0)} + \hat{H}^{(0)}\Delta\psi + \hat{H}^{(1)}\Delta\psi = E^{(0)}\psi^{(0)} + \Delta E\psi^{(0)} + E^{(0)}\Delta\psi + \Delta E\Delta\psi$$

Recall that $\hat{H}^{(0)}\psi^{(0)} = E^{(0)}\psi^{(0)}$. If we neglect the last terms on each side of the above equation because they are the product of two small terms, Equation 4 becomes

$$\hat{H}^{(0)}\Delta\psi + \hat{H}^{(1)}\psi^{(0)} = E^{(0)}\Delta\psi + \Delta E\psi^{(0)} \tag{5}$$

which we can write as

$$\left[\hat{H}^{(0)} - E^{(0)}\right]\Delta\psi + \hat{H}^{(1)}\psi^{(0)} = \Delta E\psi^{(0)}$$

Multiplying the above expression by $\psi^{(0)*}$ and integrating gives Equation 6:

$$\int \psi^{(0)*}[\hat{H}^{(0)} - E^{(0)}]\Delta\psi\, d\tau + \int \psi^{(0)*}\hat{H}^{(1)}\psi^{(0)}\, d\tau = \Delta E\int \psi^{(0)*}\psi^{(0)}\, d\tau \tag{6}$$

We now use the fact that $\hat{H}^{(0)} - E^{(0)}$ is Hermitian. Recall from Chapter 4 that a Hermitian operator $\hat{A}$ satisfies the equation

$$\int d\tau f^* \hat{A}g = \int d\tau g \hat{A}^* f^* \tag{4.31}$$

Thus

$$\int \psi^{(0)*}[\hat{H}^{(0)} - E^{(0)}]\Delta\psi\, d\tau = \int \Delta\psi\{[\hat{H}^{(0)} - E^{(0)}]\psi^{(0)}\}^*\, d\tau$$

$$= \int \{[\hat{H}^{(0)} - E^{(0)}]\psi^{(0)}\}^*\Delta\psi\, d\tau$$

The integrand vanishes because $\hat{H}^{(0)}\psi^{(0)} - E^{(0)}\psi^{(0)} = 0$. Equation 6 then becomes

$$\Delta E = \int \psi^{(0)*}\hat{H}^{(1)}\psi^{(0)}\, d\tau \tag{7}$$

7–20. Identify $\hat{H}^{(0)}$, $\hat{H}^{(1)}$, $\psi^{(0)}$, and $E^{(0)}$ for the following problems:

a. An oscillator governed by the potential

$$V(x) = \frac{k}{2}x^2 + \frac{\gamma}{6}x^3 + \frac{b}{24}x^4$$

b. A particle constrained to move in the region $0 \le x \le a$ with the potential

$$V(x) = 0 \qquad 0 < x < \frac{a}{2}$$

$$= b \qquad \frac{a}{2} < x < a$$

c. A helium atom

d. A hydrogen atom in an electric field of strength $\mathcal{E}$. The Hamiltonian operator for this system is

$$\hat{H} = -\frac{\hbar^2}{2m_e}\nabla^2 - \frac{e^2}{4\pi\varepsilon_0 r} + e\mathcal{E}r\cos\theta$$

e. A rigid rotator with a dipole moment μ in an electric field of strength $\mathcal{E}$. The Hamiltonian operator for this system is

$$\hat{H} = -\frac{\hbar^2}{2I}\nabla^2 + \mu\mathcal{E}\cos\theta$$

where ∇^2 is given by Equation 6.3.

---

a. The first term in this potential is the harmonic oscillator potential, so we will choose a harmonic oscillator as the reference system. Thus

$$\hat{H}^{(0)} = -\frac{\hbar^2}{2\mu}\frac{d^2}{dx^2} + \frac{1}{2}kx^2$$

$$\hat{H}^{(1)} = \frac{\gamma}{6}x^3 + \frac{b}{24}x^4$$

$$\psi^{(0)} = N_v H_v\left(\alpha^{1/2}x\right)e^{-\alpha x^2/2}$$

$$E^{(0)} = h\nu\left(v + \frac{1}{2}\right) \qquad v = 0, 1, 2, \ldots$$

where we have used Equations 5.26, 5.35, and 5.27 to specify $\hat{H}^{(0)}$, $\psi^{(0)}$, and $E^{(0)}$, respectively.

b. This particle is confined to the region $0 \le x \le a$, so we will choose a particle in a box as a reference system. Thus

$$\hat{H}^{(0)} = -\frac{\hbar^2}{2m}\frac{d^2}{dx^2} \qquad 0 \le x \le a$$

$$\hat{H}^{(1)} = 0 \qquad 0 < x < \frac{a}{2}$$

$$= b \qquad \frac{a}{2} < x < a$$

$$\psi^{(0)} = \left(\frac{2}{a}\right)^{1/2}\sin\frac{n\pi x}{a} \qquad n = 1, 2, \ldots \quad 0 \le x \le a$$

$$E^{(0)} = \frac{n^2 h^2}{8ma^2} \qquad n = 1, 2, \ldots$$

where we have used Equations 3.14, 3.27, and 3.21 to specify $\hat{H}^{(0)}$, $\psi^{(0)}$, and $E^{(0)}$, respectively.

c. If we neglect the interelectronic repulsion in the Hamiltonian operator of the helium atom, then we have

$$\hat{H}_{\text{He}}^{(0)} = \hat{H}_{\text{H}}(1) + \hat{H}_{\text{H}}(2)$$

where $\hat{H}_{\text{H}}(j)$ is a hydrogen-like Hamiltonian operator. Then

$$\hat{H}^{(0)} = \hat{H}_{\text{H}}(1) + \hat{H}_{\text{H}}(2)$$

$$\hat{H}^{(1)} = \frac{e^2}{4\pi\varepsilon_0 r_{12}}$$

$$\psi^{(0)} = \psi_{nlm}(\mathbf{r}_1)\psi_{n'l'm'}(\mathbf{r}_2)$$

$$E^{(0)} = -\frac{Z^2\mu e^4}{8\varepsilon_0^2 h^2 n_1^2} - \frac{Z^2\mu e^4}{8\varepsilon_0^2 h^2 n_2^2}$$

where $Z = 2$ for helium, the $\psi_{nlm}(\mathbf{r}_j)$ are given in Table 6.6 and the $\hat{H}_{\text{H}}(j)$ are found using Equation 6.2. We use Equation 6.44 for $E^{(0)}$.

d. The natural reference system to use is that of a hydrogen atom in the absence of an electric field. We then have

$$\hat{H}^{(0)} = -\frac{\hbar^2}{2\mu}\nabla^2 - \frac{e^2}{4\pi\varepsilon_0 r}$$

$$\hat{H}^{(1)} = e\mathcal{E}r\cos\theta$$

$$\psi^{(0)} = \psi_{nlm}(r, \theta, \phi)$$

$$E^{(0)} = -\frac{\mu e^4}{8\varepsilon_0^2 h^2 n^2}$$

where the $\psi_{nlm}(\mathbf{r}_j)$ are given in Table 6.6. We use Equation 6.44 for $E^{(0)}$ and Equation 6.2 for $\hat{H}^{(0)}$.

e. The natural reference system is that of a rigid rotator in the absence of an electric field. We then have

$$\hat{H}^{(0)} = -\frac{\hbar^2}{2I}\nabla^2$$

$$\hat{H}^{(1)} = \mu\mathcal{E}\cos\theta$$

$$\psi^{(0)} = Y_l^m(\theta, \phi)$$

$$E^{(0)} = -\frac{\hbar^2}{2I}J(J+1)$$

where the $Y_l^m(\theta, \phi)$ are given in Table 6.3. We use Equation 5.57 for $E^{(0)}$, Equation 5.53 for $\psi^{(0)}$, and Equation 5.48 for $\hat{H}^{(0)}$.

7–21. Using a harmonic oscillator as the unperturbed problem, calculate the first-order correction to the energy of the $v = 0$ level for the system described in Problem 7–20(a).

For $v = 0$, we use Table 5.3 to find

$$\psi_0(x) = \left(\frac{\alpha}{\pi}\right)^{1/4} e^{-\alpha x^2/2}$$

Now use Equation 7.47 and substitute the quantities found in Problem 7–20(a):

$$\Delta E = \int \psi^{(0)*} \hat{H}^{(1)} \psi^{(0)} d\tau$$

$$= \int_{-\infty}^{\infty} dx\, \psi_0^*(x)\left(\frac{\gamma}{6}x^3 + \frac{b}{24}x^4\right)\psi_0(x)$$

$$= \left(\frac{\alpha}{\pi}\right)^{1/2}\int_{-\infty}^{\infty} dx\, e^{-\alpha x^2}\left(\frac{\gamma}{6}x^3 + \frac{b}{24}x^4\right)$$

$$= 0 + \frac{b}{12}\left(\frac{\alpha}{\pi}\right)^{1/2}\int_0^{\infty} dx\, x^4 e^{-\alpha x^2}$$

$$= \frac{b}{32\alpha^2}$$

where we have used the fact that $x^n e^{-cx^2}$ is an odd function if n is odd and an even function if n is even.

7–22. Using a particle in a box as the unperturbed problem, calculate the first-order correction to the ground-state energy for the system described in Problem 7–20(b).

Again we use Equation 7.47 and substitute the parameters found in Problem 7–20(b):

$$\Delta E = \int \psi^{(0)*} \hat{H}^{(1)} \psi^{(0)} d\tau$$

$$= \left(\frac{2b}{a}\right) \int_{a/2}^{a} dx \sin^2 \frac{\pi x}{a}$$

$$= \left(\frac{2b}{a}\right) \frac{x}{2} - \frac{a}{4\pi} \sin \frac{2\pi x}{a} \Big|_{a/2}^{a}$$

$$= \left(\frac{2b}{a}\right) \left(\frac{a}{2} - \frac{a}{4}\right) = \frac{b}{2}$$

7–23. Using the result of Problem 7–20(d), calculate the first-order correction to the ground-state energy of a hydrogen atom in an external electric field of strength $\mathcal{E}$.

From Table 6.6, ψ_{nlm} for the ground state of a hydrogen atom is

$$\psi_{100} = \frac{1}{\sqrt{\pi}} \left(\frac{1}{Z_0}\right)^{3/2} e^{-r/a_0}$$

Now use Equation 7.47 and substitute the parameters found in Problem 7–20(d) to obtain

$$\Delta E = \int \psi^{(0)*} \hat{H}^{(1)} \psi^{(0)} d\tau$$

$$= \frac{e\mathcal{E}}{\pi} \left(\frac{1}{Z_0}\right)^3 \int_0^{\infty} dr r^3 e^{-r/a_0} \int_0^{2\pi} d\phi \int_0^{\pi} d\theta \sin\theta \cos\theta$$

The integral over θ is

$$\int_0^{\pi} d\theta \sin\theta \cos\theta = \int_{-1}^{1} dx x = 0$$

so $\Delta E = 0$.

7–24. Calculate the first-order correction to the energy of a particle constrained to move within the region $0 \le x \le a$ in the potential

$$V(x) = V_0 x \qquad\qquad 0 \le x \le \frac{a}{2}$$

$$= V_0(a - x) \qquad\qquad \frac{a}{2} \le x \le a$$

where V_0 is a constant.

Use Equation 7.47, $\psi^{(0)}$ from 7–20(b), and the potential $V(x)$ as $\hat{H}^{(1)}$ to obtain

$$\Delta E = \int \psi^{(0)*} \hat{H}^{(1)} \psi^{(0)} d\tau$$

$$= \frac{2V_0}{a} \int_0^{a/2} dx x \sin^2 \frac{n\pi x}{a} + \frac{2V_0}{a} \int_{a/2}^{a} dx(a - x) \sin^2 \frac{n\pi x}{a}$$

$$= \frac{2V_0}{a} \left(\frac{x^2}{4} - \frac{ax \sin \frac{2n\pi x}{a}}{4n\pi} - \frac{a^2 \cos \frac{2n\pi x}{a}}{8n^2\pi^2}\right) \Big|_0^{a/2} + 2V_0 \left(\frac{x}{2} - \frac{a}{4n\pi} \sin \frac{2\pi n x}{a}\right) \Big|_{a/2}^{a}$$

$$-\frac{2V_0}{a}\left(\frac{x^2}{4}-\frac{ax\sin\frac{2n\pi x}{a}}{4n\pi}-\frac{a^2\cos\frac{2n\pi x}{a}}{8n^2\pi^2}\right)\Bigg|_{a/2}^{a}$$

$$=\frac{2V_0}{a}\left(\frac{a^2}{16}-\frac{a^2\cos n\pi}{8n^2\pi^2}+\frac{a^2}{8n^2\pi^2}\right)+2V_0\left(\frac{a}{2}-\frac{a}{4}\right)-\frac{2V_0}{a}\left(\frac{a^2}{4}-\frac{a^2}{8n^2\pi^2}-\frac{a^2}{16}-\frac{a^2\pi}{8n^2\pi^2}\right)$$

$$=V_0 a\left(\frac{1}{4}+\frac{1-\cos n\pi}{2n^2\pi^2}\right)$$

7–25. Use first-order perturbation theory to calculate the first-order correction to the ground-state energy of a quartic oscillator whose potential energy is

$$V(x) = cx^4$$

In this case, use a harmonic oscillator as the unperturbed system. What is the perturbing potential?

The Hamiltonian operator is

$$\hat{H} = -\frac{\hbar^2}{2\mu}\frac{d^2}{dx^2} + cx^4$$

To use a harmonic oscillator as the reference system, add and substract $kx^2/2$ from $\hat{H}$:

$$\hat{H} = \underbrace{-\frac{\hbar^2}{2\mu}\frac{d^2}{dx^2} + \frac{kx^2}{2}}_{\hat{H}^0} + \underbrace{cx^4 - \frac{kx^2}{2}}_{\hat{H}^1}$$

To get perturbed $\hat{H}$, add and substract $\frac{kx^2}{2}$

so

$$\hat{H}^{(1)} = cx^4 - \frac{kx^2}{2}$$

Now use Equation 7.47:

$$\Delta E = \int \psi^{(0)*}\hat{H}^{(1)}\psi^{(0)}d\tau$$

where $\psi^0 = \sqrt[4]{\frac{\alpha}{\pi}}\,e^{-\alpha x^2/2}$

$$= \left(\frac{\alpha}{\pi}\right)^{1/2}\int_{-\infty}^{\infty}dx\,e^{-\alpha x^2}\left(cx^4-\frac{kx^2}{2}\right)$$

$$= \left(\frac{\alpha}{\pi}\right)^{1/2}2\left[\frac{3c}{8\alpha^2}\left(\frac{\pi}{\alpha}\right)^{1/2}-\frac{k}{8\alpha}\left(\frac{\pi}{\alpha}\right)^{1/2}\right]$$

$$= \frac{3c}{4\alpha^2}-\frac{k}{4\alpha} \qquad \text{α is not variational α}$$

7–26. Use a trial function

$$\phi = c_1 x(a-x) + c_2 x^2(a-x)^2$$

for a particle in a one-dimensional box. For simplicity, let $a = 1$, which amounts to measuring all distances in units of a. Show that

$$H_{11} = \frac{\hbar^2}{6m} \qquad\qquad S_{11} = \frac{1}{30}$$

$$H_{12} = H_{21} = \frac{\hbar^2}{30m} \qquad\qquad S_{12} = S_{21} = \frac{1}{140}$$

$$H_{22} = \frac{\hbar^2}{105m} \qquad\qquad S_{22} = \frac{1}{630}$$

For a particle in a box,

$$\hat{H} = -\frac{\hbar^2}{2m}\frac{d^2}{dx^2}$$

The two components of our trial function are

$$\phi_1 = x(a - x) \qquad\qquad \phi_2 = x^2(a - x)^2$$

We then have

$$\hat{H}\phi_1 = \frac{\hbar^2}{2m} \qquad\qquad \hat{H}\phi_2 = -\frac{\hbar^2}{m}\left(a^2 - 6ax + 6x^2\right)$$

We now solve for the H_{ij} and S_{ij}, using the integral

$$\int_0^1 x^m(1 - x)^n dx = \frac{m!n!}{(m + n + 1)!}$$

Recall that we let $a = 1$, so

$$H_{11} = \frac{\hbar^2}{m}\int_0^1 x(1 - x)dx = \frac{\hbar^2}{m}\frac{1}{3!} = \frac{\hbar^2}{6m}$$

$$H_{12} = -\frac{\hbar^2}{m}\int_0^1 x(1 - x)(1 - 6x + 6x^2)dx$$

$$= -\frac{\hbar^2}{m}\left(\frac{1}{6} - 6\frac{1}{12} + 6\frac{3!}{5!}\right) = \frac{\hbar^2}{30m}$$

$$H_{21} = \frac{\hbar^2}{m}\int_0^1 x^2(1 - x)^2 dx = \frac{\hbar^2}{m}\frac{2\cdot2}{5!} = \frac{\hbar^2}{30m}$$

$$H_{22} = -\frac{\hbar^2}{m}\int_0^1 x^2(1 - x)^2(1 - 6x + 6x^2)dx$$

$$= -\frac{\hbar^2}{m}\left(\frac{1}{30} - 6\frac{3!2!}{6!} + 6\frac{4!2!}{7!}\right) = \frac{\hbar^2}{105m}$$

$$S_{11} = \int_0^1 x^2(1 - x)^2 dx = \frac{2\cdot2}{5!} = \frac{1}{30}$$

$$S_{12} = S_{21} = \int_0^1 x^3(1 - x)^3 dx = \frac{6\cdot6}{7!} = \frac{1}{140}$$

$$S_{22} = \int_0^1 x^4(1 - x)^4 dx = \frac{4!4!}{9!} = \frac{1}{630}$$

as stated in the problem.

7–27. In Example 5–2, we introduced the Morse potential

$$V(x) = D(1 - e^{-\beta x})^2$$

as a description of the intramolecular potential energy of a diatomic molecule. The constants D and β are different for each molecule; for H_2, $D = 7.61 \times 10^{-19}$ J and $\beta = 0.0193$ pm^{-1}. First expand the Morse potential in a power series about x. (*Hint*: Use the expansion $e^x = 1 + x + \frac{x^2}{2} + \frac{x^3}{6} + \cdots$.) What is the Hamiltonian operator for the Morse potential? Show that the Hamiltonian operator can be written in the form

$$\hat{H} = -\frac{\hbar^2}{2\mu}\frac{d^2}{dx^2} + ax^2 + bx^3 + cx^4 + \cdots \tag{1}$$

How are the constants a, b, and c related to the constants D and β? What part of the Hamiltonian operator would you associate with $\hat{H}^{(0)}$, and what are the functions $\psi_n^{(0)}$ and energies $E_n^{(0)}$? Use perturbation theory to evaluate the first-order corrections to the energy of the first three states that arise from the cubic and quartic terms. Using these results, how different are the first two energy levels of H_2 if its intramolecular potential is described by a harmonic oscillator potential or the quartic expansion of the Morse potential (see Equation 1)?

Use the power series of e^x given in the hint to expand the Morse potential:

$$
\begin{aligned}
V(x) &= D(1 - e^{-\beta x})^2 \\
&= D\left[1 - \left(1 - \beta x + \frac{\beta^2 x^2}{2} - \frac{\beta^3 x^3}{6} + O(x^4)\right)\right]^2 \\
&= D\left[\beta x - \frac{\beta^2 x^2}{2} + \frac{\beta^3 x^3}{6} + O(x^4)\right]^2 \\
&= D\beta^2 x^2\left[1 - \frac{\beta x}{2} + \frac{\beta^2 x^2}{6} + O(x^3)\right]^2 \\
&= D\beta^2 x^2\left[1 - \beta x + \frac{\beta^2 x^2}{3} + \frac{\beta^2 x^2}{4} + O(x^3)\right] \\
&= D\beta^2 x^2 - D\beta^3 x^3 + \frac{7}{12}D\beta^4 x^4 + O(x^5)
\end{aligned}
$$

The Hamiltonian operator for the system is

$$
\begin{aligned}
\hat{H} &= -\frac{\hbar^2}{2m}\frac{d^2}{dx^2} + V \\
&= -\frac{\hbar^2}{2m}\frac{d^2}{dx^2} + ax^2 + bx^3 + cx^4 + \ldots
\end{aligned}
$$

where $a = D\beta^2$, $b = -D\beta^3$, and $c = 7D\beta^4/12$. If we use a harmonic oscillator Hamiltonian operator for $\hat{H}^{(0)}$, then [Problem 7–20(a)],

$$\hat{H}^{(0)} = -\frac{\hbar^2}{2\mu}\frac{d^2}{dx^2} + ax^2$$

$$\hat{H}^{(1)} = bx^3 + cx^4$$

$$\psi_v^{(0)} = N_v H_v\left(\alpha^{1/2}x\right)e^{-\alpha x^2/2}$$

$$E_v^{(0)} = hv\left(v + \frac{1}{2}\right) \qquad v = 0, 1, 2, \ldots$$

For the first three states,

$$\psi_0^{(0)}(x) = \left(\frac{\alpha}{\pi}\right)^{1/4} e^{-\alpha x^2/2}$$

$$\psi_1^{(0)}(x) = \left(\frac{4\alpha^3}{\pi}\right)^{1/4} x e^{-\alpha x^2/2}$$

$$\psi_2^{(0)}(x) = \left(\frac{\alpha}{4\pi}\right)^{1/4} (2\alpha x^2 - 1) e^{-\alpha x^2/2}$$

Using perturbation theory (Equation 7.47 and 7.48),

$$E_0 = E_0^{(0)} + \int d\tau \, \psi^{(0)*} \hat{H}^{(1)} \psi^{(0)}$$

$$= \frac{h\nu}{2} + b\left(\frac{\alpha}{\pi}\right)^{1/2} \int_{-\infty}^{\infty} dx \, x^3 e^{-\alpha x^2} + c\left(\frac{\alpha}{\pi}\right)^{1/2} \int_{-\infty}^{\infty} dx \, x^4 e^{-\alpha x^2}$$

[handwritten: SHOULD BE h, NOT ħ — pointing at the h\nu/2 term]

$$= \frac{h\nu}{2} + 0 + 2c\left(\frac{\alpha}{\pi}\right)^{1/2} \int_{0}^{\infty} dx \, x^4 e^{-\alpha x^2}$$

$$= \frac{h\nu}{2} + \frac{3c}{4\alpha^2}$$

$$E_1 = E_1^{(0)} + \int d\tau \, \psi^{(0)*} \hat{H}^{(1)} \psi^{(0)}$$

$$= \frac{3h\nu}{2} + b\left(\frac{4\alpha^3}{\pi}\right)^{1/2} \int_{-\infty}^{\infty} dx \, x^5 e^{-\alpha x^2} + c\left(\frac{4\alpha^3}{\pi}\right)^{1/2} \int_{-\infty}^{\infty} dx \, x^6 e^{-\alpha x^2}$$

$$= \frac{3h\nu}{2} + 0 + 2c\left(\frac{4\alpha^3}{\pi}\right)^{1/2} \int_{0}^{\infty} dx \, x^6 e^{-\alpha x^2}$$

$$= \frac{3h\nu}{2} + \frac{15c}{4\alpha^2}$$

$$E_2 = E_2^{(0)} + \int d\tau \, \psi^{(0)*} \hat{H}^{(1)} \psi^{(0)}$$

$$= \frac{5h\nu}{2} + b\left(\frac{\alpha}{4\pi}\right)^{1/2} \int_{-\infty}^{\infty} dx \, (2\alpha x^2 - 1)^2 x^3 e^{-\alpha x^2}$$

$$+ c\left(\frac{\alpha}{4\pi}\right)^{1/2} \int_{-\infty}^{\infty} dx \, (2\alpha x^2 - 1)^2 x^4 e^{-\alpha x^2}$$

$$= \frac{5h\nu}{2} + 0 + 2c\left(\frac{\alpha}{4\pi}\right)^{1/2} \int_{0}^{\infty} dx \, (2\alpha x^2 - 1)^2 x^4 e^{-\alpha x^2}$$

$$= \frac{5h\nu}{2} + \frac{39c}{4\alpha^2}$$

7–28. In this problem, we will solve the Schrödinger equation for the ground-state wave function and energy of a particle in a spherical box of radius a. The Schrödinger equation is given by Equation 6.43 with $l = 0$ (ground state) and without the $e^2/4\pi\varepsilon_0 r$ term:

$$-\frac{\hbar^2}{2mr^2}\frac{d}{dr}\left(r^2\frac{d\psi}{dr}\right) = E\psi$$

Substitute $u = r\psi$ into this equation to get

$$\frac{d^2u}{dr^2} + \frac{2mE}{\hbar^2}u = 0$$

The general solution to this equation is

$$u(r) = A \cos \alpha r + B \sin \alpha r$$

or

$$\psi(r) = \frac{A \cos \alpha r}{r} + \frac{B \sin \alpha r}{r}$$

where $\alpha = (2mE/\hbar^2)^{1/2}$. Which of these terms is finite at $r = 0$? Now use the fact that $\psi(a) = 0$ to prove that

$$\alpha a = \pi$$

for the ground state, or that the ground-state energy is

$$E = \frac{\pi^2 \hbar^2}{2ma^2}$$

Show that the normalized ground-state wave function is

$$\psi(r) = (2\pi a)^{-1/2} \frac{\sin \pi r/a}{r}$$

First find the first and second derivatives of ψ with respect to r in terms of u:

$$\psi = \frac{u}{r} \qquad \frac{d\psi}{dr} = -\frac{u}{r^2} + \frac{1}{r}\frac{du}{dr} \qquad \frac{d^2\psi}{dr^2} = \frac{2u}{r^3} - \frac{2}{r^2}\frac{du}{dr} + \frac{1}{r}\frac{d^2u}{dr^2}$$

We can now rearrange the Schrödinger equation as follows:

$$-\frac{\hbar^2}{2mr^2}\frac{d}{dr}\left(r^2\frac{d\psi}{dr}\right) = E\psi$$

$$-\frac{\hbar^2}{2m}\left[\frac{d^2\psi}{dr^2} + \left(\frac{2}{r}\right)\frac{d\psi}{dr}\right] = E\psi$$

$$-\frac{\hbar^2}{2m}\left[\frac{2u}{r^3} - \frac{2}{r^2}\frac{du}{dr} + \frac{1}{r}\frac{d^2u}{dr^2} + \frac{2}{r}\left(-\frac{u}{r^2} + \frac{1}{r}\frac{du}{dr}\right)\right] = E\left(\frac{u}{r}\right)$$

$$-\frac{\hbar^2}{2m}\frac{d^2u}{dr^2} = Eu$$

$$\frac{d^2u}{dr^2} + \frac{2mE}{\hbar^2}u = 0$$

Recall from Example 2–4 that the general solution to this equation is

$$u(r) = A \cos \alpha r + B \sin \alpha r \qquad \alpha = \left(\frac{2mE}{\hbar^2}\right)^{1/2}$$

$$\psi(r) = A\frac{\cos \alpha r}{r} + B\frac{\sin \alpha r}{r}$$

To decide which of these terms is finite at $r = 0$, examine the series expansions of $\cos \alpha r$ and $\sin \alpha r$:

$$\cos \alpha r = 1 - \frac{(\alpha r)^2}{2!} + \frac{(\alpha r)^4}{4!} + O(r^6)$$

$$\sin \alpha r = \alpha r - \frac{(\alpha r)^3}{3!} + \frac{(\alpha r)^5}{5!} + O(r^7)$$

Only the sine term will be finite at $r = 0$, because $\sin \alpha r$ and r will both approach zero as $r \to 0$. We can then write $\psi(r)$ as

$$\psi(r) = B \frac{\sin \alpha r}{r}$$

Because $\psi(a) = 0$,

$$0 = B \frac{\sin \alpha a}{a}$$
$$\alpha a = n\pi$$

or, for the ground state, $\alpha a = \pi$. Now, using the definition of α given in the problem, we find

$$\alpha a = \pi$$
$$\left(\frac{2mE_0}{\hbar^2}\right)^{1/2} a = \pi$$
$$\frac{2mE_0 a^2}{\hbar^2} = \pi^2$$
$$E_0 = \frac{\pi^2 \hbar^2}{2ma^2}$$

Now we normalize ψ by requiring

$$\int d\tau \psi^* \psi = 1$$
$$4\pi \int_0^a dr r^2 \psi^2(r) = 1$$
$$4\pi B^2 \int_0^a dr \sin^2 \alpha r = 1$$
$$4\pi B^2 \left(\frac{a}{2}\right) = 1$$
$$B = \left(\frac{1}{2\pi a}\right)^{1/2}$$

So the normalized ground-state wave function is

$$\psi(r) = (2\pi a)^{-1/2} \frac{\sin(\pi r/a)}{r}$$

7–29. In this problem, (see also Problem 4–38) we calculate the ground-state energy for the potential shown in Problem 7–12. The Schrödinger equation for this system is

$$-\frac{\hbar^2}{2m} \frac{d^2\psi}{dx^2} + V_0 \psi = E\psi \qquad -\infty < x < -a$$
$$-\frac{\hbar^2}{2m} \frac{d^2\psi}{dx^2} = E\psi \qquad -a < x < a$$
$$-\frac{\hbar^2}{2m} \frac{d^2\psi}{dx^2} + V_0 \psi = E\psi \qquad a < x < \infty$$

Label the three regions 1, 2, and 3. For the case $E < V_0$, show that

$$\psi_1(x) = Ae^{\beta x} + Be^{-\beta x}$$

$$\psi_2(x) = C\sin\alpha x + D\cos\alpha x$$

$$\psi_3(x) = Ee^{\beta x} + Fe^{-\beta x}$$

where $\beta = [2m(V_0 - E)/\hbar^2]^{1/2}$ is real and $\alpha = (2mE/\hbar^2)^{1/2}$. If $\psi_1(x)$ is to be finite as $x \to -\infty$ and $\psi_3(x)$ be finite as $x \to \infty$, we must have $B = 0$ and $E = 0$. Now there are four constants (A, C, D, and F) to be determined by the four boundary conditions

$$\psi_1(-a) = \psi_2(-a) \qquad \frac{d\psi_1}{dx}\bigg|_{x=-a} = \frac{d\psi_2}{dx}\bigg|_{x=-a}$$

$$\psi_2(a) = \psi_3(a) \qquad \frac{d\psi_2}{dx}\bigg|_{x=a} = \frac{d\psi_3}{dx}\bigg|_{x=a}$$

Before we go into all this algebra, let's remember that we are interested only in the ground-state energy. In this case, we expect $\psi_2(x)$ to be a cosine term because $\cos\alpha x$ has no nodes in region 2, whereas $\sin\alpha x$ does. Therefore, we will set $C = 0$. Show that the four boundary conditions give

$$Ae^{-\beta a} = D\cos\alpha a \qquad A\beta e^{-\beta a} = D\alpha\sin\alpha a$$

$$D\cos\alpha a = Fe^{-\beta a} \qquad -D\alpha\sin\alpha a = -F\beta e^{-\beta a}$$

These equations give $A = F$. Now divide $A\beta e^{-\beta a} = D\alpha\sin\alpha a$ by $Ae^{-\beta a} = D\cos\alpha a$ to get

$$\beta = \alpha\tan\alpha a$$

Now show that

$$\alpha^2 + \beta^2 = \frac{2mV_0}{\hbar^2}$$

and so

$$\alpha^2 + \alpha^2\tan^2\alpha a = \frac{2mV_0}{\hbar^2}$$

Multiply through by a^2 to get

$$\eta^2(1 + \tan^2\eta) = \frac{2mV_0 a^2}{\hbar^2} = \alpha$$

where $\eta = \alpha a$. Solve this equation numerically for η when $\alpha = 2mV_0 a^2/\hbar^2 = 4$ and 12 to verify the exact energies given in Problem 7–12.

The functions ψ_1, ψ_2, and ψ_3 are determined using the approach discussed in Chapter 4 (Problems 4-33 to 4-38).

Region 1 ($-\infty < x < -a$):

$$-\frac{\hbar^2}{2m}\frac{d^2\psi_1}{dx^2} + V_0\psi_1 = E\psi_1$$

$$\psi_1(x) = Ae^{\beta x} + Be^{-\beta x} \qquad \beta = \left[\frac{2m(E - V_0)}{\hbar^2}\right]^{1/2}$$

Region 2 $(-a < x < a)$:

$$-\frac{\hbar^2}{2m}\frac{d^2\psi_2}{dx^2} = E\psi_2$$

$$\psi_2(x) = C\sin\alpha x + D\cos\alpha x \qquad \alpha = \left(\frac{2mE}{\hbar^2}\right)^{1/2}$$

Region 3 $(a < x < \infty)$:

$$-\frac{\hbar^2}{2m}\frac{d^2\psi_3}{dx^2} + V_0\psi_3 = E\psi_3$$

$$\psi_3(x) = Ee^{\beta x} + Fe^{-\beta x} \qquad \beta = \left[\frac{2m(E - V_0)}{\hbar^2}\right]^{1/2}$$

As explained in the problem text, $B = E = 0$. We also set $C = 0$ because $\psi_2(x)$ should be a cosine term (as explained in the text of the problem). We then have the set of equations

$$\psi_1(x) = Ae^{\beta x} \qquad \psi_2(x) = D\cos\alpha x \qquad \psi_3(x) = Fe^{-\beta x}$$

Firstly, we impose the boundary conditions $\psi(-a) = 0$ and $d\psi/dx = 0$ at $x = -a$:

$$\psi_1(-a) = \psi_2(-a) \qquad \left.\frac{d\psi_1}{dx}\right|_{x=-a} = \left.\frac{d\psi_2}{dx}\right|_{x=-a}$$

$$Ae^{-\beta a} = D\cos\alpha a \qquad A\beta e^{-\beta a} = -D\alpha\sin(-\alpha a) = D\alpha\sin\alpha a$$

Now we impose the boundary conditions $\psi(a) = 0$ and $d\psi/dx = 0$ at $x = a$:

$$\psi_2(a) = \psi_3(a) \qquad \left.\frac{d\psi_2}{dx}\right|_{x=a} = \left.\frac{d\psi_3}{dx}\right|_{x=a}$$

$$D\cos\alpha a = Fe^{-\beta a} \qquad -D\sin\alpha a = -F\beta e^{-\beta a}$$

Following the suggestions in the problem, we find that

$$\frac{A\beta e^{-\beta a}}{Ae^{-\beta a}} = \frac{D\alpha\sin\alpha a}{D\cos\alpha a}$$
$$\beta = \alpha\tan\alpha a$$

From the definitions of α and β,

$$\alpha^2 + \beta^2 = \frac{2mE}{\hbar^2} + \frac{2m(V_0 - E)}{\hbar^2} = \frac{2mV_0}{\hbar^2}$$

Then, since $\beta^2 = \alpha^2\tan^2\alpha a$,

$$\alpha^2 + \alpha^2\tan^2\alpha a = \frac{2mV_0}{\hbar^2}$$

$$\eta^2(1 + \tan^2\eta) = \frac{2mV_0a^2}{\hbar^2} \qquad \eta = \alpha a$$

We can graph $f(\eta) = \eta^2(1 + \tan^2 \eta)$ vs. η for $\alpha = 4$ and $\alpha = 12$ to find a numerical value of η:

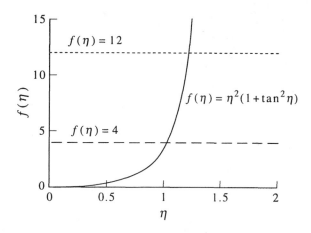

We find that $\alpha = 4$ at $\eta = 1.0299$ and $\alpha = 12$ at $\eta = 1.213$. Then, for $\alpha = 4$,

$$\frac{2mEa^2}{\hbar^2} = (1.0299)^2 \qquad E = 0.5303\frac{\hbar^2}{ma^2}$$

Likewise, for $\alpha = 12$, $E = 0.7357\dfrac{\hbar^2}{ma^2}$.

7–30. In applying first-order perturbation theory to the helium atom, we must evaluate the integral (Equation 7.50)

$$E^{(1)} = \frac{e^2}{4\pi\varepsilon_0} \int \int d\mathbf{r}_1 d\mathbf{r}_2 \psi_{1s}^*(\mathbf{r}_1)\psi_{1s}^*(\mathbf{r}_2)\frac{1}{r_{12}}\psi_{1s}(\mathbf{r}_1)\psi_{1s}(\mathbf{r}_2)$$

where

$$\psi_{1s}(\mathbf{r}_j) = \left(\frac{Z^3}{a_0^3\pi}\right)^{1/2} e^{-Zr_j/a_0}$$

and $Z = 2$ for the helium atom. This same integral occurs in a variational treatment of helium, where in that case the value of Z is left arbitrary. This problem proves that

$$E^{(1)} = \frac{5Z}{8}\left(\frac{m_e e^4}{16\pi^2\varepsilon_0^2\hbar^2}\right)$$

Let $\mathbf{r}_1$ and $\mathbf{r}_2$ be the radius vectors of electron 1 and 2, respectively, and let θ be the angle between these two vectors. Now this is generally *not* the θ of spherical coordinates, but if we choose one of the radius vectors, say $\mathbf{r}_1$, to be the z axis, then the two θ's are the same. Using the law of cosines,

$$r_{12} = (r_1^2 + r_2^2 - 2r_1 r_2 \cos\theta)^{1/2}$$

show that $E^{(1)}$ becomes

$$E^{(1)} = \frac{e^2}{4\pi\varepsilon_0}\frac{Z^6}{a_0^6\pi^2}\int_0^\infty dr_1 e^{-2Zr_1/a_0}4\pi r_1^2 \int_0^\infty dr_2 e^{-2Zr_2/a_0}r_2^2$$
$$\times \int_0^{2\pi} d\phi \int_0^\pi \frac{d\theta \sin\theta}{(r_1^2 + r_2^2 - 2r_1 r_2 \cos\theta)^{1/2}}$$

Letting $x = \cos\theta$, show that the integral over θ is

$$\int_0^\pi \frac{d\theta\,\sin\theta}{(r_1^2 + r_2^2 - 2r_1 r_2 \cos\theta)^{1/2}} = \int_{-1}^1 \frac{dx}{(r_1^2 + r_2^2 - 2r_1 r_2 x)^{1/2}}$$

$$= \frac{2}{r_1} \qquad r_1 > r_2$$

$$= \frac{2}{r_2} \qquad r_1 < r_2$$

Substituting this result into $E^{(1)}$, show that

$$E^{(1)} = \frac{e^2}{4\pi\varepsilon_0}\frac{16Z^6}{a_0^6}\int_0^\infty dr_1 e^{-2Zr_1/a_0}r_1^2\left(\frac{1}{r_1}\int_0^{r_1} dr_2 e^{-2Zr_2/a_0}r_2^2\right.$$

$$\left. + \int_{r_1}^\infty dr_2 e^{-2Zr_2/a_0}r_2\right)$$

$$= \frac{e^2}{4\pi\varepsilon_0}\frac{4Z^3}{a_0^3}\int_0^\infty dr_1 e^{-2Zr_1/a_0}r_1^2\left[\frac{1}{r_1} - e^{-2Zr_1/a_0}\left(\frac{Z}{a_0} + \frac{1}{r_1}\right)\right]$$

$$= \frac{5}{8}Z\left(\frac{e^2}{4\pi\varepsilon_0 a_0}\right) = \frac{5}{8}Z\left(\frac{m_e e^4}{16\pi^2\varepsilon_0^2\hbar^2}\right)$$

Show that the energy through first order is

$$E^{(0)} + E^{(1)} = \left(-Z^2 + \frac{5}{8}Z\right)\left(\frac{m_e e^4}{16\pi^2\varepsilon_0^2\hbar^2}\right) = -\frac{11}{4}\left(\frac{m_e e^4}{16\pi^2\varepsilon_0^2\hbar^2}\right)$$

$$= -2.75\left(\frac{m_e e^4}{16\pi^2\varepsilon_0^2\hbar^2}\right)$$

compared with the exact result, $E_{\text{exact}} = -2.9037(m_e e^4/16\pi^2\varepsilon_0^2\hbar^2)$.

Choosing $\mathbf{r}_1$ to be the z-axis allows us to write

$$r_{12} = (r_1^2 + r_2^2 - 2r_1 r_2 \cos\theta_1)^{1/2}$$

where we use the law of cosines to find r_{12}. Now

$$E^{(1)} = \frac{e^2}{4\pi\varepsilon_0}\int_0^{2\pi} d\phi_2 \int_0^\pi d\theta_2 \sin\theta_2 \int_0^{2\pi} d\phi_1 \int_0^\infty dr_1 r_1^2 \psi_1^2 \int_0^\infty dr_2 r_2^2 \psi_2^2$$

$$\times \int_0^\pi \frac{d\theta_1 \sin\theta_1}{(r_1^2 + r_2^2 - 2r_1 r_2 \cos\theta_1)^{1/2}}$$

$$= \frac{e^2}{4\pi\varepsilon_0}8\pi^2\left(\frac{Z^3}{a_0^3\pi}\right)^2\int_0^\infty dr_1 r_1^2 e^{-2Zr_1/a_0}\int_0^\infty dr_2 r_2^2 e^{-2Zr_2/a_0}$$

$$\times \int_0^\pi \frac{d\theta_1 \sin\theta_1}{\sqrt{(r_1^2 + r_2^2 - 2r_1 r_2 \cos\theta_1)}}$$

Let $x = \cos\theta_1$. Then $dx = -\sin\theta_1 d\theta_1$, and the limits on θ_1 of $(0, \pi)$ become the limits on x of $(1, -1)$. Then

$$\int_0^\pi \frac{d\theta_1 \sin\theta_1}{(r_1^2 + r_2^2 - 2r_1r_2\cos\theta_1)^{1/2}} = \int_{-1}^1 \frac{dx}{(r_1^2 + r_2^2 - 2r_1r_2x)^{1/2}}$$

$$= -\left. \frac{(r_1^2 + r_2^2 - 2r_1r_2x)^{1/2}}{r_1r_2} \right|_{-1}^1$$

$$= -\frac{(r_1^2 + r_2^2 - 2r_1r_2)^{1/2}}{r_1r_2} + \frac{(r_1^2 + r_2^2 + 2r_1r_2)^{1/2}}{r_1r_2}$$

The first term in this expression is $\pm(r_1 - r_2)/r_1r_2$ and the second term is equal to $\pm(r_1 + r_2)/r_1r_2$. Arbitrarily pick $r_1 < r_2$. Recall that E is an intrinsically positive quantity, so only positive solutions are physically acceptable as the solution of this integrand. Then

$$\int_{-1}^1 \frac{dx}{(r_1^2 + r_2^2 - 2r_1r_2x)^{1/2}} = \frac{r_2 - r_1 + r_1 + r_2}{r_1r_2} = \frac{2r_2}{r_1r_2} = \frac{2}{r_1}$$

Likewise, if $r_1 > r_2$,

$$\int_{-1}^1 \frac{dx}{(r_1^2 + r_2^2 - 2r_1r_2x)^{1/2}} = \frac{r_1 - r_2 + r_1 + r_2}{r_1r_2} = \frac{2r_1}{r_1r_2} = \frac{2}{r_2}$$

Substituting this result into $E^{(1)}$ gives

$$E^{(1)} = \frac{e^2}{4\pi\varepsilon_0} \frac{8Z^6}{a_0^6} \int_0^\infty dr_1 r_1^2 e^{-2Zr_1/a_0} \left(\frac{2}{r_1} \int_0^{r_1} dr_2 r_2^2 e^{-2Zr_2/a_0} + 2\int_{r_1}^\infty dr_2 r_2 e^{-2Zr_2/a_0} \right)$$

$$= \frac{e^2}{4\pi\varepsilon_0} \frac{16Z^6}{a_0^6} \int_0^\infty dr_1 r_1^2 e^{-2Zr_1/a_0} \left(\frac{1}{r_1} \int_0^{r_1} dr_2 r_2^2 e^{-2Zr_2/a_0} + \int_{r_1}^\infty dr_2 r_2 e^{-2Zr_2/a_0} \right)$$

Now

$$\int_{r_1}^\infty dr_2 r_2 e^{-2Zr_2/a_0} = \left. \frac{a_0^2 e^{-2Zr_2/a_0}}{4Z^2} \left(\frac{-2Zr_2}{a_0} - 1 \right) \right|_{r_1}^\infty = -\frac{a_0^2 e^{-2Zr_1/a_0}}{4Z^2} \left(\frac{-2Zr_1}{a_0} - 1 \right)$$

and

$$\int_0^{r_1} e^{-2Zr_2/a_0} r_2^2 dr_2 = \left. \frac{r_2^2 a_0 e^{-2Zr_2/a_0}}{-2Z} \right|_0^{r_1} - \frac{2a_0}{-2Z} \int_0^{r_1} e^{-2Zr_2/a_0} r_2 dr_2$$

$$= \frac{r_1^2 a_0 e^{-2Zr_1/a_0}}{-2Z} + \left. \frac{a_0^3 e^{-2Zr_2/a_0}}{4Z^3} \left(\frac{-2Zr_2}{a_0} - 1 \right) \right|_0^{r_1}$$

$$= \frac{r_1^2 a_0 e^{-2Zr_1/a_0}}{-2Z} + \frac{a_0^3 e^{-2Zr_1/a_0}}{4Z^3} \left(\frac{-2Zr_1}{a_0} - 1 \right) - \frac{a_0^3}{4Z^3}(-1)$$

Then

$$\frac{1}{r_1} \int_0^{r_1} dr_2 r_2^2 e^{-2Zr_2/a_0} + \int_{r_1}^\infty dr_2 r_2 e^{-2Zr_2/a_0} = \frac{a_0^2 e^{-2Zr_1/a_0}}{4Z^2} - \frac{a_0^2 e^{-2Zr_1/a_0}}{2Z^2}$$

$$- \frac{a_0^3 e^{-2Zr_1/a_0}}{4Z^3 r_1} + \frac{a_0^3}{4Z^3 r_1}$$

$$= \frac{a_0^3}{4Z^3} \left[\frac{1}{r_1} - e^{-2Zr_1/a_0} \left(\frac{Z}{a_0} + \frac{1}{r_1} \right) \right]$$

and our expression for $E^{(1)}$ becomes

$$
\begin{aligned}
E^{(1)} &= \frac{e^2}{4\pi\varepsilon_0} \frac{4Z^3}{a_0^3} \int_0^\infty dr_1 r_1^2 e^{-2Zr_1/a_0} \left[\frac{1}{r_1} - e^{-2Zr_1/a_0}\left(\frac{Z}{a_0} + \frac{1}{r_1} \right) \right] \\
&= \frac{e^2}{4\pi\varepsilon_0} \frac{4Z^3}{a_0^3} \left[\frac{a_0^2}{4Z^2} - \frac{Z}{a_0}\frac{2!a_0^3}{(4Z)^3} - \frac{a_0^2}{16Z^2} \right] \\
&= \frac{e^2}{4\pi\varepsilon_0} \frac{4Z^3}{a_0^3} \left(\frac{5a_0^2}{32Z^2} \right) \\
&= \frac{5Z}{8a_0}\frac{e^2}{4\pi\varepsilon_0} = \frac{5Z}{8}\left(\frac{m_e e^4}{16\pi^2\varepsilon_0^2\hbar^2} \right)
\end{aligned}
$$

From the discussion of the hydrogen atom in Chapter 6, $E^{(0)}$ for this system is

$$
E^{(0)} = -\frac{Z^2 m_e e^4}{16\pi^2\varepsilon_0^2\hbar^2} \tag{7.15}
$$

so

$$
\begin{aligned}
E^{(0)} + E^{(1)} &= \left(-Z^2 + \frac{5}{8}Z \right)\left(\frac{m_e e^4}{16\pi^2\varepsilon_0^2\hbar^2} \right) = -\frac{11}{4}\left(\frac{m_e e^4}{16\pi^2\varepsilon_0^2\hbar^2} \right) \\
&= -2.75\left(\frac{m_e e^4}{16\pi^2\varepsilon_0^2\hbar^2} \right)
\end{aligned}
$$

This is about 5% larger than the exact value.

7–31. In Problem 7–30 we evaluated the integral that occurs in the first-order perturbation theory treatment of helium (see Equation 7.50). In this problem we will evaluate the integral by another method, one that uses an expansion for $1/r_{12}$ that is useful in many applications. We can write $1/r_{12}$ as an expansion in terms of spherical harmonics

$$
\frac{1}{r_{12}} = \frac{1}{|\mathbf{r}_1 - \mathbf{r}_2|} = \sum_{l=0}^{\infty} \sum_{m=-l}^{+l} \frac{4\pi}{2l+1} \frac{r_<^l}{r_>^{l+1}} Y_l^m(\theta_1, \phi_1) Y_l^{m*}(\theta_2, \phi_2)
$$

where θ_i and ϕ_i are the angles that describe $\mathbf{r}_i$ in a spherical coordinate system and $r_<$ and $r_>$ are, respectively, the smaller and larger values of r_1 and r_2. In other words, if $r_1 < r_2$, then $r_< = r_1$ and $r_> = r_2$. Substitute $\psi_{1s}(r_i) = (Z^3/a_0^3\pi)^{1/2} e^{-Zr_i/a_0}$, and the above expansion for $1/r_{12}$ into Equation 7.50, integrate over the angles, and show that all the terms except for the $l = 0$, $m = 0$ term vanish. Show that

$$
E^{(1)} = \frac{e^2}{4\pi\varepsilon_0} \frac{16Z^6}{a_0^6} \int_0^\infty dr_1 r_1^2 e^{-2Zr_1/a_0} \int_0^\infty dr_2 r_2^2 \frac{e^{-2Zr_2/a_0}}{r_>}
$$

Now show that

$$
\begin{aligned}
E^{(1)} &= \frac{e^2}{4\pi\varepsilon_0} \frac{16Z^6}{a_0^6} \int_0^\infty dr_1 r_1 e^{-2Zr_1/a_0} \int_0^{r_1} dr_2 r_2^2 e^{-2Zr_2/a_0} \\
&\quad + \frac{e^2}{4\pi\varepsilon_0} \frac{16Z^6}{a_0^6} \int_0^\infty dr_1 r_1^2 e^{-2Zr_1/a_0} \int_{r_1}^\infty dr_2 r_2 e^{-2Zr_2/a_0} \\
&= -\frac{e^2}{4\pi\varepsilon_0} \frac{4Z^3}{a_0^3} \int_0^\infty dr_1 r_1 e^{-2Zr_1/a_0} \left[e^{-2Zr_1/a_0}\left(\frac{2Z^2 r_1^2}{a_0^2} + \frac{2Zr_1}{a_0} + 1 \right) - 1 \right] \\
&\quad + \frac{e^2}{4\pi\varepsilon_0} \frac{4Z^4}{a_0^4} \int_0^\infty dr_1 r_1^2 e^{-2Zr_1/a_0} \left[e^{-2Zr_1/a_0}\left(\frac{2Zr_1}{a_0} + 1 \right) \right]
\end{aligned}
$$

$$= -\frac{e^2}{4\pi\varepsilon_0}\frac{4Z^6}{a_0^6}\int_0^\infty dr_1 e^{-4Zr_1/a_0}\left[\frac{r_1^2 a_0^2}{Z^2} + \frac{r_1 a_0^3}{Z^3}\right]$$

$$+ \frac{e^2}{4\pi\varepsilon_0}\frac{4Z^3}{a_0^3}\int_0^\infty dr_1 r_1 e^{-2Zr_1/a_0}$$

$$= \frac{5}{8}Z\left(\frac{e^2}{4\pi\varepsilon_0 a_0}\right)$$

as in Problem 7–30.

We take Equation 7.50 and substitute the given ψ_{1s} and $1/r_{12}$:

$$E^{(1)} = \frac{e^2}{4\pi\varepsilon_0}\int\int d\mathbf{r}_1 d\mathbf{r}_2 \psi_{1s}^*(\mathbf{r}_1)\psi_{1s}^*(\mathbf{r}_2)\frac{1}{r_{12}}\psi_{1s}(\mathbf{r}_1)\psi_{1s}(\mathbf{r}_2)$$

$$= \frac{Z^6}{a_0^6\pi^2}\int_0^\infty dr_1 r_1^2 e^{-2Zr_1/a_0}\int_0^\infty dr_2 r_2^2 e^{-2Zr_2/a_0}\int_0^{2\pi} d\phi_1\int_0^{2\pi}d\phi_2$$

$$\times \int_0^\pi d\theta_1 \sin\theta_1\int_0^\pi d\theta_2\sin\theta_2\left[\sum_{l=0}^\infty\sum_{m=-l}^{+l}\frac{4\pi}{2l+1}\frac{r_<^l}{r_>^{l+1}}Y_l^m(\theta_1,\phi_1)Y_l^{m*}(\theta_2,\phi_2)\right]$$

We can write the integrals over the angles as

$$\int_0^{2\pi}d\phi_1\int_0^\pi d\theta_1\sin\theta_1 Y_l^m(\theta_1,\phi_1)\int_0^{2\pi}d\phi_2\int_0^\pi d\theta_2\sin\theta_2 Y_l^{m*}(\theta_2,\phi_2)$$

Recalling that $Y_0^{0*} = (4\pi)^{-1/2}$, we can rewrite this integral as

$$4\pi\int_0^{2\pi}d\phi_1\int_0^\pi d\theta_1\sin\theta_1 Y_l^m(\theta_1,\phi_1)Y_0^{0*}(\theta_1,\phi_1)\int_0^{2\pi}d\phi_2\int_0^\pi d\theta_2\sin\theta_2 Y_0^0(\theta_2,\phi_2)Y_l^{m*}(\theta_2,\phi_2)$$

Recall from our previous discussion of the spherical harmonics (Chapter 6) that

$$\int_0^{2\pi}d\phi\int_0^\pi d\theta\sin\theta\, Y_l^{m*}(\theta,\phi)Y_n^k(\theta,\phi) = \delta_{nl}\delta_{mk} \tag{6.31}$$

Thus, all terms containing spherical harmonics vanish unless $l = m = 0$. Returning to our expression for $E^{(1)}$, we now have

$$E^{(1)} = \frac{e^2}{4\pi\varepsilon_0}\frac{Z^6}{a_0^6\pi^2}\int_0^\infty dr_1 r_1^2 e^{-2Zr_1/a_0}\int_0^\infty dr_2 r_2^2 e^{-2Zr_2/a_0}4\pi\frac{1}{r_>}4\pi$$

$$= \frac{e^2}{4\pi\varepsilon_0}\frac{16Z^6}{a_0^6}\int_0^\infty dr_1 r_1^2 e^{-2Zr_1/a_0}\int_0^\infty dr_2 r_2^2 e^{-2Zr_2/a_0}\frac{1}{r_>}$$

Now we can examine the two cases $r_2 < r_1$ and $r_2 > r_1$ separately. When $r_2 < r_1$, $r_> = r_1$, and when $r_2 > r_1$, $r_> = r_2$. We can separate the integral over r_2 into two parts and find

$$E^{(1)} = \frac{e^2}{4\pi\varepsilon_0}\frac{16Z^6}{a_0^6}\int_0^\infty dr_1 r_1 e^{-2Zr_1/a_0}\int_0^{r_1}dr_2 r_2^2 e^{-2Zr_2/a_0}$$

$$+ \frac{e^2}{4\pi\varepsilon_0}\frac{16Z^6}{a_0^6}\int_0^\infty dr_1 r_1^2 e^{-2Zr_1/a_0}\int_{r_1}^\infty dr_2 r_2 e^{-2Zr_2/a_0}$$

$$= \frac{e^2}{4\pi\varepsilon_0}\frac{4Z^3}{a_0^3}\int_0^\infty dr_1 r_1 e^{-2Zr_1/a_0}\left[1 - e^{-2Zr_1/a_0}\left(\frac{2Z^2 r_1^2}{a_0^2} + \frac{2Zr_1}{a_0} + 1\right)\right]$$

$$+ \frac{e^2}{4\pi\varepsilon_0}\frac{4Z^4}{a_0^4}\int_0^\infty dr_1 r_1^2 e^{-2Zr_1/a_0}\left[e^{-2Zr_1/a_0}\left(\frac{2Zr_1}{a_0} + 1\right)\right]$$

(These integrals have been solved in detail in Problem 7–30.) To evaluate the resulting integrals over r_1, we can let $x = 2Zr_1/a_0$:

$$E^{(1)} = \frac{e^2}{4\pi\varepsilon_0}\frac{Z}{a_0}\int_0^\infty dx\, x e^{-x}\left[1 - e^{-x}\left(\frac{x^2}{2} + x + 1\right)\right]$$

$$+ \frac{e^2}{4\pi\varepsilon_0}\frac{Z}{2a_0}\int_0^\infty dx\, x^2 e^{-x}\left[e^{-x}(x+1)\right]$$

$$= \frac{e^2}{4\pi\varepsilon_0}\frac{Z}{a_0}\int_0^\infty dx\, x e^{-x} - \frac{e^2}{4\pi\varepsilon_0}\frac{Z}{2a_0}\int_0^\infty dx\, e^{-2x}(x^2 + 2x)$$

$$= \frac{e^2}{4\pi\varepsilon_0}\left[\frac{Z}{a_0} - \frac{Z}{2a_0}\left(\frac{2!}{8} + \frac{2}{4}\right)\right] = \frac{5Z}{8}\frac{e^2}{4\pi\varepsilon_0 a_0}$$

which is the same result as that found in Problem 7–30.

7–32. This problem fills in the steps of the variational treatment of helium. We use a trial function of the form

$$\phi(\mathbf{r}_1, \mathbf{r}_2) = \frac{Z^3}{a_0^3\pi}e^{-Z(r_1+r_2)/a_0}$$

with Z as an adjustable parameter. The Hamiltonian operator of the helium atom is

$$\hat{H} = -\frac{\hbar^2}{2m_e}\nabla_1^2 - \frac{\hbar^2}{2m_e}\nabla_2^2 - \frac{2e^2}{4\pi\varepsilon_0 r_1} - \frac{2e^2}{4\pi\varepsilon_0 r_2} + \frac{e^2}{4\pi\varepsilon_0 r_{12}}$$

We now evaluate

$$E(Z) = \int d\mathbf{r}_1 d\mathbf{r}_2 \phi^* \hat{H}\phi$$

The evaluation of this integral is greatly simplified if you recall that $\psi(r_j) = (Z^3/a_0^3\pi)^{1/2}e^{-Zr_j/a_0}$ is an eigenfunction of a hydrogenlike Hamiltonian operator, one for which the nucleus has a charge Z. Show that the helium atom Hamiltonian operator can be written as

$$\hat{H} = -\frac{\hbar^2}{2m_e}\nabla_1^2 - \frac{Ze^2}{4\pi\varepsilon_0 r_1} - \frac{\hbar^2}{2m_e}\nabla_2^2 - \frac{Ze^2}{4\pi\varepsilon_0 r_2} + \frac{(Z-2)e^2}{4\pi\varepsilon_0 r_1} + \frac{(Z-2)e^2}{4\pi\varepsilon_0 r_2} + \frac{e^2}{4\pi\varepsilon_0 r_{12}}$$

where

$$\left(-\frac{\hbar^2}{2m_e}\nabla^2 - \frac{Ze^2}{4\pi\varepsilon_0 r}\right)\left(\frac{Z^3}{a_0^3\pi}\right)^{1/2}e^{-Zr/a_0} = -\frac{\hbar^2 Z^2}{2m_e a_0^2}\left(\frac{Z^3}{a_0^3\pi}\right)^{1/2}e^{-Zr/a_0}$$

Show that

$$E(Z) = \frac{Z^6}{a_0^6\pi^2}\int\int d\mathbf{r}_1 d\mathbf{r}_2 e^{-Z(r_1+r_2)/a_0}\left[-\frac{Z^2 e^2}{8\pi\varepsilon_0 a_0} - \frac{Z^2 e^2}{8\pi\varepsilon_0 a_0} + \frac{(Z-2)e^2}{4\pi\varepsilon_0 r_1}\right.$$

$$\left. + \frac{(Z-2)e^2}{4\pi\varepsilon_0 r_2} + \frac{e^2}{4\pi\varepsilon_0 r_{12}}\right]e^{-Z(r_1+r_2)/a_0}$$

The last integral is evaluated in Problem 7–30 or 7–31 and the others are elementary. Therefore, $E(Z)$, in units of $(m_e e^4/16\pi^2\varepsilon_0^2\hbar^2)$ is given by

$$E(Z) = -Z^2 + 2(Z-2)\frac{Z^3}{\pi}\int d\mathbf{r}\frac{e^{-2Zr}}{r} + \frac{5}{8}Z$$

$$= -Z^2 + 2(Z-2)Z + \frac{5}{8}Z$$

$$= Z^2 - \frac{27}{8}Z$$

Now minimize E with respect to Z and show that

$$E = -\left(\frac{27}{16}\right)^2 = -2.8477$$

in units of $m_e e^4 / 16\pi^2 \varepsilon_0^2 \hbar^2$. Interpret the value of Z that minimizes E.

To get from

$$\hat{H} = -\frac{\hbar^2}{2m_e}\nabla_1^2 - \frac{\hbar^2}{2m_e}\nabla_2^2 - \frac{2e^2}{4\pi\varepsilon_0 r_1} - \frac{2e^2}{4\pi\varepsilon_0 r_2} + \frac{e^2}{4\pi\varepsilon_0 r_{12}}$$

to

$$\hat{H} = -\frac{\hbar^2}{2m_e}\nabla_1^2 - \frac{Ze^2}{4\pi\varepsilon_0 r_1} - \frac{\hbar^2}{2m_e}\nabla_2^2 - \frac{Ze^2}{4\pi\varepsilon_0 r_2} + \frac{(Z-2)e^2}{4\pi\varepsilon_0 r_1} + \frac{(Z-2)e^2}{4\pi\varepsilon_0 r_2} + \frac{e^2}{4\pi\varepsilon_0 r_{12}}$$

simply add and subtract the quantity $\dfrac{Ze^2}{4\pi\varepsilon_0 r_1} + \dfrac{Ze^2}{4\pi\varepsilon_0 r_2}$. The Hamiltonian operator for a hydrogenlike atom can be written as

$$\hat{H}_{\mathrm{H}} = -\frac{\hbar^2}{2m_e}\nabla^2 - \frac{Ze^2}{4\pi\varepsilon_0 r}$$

When the Hamiltonian operator acts on the hydrogenlike wave function $\psi(r_j)$ given in the problem, we find

$$\left(-\frac{\hbar^2}{2m_e}\nabla^2 - \frac{Ze^2}{4\pi\varepsilon_0 r}\right)\left(\frac{Z^3}{a_0^3\pi}\right)^{1/2} e^{-Zr/a_0} = -\frac{\hbar^2 Z^2}{2m_e a_0^2}\left(\frac{Z^3}{a_0^3\pi}\right)^{1/2} e^{-Zr/a_0}$$

where we have taken the ground-state energy from Problem 6–34 and converted it into the units used (as in Problem 6–19). Now

$$\left(-\frac{\hbar^2}{2m_e}\nabla_1^2 - \frac{Ze^2}{4\pi\varepsilon_0 r_1} - \frac{\hbar^2}{2m_e}\nabla_2^2 - \frac{Ze^2}{4\pi\varepsilon_0 r_2}\right)\phi(r_1, r_2) = \left(-\frac{\hbar^2 Z^2}{m_e a_0^2}\right)\phi(r_1, r_2)$$

so (substituting for a_0) we can write

$$E(Z) = \int\int d\mathbf{r}_1 d\mathbf{r}_2 \phi^* \hat{H}\phi$$

$$= \int\int d\mathbf{r}_1 d\mathbf{r}_2 \phi^*\left[\left(-\frac{e^2 Z^2}{4\pi\varepsilon_0}\right) + \frac{(Z-2)e^2}{4\pi\varepsilon_0 r_1} + \frac{(Z-2)e^2}{4\pi\varepsilon_0 r_2} + \frac{e^2}{4\pi\varepsilon_0 r_{12}}\right]\phi$$

$$= \frac{Z^6}{a_0^6\pi^2}\int\int d\mathbf{r}_1 d\mathbf{r}_2\left[\left(-\frac{e^2 Z^2}{4\pi\varepsilon_0}\right) + \frac{(Z-2)e^2}{4\pi\varepsilon_0 r_1} + \frac{(Z-2)e^2}{4\pi\varepsilon_0 r_2} + \frac{e^2}{4\pi\varepsilon_0 r_{12}}\right]e^{-2Z(r_1+r_2)/a_0}$$

This can be broken down into four integrals:

$$\frac{Z^6}{a_0^6\pi^2}\int\int d\mathbf{r}_1 d\mathbf{r}_2\left(-\frac{e^2 Z^2}{4\pi\varepsilon_0}\right)e^{-2Z(r_1+r_2)/a_0} = -Z^2\left(\frac{m_e e^4}{16\pi^2\varepsilon_0^2\hbar^2}\right)$$

$$\frac{Z^6}{a_0^6\pi^2}\int d\mathbf{r}_1 \frac{(Z-2)e^2}{4\pi\varepsilon_0 r_1}e^{-2Zr_1/a_0} = Z(Z-2)\left(\frac{m_e e^4}{16\pi^2\varepsilon_0^2\hbar^2}\right)$$

$$\frac{Z^6}{a_0^6\pi^2}\int d\mathbf{r}_2 \frac{(Z-2)e^2}{4\pi\varepsilon_0 r_2}e^{-2Zr_2/a_0} = Z(Z-2)\left(\frac{m_e e^4}{16\pi^2\varepsilon_0^2\hbar^2}\right)$$

$$\frac{Z^6}{a_0^6\pi^2}\int\int d\mathbf{r}_1 d\mathbf{r}_2 \frac{e^2}{4\pi\varepsilon_0 r_{12}}e^{-2Z(r_1+r_2)/a_0} = \frac{5Z}{8}\left(\frac{m_e e^4}{16\pi^2\varepsilon_0^2\hbar^2}\right)$$

Therefore, in units of $\dfrac{m_e e^4}{16\pi^2 \varepsilon_0^2 \hbar^2}$,

$$E(Z) = -Z^2 + 2Z(Z-2) + \frac{5Z}{8} = Z^2 - \frac{27Z}{8}$$

Minimizing $E(Z)$, we find

$$\frac{dE}{dZ} = 0 = 2Z_{min} - \frac{27}{8}$$

$$Z_{min} = \frac{27}{16}$$

$$E_{min} = \left(\frac{27}{16}\right)^2 - \frac{27}{8}\left(\frac{27}{16}\right)$$

$$= -\left(\frac{27}{16}\right)^2 = -2.8477$$

in units of $m_e e^4 / 16\pi^2 \varepsilon_0^2 \hbar^2$. The value of Z that minimizes E is less than 2 (the nuclear charge of helium) because one electron shields the other from the nucleus.

CHAPTER 8

Multielectron Atoms

PROBLEMS AND SOLUTIONS

8–1. Show that the atomic unit of energy can be written as

$$E_h = \frac{\hbar^2}{m_e a_0^2} = \frac{e^2}{4\pi\varepsilon_0 a_0} = \frac{m_e e^4}{16\pi^2\varepsilon_0^2\hbar^2}$$

Equations 8.2 and 8.3 give

$$a_0 = \frac{4\pi\varepsilon_0\hbar^2}{m_e e^4} \qquad \text{and} \qquad E_h = \frac{m_e e^4}{16\pi^2\varepsilon_0^2\hbar^2}$$

and substituting into Equation 8.2 into Equation 8.3 gives

$$E_h = \frac{\hbar^2}{m_e a_0^2} = \frac{e^2}{4\pi\varepsilon_0 a_0} = \frac{m_e e^4}{16\pi^2\varepsilon_0^2\hbar^2}$$

8–2. Show that the energy of a helium ion in atomic units is $-2E_h$.

The ground state energy of a helium atom is

$$E = -\frac{m_e e^4 Z^2}{32\pi^2\varepsilon_0^2\hbar^2}$$

If we let $m_e = 1$, $e = 1$, $4\pi\varepsilon_0 = 1$, and $\hbar = 1$, then we obtain

$$E = -\frac{Z^2}{2}E_h = -2E_h$$

where E_h is the atomic unit of energy and we take $Z = 2$.

8–3. The electric potential energy at a distance r from a charge q is

$$V = \frac{q}{4\pi\varepsilon_0 r}$$

Show that the atomic unit of potential energy is the potential energy at a distance of one Bohr radius from a proton (see Table 8.1).

In atomic units, $e = 1$, $4\pi\varepsilon_0 = 1$, and $a_0 = 1$. The potential at a distance of one Bohr radius from a proton is

$$V = \frac{e}{4\pi\varepsilon_0 a_0}$$

so $V = 1$ in atomic units.

8–4. Show that the speed of an electron in the first Bohr orbit is $e^2/4\pi\varepsilon_0\hbar = 2.188 \times 10^6$ m·s^{-1}. This speed is the unit of speed in atomic units.

Use Equation 1.16 and take $r = a_0$ and $n = 1$ to obtain

$$v = \frac{\hbar}{m_e a_0} = \frac{\hbar}{m_e}\left(\frac{m_e e^2}{4\pi\varepsilon_0\hbar^2}\right) = \frac{e^2}{4\pi\varepsilon_0\hbar}$$

$$= \frac{\left(1.6022 \times 10^{-19}\text{ C}\right)^2}{\left(1.1127 \times 10^{-10}\text{ C}^2\cdot\text{J}^{-1}\cdot\text{m}^{-1}\right)\left(1.0546 \times 10^{-34}\text{ J}\cdot\text{s}\right)}$$

$$= 2.1877 \times 10^6\text{ m}\cdot\text{s}^{-1}$$

8–5. Show that the speed of light is equal to 137 in atomic units.

In the previous problem we found the conversion factor from SI units to atomic units: 2.188×10^6 m·s^{-1}. Thus

$$c = \frac{2.998 \times 10^8\text{m}\cdot\text{s}^{-1}}{2.188 \times 10^6\text{m}\cdot\text{s}^{-1}} = 137$$

8–6. Another way to introduce atomic units is to express mass as multiples of m_e, the mass of an electron (instead of kg); charge as multiples of e, the protonic charge (instead of C); angular momentum as multiples of $\hbar$ (instead of in J·s = kg·m^2·s^{-1}); and permittivity as multiples of $4\pi\varepsilon_0$ (instead of in C^2·s^2·kg^{-1}·m^{-3}). This conversion can be achieved in all of our equations by letting $m_e = e = \hbar = 4\pi\varepsilon_0 = 1$. Show that this procedure is consistent with the definition of atomic units used in the chapter.

All the atomic units in Table 8.1 are clearly equal to one when we let $m_e = e = \hbar = 4\pi\varepsilon_0 = 1$ with the exception of E_h and a_0. For these two units, we use the definitions in Table 8.1 and substitute in $m_e = e = \hbar = 4\pi\varepsilon_0 = 1$ to find

$$E_h = \frac{e^2}{4\pi\varepsilon_0 a_0} = \frac{1^2}{1\cdot 1} = 1$$

$$a_0 = \frac{4\pi\varepsilon_0\hbar^2}{m_e e^2} = \frac{1\cdot 1^2}{1\cdot 1^2} = 1$$

8–7. Derive Equation 8.5 from Equation 8.4. Be sure to remember that ∇^2 has units of (distance)$^{-2}$.

Because we showed that setting $m_e = e = \hbar = 4\pi\varepsilon_0 = 1$ is consistent with the definition of atomic units, we can rewrite Equation 8.4 as follows:

$$\hat{H} = -\frac{\hbar^2}{2m_e}\nabla_1^2 - \frac{\hbar^2}{2m_e}\nabla_2^2 - \frac{2e^2}{4\pi\varepsilon_0 r_1} - \frac{2e^2}{4\pi\varepsilon_0 r_2} + \frac{e^2}{4\pi\varepsilon_0 r_{12}}$$

$$= -\frac{1}{2}\nabla_1^2 - \frac{1}{2}\nabla_2^2 - \frac{2}{r_1} - \frac{2}{r_2} + \frac{1}{r_{12}}$$

8–8. Show that the normalization constant for the radial part of Slater orbitals is $(2\zeta)^{n+\frac{1}{2}}/[(2n)!]^{1/2}$.

From Equation 8.12, the radial part of a Slater orbital is $r^{n-1}e^{-\zeta r}$. We can find the normalization constant c by solving the equation

$$c^2 \int_0^\infty r^{2n-2}e^{-2\zeta r}r^2 dr = c^2 \frac{(2n)!}{(2\zeta)^{(2n+1)}} = 1$$

and so

$$c = \frac{(2\zeta)^{n+\frac{1}{2}}}{\sqrt{(2n)!}}$$

8–9. Use Equation 8.12 to write out the normalized $1s$, $2s$, and $2p$ Slater orbitals. How do they differ from the hydrogenlike orbitals?

The Slater orbitals are given by

$$S_{nlm}(r, \theta, \phi) = N_{nl}r^{n-1}e^{-\zeta r}Y_l^m(\theta, \phi) \tag{8.12}$$

where $N_{nl} = (2\zeta)^{n+\frac{1}{2}}/[(2n)!]^{1/2}$.

Orbital	n	l	m	S_{nlm}
$1s$	1	0	0	$2\zeta^{3/2}e^{-\zeta r}/(4\pi)^{1/2}$
$2s$	2	0	0	$(2\zeta)^{5/2}e^{-\zeta r}/(96\pi)^{1/2}$
$2p$	2	1	0	$3^{1/2}(2\zeta)^{5/2}re^{-\zeta r}\cos\theta/(96\pi)^{1/2}$
$2p$	2	1	1	$3^{1/2}(2\zeta)^{5/2}re^{-\zeta r+i\phi}\sin\theta/(192\pi)^{1/2}$
$2p$	2	1	-1	$3^{1/2}(2\zeta)^{5/2}re^{-\zeta r-i\phi}\sin\theta/(192\pi)^{1/2}$

The hydrogenlike orbitals can be represented by the general function $\psi = N_{nl}R_{nl}(r)Y_l^m(\theta, \phi)$. The Slater orbitals are not orthogonal to one other, while the hydrogenlike orbitals are. Slater orbitals also have no radial nodes.

8–10. Substitute Equation 8.5 for $\hat{H}$ into

$$E = \int\int d\mathbf{r}_1 d\mathbf{r}_2 \phi^*(\mathbf{r}_1)\phi^*(\mathbf{r}_2)\hat{H}\phi(\mathbf{r}_1)\phi(\mathbf{r}_2)$$

and show that

$$E = I_1 + I_2 + J_{12}$$

where

$$I_j = \int d\mathbf{r}_j \phi^*(\mathbf{r}_j) \left[-\frac{1}{2}\nabla_j^2 - \frac{Z}{r_j} \right] \phi(\mathbf{r}_j)$$

and

$$J_{12} = \int\int d\mathbf{r}_1 d\mathbf{r}_2 \phi^*(\mathbf{r}_1)\phi(\mathbf{r}_1)\frac{1}{r_{12}}\phi^*(\mathbf{r}_2)\phi(\mathbf{r}_2)$$

Why is J_{12} called a Coulomb integral?

Equation 8.5 is the Hamiltonian operator for a helium atom. To generalize to an atom with atomic number Z we substitute Z for 2 in this equation. When we do, the energy E is given by

$$
\begin{aligned}
E &= \int\int d\mathbf{r}_1 d\mathbf{r}_2 \phi^*(\mathbf{r}_1)\phi^*(\mathbf{r}_2) \left[-\frac{1}{2}\nabla_1^2 - \frac{1}{2}\nabla_2^2 - \frac{Z}{r_1} - \frac{Z}{r_2} + \frac{1}{r_{12}} \right] \phi(\mathbf{r}_1)\phi(\mathbf{r}_2) \\
&= \int d\mathbf{r}_1 \phi^*(\mathbf{r}_1) \left[-\frac{1}{2}\nabla_1^2 - \frac{Z}{r_1} \right] \phi(\mathbf{r}_1) \int d\mathbf{r}_2 \phi^*(\mathbf{r}_2)\phi(\mathbf{r}_2) \\
&\quad + \int d\mathbf{r}_2 \phi^*(\mathbf{r}_2) \left[-\frac{1}{2}\nabla_2^2 - \frac{Z}{r_2} \right] \phi(\mathbf{r}_2) \int d\mathbf{r}_1 \phi^*(\mathbf{r}_1)\phi(\mathbf{r}_1) \\
&\quad + \int\int d\mathbf{r}_1 d\mathbf{r}_2 \phi^*(\mathbf{r}_1)\phi^*(\mathbf{r}_2)\frac{1}{r_{12}}\phi(\mathbf{r}_1)\phi(\mathbf{r}_2)
\end{aligned}
$$

Recall that $\phi(\mathbf{r}_1)$ and $\phi(\mathbf{r}_2)$ are normalized, and so the integrals of $\phi^*(\mathbf{r}_j)\phi(\mathbf{r}_j)$ are equal to one. Substituting the given definitions for I_j and J_{12} gives

$$E = I_1 + I_2 + J_{12}$$

The integral J_{12} is called a Coulomb integral because it is due to the coulombic interaction between the two electron charge densities.

8–11. In this problem we will examine the physical significance of the eigenvalue ϵ in Equation 8.20. The quantity ϵ is called the orbital energy. Using Equation 8.19 for $\hat{H}_1^{\text{eff}}(\mathbf{r}_1)$, multiply Equation 8.20 from the left by $\phi^*(\mathbf{r}_1)$ and integrate to obtain

$$\epsilon_1 = I_1 + J_{12}$$

where I_1 and J_{12} are defined in the previous problem. Show that the total energy of a helium atom $E = I_1 + I_2 + J_{12}$ is *not* the sum of its orbital energies. In fact, show that

$$\epsilon_1 = E - I_2 \tag{1}$$

But according to the definition of I_2 in Problem 8–10, I_2 is the energy of a helium ion, calculated with the helium Hartree-Fock orbital $\phi(r)$. Thus, Equation 1 suggests that the orbital energy ϵ_1 is an approximation to the ionization energy of a helium atom or that

$$\text{IE} \approx -\epsilon_1 \qquad \text{(Koopmans' approximation)}$$

Even within the Hartree-Fock approximation, Koopmans' approximation is based upon the approximation that the same orbitals can be used to calculate the energy of the neutral atom and the energy of the ion. The value of $-\epsilon_1$ obtained by Clementi (see Table 8.2) is $0.91796E_{\text{h}}$, compared with the experimental value of $0.904E_{\text{h}}$.

Use the Hamiltonian operator given by Equation 8.19 (using Z in place of 2, as in the previous problem) and multiply Equation 8.20 from the left by $\phi^*(\mathbf{r}_1)$ to obtain

$$\int d\mathbf{r}_1 \phi^*(\mathbf{r}_1)\left[-\frac{1}{2}\nabla_1^2 - \frac{Z}{r_1} + V_1^{\text{eff}}(\mathbf{r}_1)\right]\phi(\mathbf{r}_1) = \epsilon_1 \int \phi^*(\mathbf{r}_1)\phi(\mathbf{r}_1)d\mathbf{r}_1$$

But

$$\int d\mathbf{r}_1 \phi^*(\mathbf{r}_1) V_1^{\text{eff}}(\mathbf{r}_1)\phi_1(\mathbf{r}_1) = J_{12}$$

so we have

$$I_1 + J_{12} = \epsilon_1$$

Because the total energy of a helium atom is $E = I_1 + I_2 + J_{12}$, it follows that $\epsilon_1 = E - I_2$.

8–12. Show that the two-term helium Hartree-Fock orbital

$$\phi(r) = 0.81839e^{-1.44608r} + 0.52072e^{-2.86222r}$$

is normalized.

For a normalized orbital $\phi(r)$, $\int d\tau \phi^*(r)\phi(r) = 1$. Using $\phi(r)$ given in the problem,

$$\int d\tau \phi^*(r)\phi(r) = \int_0^{2\pi} d\phi \int_0^{\pi} d\theta \sin\theta \int_0^{\infty} dr\, r^2 \left(0.81839e^{-1.44608r} + 0.52072e^{-2.86222r}\right)^2$$

$$= 4\pi \int_0^{\infty} dr\, r^2 \left[(0.81839)^2 e^{-2.89216r} + 2(0.81839)(0.52072)e^{-4.3083r}\right.$$
$$\left. + (0.52072)^2 e^{-5.72444r}\right]$$

$$= 4\pi \left\{(0.81839)^2 \left[\frac{2}{(2.89216)^3}\right] + 2(0.81839)(0.52072)\left[\frac{2}{(4.3083)^3}\right]\right.$$
$$\left. + (0.52072)^2 \left[\frac{2}{(5.72444)^3}\right]\right\}$$

$$= 1.00001$$

8–13. The normalized variational helium orbital we determined in Chapter 7 is

$$\phi(r) = 1.2368e^{-27r/16}$$

A two-term Hartree-Fock orbital is given in Problem 8–12:

$$\phi(r) = 0.81839e^{-1.44608r} + 0.52072e^{-2.86222r}$$

and a five-term orbital given on page 283 is

$$\phi(r) = 0.75738e^{-1.4300r} + 0.43658e^{-2.4415r} + 0.17295e^{-4.0996r}$$
$$-0.02730e^{-6.4843r} + 0.06675e^{-7.978r}$$

Plot these orbitals on the same graph and compare them.

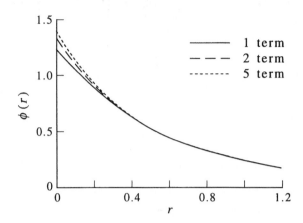

These expressions are nearly identical for $r \geq 0.3$.

8–14. Given that $\Psi(1, 2) = 1s\alpha(1)1s\beta(2) - 1s\alpha(2)1s\beta(1)$, prove that

$$\int d\tau_1 d\tau_2 \Psi^*(1, 2)\Psi(1, 2) = 2$$

if the spatial part is normalized.

$$\int d\tau_1 d\tau_2 \Psi^*(1, 2)\Psi(1, 2) = \int\int d\tau_1 d\tau_2 \left[1s\alpha^*(1)1s\beta^*(2) - 1s\alpha^*(2)1s\beta^*(1)\right]$$
$$\times \left[1s\alpha(1)1s\beta(2) - 1s\alpha(2)1s\beta(1)\right]$$
$$= \int\int d\tau_1 d\tau_2 \, 1s\alpha^*(1)1s\beta^*(2)1s\alpha(1)1s\beta(2)$$
$$- \int\int d\tau_1 d\tau_2 \, 1s\alpha^*(1)1s\beta^*(2)1s\alpha(2)1s\beta(1)$$
$$- \int\int d\tau_1 d\tau_2 \, 1s\alpha^*(2)1s\beta^*(1)1s\alpha(1)1s\beta(2)$$
$$+ \int\int d\tau_1 d\tau_2 \, 1s\alpha^*(2)1s\beta^*(1)1s\alpha(2)1s\beta(1)$$
$$= 1 - 0 - 0 + 1 = 2$$

where we have used Equations 8.27.

8–15. Show that the spin integral in Equation 8.40 is equal to 2.

The spin integral in Equation 8.40 is

$$\int\int \left[\alpha^*(\sigma_1)\beta^*(\sigma_2) - \alpha^*(\sigma_2)\beta^*(\sigma_1)\right]\left[\alpha(\sigma_1)\beta(\sigma_2) - \alpha(\sigma_2)\beta(\sigma_1)\right] d\sigma_1 d\sigma_2$$

This expression can be broken into four integrals found in Example 8–3. The sum of these four integrals is the value of the spin integral and is equal to 2.

8–16. Why is it impossible to distinguish the two electrons in a helium atom, but not the two electrons in separated hydrogen atoms? Do you think the electrons are distinguishable in the diatomic H_2 molecule? Explain your reasoning.

For two electrons in separated hydrogen atoms, the electrons are associated with two different nuclei and are independent of one another. There is no potential energy term that couples the two electrons and so the Hamiltonian operator of the system is separable. The two electrons in a helium atom cannot be distinguished because the Hamiltonian operator cannot be separated into a sum of two terms that are functions of only one electron each, and the wavefunction of this two electron system must be antisymmetric under the interchange of the two electrons.

The electrons in the diatomic H_2 molecule are also indistinguishable. We will learn in Chapter 9 that the Pauli exclusion principle also applies to molecular orbitals.

8–17. Why is the angular dependence of multielectron atomic wave functions in the Hartree-Fock approximation the same as for hydrogen atomic wave functions?

The Hamiltonian operator used in the Hartree-Fock approximation depends only on r (Equation 8.19), and so the angular dependence of the wave functions of a multielectron atom in the Hartee-Fock approximation is the same as in the hydrogen atom.

8–18. Why is the radial dependence of multielectron atomic wave functions in the Hartree-Fock approximation different from the radial dependence of hydrogen atomic wave functions?

The radial dependence of the effective Hamiltonian operator differs from the Hamiltonian operator of a hydrogen atom.

8–19. Show that the atomic determinantal wave function

$$\psi = \frac{1}{\sqrt{2}} \begin{vmatrix} 1s\alpha(1) & 1s\beta(1) \\ 1s\alpha(2) & 1s\beta(2) \end{vmatrix}$$

is normalized if the $1s$ orbitals are normalized.

Expand the determinant given to get

$$\psi = \frac{1}{\sqrt{2}} [1s\alpha(1)1s\beta(2) - 1s\alpha(2)1s\beta(1)]$$

Then

$$\int d\tau\, \psi^*\psi = \frac{1}{2} \int d\tau\, [1s\alpha(1)1s\beta(2) - 1s\alpha(2)1s\beta(1)]^2$$

$$= \frac{1}{2}(1 + 0 + 0 + 1) = \frac{1}{2}(2) = 1$$

where we have used the information in the solution of Problem 8–14 to evaluate the integral.

8–20. Show that the two-electron determinantal wave function in Problem 8–19 factors into a spatial part and a spin part.

$$\psi = \frac{1}{\sqrt{2}}[1s(1)\alpha(1)1s(2)\beta(2) - 1s(2)\alpha(2)1s(1)\beta(1)]$$

$$= 1s(1)1s(2)\left\{\frac{1}{\sqrt{2}}[\alpha(1)\beta(2) - \alpha(2)\beta(1)]\right\}$$

8–21. Argue that the normalization constant of an $N \times N$ Slater determinant of orthonormal spin orbitals is $1/\sqrt{N!}$.

The expansion of an $N \times N$ determinant yields $N!$ terms, and $(N!)^2$ terms when the wave function is squared. However, the only terms in the square of the wave function that yield a non-zero integral when integrated over all the electron coordinates are the $N!$ terms. All the cross products yield a zero integral because of the orthogonality of at least one spin orbital. Because the spin orbitals are normalized, the normalization constant is $1/N!$.

8–22. The total z component of the spin angular momentum operator for an N-electron system is

$$\hat{S}_{z,\text{total}} = \sum_{j=1}^{N} \hat{S}_{zj}$$

Show that both

$$\psi = \frac{1}{\sqrt{2}}\begin{vmatrix} 1s\alpha(1) & 1s\beta(1) \\ 1s\alpha(2) & 1s\beta(2) \end{vmatrix}$$

and

$$\psi = \frac{1}{\sqrt{3!}}\begin{vmatrix} 1s\alpha(1) & 1s\beta(1) & 2s\alpha(1) \\ 1s\alpha(2) & 1s\beta(2) & 2s\alpha(2) \\ 1s\alpha(3) & 1s\beta(3) & 2s\alpha(3) \end{vmatrix}$$

are eigenfunctions of $\hat{S}_{z,\text{total}}$. What are the eigenvalues in each case?

For the first wave function given,

$$\hat{S}_{z,\text{total}}\psi = \frac{1}{\sqrt{2}}\left(\hat{S}_{z1} + \hat{S}_{z2}\right)[1s\alpha(1)1s\beta(2) - 1s\beta(1)1s\alpha(2)]$$

$$= \frac{1}{\sqrt{2}}\left\{\left[\frac{\hbar}{2} - \frac{\hbar}{2}\right]1s\alpha(1)1s\beta(2) - \left[-\frac{\hbar}{2} + \frac{\hbar}{2}\right]1s\beta(1)1s\alpha(2)\right\}$$

$$= 0[1s\alpha(1)1s\beta(2) - 1s\beta(1)1s\alpha(2)]$$

Thus ψ is an eigenfunction of $\hat{S}_{z,\text{total}}$ with eigenvalue zero. The expansion of the second determinant given in the problem involves six terms. A typical term is the diagonal product,

$$\left(\hat{S}_{z1} + \hat{S}_{z2} + \hat{S}_{z3}\right)1s\alpha(1)1s\beta(2)2s\alpha(3) = \left(\frac{\hbar}{2} - \frac{\hbar}{2} + \frac{\hbar}{2}\right)1s\alpha(1)1s\beta(2)2s\alpha(3)$$

Of the six terms in the expansion of the 3×3 determinant, three are positive and three are negative. Evaluation of each of the six terms involved in the expansion of the determinant produces coefficients of $\hbar/2$ (multiplying the positive terms) and $-\hbar/2$ (multiplying the negative terms). Thus ψ turns out to be an eigenfunction of $\hat{S}_{z,\text{total}}$ with eigenvalue $\hbar/2$.

8–23. Consider the determinantal atomic wave function

$$\Psi(1, 2) = \frac{1}{\sqrt{2}} \begin{vmatrix} \psi_{211}\alpha(1) & \psi_{21-1}\beta(1) \\ \psi_{211}\alpha(2) & \psi_{21-1}\beta(2) \end{vmatrix}$$

where $\psi_{21\pm1}$ is a hydrogenlike wave function. Show that $\Psi(1, 2)$ is an eigenfunction of

$$\hat{L}_{z,\text{total}} = \hat{L}_{z1} + \hat{L}_{z2}$$

and

$$S_{z,\text{total}} = \hat{S}_{z1} + \hat{S}_{z2}$$

What are the eigenvalues?

Use the identities for hydrogenlike wave functions (Equations 8.24 and 8.22)

$$\hat{L}_z \psi_{nlm} = m\hbar \psi_{nlm} \qquad \hat{S}_z \alpha = \frac{\hbar}{2}\alpha \qquad \hat{S}_z \beta = -\frac{\hbar}{2}\beta$$

Now

$$\hat{L}_{z,\text{total}}\Psi = \frac{1}{\sqrt{2}} \left(\hat{L}_{z1} + \hat{L}_{z2} \right) \left[\psi_{211}\alpha(1)\psi_{21-1}\beta(2) - \psi_{211}\alpha(2)\psi_{21-1}\beta(1) \right]$$

$$= \frac{1}{\sqrt{2}} \left[(\hbar - \hbar)\psi_{211}\alpha(1)\psi_{21-1}\beta(2) - (-\hbar + \hbar)\psi_{211}\alpha(2)\psi_{21-1}\beta(1) \right]$$

$$= 0$$

Therefore Ψ is an eigenfunction of $\hat{L}_{z,\text{total}}$ with eigenvalue zero. Similarly,

$$\hat{S}_{z,\text{total}}\Psi = \frac{1}{\sqrt{2}} \left(\hat{S}_{z1} + \hat{S}_{z2} \right) \left[\psi_{211}\alpha(1)\psi_{21-1}\beta(2) - \psi_{211}\alpha(2)\psi_{21-1}\beta(1) \right]$$

$$= \frac{1}{\sqrt{2}} \left[\left(\frac{\hbar}{2} - \frac{\hbar}{2} \right) \psi_{211}\alpha(1)\psi_{21-1}\beta(2) - \left(-\frac{\hbar}{2} + \frac{\hbar}{2} \right) \psi_{211}\alpha(2)\psi_{21-1}\beta(1) \right]$$

$$= 0$$

Ψ is also an eigenfunction of $\hat{S}_{z,\text{total}}$ with eigenvalue zero.

8–24. For a two-electron system, there are four possible spin functions:

1. $\alpha(1)\alpha(2)$ **2.** $\beta(1)\alpha(2)$ **3.** $\alpha(1)\beta(2)$ **4.** $\beta(1)\beta(2)$

The concept of indistinguishability forces us to consider only linear combinations of 2 and 3,

$$\psi_{\pm} = \frac{1}{\sqrt{2}}[\alpha(1)\beta(2) \pm \beta(1)\alpha(2)]$$

instead of 2 and 3 separately. Show that of the four acceptable spin functions, 1, 4, and $\psi_{\pm}$, three are symmetric and one is antisymmetric.

Now for a two-electron system, we can combine spatial wave functions with spin functions. Show that this combination leads to only four allowable combinations:

$$[\psi(1)\phi(2) + \psi(2)\phi(1)]\frac{1}{\sqrt{2}}[\alpha(1)\beta(2) - \alpha(2)\beta(1)]$$

$$[\psi(1)\phi(2) - \psi(2)\phi(1)][\alpha(1)\alpha(2)]$$

$$[\psi(1)\phi(2) - \psi(2)\phi(1)][\beta(1)\beta(2)]$$

and

$$[\psi(1)\phi(2) - \psi(2)\phi(1)]\frac{1}{\sqrt{2}}[\alpha(1)\beta(2) + \alpha(2)\beta(1)]$$

where ψ and ϕ are two spatial wave functions. Show that $M_S = m_{s_1} + m_{s_2} = 0$ for the first of these and that $M_S = 1$, -1, and 0 (in atomic units) for the next three, respectively.

Consider the first excited state of a helium atom, in which $\psi = 1s$ and $\phi = 2s$. The first of the four wave functions above, with the symmetric spatial part, will give a higher energy than the remaining three, which form a degenerate set of three. The first state is a singlet state and the degenerate set of three represents a triplet state. Because M_S equals zero and only zero for the singlet state, the singlet state corresponds to $S = 0$. The other three, with $M_S = \pm 1$, 0, corresponds to $S = 1$. Note that the degeneracy is $2S + 1$ in each case.

Putting all this information into a more mathematical form, given that $\hat{S}_{\text{total}} = \hat{S}_1 + \hat{S}_2$, we can show that (Problem 8–53)

$$\hat{S}_{\text{total}}^2[\alpha(1)\beta(2) - \alpha(2)\beta(1)] = 0$$

corresponding to $S = 0$, and that

$$\hat{S}_{\text{total}}^2 \begin{bmatrix} \alpha(1)\alpha(2) \\ \frac{1}{\sqrt{2}}[\alpha(1)\beta(2) + \alpha(2)\beta(1)] \\ \beta(1)\beta(2) \end{bmatrix} = 2\hbar^2 \begin{bmatrix} \alpha(1)\alpha(2) \\ \frac{1}{\sqrt{2}}[\alpha(1)\beta(2) + \alpha(2)\beta(1)] \\ \beta(1)\beta(2) \end{bmatrix}$$

corresponding to $S = 1$.

When we interchange the two electrons, the only function which changes sign is ψ_-. By definition, then, the three symmetric spin functions are $\alpha(1)\alpha(2)$, $\beta(1)\beta(2)$, and ψ_+, and the antisymmetric wave function is ψ_-.

Allowable combinations of spatial wave functions and spin functions are those for which the overall product of the functions is antisymmetric under the interchange of the two electrons. For this to be the case, we must combine an antisymmetric spatial wave function with the symmetric spin functions and a symmetric spatial wave function with the antisymmetric spin function. There are then four allowable combinations

$$[\psi(1)\phi(2) + \psi(2)\phi(1)]\frac{1}{\sqrt{2}}[\alpha(1)\beta(2) - \alpha(2)\beta(1)]$$

$$[\psi(1)\phi(2) - \psi(2)\phi(1)][\alpha(1)\alpha(2)]$$

$$[\psi(1)\phi(2) - \psi(2)\phi(1)][\beta(1)\beta(2)]$$

and

$$[\psi(1)\phi(2) - \psi(2)\phi(1)]\frac{1}{\sqrt{2}}[\alpha(1)\beta(2) + \alpha(2)\beta(1)]$$

When we operate on these functions with $\hat{S}_{z,\text{total}}$ (as we did in the preceding two problems), we find that the eigenvalues of these four functions are 0, $\hbar$, $-\hbar$, and 0, respectively, or 0, 1, -1, and 0 in atomic units (recall that the atomic unit of angular momentum is $\hbar$). Thus $M_S = 0$ for the first function, and $M_S = 1$, -1, and 0 for the next three.

8–25. Consider a helium atom in an excited state in which one of its $1s$ electrons is raised to the $2s$ level, so that its electron configuration is $1s2s$. Argue that because the two orbitals are different, there are four possible determinantal wave functions for this system:

$$\phi_1 = \frac{1}{\sqrt{2}} \begin{vmatrix} 1s\alpha(1) & 2s\alpha(1) \\ 1s\alpha(2) & 2s\alpha(2) \end{vmatrix}$$

$$\phi_2 = \frac{1}{\sqrt{2}} \begin{vmatrix} 1s\beta(1) & 2s\beta(1) \\ 1s\beta(2) & 2s\beta(2) \end{vmatrix}$$

$$\phi_3 = \frac{1}{\sqrt{2}} \begin{vmatrix} 1s\alpha(1) & 2s\beta(1) \\ 1s\alpha(2) & 2s\beta(2) \end{vmatrix}$$

$$\phi_4 = \frac{1}{\sqrt{2}} \begin{vmatrix} 1s\beta(1) & 2s\alpha(1) \\ 1s\beta(2) & 2s\alpha(2) \end{vmatrix}$$

To calculate the energy of the $1s2s$ configuration, assume the variational function

$$\psi = c_1\phi_1 + c_2\phi_2 + c_3\phi_3 + c_4\phi_4$$

Show that the secular equation associated with this linear combination trial function is (this is the only lengthy part of this problem and at least you have the answer in front of you; be sure to remember that the $1s$ and $2s$ orbitals here are eigenfunctions of the hydrogenlike Hamiltonian operator)

$$\begin{vmatrix} E_0 + J - K - E & 0 & 0 & 0 \\ 0 & E_0 + J - K - E & 0 & 0 \\ 0 & 0 & E_0 + J - E & -K \\ 0 & 0 & -K & E_0 + J - E \end{vmatrix} = 0$$

where

$$J = \int\int d\tau_1 d\tau_2 \, 1s(1)1s(1)\left(\frac{1}{r_{12}}\right)2s(2)2s(2)$$

$$K = \int\int d\tau_1 d\tau_2 \, 1s(1)2s(1)\left(\frac{1}{r_{12}}\right)1s(2)2s(2)$$

and E_0 is the energy without the $1/r_{12}$ term in the helium atom Hamiltonian operator. Show that

$$E_0 = -\frac{5}{2}E_{\text{h}}$$

Explain why J is called an atomic Coulombic integral and K is called an atomic exchange integral.

Even though the above secular determinant is 4×4 and appears to give a fourth-degree polynomial in E, note that it really consists of two 1×1 blocks and a 2×2 block. Show that this symmetry in the determinant reduces the determinantal equation to

$$(E_0 + J - K - E)^2 \begin{vmatrix} E_0 + J - E & -K \\ -K & E_0 + J - E \end{vmatrix} = 0$$

and that this equation gives the four roots

$$E = E_0 + J - K \qquad \text{(twice)}$$
$$= E_0 + J \pm K$$

Show that the wave function corresponding to the positive sign in E in the $E_0 + J \pm K$ is

$$\psi_3 = \frac{1}{\sqrt{2}}(\phi_3 - \phi_4)$$

and that corresponding to the negative sign in $E_0 + J \pm K$ is

$$\psi_4 = \frac{1}{\sqrt{2}}(\phi_3 + \phi_4)$$

Now show that both ψ_3 and ψ_4 can be factored into a spatial part and a spin part, even though ϕ_3 and ϕ_4 separately cannot. Furthermore, let

$$\psi_1 = \phi_1 \quad \text{and} \quad \psi_2 = \phi_2$$

and show that both of these can be factored also. Using the argument given in Problem 8–24, group these four wave functions (ψ_1, ψ_2, ψ_3, ψ_4) into a singlet state and a triplet state.

Now calculate the energy of the singlet and triplet states in terms of E_0, J, and K. Argue that $J > 0$. Given that $K > 0$ also, does the singlet state or the triplet state have the lower energy? The values of J and K when hydrogenlike wave functions with $Z = 2$ are used are $J = 34/81 E_h$ and $K = 32/(27)^2 E_h$. Using the ground-state wave function

$$\phi = \frac{1}{\sqrt{2}} \begin{vmatrix} 1s\alpha(1) & 1s\beta(1) \\ 1s\alpha(2) & 1s\beta(2) \end{vmatrix}$$

show that the first-order perturbation theory result is $E = -11/4 \, E_h$ if hydrogenlike wave functions with $Z = 2$ are used. Use this value of E to calculate the energy difference between the ground state and the first excited singlet state and the first triplet state of helium. The experimental values of these energy differences are 159 700 cm^{-1} and 166 200 cm^{-1}, respectively (cf. Figure 8.5).

In Problem 8–24 we examined the four possible determinantal wave functions for a system with two spatial wave functions (such as the helium atom in a $1s2s$ electron configuration). Writing these wave functions in determinantal form and substituting $1s$ and $2s$ for ϕ and ψ gives the ϕ_j listed in this problem. Now we must find the secular equation associated with

$$\psi = c_1\phi_1 + c_2\phi_2 + c_3\phi_3 + c_4\phi_4$$

We use Equation 7.40,

$$\begin{vmatrix} H_{11} - ES_{11} & H_{12} - ES_{12} & H_{13} - ES_{13} & H_{14} - ES_{14} \\ H_{21} - ES_{21} & H_{22} - ES_{22} & H_{23} - ES_{23} & H_{24} - ES_{24} \\ H_{31} - ES_{31} & H_{32} - ES_{32} & H_{33} - ES_{33} & H_{34} - ES_{34} \\ H_{41} - ES_{41} & H_{42} - ES_{42} & H_{43} - ES_{43} & H_{44} - ES_{44} \end{vmatrix} = 0$$

To solve this equation, we must find the H_{ij} and S_{ij}. We can begin with the H_{ij}. For example, consider $H_{11} = \int \int d\tau_1 d\tau_2 \phi_1 \hat{H} \phi_1$. We need to evaluate the expression

$$\hat{H}\phi_1 = \left[-\frac{1}{2}\nabla_1^2 - \frac{2}{r_1} - \frac{1}{2}\nabla_2^2 - \frac{2}{r_2} + \frac{1}{r_{12}} \right] \frac{1}{\sqrt{2}} [1s\alpha(1)2s\alpha(2) - 1s\alpha(2)2s\alpha(1)]$$

$$= \left[-\frac{Z^2}{2} - \frac{Z^2}{8} + \frac{1}{r_{12}} \right] \frac{1}{\sqrt{2}} [1s\alpha(1)2s\alpha(2) - 1s\alpha(2)2s\alpha(1)]$$

where we have used the fact that the eigenvalue associated with a $1s$ orbital is $-Z^2/2$ and that associated with a $2s$ orbital is $-Z^2/8$. Because we are examining the helium atom, $Z = 2$, so

$$-\frac{Z^2}{2} - \frac{Z^2}{8} = -\frac{5}{2} E_\mathrm{h} = E_0$$

Therefore

$$H_{11} = \frac{1}{2} \int \int d\tau_1 d\tau_2 \, [1s\alpha^*(1)2s\alpha^*(2) - 1s\alpha^*(2)2s\alpha^*(1)] \left(E_0 + \frac{1}{r_{12}} \right)$$

$$\times [1s\alpha(1)2s\alpha(2) - 1s\alpha(2)2s\alpha(1)]$$

$$= \frac{1}{2} \left[E_0(1 - 0 - 0 + 1) + (J - K - K + J) \right]$$

$$= E_0 + J - K$$

The evaluation of H_{22} is similar to that of H_{11} (simply replace α by β everywhere in the expression for H_{11}). We then find

$$H_{22} = E_0 + J - K$$

To evaluate H_{33}, we evaluate the expression

$$H_{33} = \frac{1}{2} \int \int d\tau_1 d\tau_2 \, [1s\alpha^*(1)2s\beta^*(2) - 1s\alpha^*(2)2s\beta^*(1)] \left(E_0 + \frac{1}{r_{12}} \right)$$

$$\times [1s\alpha(1)2s\beta(2) - 1s\alpha(2)2s\beta(1)]$$

$$= \frac{1}{2} \left[E_0(1 - 0 - 0 + 1) + (J - 0 - 0 + J) \right]$$

$$= E_0 + J$$

The evaluation of H_{44} is similar to that of H_{33} (simply interchange α and β everywhere in the expression for H_{33}). This gives

$$H_{44} = E_0 + J$$

The evaluation of H_{34} (which is equivalent to H_{43}) is done as follows:

$$H_{34} = \frac{1}{2} \int \int d\tau_1 d\tau_2 \, [1s\alpha^*(1)2s\beta^*(2) - 1s\alpha^*(2)2s\beta^*(1)] \left(E_0 + \frac{1}{r_{12}} \right)$$

$$\times [1s\beta(1)2s\alpha(2) - 1s\beta(2)2s\alpha(1)]$$

$$= \frac{1}{2} \left[E_0(0 - 0 - 0 + 0) + (0 - K - K + 0) \right]$$

$$= -K$$

The quantities H_{12}, H_{13}, H_{14}, H_{23} and H_{24} vanish because integration over the corresponding spin functions is zero. (Notice that $H_{ij} = H_{ji}$.) The evaluation of the S_{ij} terms involves manipulations like those done in Problems 8–14 and 8–19. All of the off-diagonal S_{ij} vanish because of integration over the spin functions, and the diagonal S_{jj} terms are 1 for $j = 1, 2, 3,$ and 4.

The integral J is called an atomic Coulombic integral because it expresses the coulombic forces acting between the two electrons, and K is called an atomic exchange integral because it results from the exchange of the electrons between the orbitals. We now substitute the above results into Equation 7.40 to obtain the secular determinantal equation given in this problem. We expand this determinant to find

$$(E_0 + J - K - E) \begin{vmatrix} E_0 + J - K - E & 0 & 0 \\ 0 & E_0 + J - E & -K \\ 0 & -K & E_0 + J - E \end{vmatrix} = 0$$

$$(E_0 + J - K - E)^2 \begin{vmatrix} E_0 + J - E & -K \\ -K & E_0 + J - E \end{vmatrix} = 0$$

This gives

$$(E_0 + J - K - E)^2 \left[(E_0 + J - E)^2 - K^2 \right] = 0$$

We solve this equation to find

$$E = E_0 + J - K \qquad \text{or} \qquad E = E_0 + J \pm K$$

where (as noted in the problem) the solution $E_0 + J - K$ occurs twice. The 2×2 block in the secular determinant corresponds to the algebraic equations

$$c_3(E_0 + J - E) - c_4 K = 0$$

$$-c_3 K + c_4(E_0 + J - E) = 0$$

Substituting the solution $E = E_0 + J - K$ into either of these equations yields $c_3 = c_4$. Thus, the wave function corresponding to this solution is

$$\psi_3 = \frac{1}{\sqrt{2}}(\phi_3 - \phi_4)$$

where $1/\sqrt{2}$ is a normalization constant. Likewise, substituting the solution $E = E_0 + J + K$ into either of these equations yields $c_3 = -c_4$ and the wave function corresponding to this solution is

$$\psi_4 = \frac{1}{\sqrt{2}}(\phi_3 + \phi_4)$$

where $1/\sqrt{2}$ is a normalization constant.

Now

$$\psi_4 = \frac{1}{2} \left[1s\alpha(1)2s\beta(2) - 1s\alpha(2)2s\beta(1) + 1s\beta(1)2s\alpha(2) - 1s\beta(2)2s\alpha(1) \right]$$

$$= \frac{1}{2} \left[1s(1)2s(2) - 1s(2)2s(1) \right] \left[\alpha(1)\beta(2) + \beta(1)\alpha(2) \right]$$

$$\psi_3 = \frac{1}{2} \left[1s\alpha(1)2s\beta(2) - 1s\alpha(2)2s\beta(1) - 1s\beta(1)2s\alpha(2) + 1s\beta(2)2s\alpha(1) \right]$$

$$= \frac{1}{2} \left[1s(1)2s(2) + 1s(2)2s(1) \right] \left[\alpha(1)\beta(2) - \beta(1)\alpha(2) \right]$$

$$\psi_1 = \phi_1 = \frac{1}{\sqrt{2}} \left[1s\alpha(1)2s\alpha(2) - 1s\alpha(2)2s\alpha(1) \right]$$

$$= \frac{1}{\sqrt{2}} \left[1s(1)2s(2) - 1s(2)2s(1) \right] \left[\alpha(1)\alpha(2) \right]$$

$$\psi_2 = \phi_2 = \frac{1}{\sqrt{2}} \left[1s\beta(1)2s\beta(2) - 1s\beta(2)2s\beta(1) \right]$$

$$= \frac{1}{\sqrt{2}} \left[1s(1)2s(2) - 1s(2)2s(1) \right] \left[\beta(1)\beta(2) \right]$$

Referring to the results of Problem 8–24, we can see that ψ_4 is a singlet state and the ψ_1, ψ_2, and ψ_3 constitute a triplet state. The energy of the triplet state can be calculated using ψ_1, ψ_2, or ψ_3:

$$
\begin{aligned}
E_{\text{triplet}} &= \int\int d\tau_1 d\tau_2 \psi_1^* \hat{H} \psi_1 \\
&= \frac{1}{2}\int\int d\tau_1 d\tau_2 [1s(1)2s(2) - 1s(2)2s(1)][\alpha^*(1)\alpha^*(2)]\left(E_0 + \frac{1}{r_{12}}\right) \\
&\quad \times [1s(1)2s(2) - 1s(2)2s(1)][\alpha(1)\alpha(2)] \\
&= \frac{1}{2}\left[E_0(1 - 0 - 0 + 1) + (J - K - K + J)\right] \\
&= E_0 + J - K
\end{aligned}
$$

The energy of the singlet state is calculated in the same way and is

$$E_{\text{singlet}} = E_0 + J + K$$

The integrand in J is positive at all points, and so $J > 0$. If $K > 0$ also, then the energy of the triplet state is less than that of the singlet state. We calculated the value of E as $-11/4$ in Problem 7–30 using first-order perturbation theory. Using the given values of J and K, we find that

$$
\begin{aligned}
E(\text{excited triplet}) &= E_0 + J - K \\
&= -\frac{5}{2} + \frac{34}{81} - \frac{32}{(27)^2} \\
&= -2.1241 \\
E(\text{excited singlet}) &= E_0 + J + K \\
&= -\frac{5}{2} + \frac{34}{81} + \frac{32}{(27)^2} \\
&= -2.0364
\end{aligned}
$$

Therefore, we have

$$\Delta E(\text{triplet} \rightarrow \text{ground state}) = -2.1241 + 2.7500 = 0.6643 E_{\text{h}} = 126\,500 \text{ cm}^{-1}$$

and

$$\Delta E(\text{singlet} \rightarrow \text{ground state}) = -2.0857 + 2.7500 = 0.71365 E_{\text{h}} = 156\,630 \text{ cm}^{-1}$$

8–26. Determine the term symbols associated with an np^1 electron configuration. Show that these term symbols are the same as for an np^5 electron configuration. Which term symbol represents the ground state?

For an np^1 electron configuration, there are six entries in a table of possible sets of m_l and m_s. We have (using Equations 8.49, 8.50, and 8.52)

m_l	m_s	M_L	M_S	M_J
1	$+\frac{1}{2}$	1	$+\frac{1}{2}$	$+\frac{3}{2}$
1	$-\frac{1}{2}$	1	$-\frac{1}{2}$	$+\frac{1}{2}$
0	$+\frac{1}{2}$	0	$+\frac{1}{2}$	$+\frac{1}{2}$
0	$-\frac{1}{2}$	0	$-\frac{1}{2}$	$-\frac{1}{2}$
-1	$+\frac{1}{2}$	-1	$+\frac{1}{2}$	$-\frac{1}{2}$
-1	$-\frac{1}{2}$	-1	$-\frac{1}{2}$	$-\frac{3}{2}$

The M and M_S values given here correspond to a ^2P state, and the values of M_J correspond to a value of J of either 1/2 or 3/2. Thus, the term symbols associated with an np^1 electron configuration are $^2P_{3/2}$ and $^2P_{1/2}$. The ground state is determined by using Hund's rules; by Rule 3, the most stable state (and therefore the ground state) is $^2P_{1/2}$.

An np^5 configuration can be thought of as an np^1 configuration because two of the np orbitals are filled and so M_S and M_L are determined by the remaining half-filled p-orbital. Therefore, the term symbols associated with the np^5 configuration will be the same as those for an np^1 configuration. From Equation 8.53 we also see that the number of sets of m_{il} and m_{is} remains the same for an np^5 configuration as for an np^1 configuration.

8–27. Show that the term symbols for an np^4 electron configuration are the same as for an np^2 electron configuration.

There are fifteen entries in a table of possible m_l and m_s values for both an np^4 and np^2 electron configuration. In deducing the term symbols for an np^4 configuration, we realize that one np orbital must be filled and so M_L and M_S will be determined by the remaining two np electrons. In other words, there are two electron "holes" to fill, and so the same table of entries results as for an np^2 electron configuration, where there are two electrons to place.

8–28. Show that the number of sets of magnetic quantum numbers (m_l) and spin quantum numbers (m_s) associated with any term symbol is equal to $(2L + 1)(2S + 1)$. Apply this result to the np^2 case discussed in Section 8–9, and show that the term symbols ^1S, ^3P, and ^1D account for all the possible sets of magnetic quantum numbers and spin quantum numbers.

Each value of M_L associated with a value of L gives $2S + 1$ entries, and there are $2L + 1$ values of M_L for each value of L. The total number of entries for each term symbol, then, excluding the J subscript, is $(2L + 1)(2S + 1)$. In the previous problem, we found that there were fifteen entries in a table of m_l and m_s values for an np^2 electron configuration, so if we have accounted for all fifteen we have accounted for all possible sets of quantum numbers. Applying the result of this problem to the np^2 configuration, we find

$$^1S \qquad ^3P \qquad ^1D$$
$$(1 \times 1) + (3 \times 3) + (1 \times 5) = 15$$

These three term symbols account for all possible sets of m_l and m_s.

8–29. Calculate the number of sets of magnetic quantum numbers (m_l) and spin quantum numbers (m_s) for an nd^8 electron configuration. Prove that the term symbols ^1S, ^1D, ^3P, ^3F, and ^1G account for all possible term symbols.

There are a total of

$$\frac{G!}{N!(G-N)!} = \frac{10!}{8!2!} = 45 \text{ entries} \tag{8.53}$$

in a table of possible values of m_l and m_s. Proceeding as we did in Problem 8–28, we find

$$\begin{array}{ccccc} ^1S & ^1D & ^3P & ^3F & ^1G \\ \end{array}$$
$$(1 \times 1) + (1 \times 5) + (3 \times 3) + (3 \times 7) + (1 \times 9) = 45$$

so these five term symbols account for all possible sets of m_l and m_s.

8–30. Determine the term symbols for the electron configuration $nsnp$. Which term symbol corresponds to the lowest energy?

There are 2 possible sets of m_l and m_s for the ns electron and 6 possible sets of m_l and m_s for the np electron, so there are $2 \times 6 = 12$ possible sets of m_l and m_s for this system. We can denote values for the electron in the ns orbital as m_{1j} and those for the electron in the np orbital as m_{2j}. The allowed values are then

	m_{1l}	m_{1s}	m_{2l}	m_{2s}	M_L	M_S	M_J
1.	0	$+\frac{1}{2}$	1	$+\frac{1}{2}$	1	1	2
2.	0	$-\frac{1}{2}$	1	$+\frac{1}{2}$	1	0	1
3.	0	$+\frac{1}{2}$	1	$-\frac{1}{2}$	1	0	1
4.	0	$-\frac{1}{2}$	1	$-\frac{1}{2}$	1	-1	0
5.	0	$+\frac{1}{2}$	0	$+\frac{1}{2}$	0	1	1
6.	0	$-\frac{1}{2}$	0	$+\frac{1}{2}$	0	0	0
7.	0	$+\frac{1}{2}$	0	$-\frac{1}{2}$	0	0	0
8.	0	$-\frac{1}{2}$	0	$-\frac{1}{2}$	0	-1	-1
9.	0	$+\frac{1}{2}$	-1	$+\frac{1}{2}$	-1	1	0
10.	0	$-\frac{1}{2}$	-1	$+\frac{1}{2}$	-1	0	-1
11.	0	$+\frac{1}{2}$	-1	$-\frac{1}{2}$	-1	0	-1
12.	0	$-\frac{1}{2}$	-1	$-\frac{1}{2}$	-1	-1	-2

Entries 1, 2, 4, 5, 6, 8, 9, 10, and 12 correspond to $L = 1$ and $S = 1$, or a ^3P term symbol, and entries 3, 7, and 11 correspond to $L = 1$ and $S = 0$, which is a ^1P term symbol. The values of J can be derived from the table or by using Equation 8.54. The final result gives the term symbols

$$\begin{array}{cccc} ^3P_2 & ^3P_1 & ^3P_0 & ^1P_1 \\ (L+S) & (L+S-1) & (|L-S|) & (L+S) \end{array}$$

According to Hund's rules, the ground state is $^3P_0$.

8–31. How many sets of magnetic quantum numbers (m_l) and spin quantum numbers (m_s) are there for an $nsnd$ electron configuration? What are the term symbols? Which term symbol corresponds to the lowest energy?

There are 2 possible sets of m_l and m_s for the ns electron and 10 possible sets of m_l and m_s for the nd electron, so there are $2 \times 10 = 20$ possible sets of m_l and m_s for this system. The determination of the term symbols can be carried out as in Problem 8–30. The term symbols corresponding to the possible sets of m_l and m_s values are ^3D and ^1D. The values of J are determined using Equation 8.54, and the final result is

$$\begin{array}{cccc} ^3D_3 & ^3D_2 & ^3D_1 & ^1D_2 \\ (L+S) & (L+S-1) & (|L-S|) & (L+S) \end{array}$$

According to Hund's rules, the ground state is $^3D_1$.

8–32. The term symbols for an nd^2 electron configuration are ^1S, ^1D, ^1G, ^3P, and ^3F. Calculate the values of J associated with each of these term symbols. Which term symbol represents the ground state?

We can use Equation 8.54 to find the values of J for each term symbol.

Term Symbol	L	S	J	Full Term Symbol
^1S	0	0	0	$^1S_0$
^1D	2	0	2	$^1D_2$
^1G	4	0	4	$^1G_4$
^3P	1	1	2, 1, 0	$^3P_2, {}^3P_1, {}^3P_0$
^3F	3	1	4, 3, 2	$^3F_4, {}^3F_3, {}^3F_2$

By Hund's rules, the ground state is represented by the term symbol $^3F_2$.

8–33. The term symbols for an np^3 electron configuration are ^2P, ^2D, and ^4S. Calculate the values of J associated with each of these term symbols. Which term symbol represents the ground state?

We use Equation 8.54 to find the values of J for each term symbol.

Term Symbol	L	S	J	Full Term Symbol
^2P	1	$\frac{1}{2}$	$\frac{3}{2}, \frac{1}{2}$	$^2P_{3/2}, {}^2P_{1/2}$
^2D	2	$\frac{1}{2}$	$\frac{5}{2}, \frac{3}{2}$	$^2D_{5/2}, {}^2D_{3/2}$
^4S	0	$\frac{3}{2}$	$\frac{3}{2}$	$^4S_{3/2}$

By Hund's rules, the ground state is represented by the term symbol $^4S_{3/2}$.

8–34. Determine the electron configuration of a magnesium atom in its ground state, and its ground-state term symbol.

The ground state electron configuration of a magnesium atom is (from general chemistry) $1s^2 2s^2 2p^6 3s^2$, or $[\text{Ne}]3s^2$. We learned in our initial discussion of term symbols that ns^2 electron configurations correspond to the term symbol $^1S_0$, so the term symbol for atomic magnesium in the ground state is $^1S_0$.

8–35. Given that the electron configuration of a zirconium atom is $[\text{Kr}](4d)^2(5s)^2$, determine the ground-state term symbol for Zr.

The term-symbol for zirconium is determined by the nd^2 electrons. The ground state of an nd^2 electron configuration is $^3F_2$ (Problem 8–32).

8–36. Given that the electronic configuration of a palladium atom is $[Kr](4d)^{10}$, determine the ground-state term symbol for Pd.

Because all of the subshells of palladium are filled, the term symbol is $^1S_0$.

8–37. Consider the $1s2p$ electron configuration for helium. Determine the states (term symbols) that correspond to this electron configuration. Determine the degeneracies of each state. What will happen if you include the effect of spin orbit coupling?

This is a problem for which we did the general case (an $nsnp$ configuration) in Problem 8–30. The states corresponding to this electron configuration and their degeneracies are then

$$^3P_2 \quad ^3P_1 \quad ^3P_0 \quad ^1P_1$$
$$\;\;5 \qquad\; 3 \qquad\; 1 \qquad\; 3$$

According to Hund's rules, the ground state is $^3P_0$. Including the effect of spin-orbit coupling removes the degeneracy of the electronic states, and so spin-orbit coupling splits the lines in an atomic spectrum.

8–38. Use Table 8.5 to calculate the separation of the doublets that occur in the Lyman series of atomic hydrogen.

The Lyman series of atomic hydrogen consists of $n \to 1$ transitions. Since the $n = 1$ term symbol is $1s^2S_{1/2}$, and the selection rules (Equation 8.57) specify that $\Delta L = \pm 1$, we consider only P to S transitions. The separation of the doublets can then be obtained using the values in Table 8.5:

$$E(2p^2P_{3/2}) - E(2p^2P_{1/2}) = (82\,259.272 - 82\,258.917)\ \text{cm}^{-1}$$
$$= 0.355\ \text{cm}^{-1}$$
$$E(3p^2P_{3/2}) - E(3p^2P_{1/2}) = (97\,492.306 - 97\,492.198)\ \text{cm}^{-1}$$
$$= 0.108\ \text{cm}^{-1}$$
$$E(4p^2P_{3/2}) - E(4p^2P_{1/2}) = (102\,823.881 - 102\,823.835)\ \text{cm}^{-1}$$
$$= 0.046\ \text{cm}^{-1}$$

for the $2p$, $3p$, and $4p$ doublets respectively.

8–39. Use Table 8.6 to calculate the wavelength of the $4f\,^2F \to 3d\,^2D$ transition in atomic sodium and compare your result with that given in Figure 8.4. Be sure to use the relation $\lambda_{vac} = 1.00029\lambda_{air}$ (see Example 8–10).

From Table 8.6, we have

$$\Delta E = E(4f^2F) - E(3d^2D)$$
$$= 34\,588.6\ \text{cm}^{-1} - 29\,172.9\ \text{cm}^{-1}$$
$$= 5\,415.7\ \text{cm}^{-1}$$

Therefore, the wavelength of this transition in vacuum is

$$\lambda_{vac} = \frac{1}{\Delta E} = 1.8465 \times 10^{-4} \text{ cm} = 18\,465 \text{ Å}$$

Using the relation given in the problem, we find

$$\lambda_{air} = \frac{\lambda_{vac}}{1.00029} = 18\,459 \text{ Å}$$

in excellent agreement with Figure 8.4.

8–40. The orbital designations s, p, d, and f come from an analysis of the spectrum of atomic sodium. The series of lines due to $ns\,^2S \to 3p\,^2P$ transitions is called the *sharp* (s) series; the series due to $np\,^2P \to 3s\,^2S$ transitions is called the *principal* (p) series; the series due to $nd\,^2D \to 3p\,^2P$ transitions is called the *diffuse* (d) series; and the series due to $nf\,^2F \to 3d\,^2D$ transitions is called the *fundamental* (f) series. Identify each of these series in Figure 8.4, and tabulate the wavelengths of the first few lines in each series.

Identification of these series in the figure is easily done. The wavelengths found from the figure are

sharp series	principal series	diffuse series	fundamental series
11 404 Å	5 895.9 Å	8 194.8 Å	18 459 Å
11 382 Å	5 889.9 Å	8 183.3 Å	
6 160.7 Å	3 302.9 Å	5 688.2 Å	12 678 Å
6 154.2 Å	3 302.3 Å	5 682.7 Å	
5 153.6 Å	2 853.0 Å	4 982.9 Å	
5 149.1 Å	2 852.8 Å	4 978.6 Å	

8–41. Problem 8–40 defines the sharp, principal, diffuse, and fundamental series in the spectrum of atomic sodium. Use Table 8.6 to calculate the wavelengths of the first few lines in each series and compare your results with those in Figure 8.4. Be sure to use the relation $\lambda_{vac} = 1.00029\lambda_{air}$ (see Example 8–10).

Using Table 8.6, we find (doing the problem in the same manner as Problem 8–39, with all measurements in wavenumbers)

sharp series	principal series	diffuse series	fundamental series
8 766.48	16 956.183	12 199.5	5 415.7
8 783.68	16 973.379	12 216.7	
16 227.317	30 266.88	17 575.4	7 884.7
16 244.513	30 272.51	17 592.6	
19 399.268	35 040.27	20 063.4	
19 416.464	35 042.79	20 080.6	

Converting these results to λ_{air} by using the formula $1.00029\lambda_{air}\tilde{\nu}_{vac} = 1$, we find

sharp series	principal series	diffuse series	fundamental series
11 404 Å	5 895.8 Å	8 194.7 Å	18 459 Å
11 381 Å	5 889.9 Å	8 183.1 Å	
6 160.7 Å	3 303.0 Å	5 688.1 Å	12 679 Å
6 154.1 Å	3 302.4 Å	5 682.6 Å	
5 153.3 Å	2 853.0 Å	4 982.8 Å	
5 148.8 Å	2 852.8 Å	4 978.5 Å	

in very good agreement with the results from Figure 8.4 tabulated in Problem 8–40.

8–42. In this problem, we will derive an explicit expression for $V^{eff}(\mathbf{r}_1)$ given by Equation 8.18 using $\phi(\mathbf{r})$ of the form $(Z^3/\pi)^{1/2}e^{-Zr}$. (We have essentially done this problem in Problem 7–30.)

$$V^{eff}(\mathbf{r}_1) = \frac{Z^3}{\pi}\int d\mathbf{r}_2\frac{e^{-2Zr_2}}{r_{12}}$$

As in Problem 7–30, we use the law of cosines to write

$$r_{12} = (r_1^2 + r_2^2 - 2r_1r_2\cos\theta)^{1/2}$$

and so V^{eff} becomes

$$V^{eff}(r_1) = \frac{Z^3}{\pi}\int_0^\infty dr_2 e^{-2Zr_2}r_2^2\int_0^{2\pi}d\phi\int_0^\pi\frac{d\theta\sin\theta}{(r_1^2 + r_2^2 - 2r_1r_2\cos\theta)^{1/2}}$$

Problem 7–30 asks you to show that the integral over θ is equal to $2/r_1$ if $r_1 > r_2$ and equal to $2/r_2$ if $r_1 < r_2$. Thus, we have

$$V^{eff}(r_1) = 4Z^3\left[\frac{1}{r_1}\int_0^{r_1}e^{-Zr_2}r_2^2dr_2 + \int_{r_1}^\infty e^{-2Zr_2}r_2dr_2\right]$$

Now show that

$$V^{eff}(r_1) = \frac{1}{r_1} - e^{-2Zr_1}\left(Z + \frac{1}{r_1}\right)$$

In this problem, we are using atomic units. Use the results in the solution to Problem 7–30 and our knowledge of spherical coordinates to write

$$V^{eff}(r_1, \theta_1, \phi_1) = \frac{Z^3}{\pi}\int_0^{2\pi}d\phi_2\int_0^\infty e^{-2Zr_2}r_2^2dr_2\int_0^\pi\frac{\sin\theta_2 d\theta_2}{(r_1^2 + r_2^2 - 2r_1r_2\cos\theta_2)^{1/2}}$$

$$= 2Z^3\left(\frac{2}{r_1}\int_0^{r_1}e^{-Zr_2}r_2^2dr_2 + 2\int_{r_1}^\infty e^{-2Zr_2}r_2dr_2\right)$$

$$= 4Z^3\left(\frac{1}{r_1}\int_0^{r_1}e^{-Zr_2}r_2^2dr_2 + \int_{r_1}^\infty e^{-2Zr_2}r_2dr_2\right) \qquad (1)$$

We solved these integrals in Problem 7–30 and found that

$$\frac{1}{r_1}\int_0^{r_1}dr_2 r_2^2 e^{-2Zr_2/a_0} + \int_{r_1}^{\infty}dr_2 r_2 e^{-2Zr_2/a_0} = \frac{a_0^3}{4Z^3}\left[\frac{1}{r_1} - e^{-2Zr_1/a_0}\left(\frac{Z}{a_0} + \frac{1}{r_1}\right)\right]$$

or (in atomic units)

$$\frac{1}{r_1}\int_0^{r_1}dr_2 r_2^2 e^{-2Zr_2} + \int_{r_1}^{\infty}dr_2 r_2 e^{-2Zr_2} = \frac{1}{4Z^3}\left[\frac{1}{r_1} - e^{-2Zr_1}\left(Z + \frac{1}{r_1}\right)\right] \qquad (2)$$

Substituting Equation 2 into Equation 1 gives

$$V^{\text{eff}}(r_1) = \frac{4Z^3}{4Z^3}\left[\frac{1}{r_1} - e^{-2Zr_1}\left(Z + \frac{1}{r_1}\right)\right]$$

$$= \frac{1}{r_1} - e^{-2Zr_1}\left(Z + \frac{1}{r_1}\right)$$

8–43. Repeat Problem 8–42 using the expansion of $1/r_{12}$ given in Problem 7–31.

Again, we use atomic units in this problem. Use the expansion of $1/r_{12}$ given in Problem 7–31 to write

$$V^{\text{eff}}(r_1) = \frac{Z^3}{\pi}\int_0^{2\pi}d\phi_2 \int_0^{\pi}\sin\theta_2 d\theta_2 \sum_{l=0}^{\infty}\sum_{m=-l}^{l}\left(\frac{1}{2l+1}\right)Y_l^m(\theta_1,\phi_1)Y_l^{m*}(\theta_2,\phi_2)\int_0^{\infty}r_2^2 dr_2 \frac{r_<^l}{r_>^{l+1}}e^{-2r_2 Z}$$

As we found in Problem 7–31, the angular integration requires that $l = 0$ and $m = 0$, so that

$$V^{\text{eff}}(r_1) = \frac{Z^3(4\pi)}{\pi}\left[\frac{1}{r_1}\int_0^{r_1}dr_2 r_2^2 e^{-2r_2 Z} + \int_{r_1}^{\infty}dr_2 r_2 e^{-2r_2 Z}\right]$$

$$= 4Z^3\left[\frac{1}{r_1}\int_0^{r_1}dr_2 r_2^2 e^{-2r_2 Z} + \int_{r_1}^{\infty}dr_2 r_2 e^{-2r_2 Z}\right]$$

$$= \frac{1}{r_1} - e^{-2Zr_1}\left(Z + \frac{1}{r_1}\right)$$

The last equality uses the results of Problem 8–42 and the evaluation of the spherical harmonics was done in Problem 7–31.

Problems 8–44 through 8–48 address the energy levels of one electron atoms that include the effect of spin-orbit coupling.

8–44. Show that $\hat{\mathbf{L}} \cdot \hat{\mathbf{S}} = \frac{1}{2}(\hat{J}^2 - \hat{L}^2 - \hat{S}^2)$.

Using the definition $\hat{\mathbf{J}} = \hat{\mathbf{L}} + \hat{\mathbf{S}}$, we have

$$\hat{\mathbf{J}} \cdot \hat{\mathbf{J}} = \hat{J}^2 = (\hat{\mathbf{L}} + \hat{\mathbf{S}}) \cdot (\hat{\mathbf{L}} + \hat{\mathbf{S}})$$

$$= \hat{L}^2 + 2\hat{\mathbf{L}} \cdot \hat{\mathbf{S}} + \hat{S}^2$$

because $\hat{\mathbf{L}}$ and $\hat{\mathbf{S}}$ commute. This gives

$$\hat{\mathbf{L}} \cdot \hat{\mathbf{S}} = \frac{1}{2}(\hat{J}^2 - \hat{L}^2 - \hat{S}^2)$$

8–45. Show that $[\hat{H}, \hat{L}^2] = [\hat{H}, \hat{S}^2] = [\hat{H}, \hat{J}^2] = 0$, where $\hat{H}$ is the Hamiltonian operator of a hydrogen atom. *Hint*: Use the result of Problem 8–44 and operate on a function that is a product of a spatial part and a spin part.

First we show that $\hat{L}_z$ and $\hat{L}_z^2$ commute with $\hat{H}$, and then appeal to the fact that $\hat{H}$ is isotropic (the same in the x-, y-, and z-directions) to complete the proof. Using the spherical coordinate representation for $\hat{L}_z$,

$$\hat{L}_z = -i\frac{\partial}{\partial \phi}$$

we have

$$\begin{aligned}
[\hat{L}_z, \hat{H}]f &= \hat{L}_z \hat{H} f - \hat{H} \hat{L}_z f \\
&= -i\frac{\partial}{\partial \phi}\left(-\frac{1}{2}\nabla^2 + \frac{1}{r}\right)f + i\left(-\frac{1}{2}\nabla^2 + \frac{1}{r}\right)\frac{\partial f}{\partial \phi} \\
&= -i\left(-\frac{1}{2}\nabla^2 + \frac{1}{r}\right)\frac{\partial f}{\partial \phi} + i\left(-\frac{1}{2}\nabla^2 + \frac{1}{r}\right)\frac{\partial f}{\partial \phi} = 0
\end{aligned}$$

where we have used the fact the fact that $\partial/\partial\phi$ commutes with ∇^2 because the only ϕ dependence in ∇^2 occurs in $\partial^2/\partial\phi^2$. The same reasoning applies to $\partial^2/\partial\phi^2$, so we have

$$[\hat{L}_z^2, \hat{H}] = 0$$

Appealing to the isotropic property of $\hat{H}$, we have

$$[\hat{\mathbf{L}}, \hat{H}] = 0 \qquad \text{and} \qquad [\hat{L}^2, \hat{H}] = 0$$

Now, using the fact that $\hat{\mathbf{S}}$ and $\hat{S}^2$ operate only on the spin part of a wavefunction and $\hat{H}$ does not depend upon spin, we also have

$$[\hat{\mathbf{S}}, \hat{H}] = 0 \qquad \text{and} \qquad [\hat{S}^2, \hat{H}] = 0$$

Using the result of Problem 8–44, we see that

$$[\hat{L}^2, \hat{H}] = [\hat{S}^2, \hat{H}] = [\hat{J}^2, \hat{H}] = 0$$

8–46. Because of the coupling of the spin and orbital angular momenta of the electron, the Hamiltonian operator for a hydrogenlike atom becomes

$$\hat{H} = \hat{H}^{(0)} + \hat{H}_{so}^{(1)}$$

where (in atomic units)

$$\hat{H}^{(0)} = -\frac{1}{2}\nabla^2 - \frac{Z}{r}$$

and

$$\hat{H}_{so}^{(1)} = \frac{Z}{2(137)^2}\frac{1}{r^3}\hat{\mathbf{l}}\cdot\hat{\mathbf{s}}$$

We will now use first-order perturbation theory to evaluate the first-order correction to the energy. Recall from Chapter 7 that

$$E_n^{(1)} = \int \psi_n^{(0)*}\hat{H}_{so}^{(1)}\psi_n^{(0)}d\tau$$

Using the result of Problem 8–44, show that

$$E_n^{(1)} = \frac{Z}{2(137)^2} \int \psi_n^{(0)*} \frac{1}{r^3} (\hat{\mathbf{l}} \cdot \hat{\mathbf{s}}) \psi_n^{(0)} d\tau$$

$$= \frac{1}{2} \{ j(j+1) - l(l+1) - s(s+1) \} \frac{Z}{2(137)^2} \left\langle \frac{1}{r^3} \right\rangle \tag{1}$$

where

$$\left\langle \frac{1}{r^3} \right\rangle = \int \psi_n^{(0)*} \left(\frac{1}{r^3} \right) \psi_n^{(0)} d\tau \tag{2}$$

Problem 6–41 shows that

$$\left\langle \frac{1}{r^3} \right\rangle = \frac{Z^3}{n^3 l(l+1)(l+\frac{1}{2})} \tag{3}$$

Now combine Equations 1 through 3 to obtain

$$E_n^{(1)} / E_h = \frac{Z^4}{2(137)^2 n^3} \frac{\{j(j+1) - l(l+1) - s(s+1)\}}{2l(l+1)(l+\frac{1}{2})}$$

For $Z = 1$ and $l = 1$, what is the order of magnitude (in cm^{-1}) for the spin-orbit splitting between the two states as a function of n? (*Hint:* For a hydrogen atom $s = 1/2$ and j can be only $l \pm 1/2$.) Recall also that $1 E_h = 2.195 \times 10^5$ cm^{-1}. How does this energy compare with the energy separation between the energies for different values of n?

We know that (Problem 8–44) $\hat{\mathbf{L}} \cdot \hat{\mathbf{S}} = \frac{1}{2}(\hat{J}^2 - \hat{L}^2 - \hat{S}^2)$. In atomic units, this equation reduces to

$$\hat{\mathbf{L}} \cdot \hat{\mathbf{S}} = j(j+1) - l(l+1) - s(s+1)$$

Then

$$E_n^{(1)} = \int \psi_n^{(0)*} \hat{H}_{so}^{(1)} \psi_n^{(0)} d\tau$$

$$= \frac{Z}{2(137)^2} [j(j+1) - l(l+1) - s(s+1)] \int \psi_n^{(0)*} \frac{1}{r^3} \psi_n^{(0)} d\tau$$

$$= \frac{Z}{2(137)^2} [j(j+1) - l(l+1) - s(s+1)] \left\langle \frac{1}{r^3} \right\rangle$$

$$= \frac{Z^4}{2(137)^2 n^3} \frac{[j(j+1) - l(l+1) - s(s+1)]}{2l(l+1)(l+\frac{1}{2})}$$

The spin-orbit splitting is the difference between the state for which $j = 3/2, l = 1$, and $s = 1/2$ and the state for which $j = 1/2, l = 1$, and $s = 1/2$. Substituting into the equation above gives

$$\Delta E_n^{(1)} = E_n^{(1)}(j = \tfrac{3}{2}) - E_n^{(1)}(j = \tfrac{1}{2})$$

$$= \frac{1}{2(137)^2 n^3} \left[\frac{\left(\frac{3}{2}\right)\left(\frac{5}{2}\right) - 2 - \left(\frac{1}{2}\right)\left(\frac{3}{2}\right) - \left(\frac{1}{2}\right)\left(\frac{3}{2}\right) + 2 + \left(\frac{1}{2}\right)\left(\frac{3}{2}\right)}{(2)(2)\frac{3}{2}} \right] E_h$$

$$= \frac{1}{4(137)^2 n^3} E_h = \frac{2.9}{n^3} \text{ cm}^{-1}$$

E_h is much larger.

8–47. The two term symbols corresponding to the ns^2np^5 valence electron configuration of the halogens are $^2P_{1/2}$ and $^2P_{3/2}$. Which is the term symbol for the ground state? The energy difference between these two states for the different halogens is given below

Halogen	$[E(^2P_{1/2}) - E(^2P_{3/2})]/\text{cm}^{-1}$
F	404
Cl	880
Br	3685
I	7600

Suggest an explanation for this trend.

By Hund's rules, the ground state term symbol is $^2P_{3/2}$. The energy for this transition increases as we go down this column of the periodic table because the atomic number Z increases and $E_n^{(1)} \propto Z^4$ (Problem 8–46).

8–48. The photoionization spectra of the noble gases argon and krypton each show two closely spaced lines that correspond to the ionization of an electron from a $2p$ orbital. Explain why there are two closely spaced lines. (Assume the resulting ion is in its ground electronic state.)

There are two closely spaced lines because of spin-orbit, because the energyf the state with $m_s = +\frac{1}{2}$ is slightly different than that for $m_s = -\frac{1}{2}$. Therefore, one line corresponds to the ionization of an electron with spin of $+\frac{1}{2}$ and one to the ionization of an electron with spin of $-\frac{1}{2}$.

The spin operators satisfy the same general equations that we developed for the angular momentum operators in Problems 6–48 through 6–56. Problems 8–49 through 8–53 review these results.

8–49. The spin operators, $\hat{S}_x$, $\hat{S}_y$, and $\hat{S}_z$, like all angular momentum operators, obey the commutation relations (Problem 6–13)

$$[\hat{S}_x, \hat{S}_y] = i\hbar\hat{S}_z \quad [\hat{S}_y, \hat{S}_z] = i\hbar\hat{S}_x \quad [\hat{S}_z, \hat{S}_x] = i\hbar\hat{S}_y$$

Define the (non-Hermitian) operators

$$\hat{S}_+ = \hat{S}_x + i\hat{S}_y \qquad \hat{S}_- = \hat{S}_x - i\hat{S}_y \tag{1}$$

and show that

$$[\hat{S}_z, \hat{S}_+] = \hbar\hat{S}_+ \tag{2}$$

and

$$[\hat{S}_z, \hat{S}_-] = -\hbar\hat{S}_- \tag{3}$$

Now show that

$$\hat{S}_+\hat{S}_- = \hat{S}^2 - \hat{S}_z^2 + \hbar\hat{S}_z$$

and that

$$\hat{S}_-\hat{S}_+ = \hat{S}^2 - \hat{S}_z^2 - \hbar\hat{S}_z$$

where

$$\hat{S}^2 = \hat{S}_x^2 + \hat{S}_y^2 + \hat{S}_z^2$$

We know that (as in Problem 6–13)

$$[\hat{S}_x, \hat{S}_y] = ih\hat{S}_z \qquad [\hat{S}_y, \hat{S}_z] = ih\hat{S}_x \qquad [\hat{S}_z, \hat{S}_x] = ih\hat{S}_y$$

Now

$$
\begin{aligned}
\left[\hat{S}_z, \hat{S}_+\right] &= \hat{S}_z\hat{S}_+ - \hat{S}_+\hat{S}_z \\
&= \hat{S}_z\left(\hat{S}_x + i\hat{S}_y\right) - \hat{S}_x\hat{S}_z - i\hat{S}_y\hat{S}_z \\
&= \hat{S}_z\hat{S}_x + i\hat{S}_z\hat{S}_y - \hat{S}_x\hat{S}_z - i\hat{S}_y\hat{S}_z \\
&= (\hat{S}_z\hat{S}_x - \hat{S}_x\hat{S}_z) - i(\hat{S}_y\hat{S}_z - \hat{S}_z\hat{S}_y) \\
&= \left[\hat{S}_z, \hat{S}_x\right] - i\left[\hat{S}_y, \hat{S}_z\right] \\
&= ih\hat{S}_y - i(ih\hat{S}_x) = h(\hat{S}_x + i\hat{S}_y) = h\hat{S}_+
\end{aligned}
$$

$$
\begin{aligned}
\left[\hat{S}_z, \hat{S}_-\right] &= \hat{S}_z\hat{S}_- - \hat{S}_-\hat{S}_z \\
&= \hat{S}_z\left(\hat{S}_x - i\hat{S}_y\right) - \hat{S}_x\hat{S}_z + i\hat{S}_y\hat{S}_z \\
&= \hat{S}_z\hat{S}_x - i\hat{S}_z\hat{S}_y - \hat{S}_x\hat{S}_z + i\hat{S}_y\hat{S}_z \\
&= (\hat{S}_z\hat{S}_x - \hat{S}_x\hat{S}_z) + i(\hat{S}_y\hat{S}_z - \hat{S}_z\hat{S}_y) \\
&= \left[\hat{S}_z, \hat{S}_x\right] + i\left[\hat{S}_y, \hat{S}_z\right] \\
&= ih\hat{S}_y + i(ih\hat{S}_x) = h(-\hat{S}_x + i\hat{S}_y) = -h\hat{S}_-
\end{aligned}
$$

and

$$
\begin{aligned}
\hat{S}_-\hat{S}_+ &= (\hat{S}_x - i\hat{S}_y)(\hat{S}_x + i\hat{S}y) \\
&= \hat{S}_x^2 + \hat{S}_y^2 + i\hat{S}_x\hat{S}_y - i\hat{S}_y\hat{S}_x = \hat{S}^2 - \hat{S}_z^2 + i\left[\hat{S}_x, \hat{S}_y\right] \\
&= \hat{S}^2 - \hat{S}_z^2 - h\hat{S}_z \\
\hat{S}_+\hat{S}_- &= (\hat{S}_x + i\hat{S}_y)(\hat{S}_x - i\hat{S}_y) \\
&= \hat{S}_x^2 + \hat{S}_y^2 + i\left[\hat{S}_y, \hat{S}_x\right] = \hat{S}^2 - \hat{S}_z^2 + h\hat{S}_z
\end{aligned}
$$

8–50. Use Equation 2 from Problem 8–49 and the fact that $\hat{S}_z\beta = -\frac{h}{2}\beta$ to show that

$$\hat{S}_z\hat{S}_+\beta = \hat{S}_+\left(-\frac{h}{2}\beta + h\beta\right) = \frac{h}{2}\hat{S}_+\beta$$

Because $\hat{S}_z\alpha = \frac{h}{2}\alpha$, this result shows that

$$\hat{S}_+\beta \propto \alpha = c\alpha$$

where c is a proportionality constant. The following problem shows that $c = h$, so that we have

$$\hat{S}_+\beta = h\alpha$$

Now use Equation 3 from Problem 8–49 and the fact that $\hat{S}_z\alpha = \frac{\hbar}{2}\alpha$ to show that

$$\hat{S}_-\alpha = c\beta \tag{1}$$

where c is a proportionality constant. The following problem shows that $c = \hbar$, so that we have

$$\hat{S}_+\beta = \hbar\alpha \quad \text{and} \quad \hat{S}_-\alpha = \hbar\beta \tag{2}$$

Notice that $\hat{S}_+$ "raises" the spin function from β to α, whereas $\hat{S}_-$ "lowers" the spin function from α to β. The two operators $\hat{S}_+$ and $\hat{S}_-$ are called raising and lowering operators, respectively. Now use Equation 2 to show that

$$\hat{S}_x\alpha = \frac{\hbar}{2}\beta \qquad \hat{S}_y\alpha = \frac{i\hbar}{2}\beta$$

$$\hat{S}_x\beta = \frac{\hbar}{2}\alpha \qquad \hat{S}_y\beta = -\frac{i\hbar}{2}\alpha$$

Equation 2 from Problem 8–49 states that $\hat{S}_z\hat{S}_+\beta - \hat{S}_+\hat{S}_z\beta = \hbar\hat{S}_+\beta$, so

$$
\begin{aligned}
\hat{S}_z\hat{S}_+\beta &= \hat{S}_+\hat{S}_z\beta + \hbar\hat{S}_+\beta \\
&= \hat{S}_+\left(-\frac{\hbar}{2}\beta + \hbar\beta\right) = \hat{S}_+\left(\frac{\hbar}{2}\beta\right) \\
&= \frac{\hbar}{2}\hat{S}_+\beta
\end{aligned}
$$

where we used the equality $\hat{S}_z\beta = -\frac{\hbar}{2}\beta$. If we think of $\hat{S}_+\beta$ as a function on which $\hat{S}_z$ acts, from our previous result and the fact that $\hat{S}_z\alpha = \frac{\hbar}{2}\alpha$ we see that

$$\hat{S}_+\beta \propto \alpha = c\alpha = \hbar\alpha$$

where we have used the equality $c = \hbar$ shown in the following problem. Likewise, using Equation 3 from Problem 8–49,

$$
\begin{aligned}
\hat{S}_z\hat{S}_-\alpha - \hat{S}_-\hat{S}_z\alpha &= -\hbar\hat{S}_-\alpha \\
\hat{S}_z\hat{S}_-\alpha &= \hat{S}_-\hat{S}_z\alpha - \hbar\hat{S}_-\alpha \\
&= \hat{S}_-\left(\frac{\hbar}{2}\alpha - \hbar\alpha\right) = \hat{S}_-\left(-\frac{\hbar}{2}\alpha\right) \\
&= -\frac{\hbar}{2}\hat{S}_-\alpha \\
\hat{S}_-\alpha \propto c\beta &= \hbar\beta
\end{aligned}
$$

Now use the definitions of $\hat{S}_+$ and $\hat{S}_-$ and the results of this problem to find how $\hat{S}_x$ and $\hat{S}_y$ operate on α and β (see Problems 5–38 through 5–43). As in our results in Chapter 5 with $\hat{a}_+$ and $\hat{a}_-$, let $\hat{S}_-\beta = 0$ and $\hat{S}_+\alpha = 0$. Start with

$$\hat{S}_+ = \hat{S}_x + i\hat{S}_y, \qquad \hat{S}_- = \hat{S}_x - i\hat{S}_y$$

which gives

$$
\begin{aligned}
\hat{S}_+ + \hat{S}_- &= 2\hat{S}_x \\
\hat{S}_+\alpha + \hat{S}_-\alpha &= 2\hat{S}_x\alpha \\
0 + \hbar\beta &= 2\hat{S}_x\alpha \\
\hat{S}_x\alpha &= \frac{\hbar}{2}\beta
\end{aligned}
$$

Similarly,

$$\hat{S}_+ - \hat{S}_- = 2i\hat{S}_y$$

$$\hat{S}_+\alpha - \hat{S}_-\alpha = 2i\hat{S}_y\alpha$$

$$0 - \hbar\beta = 2i\hat{S}_y\alpha$$

$$\hat{S}_y\alpha = -\frac{\hbar}{2i}\beta = \frac{\hbar i}{2}\beta$$

Likewise, $\hat{S}_x\beta = \frac{\hbar}{2}\alpha$ and $\hat{S}_y\beta = -\frac{i\hbar}{2}\alpha$.

8–51. This problem shows that the proportionality constant c in

$$\hat{S}_+\beta = c\alpha \quad \text{or} \quad \hat{S}_-\alpha = c\beta$$

is equal to $\hbar$. Start with

$$\int \alpha^*\alpha d\tau = 1 = \frac{1}{|c|^2}\int (\hat{S}_+\beta)^*(\hat{S}_+\beta)d\tau$$

Let $\hat{S}_+ = \hat{S}_x + i\hat{S}_y$ in the second factor in the above integral and use the fact that $\hat{S}_x$ and $\hat{S}_y$ are Hermitian to get

$$\int (\hat{S}_x\hat{S}_+\beta)^*\beta d\tau + i\int (\hat{S}_y\hat{S}_+\beta)^*\beta d\tau = |c|^2$$

Now take the complex conjugate of both sides to get

$$\int \beta^*\hat{S}_x\hat{S}_+\beta d\tau - i\int \beta^*\hat{S}_y\hat{S}_+\beta d\tau = |c|^2$$

$$= \int \beta^*\hat{S}_-\hat{S}_+\beta d\tau$$

Now use the result in Problem 8–49 to show that

$$|c|^2 = \int \beta^*\hat{S}_-\hat{S}_+\beta d\tau = \int \beta^*(\hat{S}^2 - \hat{S}_z^2 - \hbar\hat{S}_z)\beta d\tau$$

$$= \int \beta^*\left(\frac{3}{4}\hbar^2 - \frac{1}{4}\hbar^2 + \frac{\hbar^2}{2}\right)\beta d\tau = \hbar^2$$

or that $c = \hbar$.

We know from Problem 8–50 that $\hat{S}_+\beta = c\alpha$, and we can assume that α is normalized. Then, following the instructions in the problem text,

$$\int \alpha^*\alpha d\tau = 1 = \frac{1}{|c|^2}\int (\hat{S}_+\beta)^*(\hat{S}_+\beta)d\tau$$

$$|c|^2 = \int (\hat{S}_+\beta)^*(\hat{S}_x\beta + i\hat{S}_y\beta)d\tau$$

$$= \int (\hat{S}_x\hat{S}_+\beta)^*\beta d\tau + i\int (\hat{S}_y\hat{S}_+\beta)^*\beta d\tau$$

$$|c|^2 = \int (\hat{S}_x\hat{S}_+\beta)^*\beta d\tau - i\int (\hat{S}_y\hat{S}_+\beta)^*\beta d\tau$$

$$= \int \beta^*\hat{S}_-\hat{S}_+\beta d\tau$$

where we have taken the complex conjugate of this equation and used the Hermitian properties of the operators $\hat{S}_x$ and $\hat{S}_y$. Now use the equation $\hat{S}_-\hat{S}_+ = \hat{S}^2 - \hat{S}_z^2 - \hbar\hat{S}_z$ from Problem 8–49 to find

$$|c|^2 = \int \beta^* \hat{S}_-\hat{S}_+ \beta \, d\tau = \int \beta^* (\hat{S}^2 - \hat{S}_z^2 - \hbar\hat{S}_z)\beta \, d\tau$$

$$= \int \beta^* \left(\frac{3}{4}\hbar^2 - \frac{1}{4}\hbar^2 + \frac{1}{2}\hbar^2 \right) \beta \, d\tau = \hbar^2$$

$$c = \hbar$$

where we have been given $\hat{S}_z\beta = -\frac{\hbar}{2}\beta$ in Problem 8–50 and will show $\hat{S}^2\beta = \frac{3}{4}\hbar^2\beta$ in the following problem.

8–52. Use the result of Problem 8–50 along with the equations $\hat{S}_z\alpha = \frac{\hbar}{2}\alpha$ and $\hat{S}_z\beta = -\frac{\hbar}{2}\beta$ to show that

$$\hat{S}^2\alpha = \tfrac{3}{4}\hbar^2\alpha = \tfrac{1}{2}\left(\tfrac{1}{2}+1\right)\hbar^2\alpha$$

and

$$\hat{S}^2\beta = \tfrac{3}{4}\hbar^2\beta = \tfrac{1}{2}\left(\tfrac{1}{2}+1\right)\hbar^2\beta$$

From Problem 8–49, we have the equation

$$\hat{S}^2 = \hat{S}_+\hat{S}_- + \hat{S}_z^2 - \hbar\hat{S}_z$$

$$\hat{S}^2\alpha = \hat{S}_+\hat{S}_-\alpha + \hat{S}_z^2\alpha - \hbar\hat{S}_z\alpha$$

$$= \hat{S}_+(\hbar\beta) + \hat{S}_z\left(\frac{\hbar\alpha}{2} - \frac{\hbar^2\alpha}{2}\right)$$

$$= \hbar^2\alpha + \frac{\hbar^2}{4}\alpha - \frac{\hbar^2}{2}\alpha = \frac{3\hbar^2}{4}\alpha$$

$$= \frac{1}{2}\left(\frac{1}{2}+1\right)\hbar^2\alpha = \frac{3}{4}\hbar^2\alpha$$

Likewise (again from Problem 8–49)

$$\hat{S}^2 = \hat{S}_-\hat{S}_+ + \hat{S}_z^2 + \hbar\hat{S}_z$$

$$\hat{S}^2\beta = \hat{S}_-\hat{S}_+\beta + \hat{S}_z^2\beta + \hbar\hat{S}_z\beta$$

$$= \hat{S}_-(\hbar\alpha) + \hat{S}_z\left(-\frac{\hbar}{2}\beta\right) - \frac{\hbar^2}{2}\beta$$

$$= \hbar^2\beta + \frac{\hbar^2}{4}\beta - \frac{\hbar^2}{2}\beta = \frac{3\hbar^2}{4}\beta$$

$$= \frac{1}{2}\left(\frac{1}{2}+1\right)\hbar^2\beta = \frac{3}{4}\hbar^2\beta$$

8–53. In this problem, we will use the results of Problems 8–50 and 8–52 to verify the statements at the end of Problem 8–24. Because $\hat{S}_{\text{total}} = \hat{S}_1 + \hat{S}_2$, we have

$$\hat{S}^2_{\text{total}} = (\hat{\mathbf{S}}_1 + \hat{\mathbf{S}}_2) \cdot (\hat{\mathbf{S}}_1 + \hat{\mathbf{S}}_2) = \hat{S}_1^2 + \hat{S}_2^2 + 2\hat{\mathbf{S}}_1 \cdot \hat{\mathbf{S}}_2$$

$$= \hat{S}_1^2 + \hat{S}_2^2 + 2(\hat{S}_{x1}\hat{S}_{x2} + \hat{S}_{y1}\hat{S}_{y2} + \hat{S}_{z1}\hat{S}_{z2})$$

Now show that

$$\hat{S}^2_{total}\alpha(1)\alpha(2) = \alpha(2)\hat{S}^2_1\alpha(1) + \alpha(1)\hat{S}^2_2\alpha(2) + 2\hat{S}_{x1}\alpha(1)\hat{S}_{x2}\alpha(2)$$

$$+2\hat{S}_{y1}\alpha(1)\hat{S}_{y2}\alpha(2) + 2\hat{S}_{z1}\alpha(1)\hat{S}_{z2}\alpha(2)$$

$$= 2\hbar^2\alpha(1)\alpha(2)$$

Similarly, show that

$$\hat{S}^2_{total}\beta(1)\beta(2) = 2\hbar^2\beta(1)\beta(2)$$

$$\hat{S}^2_{total}[\alpha(1)\beta(2) + \beta(1)\alpha(2)] = 2\hbar^2[\alpha(1)\beta(2) + \beta(1)\alpha(2)]$$

and

$$\hat{S}^2_{total}[\alpha(1)\beta(2) - \beta(1)\alpha(2)] = 0$$

$$\hat{S}_{total} = \hat{S}_1 + \hat{S}_2$$
$$\hat{S}^2_{total} = \hat{S}^2_1 + \hat{S}^2_2 + 2\hat{S}_1 \cdot \hat{S}_2 = \hat{S}^2_1 + \hat{S}^2_2 + 2(\hat{S}_{x1}\hat{S}_{x2} + \hat{S}_{y1}\hat{S}_{y2} + \hat{S}_{z1}\hat{S}_{z2})$$

Now, we evaluate $\hat{S}^2_{total}\alpha(1)\alpha(2)$,

$$\hat{S}^2_{total}\alpha(1)\alpha(2) = \hat{S}^2_1\alpha(1)\alpha(2) + \hat{S}^2_2\alpha(1)\alpha(2)$$

$$+2[\hat{S}_{x1}\alpha(1)\hat{S}_{x2}\alpha(2) + \hat{S}_{y1}\alpha(1)\hat{S}_{y2}\alpha(2) + \hat{S}_{z1}\alpha(1)\hat{S}_{z2}\alpha(2)]$$

$$= \alpha(2)\hat{S}^2_1\alpha(1) + \alpha(1)\hat{S}^2_2\alpha(2)$$

$$+2\left[\frac{\hbar}{2}\beta(1)\frac{\hbar}{2}\beta(2) + \frac{i\hbar}{2}\beta(1)\frac{i\hbar}{2}\beta(2) + \frac{\hbar}{2}\alpha(1)\frac{\hbar}{2}\alpha(2)\right]$$

$$= \frac{3\hbar^2}{4}\alpha(1)\alpha(2) + \frac{3\hbar^2}{4}\alpha(1)\alpha(2) + \frac{\hbar^2}{2}\alpha(1)\alpha(2)$$

$$= 2\hbar^2\alpha(1)\alpha(2)$$

For $\hat{S}^2_{total}\beta(1)\beta(2)$, we find

$$\hat{S}^2_{total}\beta(1)\beta(2) = \hat{S}^2_1\beta(1)\beta(2) + \hat{S}^2_2\beta(1)\beta(2)$$

$$+2[\hat{S}_{x1}\beta(1)\hat{S}_{x2}\beta(2) + \hat{S}_{y1}\beta(1)\hat{S}_{y2}\beta(2) + \hat{S}_{z1}\beta(1)\hat{S}_{z2}\beta(2)]$$

$$= \beta(2)\hat{S}^2_1\beta(1) + \beta(1)\hat{S}^2_2\beta(2)$$

$$+2\left[\frac{\hbar}{2}\alpha(1)\frac{\hbar}{2}\alpha(2) + (-1)^2\frac{i\hbar}{2}\alpha(1)\frac{i\hbar}{2}\alpha(2) + (-1)^2\frac{\hbar}{2}\beta(1)\frac{\hbar}{2}\beta(2)\right]$$

$$= \frac{3\hbar^2}{4}\beta(1)\beta(2) + \frac{3\hbar^2}{4}\beta(1)\beta(2) + \frac{\hbar^2}{2}\beta(1)\beta(2)$$

$$= 2\hbar^2\beta(1)\beta(2)$$

and for $\hat{S}^2_{total}[\alpha(1)\beta(2) + \beta(1)\alpha(2)]$, we obtain

$$\hat{S}^2_{total} = \beta(2)\hat{S}^2_1\alpha(1) + \alpha(2)\hat{S}^2_1\beta(1) + \alpha(1)\hat{S}^2_2\beta(2) + \beta(1)\hat{S}^2_2\alpha(2) + 2[\hat{S}_{x1}\alpha(1)\hat{S}_{x2}\beta(2)$$

$$+\hat{S}_{x1}\beta(1)\hat{S}_{x2}\alpha(2) + \hat{S}_{y1}\alpha(1)\hat{S}_{y2}\beta(2) + \hat{S}_{y1}\beta(1)\hat{S}_{y2}\alpha(2)$$

$$+\hat{S}_{z1}\alpha(1)\hat{S}_{z2}\beta(2) + \hat{S}_{z1}\beta(1)\hat{S}_{z2}\alpha(2)]$$

$$= \frac{3\hbar^2}{2}[\alpha(1)\beta(2) + \beta(1)\alpha(2)] + 2\left\{\frac{\hbar^2}{4}[\beta(1)\alpha(2) + \beta(2)\alpha(1)]\right.$$

$$+ \frac{\hbar^2}{4}[\beta(1)\alpha(2) + \beta(2)\alpha(1)] - \frac{\hbar^2}{4}[\beta(1)\alpha(2) + \beta(2)\alpha(1)]\right\}$$

$$= 2\hbar^2[\beta(1)\alpha(2) + \beta(2)\alpha(1)]$$

And finally, for $\hat{S}^2_{\text{total}}[\alpha(1)\beta(2) - \beta(1)\alpha(2)]$,

$$\hat{S}^2_{\text{total}} = \beta(2)\hat{S}^2_1\alpha(1) - \alpha(2)\hat{S}^2_1\beta(1) + \alpha(1)\hat{S}^2_2\beta(2) - \beta(1)\hat{S}^2_2\alpha(2)$$

$$+ 2[\hat{S}_{x1}\alpha(1)\hat{S}_{x2}\beta(2) - \hat{S}_{x1}\beta(1)\hat{S}_{x2}\alpha(2) + \hat{S}_{y1}\alpha(1)\hat{S}_{y2}\beta(2)$$

$$- \hat{S}_{y1}\beta(1)\hat{S}_{y2}\alpha(2) + \hat{S}_{z1}\alpha(1)\hat{S}_{z2}\beta(2) - \hat{S}_{z1}\beta(1)\hat{S}_{z2}\alpha(2)]$$

$$= \frac{3\hbar^2}{2}[\alpha(1)\beta(2) - \beta(1)\alpha(2)] + 2\left\{-\frac{\hbar^2}{4}[\alpha(1)\beta(2) - \beta(1)\alpha(2)]\right.$$

$$\left. - \frac{\hbar^2}{4}[\alpha(1)\beta(2) - \beta(1)\alpha(2)] - \frac{\hbar^2}{4}[\alpha(1)\beta(2) - \beta(1)\alpha(2)]\right\}$$

$$= 0$$

8–54. We discussed the Hartree-Fock method for a helium atom in Section 8–3, but the application of the Hartree-Fock method to atoms that contain more than two electrons introduces new terms because of the determinantal nature of the wave functions. For simplicity, we shall consider only closed-shell systems, in which the wave functions are represented by N doubly occupied spatial orbitals. The Hamiltonian operator for a $2N$-electron atoms is

$$\hat{H} = -\frac{1}{2}\sum_{j=1}^{2N}\nabla_j^2 - \sum_{j=1}^{2N}\frac{Z}{r_j} + \sum_{i=1}^{2N}\sum_{j>i}^{2N}\frac{1}{r_{ij}} \tag{1}$$

and the energy is given by

$$E = \int d\mathbf{r}_1 d\sigma_1 \cdots d\mathbf{r}_{2N} d\sigma_{2N} \Psi^*(1, 2, \ldots, 2N)\hat{H}\Psi(1, 2, \ldots, 2N) \tag{2}$$

Show that if Equation 1 and Equation 8.44 (with N replaced by $2N$ in this case) are substituted into Equation 2, then you obtain

$$E = 2\sum_{j=1}^{N} I_j + \sum_{i=1}^{N}\sum_{j=1}^{N}(2J_{ij} - K_{ij}) \tag{3}$$

where

$$I_j = \int d\mathbf{r}_j \phi_j^*(\mathbf{r}_j)\left[-\frac{1}{2}\nabla_j^2 - \frac{Z}{r_j}\right]\phi_j(\mathbf{r}_j) \tag{4}$$

$$J_{ij} = \int\int d\mathbf{r}_1 d\mathbf{r}_2 \phi_i^*(\mathbf{r}_1)\phi_i(\mathbf{r}_1)\frac{1}{r_{12}}\phi_j^*(\mathbf{r}_2)\phi_j(\mathbf{r}_2) \tag{5}$$

$$K_{ij} = \int\int d\mathbf{r}_1 d\mathbf{r}_2 \phi_i^*(\mathbf{r}_1)\phi_i(\mathbf{r}_2)\frac{1}{r_{12}}\phi_j^*(\mathbf{r}_2)\phi_j(\mathbf{r}_1) \tag{6}$$

Can you explain why the J_{ij} integrals are called *Coulomb integrals* and the K_{ij} integrals are called *exchange integrals* (if $i \neq j$)? Show that Equation 3 for a helium atom is the same as that given in Problem 8–10.

This problem is more difficult than most of the others in this chapter. First, replacing N by $2N$ in Equation 8.44 gives the determinantal wave function

$$\psi(1, 2, \ldots, 2N) = \frac{1}{\sqrt{(2N)!}} \begin{vmatrix} u_1(1) & u_2(1) & \cdots & u_{2N}(1) \\ u_1(2) & u_2(2) & \cdots & u_{2N}(2) \\ \vdots & \vdots & \vdots & \vdots \\ u_1(2N) & u_2(2N) & \cdots & u_{2N}(2N) \end{vmatrix}$$

Now realize that, because of its determinantal nature, $\psi(1, 2, \ldots, 2N)$ can be written as

$$\psi(1, 2, \ldots, 2N) = \frac{1}{\sqrt{(2N)!}} \sum_P \epsilon_P u_1(\hat{P}1)u_2(\hat{P}2)\ldots u_{2N}(\hat{P}2N) \tag{7}$$

where $\hat{P}$ represents a permutation of the numbers $(1, 2, 3, \ldots, 2N)$, $\epsilon_P = +1$ if the permutation $(\hat{P}1, \hat{P}2, \ldots, \hat{P}2N)$ differs from $(1, 2, \ldots, 2N)$ by an even number of interchanges of pairs of numbers, and $\epsilon_P = -1$ if the permutation $(\hat{P}1, \hat{P}2, \ldots, \hat{P}2N)$ differs from $(1, 2, \ldots, 2N)$ by an odd number of interchanges. An equivalent way of writing Equation 7 is

$$\psi(1, 2, \ldots, 2N) = \frac{1}{\sqrt{(2N)!}} \sum_P \epsilon_P u_{P1}(1)u_{P2}(2)\ldots u_{P2N}(2N) \tag{8}$$

In other words, we can interchange the labels of the spin-orbitals, rather than the electrons. We will be using both expressions for ψ in this problem. Now let us show that ψ is normalized. Using Equation 7, we can write

$$\int \cdots \int d\tau_1 d\tau_2 \ldots d\tau_{2N} \psi^* \psi = \frac{1}{(2N)!} \sum_P \epsilon_P \sum_Q \epsilon_Q \int \cdots \int d\tau_1 d\tau_2 \ldots d\tau_{2N} u_1^*(\hat{Q}1)$$

$$\times u_2^*(\hat{Q}2)\ldots u_{2N}^*(\hat{Q}2N) u_1(\hat{P}1) u_2(\hat{P}2)\ldots u_{2N}(\hat{P}2N)$$

These integrals will vanish unless $\hat{Q}1 = P1$, $\hat{Q}2 = \hat{P}2$, and, in general, $\hat{Q}n = \hat{P}n$, where n can have any integer value between 1 and $2N$. This means that the only permutation $\hat{Q}$ that will lead to a non-zero result is when $\hat{Q} = \hat{P}$. This leaves

$$\int \cdots \int d\tau_1 d\tau_2 \ldots d\tau_{2N} \psi^* \psi = \frac{1}{(2N)!} \sum_P \epsilon_P^2 \int \cdots \int d\tau_1 d\tau_2 \ldots d\tau_{2N} u_1^*(\hat{P}1)u_1(\hat{P}1)u_2^*(\hat{P}2)u_2(\hat{P}2)$$

$$\times \ldots u_{2N}^*(\hat{P}2N)u_{2N}(\hat{P}2N)$$

where $\epsilon_P^2 = 1$. Because the integration variables are dummy variables, all these integrals are equal, and equal to unity because the individual spin-orbitals are normalized. There are $(2N)!$ permutations, and so

$$\int \cdots \int d\tau_1 d\tau_2 \ldots d\tau_{2N} \psi^* \psi = \frac{1}{(2N)!}(2N)! = 1$$

Now we consider integrals of the form

$$A_1 = \int \cdots \int d\tau_1 d\tau_2 \ldots d\tau_{2N} \psi^* \left[\sum_{j=1}^{2N} \hat{H}^{(0)}(j) \right] \psi$$

where $\hat{H}^{(0)}(j)$ is a one-electron operator, meaning that it operates on only the coordinates of electron j. We can express this operator as

$$\sum_{j=1}^{2N} \hat{H}^{(0)}(j) = \sum_{j=1}^{2N} \left[-\frac{1}{2}\nabla_j^2 - \frac{Z}{r_j} \right]$$

Using Equation 7 for ψ, we write

$$A_1 = \frac{1}{(2N)!} \sum_P \epsilon_P \sum_Q \epsilon_Q \int \cdots \int d\tau_1 \ldots d\tau_{2N} u_1^*(\hat{Q}1) u_2^*(\hat{Q}2)$$

$$\times \ldots u_{2N}^*(\hat{Q}2N) \left[\sum_{j=1}^{2N} \hat{H}^{(0)}(j) \right] u_1(\hat{P}1) u_2(\hat{P}2) \ldots u_{2N}(\hat{P}2N)$$

As before, all $(2N)!$ permutations lead to the same integral, and so

$$A_1 = \sum_{j=1}^{2N} \int u_j^*(j) \hat{H}^{(0)}(j) u_j(j) d\tau_j$$

Finally, we consider integrals of the type

$$A_2 = \int \cdots \int d\tau_1 \ldots d\tau_{2N} \psi^* \left[\sum_{i=1}^{2N} \sum_{i<j} \frac{1}{r_{ij}} \right] \psi$$

Using Equation 8 for ψ, we write

$$A_2 = \frac{1}{(2N)!} \sum_P \epsilon_P \sum_Q \epsilon_Q \int \cdots \int d\tau_1 \ldots d\tau_{2N} u_{Q1}^*(1) u_{Q2}^*(2) \ldots u_{Q2N}^*(2N)$$

$$\times \left[\sum_{i=1}^{2N} \sum_{i<j} \frac{1}{r_{ij}} \right] u_{P1}(1) u_{P2}(2) \ldots u_{P2N}(2N)$$

These integrals will vanish unless the subscripts of all the spin-orbitals *except* those for electrons i and j match. There will be two non-vanishing terms in the summation of all Q permutations:

$$A_2 = \frac{1}{(2N)!} \sum_P \left\{ \epsilon_P \sum_{i=1}^{2N} \sum_{i<j} \int \int d\tau_i d\tau_j u_{P1}^*(i) u_{P2}^*(j) \frac{1}{r_{ij}} u_{P1}(i) u_{P2}^*(j) \right.$$

$$\left. - \epsilon_P \sum_{i=1}^{2N} \sum_{i<j} \int \int d\tau_i d\tau_j u_{P2}^*(i) u_{P1}^*(j) \frac{1}{r_{ij}} u_{P1}(i) u_{P2}(j) \right\}$$

Note that $Q = P$ for the term with the positive sign and that Q and P differ by the interchange of one pair of subscripts for the term with the negative sign. As before, all $(2N)!$ permutations lead to the same quantities, and so

$$A_2 = \sum_{i=1}^{2N} \sum_{i<j} \int \int d\tau_1 d\tau_2 u_i^*(1) u_j^*(2) \frac{1}{r_{12}} u_i(1) u_j(2)$$

$$- \sum_{i=1}^{2N} \sum_{i<j} \int \int d\tau_1 d\tau_2 u_i^*(1) u_j^*(2) \frac{1}{r_{12}} u_i(2) u_j(1)$$

$$= \sum_{i=1}^{2N} \sum_{i<j} J_{ij}' - \sum_{i=1}^{2N} \sum_{i<j} K_{ij}'$$

Note that the integrals here involve spin orbitals, while the integrals expressed by Equations 5 and 6 involve only spatial orbitals. Combining our results for A_1 and A_2 and substituting into Equation 1, we find

$$E = \int \cdots \int d\tau_1 \ldots d\tau_{2N} \psi^* \hat{H} \psi$$

$$= \int \cdots \int d\tau_1 d\tau_2 \ldots d\tau_{2N} \psi^* \left[\sum_j \hat{H}^{(0)}(j) \right] \psi$$

$$+ \int \cdots \int d\tau_1 \ldots d\tau_{2N} \psi^* \left[\sum_{i=1}^{2N} \sum_{i<j} \frac{1}{r_{ij}} \right] \psi$$

$$= A_1 + A_2$$

$$= \sum_{j=1}^{2N} \int u_j^*(j) \hat{H}^{(0)}(j) u_j(j) d\tau_j + \sum_{i=1}^{2N} \sum_{i<j} J'_{ij} - \sum_{i=1}^{2N} \sum_{i<j} K'_{ij}$$

$$= \sum_{j=1}^{2N} I_j + \sum_{i=1}^{2N} \sum_{i<j} J'_{ij} - \sum_{i=1}^{2N} \sum_{i<j} K'_{ij}$$

If the Slater orbital is of the form in Equation 8.44, having N doubly occupied orbitals, then $I_1 = I_2$, $I_3 = I_4$, and, in general, $I_{2n-1} = I_{2n}$, where n goes from 1 to N. Furthermore,

$$J'_{12} = J_{11} \qquad J'_{13} = J_{12} \qquad J'_{14} = J_{12}$$
$$J'_{23} = J_{12} \qquad J'_{24} = J_{22}$$

and so forth. Also,

$$K'_{12} = 0 \qquad K'_{13} = K_{12} \qquad K'_{14} = 0$$

and so forth, giving

$$E = 2 \sum_{j=1}^{N} I_{2j} + \sum_{i=1}^{N} \sum_{j=1}^{N} \left(2J_{ij} - K_{ij} \right)$$

where the summations are over spatial orbitals. If the relationships between the J'_{ij} and the J_{ij}, and those between the K'_{ij} and the K_{ij}, are unclear, take the beryllium atom as a concrete example. The integrands of the J_{ij} integrals are of the form

$$\phi_i^*(\mathbf{r}_1)\phi_i(\mathbf{r}_1) \frac{1}{r_{12}} \phi_j^*(\mathbf{r}_2)\phi_j(\mathbf{r}_2) d\mathbf{r}_1 d\mathbf{r}_2$$

which can be interpreted as the electrostatic coulombic interaction between two charges of magnitudes $\phi_i^*(\mathbf{r}_1)\phi_i(\mathbf{r}_1)d\mathbf{r}_1$ and $\phi_j^*(\mathbf{r}_2)\phi_j(\mathbf{r}_2)d\mathbf{r}_2$ separated by a distance r_{12}. The integrands of the K_{ij} do not have a similar classical interpretation. The K_{ij} may be obtained from the J_{ij} by exchanging $\mathbf{r}_1$ and $\mathbf{r}_2$ in either $\phi_i^*(\mathbf{r}_1)\phi_j^*(\mathbf{r}_2)$ or $\phi_i(\mathbf{r}_1)\phi_j(\mathbf{r}_2)$.

CHAPTER 9

The Chemical Bond: Diatomic Molecules

PROBLEMS AND SOLUTIONS

9–1. Express the Hamiltonian operator for a hydrogen molecule in atomic units.

The Hamiltonian operator for a hydrogen molecule is given in SI units by Equation 9.2. Let $m_e = e = \hbar = 4\pi\varepsilon_0 = 1$ to convert SI units to atomic units (Problem 8–6). Then Equation 9.2 becomes

$$-\frac{1}{2}\left(\nabla_1^2 + \nabla_2^2\right) - \frac{1}{r_{1A}} - \frac{1}{r_{1B}} - \frac{1}{r_{2A}} - \frac{1}{r_{2B}} + \frac{1}{r_{12}} + \frac{1}{R} \qquad (9.3)$$

9–2. Plot the product $1s_A 1s_B$ along the internuclear axis for several values of R.

Consider a plane that contains the two nuclei and let x be the coordinate along the internuclear axis. If the nuclei are located at $x = 0$ and $x = R$, then the $1s_A$ and $1s_B$ orbitals can be expressed as (see, for example, Equation 8.9)

$$1s_A = \pi^{-1/2}e^{-|x|} \qquad 1s_B = \pi^{-1/2}e^{-|x-R|}$$

Then

$$1s_A 1s_B = \frac{1}{\pi}e^{-|x|}e^{-|x-R|}$$

This formula is plotted here for three different values of R.

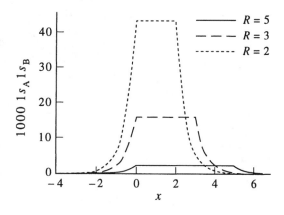

9–3. The overlap integral, Equation 9.10, and other integrals that arise in two-center systems like H_2 are called *two-center integrals*. Two-center integrals are most easily evaluated by using a coordinate system called *elliptic coordinates*. In this coordinate system (Figure 9.23), there are two fixed points separated by a distance R.

A point P is given by the three coordinates

$$\lambda = \frac{r_A + r_B}{R}$$

$$\mu = \frac{r_A - r_B}{R}$$

and the angle ϕ, which is the angle that the $(r_A,\ r_B,\ R)$ triangle makes about the interfocal axis. The differential volume element in elliptic coordinates is

$$d\mathbf{r} = \frac{R^3}{8}(\lambda^2 - \mu^2)d\lambda d\mu d\phi$$

Given the above definitions of λ, μ, and ϕ, show that

$$1 \le \lambda < \infty$$

$$-1 \le \mu \le 1$$

and

$$0 \le \phi \le 2\pi$$

Now use elliptic coordinates to evaluate the overlap integral, Equation 9.10,

$$S = \int d\mathbf{r}\, 1s_A 1s_B$$

$$= \frac{Z^3}{\pi} \int d\mathbf{r} e^{-Zr_A} e^{-Zr_B}$$

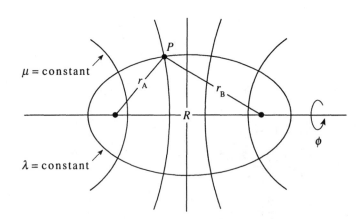

FIGURE 9.23
Elliptic coordinates are the natural coordinates for evaluating two-center integrals such as the overlap integral, Equation 9.10.

From the figure, $r_A + r_B$ can never be less than R, so $1 \leq \lambda < \infty$. Likewise, $r_A - r_B$ can never be of a magnitude greater than R, so $-1 \leq \mu \leq 1$. The variable ϕ can undergo one full revolution, so $0 \leq \phi \leq 2\pi$. We now evaluate Equation 9.10:

$$
\begin{aligned}
S &= \frac{Z^3}{\pi} \int d\mathbf{r}\, e^{-Zr_A} e^{-Zr_B} \\
&= \frac{Z^3}{\pi} \int_0^{2\pi} d\phi \int_1^\infty d\lambda \int_{-1}^1 d\mu\, \frac{R^3}{8} \left(\lambda^2 - \mu^2 \right) e^{-Z(r_A + r_B)} \\
&= \frac{R^3 Z^3}{4} \int_1^\infty d\lambda \int_{-1}^1 d\mu\, \left(\lambda^2 - \mu^2 \right) e^{-ZR\lambda} \\
&= \frac{R^3 Z^3}{4} \int_1^\infty d\lambda\, e^{-ZR\lambda} \int_{-1}^1 d\mu\, \left(\lambda^2 - \mu^2 \right) \\
&= \frac{R^3 Z^3}{4} \int_1^\infty d\lambda\, e^{-ZR\lambda} \left(2\lambda^2 - \frac{2}{3} \right) \\
&= \frac{R^3 Z^3}{2} \left[\left(\frac{1}{ZR} + \frac{2}{Z^2 R^2} + \frac{2}{Z^3 R^3} \right) e^{-ZR} - \frac{1}{3ZR} e^{-ZR} \right] \\
&= e^{-ZR} \left(1 + ZR + \frac{Z^2 R^2}{3} \right)
\end{aligned}
$$

9–4. Determine the normalized wave function for $\psi_- = c_1 (1s_A - 1s_B)$.

We follow the procedure used in Example 9–1.

$$
\begin{aligned}
1 &= \int \psi_-^* \psi_-\, d\mathbf{r} \\
&= c_1^2 \int d\mathbf{r}\, \left(1s_A^* - 1s_B^* \right) \left(1s_A - 1s_B \right) \\
&= c_1^2 \left(\int d\mathbf{r}\, 1s_A^* 1s_A - \int d\mathbf{r}\, 1s_A^* 1s_B - \int d\mathbf{r}\, 1s_B^* 1s_A + \int d\mathbf{r}\, 1s_B^* 1s_B \right) \\
&= c_1^2 \left(1 - S - S + 1 \right) \\
c_1^2 &= (2 - 2S)^{-1} \\
c_1 &= [2(1 - S)]^{-1/2}
\end{aligned}
$$

So the normalized wavefunction is

$$
\psi_- = \frac{1}{\sqrt{2(1 - S)}} (1s_A - 1s_B)
$$

9–5. Repeat the calculation in Section 9–3 for $\psi_- = (1s_A - 1s_B)$.

We follow the calculations in Section 9–3, using ψ_- instead of ψ_+.

$$
E_- = \frac{\int d\mathbf{r}\, \psi_-^* \hat{H} \psi_-}{\int d\mathbf{r}\, \psi_-^* \psi_-}
$$

$$
\int d\mathbf{r}\, \psi_-^* \psi_- = \int d\mathbf{r}\, 1s_A^* 1s_A - \int d\mathbf{r}\, 1s_A^* 1s_B - \int d\mathbf{r}\, 1s_B^* 1s_A + \int d\mathbf{r}\, 1s_B^* 1s_B
$$

$$
= 1 - S - S + 1 = 2(1 - S)
$$

9–6. Use the elliptic coordinate system of Problem 9–3 to derive analytic expressions for S, J, and K for the simple molecular-orbital treatment of H_2^+.

We have already derived an analytic expression for S in Problem 9–3. We now use Equations 9.19 and 9.20 and the coordinate system of Problem 9-3 to find analytic expressions for J and K:

$$
\begin{aligned}
J &= \frac{1}{R} - \int d\mathbf{r}\, 1s_A^* \left(\frac{1}{r_B}\right) 1s_A \\
&= \frac{1}{R} - \frac{R^3}{8\pi} \int_0^{2\pi} d\phi \int_1^\infty d\lambda \int_{-1}^1 d\mu\, (\lambda^2 - \mu^2) e^{-R(\lambda+\mu)} \left[\frac{2}{R(\lambda-\mu)}\right] \\
&= \frac{1}{R} - \frac{R^2}{2} \int_1^\infty d\lambda\, e^{-R\lambda} \int_{-1}^1 d\mu\, (\lambda + \mu) e^{-R\mu} \\
&= \frac{1}{R} - \frac{R^2}{2} \int_1^\infty d\lambda\, e^{-R\lambda} \left[\lambda \left(\frac{e^R - e^{-R}}{R}\right) + \frac{e^R(1-R) - e^{-R}(1+R)}{R^2}\right] \\
&= \frac{1}{R} - \frac{R^2}{2} \left\{\left(\frac{e^R - e^{-R}}{R}\right)\left(\frac{R+1}{R^2}\right) e^{-R} + \left[\frac{e^R(1-R) - e^{-R}(R+1)}{R^2}\right] \frac{e^{-R}}{R}\right\} \\
&= \frac{1}{R} - \frac{1}{2R} \left[2 - 2e^{-2R}(1+R)\right] = e^{-2R}\left(1 + \frac{1}{R}\right)
\end{aligned}
$$

and

$$
\begin{aligned}
K &= \frac{S}{R} - \int \frac{d\mathbf{r}\, 1s_B^* 1s_A}{r_B} \\
&= \frac{S}{R} - \frac{R^3}{4} \int_1^\infty d\lambda \int_{-1}^1 d\mu\, (\lambda^2 - \mu^2) \frac{2e^{-R\lambda}}{R(\lambda - \mu)} \\
&= \frac{S}{R} - \frac{R^2}{2} \int_1^\infty d\lambda\, e^{-R\lambda} \int_{-1}^1 d\mu\, (\lambda + \mu) \\
&= \frac{S}{R} - R^2 \int_1^\infty d\lambda\, \lambda e^{-R\lambda} \\
&= \frac{S}{R} - R^2 \left[\frac{e^{-R}(R+1)}{R^2}\right] = \frac{S}{R} - e^{-R}(R+1)
\end{aligned}
$$

9–7. Plot ψ_b and ψ_a given by Equations 9.27 and 9.28 for several values of R along the internuclear axis.

We will take the bond length to be R and let atom a be located at $x = 0$ and atom b be located at $x = R$. Here we plot ψ_b and ψ_a for several values of R:

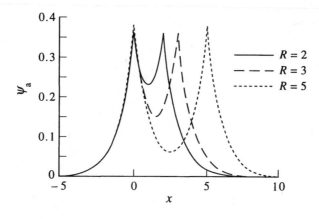

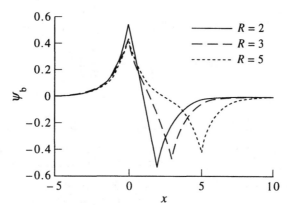

9–8. Show that

$$H_{AA} = H_{BB} = -\frac{1}{2} + J$$

and that

$$H_{AB} = -\frac{S}{2} + K$$

in the simple molecular-orbital treatment of H_2^+. The quantities J and K are given by Equations 9.23 and 9.24, respectively.

In atomic units, the Hamiltonian operator for H_2^+ is (Equation 9.4)

$$\hat{H} = -\frac{1}{2}\nabla^2 - \frac{1}{r_A} - \frac{1}{r_B} + \frac{1}{R}$$

We then have for H_{AA}

$$H_{AA} = \int d\mathbf{r}\, 1s_A \hat{H} 1s_A$$

Using Equation 9.15, we can rewrite the above equation as

$$
\begin{aligned}
H_{AA} &= \int d\mathbf{r}\, 1s_A \left(-\frac{1}{2}\nabla^2 - \frac{1}{r_A} - \frac{1}{r_B} + \frac{1}{R} \right) 1s_A \\
&= \int d\mathbf{r}\, 1s_A \left(E_{1s} - \frac{1}{r_B} + \frac{1}{R} \right) 1s_A \\
&= \int d\mathbf{r}\, 1s_A \left(-\frac{1}{2} - \frac{1}{r_B} + \frac{1}{R} \right) 1s_A = -\frac{1}{2} + J
\end{aligned}
$$

$H_{BB} = H_{AA}$, because

$$\left(-\frac{1}{2}\nabla^2 - \frac{1}{r_B}\right) 1s_B = E_{1s} 1s_B \qquad (9.16)$$

and

$$\int d\mathbf{r} \, 1s_B \left(E_{1s} - \frac{1}{r_A} + \frac{1}{R}\right) 1s_B = \int d\mathbf{r} \, 1s_A \left(E_{1s} - \frac{1}{r_B} + \frac{1}{R}\right) 1s_A$$

Finally

$$H_{AB} = \int d\mathbf{r} \, 1s_A \left(-\frac{1}{2}\nabla^2 - \frac{1}{r_A} - \frac{1}{r_B} + \frac{1}{R}\right) 1s_B$$

$$= \int d\mathbf{r}_2 \, 1s_A \left(E_{1s} - \frac{1}{r_A} + \frac{1}{R}\right) 1s_B$$

$$= \int d\mathbf{r}_2 \, 1s_A \left(-\frac{1}{2} - \frac{1}{r_A} + \frac{1}{R}\right) 1s_B = -\frac{S}{2} + K$$

9–9. Show explicitly that an s orbital on one hydrogen atom and a p_x orbital on another have zero overlap. Use the $2s$ and $2p_x$ wave functions given in Table 6.6 to set up the overlap integral. Take the z axis to lie along the internuclear axis. *Hint*: You need not evaluate any integrals, but simply show that the overlap integral can be separated into two parts that exactly cancel one another.

Table 6.6 gives the wave functions

$$\psi_{2s} = \frac{1}{4\sqrt{2\pi}} \left(\frac{1}{a_0}\right)^{3/2} \left(2 - \frac{r}{a_0}\right) e^{-r/2a_0}$$

$$\psi_{2p_z} = \frac{1}{4\sqrt{2\pi}} \left(\frac{1}{a_0}\right)^{3/2} \frac{r}{a_0} e^{-r/2a_0} \cos\theta$$

The overlap integral S is (Equation 9.10)

$$S = \int d\mathbf{r} \, \psi_{2s} \psi_{2p_z}$$

$$= C \int_0^\pi \cos\theta \, d\theta \int_0^\infty f(r) \, dr$$

$$= C \int_0^{\pi/2} \cos\theta \, d\theta \int_0^\infty f(r) \, dr + C \int_{\pi/2}^\pi \cos\theta \, d\theta \int_0^\infty f(r) \, dr$$

$$= C \int_0^\infty f(r) \, dr - C \int_0^\infty f(r) \, dr = 0$$

9–10. Show that $\Delta E_- = (J - K)/(1 - S)$ for the antibonding orbital ψ_- of H_2^+.

We replace ϕ_+ with ϕ_- in Equation 9.7 to obtain

$$E_- = \frac{\int d\mathbf{r}\phi_-^* \hat{H}\phi_-}{\int d\mathbf{r}\phi_-^* \phi_-}$$

From Problem 9–5, $\int d\mathbf{r} \, \psi_-^* \psi_- = 2(1 - S)$. To evaluate E_-, we first must evaluate the integral $\int d\mathbf{r} \, \psi_-^* \hat{H} \psi_-$:

$$
\begin{aligned}
\int d\mathbf{r} \, \psi_-^* \hat{H} \psi_- &= \int d\mathbf{r} \, (1s_A^* - 1s_B^*) \hat{H} (1s_A - 1s_B) \\
&= \int d\mathbf{r} \, (1s_A^* - 1s_B^*) \left(E_{1s} - \frac{1}{r_B} + \frac{1}{R} \right) 1s_A - \int d\mathbf{r} \, (1s_A^* - 1s_B^*) \left(E_{1s} - \frac{1}{r_A} + \frac{1}{R} \right) 1s_B \\
&= 2E_{1s}(1 - S) + \int d\mathbf{r} \, 1s_A^* \left(-\frac{1}{r_B} + \frac{1}{R} \right) 1s_A - \int d\mathbf{r} \, 1s_B^* \left(-\frac{1}{r_B} + \frac{1}{R} \right) 1s_A \\
&\quad - \int d\mathbf{r} \, 1s_A^* \left(-\frac{1}{r_A} + \frac{1}{R} \right) 1s_B + \int d\mathbf{r} \, 1s_B^* \left(-\frac{1}{r_A} + \frac{1}{R} \right) 1s_B \\
&= 2E_{1s}(1 - S) + 2J - 2K
\end{aligned}
$$

So

$$
\begin{aligned}
E_- &= \frac{\int d\mathbf{r} \, \psi_-^* \hat{H} \psi_-}{\int d\mathbf{r} \, \psi_-^* \psi_-} = \frac{2E_{1s}(1 - S) + 2J - 2K}{2(1 - S)} \\
&= E_{1s} + \frac{J - K}{1 - S}
\end{aligned}
$$

and

$$
\Delta E_- = E_- - E_{1s} = \frac{J - K}{1 - S}
$$

9–11. Show that ψ given by Equation 9.29 is an eigenfunction of $\hat{S}_z = \hat{S}_{z1} + \hat{S}_{z2}$ with $S_z = 0$.

$$
\psi = \psi_b(1)\psi_b(2) \left\{ \frac{1}{\sqrt{2}} [\alpha(1)\beta(2) - \alpha(2)\beta(1)] \right\} \tag{9.29}
$$

$\hat{S}_z$ operates on the spin part of ψ, so we wish to prove that

$$
\hat{S}_z [\alpha(1)\beta(2) - \alpha(2)\beta(1)] = S_z [\alpha(1)\beta(2) - \alpha(2)\beta(1)] = 0
$$

Begin with

$$
\left(\hat{S}_{z1} + \hat{S}_{z2} \right) [\alpha(1)\beta(2) - \alpha(2)\beta(1)] \overset{?}{=} 0
$$

$$
\hat{S}_{z1}\alpha(1)\hat{S}_{z2}\beta(2) - \hat{S}_{z2}\alpha(2)\hat{S}_{z1}\beta(1) \overset{?}{=} 0
$$

Recall (Equation 8.24) that (in atomic units)

$$
\hat{S}_z \alpha = \frac{1}{2}\alpha \qquad \hat{S}_z \beta = -\frac{1}{2}\beta
$$

Then

$$
\hat{S}_{z1}\alpha(1)\hat{S}_{z2}\beta(2) - \hat{S}_{z2}\alpha(2)\hat{S}_{z1}\beta(1) \overset{?}{=} 0
$$

$$
-\frac{1}{4}\alpha\beta + \frac{1}{4}\alpha\beta \overset{?}{=} 0
$$

$$
0 = 0
$$

9–12. Use molecular-orbital theory to explain why the dissociation energy of N_2 is greater than that of N_2^+, but the dissociation energy of O_2^+ is greater than that of O_2.

The electron configurations of N_2 and N_2^+ are (following the rules developed in Section 9–9)

$$N_2 \quad KK(\sigma 2s)^2(\sigma^* 2s)^2(\pi 2p)^4(\sigma 2p_z)^2$$
$$N_2^+ \quad KK(\sigma 2s)^2(\sigma^* 2s)^2(\pi 2p)^4(\sigma 2p_z)^1$$

Equation 9.33 gives a bond order for N_2 of 3 and a bond order of N_2^+ of $2\frac{1}{2}$. Thus, we expect a greater dissociation energy for N_2 than for N_2^+. Now consider O_2^+ and O_2:

$$O_2 \quad KK(\sigma 2s)^2(\sigma^* 2s)^2(\pi 2p)^4(\sigma 2p_z)^2(\pi^* 2p)^2$$
$$O_2^+ \quad KK(\sigma 2s)^2(\sigma^* 2s)^2(\pi 2p)^4(\sigma 2p_z)^2(\pi^* 2p)^1$$

The bond order of O_2^+ is $2\frac{1}{2}$ and the bond order of O_2 is 2. Because O_2^+ has one less antibonding electron (and hence a greater bond order) than O_2, we expect O_2^+ to have a greater dissociation energy than O_2.

9–13. Discuss the bond properties of F_2 and F_2^+ using molecular-orbital theory.

The electron configurations of F_2 and F_2^+ are

$$F_2 \quad KK(\sigma 2s)^2(\sigma^* 2s)^2(\pi 2p)^4(\sigma 2p_z)^2(\pi^* 2p)^4$$
$$F_2^+ \quad KK(\sigma 2s)^2(\sigma^* 2s)^2(\pi 2p)^4(\sigma 2p_z)^2(\pi^* 2p)^3$$

The molecule F_2^+ has one less antibonding electron than F_2, and so it has a greater bond order and a shorter bond than F_2.

9–14. Predict the relative stabilities of the species N_2, N_2^+, and N_2^-.

The electron configurations of N_2, N_2^+, and N_2^- are

$$N_2^+ \quad KK(\sigma 2s)^2(\sigma^* 2s)^2(\pi 2p)^4(\sigma 2p_z)^1$$
$$N_2 \quad KK(\sigma 2s)^2(\sigma^* 2s)^2(\pi 2p)^4(\sigma 2p_z)^2$$
$$N_2^- \quad KK(\sigma 2s)^2(\sigma^* 2s)^2(\pi 2p)^4(\sigma 2p_z)^2(\pi^* 2p)^1$$

The bond orders are then $2\frac{1}{2}$ for N_2^+, 3 for N_2, and $2\frac{1}{2}$ for N_2^-. The relative stabilities should therefore go as

$$N_2 \geq N_2^+ \approx N_2^-$$

9–15. Predict the relative bond strengths and bond lengths of diatomic carbon, C_2, and its negative ion, C_2^-.

The electron configurations of C_2 and C_2^- are

$$C_2 \quad KK(\sigma 2s)^2(\sigma^* 2s)^2(\pi 2p)^4$$
$$C_2^- \quad KK(\sigma 2s)^2(\sigma^* 2s)^2(\pi 2p)^4(\sigma 2p_z)^1$$

Since C_2^- has one more bonding electron than C_2, it has a greater bond strength and shorter bond length than C_2.

9–16. Write out the ground-state molecular-orbital electron configurations for Na_2 through Ar_2. Would you predict a stable Mg_2 molecule?

(L represents the filled $n = 2$ shell.)

$$Na_2 \quad KKLL(\sigma 3s)^2$$
$$Mg_2 \quad KKLL(\sigma 3s)^2(\sigma^*3s)^2$$
$$Al_2 \quad KKLL(\sigma 3s)^2(\sigma^*3s)^2(\pi 3p)^2$$
$$Si_2 \quad KKLL(\sigma 3s)^2(\sigma^*3s)^2(\pi 3p)^4$$
$$P_2 \quad KKLL(\sigma 3s)^2(\sigma^*3s)^2(\pi 3p)^4(\sigma 3p_z)^2$$
$$S_2 \quad KKLL(\sigma 3s)^2(\sigma^*3s)^2(\pi 3p)^4(\sigma 3p_z)^2(\pi^*3p)^2$$
$$Cl_2 \quad KKLL(\sigma 3s)^2(\sigma^*3s)^2(\pi 3p)^4(\sigma 3p_z)^2(\pi^*3p)^4$$
$$Ar_2 \quad KKLL(\sigma 3s)^2(\sigma^*3s)^2(\pi 3p)^4(\sigma 3p_z)^2(\pi^*3p)^4(\sigma^*3p_z)^2$$

The bond order of Mg_2 is 0, so we would not expect this molecule to be stable.

9–17. Determine the ground-state electron configuration of NO^+ and NO. Compare the bond orders of these two species.

The electron configurations of these two molecules are

$$NO^+ \quad KK(\sigma 2s)^2(\sigma^*2s)^2(\pi 2p)^4(\sigma 2p_z)^2$$
$$NO \quad KK(\sigma 2s)^2(\sigma^*2s)^2(\pi 2p)^4(\sigma 2p_z)^2(\pi^*2p)^1$$

The bond orders of NO^+ and NO are 3 and $2\frac{1}{2}$, respectively.

9–18. Determine the bond order of a cyanide ion.

The electron configuration of CN^- is

$$CN^- \quad KK(\sigma 2s)^2(\sigma^*2s)^2(\pi 2p)^4(\sigma 2p_z)^2$$

and so the bond order of CN^- is 3.

9–19. The force constants for the diatomic molecules B_2 through F_2 are given in the table below. Is the order what you expect? Explain.

Diatomic molecule	$k/N \cdot m^{-1}$
B_2	350
C_2	930
N_2	2260
O_2	1140
F_2	450

The force constant of a molecule is directly proportional to the bond strength of a molecule, which, in turn, is directly related to its bond order. The electron configurations of these diatomic molecules are

Molecule	Electron configuration	Bond order
B_2	$KK(\sigma 2s)^2(\sigma^* 2s)^2(\pi 2p)^2$	1
C_2	$KK(\sigma 2s)^2(\sigma^* 2s)^2(\pi 2p)^4$	2
N_2	$KK(\sigma 2s)^2(\sigma^* 2s)^2(\pi 2p)^4(\sigma 2p_z)^2$	3
O_2	$KK(\sigma 2s)^2(\sigma^* 2s)^2(\pi 2p)^4(\sigma 2p_z)^2(\pi^* 2p)^2$	2
F_2	$KK(\sigma 2s)^2(\sigma^* 2s)^2(\pi 3p)^4(\sigma 2p_z)^2(\pi^* 2p)^4$	1

Based on the bond orders, we would expect N_2 to have the largest force constant, and we expect $k(N_2) > k(C_2)$, $k(O_2) > k(B_2)$, $k(F_2)$, consistent with the above data. We cannot use the above information to order the magnitudes of the force constants for molecules of the same bond order (e.g., C_2 and O_2).

9–20. In Section 9–7, we constructed molecular orbitals for homonuclear diatomic molecules using the $n = 2$ atomic orbitals on each of the bonded atoms. In this problem, we will consider the molecular orbitals that can be constructed from the $n = 3$ atomic orbitals. These orbitals are important in describing diatomic molecules of the first row of transition metals. Once again we choose the z-axis to lie along the molecular bond. What are the designations for the $3s_A \pm 3s_B$ and $3p_A \pm 3p_B$ molecular orbitals? The $n = 3$ shell also contains a set of five $3d$ orbitals. (The shapes of the $3d$ atomic orbitals are shown in Figure 6.7.) Given that molecular orbitals with two nodal planes that contain the internuclear axis are called δ orbitals, show that ten $3d_A \pm 3d_B$ molecular orbitals consist of a bonding σ orbital, a pair of bonding π orbitals, a pair of bonding δ orbitals, and their corresponding antibonding orbitals.

Just as in the $n = 2$ case, the designations for the $3s_A \pm 3s_B$ orbitals are $\sigma_g 3s$ and $\sigma_u 3s$ and the designations for the $3p_A \pm 3p_B$ orbitals are

$$3p_{z,A} \pm 3p_{z,B} : \quad \sigma_g 3p_z \quad \sigma_u 3p_z$$
$$3p_{x,A} \pm 3p_{x,B} : \quad \pi_g 3p_x \quad \pi_u 3p_x$$
$$3p_{y,A} \pm 3p_{y,B} : \quad \pi_g 3p_y \quad \pi_u 3p_y$$

Now, through inspection of Figure 6.7, we can determine the number of nodal planes in each combination of d orbitals:

$$3d_{x^2-y^2,A} \pm 3d_{x^2-y^2,B} : \quad \pi_g 3d_{x^2-y^2} \quad \pi_u 3d_{x^2-y^2}$$
$$3d_{z^2,A} \pm 3d_{z^2,B} : \quad \sigma_g 3d_{z^2} \quad \sigma_u 3d_{z^2}$$
$$3d_{xy,A} \pm 3d_{xy,B} : \quad \pi_g 3d_{xy} \quad \pi_u 3d_{xy}$$
$$3d_{xz,A} \pm 3d_{xz,B} : \quad \delta_g 3d_{xz} \quad \delta_u 3d_{xz}$$
$$3d_{yz,A} \pm 3d_{yz,B} : \quad \delta_g 3d_{yz} \quad \delta_u 3d_{yz}$$

9–21. Determine the largest bond order for a first-row transition-metal homonuclear diatomic molecule (see Problem 9–20).

The molecule with the most bonding electrons and fewest antibonding electrons will have the largest bond order. The first-row transition metal homonuclear diatomic molecule which fits this criterion is Cr_2, with an electronic configuration of

$$KKLLMM(\sigma 4s)^2(\sigma 3d)^2(\pi 3d)^4(\delta 3d)^4$$

All the bonding orbitals derived from the atomic $3d$ orbitals in the ground state are full, and all the corresponding antibonding orbitals are empty. The bond order of Cr_2 is 6. It was first experimentally observed in molecular beam experiments, and the measured bond length agreed with molecular orbital calculations based on a bond order of 6.

9–22. Figure 9.19 plots a schematic representation of the energies of the molecular orbitals of HF. How will the energy-level diagram for the diatomic OH radical differ from that of HF? What is the highest occupied molecular orbital of OH?

In the energy-level diagram for the diatomic OH radical, the energy of the $2p_O$ orbitals is closer to that of the $1s_H$ orbital than the $2p_F$ orbitals are in HF. The highest occupied molecular orbitals of OH are the nonbonded orbitals $2p_{x,O}$ and $2p_{y,O}$.

9–23. A common light source used in photoelectron spectroscopy is a helium discharge, which generates light at 58.4 nm. A photoelectron spectrometer measures the kinetic energy of the electrons ionized when the molecule absorbs this light. What is the largest electron binding energy that can be measured using this radiation source? Explain how a measurement of the kinetic energy of the ionized electrons can be used to determine the energy of the occupied molecular orbitals of a molecule. *Hint*: Recall the photoelectron effect discussed in Chapter 1.

The energy of the source light is (Equation 1.23)

$$E = h\nu = \frac{hc}{\lambda} = 3.40 \times 10^{-18} \text{ J}$$

So 3.40×10^{-18} J is the largest electron binding energy that can be measured using this radiation source. Electrons with greater binding energies cannot be ionized from their atoms by this light.

Using Einstein's explanation of the photoelectric effect, we know that

$$\phi + \text{KE} = h\nu \tag{1.6}$$

If we can measure the kinetic energy of the ionized electrons, and also know ν, the energy absorbed by the electron, we can find ϕ, which is the energy of the molecular orbital occupied by the electron being ionized.

9–24. Using Figure 9.19, you found that the highest occupied molecular orbital for HF is a fluorine $2p$ atomic orbital. The measured ionization energy for an electron from this nonbonding molecular orbital of HF is 1550 kJ·mol^{-1}. However, the measured ionization energy of a $2p$ electron from a fluorine atom is 1795 kJ·mol^{-1}. Why is the ionization energy of an electron from the $2p$ atomic orbital on a fluorine atom greater for the fluorine atom than for the HF molecule?

The bonding electrons are unequally shared between the hydrogen and the fluorine atoms and are more localized on the most electronegative atom, which in this case is the fluorine atom. Because of this localization, the bonding pair of electrons shields the nonbonded $2p$ electron on the fluorine atom from the nucleus more than the inner shell electrons on atomic fluorine shield the $2p$ electron.

9–25. In this problem, we consider the heteronuclear diatomic molecule CO. The ionization energies of an electron from the valence atomic orbitals on the carbon atom and the oxygen atom are listed below.

Atom	Valence orbital	Ionization energy/MJ·mol^{-1}
O	$2s$	3.116
	$2p$	1.524
C	$2s$	1.872
	$2p$	1.023

Use these data to construct a molecular-orbital energy-level diagram for CO. What are the symmetry designations of the molecular orbitals of CO? What is the electron configuration of the ground state of CO? What is the bond order of CO? Is CO paramagnetic or diamagnetic?

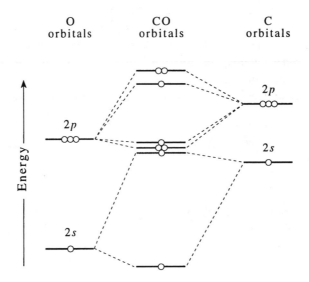

Since C and O have similar orbital energies, the bonding will be similar to that in a homonuclear diatomic molecule. In order of increasing energy, the symmetry designations of the molecular orbitals of CO are (ignoring the $1s$ orbitals) $\sigma 2s$, $\sigma^* 2s$, $\pi 2p_x$ and $\pi 2p_y$, and finally $\sigma 2p_z$. Note that we do not use the subscripts g and u because CO does not have an inversion center. The electron configuration of CO is (from Example 9–6) $KK(\sigma 2s)^2(\sigma^* 2s)^2(\pi 2p_x)^2(\pi 2p_y)^2(\sigma 2p_z)^2$, so the bond order is 3 and CO is diamagnetic because there are no unpaired electrons.

9–26. The molecule BF is isoelectronic with CO. However, the molecular orbitals for BF are different from those for CO. Unlike CO, the energy difference between the $2s$ orbitals of boron and fluorine is so large that the $2s$ orbital of boron combines with a $2p$ orbital on fluorine to make a molecular orbital. The remaining $2p$ orbitals on fluorine combine with two of the $2p$ orbitals on B to form π orbitals. The third $2p$ orbital on B is nonbonding. The energy ordering of the molecular orbitals is $\psi(2s_B + 2p_F) < \psi(2p_B - 2p_F) < \psi(2s_B - 2p_F) < \psi(2p_B + 2p_F) < \psi(2p_B)$. What are the symmetry designations of the molecular orbitals of BF? What is the electron configuration of the ground state of BF? What is the bond order of BF? Is BF diamagnetic or paramagnetic? How do the answers to these last two questions compare with those obtained for CO (Problem 9–25)?

The molecular-orbital energy-level diagram for BF resembles that for HF (Figure 9.19). The symmetry designations of the molecular orbitals of BF, in order of increasing energies, are

$2s_F$, σ_b, $\pi 2p_x$ and $\pi 2p_y$, $\pi^* 2p_x$ and $\pi^* 2p_y$, $2p_B$, and finally σ_a. The electron configuration of BF is thus

$$KK(2s_F)^2(\sigma_b)^2(\pi 2p)^4(\pi^* 2p)^2$$

The bond order is 1 and BF is paramagnetic, while CO has a bond order of 3 and is diamagnetic.

9–27. The photoelectron spectrum of O_2 exhibits two bands of 52.398 MJ·mol^{-1} and 52.311 MJ·mol^{-1} that correspond to the ionization of an oxygen $1s$ electron. Explain this observation.

These two bands correspond to the ionization of a $1s$ electron with spin $+\frac{1}{2}$ and the ionization of a $1s$ electron with spin $-\frac{1}{2}$. The different energies result from spin-orbit coupling (Problems 8–44 through 8–48).

9–28. The experimental ionization energies for a fluorine $1s$ electron from HF and F_2 are 66.981 and 67.217 MJ·mol^{-1}. Explain why these ionization energies are different even though the $1s$ electrons of the fluorine are not involved in the chemical bond.

Although the $1s$ electrons of the fluorine are not involved in the chemical bond, the bonding electrons do affect the attraction of the 1s electrons to the nucleus. In F_2, the bonding electrons are equally distributed between the two atoms, but in HF the bonding electrons are not equally distributed and are localized on the fluorine atom. This increases the shielding of the $1s$ orbital of the fluorine atom on HF relative to that on F_2. Therefore, the ionization energy of a fluorine $1s$ electron is slightly smaller for HF than for F_2.

9–29. Show that filled orbitals can be ignored in the determination of molecular term symbols.

A filled σ or σ^* orbital has two electrons and Equations 9.36 and 9.37 give $M_L = 0 + 0 = 0$ and $M_S = \frac{1}{2} - \frac{1}{2} = 0$. Likewise, for doubly degenerate π or π^* orbitals, Equations 9.36 and 9.37 give $M_L = 1 - 1 + 1 - 1 = 0$ and $M_S = \frac{1}{2} - \frac{1}{2} + \frac{1}{2} - \frac{1}{2} = 0$. Thus, a filled σ, σ^*, π, or π^* orbital does not contribute to the molecular term symbol.

9–30. Deduce the ground-state term symbols of all the diatomic molecules given in Table 9.6.

H_2^+ $(1\sigma_g)^1$ corresponds to $|M_L| = 0$ and $|M_S| = \frac{1}{2}$, or a $^2\Sigma$ term symbol. The unpaired electron occupies a molecular orbital of symmetry g and the $1\sigma_g$ wavefunction is unchanged upon reflection through a plane containing the nuclei. Therefore, the complete ground-state term symbol of H_2^+ is $^2\Sigma_g^+$.

H_2 $(1\sigma_g)^2$ corresponds to $|M_L| = 0$ and $|M_S| = 0$, or a $^1\Sigma$ term symbol. The symmetry is g and the $1\sigma_g$ wavefunction is unchanged upon reflection through a plane containing the nuclei. Therefore, the complete ground-state term symbol of H_2 is $^1\Sigma_g^+$.

He_2^+ We deduced this ground-state term symbol ($^2\Sigma_u^+$) in Example 9–11.

Li_2 $(1\sigma_g)^2(1\sigma_u)^2(2\sigma_g)^2$ corresponds to $|M_L| = 0$ and $|M_S| = 0$, or a $^1\Sigma$ term symbol. The symmetry is g and the σ orbitals remain unchanged upon reflection through a plane containing the nuclei, so the complete ground-state term symbol of Li_2 is $^1\Sigma_g^+$.

B$_2$ From Example 9–9, we know that the partial molecular term symbol is $^3\Sigma_g$. Because one of the half-filled π orbitals changes sign upon reflection through a plane containing the two nuclei, the complete ground-state term symbol of B$_2$ is $^3\Sigma_g^-$.

C$_2$ $(1\sigma_g)^2(1\sigma_u)^2(2\sigma_g)^2(2\sigma_u)^2(1\pi_u)^2(1\pi_u)^2$ corresponds to $|M_L| = 0$ and $|M_S| = 0$, or a $^1\Sigma$ term symbol. The symmetry of the molecule is g. Because the π orbital which changes sign upon reflection through a plane containing the two nuclei is filled, there is no observable difference in the molecule upon reflection through the plane and the complete ground-state term symbol of C$_2$ is $^1\Sigma_g^+$.

N$_2^+$ $(1\sigma_g)^2(1\sigma_u)^2(2\sigma_g)^2(2\sigma_u)^2(1\pi_u)^2(1\pi_u)^2(3\sigma_g)^1$ corresponds to $|M_L| = 0$ and $|M_S| = \frac{1}{2}$, or a $^2\Sigma$ term symbol. The symmetry of the molecule is g. The complete ground-state term symbol of N$_2^+$ is $^2\Sigma_g^+$, because the molecular wave function does not change when reflected through a plane containing the two nuclei.

N$_2$ $(1\sigma_g)^2(1\sigma_u)^2(2\sigma_g)^2(2\sigma_u)^2(1\pi_u)^2(1\pi_u)^2(3\sigma_g)^2$ corresponds to $|M_L| = 0$ and $|M_S| = 0$, or a $^1\Sigma$ term symbol. The symmetry of the molecule is g. The complete ground-state term symbol of N$_2$ is $^1\Sigma_g^+$, because the molecular wave function does not change when reflected through a plane containing the two nuclei.

O$_2^+$ $(1\sigma_g)^2(1\sigma_u)^2(2\sigma_g)^2(2\sigma_u)^2(3\sigma_g)^2(1\pi_u)^2(1\pi_u)^2(1\pi_g)^1$ corresponds to $|M_L| = 1$ and $|M_S| = \frac{1}{2}$, or a $^2\Pi$ term symbol. The symmetry of the molecule is g, since the only unfilled molecular orbital has symmetry g, so the complete ground-state term symbol of O$_2^+$ is $^2\Pi_g$.

O$_2$ We deduced this ground-state term symbol ($^3\Sigma_g^-$) in Example 9–10.

F$_2$ $(1\sigma_g)^2(1\sigma_u)^2(2\sigma_g)^2(2\sigma_u)^2(3\sigma_g)^2(1\pi_u)^2(1\pi_u)^2(1\pi_g)^2(1\pi_g)^2$ corresponds to $|M_L| = 0$ and $|M_S| = 0$, or a $^1\Sigma$ term symbol. The symmetry of the molecule is g and the complete ground-state term symbol of F$_2$ is $^1\Sigma_g^+$, because the molecular wave function does not change when reflected through a plane containing the two nuclei.

9–31. Determine the ground-state molecular term symbols of O$_2$, N$_2$, N$_2^+$, and O$_2^+$.

See Problem 9–30 for the ground-state molecular term symbols of these molecules.

9–32. The highest occupied molecular orbitals for an excited electronic configuration of an oxygen molecule are

$$(1\pi_g)^1(3\sigma_u)^1$$

Determine the molecular term symbols for oxygen with this electronic configuration.

This problem is similar to Example 9–12. First, set up a table of the possible momentum values:

	m_{l_1}	m_{s_1}	m_{l_2}	m_{s_2}	M_L	M_S
1.	+1	$+\frac{1}{2}$	0	$+\frac{1}{2}$	1	1
2.	+1	$+\frac{1}{2}$	0	$-\frac{1}{2}$	1	0
3.	+1	$-\frac{1}{2}$	0	$+\frac{1}{2}$	1	0
4.	+1	$-\frac{1}{2}$	0	$-\frac{1}{2}$	1	-1
5.	-1	$+\frac{1}{2}$	0	$+\frac{1}{2}$	-1	1
6.	-1	$+\frac{1}{2}$	0	$-\frac{1}{2}$	-1	0
7.	-1	$-\frac{1}{2}$	0	$+\frac{1}{2}$	-1	0
8.	-1	$-\frac{1}{2}$	0	$-\frac{1}{2}$	-1	-1

Entries 1, 2, 4, 5, 6, and 8 correspond to $|M_L| = 1$ and $S = 1$, giving a $^3\Pi$ molecular term symbol. Entries 3 and 7 correspond to $|M_L| = 1$ and $S = 0$, giving a $^1\Pi$ molecular term symbol.

9–33. Determine the values for the energies of the separated hydrogen atoms shown in Figure 9.22. Determine the energy difference of the dissociated limits.

From the legend of Figure 9.22, the energy of the ground state of each hydrogen atom is $-\frac{1}{2}E_h$. Therefore, the dissociation limit of the ground state of H_2 is $-1.0E_h$. The first excited state of H_2 also dissociates into two ground-state hydrogen atoms, so its dissociation limit is $-1.0E_h$. The dissociated limits of the $^1\Sigma_g^+$ state is given by the sum of the energies of an H_{1s} atom and H_{2s} atom, which is $-0.625E_h$.

9–34. For a set of point charges $Z_i e$ that lie along a line, we define the dipole moment (μ) of the charge distribution by

$$\mu = e\sum_i Z_i x_i$$

where e is the protonic charge and x_i is the distance of the charge $Z_i e$ from the origin. Consider the molecule LiH. A molecular-orbital calculation of LiH reveals that the bond length of this diatomic molecule is 159 pm and that there is a net charge of $+0.76e$ on the lithium atom and a net charge of $-0.76e$ on the hydrogen atom. First, determine the location of the center-of-mass of the LiH molecule. Use the center-of-mass as the origin along the x-axis and determine the dipole moment of the LiH molecule. How does your value compare with the experimental value of 19.62×10^{-30} C·m?

Li   c.m.   H

|———————x————————|———$159-x$———|

At the center of mass, $m_{Li}x = m_H(159 - x)$. Then

$$\frac{m_{Li}}{m_H}x = 159 - x$$

$$\left(1 + \frac{m_{Li}}{m_H}\right)x = 159$$

$$x = 20.2$$

where x is in picometers. Therefore, the center of mass is 20.2 pm from the Li atom. Now we can use the definition of μ to determine the dipole moment of LiH:

$$\mu = 0.76e(-20.2 \times 10^{-12} \text{ m}) - 0.76e(138.8 \times 10^{-12} \text{ m})$$
$$= (120.8 \times 10^{-12} \text{ m})e = -19.4 \times 10^{-30} \text{ C·m}$$

This answer differs by about 1% from the experimental value. (The answer is a negative number, which tells us that the dipole points toward the lithium atom, because we set up the coordinate system such that the lithium atom sits on the negative side of the origin. It is common to ignore the negative sign and tabulate μ as a positive quantity.)

9–35. Show that the value of the dipole moment μ defined in Problem 9–34 is independent of where we place the origin along the x-axis so long as the net charge of the molecule is equal to zero. Recalculate the dipole moment of LiH by placing the origin on the hydrogen atom, and compare your answer with that obtained for Problem 9–34.

Define a randomly selected origin x_0 such that $x_i = x_{\text{former, } i} - x_0$, where $x_{\text{former, } i}$ refers to the value of x_i in a previously selected coordinate system. Now substitute this value of x_i into our equation for μ:

$$\mu = \sum_i z_i x_i = \sum_i z_i (x_{\text{former, } i} - x_0)$$
$$= \sum_i z_i x_{\text{former, } i} - \sum_i z_i x_0$$
$$= \mu_{\text{former}} - \sum_i z_i x_0 = \mu_{\text{former}} - x_0 \sum_i z_i$$

If the net charge is zero, then $\sum z_i = 0$ and so $\mu = \mu_{\text{former}}$. Placing the origin of the system discussed in Problem 9–34 on the hydrogen atom, we find

$$\mu = (-0.76e)(0) + 0.76e(159 \times 10^{-12} \text{ m}) = 19.4 \times 10^{-30} \text{ C·m}$$

9–36. What would be the value of the dipole moment of LiH if its bond were purely ionic? Estimate the amount of ionic character in LiH. (See Problem 9.34.)

If the bond were purely ionic, there would be a charge of $-e$ on the hydrogen atom and $+e$ on the lithium atom. Then

$$\mu = e(-20.2 \times 10^{-12} \text{ m}) - e(138.8 \times 10^{-12} \text{ m}) = -2.55 \times 10^{-29} \text{ C·m}$$

Using the experimental result given in Problem 9–34, we find

$$\% \text{ ionic character Li} = \frac{19.4 \times 10^{-30} \text{ C·m}}{25.5 \times 10^{-30} \text{ C·m}} \times 100\% = 76.0\%$$

9–37. A dipole moment is actually a vector quantity defined by

$$\mu = e \sum_i Z_i \mathbf{r}_i$$

where $\mathbf{r}_i$ is a vector from some origin to the charge $Z_i e$. Show that μ is independent of where we take the origin if the net charge on the molecule is zero.

We can write the dipole moment above as

$$\mu = \sum_a z_a \mathbf{r}_a = \mathbf{i} \sum_a z_a r_{i,a} + \mathbf{j} \sum_a z_a r_{j,a} + \mathbf{k} \sum_a z_a r_{k,a}$$

which can then be separated into the three equations

$$\mu_i = \sum_a z_a r_{i,a} \qquad \mu_j = \sum_a z_a r_{j,a} \qquad \mu_k = \sum_a z_a r_{k,a}$$

We have shown in Problem 9–35 that each of these vector components of μ is independent of origin location if the net charge on the molecule is zero. Therefore, μ is independent of where we locate the origin under these conditions.

9–38. The dipole moment of HCl is 3.697×10^{-30} C·m. The bond length of HCl is 127.5 pm. If HCl is modeled as two point charges separated by its bond length, then what are the net charges on the H and Cl atom?

Use the chlorine atom as the origin. Then

$$\mu_{HCl} = 0 + z_H (127.5 \times 10^{-12} \text{ m})$$

$$\frac{3.697 \times 10^{-30} \text{ C·m}}{127.5 \times 10^{-12} \text{ m}} = z_H$$

$$0.181e = z_H$$

Because the net charge on the molecule is 0, the charge on the chlorine atom must be $-0.181e$.

9–39. Use the data in the table below to compute the fractional charges on the hydrogen atom and halide atom for the hydrogen halides. Is your finding in agreement with the order of the electronegativities of the halogen atoms, F > Cl > Br > I?

	R_e/pm	$\mu/10^{-30}$ C·m
HF	91.7	6.37
HCl	127.5	3.44
HBr	141.4	2.64
HI	160.9	1.40

Basing the origin at the halide atom, we can find the charges on the halide atom and hydrogen atom for each molecule as we did in the previous problem:

		Charge on H	Charge on halide
HF	0.43e	+0.43e	−0.43e
HCl	0.17e	+0.17e	−0.17e
HBr	0.12e	+0.12e	−0.12e
HI	0.054e	+0.054e	−0.054e

These data show that the ionic character of the bond increases with increasing electronegativity of the halide atom, as expected.

9–40. When we built up the molecular orbitals for diatomic molecules, we combined only those orbitals with the same energy because we said that only those with similar energies mix well. This problem is meant to illustrate this idea. Consider two atomic orbitals χ_A and χ_B. Show that a linear combination of these orbitals leads to the secular determinant

$$\begin{vmatrix} \alpha_A - E & \beta - ES \\ \beta - ES & S\alpha_B - E \end{vmatrix} = 0$$

where

$$\alpha_A = \int \chi_A h^{\text{eff}} \chi_A d\tau$$

$$\alpha_B = \int \chi_B h^{\text{eff}} \chi_B d\tau$$

$$\beta = \int \chi_B h^{\text{eff}} \chi_A d\tau = \int \chi_A h^{\text{eff}} \chi_B d\tau$$

$$S = \int \chi_A \chi_B d\tau$$

where h^{eff} is some effective one-electron Hamiltonian operator for the electron that occupies the molecular orbital ϕ. Show that this secular determinant expands to give

$$(1 - S^2)E^2 + [2\beta S - \alpha_A - \alpha_B]E + \alpha_A \alpha_B - \beta^2 = 0$$

It is usually a satisfactory first approximation to neglect S. Doing this, show that

$$E_\pm = \frac{\alpha_A + \alpha_B \pm [(\alpha_A - \alpha_B)^2 + 4\beta^2]^{1/2}}{2}$$

Now if χ_A and χ_B have the same energy, show that $\alpha_A = \alpha_B = \alpha$ and that

$$E_\pm = \alpha \pm \beta$$

giving one level of β units below α and one level of β units above α; that is, one level of β units more stable than the isolated orbital energy and one level of β units less stable. Now investigate the case in which $\alpha_A \neq \alpha_B$, say $\alpha_A > \alpha_B$. Show that

$$E_\pm = \frac{\alpha_A + \alpha_B}{2} \pm \frac{\alpha_A - \alpha_B}{2}\left[1 + \frac{4\beta^2}{(\alpha_A - \alpha_B)^2}\right]^{1/2}$$

$$= \frac{\alpha_A + \alpha_B}{2} \pm \frac{\alpha_A - \alpha_B}{2}\left[1 + \frac{2\beta^2}{(\alpha_A - \alpha_B)^2} - \frac{2\beta^4}{(\alpha_A - \alpha_B)^4} + \cdots\right]$$

$$E_\pm = \frac{\alpha_A + \alpha_B}{2} \pm \frac{\alpha_A - \alpha_B}{2} \pm \frac{\beta^2}{\alpha_A - \alpha_B} + \cdots$$

where we have assumed that $\beta^2 < (\alpha_A - \alpha_B)^2$ and have used the expansion

$$(1 + x)^{1/2} = 1 + \frac{x}{2} - \frac{x^2}{8} + \cdots$$

Show that

$$E_+ = \alpha_A + \frac{\beta^2}{\alpha_A - \alpha_B} + \cdots$$

$$E_- = \alpha_B - \frac{\beta^2}{\alpha_A - \alpha_B} + \cdots$$

Using this result, discuss the stabilization-destabilization of α_A and α_B versus the case above in which $\alpha_A = \alpha_B$. For simplicity, assume that $\alpha_A - \alpha_B$ is large.

If we let $\psi = c_1 \chi_A + \chi_B$, then (assuming these functions are normalized and following Equation 7.37)

$$\begin{vmatrix} \alpha_A - E & \beta - ES \\ \beta - ES & \alpha_A - E \end{vmatrix} = 0$$

and so

$$0 = (\alpha_A - E)(\alpha_B - E) - (\beta - ES)^2$$
$$= \alpha_A \alpha_B - E\alpha_A - E\alpha_B + E^2 - \beta^2 + 2\beta ES - E^2 S^2$$
$$= (1 - S^2)E^2 + (2\beta S - \alpha_A - \alpha_B)E + \alpha_A \alpha_B - \beta^2$$

Using the quadratic equation, we find that, neglecting S, E can be expressed by

$$E_{\pm} = \frac{\alpha_A + \alpha_B \pm [(\alpha_A - \alpha_B)^2 + 4\beta^2]^{1/2}}{2}$$

If χ_A and χ_B are associated with the same energy, $\alpha_A = \alpha_B = \alpha$, and $E_{\pm}$ becomes

$$E_{\pm} = \alpha \pm \beta$$

If $\alpha_A < \alpha_B$, $E_{\pm}$ becomes

$$E_{\pm} = \frac{\alpha_A + \alpha_B}{2} \pm \frac{[(\alpha_A - \alpha_B)^2 + 4\beta^2]^{1/2}}{2}$$
$$= \frac{\alpha_A + \alpha_B}{2} \pm \frac{\alpha_A - \alpha_B}{2}\left[1 + \frac{4\beta^2}{(\alpha_A - \alpha_B)^2}\right]$$
$$= \frac{\alpha_A + \alpha_B}{2} \pm \frac{\alpha_A - \alpha_B}{2}\left[1 + \frac{2\beta^2}{(\alpha_A - \alpha_B)^2} - \frac{2\beta^4}{(\alpha_A - \alpha_B)^4} + O(\beta^6)\right]$$
$$= \frac{\alpha_A + \alpha_B}{2} \pm \frac{\alpha_A - \alpha_B}{2} \pm \frac{\beta^2}{\alpha_A - \alpha_B} - O(\beta^4)$$

where we have made the assumptions given in the problem. Separating out E_+ and E_-, we find

$$E_+ = \frac{\alpha_A + \alpha_B}{2} + \frac{\alpha_A - \alpha_B}{2} + \frac{\beta^2}{\alpha_A - \alpha_B} - O(\beta^4) = \alpha_A + \frac{\beta^2}{\alpha_A - \alpha_B} - O(\beta^4)$$

$$E_- = \frac{\alpha_A + \alpha_B}{2} - \frac{\alpha_A - \alpha_B}{2} - \frac{\beta^2}{\alpha_A - \alpha_B} - O(\beta^4) = \alpha_B - \frac{\beta^2}{\alpha_A - \alpha_B} - O(\beta^4)$$

In this case, the energy is less stabilized and less destabilized than it was in the case where $\alpha_A = \alpha_B$. To see this, observe the case where $(\alpha_A - \alpha_B) \to \infty$. Then the stabilization/destabilization term in $E_{\pm}$ goes to zero and the two energy levels α_A and α_B remain. The smaller the difference between α_A and α_B, the greater the amount of stabilization and destabilization of energy levels that occurs.

9–41. In the Born-Oppenheimer approximation, we assume that because the nuclei are so much more massive than the electrons, the electrons can adjust essentially instantaneously to any nuclear motion, and hence we have a unique and well-defined energy, $E(R)$, at each internuclear separation R. Under this same approximation, $E(R)$ is the internuclear potential and so is the potential field in which the nuclei vibrate. Argue, then, that under the Born-Oppenheimer approximation, the force constant is independent of isotopic substitution. Using the above ideas, and given that the dissociation energy for H_2 is $D_0 = 432.1 \text{ kJ·mol}^{-1}$ and that the fundamental vibrational frequency

ν is 1.319×10^{14} s^{-1}, calculate D_0 and ν for deuterium, D_2. Realize that the observed dissociation energy is given by

$$D_0 = D_e - \frac{1}{2}h\nu$$

where D_e is the value of $E(R)$ at R_e.

The physical interpretation of the force constant is that it quantifies the curvature of the internuclear potential at its minimum. The function $E(R)$ is single valued, and the value of $E(R)$ is independent of isotopic substitution because the force constant is independent of isotopic substitution. Equation 5.33 gives

$$\nu = \frac{1}{2\pi}\left(\frac{k}{\mu}\right)^{1/2}$$

So we can write

$$\frac{\nu_{D_2}}{\nu_{H_2}} = \left(\frac{\mu_{H_2}}{\mu_{D_2}}\right)^{1/2} = \frac{(0.5005)^{1/2}}{1} = 0.7075$$

Then

$$\nu_{D_2} = 0.7075\nu_{H_2} = 0.7075(1.319 \times 10^{14}\ \text{s}^{-1})$$
$$= 9.332 \times 10^{13}\ \text{s}^{-1}$$

The value of D_e is independent of isotropic substitution, so

$$D_e = D_{0,H_2} + \frac{1}{2}h\nu_{H_2} = D_{0,D_2} + \frac{1}{2}h\nu_{D_2}$$

Then

$$D_{0,D_2} = (432.1\text{kJ}\cdot\text{mol}^{-1})\left(\frac{1}{6.022 \times 10^{23}\text{mol}^{-1}}\right) + \frac{1}{2}h(\nu_{H_2} - 0.7075\nu_{H_2})$$
$$= 7.303 \times 10^{-19}\text{J} = 439.8\ \text{kJ}\cdot\text{mol}^{-1}$$

9–42. In this problem, we evaluate the overlap integral (Equation 9.10) using spherical coordinates centered on atom A. The integral to evaluate is (Problem 9–3)

$$S(R) = \frac{1}{\pi}\int d\mathbf{r}_A e^{-r_A}e^{-r_B}$$
$$= \frac{1}{\pi}\int_0^\infty dr_A e^{-r_A}r_A^2 \int_0^{2\pi} d\phi \int_0^\pi d\theta \sin\theta e^{-r_B}$$

where r_A, r_B, and θ are shown in the figure.

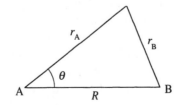

To evaluate the above integral, we must express r_B in terms of r_A, θ, and ϕ. We can do this using the law of cosines

$$r_B = (r_A^2 + R^2 - 2r_A R \cos\theta)^{1/2}$$

So the first integral we must consider is

$$I_\theta = \int_0^\pi e^{-(r_A^2 + R^2 - 2r_A R \cos\theta)^{1/2}} \sin\theta \, d\theta$$

Let $\cos\theta = x$ to get

$$\int_{-1}^1 e^{-(r_A^2 + R^2 - 2r_A Rx)^{1/2}} dx$$

Now let $u = (r_A^2 + R^2 - 2r_A Rx)^{1/2}$ and show that

$$dx = -\frac{u\,du}{r_A R}$$

Show that the limits of the integration over u are $u = r_A + R$ when $x = -1$ and $u = |R - r_A|$ when $x = 1$. Then show that

$$I_\theta = \frac{1}{r_A R} \left[e^{-(R-r_A)}(R + 1 - r_A) - e^{-(R+r_A)}(R + 1 + r_A) \right] \quad r_A < R$$

$$= \frac{1}{r_A R} \left[e^{-(r_A - R)}(r_A - R + 1) - e^{-(R+r_A)}(R + 1 + r_A) \right] \quad r_A > R$$

Now substitute this result into $S(R)$ above to get

$$S(R) = e^{-R} \left(1 + R + \frac{R^2}{3} \right)$$

Compare the length of this problem to Problem 9–3.

Following the method outlined in the problem we find that

$$I_\theta = \int_0^\pi e^{-(r_A^2 + R^2 - 2r_A R \cos\theta)^{1/2}} \sin\theta \, d\theta$$

$$= \int_{-1}^1 e^{-(r_A^2 + R^2 - 2r_A Rx)^{1/2}} dx$$

Let $u = (r_A^2 + R^2 - 2r_A Rx)^{1/2}$. Note that, by definition, $u > 0$. Now $2u\,du = -2r_A R\,dx$, or

$$dx = -\frac{u\,du}{r_A R}$$

When $x = 1$,

$$u = (r_A^2 - 2r_A R + R^2)^{1/2}$$
$$= |r_A - R|$$

Likewise, when $x = -1$,

$$u = (r_A^2 + 2r_A R + R^2)^{1/2}$$
$$= r_A + R$$

This gives us the limits of integration for the integral I_θ over u. We know that $(r_A + R) > |r_A - R|$, so we write

$$I_\theta = \frac{1}{r_A R} \int_{|R - r_A|}^{r_A + R} u e^{-u}\, du$$

$$= \frac{1}{r_A R} \int_{R - r_A}^{r_A + R} u e^{-u}\, du \qquad R > r_A$$

$$= \frac{1}{r_A R} \int_{r_A - R}^{r_A + R} u e^{-u}\, du \qquad R < r_A$$

From a table of integrals, $\int x e^{ax}\, dx = a^{-2} e^{ax}(ax - 1) + C$, so we can write

$$I_\theta = \frac{1}{r_A R}\left[e^{-(R + r_A)}(-R - r_A - 1) - e^{-(R - r_A)}(-R + r_A - 1)\right] \qquad r_A < R$$

$$= \frac{1}{r_A R}\left[e^{-(R - r_A)}(R + 1 - r_A) - e^{-(R + r_A)}(R + 1 + r_A)\right] \qquad r_A < R$$

$$= \frac{1}{r_A R}\left[e^{-(R + r_A)}(-R - r_A - 1) - e^{-(r_A - R)}(R - r_A - 1)\right] \qquad r_A > R$$

$$= \frac{1}{r_A R}\left[e^{-(r_A - R)}(r_A - R + 1) - e^{-(R + r_A)}(R + 1 + r_A)\right] \qquad r_A > R$$

Now, substituting into $S(R)$ gives

$$S(R) = 2 \int_0^R dr_A\, e^{-r_A} \frac{r_A}{R}\left[e^{-(R - r_A)}(R + 1 - r_A) - e^{-(R + r_A)}(R + 1 + r_A)\right]$$

$$+ 2 \int_R^\infty dr_A\, e^{-r_A} \frac{r_A}{R}\left[e^{-(r_A - R)}(r_A - R + 1) - e^{-(R + r_A)}(R + 1 + r_A)\right]$$

$$= \frac{2 e^{-R}}{R} \int_0^R dr_A\, r_A (R + 1 - r_A) + \frac{2 e^R}{R} \int_R^\infty dr_A\, e^{-2 r_A} r_A (r_A - R + 1)$$

$$= e^{-R}\left(\frac{R^2}{3} + R \right) - \frac{2 e^R}{R}\left[-\frac{R^2}{2} e^{-2R} + \frac{e^{-2R}}{4}(-2R - 1) \right.$$

$$\left. - \frac{R e^{-2R}}{4}(-2R - 1) + \frac{e^{-2R}}{4}(-2R - 1) \right] - e^{-R}\left(\frac{1}{2} + \frac{1}{R} \right)$$

$$= e^{-R}\left(\frac{R^2}{3} + R + R + 1 + \frac{1}{2R} - R - \frac{1}{2} + 1 + \frac{1}{2R} - \frac{1}{2} - \frac{1}{R} \right)$$

$$= e^{-R}\left(1 + R + \frac{R^2}{3} \right)$$

This is a much more lengthy procedure than that used in Problem 9–3.

9–43. Let's use the method that we developed in Problem 9–42 to evaluate the Coulomb integral, J, given by Equation 9.19. Let

$$I = -\int \frac{d\mathbf{r}\, 1 s_A^* 1 s_A}{r_B} = -\frac{1}{\pi} \int d\mathbf{r} \frac{e^{-2 r_A}}{(r_A^2 + R^2 - 2 r_A R \cos\theta)^{1/2}}$$

$$= -\frac{1}{\pi} \int_0^\infty dr_A\, r_A^2 e^{-2 r_A} \int_0^{2\pi} d\phi \int_0^\pi \frac{d\theta \sin\theta}{(r_A^2 + R^2 - 2 r_A R \cos\theta)^{1/2}}$$

Using the approach of Problem 9–42, let $\cos\theta = x$ and $u = (r_A^2 + R^2 - 2r_A Rx)^{1/2}$ to show that

$$I = \frac{2}{R}\int_0^\infty dr_A r_A e^{-2r_A}\int_{R+r_A}^{|R-r_A|} du = \frac{2}{R}\int_0^\infty dr_A r_A e^{-2r_A}[|R - r_A| - (R + r_A)]$$

$$= e^{-2R}\left(1 + \frac{1}{R}\right) - \frac{1}{R}$$

and that the Coulomb integral, J, is given by

$$J = e^{-2R}\left(1 + \frac{1}{R}\right)$$

Hint: You need to use the integrals

$$\int xe^{ax}dx = e^{ax}\left(\frac{x}{a} - \frac{1}{a^2}\right)$$

and

$$\int x^2 a^{ax}dx = e^{ax}\left(\frac{x^2}{a} - \frac{2x}{a^2} + \frac{2}{a^3}\right)$$

The integral over θ (Problem 9–42) is

$$I_\theta = \int_0^\pi \frac{d\theta \sin\theta}{(r_A^2 + R^2 - 2r_A R\cos\theta)^{1/2}} = \int_{-1}^1 \frac{dx}{(r_A^2 + R^2 - 2r_A Rx)^{1/2}}$$

$$= -\int_{R+r_A}^{|R-r_A|}\frac{u\,du}{r_A Ru} = -\frac{1}{r_A R}\int_{R+r_A}^{|R-r_A|} du$$

$$= -\frac{1}{r_A R}[|R - r_A| - (R + r_A)]$$

Substituting into I, we find

$$I = 2\int_0^\infty dr_A r_A^2 e^{-2r_A}\frac{1}{r_A R}[|R - r_A| - (R + r_A)]$$

$$= \frac{2}{R}\int_0^\infty dr_A e^{-2r_A}r_A[|R - r_A| - (R + r_A)]$$

$$= \frac{2}{R}\int_0^R dxe^{-2x}x(R - x) + \frac{2}{R}\int_R^\infty dxe^{-2x}x(x - R) - \frac{2}{R}\int_0^\infty dxe^{-2x}x(R + x)$$

$$= 2\left|_0^R \frac{e^{-2x}}{4}(-2x - 1)\right| - \frac{2}{R}\left|_0^R e^{-2x}\left(-\frac{x^2}{2} - \frac{2x}{4} - \frac{2}{8}\right)\right| + \frac{2}{R}\left|_R^\infty e^{-2x}\left(-\frac{x^2}{2} - \frac{2x}{4} - \frac{2}{8}\right)\right|$$

$$- 2\left|_R^\infty e^{-2x}\left(-\frac{x}{2} - \frac{1}{4}\right)\right| - \frac{2}{R}\left(\frac{R}{4} + \frac{2}{8}\right)$$

$$= e^{-2R}\left(-R - \frac{1}{2}\right) + \frac{1}{2} - e^{-2R}\left(-R - 1 - \frac{1}{2R}\right) - \frac{1}{2R}$$

$$+ e^{-2R}\left(R + 1 + \frac{1}{2R}\right) - e^{-2R}\left(R + \frac{1}{2}\right) - \frac{1}{2} - \frac{1}{2R}$$

$$= e^{-2R}\left(\frac{1}{2} + \frac{1}{2R} + \frac{1}{2} + \frac{1}{2R}\right) - \frac{1}{R}$$

$$= e^{-2R}\left(1 + \frac{1}{R}\right) - \frac{1}{R}$$

Finally, from Equation 9.19,

$$J = I + \frac{1}{R} = e^{-2R}\left(1 + \frac{1}{R}\right)$$

CHAPTER 10

Bonding in Polyatomic Molecules

PROBLEMS AND SOLUTIONS

10–1. Show that $\psi_{sp} = \frac{1}{\sqrt{2}}(2s \pm 2p_z)$ is normalized.

We evaluate

$$\int d\tau \, \psi_{sp}^* \psi_{sp} = \frac{1}{2} \int d\tau \, (2s \pm 2p_z)^*(2s \pm 2p_z)$$

$$= \frac{1}{2} \int d\tau \, \left(2s^*2s \pm 2s^*2p_z \pm 2p_z^*2s \pm 2p_z^*2p_z\right)$$

$$= \frac{1}{2}(1 \pm 0 \pm 0 + 1) = \frac{1}{2}(2) = 1$$

where we have used the fact that the hydrogenlike orbitals are orthonormal to each other.

10–2. Show that the three sp^2 hybrid orbitals given by Equations 10.3 through 10.5 are normalized.

Using

$$\psi_1 = \frac{1}{\sqrt{3}}2s + \sqrt{\frac{2}{3}}2p_z \tag{10.3}$$

we find

$$\int d\tau \, \psi_1^* \psi_1 = \int d\tau \, \left(\frac{1}{\sqrt{3}}2s^* + \sqrt{\frac{2}{3}}2p_z^*\right)\left(\frac{1}{\sqrt{3}}2s + \sqrt{\frac{2}{3}}2p_z\right)$$

$$= \frac{1}{3}\int d\tau \, 2s^*2s + \frac{\sqrt{2}}{3}\int d\tau \, (2s^*2p_z + 2p_z^*2s) + \frac{2}{3}\int d\tau \, 2p_z^*2p_z$$

$$= \frac{1}{3} + \frac{2}{3} = 1$$

and using

$$\psi_2 = \frac{1}{\sqrt{3}}2s - \frac{1}{\sqrt{6}}2p_z + \frac{1}{\sqrt{2}}2p_x \tag{10.4}$$

we find

$$\int d\tau \, \psi_2^* \psi_2 = \int d\tau \, \left(\frac{1}{\sqrt{3}}2s^* - \frac{1}{\sqrt{6}}2p_z^* + \frac{1}{\sqrt{2}}2p_x^*\right)\left(\frac{1}{\sqrt{3}}2s - \frac{1}{\sqrt{6}}2p_z + \frac{1}{\sqrt{2}}2p_x\right)$$

$$= \frac{1}{3}\int d\tau \, 2s^*2s + \frac{1}{6}\int d\tau \, 2p_z^*2p_z + \frac{1}{2}\int d\tau \, 2p_x^*2p_x$$

$$= \frac{1}{3} + \frac{1}{6} + \frac{1}{2} = 1$$

We have not written the integrals involving the products of different atomic orbitals, because their integrals are zero due to the orthogonality of the atomic orbitals. Likewise, using

$$\psi_3 = \frac{1}{\sqrt{3}}2s - \frac{1}{\sqrt{6}}2p_z - \frac{1}{\sqrt{2}}2p_x \qquad (10.4)$$

gives

$$\int d\tau \psi_3^* \psi_3 = \int d\tau \left(\frac{1}{\sqrt{3}}2s^* - \frac{1}{\sqrt{6}}2p_z^* - \frac{1}{\sqrt{2}}2p_x^* \right) \left(\frac{1}{\sqrt{3}}2s - \frac{1}{\sqrt{6}}2p_z - \frac{1}{\sqrt{2}}2p_x \right)$$

$$= \frac{1}{3} \int d\tau 2s^* 2s + \frac{1}{6} \int d\tau 2p_z^* 2p_z + \frac{1}{2} \int d\tau 2p_x^* 2p_x$$

$$= \frac{1}{3} + \frac{1}{6} + \frac{1}{2} = 1$$

10–3. Prove that the three sp^2 hybrid orbitals given by Equations 10.3 through 10.5 are directed at angles of 120° with respect to one another. (See Example 10–4.)

Because the s orbital is spherically symmetric, the directionality of the sp^2 hybrid orbitals will be determined by their p-orbital character. The p-orbital characters of the sp^2 hybrid orbitals are given by

$$\psi_1 = \sqrt{\frac{2}{3}}2p_z \qquad \psi_2 = -\frac{1}{\sqrt{6}}2p_z + \frac{1}{\sqrt{2}}2p_x \qquad \psi_3 = -\frac{1}{\sqrt{6}}2p_z - \frac{1}{\sqrt{2}}2p_x$$

We can represent these combinations of p-orbitals as vectors with the appropriate projections on the x- and z-axes.

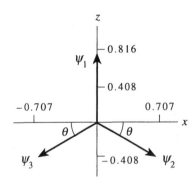

We can evaluate θ by noting that

$$\tan \theta = \frac{\frac{1}{\sqrt{6}}}{\frac{1}{\sqrt{2}}} = \frac{0.408}{0.707}$$

or $\theta = 30°$. The angle between ψ_1 and ψ_2 (and that between ψ_2 and ψ_3) is $90° + 30° = 120°$. The angle between ψ_2 and ψ_3 must therefore be $360° - 2(120°) = 120°$.

10–4. Represent the three sp^2 hybrid orbitals given by Equations 10.3 through 10.5 as vectors, where the coefficient of $2p_x$ is the x component and the coefficient of $2p_z$ is the z component. Now determine the angles between the hybrid orbitals using the formula for the dot product of two vectors. (Don't include the $2s$ orbital because it is spherically symmetric and so has no directionality.)

Excluding the 2s orbitals, the hybrid orbitals can be expressed in vector notation as

$$\psi_1 = \sqrt{\frac{2}{3}}\mathbf{k} \qquad \psi_2 = \frac{1}{\sqrt{2}}\mathbf{i} - \frac{1}{\sqrt{6}}\mathbf{k} \qquad \psi_3 = -\frac{1}{\sqrt{2}}\mathbf{i} - \frac{1}{\sqrt{6}}\mathbf{k}$$

Recall from MathChapter C that

$$\mathbf{A}\cdot\mathbf{B} = |\mathbf{A}|\,|\mathbf{B}|\cos\theta = A_x B_x + A_y B_y + A_z B_z$$

Substituting in the definition of $|\mathbf{A}|$, we can write this last equation as

$$\left(\sqrt{A_x^2 + A_y^2 + A_z^2}\right)\left(\sqrt{B_x^2 + B_y^2 + B_z^2}\right)\cos\theta = A_x B_x + A_y B_y + A_z B_z$$

We now use this expression to calculate the value of θ for the hybrid orbitals ψ_1 and ψ_2:

$$\theta = \cos^{-1}\left[\frac{0 + \left(\sqrt{\frac{2}{3}}\right)\left(-\frac{1}{\sqrt{6}}\right)}{\left(\sqrt{\frac{2}{3}}\right)\left(\sqrt{\frac{2}{3}}\right)}\right] = \cos^{-1}\left(-\frac{1}{2}\right)$$

$$= 120°$$

Likewise, for ψ_1 and ψ_3 θ is

$$\theta = \cos^{-1}\left[\frac{\left(\sqrt{\frac{2}{3}}\right)\left(-\frac{1}{\sqrt{6}}\right) + 0}{\left(\sqrt{\frac{2}{3}}\right)\left(\sqrt{\frac{2}{3}}\right)}\right] = \cos^{-1}\left(-\frac{1}{2}\right)$$

$$= 120°$$

and for ψ_2 and ψ_3 θ is

$$\theta = \cos^{-1}\left[\frac{-\frac{1}{2} + \frac{1}{6}}{\left(\sqrt{\frac{2}{3}}\right)\left(\sqrt{\frac{2}{3}}\right)}\right] = \cos^{-1}\left(-\frac{1}{2}\right)$$

$$= 120°$$

10–5. The following three orbitals are equivalent to the three sp^2 hybrid orbitals given by Equations 10.3 through 10.5

$$\phi_1 = \left(\frac{1}{3}\right)^{1/2} 2s - \left(\frac{1}{3}\right)^{1/2} 2p_x + \left(\frac{1}{3}\right)^{1/2} 2p_z$$

$$\phi_2 = \left(\frac{1}{3}\right)^{1/2} 2s + \frac{1}{2}(1 + 3^{-1/2})2p_x + \frac{1}{2}(1 - 3^{-1/2})2p_z$$

$$\phi_3 = \left(\frac{1}{3}\right)^{1/2} 2s + \frac{1}{2}(-1 + 3^{-1/2})2p_x - \frac{1}{2}(1 + 3^{-1/2})2p_z$$

First show that these orbitals are normalized. Now use the method introduced in Problem 10–4 to show that the angles between these orbitals are 120°. (These orbitals are the orbitals given by Equations 10.3 through 10.5 rotated by 45°.)

We show that the orbitals are normalized by proving that $\int \phi_j^* \phi_j = 1$.

$$\int \phi_1^* \phi_1 = \frac{1}{3} + \frac{1}{3} + \frac{1}{3} = 1$$

$$\int \phi_2^* \phi_2 = \frac{1}{3} + \frac{1}{4}\left(1 + 3^{-1/2}\right)^2 + \frac{1}{4}\left(1 - 3^{-1/2}\right)^2$$

$$= \frac{1}{3} + \frac{1}{4}\left(1 + \frac{2}{\sqrt{3}} + \frac{1}{3}\right) + \frac{1}{4}\left(1 - \frac{2}{\sqrt{3}} + \frac{1}{3}\right) = 1$$

$$\int \phi_3^* \phi_3 = \frac{1}{3} + \frac{1}{4}\left(-1 + 3^{-1/2}\right)^2 + \frac{1}{4}\left(1 + 3^{-1/2}\right)^2$$

$$= \frac{1}{3} + \frac{1}{4}\left(1 - \frac{2}{\sqrt{3}} + \frac{1}{3}\right) + \frac{1}{4}\left(1 + \frac{2}{\sqrt{3}} + \frac{1}{3}\right) = 1$$

The directionality of these orbitals can be expressed in vector notation as (see the solution to Problem 10–4)

$$\phi_1 = -\sqrt{\frac{1}{3}}\mathbf{i} + \sqrt{\frac{1}{3}}\mathbf{k} \qquad \phi_2 = \frac{1}{2}(1 + 3^{-1/2})\mathbf{i} + \frac{1}{2}(1 - 3^{-1/2})\mathbf{k}$$

$$\phi_3 = \frac{1}{2}(-1 + 3^{-1/2})\mathbf{i} - \frac{1}{2}(1 - 3^{-1/2})\mathbf{k}$$

To facilitate solving for the angle θ between these vectors, we first calculate the dot products of the vectors:

$$\phi_1 \cdot \phi_1 = \left(\frac{1}{3} + \frac{1}{3}\right)^{1/2} = \left(\frac{2}{3}\right)^{1/2}$$

$$\phi_2 \cdot \phi_2 = \left[\frac{1}{4}\left(1 + \frac{2}{\sqrt{3}} + \frac{1}{3}\right) + \frac{1}{4}\left(1 - \frac{2}{\sqrt{3}} + \frac{1}{3}\right)\right]^{1/2} = \left(\frac{2}{3}\right)^{1/2}$$

$$\phi_3 \cdot \phi_3 = \left[\frac{1}{4}\left(1 - \frac{2}{\sqrt{3}} + \frac{1}{3}\right) + \frac{1}{4}\left(1 + \frac{2}{\sqrt{3}} + \frac{1}{3}\right)\right]^{1/2} = \left(\frac{2}{3}\right)^{1/2}$$

$$\phi_1 \cdot \phi_2 = -\frac{1}{2}\left(\frac{1}{\sqrt{3}} + \frac{1}{3}\right) + \frac{1}{2}\left(\frac{1}{\sqrt{3}} - \frac{1}{3}\right) = -\frac{1}{3}$$

$$\phi_1 \cdot \phi_3 = -\frac{1}{2}\left(-\frac{1}{\sqrt{3}} + \frac{1}{3}\right) - \frac{1}{2}\left(\frac{1}{\sqrt{3}} + \frac{1}{3}\right) = -\frac{1}{3}$$

$$\phi_2 \cdot \phi_3 = \frac{1}{4}\left(-1 + \frac{1}{3}\right) - \frac{1}{4}\left(1 - \frac{1}{3}\right) = -\frac{1}{3}$$

Because the dot products of the $\phi_j \cdot \phi_j$ are the same for all j, and the dot products of the $\phi_i \cdot \phi_j$ are the same for all $i \neq j$, θ will be the same between all pairs of vectors. To calculate a numerical value of θ, we pick the pair of vectors ϕ_1 and ϕ_2 to find that

$$(\phi_1 \cdot \phi_1)(\phi_2 \cdot \phi_2)\cos\theta = \phi_1 \cdot \phi_2$$

$$\cos\theta = \frac{3}{2}\left(-\frac{1}{3}\right) = -\frac{1}{2}$$

$$\theta = \cos^{-1}\left(-\frac{1}{2}\right) = 120°$$

10–6. Given that one sp hybrid orbital is

$$\xi_1 = \frac{1}{\sqrt{2}}(2s + 2p_z)$$

construct a second one by requiring that it be normalized and orthogonal to ξ_1.

We require that ξ_2 satisfy the conditions $\int d\tau \xi_2^* \xi_2 = 1$ and $\int d\tau \xi_1^* \xi_2 = 0$. Because ξ_2 is an sp hybrid orbital, we can write $\xi_2 = c_1 2s + c_2 2p_z$. Now we use the above conditions to find c_1 and c_2:

$$0 = \int d\tau \frac{1}{\sqrt{2}}(2s^* + 2p_z^*)\xi_2$$

$$= \frac{1}{\sqrt{2}} \int d\tau 2s^* \xi_2 + \frac{1}{\sqrt{2}} \int d\tau 2p_z^* \xi_2$$

$$= \frac{c_1}{\sqrt{2}} \int d\tau 2s^* 2s + \frac{c_2}{\sqrt{2}} \int d\tau 2s^* 2p_z + \frac{c_1}{\sqrt{2}} \int d\tau 2p_z^* 2s + \frac{c_2}{\sqrt{2}} \int d\tau 2p_z^* 2p_z$$

$$0 = \frac{c_1}{\sqrt{2}} + \frac{c_2}{\sqrt{2}}$$

So $c_1 = -c_2$. Now

$$1 = \int d\tau \xi_2^* \xi_2$$

$$= \int d\tau \left(c_1 2s^* - c_1 2p_z^* \right) \left(c_1 2s - c_2 2p_z \right)$$

$$= c_1^2 \int d\tau 2s^* 2s - c_1^2 \int d\tau \left(2s^* 2p_z + 2p_z^* 2s \right) + c_1^2 \int d\tau 2p_z^* 2p_z$$

$$1 = 2c_1^2$$

and so $c_1 = \pm 1/\sqrt{2}$. Either value of c_1 is correct, because in either case ξ_2 is normalized and orthogonal to ξ_1. We arbitrarily choose $c_1 = 1/\sqrt{2}$ to write

$$\xi_2 = \frac{1}{\sqrt{2}} 2s - \frac{1}{\sqrt{2}} 2p_z$$

10–7. The relation between a tetrahedron and a cube is shown in the following figure:

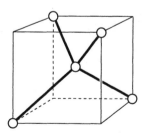

Use this figure to show that the bond angles in a regular tetrahedron are 109.47°. (*Hint:* If we let the edge of the cube be of length a, then the diagonal on a face of the cube has a length $\sqrt{2}\,a$, by the Pythagorean theorem. The distance from the center of the cube to a face is equal to $a/2$. Using this information, determine the tetrahedral angle.)

Use the following figure, where the apex of the triangle represents the center of the cube and the base represents the diagonal of a face.

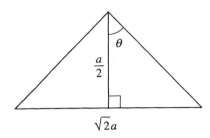

The length of the base of this triangle is $\sqrt{2}a$ because it is the diagonal of a face. The quantity $\tan\theta$ is given by

$$\tan\theta = \frac{\sqrt{2}a/2}{a/2} = \sqrt{2}$$

or $\theta = 54.736°$. The tetrahedral bond angle is equal to $2\theta = 109.47°$.

10–8. Show that the sp^3 hybrid orbitals given by Equations 10.6 through 10.9 are orthonormal.

The sp^3 hybrid orbitals are given by (Equations 10.6 through 10.9)

$$\psi_1 = \frac{1}{2}\left(2s + 2p_x + 2p_y + 2p_z\right)$$

$$\psi_2 = \frac{1}{2}\left(2s - 2p_x - 2p_y + 2p_z\right)$$

$$\psi_3 = \frac{1}{2}\left(2s + 2p_x - 2p_y - 2p_z\right)$$

$$\psi_4 = \frac{1}{2}\left(2s - 2p_x + 2p_y - 2p_z\right)$$

For orthonormal wave functions, $\int \psi_i^* \psi_j = \delta_{ij}$. Because the $2s$, $2p_x$, $2p_y$, and $2p_z$ orbitals are orthogonal, for all j we find

$$\int d\tau\,\psi_j^*\psi_j = \frac{1}{4}\int d\tau\,2s^*2s + \frac{1}{4}\int d\tau\,2p_x^*2p_x + \frac{1}{4}\int d\tau\,2p_y^*2p_y + \frac{1}{4}\int d\tau\,2p_z^*2p_z$$

$$= \frac{1}{4} + \frac{1}{4} + \frac{1}{4} + \frac{1}{4} = 1$$

Now consider the case where $i \neq j$. Specifically, let $i = 1$ and $j = 2$. Then

$$\int d\tau\,\psi_1^*\psi_2 = \frac{1}{4}\int d\tau\left(2s^* + 2p_x^* + 2p_y^* + 2p_z^*\right)\left(2s - 2p_x - 2p_y + 2p_z\right)$$

$$= \frac{1}{4} - \frac{1}{4} - \frac{1}{4} + \frac{1}{4} = 0$$

For each possible combination of $\int \psi_i^*\psi_j$, $i \neq j$, the integral reduces to the sum of four integrals, each of which has a value of 1/4. Two of these four terms are positive and two negative, so their sum is always zero. Thus the sp^3 hybrid orbitals are orthonormal.

10–9. Using the vector approach described in Problem 10–4, show that the cosine of the angle between the sp^3 hybrid orbitals given by Equations 10.6 through 10.9 is $-1/3$. What is the angle equal to?

Again, because the s orbitals are spherically symmetric, the directionality of the hybrid orbitals is due to the p orbitals. Using vector notation, we then write

$$\psi_1 = \frac{1}{2}\mathbf{i} + \frac{1}{2}\mathbf{j} + \frac{1}{2}\mathbf{k} \qquad \psi_2 = -\frac{1}{2}\mathbf{i} - \frac{1}{2}\mathbf{j} + \frac{1}{2}\mathbf{k}$$

$$\psi_3 = \frac{1}{2}\mathbf{i} - \frac{1}{2}\mathbf{j} - \frac{1}{2}\mathbf{k} \qquad \psi_4 = -\frac{1}{2}\mathbf{i} + \frac{1}{2}\mathbf{j} - \frac{1}{2}\mathbf{k}$$

From the solution to Problem 10–4,

$$\left(\sqrt{A_x^2 + A_y^2 + A_z^2}\right)\left(\sqrt{B_x^2 + B_y^2 + B_z^2}\right)\cos\theta = A_x B_x + A_y B_y + A_z B_z$$

The angles between any pair of hybrid orbitals is the same, so choosing ψ_1 and ψ_2, we have

$$\left(\frac{3}{4}\right)^{1/2}\left(\frac{3}{4}\right)^{1/2}\cos\theta = -\frac{1}{4} - \frac{1}{4} + \frac{1}{4}$$

$$\cos\theta = -\frac{1}{3}$$

$$\theta = 109.5°$$

10–10. The sp^3 hybrid orbitals given by Equations 10.6 through 10.9 are symmetric but not unique. We construct an equivalent set in this problem. We can write the four sp^3 hybrid orbitals on the carbon atom as

$$\begin{aligned}
\xi_1 &= a_1 2s + b_1 2p_x + c_1 2p_y + d_1 2p_z \\
\xi_2 &= a_2 2s + b_2 2p_x + c_2 2p_y + d_2 2p_z \\
\xi_3 &= a_3 2s + b_3 2p_x + c_3 2p_y + d_3 2p_z \\
\xi_4 &= a_4 2s + b_4 2p_x + c_4 2p_y + d_4 2p_z
\end{aligned} \qquad (1)$$

By requiring these four hybrid orbitals to be equivalent, we have that $a_1 = a_2 = a_3 = a_4$. Because there is one $2s$ orbital distributed among four equivalent hybrid orbitals, we also say that $a_1^2 + a_2^2 + a_3^2 + a_4^2 = 1$. Thus, we have that $a_1 = a_2 = a_3 = a_4 = 1/\sqrt{4}$. Without loss of generality, we take one of the hybrid orbitals to be directed along the positive z axis. Because the $2p_x$ and $2p_y$ orbitals are directed along only the x and y axes, respectively, then b and c are zero in this orbital. If we let this orbital be ξ_1, then

$$\xi_1 = \frac{1}{\sqrt{4}}2s + d_1 2p_z$$

By requiring that ξ_1 be normalized, show that

$$\xi_1 = \frac{1}{\sqrt{4}}2s + \sqrt{\frac{3}{4}}2p_z \qquad (2)$$

Equation 2 is the first of our four sp^3 hybrid orbitals. Without any loss of generality, take the second hybrid orbital to lie in the x-z plane, so that

$$\xi_2 = \frac{1}{\sqrt{4}}2s + b_2 2p_x + d_2 2p_z \qquad (3)$$

Show that if we require ξ_2 to be normalized and orthogonal to ξ_1, then

$$\xi_2 = \frac{1}{\sqrt{4}}2s + \sqrt{\frac{2}{3}}2p_x - \frac{1}{\sqrt{12}}2p_z$$

Show that the angle between ξ_1 and ξ_2 is 109.47°. Now determine ξ_3 such that it is normalized and orthogonal to ξ_1 and ξ_2. Last, determine ξ_4.

Let $\xi_1 = \frac{1}{\sqrt{4}}2s + d_1 2p_z$. For ξ_1 to be normalized,

$$\int d\tau \xi_1^* \xi_1 = \frac{1}{4}\int d\tau 2s^* 2s + d_1^2 \int d\tau 2p_z^* 2p_z + 0$$

$$1 = \frac{1}{4} + d_1^2$$

$$d_1 = \pm\sqrt{\frac{3}{4}}$$

Take $d_1 = \sqrt{\frac{3}{4}}$ to find

$$\xi_1 = \frac{1}{\sqrt{4}}2s + \sqrt{\frac{3}{4}}2p_z$$

Note that the hybrid orbital with $d_1 = -\sqrt{\frac{3}{4}}$ would also give a normalized orbital. Now take

$$\xi_2 = \frac{1}{\sqrt{4}}2s + b_2 2p_x + d_2 2p_z$$

We require that this hybrid orbital be orthogonal to ξ_1 and normalized, so

$$\int d\tau \xi_1^* \xi_2 = \frac{1}{4}\int d\tau 2s^* 2s + \sqrt{\frac{3}{4}}d_2 \int d\tau 2p_z^* 2p_z$$

$$0 = \frac{1}{4} + \sqrt{\frac{3}{4}}d_2$$

$$d_2 = -\frac{1}{\sqrt{12}}$$

and

$$\int d\tau \xi_2^* \xi_2 = \frac{1}{4}\int d\tau 2s^* 2s + b_2^2 \int d\tau 2p_x^* 2p_x + \frac{1}{12}\int d\tau 2p_z^* 2p_z$$

$$1 = \frac{1}{3} + b_2^2$$

$$b_2 = \pm\sqrt{\frac{2}{3}}$$

Once again we arbitrarily take the positive root, $b = \sqrt{\frac{2}{3}}$, and so

$$\xi_2 = \frac{1}{\sqrt{4}}2s + \sqrt{\frac{2}{3}}2p_x - \frac{1}{\sqrt{12}}2p_z$$

Now use the equation from the solution to Problem 10–4 to find the angle between ψ_1 and ψ_2:

$$\left(\sqrt{A_x^2 + A_y^2 + A_z^2}\right)\left(\sqrt{B_x^2 + B_y^2 + B_z^2}\right)\cos\theta = A_x B_x + A_y B_y + A_z B_z$$

$$\left(\sqrt{\frac{2}{3} + \frac{1}{12}}\right)\left(\sqrt{\frac{3}{4}}\right)\cos\theta = -\frac{1}{\sqrt{12}}\left(\sqrt{\frac{3}{4}}\right)$$

$$\cos\theta = -\frac{1}{\sqrt{12}}\left(\sqrt{\frac{12}{9}}\right) = -\frac{1}{3}$$

$$\theta = 109.47°$$

Now find ξ_3 so that it is orthogonal to ξ_1 and ξ_2 and normalized. Use ξ_3 from Equations 1:

$$\int d\tau\,\xi_1^*\xi_3 = \frac{1}{4}\int d\tau\,2s^*2s + d_3\sqrt{\frac{3}{4}}\int d\tau\,2p_z^*2p_z$$

$$0 = \frac{1}{4} + d_3\sqrt{\frac{3}{4}}$$

$$d_3 = -\frac{1}{\sqrt{12}}$$

$$\int d\tau\,\xi_2^*\xi_3 = \frac{1}{4}\int d\tau\,2s^*2s + \sqrt{\frac{2}{3}}b_3\int d\tau\,2p_x^*2p_x + \frac{1}{12}\int d\tau\,2p_z^*2p_z$$

$$0 = \frac{1}{4} + \sqrt{\frac{2}{3}}b_3 + \frac{1}{12}$$

$$b_3 = -\frac{1}{\sqrt{6}}$$

$$\int d\tau\,\xi_3^*\xi_3 = \frac{1}{4}\int d\tau\,2s^*2s + \frac{1}{6}\int d\tau\,2p_x^*2p_x + c_3^2\int d\tau\,2p_y^*2p_y + \frac{1}{12}\int d\tau\,2p_z^*2p_z$$

$$1 = \frac{1}{4} + \frac{1}{6} + c_3^2 + \frac{1}{12}$$

$$c_3 = \frac{1}{\sqrt{2}}$$

so

$$\xi_3 = \frac{1}{\sqrt{4}}2s - \frac{1}{\sqrt{6}}2p_x + \frac{1}{\sqrt{2}}2p_y - \frac{1}{\sqrt{12}}2p_z$$

The final hybrid orbital, ξ_4, must be normalized and orthogonal to ξ_1, ξ_2, and ξ_3. Proceeding as above, we find

$$\xi_4 = \frac{1}{\sqrt{4}}2s - \frac{1}{\sqrt{6}}2p_x - \frac{1}{\sqrt{2}}2p_y - \frac{1}{\sqrt{12}}2p_z$$

10–11. Calculate the bond angle between ψ_1 and ψ_2 in Example 10–4 using the vector approach described in Problem 10–4. (Remember not to use the 2s part of ψ_1 and ψ_2.)

Write the p-orbital contribution to the hybrid orbitals given by Equations 10.12 and 10.13 as

$$\psi_1 = 0.71\mathbf{j} + 0.55\mathbf{k}$$

$$\psi_2 = -0.71\mathbf{j} + 0.55\mathbf{k}$$

Now use the equation found in the solution to Problem 10–4 to find the angle between the orbitals:

$$\left(\sqrt{A_x^2 + A_y^2 + A_z^2}\right)\left(\sqrt{B_x^2 + B_y^2 + B_z^2}\right)\cos\theta = A_x B_x + A_y B_y + A_z B_z$$

$$\left(0.71^2 + 0.55^2\right)^{1/2}\left(0.71^2 + 0.55^2\right)^{1/2}\cos\theta = -0.71^2 + 0.55^2$$

$$\cos\theta = -0.25$$

$$\theta = 104.5°$$

10–12. Using the coordinate system shown below for a water molecule,

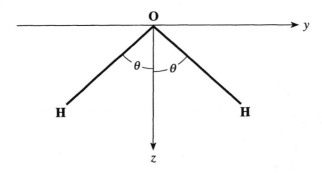

show that we can write the two bonding hybrid atomic orbitals on the oxygen atom as

$$\psi_1 = N[\gamma 2s + (\sin\theta)2p_y + (\cos\theta)2p_z]$$

and

$$\psi_2 = N[\gamma 2s - (\sin\theta)2p_y + (\cos\theta)2p_z]$$

where γ is a constant and N is the normalization constant. Now use the fact that these orbitals must be orthogonal to show that

$$\cos^2\theta - \sin^2\theta = \cos 2\theta = -\gamma^2$$

Finally, given that the H–O–H bond angle of water is 104.5°, determine the orthonormal hybrid orbitals ψ_1 and ψ_2 (see Equations 10.12 and 10.13).

The water molecule sits in the yz-plane and so the bonding hybrid atomic orbitals on oxygen will be a linear combination of the oxygen $2s$, $2p_y$, and $2p_z$ orbitals (the $2p_x$ orbital is perpendicular to the plane of the molecule). Thus

$$\psi_{bonding} = N\left(\gamma 2s + c_1 2p_y + c_2 2p_z\right)$$

where γ, N, c_1 and c_2 are constants. The directionality of the O–H bonds arises from the contributions of the $2p$ orbitals, and so we can use the orientation shown in the figure to determine c_1 and c_2. Both O–H bonds have a projection on the z-axis of $\cos\theta$. The projection of the O–H bonds on the y-axis is given by $\pm\sin\theta$. Therefore, $c_1 = \pm\sin\theta$ and $c_2 = \cos\theta$, giving

$$\psi_1 = N[\gamma 2s + (\sin\theta)2p_y + (\cos\theta)2p_z]$$

and

$$\psi_2 = N[\gamma 2s - (\sin\theta)2p_y + (\cos\theta)2p_z]$$

The two bonding hybrid molecular orbitals are orthogonal, so

$$0 = \int d\tau \psi_1^* \psi_2$$

$$0 = N^2\gamma^2 \int d\tau 2s^* 2s + N^2 \cos^2\theta \int d\tau 2p_z^* 2p_z - N^2 \sin^2\theta \int d\tau 2p_y^* 2p_y$$

$$0 = N^2 \left(\gamma^2 + \cos^2\theta - \sin^2\theta \right)$$

or

$$-\gamma^2 = \cos^2\theta - \sin^2\theta = \cos 2\theta$$

For a water molecule $2\theta = 104.5°$, so

$$-\gamma^2 = \cos(104.5°) = -0.2504$$

$$\gamma = 0.5004$$

Now $\theta = 104.5°/2 = 52.25°$, so $\cos\theta = 0.6122$ and $\sin\theta = .7907$. Using the equations for the orbitals given in the problem and requiring these orbitals to be normalized gives

$$1 = \int d\tau \psi_1^* \psi_1$$

$$1 = N^2\gamma^2 + N^2 \cos^2\theta + N^2 \sin^2\theta$$

$$N^{-2} = (0.5004)^2 + (0.6122)^2 + (0.7907)^2$$

$$N = \pm 0.8943$$

The choice of sign of N is arbitrary. We choose the positive root and write

$$\psi_1 = 0.8943 \left[0.5004(2s) + 0.7907(2p_y) + 0.6122(2p_z) \right]$$

$$\psi_2 = 0.8943 \left[0.5004(2s) - 0.7907(2p_y) + 0.6122(2p_z) \right]$$

10–13. In Problem 10–12, you found two bonding hybrid orbitals for the oxygen atom of a water molecule. In this problem, we will find the two equivalent lone-pair orbitals. Starting with the results of Problem 10–12, show that the third sp^2 hybrid orbital is given by

$$\psi_3 = 0.77 \cdot 2s - 0.64 \cdot 2p_z$$

At this point the lone pair orbitals are given by ψ_3 and the oxygen $2p_x$ orbital. Construct two equivalent lone pair orbitals by taking the appropriate linear combinations of ψ_3 and the $2p_x$ orbital. Which pair of orbitals, ψ_3 and the $2p_x$ orbital or your set of equivalent orbitals, is the correct description of the lone-pair orbitals for a water molecule? Explain your reasoning.

In Problem 10–12 we found that

$$\psi_1 = 0.8943 \left[0.5004(2s) + 0.7907(2p_y) + 0.6122(2p_z) \right]$$

$$\psi_2 = 0.8943 \left[0.5004(2s) - 0.7907(2p_y) + 0.6122(2p_z) \right]$$

We wish to construct $\psi_3 = a_3 2s + b_3 2p_y + c_3 2p_z$ such that it is orthogonal to both ψ_1 and ψ_2. We then require that

$$0 = \int d\tau\, \psi_1^* \psi_3$$

$$0 = 0.4475a_3 + 0.7071b_3 + 0.5475c_3 \tag{1}$$

and

$$0 = \int d\tau\, \psi_2^* \psi_3$$

$$0 = 0.4475a_3 - 0.7071b_3 + 0.5475c_3 \tag{2}$$

Subtracting Equation 1 from Equation 2 gives $b_3 = 0$. The function ψ_3 must be normalized, so

$$1 = \int d\tau\, \psi_3^* \psi_3$$

$$1 = a_3^2 + c_3^2$$

$$a_3 = \sqrt{1 - c_3^2}$$

Substituting this into Equation 1 gives

$$0 = 0.4475(1 - c_3^2)^{1/2} + 0.5475c_3$$

$$(1 - c_3^2)^{1/2} = -\frac{0.5475}{0.4475}c_3$$

$$1 - c_3^2 = \left(-\frac{0.5475}{0.4475}\right)^2 c_3^2$$

$$c_3^2 = 0.4005$$

$$c_3 = \pm 0.6328$$

Therefore

$$a_3 = \sqrt{1 - c_3^2} = 0.7743$$

We must select c_3 such that $\int d\tau\, \psi_1^* \psi_3 = 0$, so $c_3 = -0.6328$, giving

$$\psi_3 = 0.77(2s) - 0.63(2p_z)$$

Now we construct two new equivalent lone pair orbitals, ψ_{l1} and ψ_{l2}, by taking linear combinations of ψ_3 and the $2p_x$ orbital:

$$\psi_{l1} = c_1\psi_3 + c_2 2p_x = 0.77c_1 2s - 0.63c_1 2p_z + c_2 2p_x$$

$$\psi_{l2} = c_3\psi_3 + c_4 2p_x = 0.77c_3 2s - 0.63c_3 2p_z + c_4 2p_x$$

Since ψ_{l1} and ψ_{l2} are equivalent, $c_1 = c_3$. Recall that the coefficient of the $2s$ orbital component of ψ_1 and ψ_2 is $(0.8943)(0.5004) = 0.4475$. Because the one $2s$ orbital is distributed among ψ_1, ψ_2, ψ_{l1}, and ψ_{l2}, we have

$$1 = 2(0.4475)^2 + (c_1^2 + c_3^2)(0.77)^2$$

$$1.011 = c_1^2 + c_3^2 = 2c_1^2$$

$$0.7110 = c_1 = c_3$$

where we have arbitrarily picked the positive root for both c_1 and c_3. Now we solve for c_2 and c_4 using the normalization conditions:

$$1 = \int d\tau \psi_{l1}^* \psi_{l1} = 0.54^2 + 0.44^2 + c_2^2$$
$$c_2 = 0.717 = 0.72$$

and

$$1 = \int d\tau \psi_{l2}^* \psi_{l2} = 0.54^2 + 0.44^2 + c_4^2$$
$$c_4 = -0.717 = -0.72$$

Our final solution for the equivalent lone pair wave functions is then

$$\psi_{l1} = 0.54(2s) - 0.44(2p_z) + 0.72(2p_x)$$
$$\psi_{l2} = 0.54(2s) - 0.44(2p_z) - 0.72(2p_x)$$

Either set of orbitals is correct. The hydrogenlike $2s$ and $2p$ orbitals have the same energy. All possible normalized linear combinations of the atomic orbitals also satisfy the Schrödinger equation for a hydrogenlike atom and all these linear combinations have the same energies.

10–14. Figure 10.9 shows a schematic representation of the various molecular orbitals for a linear AH_2 molecule. We could draw similar pictures for the molecular orbitals of a linear XY_2 molecule. For example, the $3\sigma_g$ and $4\sigma_g$ molecular orbitals can be represented as

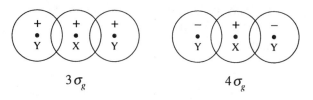

Draw a schematic representation of the $2\sigma_u$, $1\pi_u$, $2\pi_u$, and $1\pi_g$ orbitals.

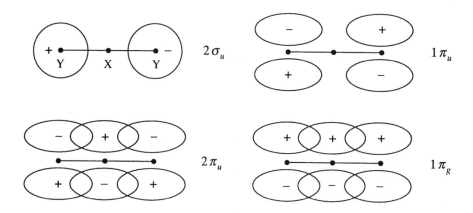

10–15. Explain why the energies of the $3\sigma_g$ and $2\sigma_u$ orbitals for an XY_2 molecule are insensitive to small changes in the bond angle.

For a constant X–Y bond length, small changes in the bond angle do not affect the overlap of the $2s$ orbitals in the three atoms. (The $2s$ orbitals are spherically symmetric.) Therefore, the energies associated with linear combinations of these three $2s$ orbitals are insensitive to small changes in bond angle.

10–16. Explain why the doubly degenerate $1\pi_u$ orbitals for a linear XY_2 molecule do not remain degenerate when the molecule is bent.

Take the z-axis to lie along the bonds of the linear molecule and let the molecule bend in the xz-plane. Then one of the $1\pi_u$ orbitals lies in the plane of the molecule and the other lies in a plane that is perpendicular to the plane of the molecule. For a linear molecule, the overlap between the $2p_x$ atomic orbitals (one of the $1\pi_u$ orbitals) and the $2p_y$ atomic orbitals (the other $1\pi_u$ orbital) are identical. Hence these molecular orbitals are degenerate. As the molecule bends, the overlap between the $2p_x$ orbitals on the three atoms is affected differently than the overlap between the $2p_y$ atomic orbitals. Therefore, the energy associated with the two $1\pi_u$ molecular orbitals depends on the bond angle.

10–17. Explain why the $3\sigma_u$ molecular orbital of a linear XY_2 molecule increases in energy as the molecule bends. (*Hint*: The $3\sigma_u$ molecular orbital is a linear combination of the $2p_z$ orbitals from each atom.)

The $3\sigma_u$ orbital is a bonding orbital formed from a linear combination of the $2p_z$ orbitals from each atom. (Recall that we have taken the z-axis to lie along the molecular bond.) When the molecule bends, the overlap between the $2p_z$ orbitals decreases and the energy associated with this linear combination of atomic orbitals increases.

10–18. Use Figure 10.25 to predict whether the following molecules are linear or bent:

  a. CO_2 **b.** CO_2^+ **c.** CF_2

---

a. CO_2 has $4 + 6 + 6 = 16$ valence electrons. The possible electron configurations of the linear and bent structures are

$$\text{Linear:} \quad 3\sigma_g^2 2\sigma_u^2 4\sigma_g^2 3\sigma_u^2 1\pi_u^2 1\pi_u^2 1\pi_g^2 1\pi_g^2$$

$$\text{Bent:} \quad 3a_1^2 2b_2^2 4a_1^2 3b_2^2 1b_1^2 5a_1^2 1a_2^2 4b_2^2$$

$$3a_1^2 2b_2^2 4a_1^2 3b_2^2 1b_1^2 5a_1^2 6a_1^2 1a_2^2$$

From the diagram we see that the linear structure is lowest in total energy.

b. CO_2^+ has 15 valence electrons and the lowest energy structure is linear.

c. CF_2 has 18 valence electrons and the lowest energy structure (using the electron configurations from part (a) of Problem 10–19) is bent.

10–19. Use Figure 10.25 to predict whether the following molecules are linear or bent:

  a. OF_2 **b.** NO_2^+ **c.** CN_2

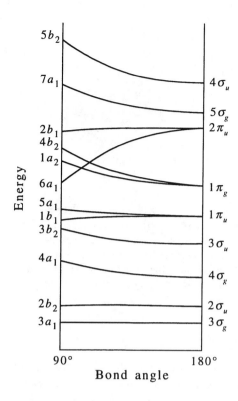

FIGURE 10.25
The Walsh correlation diagram for the valence electrons of a XY_2 molecule. The right side of the diagram gives the energy ordering of the molecular orbitals for an Y–X–Y bond angle of 180°. The left side gives the energy ordering of the molecular orbitals for an Y–X–Y bond angle of 90°. The solid lines tell us how the energies of the molecular orbitals depend upon Y–X–Y bond angles between 90° and 180°. The $1\sigma_g$, $2\sigma_g$, and $1\sigma_u$ orbitals correspond to the core $1s$ electrons on the bonded atoms and are not shown in the figure.

---

a. OF_2 has $6 + 7 + 7 = 20$ valence electrons. The possible electron configurations of the linear and bent structures are

$$\text{Linear:} \quad 3\sigma_g^2 2\sigma_u^2 4\sigma_g^2 3\sigma_u^2 1\pi_u^2 1\pi_u^2 1\pi_g^2 1\pi_g^2 2\pi_u^2 2\pi_u^2$$

$$\text{Bent:} \quad 3a_1^2 2b_2^2 4a_1^2 3b_2^2 1b_1^2 5a_1^2 1a_2^2 4b_2^2 6a_1^2 2b_1^2$$

$$3a_1^2 2b_2^2 4a_1^2 3b_2^2 1b_1^2 5a_1^2 1a_2^2 6a_1^2 4b_2^2 2b_1^2$$

$$3a_1^2 2b_2^2 4a_1^2 3b_2^2 1b_1^2 5a_1^2 6a_1^2 1a_2^2 4b_2^2 2b_1^2$$

From the energy-level diagram, we see that the bent structure is lowest in total energy.

b. NO_2^+ has 19 valence electrons and is bent.

c. CN_2 has 14 valence electrons and (using the electron configurations from part (a) of the previous problem) is linear.

10–20. Walsh correlation diagrams can be used to predict the shapes of polyatomic molecules that contain more than three atoms. In this and the following three problems we consider molecules that have the general formula XH_3. We will restrict our discussion to XH_3 molecules, where all the H–X–H bond angles are the same. If the molecule is planar, then the H–X–H bond angle is 120°. A nonplanar XH_3 molecule, then, has an H–X–H bond angle that is less than 120°. Figure 10.26 shows the Walsh correlation diagram that describes how the energies of the molecular orbitals for an XH_3 molecule change as a function of the H–X–H bond angle. Note that because XH_3 is not linear, the labels used to describe the orbitals on the two sides of the correlation diagram do not have designations such as σ and π. We see that the lowest-energy molecular orbital is insensitive to the H–X–H bond angle. Which atomic orbital(s) contribute to the lowest-energy molecular orbital? Explain why the energy of this molecular orbital is insensitive to changes in the H–X–H bond angle.

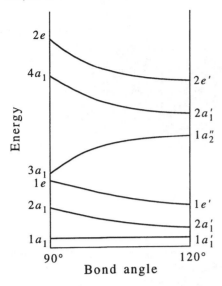

FIGURE 10.26
The Walsh correlation diagram for XH_3 molecules. The right side of the diagram gives the energy ordering of the molecular orbitals for an H–X–H bond angle of 120°. The left side gives the energy ordering of the molecular orbitals for an H–X–H bond angle of 90°. The solid lines tell us how the energies of the molecular orbitals depend upon H–X–H bond angles between 90° and 120°.

The lowest energy molecular orbital is effectively equal to the $1s_x$ orbital. This is a core atomic orbital and is not involved in the bonding. (Recall that bonding orbitals involve the valence shell electrons, which for the X atom are in the orbitals with the highest n quantum number, $n \geq 2$.)

10–21. Consider the Walsh correlation diagram given in Figure 10.26. The $2a_1'$ molecular orbital of the planar XH_3 molecule is a linear combination of the $2p$ orbital on X that lies in the molecular plane and the $1s$ orbital on each hydrogen atom. Why does the energy of this molecular orbital increase as the H–X–H bond angle decreases from 120° to 90°?

As the molecule bends out of the plane, the overlap $2p$ orbital on X (directed along the internuclear axis) and the $1s$ orbital on the hydrogen atom decreases. Thus, the energy associated with this particular linear combination of atomic orbitals increases.

10–22. Orbitals designated by the letter "e" in a Walsh correlation diagram are doubly degenerate. Which atomic orbitals can contribute to the $1e'$ molecular orbitals of the planar XH_3 molecule?

We first note that the $1e'$ orbitals remain degenerate as the molecule bends out of the plane. Only molecular orbitals comprised of atomic orbitals that are symmetric with respect to the plane containing the planar XH_3 molecule can remain degenerate as the molecule bends. Therefore, if we take the plane of the molecule to be the xy-plane, the atomic orbitals that contribute to the $1e'$ orbitals are the $2s$, $2p_x$ and $2p_y$ on X and the $1s$ orbitals on each hydrogen atom.

10–23. Use the Walsh correlation diagram in Figure 10.26 to determine which of the following molecules are planar: (a) BH_3, (b) CH_3, (c) CH_3^-, and (d) NH_3. (Orbitals designated by the letter "e" are doubly degenerate.)

The $1a_1'$ orbital is filled by the $1s$ electrons of the X atom in all cases, so we begin with the orbital with the next highest energy in the ensuing discussion.

a. BH_3 has six total valence electrons. The first three orbitals are therefore filled, so the molecule is planar (with $120°$ as the H–B–H bond angle).

b. CH_3 has seven total valence electrons. The first three orbitals are filled and one electron occupies the $1a_2''$ orbital. The energy of the $1a''$ orbital decreases more steeply than the lower energy orbitals increase as the molecule bends out of the plane, so the molecule is pyramidal.

c. CH_3^- has eight total valence electrons. The first four orbitals are filled, and the molecule is pyramidal.

d. NH_3 also has eight total valence electrons (like CH_3^-) and the molecule is pyramidal.

10–24. Show that the π molecular orbital corresponding to the energy $E = \alpha - \beta$ for ethene is $\psi_\pi = \frac{1}{\sqrt{2}}(2p_{zA} - 2p_{zB})$.

Equation 10.15 gives the bonding Hückel molecular orbitals for ethene as

$$\psi_\pi = c_1 2p_{zA} + c_2 2p_{zB}$$

where (Example 10–6)

$$c_1(\alpha - E) + c_2\beta = 0$$
$$c_1\beta + c_2(\alpha - E) = 0$$

Substituting $E = \alpha - \beta$ into these expressions gives

$$\beta c_1 + \beta c_2 = 0$$
$$c_1 = -c_2$$

Then

$$\psi_\pi = c_1(2p_{zA} - 2p_{zB})$$

To find c_1 we require that ψ_π be normalized, or

$$c_1^2(1 + 2S + 1) = 1$$

where $S = 0$ (an assumption of the Hückel theory). Then $c_1 = 1/\sqrt{2}$ and so

$$\psi_\pi = \frac{1}{\sqrt{2}}(2p_{z,A} - 2p_{z,B})$$

10–25. Generalize our Hückel molecular-orbital treatment of ethene to include overlap of $2p_{zA}$ and $2p_{zB}$. Determine the energies and the wave functions.

Including the overlap of the $2p_z$ orbitals means that we no longer set S_{12} and S_{21} to zero; however, $S_{12} = S_{21} = S$. The Hückel secular determinant then becomes

$$\begin{vmatrix} \alpha - E & \beta - ES \\ \beta - ES & \alpha - E \end{vmatrix} = 0$$

Expanding the determinant gives

$$(\alpha - E)^2 - (\beta - ES_{12})^2 = 0$$

which gives

$$\alpha - E = \pm(\beta - ES)$$

or

$$E = \frac{\alpha \pm \beta}{1 \pm S}$$

Now we can substitute the allowed energies into the linear algebraic equations for c_1 and c_2 (Equation 10.16) to find the wave functions for the energies:

$$c_1(\alpha - E) + c_2(\beta - ES_{12}) = 0$$
$$c_1(\beta - ES_{12}) + c_2(\alpha - E) = 0$$

Substituting $E = (\alpha + \beta)/(1 + S)$ into the above equations gives only one independent equation:

$$c_1\left(\frac{\alpha S - \beta}{1 + S}\right) + c_2\left(\frac{\beta - \alpha S}{1 + S}\right) = 0$$

or $c_1 = c_2$. Therefore

$$\psi_+ = \frac{1}{\sqrt{2(1 + S)}}\left(2p_{zA} + 2p_{zB}\right)$$

Similarly,

$$\psi_- = \frac{1}{\sqrt{2(1 - S)}}\left(2p_{zA} - 2p_{zB}\right)$$

10–26. Show that the four molecular orbitals for butadiene (Equation 10.18),

$$\psi_i = \sum_{j=1}^{4} c_{ij} 2p_{zj}$$

lead to the secular determinant given by Equation 10.19.

The four molecular orbitals for butadiene are

$$\psi_i = c_{i1} 2p_{z1} + c_{i2} 2p_{z2} + c_{i3} 2p_{z3} + c_{i4} 2p_{z4}$$

In Section 7–2 we learned how to find a secular determinant for a linear combination of N functions (Equation 7.40). Since there are four functions contributing to each ψ_i, the secular determinant is

$$\begin{vmatrix} H_{11} - ES_{11} & H_{12} - ES_{12} & H_{13} - ES_{13} & H_{14} - ES_{14} \\ H_{12} - ES_{12} & H_{22} - ES_{22} & H_{23} - ES_{23} & H_{24} - ES_{24} \\ H_{13} - ES_{13} & H_{23} - ES_{23} & H_{33} - ES_{33} & H_{34} - ES_{34} \\ H_{14} - ES_{14} & H_{24} - ES_{24} & H_{34} - ES_{34} & H_{44} - ES_{44} \end{vmatrix} = 0$$

Because our orbitals are Hermitian, we know that $H_{ij} = H_{ji}$ and $S_{ij} = S_{ji}$. Using Figure 10.21 for the numbering of the carbon atoms, we have (using the Hückel approximations)

$$H_{11} = H_{22} = H_{33} = H_{44} = \alpha$$
$$H_{12} = H_{23} = H_{34} = \beta$$
$$H_{ij} = 0 \qquad\qquad |i - j| \geq 2$$
$$S_{11} = S_{22} = S_{33} = S_{44} = 1$$
$$S_{ij} = 0 \qquad\qquad |i - j| \neq j$$

10–27. Show that

$$\begin{vmatrix} x & 1 & 0 & 0 \\ 1 & x & 1 & 0 \\ 0 & 1 & x & 1 \\ 0 & 0 & 1 & x \end{vmatrix} = 0$$

gives the algebraic equation

$$x^4 - 3x^2 + 1 = 0$$

We expand this 4×4 determinant as described in MathChapter E. Expanding along the first row gives

$$x \begin{vmatrix} x & 1 & 0 \\ 1 & x & 1 \\ 0 & 1 & x \end{vmatrix} - \begin{vmatrix} 1 & 1 & 0 \\ 0 & x & 1 \\ 0 & 1 & x \end{vmatrix} = 0$$

Expanding the first 3×3 determinant along its first row and the second along its first column gives

$$x^2 \begin{vmatrix} x & 1 \\ 1 & x \end{vmatrix} - x \begin{vmatrix} 1 & 1 \\ 0 & x \end{vmatrix} - \begin{vmatrix} x & 1 \\ 1 & x \end{vmatrix} = 0$$

and expanding the above 2×2 determinants gives

$$x^2(x^2 - 1) - x(x) - (x^2 - 1) = 0$$
$$(x^2 - 1)(x^2 - 1) - x^2 = 0$$
$$x^4 - 3x^2 + 1 = 0$$

10–28. Show that the four π molecular orbitals for butadiene are given by Equations 10.26.

From the text, the four energies for butadiene are given by

$$E_1 = \alpha + 1.618\beta$$
$$E_2 = \alpha + 0.618\beta$$
$$E_3 = \alpha - 0.618\beta$$
$$E_4 = \alpha - 1.618\beta$$

and the general form of the four corresponding molecular orbitals is given by

$$\psi = c_1 2p_{z1} + c_2 2p_{z2} + c_3 2p_{z3} + c_4 2p_{z4}$$

Generalizing the procedure introduced in Example 10–6, we see that the relationships among the coefficients c_1, c_2, c_3, and c_4 are

$$c_1(\alpha - E) + c_2\beta = 0 \tag{1}$$

$$c_1\beta + c_2(\alpha - E) + c_3\beta = 0 \tag{2}$$

$$c_2\beta + c_3(\alpha - E) + c_4\beta = 0 \tag{3}$$

$$c_3\beta + c_4(\alpha - E) = 0 \tag{4}$$

We now use these four equations to solve for the four unknowns. Solve Equation 1 for c_1 and Equation 4 for c_4 to find

$$c_1 = -c_2\left(\frac{\beta}{\alpha - E}\right) \tag{5}$$

$$c_4 = -c_3\left(\frac{\beta}{\alpha - E}\right) \tag{6}$$

Substitution of Equation 5 into Equation 2 gives

$$c_2\left[(\alpha - E)^2 - \beta^2\right] + c_3\beta(\alpha - E) = 0$$

For the case where $E = E_1 = \alpha + 1.618\beta$, we have

$$c_2\left[(-1.618\beta)^2 - \beta^2\right] + c_3\beta(-1.618\beta) = 0$$

$$1.618c_2 - 1.618c_3 = 0$$

$$c_2 = c_3$$

If $c_2 = c_3$, then $c_1 = c_4 = 0.618c_2$ (from Equations 5 and 6), and the wavefunction is

$$\psi_1 = c_2(0.618 2p_{z1} + 2p_{z2} + 2p_{z3} + 0.618 2p_{z4})$$

The normalization condition on ψ_1 gives

$$c_2^2\left[(0.618)^2 + 1 + 1 + (0.618)^2\right] = 1$$

$$c_2 = 0.6015$$

and so

$$\psi_1 = 0.3717 \cdot 2p_{z1} + 0.6015 \cdot 2p_{z2} + 0.6015 \cdot 2p_{z3} + 0.3717 \cdot 2p_{z4}$$

The calculation of ψ_2, ψ_3, and ψ_4 (corresponding to E_2, E_3, and E_4) is done in a similar manner. The results are given by Equations 10.26.

10–29. Derive the Hückel theory secular determinant for benzene (see Equation 10.27).

Benzene has a 6×6 determinant, where the ijth component is given by $H_{ij} - ES_{ij}$. The Hückel approximation gives $H_{jj} = \alpha$, $H_{ij} = \beta$ for neighboring atoms, $H_{ij} = 0$ for distant atoms, and $S_{ij} = \delta$.

Because benzene is a cyclic molecule, carbon 6 is adjacent to carbon 1, and so $H_{16} = H_{61} = \beta$. The Hückel secular determinant is

$$\begin{vmatrix} \alpha - E & \beta & 0 & 0 & 0 & \beta \\ \beta & \alpha - E & \beta & 0 & 0 & 0 \\ 0 & \beta & \alpha - E & \beta & 0 & 0 \\ 0 & 0 & \beta & \alpha - E & \beta & 0 \\ 0 & 0 & 0 & \beta & \alpha - E & \beta \\ \beta & 0 & 0 & 0 & \beta & \alpha - E \end{vmatrix} = 0$$

10–30. Calculate the Hückel π-electron energies of cyclobutadiene. What do Hund's rules say about the ground state of cyclobutadiene? Compare the stability of cyclobutadiene with that of two isolated ethylene molecules.

The structure of cyclobutadiene is

Letting $x = (\alpha - E)/\beta$, the Hückel determinantal equation is given by

$$\begin{vmatrix} x & 1 & 0 & 1 \\ 1 & x & 1 & 0 \\ 0 & 1 & x & 1 \\ 1 & 0 & 1 & x \end{vmatrix} = 0$$

Expanding the determinant gives

$$x \begin{vmatrix} x & 1 & 0 \\ 1 & x & 1 \\ 0 & 1 & x \end{vmatrix} - \begin{vmatrix} 1 & 1 & 0 \\ 0 & x & 1 \\ 1 & 1 & x \end{vmatrix} - \begin{vmatrix} 1 & x & 1 \\ 0 & 1 & x \\ 1 & 0 & 1 \end{vmatrix} = 0$$

and expanding the above determinants gives

$$x(x^3 - 2x) - (x^2 + 1 - 1) - (1 + x^2 - 1) = 0$$
$$x^4 - 4x^2 = 0$$
$$x = 2, 0, 0, -2$$

Because $x = (\alpha - E)/\beta$, the four π-electron energies of cyclobutadiene are

$$E = \alpha - 2\beta \qquad E = \alpha \qquad E = \alpha \qquad E = \alpha + 2\beta$$

There are four π electrons, and so the two lowest energy levels will be occupied and

$$E_\pi = 2(\alpha + 2\beta) + 2\alpha = 4\alpha + 4\beta$$

The second energy level ($E = \alpha$) is doubly degenerate. We need to place two electrons in these orbitals, and according to Hund's rules, each orbital will contain one electron and these electrons will have the same spin. Therefore, the ground state of cyclobutadiene should be a triplet state. We showed in the text that the energy of the π orbital in ethene is $2\alpha + 2\beta$, so

$$E_{\text{deloc}} = E_{\pi}(\text{cyclobutadiene}) - 2E_{\pi}(\text{ethene}) = 0$$

Cyclobutadiene has the same stability as two isolated ethene molecules.

10–31. Calculate the Hückel π-electron energy of trimethylenemethane:

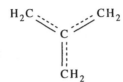

Compare the π-electron energy of trimethylenemethane with that of two isolated ethene molecules.

Let the central carbon atom be carbon 4. Then the Hückel secular determinant is

$$\begin{vmatrix} x & 0 & 0 & 1 \\ 0 & x & 0 & 1 \\ 0 & 0 & x & 1 \\ 1 & 1 & 1 & x \end{vmatrix} = 0$$

$$x\begin{vmatrix} x & 0 & 1 \\ 0 & x & 1 \\ 1 & 1 & x \end{vmatrix} - \begin{vmatrix} 0 & x & 0 \\ 0 & 0 & x \\ 1 & 1 & 1 \end{vmatrix} = 0$$

$$x(x^3 - 2x) - x^2 = 0$$
$$x^4 - 3x^2 = 0$$
$$x = \sqrt{3}, 0, 0, -\sqrt{3}$$

Because $x = (\alpha - E)/\beta$, the four π-electron energies of trimethylenemethane are

$$E = \alpha - \sqrt{3}\beta \qquad E = \alpha \qquad E = \alpha \qquad E = \alpha + \sqrt{3}\beta$$

There are four π electrons, so the two lowest energy levels are filled. The π-electron energy is then

$$E_{\pi} = 2(\alpha + \sqrt{3}\beta) + 2\alpha = 4\alpha + 2\sqrt{3}\beta$$

The energy of two isolated ethene molecules is $4\alpha + 4\beta$, so

$$E_{\text{deloc}} = E_{\pi}(\text{trimethylenemethane}) - 2E_{\pi}(\text{ethene})$$
$$= -0.5359\beta$$

10–32. Calculate the π-electronic energy levels and the total π-electron energy of bicyclobutadiene:

Let the top carbon in the figure be carbon 1 and number the other carbons clockwise. The Hückel secular determinant is then given by

$$\begin{vmatrix} x & 1 & 1 & 1 \\ 1 & x & 0 & 1 \\ 1 & 0 & x & 1 \\ 1 & 1 & 1 & x \end{vmatrix} = 0$$

$$-\begin{vmatrix} 1 & 1 & 1 \\ 0 & x & 1 \\ 1 & 1 & x \end{vmatrix} + x\begin{vmatrix} x & 1 & 1 \\ 1 & x & 1 \\ 1 & 1 & x \end{vmatrix} + \begin{vmatrix} x & 1 & 1 \\ 1 & 0 & x \\ 1 & 1 & 1 \end{vmatrix} = 0$$

$$-(x^2 + 1 - x - 1) + x(x^3 + 2 - 3x) + (x + 1 - 1 - x^2) = 0$$

$$x^4 - 5x^2 + 4x = 0$$

One root to this equation is $x = 0$, leaving the cubic equation

$$x^3 - 5x + 4 = 0$$

to be solved. We find the root $x = 1$ by inspection. Factor $x - 1$ from the cubic equation above to get

$$(x - 1)(x^2 + x - 4) = 0$$

The two roots of the quadratic equation are

$$x = \frac{-1 \pm \sqrt{17}}{2}$$

and so the four roots of this determinantal equation are $x = 1, 0, -\frac{1}{2} \pm \frac{1}{2}\sqrt{17}$. Because $x = (\alpha - E)/\beta$, the four π-electron energies of bicyclobutadiene are

$$E = \alpha - 1.562\beta \qquad E = \alpha - \beta \qquad E = \alpha \qquad E = \alpha + 2.562\beta$$

There are four π electrons, so the lowest energy levels are filled. The π-electron energy is

$$E_\pi = 2(\alpha + 2.562\beta) + 2\alpha = 4\alpha + 5.124\beta$$

The energy of two isolated ethene molecules is $4\alpha + 4\beta$, so

$$E_{\text{deloc}} = E_\pi(\text{bicyclobutadiene}) - 2E_\pi(\text{ethene})$$

$$= 1.124\beta$$

10–33. Show that the Hückel molecular orbitals of benzene given in Equations 10.31 are orthonormal.

In the Hückel approximation, the overlap integral of $2p$ orbitals on different carbon atoms is zero. Therefore, the values of $\int \psi_i^* \psi_i d\tau$, where ψ_i is a molecular orbital for benzene given in Equations 10.31, are simply the sums of the squares of the coefficients of the $2p_{zi}$ atomic orbitals for $i = 1$ to 6. For ψ_1 and ψ_6, we have

$$\left(\frac{1}{6} + \frac{1}{6} + \frac{1}{6} + \frac{1}{6} + \frac{1}{6} + \frac{1}{6}\right) = 1$$

For ψ_2 and ψ_4, we have

$$\left(\frac{1}{4} + \frac{1}{4} + \frac{1}{4} + \frac{1}{4}\right) = 1$$

And for ψ_3 and ψ_5, we have

$$\left(\frac{1}{3} + \frac{1}{12} + \frac{1}{12} + \frac{1}{3} + \frac{1}{12} + \frac{1}{12}\right) = 1$$

In evaluating $\int \phi_i \phi_j d\tau$, where $i \neq j$, we also realize that the overlap integrals between $2p_{zj}$ orbitals on different carbon atoms are zero. Therefore, for ψ_1 and ψ_2 we have

$$\int d\tau \psi_1 \psi_2 = \left(\frac{1}{\sqrt{24}} + \frac{1}{\sqrt{24}} - \frac{1}{\sqrt{24}} - \frac{1}{\sqrt{24}}\right) = 0$$

and for ψ_1 and ψ_3 we have

$$\int d\tau \psi_1 \psi_3 = \left(\frac{1}{\sqrt{18}} + \frac{1}{2\sqrt{18}} - \frac{1}{2\sqrt{18}} - \frac{1}{\sqrt{18}} - \frac{1}{2\sqrt{18}} + \frac{1}{2\sqrt{18}}\right) = 0$$

The remaining pairs of orbitals can similarly be shown to be orthogonal.

10–34. Set up, but do not try to solve, the Hückel molecular-orbital theory determinantal equation for naphthalene, $C_{10}H_8$.

The structure of naphthalene is shown below.

Using the numbering of carbon atoms shown, the Hückel determinantal equation is

$$\begin{vmatrix} x & 0 & 0 & 0 & 0 & 0 & 0 & 1 & 1 & 0 \\ 0 & x & 1 & 0 & 0 & 0 & 0 & 0 & 1 & 0 \\ 0 & 1 & x & 1 & 0 & 0 & 0 & 0 & 0 & 0 \\ 0 & 0 & 1 & x & 1 & 0 & 0 & 0 & 0 & 0 \\ 0 & 0 & 0 & 1 & x & 0 & 0 & 0 & 0 & 1 \\ 0 & 0 & 0 & 0 & 0 & x & 1 & 0 & 0 & 1 \\ 0 & 0 & 0 & 0 & 0 & 1 & x & 1 & 0 & 0 \\ 1 & 0 & 0 & 0 & 0 & 0 & 1 & x & 0 & 0 \\ 1 & 1 & 0 & 0 & 0 & 0 & 0 & 0 & x & 1 \\ 0 & 0 & 0 & 0 & 1 & 1 & 0 & 0 & 1 & x \end{vmatrix} = 0$$

10–35. A Hückel calculation for naphthalene, $C_{10}H_8$, gives the molecular-orbital energy levels $E_i = \alpha + m_i\beta$, where the 10 values of m_i are 2.3028, 1.6180, 1.3029, 1.0000, 0.6180, −0.6180, −1.0000, −1.3029, −1.6180, and −2.3028. Calculate the ground-state π-electron energy of naphthalene.

There are 10 π electrons in naphthalene, and so the five lowest energy levels are filled. The π-electron energy is then given by

$$E_\pi = 2(\alpha + 2.3028\beta) + 2(\alpha + 1.6180\beta) + 2(\alpha + 1.3029\beta)$$
$$+ 2(\alpha + \beta) + 2(\alpha + 0.6180\beta)$$
$$= 10\alpha + 13.68\beta$$

10–36. The total π-electron energy of naphthalene (Problem 10–35) is

$$E_n = 10\alpha + 13.68\beta$$

Calculate the delocalization energy of naphthalene.

To calculate the delocalization energy of naphthalene, we compare the π-electron energy of naphthalene to that of five ethene molecules. Five ethene molecules have a total π-electron energy of $10\alpha + 10\beta$, and so

$$E_{deloc} = 10\alpha + 13.68\beta - (10\alpha + 10\beta)$$
$$= 3.68\beta$$

10–37. Using Hückel molecular-orbital theory, determine whether the linear state (H—H—H⁺) or the triangular state

$$\left[\begin{array}{c} H \\ \diagup \diagdown \\ H\text{——}H \end{array} \right]^+$$

of H_3^+ is the more stable state. Repeat the calculation for H_3 and H_3^-.

For the triangular state, the Hückel determinantal equation is

$$\begin{vmatrix} x & 1 & 1 \\ 1 & x & 1 \\ 1 & 1 & x \end{vmatrix} = x^3 - 3x + 2 = 0$$

The three roots of this equation are $x = 1, 1, -2$, and so the three energies are

$$E = \alpha - \beta \qquad E = \alpha - \beta \qquad E = \alpha + 2\beta$$

The energy of H_3^+ (a two electron molecule) is

$$E_{H_3^+} = 2(\alpha + 2\beta) = 2\alpha + 4\beta$$

The energy of H_3 (a three electron molecule) is

$$E_{H_3} = 2(\alpha + 2\beta) + (\alpha - \beta) = 3\alpha + 3\beta$$

and the energy of H_3^- (a four electron molecule) is

$$E_{H_3^-} = 2(\alpha + 2\beta) + 2(\alpha - \beta) = 4\alpha + 2\beta$$

Note that H_3^- will have a triplet ground state, since the energy level $\alpha - \beta$ is doubly degenerate. For the linear molecules, the Hückel determinantal equation is

$$\begin{vmatrix} x & 1 & 0 \\ 1 & x & 1 \\ 0 & 1 & x \end{vmatrix} = x^3 - 2x = 0$$

The three roots of this equation are $x = 0, \sqrt{2}, -\sqrt{2}$, and so the three energies are

$$E = \alpha - \sqrt{2}\beta \qquad E = \alpha - \beta \qquad E = \alpha + 2\beta$$

The energies of linear H_3^+, H_3, and H_3^- are

$$E_{H_3^+} = 2(\alpha + \sqrt{2}\beta) = 2\alpha + 2\sqrt{2}\beta$$

$$E_{H_3} = 2(\alpha + \sqrt{2}\beta) + (\alpha) = 3\alpha + 2\sqrt{2}\beta$$

$$E_{H_3^-} = 2(\alpha + 2\beta) + 2(\alpha) = 4\alpha + 2\sqrt{2}\beta$$

The triangular geometry is more stable for H_3^+, the linear geometry is more stable for H_3^-, and the triangular geometry is slightly more stable for H_3.

10–38. Set up a Hückel theory secular determinant for pyridine.

The Hückel theory secular determinant is

$$
\begin{vmatrix}
\alpha_N - E & \beta_{CN} & 0 & 0 & 0 & \beta_{CN} \\
\beta_{CN} & \alpha_C - E & \beta_{CC} & 0 & 0 & 0 \\
0 & \beta_{CC} & \alpha_C - E & \beta_{CC} & 0 & 0 \\
0 & 0 & \beta_{CC} & \alpha_C - E & \beta_{CC} & 0 \\
0 & 0 & 0 & \beta_{CC} & \alpha_C - E & \beta_{CC} \\
\beta_{CN} & 0 & 0 & 0 & \beta_{CC} & \alpha_C - E
\end{vmatrix} = 0
$$

where we have introduced two different types of β's to account for the fact that the integrals can involve p-orbitals on different atoms (N and C).

10–39. The coefficients in Hückel molecular orbitals can be used to calculate charge distribution and bond orders. We will use butadiene as a concrete example. The molecular orbitals of butadiene can be expressed as

$$
\psi_i = \sum_{j=1}^{4} c_{ij} 2p_{zj}
$$

where the c_{ij} are determined by the set of linear algebraic equations that lead to the secular determinantal equation. The resulting molecular orbitals for butadiene are given by Equations 10.26:

$$
\psi_1 = 0.3717\, 2p_{z1} + 0.6015\, 2p_{z2} + 0.6015\, 2p_{z3} + 0.3717\, 2p_{z4}
$$

$$
\psi_2 = 0.6015\, 2p_{z1} + 0.3717\, 2p_{z2} - 0.3717\, 2p_{z3} - 0.6015\, 2p_{z4}
$$

$$
\psi_3 = 0.6015\, 2p_{z1} - 0.3717\, 2p_{z2} - 0.3717\, 2p_{z3} + 0.6015\, 2p_{z4}
$$

$$
\psi_4 = 0.3717\, 2p_{z1} - 0.6015\, 2p_{z2} + 0.6015\, 2p_{z3} - 0.3717\, 2p_{z4}
$$

These molecular orbitals are presented schematically in Figure 10.23. Because we have set $S_{ij} = \delta_{ij}$ in Equation 10.19, we have in effect assumed that the $2p_z$'s are orthonormal. Using this fact, show that the c_{ij} satisfy

$$
\sum_{j=1}^{4} c_{ij}^2 = 1 \qquad i = 1,\ 2,\ 3,\ 4 \tag{1}
$$

Equation 1 allows us to interpret c_{ij}^2 as the fractional π-electronic charge on the jth carbon atom due to an electron in the ith molecular orbital. Thus, the total π-electron charge on the jth carbon atom is

$$
q_j = \sum_i n_i c_{ij}^2 \tag{2}
$$

where n_i is the number of electrons in the ith molecular orbital. Show that

$$
q_1 = 2c_{11}^2 + 2c_{21}^2 + 0c_{31}^2 + 0c_{41}^2
$$

$$
= 2(0.3717)^2 + 2(0.6015)^2
$$

$$
= 1.000
$$

for butadiene. Show that the other q's are also equal to unity, indicating that the π electrons in butadiene are uniformly distributed over the molecule.

For all four ψ_j of butadiene, Equation 1 gives

$$\sum_{j=1}^{4} c_{ij}^2 = 2(0.3717)^2 + 2(0.6015)^2 = 1$$

For q_1,

$$
\begin{aligned}
q_1 &= 2c_{11}^2 + 2c_{21}^2 + 0c_{31}^2 + 0c_{41}^2 \\
&= 2(0.3717)^2 + 2(0.6015)^2 \\
&= 1.000
\end{aligned}
$$

For q_2,

$$
\begin{aligned}
q_2 &= 2c_{12}^2 + 2c_{22}^2 + 0c_{32}^2 + 0c_{42}^2 \\
&= 2(0.6015)^2 + 2(0.3717)^2 \\
&= 1.000
\end{aligned}
$$

For q_3,

$$
\begin{aligned}
q_3 &= 2c_{13}^2 + 2c_{23}^2 + 0c_{33}^2 + 0c_{43}^2 \\
&= 2(0.6015)^2 + 2(-0.3717)^2 \\
&= 1.000
\end{aligned}
$$

For q_4,

$$
\begin{aligned}
q_4 &= 2c_{14}^2 + 2c_{24}^2 + 0c_{34}^2 + 0c_{44}^2 \\
&= 2(0.3717)^2 + 2(-0.6015)^2 \\
&= 1.000
\end{aligned}
$$

10–40. Another interesting quantity that can be defined in terms of the c_{ij} in Problem 10–39 is the π-bond order. We can interpret the product $c_{ir}c_{is}$ as the π-electron charge in the ith molecular orbital between the adjacent carbon atoms r and s. We define the π-bond order between the adjacent carbon atoms r and s by

$$P_{rs}^{\pi} = \sum_i n_i c_{ir} c_{is} \tag{1}$$

where n_i is the number of π electrons in the ith molecular orbital. Show that

$$P_{12}^{\pi} = 0.8942$$

and

$$P_{23}^{\pi} = 0.4473$$

for butadiene. Clearly, $P_{12}^{\pi} = P_{34}^{\pi}$ by symmetry. If we recall that there is a σ bond between each carbon atom, then we can define a total bond order

$$P_{rs}^{\text{total}} = 1 + P_{rs}^{\pi} \tag{2}$$

where the first term on the right side is due to the σ bond between atoms r and s. For butadiene, show that

$$P_{12}^{total} = P_{34}^{total} = 1.894$$

$$P_{23}^{total} = 1.447 \tag{3}$$

Equations 3 are in excellent agreement with the experimental observations involving the reactivity of these bonds in butadiene.

We use the values of c_{ij} from Problem 10–39. Then

$$P_{rs}^{\pi} = \sum_i n_i c_{ir} c_{is}$$

$$P_{12}^{\pi} = n_1 c_{11} c_{12} + n_2 c_{21} c_{22} + n_3 c_{31} c_{32} + n_4 c_{41} c_{42}$$

$$= 2c_{11}c_{12} + 2c_{21}c_{22} = 2(0.3717)(0.6015) + 2(0.6015)(0.3717)$$

$$= 0.8942$$

$$P_{23}^{\pi} = n_1 c_{12} c_{13} + n_2 c_{22} c_{23} + n_3 c_{32} c_{33} + n_4 c_{42} c_{43}$$

$$= 2c_{12}c_{13} + 2c_{22}c_{23} = 2(0.6015)(0.6015) + 2(0.3717)(-0.3717)$$

$$= 0.4473$$

Using the definition $P_{rs}^{total} = 1 + P_{rs}^{\pi}$, we find

$$P_{12}^{total} = P_{34}^{total} = 1.894$$

$$P_{23}^{total} = 1.447$$

10–41. Calculate the delocalization energy, the charge on each carbon atom, and the bond orders for the allyl radical, cation, and anion. Sketch the molecular orbitals for the allyl system.

The structure of the allyl system is

Using the above numbering, the Hückel determinantal equation for the allyl system is

$$\begin{vmatrix} x & 1 & 0 \\ 1 & x & 1 \\ 0 & 1 & x \end{vmatrix} = x^3 - 2x = 0$$

The roots of this equation are $\sqrt{2}, 0, -\sqrt{2}$ and so the energies are

$$E = \alpha - \sqrt{2}\beta \qquad E = \alpha \qquad E = \alpha + \sqrt{2}\beta$$

For the allyl radical (with 3 π electrons)

$$E_\pi = 2(\alpha + \sqrt{2}\beta) + \alpha = 3\alpha + 2\sqrt{2}\beta$$

$$E_{deloc} = (3\alpha + 2\sqrt{2}\beta) - (3\alpha + 2\beta) = 0.828\beta$$

where we have subtracted $3\alpha + 2\beta$ from E_π because the three localized π electrons can be thought of as the sum of one ethene molecule (of energy $2\alpha + 2\beta$) and a single carbon atom (of energy α). For the allyl carbonium ion (2 π electrons)

$$E_\pi = 2(\alpha + \sqrt{2}\beta) = 2\alpha + 2\sqrt{2}\beta$$

$$E_{deloc} = E_\pi - 2(\alpha + \beta) = 0.828\beta$$

For the allyl carbanion (4 π electrons)

$$E_\pi = 2(\alpha + \sqrt{2}\beta) + 2\alpha = 4\alpha + 2\sqrt{2}\beta$$

$$E_{deloc} = E_\pi - 2(\alpha + \beta) - 2\alpha = 0.828\beta$$

In order to calculate the charges on each carbon atom and the bond orders, we must determine the molecular orbitals associated with each value of E. Following the procedure in Example 10–6 (or Problem 10–28), $\psi_\pi = c_1 2p_{z1} + c_2 2p_{z2} + c_3 2p_{z3}$, where

$$c_1(\alpha - E) + c_2\beta = 0$$

$$c_1\beta + c_2(\alpha - E) + c_3\beta = 0$$

$$c_2\beta + c_3(\alpha - E) = 0$$

Substitute the values of E into these equations to find the c_j. For $E = \alpha + \sqrt{2}\beta$, we have

$$-\sqrt{2}c_1 + c_2 = 0$$

$$c_1 - \sqrt{2}c_2 + c_3 = 0$$

$$c_2 - \sqrt{2}c_3 = 0$$

from which we obtain

$$c_1 = c_3 = \frac{1}{\sqrt{2}}c_2$$

Thus,

$$\psi_1 = c_2\left(\frac{1}{\sqrt{2}}2p_{z1} + 2p_{z2} + \frac{1}{\sqrt{2}}2p_{z3}\right)$$

Normalization gives

$$c_2^2\left(\frac{1}{2} + 1 + \frac{1}{2}\right) = 1$$

$$c_2 = \frac{1}{\sqrt{2}}$$

Therefore,

$$\psi_1 = \frac{1}{2}2p_{z1} + \frac{1}{\sqrt{2}}2p_{z2} + \frac{1}{2}2p_{z3}$$

Now, for $E = \alpha$, we have

$$c_2 = 0$$
$$c_1 + c_3 = 0$$
$$c_2 = 0$$

from which we obtain

$$\psi_2 = c_1 \left(2p_{z1} - 2p_{z3} \right)$$

Requiring that ψ_2 be normalized gives

$$c_1^2(1 + 1) = 1$$
$$c_1 = \frac{1}{\sqrt{2}}$$

So

$$\psi_2 = \frac{1}{\sqrt{2}} 2p_{z1} - \frac{1}{\sqrt{2}} 2p_{z3}$$

Finally, for $E = \alpha - \sqrt{2}\beta$, we have

$$\sqrt{2}c_1 + c_2 = 0$$
$$c_1 + \sqrt{2}c_2 + c_3 = 0$$
$$c_2 + \sqrt{2}c_3 = 0$$

from which we obtain

$$c_1 = c_3 = -\frac{1}{\sqrt{2}} c_2$$

We then find

$$\psi_3 = c_2 \left(\frac{1}{\sqrt{2}} 2p_{z1} - 2p_{z2} + \frac{1}{\sqrt{2}} 2p_{z3} \right)$$

Normalization gives

$$c_2^2 \left(\frac{1}{2} + 1 + \frac{1}{2} \right) = 1$$
$$c_2 = \frac{1}{\sqrt{2}}$$

We then write

$$\psi_3 = \frac{1}{2} 2p_{z1} - \frac{1}{\sqrt{2}} 2p_{z2} + \frac{1}{2} 2p_{z3}$$

The charge on each carbon atom is given by Equation 2 of Problem 10–39. The π-bond order can be evaluated using Equation 1 of Problem 10–40. For the allyl radical

$$q_j = \sum_i n_i c_{ij}^2$$

$$q_1 = 2 \left(\frac{1}{2} \right)^2 + 1 \left(\frac{1}{2} \right) = 1$$

$$q_2 = 2 \left(\frac{1}{2} \right) = 1$$

$$q_3 = 2 \left(\frac{1}{2} \right)^2 + 1 \left(\frac{1}{2} \right) = 1$$

and

$$P_{rs}^{\pi} = \sum_i n_i c_{ir} c_{is}$$

$$P_{12}^{\pi} = 2\left(\frac{\sqrt{2}}{4}\right) = 0.707$$

$$P_{23}^{\pi} = 2\left(\frac{\sqrt{2}}{4}\right) = 0.707$$

For the allyl carbonium ion,

$$q_1 = 2\left(\frac{1}{2}\right)^2 = \frac{1}{2}$$

$$q_2 = 2\left(\frac{1}{2}\right) = 1$$

$$q_3 = 2\left(\frac{1}{2}\right)^2 = \frac{1}{2}$$

$$P_{12}^{\pi} = 2\left(\frac{\sqrt{2}}{4}\right) = 0.707$$

$$P_{23}^{\pi} = 2\left(\frac{\sqrt{2}}{4}\right) = 0.707$$

and for the allyl carbanion,

$$q_1 = 2\left(\frac{1}{2}\right)^2 + 2\left(\frac{1}{2}\right) = \frac{3}{2}$$

$$q_2 = 2\left(\frac{1}{2}\right) = 1$$

$$q_3 = 2\left(\frac{1}{2}\right)^2 + 2\left(\frac{1}{2}\right) = \frac{3}{2}$$

$$P_{12}^{\pi} = 2\left(\frac{\sqrt{2}}{4}\right) = 0.707$$

$$P_{23}^{\pi} = 2\left(\frac{\sqrt{2}}{4}\right) = 0.707$$

These results are summarized in the table below.

	E_{deloc}	q_1	q_2	q_3	P_{12}^{π}	P_{23}^{π}
radical	0.828β	1	1	1	0.707	0.707
carbonium	0.828β	$\frac{1}{2}$	1	$\frac{1}{2}$	0.707	0.707
carbanion	0.828β	$\frac{3}{2}$	1	$\frac{3}{2}$	0.707	0.707

10–42. Calculate the π-electronic charge on each carbon atom and the total bond orders in benzene. Comment on the result.

Using Equation 2 from Problem 10–39, we find the total π-electronic charge on the nth carbon atom to be

$$q_n = 2(c_{1n}^2 + c_{2n}^2 + c_{3n}^2)$$

Therefore, using the molecular orbitals given by Equations 10.31,

$$q_1 = 2\left(\frac{1}{6} + \frac{1}{3}\right) = 1$$

$$q_2 = 2\left(\frac{1}{6} + \frac{1}{4} + \frac{1}{12}\right) = 1$$

$$q_3 = 2\left(\frac{1}{6} + \frac{1}{4} + \frac{1}{12}\right) = 1$$

$$q_4 = 2\left(\frac{1}{6} + \frac{1}{3}\right) = 1$$

$$q_5 = 2\left(\frac{1}{6} + \frac{1}{4} + \frac{1}{12}\right) = 1$$

$$q_6 = 2\left(\frac{1}{6} + \frac{1}{4} + \frac{1}{12}\right) = 1$$

Using Equation 1 from Problem 10–40 we find the π-bond orders to be

$$P_{rs}^\pi = 2(c_{1r}c_{1s} + c_{2r}c_{2s} + c_{3r}c_{3s})$$

Therefore,

$$P_{12}^\pi = 2\left(\frac{1}{6} + \frac{1}{6}\right) = \frac{2}{3}$$

$$P_{23}^\pi = 2\left(\frac{1}{6} + \frac{1}{4} - \frac{1}{12}\right) = \frac{2}{3}$$

$$P_{34}^\pi = 2\left(\frac{1}{6} + \frac{1}{6}\right) = \frac{2}{3}$$

$$P_{45}^\pi = 2\left(\frac{1}{6} + \frac{1}{6}\right) = \frac{2}{3}$$

$$P_{56}^\pi = 2\left(\frac{1}{6} + \frac{1}{4} - \frac{1}{12}\right) = \frac{2}{3}$$

$$P_{61}^\pi = 2\left(\frac{1}{6} + \frac{1}{6}\right) = \frac{2}{3}$$

All of the carbon atoms in benzene are equivalent.

10–43. Because of the symmetry inherent in the Hückel theory secular determinants of linear and cyclic conjugated polyenes, we can write mathematical formulas for the energy levels for an arbitrary number of carbon atoms in the system (for present purposes, we consider cyclic polyenes with only an even number of carbon atoms). The formula for linear chains is

$$E_n = \alpha + 2\beta \cos \frac{\pi n}{N + 1} \qquad n = 1, 2, \ldots, N$$

and the formula for cyclic chains with N even is

$$E_n = \alpha + 2\beta \cos \frac{2\pi n}{N} \qquad n = 0, \pm 1, \ldots, \pm\left(\frac{N}{2} - 1\right), \frac{N}{2}$$

where α and β are as defined in the text and N is the number of carbon atoms in the conjugated π system. (a) Use these formulas to verify the results given in the chapter for butadiene and benzene. (b) Now use these formulas to predict energy levels for linear hexatriene (C_6H_8) and octatetraene (C_8H_{10}). How does the delocalization energy of these molecules per carbon atom vary as the chains grow in length? (c) Compare the results for hexatriene and benzene. Which molecule has a greater delocalization energy? Why?

---

a. For butadiene, $n = 4$. The formula for linear polyenes gives the energies

$$E_1 = \alpha + 2\beta \cos \frac{\pi}{5} \qquad E_2 = \alpha + 2\beta \cos \frac{2\pi}{5}$$

$$E_3 = \alpha + 2\beta \cos \frac{3\pi}{5} \qquad E_4 = \alpha + 2\beta \cos \frac{4\pi}{5}$$

or

$$E_1 = \alpha + 1.618\beta \qquad E_2 = \alpha + 0.618\beta$$
$$E_3 = \alpha - 0.618\beta \qquad E_4 = \alpha - 1.618\beta$$

in agreement with the values given in the chapter. For benzene, $n = 6$ and the formula for conjugated polyenes gives the energies

$$E_0 = \alpha + 2\beta \cos(0) \qquad E_1 = \alpha + 2\beta \cos\left(\frac{2\pi}{6}\right)$$

$$E_2 = \alpha + 2\beta \cos\left(-\frac{2\pi}{6}\right) \qquad E_3 = \alpha + 2\beta \cos\left(\frac{4\pi}{6}\right)$$

$$E_4 = \alpha + 2\beta \cos\left(-\frac{4\pi}{6}\right) \qquad E_5 = \alpha + 2\beta \cos \pi$$

in agreement with the values given in the chapter,

$$E_0 = \alpha + 2\beta \qquad E_1 = \alpha + \beta \qquad E_2 = \alpha + \beta$$
$$E_3 = \alpha - \beta \qquad E_4 = \alpha - \beta \qquad E_5 = \alpha - 2\beta$$

b. Substituting into the formula, we find for hexatriene ($n = 6$)

$$E_1 = \alpha + 1.802\beta \qquad E_2 = \alpha + 1.247\beta \qquad E_3 = \alpha + 0.4450\beta$$
$$E_4 = \alpha - 0.4450\beta \qquad E_5 = \alpha - 1.247\beta \qquad E_6 = \alpha - 1.802\beta$$

and for octatriene ($n = 8$)

$$E_1 = \alpha + 1.879\beta \qquad E_2 = \alpha + 1.532\beta$$
$$E_3 = \alpha + \beta \qquad E_4 = \alpha + 0.3473\beta$$
$$E_5 = \alpha - 0.3473\beta \qquad E_6 = \alpha - \beta$$
$$E_7 = \alpha - 1.532\beta \qquad E_8 = \alpha - 1.879\beta$$

The delocalization energies of these molecules increases as the chains grow: for hexatriene $E_{deloc} = 0.9880\beta$ and for octatriene $E_{deloc} = 1.5166\beta$. Per carbon atom, the energy of delocalization (0.1647β per carbon atom for hexatriene and 0.1896β per carbon atom for octatriene) also increases as the number of carbon atoms on the chain grows.

c. Benzene is more stable than hexatriene, because its cyclic structure allows for more delocalization than the corresponding linear structure.

10–44. The problem of a linear conjugated polyene of N carbon atoms can be solved in general. The energies E_j and the coefficients of the atomic orbitals in the jth molecular orbital are given by

$$E_j = \alpha + 2\beta \cos \frac{j\pi}{N+1} \qquad j = 1, 2, 3, \ldots, N$$

and

$$c_{jk} = \left(\frac{2}{N+1}\right)^{1/2} \sin \frac{jk\pi}{N+1} \qquad k = 1, 2, 3, \ldots, N$$

Derive the energy levels and the wave functions for butadiene using these formulas.

The formula for the energy levels E_j is the same as the formula for E_n in the previous problem, where we found the energy levels of butadiene in part (a). Using the second expression in the problem, we find

$$c_{j1} = \left(\frac{2}{5}\right)^{1/2} \sin \frac{j\pi}{5} \qquad c_{j2} = \left(\frac{2}{5}\right)^{1/2} \sin \frac{2j\pi}{5}$$

$$c_{j3} = \left(\frac{2}{5}\right)^{1/2} \sin \frac{3j\pi}{5} \qquad c_{j4} = \left(\frac{2}{5}\right)^{1/2} \sin \frac{4j\pi}{5}$$

and so

$$\psi_1 = 0.37172 p_{z1} + 0.60152 p_{z2} + 0.60152 p_{z3} + 0.37172 p_{z4}$$
$$\psi_2 = 0.60152 p_{z1} + 0.37172 p_{z2} - 0.37172 p_{z3} - 0.60152 p_{z4}$$
$$\psi_3 = 0.60152 p_{z1} - 0.37172 p_{z2} - 0.37172 p_{z3} + 0.60152 p_{z4}$$
$$\psi_4 = 0.37172 p_{z1} - 0.60152 p_{z2} + 0.60152 p_{z3} - 0.37172 p_{z4}$$

10–45. We can calculate the electronic states of a hypothetical one-dimensional solid by modeling the solid as a one-dimensional array of atoms with one orbital per atom, and using Hückel theory to calculate the allowed energies. Use the formula for E_j in Problem 10–44 to show that energies will form essentially a continuous band of width 4β. *Hint:* Calculate $E_1 - E_N$ and let N be very large so that you can use $\cos x \approx 1 - x^2/2 + \cdots$.

Using the equation for E_j in Problem 10–44, we have

$$E_1 = \alpha + 2\beta \cos \frac{\pi}{N+1}$$

$$E_N = \alpha + 2\beta \cos \frac{N\pi}{N+1}$$

Subtracting these equations gives

$$E_1 - E_N = 2\beta \left(\cos \frac{\pi}{N+1} - \cos \frac{N\pi}{N+1} \right)$$

As $N \to \infty$,

$$\cos \frac{\pi}{N+1} - \cos \frac{N\pi}{N+1} \to 2$$

and $E_1 - E_N \rightarrow 4\beta$. Therefore, the width of the energy band approaches 4β for large values of N.

10–46. The band of electronic energies that we calculated in Problem 10–45 can accomodate N pairs of electrons of opposite spins, or a total of $2N$ electrons. If each atom contributes one electron (as in the case of a polyene), the band is occupied by a total of N electrons. Using some ideas you may have learned in general chemistry, would you expect such a system to be a conductor or an insulator?

The N energy levels are close together, and only half of them are occupied in the ground state. Therefore, when N is large, we would expect this system to be a conductor (since relatively little energy is required to excite an electron).

10–47. The dipole moment of a polyatomic molecule is defined by

$$\boldsymbol{\mu} = e \sum_j z_j \mathbf{r}_j$$

where $z_j e$ is the magnitude of a charge located at the point given by $\mathbf{r}_j$. Show that the value of $\boldsymbol{\mu}$ is independent of the origin chosen for $\mathbf{r}_j$ if the net charge is zero. Show that $\boldsymbol{\mu} = 0$ for SO_3 (trigonal planar), CCl_4 (tetrahedral), SF_6 (octahedral), XeF_4 (square planar), and PF_5 (trigonal bipyramidal).

See the solution to Problem 9–37 for the first part of this problem. For SO_3, we can take the coordinate system

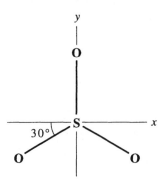

Since the molecule is in the xy-plane, $\mu_z = 0$. We can express the magnitude of the charge on the oxygen atoms as $z_O e$ and the length of the sulphur–oxygen bond as l, so

$$\mu_x = z_O el \cos 30° - z_O el \cos 30° = 0$$
$$\mu_y = -2z_O el \sin 30° + z_O el = 0$$

For CCl_4, we can refer to the figure given in the text of Problem 10–7, taking the center of the cube as the origin. We can express the magnitude of the charge on the chlorine atoms as z_{Cl} and the length of the carbon–chlorine bond as l. We then find

$$\mu_x = \mu_y = \mu_z = 2z_{Cl} el \cos\left(\frac{109.5°}{2}\right) - 2z_{Cl} el \cos\left(\frac{109.5°}{2}\right) = 0$$

The bonds in the molecules XeF_4 and SF_6 consist solely of pairs of equivalent bonds pointing directly away from each other. The net charges on these molecules are therefore zero. The molecule PF_5 has a trigonal bipyramidal structure. The net charge on the trigonal plane is zero (as we have shown for SO_3). The two remaining bonds in PF_5 are equivalent and point directly opposite each other, so the net charge of the molecule is zero.

CHAPTER 11

Computational Quantum Chemistry

PROBLEMS AND SOLUTIONS

11–1. Show that a three-dimensional Gaussian function centered at $\mathbf{r}_0 = x_0\mathbf{i} + y_0\mathbf{j} + z_0\mathbf{k}$ is a product of three one-dimensional Gaussian functions centered on x_0, y_0, and z_0.

$$
\begin{aligned}
e^{-a(\mathbf{r}-\mathbf{r}_0)^2} &= e^{-a[(x-x_0)\mathbf{i}+(y-y_0)\mathbf{j}+(z-z_0)\mathbf{k}]^2} \\
&= e^{-a[(x-x_0)^2+(y-y_0)^2+(z-z_0)^2]} \\
&= e^{-a(x-x_0)^2}e^{-a(y-y_0)^2}e^{-a(z-z_0)^2}
\end{aligned}
$$

11–2. Show that

$$
\int_{-\infty}^{\infty} e^{-(x-x_0)^2}dx = \int_{-\infty}^{\infty} e^{-x^2}dx = 2\int_0^{\infty} e^{-x^2}dx = \pi^{1/2}
$$

The first equality is true because $dx = d(x - x_0)$, since x_0 is a constant. The second equality is true because a Gaussian function is even. We then find

$$
2\int_0^{\infty} e^{-x^2}dx = 2\left(\frac{\pi^{1/2}}{2}\right) = \pi^{1/2}
$$

11–3. The Gaussian integral

$$
I_0 = \int_0^{\infty} e^{-ax^2}dx
$$

can be evaluated by a trick. First write

$$
I_0^2 = \int_0^{\infty} dx\, e^{-ax^2} \int_0^{\infty} dy\, e^{-ay^2} = \int_0^{\infty}\int_0^{\infty} dx\, dy\, e^{-a(x^2+y^2)}
$$

Now convert the integration variables from Cartesian coordinates to polar coordinates and show that

$$
I_0 = \frac{1}{2}\left(\frac{\pi}{a}\right)^{1/2}
$$

We first write

$$I_0^2 = \int_0^\infty dx e^{-ax^2} \int_0^\infty dy e^{-ay^2} = \int_0^\infty \int_0^\infty dx dy e^{-a(x^2+y^2)}$$

In polar coordinates, $x^2 + y^2 = r^2$ and $dxdy = rdrd\theta$. The limits of integration in polar coordinates corresponding to the given limits of integration in Cartesian coordinates are $0 \le r < \infty$ and $0 \le \theta \le \pi/2$. Therefore,

$$I_0^2 = \int_0^\infty \int_0^{\pi/2} rdrd\theta e^{-ar^2} = \frac{\pi}{2}\int_0^\infty rdr e^{-ar^2}$$

$$= \frac{\pi}{2}\left(-\frac{1}{2a}e^{-ar^2}\right)\Big|_0^\infty = \frac{\pi}{4a}$$

$$I_0 = \left(\frac{\pi}{4a}\right)^{1/2} = \frac{1}{2}\left(\frac{\pi}{a}\right)^{1/2}$$

11–4. Show that the integral

$$I_{2n} = \int_0^\infty x^{2n}e^{-ax^2}dx$$

can be obtained from I_0 in Problem 11–3 by differentiating n times with respect to a. Using the result of Problem 11–3, show that

$$I_{2n} = \frac{1 \cdot 3 \cdot 5 \cdots (2n-1)}{2(2a)^n}\left(\frac{\pi}{a}\right)^{1/2}$$

We first determine the first few derivatives of I_0 with respect to a:

$$I_0 = \int_0^\infty e^{-ax^2}dx$$

$$\frac{dI_0}{da} = \int_0^\infty -x^2 e^{-ax^2}dx = -\int_0^\infty x^2 e^{-ax^2}dx$$

$$\frac{d^2 I_0}{da^2} = \int_0^\infty x^{(2\cdot2)}e^{-ax^2}dx = \int_0^\infty x^4 e^{-ax^2}dx$$

For the first few I_{2n}, we can use the result of Problem 11–3 to find that

$$I_0 = \int_0^\infty e^{-ax^2}dx = \frac{1}{2}\left(\frac{\pi}{a}\right)^{1/2}$$

$$I_2 = \int_0^\infty x^2 e^{-ax^2}dx = -\frac{dI_0}{da}$$

$$= -\frac{\pi^{-1/2}}{2}\left(-\frac{1}{2}a^{-3/2}\right) = \frac{1}{2(2a)}\left(\frac{\pi}{a}\right)^{1/2}$$

$$I_4 = \int_0^\infty x^4 e^{-ax^2}dx = \frac{d^2 I_0}{da^2} = -\frac{dI_2}{da}$$

$$= -\frac{\pi^{1/2}}{4}\left(-\frac{3}{2}a^{-5/2}\right) = \frac{3}{2(2a)^2}\left(\frac{\pi}{a}\right)^{1/2}$$

$$I_6 = \int_0^\infty x^6 e^{-ax^2}dx = -\frac{d^3 I_0}{da^3} = -\frac{dI_4}{da}$$

$$= -\frac{3\pi^{1/2}}{8}\left(-\frac{5}{2}a^{-7/2}\right) = \frac{3 \cdot 5}{2(2a)^3}\left(\frac{\pi}{a}\right)^{1/2}$$

and, in general,

$$I_{2n} = (-1)^n \frac{d^n I_0}{da^n}$$

$$I_{2n} = \frac{1 \cdot 3 \cdot 5 \cdots (2n - 1)}{2(2a)^n} \left(\frac{\pi}{a}\right)^{1/2}$$

11–5. Show that the Gaussian function

$$\phi(r) = \left(\frac{2\alpha}{\pi}\right)^{3/4} e^{-\alpha r^2}$$

is normalized.

$$\int \phi^* \phi \, d\tau = 4\pi \int_0^\infty dr \, r^2 e^{-2\alpha r^2} \left(\frac{2\alpha}{\pi}\right)^{3/2}$$

$$= 4\pi \left(\frac{2\alpha}{\pi}\right)^{3/2} \int_0^\infty dr \, r^2 e^{-2\alpha r^2}$$

$$= 4\pi \left(\frac{2\alpha}{\pi}\right)^{3/2} \left[\frac{1}{4(2\alpha)} \left(\frac{\pi}{2\alpha}\right)^{1/2}\right]$$

$$= 4\pi \left(\frac{2\alpha}{\pi}\right) \left[\frac{1}{4(2\alpha)}\right] = 1$$

11–6. Show that the product of a (not normalized) Gaussian function centered at $\mathbf{R}_A$ and one centered at $\mathbf{R}_B$, i.e.

$$\phi_1 = e^{-\alpha|\mathbf{r}-\mathbf{R}_A|^2} \qquad \text{and} \qquad \phi_2 = e^{-\beta|\mathbf{r}-\mathbf{R}_B|^2}$$

is a Gaussian function centered at

$$\mathbf{R}_p = \frac{\alpha \mathbf{R}_A + \beta \mathbf{R}_B}{\alpha + \beta}$$

For simplicity, work in one dimension and appeal to Problem 11–1 to argue that it is true in three dimensions.

In one dimension, these equations become

$$\phi_1 = e^{-\alpha(x-x_A)^2} \qquad \text{and} \qquad \phi_2 = e^{-\beta(x-x_B)^2}$$

and the Gaussian function is centered at

$$x_p = \frac{\alpha x_A + \beta x_B}{\alpha + \beta}$$

The easiest way to do this problem is to show that $\phi_1\phi_2$ can be written as a Gaussian function centered at x_p. The product $\phi_1\phi_2$ is

$$\begin{aligned}
\phi_1\phi_2 &= \exp\left[-\alpha(x-x_A)^2 - \beta(x-x_B)^2\right] \\
&= \exp\left(-\alpha x^2 + 2x_A x\alpha - \alpha x_A^2 - \beta x^2 + 2\beta x x_B - \beta x_B^2\right) \\
&= \exp\left[-(\alpha x_A^2 + \beta x_B^2)\right]\exp\left[-(\alpha+\beta)x^2 + 2x(x_A\alpha + x_B\beta)\right] \\
&= \exp\left[-(\alpha x_A^2 + \beta x_B^2)\right]\exp\left\{-(\alpha+\beta)\left[x^2 - 2x\frac{\alpha x_A + \beta x_B}{\alpha+\beta} + \left(\frac{\alpha x_A + \beta x_B}{\alpha+\beta}\right)^2\right]\right\} \\
&\quad\quad \times \exp\left[(\alpha+\beta)\left(\frac{\alpha x_A + \beta x_B}{\alpha+\beta}\right)^2\right] \\
&= \exp\left[-\frac{\alpha\beta(x_A - x_B)^2}{\alpha+\beta}\right]\exp\left[-(\alpha+\beta)(x-x_p)^2\right]
\end{aligned}$$

This is a Gaussian function centered at x_p. We can find three one-dimensional functions like this in x, y, and z centered at x_p, y_p, and z_p; their product would be the three-dimensional function sought (as shown in Problem 11–1).

11–7. Show explicitly that if

$$\phi_{1s}(\alpha, \mathbf{r} - \mathbf{R}_A) = \left(\frac{2\alpha}{\pi}\right)^{3/4} e^{-\alpha|\mathbf{r}-\mathbf{R}_A|^2}$$

and

$$\phi_{1s}(\beta, \mathbf{r} - \mathbf{R}_B) = \left(\frac{2\beta}{\pi}\right)^{3/4} e^{-\beta|\mathbf{r}-\mathbf{R}_B|^2}$$

are normalized Gaussian $1s$ functions, then

$$\phi_{1s}(\alpha, \mathbf{r} - \mathbf{R}_A)\phi_{1s}(\beta, \mathbf{r} - \mathbf{R}_B) = K_{AB}\phi_{1s}(p, \mathbf{r} - \mathbf{R}_p)$$

where $p = \alpha + \beta$, $\mathbf{R}_p = (\alpha\mathbf{R}_A + \beta\mathbf{R}_B)/(\alpha+\beta)$ (see Problem 11–6), and

$$K_{AB} = \left[\frac{2\alpha\beta}{(\alpha+\beta)\pi}\right]^{3/4} e^{-\frac{\alpha\beta}{\alpha+\beta}|\mathbf{R}_A - \mathbf{R}_B|^2}$$

Using the result of Problem 11–6,

$$\phi_1(\alpha, \mathbf{r} - \mathbf{R}_A)\phi_2(\beta, \mathbf{r} - \mathbf{R}_B) = \left(\frac{4\alpha\beta}{\pi^2}\right)^{3/4} e^{-\alpha\beta|\mathbf{R}_A - \mathbf{R}_B|^2/(\alpha+\beta)} e^{-(\alpha+\beta)(\mathbf{r}-\mathbf{R}_p)^2} \tag{1}$$

The normalization constant of $e^{-(\alpha+\beta)(\mathbf{r}-\mathbf{R}_p)^2}$ is A, where

$$\begin{aligned}
1 &= A^2 \int_{-\infty}^{\infty}\int_{-\infty}^{\infty}\int_{-\infty}^{\infty} e^{-(\alpha+\beta)[(x-x_p)^2 + (y-y_p)^2 + (z-z_p)^2]} \\
&= A^2\left[2\int_0^{\infty} e^{-2(\alpha+\beta)u^2}du\right]^3 = A^2\left[\frac{\pi}{2(\alpha+\beta)}\right]^{3/2} \\
A &= \left[\frac{2(\alpha+\beta)}{\pi}\right]^{3/4}
\end{aligned}$$

Therefore we can write a normalized Gaussian $1s$ function in p, where $p = \alpha + \beta$:

$$\phi_{1s}(\alpha + \beta, \mathbf{r} - \mathbf{R}_p) = \left[\frac{2(\alpha + \beta)}{\pi}\right]^{3/4} e^{-(\alpha + \beta)(\mathbf{r} - \mathbf{R}_p)^2}$$

Substituting into Equation 1 gives

$$\phi_{1s}(\alpha, \mathbf{r} - \mathbf{R}_A)\phi_{1s}(\beta, \mathbf{r} - \mathbf{R}_B) = \left(\frac{4\alpha\beta}{\pi^2}\right)^{3/4}\left[\frac{\pi}{2(\alpha + \beta)}\right]^{3/4} e^{-\alpha\beta|\mathbf{R}_A - \mathbf{R}_B|^2/(\alpha + \beta)}\phi_{1s}(p, \mathbf{r} - \mathbf{R}_p)$$

$$= \left[\frac{2\alpha\beta}{(\alpha + \beta)\pi}\right]^{3/4} e^{-\alpha\beta|\mathbf{R}_A - \mathbf{R}_B|^2/(\alpha + \beta)}\phi_{1s}(p, \mathbf{r} - \mathbf{R}_p)$$

$$= K_{AB}\phi_{1s}(p, \mathbf{r} - \mathbf{R}_p))$$

where

$$K_{AB} = \left[\frac{2\alpha\beta}{(\alpha + \beta)\pi}\right]^{3/4} e^{-\frac{\alpha\beta}{\alpha + \beta}|\mathbf{R}_A - \mathbf{R}_B|^2}$$

11–8. Plot the product of the two (unnormalized) Gaussian functions

$$\phi_1 = e^{-2(x-1)^2} \qquad \text{and} \qquad \phi_2 = e^{-3(x-2)^2}$$

Interpret the result.

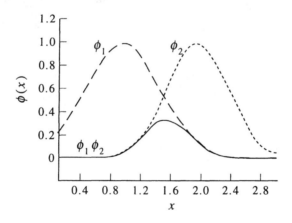

The product of these functions (shown by the solid line in the figure) is a Gaussian function centered at 1.5.

11–9. Using the result of Problem 11–7, show that the overlap integral of the two normalized Gaussian functions

$$\phi_{1s} = \left(\frac{2\alpha}{\pi}\right)^{3/4} e^{-\alpha|\mathbf{r} - \mathbf{R}_A|^2} \quad \text{and} \quad \phi_{1s} = \left(\frac{2\beta}{\pi}\right)^{3/4} e^{-\beta|\mathbf{r} - \mathbf{R}_B|^2}$$

is

$$S(|\mathbf{R}_A - \mathbf{R}_B|) = \left[\frac{4\alpha\beta}{(\alpha + \beta)^2}\right]^{3/4} e^{-\frac{\alpha\beta|\mathbf{R}_A - \mathbf{R}_B|^2}{\alpha + \beta}}$$

Plot this result as a function of $|\mathbf{R}_A - \mathbf{R}_B|$.

Recall from Problem 11–7 that we can write

$$\phi_{1s}(p, \mathbf{r} - \mathbf{R}_p) = \left[\frac{2(\alpha + \beta)}{\pi}\right]^{3/4} e^{-(\alpha+\beta)(\mathbf{r}-\mathbf{R}_p)^2}$$

where p and $\mathbf{R}_p$ are defined in Problem 11–7. Then (using the result of Problem 11–7)

$$\begin{aligned}
S(|\mathbf{R}_A - \mathbf{R}_B|) &= \int \phi_{1s}(\alpha, \mathbf{r} - \mathbf{R}_A)\phi_{1s}(\beta, \mathbf{r} - \mathbf{R}_B)d\mathbf{r} \\
&= \int K_{AB}\phi_{1s}(p, \mathbf{r} - \mathbf{R}_p) = K_{AB} \int \phi_{1s}(p, \mathbf{r} - \mathbf{R}_p)d\mathbf{r} \\
&= K_{AB} \left[\frac{2(\alpha + \beta)}{\pi}\right]^{3/4} \int e^{-(\alpha+\beta)(\mathbf{r}-\mathbf{r}_p)^2}d\mathbf{r}
\end{aligned}$$

The integral is given by

$$\int_{-\infty}^{\infty}\int_{-\infty}^{\infty}\int_{-\infty}^{\infty} dxdydz\, e^{-(\alpha+\beta)(x^2+y^2+z^2)} = \left[2\int_0^{\infty} e^{-(\alpha+\beta)u^2}du\right]^3 = \left(\frac{\pi}{\alpha + \beta}\right)^{3/2}$$

so

$$\begin{aligned}
S &= \left[\frac{2\alpha\beta}{(\alpha + \beta)\pi}\right]^{3/4} e^{-\alpha\beta|\mathbf{R}_A-\mathbf{R}_B|^2/(\alpha+\beta)} \left[\frac{2(\alpha + \beta)}{\pi}\right]^{3/4} \left(\frac{\pi}{\alpha + \beta}\right)^{3/2} \\
&= \left[\frac{4\alpha\beta}{(\alpha + \beta)^2}\right]^{3/4} e^{-\alpha\beta|\mathbf{R}_A-\mathbf{R}_B|^2/(\alpha+\beta)}
\end{aligned}$$

The overlap function $S(\mathbf{R}_A - \mathbf{R}_B)$ decays as a Gaussian function of the distance between the two centers.

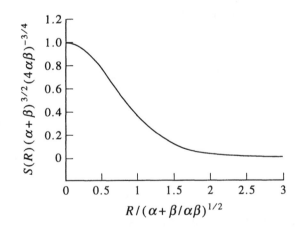

11–10. One criterion for the best possible "fit" of a Gaussian function to a Slater orbital is a fit that minimizes the integral of the square of their difference. For example, we can find the optimal value of α in $\phi_{1s}^{GF}(r, \alpha)$ by minimizing

$$I = \int d\mathbf{r}[\phi_{1s}^{STO}(r, 1.00) - \phi_{1s}^{GF}(r, \alpha)]^2$$

with respect to α. If the two functions $\phi_{1s}^{STO}(r, 1.00)$ and $\phi_{1s}^{GF}(r, \alpha)$ are normalized, show that minimizing I is equivalent to maximizing the overlap integral of $\phi_{1s}^{STO}(r, 1.00)$ and $\phi_{1s}^{GF}(r, \alpha)$:

$$S = \int d\mathbf{r}\, \phi_{1s}^{STO}(r, 1.00)\phi_{1s}^{GF}(r, \alpha)$$

$$I = \int d\mathbf{r}[\phi_{1s}^{STO}(r, 1.00) - \phi_{1s}^{GF}(r, \alpha)]^2$$

$$= \int d\mathbf{r}\left[\phi_{1s}^{STO}(r, 1.00)\right]^2 - 2\int d\mathbf{r}\, \phi_{1s}^{STO}(r, 1.00)\phi_{1s}^{GF}(r, \alpha) + \int d\mathbf{r}\left[\phi_{1s}^{GF}(r, \alpha)\right]^2$$

$$= 2 - 2\int d\mathbf{r}\, \phi_{1s}^{STO}(r, 1.00)\phi_{1s}^{GF}(r, \alpha) = 2 - 2S$$

where we have used the fact that both functions are normalized to write the last equality. Minimizing I is then equivalent to maximizing S.

11–11. Show that S in Problem 11–10 is given by

$$S = 4\pi^{1/2}\left(\frac{2\alpha}{\pi}\right)^{3/4}\int_0^\infty r^2 e^{-r}e^{-\alpha r^2}\, dr$$

Using a numerical integration computer program such as Mathematica or MathCad, show that the following results are correct:

α	S
0.10	0.8642
0.15	0.9367
0.20	0.9673
0.25	0.9776
0.30	0.9772
0.35	0.9706
0.40	0.9606

These numbers show that the maximum occurs around $\alpha = 0.25$. A more detailed calculation would show that the maximum actually occurs at $\alpha = 0.27095$. Thus, the normalized Gaussian $1s$ function $\phi_{1s}^{GF}(r, 0.2709)$ is an optimal fit to the $1s$ Slater orbital $\phi_{1s}^{STO}(r, 1.00)$.

We substitute Equations 11.6 and 11.7 into the expression for S:

$$S = \int d\mathbf{r}\, \phi_{1s}^{STO}(r, \zeta)\phi_{1s}^{GF}(r, \alpha)$$

$$= \int \left(\frac{\zeta^3}{\pi}\right)^{1/2} e^{-\zeta r}\left(\frac{2\alpha}{\pi}\right)^{3/4} e^{-\alpha r^2}\, d\tau$$

Taking $\zeta = 1.00$ gives

$$S = \left(\frac{1}{\pi}\right)^{1/2}\left(\frac{2\alpha}{\pi}\right)^{3/4}\int_0^\pi \sin\theta\, d\theta \int_0^{2\pi} d\phi \int_0^\infty r^2\, dr\, e^{-r}e^{-\alpha r^2}$$

$$= 4\pi^{1/2}\left(\frac{2\alpha}{\pi}\right)^{3/4}\int_0^\infty r^2\, dr\, e^{-r}e^{-\alpha r^2}$$

11–12. Compare $\phi_{1s}^{STO}(r, 1.00)$ and $\phi_{1s}^{GF}(r, 0.27095)$ graphically by plotting them on the same graph.

From Equations 11.6 and 11.7,

$$\phi_{1s}^{STO}(r, 1.00) = \pi^{-1/2}e^{-r} \qquad \phi_{1s}^{GF}(r, 0.27095) = \left(\frac{2(0.27095)}{\pi}\right)^{3/4} e^{-0.27095r^2}$$

We can now plot these functions versus r:

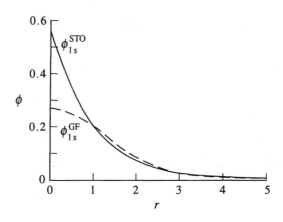

11–13. In Problems 11–11 and 11–12, we discussed a one-term Gaussian fit to a $1s$ Slater orbital $\phi_{1s}^{STO}(r, 1.00)$. Can we use the result of Problem 11–11 to find the optimal Gaussian fit to a $1s$ Slater orbital with a different orbital exponent, $\phi_{1s}^{GF}(r, \zeta)$? The answer is "yes." To see how, start with the overlap integral of $\phi_{1s}^{STO}(r, \zeta)$ and $\phi_{1s}^{GF}(r, \beta)$:

$$S = 4\pi^{1/2}\left(\frac{2\beta}{\pi}\right)^{3/4}\zeta^{3/2}\int_0^\infty r^2 e^{-\zeta r}e^{-\beta r^2}\,dr$$

Now let $u = \zeta r$ to get

$$S = 4\pi^{1/2}\left(\frac{2\beta/\zeta^2}{\pi}\right)^{3/4}\int_0^\infty u^2 e^{-u}e^{-(\beta/\zeta^2)u^2}\,du$$

Compare this result for S with that in Problem 11–11 to show that $\beta = \alpha\zeta^2$ or, in more detailed notation,

$$\alpha(\zeta = \zeta) = \alpha(\zeta = 1.00) \times \zeta^2$$

If we let $u = \zeta r$, $r = u/\zeta$ and $dr = du/\zeta$. Then we can write S as

$$\begin{aligned}
S &= 4\pi^{1/2}\left(\frac{2\beta}{\pi}\right)^{3/4}\zeta^{3/2}\int_0^\infty r^2 e^{-\zeta r}e^{-\beta r^2}\,dr \\
&= 4\pi^{1/2}\left(\frac{2\beta}{\pi}\right)^{3/4}\zeta^{3/2}\int_0^\infty \frac{u^2}{\zeta^2}e^{-u}e^{-\beta u^2/\zeta^2}\frac{du}{\zeta} \\
&= 4\pi^{1/2}\left(\frac{2\beta}{\pi}\right)^{3/4}\frac{\zeta^{3/2}}{\zeta^3}\int_0^\infty u^2 e^{-u}e^{-(\beta/\zeta^2)u^2}\,du
\end{aligned}$$

This result is equivalent to that found in Problem 11–11 if

$$\frac{\beta}{\zeta^2} = \alpha \qquad \text{or} \qquad \beta = \alpha\zeta^2$$

Therefore,

$$\alpha(\zeta = \zeta) = \alpha(\zeta = 1.00) \times \zeta^2$$

11–14. Use the result of Problem 11–13 to verify the value of α used in Equation 11.5 and Figure 11.1.

In Equation 11.5 and Figure 11.1 $\alpha = 0.4166$ and $\zeta = 1.24$. We substitute this value of ζ into the result of Problem 11–13 and use the result of Problem 11–11 for $\alpha(\zeta = 1.00)$ to find

$$\alpha(\zeta = 1.24) = (0.27095) \times (1.24)^2 = 0.4166$$

which is the value of α used in Equation 11.5.

11–15. Because of the scaling law developed in Problem 11–13, Gaussian fits are usually made with respect to a Slater orbital with $\zeta = 1.00$ and then the various Gaussian exponents are scaled according to $\alpha(\zeta = \zeta) = \alpha(\zeta = 1.00) \times \zeta^2$. Given the fit

$$\phi_{1s}^{STO-3G}(r, 1.0000) = 0.4446\phi_{1s}^{GF}(r, 0.10982)$$
$$+ 0.5353\phi_{1s}^{GF}(r, 0.40578)$$
$$+ 0.1543\phi_{1s}^{GF}(r, 2.2277)$$

verify Equation 11.8.

Equation 11.8 states that, for $\zeta = 1.24$,

$$\phi_{1s}(r) = \sum_{i=1}^{3} d_{1si}\phi_{1s}^{GF}(r, \alpha_{1si})$$
$$= 0.4446\phi_{1s}^{GF}(r, 0.1688) + 0.5353\phi_{1s}^{GF}(r, 0.6239) + 0.1543\phi_{1s}^{GF}(r, 3.425)$$

The fit given is for $\zeta = 1.0000$. Using the scaling factor,

$$\alpha_1(\zeta = 1.24) = (0.10982)(1.24)^2 = 0.1688$$
$$\alpha_2(\zeta = 1.24) = (0.40578)(1.24)^2 = 0.6239$$
$$\alpha_3(\zeta = 1.24) = (2.2277)(1.24)^2 = 3.425$$

which are the coefficients used in Equation 11.8.

11–16. The Gaussian function exponents and expansion coefficients for the valence shell orbitals of chlorine are as follows:

$\alpha_{3si} = \alpha_{3pi}$	d_{3si}	$\alpha'_{3s} = \alpha'_{3p}$	d_{3p}
3.18649	−2.51830	1.42657	−1.42993
1.19427	6.15890		3.23572
4.20377	1.06018		7.43507

Write the expression for the Gaussian functions corresponding to the $3s$ and $3p$ atomic orbitals of chlorine. Plot the function for the $3s$ orbital for several values of the expansion coefficient for the α'_{3s} term.

$$\phi_{3s} = \sum_{i=1}^{3} d_{3si} \phi_{3si}^{GF}(r, \alpha_{3si}) + d'_{3s} \phi_{3s}^{GF}(r, \alpha'_{3s})$$

$$= -2.51830 \left[\frac{2(3.18649)}{\pi} \right]^{3/4} e^{-3.18649r^2}$$

$$+6.15890 \left[\frac{2(1.19427)}{\pi} \right]^{3/4} e^{-1.19427r^2}$$

$$+1.06018 \left[\frac{2(4.20377)}{\pi} \right]^{3/4} e^{-4.20377r^2}$$

$$+d'_{3s} \left[\frac{2(1.42657)}{\pi} \right]^{3/4} e^{-1.42657r^2}$$

$$\phi_{3p} = \sum_{i=1}^{3} d_{3pi} \phi_{3pi}^{GF}(r, \alpha_{3pi}) + d'_{3p} \phi_{3p}^{GF}(r, \alpha'_{3p})$$

$$= -1.42993 \left[\frac{2(3.18649)}{\pi} \right]^{3/4} e^{-3.18649r^2}$$

$$+3.23572 \left[\frac{2(1.19427)}{\pi} \right]^{3/4} e^{-1.19427r^2}$$

$$+7.43507 \left[\frac{2(4.20377)}{\pi} \right]^{3/4} e^{-4.20377r^2}$$

$$+d'_{3p} \left[\frac{2(1.42657)}{\pi} \right]^{3/4} e^{-1.42657r^2}$$

Below we plot ϕ_{3s} for several values of d'_{3s}:

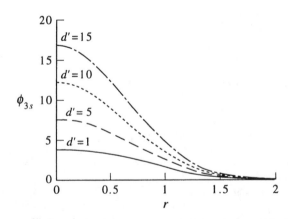

11–17. The input file to a computational quantum chemistry program must specify the coordinates of the atoms that comprise the molecule. Determine a set of Cartesian coordinates of the atoms in the molecule CH_4. The HCH bond angle is $109.5°$ and the C–H bond length is 109.1 pm. (*Hint:* Use the figure in Problem 10–7.)

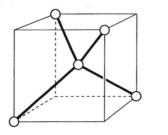
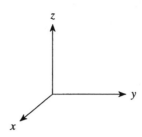

This figure (from Problem 10–7) represents a tetrahedral molecule, so we can allow the central atom to be a carbon atom and the atoms at the four vertices to be hydrogen atoms and thus represent methane. If we assign the origin of our coordinate system to be the carbon atom and allow the length of one edge of the cube to be $2a$, the coordinates of the four hydrogen atoms are (a, a, a), $(-a, -a, a)$, $(a, -a, -a)$, and $(-a, -a, -a)$. As we found in Problem 10–7, the bond length is $\sqrt{3}/2$ times the length of one edge of the cube, so

$$\frac{109.1 \text{ pm}}{\sqrt{3}} = 63.00 \text{ pm} = a$$

We now have a set of Cartesian coordinates assigned to all atoms in the molecule CH_4.

11–18. The input file to a computational quantum chemistry program must specify the coordinates of the atoms that comprise the molecule. Determine a set of Cartesian coordinates of the atoms in the molecule CH_3Cl. The HCH bond angle is $110.0°$ and the C–H and C–Cl bond lengths are 109.6 and 178.1 pm, respectively. (*Hint*: locate the origin at the carbon atom.)

Remember that only a rough estimate of the coordinates of the atoms is needed, since the molecular geometry is optimized as part of the computational program. Since this is the case, we can use essentially the same coordinates for CH_3Cl as those we found in the previous problem for CH_4, substituting a chlorine atom for one of the hydrogen atoms. Since the chlorine atom is farther from the carbon atom than the hydrogen atom it replaces would be, we might take that into account in determining a set of Cartesian coordinates. One possible set is shown below.

	x/pm	y/pm	z/pm
C	0	0	0
H	63	63	63
H	−63	−63	63
H	63	−63	−63
Cl	−100	−100	−100

11–19. The calculated vibrational frequencies and bond lengths for three diatomic molecules are listed below.

Molecule	Calculated values (6-31G*)	
	Frequency/cm^{-1}	R_e/pm
H_2	4647	73.2
CO	2438	111.4
N_2	2763	107.9

Determine the force constants that correspond to these vibrational frequencies. How do these values compare with the data in Table 5.1? How do the calculated bond lengths compare with the experimental values (also Table 5.1)? Why do you think the bond-length calculations show a higher accuracy than the vibrational-frequency calculations?

For a harmonic oscillator,

$$\tilde{\nu} = \frac{1}{2\pi c} \left(\frac{k}{\mu} \right)^{1/2} \tag{5.29}$$

$$(2\pi c \tilde{\nu})^2 \mu = k$$

Using the formula

$$\mu = \frac{m_1 m_2}{m_1 + m_2}$$

we can calculate the reduced masses of the three molecules listed above as $\mu_{H_2} = 8.38 \times 10^{-28}$ kg, $\mu_{CO} = 1.140 \times 10^{-26}$ kg, and $\mu_{N_2} = 1.163 \times 10^{-26}$ kg. Then

$$k_{H_2} = \left[2\pi (2.998 \times 10^8 \text{ m·s}^{-1})(464\,700 \text{ m}^{-1}) \right]^2 (8.38 \times 10^{-28} \text{ kg})$$

$$= 642 \text{ kg·s}^{-2} = 642 \text{ N·m}^{-1}$$

$$k_{CO} = \left[2\pi (2.998 \times 10^8 \text{ m·s}^{-1})(243\,800 \text{ m}^{-1}) \right]^2 (1.140 \times 10^{-26} \text{ kg})$$

$$= 2404 \text{ kg·s}^{-2} = 2404 \text{ N·m}^{-1}$$

$$k_{NO} = \left[2\pi (2.998 \times 10^8 \text{ m·s}^{-1})(276\,300 \text{ m}^{-1}) \right]^2 (1.163 \times 10^{-26} \text{ kg})$$

$$= 3150 \text{ kg·s}^{-2} = 3150 \text{ N·m}^{-1}$$

A table combining experimental results from Table 5.1, calculated results from the problem text, and the approximate percentage difference between them is presented below:

	k_{exp}/N·m^{-1}	k_{calc}/N·m^{-1}	Difference	r_{exp}/pm	r_{calc}/pm	Difference
H$_2$	510	642	25%	74.2	73	1.6%
CO	1857	2404	29%	112.8	111.4	1.2%
NO	1550	3150	40%	115.1	107.9	6.2%

To accurately account for the shape of a potential energy surface near its minimum requires a large basis set. The minimum istelf, however, can usually be found using a comparatively small basis set. Because k is sensitive to the curvature of the potential, accurate calculations of k require larger basis sets than the 6-31G* basis set.

11–20. Normalize the following Gaussian functions

a. $\phi(r) = xe^{-\alpha r^2}$

b. $\phi(r) = x^2 e^{-\alpha r^2}$

In both cases, we write $\phi(r)$ in spherical coordinates (MathChapter D) and then apply the normalization condition to the normalized function $A\phi(r)$.

a. The normalization condition is

$$\int d\tau A^2 x^2 e^{-2\alpha r^2} = 1$$

where A is the normalization constant. In spherical coordinates,

$$1 = \int d\tau A^2 r^2 \sin^2\theta \cos^2\phi e^{-2\alpha r^2}$$

$$= A^2 \int_0^\infty dr r^4 e^{-2\alpha r^2} \int_0^\pi d\theta \sin^3\theta \int_0^{2\pi} d\phi \cos^2\phi$$

$$\frac{1}{A^2} = \frac{3}{8(2\alpha)^2} \left(\frac{\pi}{2\alpha}\right)^{1/2} \left(\frac{4}{3}\right)(\pi)$$

$$= \frac{\pi^{3/2}}{2^{7/2}\alpha^{5/2}}$$

so

$$A = \left(\frac{128\alpha^5}{\pi^3}\right)^{1/4}$$

b. The normalization condition is

$$\int d\tau A^2 x^4 e^{-2\alpha r^2} = 1$$

where A is the normalization constant. In spherical coordinates,

$$1 = \int d\tau A^2 r^4 \sin^4\theta \cos^4\phi e^{-2\alpha r^2}$$

$$= A^2 \int_0^\infty dr r^6 e^{-2\alpha r^2} \int_0^\pi d\theta \sin^5\theta \int_0^{2\pi} d\phi \cos^4\phi$$

$$\frac{1}{A^2} = \frac{15}{16(2\alpha)^3} \left(\frac{\pi}{2\alpha}\right)^{1/2} \left(\frac{16}{15}\right)\left(\frac{3\pi}{4}\right)$$

$$= \frac{3\pi^{3/2}}{4(2\alpha)^{7/2}}$$

so

$$A = \left(\frac{2048\alpha^7}{9\pi^3}\right)^{1/4}$$

11–21. Which hydrogen atomic orbital corresponds to the following normalized Gaussian orbital?

$$G(x, y, z; \alpha) = \left(\frac{128\alpha^5}{\pi^3}\right)^{1/4} y e^{-\alpha r^2}$$

How many radial and angular nodes does the above function have? Is this result what you would expect for the corresponding hydrogenic function?

Recall that a Gaussian orbital has the form

$$G_{nlm}(r, \theta, \phi) = N_n r^{n-1} e^{-\alpha r^2} Y_l^m(\theta, \phi) \tag{11.7}$$

Because $G(\mathbf{r}, \alpha)$ has a y ($r\sin\theta\sin\phi$) component, $n = 2$ and $Y_l^m = Y_1^{-1}$, corresponding to a $2p_y$ orbital. There is one nodal plane ($y = 0$) in the Gaussian orbital ($y = 0$), just as is true for the $2p_y$ hydrogenic orbital. There are no radial nodes in either orbital.

11–22. Using Equations 6.62 for the spherical harmonic components of ϕ_{2p_x} and ϕ_{2p_y}, show that the Slater orbitals for the $2p_x$, $2p_y$, and $2p_z$ orbitals are given by the formulas in Example 11–4. Recall that the $2p_x$ and $2p_y$ orbitals are given by Equations 6.62.

We can use the $2p_x$ and $2p_y$ orbitals given by Equations 6.62 in place of Y_1^1 and Y_1^{-1}:

$$p_x = \frac{1}{\sqrt{2}} \left(Y_1^1 + Y_1^{-1} \right) = \left(\frac{3}{4\pi} \right)^{1/2} \sin\theta \cos\phi$$

$$p_y = \frac{1}{\sqrt{2}} \left(Y_1^1 - Y_1^{-1} \right) = \left(\frac{3}{4\pi} \right)^{1/2} \sin\theta \sin\phi$$

For $n = 2$, the Slater-type orbitals become (from Equation 11.2)

$$\phi_{nlm}^{STO} = \frac{(2\zeta)^{n+1/2}}{[(2n)!]^{1/2}} r^{n-1} e^{-\zeta r} Y_l^m(\theta, \phi)$$

$$\phi_{2p_x}^{STO} = \left(\frac{2^5 \zeta^5}{24} \right)^{1/2} r e^{-\zeta r} \left(\frac{3}{4\pi} \right)^{1/2} \sin\theta \cos\phi$$

$$= \left(\frac{\zeta^5}{\pi} \right)^{1/2} r \sin\theta \cos\phi\, e^{-\zeta r} = \left(\frac{\zeta^5}{\pi} \right)^{1/2} x e^{-\zeta r}$$

$$\phi_{2p_y}^{STO} = \left(\frac{2^5 \zeta^5}{24} \right)^{1/2} r e^{-\zeta r} \left(\frac{3}{4\pi} \right)^{1/2} \sin\theta \sin\phi$$

$$= \left(\frac{\zeta^5}{\pi} \right)^{1/2} r \sin\theta \sin\phi\, e^{-\zeta r} = \left(\frac{\zeta^5}{\pi} \right)^{1/2} y e^{-\zeta r}$$

$$\phi_{2p_z}^{STO} = \left(\frac{2^5 \zeta^5}{24} \right)^{1/2} r e^{-\zeta r} \left(\frac{3}{4\pi} \right)^{1/2} \cos\theta$$

$$= \left(\frac{\zeta^5}{\pi} \right)^{1/2} r \cos\theta\, e^{-\zeta r} = \left(\frac{\zeta^5}{\pi} \right)^{1/2} z e^{-\zeta r}$$

These are the equations found in Example 11–4.

11–23. Consider the normalized functions

$$G_1(x, y, z; \alpha) = \left(\frac{2048\alpha^7}{9\pi^3} \right)^{1/4} x^2 e^{-\alpha r^2}$$

$$G_2(x, y, z; \alpha) = \left(\frac{2048\alpha^7}{9\pi^3} \right)^{1/4} y^2 e^{-\alpha r^2}$$

$$G_3(x, y, z; \alpha) = \left(\frac{2048\alpha^7}{9\pi^3} \right)^{1/4} z^2 e^{-\alpha r^2}$$

Which hydrogen atomic orbital corresponds to the linear combination $G_1(x, y, z; \alpha) - G_2(x, y, z; \alpha)$?

$$G_1(\mathbf{r}, \alpha) - G_2(\mathbf{r}, \alpha) = \left(\frac{2048\alpha^7}{9\pi^3} \right)^{1/4} (x^2 - y^2) e^{-\alpha r^2}$$

This corresponds to the $3d_{x^2-y^2}$ hydrogen atomic orbital.

11–24. What is meant by the phrase "triple-zeta basis set"?

A triple zeta basis set is one in which each atomic orbital is expressed as a sum of three Slater-type orbitals:

$$\phi(r) = \phi^{STO}(r, \zeta_1) + d_1 \phi^{STO}(r, \zeta_2) + d_2 \phi^{STO}(r, \zeta_3)$$

11–25. Part of the output of most computational programs is a list of numbers that comprise what is called Mulliken Population Analysis. This list assigns a net charge to each atom in the molecule. The value of this net charge is the difference between the charge of the isolated atom, Z, and the calculated charge on the bonded atom, q. Thus if $Z - q > 0$, the atom is assigned a net positive charge and if $Z - q < 0$, the atom is assigned a net negative charge. What would be the sum of the Mulliken Populations for the molecules H_2CO, CO_3^{2-}, and NH_4^+?

The sum of the Mulliken Populations for a molecule will be the net charge on the molecule. Therefore, the sum for H_2CO is 0, the sum for CO_3^{2-} is -2, and the sum for NH_4^+ is $+1$.

11–26. In this problem, we show that the Mulliken Populations (Problem 11–25) can be used to calculate the molecular dipole moment. Consider the formaldehyde molecule, H_2CO. The calculated bond lengths for the CO and CH bonds are 121.7 pm and 110.0 pm, respectively, and the optimized H–C–H bond angle was found to be 114.5°. Use this information along with the Mulliken Population Analysis shown below

to calculate the dipole moment of formaldehyde. The experimentally determined values for the bond lengths and bond angles are $R_{CO} = 120.8$ pm, $R_{CH} = 111.6$ pm and $\angle(HCH) = 116.5°$. What is the value of the dipole moment if you combine the experimental geometry and the calculated Mulliken Populations? How do your calculated dipole moments compare with the experimental value of 7.8×10^{-30} C·m?

We can use the following coordinate system:

When we use the calculated bond angle (57.25°), we find

$$\mu = e \sum_i z_i \mathbf{r}_i$$
$$= 2(0.0566e)(110.0 \times 10^{-12} \text{ m}) \cos(57.25°) - (-0.1879e)(121.7 \times 10^{-12} \text{ m})$$
$$= 4.743 \times 10^{-30} \text{ C} \cdot \text{m}$$

With the experimentally observed bond angle, we find $\mu = 4.702 \times 10^{-30}$ C·m. In both cases, this is about 40% below the experimental dipole moment.

11–27. The experimentally determined dipole moment of CO is 3.66×10^{-31} C·m, with the oxygen atom being positively charged. The Mulliken Populations from Hartree-Fock calculations using the STO-3G or the 6-31G* basis sets predict a dipole moment of 5.67×10^{-31} C·m and 1.30×10^{-30} C·m, respectively, and pointing in the opposite direction of the experimental results. The experimental and two calculated bond lengths are 112.8 pm, 114.6 pm, and 111.4 pm, respectively. Why do you think the bond-length calculation is significantly more accurate than the dipole-moment calculation?

The dipole moment requires an accurate knowledge of the electron density at each center. This requires accurate descriptions of molecular orbitals, which become more accurate representations of the electron densities as the size of the basis set used increases. The 6-31G* basis set is larger and more flexible than the STO-3G basis set and so gives a better result. It is not large enough to give an accurate dipole moment for CO, however.

11–28. The orbital energies calculated for formaldehyde using STO-3G and 3-21G basis sets are given below.

Orbital	STO-3G energy/E_h	3-21G energy/E_h
$1a_1$	−20.3127	−20.4856
$2a_1$	−11.1250	−11.2866
$3a_1$	−1.3373	−1.4117
$4a_1$	−0.8079	−0.8661
$1b_2$	−0.6329	−0.6924
$5a_1$	−0.5455	−0.6345
$1b_1$	−0.4431	−0.5234
$2b_2$	−0.3545	−0.4330
$2b_1$	0.2819	0.1486
$6a_1$	0.6291	0.2718
$3b_2$	0.7346	0.3653
$7a_1$	0.9126	0.4512

Determine the ground-state electronic configuration of formaldehyde. The photoelectron spectrum of formaldehyde is shown below.

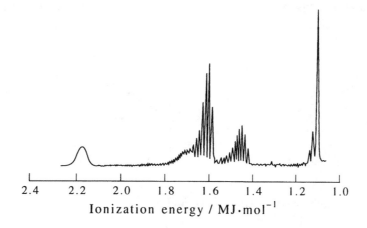

$$\text{Ionization energy / MJ} \cdot \text{mol}^{-1}$$

Assign the bands. Which calculated set of energies shows the best agreement with the photoelectron spectrum? Why is there such a large energy separation between the $1a_1$ and $2a_1$ orbitals? Predict the ionization energy and electron affinity of formaldehyde for each calculated set of energy levels. How do these compare with the experimental values?

There are sixteen electrons in formaldehyde (six from carbon, one from each hydrogen, and eight from oxygen). This gives a ground-state electronic configuration of

$$1a_1^2 2a_1^2 3a_1^2 4a_1^2 1b_2^2 5a_1^2 1b_1^2 2b_2^2$$

The band at approximately 1.5 MJ·mol^{-1} corresponds to the $2b_2$ electrons, the bands at 1.4 and 1.6 MJ·mol^{-1} arise from the ionization of electrons in the $1b_1$ and $5a_1$ orbitals, and the band at 1.7 MJ·mol^{-1} corresponds to the $1b_2$ electrons. The broad peak at about 2.2 MJ·mol^{-1} is probably due to the electrons in the $4a_1$ orbital. There is a large energy separation between the $1a_1$ and $2a_1$ orbitals because the $1a_1$ electrons are much closer to the nuclei than the $2a_1$ electrons are. The set of calculated energies showing the best agreement with the spectrum is the 3-21G set. Using this basis set, we can find

$$\text{IE} = -E_{2b_2} = 0.4330 E_{\text{h}} = 1.136 \text{ MJ} \cdot \text{mol}^{-1}$$

and

$$\text{EA} = -E_{2b_1} = -0.1486 E_{\text{h}} = -0.390 \text{ MJ} \cdot \text{mol}^{-1}$$

11–29. The units of dipole moment given by Gaussian 94 are called debyes (D), after the Dutch-American chemist, Peter Debye, who was awarded the Nobel Prize for chemistry in 1936 for his work on dipole moments. One debye is equal to 10^{-18} esu · cm where esu (*electrostatic units*) is a non-SI unit for electric charge. Given that the protonic charge is 4.803×10^{-10} esu, show that the conversion factor between debyes and C · m (coulomb · meters) is $1 \text{ D} = 3.33 \times 10^{-30}$ C · m.

$$1 \text{ D} = 1 \times 10^{-18} \text{ esu} \cdot \text{cm} \left(\frac{1.6022 \times 10^{-19} \text{ C}}{4.803 \times 10^{-10} \text{ esu}} \right) \left(\frac{1 \text{ m}}{100 \text{ cm}} \right) = 3.33 \times 10^{-30} \text{ C} \cdot \text{m}$$

11–30. Using the geometry and the charges given in Table 11.8, verify the value of the dipole moment of water.

We can calculate the dipole moment from the values in the table as we did in Problem 11–26. We will use the coordinate system

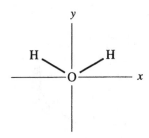

When we use the calculated bond angle, we find

$$\mu = e \sum z_i \mathbf{r}_i$$
$$= e \left\{ (94.7 \times 10^{-12} \text{ m}) \left[(0.41)(\sin 52.75°)\mathbf{i} + (0.41)(\cos(52.75°)\mathbf{j}] \right.$$
$$\left. + (94.7 \times 10^{-12} \text{ m}) [-(0.41)(\sin 52.75°) \mathbf{i} + (0.41)(\cos 52.75°)\mathbf{j}] \right\}$$
$$= 7.53 \times 10^{-30} \text{ C·m} = 2.3 \text{ D}$$

11–31. Using the geometry and the charges given in Table 11.9, verify the value of the dipole moment of ammonia.

We can calculate the dipole moment from the values in the table as we did in Problem 11–26. We will center the coordinate system at the nitrogen atom and allow the hydrogens to extend upward in the direction of the z-axis. The projection onto the xy-plane is shown in part (a) of the figure below:

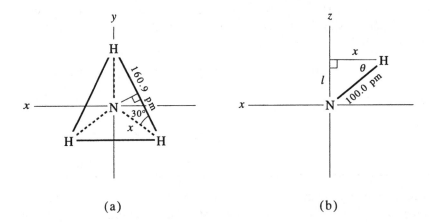

(a)   (b)

The distance on the xy-plane between the nitrogen atom and any of the hydrogen atoms is given by

$$x = \frac{160.9 \text{ pm}}{2} \left(\frac{1}{\cos 30°} \right) = 92.90 \text{ pm}$$

We now use this result to find the distance l and angle θ shown in part (b) of the figure:

$$l = \sqrt{(100.0 \text{ pm})^2 - (92.90 \text{ pm})^2} = 37.02 \text{ pm}$$
$$\theta = \cos^{-1} \left(\frac{92.90 \text{ pm}}{100.0 \text{ pm}} \right) = 21.72°$$

Finally, we calculate the dipole moment of ammonia:

$$\mu = e \sum z_i \mathbf{r}_i$$

$$= e\big[(0.37)(\cos 30°)(92.90 \text{ pm})\mathbf{i} - (0.37)(\cos 30°)(92.90 \text{ pm})\mathbf{i}$$

$$+(0.37)(92.90 \text{ pm})\mathbf{j} - 2(0.37)(\sin 30°)(92.90 \text{ pm})\mathbf{j} + 3(0.37)(37.02 \text{ pm})\mathbf{k}\big]$$

$$= 6.58 \times 10^{-30} \text{ C·m} = 2.0 \text{ D}$$

Matrices

PROBLEMS AND SOLUTIONS

F–1. Given the two matrices

$$A = \begin{pmatrix} 1 & 0 & -1 \\ -1 & 2 & 0 \\ 0 & 1 & 1 \end{pmatrix} \quad \text{and} \quad B = \begin{pmatrix} -1 & 1 & 0 \\ 3 & 0 & 2 \\ 1 & 1 & 1 \end{pmatrix}$$

form the matrices $C = 2A - 3B$ and $D = 6B - A$.

a. $C = 2A - 3B$

$$C = \begin{pmatrix} 2 & 0 & -2 \\ -2 & 4 & 0 \\ 0 & 2 & 2 \end{pmatrix} - \begin{pmatrix} -3 & 3 & 0 \\ 9 & 0 & 6 \\ 3 & 3 & 3 \end{pmatrix} = \begin{pmatrix} 5 & -3 & -2 \\ -11 & 4 & -6 \\ -3 & -1 & -1 \end{pmatrix}$$

b. $D = 6B - A$

$$D = \begin{pmatrix} -6 & 6 & 0 \\ 18 & 0 & 12 \\ 6 & 6 & 6 \end{pmatrix} - \begin{pmatrix} 1 & 0 & -1 \\ -1 & 2 & 0 \\ 0 & 1 & 1 \end{pmatrix} = \begin{pmatrix} -7 & 6 & 1 \\ 19 & -2 & 12 \\ 6 & 5 & 5 \end{pmatrix}$$

F–2. Given the three matrices

$$A = \frac{1}{2}\begin{pmatrix} 0 & 1 \\ 1 & 0 \end{pmatrix} \quad B = \frac{1}{2}\begin{pmatrix} 0 & -i \\ i & 0 \end{pmatrix} \quad C = \frac{1}{2}\begin{pmatrix} 1 & 0 \\ 0 & -1 \end{pmatrix}$$

show that $A^2 + B^2 + C^2 = \frac{3}{4}I$, where I is a unit matrix. Also show that

$$AB - BA = iC$$

$$BC - CB = iA$$

$$CA - AC = iB$$

a.

$$A^2 + B^2 + C^2 = \frac{1}{4}\begin{pmatrix} 0 & 1 \\ 1 & 0 \end{pmatrix}\begin{pmatrix} 0 & 1 \\ 1 & 0 \end{pmatrix} + \frac{1}{4}\begin{pmatrix} 0 & -i \\ i & 0 \end{pmatrix}\begin{pmatrix} 0 & -i \\ i & 0 \end{pmatrix}$$

$$+ \frac{1}{4}\begin{pmatrix} 1 & 0 \\ 0 & -1 \end{pmatrix}\begin{pmatrix} 1 & 0 \\ 0 & -1 \end{pmatrix}$$

$$= \frac{1}{4}\begin{pmatrix} 1 & 0 \\ 0 & 1 \end{pmatrix} + \frac{1}{4}\begin{pmatrix} 1 & 0 \\ 0 & 1 \end{pmatrix} + \frac{1}{4}\begin{pmatrix} 1 & 0 \\ 0 & 1 \end{pmatrix} = \frac{3}{4}\mathbf{I}$$

b.

$$AB - BA = \frac{1}{4}\begin{pmatrix} 0 & 1 \\ 1 & 0 \end{pmatrix}\begin{pmatrix} 0 & -i \\ i & 0 \end{pmatrix} - \frac{1}{4}\begin{pmatrix} 0 & -i \\ i & 0 \end{pmatrix}\begin{pmatrix} 0 & 1 \\ 1 & 0 \end{pmatrix}$$

$$= \frac{1}{4}\begin{pmatrix} i & 0 \\ 0 & -i \end{pmatrix} - \frac{1}{4}\begin{pmatrix} -i & 0 \\ 0 & i \end{pmatrix} = \frac{i}{2}\begin{pmatrix} 1 & 0 \\ 0 & -1 \end{pmatrix} = iC$$

c.

$$BC - CB = \frac{1}{4}\begin{pmatrix} 0 & -i \\ i & 0 \end{pmatrix}\begin{pmatrix} 1 & 0 \\ 0 & -1 \end{pmatrix} - \frac{1}{4}\begin{pmatrix} 1 & 0 \\ 0 & -1 \end{pmatrix}\begin{pmatrix} 0 & -i \\ i & 0 \end{pmatrix}$$

$$= \frac{1}{4}\begin{pmatrix} 0 & i \\ i & 0 \end{pmatrix} - \frac{1}{4}\begin{pmatrix} 0 & -i \\ -i & 0 \end{pmatrix} = \frac{i}{2}\begin{pmatrix} 0 & 1 \\ 1 & 0 \end{pmatrix} = iA$$

d.

$$CA - AC = \frac{1}{4}\begin{pmatrix} 1 & 0 \\ 0 & -1 \end{pmatrix}\begin{pmatrix} 0 & 1 \\ 1 & 0 \end{pmatrix} - \frac{1}{4}\begin{pmatrix} 0 & 1 \\ 1 & 0 \end{pmatrix}\begin{pmatrix} 1 & 0 \\ 0 & -1 \end{pmatrix}$$

$$= \frac{1}{4}\begin{pmatrix} 0 & 1 \\ -1 & 0 \end{pmatrix} - \frac{1}{4}\begin{pmatrix} 0 & -1 \\ 1 & 0 \end{pmatrix} = \frac{i}{2}\begin{pmatrix} 0 & -i \\ i & 0 \end{pmatrix} = iB$$

F–3. Given the matrices

$$A = \frac{1}{\sqrt{2}}\begin{pmatrix} 0 & 1 & 0 \\ 1 & 0 & 1 \\ 0 & 1 & 0 \end{pmatrix} \quad B = \frac{1}{\sqrt{2}}\begin{pmatrix} 0 & -i & 0 \\ i & 0 & -i \\ 0 & i & 0 \end{pmatrix} \quad C = \begin{pmatrix} 1 & 0 & 0 \\ 0 & 0 & 0 \\ 0 & 0 & -1 \end{pmatrix}$$

show that

$$AB - BA = iC$$

$$BC - CB = iA$$

$$CA - AC = iB$$

and

$$A^2 + B^2 + C^2 = 2I$$

where I is a unit matrix.

---

a.

$$AB - BA = \frac{1}{2}\begin{pmatrix} 0 & 1 & 0 \\ 1 & 0 & 1 \\ 0 & 1 & 0 \end{pmatrix}\begin{pmatrix} 0 & -i & 0 \\ i & 0 & -i \\ 0 & i & 0 \end{pmatrix} - \frac{1}{2}\begin{pmatrix} 0 & -i & 0 \\ i & 0 & -i \\ 0 & i & 0 \end{pmatrix}\begin{pmatrix} 0 & 1 & 0 \\ 1 & 0 & 1 \\ 0 & 1 & 0 \end{pmatrix}$$

$$= \frac{1}{2}\begin{pmatrix} i & 0 & -i \\ 0 & 0 & 0 \\ i & 0 & -i \end{pmatrix} - \frac{1}{2}\begin{pmatrix} -i & 0 & -i \\ 0 & 0 & 0 \\ i & 0 & i \end{pmatrix} = \frac{1}{2}\begin{pmatrix} 2i & 0 & 0 \\ 0 & 0 & 0 \\ 0 & 0 & -2i \end{pmatrix}$$

$$= i\begin{pmatrix} 1 & 0 & 0 \\ 0 & 0 & 0 \\ 0 & 0 & -1 \end{pmatrix} = iC$$

b.

$$BC - CB = \frac{1}{\sqrt{2}}\begin{pmatrix} 0 & -i & 0 \\ i & 0 & -i \\ 0 & i & 0 \end{pmatrix}\begin{pmatrix} 1 & 0 & 0 \\ 0 & 0 & 0 \\ 0 & 0 & -1 \end{pmatrix} - \frac{1}{\sqrt{2}}\begin{pmatrix} 1 & 0 & 0 \\ 0 & 0 & 0 \\ 0 & 0 & -1 \end{pmatrix}\begin{pmatrix} 0 & -i & 0 \\ i & 0 & -i \\ 0 & i & 0 \end{pmatrix}$$

$$= \frac{1}{\sqrt{2}}\begin{pmatrix} 0 & 0 & 0 \\ i & 0 & i \\ 0 & 0 & 0 \end{pmatrix} - \frac{1}{\sqrt{2}}\begin{pmatrix} 0 & -i & 0 \\ 0 & 0 & 0 \\ 0 & -i & 0 \end{pmatrix} = \frac{1}{\sqrt{2}}\begin{pmatrix} 0 & i & 0 \\ i & 0 & i \\ 0 & i & 0 \end{pmatrix}$$

$$= \frac{i}{\sqrt{2}}\begin{pmatrix} 0 & 1 & 0 \\ 1 & 0 & 1 \\ 0 & 1 & 0 \end{pmatrix} = iA$$

c.

$$CA - AC = \frac{1}{\sqrt{2}}\begin{pmatrix} 1 & 0 & 0 \\ 0 & 0 & 0 \\ 0 & 0 & -1 \end{pmatrix}\begin{pmatrix} 0 & 1 & 0 \\ 1 & 0 & 1 \\ 0 & 1 & 0 \end{pmatrix} - \frac{1}{\sqrt{2}}\begin{pmatrix} 0 & 1 & 0 \\ 1 & 0 & 1 \\ 0 & 1 & 0 \end{pmatrix}\begin{pmatrix} 1 & 0 & 0 \\ 0 & 0 & 0 \\ 0 & 0 & -1 \end{pmatrix}$$

$$= \frac{1}{\sqrt{2}}\begin{pmatrix} 0 & 1 & 0 \\ 0 & 0 & 0 \\ 0 & -1 & 0 \end{pmatrix} - \frac{1}{\sqrt{2}}\begin{pmatrix} 0 & 0 & 0 \\ 1 & 0 & -1 \\ 0 & 0 & 0 \end{pmatrix} = \frac{1}{\sqrt{2}}\begin{pmatrix} 0 & 1 & 0 \\ -1 & 0 & 1 \\ 0 & -1 & 0 \end{pmatrix}$$

$$= \frac{i}{\sqrt{2}}\begin{pmatrix} 0 & -i & 0 \\ i & 0 & -i \\ 0 & i & 0 \end{pmatrix} = iB$$

d.

$$A^2 + B^2 + C^2 = \frac{1}{2}\begin{pmatrix} 0 & 1 & 0 \\ 1 & 0 & 1 \\ 0 & 1 & 0 \end{pmatrix}\begin{pmatrix} 0 & 1 & 0 \\ 1 & 0 & 1 \\ 0 & 1 & 0 \end{pmatrix} + \frac{1}{2}\begin{pmatrix} 0 & -i & 0 \\ i & 0 & -i \\ 0 & i & 0 \end{pmatrix}\begin{pmatrix} 0 & -i & 0 \\ i & 0 & -i \\ 0 & i & 0 \end{pmatrix}$$

$$+ \begin{pmatrix} 1 & 0 & 0 \\ 0 & 0 & 0 \\ 0 & 0 & -1 \end{pmatrix}\begin{pmatrix} 1 & 0 & 0 \\ 0 & 0 & 0 \\ 0 & 0 & -1 \end{pmatrix}$$

$$= \frac{1}{2}\begin{pmatrix} 1 & 0 & 1 \\ 0 & 2 & 0 \\ 1 & 0 & 1 \end{pmatrix} + \frac{1}{2}\begin{pmatrix} 1 & 0 & -1 \\ 0 & 2 & 0 \\ -1 & 0 & 1 \end{pmatrix} + \begin{pmatrix} 1 & 0 & 0 \\ 0 & 0 & 0 \\ 0 & 0 & 1 \end{pmatrix}$$

$$= \frac{1}{2}\begin{pmatrix} 4 & 0 & 0 \\ 0 & 4 & 0 \\ 0 & 0 & 4 \end{pmatrix} = 2I$$

F–4. Do you see any similarity between the results of Problems F–2 and F–3 and the commutation relations involving the components of angular momentum?

Yes - the commutation relations for these sets of matrices behave like the components of angular momentum.

F–5. A three-dimensional rotation about the z axis can be represented by the matrix

$$R = \begin{pmatrix} \cos\theta & -\sin\theta & 0 \\ \sin\theta & \cos\theta & 0 \\ 0 & 0 & 1 \end{pmatrix}$$

Show that

$$\det R = |R| = 1$$

Also show that

$$R^{-1} = R(-\theta) = \begin{pmatrix} \cos\theta & \sin\theta & 0 \\ -\sin\theta & \cos\theta & 0 \\ 0 & 0 & 1 \end{pmatrix}$$

By definition, $\det R = |R|$. To evaluate $|R|$, we expand along the third column to obtain

$$|R| = \begin{vmatrix} \cos\theta & -\sin\theta & 0 \\ \sin\theta & \cos\theta & 0 \\ 0 & 0 & 1 \end{vmatrix} = 1\begin{vmatrix} \cos\theta & -\sin\theta \\ \sin\theta & \cos\theta \end{vmatrix}$$

$$= \cos^2\theta + \sin^2\theta = 1$$

The function $\sin\theta$ is an odd function, so $\sin(-\theta) = -\sin\theta$. The function $\cos\theta$ is an even function, so $\cos(-\theta) = \cos\theta$. Therefore,

$$R(-\theta) = \begin{pmatrix} \cos\theta & \sin\theta & 0 \\ -\sin\theta & \cos\theta & 0 \\ 0 & 0 & 1 \end{pmatrix}$$

Now consider the product

$$R(-\theta)R(\theta) = \begin{pmatrix} \cos\theta & \sin\theta & 0 \\ -\sin\theta & \cos\theta & 0 \\ 0 & 0 & 1 \end{pmatrix} \begin{pmatrix} \cos\theta & -\sin\theta & 0 \\ \sin\theta & \cos\theta & 0 \\ 0 & 0 & 1 \end{pmatrix} = \begin{pmatrix} 1 & 0 & 0 \\ 0 & 1 & 0 \\ 0 & 0 & 1 \end{pmatrix} = I$$

Therefore, $R(-\theta) = R^{-1}(\theta)$.

F–6. The *transpose* of a matrix A, which we denote by $\tilde{A}$, is formed by replacing the first row of A by its first column, its second row by its second column, etc. Show that this procedure is equivalent to the relation $\tilde{a}_{ij} = a_{ji}$. Show that the transpose of the matrix R given in Problem F–5 is

$$\tilde{R} = \begin{pmatrix} \cos\theta & \sin\theta & 0 \\ -\sin\theta & \cos\theta & 0 \\ 0 & 0 & 1 \end{pmatrix}$$

Note that $\tilde{R} = R^{-1}$. When $\tilde{R} = R^{-1}$, the matrix R is said to be *orthogonal*.

The element a_{ij} is defined to be the element in the ith row and jth column. If we interchange the rows and columns of a matrix, this operation will place the original element a_{ij} into the jth row and ith column. In other words, $\tilde{a}_{ij} = a_{ji}$. Finding the transpose of R (Problem F–5) involves replacing the rows by the corresponding columns, so

$$\tilde{R} = \begin{pmatrix} \cos\theta & \sin\theta & 0 \\ -\sin\theta & \cos\theta & 0 \\ 0 & 0 & 1 \end{pmatrix}$$

F–7. Given the matrices

$$C_3 = \begin{pmatrix} -\frac{1}{2} & -\frac{\sqrt{3}}{2} \\ \frac{\sqrt{3}}{2} & -\frac{1}{2} \end{pmatrix} \qquad \sigma_v = \begin{pmatrix} 1 & 0 \\ 0 & -1 \end{pmatrix}$$

$$\sigma_v' = \begin{pmatrix} -\frac{1}{2} & \frac{\sqrt{3}}{2} \\ \frac{\sqrt{3}}{2} & \frac{1}{2} \end{pmatrix} \qquad \sigma_v'' = \begin{pmatrix} -\frac{1}{2} & -\frac{\sqrt{3}}{2} \\ -\frac{\sqrt{3}}{2} & \frac{1}{2} \end{pmatrix}$$

show that

$$\sigma_v C_3 = \sigma_v'' \qquad C_3 \sigma_v = \sigma_v'$$

$$\sigma_v'' \sigma_v' = C_3 \qquad C_3 \sigma_v'' = \sigma_v$$

Calculate the determinant associated with each matrix. Calculate the trace of each matrix.

---

a.

$$\sigma_v C_3 = \begin{pmatrix} 1 & 0 \\ 0 & -1 \end{pmatrix} \begin{pmatrix} -\frac{1}{2} & -\frac{\sqrt{3}}{2} \\ \frac{\sqrt{3}}{2} & -\frac{1}{2} \end{pmatrix} = \begin{pmatrix} -\frac{1}{2} & -\frac{\sqrt{3}}{2} \\ -\frac{\sqrt{3}}{2} & \frac{1}{2} \end{pmatrix} = \sigma_v''$$

b.

$$C_3 \sigma_v = \begin{pmatrix} -\frac{1}{2} & -\frac{\sqrt{3}}{2} \\ \frac{\sqrt{3}}{2} & -\frac{1}{2} \end{pmatrix} \begin{pmatrix} 1 & 0 \\ 0 & -1 \end{pmatrix} = \begin{pmatrix} -\frac{1}{2} & \frac{\sqrt{3}}{2} \\ \frac{\sqrt{3}}{2} & \frac{1}{2} \end{pmatrix} = \sigma_v'$$

c.
$$\sigma_v''\sigma_v' = \begin{pmatrix} -\frac{1}{2} & -\frac{\sqrt{3}}{2} \\ -\frac{\sqrt{3}}{2} & \frac{1}{2} \end{pmatrix}\begin{pmatrix} -\frac{1}{2} & \frac{\sqrt{3}}{2} \\ \frac{\sqrt{3}}{2} & \frac{1}{2} \end{pmatrix} = \begin{pmatrix} -\frac{1}{2} & -\frac{\sqrt{3}}{2} \\ \frac{\sqrt{3}}{2} & -\frac{1}{2} \end{pmatrix} = C_3$$

d.
$$C_3\sigma_v'' = \begin{pmatrix} -\frac{1}{2} & -\frac{\sqrt{3}}{2} \\ \frac{\sqrt{3}}{2} & -\frac{1}{2} \end{pmatrix}\begin{pmatrix} -\frac{1}{2} & -\frac{\sqrt{3}}{2} \\ -\frac{\sqrt{3}}{2} & \frac{1}{2} \end{pmatrix} = \begin{pmatrix} 1 & 0 \\ 0 & -1 \end{pmatrix} = \sigma_v$$

Let Tr A be the trace of matrix A. Then

$$|C_3| = \frac{1}{4} + \frac{3}{4} = 1 \qquad \text{Tr } C_3 = -1$$

$$|\sigma_v| = -1 \qquad\qquad \text{Tr } \sigma_v = 0$$

$$|\sigma_v'| = -\frac{1}{4} - \frac{3}{4} = -1 \qquad \text{Tr } \sigma_v' = 0$$

$$|\sigma_v''| = -\frac{1}{4} - \frac{3}{4} = -1 \qquad \text{Tr } \sigma_v'' = 0$$

F–8. Which of the matrices in Problem F–7 are orthogonal (see Problem F–6)?

A matrix R is orthogonal if $\tilde{R} = R^{-1}$. In other words, R is orthogonal if $R^{-1}R = \tilde{R}R = I$ where I is the identity matrix.

a.
$$C_3 = \begin{pmatrix} -\frac{1}{2} & -\frac{\sqrt{3}}{2} \\ \frac{\sqrt{3}}{2} & -\frac{1}{2} \end{pmatrix} \qquad \tilde{C}_3 = \begin{pmatrix} -\frac{1}{2} & \frac{\sqrt{3}}{2} \\ -\frac{\sqrt{3}}{2} & -\frac{1}{2} \end{pmatrix}$$

$$\tilde{C}_3 C_3 = \begin{pmatrix} -\frac{1}{2} & \frac{\sqrt{3}}{2} \\ -\frac{\sqrt{3}}{2} & -\frac{1}{2} \end{pmatrix}\begin{pmatrix} -\frac{1}{2} & -\frac{\sqrt{3}}{2} \\ \frac{\sqrt{3}}{2} & -\frac{1}{2} \end{pmatrix} = \begin{pmatrix} 1 & 0 \\ 0 & 1 \end{pmatrix}$$

C_3 is orthogonal.

b.
$$\sigma_v = \begin{pmatrix} 1 & 0 \\ 0 & -1 \end{pmatrix} \qquad \tilde{\sigma}_v = \begin{pmatrix} 1 & 0 \\ 0 & -1 \end{pmatrix}$$

$$\tilde{\sigma}_v\sigma_v = \begin{pmatrix} 1 & 0 \\ 0 & -1 \end{pmatrix}\begin{pmatrix} 1 & 0 \\ 0 & -1 \end{pmatrix} = \begin{pmatrix} 1 & 0 \\ 0 & 1 \end{pmatrix}$$

σ_v is orthogonal.

c.
$$\sigma_v' = \begin{pmatrix} -\frac{1}{2} & \frac{\sqrt{3}}{2} \\ \frac{\sqrt{3}}{2} & \frac{1}{2} \end{pmatrix} \qquad \tilde{\sigma}_v' = \begin{pmatrix} -\frac{1}{2} & \frac{\sqrt{3}}{2} \\ \frac{\sqrt{3}}{2} & \frac{1}{2} \end{pmatrix}$$

$$\tilde{\sigma}_v'\sigma_v' = \begin{pmatrix} -\frac{1}{2} & \frac{\sqrt{3}}{2} \\ \frac{\sqrt{3}}{2} & \frac{1}{2} \end{pmatrix}\begin{pmatrix} -\frac{1}{2} & \frac{\sqrt{3}}{2} \\ \frac{\sqrt{3}}{2} & \frac{1}{2} \end{pmatrix} = \begin{pmatrix} 1 & 0 \\ 0 & 1 \end{pmatrix}$$

σ_v' is orthogonal.

d.

$$\sigma_v'' = \begin{pmatrix} -\frac{1}{2} & -\frac{\sqrt{3}}{2} \\ -\frac{\sqrt{3}}{2} & \frac{1}{2} \end{pmatrix} \qquad \tilde{\sigma}_v'' = \begin{pmatrix} -\frac{1}{2} & -\frac{\sqrt{3}}{2} \\ -\frac{\sqrt{3}}{2} & \frac{1}{2} \end{pmatrix}$$

$$\tilde{\sigma}_v''\sigma_v'' = \begin{pmatrix} -\frac{1}{2} & -\frac{\sqrt{3}}{2} \\ -\frac{\sqrt{3}}{2} & \frac{1}{2} \end{pmatrix}\begin{pmatrix} -\frac{1}{2} & -\frac{\sqrt{3}}{2} \\ -\frac{\sqrt{3}}{2} & \frac{1}{2} \end{pmatrix} = \begin{pmatrix} 1 & 0 \\ 0 & 1 \end{pmatrix}$$

σ_v'' is orthogonal.

All four matrices are orthogonal.

F–9. The inverse of a matrix A can be found by using the following procedure:

a. Replace each element of A by its cofactor in the corresponding determinant (see MathChapter E for a definition of a cofactor).

b. Take the transpose of the matrix obtained in step 1.

c. Divide each element of the matrix obtained in Step 2 by the determinant of A.

For example, if

$$A = \begin{pmatrix} 1 & 2 \\ 3 & 4 \end{pmatrix}$$

then det A = −2 and

$$A^{-1} = -\frac{1}{2}\begin{pmatrix} 4 & -2 \\ -3 & 1 \end{pmatrix}$$

Show that $AA^{-1} = A^{-1}A = I$. Use the above procedure to find the inverse of

$$A = \begin{pmatrix} \frac{1}{2} & \frac{1}{\sqrt{2}} \\ \frac{1}{\sqrt{2}} & 0 \end{pmatrix} \quad \text{and} \quad A = \begin{pmatrix} 0 & 2 & 3 \\ 1 & 1 & 1 \\ 2 & 0 & 1 \end{pmatrix}$$

Recall from MathChapter E that the cofactor, A_{ij}, of an element a_{ij} is a $(n-1) \times (n-1)$ determinant obtained by deleting the ith row and the jth column, multiplied by $(-1)^{i+j}$.

a. For the first matrix given, det A = $-\frac{1}{2}$. The cofactors are $A_{11} = 0$, $A_{12} = -\frac{1}{\sqrt{2}}$, $A_{21} = -\frac{1}{\sqrt{2}}$, and $A_{22} = \frac{1}{2}$. Transpose these elements and divide by the determinant to form the matrix A^{-1}:

$$A^{-1} = -2\begin{pmatrix} 0 & -\frac{1}{\sqrt{2}} \\ -\frac{1}{\sqrt{2}} & \frac{1}{2} \end{pmatrix} = \begin{pmatrix} 0 & \sqrt{2} \\ \sqrt{2} & -1 \end{pmatrix}$$

b.

$$\det A = -2\begin{vmatrix} 1 & 1 \\ 2 & 1 \end{vmatrix} + 3\begin{vmatrix} 1 & 1 \\ 2 & 0 \end{vmatrix} = -2(-1) + 3(-2) = -4$$

The cofactors are

$$\begin{array}{lll}
A_{11} = 1 & A_{12} = 1 & A_{13} = -2 \\
A_{21} = -2 & A_{22} = -6 & A_{23} = 4 \\
A_{31} = -1 & A_{32} = 3 & A_{33} = -2
\end{array}$$

So

$$A^{-1} = -\frac{1}{4} \begin{pmatrix} 1 & -2 & -1 \\ 1 & -6 & 3 \\ -2 & 4 & -2 \end{pmatrix}$$

F–10. Recall that a singular matrix is one whose determinant is equal to zero. Referring to the procedure in Problem F–9, do you see why a singular matrix has no inverse?

To find the inverse of a matrix, we first construct a new matrix and then divide each element of that matrix by the determinant of the original matrix. If the determinant of a matrix is equal to zero (that is, if the matrix is singular), then this division process is undefined and we cannot obtain an inverse.

F–11. Consider the simultaneous algebraic equations

$$x + y = 3$$

$$4x - 3y = 5$$

Show that this pair of equations can be written in the matrix form

$$A\mathbf{x} = \mathbf{c} \tag{1}$$

where

$$\mathbf{x} = \begin{pmatrix} x \\ y \end{pmatrix} \qquad \mathbf{c} = \begin{pmatrix} 3 \\ 5 \end{pmatrix} \quad \text{and} \quad A = \begin{pmatrix} 1 & 1 \\ 4 & -3 \end{pmatrix}$$

Now multiply Equation 1 from the left by A^{-1} to obtain

$$\mathbf{x} = A^{-1}\mathbf{c} \tag{2}$$

Now show that

$$A^{-1} = -\frac{1}{7} \begin{pmatrix} -3 & -1 \\ -4 & 1 \end{pmatrix}$$

and that

$$\mathbf{x} = -\frac{1}{7} \begin{pmatrix} -3 & -1 \\ -4 & 1 \end{pmatrix} \begin{pmatrix} 3 \\ 5 \end{pmatrix} = \begin{pmatrix} 2 \\ 1 \end{pmatrix}$$

or that $x = 2$ and $y = 1$. Do you see how this procedure generalizes to any number of simultaneous equations?

The equation $A\mathbf{x} = \mathbf{c}$ can be expressed as

$$\begin{pmatrix} 1 & 1 \\ 4 & -3 \end{pmatrix} \begin{pmatrix} x \\ y \end{pmatrix} = \begin{pmatrix} 3 \\ 5 \end{pmatrix}$$

Multiplying the two matrices on the left side and equating to the matrix elements on the right side gives

$$x + y = 3$$

$$4x - 3y = 5$$

which are the original two equations. Multiplying $A\mathbf{x} = \mathbf{c}$ from the left by A^{-1} gives

$$A^{-1}A\mathbf{x} = A^{-1}\mathbf{c}$$

$$I\mathbf{x} = A^{-1}\mathbf{c}$$

$$\mathbf{x} = A^{-1}\mathbf{c}$$

Using the approach in Problem F–9, we find that det $A = -7$. The cofactors are $A_{11} = -3$, $A_{12} = -4$, $A_{21} = -1$, and $A_{22} = 1$ and so the inverse is

$$A^{-1} = -\frac{1}{7}\begin{pmatrix} -3 & -1 \\ -4 & 1 \end{pmatrix}$$

Therefore,

$$\mathbf{x} = -\frac{1}{7}\begin{pmatrix} -3 & -1 \\ -4 & 1 \end{pmatrix}\begin{pmatrix} 3 \\ 5 \end{pmatrix} = \begin{pmatrix} 2 \\ 1 \end{pmatrix}$$

This procedure is easily generalized.

F–12. Solve the following simultaneous algebraic equations by the matrix inverse method developed in Problem F–11:

$$x + y - z = 1$$

$$2x - 2y + z = 6$$

$$x + 3z = 0$$

First show that

$$A^{-1} = \frac{1}{13}\begin{pmatrix} 6 & 3 & 1 \\ 5 & -4 & 3 \\ -2 & -1 & 4 \end{pmatrix}$$

and evaluate $\mathbf{x} = a^{-1}\mathbf{c}$.

The matrix A is given by

$$A = \begin{pmatrix} 1 & 1 & -1 \\ 2 & -2 & 1 \\ 1 & 0 & 3 \end{pmatrix}$$

We can find its inverse, A^{-1}, using the procedure from Problem F–9. The determinant of A is

$$\det A = -1\begin{vmatrix} 2 & 1 \\ 1 & 3 \end{vmatrix} - 2\begin{vmatrix} 1 & -1 \\ 1 & 3 \end{vmatrix} = -1(5) - 2(4) = -13$$

The cofactors are

$$
\begin{array}{lll}
A_{11} = -6 & A_{12} = -5 & A_{13} = 2 \\
A_{21} = -3 & A_{22} = 4 & A_{23} = 1 \\
A_{31} = -1 & A_{32} = -3 & A_{33} = -4
\end{array}
$$

So

$$A^{-1} = -\frac{1}{13}\begin{pmatrix} -6 & -3 & -1 \\ -5 & 4 & -3 \\ 2 & 1 & -4 \end{pmatrix} \tag{1}$$

Then

$$\begin{pmatrix} x \\ y \\ z \end{pmatrix} = A^{-1} \begin{pmatrix} 1 \\ 6 \\ 0 \end{pmatrix} \tag{2}$$

Substituting Equation 1 into Equation 2 gives

$$\begin{pmatrix} x \\ y \\ z \end{pmatrix} = -\frac{1}{13} \begin{pmatrix} -6 & -3 & -1 \\ -5 & 4 & -3 \\ 2 & 1 & -4 \end{pmatrix} \begin{pmatrix} 1 \\ 6 \\ 0 \end{pmatrix} = -\frac{1}{13} \begin{pmatrix} -24 \\ 19 \\ 8 \end{pmatrix}$$

so $x = 24/13$, $y = -19/13$, and $z = -8/13$.

CHAPTER 12

Group Theory: the Exploitation of Symmetry

12–1. Neglecting overlap, show that ϕ_1 and ϕ_2 given by Equations 12.3 are orthonormal to the other four molecular orbitals.

Recall that the overlap integral S_{ij} is 0 in the Hückel approximation if $i \neq j$. Then

$$\int \phi_1^* \phi_1 d\tau = \frac{1}{6}(1 + 1 + 1 + 1 + 1 + 1) = 1$$

$$\int \phi_2^* \phi_2 d\tau = \frac{1}{6}(1 + 1 + 1 + 1 + 1 + 1) = 1$$

so ϕ_1 and ϕ_2 are both normalized. Now consider

$$\int \phi_1^* \phi_2 d\tau = \frac{1}{6}(1 - 1 + 1 - 1 + 1 - 1) = 0$$

$$\int \phi_1^* \phi_3 d\tau = \frac{1}{\sqrt{72}}(2 + 1 - 1 - 2 - 1 + 1) = 0$$

$$\int \phi_2^* \phi_3 d\tau = \frac{1}{\sqrt{72}}(2 - 1 - 1 + 2 - 1 - 1) = 0$$

$$\int \phi_1^* \phi_4 d\tau = \frac{1}{12}(1 + 2 + 1 - 1 - 2 - 1) = 0$$

$$\int \phi_2^* \phi_4 d\tau = \frac{1}{12}(1 - 2 + 1 + 1 - 2 + 1) = 0$$

$$\int \phi_1^* \phi_5 d\tau = \frac{1}{12}(2 - 1 - 1 + 2 - 1 - 1) = 0$$

$$\int \phi_2^* \phi_5 d\tau = \frac{1}{12}(2 + 1 - 1 - 2 - 1 + 1) = 0$$

$$\int \phi_1^* \phi_6 d\tau = \frac{1}{12}(-1 + 2 - 1 - 1 + 2 - 1) = 0$$

$$\int \phi_2^* \phi_6 d\tau = \frac{1}{12}(-1 - 2 - 1 + 1 + 2 + 1) = 0$$

We see that ϕ_1 and ϕ_2 are orthogonal to ϕ_3, ϕ_4, ϕ_5, and ϕ_6.

12–2. Using the six molecular orbitals given by Equations 12.3, verify that $H_{11} = \alpha + 2\beta$, $H_{22} = \alpha - 2\beta$, $H_{12} = H_{13} = H_{14} = H_{15} = H_{16} = 0$ (see Equation 12.4).

Recall that the Hückel molecular-orbital theory sets the Coulomb integrals to be α, the overlap integrals to be $S_{ij} = \delta_{ij}$, and the resonance integrals involving the $2p_z$ orbitals of nearest-neighbor carbon atoms to be β (Section 10–5). Also recall that the ψ_j's that comprise the ϕ_j molecular orbitals are the $2p_z$ orbitals centered on the individual carbon atoms. Thus

$$H_{11} = \int \phi_1^* \hat{H} \phi_1 d\tau = \frac{1}{6} \left(\sum_{j=1}^{6} \int \psi_j^* \hat{H} \psi_j d\tau + \sum_{\substack{i=1 \\ i \neq j}}^{6} \sum_{j=1}^{6} \int \psi_i^* \hat{H} \psi_j d\tau \right)$$

$$= \frac{1}{6} (6\alpha + 12\beta) = \alpha + 2\beta$$

Likewise,

$$H_{22} = \int \phi_2^* \hat{H} \phi_2 = \frac{1}{6} \left(\sum_{j=1}^{6} \int \psi_j^* \hat{H} \psi_j d\tau - \sum_{\substack{i=1 \\ i \neq j}}^{6} \sum_{j=1}^{6} \int \psi_i^* \hat{H} \psi_j d\tau \right)$$

$$= \frac{1}{6} (6\alpha - 12\beta) = \alpha - 2\beta$$

where the $(-)$ sign in front of the second sum occurs because the ψ_j on adjacent carbon atoms have opposite signs. Now consider H_{12} and H_{13}:

$$H_{12} = \int \phi_1^* \hat{H} \phi_2 d\tau = \frac{1}{6} [(\alpha - \alpha + \alpha - \alpha + \alpha - \alpha) + (-2\beta + 2\beta - 2\beta + 2\beta - 2\beta + 2\beta)] = 0$$

$$H_{13} = \int \phi_1^* \hat{H} \phi_3 d\tau = \frac{1}{\sqrt{72}} [(2\alpha + \alpha - \alpha - 2\alpha - \alpha + \alpha) + (4\beta + 2\beta - 2\beta - 4\beta - 2\beta + 2\beta)] = 0$$

We can show that $H_{14} = H_{15} = H_{16} = 0$ in the same manner.

12–3. List the various symmetry elements for the trigonal planar molecule SO_3.

As is true for all molecules, SO_3 has the identity element E. There is a C_3 axis perpendicular to the plane of the molecule centered on the sulphur atom, and three C_2 axes in the plane of the molecule along each of the S–O bonds. The plane of symmmetry of the molecule is a σ_h plane (perpendicular to the principal C_3 axis). There are three σ_v planes, all of which contain the C_3 axis and one of the S–O bonds. Finally, there is a three-fold rotation-reflection axis, S_3, coincident with C_3.

12–4. Verify that a methane molecule has the symmetry elements given in Table 12.2.

The point group of methane is $\mathbf{T}_d$. The identity element is given. There are C_3 axes coinciding with each of the C–H bonds. The three C_2 and S_4 axes are represented in Figure 12.3. Finally, each H–C–H unit lies in a σ_d plane, so there are six σ_d planes in methane.

12–5. Verify that a benzene molecule has the symmetry elements given in Table 12.2.

The point group of benzene is $\mathbf{D}_{6h}$. The C_6 and S_6 axes pass through the center of the benzene ring and are perpendicular to the plane of the ring. There are three C_2 axes perpendicular to the C_6 axis that each bisect opposing C–C bonds. There are three C_2' axes perpendicular to the C_6 axis (shown in Figure 12.5). The inversion center i is located at the center of the benzene ring. The plane of the molecule is a σ_h plane. There are three σ_v planes that each contain the C_6 axis and one of the three C_2 axes. The three σ_d axes are shown in Figure 12.5. There is also the identity element.

12–6. Verify that a xenon tetrafluoride (square planar) molecule has the symmetry elements given in Table 12.2.

The point group of XeF_4 is $\mathbf{D_{4h}}$. All molecules have an identity element. The C_4 axis is perpendicular to the plane of the molecule and passes through the center of the xenon atom. The four C_2 axes lie in the plane of the molecule; two contain opposite Xe–F bonds and bisect the F–Xe–F bond angle. The inversion center i is located at the center of the xenon atom, the S_4 axis is coincident with C_4, and the plane of the molecule is a σ_h symmetry plane. Each of the two σ_v planes contains the C_4 axis and one of the C_2 axes that lie along the F–Xe–F bonds. Each of the two σ_d planes contains the C_4 axis and one of the C_2 axes that bisects the F–Xe–F bond angle.

12–7. Explain why $\hat{C}_4^3 = \hat{C}_4^{-1}$.

$\hat{C}_4^3$ represents a clockwise rotation of $270°$ and $\hat{C}_4^{-1}$ represents a counterclockwise rotation of $90°$. These are equivalent because a full rotation is given by $360°$.

12–8. Deduce the group multiplication table for the point group $\mathbf{C_{2v}}$ (see Table 12.3).

The property of the identity element ($\hat{E}\hat{X} = \hat{X}\hat{E} = \hat{X}$ for all operators $\hat{X}$) accounts for the first column and the first row of the group multiplication table, and the products $\hat{\sigma}_v\hat{C}_2$, $\hat{\sigma}_v'\hat{\sigma}_v$, and $\hat{\sigma}_v'\hat{C}_2$ are evaluated for H_2O in Section 12–3 and Example 12–3. Also, $\hat{C}_2\hat{C}_2 = \hat{E}$ (two rotations of $180°$), $\hat{\sigma}_v\hat{\sigma}_v = \hat{E}$, and $\hat{\sigma}_v'\hat{\sigma}_v' = \hat{E}$ (subsequent reflections in one plane of symmetry). We now have the table

$\mathbf{C_{2v}}$	$\hat{E}$	$\hat{C}_2$	$\hat{\sigma}_v$	$\hat{\sigma}_v'$
$\hat{E}$	$\hat{E}$	$\hat{C}_2$	$\hat{\sigma}_v$	$\hat{\sigma}_v'$
$\hat{C}_2$	$\hat{C}_2$	$\hat{E}$		
$\hat{\sigma}_v$	$\hat{\sigma}_v$	$\hat{\sigma}_v'$	$\hat{E}$	
$\hat{\sigma}_v'$	$\hat{\sigma}_v'$	$\hat{\sigma}_v$	$\hat{C}_2$	$\hat{E}$

An important property of group multiplication tables is that all the entries in any row or column must be different. This fact allows us to easily complete the table as in Table 12.3. The missing entry in the third row must be $\hat{C}_2$ and the missing entry in the third column must be $\hat{\sigma}_v'$, thereby giving $\hat{\sigma}_v$ for the final element (second row, fourth column).

12–9. Determine the order of the $\mathbf{D_{4h}}$ point group (see Table 12.2).

The symmetry elements of this group are E, C_4, $4C_2$, i, S_4, σ_h, $2\sigma_v$, and $2\sigma_d$. The C_4 element has three operators associated with it ($\hat{C}_4$, $\hat{C}_4^2$, $\hat{C}_4^3$) and the S_4 element has two operators associated with it ($\hat{S}_4$ and $\hat{S}_4^2$); all the remaining symmetry elements are associated with one operator. Therefore, the order of this point group is 16.

12–10. Determine the order of the $\mathbf{D}_{6h}$ point group (see Table 12.2).

The symmetry elements of this group are E, C_6, $3C_2$, $3C_2'$, i, S_6, σ_h, $3\sigma_v$, and $3\sigma_d$. The C_6 element has five operators associated with it ($\hat{C}_6$, $\hat{C}_6^2$, $\hat{C}_6^3$, $\hat{C}_6^4$, and $\hat{C}_6^5$) and the S_6 element has four operators associated with it ($\hat{S}_6$, $\hat{S}_6^2$, $\hat{S}_6^4$, and $\hat{S}_6^5$); the remaining symmetry operators are associated with only one operator. Consequently, the order of this point group is 24.

12–11. Evaluate the products $\hat{\sigma}_v\hat{\sigma}_v'$, $\hat{C}_2\hat{\sigma}_v$, and $\hat{C}_2\hat{\sigma}_v'$ for a $\mathbf{C}_{2v}$ point group (see Table 12.3).

We can use the geometry defined in Example 12–3.

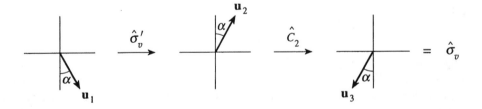

12–12. Evaluate the products $\hat{C}_3\hat{\sigma}_v$ and $\hat{C}_3\hat{\sigma}_v''$ for a $\mathbf{C}_{3v}$ point group (see Table 12.4).

We can use the geometry defined in Example 12–4.

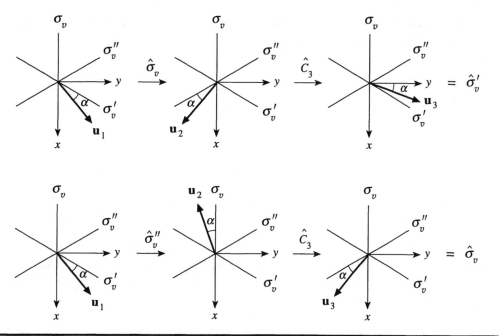

12–13. Show that Equation 12.7 is valid for the point groups given in Tables 12.9 through 12.14.

$$\sum_{j=1}^{N} d_j^2 = h \tag{12.7}$$

The point groups for which we need to show Equation 12.7 are $\mathbf{C}_{3v}$, $\mathbf{C}_{2h}$, $\mathbf{D}_{3h}$, $\mathbf{D}_{4h}$, $\mathbf{D}_{6h}$, and $\mathbf{T}_d$. The character tables for these groups are given by Tables 12.8 ($\mathbf{C}_{3v}$), 12.10 ($\mathbf{C}_{2h}$), 12.11 ($\mathbf{D}_{3h}$), 12.12 ($\mathbf{D}_{4h}$), 12.13 ($\mathbf{D}_{6h}$), and 12.14 ($\mathbf{T}_d$). From the number of symmetry elements listed in each table, we find the orders of the point groups to be 4, 4, 12, 16, 24, and 24, respectively. The dimensionality d of the irreducible representations of each group is given by the character tables; substituting these values into Equation 12.7 gives

$$
\begin{aligned}
\mathbf{C}_{3v} &: \quad 1^2 + 1^2 + 2^2 = 4 \\
\mathbf{C}_{2h} &: \quad 1^2 + 1^2 + 1^2 + 1^2 = 4 \\
\mathbf{D}_{3h} &: \quad 1^2 + 1^2 + 2^2 + 1^2 + 1^2 + 2^2 = 12 \\
\mathbf{D}_{4h} &: \quad 1^2 + 1^2 + 1^2 + 1^2 + 2^2 + 1^2 + 1^2 + 1^2 + 1^2 + 2^2 = 16 \\
\mathbf{D}_{6h} &: \quad 1^2 + 1^2 + 1^2 + 1^2 + 2^2 + 2^2 + 1^2 + 1^2 + 1^2 + 1^2 + 2^2 + 2^2 = 24 \\
\mathbf{T}_d &: \quad 1^2 + 1^2 + 2^2 + 3^2 + 3^2 = 24
\end{aligned}
$$

in agreement with the values of the orders of each group.

12–14. Show that the 2×2 matrices given in Table 12.6 are a representation for the $\mathbf{C}_{3v}$ point group.

We can show this by demonstrating that the representations given in Table 12.6 obey the $\mathbf{C}_{3v}$ group multiplication table (Table 12.4). We calculate two products below to illustrate that the matrices obey the multiplication table. The remaining products are easily evaluated.

$$
\hat{C}_3 \hat{\sigma}_v = \begin{pmatrix} -\frac{1}{2} & -\frac{\sqrt{3}}{2} \\ \frac{\sqrt{3}}{2} & -\frac{1}{2} \end{pmatrix} \begin{pmatrix} 1 & 0 \\ 0 & -1 \end{pmatrix} = \begin{pmatrix} -\frac{1}{2} & \frac{\sqrt{3}}{2} \\ \frac{\sqrt{3}}{2} & \frac{1}{2} \end{pmatrix} = \hat{\sigma}_v'
$$

$$
\hat{\sigma}_v'' \hat{\sigma}_v' = \begin{pmatrix} -\frac{1}{2} & -\frac{\sqrt{3}}{2} \\ -\frac{\sqrt{3}}{2} & \frac{1}{2} \end{pmatrix} \begin{pmatrix} -\frac{1}{2} & \frac{\sqrt{3}}{2} \\ \frac{\sqrt{3}}{2} & \frac{1}{2} \end{pmatrix} = \begin{pmatrix} -\frac{1}{2} & -\frac{\sqrt{3}}{2} \\ \frac{\sqrt{3}}{2} & -\frac{1}{2} \end{pmatrix} = \hat{C}_3
$$

12–15. In Section 12–4, we derived matrix representations for various symmetry operators. Starting with an arbitrary vector $\mathbf{u}$, where $\mathbf{u} = u_x\mathbf{i} + u_y\mathbf{j} + u_z\mathbf{k}$, show that the matrix representation for a counterclockwise rotation about the z-axis by an angle α, $\hat{C}_{360/\alpha}$, is given by

$$\begin{pmatrix} \cos\alpha & -\sin\alpha & 0 \\ \sin\alpha & \cos\alpha & 0 \\ 0 & 0 & 1 \end{pmatrix}$$

Show that the corresponding matrix for rotation-reflection $\hat{S}_{360/\alpha}$ about the z-axis by an angle α is

$$\begin{pmatrix} \cos\alpha & -\sin\alpha & 0 \\ \sin\alpha & \cos\alpha & 0 \\ 0 & 0 & -1 \end{pmatrix}$$

Consider a vector $\mathbf{u}$ with components (in spherical coordinates) of $u_x = r\sin\theta\cos\phi$, $u_y = r\sin\theta\sin\phi$, and $u_z = r\cos\theta$. Operating on this vector by the matrix representation of $\hat{C}_{360/\alpha}$ given above yields

$$\begin{pmatrix} \cos\alpha & -\sin\alpha & 0 \\ \sin\alpha & \cos\alpha & 0 \\ 0 & 0 & 1 \end{pmatrix} \begin{pmatrix} r\sin\theta\cos\phi \\ r\sin\theta\sin\phi \\ r\cos\theta \end{pmatrix} = \begin{pmatrix} r\cos\alpha\sin\theta\cos\phi - r\sin\alpha\sin\theta\sin\phi \\ r\sin\alpha\sin\theta\cos\phi + r\cos\alpha\sin\theta\sin\phi \\ r\cos\theta \end{pmatrix}$$

$$= \begin{pmatrix} r\sin\theta(\cos\alpha\cos\phi - \sin\alpha\sin\phi) \\ r\sin\theta(\sin\alpha\cos\phi + \cos\alpha\cos\phi) \\ r\cos\theta \end{pmatrix}$$

$$= \begin{pmatrix} r\sin\theta\cos(\alpha+\phi) \\ r\sin\theta\sin(\alpha+\phi) \\ r\cos\theta \end{pmatrix}$$

The new coordinates of $\mathbf{u}$ are $u_x = r\sin\theta\cos(\alpha+\phi)$, $u_y = r\sin\theta\sin(\alpha+\phi)$, and $u_z = r\cos\theta$, indicating that the vector has been rotated about the z-axis by an angle α. The improper rotation $\hat{S}_{360/\alpha}$ is a rotation by $360/\alpha$ followed by a reflection through a plane perpendicular to the rotation axis, so

$$\hat{S}_{360/\alpha} = \begin{pmatrix} 1 & 0 & 0 \\ 0 & 1 & 0 \\ 0 & 0 & -1 \end{pmatrix} \begin{pmatrix} \cos\alpha & -\sin\alpha & 0 \\ \sin\alpha & \cos\alpha & 0 \\ 0 & 0 & 1 \end{pmatrix} = \begin{pmatrix} \cos\alpha & -\sin\alpha & 0 \\ \sin\alpha & \cos\alpha & 0 \\ 0 & 0 & -1 \end{pmatrix}$$

12–16. Show that u_x forms a basis for the irreducible representation B_1 of the point group $\mathbf{C}_{2v}$.

See Example 12–9.

12–17. Show that R_x forms a basis for the irreducible representation B_2 of the point group $\mathbf{C}_{2v}$.

We do this problem in the same way as Example 12–10.

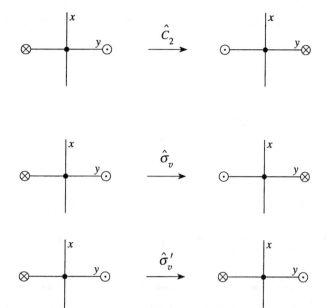

First of all, $\hat{E}R_x = (1)R_x$. We see from the above drawings that $\hat{\sigma}_v'R_x = (1)R_x$ and $\hat{C}_2R_x = \hat{\sigma}_vR_x = (-1)R_x$, and therefore R_x forms a basis for the B_2 irreducible representation of the point group $\mathbf{C}_{2v}$.

12–18. Show that (u_x, u_y) forms a joint basis for the irreducible representation E of the point group $\mathbf{C}_{3v}$.

The operations of the irreducible representation E transform u_x and u_y into linear combinations of u_x and u_y (Section 12–5); therefore, u_x and u_y form a basis for this irreducible representation.

12–19. Show that the rows of the character table of $\mathbf{C}_{2h}$ satisfy Equation 12.20.

The orthogonality condition is

$$\sum_{\substack{\text{classes}}} n(\hat{R})\chi_i(\hat{R})\chi_j(\hat{R}) = h\delta_{ij} \tag{12.20}$$

The character table of $\mathbf{C}_{2h}$ is given in Table 12.10. For the point group $\mathbf{C}_{2h}$, $h = 4$. First consider the cases where $i = j$, where the sum should give a value of 4:

$$\sum_{\hat{R}} \chi_{A_g}(\hat{R})\chi_{A_g}(\hat{R}) = 1 + 1 + 1 + 1 = 4$$

$$\sum_{\hat{R}} \chi_{B_g}(\hat{R})\chi_{B_g}(\hat{R}) = 1 + 1 + 1 + 1 = 4$$

$$\sum_{\hat{R}} \chi_{A_u}(\hat{R})\chi_{A_u}(\hat{R}) = 1 + 1 + 1 + 1 = 4$$

$$\sum_{\hat{R}} \chi_{B_u}(\hat{R})\chi_{B_u}(\hat{R}) = 1 + 1 + 1 + 1 = 4$$

Now consider the cases where $i \neq j$, where the sum should give a value of 0:

$$\sum_{\hat{R}} \chi_{A_g}(\hat{R})\chi_{B_g}(\hat{R}) = 1 - 1 + 1 - 1 = 0$$

$$\sum_{\hat{R}} \chi_{A_g}(\hat{R})\chi_{A_u}(\hat{R}) = 1 + 1 - 1 - 1 = 0$$

$$\sum_{\hat{R}} \chi_{A_g}(\hat{R})\chi_{B_u}(\hat{R}) = 1 - 1 - 1 + 1 = 0$$

$$\sum_{\hat{R}} \chi_{B_g}(\hat{R})\chi_{A_u}(\hat{R}) = 1 - 1 - 1 + 1 = 0$$

$$\sum_{\hat{R}} \chi_{B_g}(\hat{R})\chi_{B_u}(\hat{R}) = 1 + 1 - 1 - 1 = 0$$

$$\sum_{\hat{R}} \chi_{A_u}(\hat{R})\chi_{B_u}(\hat{R}) = 1 - 1 + 1 - 1 = 0$$

12–20. Show that the rows of the character table of $\mathbf{D}_{3h}$ satisfy Equation 12.20.

The orthogonality condition is

$$\sum_{\text{classes}} n(\hat{R})\chi_i(\hat{R})\chi_j(\hat{R}) = h\delta_{ij} \qquad (12.20)$$

The character table of $\mathbf{D}_{3h}$ is given in Table 12.11, and $h = 12$ for the $\mathbf{D}_{3h}$ point group. Showing that the rows satsify Equation 12.20 requires the evaluation of 21 different sums. We will show four of these to illustrate the point, as the rest are easily evaluated. When $i = j$, the sum should give a value of 12:

$$\sum_{\hat{R}} \chi_{A_1'}(\hat{R})\chi_{A_1'}(\hat{R}) = 1 + 2(1) + 3(1) + 1 + 2(1) + 3(1) = 12$$

When $i \neq j$, the sum should give a value of 0:

$$\sum_{\hat{R}} \chi_{A_1'}(\hat{R})\chi_{A_2'}(\hat{R}) = 1 + 2(1) + 3(-1) + 1 + 2(1) + 3(-1) = 0$$

$$\sum_{\hat{R}} \chi_{A_1'}(\hat{R})\chi_{E'}(\hat{R}) = 2 + 2(-1) + 3(0) + 2 + 2(-1) + 3(0) = 0$$

$$\sum_{\hat{R}} \chi_{E'}(\hat{R})\chi_{E''}(\hat{R}) = 4 + 2(1) + 3(0) - 4 + 2(-1) + 3(0) = 0$$

12–21. Suppose the characters of a reducible representation of the $\mathbf{T}_d$ point group are $\chi(\hat{E}) = 17$, $\chi(\hat{C}_3) = 2$, $\chi(\hat{C}_2) = 5$, $\chi(\hat{S}_4) = -3$, and $\chi(\hat{\sigma}_d) = -5$, or $\Gamma = 17\ 2\ 5\ -3\ -5$. Determine how many times each irreducible representation of $\mathbf{T}_d$ is contained in Γ.

Using Equation 12.23 and summing over classes, we have

$$a_i = \frac{1}{h} \sum_{\text{classes}} n(\hat{R})\chi(\hat{R})\chi_i(\hat{R})$$

$$a_{A_1} = \frac{1}{24}[17 + 8(2) + 3(5) + 6(-3) + 6(-5)] = 0$$

$$a_{A_2} = \frac{1}{24}[17 + 8(2) + 3(5) - 6(-3) - 6(-5)] = 4$$

$$a_E = \frac{1}{24}[2(17) - 8(2) + 3(2)(5) + 0 + 0] = 2$$

$$a_{T_1} = \frac{1}{24}[3(17) + 0 - 3(5) + 6(-3) - 6(-5)] = 2$$

$$a_{T_2} = \frac{1}{24}[3(17) + 0 - 3(5) - 6(-3) + 6(-5)] = 1$$

Then

$$\Gamma = 4A_2 + 2E + 2T_1 + T_2$$

12–22. Suppose the characters of a reducible representation of the C_{2v} point group are $\Gamma = 27 \ -1 \ 1 \ 5$. Determine how many times each irreducible representation of C_{2v} is contained in Γ.

We use Equation 12.23 and sum over classes.

$$a_{A_1} = \frac{1}{4}[27 - 1 + 1 + 5] = 8$$

$$a_{A_2} = \frac{1}{4}[27 - 1 - 1 - 5] = 5$$

$$a_{B_1} = \frac{1}{4}[27 + 1 + 1 - 5] = 6$$

$$a_{B_2} = \frac{1}{4}[27 + 1 - 1 + 5] = 8$$

So

$$\Gamma = 8A_1 + 5A_2 + 6B_1 + 8B_2$$

12–23. Suppose the characters of a reducible representation of the D_{3h} point group are $\Gamma = 12 \ 0 \ -2 \ 4 \ -2 \ 2$. Determine how many times each irreducible representation of D_{3h} is contained in Γ.

We use Equation 12.23 and sum over classes.

$$a_{A_1'} = \frac{1}{12}[12 + 2(0) + 3(-2) + 4 + 2(-2) + 3(2)] = 1$$

$$a_{A_2'} = \frac{1}{12}[12 + 2(0) - 3(-2) + 4 + 2(-2) - 3(2)] = 1$$

$$a_{E'} = \frac{1}{12}[2(12) + 0 + 0 + 2(4) - 2(-2) + 0] = 3$$

$$a_{A_1''} = \frac{1}{12}[12 + 2(0) + 3(-2) - 4 - 2(-2) - 3(2)] = 0$$

$$a_{A_2''} = \frac{1}{12}[12 + 2(0) - 3(-2) - 4 - 2(-2) + 3(2)] = 2$$

$$a_{E''} = \frac{1}{12}[2(12) + 0 + 0 - 2(4) + 2(-2) + 0] = 1$$

$$\Gamma = A_1' + A_2' + 3E' + 2A_2'' + E''$$

12–24. In Example 12–14, we showed that the overlap integral involving $2p_{xN}$ and $1s_{H_A} + 1s_{H_B} + 1s_{H_C}$ in the NH_3 molecule is equal to zero. Is this necessarily true for the $2p_{zN}$ rather than the $2p_{xN}$ orbital?

No. NH_3 belongs to the C_{3v} point group (Example 12–14) and in that example we showed that the linear combination of the $1s$ orbitals on the three hydrogens belongs to the A_1 representation. The $2p_{zN}$ orbital transforms as z, which also belongs to A_1. Therefore, the integrand of the overlap integral between the $2p_{zN}$ orbital and the linear combination $1s_{H_A} + 1s_{H_B} + 1s_{H_C}$ is totally symmetric and is not zero by symmetry.

12–25. Show that the molecular orbital ϕ_2 given by Equations 12.3 belongs to the irreducible representation B_{2g}.

$$\phi_2 = \frac{1}{\sqrt{6}} \left(\psi_1 - \psi_2 + \psi_3 - \psi_4 + \psi_5 - \psi_6 \right) \tag{12.3}$$

We will use the following diagram to visualize the effects of the symmetry operators on ϕ_2. Recall that benzene belongs to the D_{6h} point group.

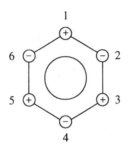

Consider the operator $\hat{C}_6$. This rotates the benzene molecule counterclockwise by 60°. Therefore $\psi_1 \to \psi_6$, $\psi_2 \to \psi_1$, etc., giving

$$\hat{C}_6 \phi_2 = \frac{1}{\sqrt{6}} \left(\psi_6 - \psi_1 + \psi_2 - \psi_3 + \psi_4 - \psi_5 \right) = (-1)\phi_2$$

so the associated character is -1. Proceeding in a similar fashion, we show that the following operators have the associated characters

$$
\begin{array}{lll}
\hat{E} \to +1 & \hat{C}_2' \to -1 & \hat{S}_6 \to +1 \\
\hat{C}_6 \to -1 & \hat{C}_2'' \to +1 & \hat{\sigma}_h \to -1 \\
\hat{C}_3 \to +1 & \hat{i} \to +1 & \hat{\sigma}_d \to -1 \\
\hat{C}_2 \to -1 & \hat{S}_3 \to -1 & \hat{\sigma}_v \to +1
\end{array}
$$

This set of characters is equivalent to the B_{2g} representation.

12–26. Because the benzene molecular orbitals ϕ_3 and ϕ_4 in Equations 12.3 belong to a two-dimensional irreducible representation (E_{1g}), they are not unique. Any two linear combinations of ϕ_3 and ϕ_4 will also form a basis for E_{1g}. Consider

$$\phi_3' = \phi_3 = \frac{1}{\sqrt{12}} (2\psi_1 + \psi_2 - \psi_3 - 2\psi_4 - \psi_5 + \psi_6)$$

and

$$\phi_4' = \frac{1}{2}(\psi_2 + \psi_3 - \psi_5 - \psi_6)$$

First show that

$$\phi_4' = \frac{\sqrt{12}}{6}(2\phi_4 - \phi_3)$$

Now show that ϕ_4' is normalized. (Realize that ϕ_3 and ϕ_4 are not necessarily orthogonal because they are degenerate (see Problem 4–29).) Evaluate the 2×2 block of the secular determinant corresponding to E_{1g} and show that the final value of the energy is the same as that given in Equation 12.4.

Working from the given values of ϕ_3' and ϕ_4' and the definition of ϕ_4 in Equation 12.3, we see that

$$2\phi_4 - \phi_3 = 2\phi_4 - \phi_3'$$

$$= 2\phi_4 - \frac{1}{\sqrt{12}}(2\psi_1 + \psi_2 - \psi_3 - 2\psi_4 - \psi_5 + \psi_6)$$

$$= \frac{1}{\sqrt{12}}[2(\psi_1 + 2\psi_2 + \psi_3 - \psi_4 - 2\psi_5 - \psi_6)$$

$$- (2\psi_1 + \psi_2 - \psi_3 - 2\psi_4 - \psi_5 + \psi_6)]$$

$$= \frac{1}{\sqrt{12}}(3\psi_2 + 3\psi_3 - 3\psi_5 - 3\psi_6)$$

$$= \frac{3}{\sqrt{12}}(\psi_2 + \psi_3 - \psi_5 - \psi_6) = \frac{6}{\sqrt{12}}\phi_4'$$

$$\phi_4' = \frac{\sqrt{12}}{6}(2\phi_4 - \phi_3)$$

To show that ϕ_4' is normalized, we evaluate

$$\int \phi_4'^* \phi_4' d\tau = \frac{12}{36}\left[4\int \phi_4^* \phi_4 d\tau - 2\int \phi_4^* \phi_3 d\tau - 2\int \phi_3^* \phi_4 d\tau + \int \phi_3^* \phi_3 d\tau\right]$$

The wavefunctions ϕ_4 and ϕ_3 are normalized, so

$$\int \phi_4'^* \phi_4' d\tau = \frac{12}{36}\left[5 - 2\int \phi_4^* \phi_3 d\tau - 2\int \phi_3^* \phi_4 d\tau\right]$$

Recall that the Hückel theory assumes $S_{ij} = 1$ for $i = j$ and 0 for $i \neq j$. The overlap integrals are then

$$\int \phi_4^* \phi_3 d\tau = \int \phi_3^* \phi_4 d\tau = \frac{1}{2}$$

so

$$\int \phi_4'^* \phi_4' d\tau = \frac{12}{36}(5 - 1 - 1) = 1$$

To evaluate the 2×2 block of the secular determinant, we need to evaluate the Hamiltonian and overlap integrals for ϕ_3' and ϕ_4' using the Hückel approximation. This gives

$$H_{33}' = \int \phi_3' \hat{H} \phi_3' d\tau = \frac{1}{12}[12\alpha + 12\beta] = \alpha + \beta$$

$$H_{44}' = \int \phi_4' \hat{H} \phi_4' d\tau = \frac{1}{4}[4\alpha + 4\beta] = \alpha + \beta$$

$$H'_{34} = \int \phi'_3 \hat{H} \phi'_4 d\tau = \frac{1}{\sqrt{48}} [2\alpha - 2\alpha + 6\beta - 6\beta] = 0$$

$$S'_{33} = S'_{44} = 1$$

$$S'_{34} = \frac{\sqrt{12}}{6} \int (2\phi_4 - \phi_3)\phi_3 d\tau = \frac{\sqrt{12}}{6}\left(\frac{2}{2} - 1\right) = 0$$

The 2×2 block of the secular determinant is then

$$\begin{vmatrix} \alpha + \beta - E & 0 \\ 0 & \alpha + \beta - E \end{vmatrix} = \beta^2 \begin{vmatrix} x+1 & 0 \\ 0 & x+1 \end{vmatrix} = 0$$

where $x = (\alpha - E)/\beta$. Expanding and solving for x gives

$$(x+1)^2 = 0 \quad \longrightarrow \quad x = \pm 1$$

which are two of the roots that we found from Equation 12.4.

12–27. Arrange the benzene molecular orbitals given by Equations 12.3 in the order of the number of nodal planes perpendicular to the plane of the molecule. Label the molecular orbitals according to the irreducible representation (see Example 10–8).

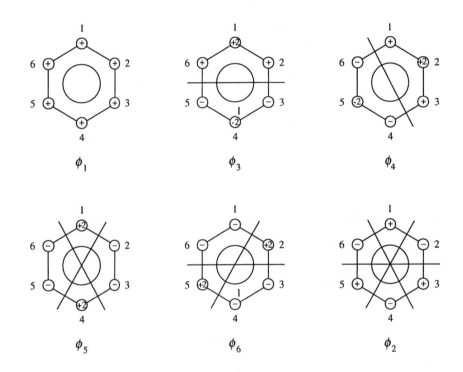

The nodal planes are shown in the pictures. The molecular orbital ϕ_1 has no nodal planes and belongs to A_{2u}, ϕ_2 has 3 nodal planes and belongs to B_{2g}, ϕ_3 and ϕ_4 have 1 nodal plane and belong to E_{1g}, and ϕ_5 and ϕ_6 have 2 nodal planes and belong to E_{2u}.

12–28. Using the symmetry orbitals for butadiene given by Equations 12.33, show that the Hückel theory secular determinant is given by Equation 12.34.

The symmetry orbitals given by Equation 12.23 are

$$\phi_1 = \frac{1}{\sqrt{2}}(\psi_1 - \psi_4) \quad \phi_3 = \frac{1}{\sqrt{2}}(\psi_1 + \psi_4)$$

$$\phi_2 = \frac{1}{\sqrt{2}}(\psi_2 - \psi_3) \quad \phi_4 = \frac{1}{\sqrt{2}}(\psi_2 + \psi_3)$$

(The $1/\sqrt{2}$ normalization factor derives from the fact that the ψ_j are orthonormal.) To construct the Hückel theory secular determinant, we must evaluate the various elements H_{ij} and S_{ij}, subject to the Hückel assumptions:

$$H_{11} = \frac{1}{2}\int (\psi_1 - \psi_4)\hat{H}(\psi_1 - \psi_4)d\tau = \frac{1}{2}\left(H_{11} + H_{44} - 0\right) = \frac{1}{2}(2\alpha) = \alpha$$

$$H_{22} = \frac{1}{2}\int (\psi_2 - \psi_3)\hat{H}(\psi_2 - \psi_3)d\tau = \frac{1}{2}\left(H_{22} + H_{33} - 2H_{23}\right) = \frac{1}{2}(2\alpha - 2\beta) = \alpha - \beta$$

$$H_{33} = \frac{1}{2}\int (\psi_1 + \psi_4)\hat{H}(\psi_1 + \psi_4)d\tau = \alpha$$

$$H_{44} = \frac{1}{2}\int (\psi_2 + \psi_3)\hat{H}(\psi_2 + \psi_3)d\tau = \alpha + \beta$$

$$S_{11} = S_{22} = S_{33} = S_{44} = 1$$

$$S_{12} = S_{14} = S_{23} = S_{34} = 0$$

Likewise,

$$S_{24} = \tfrac{1}{2}(1 - 1) = 0 \qquad S_{13} = \tfrac{1}{2}(1 - 1) = 0$$
$$H_{12} = \tfrac{1}{2}(\beta + \beta) = \beta \qquad H_{13} = \tfrac{1}{2}(\alpha - \alpha) = 0$$
$$H_{14} = \tfrac{1}{2}(\beta - \beta) = 0 \qquad H_{23} = \tfrac{1}{2}(\beta - \beta) = 0$$
$$H_{24} = \tfrac{1}{2}(\alpha - \alpha + \beta - \beta) = 0 \quad H_{34} = \tfrac{1}{2}(\beta + \beta) = \beta$$

This gives a blocked secular determinant of

$$\begin{vmatrix} \alpha - E & \beta & 0 & 0 \\ \beta & \alpha - \beta - E & 0 & 0 \\ 0 & 0 & \alpha - E & \beta \\ 0 & 0 & \beta & \alpha + \beta - E \end{vmatrix} = \beta^4 \begin{vmatrix} x & 1 & 0 & 0 \\ 1 & x - 1 & 0 & 0 \\ 0 & 0 & x & 1 \\ 0 & 0 & 1 & x + 1 \end{vmatrix} = 0$$

$$(x^2 - x - 1)(x^2 + x - 1) = 0$$

which is Equation 12.34.

12–29. Show that if we used a $2p_z$ orbital on each carbon atom as the basis for a (reducible) representation for benzene, ($\mathbf{D}_{6h}$) then $\Gamma = 6\ 0\ 0\ 0\ -2\ 0\ 0\ 0\ 0\ -6\ 0\ 2$. Reduce Γ into its component irreducible representations. What does your answer tell you about the expected Hückel secular determinant?

The operator $\hat{E}$ leaves all six $2p_z$ orbitals unchanged (a reducible character of 6); $\hat{C}_2'$ moves four of the six p-orbitals from one atom to another atom (contributing 0 to the reducible character) and inverts the two remaining p-orbitals (contributing -2 to the reducible character); $\hat{\sigma}_h$ inverts all six p-orbitals (a reducible character of -6); and $\hat{\sigma}_v$ leaves two p-orbitals unchanged (contributing 2 to the reducible character) and moves four p-orbitals between atoms (contributing 0 to the reducible character). All of the other operators move the orbitals from one atom to another, and so have a reducible character of 0. This gives the representation Γ above. Using Equation 12.23, we find that the only nonzero values of a_i occur for

$$a_{A_{2u}} = \frac{1}{24}(6+6+6+6) = 1$$

$$a_{B_{2g}} = \frac{1}{24}(6+6+6+6) = 1$$

$$a_{E_{1g}} = \frac{1}{24}(12+0+12+0) = 1$$

$$a_{E_{2u}} = \frac{1}{24}(12+0+12+0) = 1$$

This result shows that the secular determinant can be written in diagonal form, which when expanded will be the product of two 2×2 determinants (corresponding to the E_{1g} and E_{2u} representations) and two 1×1 determinants (corresponding to the A_{2u} and B_{2g} representations).

12–30. Show that if we used a $2p_z$ orbital on each carbon atom as the basis for a (reducible) representation for cyclobutadiene ($\mathbf{D}_{4h}$), then $\Gamma = 4\ 0\ 0\ 0\ -2\ 0\ 0\ -4\ 0\ 2$. Reduce Γ into its component irreducible representations. What does your answer tell you about the expected Hückel secular determinant?

The operator $\hat{E}$ leaves all four $2p_z$ orbitals unchanged (a reducible character of 4); $\hat{C}_2'$ inverts two p-orbitals and moves two of them (a reducible character of -2); $\hat{\sigma}_h$ inverts all four (a reducible character of -4); and $\hat{\sigma}_v$ leaves two orbitals unchanged and moves two of them (a reducible character of 2). All of the other operators move the orbitals from one atom to another (a reducible character of 0). This gives the representation Γ above. Using Equation 12.23, we find that

$$a_{A_{1g}} = \tfrac{1}{16}(4-4-4+4) = 0 \qquad a_{A_{2g}} = \tfrac{1}{16}(4+4-4-4) = 0$$

$$a_{B_{1g}} = \tfrac{1}{16}(4+4-4-4) = 0 \qquad a_{B_{2g}} = \tfrac{1}{16}(4-4-4+4) = 0$$

$$a_{E_g} = \tfrac{1}{16}(8+0+8) = 1 \qquad a_{A_{1u}} = \tfrac{1}{16}(4-4+4-4) = 0$$

$$a_{A_{2u}} = \tfrac{1}{16}(4+4+4+4) = 1 \qquad a_{B_{1u}} = \tfrac{1}{16}(4+4+4+4) = 1$$

$$a_{B_{2u}} = \tfrac{1}{16}(4-4+4-4) = 0 \qquad a_{E_u} = \tfrac{1}{16}(8+0-8+0) = 0$$

This result shows that the secular determinant can be written in block diagonal form, which, when expanded, will be the product of two 1×1 determinants (corresponding to the A_{2u} and B_{1u} representations) and one 2×2 determinant (corresponding to the E_g representation).

12–31. Consider an allyl anion, $CH_2CHCH_2^-$, which belongs to the point group $\mathbf{C}_{2v}$. Show that if we use ψ_1, ψ_2, and ψ_3 ($2p_z$ on each carbon atom) to calculate the Hückel secular determinant, then we obtain

$$\begin{vmatrix} \alpha - E & \beta & 0 \\ \beta & \alpha - E & \beta \\ 0 & \beta & \alpha - E \end{vmatrix} = \begin{vmatrix} x & 1 & 0 \\ 1 & x & 1 \\ 0 & 1 & x \end{vmatrix} = 0$$

or $x^3 - 2x = 0$, or $x = 0$, $\pm\sqrt{2}$. Now show that if we use the ψ_j as the basis for a (reducible) representation for the allyl anion, then $\Gamma = 3\ -1\ 1\ -3$. Now show that $\Gamma = A_2 + 2B_1$. What does this say about the expected Hückel secular determinant? Now use the generating operator, Equation 12.32, to derive three symmetry orbitals for the allyl anion. Normalize them and use them to calculate the Hückel secular determinantal equation and solve for the π-electron energies.

We apply Hückel theory to show that the secular determinant is (Problem 10–41)

$$\begin{vmatrix} \alpha - E & \beta & 0 \\ \beta & \alpha - E & \beta \\ 0 & \beta & \alpha - E \end{vmatrix} = \beta^3 \begin{vmatrix} x & 1 & 0 \\ 1 & x & 1 \\ 0 & 1 & x \end{vmatrix} = 0$$

Expanding gives $x^3 - 2x = 0$, and solving this equation gives $x = 0, \pm\sqrt{2}$. There are four operators, $\hat{E}$, $\hat{C}_2$, $\hat{\sigma}_v$, and $\hat{\sigma}'_v$, for the C_{2v} point group. The operator $\hat{E}$ leaves all three orbitals unchanged (a reducible character of 3); $\hat{C}_2$ inverts one of the $2p$-orbitals and moves two of them (a reducible character of -1); $\hat{\sigma}_v$ leaves one $2p$-orbital unchanged and moves two of them (a reducible character of 1); and $\hat{\sigma}'_v$ inverts all three $2p$-orbitals (a reducible character of -3). Therefore, $\Gamma = 3 \ -1 \ 1 \ -3$. We can use Equation 12.23 to find the irreducible representations

$$a_{A_1} = \frac{1}{4}(3 - 1 + 1 - 3) = 0$$

$$a_{A_2} = \frac{1}{4}(3 - 1 - 1 + 3) = 1$$

$$a_{B_1} = \frac{1}{4}(3 + 1 + 1 + 3) = 2$$

$$a_{B_2} = \frac{1}{4}(3 + 1 - 1 - 3) = 0$$

We can therefore write $\Gamma = A_2 + 2B_1$. This result shows that the secular determinant can be written in block diagonal form; a 1×1 block corresponding to the A_2 representation and a 2×2 block corresponding to the doubly degenerate wavefunction that transforms as B_1. Therefore, we need only use generating operators $\hat{P}_{A_2}$ and $\hat{P}_{B_1}$ to find three symmetry orbitals for the allyl anion:

$$\hat{P}_{A_2}\psi_2 = \frac{1}{4}(\psi_1 - \psi_3 - \psi_3 + \psi_1) \propto \psi_1 - \psi_3$$

$$\hat{P}_{B_1}\psi_1 = \frac{1}{4}(\psi_1 + \psi_3 + \psi_3 + \psi_1) \propto \psi_1 + \psi_3$$

$$\hat{P}_{B_1}\psi_2 = \frac{1}{4}(\psi_2 + \psi_2 + \psi_2 + \psi_2) = \psi_2$$

The three normalized symmetry orbitals are

$$\phi_1 = \frac{1}{\sqrt{2}}(\psi_1 - \psi_3) \qquad \phi_2 = \psi_2 \qquad \phi_3 = \frac{1}{\sqrt{2}}(\psi_1 + \psi_3)$$

Using these three orbitals,

$$H_{11} = \tfrac{1}{2}(2\alpha) = \alpha \qquad H_{22} = \alpha$$

$$H_{33} = \tfrac{1}{2}(2\alpha) = \alpha \qquad H_{13} = \tfrac{1}{2}(\alpha - \alpha) = 0$$

$$H_{12} = \tfrac{1}{2}(\beta - \beta) = 0 \qquad H_{23} = \frac{1}{\sqrt{2}}(2\beta) = \sqrt{2}\beta$$

$$S_{11} = S_{22} = S_{33} = 1 \qquad S_{12} = S_{23} = 0$$

$$S_{13} = \tfrac{1}{2}(1 - 1) = 0$$

This gives the secular determinant

$$\begin{vmatrix} \alpha - E & 0 & 0 \\ 0 & \alpha - E & \sqrt{2}\beta \\ 0 & \sqrt{2}\beta & \alpha - E \end{vmatrix} = \beta \begin{vmatrix} x & 0 & 0 \\ 0 & x & \sqrt{2} \\ 0 & \sqrt{2} & x \end{vmatrix} = 0$$

which gives $x = 0, \pm\sqrt{2}$ and energies of

$$E = \alpha - \sqrt{2}\beta \qquad E = \alpha \qquad E = \alpha + \sqrt{2}\beta$$

Note that the secular determinant has the block diagonal form predicted above.

12–32. Apply the analysis we use in Example 12–15 to a minimal basis set for NH_3.

There are eight orbitals to consider: $1s_{H_1}$, $1s_{H_2}$, $1s_{H_3}$, $1s_N$, $2s_N$, $2p_{xN}$, $2p_{yN}$, and $2p_{zN}$. NH_3 has the symmetry properties of the point group C_{3v}, so we consider the effects of the operators of this point group. The operator $\hat{E}$ leaves all eight orbitals unchanged. $\hat{C}_3$ leaves only the $1s_N$ and $2s_N$ orbitals unchanged, and so $\chi(\hat{C}_3) = 2$. $\hat{C}_3^2$ has the same effect (the direction of rotation of the molecule is arbitrary). The three $\hat{\sigma}_v$ operators leave four orbitals unchanged and two of the $2p$ orbitals move to other atoms, so $\chi(\hat{\sigma}_v) = \chi(\hat{\sigma}_v') = \chi(\hat{\sigma}_v'') = 4$. We therefore find

C_{3v}	$\hat{E}$	$\hat{C}_3$	$\hat{C}_3^2$	$\hat{\sigma}_v$	$\hat{\sigma}_v'$	$\hat{\sigma}_v''$
Γ	8	2	2	4	4	4

We now use Equation 12.23 to find

$$a_{A_1} = \frac{1}{6}(8 + 2 + 2 + 4 + 4 + 4) = 4$$

$$a_{A_2} = \frac{1}{6}(8 + 2 + 2 - 4 - 4 - 4) = 0$$

$$a_E = \frac{1}{6}(16 - 2 - 2) = 2$$

so $\Gamma = 4A_1 + 2E$. The original 8×8 secular determinant can be written in block diagonal form, consisting of one 4×4 block and two 2×2 blocks.

12–33. Just as we have orthogonality conditions for the characters of irreducible representations, there are also orthogonality conditions of their matrix elements. For example, if $\Gamma_i(\hat{R})_{mn}$ denotes the mn matrix element of the matrix of the ith irreducible representation, then

$$\sum_{\hat{R}} \Gamma_i(\hat{R})_{mn} \Gamma_j(\hat{R})_{m'n'} = \frac{h}{d_i} \delta_{ij} \delta_{mm'} \delta_{nn'}$$

This rather complicated looking equation is called the *great orthogonality theorem*. Show how this equation applies to the elements of the matrices in Table 12.6.

We will choose several examples to illustrate how this equation applies to the elements in Table 12.6. First, consider the squares of the elements of the 2×2 matrices in Table 12.6 ($i = j = E$, $m = m'$, and $n = n'$. In this case the sum should be $h/l = 6/2 = 3$.

$$\sum_{\hat{R}} [\Gamma_E(\hat{R})_{11}]^2 = 1 + \frac{1}{4} + \frac{1}{4} + 1 + \frac{1}{4} + \frac{1}{4} = 3$$

$$\sum_{\hat{R}} [\Gamma_E(\hat{R})_{12}]^2 = 0 + \frac{3}{4} + \frac{3}{4} + 0 + \frac{3}{4} + \frac{3}{4} = 3$$

$$\sum_{\hat{R}} [\Gamma_E(\hat{R})_{21}]^2 = 0 + \frac{3}{4} + \frac{3}{4} + 0 + \frac{3}{4} + \frac{3}{4} = 3$$

$$\sum_{\hat{R}} [\Gamma_E(\hat{R})_{11}]^2 = 1 + \frac{1}{4} + \frac{1}{4} + 1 + \frac{1}{4} + \frac{1}{4} = 3$$

Now consider the products of the same elements (letting $i = j = E$). In this case, either $m \neq m'$ or $n \neq n'$, and the sum should be zero.

$$\sum_{\hat{R}} \Gamma_E(\hat{R})_{11} \Gamma_E(\hat{R})_{12} = 0 + \frac{\sqrt{3}}{4} - \frac{\sqrt{3}}{4} + 0 - \frac{\sqrt{3}}{4} + \frac{\sqrt{3}}{4} = 0$$

$$\sum_{\hat{R}} \Gamma_E(\hat{R})_{11} \Gamma_E(\hat{R})_{22} = 1 + \frac{1}{4} + \frac{1}{4} - 1 - \frac{1}{4} - \frac{1}{4} = 0$$

Finally, consider the products of A_2 with E, where $i \neq j$ and the sum should be zero.

$$\sum_{\hat{R}} \Gamma_{A_2}(\hat{R})_{11} \Gamma_E(\hat{R})_{11} = 1 - \frac{1}{2} - \frac{1}{2} - 1 + \frac{1}{2} + \frac{1}{2} = 0$$

$$\sum_{\hat{R}} \Gamma_{A_2}(\hat{R})_{11} \Gamma_E(\hat{R})_{12} = 0 - \frac{\sqrt{3}}{2} + \frac{\sqrt{3}}{2} - 0 - \frac{\sqrt{3}}{2} + \frac{\sqrt{3}}{2} = 0$$

The other products where $i \neq j$ can easily be shown to be zero.

12–34.

a. Let $i = j$, $m = n$, and $m' = n'$ in the great orthogonality theorem (Problem 12–33) and sum over n and n' to derive Equation 12.18.

b. Let $m = n$, $m' = n'$ and sum over n and n' to derive Equation 12.14.

c. Combine these results to derive Equation 12.20.

---

a. Recall that $\chi_j(\hat{R})$ is defined as the character of the jth irreducible representation of $\hat{R}$, which, in terms of matrix elements, is given by

$$\chi_i(\hat{R}) = \sum_m \Gamma_i(\hat{R})_{mm}$$

We now use the great orthogonality theorem to find Equation 12.18:

$$\sum_{\hat{R}} \Gamma_i(\hat{R})_{mn} \Gamma_j(\hat{R})_{m'n'} = \frac{h}{l_i} \delta_{ij} \delta_{mm'} \delta_{nn'}$$

Let $i = j$, $m = n$, and $m' = n'$. Then

$$\sum_{\hat{R}} \Gamma_i(\hat{R})_{nn} \Gamma_i(\hat{R})_{n'n'} = \frac{h}{l_i} \delta_{nn'}$$

$$\sum_{\hat{R}} \sum_n \Gamma_i(\hat{R})_{nn} \sum_{n'} \Gamma_i(\hat{R})_{n'n'} = \frac{h}{l_i} \delta_{nn'}$$

$$\sum_{\hat{R}} [\chi_i(\hat{R})]^2 = \frac{h}{l_i} = h$$

(Note that $l_i = 1$ for all elements in Tables 12.7 through 12.14.)

b. Use the great orthogonality theorem to find Equation 12.14:

$$\sum_{\hat{R}} \Gamma_i(\hat{R})_{mn} \Gamma_j(\hat{R})_{m'n'} = \frac{h}{l_i} \delta_{ij} \delta_{mm'} \delta_{nn'}$$

Let $i \neq j$, $m = n$, and $m' = n'$. Then

$$\sum_{\hat{R}} \Gamma_i(\hat{R})_{nn} \Gamma_j(\hat{R})_{n'n'} = \frac{h}{l_i} \delta_{ij} \delta_{nn'}$$

$$\sum_{\hat{R}} \sum_{n} \Gamma_i(\hat{R})_{nn} \sum_{n'} \Gamma_j(\hat{R})_{n'n'} = \frac{h}{l_i} \delta_{ij} \delta_{nn'}$$

$$\sum_{\hat{R}} \chi_i(\hat{R}) \chi_j(\hat{R}) = \frac{h}{l_i} \delta_{ij} = h \delta_{ij}$$

$$\sum_{\hat{R}} \chi_i(\hat{R}) \chi_j(\hat{R}) = 0 \qquad i \neq j$$

c. Combining the results of parts (a) and (b),

$$\sum_{\hat{R}} \chi_i(\hat{R}) \chi_i(\hat{R}) = \sum_{m} \frac{h}{l_i} = h$$

and

$$\sum_{\hat{R}} \chi_i(\hat{R}) \chi_j(\hat{R}) = 0 \qquad i \neq j$$

gives

$$\sum_{\hat{R}} \chi_i(\hat{R}) \chi_j(\hat{R}) = h \delta_{ij}$$

12–35. Consider the point group $\mathbf{C}_s$, which contains only the symmetry elements E and σ. Determine the character table for $\mathbf{C}_s$. (The molecule NOCl belongs to this point group.)

There are two symmetry elements, and so, by Equation 12.7, the $\mathbf{C}_s$ character table consists of two one-dimensional representations. All groups contain a totally symmetric representation, so the first row of the character table must be 1 1. The second row must be orthogonal to the first, so we find

$\mathbf{C}_s$	$\hat{E}$	$\hat{\sigma}$
	1	1
	1	−1

12–36. Consider the simple point group $\mathbf{C}_s$ whose character table is

$\mathbf{C}_s$	$\hat{E}$	$\hat{\sigma}$
A'	1	1
A''	1	−1

where $\hat{\sigma}$ represents reflection through the y axis in a two-dimensional x, y Cartesian coordinate system. Show that the bases for this point group are even and odd functions of x over a symmetric interval, $-a \leq x \leq a$. Now use group theory to show that

$$\int_{-a}^{a} f_{\text{even}}(x) f_{\text{odd}}(x) dx = 0$$

From the definition of even and odd functions,

$$\hat{\sigma} f_{\text{even}} = f_{\text{even}} \qquad \hat{\sigma} f_{\text{odd}} = -f_{\text{odd}}$$

Thus f_{even} belongs to A' and f_{odd} belongs to A'', so these types of functions serve as the bases for this point group. Equation 12.26 requires that $S_{ij} = \hat{R} S_{ij}$ for all symmetry elements of the group. We then have

$$S_{ij} = \int_{-a}^{a} f_{\text{even}}(x) f_{\text{odd}}(x) dx$$

and

$$\hat{\sigma} S_{ij} = \int_{-a}^{a} \hat{\sigma} f_{\text{even}}(x) \hat{\sigma} f_{\text{odd}}(x) dx$$

$$= -\int_{-a}^{a} f_{\text{even}}(x) f_{\text{odd}}(x) dx$$

In order for Equation 12.26 to hold,

$$\int_{-a}^{a} f_{\text{even}}(x) f_{\text{odd}}(x) dx = 0$$

12–37. We calculated the π-electron energy of a trimethylenemethane molecule in Problem 10–31. Derive the symmetry orbitals for the π orbitals by applying the generating operator, Equation 12.32, to the atomic $2p_z$ orbital on each carbon atom. Identify the irreducible representation to which each resulting symmetry orbital belongs. Derive the Hückel secular determinant corresponding to these symmetry orbitals and compare it to the one that you obtained in Problem 10–31. Compare the π-electron energies.

The point group of the molecule trimethylenemethane is $\mathbf{D}_{3h}$.

We use the schematic diagram above to picture the effects of the operations on the molecule. Applying the symmetry operations to the $2p_z$ orbitals on each carbon atom gives the following reducible representation:

$\mathbf{D}_{3h}$	$\hat{E}$	$2\hat{C}_3$	$3\hat{C}_2$	$\hat{\sigma}_h$	$2\hat{S}_3$	$3\hat{\sigma}_v$
Γ	4	1	-2	-4	-1	2

We now use Equation 12.23 to reduce this representation

$$a_{A_1'} = \frac{1}{12}(4 + 2 - 6 - 4 - 2 + 6) = 0$$

$$a_{A_2'} = \frac{1}{12}(4 + 2 + 6 - 4 - 2 - 6) = 0$$

$$a_{E'} = \frac{1}{12}(8 - 2 + 0 - 8 + 2 + 0) = 0$$

$$a_{A_1''} = \frac{1}{12}(4 + 2 - 6 + 4 + 2 - 6) = 0$$

$$a_{A_2''} = \frac{1}{12}(4 + 2 + 6 + 4 + 2 + 6) = 2$$

$$a_{E''} = \frac{1}{12}(8 - 2 + 0 + 8 - 2 + 0) = 1$$

and find that $\Gamma = 2A_2'' + E''$. Therefore, we need only use the generating operators $\hat{P}_{A_2''}$ and $\hat{P}_{E''}$ to find the symmetry orbitals for trimethylenemethane:

$$\hat{P}_{A_2''}\psi_1 = \frac{1}{12}(\psi_1 + \psi_1 + \psi_1 + 3\psi_1 + \psi_1 + 2\psi_1 + 3\psi_1) = \psi_1$$

$$\hat{P}_{A_2''}\psi_2 = \frac{1}{12}(\psi_2 + \psi_3 + \psi_4 + \psi_2 + \psi_4 + \psi_3 + \psi_2 + \psi_3 + \psi_4 + \psi_2 + \psi_3 + \psi_4) \propto \psi_2 + \psi_3 + \psi_4$$

$$\hat{P}_{E''}\psi_1 = \frac{2}{12}(2\psi_1 - 2\psi_1 + 0 + 2\psi_1 - 2\psi_1 + 0) = 0$$

$$\hat{P}_{E''}\psi_2 = \frac{2}{12}(2\psi_2 - \psi_3 - \psi_4 + 0 + 2\psi_2 - \psi_3 - \psi_4 + 0) \propto 2\psi_2 - \psi_3 - \psi_4$$

$$\hat{P}_{E''}\psi_3 = \frac{2}{12}(2\psi_3 - \psi_2 - \psi_4 + 0 + 2\psi_3 - \psi_2 - \psi_4 + 0) \propto 2\psi_3 - \psi_2 - \psi_4$$

$$\hat{P}_{E''}\psi_4 = \frac{2}{12}(2\psi_4 - \psi_2 - \psi_3 + 0 + 2\psi_4 - \psi_2 - \psi_3 + 0) \propto 2\psi_4 - \psi_2 - \psi_3$$

Only two of the wave functions found using the $\hat{P}_{E''}$ are linearly independent. This gives us a total of four symmetry orbitals. The normalized symmetry orbitals are

$$\phi_1(A_2'') = \psi_1 \qquad\qquad \phi_2(A_2'') = \frac{1}{\sqrt{3}}(\psi_2 + \psi_3 + \psi_4)$$

$$\phi_3(E'') = \frac{1}{\sqrt{6}}(2\psi_2 - \psi_3 - \psi_4) \qquad \phi_4(E'') = \frac{1}{\sqrt{6}}(2\psi_3 - \psi_2 - \psi_4)$$

The elements of the corresponding Hückel secular determinants are

$$H_{11} = \int \psi_1 \hat{H}\psi_1 d\tau = \alpha$$

$$H_{12} = \int \psi_1 \hat{H}\frac{1}{\sqrt{3}}(\psi_2 + \psi_3 + \psi_4)\,d\tau = \sqrt{3}\beta$$

$$H_{22} = H_{33} = H_{34} = \alpha$$

$$H_{13} = H_{14} = \frac{1}{\sqrt{6}}(2\beta - \beta - \beta) = 0$$

$$H_{23} = H_{24} = \frac{1}{3\sqrt{2}}(2\alpha - \alpha - \alpha) = 0$$

$$H_{34} = \frac{1}{6}(-2\alpha - 2\alpha + \alpha) = -\frac{1}{2}\alpha$$

$$S_{11} = S_{22} = S_{33} = S_{44} = 1$$

$$S_{12} = S_{13} = S_{14} = 0$$

$$S_{23} = S_{24} = 0$$

$$S_{34} = -\frac{1}{2}$$

which leads to the determinantal equation

$$\begin{vmatrix} x & \sqrt{3} & 0 & 0 \\ \sqrt{3} & x & 0 & 0 \\ 0 & 0 & x & -\frac{x}{2} \\ 0 & 0 & -\frac{x}{2} & x \end{vmatrix} = \begin{vmatrix} x & \sqrt{3} \\ \sqrt{3} & x \end{vmatrix}\frac{3x^2}{4} = 0$$

Expanding this result gives

$$(x^2 - 3)\frac{3x^2}{4} = 0$$

Solving this last equation for x gives $x = 0, \pm\sqrt{3}$, which are the solutions that we found in Problem 10–31. The π-electron energy is then

$$E_\pi = 2(\alpha + \sqrt{3}\beta) + 2\alpha = 4\alpha + 2\sqrt{3}\beta$$

Both approaches give the same π-electron energies.

12–38. We calculated the π-electron energy of a bicyclobutadiene molecule in Problem 10–32. Using the point group $\mathbf{C}_{2h}$, derive the symmetry orbitals for the π orbitals by applying the generating operator, Equation 12.32, to the atomic $2p_z$ orbital on each carbon atom. Identify the irreducible representation to which each resulting symmetry orbital belongs. Derive the Hückel secular determinant corresponding to these symmetry orbitals and compare it to the one that you obtained in Problem 10–32. Compare the π-electron energies.

The point group of bicyclobutadiene is $\mathbf{C}_{2h}$.

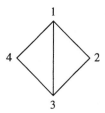

We use the schematic diagram above to picture the effects of the operations on the molecule. Applying the symmetry operations to the $2p_z$ orbitals on each carbon atom gives the following reducible representation:

$\mathbf{D}_{3h}$	$\hat{E}$	$\hat{C}_2$	$\hat{i}$	$\hat{\sigma}_h$
Γ	4	0	0	−4

We now use Equation 12.23 to reduce this representation:

$$a_{A_g} = \frac{1}{4}(4 + 0 + 0 - 4) = 0$$

$$a_{B_g} = \frac{1}{4}(4 + 0 + 0 + 4) = 2$$

$$a_{A_u} = \frac{1}{4}(4 + 0 + 0 + 4) = 2$$

$$a_{B_u} = \frac{1}{4}(4 + 0 + 0 - 4) = 0$$

and find that $\Gamma = 2B_g + 2A_u$. Therefore, we need only to use the generating operators $\hat{P}_{B_g}$ and $\hat{P}_{A_u}$ to find the symmetry orbitals for bicyclobutadiene:

$$\hat{P}_{B_g}\psi_1 = \frac{1}{4}(\psi_1 - \psi_4 - \psi_4 + \psi_1) \propto \psi_1 - \psi_4$$

$$\hat{P}_{B_g}\psi_2 = \frac{1}{4}(\psi_2 - \psi_3 - \psi_3 + \psi_2) \propto \psi_2 - \psi_3$$

$$\hat{P}_{A_u}\psi_1 = \frac{1}{4}\left(\psi_1 + \psi_4 + \psi_4 + \psi_1\right) \propto \psi_1 + \psi_4$$

$$\hat{P}_{A_u}\psi_2 = \frac{1}{4}\left(\psi_2 + \psi_3 + \psi_3 + \psi_2\right) \propto \psi_2 + \psi_3$$

This gives us four symmetry orbitals. The normalized symmetry orbitals are

$$\phi_1(B_g) = \frac{1}{\sqrt{2}}\left(\psi_1 - \psi_4\right) \qquad \phi_2(B_g) = \frac{1}{\sqrt{2}}\left(\psi_2 - \psi_3\right)$$

$$\phi_3(A_u) = \frac{1}{\sqrt{2}}\left(\psi_1 + \psi_4\right) \qquad \phi_4(A_u) = \frac{1}{\sqrt{2}}\left(\psi_2 + \psi_3\right)$$

We then have

$$H_{11} = \alpha - \beta \qquad H_{12} = 0 \qquad H_{22} = H_{44} = \alpha$$

$$H_{13} = H_{14} = 0 \qquad H_{23} = H_{24} = 0 \qquad H_{34} = 2\beta$$

and

$$H_{33} = \alpha + \beta \qquad S_{11} = S_{22} = S_{33} = S_{44} = 1$$

$$S_{12} = S_{13} = S_{14} = 0 \qquad S_{23} = S_{24} = S_{34} = 0$$

which leads to the determinantal equation

$$\begin{vmatrix} x-1 & 0 & 0 & 0 \\ 0 & x & 0 & 0 \\ 0 & 0 & x+1 & 2 \\ 0 & 0 & 2 & x \end{vmatrix} = x(x-1)\begin{vmatrix} x+1 & 2 \\ 2 & x \end{vmatrix} = 0$$

Expanding this result gives

$$x(x-1)(x^2 + x - 4) = x^4 - 5x^2 + 4x = 0$$

This is the same equation as that found in Problem 10–32. The π-electron energy is then

$$E_\pi = 2(\alpha + 2.562\beta) + 2\alpha = 4\alpha + 5.124\beta$$

Both approaches give the same π-electron energies.

12–39. Use the generating operator, Equation 12.32, to derive the symmetry orbitals for the π orbitals of the (bent) allyl radical ($C_3H_5\cdot$) from a basis set consisting of a $2p_z$ orbital on each carbon atom. (Assume the three carbon atoms lie in the x–y plane.) Now create a set of orthonormal molecular orbitals from these symmetry orbitals. Sketch each orbital. How do your results compare with the π orbitals predicted by Hückel theory (see Problem 10–47)?

The (bent) allyl radical $C_3H_5\cdot$ belongs to the C_{2v} point group. Applying the symmetry operators to the $2p_z$ orbital on each carbon atom gives the following reducible representation:

C_{2v}	$\hat{E}$	$\hat{C}_2$	$\hat{\sigma}_v$	$\hat{\sigma}_v'$
Γ	3	−1	1	−3

Using Equation 12.23, this reducible representation corresponds to $\Gamma = A_2 + 2B_1$. Using the generating operators for A_2 and B_1 gives

$$\phi_1(A_2) \propto \psi_1 - \psi_3 \qquad \phi_2(B_1) \propto \psi_1 + \psi_3 \qquad \phi_3(B_1) \propto \psi_2$$

and the normalized symmetry orbitals are (Problem 12–21)

$$\phi_1 = \frac{1}{\sqrt{2}}(\psi_1 - \psi_3) \qquad \phi_2 = \psi_2 \qquad \phi_3 = \frac{1}{\sqrt{2}}(\psi_1 + \psi_3)$$

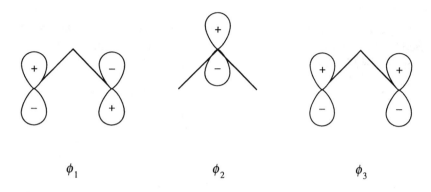

$$\phi_1 \qquad\qquad\qquad \phi_2 \qquad\qquad\qquad \phi_3$$

These correspond to the π orbitals predicted by Hückel theory.

The following four problems illustrate the application of group theory to the formation of hybrid orbitals.

12–40. Consider a trigonal planar molecule XY_3 whose point group is $\mathbf{D}_{3h}$. All three bonds are unmoved by the operation of $\hat{E}$; all three are moved by the operation of $\hat{C}_3$; one is unmoved by the operation of $\hat{C}_2$; all three are unmoved by the operation of $\hat{\sigma}_h$; all are moved by the operation of $\hat{S}_3$; and one is unmoved by the operation of $\hat{\sigma}_v$. This result leads to the reducible representation $\Gamma = 3\ 0\ 1\ 3\ 0\ 1$. Now show that $\Gamma = A_1' + E'$. Argue now that this result suggests that hybrid orbitals with $\mathbf{D}_{3h}$ symmetry can be formed from an s orbital and the p_x and p_y orbitals (or the $d_{x^2-y^2}$ and d_{z^2} orbitals) to give sp^2 (or sd^2) hybrid orbitals.

We can use Equation 12.23 to reduce the representation given in the problem text:

$$a_{A_1'} = \frac{1}{12}(3 + 0 + 3 + 3 + 0 + 3) = 1$$

$$a_{A_2'} = \frac{1}{12}(3 + 0 - 3 + 3 + 0 - 3) = 0$$

$$a_{E'} = \frac{1}{12}(6 + 0 + 0 + 6 + 0 + 0) = 1$$

$$a_{A_1''} = \frac{1}{12}(3 + 0 + 3 - 3 + 0 - 3) = 0$$

$$a_{A_2''} = \frac{1}{12}(3 + 0 - 3 - 3 + 0 + 3) = 0$$

$$a_{E''} = \frac{1}{12}(6 + 0 + 0 - 6 + 0 + 0) = 0$$

So we find the irreducible representation $\Gamma = A_1' + E'$. Looking at the rightmost column of Table 12.11, we see that the $d_{x^2-y^2}$ and d_{z^2} orbitals can combine with an s orbital to form sd^2 hybrid orbitals; likewise, looking at the next-to-rightmost column of Table 12.11 shows that the p_x and p_y orbitals can combine with an s orbital to form sp^2 hybrid orbitals.

12–41. Consider a tetrahedral molecule XY_4 whose point group is $\mathbf{T}_d$. Using the procedure introduced in Problem 12–40, show that $\Gamma = 4\ 1\ 0\ 0\ 2$, which reduces to $\Gamma = A_1 + T_2$. Now argue that hybrid orbitals with $\mathbf{T}_d$ symmetry can be formed from an s orbital and the p_x, p_y and p_z orbitals (or the d_{xy}, d_{xz}, and d_{yz} orbitals) to give sp^3 (or sd^3) hybrid orbitals.

All four bonds are unmoved by the operation of $\hat{E}$; one is unmoved by the operation of $\hat{C}_3$; all four are moved by the operation of $\hat{C}_2$; all four are moved by the operation of $\hat{S}_4$; and two are unmoved by the operation of $\hat{\sigma}_d$. This result leads to the reducible representation $\Gamma = 4\ 1\ 0\ 0\ 2$. We can now use Equation 12.23 to reduce this representation:

$$a_{A_1} = \frac{1}{24}(4 + 8 + 0 + 0 + 12) = 1$$

$$a_{A_2} = \frac{1}{24}(4 + 8 + 0 + 0 - 12) = 0$$

$$a_E = \frac{1}{24}(8 - 8 + 0 + 0 + 0) = 0$$

$$a_{T_1} = \frac{1}{24}(12 + 0 + 0 + 0 - 12) = 0$$

$$a_{T_2} = \frac{1}{24}(12 + 0 + 0 + 0 + 12) = 1$$

So we find the irreducible representation $\Gamma = A_1 + T_2$. Looking at the rightmost column of Table 12.14, we see that the d_{xy}, d_{xz} and d_{yz} orbitals can combine with an s orbital to form sd^3 hybrid orbitals; likewise, looking at the next-to-rightmost column of Table 12.14 shows that the p_x, p_y and p_z orbitals can combine with an s orbital to form sp^3 hybrid orbitals.

12–42. Consider a square planar molecule XY_4 whose point group is $\mathbf{D}_{4h}$. Using the procedure introduced in Problem 12–40, show that $\Gamma = 4\ 0\ 0\ 2\ 0\ 0\ 0\ 4\ 2\ 0$, which reduces to $\Gamma = A_{1g} + B_{1g} + E_u$. Now argue that hybrid orbitals $\mathbf{D}_{4h}$ symmetry can be formed from a s orbital, a $d_{x^2-y^2}$ orbital, and the p_x and p_y orbitals to give sdp^2 hybrid orbitals.

All four bonds are unmoved by the operation of $\hat{E}$; all four are moved by the operation of $\hat{C}_4$; all four are moved by the operation of $\hat{C}_2$; two bonds are unmoved by the operation of $\hat{C}_2'$; all four are moved by the operation of $\hat{C}_2''$, $\hat{i}$, and $\hat{S}_4$; all four are unmoved by the operation of $\hat{\sigma}_h$; two bonds are unmoved by the operation of $\hat{\sigma}_v$; and all four are moved by the operation of $\hat{\sigma}_d$. This result leads to the reducible representation $\Gamma = 4\ 0\ 0\ 2\ 0\ 0\ 0\ 4\ 2\ 0$. We can now use Equation 12.23 to reduce this representation:

$$a_{A_{1g}} = \frac{1}{16}(4 + 4 + 4 + 4) = 1$$

$$a_{A_{2g}} = \frac{1}{16}(4 - 4 + 4 - 4) = 0$$

$$a_{B_{1g}} = \frac{1}{16}(4 + 4 + 4 + 4) = 1$$

$$a_{B_{2g}} = \frac{1}{16}(4 - 4 + 4 - 4) = 0$$

$$a_{E_g} = \frac{1}{16}(8 - 8) = 0$$

$$a_{A_{1u}} = \frac{1}{16}(4 + 4 - 4 - 4) = 0$$

$$a_{A_{2u}} = \frac{1}{16}(4 - 4 - 4 + 4) = 0$$

$$a_{B_{1u}} = \frac{1}{16}(4 + 4 - 4 - 4) = 0$$

$$a_{B_{2u}} = \frac{1}{16}(4 - 4 - 4 + 4) = 0$$

$$a_{E_u} = \frac{1}{16}(8 + 8) = 1$$

So we find the irreducible representation $\Gamma = A_{1g} + B_{1g} + E_u$. Looking at the rightmost and next-to-rightmost columns of Table 12.12, we see that the $d_{x^2-y^2}$, p_x and p_y orbitals can combine with an s orbital to form sp^2d hybrid orbitals.

12–43. Consider a trigonal bipyramidal molecule XY_5 whose point group is D_{3h}. Using the procedure introduced in Problem 12–40, show that $\Gamma = 5\ 2\ 1\ 3\ 0\ 3$, and that $\Gamma = 2A_1' + A_2'' + E'$. Now argue that hybrid orbitals with D_{3h} symmetry can be formed from an s orbital, a d_{z^2} orbital, a p_z orbital, and p_x and p_y orbitals to give spd^3 hybrid orbitals.

All five bonds are unmoved by the operation of $\hat{E}$; two bonds are unmoved by the operation of $\hat{C}_3$; only one bond is unmoved by the operation of $\hat{C}_2$; three bonds are unmoved by the operation of $\hat{\sigma}_h$; all five are moved by the operation of $\hat{S}_3$; and three are unmoved by the operation of $\hat{\sigma}_v$. This result leads to the reducible representation $\Gamma = 5\ 2\ 1\ 3\ 0\ 3$. We can now use Equation 12.23 to reduce this representation:

$$a_{A_1'} = \frac{1}{12}(5 + 4 + 3 + 3 + 9) = 2$$

$$a_{A_2'} = \frac{1}{12}(5 + 4 - 3 + 3 - 9) = 0$$

$$a_{E'} = \frac{1}{12}(10 - 4 + 6) = 1$$

$$a_{A_1''} = \frac{1}{12}(5 + 4 + 3 - 3 - 9) = 0$$

$$a_{A_2''} = \frac{1}{12}(5 + 4 - 3 - 3 + 9) = 1$$

$$a_{E''} = \frac{1}{12}(10 - 4 - 6) = 0$$

So we find the irreducible representation $\Gamma = 2A_1' + A_2'' + E'$. Looking at the rightmost and next-to-rightmost columns of Table 12.11, we see that the d_{z^2}, p_x, p_y and p_z orbitals can combine with an s orbital to form sp^3d hybrid orbitals.

13–1. The spacing between the lines in the microwave spectrum of $H^{35}Cl$ is 6.350×10^{11} Hz. Calculate the bond length of $H^{35}Cl$.

Using Equation 13.6 for E_J, we find

$$\Delta E = E_{J+1} - E_J = \frac{\hbar^2}{2I}(J+1)(J+2) - \frac{\hbar^2}{2I}J(J+1)$$

$$= \frac{\hbar^2}{2I}[(J+1)(J+2-J)]$$

$$= \frac{\hbar^2}{I}(J+1)$$

Substituting the last result into Equation 13.1 gives

$$\nu = \frac{\Delta E}{h} = \frac{h}{4\pi^2 I}(J+1) \tag{1}$$

The spectrum consists of lines of frequencies ν, separated from each other by $h/4\pi^2 I$. Then

$$6.350 \times 10^{11} \text{ s}^{-1} = \frac{h}{4\pi^2 I}$$

$$I = \frac{6.626 \times 10^{-34} \text{ J·s}}{4\pi^2 \left(6.350 \times 10^{11} \text{ s}^{-1}\right)} = 2.643 \times 10^{-47} \text{ kg·m}^2$$

The reduced mass of $H^{35}Cl$ is

$$\mu = \frac{(1.0079)(34.969)}{(34.969 + 1.0079)} \text{ amu} = 0.97966 \text{ amu}$$

For a diatomic molecule,

$$I = \mu R_e^2$$

$$R_e = \left(\frac{I}{\mu}\right)^{1/2}$$

$$= \left[\frac{2.643 \times 10^{-47} \text{ kg·m}^2}{(0.97966 \text{ amu]} \left(1.661 \times 10^{-27} \text{ kg·amu}^{-1}\right)}\right]^{1/2}$$

$$= 1.275 \times 10^{-10} \text{ m} = 127.5 \text{ pm}$$

13–2. The microwave spectrum of $^{39}K^{127}I$ consists of a series of lines whose spacing is almost constant at 3634 MHz. Calculate the bond length of $^{39}K^{127}I$.

We use the same method as in Problem 13–1. From Equation 1 of Problem 13–1,

$$3\,634 \times 10^6 \text{ s}^{-1} = \frac{h}{4\pi^2 I}$$

$$I = \frac{6.626 \times 10^{-34} \text{ J} \cdot \text{s}}{4\pi^2 \left(3.634 \times 10^9 \text{ s}^{-1}\right)} = 4.619 \times 10^{-45} \text{ kg} \cdot \text{m}^2$$

The reduced mass of $^{39}K^{127}I$ is

$$\mu = \frac{(38.964)(126.90)}{(38.964 + 126.90)} \text{ amu} = 29.811 \text{ amu}$$

From the definition of I,

$$I = \mu R_e^2$$

$$R_e = \left(\frac{I}{\mu}\right)^{1/2}$$

$$= \left(\frac{4.619 \times 10^{-45} \text{ kg} \cdot \text{m}^2}{(29.811 \text{ amu}) \left(1.661 \times 10^{-27} \text{ kg} \cdot \text{amu}^{-1}\right)}\right)^{1/2}$$

$$= 3.055 \times 10^{-10} \text{ m} = 305.5 \text{ pm}$$

13–3. The equilibrium internuclear distance of $H^{127}I$ is 160.4 pm. Calculate the value of B in wave numbers and megahertz.

The reduced mass of $H^{127}I$ is

$$\mu = \frac{(1.0079)(126.90)}{(1.0079 + 126.90)} \text{ amu} = 0.99996 \text{ amu}$$

The moment of inertia for $H^{127}I$ is

$$I = \mu r^2$$

$$= (0.99996 \text{ amu}) \left(1.661 \times 10^{-27} \text{ kg} \cdot \text{amu}^{-1}\right) \left(1.604 \times 10^{-10} \text{ m}\right)^2$$

$$= 4.272 \times 10^{-47} \text{ kg} \cdot \text{m}^2$$

We can use Equation 13.9 for B, since B and $\tilde{B}$ are related by $c\tilde{B} = B$:

$$B = c\tilde{B} = \frac{h}{8\pi^2 I}$$

$$= \frac{6.626 \times 10^{-34} \text{ J} \cdot \text{s}}{8\pi^2 \left(4.272 \times 10^{-47} \text{ kg} \cdot \text{m}^2\right)}$$

$$= 1.964 \times 10^{11} \text{ s}^{-1} = 1.964 \times 10^5 \text{ MHz}$$

or 6.552 cm^{-1}.

13–4. Assuming the rotation of a diatomic molecule in the $J = 10$ state may be approximated by classical mechanics, calculate how many revolutions per second $^{23}Na^{35}Cl$ makes in the $J = 10$ rotational state. The rotational constant of $^{23}Na^{35}Cl$ is 6500 MHz.

The energy of a classical rotator is

$$K = \frac{1}{2}I\omega^2$$

The quantum-mechanical energy is given by

$$E = hBJ(J+1)$$

where $B = h/8\pi^2 I$. Equating K with E gives

$$\frac{I\omega^2}{2} = hBJ(J+1)$$

$$\omega^2 = \frac{2hBJ(J+1)}{I} = [2hBJ(J+1)]\left(\frac{h}{8\pi^2 B}\right)^{-1}$$

$$\omega = 4\pi B[J(J+1)]^{1/2} = 4\pi(6500 \times 10^6 \text{ s}^{-1})(110)^{1/2}$$

$$= 8.57 \times 10^{11} \text{ radian·s}^{-1} = 1.36 \times 10^{11} \text{ revolution·s}^{-1}$$

13–5. The results we derived for a rigid rotator apply to linear polyatomic molecules as well as to diatomic molecules. Given that the moment of inertia I for $H^{12}C^{14}N$ is 1.89×10^{-46} kg·m^2 (cf. Problem 13–6), predict the microwave spectrum of $H^{12}C^{14}N$.

The microwave spectrum of $H^{12}C^{14}N$ will be a series of equally spaced lines separated by $2\tilde{B}$ (see Equations 13.12 and 13.13). Substituting into Equation 13.9,

$$2\tilde{B} = \frac{h}{4\pi^2 cI}$$

$$= \frac{6.626 \times 10^{-34} \text{ J·s}}{4\pi^2(2.998 \times 10^8 \text{ m·s}^{-1})(1.89 \times 10^{-46} \text{ kg·m}^2)}$$

$$= 296 \text{ m}^{-1} = 2.96 \text{ cm}^{-1}$$

13–6. This problem involves the calculation of the moment of inertia of a linear triatomic molecule such as $H^{12}C^{14}N$ (see Problem 13–5). The moment of inertia of a linear molecule is

$$I = \sum_j m_j d_j^2$$

where d_j is the distance of the jth mass from the center of mass. Thus, the moment of inertia of $H^{12}C^{14}N$ is

$$I = m_H d_H^2 + m_C d_C^2 + m_N d_N^2 \tag{1}$$

Show that Equation 1 can be written as

$$I = \frac{m_H m_C R_{HC}^2 + m_H m_N R_{HN}^2 + m_C m_N R_{CN}^2}{m_H + m_C + m_N}$$

where the R's are the various internuclear distances. Given that $R_{HC} = 106.8$ pm and $R_{CN} = 115.6$ pm, calculate the value of I and compare the result with that given in Problem 13–5.

The easiest (and best) way to do this problem is to work backwards. Let $M = m_H + m_C + m_N$. Then the desired equation becomes

$$
\begin{aligned}
MI &= m_H m_C R_{HC}^2 + m_H m_N R_{HN}^2 + m_C m_N R_{CN}^2 \\
&= m_H m_C \left(d_H - d_C \right)^2 + m_H m_N \left(d_H - d_N \right)^2 + m_C m_N \left(d_C - d_N \right)^2 \\
&= m_H m_C d_H^2 - 2 m_H m_C d_H d_C + m_H m_C d_C^2 + m_H m_N d_H^2 - 2 m_H m_N d_H d_N \\
&\quad + m_H m_N d_N^2 + m_C m_N d_C^2 - 2 m_C m_N d_C d_N + m_C m_N d_N^2
\end{aligned}
$$

Now add and subtract $m_H d_H^2 + m_C d_C^2 + m_N d_N^2$ from the right side of this equation to obtain

$$
\begin{aligned}
MI &= M m_H d_H^2 + M m_C d_C^2 + M m_N d_N^2 \\
&\quad - \left(m_H d_H^2 + m_C d_C^2 + m_N d_N^2 + 2 m_H m_C d_H d_C + 2 m_H m_N d_H d_N + 2 m_C m_N d_C d_N \right) \\
&= M m_H d_H^2 + M m_C d_C^2 + M m_N d_N^2 - \left[m_H d_H + m_C d_C + m_N d_N \right]^2
\end{aligned}
$$

The term in brackets is equal to zero by the definition of the center of mass, so we have

$$
I = m_H d_H^2 + m_C d_C^2 + m_N d_N^2
$$

which is Equation 1. Numerically, for $H^{12}C^{14}N$,

$$
\begin{aligned}
I &= \frac{(1.0079 \text{ amu})(12.000 \text{ amu})(106.8 \text{ pm})^2}{1.0079 \text{ amu} + 12.000 \text{ amu} + 14.003 \text{ amu}} \\
&\quad + \frac{(1.0079 \text{ amu})(14.003 \text{ amu})(106.8 \text{ pm} + 115.6 \text{ pm})^2}{1.0079 \text{ amu} + 12.000 \text{ amu} + 14.003 \text{ amu}} \\
&\quad + \frac{(12.000 \text{ amu})(14.003 \text{ amu})(115.6 \text{ pm})^2}{1.0079 \text{ amu} + 12.000 \text{ amu} + 14.003 \text{ amu}} \\
&= \frac{3.082 \times 10^6 \text{ amu}^2 \cdot \text{pm}^2}{27.011 \text{ amu}} = 1.141 \times 10^5 \text{ amu} \cdot \text{pm}^2 \\
&= 1.141 \times 10^5 \text{ amu} \cdot \text{pm}^2 \left(\frac{1.661 \times 10^{-27} \text{ kg}}{\text{amu}} \right) \left(\frac{10^{-12} \text{ m}}{1 \text{ pm}} \right)^2 \\
&= 1.894 \times 10^{-46} \text{ kg} \cdot \text{m}^2
\end{aligned}
$$

This is the same as the value given in Problem 13–5.

13–7. The far infrared spectrum of $^{39}K^{35}Cl$ has an intense line at 278.0 cm^{-1}. Calculate the force constant and the period of vibration of $^{39}K^{35}Cl$.

The reduced mass of $^{39}K^{35}Cl$ is

$$
\mu = \frac{(38.964)(34.969)}{(38.964 + 34.969)} \text{ amu} = 18.429 \text{ amu}
$$

Use Equation 13.5:

$$
\tilde{\nu} = \frac{1}{2\pi c} \left(\frac{k}{\mu} \right)^{1/2}
$$

$$
k = (2\pi c \tilde{\nu})^2 \mu
$$

$$= \left[2\pi(2.998 \times 10^8 \text{ m}\cdot\text{s}^{-1})(278.0 \times 10^2 \text{ m}^{-1})\right]^2$$
$$\times (18.429 \text{ amu}) \left(1.661 \times 10^{-27} \text{ kg}\cdot\text{amu}^{-1}\right)$$
$$= 83.92 \text{ N}\cdot\text{m}^{-1}$$

The period of vibration is

$$T = \frac{1}{\nu} = \frac{1}{c\tilde{\nu}} = 1.20 \times 10^{-13} \text{ s}$$

13–8. The force constant of $^{79}\text{Br}^{79}\text{Br}$ is $240 \text{ N}\cdot\text{m}^{-1}$. Calculate the fundamental vibrational frequency and the zero-point energy of $^{79}\text{Br}_2$.

The reduced mass of $^{79}\text{Br}^{79}\text{Br}$ is

$$\mu = \frac{(78.92)(78.92)}{(78.92 + 78.92)} \text{ amu} = 39.46 \text{ amu}$$

Now we use Equation 13.5:

$$\nu = \frac{1}{2\pi} \left(\frac{k}{\mu}\right)^{1/2}$$

$$= \frac{1}{2\pi} \left[\frac{240 \text{ N}\cdot\text{m}^{-1}}{(39.46 \text{ amu}) \left(1.661 \times 10^{-27} \text{ kg}\cdot\text{amu}^{-1}\right)}\right]^{1/2}$$

$$= 9.63 \times 10^{12} \text{ s}^{-1}$$

$$\tilde{\nu} = \frac{\nu}{c} = 321 \text{ cm}^{-1}$$

From Equation 13.2,

$$E_0 = \tfrac{1}{2}h\nu = 3.19 \times 10^{-21} \text{ J}$$

13–9. Prove that

$$\langle x^2 \rangle = \frac{\hbar}{2(\mu k)^{1/2}}$$

for the ground state of a harmonic oscillator. Use this equation to calculate the root-mean-square amplitude of $^{14}\text{N}_2$ in its ground state. Compare your result to the bond length. Use $k = 2260 \text{ N}\cdot\text{m}^{-1}$ for $^{14}\text{N}_2$.

We proved this result in Problem 5–26. The reduced mass of $^{14}\text{N}_2$ is

$$\mu = \frac{(14.003)(14.003)}{(14.003 + 14.003)} \text{ amu} = 7.0015 \text{ amu}$$

For $^{14}\text{N}_2$,

$$x_{\text{rms}} = \left(\langle x^2 \rangle\right)^{1/2} = \left[\frac{\hbar}{2(\mu k)^{1/2}}\right]^{1/2}$$

$$= \left\{\frac{1.055 \times 10^{-34} \text{ J}\cdot\text{s}}{2\left[(7.0015 \text{ amu}) \left(1.661 \times 10^{-27} \text{ kg}\cdot\text{amu}^{-1}\right) \left(2260 \text{ N}\cdot\text{m}^{-1}\right)\right]^{1/2}}\right\}^{1/2}$$

$$= 3.21 \times 10^{-12} \text{ m} = 3.21 \text{ pm}$$

The bond length of $^{14}N_2$ is 109.77 pm, so x_{rms} is 3% of the bond length.

13–10. Derive Equations 13.15 and 13.16.

We begin with the equation

$$\tilde{E}_{v,J} = \tilde{\nu}\left(v + \tfrac{1}{2}\right) + \tilde{B}_v J(J+1) \tag{13.14}$$

For the R branches of the $v = 0 \to 1$ transition, $\Delta J = +1$, so

$$\begin{aligned}
\tilde{\nu}_R(\Delta J = +1) &= \tilde{E}_{1,J+1} - \tilde{E}_{0,J}\\
&= \tfrac{3}{2}\tilde{\nu} + \tilde{B}_1(J+1)(J+2) - \tfrac{1}{2}\tilde{\nu} - \tilde{B}_0 J(J+1)\\
&= \tilde{\nu} + \tilde{B}_1(J^2 + 3J + 2) - \tilde{B}_0(J^2 + J)\\
&= \tilde{\nu} + 2\tilde{B}_1 + (3\tilde{B}_1 - \tilde{B}_0)J + (\tilde{B}_1 - \tilde{B}_0)J^2
\end{aligned}$$

For the P branches, $\Delta J = -1$, and

$$\begin{aligned}
\tilde{\nu}_P(\Delta J = -1) &= \tilde{E}_{1,J-1} - \tilde{E}_{0,J}\\
&= \tfrac{3}{2}\tilde{\nu} + \tilde{B}_1(J-1)(J) - \tfrac{1}{2}\tilde{\nu} - \tilde{B}_0 J(J+1)\\
&= \tilde{\nu} + \tilde{B}_1(J^2 - J) - \tilde{B}_0(J^2 + J)\\
&= \tilde{\nu} - (\tilde{B}_1 + \tilde{B}_0)J + (\tilde{B}_1 - \tilde{B}_0)J^2
\end{aligned}$$

13–11. Given that $B = 58\,000$ MHz and $\tilde{\nu} = 2160.0$ cm^{-1} for CO, calculate the frequencies of the first few lines of the R and P branches in the vibration-rotation spectrum of CO.

We use Equation 13.12 for the R branch and Equation 13.13 for the P branch (recall that $c\tilde{B} = B$) to obtain the general results

$$\begin{aligned}
\tilde{\nu}_R &= \tilde{\nu} + 2\tilde{B}(J+1)\\
&= 2160.0\ \text{cm}^{-1} + 2\left(\frac{58\,000 \times 10^6\ \text{s}^{-1}}{2.998 \times 10^{10}\ \text{cm}\cdot\text{s}^{-1}}\right)(J+1)\\
&= 2160.0\ \text{cm}^{-1} + (3.87\ \text{cm}^{-1})(J+1)
\end{aligned}$$

and

$$\begin{aligned}
\tilde{\nu}_P &= \tilde{\nu} - 2\tilde{B}J\\
&= 2160.0\ \text{cm}^{-1} - 2\left(\frac{58\,000 \times 10^6\ \text{s}^{-1}}{2.998 \times 10^{10}\ \text{cm}\cdot\text{s}^{-1}}\right)J\\
&= 2160.0\ \text{cm}^{-1} - (3.87\ \text{cm}^{-1})J
\end{aligned}$$

13–12. Given that $R_e = 156.0$ pm and $k = 250.0$ N·m^{-1} for ^{7}Li^{19}F, use the rigid rotator-harmonic oscillator approximation to construct to scale an energy-level diagram for the first five rotational levels in the $v = 0$ and $v = 1$ vibrational states. Indicate the allowed transitions in an absorption experiment, and calculate the frequencies of the first few lines in the R and P branches of the vibration-rotation spectrum of ^{6}Li^{19}F.

To find the lines in the R and P branches of the spectrum, we will need to find $\tilde{\nu}$ and $\tilde{B}$ and use Equations 13.12 and 13.13. The reduced mass of ^{6}LiF is

$$\mu = \frac{(6.015)(18.998)}{(6.015 + 18.998)} \text{ amu} = 4.569 \text{ amu}$$

First, find the value of I:

$$I = \mu R_e^2$$
$$= (4.569 \text{ amu}) (1.661 \times 10^{-27} \text{ kg·amu}^{-1})(156.0 \times 10^{-12} \text{ m})^2$$
$$= 1.846 \times 10^{-46} \text{ kg·m}^2$$

Now find $\tilde{B}$ (Equation 13.9) and $\tilde{\nu}$ (Equation 13.5) :

$$\tilde{B} = \frac{h}{8\pi^2 c I}$$
$$= \frac{6.626 \times 10^{-34} \text{ J·s}}{8\pi^2 (2.998 \times 10^8 \text{ m·s}^{-1})(1.846 \times 10^{-46} \text{ kg·m}^2)}$$
$$= 151.6 \text{ m}^{-1} = 1.516 \text{ cm}^{-1}$$
$$\tilde{\nu} = \frac{1}{2\pi c} \left(\frac{k}{\mu} \right)^{1/2}$$
$$= \frac{1}{2\pi (2.998 \times 10^{10} \text{ cm·s}^{-1})} \left[\frac{250.0 \text{ N·m}^{-1}}{(4.569 \text{ amu}) (1.661 \times 10^{-27} \text{kg·amu}^{-1})} \right]^{1/2}$$
$$= 963.7 \text{ cm}^{-1}$$

To construct the energy level diagram, we use Equation 13.10:

$$\tilde{E}_{v,J} = (v + \tfrac{1}{2})\tilde{\nu} + \tilde{B} J(J + 1)$$
$$\tilde{E}_{0,J} = 481.9 \text{ cm}^{-1} + (1.516 \text{ cm}^{-1}) J(J + 1)$$
$$\tilde{E}_{1,J} = 1445 \text{ cm}^{-1} + (1.516 \text{ cm}^{-1}) J(J + 1)$$

We can now construct a table of values to use in the energy-level diagram:

J	$\tilde{E}_{0,J}/\text{cm}^{-1}$	$\tilde{E}_{1,J}/\text{cm}^{-1}$
0	481.9	1445
1	484.9	1448
2	490.9	1454
3	500.0	1463
4	512.1	1475

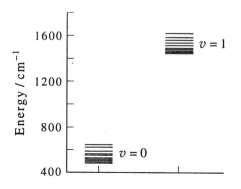

The selection rules $\Delta v = +1$ and $\Delta J = \pm 1$ (Equations 13.11) determine the allowed transitions. The frequencies of the lines in the R and P branches are (Equations 13.12 and 13.13)

$$\tilde{\nu}_R = \tilde{\nu} + 2\tilde{B}(J + 1)$$
$$= 963.7 \text{ cm}^{-1} + (1.516 \text{ cm}^{-1})(J + 1) \qquad J = 0, 1, 2, \ldots$$
$$\tilde{\nu}_P = \tilde{\nu} - 2\tilde{B}J$$
$$= 963.7 \text{ cm}^{-1} - (1.516 \text{ cm}^{-1})J \qquad J = 0, 1, 2, \ldots$$

The first few allowed transitions in the R branch are

Transition ($J'' \to J'$)	$\tilde{\nu}/\text{cm}^{-1}$
$0 \to 1$	966.7
$1 \to 2$	969.7
$2 \to 3$	972.8
$3 \to 4$	975.8

and the first few allowed transitions in the P branch are

Transition ($J'' \to J'$)	$\tilde{\nu}/\text{cm}^{-1}$
$1 \to 0$	960.7
$2 \to 1$	957.6
$3 \to 2$	954.6
$4 \to 3$	951.6

13–13. Using the values of $\tilde{\nu}_e$, $\tilde{x}_e\tilde{\nu}_e$, $\tilde{B}_e$, and $\tilde{\alpha}_e$ given in Table 13.2, construct to scale an energy-level diagram for the first five rotational levels in the $v = 0$ and $v = 1$ vibrational states for H^{35}Cl. Indicate the allowed transitions in an absorption experiment, and calculate the frequencies of the first few lines in the R and P branches.

From the table,

$$\tilde{\nu}_e = 2990.946 \text{ cm}^{-1} \qquad \tilde{B}_e = 10.5934 \text{ cm}^{-1}$$

$$\tilde{x}_e\tilde{\nu}_e = 52.819 \text{ cm}^{-1} \qquad \tilde{\alpha}_e = 0.3072 \text{ cm}^{-1}$$

Using Equations 13.10, 13.17, and 13.21, we obtain

$$\tilde{E}_{v,J} = \tilde{\nu}(v + \tfrac{1}{2}) + \tilde{B}_v J(J + 1)$$
$$= \tilde{\nu}_e(v + \tfrac{1}{2}) - \tilde{x}_e\tilde{\nu}_e(v + \tfrac{1}{2})^2 + \left[\tilde{B}_e - \tilde{\alpha}_e(v + \tfrac{1}{2})\right] J(J + 1)$$

So

$$\tilde{E}_{0,J} = \tfrac{1}{2}\tilde{\nu}_e - \tfrac{1}{4}\tilde{x}_e\tilde{\nu}_e + (\tilde{B}_e - \tfrac{1}{2}\tilde{\alpha}_e)J(J + 1)$$
$$= 1482.268 \text{ cm}^{-1} + (10.4398 \text{ cm}^{-1})J(J + 1)$$
$$\tilde{E}_{1,J} = \tfrac{3}{2}\tilde{\nu}_e - \tfrac{9}{4}\tilde{x}_e\tilde{\nu}_e + (\tilde{B}_e - \tfrac{3}{2}\tilde{\alpha}_e)J(J + 1)$$
$$= 4367.576 \text{ cm}^{-1} + (10.1326 \text{ cm}^{-1})J(J + 1)$$

We can use the expressions for $\tilde{E}_{0,J}$ and $\tilde{E}_{1,J}$ to construct a table of energy values for different values of J.

J	$\tilde{E}_{0,J}/\text{cm}^{-1}$	$\tilde{E}_{1,J}/\text{cm}^{-1}$
0	1482.268	4367.576
1	1503.148	4387.841
2	1544.907	4428.372
3	1607.546	4489.167
4	1691.064	4570.228

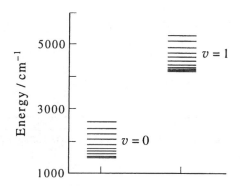

We now subtract the appropriate energy levels subject to the selection rules $\Delta v = +1$ and $\Delta J = \pm 1$ to obtain the frequencies of the first few lines in the P and R branches of the spectrum:

$$\tilde{\nu}_R(0 \to 1) = 2905.573 \text{ cm}^{-1} \qquad \tilde{\nu}_P(1 \to 0) = 2864.428 \text{ cm}^{-1}$$
$$\tilde{\nu}_R(1 \to 2) = 2925.222 \text{ cm}^{-1} \qquad \tilde{\nu}_P(2 \to 1) = 2842.934 \text{ cm}^{-1}$$
$$\tilde{\nu}_R(2 \to 3) = 2944.260 \text{ cm}^{-1} \qquad \tilde{\nu}_P(3 \to 2) = 2820.826 \text{ cm}^{-1}$$
$$\tilde{\nu}_R(3 \to 4) = 2962.682 \text{ cm}^{-1} \qquad \tilde{\nu}_P(4 \to 3) = 2798.103 \text{ cm}^{-1}$$

13–14. The following data are obtained for the vibration-rotation spectrum of $H^{79}Br$. Determine $\tilde{B}_0$, $\tilde{B}_1$, $\tilde{B}_e$, and $\tilde{\alpha}_e$ from these data.

Line	Frequency/cm^{-1}
$R(0)$	2642.60
$R(1)$	2658.36
$P(1)$	2609.67
$P(2)$	2592.51

Following the method of Example 13.4, we substitute into Equations 13.15 and 13.16 to find

$$\tilde{R}(0) = 2642.60 \text{ cm}^{-1} = \tilde{\nu} + 2\tilde{B}_1$$
$$\tilde{R}(1) = 2658.36 \text{ cm}^{-1} = \tilde{\nu} + 6\tilde{B}_1 - 2\tilde{B}_0$$
$$\tilde{P}(1) = 2609.67 \text{ cm}^{-1} = \tilde{\nu} - 2\tilde{B}_0$$
$$\tilde{P}(2) = 2592.51 \text{ cm}^{-1} = \tilde{\nu} - 2\tilde{B}_1 - 6\tilde{B}_0$$

Subtracting $\tilde{P}(1)$ from $\tilde{R}(1)$ gives

$$\tilde{R}(1) - \tilde{P}(1) = 48.69 \text{ cm}^{-1} = 6\tilde{B}_1$$

so that $\tilde{B}_1 = 8.12 \text{ cm}^{-1}$. Likewise, subtracting $\tilde{P}(2)$ from $\tilde{R}(0)$ gives

$$\tilde{R}(0) - \tilde{P}(2) = 50.09 \text{ cm}^{-1} = 6\tilde{B}_0$$

so $\tilde{B}_0 = 8.35 \text{ cm}^{-1}$.

We now use Equation 13.17 to write $\tilde{B}_0$ and $\tilde{B}_1$ in terms of $\tilde{B}_e$ and $\tilde{\alpha}_e$:

$$\tilde{B}_0 = \tilde{B}_e - \tfrac{1}{2}\tilde{\alpha}_e = 8.35 \text{ cm}^{-1}$$
$$\tilde{B}_1 = \tilde{B}_e - \tfrac{3}{2}\tilde{\alpha}_e = 8.12 \text{ cm}^{-1}$$

Solving for $\tilde{B}_e$ and $\tilde{\alpha}_e$ gives

$$\tilde{\alpha}_e = \tilde{B}_0 - \tilde{B}_1 = 0.23 \text{ cm}^{-1}$$
$$2\tilde{B}_e = 3\tilde{B}_0 - \tilde{B}_1 = 16.93 \text{ cm}^{-1}$$
$$\tilde{B}_e = 8.47 \text{ cm}^{-1}$$

13–15. The following lines were observed in the microwave absorption spectrum of $H^{127}I$ and $D^{127}I$ between 60 cm^{-1} and 90 cm^{-1}.

$$\tilde{\nu}/\text{cm}^{-1}$$

$H^{127}I$	64.275	77.130	89.985	
$D^{127}I$	65.070	71.577	78.084	84.591

Use the rigid-rotator approximation to determine the values of $\tilde{B}$, I, and R_e for each molecule. Do your results for the bond length agree with what you would expect based upon the Born-Oppenheimer approximation? Take the mass of ^{127}I to be 126.904 amu and the mass of D to be 2.014 amu.

In the rigid-rotator approximation, the spacing between the lines in a microwave absorption spectrum is $2\tilde{B}$ (Equation 13.19). The spacing between the lines given for $H^{127}I$ is 12.855 cm^{-1}, and the spacing between the lines given for $D^{127}I$ is 6.507 cm^{-1}. Therefore,

$$\tilde{B}_{H^{127}I} = \tfrac{1}{2}(12.855 \text{ cm}^{-1}) = 6.428 \text{ cm}^{-1}$$

and

$$\tilde{B}_{D^{127}I} = \tfrac{1}{2}(6.507 \text{ cm}^{-1}) = 3.254 \text{ cm}^{-1}$$

We now use Equation 13.9 to find I for both molecules:

$$I = \frac{h}{8\pi^2 c \tilde{B}}$$

$$I_{HI} = \frac{6.626 \times 10^{-34} \text{ J·s}}{8\pi^2 (2.998 \times 10^{10} \text{ cm·s}^{-1})(6.428 \text{ cm}^{-1})} = 4.355 \times 10^{-47} \text{ kg·m}^2$$

$$I_{DI} = \frac{6.626 \times 10^{-34} \text{ J·s}}{8\pi^2 (2.998 \times 10^{10} \text{ cm·s}^{-1})(3.254 \text{ cm}^{-1})} = 8.604 \times 10^{-47} \text{ kg·m}^2$$

Now we use the fact that $I = \mu R_e^2$ for a diatomic molecule to find R_e:

$$I = \mu R_e^2$$

$$R_e = \left(\frac{I}{\mu}\right)^{1/2}$$

$$R_{e,\mathrm{HI}} = \left[\frac{4.355 \times 10^{-47} \ \mathrm{kg \cdot m^2}}{\left(\frac{1.008 \times 126.904}{127.912} \ \mathrm{amu} \right) \left(1.661 \times 10^{-27} \ \mathrm{kg \cdot amu^{-1}} \right)} \right]^{1/2}$$

$$= 1.619 \times 10^{-10} \ \mathrm{m} = 161.9 \ \mathrm{pm}$$

$$R_{e,\mathrm{DI}} = \left[\frac{8.604 \times 10^{-47} \ \mathrm{kg \cdot m^2}}{\left(\frac{2.014 \times 126.904}{128.917} \ \mathrm{amu} \right) \left(1.661 \times 10^{-27} \ \mathrm{kg \cdot amu^{-1}} \right)} \right]^{1/2}$$

$$= 1.617 \times 10^{-10} \ \mathrm{m} = 161.7 \ \mathrm{pm}$$

These values differ by approximately 0.1%. In the Born-Oppenheimer approximation, the bond length is independent of the isotope of the atoms, in agreement with the above calculations.

13–16. The following spectroscopic constants were determined for pure samples of $^{74}\mathrm{Ge}^{32}\mathrm{S}$ and $^{72}\mathrm{Ge}^{32}\mathrm{S}$:

Molecule	B_e/MHz	α_e/MHz	D/kHz	$R_e(v=0)$/pm
$^{74}\mathrm{Ge}^{32}\mathrm{S}$	5593.08	22.44	2.349	0.201 20
$^{72}\mathrm{Ge}^{32}\mathrm{S}$	5640.06	22.74	2.388	0.201 20

Determine the frequency of the $J = 0$ to $J = 1$ transition for $^{74}\mathrm{Ge}^{32}\mathrm{S}$ and $^{72}\mathrm{Ge}^{32}\mathrm{S}$ in their ground vibrational states. The width of a microwave absorption line is on the order of 1 kHz. Could you distinguish a pure sample of $^{74}\mathrm{Ge}^{32}\mathrm{S}$ from a 50/50 mixture of $^{74}\mathrm{Ge}^{32}\mathrm{S}$ and $^{72}\mathrm{Ge}^{32}\mathrm{S}$ using microwave spectroscopy?

To convert the values given in the above table to wave numbers, divide by c. Then substitute into a combination of Equations 13.17 and 13.19, with $v = 0$:

$$\tilde{\nu} = 2 \left[\tilde{B}_e - \tilde{\alpha}_e \left(\tfrac{1}{2} \right) \right] (J + 1) - 4\tilde{D}(J + 1)^3$$

Molecule	$(\tilde{\nu}, J = 0)$/cm^{-1}	$(\tilde{\nu}, J = 1)$/cm^{-1}	$\Delta\tilde{\nu}$/cm^{-1}
$^{74}\mathrm{Ge}^{32}\mathrm{S}$	0.372 381	0.744 761	0.372 379
$^{72}\mathrm{Ge}^{32}\mathrm{S}$	0.375 505	0.751 009	0.375 504

Now, the width of the absorption line is on the order of 1000 s^{-1}, which corresponds to an energy of approximately 3×10^{-8} cm^{-1}. This width is less than the difference between the absorption lines of $^{74}\mathrm{Ge}^{32}\mathrm{S}$ and $^{72}\mathrm{Ge}^{32}\mathrm{S}$ and so a 50/50 mixture can be distinguished from a pure sample using microwave spectroscopy.

13–17. The frequencies of the rotational transitions in the nonrigid-rotator approximation are given by Equation 13.19. Show how both $\tilde{B}$ and $\tilde{D}$ may be obtained by curve fitting $\tilde{\nu}$ to Equation 13.19. Use this method and the data in Table 13.3 to determine both $\tilde{B}$ and $\tilde{D}$ for H^{35}Cl.

$$\tilde{\nu} = 2\tilde{B}(J + 1) - 4\tilde{D}(J + 1)^3 \qquad (13.19)$$

The data from Table 13.3 are plotted in the following figure. (Compare the accuracy of this data to that in the following problem.)

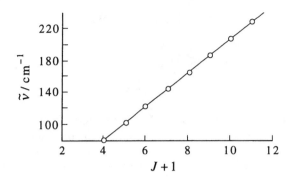

The best-fit line to this data gives $2\tilde{B} = 20.81$ cm^{-1} and $-4\tilde{D} = -1.772 \times 10^{-3}$ cm^{-1}. Therefore, $\tilde{B} = 10.40$ cm^{-1} and $\tilde{D} = 4.43 \times 10^{-4}$ cm^{-1}.

13–18. The following data are obtained in the microwave spectrum of $^{12}C^{16}O$. Use the method of Problem 13–17 to determine the values of $\tilde{B}$ and $\tilde{D}$ from these data.

Transitions	Frequency/cm^{-1}
$0 \rightarrow 1$	3.845 40
$1 \rightarrow 2$	7.690 60
$2 \rightarrow 3$	11.535 50
$3 \rightarrow 4$	15.379 90
$4 \rightarrow 5$	19.223 80
$5 \rightarrow 6$	23.066 85

Using the method of Problem 13–17, we plot $\tilde{v}$ vs. $J + 1$.

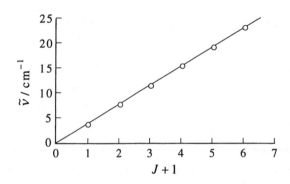

The best-fit line to this data gives $2\tilde{B} = 3.8454$ cm^{-1} and $-4\tilde{D} = -2.5547 \times 10^{-5}$ cm^{-1}. Therefore, $\tilde{B} = 1.9227$ cm^{-1} and $\tilde{D} = 6.387 \times 10^{-6}$ cm^{-1}.

13–19. Using the parameters given in Table 13.2, calculate the frequencies (in cm^{-1}) of the $0 \rightarrow 1$, $1 \rightarrow 2$, $2 \rightarrow 3$, and $3 \rightarrow 4$ rotational transitions in the ground vibrational state of $H^{35}Cl$ in the nonrigid-rotator approximation.

We substitute $\tilde{B} = 10.5934 \text{ cm}^{-1}$ and $\tilde{D} = 5.319 \times 10^{-4} \text{ cm}^{-1}$ (Table 13.2) into Equation 13.19 to find

$$\tilde{\nu}_{\text{calc}}/\text{cm}^{-1} = 21.1868(J + 1) - 0.002128(J + 1)^3$$

Transition	$\tilde{\nu}_{\text{calc}}/\text{cm}^{-1}$
$0 \rightarrow 1$	21.1847
$1 \rightarrow 2$	42.3566
$2 \rightarrow 3$	63.5030
$3 \rightarrow 4$	84.6110

13–20. The vibrational term of a diatomic molecule is given by

$$G(v) = \left(v + \tfrac{1}{2}\right)\tilde{\nu}_e - \left(v + \tfrac{1}{2}\right)^2 \tilde{x}_e \tilde{\nu}_e$$

where v is the vibrational quantum number. Show that the spacing between the adjacent levels ΔG is given by

$$\Delta G = G(v + 1) - G(v) = \tilde{\nu}_e\{1 - 2\tilde{x}_e(v + 1)\} \tag{1}$$

The diatomic molecule dissociates in the limit that $\Delta G \rightarrow 0$. Show that the maximum vibrational quantum number, $v_{\max}$, is given by

$$v_{\max} = \frac{1}{2\tilde{x}_e} - 1$$

Use this result to show that the dissociation energy $\tilde{D}_e$ of the diatomic molecule can be written as

$$\tilde{D}_e = \frac{\tilde{\nu}_e(1 - \tilde{x}_e^2)}{4\tilde{x}_e} \approx \frac{\tilde{\nu}_e}{4\tilde{x}_e} \tag{2}$$

Referring to Equation 1, explain how the constants $\tilde{\nu}_e$ and $\tilde{x}_e$ can be evaluated from a plot of ΔG versus $v + 1$. This type of plot is called a *Birge-Sponer plot*. Once the values of $\tilde{\nu}_e$ and $\tilde{x}_e$ are known, Equation 2 can be used to determine the dissociation energy of the molecule. Use the following experimental data for H_2 to calculate the dissociation energy, $\tilde{D}_e$.

v	$G(v)/\text{cm}^{-1}$	v	$G(v)/\text{cm}^{-1}$
0	4161.12	7	26 830.97
1	8087.11	8	29 123.93
2	11 782.35	9	31 150.19
3	15 250.36	10	32 886.85
4	18 497.92	11	34 301.83
5	21 505.65	12	35 351.01
6	24 287.83	13	35 972.97

Explain why your Birge-Sponer plot is not linear for high values of v. How does the value of $\tilde{D}_e$ obtained from the Birge-Sponer analysis compare with the experimental value of $38\,269.48 \text{ cm}^{-1}$?

Use the expression given in the problem to find Equation 1:

$$\begin{aligned}
\Delta G &= G(v + 1) - G(v) \\
&= \left(v + \tfrac{3}{2}\right)\tilde{\nu}_e - \left(v + \tfrac{3}{2}\right)^2 \tilde{x}_e \tilde{\nu}_e - \left(v + \tfrac{1}{2}\right)\tilde{\nu}_e + \left(v + \tfrac{1}{2}\right)^2 \tilde{x}_e \tilde{\nu}_e \\
&= \tilde{\nu}_e - \tilde{x}_e \tilde{\nu}_e (2v + 2) = \tilde{\nu}_e \left[1 - 2\tilde{x}_e(v + 1)\right]
\end{aligned}$$

In the limit $\Delta G \to 0$, $v \to v_{max}$. Solving for v_{max} gives

$$0 = \tilde{\nu}_e \left[1 - 2\tilde{x}_e(v_{max} + 1) \right]$$

$$2\tilde{x}_e(v_{max} + 1) = 1$$

$$v_{max} = \frac{1}{2\tilde{x}_e} - 1$$

The molecule dissociates in the limit $\Delta G \to 0$, so the dissociation energy is

$$\tilde{D}_e = G(v_{max}) = \left(\frac{1}{2\tilde{x}_e} - \frac{1}{2} \right) \tilde{\nu}_e - \left(\frac{1}{2\tilde{x}_e} - \frac{1}{2} \right)^2 \tilde{x}_e \tilde{\nu}_e$$

$$= \frac{\tilde{\nu}_e}{2\tilde{x}_e}(1 - \tilde{x}_e) - \frac{\tilde{\nu}_e}{4\tilde{x}_e}(1 - \tilde{x}_e)^2$$

$$= \frac{\tilde{\nu}_e}{4\tilde{x}_e} \left(2 - 2\tilde{x}_e - 1 + 2\tilde{x}_e - \tilde{x}_e^2 \right)$$

$$= \frac{\tilde{\nu}_e}{4\tilde{x}_e}(1 - \tilde{x}_e^2) \approx \frac{\tilde{\nu}_e}{4\tilde{x}_e}$$

because $\tilde{x}_e$ is very small compared to one. We can expand Equation 1 to write

$$\Delta G = \tilde{\nu}_e - 2\tilde{x}_e\tilde{\nu}_e(v + 1)$$

Therefore, a plot of ΔG vs. $(v + 1)$ will have an intercept of $\tilde{\nu}_e$ and a slope of $-2\tilde{x}_e\tilde{\nu}_e$. The experimental data points for H_2 are plotted below:

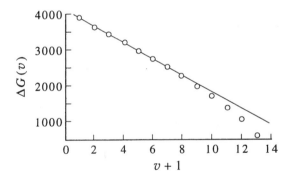

Using the best linear fit to the first eight data points gives $\tilde{\nu}_e = 4164.4$ cm^{-1} and $-2\tilde{x}_e\tilde{\nu}_e = -232.01$ cm^{-1}. Then $\tilde{x}_e = 0.0279$, and so

$$\tilde{D}_e = \frac{\tilde{\nu}_e}{4\tilde{x}_e} = 37\,400 \text{ cm}^{-1}$$

The Birge-Sponer plot is not linear for large values of v because the potential curve is not well described by the anharmonic potential energy function $G(v)$ given in the statement of the problem.

13–21. An analysis of the vibrational spectrum of the ground-state homonuclear diatomic molecule C_2 gives $\tilde{\nu}_e = 1854.71$ cm^{-1} and $\tilde{\nu}_e\tilde{x}_e = 13.34$ cm^{-1}. Suggest an experimental method that can be used to determine these spectroscopic parameters. Use the expression derived in Problem 13–20 to determine the number of bound vibrational levels for the ground state of C_2.

The molecule C_2 does not have a dipole moment, and so we cannot record an infrared absorption spectrum. We can determine $\tilde{\nu}_e$ and $\tilde{\nu}_e\tilde{x}_e$ from an emission spectrum, where the lines correspond to transitions between a specific vibrational state in the upper electronic state and the various vibrational states in the ground electronic state. The number of bound vibrational levels, v_{max}, is given by (see Problem 13–20)

$$v_{max} = \frac{1}{2\tilde{x}_e} - 1 = \frac{1854.71 \text{ cm}^{-1}}{2(13.34 \text{ cm}^{-1})} - 1 = 68.5$$

There are 68 bound vibrational levels for the ground state of C_2.

13–22. A simple function that is a good representation of an internuclear potential is the Morse potential,

$$U(q) = D_e(1 - e^{-\beta q})^2$$

where q is $R - R_e$. Show that the force constant calculated for a Morse potential is given by

$$k = 2D_e\beta^2$$

Given that $D_e = 7.31 \times 10^{-19}$ J·molecule^{-1} and $\beta = 1.83 \times 10^{10}$ m^{-1} for HCl, calculate the value of k.

We can expand $U(q)$ using a Maclaurin series:

$$U(q) = D_e(1 - e^{-\beta q})^2$$
$$= D_e\left\{1 - \left[1 - \beta q + \frac{\beta^2 q^2}{2} + O(q^3)\right]\right\}^2$$
$$= D_e\left[\beta^2 q^2 + O(q^3)\right]$$

Equating this result to $U(q) = kq^2/2$ (as in Equation 5.11) gives

$$k = 2D_e\beta^2$$

For HCl,

$$k = 2\left(7.31 \times 10^{-19} \text{ J·molecule}^{-1}\right)\left(1.83 \times 10^{10} \text{ m}^{-1}\right)^2 = 490 \text{ N·m}^{-1}$$

13–23. The Morse potential is presented in Problem 13–22. Given that $D_e = 7.33 \times 10^{-19}$ J·molecule^{-1}, $\tilde{v}_e = 1580.0$ cm^{-1}, and $R_e = 121$ pm for $^{16}O_2$, plot a Morse potential for $^{16}O_2$. Plot the corresponding harmonic-oscillator potential on the same graph.

$$\tilde{v}_e = \frac{1}{2\pi c}\left(\frac{k}{\mu}\right)^{1/2} \tag{13.5}$$

We can find k from the parameters given in the problem. Solving for k gives

$$k = \left(2\pi c\tilde{v}_e\right)^2 \mu$$
$$= \left[2\pi(2.998 \times 10^{10} \text{ cm·s}^{-1})(1580.0 \text{ cm}^{-1})\right]^2 (7.9975 \text{ amu})(1.661 \times 10^{-27} \text{ kg·amu}^{-1})$$
$$= 1176.3 \text{ N·m}^{-1}$$

From Problem 13–22, we know that $k = 2D_e\beta^2$, so

$$\beta = \left(\frac{k}{2D_e}\right)^{1/2} = \left[\frac{1176.3\ \text{N}\cdot\text{m}^{-1}}{2\left(7.33 \times 10^{-19}\ \text{J}\right)}\right]^{1/2}$$

$$= 2.83 \times 10^{10}\ \text{m}^{-1}$$

Recall (Equation 5.11) that the potential energy of a harmonic oscillator is $U(q) = kq^2/2$. Now graph the Morse potential and harmonic-oscillator potential for $^{16}\text{O}_2$:

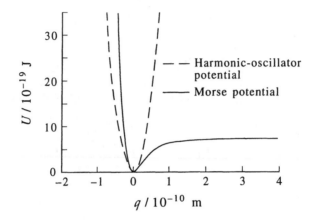

The Morse potential is a more realistic approximation of the behavior of a molecule.

13–24. The fundamental line in the infrared spectrum of $^{12}\text{C}^{16}\text{O}$ occurs at 2143.0 cm^{-1}, and the first overtone occurs at 4260.0 cm^{-1}. Calculate the values of $\tilde{\nu}_e$ and $\tilde{x}_e\tilde{\nu}_e$ for $^{12}\text{C}^{16}\text{O}$.

We do this problem in the manner of Example 13–5, where we derived equations for the frequency of the fundamental and first overtone.

Fundamental: $\tilde{\nu}_{obs} = \tilde{\nu}_e - 2\tilde{x}_e\tilde{\nu}_e = 2143.0\ \text{cm}^{-1}$

First overtone: $\tilde{\nu}_{obs} = 2\tilde{\nu}_e - 6\tilde{x}_e\tilde{\nu}_e = 4260.0\ \text{cm}^{-1}$

Multiply the fundamental frequency by 3 and subtract the overtone to get

$$\tilde{\nu}_e = 3(2143.0\ \text{cm}^{-1}) - 4260.0\ \text{cm}^{-1} = 2169\ \text{cm}^{-1}$$

Multiply the fundamental frequency by 2 and subtract from the overtone to get

$$2\tilde{x}_e\tilde{\nu}_e = 26.0\ \text{cm}^{-1}$$

or

$$\tilde{x}_e\tilde{\nu}_e = 13.0\ \text{cm}^{-1}$$

13–25. Using the parameters given in Table 13.2, calculate the fundamental and the first three overtones of H^{79}Br.

Using Equation 13.22, we find

Fundamental: $\quad \tilde{\nu}_{obs} = \tilde{\nu}_e - 2\tilde{x}_e\tilde{\nu}_e$

$\qquad\qquad\qquad = 2648.975 \text{ cm}^{-1} - 2(45.218 \text{ cm}^{-1})$

$\qquad\qquad\qquad = 2558.539 \text{ cm}^{-1}$

First overtone: $\quad \tilde{\nu}_{obs} = 2\tilde{\nu}_e - 6\tilde{x}_e\tilde{\nu}_e$

$\qquad\qquad\qquad = 2(2648.975 \text{ cm}^{-1}) - 6(45.218 \text{ cm}^{-1})$

$\qquad\qquad\qquad = 5026.642 \text{ cm}^{-1}$

Second overtone: $\quad \tilde{\nu}_{obs} = 3\tilde{\nu}_e - 12\tilde{x}_e\tilde{\nu}_e$

$\qquad\qquad\qquad = 3(2648.975 \text{ cm}^{-1}) - 12(45.218 \text{ cm}^{-1})$

$\qquad\qquad\qquad = 7404.309 \text{ cm}^{-1}$

Third overtone: $\quad \tilde{\nu}_{obs} = 4\tilde{\nu}_e - 20\tilde{x}_e\tilde{\nu}_e$

$\qquad\qquad\qquad = 4(2648.975 \text{ cm}^{-1}) - 20(45.218 \text{ cm}^{-1})$

$\qquad\qquad\qquad = 9691.54 \text{ cm}^{-1}$

13–26. The frequencies of the vibrational transitions in the anharmonic-oscillator approximation are given by Equation 13.22. Show how the values of both $\tilde{\nu}_e$ and $\tilde{x}_e\tilde{\nu}_e$ may be obtained by plotting $\tilde{\nu}_{obs}/v$ versus $(v + 1)$. Use this method and the data in Table 13.4 to determine the values $\tilde{\nu}_e$ and $\tilde{x}_e\tilde{\nu}_e$ for H^{35}Cl.

$$\tilde{\nu}_{obs} = \tilde{\nu}_e v - \tilde{x}_e\tilde{\nu}_e v(v + 1) \qquad (13.22)$$

Divide by v to obtain

$$\frac{\tilde{\nu}_{obs}}{v} = \tilde{\nu}_e - \tilde{x}_e\tilde{\nu}_e(v + 1)$$

If we plot $\tilde{\nu}_{obs}/v$ versus $(v + 1)$, the slope of the line will be $-\tilde{x}_e\tilde{\nu}_e$ and the intercept will be $\tilde{\nu}_e$. Using the data in Table 13.4, we obtain the following plot:

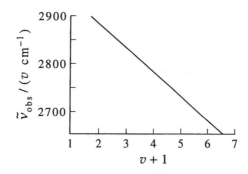

The best fit line to the data has an intercept of $\tilde{\nu}_e = 2989 \text{ cm}^{-1}$ and a slope of -51.6 cm^{-1}, so $\tilde{\nu}_e\tilde{x}_e = 51.6 \text{ cm}^{-1}$.

13–27. The following data are obtained from the infrared spectrum of $^{127}I^{35}Cl$. Using the method of Problem 13–26, determine the values of $\tilde{\nu}_e$ and $\tilde{x}_e\tilde{\nu}_e$ from these data.

Transitions	Frequency/cm^{-1}
$0 \rightarrow 1$	381.20
$0 \rightarrow 2$	759.60
$0 \rightarrow 3$	1135.00
$0 \rightarrow 4$	1507.40
$0 \rightarrow 5$	1877.00

We plot $\tilde{\nu}_{obs}/v$ vs. $(v+1)$.

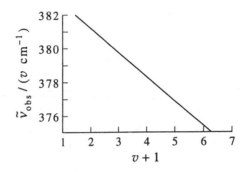

The best fit line to the data has an intercept of $\tilde{\nu}_e = 384.1$ cm^{-1} and a slope of -1.45 cm^{-1}, so $\tilde{\nu}_e\tilde{x}_e = 1.45$ cm^{-1}.

13–28. The values of $\tilde{\nu}_e$ and $\tilde{x}_e\tilde{\nu}_e$ of $^{12}C^{16}O$ are 2169.81 cm^{-1} and 13.29 cm^{-1} in the ground electronic state and 1514.10 cm^{-1} and 17.40 cm^{-1} in the first excited electronic state. If the $0 \rightarrow 0$ vibronic transition occurs at $6.475\,15 \times 10^4$ cm^{-1}, calculate the value of $\tilde{T}_e = \tilde{\nu}'_{el} - \tilde{\nu}''_{el}$, the energy difference between the minima of the potential curves of the two electronic states.

For the $0 \rightarrow 0$ vibronic transition, $v' = v'' = 0$ in Equation 13.24 and so

$$\tilde{\nu}_{0,0} = \tilde{T}_e + (\tfrac{1}{2}\tilde{\nu}'_e - \tfrac{1}{4}\tilde{x}'_e\tilde{\nu}'_e) - (\tfrac{1}{2}\tilde{\nu}''_e - \tfrac{1}{4}\tilde{x}''_e\tilde{\nu}''_e)$$

$$64\,751.5 \text{ cm}^{-1} = \tilde{T}_e + \left[\tfrac{1}{2}(1514.10 \text{ cm}^{-1}) - \tfrac{1}{4}(17.40 \text{ cm}^{-1})\right]$$
$$- \left[\tfrac{1}{2}(2169.81 \text{ cm}^{-1}) - \tfrac{1}{4}(13.29 \text{ cm}^{-1})\right]$$

$$65\,080.4 \text{ cm}^{-1} = \tilde{T}_e$$

13–29. Given the following parameters for $^{12}C^{16}O$: $\tilde{T}_e = 6.508\,043 \times 10^4$ cm^{-1}, $\tilde{\nu}'_e = 1514.10$ cm^{-1}, $\tilde{x}'_e\tilde{\nu}'_e = 17.40$ cm^{-1}, $\tilde{\nu}''_e = 2169.81$ cm^{-1}, and $\tilde{x}''_e\tilde{\nu}''_e = 13.29$ cm^{-1}, construct to scale an energy-level diagram of the first two electronic states, showing the first four vibrational states in each electronic state. Indicate the allowed transitions from $v'' = 0$, and calculate the frequencies of these transitions. Also, calculate the zero-point vibrational energy in each electronic state.

Neglecting rotational energies, Equation 13.23 becomes

$$\tilde{E}_v = \tilde{\nu}_{\text{el}} + \tilde{\nu}_e(v + \tfrac{1}{2}) - \tilde{x}_e\tilde{\nu}_e(v + \tfrac{1}{2})^2$$

Setting the energy at the minimum of the lower electronic potential energy curve equal to zero, we have

$$\tilde{E}_v'' = 0 + \tilde{\nu}_e''(v + \tfrac{1}{2}) - \tilde{x}_e''\tilde{\nu}_e''(v + \tfrac{1}{2})^2$$

$$\tilde{E}_v' = \tilde{T}_e + \tilde{\nu}_e'(v + \tfrac{1}{2}) - \tilde{x}_e'\tilde{\nu}_e'(v + \tfrac{1}{2})^2$$

$$\tilde{E}_v'' = 2169.81 \text{ cm}^{-1}(v + \tfrac{1}{2}) - 13.29 \text{ cm}^{-1}(v + \tfrac{1}{2})^2$$

$$\tilde{E}_v' = 65\,080.43 \text{ cm}^{-1} + 1514.10 \text{ cm}^{-1}(v + \tfrac{1}{2}) - 17.40 \text{ cm}^{-1}(v + \tfrac{1}{2})^2$$

We can now make a table of the energies of the various vibrational states:

v	$\tilde{E}_v''/\text{cm}^{-1}$	$\tilde{E}_v'/\text{cm}^{-1}$
0	1081.58	65 833.13
1	3224.81	67 312.43
2	5341.63	68 756.93
3	7431.53	70 166.63

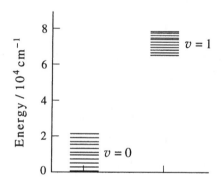

We can also create a table of allowed transitions from $v'' = 0$ and their frequencies:

$v'' \to v'$	$\Delta E/\text{cm}^{-1}$
$0 \to 0$	64 751.55 cm^{-1}
$0 \to 1$	66 230.85 cm^{-1}
$0 \to 2$	67 675.35 cm^{-1}
$0 \to 3$	69 085.05 cm^{-1}

13–30. An analysis of the rotational spectrum of $^{12}\text{C}^{32}\text{S}$ gives the following results:

v	0	1	2	3
$\tilde{B}_v/\text{cm}^{-1}$	0.817 08	0.811 16	0.805 24	0.799 32

Determine the values of $\tilde{B}_e$ and $\tilde{\alpha}_e$ from these data.

$$\tilde{B}_v = \tilde{B}_e - \tilde{\alpha}_e(v + \tfrac{1}{2}) \tag{13.17}$$

If we subtract $\tilde{B}_v$ from $\tilde{B}_{v+1}$, we find that

$$\tilde{B}_{v+1} - \tilde{B}_v = -\tilde{\alpha}_e$$

Using the values given in the problem, we then find

v	0	1	2
$\tilde{\alpha}_e$/cm^{-1}	0.00592	0.00592	0.00590

Take $\tilde{\alpha}_e = 0.00592$ cm^{-1} and substitute into Equation 13.17 to find

$$\begin{aligned}
\tilde{B}_e &= \tilde{B}_v + (0.00592 \text{ cm}^{-1})(v + \tfrac{1}{2}) \\
&= 0.81708 \text{ cm}^{-1} + (0.00592 \text{ cm}^{-1})(\tfrac{1}{2}) \\
&= 0.82004 \text{ cm}^{-1}
\end{aligned}$$

where we have let $v = 0$. We obtain the same value of $\tilde{B}_e$ for $v = 1, 2$, and 3.

13–31. The frequencies of the first few vibronic transitions to an excited state of BeO are as follows:

Vibronic transitions	$0 \to 2$	$0 \to 3$	$0 \to 4$	$0 \to 5$
$\tilde{\nu}_{obs}$/cm^{-1}	12 569.95	13 648.43	14 710.85	15 757.50

Use these data to calculate the values of $\tilde{\nu}_e$ and $\tilde{x}_e\tilde{\nu}_e$ for the excited state of BeO.

We use Equation 13.24, with $v'' = 0$:

$$\tilde{\nu}_{obs} = \tilde{T}_e + (\tfrac{1}{2}\tilde{\nu}'_e - \tfrac{1}{4}\tilde{x}'_e\tilde{\nu}'_e) - (\tfrac{1}{2}\tilde{\nu}''_e - \tfrac{1}{4}\tilde{x}''_e\tilde{\nu}''_e) + \tilde{\nu}'_e v' - \tilde{x}'_e\tilde{\nu}'_e v'(v' + 1) \qquad (1)$$

We can rewrite this equation in the form

$$\tilde{\nu}_{obs} = A v'^2 + B v' + C$$

where $A = \tilde{x}'_e\tilde{\nu}'_e$, $B = \tilde{\nu}'_e - \tilde{x}'_e\tilde{\nu}'_e$, and C is the sum of the first three terms in Equation 1. The best fit of the quadratic equation to the experimental data gives $\tilde{\nu}_{obs} = -8.0v'^2 + 1118.2v' + 10\,365$. Therefore,

$$\begin{aligned}
\tilde{\nu}'_e - \tilde{x}'_e\tilde{\nu}'_e &= 1118.2 \text{ cm}^{-1} \\
\tilde{x}'_e\tilde{\nu}'_e &= 8.0 \text{ cm}^{-1} \\
\tilde{\nu}'_e &= 1126.2 \text{ cm}^{-1}
\end{aligned}$$

13–32. The frequencies of the first few vibronic transitions to an excited state of $^7\text{Li}_2$ are as follows:

Vibronic transitions	$0 \to 0$	$0 \to 1$	$0 \to 2$	$0 \to 3$	$0 \to 4$	$0 \to 5$
$\tilde{\nu}_{obs}$/cm^{-1}	14 020	14 279	14 541	14 805	15 074	15 345

Use these data to calculate the values of $\tilde{\nu}_e$ and $\tilde{x}_e\tilde{\nu}_e$ for the excited state of $^7\text{Li}_2$.

We use the same technique as in Problem 13–31. The best-fit quadratic equation to the data is $\tilde{\nu}_{obs} = -0.04v'^2 + 267.72v' + 14\,005$. Thus, we have

$$\tilde{\nu}_e' - \tilde{x}_e'\tilde{\nu}_e' = 267.72 \text{ cm}^{-1}$$

$$\tilde{x}_e'\tilde{\nu}_e' = 0.04 \text{ cm}^{-1}$$

$$\tilde{\nu}_e' = 267.76 \text{ cm}^{-1}$$

13–33. Determine the number of translational, rotational, and vibrational degrees of freedom in

a. CH_3Cl b. OCS c. C_6H_6 d. H_2CO

The total number of degrees of freedom is $3N$, where N is the number of atoms in the molecule. All molecules have three translational degrees of freedom. A nonlinear molecule has three rotational degrees of freedom and a linear molecule has two rotational degrees of freedom. A linear molecule has $3N - 5$ vibrational degrees of freedom and a nonlinear molecule has $3N - 6$ vibrational degrees of freedom.

a. 3 translational, 3 rotational, 9 vibrational

b. 3 translational, 2 rotational, 4 vibrational

c. 3 translational, 3 rotational, 30 vibrational

d. 3 translational, 3 rotational, 6 vibrational

13–34. Determine which of the following molecules will exhibit a microwave rotational absorption spectrum: H_2, HCl, CH_4, CH_3I, H_2O, and SF_6.

The molecules HCl, CH_3I, and H_2O exhibit a microwave rotational absorption spectrum. All the other molecules do not have a permanent dipole moment and so do not have a microwave absorption spectrum.

13–35. Classify each of the following molecules as a spherical, a symmetric, or an asymmetric top: CH_3Cl, CCl_4, SO_2, and SiH_4.

CH_3Cl: symmetric top
CCl_4: spherical top
SO_2: asymmetric top
SiH_4: spherical top

13–36. Classify each of the following molecules as either a prolate or an oblate symmetric top: FCH_3, $HCCl_3$, PF_3, and CH_3CCH.

FCH_3: prolate symmetric top
$HCCl_3$: oblate symmetric top
PF_3: oblate symmetric top
CH_3CCH: prolate symmetric top

13–37. Show that the components of the moment of inertia of the trigonal planar molecule shown below are $I_{xx} = I_{yy} = 3m/2$ and $I_{zz} = 3m$ if all the masses are m units, all the bond lengths are unit length, and all the bond angles are 120°.

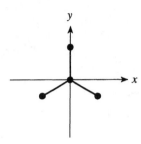

The center of mass sits at the origin, so

$$I_{xx} = \sum m_j y_j^2 = m(1)^2 + 2m(\sin^2 30°) = \frac{3}{2}m$$

$$I_{yy} = \sum m_j x_j^2 = m(0)^2 + 2m(\cos^2 30°) = \frac{3}{2}m$$

$$I_{zz} = \sum m_j x_j^2 + \sum m_j y_j^2 = 3m$$

13–38. This problem illustrates how the principal moments of inertia can be obtained as an eigenvalue problem. We will work in two dimensions for simplicity. Consider the "molecule" represented below,

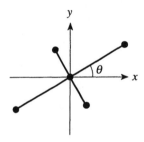

where all the masses are unit masses and the long and short bond lengths are 2 and 1, respectively. Show that

$$I_{xx} = 2\cos^2 \theta + 8\sin^2 \theta$$

$$I_{yy} = 8\cos^2 \theta + 2\sin^2 \theta$$

$$I_{xy} = -6\cos \theta \sin \theta$$

The fact that $I_{xy} \neq 0$ indicates that these I_{ij} are not the principal moments of inertia. Now solve the secular determinantal equation for λ

$$\begin{vmatrix} I_{xx} - \lambda & I_{xy} \\ I_{xy} & I_{yy} - \lambda \end{vmatrix} = 0$$

and compare your result with the values of I_{xx} and I_{yy} that you would obtain if you align the "molecule" and the coordinate system such that $\theta = 90°$. What does this comparison tell you? What are the values of I_{xx} and I_{yy} if $\theta = 0°$?

We again use trigonometric functions to find the x and y components of the direction vectors.

$$I_{xx} = \sum m_j y_j^2 = 2(1)(2\sin\theta)^2 + 2(1)\left[\sin\left(\tfrac{\pi}{2} - \theta\right)\right]^2$$
$$= 8\sin^2\theta + 2\cos^2\theta$$
$$I_{yy} = \sum m_j x_j^2 = 2(1)(2\cos\theta)^2 + 2(1)\left[\cos\left(\tfrac{\pi}{2} - \theta\right)\right]^2$$
$$= 8\cos^2\theta + 2\sin^2\theta$$
$$I_{xy} = -\sum m_j x_j y_j$$
$$= -(2\cos\theta)(2\sin\theta) - 2\left[\cos(\pi+\theta)\right]\left[\sin(\pi+\theta)\right]$$
$$- \sin(\tfrac{\pi}{2}-\theta)\cos(\tfrac{\pi}{2}-\theta) - \cos(\tfrac{\pi}{2}+\theta)\sin(\tfrac{\pi}{2}+\theta)$$
$$= -4\cos\theta\sin\theta - 2\cos\theta\sin\theta - \sin\theta\cos\theta + \sin\theta\cos\theta$$
$$= -6\cos\theta\sin\theta$$

Now the secular determinantal equation becomes

$$\begin{vmatrix} 8\sin^2\theta + 2\cos^2\theta - \lambda & -6\cos\theta\sin\theta \\ -6\cos\theta\sin\theta & 8\cos^2\theta + 2\sin^2\theta - \lambda \end{vmatrix} = 0$$

Expanding this determinant gives

$$0 = 64\cos^2\theta\sin^2\theta + 16\cos^4\theta + 16\sin^4\theta + 4\sin^2\theta\cos^2\theta$$
$$- \lambda\left(8\cos^2\theta + 2\sin^2\theta + 8\sin^2\theta + 2\cos^2\theta\right) + \lambda^2 - 36\cos^2\theta\sin^2\theta$$
$$= \lambda^2 - \lambda(10)(\sin^2\theta + \cos^2\theta) + 16(\cos^2\theta + \sin^2\theta)^2$$
$$= \lambda^2 - 10\lambda + 16$$
$$\lambda = 5 \pm \frac{\sqrt{100 - 64}}{2} = 5 \pm 3$$
$$\lambda = 8 \text{ or } 2$$

If we align the molecule such that $\theta = 90°$, then $\sin^2\theta = 1$ and $\cos^2\theta = 0$, so $I_{xx} = 8$, $I_{yy} = 2$, and $I_{xy} = 0$. If $\theta = 0°$, $I_{xx} = 2$ and $I_{yy} = 8$ (I_{xy} will still equal zero). This tells us that the coordinate system chosen does not affect the values of the principal moments of inertia.

13–39. Sketch an energy-level diagram for a prolate symmetric top and an oblate symmetric top. How do they differ? Indicate some of the allowed transitions in each case.

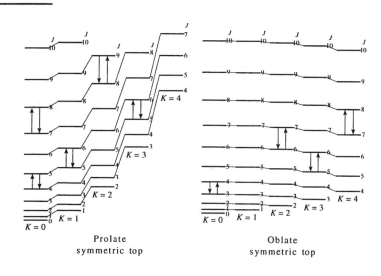

Prolate
symmetric top

Oblate
symmetric top

The energies of the J levels increase as K increases for the prolate symmetric top (represented by (a) in the diagram), but decrease as K increases for the oblate symmetric top (b). Some of the allowed transitions are indicated in each diagram.

13–40. Derive Equation 13.57 from Equation 13.55.

Because $E_2 > E_1$, the second term in Equation 13.55 is dominant and so

$$a_2(t) \propto \frac{1 - \exp\left[i(E_2 - E_1 - h\nu)t/\hbar\right]}{E_2 - E_1 - h\nu}$$

$$a_2^*(t) \propto \frac{1 - \exp\left[-i(E_2 - E_1 - h\nu)t/\hbar\right]}{E_2 - E_1 - h\nu}$$

$$a_2(t)a_2^*(t) \propto \frac{1 - \exp\left[i(E_2 - E_1 - h\nu)t/\hbar\right] - \exp\left[-i(E_2 - E_1 - h\nu)t/\hbar\right] + 1}{(E_2 - E_1 - h\nu)^2}$$

$$\propto \frac{2 - 2\cos\left[(E_2 - E_1 - h\nu)t/\hbar\right]}{(E_2 - E_1 - h\nu)^2}$$

$$\propto \frac{2\left\{2\sin^2\left[(E_2 - E_1 - h\nu)t/2\hbar\right]\right\}}{(E_2 - E_1 - h\nu)^2}$$

$$\propto \frac{\sin^2\left[(E_2 - E_1 - h\nu)t/2\hbar\right]}{(E_2 - E_1 - h\nu)^2} = \frac{\sin^2\left[(E_2 - E_1 - \hbar\omega)t/2\hbar\right]}{(E_2 - E_1 - \hbar\omega)^2}$$

13–41. Show that the first few associated Legendre functions satisfy the recursion formula given by Equation 13.62.

$$(2J + 1)x P_J^{|M|}(x) = (J - |M| + 1) P_{J+1}^{|M|}(x) + J + |M|) P_{J-1}^{|M|}(x) \tag{13.62}$$

The associated Legendre functions are given in Table 6.2. For $J = 1$, $|M| = 0$,

$$3x P_1^0(x) \overset{?}{=} 2 P_2^0(x) + P_0^0(x)$$

$$3x(x) \overset{?}{=} 2\left[\tfrac{1}{2}(3x^2 - 1)\right] + 1$$

$$3x^2 = 3x^2 - 1 + 1$$

For $J = 1$, $|M| = 1$,

$$3x P_1^1(x) \overset{?}{=} P_2^1(x) + P_0^1(x)$$

$$3x(1 - x^2)^{1/2} = 3x(1 - x^2)^{1/2}$$

(Remember that the superscript on P cannot be greater than the subscript.) For $J = 2$, $|M| = 0$,

$$5x P_2^0(x) \overset{?}{=} 3 P_3^0(x) + 2 P_1^0(x)$$

$$5x\left[\tfrac{1}{2}(3x^2 - 1)\right] \overset{?}{=} 3\left[\tfrac{1}{2}(5x^3 - 3x)\right] + 2x$$

$$\frac{15}{2}x^3 - \frac{5}{2}x \overset{?}{=} \frac{15}{2}x^3 - \frac{9}{2}x + 2x$$

$$\frac{15}{2}x^3 - \frac{5}{2}x = \frac{15}{2}x^3 - \frac{5}{2}x$$

For $J = 2, |M| = 1$,

$$5x P_2^1(x) \stackrel{?}{=} 2P_3^1(x) + 3P_1^1(x)$$

$$5x\left[3x(1-x^2)^{1/2}\right] \stackrel{?}{=} 2\left[\tfrac{3}{2}(5x^2-1)(1-x^2)^{1/2}\right] + 3(1-x^2)^{1/2}$$

$$15x^2(1-x^2)^{1/2} \stackrel{?}{=} 15x^2(1-x^2)^{1/2} - 3(1-x^2)^{1/2} + 3(1-x^2)^{1/2}$$

$$15x^2(1-x^2)^{1/2} = 15x^2(1-x^2)^{1/2}$$

13–42. Calculate the ratio of the dipole transition moments for the $0 \to 1$ and $1 \to 2$ rotational transitions in the rigid-rotator approximation.

For the $0 \to 1$ transition (Example 13–12),

$$I_{0\to 1} = \frac{1}{\sqrt{3}}$$

For the $1 \to 2$ transition,

$$
\begin{aligned}
I_{1\to 2} &= \int_0^{2\pi} d\phi \int_0^\pi d\theta \sin\theta \cos\theta\, Y_2^0(\theta,\phi) Y_1^0(\theta,\phi) \\
&= 2\pi \int_0^\pi d\theta \sin\theta \cos\theta \left(\frac{5}{16\pi}\right)^{1/2}(3\cos^2 - 1)\left(\frac{3}{4\pi}\right)^{1/2}\cos\theta \\
&= \frac{\sqrt{15}}{4}\int_{-1}^1 dx\, x^2(3x^2 - 1) = \frac{2}{\sqrt{15}}
\end{aligned}
$$

The ratio of the dipole transition moments for these two rotational transitions is, therefore,

$$\frac{I_{0\to 1}}{I_{1\to 2}} = \frac{\sqrt{15}}{2\sqrt{3}} = \frac{\sqrt{5}}{2}$$

13–43. Calculate the ratio of the dipole transition moments for the $0 \to 1$ and $1 \to 2$ vibrational transitions in the harmonic-oscillator approximation.

For the $0 \to 1$ transition (Example 13–13),

$$I_{0\to 1} \propto \int_{-\infty}^\infty \psi_1(\xi)\xi\psi_0(\xi)d\xi$$

$$\propto \left(\frac{2\alpha}{\pi}\right)^{1/2}\int_{-\infty}^\infty \xi^2 e^{-\xi^2}d\xi$$

$$\propto \frac{\alpha^{1/2}}{\sqrt{2}}$$

For the $1 \to 2$ transition,

$$I_{1\to 2} \propto \int_{-\infty}^\infty \psi_1(\xi)\xi\psi_0(\xi)d\xi$$

$$\propto \left(\frac{\alpha}{\pi}\right)^{1/2}\int_{-\infty}^\infty \xi^2(2\xi^2 - 1)e^{-\xi^2}d\xi$$

$$\propto \alpha^{1/2}$$

The ratio of the two transitions is then

$$\frac{I_{0\to1}}{I_{1\to2}} = \frac{\alpha^{1/2}(\sqrt{2})^{-1}}{\alpha^{1/2}} = \frac{1}{\sqrt{2}}$$

13-44. Use Table 13.7 to determine the 12-dimensional reducible representation for the vibrational motion of NH_3. Use this result to determine the symmetries and the infrared activity of the normal coordinates of NH_3.

The point group of NH_3 is $\mathbf{C}_{3v}$ (Table 12.2). Recall from the character table (Table 12.9) that this group has the operators $\hat{E}$, $2\hat{C}_3$, and $3\hat{\sigma}_v$. $\hat{E}$ leaves all of the atoms unmoved, the $\hat{C}_3$ leave the nitrogen atom unmoved, and the $\hat{\sigma}_v$ planes leave the nitrogen atom and one hydrogen atom unmoved. Using Table 13.8, we find

	$\hat{E}$	$\hat{C}_3$	$\hat{C}_3^2$	$\hat{\sigma}_v$	$\hat{\sigma}_v'$	$\hat{\sigma}_v''$
Γ	12	0	0	2	2	2

The quantity $h = 6$ for the $\mathbf{C}_{3v}$ point group, so Equation 12.23 becomes

$$a_{A_1} = \frac{1}{6}[12 + 0 + 6] = 3$$

$$a_{A_2} = \frac{1}{6}[12 + 0 - 6] = 1$$

$$a_E = \frac{1}{6}[24 + 0 + 0] = 4$$

where we have used Table 12.9 to determine the values of χ_i. We thus have

$$\Gamma_{3N} = 3A_1 + A_2 + 4E$$

The character table shows that x and y are jointly represented by E, that z is represented by A_1, that R_z is represented by A_2, and that R_x and R_y are jointly represented by E. Subtracting these translational and rotational degrees of freedom from Γ_{3N} gives us

$$\Gamma_{vib} = 2A_1 + 2E$$

Recall that we have infrared activity only if Q_j belongs to the same irreducible representation as x, y, or z, which are represented by E and A_1. In this case, all of the vibrational modes in Γ_{vib} are infrared active.

13-45. Use Table 13.7 to determine the 15-dimensional reducible representation for the vibrational motion of CH_2Cl_2. Use this result to determine the symmetries and the infrared activity of the normal coordinates of CH_2Cl_2.

The point group of CH_2Cl_2 is $\mathbf{C}_{2v}$ (Table 12.2). Recall from the character table (Table 12.7) that this group has the operators $\hat{E}$, $\hat{C}_2$, and $2\hat{\sigma}_v$. $\hat{E}$ leaves all of the atoms unmoved, $\hat{C}_2$ leaves the carbon atom unmoved, and the $\hat{\sigma}_v$ planes leave the carbon atom and two other atoms unmoved. Using Table 13.8, we find

	$\hat{E}$	$\hat{C}_2$	$\hat{\sigma}_v$	$\hat{\sigma}_v'$
Γ	15	−1	3	3

The quantity $h = 4$ for the C_{2v} point group, so Equation 12.23 becomes

$$a_{A_1} = \frac{1}{4}[15 - 1 + 3 + 3] = 5$$

$$a_{A_2} = \frac{1}{4}[15 - 1 - 3 - 3] = 2$$

$$a_{B_1} = \frac{1}{4}[15 + 1 + 3 - 3] = 4$$

$$a_{B_2} = \frac{1}{4}[15 + 1 - 3 + 3] = 4$$

where we have used Table 12.7 to determine the values of χ_i. We thus have

$$\Gamma_{3N} = 5A_1 + 2A_2 + 4B_1 + 4B_2$$

The character table shows that x transforms as B_1, y as B_2, z as A_1, R_z as A_2, R_y as B_1, and R_x as B_2. Subtracting these translational and rotational degrees of freedom from Γ_{3N} gives us

$$\Gamma_{\text{vib}} = 4A_1 + A_2 + 2B_1 + 2B_2$$

Recall that we have infrared activity only if the representation of the normal mode transforms as x, y, or z, which are represented by B_1, B_2, and A_1. Therefore, the A_2 vibrational mode is infrared inactive, and the other vibrational modes are infrared active.

13–46. Use Table 13.7 to determine the 18-dimensional reducible representation for the vibrational motion of *trans*-dichloroethene. Use this result to determine the symmetries and the infrared activity of the normal coordinates of *trans*-dichloroethene.

The point group of *trans*-dichloroethene is C_{2h} (Table 12.2). Recall from the character table (Table 12.10) that this group has the operators $\hat{E}$, $\hat{C}_2$, $\hat{i}$, and $\hat{\sigma}_h$. $\hat{E}$ leaves all of the atoms unmoved, $\hat{C}_2$ and $\hat{i}$ move all the atoms, and $\hat{\sigma}_h$ leaves all the atoms unmoved. Using Table 13.8, we find

	$\hat{E}$	$\hat{C}_2$	$\hat{i}$	$\hat{\sigma}_h$
Γ	18	0	0	6

The quantity $h = 4$ for the C_{2h} point group, so Equation 12.23 becomes

$$a_{A_g} = \frac{1}{4}[18 + 6] = 6$$

$$a_{B_g} = \frac{1}{4}[18 - 6] = 3$$

$$a_{A_u} = \frac{1}{4}[18 - 6] = 3$$

$$a_{B_u} = \frac{1}{4}[18 + 6] = 6$$

where we have used Table 12.10 to determine the values of χ_i. We thus have

$$\Gamma_{3N} = 6A_g + 3B_g + 3A_u + 6B_u$$

The character table shows that x and y are independently represented by B_u, z by A_u, R_z by A_g, and R_x and R_y independently by B_g. Subtracting these translational and rotational degrees of freedom from Γ_{3N} gives us

$$\Gamma_{\text{vib}} = 5A_g + B_g + 2A_u + 4B_u$$

Recall that we have infrared activity only if the representation of the normal mode transforms as x, y, or z, which are represented by B_u and A_u. Therefore, the A_g and B_g vibrational modes are infrared inactive, and the other vibrational modes are infrared active.

13–47. Use Table 13.7 to determine the 15-dimensional reducible representation for the vibrational motion of XeF$_4$ (square planar). Use this result to determine the symmetries and the infrared activity of the normal coordinates of XeF$_4$.

The point group of XeF$_4$ is $\mathbf{D}_{4h}$ (Table 12.2). Recall from the character table (Table 12.12) that this group has the operators $\hat{E}$, $2\hat{C}_4$, $\hat{C}_2$, $2\hat{C}_2'$, $2\hat{C}_2''$, $\hat{\imath}$, $2\hat{S}_4$, $\hat{\sigma}_h$, $2\hat{\sigma}_v$, and $2\hat{\sigma}_d$. $\hat{E}$ and $\hat{\sigma}_h$ leave all of the atoms unmoved, the $\hat{C}_4$, $\hat{C}_2$, $\hat{C}_2''$, $\hat{\imath}$, $\hat{S}_4$, and $\hat{\sigma}_d$ leave only Xe unmoved. The $\hat{C}_2'$ each leave Xe and two F unmoved, as do the two $\hat{\sigma}_v$. Using Table 13.8, we find

	$\hat{E}$	$2\hat{C}_4$	$\hat{C}_2$	$2\hat{C}_2'$	$2\hat{C}_2''$	$\hat{\imath}$	$2\hat{S}_4$	$\hat{\sigma}_h$	$2\hat{\sigma}_v$	$2\hat{\sigma}_d$
Γ	15	1	-1	-3	-1	-3	-1	5	3	1

The quantity $h = 16$ for the $\mathbf{D}_{4h}$ point group, so Equation 12.23 becomes

$$a_{A_{1g}} = \frac{1}{16}[15 + 2 - 1 - 6 - 2 - 3 - 2 + 5 + 6 + 2] = 1$$

$$a_{A_{2g}} = \frac{1}{16}[15 + 2 - 1 + 6 + 2 - 3 - 2 + 5 - 6 - 2] = 1$$

$$a_{B_{1g}} = \frac{1}{16}[15 - 2 - 1 - 6 + 2 - 3 + 2 + 5 + 6 - 2] = 1$$

$$a_{B_{2g}} = \frac{1}{16}[15 - 2 - 1 + 6 - 2 - 3 + 2 + 5 - 6 + 2] = 1$$

$$a_{E_g} = \frac{1}{16}[30 + 0 + 2 + 0 + 0 - 6 + 0 - 10 + 0 + 0] = 1$$

$$a_{A_{1u}} = \frac{1}{16}[15 + 2 - 1 - 6 - 2 + 3 + 2 - 5 - 6 - 2] = 0$$

$$a_{A_{2u}} = \frac{1}{16}[15 + 2 - 1 + 6 + 2 + 3 + 2 - 5 + 6 + 2] = 2$$

$$a_{B_{1u}} = \frac{1}{16}[15 - 2 - 1 - 6 + 2 + 3 - 2 - 5 - 6 + 2] = 0$$

$$a_{B_{2u}} = \frac{1}{16}[15 - 2 - 1 + 6 - 2 + 3 - 2 - 5 + 6 - 2] = 1$$

$$a_{E_u} = \frac{1}{16}[30 + 0 + 2 + 0 + 0 + 6 + 0 + 10 + 0 + 0] = 3$$

where we have used Table 12.12 to determine the values of χ_i. We thus have

$$\Gamma_{3N} = A_{1g} + A_{2g} + B_{1g} + B_{2g} + E_g + 2A_{2u} + B_{2u} + 3E_u$$

The character table shows that x and y are jointly represented by E_u, z by A_{2u}, R_z by A_{2g}, and R_x and R_y jointly by E_g. Subtracting these translational and rotational degrees of freedom from Γ_{3N} gives us

$$\Gamma_{\text{vib}} = A_{1g} + B_{1g} + B_{2g} + A_{2u} + B_{2u} + 2E_u$$

Recall that we have infrared activity only if the normal mode transforms as as x, y, or z, which are represented by E_u and A_{2u}. Therefore, the A_{1g}, B_{1g}, B_{2g}, and B_{2u} vibrational modes are infrared inactive, and the other vibrational modes are infrared active.

13–48. Use Table 13.7 to determine the 15-dimensional reducible representation for the vibrational motion of CH_4. Use this result to determine the symmetries and the infrared activity of the normal coordinates of CH_4.

The point group of CH_4 is T_d (Table 12.2). Recall from the character table (Table 12.14) that this group has the operators $\hat{E}$, $8\hat{C}_3$, $3\hat{C}_2$, $6\hat{S}_4$, and $6\hat{\sigma}_d$. $\hat{E}$ leaves all of the atoms unmoved, all atoms are moved by $\hat{C}_3$, $\hat{C}_2$ and $\hat{S}_4$ leave only the central carbon atom unmoved, and $\hat{\sigma}_d$ leaves the central carbon atom and two of the hydrogen atoms unmoved. Using Table 13.8, we find

	$\hat{E}$	$8\hat{C}_3$	$3\hat{C}_2$	$6\hat{S}_4$	$6\hat{\sigma}_d$
Γ	15	0	-1	-1	3

The quantity $h = 24$ for the T_d point group, so Equation 12.23 becomes

$$a_{A_1} = \frac{1}{24}[15 - 3 - 6 + 18] = 1$$

$$a_{A_2} = \frac{1}{24}[15 - 3 + 6 - 18] = 0$$

$$a_E = \frac{1}{24}[30 - 6] = 1$$

$$a_{T_1} = \frac{1}{24}[45 + 3 - 6 - 18] = 1$$

$$a_{T_2} = \frac{1}{24}[45 + 3 + 6 + 18] = 3$$

where we have used Table 12.14 to determine the values of χ_i. We thus have

$$\Gamma_{3N} = A_1 + E + T_1 + 3T_2$$

The character table shows that x, y, and z are jointly represented by T_2 and R_x, R_y, and R_z are jointly represented by T_1. Subtracting these translational and rotational degrees of freedom from Γ_{3N} gives us

$$\Gamma_{vib} = A_1 + E + 2T_2$$

Recall that we have infrared activity only if the normal mode transforms as x, y, or z, which are represented by T_2. The A_1 and E vibrational modes are infrared inactive and the T_2 vibrational mode is infrared active.

13–49. Consider a molecule with a dipole moment μ in an electric field $\mathbf{E}$. We picture the dipole moment as a positive charge and a negative charge of magnitude q separated by a vector $\mathbf{l}$.

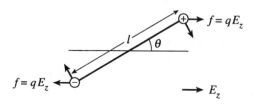

The field $\mathbf{E}$ causes the dipole to rotate into a direction parallel to $\mathbf{E}$. Therefore, work is required to rotate the dipole to an angle θ to $\mathbf{E}$. The force causing the molecule to rotate is actually a torque (torque is the angular analog of force) and is given by $l/2$ times the force perpendicular to $\mathbf{l}$ at each end of the vector $\mathbf{l}$. Show that this torque is equal to $\mu E \sin \theta$ and that the energy required to rotate the dipole from some initial angle θ_0 to some arbitrary angle θ is

$$V = \int_{\theta_0}^{\theta} \mu E \sin \theta' d\theta'$$

Given that θ_0 is customarily taken to be $\pi/2$, show that

$$V = -\mu E \cos \theta = -\boldsymbol{\mu} \cdot \mathbf{E}$$

The magnetic analog of this result will be given by Equation 14.10.

The component of the force perpendicular to $\mathbf{l}$ is $qE \sin \theta$, so the torque at each end of the dipole is given by $lqE \sin \theta / 2$. The total torque is twice this amount, and so the total torque is $lqE \sin \theta$ or $\mu E \sin \theta$ (recall that $\mu = lq$). To find the energy required to rotate the dipole, we must integrate over the torque:

$$
\begin{aligned}
V &= \int_{\theta_0}^{\theta} \mu E \sin \theta' d\theta' \\
&= \mu E \int_{\pi/2}^{\theta} \sin \theta' d\theta' \\
&= \mu E (-\cos \theta') \Big|_{\pi/2}^{\theta} = -\mu E \cos \theta \\
&= -\boldsymbol{\mu} \cdot \mathbf{E}
\end{aligned}
$$

(This last equality is from the definition of dot product in MathChapter C.)

13–50. The observed vibrational-rotational lines for the $v = 0$ to $v = 1$ transition of $^{12}\mathrm{C}^{16}\mathrm{O}(\mathrm{g})$ are listed below. Determine $\tilde{B}_0$, $\tilde{B}_1$, $\tilde{B}_e$, $\tilde{\alpha}_e$, $\tilde{I}_e$, and r_e.

2238.89	2215.66	2189.84	2161.83	2127.61	2094.69	2059.79
2236.06	2212.46	2186.47	2158.13	2123.62	2090.56	2055.31
2233.34	2209.31	2183.14	2154.44	2119.64	2086.27	2050.72
2230.49	2206.19	2179.57	2150.83	2115.56	2081.95	2046.14
2227.55	2202.96	2176.12	2147.05	2111.48	2077.57	
2224.63	2199.77	2172.63	2139.32	2107.33	2073.19	
2221.56	2196.53	2169.05	2135.48	2103.12	2068.69	
2218.67	2193.19	2165.44	2131.49	2099.01	2064.34	

[*Hint:* Recall that the transition $(v'' = 0, J'' = 0) \rightarrow (v'' = 1, J'' = 0)$ is forbidden.]

We must first establish where the R and P branches are in this spectrum. The $(v'' = 0, J'' = 0) \rightarrow (v'' = 1, J'' = 0)$ transition is forbidden and so we expect the energy difference between $R(0)$ and $P(1)$ to be approximately twice the difference between successive lines in the R and P branches. Examining the data in the table, we find that the gap between the lines at 2147.05 cm^{-1} and 2139.32 cm^{-1} is approximately twice as large as the gap between any other neighboring pair of lines. So the first line in the R branch (for $J = 0$) occurs at 2147.05 cm^{-1}, and the first line in the P branch (for $J = 1$) occurs at 2139.32 cm^{-1}.

We now combine Equations 13.15 and 13.16

$$\tilde{\nu}_R(J \to J+1) = \tilde{\nu} + 2\tilde{B}_1 + (3\tilde{B}_1 - \tilde{B}_0)J + (\tilde{B}_1 - \tilde{B}_0)J^2 \tag{13.15}$$

$$\tilde{\nu}_P(J \to J-1) = \tilde{\nu} - (\tilde{B}_1 + \tilde{B}_0)J + (\tilde{B}_1 - \tilde{B}_0)J^2 \tag{13.16}$$

to find the following equations for $\tilde{B}_1$ and $\tilde{B}_0$:

$$\begin{aligned}
\Delta\tilde{\nu}_1 &= \tilde{\nu}_R(J \to J+1) - \tilde{\nu}_P(J \to J-1) \\
&= 2\tilde{B}_1 + (4\tilde{B}_1)J \\
&= 2\tilde{B}_1(2J+1) \\
\Delta\tilde{\nu}_2 &= \tilde{\nu}_R(J \to J+1) - \tilde{\nu}_P(J+2 \to J+1) \\
&= 2\tilde{B}_1(2J+1) + 2(\tilde{B}_1 + \tilde{B}_0) - 4J(\tilde{B}_1 - \tilde{B}_0) - 4(\tilde{B}_1 - \tilde{B}_0) \\
&= 2\tilde{B}_0(3+2J)
\end{aligned}$$

These equations show that a plot of $\Delta\tilde{\nu}_1$ versus $2(2J+1)$ has a slope of $\tilde{B}_1$, and a plot of $\Delta\tilde{\nu}_2$ versus $2(2J+3)$ has a slope of $\tilde{B}_0$. These two plots are shown below.

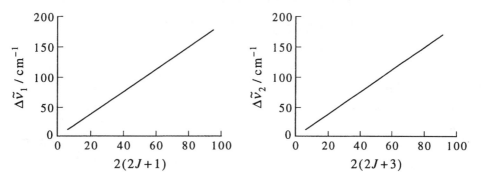

From the slopes of the best fit lines to the data, we find that $\tilde{B}_0 = 1.9163 \text{ cm}^{-1}$ and $\tilde{B}_1 = 1.8986 \text{ cm}^{-1}$. To find $\tilde{B}_e$ and $\tilde{\alpha}_e$, we use Equation 13.17:

$$\tilde{B}_v = \tilde{B}_e - \tilde{\alpha}_e(v + \tfrac{1}{2})$$

$$\tilde{B}_0 = \tilde{B}_e - \tfrac{1}{2}\tilde{\alpha}_e$$

$$\tilde{B}_1 = \tilde{B}_e - \tfrac{3}{2}\tilde{\alpha}_e$$

$$\tilde{B}_0 - \tilde{B}_1 = \tilde{\alpha}_e$$

$$1.9163 \text{ cm}^{-1} - 1.8986 \text{ cm}^{-1} = 0.0177 \text{ cm}^{-1} = \tilde{\alpha}_e$$

$$1.9163 \text{ cm}^{-1} = \tilde{B}_e - \tfrac{1}{2}(0.0177 \text{ cm}^{-1})$$

$$1.92515 \text{ cm}^{-1} = \tilde{B}_e$$

We can now use Equation 13.9 to find cI and then use the definition of the moment of the inertia to find R_e.

$$\mu = \frac{(12.00 \text{ amu})(15.99 \text{ amu})}{27.99 \text{ amu}}(1.661 \times 10^{-27} \text{ kg·amu}^{-1}) = 1.139 \times 10^{-26} \text{ kg}$$

$$\tilde{B}_e = \frac{h}{8\pi^2 c I}$$

$$I = \frac{6.626 \times 10^{-34} \text{ J·s}}{8\pi^2 (2.998 \times 10^8 \text{ m·s}^{-1})(192.515 \text{ m}^{-1})} = 1.454 \times 10^{-46} \text{ kg·m}^2$$

$$I = \mu R_e^2$$

$$R_e = \left(\frac{I_e}{\mu}\right)^{1/2} = 113.0 \text{ pm}$$

13–51. This problem is a three-dimensional version of Problem 13–41. The rotational spectrum of a polyatomic molecule can be predicted once the values of $\tilde{A}$, $\tilde{B}$, and $\tilde{C}$ are known. These, in turn, can be calculated from the principal moments of inertia I_A, I_B, and I_C. In this problem, we show how I_A, I_B, and I_C can be determined from the molecular geometry. We set up an arbitrarily oriented coordinate system, whose origin sits at the center-of-mass of the molecule, and determine the moments of inertia I_{xx}, I_{xy}, I_{xz}, I_{yy}, I_{yz}, and I_{zz}. The principal moments of inertia are the solution to the secular determinantal equation

$$\begin{vmatrix} I_{xx} - \lambda & I_{xy} & I_{xz} \\ I_{xy} & I_{yy} - \lambda & I_{yz} \\ I_{xz} & I_{yz} & I_{zz} - \lambda \end{vmatrix} = 0$$

The assignment for the subscripts A, B, and C to the three roots of this determinant are done according to the convention $I_A \le I_B \le I_C$. Use this approach to find the principal moments of inertia for the planar formate radical, HCO_2, given the following geometry: The H–C bond length is 109.7 pm, the C=O bond length is 120.2 pm, and the C–O bond length is 134.3 pm.

Take the origin to sit on the carbon atom and the molecule to sit in the xy-plane, as shown below:

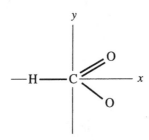

We can use straightforward trigonometry to find the following Cartesian coordinates for the atoms given the geometry in the problem.

	C	H	O (carbonyl)	O
x	0.0	−109.7 pm	43.1 pm	75.5 pm
y	0.0	0.0	112.2 pm	−111.1 pm
z	0.0	0.0	0.0	0.0

The center of mass is then located at

$$x_{cm} = \frac{1}{M} \sum_i m_i x_i = \frac{1}{45.0 \text{ amu}} [(1.008)(-109.7) + (15.999)(43.1 + 75.5)] \text{ amu·pm}$$

$$= 39.7 \text{ pm}$$

$$y_{cm} = \frac{1}{45.0 \text{ amu}} [(15.999)(112.2 - 111.1)] \text{ amu·pm}$$

$$= 0.39 \text{ pm}$$

$$z_{cm} = \frac{1}{45.0 \text{ amu}} (0) = 0 \text{ pm}$$

The coordinates of the atoms relative to the center of mass are then

	C	H	O (carbonyl)	O
x	−39.7 pm	−149.4 pm	3.4 pm	40.2 pm
y	−0.4 pm	−0.4 pm	111.8 pm	−111.5 pm
z	0.0	0.0	0.0	0.0

We now find the components of the moment of inertia.

$$I_{xx} = \sum_i m_i(y_i^2 + z_i^2)$$

$$= \{(12.01)(-0.4)^2 + (1.008)(-0.4)^2 + 15.99[(111.8)^2 + (-111.5)^2]\} \text{ amu·pm}^2$$

$$= 3.97 \times 10^5 \text{ pm}^2 \cdot \text{g·mol}^{-1}$$

$$I_{yy} = \sum_i m_i(x_i^2 + z_i^2)$$

$$= \{(12.01)(-39.7)^2 + (1.008)(-149.4)^2 + 15.99[(3.4)^2 + (40.2)^2]\} \text{ amu·pm}^2$$

$$= 6.74 \times 10^4 \text{ pm}^2 \cdot \text{g·mol}^{-1}$$

$$I_{zz} = I_{xx} + I_{yy} = 4.64 \times 10^5 \text{ pm}^2 \cdot \text{g·mol}^{-1}$$

$$I_{xz} = I_{zx} = I_{yz} = I_{zy} = 0$$

$$I_{xy} = I_{yx} = -\sum_i m_i x_i y_i$$

$$= -\{(12.01)(-39.7)(-0.4) + (1.008)(-149.4)(-0.4)$$

$$+ 15.99[(3.4)(111.8) + (40.2)(-111.5)]\} \text{ amu·pm}^2$$

$$= 6.53 \times 10^4 \text{ pm}^2 \cdot \text{g·mol}^{-1}$$

We then have the determinantal equation

$$\begin{vmatrix} 3.97 \times 10^5 - \lambda & 6.53 \times 10^4 & 0 \\ 6.53 \times 10^4 & 6.74 \times 10^4 - \lambda & 0 \\ 0 & 0 & 4.64 \times 10^5 \end{vmatrix} = 0$$

One of the principal moments of inertia is $4.64 \times 10^5 \text{ pm}^2 \cdot \text{g·mol}^{-1}$. To find the other two we need to solve the 2×2 determinantal equation

$$\begin{vmatrix} 3.97 \times 10^5 - \lambda & 6.53 \times 10^4 \\ 6.53 \times 10^4 & 6.74 \times 10^4 - \lambda \end{vmatrix} = 0$$

Expanding this determinant gives

$$\lambda^2 - 4.64 \times 10^5 \lambda + 2.25 \times 10^{10} = 0$$

or

$$\lambda = \frac{4.64 \times 10^5 \pm \sqrt{(4.64 \times 10^5)^2 - 4(2.25 \times 10^{10})}}{2} = \frac{4.64 \times 10^5 \pm 3.54 \times 10^5}{2}$$

The other two moments of intertia are $5.50 \times 10^4 \text{ pm}^2 \cdot \text{g·mol}^{-1}$ and $4.09 \times 10^5 \text{ pm}^2 \cdot \text{g·mol}^{-1}$.

CHAPTER 14

Nuclear Magnetic Resonance Spectroscopy

PROBLEMS AND SOLUTIONS

14–1. Show how Equation 14.7 reduces to Equation 14.6 for a circular orbit.

$$\mu = \frac{q(\mathbf{r} \times \mathbf{v})}{2} = \frac{qrv\sin\theta}{2} \tag{14.7}$$

In a circular orbit, motion is always perpendicular to the radius, so $\sin\theta = \sin 90° = 1$, and the equation above reduces to

$$\mu = \frac{qrv}{2} \tag{14.6}$$

14–2. What magnetic field strength must be applied for C–13 spin transitions to occur at 90.0 MHz?

From Table 14.1, for ^{13}C, $\gamma = 6.7283 \times 10^7$ rad·T^{-1}·s^{-1}. We now use Equation 14.18 to find B_z:

$$B_z = \frac{2\pi\nu}{\gamma} = \frac{2\pi(90.0 \times 10^6 \text{ s}^{-1})}{6.7283 \times 10^7 \text{ rad·T}^{-1}\text{·s}^{-1}} = 8.40 \text{ T}$$

14–3. What magnetic field strength must be applied for proton spin transitions to occur at 270.0 MHz?

The value of γ for ^1H (Table 14.1) is 26.7522×10^7 rad·T^{-1}·s^{-1}, and so (Equation 14.18)

$$B_z = \frac{2\pi\nu}{\gamma} = \frac{2\pi(270.0 \times 10^6 \text{ s}^{-1})}{26.7522 \times 10^7 \text{ rad·T}^{-1}\text{·s}^{-1}} = 6.341 \text{ T}$$

14–4. Calculate the magnetic field strength necessary to observe resonances of the nuclei given in Table 14.1 using a 300-MHz NMR spectrometer.

$$B_z = \frac{2\pi\nu}{\gamma} = \frac{(2\pi \text{ rad})(300 \times 10^6 \text{ s}^{-1})}{\gamma} \tag{14.18}$$

Nucleus	$\gamma/10^7$ rad·T^{-1}·s^{-1}	B_z/T
^1H	26.7522	7.05
^2H	4.1066	45.9
^{13}C	6.7283	28.0
^{14}N	1.9338	97.5
^{31}P	10.841	17.4

14–5. It turns out that a proton chemical shift of 2.2 ppm corresponds to a frequency range of 1100 Hz on a certain NMR instrument. Determine the magnetic field strength of this instrument.

We can determine $\nu_{spectrometer}$ using Equation 14.23:

$$\delta = \frac{\nu_H - \nu_{TMS}}{\nu_{spectrometer}} \times 10^6$$

or

$$\nu_{spectrometer} = \frac{1100 \text{ Hz}}{2.2 \times 10^{-6}} = 500 \text{ MHz}$$

The corresponding magnetic field strength (in Teslas) can be found using Equation 14.18:

$$\nu = \frac{\gamma B_z}{2\pi}$$

$$500 \times 10^6 \text{ s}^{-1} = \frac{(26.7522 \times 10^7 \text{ rad}\cdot\text{T}^{-1}\cdot\text{s}^{-1}) B_z}{2\pi}$$

$$B_z = 12 \text{ T}$$

14–6. Show that a chemical shift range of 8.0 ppm corresponds to a frequency range of 480 Hz on a 60-MHz instrument. What is the frequency range on a 270-MHz instrument?

$$\delta = \frac{\nu_H - \nu_{TMS}}{\nu_{spectrometer}} \times 10^6 \tag{14.23}$$

On a 60-MHz instrument, a chemical shift of 8.0 ppm corresponds to a frequency range of

$$\nu_H - \nu_{TMS} = \left(8.0 \times 10^{-6}\right)\left(60 \times 10^6 \text{ Hz}\right) = 480 \text{ Hz}$$

On a 270-MHz instrument, the same shift corresponds to the frequency range of

$$\nu_H - \nu_{TMS} = \left(8.0 \times 10^{-6}\right)\left(270 \times 10^6 \text{ Hz}\right) = 2200 \text{ Hz}$$

For the solutions to Problems 14–7 through 14–14, approximate chemical shifts are given in units of δ.

14–7. Show that the top and bottom scales in Figure 14.6 are consistent.

The bottom scale shows that $\delta = 2.2$ ppm for the protons in CH_3I. Using Equation 14.24, we have

$$\nu_H - \nu_{TMS} = \delta\nu_{spectrometer} \times 10^{-6} = (2.2)(60 \text{ MHz}) \times 10^{-6} = 130 \text{ Hz}$$

which is consistent with the value for this peak in Figure 14.6.

14–8. Use Equation 14.21 to show that $B_{TMS} - B_H$ is directly proportional to δ_H, in analogy with Equation 14.23. Interpret this result.

From Equation 14.21, we have

$$B_{TMS} \propto \nu_{TMS} \qquad B_H \propto \nu_H$$

Then

$$B_{TMS} - B_H \propto \nu_{TMS} - \nu_H \propto \delta_H$$

The quantity $B_{TMS} - B_H$ increases as δ_H decreases, consistent with the fact that the strength of the magnetic field increases from left to right in a NMR spectrum.

14–9. Make a rough sketch of what you think the NMR spectrum of methyl acetate looks like.

The molecular formula for methyl acetate is $H_3CCOOCH_3$. Using the trends given in Section 14–5, we expect to observe two lines in the NMR spectra:

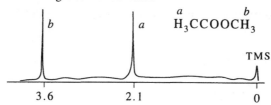

The actual spectra is shown in Example 14–5.

14–10. Make rough sketches of what you think the NMR spectra of the two isomers dimethyl ether and ethanol look like and compare the two.

Using the trends given in Section 14–5, we draw

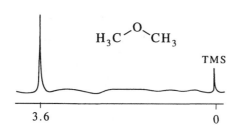
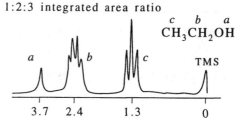

These two molecules have clearly distinguishable NMR spectra. There is no spin-spin coupling in dimethyl ether, and so no multiplets are observed in the spectrum of this molecule. However, there are multiplets in the spectrum of ethanol, since spin-spin coupling occurs.

14–11. Make a rough sketch of what you think the NMR spectrum of diethyl ether looks like.

Using the trends given in Section 14–5, we draw

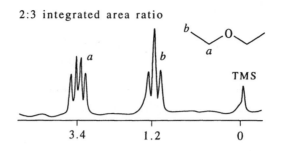

2:3 integrated area ratio

14–12. Make a rough sketch of what you think the NMR spectrum of 3-pentanone looks like.

Using the trends given in Section 14–5, we obtain

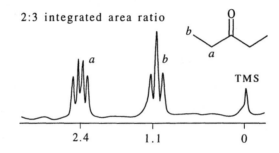

2:3 integrated area ratio

14–13. Make a rough sketch of what you think the NMR spectrum of methyl propanoate looks like.

Using the trends given in Section 14–5, we draw

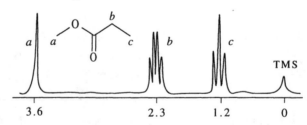

3:2:3 integrated area ratio

14–14. Make a rough sketch of what you think the NMR spectrum of ethyl acetate looks like.

Using the trends given in Section 14–5, we draw

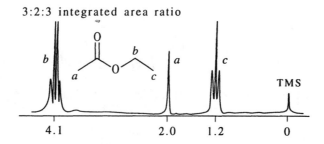

3:2:3 integrated area ratio

$$\hat{H} = -\gamma B_0(1 - \sigma_1)\hat{I}_{z1} - \gamma B_0(1 - \sigma_2)\hat{I}_{z2} + \frac{hJ_{12}}{\hbar^2}\hat{\mathbf{I}}_1 \cdot \hat{\mathbf{I}}_2 \qquad (14.27)$$

14–15. Show that Equation 14.27 has units of joules.

$$\hat{H} = -\gamma B_0(1 - \sigma_1)\hat{I}_{z1} - \gamma B_0(1 - \sigma_2)\hat{I}_{z2} + \frac{hJ_{12}}{\hbar^2}\hat{\mathbf{I}}_1 \cdot \hat{\mathbf{I}}_2 \qquad (14.27)$$

The quantity σ is unitless, B_0 has units of T, γ has units of rad·T^{-1}·s^{-1}, and $\hat{I}$ has units of J·s·rad^{-1} ($\hat{I}\alpha \propto \hbar\alpha$). The units of Equation 14.27 are then

$$\text{Units} = \left(\frac{\text{rad}}{\text{T·s}}\right)(\text{T})\left(\frac{\text{J·s}}{\text{rad}}\right) - \left(\frac{\text{rad}}{\text{T·s}}\right)(\text{T})\left(\frac{\text{J·s}}{\text{rad}}\right) + \left(\frac{1}{\text{J·s}}\right)\left(\frac{1}{\text{s}}\right)(\text{J·s})^2 = \text{J}$$

14–16. Verify Equations 14.36 and 14.37.

Use the unperturbed wave functions in Equation 14.30, the Hamiltonian operator given in Equation 14.28, and the equivalence given by Equation 14.32:

$$\begin{aligned}
\hat{H}^{(0)}\psi_2 &= \hat{H}^{(0)}\beta(1)\alpha(2) \\
&= -\gamma B_0(1 - \sigma_1)\hat{I}_{z1}\beta(1)\alpha(2) - \gamma B_0(1 - \sigma_2)\hat{I}_{z2}\beta(1)\alpha(2) \\
&= \tfrac{1}{2}\gamma B_0\hbar(1 - \sigma_1)\beta(1)\alpha(2) - \tfrac{1}{2}\hbar\gamma B_0(1 - \sigma_2)\beta(1)\alpha(2) \\
&= -\tfrac{1}{2}\hbar\gamma B_0(\sigma_1 - \sigma_2)\psi_2
\end{aligned}$$

So

$$E_2^{(0)} = -\tfrac{1}{2}\hbar\gamma B_0(\sigma_1 - \sigma_2)$$

Likewise, to verify Equation 14.37, we write

$$\begin{aligned}
\hat{H}^{(0)}\psi_4 &= \hat{H}^{(0)}\beta(1)\beta(2) \\
&= \tfrac{1}{2}\hbar\gamma B_0(1 - \sigma_1)\beta(1)\beta(2) + \tfrac{1}{2}\hbar\gamma B_0(1 - \sigma_2)\beta(1)\beta(2) \\
&= \tfrac{1}{2}\hbar\gamma B_0(1 - \sigma_1)\beta(1)\beta(2) + \tfrac{1}{2}\hbar\gamma B_0(1 - \sigma_2)\beta(1)\beta(2) \\
&= \hbar\gamma B_0\left(1 - \frac{\sigma_1 + \sigma_2}{2}\right)\psi_4
\end{aligned}$$

and

$$E_4^{(0)} = \hbar\gamma B_0\left(1 - \frac{\sigma_1 + \sigma_2}{2}\right)$$

14–17. Verify Equations 14.41 and 14.42.

As in Equation 14.40,

$$H_{z,ii} = \frac{h J_{12}}{\hbar^2} \int \psi_i^* \hat{I}_{z1} \hat{I}_{z2} \psi_i d\tau_1 d\tau_2$$

Now use the unperturbed wave functions from Equation 14.30 to find $H_{z,ii}$ for $i = 2, 3,$ and 4.

$$H_{z,22} = \frac{h J_{12}}{\hbar^2} \int \beta^*(1)\alpha^*(2) \hat{I}_{z1} \hat{I}_{z2} \beta(1)\alpha(2) d\tau_1 d\tau_2$$

$$= \frac{h J_{12}}{\hbar^2} \left(-\frac{\hbar^2}{4}\right) \int \beta^*(1)\beta(1) d\tau_1 \int \alpha^*(2)\alpha(2) d\tau_2 = -\frac{h J_{12}}{4}$$

$$H_{z,33} = \frac{h J_{12}}{\hbar^2} \int \alpha^*(1)\beta^*(2) \hat{I}_{z1} \hat{I}_{z2} \alpha(1)\beta(2) d\tau_1 d\tau_2$$

$$= \frac{h J_{12}}{\hbar^2} \left(-\frac{\hbar^2}{4}\right) \int \alpha^*(1)\alpha(1) d\tau_1 \int \beta^*(2)\beta(2) d\tau_2 = -\frac{h J_{12}}{4}$$

$$H_{z,44} = \frac{h J_{12}}{\hbar^2} \int \beta^*(1)\beta^*(2) \hat{I}_{z1} \hat{I}_{z2} \beta(1)\beta(2) d\tau_1 d\tau_2$$

$$= \frac{h J_{12}}{\hbar^2} \left(\frac{\hbar^2}{4}\right) \int \beta^*(1)\beta(1) d\tau_1 \int \beta^*(2)\beta(2) d\tau_2 = \frac{h J_{12}}{4}$$

14–18. The nuclear spin operators, $\hat{I}_x$, $\hat{I}_y$, and $\hat{I}_z$, like all angular momentum operators, obey the commutation relations (Problem 6–13)

$$[\hat{I}_x, \hat{I}_y] = i\hbar \hat{I}_z, \quad [\hat{I}_y, \hat{I}_z] = i\hbar \hat{I}_x, \quad \text{and} \quad [\hat{I}_z, \hat{I}_x] = i\hbar \hat{I}_y$$

Define the (non-Hermitian) operators

$$\hat{I}_+ = \hat{I}_x + i\hat{I}_y \quad \text{and} \quad \hat{I}_- = \hat{I}_x - i\hat{I}_y \tag{1}$$

and show that

$$\hat{I}_z \hat{I}_+ = \hat{I}_+ \hat{I}_z + \hbar \hat{I}_+ \tag{2}$$

and

$$\hat{I}_z \hat{I}_- = \hat{I}_- \hat{I}_z - \hbar \hat{I}_- \tag{3}$$

First write the commutation relations as

$$\hat{I}_x \hat{I}_y - \hat{I}_y \hat{I}_x = i\hbar \hat{I}_z \quad \hat{I}_y \hat{I}_z - \hat{I}_z \hat{I}_y = i\hbar \hat{I}_x \quad \hat{I}_z \hat{I}_x - \hat{I}_x \hat{I}_z = i\hbar \hat{I}_y$$

Then

$$\hat{I}_z \hat{I}_+ = \hat{I}_z \left(\hat{I}_x + i\hat{I}_y\right)$$

$$= i\hbar \hat{I}_y + \hat{I}_x \hat{I}_z + i\hat{I}_z \hat{I}_y$$

$$= i\hbar \hat{I}_y + \hat{I}_x \hat{I}_z + i\left(\hat{I}_y \hat{I}_z - i\hbar \hat{I}_x\right)$$

$$= \hat{I}_x \hat{I}_z + i\hat{I}_y \hat{I}_z + \hbar \hat{I}_x + i\hbar \hat{I}_y$$

$$= \hat{I}_+ \hat{I}_z + \hbar \hat{I}_+$$

and

$$\hat{I}_z \hat{I}_- = \hat{I}_z \left(\hat{I}_x - i\hat{I}_y \right)$$
$$= i\hbar \hat{I}_y + \hat{I}_x \hat{I}_z - i\hat{I}_z \hat{I}_y$$
$$= i\hbar \hat{I}_y + \hat{I}_x \hat{I}_z - i \left(\hat{I}_y \hat{I}_z - i\hbar \hat{I}_x \right)$$
$$= \hat{I}_- \hat{I}_z - \hbar \hat{I}_-$$

14–19. Using the definitions of $\hat{I}_+$ and $\hat{I}_-$ from the previous problem, show that

$$\hat{I}_+ \hat{I}_- = \hat{I}^2 - \hat{I}_z^2 + \hbar \hat{I}_z$$

and that

$$\hat{I}_- \hat{I}_+ = \hat{I}^2 - \hat{I}_z^2 - \hbar \hat{I}_z$$

where

$$\hat{I}^2 = \hat{I}_x^2 + \hat{I}_y^2 + \hat{I}_z^2$$

Using the definitions from the previous problem,

$$\hat{I}_+ \hat{I}_- = \hat{I}_x^2 + i\hat{I}_y \hat{I}_x - i\hat{I}_x \hat{I}_y + \hat{I}_y^2$$

Now $\hat{I}^2 - \hat{I}_z^2 = \hat{I}_x^2 + \hat{I}_y^2$, so

$$\hat{I}_+ \hat{I}_- = \hat{I}^2 - \hat{I}_z^2 + i\hat{I}_y \hat{I}_x - i\hat{I}_x \hat{I}_y$$
$$= \hat{I}^2 - \hat{I}_z^2 - i \left(i\hbar \hat{I}_z \right)$$
$$= \hat{I}^2 - \hat{I}_z^2 + \hbar \hat{I}_z$$

Likewise,

$$\hat{I}_- \hat{I}_+ = \hat{I}_x^2 + i\hat{I}_x \hat{I}_y - i\hat{I}_y \hat{I}_x + \hat{I}_y^2$$
$$= \hat{I}^2 - \hat{I}_z^2 + i(i\hbar \hat{I}_z)$$
$$= \hat{I}^2 - \hat{I}_z^2 - \hbar \hat{I}_z$$

14–20. Use Equation 2 from Problem 14–18 and the fact that $\hat{I}_z \beta = -\frac{\hbar}{2}\beta$ to show that

$$\hat{I}_z \hat{I}_+ \beta = \hat{I}_+ \left(-\frac{\hbar}{2}\beta + \hbar\beta \right) = \frac{\hbar}{2}\hat{I}_+ \beta$$

Because $\hat{I}_z \alpha = \frac{\hbar}{2}\alpha$, this result shows that

$$\hat{I}_+ \beta \propto \alpha = c\alpha$$

where c is a proportionality constant. The following problem shows that $c = \hbar$, so we have

$$\hat{I}_+ \beta = \hbar\alpha \tag{1}$$

Now use Equation 3 from Problem 14–18 and the fact that $\hat{I}_z\alpha = \frac{\hbar}{2}\alpha$ to show that

$$\hat{I}_-\alpha = c\beta$$

where c is a proportionality constant. The following problem shows that $c = \hbar$, so we have

$$\hat{I}_-\alpha = \hbar\beta \tag{2}$$

Notice that $\hat{I}_+$ "raises" the spin function from β to α, whereas $\hat{I}_-$ "lowers" the spin function from α to β. The two operators $\hat{I}_+$ and $\hat{I}_-$ are called raising and lowering operators, respectively. Now argue that a consequence of the raising and lowering properties of $\hat{I}_+$ and $\hat{I}_-$ is that

$$\hat{I}_+\alpha = 0 \quad\text{and}\quad \hat{I}_-\beta = 0 \tag{3}$$

Now use Equations 1, 2, and 3 to show that

$$\hat{I}_x\alpha = \frac{\hbar}{2}\beta \qquad \hat{I}_y\alpha = \frac{i\hbar}{2}\beta$$

$$\hat{I}_x\beta = \frac{\hbar}{2}\alpha \qquad \hat{I}_y\beta = -\frac{i\hbar}{2}\alpha$$

Using Equation 2 from Problem 14–18 and $\hat{I}_z\beta = -\frac{\hbar}{2}\beta$, we find

$$\hat{I}_z\hat{I}_+\beta = \hat{I}_+\hat{I}_z\beta + \hbar\hat{I}_+\beta$$

$$= \hat{I}_+\left(-\frac{\hbar}{2}\beta\right) + \hbar\hat{I}_+\beta$$

$$= \hat{I}_+\left(-\frac{\hbar}{2}\beta + \hbar\beta\right)$$

$$= \frac{\hbar}{2}\hat{I}_+\beta$$

As explained in the problem, we find $\hat{I}_+\beta = \hbar\alpha$. We can similarly use Equation 3 from Problem 14–18 and $\hat{I}_z\alpha = \frac{\hbar}{2}\alpha$ to find

$$\hat{I}_z\hat{I}_-\alpha = \hat{I}_-\hat{I}_z\alpha - \hbar\hat{I}_-\alpha$$

$$= \hat{I}_-\left(\frac{\hbar}{2}\alpha\right) - \hbar\hat{I}_-\beta$$

$$= \hat{I}_-\left(\frac{\hbar}{2}\alpha - \hbar\alpha\right)$$

$$= -\frac{\hbar}{2}\hat{I}_-\alpha$$

and we take $\hat{I}_-\alpha = \hbar\beta$. Now, since α and β correspond to the two possible spin states $+1/2$ and $-1/2$, respectively, α cannot be raised and β cannot be lowered. Therefore,

$$I_+\alpha = 0 \quad\text{and}\quad I_-\beta = 0 \tag{3}$$

Recall from Problem 14–18 that

$$\hat{I}_+ = \hat{I}_x + i\hat{I}_y \qquad \hat{I}_- = \hat{I}_x - i\hat{I}_y$$

Then, using Equations 2 and 3, we have

$$\hat{I}_+\alpha = \hat{I}_x\alpha + i\hat{I}_y\alpha = 0 \tag{4}$$

$$\hat{I}_-\alpha = \hat{I}_x\alpha - i\hat{I}_y\alpha = \hbar\beta \tag{5}$$

Adding Equations 4 and 5 gives

$$2\hat{I}_x\alpha = \hbar\beta$$

$$\hat{I}_x\alpha = \frac{\hbar}{2}\beta$$

Substituting this result into Equation 5 gives

$$\frac{\hbar}{2}\beta - i\hat{I}_y\alpha = \hbar\beta$$

or

$$\hat{I}_y\alpha = -\frac{\hbar}{2i}\beta = \frac{i\hbar}{2}\beta$$

Using Equations 1 and 3, we have

$$\hat{I}_+\beta = \hat{I}_x\beta + i\hat{I}_y\beta = \hbar\alpha \tag{6}$$

$$\hat{I}_-\beta = \hat{I}_x\beta - i\hat{I}_y\beta = 0 \tag{7}$$

Adding Equations 6 and 7 gives

$$2\hat{I}_x\beta = \hbar\alpha$$

$$\hat{I}_x\beta = \frac{\hbar}{2}\alpha$$

and substituting this result into Equation 7 gives

$$\frac{\hbar}{2}\alpha - i\hat{I}_y\beta = 0$$

or

$$\hat{I}_y\beta = \frac{\hbar}{2i}\alpha = -\frac{i\hbar}{2}\alpha$$

14–21. This problem shows that the proportionality constant c in

$$\hat{I}_+\beta = c\alpha \quad \text{or} \quad \hat{I}_-\alpha = c\beta$$

is equal to $\hbar$. Start with

$$\int \alpha^*\alpha\, d\tau = 1 = \frac{1}{c^2}\int (\hat{I}_+\beta)^*(\hat{I}_+\beta)\, d\tau$$

Let $\hat{I}_+ = \hat{I}_x + i\hat{I}_y$ in the second factor in the above integral and use the fact that $\hat{I}_x$ and $\hat{I}_y$ are Hermitian to get

$$\int (\hat{I}_x\hat{I}_+\beta)^*\beta\, d\tau + i\int (\hat{I}_y\hat{I}_+\beta)^*\beta\, d\tau = c^2$$

Now take the complex conjugate of both sides to get

$$\int \beta^*\hat{I}_x\hat{I}_+\beta\, d\tau - i\int \beta^*\hat{I}_y\hat{I}_+\beta\, d\tau = c^2$$

$$= \int \beta^*\hat{I}_-\hat{I}_+\beta\, d\tau$$

Now use the result in Problem 14–19 to show that

$$c^2 = \int \beta^* \hat{I}_- \hat{I}_+ \beta d\tau = \int \beta^* (\hat{I}^2 - \hat{I}_z^2 - \hbar \hat{I}_z) \beta d\tau$$

$$= \int \beta^* \left(\frac{3}{4}\hbar^2 - \frac{1}{4}\hbar^2 + \frac{\hbar^2}{2} \right) \beta d\tau = \hbar^2$$

or that $c = \hbar$.

Recall that for a Hermitian operator $\hat{A}$,

$$\int f^*(x)\hat{A}g(x)dx = \int g(x)\hat{A}^* f^*(x)dx \tag{4.31}$$

Begin with the expression

$$\int \alpha^* \alpha d\tau = 1 = \frac{1}{c^2} \int (\hat{I}_+ \beta)^* (\hat{I}_+ \beta) d\tau$$

Solving for c^2 gives

$$c^2 = \int (\hat{I}_+ \beta)^* (\hat{I}_+ \beta) d\tau = \int (\hat{I}_+ \beta)^* (\hat{I}_x \beta + i\hat{I}_y \beta) d\tau$$

$$= \int (\hat{I}_+ \beta)^* \hat{I}_x \beta d\tau + i \int (\hat{I}_+ \beta)^* \hat{I}_y \beta d\tau$$

We can use the fact that $\hat{I}_x$ and $\hat{I}_y$ are Hermitian to write this as

$$c^2 = \int \beta \hat{I}_x^* (\hat{I}_+ \beta)^* d\tau + i \int \beta \hat{I}_y^* (\hat{I}_+ \beta)^* d\tau$$

Take the complex conjugate of both sides of the last equation to find

$$c^2 = \int \beta^* \hat{I}_x \hat{I}_+ \beta d\tau - i \int \beta^* \hat{I}_y \hat{I}_+ \beta d\tau$$

$$= \int \beta^* (\hat{I}_x - i\hat{I}_y)\hat{I}_+ \beta d\tau = \int \beta^* \hat{I}_- \hat{I}_+ \beta d\tau$$

Substituting $\hat{I}_- \hat{I}_+ = \hat{I}^2 - \hat{I}_z^2 - \hbar \hat{I}_z$ (Problem 14–19), we obtain

$$c^2 = \int \beta^* \hat{I}_- \hat{I}_+ \beta d\tau = \int \beta^* (\hat{I}^2 - \hat{I}_z^2 - \hbar \hat{I}_z) \beta d\tau$$

$$= \int \beta^* \left(\frac{3\hbar^2}{4} - \frac{\hbar^2}{4} + \frac{\hbar^2}{2} \right) \beta d\tau = \hbar^2$$

where we have used Equations 14.3 to evaluate the various terms involving $\hat{I}^2$ and $\hat{I}_z$. Therefore, $c = \hbar$.

14–22. Show that

$$H_{y,11} = \frac{hJ_{12}}{\hbar^2} \iint d\tau_1 d\tau_2 \alpha^*(1)\alpha^*(2)\hat{I}_{y1}\hat{I}_{y2}\alpha(1)\alpha(2)$$

$$= 0$$

and more generally that

$$H_{x,jj} = H_{y,jj} = 0 \qquad j = 1, 2, 3, 4$$

where $j = 1, 2, 3, 4$ refer to the four spin functions given by Equations 14.30.

$$H_{y,11} = \frac{hJ_{12}}{\hbar^2} \int\int d\tau_1 d\tau_2 \alpha^*(1)\alpha^*(2)\hat{I}_{y1}\hat{I}_{y2}\alpha(1)\alpha(2)$$

$$= \frac{hJ_{12}}{\hbar^2} \int\int d\tau_1 d\tau_2 \alpha^*(1)\alpha^*(2)\left[\hat{I}_{y1}\alpha(1)\right]\left[\hat{I}_{y2}\alpha(2)\right]$$

$$= \frac{hJ_{12}}{\hbar^2} \int\int d\tau_1 d\tau_2 \alpha^*(1)\alpha^*(2)\left[\frac{i\hbar}{2}\beta(1)\right]\left[\frac{i\hbar}{2}\beta(2)\right]$$

$$= -\frac{hJ_{12}}{4} \int\int d\tau_1 d\tau_2 \alpha^*(1)\alpha^*(2)\beta(1)\beta(2) = 0$$

because α and β are orthogonal. When $\hat{I}_z$ acts on a spin function, it returns a constant times the same spin function. However, when $\hat{I}_x$ and $\hat{I}_y$ act on a spin function, they return a constant times a different spin function (Problem 14–19). Because the set of spin functions is orthogonal, the integrals $H_{x,jj} = H_{y,jj}$ will all contain the product $\alpha\beta$ in the integrand and hence be zero (for $j = 1, 2, 3,$ and 4).

14–23. Verify Equations 14.44.

$$E_j = E_j^{(0)} + \int d\tau_1 d\tau_2 \psi_j^* \hat{H}^{(1)}\psi_j = E_j^{(0)} + H_{jj}^{(1)} \tag{14.31}$$

We can separate the x, y, and z-components of $\hat{H}^{(1)}$ to write this as

$$E_j = E_j^{(0)} + \hat{H}_{x,jj}^{(1)} + \hat{H}_{y,jj}^{(1)} + \hat{H}_{z,jj}^{(1)}$$

The $E_j^{(0)}$ expressions are given in Equations 14.34 through 14.37, and the expressions for the $H_{z,jj}$ are given by Equations 14.40 through 14.42. We just found that $H_{x,jj} = H_{y,jj} = 0$ (Problem 14–22), so

$$E_j = E_j^{(0)} + H_{z,jj}^{(1)}$$

which are given by Equations 14.44.

14–24. Verify Equations 14.46.

Using the relationship $E = h\nu$, we have

$$E_{1\to2} = E_2 - E_1$$
$$= \frac{h\nu_0}{2}(\sigma_2 - \sigma_1) - \frac{hJ_{12}}{4} + h\nu_0\left(1 - \frac{\sigma_1+\sigma_2}{2}\right) - \frac{hJ_{12}}{4}$$
$$= h\nu_0(1-\sigma_1) - \frac{hJ_{12}}{2}$$
$$\nu_{1\to2} = \nu_0(1-\sigma_1) - \frac{J_{12}}{2}$$

Likewise,

$$E_{1\to3} = E_3 - E_1$$
$$= \frac{h\nu_0}{2}(\sigma_1 - \sigma_2) - \frac{hJ_{12}}{4} + h\nu_0\left(1 - \frac{\sigma_1+\sigma_2}{2}\right) - \frac{hJ_{12}}{4}$$
$$= h\nu_0(1-\sigma_3) - \frac{hJ_{12}}{2}$$
$$\nu_{1\to3} = \nu_0(1-\sigma_2) - \frac{J_{12}}{2}$$

$$E_{2\rightarrow4} = E_4 - E_2$$

$$= h\nu_0\left(1 - \frac{\sigma_1 + \sigma_2}{2}\right) + \frac{hJ_{12}}{4} - \frac{h\nu_0}{2}(\sigma_2 - \sigma_1) + \frac{hJ_{12}}{4}$$

$$= h\nu_0(1 - \sigma_2) + \frac{hJ_{12}}{2}$$

$$\nu_{2\rightarrow4} = \nu_0(1 - \sigma_2) + \frac{J_{12}}{2}$$

$$E_{3\rightarrow4} = E_4 - E_3$$

$$= h\nu_0\left(1 - \frac{\sigma_1 + \sigma_2}{2}\right) + \frac{hJ_{12}}{4} - \frac{h\nu_0}{2}(\sigma_1 - \sigma_2) + \frac{hJ_{12}}{4}$$

$$= h\nu_0(1 - \sigma_1) + \frac{hJ_{12}}{2}$$

$$\nu_{3\rightarrow4} = \nu_0(1 - \sigma_1) + \frac{J_{12}}{2}$$

14–25. Make a sketch like Figure 14.11 for a spectrum taken at 500 MHz.

$$\text{center of doublet 1} = (130\text{ Hz})\left(\frac{500\text{ MHz}}{90\text{ MHz}}\right) = 722\text{ Hz}$$

$$\text{center of doublet 2} = (210\text{ Hz})\left(\frac{500\text{ MHz}}{90\text{ MHz}}\right) = 1167\text{ Hz}$$

The spacing within the doublets remains the same.

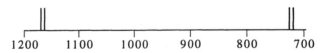

14–26. For a first-order spectrum with (Equations 14.47)

$$\nu_1^{\pm} = \nu_0(1 - \sigma_1) \pm \frac{J_{12}}{2}$$

and

$$\nu_2^{\pm} = \nu_0(1 - \sigma_2) \pm \frac{J_{12}}{2}$$

show that the centers of the doublets are separated by $\nu_0|\sigma_1 - \sigma_2|$ and that the separations of the peaks within the two doublets is J_{12}.

The separation of the peaks within the doublets is

$$\nu_j^+ - \nu_j^- = 2\left(\frac{J_{12}}{2}\right) = J_{12} \qquad j = 1, 2$$

The centers of the doublets are located at $\nu_0(1 - \sigma_1)$ and $\nu_0(1 - \sigma_2)$, so the doublets are separated by

$$\left|\nu_0(1 - \sigma_2) - \nu_0(1 - \sigma_1)\right| = \nu_0\left|\sigma_1 - \sigma_2\right|$$

14-27. Verify Equations 14.55 and 14.56.

We can verify both of these equations using the method of Example 14-7. Recall for both calculations that

$$\hat{H}^{(1)} = \frac{hJ_{12}}{\hbar^2}\hat{\mathbf{I}}_1 \cdot \hat{\mathbf{I}}_2 \tag{14.50}$$

$$E_3 = E_3^{(0)} + E_3^{(1)}$$
$$= \int\int d\tau_1 d\tau_2 \phi_3^* \hat{H}^{(0)}\phi_3 + \int\int d\tau_1 d\tau_2 \phi_3^* \hat{H}^{(1)}\phi_3 \tag{1}$$

The first integral in this equation requires that we evaluate

$$(\hat{I}_{z1} + \hat{I}_{z2})\phi_3 = \frac{1}{\sqrt{2}}(\hat{I}_{z1} + \hat{I}_{z2})\left[\alpha(1)\beta(2) + \beta(1)\alpha(2)\right]$$
$$= \frac{1}{\sqrt{2}}\left[\frac{\hbar}{2}\beta(2)\alpha(1) - \frac{\hbar}{2}\beta(2)\alpha(1) - \frac{\hbar}{2}\beta(1)\alpha(2) + \frac{\hbar}{2}\beta(1)\alpha(2)\right]$$
$$= 0$$

and so

$$E_3^{(0)} = \int\int d\tau_1 d\tau_2 0 = 0$$

The second integral in Equation 1 requires the evaluation of

$$(\hat{I}_{x1}\hat{I}_{x2} + \hat{I}_{y1}\hat{I}_{y2} + \hat{I}_{z1}\hat{I}_{z2})\left[\alpha(1)\beta(2) + \beta(1)\alpha(2)\right]$$

Using the relations in Table 14.4, we have

$$\hat{I}_{x1}\hat{I}_{x2}\alpha(1)\beta(2) = \frac{\hbar^2}{4}\beta(1)\alpha(2) \qquad \hat{I}_{x1}\hat{I}_{x2}\beta(1)\alpha(2) = \frac{\hbar^2}{4}\alpha(1)\beta(2)$$

$$\hat{I}_{y1}\hat{I}_{y2}\alpha(1)\beta(2) = \frac{\hbar^2}{4}\beta(1)\alpha(2) \qquad \hat{I}_{y1}\hat{I}_{y2}\beta(1)\alpha(2) = \frac{\hbar^2}{4}\alpha(1)\beta(2)$$

$$\hat{I}_{z1}\hat{I}_{z2}\alpha(1)\beta(2) = -\frac{\hbar^2}{4}\alpha(1)\beta(2) \qquad \hat{I}_{z1}\hat{I}_{z2}\beta(1)\alpha(2) = -\frac{\hbar^2}{4}\beta(1)\alpha(2)$$

Then

$$(\hat{I}_{x1}\hat{I}_{x2} + \hat{I}_{y1}\hat{I}_{y2} + \hat{I}_{z1}\hat{I}_{z2})\left[\alpha(1)\beta(2) + \beta(1)\alpha(2)\right] = \frac{\hbar^2}{4}\left[\beta(1)\alpha(2) + \alpha(1)\beta(2)\right]$$

Substituting this result into Equation 1 gives

$$E_3 = 0 + \int\int d\tau_1 d\tau_2 \phi_3^* \hat{H}^{(1)}\phi_3$$
$$= \int\int d\tau_1 d\tau_2 \frac{1}{2}\left[\alpha^*(1)\beta^*(2) + \beta^*(1)\alpha^*(2)\right]\left(\frac{hJ_{AA}}{\hbar^2}\right)\frac{\hbar^2}{4}\left[\beta(1)\alpha(2) + \alpha(1)\beta(2)\right]$$
$$= \left(\frac{hJ_{AA}}{8}\right)(1+1) = \frac{hJ_{AA}}{4}$$

Similarly, E_4 is given by

$$E_4 = E_4^{(0)} + E_4^{(1)}$$
$$= \int\int d\tau_1 d\tau_2 \phi_4^* \hat{H}^{(0)}\phi_4 + \int\int d\tau_1 d\tau_2 \phi_4^* \hat{H}^{(1)}\phi_4 \tag{2}$$

The first integral in this equation requires that we evaluate

$$(\hat{I}_{z1} + \hat{I}_{z2})\phi_4 = (\hat{I}_{z1} + \hat{I}_{z2})\,[\beta(1)\beta(2)] = \left[-\frac{\hbar}{2}\beta(1) - \frac{\hbar}{2}\beta(2)\right] = -\frac{\hbar}{2}\,[\beta(1)\beta(2)]$$

When we substitute this into the integral for $E_4^{(0)}$ we find that

$$E_4^{(0)} = \int \int d\tau_1 d\tau_2 \beta^*(1)\beta^*(2)\left[-\gamma B_0(1-\sigma_A)\right]\left(-\frac{\hbar}{2}\right)\beta(1)\beta(2)$$

$$= \frac{\hbar\gamma B_0}{2}(1-\sigma_A)(2) = \hbar\gamma B_0(1-\sigma_A) \tag{3}$$

The second integral requires the evaluation of

$$(\hat{I}_{x1}\hat{I}_{x2} + \hat{I}_{y1}\hat{I}_{y2} + \hat{I}_{z1}\hat{I}_{z2})\,[\beta(1)\beta(2)]$$

Using the relations in Table 14.4, we have

$$\hat{I}_{x1}\hat{I}_{x2}\beta(1)\beta(2) = \frac{\hbar^2}{4}\alpha(1)\alpha(2)$$

$$\hat{I}_{y1}\hat{I}_{y2}\beta(1)\beta(2) = -\frac{\hbar^2}{4}\alpha(1)\alpha(2)$$

$$\hat{I}_{z1}\hat{I}_{z2}\beta(1)\beta(2) = \frac{\hbar^2}{4}\beta(1)\beta(2)$$

Then

$$(\hat{I}_{x1}\hat{I}_{x2} + \hat{I}_{y1}\hat{I}_{y2} + \hat{I}_{z1}\hat{I}_{z2})\,[\beta(1)\beta(2)] = \frac{\hbar^2}{4}\beta(1)\beta(2)$$

Substituting the result and Equation 3 into Equation 2 gives

$$E_4 = \hbar\gamma B_0(1-\sigma_A) + \int \int d\tau_1 d\tau_2 \phi_4^* \hat{H}^{(1)}\phi_4$$

$$= \hbar\gamma B_0(1-\sigma_A) + \int \int d\tau_1 d\tau_2\,[\beta^*(1)\beta^*(2)]\left(\frac{hJ_{AA}}{\hbar^2}\right)\frac{\hbar^2}{4}\,[\beta(1)\beta(2)]$$

$$= \hbar\gamma B_0(1-\sigma_A) + \frac{hJ_{AA}}{4}$$

14–28. Prove that

$$H_{13} = \int \int d\tau_1 d\tau_2 \alpha^*(1)\alpha^*(2)\hat{H}\beta(1)\alpha(2) = 0$$

with $\hat{H}$ given by Equation 14.58.

$$H_{13} = \int \int d\tau_1 d\tau_2 \alpha^*(1)\alpha^*(2)\hat{H}\beta(1)\alpha(2)$$

Use the relationships given in Table 14.4 to evaluate

$$\hat{H}\beta(1)\alpha(2) = -\gamma B_0(1-\sigma_1)\left(-\frac{\hbar}{2}\right)\beta(1)\alpha(2) - \gamma B_0(1-\sigma_2)\left(\frac{\hbar}{2}\right)\beta(1)\alpha(2)$$

$$+ \frac{hJ_{12}}{\hbar^2}\left[\frac{\hbar^2}{4}\alpha(1)\beta(2) + \frac{\hbar^2}{4}\alpha(1)\beta(2) - \frac{\hbar^2}{4}\beta(1)\alpha(2)\right]$$

$$= \frac{\hbar}{2}\gamma B_0\beta(1)\alpha(2)\left[(1-\sigma_1) - (1-\sigma_2)\right] + \frac{hJ_{12}}{4}\left[2\alpha(1)\beta(2) - \beta(1)\alpha(2)\right]$$

Then

$$H_{13} = \int\int d\tau_1 d\tau_2 \alpha^*(1)\alpha^*(2) \left\{ \frac{\hbar}{2}\gamma B_0 \beta(1)\alpha(2) \left[(1-\sigma_1)-(1-\sigma_2)\right] \right.$$

$$\left. + \frac{hJ_{12}}{4}\left[2\alpha(1)\beta(2)-\beta(1)\alpha(2)\right] \right\} = 0$$

because α and β are orthonormal.

14–29. Prove that

$$H_{11} = \int\int d\tau_1 d\tau_2 \alpha^*(1)\alpha^*(2)\hat{H}\alpha(1)\alpha(2)$$

$$= -\frac{1}{2}h\nu_0(1-\sigma_1) - \frac{1}{2}h\nu_0(1-\sigma_2) + \frac{hJ_{12}}{4}$$

with $\hat{H}$ given by Equation 14.58.

$$H_{11} = \int\int d\tau_1 d\tau_2 \alpha^*(1)\alpha^*(2)\hat{H}\alpha(1)\alpha(2)$$

First, we use the relationships in Table 14.4 to evaluate

$$\hat{H}\alpha(1)\alpha(2) = -\gamma B_0(1-\sigma_1)\left(\frac{\hbar}{2}\right)\alpha(1)\alpha(2) - \gamma B_0(1-\sigma_2)\left(\frac{\hbar}{2}\right)\alpha(1)\alpha(2)$$

$$+ \frac{hJ_{12}}{\hbar^2}\left[\frac{\hbar^2}{4}\beta(1)\beta(2) - \frac{\hbar^2}{4}\beta(1)\beta(2) + \frac{\hbar^2}{4}\alpha(1)\alpha(2)\right]$$

$$= \left[-\frac{\hbar\gamma B_0}{2}(1-\sigma_1) - \frac{\hbar\gamma B_0}{2}(1-\sigma_2) + \frac{hJ_{12}}{4}\right]\alpha(1)\alpha(2)$$

Then

$$H_{11} = \int\int d\tau_1 d\tau_2 \alpha^*(1)\alpha^*(2)\left[-\frac{\hbar\gamma B_0}{2}(1-\sigma_1) - \frac{\hbar\gamma B_0}{2}(1-\sigma_2) + \frac{hJ_{12}}{4}\right]\alpha(1)\alpha(2)$$

$$= -\frac{1}{2}h\nu_0(1-\sigma_1) - \frac{1}{2}h\nu_0(1-\sigma_2) + \frac{hJ_{12}}{4}$$

using Equation 14.45 for ν_0.

14–30. Prove that

$$H_{44} = \int\int d\tau_1 d\tau_2 \beta^*(1)\beta^*(2)\hat{H}\beta(1)\beta(2)$$

$$= \frac{1}{2}h\nu_0(1-\sigma_1) + \frac{1}{2}h\nu_0(1-\sigma_2) + \frac{hJ_{12}}{4}$$

with $\hat{H}$ given by Equation 14.58.

$$H_{44} = \int\int d\tau_1 d\tau_2 \beta^*(1)\beta^*(2)\hat{H}\beta(1)\beta(2)$$

First, we use the relationships in Table 14.4 to evaluate

$$
\hat{H}\beta(1)\beta(2) = \gamma B_0(1-\sigma_1)\left(\frac{\hbar}{2}\right)\beta(1)\beta(2) + \gamma B_0(1-\sigma_2)\left(\frac{\hbar}{2}\right)\beta(1)\beta(2)
$$

$$
+ \frac{hJ_{12}}{\hbar^2}\left[\frac{\hbar^2}{4}\alpha(1)\alpha(2) - \frac{\hbar^2}{4}\alpha(1)\alpha(2) + \frac{\hbar^2}{4}\beta(1)\beta(2)\right]
$$

$$
= \left[\frac{\hbar\gamma B_0}{2}(1-\sigma_1) + \frac{\hbar\gamma B_0}{2}(1-\sigma_2) + \frac{hJ_{12}}{4}\right]\beta(1)\beta(2)
$$

Then

$$
H_{44} = \int\int d\tau_1 d\tau_2 \beta^*(1)\beta^*(2)\left[\frac{\hbar\gamma B_0}{2}(1-\sigma_1) + \frac{\hbar\gamma B_0}{2}(1-\sigma_2) + \frac{hJ_{12}}{4}\right]\beta(1)\beta(2)
$$

$$
= \frac{1}{2}h\nu_0(1-\sigma_1) + \frac{1}{2}h\nu_0(1-\sigma_2) + \frac{hJ_{12}}{4}
$$

using Equation 14.45 for ν_0.

14–31. Show that Equation 14.64 leads to Equation 14.65.

Expand the determinant in Equation 14.64:

$$
\begin{vmatrix}
-d_1 - d_2 + \frac{hJ}{4} - E & 0 & 0 & 0 \\
0 & -d_1 + d_2 - \frac{hJ}{4} - E & \frac{hJ}{2} & 0 \\
0 & \frac{hJ}{2} & d_1 - d_2 - \frac{hJ}{4} - E & 0 \\
0 & 0 & 0 & d_1 + d_2 + \frac{hJ}{4} - E
\end{vmatrix} = 0
$$

$$
\left(-d_1 - d_2 + \frac{hJ}{4} - E\right)\begin{vmatrix}
-d_1 + d_2 - \frac{hJ}{4} - E & \frac{hJ}{2} & 0 \\
\frac{hJ}{2} & d_1 - d_2 - \frac{hJ}{4} - E & 0 \\
0 & 0 & d_1 + d_2 + \frac{hJ}{4} - E
\end{vmatrix} = 0
$$

$$
\left(-d_1 - d_2 + \frac{hJ}{4} - E\right)\left(d_1 + d_2 + \frac{hJ}{4} - E\right)\begin{vmatrix}
-d_1 + d_2 - \frac{hJ}{4} - E & \frac{hJ}{2} \\
\frac{hJ}{2} & d_1 - d_2 - \frac{hJ}{4} - E
\end{vmatrix} = 0
$$

We then have one of three cases:

$$
-d_1 - d_2 + \frac{hJ}{4} - E = 0 \tag{1}
$$

$$
d_1 + d_2 + \frac{hJ}{4} - E = 0 \tag{2}
$$

or

$$
\left(-d_1 + d_2 - \frac{hJ}{4} - E\right)\left(d_1 - d_2 - \frac{hJ}{4} - E\right) - \left(\frac{hJ}{2}\right)^2 = 0 \tag{3}
$$

Solving Equation 1 for E gives

$$
E_1 = -d_1 - d_2 + \frac{hJ}{4}
$$

$$
= -\frac{1}{2}h\nu_0(1-\sigma_1) - \frac{1}{2}h\nu_0(1-\sigma_2) + \frac{hJ}{4}
$$

$$
= -h\nu_0\left(1 - \frac{\sigma_1 + \sigma_2}{2}\right) + \frac{hJ}{4}
$$

Likewise, solving Equation 2 for E gives

$$E_4 = d_1 + d_2 + \frac{hJ}{4}$$

$$= \frac{1}{2}h\nu_0(1 - \sigma_1) + \frac{1}{2}h\nu_0(1 - \sigma_2) + \frac{hJ}{4}$$

$$= h\nu_0\left(1 - \frac{\sigma_1 + \sigma_2}{2}\right) + \frac{hJ}{4}$$

We can then solve the last equation to find the other two energies (E_2 and E_3) listed in Equations 14.65.

$$0 = \left(-d_1 + d_2 - \frac{hJ}{4} - E\right)\left(d_1 - d_2 - \frac{hJ}{4} - E\right) - \left(\frac{hJ}{2}\right)^2$$

$$= -d_1^2 + 2d_1d_2 - d_2^2 + h^2J^2\left(\frac{1}{16} - \frac{1}{4}\right) + \frac{hJ}{2}E + E^2$$

$$= E^2 + \frac{hJ}{2}E - \frac{1}{4}h^2\nu_0^2\left(\sigma_2 - \sigma_1\right)^2 - \frac{3}{16}h^2J^2$$

Now use the quadratic equation to find E:

$$E = \frac{-b \pm \sqrt{b^2 - 4ac}}{2a}$$

$$= -\frac{hJ}{4} \pm \frac{h}{2}\left[\nu_0^2(\sigma_1 - \sigma_2)^2 + J^2\right]^{1/2}$$

which are E_2 and E_3 in Equations 14.65.

14–32. Sketch the splitting pattern of a two-spin system –$\overset{\shortmid}{C}$H–$\overset{\shortmid}{C}$H– for $\nu_0|\sigma_1 - \sigma_2|/J = 20, 10, 5, 2, 1, 0.10$, and 0.01.

Let $\alpha = (\nu_0/2)(2 - \sigma_1 - \sigma_2)$ and $\beta = \dfrac{\nu_0(\sigma_1 - \sigma_2)}{J}$. Then the equations in Table 14.6 become

$$\nu_{1\to2} = \alpha - \frac{J}{2} - \frac{1}{2}\left[\nu_0^2(\sigma_1 - \sigma_2)^2 + J^2\right]^{1/2}$$

$$= \alpha - \frac{J}{2} - \frac{J}{2}\left(1 + \beta^2\right)^{1/2} = \alpha - \frac{J}{2}\left[1 + \left(1 + \beta^2\right)^{1/2}\right]$$

$$\nu_{1\to3} = \alpha - \frac{J}{2} + \frac{J}{2}\left(1 + \beta^2\right)^{1/2} = \alpha - \frac{J}{2}\left[1 - \left(1 + \beta^2\right)^{1/2}\right]$$

$$\nu_{2\to4} = \alpha + \frac{J}{2} + \frac{J}{2}\left(1 + \beta^2\right)^{1/2} = \alpha + \frac{J}{2}\left[1 + \left(1 + \beta^2\right)^{1/2}\right]$$

$$\nu_{3\to4} = \alpha + \frac{J}{2} - \frac{J}{2}\left(1 + \beta^2\right)^{1/2} = \alpha + \frac{J}{2}\left[1 - \left(1 + \beta^2\right)^{1/2}\right]$$

Also from Table 14.6, we can find r in terms of β:

$$r = \left[\frac{(\Delta^2 + J^2)^{1/2} + \Delta}{(\Delta^2 + J^2)^{1/2} - \Delta}\right]^{1/2} = \left[\frac{J(\beta^2 + 1)^{1/2} + J\beta}{J(\beta^2 + 1)^{1/2} - J\beta}\right]^{1/2}$$

$$= \left[\frac{(\beta^2 + 1)^{1/2} + \beta}{(\beta^2 + 1)^{1/2} - \beta}\right]^{1/2}$$

We are given β in the problem, so we can make a table for different values of v, r, and $(r-1)^2/(r+1)^2$:

β	$v_{1\to2}$	$v_{1\to3}$	$v_{2\to4}$	$v_{3\to4}$	r	$\left(\frac{r-1}{r+1}\right)^2$
0	$\alpha - J$	$\approx \alpha$	$\alpha + J$	$\approx \alpha$	1.01	≈ 0
0.1	$\alpha - 1.002J$	$\approx \alpha$	$\alpha + 1.002J$	$\approx \alpha$	1.10	0.0025
1	$\alpha - 1.207J$	$\alpha + 0.207J$	$\alpha + 1.207J$	$\alpha - 0.207J$	2.41	0.17
2	$\alpha - 1.618J$	$\alpha + 0.618J$	$\alpha + 1.618J$	$\alpha - 0.618J$	4.24	0.38
5	$\alpha - 3.050J$	$\alpha + 2.049J$	$\alpha + 3.050J$	$\alpha - 2.049J$	10.10	0.67
10	$\alpha - 5.525J$	$\alpha + 4.525J$	$\alpha + 5.525J$	$\alpha - 4.525J$	20.05	0.82
20	$\alpha - 10.512J$	$\alpha + 9.512J$	$\alpha + 10.512J$	$\alpha - 9.512J$	40.02	0.90

Below we sketch the splitting pattern for different values of β, where $\alpha = 300$.

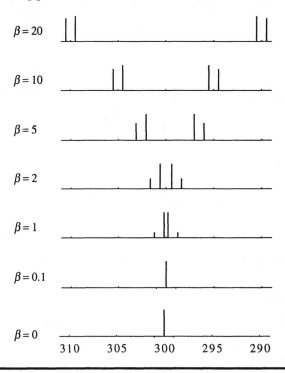

14–33. Show that a two-spin system with $J = 0$ consists of just two peaks with frequencies $v_0(1 - \sigma_1)$ and $v_0(1 - \sigma_2)$.

We can determine the frequency of the peaks using the equations from Table 14.6, with $J = 0$. We find

$$v_{1\to2} = v_{3\to4} = v_0(1 - \sigma_1)$$
$$v_{1\to3} = v_{2\to4} = v_0(1 - \sigma_2)$$

14–34. Show that

$$v_{1\to2} = \frac{v_0}{2}(2 - \sigma_1 - \sigma_2) - \frac{J}{2} - \frac{1}{2}[v_0^2(\sigma_1 - \sigma_2)^2 + J^2]^{1/2}$$

for a general two-spin system (see Table 14.6).

Use the energies given by Equations 14.65, recalling that $E = h\nu$:

$$E_2 - E_1 = -\frac{hJ}{4} - \frac{h}{2}\left[\nu_0^2(\sigma_1 - \sigma_2)^2 + J^2\right]^{1/2} + h\nu_0\left(1 - \frac{\sigma_1 + \sigma_2}{2}\right) - \frac{hJ}{4}$$

$$= -\frac{hJ}{2} - \frac{h}{2}\left[\nu_0^2(\sigma_1 - \sigma_2)^2 + J^2\right]^{1/2} + \frac{h\nu_0}{2}(2 - \sigma_1 - \sigma_2)$$

$$\frac{E_2 - E_1}{h} = \nu_{1\to2} = -\frac{J}{2} - \frac{1}{2}\left[\nu_0^2(\sigma_1 - \sigma_2)^2 + J^2\right]^{1/2} + \frac{\nu_0}{2}(2 - \sigma_1 - \sigma_2)$$

14–35. Show that the frequencies given in Table 14.6 reduce to Equations 14.66 (and also Equations 14.46) when $J \ll \nu_0(\sigma_1 - \sigma_2)$.

$$\nu_{1\to2} = \frac{\nu_0}{2}(2 - \sigma_1 - \sigma_2) - \frac{J}{2} - \frac{1}{2}\left[\nu_0^2(\sigma_1 - \sigma_2)^2 + J^2\right]^{1/2}$$

$$= \frac{\nu_0}{2}(2 - \sigma_1 - \sigma_2) - \frac{J}{2} - \frac{\nu_0(\sigma_1 - \sigma_2)}{2}\left[1 + \frac{J^2}{\nu_0^2(\sigma_1 - \sigma_2)^2}\right]^{1/2}$$

Since $J \ll \nu_0(\sigma_1 - \sigma_2)$, we can use the expansion

$$(1 + x)^{1/2} = 1 + \frac{x}{2} - \frac{x^2}{8} + \cdots$$

Keeping the terms that are linear in J gives

$$\nu_{1\to2} = \frac{\nu_0}{2}(2 - \sigma_1 - \sigma_2) - \frac{J}{2} - \frac{\nu_0(\sigma_1 - \sigma_2)}{2} - O(J^2)$$

$$= \nu_0(1 - \sigma_1) - \frac{J}{2}$$

Because $\nu_{3\to4}$ differs from $\nu_{1\to2}$ only in the sign of the $J/2$ term,

$$\nu_{3\to4} = \nu_0(1 - \sigma_1) + \frac{J}{2}$$

Likewise,

$$\nu_{1\to3} = \frac{\nu_0}{2}(2 - \sigma_1 - \sigma_2) - \frac{J}{2} + \frac{1}{2}\left[\nu_0^2(\sigma_1 - \sigma_2)^2 + J^2\right]^{1/2}$$

$$= \frac{\nu_0}{2}(2 - \sigma_1 - \sigma_2) - \frac{J}{2} + \frac{\nu_0(\sigma_1 - \sigma_2)}{2}\left[1 + \frac{J^2}{\nu_0^2(\sigma_1 - \sigma_2)^2}\right]^{1/2}$$

$$= \frac{\nu_0}{2}(2 - \sigma_1 - \sigma_2) - \frac{J}{2} + \frac{\nu_0(\sigma_1 - \sigma_2)}{2} - O(J^2)$$

$$= \nu_0(1 - \sigma_2) - \frac{J}{2}$$

Because $\nu_{2\to4}$ differs from $\nu_{1\to3}$ only in the sign of the $J/2$ term,

$$\nu_{2\to4} = \nu_0(1 - \sigma_2) + \frac{J}{2}$$

14–36. Using the results in Table 14.6, compute the spectrum of a two-spin system for $\nu_0 = 60$ MHz and 500 MHz given that $\sigma_1 - \sigma_2 = 0.12 \times 10^{-6}$ and $J = 8.0$ Hz.

See Example 14–11. Using the values given for 60 MHz (Table 14.6),

$$v_0^2(\sigma_1 - \sigma_2)^2 = (60.0 \text{ MHz})^2 (0.12 \times 10^{-6})^2 = 51.84 \text{ Hz}^2$$

$$\left[v_0^2(\sigma_1 - \sigma_2)^2 + J^2\right]^{1/2} = \left[51.84 \text{ Hz}^2 + 64.0 \text{ Hz}^2\right]^{1/2} = 10.76 \text{ Hz}$$

Therefore,

$$v_{1\to2} = 60 \text{ MHz} - 9.38 \text{ Hz}$$
$$v_{1\to3} = 60 \text{ MHz} + 1.38 \text{ Hz}$$
$$v_{2\to4} = 60 \text{ MHz} + 9.38 \text{ Hz}$$
$$v_{3\to4} = 60 \text{ MHz} - 1.38 \text{ Hz}$$

To calculate the relative intensities of the signals, we calculate r:

$$r = \left(\frac{10.76 \text{ Hz} + \sqrt{51.84} \text{ Hz}}{10.76 \text{ Hz} - \sqrt{51.84} \text{ Hz}}\right)^{1/2} = 2.25$$

$$\frac{(r-1)^2}{(r+1)^2} = 0.15$$

so the relative intensities are 0.15 to 1. For 500 MHz,

$$v_0^2(\sigma_1 - \sigma_2)^2 = (500 \text{ MHz})^2 (0.12 \times 10^{-6})^2 = 3600 \text{ Hz}^2$$

$$\left[v_0^2(\sigma_1 - \sigma_2)^2 + J^2\right]^{1/2} = \left[3600 \text{ Hz}^2 + 64.0 \text{ Hz}^2\right]^{1/2} = 60.5 \text{ Hz}$$

Therefore,

$$v_{1\to2} = 500 \text{ MHz} - 34.3 \text{ Hz}$$
$$v_{1\to3} = 500 \text{ MHz} + 26.3 \text{ Hz}$$
$$v_{2\to4} = 500 \text{ MHz} + 34.3 \text{ Hz}$$
$$v_{3\to4} = 500 \text{ MHz} - 26.3 \text{ Hz}$$

Also,

$$r = \left(\frac{60.5 \text{ Hz} + 60 \text{ Hz}}{60.5 \text{ Hz} - 60 \text{ Hz}}\right)^{1/2} = 15.52$$

$$\frac{(r-1)^2}{(r+1)^2} = 0.77$$

so the relative intensities are 0.77 to 1.

14–37. In Chapter 13, we learned that selection rules for a transition from state i to state j are governed by an integral of the form (Equation 13.52)

$$\int \psi_j^* \hat{H}^{(1)} \psi_i \, d\tau$$

where $\hat{H}^{(1)}$ is the Hamiltonian operator that causes the transitions from one state to another. In NMR spectroscopy, there are two magnetic fields to consider. There is the static field $\mathbf{B}$ that is produced by the magnets and aligns the nuclear spins of the sample. We customarily take this field to be in the z direction, and the nuclear (proton) spin states α and β are defined with respect to this field. Nuclear spin transitions occur when the spin system is irradiated with a radio-frequency field $\mathbf{B}_1 = \mathbf{B}_1^0 \cos 2\pi \nu t$. In this case,

$$\hat{H}^{(1)} = -\hat{\mu} \cdot \mathbf{B}_1 = -\gamma \hat{\mathbf{I}} \cdot \mathbf{B}_1$$

Show that the NMR selection rules are governed by integrals of the form

$$P_x = \int \psi_j^* \hat{I}_x \psi_i d\tau$$

with similar integrals involving $\hat{I}_y$ and $\hat{I}_z$. Now show that $P_x \neq 0$, $P_y \neq 0$, and $P_z = 0$, indicating that the radio-frequency field must be perpendicular to the static magnetic field.

Let us assume that we are dealing with hydrogen nuclei, in which case there are only two spin functions α and β. Therefore, either $\psi_i = \alpha$ and $\psi_j = \beta$ or $\psi_i = \beta$ and $\psi_j = \alpha$. We are given the equation for the first-order Hamiltonian $\hat{H}^{(1)}$ and the integral form governing the selection rules. Substituting into this integral, we find

$$\int \psi_j^* \hat{H}^{(1)} \psi_i d\tau = -\gamma \int \psi_j^* \hat{\mathbf{I}} \cdot \mathbf{B}_1 \psi_i d\tau$$

$$= -\gamma B_{1x} \int \psi_j^* \hat{I}_x \psi_i d\tau - \gamma B_{1y} \int \psi_j^* \hat{I}_y \psi_i d\tau - \gamma B_{1z} \int \psi_j^* \hat{I}_z \psi_i d\tau$$

We see that the selection rules are governed by the integrals

$$P_x = \int \psi_j^* \hat{I}_x \psi_i d\tau \qquad P_y = \int \psi_j^* \hat{I}_y \psi_i d\tau \qquad P_z = \int \psi_j^* \hat{I}_z \psi_i d\tau$$

Now examine the case for which $\psi_j = \beta$ and $\psi_i = \alpha$. In this case, we find

$$P_x = \frac{\hbar}{2} \int \beta^* \beta d\tau = \frac{\hbar}{2} \neq 0$$

$$P_y = \frac{i\hbar}{2} \int \beta^* \beta d\tau = \frac{i\hbar}{2} \neq 0$$

$$P_z = \frac{\hbar}{2} \int \beta^* \alpha d\tau = 0$$

Likewise, if we examine the case for which $\psi_j = \alpha$ and $\psi_i = \beta$, we find

$$P_x = \frac{\hbar}{2} \neq 0 \qquad P_y = -\frac{i\hbar}{2} \neq 0 \qquad P_z = 0$$

Because $P_z = 0$, the radio-frequency field cannot be parallel to the direction of the static magnetic field. Because $P_x \neq 0$ and $P_y \neq 0$, a transition can occur if the radio-frequency field is perpendicular to the direction of the static magnetic field.

14–38. Consider the two-spin system discussed in Section 14–6. In this case, the selection rule is governed by

$$P_x = \int d\tau_1 d\tau_2 \psi_j^* (\hat{I}_{x1} + \hat{I}_{x2}) \psi_i$$

with a similar equation for P_y. Using the notation given by Equations 14.30, show that the only allowed transitions are for $1 \rightarrow 2$, $1 \rightarrow 3$, $2 \rightarrow 4$, and $3 \rightarrow 4$.

The functions given by Equations 14.30 are

$$\psi_1 = \alpha(1)\alpha(2) \qquad \psi_3 = \beta(1)\alpha(2)$$
$$\psi_2 = \alpha(1)\beta(2) \qquad \psi_4 = \beta(1)\beta(2)$$

We will consider all $\psi_i \rightarrow \psi_j$ transitions where $j > i$. First, consider the effect of the operator combinations $\hat{I}_{x1} + \hat{I}_{x2}$ and $\hat{I}_{y1} + \hat{I}_{y2}$ on ψ_1, ψ_2, and ψ_3:

$$\left[\hat{I}_{x1} + \hat{I}_{x2}\right]\psi_1 = \frac{\hbar}{2}\left[\beta(1)\alpha(2) + \alpha(1)\beta(2)\right]$$

$$\left[\hat{I}_{y1} + \hat{I}_{y2}\right]\psi_1 = \frac{i\hbar}{2}\left[\beta(1)\alpha(2) + \alpha(1)\beta(2)\right]$$

$$\left[\hat{I}_{x1} + \hat{I}_{x2}\right]\psi_2 = \frac{\hbar}{2}\left[\beta(1)\beta(2) + \alpha(1)\alpha(2)\right]$$

$$\left[\hat{I}_{y1} + \hat{I}_{y2}\right]\psi_2 = \frac{i\hbar}{2}\left[\beta(1)\beta(2) - \alpha(1)\alpha(2)\right]$$

$$\left[\hat{I}_{x1} + \hat{I}_{x2}\right]\psi_3 = \frac{\hbar}{2}\left[\alpha(1)\alpha(2) + \beta(1)\beta(2)\right]$$

$$\left[\hat{I}_{x1} + \hat{I}_{x2}\right]\psi_3 = \frac{-i\hbar}{2}\left[\alpha(1)\alpha(2) - \beta(1)\beta(2)\right]$$

Carrying out the integration gives

$$P_{x,1\rightarrow2} = \int d\tau_1 d\tau_2 \psi_2^*(\hat{I}_{x1} + \hat{I}_{x2})\psi_1 = \frac{\hbar}{2}\int d\tau_1 d\tau_2 \alpha^*(1)\beta^*(2)\left[\beta(1)\alpha(2) + \alpha(1)\beta(2)\right] \neq 0$$

$$P_{y,1\rightarrow2} = \int d\tau_1 d\tau_2 \psi_2^*(\hat{I}_{y1} + \hat{I}_{y2})\psi_1 \neq 0$$

$$P_{x,1\rightarrow3} = \int d\tau_1 d\tau_2 \psi_3^*(\hat{I}_{x1} + \hat{I}_{x2})\psi_1 = \frac{\hbar}{2}\int d\tau_1 d\tau_2 \beta^*(1)\alpha^*(2)\left[\beta(1)\alpha(2) + \alpha(1)\beta(2)\right] \neq 0$$

$$P_{y,1\rightarrow3} = \int d\tau_1 d\tau_2 \psi_3^*(\hat{I}_{y1} + \hat{I}_{y2})\psi_1 \neq 0$$

$$P_{x,1\rightarrow4} = \int d\tau_1 d\tau_2 \psi_4^*(\hat{I}_{x1} + \hat{I}_{x2})\psi_1 = \frac{\hbar}{2}\int d\tau_1 \beta^*(1)\beta^*(2)\left[\beta(1)\alpha(2) + \alpha(1)\beta(2)\right] = 0$$

$$P_{y,1\rightarrow4} = \int d\tau_1 d\tau_2 \psi_4^*(\hat{I}_{y1} + \hat{I}_{y2})\psi_1 = 0$$

$$P_{x,2\rightarrow3} = \int d\tau_1 d\tau_2 \psi_3^*(\hat{I}_{x1} + \hat{I}_{x2})\psi_2 = \frac{\hbar}{2}\int d\tau_1 d\tau_2 \beta^*(1)\alpha^*(2)\left[\alpha(1)\alpha(2) + \beta(1)\beta(2)\right] = 0$$

$$P_{y,2\rightarrow3} = \int d\tau_1 d\tau_2 \psi_3^*(\hat{I}_{y1} + \hat{I}_{y2})\psi_2 = \frac{i\hbar}{2}\int d\tau_1 d\tau_2 \beta^*(1)\alpha^*(2)\left[\alpha(1)\alpha(2) - \beta(1)\beta(2)\right] = 0$$

$$P_{x,2\rightarrow4} = \int d\tau_1 d\tau_2 \psi_4^*(\hat{I}_{x1} + \hat{I}_{x2})\psi_2 = \frac{\hbar}{2}\int d\tau_1 \beta^*(1)\beta^*(2)\left[\beta(1)\beta(2) + \alpha(1)\alpha(2)\right] \neq 0$$

$$P_{y,2\rightarrow4} = \int d\tau_1 d\tau_2 \psi_4^*(\hat{I}_{y1} + \hat{I}_{y2})\psi_2 = \frac{i\hbar}{2}\int d\tau_1 \beta^*(1)\beta^*(2)\left[\beta(1)\beta(2) - \alpha(1)\alpha(2)\right] \neq 0$$

$$P_{x,3\rightarrow4} = \int d\tau_1 d\tau_2 \psi_4^*(\hat{I}_{x1} + \hat{I}_{x2})\psi_3 = \frac{\hbar}{2}\int d\tau_1 \beta^*(1)\beta^*(2)\left[\alpha(1)\alpha(2) + \beta(1)\beta(2)\right] \neq 0$$

$$P_{y,3\rightarrow4} = \int d\tau_1 d\tau_2 \psi_4^*(\hat{I}_{y1} + \hat{I}_{y2})\psi_3 = -\frac{i\hbar}{2}\int d\tau_1 \beta^*(1)\beta^*(2)\left[\alpha(1)\alpha(2) + \beta(1)\beta(2)\right] \neq 0$$

Transitions are possible only if $P_{i \to j} \neq 0$, so the transitions $1 \to 2$, $1 \to 3$, $2 \to 4$, and $3 \to 4$ are possible - the others are forbidden.

14–39. Using the spin functions given by Equations 14.51, show that the only allowed transitions are $1 \to 3$ and $3 \to 4$.

The functions given by Equations 14.51 are

$$\phi_1 = \psi_1 \qquad\qquad \phi_2 = \tfrac{1}{\sqrt{2}} (\psi_2 - \psi_3)$$
$$\phi_3 = \tfrac{1}{\sqrt{2}} (\psi_2 + \psi_3) \qquad \phi_4 = \psi_4$$

As in the previous problem, we will consider all $\psi_i \to \psi_j$ transitions where $j > i$. Because $\phi_1 = \psi_1$ and $\phi_4 = \psi_4$, the $1 \to 4$ transition (Problem 14–38) is forbidden. We first evaluate the relationships

$$\left[\hat{I}_{x1} + \hat{I}_{x2} \right] \phi_1 = \frac{\hbar}{2} [\beta(1)\alpha(2) + \alpha(1)\beta(2)]$$

$$\left[\hat{I}_{y1} + \hat{I}_{y2} \right] \phi_1 = \frac{i\hbar}{2} [\beta(1)\alpha(2) + \alpha(1)\beta(2)]$$

$$\left[\hat{I}_{x1} + \hat{I}_{x2} \right] \phi_2 = \frac{\hbar}{2\sqrt{2}} [\beta(1)\beta(2) + \alpha(1)\alpha(2) - \beta(1)\beta(2) - \alpha(1)\alpha(2)] = 0$$

$$\left[\hat{I}_{y1} + \hat{I}_{y2} \right] \phi_2 = \frac{i\hbar}{2\sqrt{2}} [\beta(1)\beta(2) - \alpha(1)\alpha(2) + \alpha(1)\alpha(2) - \beta(1)\beta(2)] = 0$$

$$\left[\hat{I}_{x1} + \hat{I}_{x2} \right] \phi_3 = \frac{\hbar}{\sqrt{2}} [\beta(1)\beta(2) + \alpha(1)\alpha(2)]$$

$$\left[\hat{I}_{y1} + \hat{I}_{y2} \right] \phi_3 = \frac{i\hbar}{\sqrt{2}} [\beta(1)\beta(2) + \alpha(1)\alpha(2)]$$

Because $\left[\hat{I}_{x1} + \hat{I}_{x2} \right] \phi_2 = \left[\hat{I}_{y1} + \hat{I}_{y2} \right] \phi_2 = 0$, the transitions $2 \to 3$ and $2 \to 4$ are forbidden. Using the above results,

$$P_{x,1\to2} = \int d\tau_1 d\tau_2 \phi_2^*(\hat{I}_{x1} + \hat{I}_{x2})\phi_1$$

$$= \frac{\hbar}{2\sqrt{2}} \int d\tau_1 d\tau_2 [\alpha^*(1)\beta^*(2) - \beta^*(1)\alpha^*(2)][\beta(1)\alpha(2) + \alpha(1)\beta(2)] = 0$$

$$P_{y,1\to2} = \int d\tau_1 d\tau_2 \phi_2^*(\hat{I}_{y1} + \hat{I}_{y2})\phi_1$$

$$= \frac{i\hbar}{2\sqrt{2}} \int d\tau_1 d\tau_2 [\alpha^*(1)\beta^*(2) - \beta^*(1)\alpha^*(2)][\beta(1)\alpha(2) + \alpha(1)\beta(2)] = 0$$

$$P_{x,1\to3} = \int d\tau_1 d\tau_2 \phi_3^*(\hat{I}_{x1} + \hat{I}_{x2})\phi_1$$

$$= \frac{\hbar}{2\sqrt{2}} \int d\tau_1 d\tau_2 [\alpha^*(1)\beta^*(2) + \beta^*(1)\alpha^*(2)][\beta(1)\alpha(2) + \alpha(1)\beta(2)] \neq 0$$

$$P_{y,1\to3} = \int d\tau_1 d\tau_2 \phi_3^*(\hat{I}_{y1} + \hat{I}_{y2})\phi_1$$

$$= \frac{i\hbar}{2\sqrt{2}} \int d\tau_1 d\tau_2 [\alpha^*(1)\beta^*(2) + \beta^*(1)\alpha^*(2)][\beta(1)\alpha(2) + \alpha(1)\beta(2)] \neq 0$$

$$P_{x,3\to4} = \int d\tau_1 d\tau_2 \phi_4^*(\hat{I}_{x1} + \hat{I}_{x2})\phi_3$$

$$= \frac{\hbar}{\sqrt{2}} \int d\tau_1 d\tau_2 \beta^*(1)\beta^*(2) [\beta(1)\beta(2) + \alpha(1)\alpha(2)] \neq 0$$

$$P_{y,3\to4} = \int d\tau_1 d\tau_2 \phi_4^*(\hat{I}_{y1} + \hat{I}_{y2})\phi_3$$

$$= \frac{i\hbar}{\sqrt{2}} \int d\tau_1 d\tau_2 \beta^*(1)\beta^*(2)\,[\beta(1)\beta(2) + \alpha(1)\alpha(2)] \neq 0$$

The only possible transitions are $1 \to 3$ and $3 \to 4$.

CHAPTER 15

Lasers, Laser Spectroscopy, and Photochemistry

PROBLEMS AND SOLUTIONS

15–1. The ground-state term symbol for O_2^+ is $^2\Pi_g$. The first electronic excited state has an energy of $38\,795$ cm^{-1} above that of the ground state and has a term symbol of $^2\Pi_u$. Is the radiative $^2\Pi_u \rightarrow ^2\Pi_g$ decay of the O_2^+ molecule an example of fluorescence or phosphorescence?

Fluorescence occurs when there is a radiative transition between states of the same spin multiplicity, while phosphorescence occurs when there is a radiative transition between states of different spin multiplicity. Since there is no change in spin multiplicity in the $^2\Pi_u \rightarrow ^2\Pi_g$ decay, it is an example of fluorescence.

15–2. Consider the absorption and fluorescence spectrum of a diatomic molecule for the specific case in which $R_e(S_1) > R_e(S_0)$. Using the potential energy curves shown in Figure 15.1, draw the expected absorption and fluorescence spectra of the molecule. You can assume that the molecule relaxes to $v' = 0$ before it fluoresces. Do your spectra look like the spectra in Figure 15.2? Explain.

Using the potential curves in Figure 15.1, we can see that, due to the differences in bond length for $v' = 0$ and $v'' = 0$, the 0,0-transition and perhaps some of the $v' = 0$ to $v'' = n$ for small n will be missing if we compare this spectra to the spectra in Figure 15.2.

Absorption   Fluorescence

15–3. In Section 15–2, the spectral radiant energy density was expressed in terms of the frequency of the electromagnetic radiation. We could have chosen to express the spectral radiant energy density in terms of the wave number or wavelength of the electromagnetic radiation. Recall that the units of $\rho_v(v)$ are J·m^{-3}·s. Show that the units of $\rho_{\tilde{v}}(\tilde{v})$, the spectral radiant energy density in terms of wave numbers, are J·m^{-2} and that the units of $\rho_\lambda(\lambda)$, the spectral radiant energy density in terms of wavelength, are J·m^{-4}. What are the units of the Einstein B coefficient if we use $\rho_{\tilde{v}}(\tilde{v})$ to describe the spectral radiant energy density? What are the units of the Einstein B coefficient if we use $\rho_\lambda(\lambda)$ to describe the spectral radiant energy density?

The spectral radiant energy density was defined as $\rho_\nu(\nu) = d\rho/d\nu$. Replacing ν with $\tilde{\nu}$ or λ gives units of

$$\rho_{\tilde{\nu}}(\tilde{\nu}) = \frac{d\rho}{d\tilde{\nu}} = \frac{J \cdot m^{-3}}{m^{-1}} = J \cdot m^{-2}$$

or

$$\rho_\lambda(\lambda) = \frac{d\rho}{d\lambda} = \frac{J \cdot m^{-3}}{m} = J \cdot m^{-4}$$

From Equation 15.1, we see that

$$\text{rate} = B_{12}\rho_\nu(\nu_{12})N_1(t)$$

Replacing $\rho_\nu(\nu)$ with $\rho_{\tilde{\nu}}(\tilde{\nu})$ or $\rho_\lambda(\lambda)$ gives B_{12} units of

$$s^{-1} = B_{12}\rho_{\tilde{\nu}}(\tilde{\nu}) = B_{12}(J \cdot m^{-2})$$
$$J^{-1} \cdot m^2 \cdot s^{-1} = B_{12}$$

or

$$s^{-1} = B_{12}\rho_\lambda(\lambda) = B_{12}(J \cdot m^{-4})$$
$$J^{-1} \cdot m^4 \cdot s^{-1} = B_{12}$$

15–4. Show that Equations 15.7 and 15.11 are equivalent only if $B_{12} = B_{21}$ and $A_{21} = (8h\pi\nu_{12}^3/c^3)B_{21}$.

$$\rho_\nu(\nu_{12}) = \frac{8\pi h}{c^3} \frac{\nu_{12}^3}{e^{h\nu_{12}/k_B T} - 1} \tag{15.7}$$

$$\rho_\nu(\nu_{12}) = \frac{A_{21}}{B_{12}e^{h\nu_{12}/k_B T} - B_{21}} \tag{15.11}$$

These are equivalent if

$$\frac{8\pi h}{c^3} \frac{\nu_{12}^3}{e^{h\nu_{12}/k_B T} - 1} = \frac{A_{21}}{B_{12}e^{h\nu_{12}/k_B T} - B_{21}}$$

For this to be true, we need to be able to factor out an $(e^{h\nu_{12}/k_B T} - 1)$ from the denominator of the right-hand side, so $B_{12} = B_{21}$. Then

$$\frac{8\pi h}{c^3} \frac{\nu_{12}^3}{e^{h\nu_{12}/k_B T} - 1} = \frac{A_{21}}{B_{12}\left(e^{h\nu_{12}/k_B T} - 1\right)}$$

$$\frac{8\pi h\nu_{12}^3}{c^3} = \frac{A_{21}}{B_{12}}$$

15–5. Substitute Equation 15.16 into Equation 15.15 to prove that it is a solution to Equation 15.15.

$$\frac{dN_2(t)}{dt} = B\rho_\nu(\nu_{12})\left[N_1(t) - N_2(t)\right] - AN_2(t) \tag{15.15}$$

$$N_2(t) = \frac{B\rho_\nu(\nu_{12})N_{\text{total}}}{A + 2B\rho_\nu(\nu_{12})}\left\{1 - e^{-[A+2B\rho_\nu(\nu_{12})]t}\right\} \tag{15.16}$$

$N_1(t) + N_2(t) = N_{total}$, so we can write Equation 15.15 as

$$\frac{dN_2(t)}{dt} = B\rho_\nu(\nu_{12})\left[N_{total}(t) - 2N_2(t)\right] - AN_2(t) \tag{1}$$

Taking the derivative of Equation 15.16 gives

$$\frac{dN_2}{dt} = \frac{B\rho_\nu(\nu_{12})N_{total}}{A + 2B\rho_\nu(\nu_{12})}\left[A + 2B\rho_\nu(\nu_{12})\right]e^{-[A+2B\rho_\nu(\nu_{12})]t}$$

$$= B\rho_\nu(\nu_{12})N_{total}\,e^{-[A+2B\rho_\nu(\nu_{12})]t}$$

Substituting into the above expression from Equation 15.16, we find

$$\frac{dN_2}{dt} = -\left[A + 2B\rho_\nu(\nu_{12})\right]N_2(t) + B\rho_\nu(\nu_{12})N_{total}$$

$$= B\rho_\nu(\nu_{12})\left[N_{total} - 2N_2(t)\right] - AN_2(t)$$

which is the same as the expression we found using Equation 15.15 (Equation 1).

15–6. Use the fact that $N_1(t) + N_2(t) = N_{total}$ to write Equation 15.15 as

$$\frac{dN_2}{B\rho_\nu(\nu_{12})N_{total} - [A + 2B\rho_\nu(\nu_{12})]N_2} = dt$$

Now show that the integral of this equation gives Equation 15.16.

$$\frac{dN_2(t)}{dt} = B\rho_\nu(\nu_{12})\left[N_1(t) - N_2(t)\right] - AN_2(t) \tag{15.15}$$

We set $N_1(t) = N_{total} - N_2(t)$ to find

$$\frac{dN_2(t)}{dt} = B\rho_\nu(\nu_{12})\left[N_{total} - 2N_2(t)\right] - AN_2(t)$$

and solving for t gives

$$dt = \frac{dN_2}{B\rho_\nu(\nu_{12})N_{total} - \left[A + 2B\rho_\nu(\nu_{12})\right]N_2}$$

For simplicity, we let $\alpha = B\rho_\nu(\nu_{12})N_{total}$ and $\beta = A + 2B\rho_\nu(\nu_{12})$. We then have

$$dt = \frac{dN_2}{\alpha - \beta N_2}$$

Now we let $u = \alpha - \beta N_2$. Then $du = -\beta\, dN_2$, so $dN_2 = -\beta^{-1}du$, and the equation above becomes

$$dt = -\frac{du}{\beta u}$$

Integrating both sides of the equation gives $-\beta t = \ln u$, or $u = e^{-\beta t}$, so $\alpha - \beta N_2 = e^{-\beta t}$, or (substituting)

$$N_2 = -\frac{1}{\beta}e^{-\beta t} + \frac{\alpha}{\beta}$$

$$= -\frac{1}{A + 2B\rho_\nu(\nu_{12})}e^{-[A+2B\rho_\nu(\nu_{12})]t} + \frac{B\rho_\nu(\nu_{12})N_{total}}{A + 2B\rho_\nu(\nu_{12})}$$

$$= \frac{B\rho_\nu(\nu_{12})N_{total}}{A + 2B\rho_\nu(\nu_{12})}\left\{1 - e^{-[A+2B\rho_\nu(\nu_{12})]t}\right\} \tag{15.16}$$

15–7. Prove that Equation 15.17 implies that N_2/N_{total} is less than 1/2 because $A > 0$.

If $A = 0$, the ratio is

$$\frac{N_2(t \to \infty)}{N_{total}} = \frac{B\rho_\nu(\nu_{12})}{A + 2B\rho_\nu(\nu_{12})} = \frac{1}{2}$$

Because $A > 0$, $A + 2B\rho_\nu(\nu_{12}) > 2B\rho_\nu(\nu_{12})$ and so the ratio N_2/N_{total} must be less than 1/2.

15–8. Prove that the inequality

$$\frac{N_2}{N_{total}} < \frac{1}{2}$$

implies that N_2/N_1 is less than 1. (*Hint:* Use the fact that $1/a > 1/b$ if $a < b$.)

$$\frac{N_2}{N_{total}} < \frac{1}{2}$$

$$\frac{1}{\frac{N_1}{N_2} + 1} < \frac{1}{2}$$

This inequality implies that $N_1/N_2 > 1$ or that $N_2/N_1 < 1$.

15–9. The Einstein coefficients can also be derived using quantum mechanics. If the ground state and the excited state have a degeneracy of g_1 and g_2, respectively, the Einstein A coefficient is given by

$$A = \frac{16\pi^3\nu^3 g_1}{3\varepsilon_0 hc^3 g_2}|\mu|^2$$

where $|\mu|$ is the transition dipole moment (see Section 13–11). Now consider the $1s \to 2p$ absorption of H(g), which is observed at 121.8 nm. The radiative lifetime (see Example 15–3) of the triply degenerate excited $2p$ state of H(g) is 1.6×10^{-9} s. Determine the value of the transition dipole moment for this transition.

Using the expression given in the problem

$$A = \frac{16\pi^3\nu^3 g_1}{3\varepsilon_0 hc^3 g_2}|\mu|^2$$

we find

$$|\mu| = \left(\frac{3A\varepsilon_0 hc^3 g_2}{16\pi^3\nu^3 g_1}\right)^{1/2}$$

We are given λ, and so $\nu = c\lambda^{-1} = 2.46 \times 10^{15}$ s^{-1}. The reciprocal of A is denoted as τ_R and called the radiative lifetime (Example 15–3). The $2p$ orbital is threefold degenerate ($g_2 = 3$) and the $1s$ orbital is singly degenerate ($g_1 = 1$). We then have

$$|\mu|^2 = \frac{3A\varepsilon_0 hc^3 g_2}{16\pi^3\nu^3 g_1}$$

$$|\mu|^2 = \frac{3(1.6 \times 10^{-9}\text{ s})^{-1}(8.854 \times 10^{-12}\text{ C}^2\cdot\text{J}^{-1}\cdot\text{m}^{-1}) \times (6.626 \times 10^{-34}\text{ J}\cdot\text{s})(2.998 \times 10^8\text{ m}\cdot\text{s}^{-1})^3(3)}{16\pi^3(2.46 \times 10^{15}\text{ s}^{-1})^3(1)}$$

$$|\mu| = 1.1 \times 10^{-29}\text{ C}\cdot\text{m}$$

15–10. Use the equation given in Problem 15–9 and Equation 15.13 to derive the quantum mechanical expression for the Einstein B coefficient. Consider the $5s^1P_1 \rightarrow 3p^3P_2$ transition of neon at 632.8 nm, which is the lasing transition of most commercially available helium-neon lasers. Table 15.4 gives the Einstein A coefficient for this transition to be 6.56×10^6 s^{-1}. Determine the values of the Einstein B coefficient and the transition moment dipole for this transition. $(g_1 = g_2 = 1.)$

$$B = \frac{c^3 A}{8h\pi \nu^3} = \frac{16c^3\pi^3\nu^3 g_1}{24h^2\varepsilon_0 c^3\pi\nu^3 g_2}|\mu|^2 = \frac{2\pi^2 g_1}{3h^2\varepsilon_0 g_2}|\mu|^2$$

We are given λ, and so $\nu = c\lambda^{-1} = 4.738 \times 10^{14}$ s^{-1}. Then

$$B = \frac{c^3 A}{8h\pi\nu^3}$$

$$B = \frac{(2.998 \times 10^8 \text{ m·s}^{-1})^3 (6.56 \times 10^6 \text{ s}^{-1})}{8(6.626 \times 10^{-34} \text{ J·s})\pi(4.738 \times 10^{14} \text{ s}^{-1})^3} = 9.98 \times 10^{19} \text{ kg}^{-1}\text{·m}$$

$$|\mu|^2 = \frac{3A\varepsilon_0 hc^3 g_2}{16\pi^3\nu^3 g_1}$$

$$= \frac{3(6.56 \times 10^6 \text{ s}^{-1})(8.854 \times 10^{-12} \text{ C}^2\text{·J}^{-1}\text{·m}^{-1})}{16\pi^3(4.738 \times 10^{14} \text{ s}^{-1})^3}$$

$$\times (6.626 \times 10^{-34} \text{ J·s})(2.998 \times 10^8 \text{ m·s})^3$$

$$|\mu| = 7.68 \times 10^{-30} \text{ C·m}$$

15–11. Derive (but do not try to solve) rate equations for $N_1(t)$, $N_2(t)$, and $N_3(t)$ for the three-level system described by Figure 15.8.

First consider dN_1/dt. There are four components to the rate equation: excitation from level 1 to level 3, stimulated emission from level 3 to level 1, spontaneous emission from level 3 to level 1, and spontaneous emission from level 2 to level 1. This gives

$$\frac{dN_1}{dt} = -B_{31}\rho_\nu(\nu_{31})N_1 + B_{31}\rho_\nu(\nu_{31})N_3 + A_{31}N_3 + A_{21}N_2$$

Likewise, for dN_2/dt, we must consider spontaneous emission from level 3 to level 2, spontaneous emission from level 2 to level 1, stimulated emission from level 3 to level 2, and absorption from level 2 to level 3. We find the rate equation

$$\frac{dN_2}{dt} = A_{32}N_3 - A_{21}N_2 + B_{32}\rho_\nu(\nu_{32})N_3 - B_{32}\rho_\nu(\nu_{32})N_2$$

Finally, for dN_3/dt, we consider absorption from level 1 to level 3, stimulated emission from level 3 to level 1, spontaneous emission from level 3 to level 2, spontaneous emission from level 3 to level 1, stimulated emission from level 3 to level 2, and absorption from level 2 to level 3, to find

$$\frac{dN_3}{dt} = B_{31}\rho_\nu(\nu_{31})N_1 - B_{31}\rho_\nu(\nu_{31})N_3 - A_{32}N_3 - A_{31}N_3$$

$$- B_{32}\rho_\nu(\nu_{32})N_3 + B_{32}\rho_\nu(\nu_{32})N_2$$

15–12. Consider the nondegenerate three-level system shown in Figure 15.8. Suppose that an incident light beam of energy $h\nu = E_3 - E_1$ is turned on for a while and then turned off. Show that the subsequent decay of the E_3 level is given by

$$N_3(t) = N_3^0 e^{-(A_{32}+A_{31})t}$$

where N_3^0 is the number of atoms in state 3 at the instant the light source is turned off. What will be the observed radiative lifetime of this excited state?

After the light is turned off, no stimulated processes will occur. Then the rate equation of N_3 that we found in the previous problem becomes

$$\frac{dN_3}{dt} = -A_{32}N_3 - A_{31}N_3 = -(A_{32} + A_{31})N_3$$

or

$$N_3 = Ce^{-(A_{32}+A_{31})t}$$

At $t = 0$ (when the light is turned off), $N_3(t) = N_3^0 = C$, so

$$N_3(t) = N_3^0 e^{-(A_{32}+A_{31})t}$$

The observed radiative lifetime will be $(A_{32} + A_{31})^{-1}$. (The radiative lifetime is the reciprocal of the coefficient of t in the exponential term.)

15–13. In this problem, we will generalize the result of Problem 15–12. Consider a system that has N nondegenerate levels of energy, E_1, E_2, ..., E_N such that $E_1 < E_2 < \cdots < E_N$. Suppose that all the atoms are initially in the level of energy E_1. The system is then exposed to light of energy $h\nu = E_N - E_1$. Defining $t = 0$ to be the instant the light source is turned off, show that the decay of p_N, the population in state N, is given by

$$p_N(t) = p_N^0 e^{-\sum_{i=1}^{N-1} A_{Ni} t}$$

where p_N^0 is the population of level N at $t = 0$. Show that the radiative lifetime of level N is given by $1/\sum_{i=1}^{N-1} A_{Ni}$. Use this result and the data in Table 15.4 to evaluate the radiative lifetime of the $5s\,^1P_1$ level of neon, assuming the only radiative decay channels are to the eight levels tabulated in Table 15.4.

No stimulated processes will occur without the light source. The rate equation for $dp_N(t)/dt$ will then depend on only spontaneous emission terms to each of the $N - 1$ lower levels:

$$\frac{dp_N(t)}{dt} = -A_{N1}p_N(t) - A_{N2}p_N(t) - \cdots - A_{N(N-1)}p_N(t)$$

$$= -p_N(t) \sum_{i=1}^{N-1} A_{Ni}$$

Integrating this equation gives the solution

$$p_N(t) = Ce^{-\sum_{i=1}^{N-1} A_{Ni} t}$$

At $t = 0$, $p_N(t) = p_N^0 = C$, so

$$p_N(t) = p_N^0 e^{-\sum_{i=1}^{N-1} A_{Ni} t}$$

and a radiative lifetime of $1/\sum_{i=1}^{N-1} A_{Ni}$. The radiative lifetime of neon would be

$$\tau_{rad} = \frac{1}{(0.48 + 0.60 + 0.70 + 6.56 + 1.35 + 1.28 + 0.68 + 0.56) \times 10^6 \text{ s}^{-1}}$$
$$= 8.19 \times 10^{-8} \text{ s} = 81.9 \text{ ns}$$

15–14. The excited states of helium shown in Figure 15.13 have the electron configuration $1s2s$. Show that this electron configuration leads to a $^3S_1$ and a $^1S_0$ state. Which state has the lower energy?

Refer to Section 8–9. The maximum value of M_L is 0 and M_S can equal +1, 0, or -1. A table of microstates is then

	M_L
M_S	0
---------	-------
+1	$0^+, 0^+$
0	$0^+, 0^-; 0^-, 0^+$
−1	$0^-, 0^-$

One microstate (either $0^+, 0^-$ or $0^-, 0^+$; it makes no difference) gives $M_L = 0$ and $M_S = 0$, or ^1S, and the other three microstates give $M_L = 0$ and $M_S = +1, 0, -1$, or ^3S. Because $L = 0$ in both cases, $J = S$, so the two possible states are $^3S_1$ and $^1S_0$. In accordance with Hund's rules, $^3S_1$ is the lower-energy state.

15–15. According to Table 8.2, the ground-state energy of a helium atom is -2.904 hartrees. Use this value and the fact that the energy of a helium ion is given by $E = -Z^2/2n^2$ (in hartrees) to verify the energy of He$^+$ in Figure 15.13.

Using $E = -Z^2/2n^2$, the energy of the ground state of He$^+$ is $-2E_h$. Therefore, the energy of He$^+$ is 0.904 hartrees above that of He, or (because 1 hartree = 2.195×10^5 cm^{-1}) about 198 000 cm^{-1} above that of He, as shown in Figure 15.13.

15–16. The 3391.3 nm line in a He-Ne laser is due to the $5s\,^1P_1 \rightarrow 3p\,^3P_2$ transition. According to the *Table of Atomic Energy Levels* by Charlotte Moore, the energies of these levels are 166 658.484 cm^{-1} and 163 710.581 cm^{-1}, respectively. Calculate the wavelength of this transition. Why does your answer not come out to be 3391.3 nm? (See Example 8–10.)

$$\Delta E = 166\,658.484 \text{ cm}^{-1} - 163\,710.581 \text{ cm}^{-1} = 2947.903 \text{ cm}^{-1}$$

so

$$\lambda = \frac{1}{\tilde{\nu}} = \frac{1}{2947.903 \text{ cm}^{-1}} = 3392.2419 \text{ nm}$$

Example 8–10 reminds us that the values from the *Table of Atomic Energy Levels* are corrected for vacuum and the 3391.3 nm value calculated is for a measurement in air. Correcting, we find

$$\lambda_{air} = \frac{\lambda_{vacuum}}{1.00029} = \frac{3392.2419 \text{ nm}}{1.00029} = 3391.3 \text{ nm}$$

15–17. Using the method explained in Section 8–9, show that the states associated with a $2p^5ns$ electron configuration are $^3P_2$, $^3P_1$, $^3P_0$, and $^1P_1$.

There are six distinct ways of assigning the $2p$ electrons to their orbitals and two ways of placing the ns electron, for a total of twelve different microstates. (We can use Equation 8.53 to determine this or determine it by inspection.) We can then create a table of microstates. The maximum value of M_L is 1 and its possible values are 1, 0, and -1; the maximum value of M_S is 1 and its possible values are 1, 0, and -1.

	M_L		
M_S	-1	0	1
$+1$	$0^+, -1^+$	$0^+, 0^+$	$0^+, 1^+$
0	$0^+, -1^-; 0^-, -1^+$	$0^+, 0^-; 0^-, 0^+$	$0^+, 1^-; 0^-, 1^+$
-1	$0^-, -1^-$	$0^-, 0^-$	$0^-, 1^-$

There are nine states corresponding to $L = 1$ and $S = 1$ and three states corresponding to $L = 1$ and $S = 0$. This gives us the term symbols $^3P_2$, $^3P_1$, $^3P_0$, and $^1P_1$.

15–18. Consider the excited-state electron configuration $2p^5np$, with $n \geq 3$. How many microstates are associated with this electron configuration? The term symbols that correspond to $2p^5np$ are $^3D_3$, $^3D_2$, $^3D_1$, $^1D_2$, $^3P_2$, $^3P_1$, $^3P_0$, $^1P_1$, $^3S_1$, and $^1S_0$. Show that these term symbols account for all the microstates of the electron configuration $2p^5np$, $n \geq 3$.

We can use the fact that the total number of microstates associated with any term symbol is $(2S + 1)(2L + 1)$ (Section 8–9) to determine the number of microstates the above term symbols account for:

$$\begin{array}{cccccc} ^3D & ^1D & ^3P & ^1P & 3S & ^1S \\ (3 \times 5) & +(1 \times 5) & +(3 \times 3) & +(1 \times 3) & +(3 \times 1) & +(1 \times 1)=36 \end{array}$$

The total number of possible microstates in the electron configuration $2p^5np$ is given by (Equation 8.53)

$$\frac{G!}{N!(G-N)!} = \left(\frac{6!}{5!1!}\right)\left(\frac{6!}{1!5!}\right) = 36$$

where we have multiplied the number of possible microstates for the $2p^5$ configuration by the number of possible microstates for the np configuration.

15–19. A titanium sapphire laser operating at 780 nm produces pulses at a repetition rate of 100 MHz. If each pulse is 25 fs in duration and the average radiant power of the laser is 1.4 W, calculate the radiant power of each laser pulse. How many photons are produced by this laser in one second?

The laser produces 1.4 J of energy per second. Because there are 1.0×10^8 laser pulses per second,

$$\frac{1.4 \text{ J} \cdot \text{s}^{-1}}{1.0 \times 10^8 \text{ pulses} \cdot \text{s}^{-1}} = 1.4 \times 10^{-8} \text{ J} \cdot \text{pulse}^{-1}$$

Because each pulse is 25 fs long, the radiant energy of each laser pulse (which is measured in watts) is

$$\frac{1.4 \times 10^{-8} \text{ J} \cdot \text{pulse}^{-1}}{25 \times 10^{-15} \text{ s}} = 560 \text{ kW}$$

The radiant energy of one photon, given by Q_p, is (see Example 15–5)

$$Q_p = \frac{hc}{\lambda} = \frac{(6.626 \times 10^{-34} \text{ J} \cdot \text{s})(2.998 \times 10^8 \text{ m} \cdot \text{s}^{-1})}{780 \times 10^{-9} \text{ m}} = 2.55 \times 10^{-19} \text{ J}$$

Because the laser produces 1.4 J of energy per second, it produces

$$\frac{1.4 \text{ J}}{2.55 \times 10^{-19} \text{ J} \cdot \text{photon}^{-1}} = 5.50 \times 10^{18} \text{ photons}$$

in one second.

15–20. A typical chromium doping level of a ruby rod is 0.050% by mass. How many chromium atoms are there in a ruby rod of diameter 1.15 cm and length 15.2 cm? The density of corundum (Al_2O_3) is 4.05 g·cm^{-3}, and you can assume that the doping with chromium has no effect on the density of the solid. Now suppose all the chromium atoms are in the upper lasing level. If a laser pulse of 100 ps is generated by the simultaneous stimulated emission of all the chromium atoms, determine the radiant power of the laser pulse. (See Table 15.2.)

The volume of the rod is given by

$$V_{rod} = \pi r^2 l = \pi \left(\frac{1.15 \text{ cm}}{2}\right)^2 (15.2 \text{ cm}) = 15.8 \text{ cm}^3$$

The mass of the rod is then

$$(4.05 \text{ g} \cdot \text{cm}^{-3})(15.8 \text{ cm}^3) = 64.0 \text{ g}$$

and the mass of the contained chromium is

$$\text{mass}_{Cr} = (0.0005)(64.0 \text{ g}) = 0.032 \text{ g}$$

$$n_{Cr} = \frac{0.032 \text{ g}}{51.996 \text{ g} \cdot \text{mol}^{-1}} = 6.15 \times 10^{-4} \text{ mol}$$

which means there will be 3.70×10^{20} atoms of chromium in the rod. If all of the chromium atoms are in the upper lasing level and all undergo simultaneous stimulated emission, there will be 3.70×10^{20} photons emitted. From Table 15.2, we know that the light emitted from a chromium/ruby laser has a wavelength of 694.3 nm, so the energy produced by the stimulated emission is

$$E = nh\nu = nhc\lambda^{-1}$$

$$= (3.70 \times 10^{20})(6.626 \times 10^{-34} \text{ J} \cdot \text{s})(2.998 \times 10^8 \text{ m} \cdot \text{s}^{-1})(694.3 \times 10^{-9} \text{ m})^{-1}$$

$$= 106 \text{ J}$$

and the radiant power of the pulse, in watts, would be

$$\text{radiant power} = \frac{106 \text{ J}}{100 \times 10^{-12} \text{ s}} = 1.06 \times 10^{12} \text{ W}$$

15–21. Which laser pulse contains more photons, a 10-ns, 1.60-mJ pulse at 760 nm or a 500-ms, 1.60-mJ pulse at 532 nm?

The pulse duration does not affect the number of photons a laser pulse contains. Because the energies of both laser pulses are equivalent, and because the energy per pulse is proportional to the number of photons and inversely proportional to the wavelength of the emitted photons, the 760 nm pulse contains more photons.

15–22. Consider a flashlamp-pumped Nd^{3+}:YAG laser operating at a repetition rate of 10 Hz. Suppose the average radiant power of the flashlamp is 100 W. Determine the maximum number of photons that each laser pulse can contain using this pump source. The actual number of photons per laser pulse is 6.96×10^{17}. Determine the efficiency for converting the flashlamp output into laser output. (See Table 15.2.)

The pump energy available to each laser pulse is

$$\frac{100 \text{ J} \cdot \text{s}^{-1}}{10 \text{ pulse} \cdot \text{s}^{-1}} = 10 \text{ J} \cdot \text{pulse}^{-1}$$

A Nd^{3+}:YAG laser produces light at 1064.1 nm. The maximum number of photons that a laser pulse can contain, n, is given by

$$n = \frac{E\lambda}{hc}$$

$$= \frac{(10 \text{ J})(1064.1 \times 10^{-9} \text{ m})}{(6.626 \times 10^{-34} \text{ J} \cdot \text{s})(2.998 \times 10^8 \text{ m} \cdot \text{s}^{-1})} = 5.36 \times 10^{19} \text{ photons}$$

The efficiency of the actual laser is then

$$\text{efficiency} = \frac{6.96 \times 10^{17} \text{ photons}}{5.36 \times 10^{19} \text{ photons}} = 1.3\%$$

15–23. Chemical lasers are devices that create population inversions by a chemical reaction. One example is the HF gas laser, in which HF(g) is generated by the reaction

$$F(g) + H_2(g) \longrightarrow HF(g) + H(g)$$

The major product of this reaction is HF(g) in the excited $v = 3$ vibrational state. The reaction creates a population inversion in which $N(v)$, the number of molecules in each vibrational state, is such that $N(3) > N(v)$ for $v = 0$, 1, and 2. The output of the HF(g) laser corresponds to transitions between rotational lines of the $v = 3 \rightarrow v = 2$ ($\lambda = 2.7$–3.2 μm) transition. Why is there no lasing action from $v = 3 \rightarrow v = 1$ and $v = 3 \rightarrow v = 0$ even though there is a population inversion between these pairs of levels?

Recall from Chapter 13 (Section 13–11) that the harmonic oscillator selection rule is $\Delta v = \pm 1$. There may be weak anharmonic transitions which occur at $v = 3 \rightarrow v = 1$ and $v = 3 \rightarrow v = 0$, but the magnitude of the Einstein B coefficient will not be sufficient for lasing to occur.

15–24. A CO_2 laser operating at 9.6 μm uses an electrical power of 5.00 kW. If this laser produces 100-ns pulses at a repetition rate of 10 Hz and has an efficiency of 27%, how many photons are in each laser pulse?

The pump energy per pulse is

$$\frac{5000 \text{ J·s}^{-1}}{10 \text{ s}^{-1}} = 500 \text{ J·pulse}^{-1}$$

The laser is 27% efficient, so the radiant energy per pulse is $(500 \text{ J})(0.27) = 135$ J. The number of photons per pulse, n, is

$$n = \frac{E\lambda}{hc}$$

$$= \frac{(500 \text{ J})(0.27)(9.4 \times 10^{-6} \text{ m})}{(6.626 \times 10^{-34} \text{ J·s})(2.998 \times 10^8 \text{ m·s}^{-1})} = 6.39 \times 10^{21} \text{ photons}$$

15–25. Figure 15.10 displays the energy levels of the CO_2 laser. Given the following spectroscopic data for $CO_2(g)$, calculate the spacing between the $J' = 1 \to 0$ and $J' = 2 \to 1$ laser lines for the $001 \to 100$ vibrational transition.

Fundamental frequency$(J' = 0 \to 0)$ $100 \to 001 = 960.80$ cm^{-1}

$\tilde{B}(001) = 0.3871$ cm^{-1} $\tilde{B}(100) = 0.3902$ cm^{-1}

Why is no lasing observed at the fundamental frequency of 960.80 cm^{-1}?

We can use Equation 13.8 to find $F(J)$ of the 100 and 001 levels for $J = 1$ and 2:

$$F(J) = \tilde{B}J(J + 1)$$
$$F_{001}(1) = (0.3871 \text{ cm}^{-1})2 = 0.7742 \text{ cm}^{-1}$$
$$F_{001}(2) = (0.3871 \text{ cm}^{-1})6 = 2.3226 \text{ cm}^{-1}$$
$$F_{100}(1) = (0.3902 \text{ cm}^{-1})2 = 0.7804 \text{ cm}^{-1}$$

For the $J' = 1 \to 0$ transition, the spacing is then

$$960.80 \text{ cm}^{-1} + 0.7742 \text{ cm}^{-1} = 961.57 \text{ cm}^{-1}$$

and for the $J' = 2 \to 1$ transition, the spacing is

$$960.80 \text{ cm}^{-1} + 2.3226 \text{ cm}^{-1} - 0.7804 \text{ cm}^{-1} = 962.34 \text{ cm}^{-1}$$

There is no lasing observed at the fundamental frequency because the transition $\Delta J = 0$ is forbidden in the rigid-rotator approximation (Section 13–12).

15–26. The upper level of the $H_2(g)$ laser is the lowest excited state of the molecule, the $B^1\Sigma_u^+$ state, and the lower level is the $X^1\Sigma_g^+$ ground state. The lasing occurs between the $v' = 5$ level of the excited state and the $v'' = 12$ level of the ground state. Use the following spectroscopic data to determine the wavelength of the laser light from the $H_2(g)$ laser.

State	$\tilde{T}_e/\text{cm}^{-1}$	$\tilde{v}_e/\text{cm}^{-1}$	$\tilde{v}_e\tilde{x}_e/\text{cm}^{-1}$
$B^1\Sigma_u^+$	91 689.9	1356.9	19.93
$X^1\Sigma_g^+$	0	4401.2	121.34

A 1.0 ns pulse can be generated with a pulse radiant power of 100 kW. Calculate the radiant energy of such a laser pulse. How many photons are there in this pulse?

We can use Equation 13.21 to calculate the energy of the upper and lower lasing levels:

$$G(v) = \tilde{v}_e(v + \tfrac{1}{2}) - \tilde{x}_e\tilde{v}_e(v + \tfrac{1}{2})^2$$

$$G''(12) = (4401.2 \text{ cm}^{-1})(12.5) - (121.34 \text{ cm}^{-1})(12.5)^2$$

$$= 36\ 055.6 \text{ cm}^{-1}$$

$$G'(5) = (1356.9 \text{ cm}^{-1})(5.5) - (19.93 \text{ cm}^{-1})(5.5)^2$$

$$= 6860.1 \text{ cm}^{-1}$$

$\tilde{T}_e$ is the difference in the minima of the electronic potential energy curve in wave numbers, so the transition will have the energy

$$\tilde{v} = \tilde{T}_e + G'(5) - G''(12)$$

$$= 91\ 689.9 \text{ cm}^{-1} + 6860.1 \text{ cm}^{-1} - 36\ 055.6 \text{ cm}^{-1}$$

$$= 62\ 494.3 \text{ cm}^{-1}$$

Then $\lambda = (62\ 494.3 \text{ cm}^{-1})^{-1} = 160$ nm. The radiant energy of a laser pulse is

$$(100 \text{ kJ·s}^{-1})(1.00 \times 10^{-9} \text{ s}) = 1.0 \times 10^{-4} \text{ J}$$

and the number of photons per pulse is determined by using $E = nh\nu$:

$$n = \frac{E}{h\nu} = \frac{E\lambda}{hc}$$

$$= \frac{(1.0 \times 10^{-4} \text{ J})(160 \times 10^{-9} \text{ m})}{(6.626 \times 10^{-34} \text{ J·s})(2.998 \times 10^8 \text{ m·s}^{-1})}$$

$$= 8.06 \times 10^{13} \text{ photons}$$

15–27. In this problem, we will determine the excited-state rotational quantum numbers for the $X \rightarrow A$ absorption bands of ICl(g) that are shown in Figure 15.14. The transition is from the $v'' = 0$ state of the X state to a highly excited vibrational level of the A state ($v' = 32$). To accurately calculate the vibrational term $G(v)$ for the excited A state, we will need to include a second-order anharmonic correction to take into account the shape of the potential curve. First-order corrections will be sufficient for the ground electronic state. Extending the approach discussed in Chapter 13, we would write

$$G(v) = \tilde{v}_e(v + \tfrac{1}{2}) - \tilde{v}_e\tilde{x}_e(v + \tfrac{1}{2})^2 + \tilde{v}_e\tilde{y}_e(v + \tfrac{1}{2})^3$$

Some of the spectroscopic constants for the X ground state and the A excited state of ICl(g) are tabulated below.

State	$\tilde{T}_e$/cm^{-1}	$\tilde{v}_e$/cm^{-1}	$\tilde{v}_e\tilde{x}_e$/cm^{-1}	$\tilde{v}_e\tilde{y}_e$/cm^{-1}	$\tilde{B}_e$/cm^{-1}	$\tilde{\alpha}_e$/cm^{-1}
A	13 745.6	212.30	1.927	−0.03257	0.08389	0.00038
X	0	384.18	1.46			

Determine the value of $\tilde{v}$ corresponding to the transition $X(v'' = 0, J'' = 0) \rightarrow A(v' = 32, J' = 0)$. Given that the ground state for the lines shown in Figure 15.14 is the $v'' = 0$, $J'' = 2$ level of the X state and that the rotational term for this level is $F(2) = 0.65$ cm^{-1}, determine the closest value of J', the rotational number of the $v'' = 32$ level of the excited A state that gives the two observed spectral lines. Using your results, do you think that the individual lines between 17 299.45 and 17 299.55 cm^{-1} in Figure 15.15 can be attributed to transitions to different excited rotational states from the $X(v'' = 0, J'' = 2)$ ground state?

For the ground state, we can use Equation 13.21 (as in the previous problem):

$$G''(0) = (\tfrac{1}{2})(384.18 \text{ cm}^{-1}) - (\tfrac{1}{2})^2(1.46 \text{ cm}^{-1})$$
$$= 191.73 \text{ cm}^{-1}$$

We use Equation 13.21 as given in the problem to calculate G':

$$G'(32) = (212.30 \text{ cm}^{-1})(32 + \tfrac{1}{2}) - (1.927 \text{ cm}^{-1})(32 + \tfrac{1}{2})^2 - (0.03257 \text{ cm}^{-1})(32 + \tfrac{1}{2})^3$$
$$= 3746.29 \text{ cm}^{-1}$$

$\tilde{T}_e$ is the difference in the minima of the electronic potential energy curve in wave numbers, so the transition will have the energy

$$\tilde{v} = \tilde{T}_e + G'(32) - G''(0)$$
$$= 13 745.6 \text{ cm}^{-1} + 3746.29 \text{ cm}^{-1} - 191.73 \text{ cm}^{-1}$$
$$= 17 300.2 \text{ cm}^{-1}$$

Using Equation 13.17, we can find $\tilde{B}_v$ for the A-state, $v = 32$:

$$\tilde{B}_{32} = \tilde{B}_e - \tilde{\alpha}_e(v + \tfrac{1}{2})$$
$$= 0.08389 \text{ cm}^{-1} - (0.00038 \text{ cm}^{-1})(32 + \tfrac{1}{2}) = 0.07154 \text{ cm}^{-1}$$

The observed lines are between 17 299.45 and 17 299.55 cm^{-1}. Recall that $\tilde{E}_{v,J} = G(v) + F(J)$ (Equation 13.10), so

$$17 299.45 \text{ cm}^{-1} \approx 17 300.2 \text{ cm}^{-1} - 0.65 \text{ cm}^{-1} + F'(J)$$
$$17 299.45 \text{ cm}^{-1} \approx 17 299.5 + F'(J)$$

Therefore, $0 \approx F'(J)$. Recall that $F(J) = \tilde{B}J(J + 1)$, so the best value for J for the excited state rotational level is $J = 0$. The lines cannot be attributed to different excited rotational ground states, because the differences between their energies, 0.1 cm^{-1}, is less than the difference $F'(1) - F'(0) = 0.14$ cm^{-1}.

15–28. Hydrogen iodide decomposes to hydrogen and iodine when it is irradiated with radiation of frequency 1.45×10^{15} Hz. When 2.31 J of energy is absorbed by HI(g), 0.153 mg of HI(g) is decomposed. Calculate the quantum yield for this reaction.

First, determine the number of molecules that decompose:

$$\left[\frac{0.153 \times 10^{-3} \text{ g HI}}{(126.904 + 1.008) \text{ g} \cdot \text{mol}^{-1}} \right] (6.022 \times 10^{23} \text{ mol}^{-1}) = 7.203 \times 10^{17} \text{ molecules}$$

Now determine the number of photons absorbed (using the equation $E = nh\nu$):

$$n = \frac{E}{h\nu} = \frac{2.31 \text{ J}}{(6.626 \times 10^{-34} \text{ J} \cdot \text{s})(1.45 \times 10^{15} \text{ s}^{-1})} = 2.40 \times 10^{18} \text{ photons}$$

The quantum yield is therefore

$$\Phi = \frac{7.203 \times 10^{17}}{2.40 \times 10^{18}} = 0.30$$

15–29. Ozone decomposes to $O_2(g)$ and $O(g)$ with a quantum yield of 1.0 when it is irradiated with radiation of wavelength 300 nm. If ozone is irradiated with a power of 100 W, how long will it take for 0.020 mol of $O_3(g)$ to decompose?

The number of photons of light produced in one second is

$$n = \frac{E\lambda}{hc} = \frac{(100 \text{ J} \cdot \text{s}^{-1})(300 \times 10^{-9} \text{m})}{(6.626 \times 10^{-34} \text{ J} \cdot \text{s})(2.998 \times 10^{8} \text{ m} \cdot \text{s}^{-1})} = 1.51 \times 10^{20} \text{ photon} \cdot \text{s}^{-1}$$

Because the quantum yield is 1.0, 1.51×10^{20} molecules, or 2.51×10^{-4} moles, of ozone will decompose per second. We can now easily find how long it will take to decompose 0.020 moles of ozone:

$$\frac{0.020 \text{ mol}}{2.51 \times 10^{-4} \text{ mol} \cdot \text{s}^{-1}} = 79.8 \text{ s}$$

15–30. The quantum yield for the photosubstitution reaction

$$Cr(CO)_6 + NH_3 + h\nu \longrightarrow Cr(CO)_5NH_3 + CO$$

in octane solution at room temperature is 0.71 for a photolysis wavelength of 308 nm. How many $Cr(CO)_6$ molecules are destroyed per second when the solution is irradiated by a continuous laser with an output radiant power of 1.00 mW at 308 nm? If you wanted to produce one mole of $Cr(CO)_5NH_3$ per minute of exposure, what would the output radiant power of the laser need to be? (For both questions, assume the sample is sufficiently concentrated so that all the incident light is absorbed.)

The number of photons of light produced in one second is

$$n = \frac{(1.00 \times 10^{-3} \text{ J} \cdot \text{s}^{-1})(308 \times 10^{-9} \text{m})}{(6.626 \times 10^{-34} \text{ J} \cdot \text{s})(2.998 \times 10^{8} \text{ m} \cdot \text{s}^{-1})} = 1.55 \times 10^{15} \text{ photon} \cdot \text{s}^{-1}$$

We can then use the quantum yield to determine the number of molecules destroyed per second:

$$(0.71)(1.55 \times 10^{15} \text{ photon} \cdot \text{s}^{-1}) = 1.10 \times 10^{15} \text{ molecules destroyed} \cdot \text{s}^{-1}$$

To produce one mole per minute, we must produce

$$\left(\frac{1 \text{ mol}}{60 \text{ s}} \right) 6.022 \times 10^{23} \text{ mol}^{-1} = 1.00 \times 10^{22} \text{ molecule} \cdot \text{s}^{-1}$$

This means that we need

$$\frac{(1.00 \times 10^{22} \text{ molecule} \cdot \text{s}^{-1})}{0.71} = 1.41 \times 10^{22} \text{ photon} \cdot \text{s}^{-1}$$

or an output radiant power of

$$\text{Power} = \frac{nhc}{\lambda} = \frac{(1.41 \times 10^{22} \text{ s}^{-1})(6.626 \times 10^{-34} \text{ J} \cdot \text{s})(2.998 \times 10^8 \text{ m} \cdot \text{s}^{-1})}{308 \times 10^{-9} \text{ m}}$$
$$= 9.12 \text{ kW}$$

15–31. A mole of photons is called an *einstein*. Calculate the radiant energy of an einstein if the photons have a wavelength of 608.7 nm.

Again, use $E = nh\nu$:

$$E = \frac{nhc}{\lambda} = \frac{(6.022 \times 10^{23})(6.626 \times 10^{-34} \text{ J} \cdot \text{s})(2.998 \times 10^8 \text{ m} \cdot \text{s}^{-1})}{608.7 \times 10^{-9} \text{ m}} = 1.965 \times 10^5 \text{ J}$$

15–32. The width of the duration of an electromagnetic pulse, Δt, and the width of the frequency distribution of the pulse, $\Delta \nu$, are related by $\Delta t \Delta \nu = 1/2\pi$. Compute the width of the frequency distribution of a 10-fs laser pulse and a 1-ms laser pulse. Can you record high-resolution spectra like that shown in Figure 15.15 for ICl(g) using a tunable femtosecond laser?

$$\Delta \nu = \frac{1}{2\pi \Delta t}$$

For a 10-fs laser pulse, $\Delta t = 10$ fs and so $\Delta \nu = 1.59 \times 10^{13}$ s^{-1}. For a 1-ms laser pulse, $\Delta t = 1$ ms and $\Delta \nu = 159$ s^{-1}. The resolution of the femtosecond laser is

$$\frac{\Delta \nu}{c} = \frac{1.59 \times 10^{13} \text{ s}^{-1}}{2.998 \times 10^{10} \text{ cm} \cdot \text{s}^{-1}} = 531 \text{ cm}^{-1} = \Delta \tilde{\nu}$$

The spectrometer used to record Figure 15.16 has a spectral resolution of 0.002 cm^{-1}, so a femtosecond laser cannot be used to record high-resolution spectra of the quality shown in Figure 15.15.

15–33. In Section 15–8, we found that in the photodissociation reaction of ICN(g), 205 fs is required for the I(g) and CN(g) photofragments to separate by 400 pm (Figure 15.18). Calculate the relative velocity of the two photofragments. (*Hint*: The equilibrium bond length in the ground state is 275 pm.)

Initially, the two photofragments were 275 pm apart, so the distance travelled in 205 fs is 400 pm − 275 pm = 125 pm. The relative velocity of the photofragments is then

$$\frac{125 \times 10^{-12} \text{ m}}{205 \times 10^{-15} \text{ s}} = 610 \text{ m} \cdot \text{s}^{-1}$$

15–34. In the photolysis of ICN(g), the CN(g) fragment can be generated in several different vibrational and rotational states. At what wavelength would you set your probe laser to excite the

$v'' = 0$, $J'' = 3$ of the $X^2\Sigma^+$ ground state to the $v' = 0$, $J' = 3$ level of the $B^2\Sigma^+$ excited state? Use the following spectroscopic data.

State	$\tilde{T}_e/\text{cm}^{-1}$	$\tilde{\nu}_e/\text{cm}^{-1}$	$\tilde{\nu}_e\tilde{x}_e/\text{cm}^{-1}$	$\tilde{B}_e/\text{cm}^{-1}$	$\tilde{\alpha}_e/\text{cm}^{-1}$
$B^2\Sigma^+$	25 751.8	2164.13	20.25	1.970	0.0222
$X^2\Sigma^+$	0	2068.71	13.14	1.899	0.0174

Calculate the energy-level spacing between the $v'' = 0$, $J'' = 3$ and the $v'' = 0$, $J'' = 4$ levels. Can the formation dynamics of a single vibrational-rotational state of CN(g) be monitored by a femtosecond pump-probe experiment? (*Hint*: See Problem 15–32.)

We can use Equation 13.17 to determine $\tilde{B}_v$ for the $v' = 0$ and $v'' = 0$ levels, and then we can calculate $F'(J)$ and $F''(J)$:

$$\tilde{B}_v = \tilde{B}_e - \tilde{\alpha}_e(v + \tfrac{1}{2})$$
$$\tilde{B}_0(X) = 1.899\ \text{cm}^{-1} - \tfrac{1}{2}(0.0174\ \text{cm}^{-1}) = 1.890\ \text{cm}^{-1}$$
$$\tilde{B}_0(B) = 1.970\ \text{cm}^{-1} - \tfrac{1}{2}(0.0222\ \text{cm}^{-1}) = 1.959\ \text{cm}^{-1}$$

Equation 13.8 gives $F(J) = \tilde{B}J(J+1)$. Now $\tilde{T}_e$ is the energy difference between $v' = 0$, $J' = 0$ and $v'' = 0$, $J'' = 0$. Therefore, the energy difference between the $v'' = 0$, $J'' = 3$ of the $X^2\Sigma^+$ ground state and the $v' = 0$, $J' = 3$ level of the $B^2\Sigma^+$ excited state is

$$E = \tilde{T}_e + F'(J) - F''(J)$$
$$= 25\,751.8\ \text{cm}^{-1} + (1.959\ \text{cm}^{-1})(12) - (1.890\ \text{cm}^{-1})(12)$$
$$= 25\,752.6\ \text{cm}^{-1}$$

This means that laser light of wavelength 388 nm $(1/\tilde{\nu})$ should be used to excite this transition. The energy-level spacing between the $v'' = 0$, $J'' = 3$ and $v'' = 0$, $J'' = 4$ levels is simply

$$F''(4) - F''(3) = (1.890\ \text{cm}^{-1})(20) - (1.890\ \text{cm}^{-1})(12) = 15.12\ \text{cm}^{-1}$$

Recall from Problem 15–32 that the spectral width of the 10 fs laser pulse is 531 cm^{-1}. The width of the pulse is broader than the difference between the rotational levels we just calculated; therefore, femtosecond lasers cannot be used to monitor the formation dynamics of a single rovibrational state of CN.

15–35. The $X^1A_1 \to \tilde{A}$ electronic excitation of CH$_3$I(g) at 260 nm results in the following two competing photodissociation reactions:

$$\text{CH}_3\text{I(g)} + h\nu \longrightarrow \text{CH}_3\text{(g)} + \text{I(g)}(^2P_{3/2})$$
$$\longrightarrow \text{CH}_3\text{(g)} + \text{I}^*\text{(g)}(^2P_{1/2})$$

The energy difference between the excited $^2P_{1/2}$ state and the ground $^2P_{3/2}$ state of I(g) is 7603 cm^{-1}. The total quantum yield for dissociation is 1.00 with 31% of the excited molecules producing I*(g). Assuming that I*(g) relaxes by only radiative decay, calculate the number of photons emitted per second by a CH$_3$I(g) sample that absorbs 10% of the light generated by a 1.00-mW 260-nm laser.

The number of photons emitted by the laser is

$$n = \frac{E\lambda}{hc} = \frac{(1.00 \times 10^{-3}\ \text{J})(260 \times 10^{-9}\ \text{m})}{(6.626 \times 10^{-34}\ \text{J·s})(2.998 \times 10^8\ \text{m·s}^{-1})} = 1.31 \times 10^{15}\ \text{photons}$$

Because the sample absorbs only 10% of these photons, 1.31×10^{14} photons are absorbed by the sample, and because the quantum yield is one, 1.31×10^{14} molecules dissociate. Only 31% of these produce $I^*(g)$, so 4.06×10^{13} molecules react to produce $I^*(g)$. Each excited iodine releases only one photon in its radiative decay process, so 4.06×10^{13} photons are emitted.

15–36. The frequency of laser light can be converted using nonlinear optical materials. The most common form of frequency conversion is second harmonic generation, whereby laser light of frequency ν is converted to light at frequency 2ν. Calculate the wavelength of the second harmonic light from a Nd^{3+}:YAG laser. If the output pulse of a Nd^{3+}:YAG laser at 1064.1 nm has a radiant energy of 150.0 mJ, how many photons are contained in this pulse? Calculate the maximum number of photons that can be generated at the second harmonic. (*Hint*: Energy must be conserved.)

To determine the wavelength of the second harmonic from a Nd^{3+}:YAG laser, we can use

$$\nu_2 = 2\nu_1 = \frac{c}{\lambda_2} = \frac{2c}{\lambda_1}$$

or $\lambda_2 = \lambda_1/2$. Therefore, $\lambda_2 = 1064.1$ nm$/2 = 532.05$ nm. The 1064.1 nm pulse contains

$$n = \frac{E\lambda}{hc} = \frac{(150.0 \times 10^{-3} \text{ J})(1064.1 \times 10^{-9} \text{ m})}{(6.626 \times 10^{-34} \text{ J}\cdot\text{s})(2.998 \times 10^8 \text{ m}\cdot\text{s}^{-1})} = 8.035 \times 10^{17} \text{ photons}$$

At the second harmonic, for the same amount of energy

$$n = \frac{E\lambda}{hc} = \frac{(150.0 \times 10^{-3} \text{ J})(532.05 \times 10^{-9} \text{ m})}{(6.626 \times 10^{-34} \text{ J}\cdot\text{s})(2.998 \times 10^8 \text{ m}\cdot\text{s}^{-1})} = 4.018 \times 10^{17} \text{ photons}$$

15–37. There are nonlinear optical materials that can sum two laser beams at frequencies ν_1 and ν_2 and thereby generate light at frequency $\nu_3 = \nu_1 + \nu_2$. Suppose that part of the output from a krypton ion laser operating at 647.1 nm is used to pump a rhodamine 700 dye laser that produces laser light at 803.3 nm. The dye laser beam is then combined with the remaining output from the krypton ion laser in a nonlinear optical material that sums the two laser beams. Calculate the wavelength of the light created by the nonlinear optical material.

We can use

$$\nu_1 + \nu_2 = \frac{c}{\lambda_1} + \frac{c}{\lambda_2} = c\left(\frac{1}{\lambda_1} + \frac{1}{\lambda_2}\right) = \nu_3 = \frac{c}{\lambda_3}$$

or

$$\frac{1}{\lambda_1} + \frac{1}{\lambda_2} = \frac{1}{\lambda_3}$$

$$\frac{1}{803.3 \text{ nm}} + \frac{1}{647.1 \text{ nm}} = \frac{1}{\lambda_3}$$

giving $\lambda_3 = 358.4$ nm.

The following four problems examine how the intensity of absorption lines are quantified.

15–38. The *decadic absorbance A* of a sample is defined by $A = \log(I_0/I)$, where I_0 is the light intensity incident on the sample and I is the intensity of the light after it has passed through the sample. The decadic absorbance is proportional to c, the molar concentration of the sample, and l, the path length of the sample in meters, or in an equation

$$A = \varepsilon cl$$

where the proportionality factor ε is called the *molar absorption coefficient*. This expression is called the *Beer-Lambert law*. What are the units of A and ε? If the intensity of the transmitted light is 25.0% of that of the incident light, then what is the decadic absorbance of the sample? At 200 nm, a 1.42×10^{-3} M solution of benzene has decadic absorbance of 1.08. If the pathlength of the sample cell is 1.21×10^{-3} m, what is the value of ε? What percentage of the incident light is transmitted through this benzene sample? (It is common to express ε in the non SI units $L \cdot mol^{-1} \cdot cm^{-1}$ because l and c are commonly expressed in cm and $mol \cdot L^{-1}$, respectively. This difference in units leads to annoying factors of 10 that you need to be aware of.)

The quantity A is unitless, because it is a logarithmic quantity, and the quantity ε has units of $m^2 \cdot mol^{-1}$. If the intensity of the transmitted light is 25.0% of that of the incident light, then

$$A = \log\left(\frac{1}{0.250}\right) = 0.602$$

For benzene,

$$\varepsilon = \frac{A}{cl} = \frac{1.08}{(1.21 \times 10^{-3}\ m)(1.42 \times 10^{-3}\ mol \cdot dm^{-3})} = 629\ m^2 \cdot mol^{-1}$$

The percentage of light transmitted through the sample is

$$1.08 = \log\left(\frac{I_0}{I}\right)$$

$$12.0I = I_0$$

$$I = 0.083 I_0$$

(8.3% of the light is transmitted.)

15–39. The Beer-Lambert law (Problem 15–38) can also be written as

$$I = I_0 e^{-\sigma N l}$$

where N is the number of molecules per cubic meter and l is the pathlength of the cell in units of meters. What are the units of σ? The constant σ in this equation is called the *absorption cross section*. Derive an expression relating σ to ε, the molar absorption coefficient introduced in Problem 15–38. Determine σ for the benzene solution described in Problem 15–38.

Because the exponential term must be unitless, σ has units of m^2. Now $\ln x = 2.303 \log x$, so

$$\ln\left(\frac{I}{I_0}\right) = 2.303 \log\left(\frac{I}{I_0}\right) = -2.303 \log\left(\frac{I_0}{I}\right) = -\sigma N l$$

so

$$2.303 \varepsilon cl = \sigma Nl$$

$$2.303 \varepsilon \left(\frac{N}{N_A} \right) = \sigma N$$

$$\frac{2.303 \varepsilon}{N_A} = \sigma$$

For the benzene solution in Problem 15–39, $\varepsilon = 629 \text{ m}^2 \cdot \text{mol}^{-1}$, so

$$\sigma = \frac{2.303(629 \text{ m}^2 \cdot \text{mol}^{-1})}{6.022 \times 10^{23} \text{ mol}^{-1}} = 2.41 \times 10^{-21} \text{ m}^2$$

15–40. The Beer-Lambert law (Problem 15–38) can also be written in terms of the natural logarithm instead of the base ten logarithm:

$$A_e = \ln \frac{I_0}{I} = \kappa cl$$

In this form, the constant κ is called the *molar napierian absorption coefficient*, and A_e is called the *naperian absorbance*. What are the units of κ? Derive a relationship between κ and ε (see Problem 15–38). Determine κ for the benzene solution described in Problem 15–38.

The units of κ are the same as those of ε: $\text{m}^2 \cdot \text{mol}^{-1}$. As in the previous problem, use the fact that $\ln x = 2.303 \log x$ to find

$$\ln \frac{I}{I_0} = 2.303 \log \frac{I_0}{I} = \kappa cl$$

$$2.303 \varepsilon cl = \kappa cl$$

$$2.303 \varepsilon = \kappa$$

The quantity $\varepsilon = 629 \text{ m}^2 \cdot \text{mol}^{-1}$, so $\kappa = 1450 \text{ m}^2 \cdot \text{mol}^{-1}$.

15–41. A re-examination of the spectra in Chapter 13 reveals that the transitions observed have a line width. We define A, the integrated absorption intensity to be

$$A = \int_{-\infty}^{\infty} \kappa(\tilde{\nu}) d\tilde{\nu}$$

where $\kappa(\tilde{\nu})$ is the molar napierian absorption coefficient in terms of wavenumbers, $\tilde{\nu}$ (see Problem 15–40). What are the units of A? Now suppose that the absorption line has a Gaussian line shape, or that

$$\kappa(\tilde{\nu}) = \kappa(\tilde{\nu}_{max}) e^{-\alpha(\tilde{\nu} - \tilde{\nu}_{max})^2}$$

where α is a constant and $\tilde{\nu}_{max}$ is the maximum frequency of absorption. Plot $\kappa(\tilde{\nu})$. How is α related to $\Delta\tilde{\nu}_{1/2}$, the width of the absorption line at half of its maximum intensity? Now show that

$$A = 1.07\kappa(\tilde{\nu}_{max})\Delta\tilde{\nu}_{1/2}$$

(*Hint:* $\int_0^\infty e^{-\beta x^2}dx = (\pi/4\beta)^{1/2}$.)

The quantity κ is in units of $m^2\cdot mol^{-1}$ and we integrate over m^{-1} to find A, so A has units of $m\cdot mol^{-1}$. A plot of the Gaussian shape is shown below.

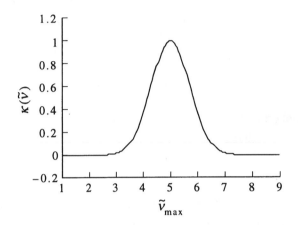

At half the maximum intensity, $\kappa(\tilde{\nu}_{1/2}) = \frac{1}{2}\kappa(\tilde{\nu}_{max})$. Substituting this value into the equation for a Gaussian lineshape gives

$$\frac{\kappa(\tilde{\nu}_{max})}{2} = \kappa(\tilde{\nu}_{max})e^{-\alpha(\tilde{\nu}_{1/2}-\tilde{\nu}_{max})^2}$$

$$\frac{1}{2} = e^{-\alpha(\tilde{\nu}_{1/2}-\tilde{\nu}_{max})^2}$$

so

$$\ln 2 = \alpha(\tilde{\nu}_{1/2} - \tilde{\nu}_{max})^2$$

$$\frac{0.833}{\alpha^{1/2}} = (\tilde{\nu}_{1/2} - \tilde{\nu}_{max})$$

The width of the absorption line is twice the difference $\tilde{\nu}_{1/2} - \tilde{\nu}_{max}$, so

$$\Delta\tilde{\nu}_{1/2} = \frac{1.66}{\alpha^{1/2}}$$

$$\alpha^{1/2} = \frac{1.66}{\Delta\tilde{\nu}_{1/2}}$$

Letting κ have the Gaussian line shape described in the problem,

$$A = \int_{-\infty}^{\infty} \kappa(\tilde{\nu}_{max})e^{-\alpha(\tilde{\nu}-\tilde{\nu}_{max})^2}d\tilde{\nu}$$

$$= \kappa(\tilde{\nu}_{max})\left(\frac{\pi}{\alpha}\right)^{1/2}$$

$$= \kappa(\tilde{\nu}_{max})\frac{\pi^{1/2}\Delta\tilde{\nu}_{1/2}}{1.66}$$

$$= 1.07\kappa(\tilde{\nu}_{max})\Delta\tilde{\nu}_{1/2}$$

Numerical Methods

PROBLEMS AND SOLUTIONS

Excel was used to create spreadsheets for the approximations in the problems below. Any spreadsheet program can be used, but programs such as Excel, where the formulas can be saved and are automatically recalculated when different values are entered, are more powerful in this case than programs which are primarily used for graphing (such as Kaleidagraph).

G–1. Solve the equation $x^5 + 2x^4 + 4x = 5$ to four significant figures for the root that lies between 0 and 1.

$$f(x) = x^5 + 2x^4 + 4x - 5$$
$$f'(x) = 5x^4 + 8x^3 + 4$$

The iterative formula for the Newton-Raphson method is

$$x_{n+1} = x_n - \frac{f(x_n)}{f'(x_n)} \tag{G.1}$$

To set up a spreadsheet for the Newton-Raphson method, let one column contain x_n, one column contain the formula for $f(x)$, and one column contain the formula for $f'(x)$. [Allow the first x_n to be input manually; thereafter, let your spreadsheet calculate the values of x_n using the equation above.] Since we wish to find the root which lies between 0 and 1, we can take x_0 to be 0.5:

n	x_n	$f(x_n)$	$f'(x_n)$
0	0.50000	−2.84375	5.3125
1	1.03529	2.62821	18.62145
2	0.894155	0.426631	12.91523
3	0.861122	0.0177315	11.85774
4	0.859627	3.4128×10^{-5}	11.81213
5	0.859624		

Thus the root is 0.8596.

G–2. Use the Newton-Raphson method to derive the iterative formula

$$x_{n+1} = \frac{1}{2}\left(x_n + \frac{A}{x_n}\right)$$

for the value of $\sqrt{A}$. This formula was discovered by a Babylonian mathematician more than 2000 years ago. Use this formula to evaluate $\sqrt{2}$ to five significant figures.

$x^2 = A$, so $x^2 - A = 0 = f(x)$ and $2x = f'(x)$. From the Newton-Raphson equation,

$$x_{n+1} = x_n - \frac{f(x_n)}{f'(x_n)}$$

$$= x_n - \frac{(x_n^2 - A)}{2x_n} = \frac{2x_n^2 - x_n^2 + A}{2x_n}$$

$$= \frac{1}{2}\left(x_n + \frac{A}{x_n}\right)$$

We know that $\sqrt{2}$ is between 1 and 2, so we can take x_0 to be 1.5. In three iterations, we find that $\sqrt{2} = 1.4142$ to five significant figures.

G–3. Use the Newton-Raphson method to solve the equation $e^{-x} + (x/5) = 1$ to four significant figures. This equation occurs in Problem 1–5.

$$f(x) = e^{-x} + \frac{x}{5} - 1 = 0$$

$$f'(x) = -e^{-x} + \frac{1}{5}$$

We select 5 as x_0 because $(5/5) - 1 = 0$ and e^{-5} is a small number. Using the spreadsheet, we find

n	x_n	$f(x_n)$	$f'(x_n)$
0	5.00000	6.73795×10^{-2}	0.19326
1	4.96514	4.143×10^{-6}	0.19302
2	4.96511		

Thus the solution to the equation is 4.965 (to four significant figures).

G–4. Consider the chemical reaction described by the equation

$$CH_4(g) + H_2O(g) \rightleftharpoons CO(g) + 3\,H_2(g)$$

at 300 K. If 1.00 atm of $CH_4(g)$ and $H_2O(g)$ are introduced into a reaction vessel, the pressures at equilibrium obey the equation

$$\frac{P_{CO}P_{H_2}^3}{P_{CH_4}P_{H_2O}} = \frac{(x)(3x)^3}{(1-x)(1-x)} = 26$$

Solve this equation for x.

$$\frac{27x^4}{1 - 2x + x^2} = 26$$

$$27x^4 = 26 - 52x + 26x^2$$

The functions we will use in the spreadsheet created for Problem G–1 are then

$$f(x) = 0 = 27x^4 - 26x^2 + 52x - 26$$
$$f'(x) = 108x^3 - 52x + 52$$

We know that x must be between 0 and 1, so we can take x_0 to be 0.5. Then

n	x_n	$f(x_n)$	$f'(x_n)$
0	0.50000	−4.81250	39.50000
1	0.62184	0.31884	45.63326
2	0.61485	0.00177	45.13101
3	0.61481		

To three significant figures, $x = 0.615$ atm.

G–5. In Chapter 16, we will solve the cubic equation

$$64x^3 + 6x^2 + 12x - 1 = 0$$

Use the Newton-Raphson method to find the only real root of this equation to five significant figures.

$$f(x) = 64x^3 + 6x^2 + 12x - 1 = 0$$
$$f'(x) = 192x^2 + 12x + 12$$

The solution must be small ($-1 \leq x \leq 1$), so let us take $x_0 = -0.5$. Then

n	x_n	$f(x_n)$	$f'(x_n)$
0	−0.50000	−13.5000	54.00000
1	−0.25000	−4.62500	21.00000
2	−0.02976	−1.35352	11.81293
3	0.084817	0.10002	14.39905
4	0.0778708	1.0539×10^{-3}	14.09871
5	0.0777961	1.17×10^{-7}	14.09558
6	0.0777961		

So $x = 7.7780 \times 10^{-2}$.

G–6. Solve the equation $x^3 - 3x + 1 = 0$ for all three of its roots to four decimal places.

$$f(x) = x^3 - 3x + 1$$
$$f'(x) = 3x^2 - 3$$

Setting $f'(x)$ equal to zero, we find that the inflection points of the equation $x^3 - 3x + 1$ are 1 and -1. We can therefore set our x_0's to 0, -1.5, and 1.5.

n	x_n	$f(x_n)$	$f'(x_n)$
0	0.00000	1.00000	-3.00000
1	0.333333	0.03704	-2.66667
2	0.347222	1.956×10^{-4}	-2.63831
3	0.347296	6×10^{-9}	-2.63816
4	0.347296		

n	x_n	$f(x_n)$	$f'(x_n)$
0	1.50000	-0.12500	3.75000
1	1.53333	0.005037	4.05333
2	1.53209	7.102×10^{-6}	4.04191
3	1.53209		

n	x_n	$f(x_n)$	$f'(x_n)$
0	-1.50000	2.12500	3.75000
1	-2.06667	-1.62696	9.81333
2	-1.90088	-0.16586	7.83998
3	-1.87972	-2.5428×10^{-3}	7.60004
4	-1.87939	-6.312×10^{-7}	7.59627
5	-1.87939		

The three roots of $x^3 - 3x + 1$ are 0.3473, 1.532, and -1.879.

G–7. In Example 16–3 we will solve the cubic equation

$$\overline{V}^3 - 0.1231\overline{V}^2 + 0.02056\overline{V} - 0.001271 = 0$$

Use the Newton-Raphson method to find the root to this equation that is near $\overline{V} = 0.1$.

Let $\overline{V} = x$:

$$f(x) = x^3 - 0.1231x^2 + 0.02056x - 0.001271$$
$$f'(x) = 3x^2 - 0.2462x + 0.02056$$

We can take $x_0 = 0.120$. Then

n	x_n	$f(x_n)$	$f'(x_n)$
0	0.120	1.1516×10^{-3}	3.4216×10^{-2}
1	0.086344	2.3021×10^{-4}	2.1668×10^{-2}
2	0.075720	1.4145×10^{-5}	1.9118×10^{-2}
3	0.074980	5.6559×10^{-8}	1.8966×10^{-2}
4	0.074977	9.0564×10^{-13}	1.8965×10^{-2}
5	0.074977		

So the root to the equation near $\overline{V} = 0.120$ is $\overline{V} = 0.074977$.

G–8. In Section 16–3 we will solve the cubic equation

$$\overline{V}^3 - 0.3664\overline{V}^2 + 0.03802\overline{V} - 0.001210 = 0$$

Use the Newton-Raphson method to show that the three roots to this equation are 0.07073, 0.07897, and 0.2167.

Let $\overline{V} = x$:

$$f(x) = x^3 - 0.3664x^2 + 0.03802x - 0.001210$$

$$f'(x) = 3x^2 - 0.7328x + 0.03802$$

We can set our x_0's to 0.069, 0.080, and 0.20, conveniently close to those given in the problem text. Then

n	x_n	$f(x_n)$	$f'(x_n)$
0	0.069	-2.5414×10^{-6}	1.7398×10^{-3}
1	0.070461	-3.3701×10^{-7}	1.2805×10^{-3}
2	0.070724	-1.0719×10^{-8}	1.1991×10^{-3}
3	0.070733	-1.2323×10^{-11}	1.1964×10^{-3}
4	0.070733		

n	x_n	$f(x_n)$	$f'(x_n)$
0	0.080	-1.36×10^{-6}	-1.4040×10^{-3}
1	0.079031	-1.1951×10^{-7}	-1.1563×10^{-3}
2	0.078928	-1.3824×10^{-9}	-1.1295×10^{-3}
3	0.078927	-1.9414×10^{-13}	-1.1292×10^{-3}
4	0.078927		

n	x_n	$f(x_n)$	$f'(x_n)$
0	0.20	-2.62×10^{-4}	1.1460×10^{-2}
1	0.22286	1.3405×10^{-4}	2.3709×10^{-2}
2	0.21721	9.4788×10^{-6}	2.0388×10^{-2}
3	0.21674	6.1550×10^{-8}	2.0124×10^{-2}
4	0.21674		

G–9. The Newton-Raphson method is not limited to polynomial equations. For example, in Problem 4–38 we solved the equation

$$\varepsilon^{1/2} \tan \varepsilon^{1/2} = (12 - \varepsilon)^{1/2}$$

for ε by plotting $\varepsilon^{1/2} \tan \varepsilon^{1/2}$ and $(12 - \varepsilon)^{1/2}$ versus ε on the same graph and noting the intersections of the two curves. We found that $\varepsilon = 1.47$ and 11.37. Solve the above equation using the Newton-Raphson method and obtain the same values of ε.

Let $\varepsilon = x$:

$$f(x) = (12 - x)^{1/2} - x^{1/2} \tan x^{1/2}$$

$$f'(x) = \frac{-1}{2(12 - x)^{1/2}} - \frac{\tan x^{1/2}}{2x^{1/2}} - \frac{\sec^2 x^{1/2}}{2}$$

We use the spreadsheet created for Problem G–1 to find that $\varepsilon = 1.4715$ and 11.372.

G–10. Use the trapezoidal approximation and Simpson's rule to evaluate

$$I = \int_0^1 \frac{dx}{1 + x^2}$$

This integral can be evaluated analytically; it is given by $\tan^{-1}(1)$, which is equal to $\pi/4$, so $I = 0.78539816$ to eight decimal places.

Set up a new spreadsheet which will use the trapezoidal approximation, and another for Simpson's rule. For Simpson's rule, your coefficients of the functions vary according to whether the variable's subscript is even or odd. I set up two columns: one for $f(x_1)$, $f(x_3)$, $f(x_5)$, ... and one for $f(x_2)$, $f(x_4)$, $f(x_6)$, ... One can then calculate $f(x_0)$ and $f(x_{2n})$ elsewhere (I used cells above my two columns) and creating an equation for the approximation to the integral then becomes trivial. The spreadsheet for the trapezoidal approximation is much the same. This spreadsheet can be used for the remainder of the problems.

n	h	I_n(trapezoidal)	I_{2n}(Simpson's rule)
10	0.1	0.7849814972	0.7853981632
50	0.02	0.7853814967	0.7853981634
100	0.01	0.7853939967	0.7853981634

G–11. Evaluate ln 2 to six decimal places by evaluating

$$\ln 2 = \int_1^2 \frac{dx}{x}$$

What must n be to assure six-digit accuracy?

To find n to six-digit accuracy, the error must be no greater than 1×10^{-6}. Then

$$E = \frac{M(b - a)h^4}{180}$$

where M is the maximum value of $f^{IV}(x)$. Differentiate $f(x)$ to find

$$f^{IV}(x) = \frac{24}{x^5}$$

The maximum value of $f^{IV}(x)$ in the interval from 1 to 2 is thus 24. Then

$$1 \times 10^{-6} = \frac{24(2 - 1)h^4}{180}$$

$$h = 0.0523$$

Recall $h = (b - a)/2n$. Thus $n = (b - a)/2h = 1/0.105 = 9.6$, and the smallest value of n needed is 10. We use this value of n to find $\ln 2 = 0.693147$.

G–12. Use Simpson's rule to evaluate

$$I = \int_0^\infty e^{-x^2}\,dx$$

and compare your result with the exact value, $\sqrt{\pi}/2$.

Use the spreadsheet created for Problem G–10 to find

$$\int_0^\infty e^{-x^2}\,dx = 0.8862269$$

We can use large values of n to find this value to greater accuracy, if need be.

G–13. The integral

$$I = \int_0^\infty \frac{x^3\,dx}{e^x - 1}$$

occurs in Problem 1–42, where we use its exact value $\pi^4/15$. Use Simpson's rule to evaluate I to six decimal places.

Use the spreadsheet created for Problem G–10 to find $I = 6.49394$.

G–14. Use a numerical software package such as *MathCad*, *Kaleidagraph*, or *Mathematica* to evaluate the integral

$$S = 4\pi^{1/2}\left(\frac{2\alpha}{\pi}\right)^{3/4}\int_0^\infty r^2 e^{-r}e^{-\alpha r^2}\,dr$$

for values of α between 0.200 and 0.300 and show that S has a maximum value at $\alpha = 0.271$ (see Problem 11–11).

Here, we use values of α calculated by *Mathematica* at intervals of 0.005 and plot S against α.

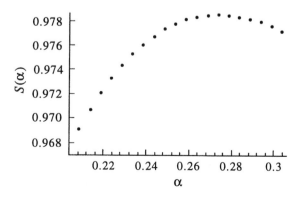

The maximum value of S is around $\alpha = 0.27$. The values of $S(\alpha)$ close to $\alpha = 0.271$ are given below.

α	0.2700	0.2705	0.2710	0.2715	0.2720
$S(\alpha)$	0.9784029	0.9784041	0.9784044	0.9784039	0.9784024

CHAPTER 16

The Properties of Gases

PROBLEMS AND SOLUTIONS

16–1. In an issue of the journal *Science* a few years ago, a research group discussed experiments in which they determined the structure of cesium iodide crystals at a pressure of 302 gigapascals (GPa). How many atmospheres and bars is this pressure?

2.98×10^6 atm, 3.02×10^6 bar

16–2. In meteorology, pressures are expressed in units of millibars (mbar). Convert 985 mbar to torr and to atmospheres.

739 torr, 0.972 atm

16–3. Calculate the value of the pressure (in atm) exerted by a 33.9-foot column of water. Take the density of water to be $1.00 \text{ g} \cdot \text{mL}^{-1}$.

We first convert the height of the column to metric units: 33.9 ft = 10.33 m. Now

$$P = \rho gh = (1.00 \text{ kg} \cdot \text{dm}^{-3})(98.067 \text{ dm} \cdot \text{s}^{-2})(103.3 \text{ dm})$$
$$= 1.013 \times 10^4 \text{ kg} \cdot \text{dm}^{-1} \cdot \text{s}^{-2}$$
$$= 1.013 \times 10^5 \text{ Pa} = 1.00 \text{ atm}$$

16–4. At which temperature are the Celsius and Farenheit temperature scales equal?

$-40°$

16–5. A travel guide says that to convert Celsius temperatures to Farenheit temperatures, double the Celsius temperature and add 30. Comment on this recipe.

This will provide a rough estimate of the temperature, decreasing in accuracy as temperature increases. (Of course, it is not valid for Celsius temperatures below zero degrees.) At room temperatures, it is accurate enough for ordinary purposes.

Actual T (°C)	Actual T (°F)	Travel T (°F)
0	32	30
10	50	50
20	68	70
30	86	90
40	104	110

16–6. Research in surface science is carried out using ultra-high vacuum chambers that can sustain pressures as low as 10^{-12} torr. How many molecules are there in a 1.00-cm³ volume inside such an apparatus at 298 K? What is the corresponding molar volume $\overline{V}$ at this pressure and temperature?

We will assume ideal gas behavior, so

$$\frac{PV}{RT} = n \tag{16.1a}$$

$$\frac{(10^{-12} \text{ torr})(1.00 \text{ cm}^3)}{(82.058 \text{ cm}^3 \cdot \text{atm} \cdot \text{mol}^{-1} \cdot \text{K}^{-1})(760 \text{ torr} \cdot \text{atm}^{-1})(298 \text{ K})} = n$$

$$5.38 \times 10^{-20} \text{ mol} = n$$

so there are 3.24×10^4 molecules in the apparatus. The molar volume is

$$\overline{V} = \frac{V}{n} = \frac{1.00 \text{ cm}^3}{5.38 \times 10^{-20} \text{ mol}} = 1.86 \times 10^{19} \text{ cm}^3 \cdot \text{mol}^{-1}$$

16–7. Use the following data for an unknown gas at 300 K to determine the molecular mass of the gas.

P/bar	0.1000	0.5000	1.000	1.01325	2.000
ρ/g·L^{-1}	0.1771	0.8909	1.796	1.820	3.652

The line of best fit of a plot of P/ρ versus ρ will have an intercept of RT/M. Plotting, we find that the intercept of this plot is 0.56558 bar·g^{-1}·dm³, and so $M = 44.10$ g·mol^{-1}.

16–8. Recall from general chemistry that Dalton's law of partial pressures says that each gas in a mixture of ideal gases acts as if the other gases were not present. Use this fact to show that the partial pressure exerted by each gas is given by

$$P_j = \left(\frac{n_j}{\sum n_j}\right) P_{\text{total}} = y_j P_{\text{total}}$$

where P_j is the partial pressure of the jth gas and y_j is its mole fraction.

The ideal gas law (Equation 16.1) gives

$$P_{total} V = n_{total} RT = \sum_j n_j RT$$

and

$$P_j V = n_j RT$$

for all component gases j. Solving each expression for RT/V and equating the results gives

$$\frac{P_{total}}{\sum_j n_j} = \frac{P_j}{n_j}$$

or

$$P_j = \frac{n_j}{\sum_j n_j} P_{total} = y_j P_{total}$$

16–9. A mixture of $H_2(g)$ and $N_2(g)$ has a density of 0.216 g·L^{-1} at 300 K and 500 torr. What is the mole fraction composition of the mixture?

The density of the mixture is 0.216 g·L^{-1}, so there are 216 g of gas present in one m³ of gas. Take the total volume of the mixture to be 1 m³. Then, using the ideal gas law (Equation 16.1), we find

$$P_{tot} V = n_{tot} RT$$

$$500 \text{ torr} \left(\frac{101\,325 \text{ Pa}}{760 \text{ torr}} \right) 1 \text{ m}^3 = n_{tot} \left(8.3145 \text{ J·mol}^{-1}\cdot \text{K}^{-1} \right) (300 \text{ K})$$

$$26.7 \text{ mol} = n_{tot}$$

There are 26.7 mol of gas per cubic meter. Let x be the number of moles of hydrogen gas. Then $n_{tot} - x$ is the number of moles of nitrogen gas. Since $M_{H_2} = 2.01588 \text{ g·mol}^{-1}$ and $M_{N_2} = 28.01348 \text{ g·mol}^{-1}$, we can write

$$216 \text{ g} = \left(28.01348 \text{ g·mol}^{-1}\right)(26.7 \text{ mol} - x \text{ mol}) + \left(2.01588 \text{ g·mol}^{-1}\right)(x \text{ mol})$$

$$26x = 532.6 \text{ g}$$

$$x = 20.5 \text{ g}$$

The mole fractions of each component of the mixture are therefore

$$y_{H_2} = \frac{n_{H_2}}{n_{tot}} = \frac{20.5 \text{ mol}}{26.7 \text{ mol}} = 0.77$$

and

$$y_{N_2} = \frac{n_{N_2}}{n_{tot}} = \frac{6.2 \text{ mol}}{26.7 \text{ mol}} = 0.23$$

16–10. One liter of $N_2(g)$ at 2.1 bar and two liters of $Ar(g)$ at 3.4 bar are mixed in a 4.0-L flask to form an ideal-gas mixture. Calculate the value of the final pressure of the mixture if the initial and final temperature of the gases are the same. Repeat this calculation if the initial temperatures of the $N_2(g)$ and $Ar(g)$ are 304 K and 402 K, respectively, and the final temperature of the mixture is 377 K. (Assume ideal-gas behavior.)

---

a. Initially, we have one liter of N_2 at 2.1 bar and two liters of Ar at 3.4 bar. We can use the ideal gas law (Equation 16.1) to find the number of moles of each gas:

$$n_{N_2} = \frac{P_{N_2} V_{N_2}}{RT} \qquad\qquad n_{Ar} = \frac{P_{Ar} V_{Ar}}{RT}$$

$$= \frac{(2.1 \times 10^5 \text{ Pa})(1 \times 10^{-3} \text{ m}^3)}{RT} \qquad = \frac{(3.4 \times 10^5 \text{ Pa})(2 \times 10^{-3} \text{ m}^3)}{RT}$$

$$= \frac{210 \text{ Pa·m}^3}{RT} \qquad\qquad = \frac{680 \text{ Pa·m}^3}{RT}$$

The total moles of gas in the final mixture is the sum of the moles of each gas in the mixture, which is $(890 \text{ Pa·m}^3)/RT$. So (Equation 16.1)

$$P = \frac{nRT}{V} = \frac{890 \text{ Pa·m}^3}{0.0040 \text{ m}^3} = 2.2 \times 10^5 \text{ Pa} = 2.2 \text{ bar}$$

b. Here, the initial temperatures of N_2 and Ar are different from each other and from the temperature of the final mixture. From above,

$$n_{total} = n_{N_2} + n_{Ar} = \frac{210 \text{ Pa·m}^3}{R(304 \text{ K})} + \frac{680 \text{ Pa·m}^3}{R(402 \text{ K})}$$

Substituting into the ideal gas law (Equation 16.1),

$$P = \left[\frac{210 \text{ Pa·m}^3}{R(304 \text{ K})} + \frac{680 \text{ Pa·m}^3}{R(402 \text{ K})} \right] \left[\frac{R(377 \text{ K})}{0.0040 \text{ m}^3} \right]$$

$$= 2.2 \times 10^5 \text{ Pa} = 2.2 \text{ bar}$$

16–11. It takes 0.3625 g of nitrogen to fill a glass container at 298.2 K and 0.0100 bar pressure. It takes 0.9175 g of an unknown homonuclear diatomic gas to fill the same bulb under the same conditions. What is this gas?

The number of moles of each gas must be the same, because P, V, and T are held constant. The number of moles of nitrogen is

$$n_{N_2} = \frac{0.3625 \text{ g}}{28.0135 \text{ g·mol}^{-1}} = 1.294 \times 10^{-2} \text{ mol}$$

The molar mass of the unknown compound must be

$$M = \frac{0.9175 \text{ g}}{1.294 \times 10^{-2} \text{ mol}} = 70.903 \text{ g·mol}^{-1}$$

The homonuclear diatomic gas must be chlorine (Cl_2).

16–12. Calculate the value of the molar gas constant in units of $dm^3 \cdot torr \cdot K^{-1} \cdot mol^{-1}$.

$$R = 8.31451 \text{ J·mol}^{-1} \cdot \text{K}^{-1}$$

$$= (8.31451 \text{ Pa·m}^3 \cdot \text{mol}^{-1} \cdot \text{K}^{-1}) \left(\frac{10 \text{ dm}}{1 \text{ m}} \right)^3 \left(\frac{760 \text{ torr}}{1.01325 \times 10^5 \text{ Pa}} \right)$$

$$= 62.3639 \text{ dm}^3 \cdot \text{torr} \cdot \text{K}^{-1} \cdot \text{mol}^{-1}$$

16–13. Use the van der Waals equation to plot the compressibility factor, Z, against P for methane for $T = 180$ K, 189 K, 190 K, 200 K, and 250 K. *Hint*: Calculate Z as a function of $\overline{V}$ and P as a function of $\overline{V}$, and then plot Z versus P.

For methane, $a = 2.3026$ dm$^6\cdot$bar$\cdot$mol^{-2} and $b = 0.043067$ dm$^3\cdot$mol^{-1}. By definition,

$$Z = \frac{P\overline{V}}{RT}$$

and the van der Waals equation of state is (Equation 16.5)

$$P = \frac{RT}{\overline{V} - b} - \frac{a}{\overline{V}^2}$$

We can create a parametric plot of Z versus P for the suggested temperatures, shown below. Note that the effect of molecular attraction becomes less important at higher temperatures, as observed in the legend of Figure 16.4.

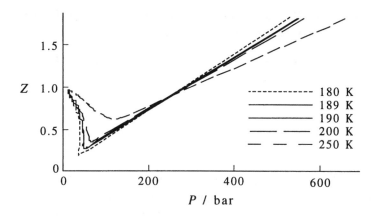

16–14. Use the Redlich-Kwong equation to plot the compressibility factor, Z, against P for methane for $T = 180$ K, 189 K, 190 K, 200 K, and 250 K. *Hint*: Calculate Z as a function of $\overline{V}$ and P as a function of $\overline{V}$, and then plot Z versus P.

For methane, $A = 32.205$ dm$^6\cdot$bar$\cdot$mol$^{-2}\cdot$K$^{1/2}$ and $B = 0.029850$ dm$^3\cdot$mol^{-1}. By definition,

$$Z = \frac{P\overline{V}}{RT}$$

and the Redlich-Kwong equation of state is (Equation 16.7)

$$P = \frac{RT}{\overline{V} - B} - \frac{A}{T^{1/2}\overline{V}(\overline{V} + B)}$$

We can create a parametric plot of Z versus P for the suggested temperatures, shown below. Note that the effect of molecular attraction becomes less important at higher temperatures, as observed in the legend of Figure 16.4.

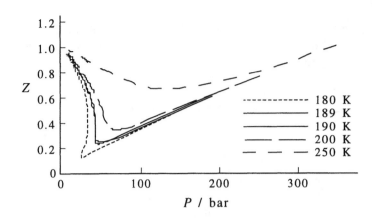

16–15. Use both the van der Waals and the Redlich-Kwong equations to calculate the molar volume of CO at 200 K and 1000 bar. Compare your result to the result you would get using the ideal-gas equation. The experimental value is $0.04009 \text{ L·mol}^{-1}$.

We can use the Newton-Raphson method (MathChapter G) to solve these cubic equations of state. We can express $f(\overline{V})$ for the van der Waals equation as (Example 16–2)

$$f(\overline{V}) = \overline{V}^3 - \left(b + \frac{RT}{P}\right)\overline{V}^2 + \frac{a}{P}\overline{V} - \frac{ab}{P}$$

and $f'(\overline{V})$ as

$$f'(\overline{V}) = 3\overline{V}^2 - 2\left(b + \frac{RT}{P}\right)\overline{V} + \frac{a}{P}$$

For CO, $a = 1.4734 \text{ dm}^6\text{·bar·mol}^{-2}$ and $b = 0.039523 \text{ dm}^3\text{·mol}^{-1}$ (Table 16.3). Then, using the Newton-Raphson method, we find that the van der Waals equation gives a result of $\overline{V} = 0.04998 \text{ dm}^3\text{·mol}^{-1}$. Likewise, we can express $f(\overline{V})$ for the Redlich-Kwong equation as (Equation 16.9)

$$f(\overline{V}) = \overline{V}^3 - \frac{RT}{P}\overline{V}^2 - \left(B^2 + \frac{BRT}{P} - \frac{A}{T^{1/2}P}\right)\overline{V} - \frac{AB}{T^{1/2}P}$$

and $f'(\overline{V})$ as

$$f'(\overline{V}) = 3\overline{V}^2 - \frac{2RT}{P}\overline{V} - \left(B^2 + \frac{BRT}{P} - \frac{A}{T^{1/2}P}\right)$$

For CO, $A = 17.208 \text{ dm}^6\text{·bar·mol}^{-2}\text{·K}^{1/2}$ and $B = 0.027394 \text{ dm}^3\text{·mol}^{-1}$ (Table 16.4). Applying the Newton-Raphson method, we find that the Redlich-Kwong equation gives a result of $\overline{V} = 0.03866 \text{ dm}^3\text{·mol}^{-1}$. Finally, the ideal gas equation gives (Equation 16.1)

$$\overline{V} = \frac{RT}{P} = \frac{(0.083145 \text{ dm}^3\text{·bar·mol}^{-1}\text{·K}^{-1})(200 \text{ K})}{1000 \text{ bar}} = 0.01663 \text{ dm}^3\text{·mol}^{-1}$$

The experimental value of $0.04009 \text{ dm}^3\text{·mol}^{-1}$ is closest to the result given by the Redlich-Kwong equation (the two values differ by about 3%).

16–16. Compare the pressures given by (a) the ideal-gas equation, (b) the van der Waals equation, (c) the Redlich-Kwong equation, and (d) the Peng-Robinson equation for propane at 400 K and $\rho = 10.62 \text{ mol·dm}^{-3}$. The experimental value is 400 bar. Take $\alpha = 9.6938 \text{ L}^2\text{·mol}^{-2}$ and $\beta = 0.05632 \text{ L·mol}^{-1}$ for the Peng-Robinson equation.

The molar volume corresponding to a density of $10.62 \, \text{mol} \cdot \text{dm}^{-3}$ is $0.09416 \, \text{dm}^3 \cdot \text{mol}^{-1}$.

a. The ideal gas equation gives a pressure of (Equation 16.1)

$$P = \frac{RT}{\overline{V}} = \frac{(0.083145 \, \text{dm}^3 \cdot \text{bar} \cdot \text{mol}^{-1} \cdot \text{K}^{-1})(400 \, \text{K})}{0.09416 \, \text{dm}^3 \cdot \text{mol}^{-1}} = 353.2 \, \text{bar}$$

b. The van der Waals equation gives a pressure of (Equation 16.5)

$$P = \frac{RT}{\overline{V} - b} - \frac{a}{\overline{V}^2}$$

For propane, $a = 9.3919 \, \text{dm}^6 \cdot \text{bar} \cdot \text{mol}^{-2}$ and $b = 0.090494 \, \text{dm}^3 \cdot \text{mol}^{-1}$ (Table 16.3). Then

$$P = \frac{(0.083145 \, \text{dm}^3 \cdot \text{bar} \cdot \text{mol}^{-1} \cdot \text{K}^{-1})(400 \, \text{K})}{0.09416 \, \text{dm}^3 \cdot \text{mol}^{-1} - 0.090494 \, \text{dm}^3 \cdot \text{mol}^{-1}} - \frac{9.3919 \, \text{dm}^6 \cdot \text{bar} \cdot \text{mol}^{-2}}{(0.09416 \, \text{dm}^3 \cdot \text{mol}^{-1})^2}$$
$$= 8008 \, \text{bar}$$

c. The Redlich-Kwong equation gives a pressure of (Equation 16.7)

$$P = \frac{RT}{\overline{V} - B} - \frac{A}{T^{1/2} \overline{V}(\overline{V} + B)}$$

For propane, $A = 183.02 \, \text{dm}^6 \cdot \text{bar} \cdot \text{mol}^{-2} \cdot \text{K}^{1/2}$ and $B = 0.062723 \, \text{dm}^3 \cdot \text{mol}^{-1}$ (Table 16.4). Then

$$P = \frac{(0.083145 \, \text{dm}^3 \cdot \text{bar} \cdot \text{mol}^{-1} \cdot {}^{-1})(400 \, \text{K})}{0.09416 \, \text{dm}^3 \cdot \text{mol}^{-1} - 0.062723 \, \text{dm}^3 \cdot \text{mol}^{-1}}$$
$$- \frac{183.02 \, \text{dm}^6 \cdot \text{bar} \cdot \text{mol}^{-2} \cdot \text{K}^{1/2}}{(400 \, \text{K})^{1/2}(0.09416 \, \text{dm}^3 \cdot \text{mol}^{-1})(0.09416 \, \text{dm}^3 \cdot \text{mol}^{-1} + 0.062723 \, \text{dm}^3 \cdot \text{mol}^{-1})}$$
$$= 438.4 \, \text{bar}$$

d. The Peng-Robinson equation gives a pressure of (Equation 16.8)

$$P = \frac{RT}{\overline{V} - \beta} - \frac{\alpha}{\overline{V}(\overline{V} + \beta) + \beta(\overline{V} - \beta)}$$

For propane, $\alpha = 9.6938 \, \text{dm}^6 \cdot \text{bar} \cdot \text{mol}^{-2}$ and $\beta = 0.05632 \, \text{dm}^3 \cdot \text{mol}^{-1}$. Then

$$P = \frac{(0.083145 \, \text{dm}^3 \cdot \text{bar} \cdot \text{mol}^{-1} \cdot {}^{-1})(400 \, \text{K})}{0.09416 \, \text{dm}^3 \cdot \text{mol}^{-1} - 0.05632 \, \text{dm}^3 \cdot \text{mol}^{-1}}$$
$$- \frac{9.6938 \, \text{dm}^6 \cdot \text{bar} \cdot \text{mol}^{-2}}{(0.09416)(0.09416 + 0.05632) \, \text{dm}^6 \cdot \text{mol}^{-2} + (0.05632)(0.09416 - 0.05632) \, \text{dm}^6 \cdot \text{mol}^{-2}}$$
$$= 284.2 \, \text{bar}$$

The Redlich-Kwong equation of state gives a pressure closest to the experimentally observed pressure (the two values differ by about 10%).

16–17. Use the van der Waals equation and the Redlich-Kwong equation to calculate the value of the pressure of one mole of ethane at $400.0 \, \text{K}$ confined to a volume of $83.26 \, \text{cm}^3$. The experimental value is 400 bar.

Here, the molar volume of ethane is $0.08326 \, \text{dm}^3 \cdot \text{mol}^{-1}$.

a. The van der Waals equation gives a pressure of (Equation 16.5)

$$P = \frac{RT}{\overline{V} - b} - \frac{a}{\overline{V}^2}$$

For ethane, $a = 5.5818 \text{ dm}^6 \cdot \text{bar} \cdot \text{mol}^{-2}$ and $b = 0.065144 \text{ dm}^3 \cdot \text{mol}^{-1}$ (Table 16.3). Then

$$P = \frac{(0.083145 \text{ dm}^3 \cdot \text{bar} \cdot \text{mol}^{-1} \cdot \text{K}^{-1})(400 \text{ K})}{0.08326 \text{ dm}^3 \cdot \text{mol}^{-1} - 0.065144 \text{ dm}^3 \cdot \text{mol}^{-1}} - \frac{5.5818 \text{ dm}^6 \cdot \text{bar} \cdot \text{mol}^{-2}}{(0.08326 \text{ dm}^3 \cdot \text{mol}^{-1})^2}$$
$$= 1031 \text{ bar}$$

b. The Redlich-Kwong equation gives a pressure of (Equation 16.7)

$$P = \frac{RT}{\overline{V} - B} - \frac{A}{T^{1/2}\overline{V}(\overline{V} + B)}$$

For ethane, $A = 98.831 \text{ dm}^6 \cdot \text{bar} \cdot \text{mol}^{-2} \cdot \text{K}^{1/2}$ and $B = 0.045153 \text{ dm}^3 \cdot \text{mol}^{-1}$ (Table 16.4). Then

$$P = \frac{(0.083145 \text{ dm}^3 \cdot \text{bar} \cdot \text{mol}^{-1} \cdot \text{K}^{-1})(400 \text{ K})}{0.08326 \text{ dm}^3 \cdot \text{mol}^{-1} - 0.045153 \text{ dm}^3 \cdot \text{mol}^{-1}}$$
$$- \frac{98.831 \text{ dm}^6 \cdot \text{bar} \cdot \text{mol}^{-2} \cdot \text{K}^{1/2}}{(400 \text{ K})^{1/2}(0.08326 \text{ dm}^3 \cdot \text{mol}^{-1})(0.08326 + 0.045153) \text{ dm}^3 \cdot \text{mol}^{-1}}$$
$$= 410.6 \text{ bar}$$

The value of P found using the Redlich-Kwong equation of state is the closest to the experimentally observed value (the two values differ by about 3%).

16–18. Use the van der Waals equation and the Redlich-Kwong equation to calculate the molar density of one mole of methane at 500 K and 500 bar. The experimental value is 10.06 mol·L^{-1}.

We can use the Newton-Raphson method (MathChapter G) to solve the cubic equations of state for $\overline{V}$, and take the reciprocal to find the molar density. We use the experimentally observed molar volume of 0.09940 dm$^3 \cdot$mol^{-1} as the starting point for the iteration. We can express $f(\overline{V})$ for the van der Waals equation as (Example 16–2)

$$f(\overline{V}) = \overline{V}^3 - \left(b + \frac{RT}{P}\right)\overline{V}^2 + \frac{a}{P}\overline{V} - \frac{ab}{P}$$

and $f'(\overline{V})$ as

$$f'(\overline{V}) = 3\overline{V}^2 - 2\left(b + \frac{RT}{P}\right)\overline{V} + \frac{a}{P}$$

For methane, $a = 2.3026 \text{ dm}^6 \cdot \text{bar} \cdot \text{mol}^{-2}$ and $b = 0.043067 \text{ dm}^3 \cdot \text{mol}^{-1}$ (Table 16.3). Then (using the Newton-Raphson method) we find that the van der Waals equation gives a result of $\overline{V} = 0.09993 \text{ dm}^3 \cdot \text{mol}^{-1}$, which corresponds to a molar density of 10.01 mol·dm^{-3}. Likewise, we can express $f(\overline{V})$ for the Redlich-Kwong equation as (Equation 16.9)

$$f(\overline{V}) = \overline{V}^3 - \frac{RT}{P}\overline{V}^2 - \left(B^2 + \frac{BRT}{P} - \frac{A}{T^{1/2}P}\right)\overline{V} - \frac{AB}{T^{1/2}P}$$

and $f'(\overline{V})$ as

$$f'(\overline{V}) = 3\overline{V}^2 - \frac{2RT}{P}\overline{V} - \left(B^2 + \frac{BRT}{P} - \frac{A}{T^{1/2}P}\right)$$

For methane, $A = 32.205$ dm$^6 \cdot$bar$\cdot$mol$^{-2} \cdot$K$^{1/2}$ and $B = 0.029850$ dm$^3 \cdot$mol^{-1} (Table 16.4). Then (using the Newton-Raphson method) we find that the Redlich-Kwong equation gives a result of $\overline{V} = 0.09729$ dm$^3 \cdot$mol^{-1}, which corresponds to a molar density of 10.28 mol$\cdot$dm^{-3}. The molar density of methane found using the van der Waals equation of state is within 0.5% of the experimentally observed value.

16–19. Use the Redlich-Kwong equation to calculate the pressure of methane at 200 K and a density of 27.41 mol$\cdot$L^{-1}. The experimental value is 1600 bar. What does the van der Waals equation give?

The molar volume of the methane is 0.03648 dm$^3 \cdot$mol^{-1}.

a. The van der Waals equation gives a pressure of (Equation 16.5)

$$P = \frac{RT}{\overline{V} - b} - \frac{a}{\overline{V}^2}$$

For methane, $a = 2.3026$ dm$^6 \cdot$bar$\cdot$mol^{-2} and $b = 0.043067$ dm$^3 \cdot$mol^{-1} (Table 16.3). Then

$$P = \frac{(0.083145 \text{ dm}^3 \cdot \text{bar} \cdot \text{mol}^{-1} \cdot \text{K}^{-1})(200 \text{ K})}{0.03648 \text{ dm}^3 \cdot \text{mol}^{-1} - 0.043067 \text{ dm}^3 \cdot \text{mol}^{-1}} - \frac{2.3026 \text{ dm}^6 \cdot \text{bar} \cdot \text{mol}^{-2}}{(0.03648 \text{ dm}^3 \cdot \text{mol}^{-1})^2}$$
$$= -4256 \text{ bar}$$

b. The Redlich-Kwong equation gives a pressure of (Equation 16.7)

$$P = \frac{RT}{\overline{V} - B} - \frac{A}{T^{1/2}\overline{V}(\overline{V} + B)}$$

For ethane, $A = 32.205$ dm$^6 \cdot$bar$\cdot$mol$^{-2} \cdot$K$^{1/2}$ and $B = 0.029850$ dm$^3 \cdot$mol^{-1} (Table 16.4). Then

$$P = \frac{(0.083145 \text{ dm}^3 \cdot \text{bar} \cdot \text{mol}^{-1} \cdot ^{-1})(200 \text{ K})}{0.03648 \text{ dm}^3 \cdot \text{mol}^{-1} - 0.029850 \text{ dm}^3 \cdot \text{mol}^{-1}}$$
$$- \frac{32.205 \text{ dm}^6 \cdot \text{bar} \cdot \text{mol}^{-2} \cdot \text{K}^{1/2}}{(200 \text{ K})^{1/2}(0.03648 \text{ dm}^3 \cdot \text{mol}^{-1})(0.03648 \text{ dm}^3 \cdot \text{mol}^{-1} + 0.029850 \text{ dm}^3 \cdot \text{mol}^{-1})}$$
$$= 1566 \text{ bar}$$

The value of P found using the Redlich-Kwong equation of state is within 2% of the experimentally observed value. The value of P found using the van der Waals equation is obviously incorrect (as it is negative). This is a good example of the problems associated with the van der Waals equation.

16–20. The pressure of propane versus density at 400 K can be fit by the expression

$$P/\text{bar} = 33.258(\rho/\text{mol}\cdot\text{L}^{-1}) - 7.5884(\rho/\text{mol}\cdot\text{L}^{-1})^2$$
$$+ 1.0306(\rho/\text{mol}\cdot\text{L}^{-1})^3 - 0.058757(\rho/\text{mol}\cdot\text{L}^{-1})^4$$
$$- 0.0033566(\rho/\text{mol}\cdot\text{L}^{-1})^5 + 0.00060696(\rho/\text{mol}\cdot\text{L}^{-1})^6$$

for $0 \le \rho/\text{mol}\cdot\text{L}^{-1} \le 12.3$. Use the van der Waals equation and the Redlich-Kwong equation to calculate the pressure for $\rho = 0$ mol$\cdot$L^{-1} up to 12.3 mol$\cdot$L^{-1}. Plot your results. How do they compare to the above expression?

The van der Waals constants for propane are (Table 16.3) $a = 9.3919 \text{ dm}^6 \cdot \text{bar} \cdot \text{mol}^{-2}$ and $b = 0.090494 \text{ dm}^3 \cdot \text{mol}^{-1}$. From Equation 16.5, we can write the pressure calculated using the van der Waals equation of state as

$$P = \frac{RT}{\rho^{-1} - b} - \frac{a}{\rho^{-2}}$$

$$= \frac{(0.083145 \text{ dm}^3 \cdot \text{bar} \cdot \text{mol}^{-1} \cdot \text{K}^{-1})(400 \text{ K})}{\rho^{-1} - 0.090494 \text{ dm}^3 \cdot \text{mol}^{-1}} - \frac{9.3919 \text{ dm}^6 \cdot \text{bar} \cdot \text{mol}^{-2}}{\rho^{-2}}$$

Likewise, the Redlich-Kwong constants for propane are (Table 16.4) $A = 183.02 \text{ dm}^6 \cdot \text{bar} \cdot \text{mol}^{-2} \cdot \text{K}^{1/2}$ and $B = 0.062723 \text{ dm}^3 \cdot \text{mol}^{-1}$. From Equation 16.7, we can write the pressure calculated using the Redlich-Kwong equation of state as

$$P = \frac{RT}{\rho^{-1} - B} - \frac{A}{T^{1/2}\rho^{-1}(\rho^{-1} + B)}$$

$$= \frac{(0.083145 \text{ dm}^3 \cdot \text{bar} \cdot \text{mol}^{-1} \cdot \text{K}^{-1})(400 \text{ K})}{\rho^{-1} - 0.062723 \text{ dm}^3 \cdot \text{mol}^{-1}}$$

$$- \frac{183.02 \text{ dm}^6 \cdot \text{bar} \cdot \text{mol}^{-2} \cdot \text{K}^{1/2}}{(400 \text{ K})^{1/2}\rho^{-1}(\rho^{-1} + 0.062723 \text{ dm}^3 \cdot \text{mol}^{-1})}$$

We plot these equations expressing pressure as a function of ρ as shown below.

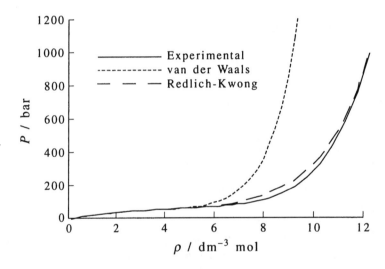

The Redlich-Kwong equation of state describes the data very well, while the van der Waals equation gives a markedly poorer approximation of the observed behavior, especially at high densities.

16–21. The Peng-Robinson equation is often superior to the Redlich-Kwong equation for temperatures near the critical temperature. Use these two equations to calculate the pressure of $CO_2(g)$ at a density of 22.0 mol·L^{-1} at 280 K [the critical temperature of $CO_2(g)$ is 304.2 K]. Use $\alpha = 4.192 \text{ bar} \cdot \text{L}^2 \cdot \text{mol}^{-2}$ and $\beta = 0.02665 \text{ L} \cdot \text{mol}^{-1}$ for the Peng-Robinson equation.

The molar volume of CO_2 is 0.04545 dm^3·mol^{-1}.

a. The Redlich-Kwong equation gives a pressure of (Equation 16.7)

$$P = \frac{RT}{\overline{V} - B} - \frac{A}{T^{1/2}\overline{V}(\overline{V} + B)}$$

For CO_2, $A = 64.597$ $dm^6 \cdot bar \cdot mol^{-2} \cdot K^{1/2}$ and $B = 0.029677$ $dm^3 \cdot mol^{-1}$ (Table 16.4). Then

$$P = \frac{(0.083145 \ dm^3 \cdot bar \cdot mol^{-1} \cdot ^{-1})(280 \ K)}{0.04545 \ dm^3 \cdot mol^{-1} - 0.029677 \ dm^3 \cdot mol^{-1}}$$
$$- \frac{64.597 \ dm^6 \cdot bar \cdot mol^{-2} \cdot K^{1/2}}{(280 \ K)^{1/2}(0.04545 \ dm^3 \cdot mol^{-1})(0.04545 + 0.029677) \ dm^3 \cdot mol^{-1}}$$
$$= 345 \ bar$$

b. The Peng-Robinson equation gives a pressure of (Equation 16.8)

$$P = \frac{RT}{\overline{V} - \beta} - \frac{\alpha}{\overline{V}(\overline{V} + \beta) + \beta(\overline{V} - \beta)}$$

For CO_2, $\alpha = 4.192$ $dm^6 \cdot bar \cdot mol^{-2}$ and $\beta = 0.02665$ $dm^3 \cdot mol^{-1}$. Then

$$P = \frac{(0.083145 \ dm^3 \cdot bar \cdot mol^{-1} \cdot ^{-1})(280 \ K)}{0.04545 \ dm^3 \cdot mol^{-1} - 0.02665 \ dm^3 \cdot mol^{-1}}$$
$$- \frac{4.192 \ dm^6 \cdot bar \cdot mol^{-2}}{[(0.04545)(0.04545 + 0.02665) + (0.02665)(0.04545 - 0.02665)] \ dm^6 \cdot mol^{-2}}$$
$$= 129 \ bar$$

The Peng-Robinson result is much closer to the experimental value than the value predicted by the Redlich-Kwong equation.

16–22. Show that the van der Waals equation for argon at $T = 142.69$ K and $P = 35.00$ atm can be written as

$$\overline{V}^3 - 0.3664 \ \overline{V}^2 + 0.03802 \ \overline{V} - 0.001210 = 0$$

where, for convenience, we have supressed the units in the coefficients. Use the Newton-Raphson method (MathChapter G) to find the three roots to this equation, and calculate the values of the density of liquid and vapor in equilibrium with each other under these conditions.

For argon, $a = 1.3307$ $dm^6 \cdot atm \cdot mol^{-2}$ and $b = 0.031830$ $dm^3 \cdot mol^{-1}$ (Table 16.3). The van der Waals equation of state can be written as (Example 16–2)

$$\overline{V}^3 - \left(b + \frac{RT}{P}\right)\overline{V}^2 + \frac{a}{P}\overline{V} - \frac{ab}{P} = 0$$

$$\overline{V}^3 - \left[0.03183 + \frac{(0.082058)(142.69)}{35.00}\right]\overline{V}^2 + \frac{1.3307}{35.00}\overline{V} - \frac{(1.3307)(0.031830)}{35.00} = 0$$

$$\overline{V}^3 - 0.3664\overline{V}^2 + 0.03802\overline{V} - 0.001210 = 0$$

where we have suppressed the units of the coefficients for convenience. (The quantity $\overline{V}$ is expressed in $dm^3 \cdot mol^{-1}$.) We apply the Newton-Raphson method, using the function

$$f(\overline{V}) = \overline{V}^3 - 0.3664\overline{V}^2 + 0.03802\overline{V} - 0.001210$$

and its derivative

$$f'(\overline{V}) = 3\overline{V}^2 - 0.7328\overline{V} + 0.03802$$

to find the three roots of this equation, 0.07893 $dm^3 \cdot mol^{-1}$, 0.07073 $dm^3 \cdot mol^{-1}$, and 0.21674 $dm^3 \cdot mol^{-1}$. The smallest root represents the molar volume of liquid argon, and the largest root represents the molar volume of the vapor. The corresponding densities are 14.14 mol·dm^{-3} and 4.614 mol·dm^{-3}, respectively.

16–23. Use the Redlich-Kwong equation and the Peng-Robinson equation to calculate the densities of the coexisting argon liquid and vapor phases at 142.69 K and 35.00 atm. Use the Redlich-Kwong constants given in Table 16.4 and take $\alpha = 1.4915$ atm·L^2·mol^{-2} and $\beta = 0.01981$ L·mol^{-1} for the Peng-Robinson equation.

---

a. For argon, $A = 16.566$ dm^6·atm·mol^{-2}·K$^{1/2}$ and $B = 0.022062$ dm^3·mol^{-1} (Table 16.4). The Redlich-Kwong equation of state can be written as (Equation 16.9)

$$0 = \overline{V}^3 - \frac{RT}{P}\overline{V}^2 - \left(B^2 + \frac{BRT}{P} - \frac{A}{T^{1/2}P}\right)\overline{V} - \frac{AB}{T^{1/2}P}$$

$$0 = \overline{V}^3 - \frac{(0.082058)(142.69)}{35.00}\overline{V}^2 - \left[(0.022062)^2 + \frac{(0.022062)(0.082058)(142.69)}{35.00} - \right.$$
$$\left. \frac{16.566}{(142.69)^{1/2}(35.00)}\right]\overline{V} - \frac{(16.566)(0.022062)}{(142.69)^{1/2}(35.00)}$$

$$0 = \overline{V}^3 - 0.3345\overline{V}^2 + 0.03176\overline{V} - 0.0008742$$

where we have suppressed the units of the coefficients for convenience. (The quantity $\overline{V}$ is expressed in dm^3·mol^{-1}.) We apply the Newton-Raphson method, using the function

$$f(\overline{V}) = \overline{V}^3 - 0.3345\overline{V}^2 + 0.03176\overline{V} - 0.0008742$$

and its derivative

$$f'(\overline{V}) = 3\overline{V}^2 - 0.6690\overline{V} + 0.03176$$

to find the three roots of this equation to be 0.04961 dm^3·mol^{-1}, 0.09074 dm^3·mol^{-1}, and 0.19419 dm^3·mol^{-1}. The smallest root represents the molar volume of liquid argon, and the largest root represents the molar volume of the vapor. The corresponding densities are 20.16 mol·dm^{-3} and 5.150 mol·dm^{-3}, respectively.

b. The Peng-Robinson equation is given as (Equation 16.8)

$$P = \frac{RT}{\overline{V} - \beta} - \frac{\alpha}{\overline{V}(\overline{V} + \beta) + \beta(\overline{V} - \beta)}$$

This can be expressed as the cubic equation in $\overline{V}$

$$0 = \overline{V}^3 + \left(\beta - \frac{RT}{P}\right)\overline{V}^2 + \left(\frac{\alpha - 3\beta^2 P - 2\beta RT}{P}\right)\overline{V} + \frac{\beta^3 P + \beta^2 RT - \alpha\beta}{P}$$

Substituting the values given in the text of the problem, we find that the Peng-Robinson equation for argon at 142.69 K and 35.00 atm becomes

$$0 = \overline{V}^3 + \left[(0.01981) - \frac{(0.082058)(142.69)}{35.00}\right]\overline{V}^2$$
$$+ \left[\frac{(1.4915) - 3(0.01981)^2(35.00) - 2(0.01981)(0.082058)(142.69)}{35.00}\right]\overline{V}$$
$$+ \frac{(0.01981)^3(35.00) + (0.01981)^2(0.082058)(142.69) - (1.4915)(0.01981)}{35.00}$$

$$= \overline{V}^3 - 0.3147\overline{V}^2 + 0.02818\overline{V} - 0.0007051$$

where we have suppressed the units of the coefficients for convenience. (The quantity $\overline{V}$ is expressed in $dm^3 \cdot mol^{-1}$.) We apply the Newton-Raphson method, using the function

$$f(\overline{V}) = \overline{V}^3 - 0.3147\overline{V}^2 + 0.02818\overline{V} - 0.0007051$$

and its derivative

$$f'(\overline{V}) = 3\overline{V}^2 - 0.6294\overline{V} + 0.02818$$

to find the three roots of this equation to be 0.04237 $dm^3 \cdot mol^{-1}$, 0.09257 $dm^3 \cdot mol^{-1}$, and 0.17979 $dm^3 \cdot mol^{-1}$. The smallest root represents the molar volume of liquid argon, and the largest root represents the molar volume of the vapor. The corresponding densities are 23.60 $mol \cdot dm^{-3}$ and 5.562 $mol \cdot dm^{-3}$, respectively.

16–24. Butane liquid and vapor coexist at 370.0 K and 14.35 bar. The densities of the liquid and vapor phases are 8.128 $mol \cdot L^{-1}$ and 0.6313 $mol \cdot L^{-1}$, respectively. Use the van der Waals equation, the Redlich-Kwong equation, and the Peng-Robinson equation to calculate these densities. Take $\alpha = 16.44$ $bar \cdot L^2 \cdot mol^{-2}$ and $\beta = 0.07245$ $L \cdot mol^{-1}$ for the Peng-Robinson equation.

---

a. For butane, $a = 13.888$ $dm^6 \cdot bar \cdot mol^{-2}$ and $b = 0.11641$ $dm^3 \cdot mol^{-1}$ (Table 16.3). The van der Waals equation of state can be written as (Example 16–2)

$$\overline{V}^3 - \left(b + \frac{RT}{P}\right)\overline{V}^2 + \frac{a}{P}\overline{V} - \frac{ab}{P} = 0$$

$$\overline{V}^3 - \left[0.11641 + \frac{(0.083145)(370.0)}{14.35}\right]\overline{V}^2 + \frac{13.888}{14.35}\overline{V} - \frac{(13.888)(0.11641)}{14.35} = 0$$

$$\overline{V}^3 - 2.2602\overline{V}^2 + 0.9678\overline{V} - 0.1127 = 0$$

where we have suppressed the units of the coefficients for convenience. (The quantity $\overline{V}$ is expressed in $dm^3 \cdot mol^{-1}$.) We apply the Newton-Raphson method, using the function

$$f(\overline{V}) = \overline{V}^3 - 2.2602\overline{V}^2 + 0.9678\overline{V} - 0.1127$$

and its derivative

$$f'(\overline{V}) = 3\overline{V}^2 - 4.5204\overline{V} + 0.9678$$

to find the three roots of this equation to be 0.20894 $dm^3 \cdot mol^{-1}$, 0.30959 $dm^3 \cdot mol^{-1}$, and 1.7417 $dm^3 \cdot mol^{-1}$. The smallest root represents the molar volume of liquid butane, and the largest root represents the molar volume of the vapor. The corresponding densities are 4.786 $mol \cdot dm^{-3}$ and 0.5741 $mol \cdot dm^{-3}$, respectively.

b. For butane, $A = 290.16$ $dm^6 \cdot bar \cdot mol^{-2} \cdot K^{1/2}$ and $B = 0.08068$ $dm^3 \cdot mol^{-1}$ (Table 16.4). The Redlich-Kwong equation of state can be written as (Equation 16.9)

$$0 = \overline{V}^3 - \frac{RT}{P}\overline{V}^2 - \left(B^2 + \frac{BRT}{P} - \frac{A}{T^{1/2}P}\right)\overline{V} - \frac{AB}{T^{1/2}P}$$

$$0 = \overline{V}^3 - \frac{(0.083145)(370.0)}{14.35}\overline{V}^2 - \left[(0.08068)^2 + \frac{(0.08068)(0.083145)(370.0)}{14.35} - \right.$$
$$\left. \frac{290.16}{(370.0)^{1/2}(14.35)}\right]\overline{V} - \frac{(290.16)(0.08068)}{(370.0)^{1/2}(14.35)}$$

$$0 = \overline{V}^3 - 2.144\overline{V}^2 + 0.8717\overline{V} - 0.08481$$

where we have suppressed the units of the coefficients for convenience. (The quantity $\overline{V}$ is expressed in $dm^3 \cdot mol^{-1}$.) We apply the Newton-Raphson method, using the function

$$f(\overline{V}) = \overline{V}^3 - 2.144\overline{V}^2 + 0.8717\overline{V} - 0.08481$$

and its derivative

$$f'(\overline{V}) = 3\overline{V}^2 - 4.288\overline{V} + 0.8717$$

to to find the three roots of this equation to be $0.14640\ dm^3 \cdot mol^{-1}$, $0.35209\ dm^3 \cdot mol^{-1}$, and $1.6453\ dm^3 \cdot mol^{-1}$. The smallest root represents the molar volume of liquid butane, and the largest root represents the molar volume of the vapor. The corresponding densities are $6.831\ mol \cdot dm^{-3}$ and $0.6078\ mol \cdot dm^{-3}$, respectively.

c. The Peng-Robinson equation can be expressed as (Problem 16–23)

$$0 = \overline{V}^3 + \left(\beta - \frac{RT}{P}\right)\overline{V}^2 + \left(\frac{\alpha - 3\beta^2 P - 2\beta RT}{P}\right)\overline{V} + \frac{\beta^3 P + \beta^2 RT - \alpha\beta}{P}$$

Substituting the values given in the text of the problem, we find that the Peng-Robinson equation for butane at 370.0 K and 14.35 bar becomes

$$0 = \overline{V}^3 + \left[(0.07245) - \frac{(0.081345)(370.0)}{14.35}\right]\overline{V}^2$$
$$+ \left[\frac{(16.44) - 3(0.07245)^2(14.35) - 2(0.07245)(0.081345)(370.0)}{14.35}\right]\overline{V}$$
$$+ \frac{(0.07245)^3(14.35) + (0.07245)^2(0.081345)(370.0) - (16.44)(0.07245)}{14.35}$$

$$= \overline{V}^3 - 2.071\overline{V}^2 + 0.8193\overline{V} - 0.07137$$

where we have suppressed the units of the coefficients for convenience. (The quantity $\overline{V}$ is expressed in $dm^3 \cdot mol^{-1}$.) We apply the Newton-Raphson method, using the function

$$f(\overline{V}) = \overline{V}^3 - 2.071\overline{V}^2 + 0.8193\overline{V} - 0.07137$$

and its derivative

$$f'(\overline{V}) = 3\overline{V}^2 - 4.142\overline{V} + 0.8193$$

to find the three roots of this equation to be $0.12322\ dm^3 \cdot mol^{-1}$, $0.36613\ dm^3 \cdot mol^{-1}$, and $1.5820\ dm^3 \cdot mol^{-1}$. The smallest root represents the molar volume of liquid butane, and the largest root represents the molar volume of the vapor. The corresponding densities are $8.116\ mol \cdot dm^{-3}$ and $0.6321\ mol \cdot dm^{-3}$, respectively.

Below is a table which summarizes the densities of liquid and vapor butane observed experimentally and calculated with the various equations of state above.

Equation used	Liquid $\rho/mol \cdot dm^{-3}$	Gas $\rho/mol \cdot dm^{-3}$
Experimental	8.128	0.6313
van der Waals	4.786	0.5741
Redlich-Kwong	6.831	0.6078
Peng-Robinson	8.116	0.6321

16–25. Another way to obtain expressions for the van der Waals constants in terms of critical parameters is to set $(\partial P/\partial \overline{V})_T$ and $(\partial^2 P/\partial \overline{V}^2)_T$ equal to zero at the critical point. Why are these quantities equal to zero at the critical point? Show that this procedure leads to Equations 16.12 and 16.13.

These values are equal to zero at the critical point because the critical point is an inflection point in a plot of P versus V at constant temperature.

$$P = \frac{RT}{\overline{V} - b} - \frac{a}{\overline{V}^2}$$

So

$$\left(\frac{\partial P}{\partial V}\right)_T = \frac{-RT}{\left(\overline{V} - b\right)^2} + \frac{2a}{\overline{V}^3} \tag{1}$$

$$\left(\frac{\partial^2 P}{\partial V^2}\right)_T = \frac{2RT}{\left(\overline{V} - b\right)^3} - \frac{6a}{\overline{V}^4} \tag{2}$$

If $(\partial P/\partial V)_T$ and $(\partial^2 P/\partial V^2)_T$ are zero at the critical point, then Equations 1 and 2 give

$$RT_c \overline{V}_c^{\,3} = 2a\left(\overline{V}_c - b\right)^2 \tag{3}$$

and

$$2RT_c \overline{V}_c^{\,4} = 6a\left(\overline{V}_c - b\right)^3 \tag{4}$$

Multiplying Equation 3 by $2\overline{V}_c$ gives

$$2RT_c \overline{V}_c^{\,4} = 4a\overline{V}_c\left(\overline{V}_c - b\right)^2$$

and then using Equation 4 yields

$$4a\overline{V}_c\left(\overline{V}_c - b\right)^2 = 6a\left(\overline{V}_c - b\right)^3$$
$$4\overline{V}_c = 6\overline{V}_c - 6b$$
$$3b = \overline{V}_c \tag{16.13a}$$

Substituting Equation 16.13a into Equation 3 gives

$$RT_c(3b)^3 = 2a(3b - b)^2$$
$$T_c = \frac{8ab^2}{27b^3 R} = \frac{8a}{27bR} \tag{16.13c}$$

Now substitute Equations 16.13a and 16.13c into the van der Waals equation to find P_c:

$$P_c = \frac{RT_c}{\overline{V}_c - b} - \frac{a}{\overline{V}_c^{\,2}} = \frac{8aR}{27bR\,(3b - b)} - \frac{a}{(3b)^2} = \frac{a}{27b^2} \tag{16.13b}$$

Equation 16.12 follows naturally from these expressions for $\overline{V}_c$, P_c, and T_c.

16–26. Show that the Redlich-Kwong equation can be written in the form

$$\overline{V}^3 - \frac{RT}{P}\overline{V}^2 - \left(B^2 + \frac{BRT}{P} - \frac{A}{PT^{1/2}}\right)\overline{V} - \frac{AB}{PT^{1/2}} = 0$$

Now compare this equation with $(\overline{V} - \overline{V}_c)^3 = 0$ to get

$$3\overline{V}_c = \frac{RT_c}{P_c} \tag{1}$$

$$3\overline{V}_c^2 = \frac{A}{P_c T_c^{1/2}} - \frac{BRT_c}{P_c} - B^2 \tag{2}$$

and

$$\overline{V}_c^3 = \frac{AB}{P_c T_c^{1/2}} \tag{3}$$

Note that Equation 1 gives

$$\frac{P_c \overline{V}_c}{RT_c} = \frac{1}{3} \tag{4}$$

Now solve Equation 3 for A and substitute the result and Equation 4 into Equation 2 to obtain

$$B^3 + 3\overline{V}_c B^2 + 3\overline{V}_c^2 B - \overline{V}_c^3 = 0 \tag{5}$$

Divide this equation by $\overline{V}_c^3$ and let $B/\overline{V}_c = x$ to get

$$x^3 + 3x^2 + 3x - 1 = 0$$

Solve this cubic equation by the Newton-Raphson method (MathChapter G) to obtain $x = 0.25992$, or

$$B = 0.25992 \overline{V}_c \tag{6}$$

Now substitute this result and Equation 4 into Equation 3 to obtain

$$A = 0.42748 \frac{R^2 T_c^{5/2}}{P_c}$$

We start with the Redlich-Kwong equation of state,

$$P = \frac{RT}{\overline{V} - B} - \frac{A}{T^{1/2}\overline{V}\left(\overline{V} + B\right)} \tag{16.7}$$

We can rewrite the above equation as

$$P\left(\overline{V} - B\right)\left(\overline{V} + B\right)T^{1/2}\overline{V} = RT^{3/2}\overline{V}\left(\overline{V} + B\right) - A\left(\overline{V} - B\right)$$
$$PT^{1/2}\overline{V}\left(\overline{V}^2 - B^2\right) = RT^{3/2}\overline{V}^2 + RT^{3/2}\overline{V}B - A\overline{V} + AB$$

We express this equation as a cubic equation in $\overline{V}$:

$$\overline{V}^3 - \frac{RT}{P}\overline{V}^2 - \left(B^2 + \frac{BRT}{P} - \frac{A}{PT^{1/2}}\right)\overline{V} - \frac{AB}{PT^{1/2}} = 0 \tag{a}$$

Expanding the equation $(\overline{V} - \overline{V}_c^3) = 0$ gives

$$\left(\overline{V} - \overline{V}_c\right)^3 = \overline{V}^3 - \overline{V}_c^3 + 3\overline{V}_c^2\overline{V} - 3\overline{V}^2\overline{V}_c = 0 \tag{b}$$

Setting the coefficients of $\overline{V}^3$, $\overline{V}^2$, $\overline{V}$, and $\overline{V}^0$ in Equations a and b equal to one another at the critical point gives

$$3\overline{V}_c = \frac{RT_c}{P_c} \tag{1}$$

$$3\overline{V}_c^2 = -B^2 - \frac{BRT_c}{P_c} + \frac{A}{P_c T_c^{1/2}} \tag{2}$$

$$\overline{V}_c^3 = \frac{AB}{P_c T_c^{1/2}} \tag{3}$$

$$\frac{P_c \overline{V}_c}{RT_c} = \frac{1}{3} \tag{4}$$

We can solve Equation 3 for A to find

$$A = \frac{\overline{V}_c^3 P_c T_c^{1/2}}{B}$$

Substituting this result into Equation 2 gives

$$3\overline{V}_c^2 = -B^2 - \frac{BRT_c \overline{V}_c}{P_c \overline{V}_c} + \frac{\overline{V}_c^3 P_c T_c^{1/2}}{B P_c T_c^{1/2}}$$

$$3\overline{V}_c^2 = -B^2 - 3B\overline{V}_c + B^{-1}\overline{V}_c^3$$

$$0 = \frac{B^3}{\overline{V}_c^3} + \frac{3B^2}{\overline{V}_c^2} + \frac{3B}{\overline{V}_c} - 1$$

$$0 = x^3 + 3x^2 + 3x - 1$$

where we set $x = B/\overline{V}_c$. We solved this cubic equation using the Newton-Raphson method in Example G–1 and found that $x = 0.25992$. Then $B = 0.25992\overline{V}_c$, and substituting into Equation 3 gives

$$\left(\frac{P_c V_c}{RT_c}\right) T_c^{1/2} \overline{V}_c^2 = \frac{AB}{RT_c}$$

$$T_c^{1/2} \overline{V}_c^2 = \frac{3AB}{RT_c}$$

$$A = \frac{T_c^{3/2} \overline{V}_c^2 R}{3B} = \frac{T_c^{3/2} \overline{V} R}{3(0.25992)}$$

$$A = \frac{P_c \overline{V}_c}{RT_c}\left(\frac{RT_c}{P_c}\right) \frac{T_c^{3/2} R}{3(0.25992)}$$

$$A = 0.42748 \frac{R^2 T_c^{5/2}}{P_c} \tag{7}$$

16–27. Use the results of the previous problem to derive Equations 16.14.

Equation 6 of Problem 16.26 gives $B = 0.25992\overline{V}_c$, and so

$$\overline{V}_c = 3.8473B \tag{16.14a}$$

Now we can use Equation 7 of Problem 16–26 to write

$$A = \frac{0.42748 R^2 T_c^{5/2}}{P_c} = 0.42748 RT_c^{3/2} \left(\frac{RT_c}{P_c \overline{V}_c}\right) \overline{V}_c$$

Substituting Equation 4 from Problem 16–26 and Equation 16.14a into the above expression gives

$$A = 3(0.42748) RT_c^{3/2}(3.8473 B)$$

$$T_c^{3/2} = \frac{A}{3(0.42748)(3.8473) RB}$$

$$T_c = 0.34504 \left(\frac{A}{BR}\right)^{2/3}$$

which is Equation 16.14c. Substitute this last result into the final equation of Problem 16–26 to find

$$P_c = \frac{0.42748 R^2 T_c^{5/2}}{A}$$

$$= \left(\frac{0.42748}{A}\right) R^2 \left[0.34504 \left(\frac{A}{BR}\right)^{2/3}\right]^{5/2}$$

$$= 0.029894 \frac{A^{2/3} R^{1/3}}{B^{5/3}}$$

which is Equation 16.14b.

16–28. Write the Peng-Robinson equation as a cubic polynomial equation in $\overline{V}$ (with the coefficient of $\overline{V}^3$ equal to one), and compare it with $(\overline{V} - \overline{V}_c)^3 = 0$ at the critical point to obtain

$$\frac{RT_c}{P_c} - \beta = 3\overline{V}_c \tag{1}$$

$$\frac{\alpha_c}{P_c} - 3\beta^2 - 2\beta \frac{RT_c}{P_c} = 3\overline{V}_c^2 \tag{2}$$

and

$$\frac{\alpha_c \beta}{P_c} - \beta^2 \frac{RT_c}{P_c} - \beta^3 = \overline{V}_c^3 \tag{3}$$

(We write α_c because α depends upon the temperature.) Now eliminate α_c/P_c between Equations 2 and 3, and then use Equation 1 for $\overline{V}_c$ to obtain

$$64\beta^3 + 6\beta^2 \frac{RT_c}{P_c} + 12\beta \left(\frac{RT_c}{P_c}\right)^2 - \left(\frac{RT_c}{P_c}\right)^3 = 0$$

Let $\beta/(RT_c/P_c) = x$ and get

$$64x^3 + 6x^2 + 12x - 1 = 0$$

Solve this equation using the Newton-Raphson method to obtain

$$\beta = 0.077796 \frac{RT_c}{P_c}$$

Substitute this result and Equation 1 into Equation 2 to obtain

$$\alpha_c = 0.45724 \frac{(RT_c)^2}{P_c}$$

Last, use Equation 1 to show that

$$\frac{P_c \overline{V}_c}{RT_c} = 0.30740$$

_____

First, we write the Peng-Robinson equation as a cubic polynomial in $\overline{V}$, as we did in Problem 16–23.

$$0 = \overline{V}^3 + \left(\beta - \frac{RT}{P} \right) \overline{V}^2 + \left(\frac{\alpha - 3\beta^2 P - 2\beta RT}{P} \right) \overline{V} + \frac{\beta^3 P + \beta^2 RT - \alpha \beta}{P} \tag{a}$$

Expanding the equation $(\overline{V} - \overline{V}_c^3) = 0$ gives

$$(\overline{V} - \overline{V}_c)^3 = \overline{V}^3 - \overline{V}_c^3 + 3\overline{V}_c^2 \overline{V} - 3\overline{V}^2 \overline{V}_c = 0 \tag{b}$$

Setting the coefficients of $\overline{V}^3$, $\overline{V}^2$, $\overline{V}$, and $\overline{V}^0$ in Equations a and b equal to one another at the critical point gives

$$\beta - \frac{RT_c}{P_c} = -3\overline{V}_c \tag{1}$$

$$\frac{\alpha_c}{P_c} - 3\beta^2 - 2\beta \frac{RT_c}{P_c} = 3\overline{V}_c^2 \tag{2}$$

$$\frac{\alpha_c \beta}{P_c} - \beta^3 - \frac{RT_c}{P_c} \beta^2 = \overline{V}_c^3 \tag{3}$$

Solving Equation 2 for α_c / P_c gives

$$\frac{\alpha_c}{P_c} = 3\overline{V}_c^2 + 3\beta^2 + 2\beta \frac{RT_c}{P_c}$$

We substitute this last result into Equation 3 to find

$$0 = \overline{V}_c^3 - \beta \left(3\overline{V}_c^2 + 3\beta^2 + 2\beta \frac{RT_c}{P_c} \right) + \beta^2 \frac{RT_c}{P_c} + \beta^3$$

$$= \overline{V}_c^3 - 2\beta^3 - \beta^2 \frac{RT_c}{P_c} - 3\beta \overline{V}_c^2$$

$$= \frac{1}{27} \left(\frac{RT_c}{P_c} - \beta \right)^3 - 2\beta^3 - \beta^2 \frac{RT_c}{P_c} - \frac{\beta}{3} \left(\frac{RT_c}{P_c} - \beta \right)^2$$

$$= \left(\frac{RT_c}{P_c} \right)^3 - 3\beta \left(\frac{RT_c}{P_c} \right)^2 + 3\beta^2 \frac{RT_c}{P_c} - \beta^3 - 54\beta^3 - 27\beta^2 \frac{RT_c}{P_c} - 9 \left(\frac{RT_c}{P_c} \right)^2 \beta + 18\beta \frac{RT_c}{P_c} - 9\beta^3$$

$$= \left(\frac{RT_c}{P_c} \right)^3 - 12\beta \left(\frac{RT_c}{P_c} \right)^2 - 6\beta^2 \left(\frac{RT_c}{P_c} \right) - 64\beta^3$$

Set $x = \beta / \left(RT_c / P_c \right)$ and the above equation becomes

$$64x^3 + 6x^2 + 12x - 1 = 0$$

Using the Newton-Raphson method (MathChapter G), we find $x = 0.077796$ and so

$$\beta = 0.077796 \frac{RT_c}{P_c}$$

We now substitute Equation 1 into Equation 2 and use the expression for β given above to write

$$
\begin{aligned}
\alpha_c &= P_c \left(3\overline{V}_c^2 + 3\beta^2 + 2\beta \frac{RT_c}{P_c} \right) \\
&= P_c \left[\frac{1}{3} \left(\frac{RT_c}{P_c} - \beta \right)^2 + 3\beta^2 + 2\beta \frac{RT_c}{P_c} \right] \\
&= P_c \left(\frac{RT_c}{P_c} \right)^2 \left[\frac{1}{3}(1 - 0.77796)^2 + 3(0.077796)^2 + 2(0.077796) \right] \\
&= 0.45724 \frac{(RT_c)^2}{P_c}
\end{aligned}
$$

Finally, substitute the expression of β given above into Equation 1 to write

$$(1 - 0.077796) \frac{RT_c}{P_c} = 3\overline{V}_c$$

$$0.92220 RT_c = 3 P_c \overline{V}_c$$

$$0.30740 = \frac{P_c \overline{V}_c}{RT_c}$$

16-29. Look up the boiling points of the gases listed in Table 16.5 and plot these values versus the critical temperatures T_c. Is there any correlation? Propose a reason to justify your conclusions from the plot.

A graph of boiling points versus critical temperatures of the gases listed in Table 16.5 is shown below. There appears to be a direct correlation between the boiling point of a gas and its critical temperature.

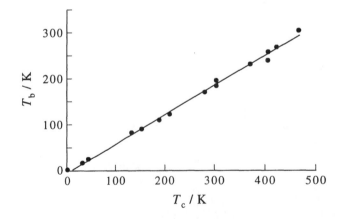

This is another illustration of the law of corresponding states: if we compare the boiling points of different gases relative to their critical temperatures, we find that all behaviors can be similarly explained (hence, the constant slope in the figure).

16–30. Show that the compressibility factor Z for the Redlich-Kwong equation can be written as in Equation 16.21.

The Redlich-Kwong equation of state is given by Equation 16.7. Thus

$$\left(\frac{\overline{V}}{RT}\right)P = \left(\frac{\overline{V}}{RT}\right)\frac{RT}{\overline{V}-B} - \frac{A}{RT^{3/2}\left(\overline{V}+B\right)}$$

or

$$Z = \frac{\overline{V}}{\overline{V}-B} - \frac{A}{RT^{3/2}\left(\overline{V}+B\right)}$$

We know from Equation 16.18 that

$$A = 0.42748\frac{R^2 T_c^{5/2}}{P_c} \qquad \text{and} \qquad B = 0.086640\frac{RT_c}{P_c}$$

We can then write Z as

$$Z = \overline{V}\left(\overline{V} - 0.086640\frac{RT_c}{P_c}\right)^{-1} - \left(0.42748\frac{R^2 T_c^{5/2}}{P_c}\right)\left[RT^{3/2}\left(\overline{V} + 0.086640\frac{RT_c}{P_c}\right)\right]^{-1}$$

In the solution to Problem 16.26, we showed that $3\overline{V}_c P_c = RT_c$, so

$$Z = \frac{\overline{V}}{\overline{V} - 0.086640\left(3\overline{V}_c\right)} - \frac{0.42748 T_c^{3/2}\left(3\overline{V}_c\right)}{T^{3/2}\left[\overline{V} + 0.086640\left(3\overline{V}_c\right)\right]}$$

$$= \frac{\overline{V}_R}{\overline{V}_R - 0.25992} - \frac{1.28244}{T_R^{3/2}\left(\overline{V}_R + 0.25992\right)}$$

16–31. Use the following data for ethane and argon at $T_R = 1.64$ to illustrate the law of corresponding states by plotting Z against $\overline{V}_R$.

Ethane ($T = 500$ K)		Argon ($T = 247$ K)	
P/bar	$\overline{V}$/L·mol^{-1}	P/atm	$\overline{V}$/L·mol^{-1}
0.500	83.076	0.500	40.506
2.00	20.723	2.00	10.106
10.00	4.105	10.00	1.999
20.00	2.028	20.00	0.9857
40.00	0.9907	40.00	0.4795
60.00	0.6461	60.00	0.3114
80.00	0.4750	80.00	0.2279
100.0	0.3734	100.0	0.1785
120.0	0.3068	120.0	0.1462
160.0	0.2265	160.0	0.1076
200.0	0.1819	200.0	0.08630
240.0	0.1548	240.0	0.07348
300.0	0.1303	300.0	0.06208
350.0	0.1175	350.0	0.05626
400.0	0.1085	400.0	0.05219
450.0	0.1019	450.0	0.04919
500.0	0.09676	500.0	0.04687
600.0	0.08937	600.0	0.04348
700.0	0.08421	700.0	0.04108

We can use Table 16.5 to find the critical molar volumes of ethane and argon $(0.1480 \; \text{dm}^3 \cdot \text{mol}^{-1}$ and $0.07530 \; \text{dm}^3 \cdot \text{mol}^{-1}$, respectively) and use the data given in the problem text in the equations

$$Z = \frac{P\overline{V}}{RT} \qquad \text{and} \qquad \overline{V}_R = \frac{\overline{V}}{\overline{V}_c}$$

to plot Z versus $\overline{V}_R$. Note that the pressures for ethane are given in units of bar, while the pressures for argon are given in units of atm.

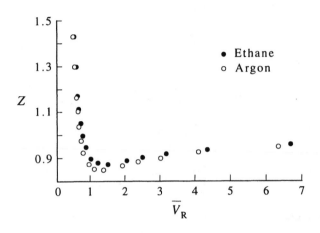

16–32. Use the data in Problem 16–31 to illustrate the law of corresponding states by plotting Z against P_R.

We can use Table 16.5 to find the critical pressures of ethane and argon (48.714 bar and 48.643 atm, respectively) and use the data given in Problem 16–31 in the equations

$$Z = \frac{P\overline{V}}{RT} \qquad \text{and} \qquad P_R = \frac{P}{P_c}$$

to plot Z versus P_R.

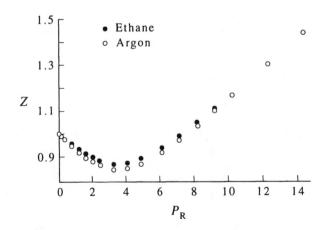

16–33. Use the data in Problem 16.31 to test the quantitative reliability of the van der Waals equation by comparing a plot of Z versus $\overline{V}_R$ from Equation 16.20 to a similar plot of the data.

We can use Equation 16.20 to express Z as a function of $\overline{V}_R$:

$$Z = \frac{\overline{V}_R}{\overline{V}_R - 1/3} - \frac{9}{8\overline{V}_R T_R}$$

We can substitute the appropriate values of $T_R = T/T_c$ (Problem 16–31) and plot Z versus $\overline{V}_R$ for both argon and ethane. The plot below shows the lines generated by applying the van der Waals equation and the actual data from Problem 16–31.

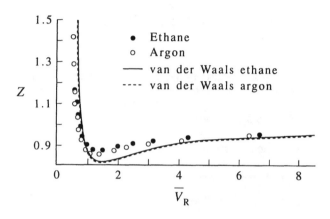

16–34. Use the data in Problem 16.31 to test the quantitative reliability of the Redlich-Kwong equation by comparing a plot of Z versus $\overline{V}_R$ from Equation 16.21 to a similar plot of the data.

We can use Equation 16.21 to express Z as a function of $\overline{V}_R$:

$$Z = \frac{\overline{V}_R}{\overline{V}_R - 0.25992} - \frac{1.2824}{T_R^{3/2}(\overline{V}_R + 0.25992)}$$

We can substitute the appropriate values of $T_R = T/T_c$ (Problem 16–31) and plot Z versus $\overline{V}_R$ for both argon and ethane. The plot below shows the lines generated by applying the Redlich-Kwong equation and the actual data from Problem 16–31.

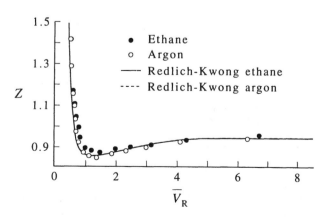

16–35. Use Figure 16.10 to estimate the molar volume of CO at 200 K and 180 bar. An accurate experimental value is 78.3 $cm^3 \cdot mol^{-1}$.

We use the critical values in Table 16.5 to write

$$T_R = \frac{T}{T_c} = \frac{200\ K}{132.85\ K} = 1.51$$

and

$$P_R = \frac{P}{P_c} = \frac{180\ bar}{34.935\ bar} = 5.15$$

From Figure 16.10,

$$Z = \frac{P\overline{V}}{RT} \approx 0.85$$

We can now find $\overline{V}$:

$$\overline{V} \approx \frac{0.85\ (0.083145\ dm^3 \cdot bar \cdot mol^{-1} \cdot K^{-1})\ (200\ K)}{180\ bar}$$

$$\overline{V} \approx 78.5\ cm^3 \cdot mol^{-1}$$

in excellent agreement with the experimental value.

16–36. Show that $B_{2V}(T) = RT\,B_{2P}(T)$ (see Equation 16.24).

We begin with Equations 16.22 and 16.23,

$$\frac{P\overline{V}}{RT} = 1 + \frac{B_{2V}(T)}{\overline{V}} + \frac{B_{3V}(T)}{\overline{V}^2} + O(\overline{V}^3)$$

and

$$\frac{P\overline{V}}{RT} = 1 + B_{2P}(T)\,P + O(P^2)$$

We now solve Equation 16.23 for P:

$$P = \frac{RT}{\overline{V}} + \frac{PRT}{\overline{V}}B_{2P}(T) + O(\overline{V}^2)$$

Substituting this expression for P into Equation 16.22 gives

$$1 + \frac{B_{2V}(T)}{\overline{V}} + O(\overline{V}^2) = 1 + B_{2P}(T)\frac{RT}{\overline{V}} + \frac{PRT}{\overline{V}}B_{2P}(T) + O(\overline{V}^2)$$

and equating the coefficients of $\overline{V}^{-1}$ on both sides of the equation gives

$$B_{2V}(T) = RT\,B_{2P}(T)$$

16–37. Use the following data for $NH_3(g)$ at 273 K to determine $B_{2P}(T)$ at 273 K.

P/bar	0.10	0.20	0.30	0.40	0.50	0.60	0.70
$(Z-1)/10^{-4}$	1.519	3.038	4.557	6.071	7.583	9.002	10.551

Ignoring terms of $O(P^2)$, we can write Equation 16.23 as

$$Z - 1 = B_{2P}(T)P$$

A plot of $(Z - 1)$ versus pressure for the data given for NH_3 at 273 K is shown below.

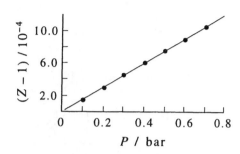

The slope of the best-fit line to the data is B_{2P} and is equal to 15.0×10^{-4} bar^{-1}.

16–38. The density of oxygen as a function of pressure at 273.15 K is listed below.

P/atm	0.2500	0.5000	0.7500	1.0000
ρ/g·dm^{-3}	0.356985	0.714154	1.071485	1.428962

Use the data to determine $B_{2V}(T)$ of oxygen. Take the atomic mass of oxygen to be 15.9994 and the value of the molar gas constant to be 8.31451 J·K^{-1}·mol^{-1} = 0.0820578 dm^3·atm·K^{-1}·mol^{-1}.

We can express the molar volume of oxygen as

$$\overline{V} = \frac{(15.9994 \text{ g·mol}^{-1})}{\rho}$$

where ρ has units of g·dm^{-3}. Using Equation 16.22 to express Z and neglecting terms of $O(\overline{V}^{-2})$, we find

$$Z - 1 = \overline{V}^{-1} B_{2V}(T)$$

or

$$\frac{P(15.994 \text{ g·mol}^{-1})}{\rho RT} - 1 = \overline{V}^{-1} B_{2V}(T)$$

A plot of $(Z - 1)$ versus $\overline{V}^{-1}$ for the data given for oxygen at 273.15 K is shown below.

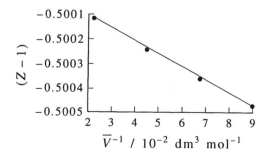

The slope of the best-fit line to the data is B_{2V} and is equal to -5.33×10^{-3} dm$^3 \cdot$mol^{-1}.

16–39. Show that the Lennard-Jones potential can be written as

$$u(r) = \varepsilon \left(\frac{r^*}{r}\right)^{12} - 2\varepsilon \left(\frac{r^*}{r}\right)^{6}$$

where r^* is the value of r at which $u(r)$ is a minimum.

The Lennard-Jones potential is (Equation 16.29)

$$u(r) = 4\varepsilon \left[\left(\frac{\sigma}{r}\right)^{12} - \left(\frac{\sigma}{r}\right)^{6} \right]$$

From Example 16.9, we know $\sigma = r^* 2^{-1/6}$, so

$$u(r) = 4\varepsilon \left[\left(\frac{r^*}{2^{1/6} r}\right)^{12} - \left(\frac{r^*}{2^{1/6} r}\right)^{6} \right]$$

$$= \frac{4\varepsilon (r^*)^{12}}{2^2 r^{12}} - \frac{4\varepsilon (r^*)^{6}}{2 r^6} = \varepsilon \left(\frac{r^*}{r}\right)^{12} - 2\varepsilon \left(\frac{r^*}{r}\right)^{6}$$

16–40. Using the Lennard-Jones parameters given in Table 16.7, compare the depth of a typical Lennard-Jones potential to the strength of a covalent bond.

The parameter ε is the depth of a Lennard-Jones potential. From Table 16.7, an average value of $\varepsilon / k_B \approx 139$ K for one molecule. So, for one mole,

$$\varepsilon \approx (139 \text{ K}) k_B N_A \approx \text{J} \approx 1 \text{ kJ}$$

In comparison, the strength of a covalent bond is on the order of 100 kJ per mole (Table 13.2).

16–41. Compare the Lennard-Jones potentials of H_2(g) and O_2(g) by plotting both on the same graph.

Shown below are plots of $u(r)$ versus r for both H_2(g) and O_2(g). Oxygen has a deeper potential well than hydrogen and the minimum of its potential curve occurs at a higher value of r than the minimum of the potential curve of hydrogen.

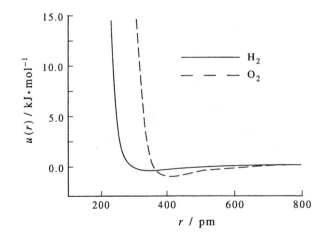

16–42. Use the data in Tables 16.5 and 16.7 to show that *roughly* $\epsilon/k_B = 0.75\, T_c$ and $b_0 = 0.7\, \overline{V}_c$. Thus, critical constants can be used as rough, first estimates of ϵ and b_0 ($= 2\pi N_A \sigma^3 /3$).

Let us select argon as a representative molecule. For Ar, $\varepsilon/k_B = 120$ K, $b_0 = 50.0$ cm$^3\cdot$mol^{-1}, $T_c = 150.95$ K, and $\overline{V}_c = 75.3$ cm$^3\cdot$mol^{-1}.

$$0.75\, T_c = 113 \text{ K} \qquad \text{compared to} \quad \varepsilon/k_B = 120 \text{ K}$$

$$0.7\, \overline{V}_c = 53 \text{ cm}^3\cdot\text{mol}^{-1} \quad \text{compared to} \qquad b_0 = 50.0 \text{ cm}^3\cdot\text{mol}^{-1}$$

The equivalencies stated in the problem text hold for argon.

16–43. Prove that the second virial coefficient calculated from a general intermolecular potential of the form

$$u(r) = (\text{energy parameter}) \times f\left(\frac{r}{\text{distance parameter}}\right)$$

rigorously obeys the law of corresponding states. Does the Lennard-Jones potential satisfy this condition?

Begin with Equation 16.25,

$$B_{2V}(T) = -2\pi N_A \int_0^\infty [e^{-u(r)/k_B T} - 1]r^2 dr$$

Let $u(r) = Ef\left(r/r_0\right)$ and $T^* = k_B T$, so that we can write $B_{2V}(T)$ as

$$B_{2V}(T^*) = -2\pi N_A \int_0^\infty [e^{-Ef\left(r/r_0\right)/T^*} - 1]r^2 dr$$

Now let $r/r_0 = r_R$, where r_R is the reduced distance variable. Then $dr = r_0 dr_R$, so we can write

$$B_{2V}(T^*) = -2\pi r_0^3 N_A \int_0^\infty [e^{-Ef(r_R)/T^*} - 1]dr_R$$

We can divide both sides by $-2\pi r_0^3 N_A$ to obtain $B_{2V}{}^*$ as a function of only reduced variables:

$$B_{2V}{}^*(T^*) = \int_0^\infty [e^{-Ef(r_R)/T^*} - 1]dr_R$$

Therefore, the functional form of $u(r)$ given in the problem text rigorously obeys the law of corresponding states. The Lennard-Jones potential can be written as (Equation 16.29)

$$u(r) = 4\varepsilon\left[\left(\frac{\sigma}{r}\right)^{12} - \left(\frac{\sigma}{r}\right)^6\right] = E[x^{12} - x^6]$$

where E is an energy parameter and x is a distance parameter ($x \sim r^{-1}$). So the Lennard-Jones potential can be written as $Ef(r)$ and so satisfies the conditions of the above general intermolecular potential.

16-44. Use the following data for argon at 300.0 K to determine the value of B_{2V}. The accepted value is -15.05 cm$^3\cdot$mol^{-1}.

P/atm	ρ/mol$\cdot$L^{-1}	P/atm	ρ/mol$\cdot$L^{-1}
0.01000	0.000406200	0.4000	0.0162535
0.02000	0.000812500	0.6000	0.0243833
0.04000	0.00162500	0.8000	0.0325150
0.06000	0.00243750	1.000	0.0406487
0.08000	0.00325000	1.500	0.0609916
0.1000	0.00406260	2.000	0.0813469
0.2000	0.00812580	3.000	0.122094

We can use Equation 16.22 to express $Z - 1$ (neglecting terms of $O(\overline{V}^{-2})$) as

$$Z - 1 = \overline{V}^{-1} B_{2V}(T) = \rho B_{2V}(T)$$

A plot of $(Z - 1)$ versus $\overline{V}^{-1}$ for the data given for oxygen at 273.15 K is shown below.

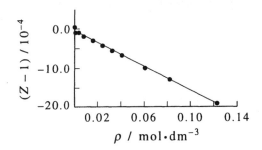

The slope of the best-fit line to the data is B_{2V} and is equal to -15.13 cm$^3\cdot$mol^{-1}.

16-45. Using Figure 16.15 and the Lennard-Jones parameters given in Table 16.7, estimate $B_{2V}(T)$ for CH$_4$(g) at 0°C.

For methane, $\epsilon/k_B = 149$ K and $2\pi\sigma^3 N_A/3 = 68.1$ cm$^3\cdot$mol^{-1} (Table 16.7). Then (by definition of T^*)

$$T^* = \frac{k_B T}{\epsilon} = \frac{273.15 \text{ K}}{149 \text{ K}} = 1.83$$

From Figure 16.15, we estimate $B_{2V}^*(T^*) \approx -0.9$. Then (also by definition)

$$B_{2V}(273.15 \text{ K}) = \frac{2\pi\sigma^3 N_A B_{2V}^*(T^*)}{3} \approx -60 \text{ cm}^3\cdot\text{mol}^{-1}$$

16-46. Show that $B_{2V}(T)$ obeys the law of corresponding states for a square-well potential with a *fixed* value of λ (in other words, if all molecules had the same value of λ).

We use Equation 16.25 to express $B_{2V}(T)$:

$$B_{2V}(T) = -2\pi N_A \int_0^\infty \left[e^{-u(r)/k_B T} - 1\right] r^2 dr$$

where, since we have a square-well potential of fixed value λ,

$$
\begin{aligned}
u(r) &= \infty && \text{if} \quad r < \sigma \\
&= -\epsilon && \text{if} \quad \sigma < r < \lambda\sigma \\
&= 0 && \text{if} \quad \lambda\sigma < r
\end{aligned}
$$

We can now integrate $B_{2V}(T)$ over the three intervals $0 < r < \sigma$, $\sigma < r < \lambda\sigma$, and $\lambda\sigma < r < \infty$:

$$
\begin{aligned}
B_{2V}(T) &= -2\pi N_A \int_0^\sigma (0-1)\, r^2 dr - 2\pi N_A \int_\sigma^{\lambda\sigma} \left[e^{\epsilon/k_B T} - 1 \right] r^2 dr + 0 \\
&= 2\pi N_A \left(\frac{\sigma^3}{3} \right) - 2\pi N_A \left(e^{\epsilon/k_B T} - 1 \right) \frac{\lambda^3\sigma^3 - \sigma^3}{3} \\
&= \frac{2\pi N_A \sigma^3}{3} \left[1 - \left(\lambda^3 - 1 \right) \left(e^{\epsilon/k_B T} - 1 \right) \right]
\end{aligned}
$$

If we divide $B_{2V}(T)$ by σ^3 and let this quantity be a reduced value of $B_{2V}(T)$, this reduced second virial coefficient will be molecule-independent and therefore satisfy the law of corresponding states.

16–47. Using the Lennard-Jones parameters in Table 16.7, show that the following second virial cofficient data satisfy the law of corresponding states.

Argon		Nitrogen		Ethane	
T/K	$B_{2V}(T)$ $/10^{-3}$ dm$^3 \cdot$mol^{-1}	T/K	$B_{2V}(T)$ $/10^{-3}$ dm$^3 \cdot$mol^{-1}	T/K	$B_{2V}(T)$ $/10^{-3}$ dm$^3 \cdot$mol^{-1}
173	−64.3	143	−79.8	311	−164.9
223	−37.8	173	−51.9	344	−132.5
273	−22.1	223	−26.4	378	−110.0
323	−11.0	273	−10.3	411	−90.4
423	+1.2	323	−0.3	444	−74.2
473	4.7	373	+6.1	478	−59.9
573	11.2	423	11.5	511	−47.4
673	15.3	473	15.3		
		573	20.6		
		673	23.5		

Find the reduced parameters of each gas by dividing T by ε/k_B and B_{2V} by $2\pi\sigma^3 N_A/3$ (Table 16.7). Below, we plot $B_{2V}^*(T)$ versus T^* for each gas.

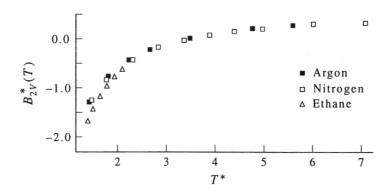

The data points for all three gases fall on the same curve, consistent with the law of corresponding states.

16–48. In Section 16–4, we expressed the van der Waals equation in reduced units by dividing P, $\overline{V}$, and T by their critical values. This suggests we can write the second virial coefficient in reduced form by dividing $B_{2V}(T)$ by $\overline{V}_c$ and T by T_c (instead of $2\pi N_A \sigma^3/3$ and ε/k as we did in Section 16–5). Reduce the second virial coefficient data given in the previous problem by using the values of $\overline{V}_c$ and T_c in Table 16.5 and show that the reduced data satisfy the law of corresponding states.

We find the reduced parameters of each gas by dividing T by T_c and B_{2V} by $\overline{V}_c$ (Table 16.5). Below, we plot $B_{2V}(T)/\overline{V}_c$ vs. T/T_c for each gas.

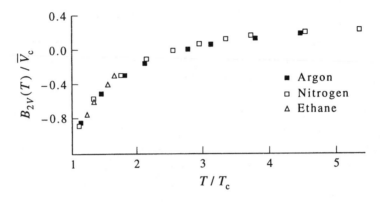

The data points for all three gases fall on the same curve, consistent with the law of corresponding states.

16–49. Listed below are experimental second virial coefficient data for argon, krypton, and xenon.

T/K	$B_{2V}(T)/10^{-3}\,\text{dm}^3\cdot\text{mol}^{-1}$ Argon	Krypton	Xenon
173.16	−63.82		
223.16	−36.79		
273.16	−22.10	−62.70	−154.75
298.16	−16.06		−130.12
323.16	−11.17	−42.78	−110.62
348.16	−7.37		−95.04
373.16	−4.14	−29.28	−82.13
398.16	−0.96		
423.16	+1.46	−18.13	−62.10
473.16	4.99	−10.75	−46.74
573.16	10.77	+0.42	−25.06
673.16	15.72	7.42	−9.56
773.16	17.76	12.70	−0.13
873.16	19.48	17.19	+7.95
973.16			14.22

Use the Lennard-Jones parameters in Table 16.7 to plot $B_{2V}^*(T^*)$, the reduced second virial coefficient, versus T^*, the reduced temperature, to illustrate the law of corresponding states.

Find the reduced parameters of each gas by dividing T by ε/k_B and B_{2V} by $2\pi\sigma^3 N_A/3$ (Table 16.7). Below, we plot $B_{2V}^*(T)$ versus T^* for each gas.

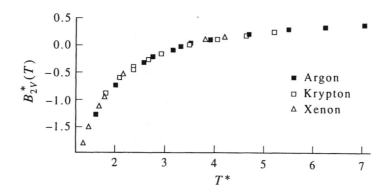

The data points for all three gases fall on the same curve, consistent with the law of corresponding states.

16–50. Use the critical temperatures and the critical molar volumes of argon, krypton, and xenon to illustrate the law of corresponding states with the data given in Problem 16–49.

We find the reduced parameters of each gas by dividing T by T_c and B_{2V} by $\overline{V}_c$ (Table 16.5). Below, we plot $B_{2V}(T)/\overline{V}_c$ vs. T/T_c for each gas.

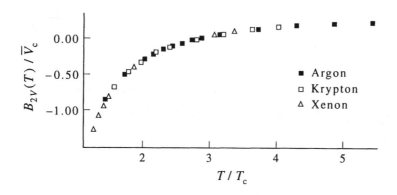

The data points for all three gases fall on the same curve, consistent with the law of corresponding states.

16–51. Evaluate $B_{2V}^*(T^*)$ in Equation 16.31 numerically from $T^* = 1.00$ to 10.0 using a packaged numerical integration program such as *MathCad* or *Mathematica*. Compare the reduced second virial coefficient data from Problem 16–49 and $B_{2V}^*(T^*)$ by plotting them all on the same graph.

Below is a table with some representative values of $B_{2V}^*(T^*)$ calculated using the numerical integration package in *Mathematica*.

T^*	$B_{2V}^*(T^*)$
1.00	-2.538081336
2.00	-0.6276252881
3.00	-0.1152339638
4.00	0.1154169217
5.00	0.2433435028
6.00	0.3229043727
7.00	0.3760884671
8.00	0.4134339539
9.00	0.4405978376
10.00	0.4608752841

The plot below shows both the curve obtained from numerically integrating $B_{2V}^*(T^*)$ and the reduced second virial coefficient data from Problem 16–49. The numerical integration is an excellent fit to the data.

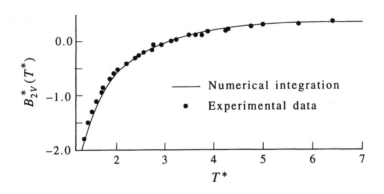

16–52. Show that the units of the right side of Equation 16.35 are energy.

$$u_{\text{induced}}(r) = -\frac{\mu_1^2 \alpha_2}{(4\pi\varepsilon_0)^2 r^6} - \frac{\mu_2^2 \alpha_1}{(4\pi\varepsilon_0)^2 r^6}$$

We know that $\alpha/4\pi\varepsilon_0$ has units of m³, $4\pi\varepsilon_0$ has units of $\text{C}\cdot\text{V}^{-1}\cdot\text{m}^{-1}$, μ has units of $\text{C}\cdot\text{m}$, and r has units of m. Thus

$$\text{units}[u_{\text{induced}}(r)] = \frac{(\text{C}\cdot\text{m})^2 \text{m}^3}{\text{C}\cdot\text{V}^{-1}\cdot\text{m}^{-1}\text{m}^6} = \text{C}\cdot\text{V} = \text{J}$$

16–53. Show that the sum of Equations 16.33, 16.35, and 16.36 gives Equation 16.37.

The sum of these three equations is

$$-\frac{\mu_1^2 \alpha_2 + \mu_2^2 \alpha_1}{(4\pi\varepsilon_0)^2 r^6} - \frac{2\mu_1^2 \mu_2^2}{(4\pi\varepsilon_0)^2 (3k_B T)}\frac{1}{r^6} - \frac{3}{2}\left(\frac{I_1 I_2}{I_1 + I_2}\right)\frac{\alpha_1 \alpha_2}{(4\pi\varepsilon_0)^2}\frac{1}{r^6}$$

For identical atoms or molecules, $I_1 = I_2 = I$, $\alpha_2 = \alpha_1 = \alpha$, and $\mu_1 = \mu_2 = \mu$, and so the sum becomes

$$\Sigma(u) = -\frac{1}{(4\pi\varepsilon_0)^2 r^6}\left(2\mu^2\alpha + \frac{2\mu^4}{3k_BT} + \frac{3I\alpha^2}{4}\right)$$

The coefficient of the r^6 term, C_6, is therefore

$$C_6 = \frac{2\mu^4}{3(4\pi\varepsilon_0)^2 k_B T} + \frac{2\alpha\mu^2}{(4\pi\varepsilon_0)^2} + \frac{3}{4}\frac{I\alpha^2}{(4\pi\varepsilon_0)^2}$$

as in Equation 16.37.

16–54. Compare the values of the coefficient of r^{-6} for $N_2(g)$ using Equation 16.37 and the Lennard-Jones parameters given in Table 16.7.

Using Equation 16.37, we find that

$$\begin{aligned}
C_6 &= \frac{2\mu^4}{3(4\pi\varepsilon_0)^2 k_B T} + \frac{2\alpha\mu^2}{(4\pi\varepsilon_0)^2} + \frac{3}{4}\frac{I\alpha^2}{(4\pi\varepsilon_0)^2}\\
&= 0 + 0 + 0.75\left(2.496 \times 10^{-18}\ \text{J}\right)\left(1.77 \times 10^{-30}\ \text{m}^3\right)^2\\
&= 5.86 \times 10^{-78}\ \text{J·m}^6
\end{aligned}$$

Using the Lennard-Jones parameters,

$$\begin{aligned}
C_6 &= 4\varepsilon\sigma^6 = 4\frac{\varepsilon}{k_B}k_B\sigma^6\\
&= 4\,(95.1\ \text{K})\,(1.381 \times 10^{-23}\ \text{J·K}^{-1})\left(370 \times 10^{-12}\ \text{m}\right)^6\\
&= 1.35 \times 10^{-77}\ \text{J·m}^6
\end{aligned}$$

The coefficient of r^{-6} obtained using the Lennard-Jones parameters is about twice that obtained using Equation 16.37.

16–55. Show that

$$B_{2V}(T) = B - \frac{A}{RT^{3/2}}$$

and

$$B_{3V}(T) = B^2 + \frac{AB}{RT^{3/2}}$$

for the Redlich-Kwong equation.

Begin with the Redlich-Kwong equation (Equation 16.7):

$$\begin{aligned}
P &= \frac{RT}{\overline{V} - B} - \frac{A}{T^{1/2}\overline{V}\left(\overline{V} + B\right)}\\
&= \frac{RT}{\overline{V}\left(1 - \frac{B}{\overline{V}}\right)} - \frac{A}{T^{1/2}\overline{V}^2\left(1 + \frac{B}{\overline{V}}\right)}
\end{aligned}$$

Expanding the fractions $1/(1 - B/\overline{V})$ and $1/(1 + B/\overline{V})$ (Equation I.3) gives

$$P = \frac{RT}{\overline{V}}\left[1 + \frac{B}{\overline{V}} + \frac{B^2}{\overline{V}^2} + O(\overline{V}^{-3})\right] - \frac{A}{T^{1/2}\overline{V}^2}\left[1 - \frac{B}{\overline{V}} - \frac{B^2}{\overline{V}^2} - O(\overline{V}^{-3})\right]$$

$$= \frac{RT}{\overline{V}} - \frac{A}{T^{1/2}\overline{V}^2} + \left(\frac{RT}{\overline{V}} + \frac{A}{T^{1/2}\overline{V}^2}\right)\left[\frac{B}{\overline{V}} - \frac{B^2}{\overline{V}^2} - O(\overline{V}^{-3})\right]$$

We then use the definition of Z to find that

$$Z = \frac{P\overline{V}}{RT} = 1 - \frac{A}{RT^{3/2}\overline{V}} + \left(1 + \frac{A}{RT^{3/2}\overline{V}}\right)\left[\frac{B}{\overline{V}} - \frac{B^2}{\overline{V}^2} - O(\overline{V}^{-3})\right]$$

We compare this with Equation 16.22,

$$Z = 1 + \frac{B_{2V}(T)}{\overline{V}} + \frac{B_{3V}(T)}{\overline{V}^2} + O(\overline{V}^3)$$

Setting the coefficients of $1/\overline{V}$ and $1/\overline{V}^2$ equal to one another gives

$$B_{2V} = B - \frac{A}{RT^{3/2}}$$

and

$$B_{3V} = B^2 + \frac{AB}{RT^{3/2}}$$

16–56. Show that the second and third virial coefficients of the Peng-Robinson equation are

$$B_{2V}(T) = \beta - \frac{\alpha}{RT}$$

and

$$B_{3V}(T) = \beta^2 + \frac{2\alpha\beta}{RT}$$

Begin with the Peng-Robinson equation (Equation 16.8):

$$P = \frac{RT}{\overline{V} - \beta} - \frac{\alpha}{\overline{V}(\overline{V} + \beta) + \beta(\overline{V} - \beta)}$$

$$= \frac{RT}{\overline{V}\left(1 - \frac{\beta}{\overline{V}}\right)} - \frac{\alpha}{\overline{V}^2}\left[\frac{1}{1 - \left(\frac{\beta^2}{\overline{V}^2} - \frac{2\beta}{\overline{V}}\right)}\right]$$

Expanding the fractions $1/(1 - \beta/\overline{V})$ and $1/(1 - \beta^2/\overline{V}^2 - 2\beta/\overline{V})$ (Equation I.3) gives

$$P = \frac{RT}{\overline{V}}\left[1 + \frac{\beta}{\overline{V}} + \frac{\beta^2}{\overline{V}^2} + O(\overline{V}^{-3})\right] - \frac{\alpha}{\overline{V}^2}\left[1 + \left(\frac{\beta^2}{\overline{V}^2} - \frac{2\beta}{\overline{V}}\right) + \left(\frac{\beta^2}{\overline{V}^2} - \frac{2\beta}{\overline{V}}\right)^2 + O(\overline{V}^{-3})\right]$$

$$= \frac{RT}{\overline{V}}\left[1 + \frac{\beta}{\overline{V}} + \frac{\beta^2}{\overline{V}^2} + O(\overline{V}^{-3})\right] - \frac{\alpha}{\overline{V}^2}\left[1 + \frac{2\beta}{\overline{V}} + O(\overline{V}^{-2})\right]$$

We then use the definition of Z to find that

$$Z = \frac{P\overline{V}}{RT} = 1 + \frac{\beta}{\overline{V}} + \frac{\beta^2}{\overline{V}^2} - \frac{\alpha}{\overline{V}RT} - \frac{2\alpha\beta}{RT\overline{V}^2} + O(\overline{V}^{-3})$$

We compare this with Equation 16.22,

$$Z = 1 + \frac{B_{2V}(T)}{\overline{V}} + \frac{B_{3V}(T)}{\overline{V}^2} + O(\overline{V}^3)$$

Setting the coefficients of $1/\overline{V}$ and $1/\overline{V}^2$ equal to one another gives

$$B_{2V} = \beta - \frac{\alpha}{RT}$$

and

$$B_{3V} = \beta^2 + \frac{2\alpha\beta}{RT}$$

16–57. The square-well parameters for krypton are $\varepsilon/k_B = 136.5$ K, $\sigma = 327.8$ pm, and $\lambda = 1.68$. Plot $B_{2V}(T)$ against T and compare your results with the data given in Problem 16–49.

From Problem 16.46, we know that, for a square-well potential,

$$
\begin{aligned}
B_{2V}(T) &= \frac{2\pi N_A \sigma^3}{3} \left[1 - \left(\lambda^3 - 1 \right) \left(e^{\varepsilon/k_B T} - 1 \right) \right] \\
&= \frac{2\pi (6.022 \times 10^{23} \text{ mol}^{-1})(327.8 \text{ pm})^3}{3} \\
&\quad \times \left\{ 1 - \left[(1.68)^3 - 1 \right] \left(e^{(136.5 \text{ K})/T} - 1 \right) \right\}
\end{aligned}
$$

The plot below shows both the square-well potential curve and the experimental data from Problem 16–49 for krypton. The square-well potential is a very good fit to the data.

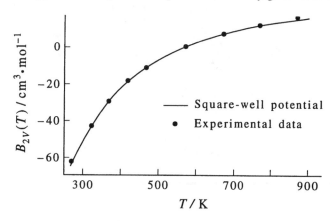

16–58. The coefficient of thermal expansion α is defined as

$$\alpha = \frac{1}{\overline{V}} \left(\frac{\partial \overline{V}}{\partial T} \right)_P$$

Show that

$$\alpha = \frac{1}{T}$$

for an ideal gas.

For an ideal gas, $P\overline{V} = RT$. Taking the partial derivative of both sides of this equation with respect to T gives

$$P\overline{V} = RT$$

$$\left(\frac{\partial \overline{V}}{\partial T}\right)_P = \frac{R}{P}$$

and so

$$\alpha = \frac{1}{\overline{V}}\left(\frac{\partial \overline{V}}{\partial T}\right)_P = \frac{R}{P\overline{V}} = \frac{1}{T}$$

16–59. The isothermal compressibility κ is defined as

$$\kappa = -\frac{1}{\overline{V}}\left(\frac{\partial \overline{V}}{\partial P}\right)_T$$

Show that

$$\kappa = \frac{1}{P}$$

for an ideal gas.

For an ideal gas, $P = RT/\overline{V}$. Taking the partial derivative of both sides of this equation with respect to P gives

$$1 = \frac{-RT}{\overline{V}^2}\left(\frac{\partial \overline{V}}{\partial P}\right)_T$$

$$-\frac{\overline{V}^2}{RT} = \left(\frac{\partial \overline{V}}{\partial P}\right)_T$$

and so

$$\kappa = -\frac{1}{\overline{V}}\left(\frac{\partial \overline{V}}{\partial P}\right)_T = \frac{\overline{V}}{RT} = \frac{1}{P}$$

Partial Derivatives

PROBLEMS AND SOLUTIONS

H–1. The isothermal compressibility, κ_T, of a substance is defined as

$$\kappa_T = -\frac{1}{V}\left(\frac{\partial V}{\partial P}\right)_T$$

Obtain an expression for the isothermal compressibility of an ideal gas.

For an ideal gas, $PV = nRT$. Taking the partial derivative of both sides of this equation with respect to P gives

$$V\left(\frac{\partial P}{\partial P}\right)_T + P\left(\frac{\partial V}{\partial P}\right)_T = nR\left(\frac{\partial T}{\partial P}\right)_T$$

$$V + P\left(\frac{\partial V}{\partial P}\right)_T = 0$$

Then

$$\left(\frac{\partial V}{\partial P}\right)_T = -\frac{V}{P}$$

$$-\frac{1}{V}\left(\frac{\partial V}{\partial P}\right)_T = \frac{1}{P}$$

$$-\frac{1}{V}\left(\frac{\partial V}{\partial P}\right)_T = \kappa_T = \frac{1}{P}$$

H–2. The coefficient of thermal expansion, α, of a substance is defined as

$$\alpha = \frac{1}{V}\left(\frac{\partial V}{\partial T}\right)_P$$

Obtain an expression for the coefficient of thermal expansion of an ideal gas.

For an ideal gas, $PV = nRT$. Taking the partial derivative of both sides of this equation with respect to T gives

$$P\left(\frac{\partial V}{\partial T}\right)_P = nR$$

$$\left(\frac{\partial V}{\partial T}\right)_P = \frac{nR}{P}$$

$$\frac{1}{V}\left(\frac{\partial V}{\partial T}\right)_P = \frac{nR}{PV}$$

or

$$\alpha = \frac{1}{T}$$

H–3. Prove that

$$\left(\frac{\partial P}{\partial V}\right)_{n,T} = \frac{1}{\left(\dfrac{\partial V}{\partial P}\right)_{n,T}}$$

for an ideal gas and for a gas whose equation of state is $P = nRT/(V - nb)$, where b is a constant. This relation is generally true and is called the reciprocal identity. Notice that the same variables must be held fixed on both sides of the identity.

For an ideal gas, $PV = nRT$. The partial of P with respect to V is

$$\left(\frac{\partial P}{\partial V}\right)_{n,T} = -\frac{nRT}{V^2} = -\frac{P}{V}$$

and the partial of V with respect to P is

$$\left(\frac{\partial V}{\partial P}\right)_{n,T} = -\frac{nRT}{P^2} = -\frac{V}{P}$$

Because

$$-\frac{P}{V} = \left(-\frac{V}{P}\right)^{-1}$$

the reciprocal identity given in the text of the problem holds for an ideal gas. Likewise, for a gas with the equation of state $P = nRT/(V - nb)$, the partial of P with respect to V is

$$\left(\frac{\partial P}{\partial V}\right)_{n,T} = -\frac{nRT}{(V - nb)^2} = -\frac{P}{V - nb}$$

and the partial of V with respect to P is

$$\left(\frac{\partial V}{\partial P}\right)_{n,T} = -\frac{nRT}{P^2} = -\frac{V - nb}{P}$$

Because

$$-\frac{P}{(V - nb)} = \left[-\frac{(V - nb)}{P}\right]^{-1}$$

the reciprocal identity also holds for a gas with equation of state $P = nRT/(V - nb)$.

H–4. Given that

$$U = kT^2 \left(\frac{\partial \ln Q}{\partial T} \right)_{N,V}$$

where

$$Q(N, V, T) = \frac{1}{N!} \left(\frac{2\pi m k_B T}{h^2} \right)^{3N/2} V^N$$

and k_B, m, and h are constants, determine U as a function of T.

We are given

$$U = kT^2 \left(\frac{\partial \ln Q}{\partial T} \right)_{N,V}$$

and

$$Q(N, V, T) = \frac{1}{N!} \left(\frac{2\pi m k_B T}{h^2} \right)^{3N/2} V^N$$

Let $K = \dfrac{V^N}{N!} \left(\dfrac{2\pi m k_B}{h^2} \right)^{3N/2}$. Then $Q = K T^{3N/2}$, and

$$\ln Q = 3N/2 \ln T + \ln K$$

Taking the partial derivative with respect to T gives

$$\left(\frac{\partial \ln Q}{\partial T} \right)_{N,V} = \frac{3N}{2T}$$

and so

$$U = kT^2 \left(\frac{\partial \ln Q}{\partial T} \right)_{N,V} = kT^2 \frac{3N}{2T} = \frac{3NkT}{2}$$

H–5. Show that the total derivative of P for the Redlich-Kwong equation,

$$P = \frac{RT}{\overline{V} - B} - \frac{A}{T^{1/2} \overline{V}(\overline{V} + B)}$$

is given by Equation H.14.

We can write (as in Equation H.11)

$$dP = \left(\frac{\partial P}{\partial T} \right)_{\overline{V}} dT + \left(\frac{\partial P}{\partial \overline{V}} \right)_T d\overline{V}$$

For a Redlich-Kwong gas,

$$\left(\frac{\partial P}{\partial T} \right)_{\overline{V}} = \frac{R}{\overline{V} - B} + \frac{A}{2T^{3/2} \overline{V}(\overline{V} + B)}$$

$$\left(\frac{\partial P}{\partial \overline{V}} \right)_T = \frac{-RT}{(\overline{V} - B)^2} + \frac{A}{T^{1/2} \overline{V}^2 (\overline{V} + B)} + \frac{A}{T^{1/2} \overline{V}(\overline{V} + B)^2}$$

Then

$$dP = \left[\frac{R}{\overline{V} - B} + \frac{A}{2T^{3/2}\overline{V}\left(\overline{V} + B\right)} \right] dT + \left[-\frac{RT}{\left(\overline{V} - B\right)^2} + \frac{A(2\overline{V} + B)}{T^{1/2}\overline{V}^2(\overline{V} + B)^2} \right] d\overline{V}$$

H–6. Show explicitly that

$$\left(\frac{\partial^2 P}{\partial \overline{V} \partial T} \right) = \left(\frac{\partial^2 P}{\partial T \partial \overline{V}} \right)$$

for the Redlich-Kwong equation (Problem H–5).

We found expressions for $(\partial P/\partial T)_{\overline{V}}$ and $(\partial P/\partial \overline{V})_T$ in Problem H–5. Differentiating $(\partial P/\partial T)_{\overline{V}}$ with respect to $\overline{V}$ and $(\partial P/\partial \overline{V})_T$ with respect to T gives

$$\frac{\partial}{\partial \overline{V}} \left(\frac{\partial P}{\partial T} \right) = \frac{-R}{\left(\overline{V} - B\right)^2} - \frac{A}{2T^{3/2}\overline{V}^2\left(\overline{V} + B\right)} - \frac{A}{2\overline{V}T^{3/2}\left(\overline{V} + B\right)^2}$$

$$\frac{\partial}{\partial T} \left(\frac{\partial P}{\partial \overline{V}} \right) = \frac{-R}{\left(\overline{V} - B\right)^2} - \frac{A}{2T^{3/2}\overline{V}^2\left(\overline{V} + B\right)} - \frac{A}{2\overline{V}T^{3/2}\left(\overline{V} + B\right)^2}$$

So

$$\frac{\partial^2 P}{\partial \overline{V} \partial T} = \frac{\partial^2 P}{\partial T \partial \overline{V}}$$

H–7. We will derive the following equation in Chapter 19:

$$\left(\frac{\partial U}{\partial V} \right)_T = T \left(\frac{\partial P}{\partial T} \right)_V - P$$

Evaluate $(\partial U/\partial V)_T$ for an ideal gas, for a van der Waals gas (Equation H.4), and for a Redlich-Kwong gas (Problem H–5).

For an ideal gas $(\partial P/\partial T)_V = R/V$, so (using the equation given in the problem)

$$\left(\frac{\partial U}{\partial V} \right)_T = T \left(\frac{R}{V} \right) - P = P - P = 0$$

For a van der Waals gas $(\partial P/\partial T)_V = R/(\overline{V} - b)$, so

$$\left(\frac{\partial U}{\partial V} \right)_T = T\frac{R}{\overline{V} - b} - P = \frac{RT}{\overline{V} - b} - \frac{RT}{\overline{V} - b} + \frac{a}{\overline{V}^2} = \frac{a}{\overline{V}^2}$$

For a Redlich-Kwong gas,

$$\left(\frac{\partial P}{\partial T} \right)_{\overline{V}} = \frac{R}{\overline{V} - B} + \frac{A}{2\overline{V}\left(\overline{V} + B\right)T^{3/2}}$$

$$\left(\frac{\partial U}{\partial V} \right)_T = T \left[\frac{R}{\overline{V} - B} + \frac{A}{2\overline{V}\left(\overline{V} + B\right)T^{3/2}} \right] - P$$

$$= \frac{RT}{\overline{V} - B} + \frac{A}{2\overline{V}\left(\overline{V} + B\right)T^{1/2}} - \frac{RT}{\overline{V} - B} + \frac{A}{\overline{V}\left(\overline{V} + B\right)T^{1/2}}$$

$$= \frac{3A}{2\overline{V}\left(\overline{V} + B\right)T^{1/2}}$$

H–8. Given that the heat capacity at constant volume is defined by

$$C_V = \left(\frac{\partial U}{\partial T}\right)_V$$

and given the expression in Problem H–7, derive the equation

$$\left(\frac{\partial C_V}{\partial V}\right)_T = T\left(\frac{\partial^2 P}{\partial T^2}\right)_V$$

We know that

$$C_V = \left(\frac{\partial U}{\partial T}\right)_V \tag{1}$$

and

$$\left(\frac{\partial U}{\partial V}\right)_T = T\left(\frac{\partial P}{\partial T}\right)_V - P \tag{2}$$

Substituting Equation 2 into Equation 1 gives

$$\left(\frac{\partial C_V}{\partial V}\right)_T = \frac{\partial}{\partial V}\left(\frac{\partial U}{\partial T}\right)_V = \frac{\partial^2 U}{\partial V \partial T} = \frac{\partial^2 U}{\partial T \partial V} = \frac{\partial}{\partial T}\left(\frac{\partial U}{\partial V}\right)_T$$

$$= \frac{\partial}{\partial T}\left[T\left(\frac{\partial P}{\partial T}\right)_V - P\right]$$

$$= \left(\frac{\partial P}{\partial T}\right)_V + T\left(\frac{\partial^2 P}{\partial T^2}\right)_V - \left(\frac{\partial P}{\partial T}\right)_V$$

$$= T\left(\frac{\partial^2 P}{\partial T^2}\right)_V$$

as stated in the text of the problem.

H–9. Use the expression in Problem H–8 to determine $(\partial C_V/\partial V)_T$ for an ideal gas, a van der Waals gas (Equation H.4), and a Redlich-Kwong gas (see Problem H–5).

The expression in Problem H–8 is

$$\left(\frac{\partial C_V}{\partial V}\right)_T = T\left(\frac{\partial^2 P}{\partial T^2}\right)_V$$

For an ideal gas, $(\partial P/\partial T)_V = R/\overline{V}$, and $(\partial P/\partial T)_V = R/(\overline{V} - b)$ for a van der Waals gas. Since neither of these expressions varies when temperature is held constant, for both ideal and van der Waals gases

$$\left(\frac{\partial C_V}{\partial V}\right)_T = 0$$

For a Redlich-Kwong gas, we know that (Problem H–7)

$$\left(\frac{\partial P}{\partial T}\right)_{\overline{V}} = \frac{R}{\overline{V} - B} + \frac{A}{2\overline{V}\left(\overline{V} + B\right)T^{3/2}}$$

Then

$$\left(\frac{\partial^2 P}{\partial T^2}\right)_V = 0 - \frac{3}{4}\left[\frac{A}{\overline{V}\left(\overline{V}+B\right)T^{5/2}}\right]$$

Substituting into the expression from Problem H–8 gives

$$\left(\frac{\partial C_V}{\partial V}\right)_T = T\left(\frac{\partial^2 P}{\partial T^2}\right)_V = -\frac{3}{4}\left[\frac{A}{\overline{V}\left(\overline{V}+B\right)T^{3/2}}\right]$$

H–10. Is

$$dV = \pi r^2 dh + 2\pi r h dr$$

an exact or inexact differential?

For an exact differential, the cross derivatives are equal. We evaluate the two derivatives

$$\left[\frac{\partial}{\partial r}\pi r^2\right]_h = 2\pi r$$

and

$$\left[\frac{\partial}{\partial h}2\pi r h\right]_r = 2\pi r$$

Because these two derivatives are equal, dV is an exact differential.

H–11. Is

$$dx = C_V(T)dT + \frac{nRT}{V}dV$$

an exact or inexact differential? The quantity $C_V(T)$ is simply an arbitrary function of T. What about dx/T?

We evaluate the two derivatives

$$\left[\frac{\partial}{\partial V}C_V(T)\right]_T = 0$$

and

$$\left[\frac{\partial}{\partial T}\frac{nRT}{V}\right]_V = \frac{nR}{V}$$

Because these derivatives are unequal, dx is an inexact differential. We can express dx/T as

$$\frac{dx}{T} = T^{-1}C_V(T)dT + \frac{nR}{V}dV$$

and evaluate the two derivatives

$$\left[\frac{\partial}{\partial V}T^{-1}C_V(T)\right]_T = 0$$

$$\left[\frac{\partial}{\partial T}\frac{nR}{V}\right]_V = 0$$

Because these derivatives are equal, dx/T is an exact differential.

H–12. Prove that

$$\frac{1}{Y}\left(\frac{\partial Y}{\partial P}\right)_{T,n} = \frac{1}{\overline{Y}}\left(\frac{\partial \overline{Y}}{\partial P}\right)_{T}$$

and that

$$\left(\frac{\partial P}{\partial \overline{Y}}\right)_{T} = n\left(\frac{\partial P}{\partial Y}\right)_{T,n}$$

where $Y = Y(P, T, n)$ is an extensive variable.

We know that $\overline{Y}n = Y$. Then

$$\frac{1}{Y}\left(\frac{\partial Y}{\partial P}\right)_{T,n} = \frac{1}{Y}\left[\frac{\partial(\overline{Y}n)}{\partial P}\right]_{T,n}$$

$$= \frac{n}{Y}\left(\frac{\partial \overline{Y}}{\partial P}\right)_{T,n}$$

Since $\overline{Y} = \overline{Y}(P, T)$, this becomes

$$\frac{1}{Y}\left(\frac{\partial Y}{\partial P}\right)_{T,n} = \frac{1}{\overline{Y}}\left(\frac{\partial \overline{Y}}{\partial P}\right)_{T}$$

We can use the reciprocal identity from Problem H–3 to write this as

$$\frac{1}{Y}\left(\frac{\partial P}{\partial \overline{Y}}\right)_{T} = \frac{1}{\overline{Y}}\left(\frac{\partial P}{\partial Y}\right)_{T,n}$$

$$\left(\frac{\partial P}{\partial \overline{Y}}\right)_{T} = n\left(\frac{\partial P}{\partial Y}\right)_{T,n}$$

H–13. Equation 16.5 gives P for the van der Waals equation as a function of $\overline{V}$ and T. Show that P expressed as a function of V, T, and n is

$$P = \frac{nRT}{V - nb} - \frac{n^2 a}{V^2} \tag{1}$$

Now evaluate $(\partial P/\partial \overline{V})_T$ from Equation 16.5 and $(\partial P/\partial V)_{T,n}$ from Equation 1 above and show that (see Problem H–12)

$$\left(\frac{\partial P}{\partial \overline{V}}\right)_{T} = n\left(\frac{\partial P}{\partial V}\right)_{T,n}$$

We begin with Equation 16.5:

$$\left(P + \frac{a}{\overline{V}^2}\right)(\overline{V} - b) = RT$$

$$\left(P + \frac{an^2}{V^2}\right) = \frac{nRT}{V - nb}$$

$$P = \frac{nRT}{V - nb} - \frac{n^2 a}{V^2} \tag{1}$$

Evaluating $(\partial P/\partial \overline{V})_T$ from Equation 16.5 gives

$$P = \frac{RT}{\overline{V} - b} - \frac{a}{\overline{V}^2}$$

$$\left(\frac{\partial P}{\partial \overline{V}}\right)_T = -\frac{RT}{(\overline{V} - b)^2} + \frac{2a}{\overline{V}^3}$$

and evaluating $(\partial P/\partial V)_T$ from Equation 1 gives

$$P = \frac{nRT}{V - nb} - \frac{n^2 a}{V^2}$$

$$\left(\frac{\partial P}{\partial V}\right)_T = -\frac{nRT}{(V - nb)^2} + \frac{2n^2 a}{V^3}$$

$$= -\frac{RT}{n(\overline{V} - b)^2} + \frac{2a}{n\overline{V}^3}$$

$$= \frac{1}{n}\left(\frac{\partial P}{\partial \overline{V}}\right)_T$$

H–14. Referring to Problem H–13, show that

$$\left(\frac{\partial P}{\partial T}\right)_{\overline{V}} = \left(\frac{\partial P}{\partial T}\right)_{V,n}$$

and generally that

$$\left[\frac{\partial y(x, \overline{V})}{\partial x}\right]_{\overline{V}} = \left[\frac{\partial y(x, n, V)}{\partial x}\right]_{V,n}$$

where y and x are intensive variables and $y(x, n, V)$ can be written as $y(x, V/n)$.

Evaluating $(\partial P/\partial T)_{\overline{V}}$ from Equation 16.5 gives

$$P = \frac{RT}{\overline{V} - b} - \frac{a}{\overline{V}^2}$$

$$\left(\frac{\partial P}{\partial T}\right)_{\overline{V}} = \frac{R}{\overline{V} - b}$$

and evaluating $(\partial P/\partial T)_V$ from Equation 1 of Problem H–13 gives

$$P = \frac{nRT}{V - nb} - \frac{n^2 a}{V^2}$$

$$\left(\frac{\partial P}{\partial T}\right)_V = \frac{nR}{V - nb} = \frac{R}{\overline{V} - b}$$

$$= \left(\frac{\partial P}{\partial T}\right)_{\overline{V}}$$

Generally, for any $y(x, n, V)$ which can be written as $y(x, \overline{V})$,

$$dy = \left[\frac{\partial y(x, \overline{V})}{\partial x}\right]_{\overline{V}} dx + \left[\frac{\partial y(x, \overline{V})}{\partial \overline{V}}\right]_x d\overline{V}$$

and

$$dy = \left[\frac{\partial y(x, n, V)}{\partial x}\right]_{V,n} dx + \left[\frac{\partial y(x, n, V)}{\partial n}\right]_{x,V} dn + \left[\frac{\partial y(x, n, V)}{\partial V}\right]_{x,n} dV$$

Equating the coefficients of dx in these two expressions gives

$$\left[\frac{\partial y(x, \overline{V})}{\partial x}\right]_{\overline{V}} = \left[\frac{\partial y(x, n, V)}{\partial x}\right]_{V,n}$$

CHAPTER 17

The Boltzmann Factor and Partition Functions

PROBLEMS AND SOLUTIONS

17–1. How would you describe an ensemble whose systems are one-liter containers of water at 25°C?

An unlimited number of one-liter containers of water in an essentially infinite heat bath at 298 K.

17–2. Show that Equation 17.8 is equivalent to $f(x + y) = f(x)f(y)$. In this problem, we will prove that $f(x) \propto e^{ax}$. First, take the logarithm of the above equation to obtain

$$\ln f(x + y) = \ln f(x) + \ln f(y)$$

Differentiate both sides with respect to x (keeping y fixed) to get

$$\left[\frac{\partial \ln f(x + y)}{\partial x}\right]_y = \frac{d \ln f(x + y)}{d(x + y)}\left[\frac{\partial(x + y)}{\partial x}\right]_y = \frac{d \ln f(x + y)}{d(x + y)}$$

$$= \frac{d \ln f(x)}{dx}$$

Now differentiate with respect to y (keeping x fixed) and show that

$$\frac{d \ln f(x)}{dx} = \frac{d \ln f(y)}{dy}$$

For this relation to be true for all x and y, each side must equal a constant, say a. Show that

$$f(x) \propto e^{ax} \qquad \text{and} \qquad f(y) \propto e^{ay}$$

Let $x = E_1 - E_2$ and $y = E_2 - E_3$. Then $x + y = E_1 - E_3$, and we can write Equation 17.8 as $f(x + y) = f(x)f(y)$. Taking the logarithm of this equation gives

$$\ln f(x + y) = \ln[f(x)f(y)]$$
$$= \ln f(x) + \ln f(y)$$

We then differentiate with respect to x and find

$$\left[\frac{\partial \ln f(x + y)}{\partial x}\right]_y = \left[\frac{\partial \ln f(x)}{\partial x}\right]_y + 0$$

$$\frac{d \ln f(x + y)}{d(x + y)}\left[\frac{\partial(x + y)}{\partial x}\right]_y = \frac{d \ln f(x)}{dx}$$

$$\frac{d \ln f(x + y)}{d(x + y)} = \frac{d \ln f(x)}{dx} \tag{1}$$

Likewise, we differentiate with respect to y and find

$$\left[\frac{\partial \ln f(x+y)}{\partial y}\right]_x = 0 + \left[\frac{\partial \ln f(y)}{\partial y}\right]_x$$

$$\frac{d \ln f(x+y)}{d(x+y)}\left[\frac{\partial(x+y)}{\partial y}\right]_x = \frac{d \ln f(y)}{dy}$$

$$\frac{d \ln f(x+y)}{d(x+y)} = \frac{d \ln f(y)}{dy} \tag{2}$$

Equations 1 and 2 are equal to one another, and equal to a constant a (as stated in the text of the problem), so

$$\frac{d \ln f(x)}{dx} = \frac{d \ln f(y)}{dy} = a$$

We can integrate $d \ln f(x) = a dx$ to find

$$\int d \ln f(x) = \int a dx$$

$$\ln f(x) \propto ax$$

$$f(x) \propto e^{ax}$$

Similarly, integrating $d \ln f(y) = a dy$ gives $f(y) \propto e^{ay}$.

17–3. Show that $a_l/a_i = e^{\beta(E_i - E_l)}$ implies that $a_j = Ce^{-\beta E_j}$.

$$a_l/a_i = e^{\beta(E_i - E_l)} = e^{-\beta E_l} e^{\beta E_i}$$

$$a_l = \left(a_i e^{\beta E_i}\right) e^{-\beta E_l} = Ce^{-\beta E_l}$$

The subscript l is arbitrary, and so this shows the desired result.

17–4. Prove to yourself that $\sum_i e^{-\beta E_i} = \sum_j e^{-\beta E_j}$.

Expanding both sums gives

$$e^{-\beta E_1} + e^{-\beta E_2} + \dots = e^{-\beta E_1} + e^{-\beta E_2} + \dots$$

17–5. Show that the partition function in Example 17–1 can be written as

$$Q(\beta, B_z) = 2\cosh\left(\frac{\beta\hbar\gamma B_z}{2}\right) = 2\cosh\left(\frac{\hbar\gamma B_z}{2k_B T}\right)$$

Use the fact that $d\cosh x/dx = \sinh x$ to show that

$$\langle E \rangle = -\frac{\hbar\gamma B_z}{2}\tanh\frac{\beta\hbar\gamma B_z}{2} = -\frac{\hbar\gamma B_z}{2}\tanh\frac{\hbar\gamma B_z}{2k_B T}$$

Recall that the definition of $\cosh u$ is

$$\cosh u = \frac{e^u + e^{-u}}{2}$$

Therefore, $2\cosh u = e^u + e^{-u}$. The partition function in Example 17–1 is

$$Q(\beta, B_z) = e^{\beta\hbar\gamma B_z/2} + e^{-\beta\hbar\gamma B_z/2} = 2\cosh\frac{\beta\hbar\gamma B_z}{2}$$

We use Equation 17.20 to write

$$\langle E\rangle = -\frac{1}{Q}\left(\frac{\partial Q}{\partial\beta}\right)_{B_z}$$

$$= -\frac{1}{2\cosh\dfrac{\beta\hbar\gamma B_z}{2}}\left(\frac{2\partial\cosh\dfrac{\beta\hbar\gamma B_z}{2}}{\partial\beta}\right)_{B_z}$$

Let $K = \hbar\gamma B_z/2$. Then we can write $\langle E\rangle$ as

$$\langle E\rangle = -\frac{1}{2\cosh(K\beta)}\left[\frac{2\partial\cosh(K\beta)}{\partial\beta}\right]_{B_z}$$

$$= \left[-\frac{1}{\cosh(K\beta)}\right]K\sinh(K\beta)$$

$$= -K\tanh(K\beta)$$

$$= -\frac{\hbar\gamma B_z}{2}\tanh\frac{\hbar\gamma B_z}{2k_BT}$$

17–6. Use either the expression for $\langle E\rangle$ in Example 17–1 or the one in Problem 17–5 to show that

$$\langle E\rangle \longrightarrow -\frac{\hbar\gamma B_z}{2}\quad\text{as}\quad T\longrightarrow 0$$

and that

$$\langle E\rangle \longrightarrow 0\quad\text{as}\quad T\longrightarrow\infty$$

A graph of $\tanh(x)$ is presented below for reference.

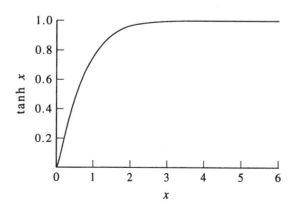

From Problem 17–5, we write

$$\langle E\rangle = -\frac{\hbar\gamma B_z}{2}\tanh\frac{\hbar\gamma B_z}{2k_BT} = -\frac{\hbar\gamma B_z}{2}\tanh\frac{\hbar\gamma B_z\beta}{2}$$

As $T \to 0$, $\beta \to \infty$, so

$$\lim_{T \to 0}\langle E \rangle = \lim_{\beta \to \infty} -\frac{\hbar \gamma B_z}{2} \tanh \frac{\hbar \gamma B_z \beta}{2} = -\frac{\hbar \gamma B_z}{2}$$

As $T \to \infty$, $\beta \to 0$, so

$$\lim_{T \to \infty} \langle E \rangle = \lim_{\beta \to 0} -\frac{\hbar \gamma B_z}{2} \tanh \frac{\hbar \gamma B_z \beta}{2} = 0$$

17–7. Generalize the results of Example 17–1 to the case of a spin-1 nucleus. Determine the low-temperature and high-temperature limits of $\langle E \rangle$.

For a spin-1 nucleus, $m_I = -1, 0,$ or 1. Substituting these values into Equation 14.16 gives

$$E_0 = 0 \quad \text{and} \quad E_{\pm 1} = \mp \hbar \gamma B_z$$

We now apply Equation 17.14 to find

$$Q(\beta, B_z) = 1 + e^{-\beta \hbar \gamma B_z} + e^{\beta \hbar \gamma B_z} = 1 + 2\cosh(\beta \hbar \gamma B_z)$$

Using Equation 17.20 for $\langle E \rangle$ gives

$$\langle E \rangle = -\left(\frac{\partial \ln Q}{\partial \beta}\right)_{B_z} = -\frac{1}{Q}\left(\frac{\partial Q}{\partial \beta}\right)_{B_z}$$

$$= -\frac{1}{1 + 2\cosh(\beta \hbar \gamma B_z)}\left[2\hbar \gamma B_z \sinh(\beta \hbar \gamma B_z)\right]$$

$$= -2\hbar \gamma B_z \frac{\sinh(\beta \hbar \gamma B_z)}{1 + 2\cosh(\beta \hbar \gamma B_z)}$$

Below, graphs of $\sinh(x)$ and $\cosh(x)$ are presented for reference.

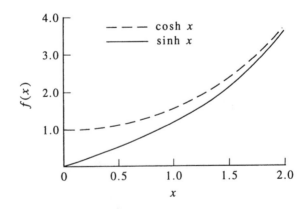

As $T \to 0$, $\beta \to \infty$, so

$$\lim_{T \to 0}\langle E \rangle = \lim_{\beta \to \infty} -2\hbar \gamma B_z \frac{\sinh(\beta \hbar \gamma B_z)}{1 + 2\cosh(\beta \hbar \gamma B_z)} = \lim_{\beta \to \infty} -\frac{\hbar \gamma B_z}{\tanh(\beta \hbar \gamma B_z)} = -\hbar \gamma B_z$$

As $T \to \infty$, $\beta \to 0$, so

$$\lim_{T \to \infty} \langle E \rangle = \lim_{\beta \to 0} -2\hbar \gamma B_z \frac{\sinh(\beta \hbar \gamma B_z)}{1 + 2\cosh(\beta \hbar \gamma B_z)} = 0$$

17–8. If N_w is the number of protons aligned with a magnetic field B_z and N_o is the number of protons opposed to the field, show that

$$\frac{N_o}{N_w} = e^{-\hbar\gamma B_z/k_B T}$$

Given that $\gamma = 26.7522 \times 10^7$ rad·T^{-1}·s^{-1} for a proton, calculate N_o/N_w as a function of temperature for a field strength of 5.0 T. At what temperature is $N_o = N_w$? Interpret this result physically.

The energy of the protons aligned with the magnetic field is $E_w = -\frac{1}{2}\hbar\gamma B_z$, and the energy of the protons aligned against the magnetic field is $E_o = \frac{1}{2}\hbar\gamma B_z$. Using Equation 17.10, we can write

$$\frac{N_o}{N_w} = \exp\left[\beta\left(-\frac{1}{2}\hbar\gamma B_z - \frac{1}{2}\hbar\gamma B_z\right)\right] = \exp\left[-\frac{\hbar\gamma B_z}{k_B T}\right]$$

For a field strength of 5.0 T,

$$\frac{N_o}{N_w} = \exp\left[\frac{-(1.05459 \times 10^{-34}\ \text{J·s})(26.7522 \times 10^7\ \text{rad·T}^{-1}\text{·s}^{-1})(5.0\ \text{T})}{(1.38066 \times 10^{-23}\ \text{J·K}^{-1})T}\right]$$

$$= \exp\left[\frac{-0.010\ \text{K}}{T}\right]$$

Let $N_o = N_w$. Then

$$\frac{N_o}{N_w} = 1 = \exp\left[\frac{-0.010\ \text{K}}{T}\right]$$

$$0 = \frac{-0.010\ \text{K}}{T}$$

and T is undefined. Therefore, we can never attain the condition $N_w = N_o$. (See Section 15–3 for a similar problem, where we find that the population of the ground state is always greater than the population of the excited state in a two-level system.)

17–9. In Section 17–3, we derived an expression for $\langle E \rangle$ for a monatomic ideal gas by applying Equation 17.20 to $Q(N, V, T)$ given by Equation 17.22. Apply Equation 17.21 to

$$Q(N, V, T) = \frac{1}{N!}\left(\frac{2\pi m k_B T}{h^2}\right)^{3N/2} V^N$$

to derive the same result. Note that this expression for $Q(N, V, T)$ is simply Equation 17.22 with β replaced by $1/k_B T$.

Begin with Equation 17.21:

$$\langle E \rangle = k_B T^2 \left(\frac{\partial \ln Q}{\partial T}\right)_{N,V}$$

We can use the partition function given in the problem to find $(\partial \ln Q/\partial T)_{N,V}$:

$$Q(N, V, T) = \frac{V^N}{N!}\left(\frac{2\pi m k_B}{h^2}\right)^{3N/2} T^{3N/2}$$

$$\ln Q = \frac{3N}{2}\ln T + \text{terms not involving } T$$

$$\left(\frac{\partial \ln Q}{\partial T}\right)_{N,V} = \frac{3N}{2T}$$

Substituting this last result into Equation 17.21 gives

$$\langle E \rangle = k_B T^2 \frac{3N}{2T} = \frac{3}{2}(Nk_B T)$$

17–10. A gas absorbed on a surface can sometimes be modelled as a two-dimensional ideal gas. We will learn in Chapter 18 that the partition function of a two-dimensional ideal gas is

$$Q(N, A, T) = \frac{1}{N!}\left(\frac{2\pi m k_B T}{h^2}\right)^N A^N$$

where A is the area of the surface. Derive an expression for $\langle E \rangle$ and compare your result with the three-dimensional result.

We can use the partition function given in the problem to find $(\partial \ln Q/\partial T)_{N,V}$:

$$Q(N, A, T) = \frac{1}{N!}\left(\frac{2\pi m k_B T}{h^2}\right)^N A^N$$

$$\ln Q = N \ln T + \text{terms not involving } T$$

$$\left(\frac{\partial \ln Q}{\partial T}\right)_{N,V} = \frac{N}{T}$$

Substituting this last result into Equation 17.21 gives

$$\langle E \rangle = \frac{Nk_B T^2}{T} = Nk_B T$$

The value for $\langle E \rangle$ for a two-dimensional ideal gas is less than the three-dimensional ideal gas result by $Nk_B T/2$. We infer that each dimension contributes $Nk_B T/2$ to the value of $\langle E \rangle$.

17–11. Although we will not do so in this book, it is possible to derive the partition function for a monatomic van der Waals gas.

$$Q(N, V, T) = \frac{1}{N!}\left(\frac{2\pi m k_B T}{h^2}\right)^{3N/2}(V - Nb)^N e^{aN^2/Vk_B T}$$

where a and b are the van der Waals constants. Derive an expression for the energy of a monatomic van der Waals gas.

We can use the partition function given in the problem to find $(\partial \ln Q/\partial T)_{N,V}$:

$$Q(N, V, T) = \frac{(V - Nb)^N}{N!}\left(\frac{2\pi m k_B}{h^2}\right)^{3N/2} e^{aN^2/Vk_B T} T^{3N/2}$$

$$\ln Q = \frac{3N}{2}\ln T + \frac{aN^2}{Vk_B T} + \text{terms not involving } T$$

$$\left(\frac{\partial \ln Q}{\partial T}\right)_{N,V} = \frac{3N}{2T} - \frac{aN^2}{Vk_B T^2}$$

Substituting this last result into Equation 17.21 gives

$$\langle E \rangle = k_B T^2\left(\frac{3N}{2T} - \frac{aN^2}{Vk_B T^2}\right) = \frac{3}{2}Nk_B T - \frac{aN^2}{V}$$

17–12. An approximate partition function for a gas of hard spheres can be obtained from the partition function of a monatomic gas by replacing V in Equation 17.22 (and the following equation) by $V - b$, where b is related to the volume of the N hard spheres. Derive expressions for the energy and the pressure of this system.

We can use the partition function specified in the problem to find

$$Q(N, V, T) = \frac{(V - b)^N}{N!} \left(\frac{2\pi m k_B}{h^2} \right)^{3N/2} T^{3N/2}$$

$$\ln Q = \frac{3N}{2} \ln T + \text{terms not involving } T$$

Substituting into Equation 17.21, we find that the energy $\langle E \rangle$ is the same as that for a monatomic ideal gas: $3Nk_B T/2$. We can use the partition function specified in the problem to find

$$Q(N, V, T) = \frac{1}{N!} \left(\frac{2\pi m k_B T}{h^2} \right)^{3N/2} (V - b)^N$$

$$\ln Q = N \ln(V - b) + \text{terms not involving } V$$

We substitute into Equation 17.32 to find

$$\langle P \rangle = k_B T \left(\frac{\partial \ln Q}{\partial V} \right)_{N,\beta} = \frac{N k_B T}{V - b}$$

17–13. Use the partition function in Problem 17–10 to calculate the heat capacity of a two-dimensional ideal gas.

In Problem 17–10, we found that $\langle E \rangle = Nk_B T$ for the given partition function. Since $\langle E \rangle = U$, we can substitute into Equation 17.25 to write

$$C_V = \left(\frac{\partial U}{\partial T} \right)_{N,V} = N k_B$$

17–14. Use the partition function for a monatomic van der Waals gas given in Problem 17–11 to calculate the heat capacity of a monatomic van der Waals gas. Compare your result with that of a monatomic ideal gas.

The partition function given in Problem 17–11 is

$$Q(N, V, \beta) = \frac{1}{N!} \left(\frac{2\pi m}{h^2 \beta} \right)^{3N/2} (V - Nb)^N e^{\beta a N^2/V}$$

In Problem 17.11, we found that $\langle E \rangle = 3Nk_B T/2 - aN^2/V$ for a monatomic van der Waals gas. Since $\langle E \rangle = U$, we can substitute into Equation 17.25 to write

$$C_V = \left(\frac{\partial U}{\partial T} \right)_{N,V} = \frac{3Nk_B}{2}$$

The heat capacity of a monatomic van der Waals gas is the same as that of a monatomic ideal gas.

17–15. Using the partition function given in Example 17–2, show that the pressure of an ideal diatomic gas obeys $PV = Nk_BT$, just as it does for a monatomic ideal gas.

We can use the partition function specified in Example 17–2 to find $(\partial \ln Q/\partial V)$:

$$Q(N, V, T) = \frac{1}{N!}\left(\frac{2\pi m}{h^2\beta}\right)^N V^N \left(\frac{8\pi^2 I}{h^2\beta}\right)^N \left(\frac{e^{-\beta h\nu/2}}{1 - e^{-\beta h\nu}}\right)^N$$

$$\ln Q = N \ln V + \text{terms independent of } V$$

$$\left(\frac{\partial \ln Q}{\partial V}\right) = \frac{N}{V}$$

Substituting this result into Equation 17.32 gives

$$\langle P \rangle = \frac{Nk_BT}{V}$$

which is the ideal gas law.

17–16. Show that if a partition function is of the form

$$Q(N, V, T) = \frac{[q(V, T)]^N}{N!}$$

and if $q(V, T) = f(T)V$ [as it does for a monatomic ideal gas (Equation 17.22) and a diatomic ideal gas (Example 17–2)], then the ideal-gas equation of state results.

If $q(V, T) = f(T)V$, then

$$Q = \frac{f(T)^N V^N}{N!}$$

We can use this partition function to find $(\partial \ln Q/\partial V)$:

$$\ln Q = \ln\left(\frac{f(T)^N}{N!}\right) + N \ln V$$

$$= N \ln V + \text{terms independent of } V$$

$$\left(\frac{\partial \ln Q}{\partial V}\right)_{N,\beta} = \frac{N}{V}$$

Substituting this result into Equation 17.32 gives

$$\langle P \rangle = \frac{Nk_BT}{V} = \frac{nRT}{V}$$

which is the ideal gas law.

17–17. Use Equation 17.27 and the value of $\tilde{\nu}$ for O_2 given in Table 5.1 to calculate the value of the molar heat capacity of $O_2(g)$ from 300 K to 1000 K (see Figure 17.3).

From Table 5.1, $\tilde{\nu} = 1556 \text{ cm}^{-1}$. Using this value, we can write

$$\frac{hc\tilde{\nu}}{k_B} = \frac{(6.626 \times 10^{-34} \text{ J·s}^{-1})(2.998 \times 10^{10} \text{ cm·s}^{-1})(1556 \text{ cm}^{-1})}{1.3807 \times 10^{-23} \text{ J·K}^{-1}} = 2240 \text{ K}$$

We can substitute this value into Equation 17.27 to obtain the expression

$$\frac{\overline{C}_V}{R} = \frac{5}{2} + \left(\frac{2240 \text{ K}}{T}\right)^2 \frac{e^{-2240 \text{ K}/T}}{(1 - e^{-2240 \text{ K}/T})^2}$$

We plot $\overline{C}_V$ versus T below.

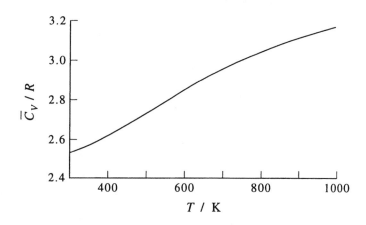

17–18. Show that the heat capacity given by Equation 17.29 in Example 17–3 obeys a law of corresponding states.

Let $h\nu/k_B = \Theta_E$. We can then write Equation 17.29 as

$$\frac{\overline{C}_V}{R} = 3\left(\frac{\Theta_E}{T}\right)^2 \frac{e^{-\Theta_E/T}}{(1 - e^{-\Theta_E/T})^2}$$

Now let $T_R = T/\Theta_E$. The above expression becomes

$$\frac{\overline{C}_V}{R} = 3\left(\frac{1}{T_R}\right)^2 \frac{e^{-1/T_R}}{(1 - e^{-1/T_R})^2}$$

This equation obeys a law of corresponding states.

17–19. Consider a system of independent, distinguishable particles that have only two quantum states with energy ε_0 (let $\varepsilon_0 = 0$) and ε_1. Show that the molar heat capacity of such a system is given by

$$\overline{C}_V = R(\beta\varepsilon)^2 \frac{e^{-\beta\varepsilon}}{(1 + e^{-\beta\varepsilon})^2}$$

and that $\overline{C}_V$ plotted against $\beta\varepsilon$ passes through a maximum value at $\beta\varepsilon$, given by the solution to $\beta\varepsilon/2 = \coth \beta\varepsilon/2$. Use a table of values of $\coth x$ (for example, the *CRC Standard Mathematical Tables*) to show that $\beta\varepsilon = 2.40$.

For the system described above, Equation 17.34 becomes

$$q = \sum_i e^{-\beta\varepsilon_i} = 1 + e^{-\beta\varepsilon_1} = 1 + e^{-\beta\varepsilon}$$

and Equation 17.43 becomes

$$\langle \varepsilon \rangle = \sum_j \varepsilon_j \frac{e^{-\beta \varepsilon_j}}{q} = \frac{\varepsilon_1 e^{-\beta \varepsilon_1}}{1 + e^{-\beta \varepsilon_1}} = \frac{\varepsilon e^{-\beta \varepsilon}}{1 + e^{-\beta \varepsilon}}$$

where we have dropped the subscript "1" on ε in both of the above expressions for simplicity. We can use Equation 17.42 to write $\langle E \rangle$ as

$$\langle E \rangle = N \langle \varepsilon \rangle = \frac{N \varepsilon e^{-\beta \varepsilon}}{1 + e^{-\beta \varepsilon}}$$

We now substitute this last expression into Equation 17.25 to find C_V:

$$C_V = \frac{\partial \langle E \rangle}{\partial T} = -\frac{1}{k_B T^2} \left(\frac{\partial \langle E \rangle}{\partial \beta} \right)$$

$$= -\frac{N}{k_B T^2} \left[\frac{-\varepsilon^2 e^{-\beta \varepsilon}}{1 + e^{-\beta \varepsilon}} + \frac{\varepsilon^2 e^{-2\beta \varepsilon}}{(1 + e^{-\beta \varepsilon})^2} \right]$$

$$= -\frac{N}{k_B T^2} \left[\frac{-\varepsilon^2 e^{-\beta \varepsilon}}{(1 + e^{-\beta \varepsilon})^2} \right] = N k_B \frac{(\beta \varepsilon)^2 e^{-\beta \varepsilon}}{(1 + e^{-\beta \varepsilon})^2}$$

$$\overline{C}_V = \frac{R(\beta \varepsilon)^2 e^{-\beta \varepsilon}}{(1 + e^{-\beta \varepsilon})^2}$$

We now wish to find the maximum value of $\overline{C}_V$. Let $\beta \varepsilon = x$. Then

$$\frac{\overline{C}_V}{R} = \frac{x^2 e^{-x}}{(1 + e^{-x})^2}$$

$$\frac{\partial (\overline{C}_V / R)}{\partial x} = \frac{2x e^{-x}}{(1 + e^{-x})^2} - \frac{x^2 e^{-x}}{(1 + e^{-x})^2} + \frac{2x^2 e^{-2x}}{(1 + e^{-x})^3}$$

At the maximum value of $\overline{C}_V$, $(\partial \overline{C}_V / \partial x) = 0$, so

$$0 = 2 - x + 2x \frac{e^{-x}}{1 + e^{-x}}$$

$$2 = x \left(1 - \frac{2e^{-x}}{1 + e^{-x}} \right) = x \left(\frac{1 - e^{-x}}{1 + e^{-x}} \right)$$

$$\frac{1 + e^{-x}}{1 - e^{-x}} = \frac{x}{2}$$

$$\frac{e^{x/2} + e^{-x/2}}{e^{x/2} - e^{-x/2}} = \frac{x}{2}$$

$$\coth \frac{x}{2} = \frac{x}{2}$$

$$x = 2.40 = \beta \varepsilon$$

and $x = \beta \varepsilon = 2.40$.

17–20. Deriving the partition function for an Einstein crystal is not difficult (see Example 17–3). Each of the N atoms of the crystal is assumed to vibrate independently about its lattice position, so that the crystal is pictured as N independent harmonic oscillators, each vibrating in three directions. The partition function of a harmonic oscillator is

$$q_{ho}(T) = \sum_{v=0}^{\infty} e^{-\beta (v + \frac{1}{2}) h \nu}$$

$$= e^{-\beta h \nu / 2} \sum_{v=0}^{\infty} e^{-\beta v h \nu}$$

This summation is easy to evaluate if you recognize it as the so-called geometric series (MathChapter I)

$$\sum_{v=0}^{\infty} x^v = \frac{1}{1-x}$$

Show that

$$q_{\text{ho}}(T) = \frac{e^{-\beta h v/2}}{1 - e^{-\beta h v}}$$

and that

$$Q = e^{-\beta U_0}\left(\frac{e^{-\beta h v/2}}{1 - e^{-\beta h v}}\right)^{3N}$$

where U_0 simply represents the zero-of-energy, where all N atoms are infinitely separated.

For a one-dimensional harmonic oscillator,

$$q_{\text{ho}}(T) = e^{-\beta h v/2}\sum_{v=0}^{\infty} e^{-\beta v h v}$$

Let $x = e^{-\beta h v}$. Then

$$q_{\text{ho}}(T) = e^{-\beta h v/2}\sum_{v=0}^{\infty} x^v = e^{-\beta h v/2}\frac{1}{1-x} = \frac{e^{-\beta h v/2}}{1 - e^{-\beta h v}}$$

A three-dimensional harmonic oscillator can be viewed as three independent one-dimensional harmonic oscillators, and so in three dimensions

$$q_{\text{3D-ho}}(T) = [q_{\text{ho}}(T)]^3 = \left(\frac{e^{-\beta h v/2}}{1 - e^{-\beta h v}}\right)^3$$

Because each particle is distinguishable (due to its position in the lattice), the system partition function can be written as

$$Q = q_{\text{3D-ho},a}(V, T)q_{\text{3D-ho},b}(V, T)q_{\text{3D-ho},c}(V, T)...\tag{17.33}$$

where $a, b, c, ...$ are independent atoms. Then

$$Q = \left(\frac{e^{-\beta h v/2}}{1 - e^{-\beta h v}}\right)^{3N} e^{-\beta U_0}$$

The $e^{-\beta U_0}$ term takes the zero-of-energy into account.

17–21. Show that

$$S = \sum_{i=1}^{2}\sum_{j=0}^{1} x^i y^j = x(1 + y) + x^2(1 + y) = (x + x^2)(1 + y)$$

by summing over j first and then over i. Now obtain the same result by writing S as a product of two separate summations.

Summing first over j and then over i, we obtain

$$S = \sum_{i=1}^{2}\sum_{j=0}^{1} x^i y^j = \sum_{i=1}^{2} x^i \left(y^0 + y^1\right) = x(1+y) + x^2(1+y) = (x+x^2)(1+y)$$

Writing S as a product of two separate summations, we find

$$S = \sum_{i=1}^{2} x^i \sum_{j=0}^{1} y^j = (x+x^2)(1+y)$$

17–22. Evaluate

$$S = \sum_{i=0}^{2}\sum_{j=0}^{1} x^{i+j}.$$

by summing over j first and then over i. Now obtain the same result by writing S as a product of two separate summations.

Summing first over j and then over i, we obtain

$$S = \sum_{i=0}^{2}\sum_{j=0}^{1} x^{i+j} = \sum_{i=0}^{2}(x^{i+0} + x^{i+1}) = 1 + x + x^2 + x + x^2 + x^3 = 1 + 2x + 2x^2 + x^3$$

Writing S as a product of two separate summations, we find

$$\sum_{i=0}^{2} x^i \sum_{j=0}^{1} x^j = (1 + x + x^2)(1 + x) = 1 + 2x + 2x^2 + x^3$$

17–23. How many terms are there in the following summations?

a. $S = \sum_{i=1}^{3}\sum_{j=1}^{2} x^i y^j$ b. $S = \sum_{i=1}^{3}\sum_{j=0}^{2} x^i y^j$ c. $S = \sum_{i=1}^{3}\sum_{j=1}^{2}\sum_{k=1}^{2} x^i y^j z^k$

---

a. 6 terms

b. 9 terms

c. 12 terms

17–24. Consider a system of two noninteracting identical fermions, each of which has states with energies ε_1, ε_2, and ε_3. How many terms are there in the unrestricted evaluation of $Q(2, V, T)$? Enumerate the allowed total energies in the summation in Equation 17.37 (see Example 17–5). How many terms occur in $Q(2, V, T)$ when the fermion restriction is taken into account?

There are nine terms in the unrestricted evaluation. The allowed total energies given by the summation in Equation 17.37 are

$$\varepsilon_1 + \varepsilon_3 = \varepsilon_3 + \varepsilon_1 \qquad \varepsilon_1 + \varepsilon_1$$
$$\varepsilon_1 + \varepsilon_2 = \varepsilon_2 + \varepsilon_1 \qquad \varepsilon_2 + \varepsilon_2$$
$$\varepsilon_2 + \varepsilon_3 = \varepsilon_3 + \varepsilon_2 \qquad \varepsilon_2 + \varepsilon_2$$

When the fermion restriction is taken into account (no two identical fermions can occupy the same single-particle energy state), the allowed total energies are $\varepsilon_1 + \varepsilon_3$, $\varepsilon_1 + \varepsilon_2$, and $\varepsilon_2 + \varepsilon_3$, and so there are only three terms in $Q(2, V, T)$.

17–25. Redo Problem 17–24 for the case of bosons instead of fermions.

In this case there are six allowed terms: the three alloweed in Problem 17–24 and the three in which the ε_j are the same ($\varepsilon_1 + \varepsilon_1$, $\varepsilon_2 + \varepsilon_2$, and $\varepsilon_3 + \varepsilon_3$).

17–26. Consider a system of three noninteracting identical fermions, each of which has states with energies ε_1, ε_2, and ε_3. How many terms are there in the unrestricted evaluation of $Q(3, V, T)$? Enumerate the allowed total energies in the summation of Equation 17.37 (see Example 17–5). How many terms occur in $Q(3, V, T)$ when the fermion restriction is taken into account?

There are 27 terms in the unrestricted evaluation. The allowed total energies given by the summation in Equation 17.37 are

$$
\begin{aligned}
1. \quad & \varepsilon_1 + \varepsilon_2 + \varepsilon_3 = \varepsilon_1 + \varepsilon_3 + \varepsilon_2 = \varepsilon_2 + \varepsilon_1 + \varepsilon_3 = \varepsilon_2 + \varepsilon_3 + \varepsilon_1 \\
& = \varepsilon_3 + \varepsilon_1 + \varepsilon_2 = \varepsilon_3 + \varepsilon_2 + \varepsilon_1 \\
2. \quad & \varepsilon_1 + \varepsilon_1 + \varepsilon_2 = \varepsilon_1 + \varepsilon_2 + \varepsilon_1 = \varepsilon_2 + \varepsilon_1 + \varepsilon_1 \\
3. \quad & \varepsilon_1 + \varepsilon_1 + \varepsilon_3 = \varepsilon_1 + \varepsilon_3 + \varepsilon_1 = \varepsilon_3 + \varepsilon_1 + \varepsilon_1 \\
4. \quad & \varepsilon_1 + \varepsilon_2 + \varepsilon_2 = \varepsilon_2 + \varepsilon_2 + \varepsilon_1 = \varepsilon_2 + \varepsilon_1 + \varepsilon_2 \\
5. \quad & \varepsilon_1 + \varepsilon_3 + \varepsilon_3 = \varepsilon_3 + \varepsilon_3 + \varepsilon_1 = \varepsilon_3 + \varepsilon_1 + \varepsilon_3 \\
6. \quad & \varepsilon_2 + \varepsilon_3 + \varepsilon_3 = \varepsilon_3 + \varepsilon_3 + \varepsilon_2 = \varepsilon_3 + \varepsilon_2 + \varepsilon_3 \\
7. \quad & \varepsilon_2 + \varepsilon_2 + \varepsilon_3 = \varepsilon_2 + \varepsilon_3 + \varepsilon_2 = \varepsilon_3 + \varepsilon_2 + \varepsilon_2 \\
8. \quad & \varepsilon_1 + \varepsilon_1 + \varepsilon_1 \\
9. \quad & \varepsilon_2 + \varepsilon_2 + \varepsilon_2 \\
10. \quad & \varepsilon_3 + \varepsilon_3 + \varepsilon_3
\end{aligned}
$$

When the fermion restriction is taken into account (no two identical fermions can occupy the same single-particle energy state), the only allowed total energy is $\varepsilon_1 + \varepsilon_2 + \varepsilon_3$, and so there is only one term in $Q(3, V, T)$.

17–27. Redo Problem 17–26 for the case of bosons instead of fermions.

In this case there are ten allowed terms, given by the total energies

$$
\begin{array}{ll}
\varepsilon_1 + \varepsilon_2 + \varepsilon_3 & \varepsilon_2 + \varepsilon_2 + \varepsilon_3 \\
\varepsilon_1 + \varepsilon_1 + \varepsilon_3 & \varepsilon_1 + \varepsilon_3 + \varepsilon_3 \\
\varepsilon_2 + \varepsilon_2 + \varepsilon_3 & \varepsilon_3 + \varepsilon_2 + \varepsilon_3 \\
\varepsilon_2 + \varepsilon_2 + \varepsilon_1 & \varepsilon_1 + \varepsilon_1 + \varepsilon_1 \\
\varepsilon_2 + \varepsilon_2 + \varepsilon_2 & \varepsilon_3 + \varepsilon_3 + \varepsilon_3
\end{array}
$$

17–28. Evaluate $(N/V)(h^2/8mk_BT)^{3/2}$ (see Table 17.1) for $O_2(g)$ at its normal boiling point, 90.20 K. Use the ideal-gas equation of state to calculate the density of $O_2(g)$ at 90.20 K.

Using the ideal-gas equation of state,

$$\frac{N}{V} = \frac{P}{k_B T} = \frac{1.013 \times 10^5 \text{ kg·m}^{-1} \cdot \text{s}^{-2}}{(1.381 \times 10^{-23} \text{ J·K}^{-1})(90.20 \text{ K})} = 8.134 \times 10^{25} \text{ kg·m}^{-3}$$

Then

$$\frac{N}{V}\left(\frac{h^2}{8mk_B T}\right)^{3/2} = 8.134 \times 10^{25} \text{ kg·m}^{-3}\left[\frac{(6.626 \times 10^{-34} \text{ J·s})^2}{8(5.315 \times 10^{-26} \text{ kg})(1.381 \times 10^{-23} \text{ J·K}^{-1})(90.20 \text{ K})}\right]^{3/2}$$

$$= 1.943 \times 10^{-6}$$

which is much less than unity.

17–29. Evaluate $(N/V)(h^2/8mk_B T)^{3/2}$ (see Table 17.1) for He(g) at its normal boiling point 4.22 K. Use the ideal-gas equation of state to calculate the density of He(g) at 4.22 K.

Using the ideal-gas equation of state,

$$\frac{N}{V} = \frac{P}{k_B T} = \frac{1.013 \times 10^5 \text{ kg·m}^{-1} \cdot \text{s}^{-2}}{(1.381 \times 10^{-23} \text{ J·K}^{-1})(4.22 \text{ K})} = 1.739 \times 10^{27} \text{ kg·m}^{-3}$$

Then

$$\frac{N}{V}\left(\frac{h^2}{8mk_B T}\right)^{3/2} = 1.739 \times 10^{27} \text{ kg·m}^{-3}\left[\frac{(6.626 \times 10^{-34} \text{ J·s})^2}{8(6.646 \times 10^{-27} \text{ kg})(1.381 \times 10^{-23} \text{ J·K}^{-1})(4.22 \text{ K})}\right]^{3/2}$$

$$= 0.0928$$

which is not much less than unity, because of the small mass of He and the fact that here we consider helium at 4.22 K.

17–30. Evaluate $(N/V)(h^2/8mk_B T)^{3/2}$ for the electrons in sodium metal at 298 K. Take the density of sodium to be 0.97 g·mL^{-1}. Compare your result with the value given in Table 17.1.

The mass of an electron is 9.11×10^{-31} kg, and the density of sodium is 2.54×10^{28} particles per cubic meter. Thus

$$\frac{N}{V}\left(\frac{h^2}{8mk_B T}\right)^{3/2} = (2.54 \times 10^{28} \text{ m}^3)\left[\frac{(6.626 \times 10^{-34} \text{ J·s})^2}{8(9.11 \times 10^{-31} \text{ kg})(1.381 \times 10^{-23} \text{ J·K}^{-1})(298 \text{ K})}\right]^{3/2} = 1420$$

which, because of the very small mass of an electron, is much greater than unity. This is essentially equivalent to the value given in Table 5.1.

17–31. Evaluate $(N/V)(h^2/8mk_B T)^{3/2}$ (see Table 17.1) for liquid hydrogen at its normal boiling point 20.3 K. The density of H_2(l) at its boiling point is 0.067 g·mL^{-1}.

The density of hydrogen gas is 2.00×10^{28} particles per cubic meter. Thus

$$\frac{N}{V}\left(\frac{h^2}{8mk_B T}\right)^{3/2} = (2.00 \times 10^{28} \text{ m}^{-3})\left[\frac{(6.626 \times 10^{-34} \text{ J·s})^2}{8(3.35 \times 10^{-27} \text{ kg})(1.381 \times 10^{-23} \text{ J·K}^{-1})(20.3 \text{ K})}\right]^{3/2}$$

$$= 0.283$$

17–32. Because the molecules in an ideal gas are independent, the partition function of a mixture of monatomic ideal gases is of the form

$$Q(N_1, N_2, V, T) = \frac{[q_1(V, T)]^{N_1}}{N_1!} \frac{[q_2(V, T)]^{N_2}}{N_2!}$$

where

$$q_j(V, T) = \left(\frac{2\pi m_j k_B T}{h^2}\right)^{3/2} V \qquad j = 1, 2$$

Show that

$$\langle E \rangle = \frac{3}{2}(N_1 + N_2)k_B T$$

and that

$$PV = (N_1 + N_2)k_B T$$

for a mixture of monatomic ideal gases.

We can use the given partition function to find $\ln Q$:

$$Q = \frac{q_1^{N_1} q_2^{N_2}}{N_1! N_2!}$$

$$\ln Q = \ln \frac{q_1^{N_1}}{N_1!} + \ln \frac{q_2^{N_2}}{N_2!} = N_1 \ln q_1 + N_2 \ln q_2 - \ln(N_2!) - \ln(N_1!)$$

Using the definition of q_j given in the problem, we can find $(\partial \ln Q/\partial \beta)_{N,V}$ and $(\partial \ln Q/\partial V)_{N,\beta}$:

$$\ln Q = -\frac{3N_1}{2} \ln \beta - \frac{3N_2}{2} \ln \beta + \text{non-}\beta\text{-related terms}$$

$$\left(\frac{\partial \ln Q}{\partial \beta}\right)_{N,V} = -\frac{3}{2}\left(\frac{N_1 + N_2}{\beta}\right)$$

$$\ln Q = N_1 \ln V + N_2 \ln V + \text{non-}V\text{-related terms}$$

$$\left(\frac{\partial \ln Q}{\partial V}\right)_{N,\beta} = \frac{N_1 + N_2}{V}$$

Now we use Equations 17.20 and 17.32 to find $\langle E \rangle$ and $\langle P \rangle$:

$$\langle E \rangle = \frac{3}{2}\left(\frac{N_1 + N_2}{\beta}\right) = \frac{3}{2}(N_1 + N_2)k_B T$$

and

$$\langle P \rangle = k_B T \left(\frac{N_1 + N_2}{V}\right)$$

$$PV = k_B T \left(N_1 + N_2\right)$$

17–33. We will learn in Chapter 18 that the rotational partition function of an asymmetric top molecule is given by

$$q_{rot}(T) = \frac{\pi^{1/2}}{\sigma} \left(\frac{8\pi^2 I_A k_B T}{h^2}\right)^{1/2} \left(\frac{8\pi^2 I_B k_B T}{h^2}\right)^{1/2} \left(\frac{8\pi^2 I_C k_B T}{h^2}\right)^{1/2}$$

where σ is a constant and I_A, I_B, and I_C are the three (distinct) moments of inertia. Show that the rotational contribution to the molar heat capacity is $\overline{C}_{V,rot} = \frac{3}{2} R$.

_____

Using the partition function in the problem text, we can write

$$\ln q_{rot} = \frac{3}{2} \ln T + \text{non-}T\text{-related terms}$$

Then we can use Equation 17.21 to write

$$\langle E \rangle_{rot} = k_B T^2 \left(\frac{\partial \ln q_{rot}}{\partial T}\right) = \frac{3}{2} k_B T$$

and Equation 17.25 to write

$$C_{V,rot} = \frac{\partial \langle E \rangle_{rot}}{\partial T} = \frac{3}{2} k_B$$

For N_A particles, $\overline{C}_{V,rot} = 3 k_B N_A / 2 = 3R/2$.

_____

17–34. The allowed energies of a harmonic oscillator are given by $\varepsilon_v = (v + \frac{1}{2})h\nu$. The corresponding partition function is given by

$$q_{vib}(T) = \sum_{v=0}^{\infty} e^{-(v+\frac{1}{2})h\nu/k_B T}$$

Let $x = e^{-h\nu/k_B T}$ and use the formula for the summation of a geometric series (Problem 17–20) to show that

$$q_{vib}(T) = \frac{e^{-h\nu/2k_B T}}{1 - e^{-h\nu/k_B T}}$$

_____

Let $x = e^{-h\nu/k_B T}$. The equation for the partition function given in the problem text then becomes

$$q_{vib}(T) = \sum_{v=0}^{\infty} x^{v+1/2} = x^{1/2} \sum_{v=0}^{\infty} x^v$$

$$= \frac{x^{1/2}}{1 - x} = \frac{e^{-h\nu/2k_B T}}{1 - e^{-h\nu/k_B T}}$$

_____

17–35. Derive an expression for the probability that a harmonic oscillator will be found in the vth state. Calculate the probability that the first few vibrational states are occupied for HCl(g) at 300 K. (See Table 5–1 and Problem 17–34.)

_____

$$\text{prob}\,(v = j) = \frac{e^{-\beta \varepsilon_{vib,j}}}{q_{vib}}$$

Using the harmonic oscillator partition function, we find that

$$\text{prob } (v = j) = \frac{e^{-\beta h v v} e^{-\beta h v/2} (1 - e^{-h\beta v})}{e^{-\beta h v/2}}$$

$$= e^{-\beta h v v} \left(1 - e^{-\beta h v}\right)$$

For HCl, $h v/k_B = 4140$ K. The probabilities of the first three vibrational states of HCl being occupied at 300 K are

$$\text{prob } (v = 0) = \left(1 - e^{-\beta h v}\right) \qquad = 0.99999898$$
$$\text{prob } (v = 1) = e^{-\beta h v} \left(1 - e^{-\beta h v}\right) = 1.01 \times 10^{-6}$$
$$\text{prob } (v = 2) = e^{-2\beta h v} \left(1 - e^{-\beta h v}\right) = 1.03 \times 10^{-12}$$

It is clear that at room temperature, the majority of the gaseous HCl molecules are in the ground vibrational state.

17-36. Show that the fraction of harmonic oscillators in the ground vibrational state is given by

$$f_0 = 1 - e^{-h v/k_B T}$$

Calculate f_0 for N_2 at 300 K, 600 K, and 1000 K (see Table 5.1).

The fraction of harmonic oscillators in the ground vibrational state is the same as the probability that a harmonic oscillator will be in the ground vibrational state (from Problem 17.35):

$$f_0 = \text{prob } (v = 0) = 1 - e^{-h v/k_B T}$$

For N_2, $h v/k_B = 3352$ K.

Temperature	f_0
300 K	1.000
600 K	0.9962
1000 K	0.9650

17-37. Use Equation 17.55 to show that the fraction of rigid rotators in the Jth rotational level is given by

$$f_J = \frac{(2J + 1)e^{-\hbar^2 J(J+1)/2Ik_B T}}{q_{rot}(T)}$$

Plot the fraction in the Jth level relative to the $J = 0$ level (f_J/f_0) against J for HCl(g) at 300 K. Take $\tilde{B} = 10.44$ cm^{-1}.

$$q_{rot} = \sum_{J=0}^{\infty} (2J + 1)e^{-\hbar^2 J(J+1)/2Ik_B T} \tag{17.55}$$

The number of rigid rotators in the jth rotational level is the $J = j$ term in the above sum. Thus, the fraction of rigid rotators in the Jth rotational level can be expressed as

$$f_J = \frac{(2J + 1)e^{-\hbar^2 J(J+1)/2Ik_B T}}{q_{rot}}$$

and

$$\frac{f_J}{f_0} = \frac{(2J+1)e^{-\hbar^2 J(J+1)/2Ik_B T}}{1}$$

For HCl(g), $B_{HCl} = 3.13 \times 10^{11}$ Hz. So $I = h/8\pi^2 B = 2.68 \times 10^{-47}$ J. Below is a plot of (f_J/f_0) vs. J for HCl at 300 K.

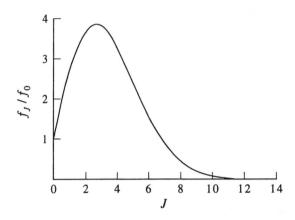

17–38. Equations 17.20 and 17.21 give the ensemble average of E, which we assert is the same as the experimentally observed value. In this problem, we will explore the standard deviation about $\langle E \rangle$ (MathChapter B). We start with either Equation 17.20 or 17.21:

$$\langle E \rangle = U = -\left(\frac{\partial \ln Q}{\partial \beta}\right)_{N,V} = k_B T^2 \left(\frac{\partial \ln Q}{\partial T}\right)_{N,V}$$

Differentiate again with respect to β or T to show that (MathChapter B)

$$\sigma_E^2 = \langle E^2 \rangle - \langle E \rangle^2 = k_B T^2 C_V$$

where C_V is the heat capacity. To explore the relative magnitude of the spread about $\langle E \rangle$, consider

$$\frac{\sigma_E}{\langle E \rangle} = \frac{(k_B T C_V)^{1/2}}{\langle E \rangle}$$

To get an idea of the size of this ratio, use the values of $\langle E \rangle$ and C_V for a (monatomic) ideal gas, namely, $\frac{3}{2} N k_B T$ and $\frac{3}{2} N k_B$, respectively, and show that $\sigma_E/\langle E \rangle$ goes as $N^{-1/2}$. What does this trend say about the likely observed deviations from the average macroscopic energy?

We use the following equalities (from Equation 17.18):

$$\langle E \rangle = \frac{\sum_j E_j e^{-\beta E_j}}{Q}$$

$$\langle E^2 \rangle = \frac{\sum_j E_j^2 e^{-\beta E_j}}{Q}$$

Now we can find σ_E^2:

$$\sigma_E^2 = \langle E^2 \rangle - \langle E \rangle^2 = \frac{\sum_j E_j^2 e^{-\beta E_j}}{Q} - \left(\frac{\sum_j E_j e^{-\beta E_j}}{Q}\right)^2$$

We can substitute the definition $Q = \sum_j e^{-\beta E_j}$ into Equation 17.18 and take the partial derivative with respect to β to find

$$\langle E \rangle = \frac{\sum_j E_j e^{-\beta E_j}}{\sum_j e^{-\beta E_j}}$$

$$\left(\frac{\partial \langle E \rangle}{\partial \beta} \right) = -\frac{\sum_j E_j^2 e^{-\beta E_j}}{\sum_j e^{-\beta E_j}} + \frac{(\sum_j E_j e^{-\beta E_j})(\sum_j E_j e^{-\beta E_j})}{\sum_j e^{-2\beta E_j}}$$

$$= -\frac{\sum_j E_j^2 e^{-\beta E_j}}{Q} + \left(\frac{\sum_j E_j e^{-\beta E_j}}{Q} \right)^2 = -\sigma_E^2$$

We define C_V as (Example 17–3)

$$C_V = \left(\frac{\partial \langle E \rangle}{\partial T} \right)_{N,V} = -\frac{1}{k_B T^2} \left(\frac{\partial \langle E \rangle}{\partial \beta} \right)_{N,V} = \frac{1}{k_B T^2} \sigma_E^2$$

So $\sigma_E^2 = k_B T^2 C_V$. Now consider the spread about $\langle E \rangle$ for a monatomic ideal gas:

$$\frac{\sigma_E}{\langle E \rangle} = \frac{(k_B T^2 C_V)^{1/2}}{\langle E \rangle}$$

$$= \frac{[k_B T^2 (\frac{3}{2} N k_B)]^{1/2}}{\frac{3}{2} N k_B T}$$

$$= \frac{1}{(\frac{3}{2})^{1/2} N^{1/2}} \sim N^{-1/2}$$

Because $N^{-1/2}$ is on the order of 10^{-12} for one mole, it is very unlikely that significant deviations from the average macroscopic energy will be observed.

17–39. Following Problem 17–38, show that the variance about the average values of a *molecular* energy is given by

$$\sigma_\varepsilon^2 = \langle \varepsilon^2 \rangle - \langle \varepsilon \rangle^2 = \frac{k_B T^2 C_V}{N}$$

and that $\sigma_\varepsilon / \langle \varepsilon \rangle$ goes as order unity. What does this result say about the deviations from the average molecular energy?

We use the following equalities:

$$q = e^{-\beta \varepsilon_j}$$

$$\langle \varepsilon \rangle = \frac{\sum_j \varepsilon_j e^{-\beta \varepsilon_j}}{q}$$

$$\langle \varepsilon^2 \rangle = \frac{\sum_j \varepsilon_j^2 e^{-\beta \varepsilon_j}}{q}$$

We know that $N \langle \varepsilon \rangle = \langle E \rangle$, where N is the number of particles with molecular energy ε. Thus

$$C_V = \left(\frac{\partial \langle E \rangle}{\partial T} \right) = N \left(\frac{\partial \langle \varepsilon \rangle}{\partial T} \right)$$

We can follow the steps in Problem 17.39, subsituting ε for E and q for Q, to find

$$\sigma_\varepsilon^2 = k_B T^2 \left(\frac{\partial \langle \varepsilon \rangle}{\partial T} \right)$$

$$= k_B T^2 C_V N^{-1}$$

For a monatomic ideal gas, $C_V N^{-1} = \frac{3}{2} k_B$. Now consider the spread about $\langle \varepsilon \rangle$ for a monatomic ideal gas:

$$\frac{\sigma_E}{\langle E \rangle} = \frac{\left(\frac{3}{2} k_B^2 T^2 \right)^{1/2}}{\langle \varepsilon \rangle}$$

$$= \frac{\left(\frac{3}{2} \right)^{1/2} k_B T}{\frac{3}{2} k_B T}$$

$$= \left(\frac{3}{2} \right)^{-1/2} \approx 0.8$$

which is of order unity. The molecular energy is much more likely to deviate from the average molecular energy than the corresponding macroscopic energy is to deviate from the average macroscopic energy.

17–40. Use the result of Problem 17–38 to show that C_V is never negative.

The quantity σ^2 must be positive, because σ is a real number and the square of any real number is always positive. Thus $k_B T^2 C_V$ must be a positive number. Both k_B and T are always positive; therefore, C_V must always be positive.

17–41. The lowest electronic states of Na(g) are tabulated below.

Term symbol	Energy/cm^{-1}	Degeneracy
$^2S_{1/2}$	0.000	2
$^2P_{1/2}$	16 956.183	2
$^2P_{3/2}$	16 973.379	4
$^2S_{1/2}$	25 739.86	2

Calculate the fraction of the atoms in each of these electronic states in a sample of Na(g) at 1000 K. Repeat this calculation for a temperature of 2500 K.

The probability of an atom being in the lth electronic state is (Equation 17.44)

$$\text{prob}_l = \frac{e^{-\varepsilon_l/k_B T}}{\sum_l e^{-\varepsilon_l/k_B T}}$$

where, for Na(g),

$$\sum_l e^{-\varepsilon_l/k_B T} = 2e^{-0/k_B T} + 2e^{-16\,956.183/k_B T} + 4e^{-16\,973.379/k_B T} + 2e^{-25\,739.86/k_B T}$$

To calculate the fraction of atoms in each electronic state, we multiply the state's degeneracy by $e^{-\varepsilon_l/k_B T}$ and divide by the summation above. (Note: $k_B = 0.69509$ cm^{-1}.)

Term Symbol	Fraction of atoms, 1000 K	Fraction of atoms, 2500 K
$^2S_{1/2}$	1.000	0.9998
$^2P_{1/2}$	2.545×10^{-11}	5.784×10^{-5}
$^2P_{3/2}$	4.966×10^{-11}	1.145×10^{-4}
$^2S_{1/2}$	8.273×10^{-17}	3.690×10^{-7}

17–42. The vibrational frequency of NaCl(g) is 159.23 cm^{-1}. Calculate the molar heat capacity of $\overline{C}_V$ at 1000 K. (See Equation 17.27.)

$$\overline{C}_V = \frac{5R}{2} + R\left(\frac{h\nu}{k_B T}\right)^2 \frac{e^{-h\nu/k_B T}}{\left(1 - e^{-h\nu/k_B T}\right)^2}$$

$\tilde{\nu}_{\text{NaCl(g)}} = 159.23$ cm^{-1}, so $h\nu/k_B T = 0.22909$, and

$$\overline{C}_V = \left[\frac{5}{2} + (0.22909)^2 \frac{e^{0.22909}}{\left(1 - e^{0.22909}\right)^2}\right] R = 3.4956R = 29.064 \text{ J}\cdot\text{mol}^{-1}\cdot\text{K}^{-1}$$

17–43. The energies and degeneracies of the two lowest electronic states of atomic iodine are listed below.

Energy/cm^{-1}	Degeneracy
0	4
7603.2	2

What temperature is required so that 2% of the atoms are in the excited state?

As in Problem 17–41, we have

$$\text{prob}_l = \frac{e^{-\varepsilon_l/k_B T}}{\sum_l e^{-\varepsilon_l/k_B T}} \tag{17.44}$$

For 2% of the atoms to be in the excited state, we can substitute into the expression above and write

$$0.02 = 2\exp\frac{-7603.2}{0.69509T}\left(4 + 2\exp\frac{-7603.2}{0.69509T}\right)^{-1}$$

$$\ln\frac{0.04}{0.98} = \frac{-7603.2}{0.69509T}$$

$$T = 3420 \text{ K}$$

Series and Limits

PROBLEMS AND SOLUTIONS

I–1. Calculate the percentage difference between e^x and $1 + x$ for $x = 0.0050, 0.0100, 0.0150, \ldots, 0.1000$.

We define percentage difference as

$$\left| \frac{e^x - (1 + x)}{e^x} \right|$$

Then we can create the table below:

x	e^x	$1 + x$	percentage difference
0.0050	1.00501	1.0050	0.0012%
0.0100	1.01005	1.0100	0.0050%
0.0150	1.01511	1.0150	0.0111%
0.0200	1.02020	1.0200	0.0197%
0.0250	1.02532	1.0250	0.0307%
0.0300	1.03045	1.0300	0.0441%
0.0350	1.03562	1.0350	0.0598%
0.0400	1.04081	1.0400	0.0779%
0.0450	1.04603	1.0450	0.0983%
0.0500	1.05127	1.0500	0.1209%
0.0550	1.05654	1.0550	0.1458%
0.0600	1.06184	1.0600	0.1729%
0.0650	1.06716	1.0650	0.2023%
0.0700	1.07251	1.0700	0.2339%
0.0750	1.07788	1.0750	0.2676%
0.0800	1.08329	1.0800	0.3034%
0.0850	1.08872	1.0850	0.3414%
0.0900	1.09417	1.0900	0.3815%
0.0950	1.09966	1.0950	0.4237%
0.1000	1.10517	1.1000	0.4679%

I–2. Calculate the percentage difference between $\ln(1 + x)$ and x for $x = 0.0050, 0.0100,$ $0.0150, \ldots, 0.1000.$

We define percentage difference as

$$\left| \frac{\ln(1 + x) - x}{\ln(1 + x)} \right|$$

Then we can create the table below:

x	$\ln(1 + x)$	percentage difference
0.0050	0.0049875	0.2498%
0.0100	0.0099503	0.4992%
0.0150	0.0148886	0.7481%
0.0200	0.0198026	0.9967%
0.0250	0.0246926	1.2449%
0.0300	0.0295588	1.4926%
0.0350	0.0344014	1.7400%
0.0400	0.0392207	1.9869%
0.0450	0.0440169	2.2335%
0.0500	0.0487902	2.4797%
0.0550	0.0535408	2.7255%
0.0600	0.0582689	2.9709%
0.0650	0.0629748	3.2159%
0.0700	0.0676586	3.4605%
0.0750	0.0723207	3.7048%
0.0800	0.0769610	3.9487%
0.0850	0.0815800	4.1922%
0.0900	0.0861777	4.4354%
0.0950	0.0907544	4.6782%
0.1000	0.0953102	4.9206%

I–3. Write out the expansion of $(1 + x)^{1/2}$ through the quadratic term.

$$(1 + x)^{1/2} = 1 + \frac{x}{2} + \frac{\frac{1}{2}(\frac{1}{2} - 1)}{2}x^2 + O(x^3)$$

$$= 1 + \frac{x}{2} - \frac{x^2}{8} + O(x^3)$$

I–4. Evaluate the series

$$S = \sum_{v=0}^{\infty} e^{-(v + \frac{1}{2})\beta h v}$$

We can express this series as

$$S = \sum_{v=0}^{\infty} e^{-(v+\frac{1}{2})\beta h\nu} = e^{-\frac{1}{2}\beta h\nu} \sum_{v=0}^{\infty} e^{-v\beta h\nu}$$

Let $e^{-\beta h\nu} = x$, so

$$S = x^{\frac{1}{2}} \sum_{v=0}^{\infty} x^{v}$$

and use Equation I.3 to write

$$S = \frac{e^{-\frac{1}{2}\beta h\nu}}{1 - e^{-\beta h\nu}}$$

I–5. Show that

$$\frac{1}{(1-x)^2} = 1 + 2x + 3x^2 + 4x^3 + \cdots$$

We know (Equation I.3)

$$\frac{1}{1-x} = 1 + x + x^2 + x^3 + x^4 + \cdots$$

Differentiating both sides of this equation with respect to x gives

$$\frac{1}{(1-x)^2} = 1 + 2x + 3x^2 + 4x^3 + \cdots$$

I–6. Evaluate the series

$$S = \frac{1}{2} + \frac{1}{4} + \frac{1}{8} + \frac{1}{16} + \cdots$$

We can write S as

$$S = \sum_{n=1}^{\infty} \left(\frac{1}{2}\right)^n = \sum_{n=0}^{\infty} \left(\frac{1}{2}\right)^n - 1$$

Using the geometric series (Equation I.3), we find

$$S = \frac{1}{1 - \frac{1}{2}} - 1 = 1$$

I–7. Use Equation I.9 to derive Equations I.10 and I.11.

Let $f(x) = \sin x$. Then

$$f'(x) = \cos x \qquad f''(x) = -\sin x \quad \text{and} \quad f''' = -\cos x$$

The Maclaurin series is

$$f(x) = f(0) + \left(\frac{df}{dx}\right)_{x=0} x + \frac{1}{2!}\left(\frac{d^2 f}{dx^2}\right)_{x=0} x^2 + \frac{1}{3!}\left(\frac{d^3 f}{dx^3}\right)_{x=0} x^3 + \cdots \tag{I.9}$$

So

$$\sin(x) = 0 + x + 0 - \frac{1}{3!}x^3 + 0 + \frac{1}{5!}x^5 + \cdots \tag{I.10}$$

Likewise, letting $f(x) = \cos x$ and applying Equation I.9 gives

$$\cos(x) = 1 + 0 - \frac{1}{2!}x^2 + 0 + \frac{1}{4!}x^4 + 0 + \cdots \tag{I.11}$$

I–8. Show that Equations I.2, I.10, and I.11 are consistent with the relation $e^{ix} = \cos x + i \sin x$.

We begin with the relation $e^{ix} = \cos x + i \sin x$. We use the expressions for $\sin x$ and $\cos x$ given in Equations I.10 and I.11 to write this relation as

$$\begin{aligned}
e^{ix} &= 1 - \frac{x^2}{2!} + \frac{x^4}{4!} - \frac{x^6}{6!} + \cdots + i\left[x - \frac{x^3}{3!} + \frac{x^5}{5!} - \frac{x^7}{7!} + \cdots\right] \\
&= 1 + ix - \frac{x^2}{2!} - \frac{ix^3}{3!} + \frac{x^4}{4!} + \frac{ix^5}{5!} - \frac{x^6}{6!} - \frac{ix^7}{7!} + \cdots \\
&= 1 + ix + \frac{i^2 x^2}{2!} + \frac{i^3 x^3}{3!} + \frac{i^4 x^4}{4!} + \cdots
\end{aligned}$$

Let $ix = z$. Then

$$e^z = 1 + z + \frac{z^2}{2!} + \frac{z^3}{3!} + \cdots$$

in accordance with Equation I.2.

I–9. In Example 17–3, we derived a simple formula for the molar heat capacity of a solid based on a model by Einstein:

$$\overline{C}_V = 3R\left(\frac{\Theta_E}{T}\right)^2 \frac{e^{-\Theta_E/T}}{(1 - e^{-\Theta_E/T})^2}$$

where R is the molar gas constant and $\Theta_E = h\nu/k_B$ is a constant, called the Einstein constant, that is characteristic of the solid. Show that this equation gives the Dulong and Petit limit ($\overline{C}_V \to 3R$) at high temperatures.

Let $x = \Theta_E/T$. Then

$$\overline{C}_V = 3Rx^2 \frac{e^{-x}}{(1 - e^{-x})^2}$$

At high temperatures, $x \to 0$. We can use a series expansion of e^x (Equation I.2) and write

$$\begin{aligned}
\overline{C}_V &= \lim_{x \to 0} 3Rx^2 \frac{e^{-x}}{(1 - e^{-x})^2} \\
&= 3R \lim_{x \to 0} \frac{1 - x + \cdots}{(1 - 1 + x + \cdots)^2} = 3R
\end{aligned}$$

I–10. Evaluate the limit of

$$f(x) = \frac{e^{-x} \sin^2 x}{x^2}$$

as $x \to 0$.

We can use Equations I.2 and I.10 to write this function as a series expansion:

$$f(x) = \frac{[1 - x + O(x^2)][x + O(x^2)]^2}{x^2} = \frac{x^2 + O(x^3)}{x^2}$$

As $x \to 0$, $f(x) \to 1$.

I–11. Evaluate the integral

$$I = \int_0^a x^2 e^{-x} \cos^2 x \, dx$$

for small values of a by expanding I in powers of a through quadratic terms.

$$
\begin{aligned}
I &= \int_0^a x^2 e^{-x} \cos^2 x \, dx \\
&= \int_0^a x^2 \left[1 - x + \frac{x^2}{2!} - \frac{x^3}{3!} + O(x^4) \right] \left[1 - \frac{x^2}{2!} + O(x^4) \right]^2 dx \\
&= \int_0^a \left[x^2 - x^3 + O(x^4) \right] \left[1 - \frac{x^2}{2!} + O(x^4) \right]^2 dx \\
&= \int_0^a \left[x^2 - x^3 + O(x^4) \right] \left[1 - x^2 + O(x^4) \right] dx \\
&= \int_0^a \left[x^2 - x^3 + O(x^4) \right] dx \\
&= \frac{a^3}{3} + O(a^4)
\end{aligned}
$$

I–12. Prove that the series for $\sin x$ converges for all values of x.

We can use the ratio test (Equation I.5) and the Maclaurin series for $\sin x$ (Equation I.10):

$$
\begin{aligned}
r &= \lim_{n \to \infty} \left| \frac{u_{n+1}}{u_n} \right| \\
&= \lim_{n \to \infty} \left| \frac{x^{2n+1}}{(2n+1)!} \frac{(2n-1)!}{x^{2n-1}} \right| \\
&= \lim_{n \to \infty} \left| \frac{x^2}{(2n+1)(2n)} \right| = 0
\end{aligned}
$$

Because $r < 1$, the series converges for all values of x.

I–13. A Maclaurin series is an expansion about the point $x = 0$. A series of the form

$$f(x) = c_0 + c_1(x - x_0) + c_2(x - x_0)^2 + \cdots$$

is an expansion about the point x_0 and is called a Taylor series. First show that $c_0 = f(x_0)$. Now differentiate both sides of the above expansion with respect to x and then let $x = x_0$ to show that $c_1 = (df/dx)_{x=x_0}$. Now show that

$$c_n = \frac{1}{n!}\left(\frac{d^n f}{dx^n}\right)_{x=x_0}$$

and so

$$f(x) = f(x_0) + \left(\frac{df}{dx}\right)_{x_0}(x - x_0) + \frac{1}{2}\left(\frac{d^2 f}{dx^2}\right)_{x_0}(x - x_0)^2 + \cdots$$

At x_0, we find

$$f(x) = c_0 + c_1(x - x_0) + c_2(x - x_0)^2 + \cdots$$
$$f(x_0) = c_0$$

Differentiating $f(x)$ with respect to x gives

$$f'(x) = c_1 + 2c_2(x - x_0) + \cdots$$
$$f'(x_0) = c_1$$

Likewise, the second derivative of $f(x)$ at x_0 is $2c_2$, the third derivative of $f(x)$ at x_0 is $3!c_3$, and, generally,

$$\left(\frac{d^n f}{dx^n}\right)_{x=x_0} = n!c_n$$

$$c_n = \frac{1}{n!}\left(\frac{d^n f}{dx^n}\right)_{x=x_0}$$

Substituting into the Taylor series for c_n, we find

$$f(x) = f(x_0) + \left(\frac{df}{dx}\right)_{x=x_0}(x - x_0) + \frac{1}{2}\left(\frac{d^2 f}{dx^2}\right)_{x=x_0}(x - x_0)^2 + \cdots$$

I–14. Show that l'Hôpital's rule amounts to forming a Taylor expansion of both the numerator and the denominator. Evaluate the limit

$$\lim_{x \to 0} \frac{\ln(1 + x) - x}{x^2}$$

both ways.

We use two Maclaurin series, one for the numerator and one for the denominator:

$$f(x) = a_0 + a_1 x + a_2 x^2 + \cdots$$

$$g(x) = b_0 + b_1 x + b_2 x^2 + \cdots$$

We can then write

$$\frac{f(x)}{g(x)} = \frac{a_0 + a_1 x + a_2 x^2 + \cdots + a_n x^n}{b_0 + b_1 x + b_2 x^2 + \cdots + b_n x^n}$$

and

$$\frac{f'(x)}{g'(x)} = \frac{a_1 + a_2 x + a_3 x^2 + \cdots + a_n x^{n-1}}{b_1 + b_2 x + b_3 x^2 + \cdots + b_n x^{n-1}}$$

L'Hôpital's rule states that if both $f(x)$ and $g(x)$ go to 0 or ∞ as x approaches 0, then

$$\lim_{x \to 0} \frac{f(x)}{g(x)} = \lim_{x \to 0} \frac{f'(x)}{g'(x)}$$

In terms of the Maclaurin series above, let us consider only terms without x. Then

$$\frac{f(x)}{g(x)} = \frac{a_0 + O(x)}{b_0 + O(x)} = \frac{a_0}{b_0}$$

If the numerator and denominator of this fraction are equal to zero, we shall consider terms up to (but not including) $O(x^2)$. Since we have already found that $a_0 = b_0 = 0$,

$$\frac{f(x)}{g(x)} = \frac{a_1 + O(x^2)}{b_1 + O(x^2)} = \frac{a_1}{b_1}$$

which is the derivative of $f(x)/g(x)$, as l'Hôpital's rule states. This process can be continued indefinitely until a non-zero numerator or denominator is found. To extend this to situations where $x \to K$, where K is a nonzero number, we need only consider the Taylor series about the point K. By l'Hôpital's rule,

$$\lim_{x \to 0} \frac{\ln(1 + x) - x}{x^2} = \lim_{x \to 0} \frac{(1 + x)^{-1} - 1}{2x}$$

$$= \lim_{x \to 0} \frac{-x}{2x(1 + x)} = \lim_{x \to 0} \frac{-1}{2(1 + x) + 2x} = -\frac{1}{2}$$

For the Maclaurin expansion,

$$f(x) = \ln(1 + x) - x = 0 + 0x - \frac{x^2}{2} + \cdots$$

$$g(x) = 0 + 0x + x^2$$

Now $f(x)/g(x) = 0/0$ and $f'(x)/g'(x) = 0/0$, but

$$\frac{f''(x)}{g''(x)} = -\frac{1}{2}$$

as above.

I–15. In Problem 18–45, we will need to sum the series

$$s_1 = \sum_{v=0}^{\infty} v x^v$$

and

$$s_2 = \sum_{v=0}^{\infty} v^2 x^v$$

To sum the first one, start with (Equation I.3)

$$s_0 = \sum_{v=0}^{\infty} x^v = \frac{1}{1 - x}$$

Differentiate with respect to x and then multiply by x to obtain

$$s_1 = \sum_{v=0}^{\infty} v x^v = x \frac{ds_0}{dx} = x \frac{d}{dx}\left(\frac{1}{1-x}\right) = \frac{x}{(1-x)^2}$$

Using the same approach, show that

$$s_2 = \sum_{v=0}^{\infty} v^2 x^v = \frac{x+x^2}{(1-x)^3}$$

Following the procedure outlined in the problem, we find

$$s_0 = \sum_{n=0}^{\infty} x^n = \frac{1}{1-x}$$

$$\frac{ds_0}{dx} = \sum_{n=0}^{\infty} n x^{n-1} = \frac{1}{(1-x)^2}$$

$$x\frac{ds_0}{dx} = \sum_{n=0}^{\infty} n x^n = \frac{x}{(1-x)^2} = s_1$$

Likewise,

$$s_1 = \sum_{n=0}^{\infty} n x^n = \frac{x}{(1-x)^2}$$

$$\frac{ds_1}{dx} = \sum_{n=0}^{\infty} n^2 x^{n-1} = \frac{1}{(1-x)^2} + \frac{2x}{(1-x)^3}$$

$$x\frac{ds_1}{dx} = \sum_{n=0}^{\infty} n^2 x^n = x\left[\frac{1-x+2x}{(1-x)^3}\right]$$

$$= \frac{x+x^2}{(1-x)^3}$$

CHAPTER 18

Partition Functions and Ideal Gases

PROBLEMS AND SOLUTIONS

18–1. Equation 18.7 shows that $\langle \varepsilon_{\text{trans}} \rangle = \frac{3}{2}k_B T$ in three dimensions, and Problem 18–3 shows that $\langle \varepsilon_{\text{trans}} \rangle = \frac{1}{2}k_B T$ in one dimension and $\frac{2}{2}k_B T$ in two dimensions. Show that typical values of translational quantum numbers at room temperature are $O(10^9)$ for $m = 10^{-26}$ kg, $a = 1$ dm, and $T = 300$ K.

The average translational energy at 300 K is on the order of $k_B T = 4.142 \times 10^{-21}$ J. Recall that

$$\langle \varepsilon \rangle = \frac{h^2 n^2}{8ma^2}$$

So

$$n^2 \approx \frac{8ma^2}{h^2}k_B T = \frac{8(10^{-26}\text{ kg})(0.1\text{ m})^2}{(6.626 \times 10^{-34}\text{ J·s})^2}(4.142 \times 10^{-21}\text{ J})$$

$$n \approx 2.75 \times 10^9$$

In two and three dimensions, $\langle \varepsilon_{\text{trans}} \rangle$ depends on the sum of the squares of the respective quantum numbers. So for comparable values of a for each dimension, n will be of $O(10^9)$.

18–2. Show that the difference between the successive terms in the summation in Equation 18.4 is very small for $m = 10^{-26}$ kg, $a = 1$ dm, and $T = 300$ K. Recall from Problem 18–1 that typical values of n are $O(10^9)$.

Equation 18.4 gives

$$q_{\text{trans}}(V, T) = \left[\sum_{n=1}^{\infty} \exp\left(\frac{-\beta h^2 n^2}{8ma^2}\right) \right]^3$$

The difference between terms in $q_{\text{trans}}(V, T)$ is then

$$e^{-A(n+1)^2} - e^{-An^2} = e^{-An^2}\left[e^{-A(2n+1)} - 1\right]$$

where $A = \beta h^2/8ma^2 \sim 10^{-19}$. Therefore,

$$e^{-A(2n+1)} - 1 \approx 10^{-10}$$

557

18–3. Show that

$$q_{\text{trans}}(a, T) = \left(\frac{2\pi m k_B T}{h^2}\right)^{1/2} a$$

in one dimension and that

$$q_{\text{trans}}(a, T) = \left(\frac{2\pi m k_B T}{h^2}\right) a^2$$

in two dimensions. Use these results to show that $\langle \varepsilon_{\text{trans}} \rangle$ has a contribution of $k_B T/2$ to its total value for each dimension.

Remember that $\int_0^\infty e^{-\alpha n^2} dn = \left(\frac{\pi}{4\alpha}\right)^{1/2}$. Then, for one dimension,

$$q_{\text{trans}}(a, T) = \int_0^\infty e^{-\beta h^2 n^2/8ma^2} dn = \left(\frac{2\pi m k_B T}{h^2}\right)^{1/2} a$$

And for two dimensions,

$$q_{\text{trans}}(a, T) = \left(\int_0^\infty e^{-\beta h^2 n^2/8ma^2} dn\right)^2 = \left(\frac{2\pi m k_B T}{h^2}\right) a^2$$

Now

$$\langle \epsilon_{\text{trans}} \rangle = k_B T^2 \left(\frac{\partial \ln q_{\text{trans}}}{\partial T}\right)_V$$

The partition function is proportional to $T^{n/2}$, where n is the dimension. So

$$\left(\frac{\partial \ln q_{\text{trans}}}{\partial T}\right)_V = \frac{n}{2T}$$

and

$$\langle \epsilon_{\text{trans}} \rangle = k_B T^2 \frac{n}{2T} = \frac{n k_B T}{2}$$

18–4. Using the data in Table 8.6, calculate the fraction of sodium atoms in the first excited state at 300 K, 1000 K, and 2000 K.

We can use the second line of Equation 18.10 to calculate the fraction of sodium atoms in the first excited state, with $g_{e1} = 2$, $g_{e2} = 2$, $g_{e3} = 4$, and $g_{e4} = 2$:

$$f_2 = \frac{2e^{-\beta \epsilon_{e2}}}{2 + 2e^{-\beta \epsilon_{e2}} + 4e^{-\beta \epsilon_{e3}} + 2e^{-\beta \epsilon_{e4}} + \ldots}$$

Using the data in Table 8.6, we find that the numerator of this fraction is

$$2 \exp\left[-\frac{16\,956.183 \text{ cm}^{-1}}{(0.6950 \text{ cm}^{-1} \cdot \text{K}^{-1})T}\right]$$

and the denominator is

$$2 + 2\exp\left[-\frac{16\,956.183\text{ cm}^{-1}}{(0.6950\text{ cm}^{-1}\cdot\text{K}^{-1})T}\right] + 4\exp\left[-\frac{16\,973.379\text{ cm}^{-1}}{(0.6950\text{ cm}^{-1}\cdot\text{K}^{-1})T}\right]$$
$$+2\exp\left[-\frac{25\,739.86\text{ cm}^{-1}}{(0.6950\text{ cm}^{-1}\cdot\text{K}^{-1})T}\right] + \cdots$$

Using these values, we find that the values of f_2 for the various temperatures are

$$f_2(T = 300\text{ K}) = 4.8 \times 10^{-36}$$
$$f_2(T = 1000\text{ K}) = 2.5 \times 10^{-11}$$
$$f_2(T = 2000\text{ K}) = 5.0 \times 10^{-6}$$

18–5. Using the data in Table 18.1, evaluate the fraction of lithium atoms in the first excited state at 300 K, 1000 K, and 2000 K.

We can use the second line of Equation 18.10 to calculate the fraction of lithium atoms in the first excited state, with $g_{e1} = 2$, $g_{e2} = 2$, $g_{e3} = 4$, and $g_{e4} = 2$:

$$f_2 = \frac{2e^{-\beta\epsilon_{e2}}}{2 + 2e^{-\beta\epsilon_{e2}} + 4e^{-\beta\epsilon_{e3}} + 2e^{-\beta\epsilon_{e4}} + \cdots}$$

Using the data in Table 8.6, we find that the numerator of this fraction is

$$2\exp\left[-\frac{14\,903.66\text{ cm}^{-1}}{(0.6950\text{ cm}^{-1}\cdot\text{K}^{-1})T}\right]$$

and the denominator is

$$2 + 2\exp\left[-\frac{14\,903.66\text{ cm}^{-1}}{(0.6950\text{ cm}^{-1}\cdot\text{K}^{-1})T}\right] + 4\exp\left[-\frac{14\,904.00\text{ cm}^{-1}}{(0.6950\text{ cm}^{-1}\cdot\text{K}^{-1})T}\right]$$
$$+2\exp\left[-\frac{27\,206.12\text{ cm}^{-1}}{(0.6950\text{ cm}^{-1}\cdot\text{K}^{-1})T}\right] + \cdots$$

Using these values, we find that the values of f_2 for the various temperatures are

$$f_2(T = 300\text{ K}) = 9.0 \times 10^{-32}$$
$$f_2(T = 1000\text{ K}) = 4.9 \times 10^{-10}$$
$$f_2(T = 2000\text{ K}) = 2.2 \times 10^{-5}$$

18–6. Show that each dimension contributes $R/2$ to the molar translational heat capacity.

In Problem 18.3, we showed that $\langle\epsilon_{\text{trans}}\rangle$ has a contribution of $k_B T/2$ from each dimension. From Chapter 17,

$$C_V = \left(\frac{\partial\langle E\rangle}{\partial T}\right)_{N,V} = N\left(\frac{\partial\langle\epsilon\rangle}{\partial T}\right)_{N,V}$$

Because

$$\left(\frac{\partial \langle \epsilon \rangle}{\partial T}\right)_{N,V} = \left[\frac{\partial (k_B T/2)}{\partial T}\right]_{N,V} = \frac{k_B}{2}$$

each dimension contributes $Nk_B/2 = R/2$ to the molar translational heat capacity.

18–7. Using the values of Θ_{vib} and D_0 in Table 18.2, calculate the vaues of D_e for CO, NO, and K_2.

We can use the definitions $D_e = D_0 + h\nu/2$ and $\Theta_{vib} = h\nu/k_B$ to write

$$D_e = D_0 + \frac{k_B \Theta_{vib}}{2} = D_0 + \frac{R\Theta_{vib}}{2}$$

$$D_e(CO) = 1070 \text{ kJ} \cdot \text{mol}^{-1} + \frac{(8.314 \times 10^{-3} \text{ kJ} \cdot \text{mol}^{-1} \cdot \text{K}^{-1})(3103 \text{ K})}{2}$$

$$= 1083 \text{ kJ} \cdot \text{mol}^{-1}$$

$$D_e(NO) = 626.8 \text{ kJ} \cdot \text{mol}^{-1} + \frac{(8.314 \times 10^{-3} \text{ kJ} \cdot \text{mol}^{-1} \cdot \text{K}^{-1})(2719 \text{ K})}{2}$$

$$= 638.1 \text{ kJ} \cdot \text{mol}^{-1}$$

$$D_e(K_2) = 53.5 \text{ kJ} \cdot \text{mol}^{-1} + \frac{(8.314 \times 10^{-3} \text{ kJ} \cdot \text{mol}^{-1} \cdot \text{K}^{-1})(133 \text{ K})}{2}$$

$$= 54.1 \text{ kJ} \cdot \text{mol}^{-1}$$

18–8. Calculate the characteristic vibrational temperature Θ_{vib} for $H_2(g)$ and $D_2(g)$ ($\tilde{\nu}_{H_2} = 4401 \text{ cm}^{-1}$ and $\tilde{\nu}_{D_2} = 3112 \text{ cm}^{-1}$).

From the definition of Θ_{vib}, we can write $\Theta_{vib} = hc\tilde{\nu}/k_B$. Then

$$\Theta_{vib}(H_2) = \frac{(6.626 \times 10^{-34} \text{ J} \cdot \text{s})(2.9979 \times 10^{10} \text{ cm} \cdot \text{s}^{-1})(4401 \text{ cm}^{-1})}{1.381 \times 10^{-23} \text{ J} \cdot \text{K}^{-1}} = 6332 \text{ K}$$

$$\Theta_{vib}(D_2) = \frac{(6.626 \times 10^{-34} \text{ J} \cdot \text{s})(2.9979 \times 10^{10} \text{ cm} \cdot \text{s}^{-1})(3112 \text{ cm}^{-1})}{1.381 \times 10^{-23} \text{ J} \cdot \text{K}^{-1}} = 4478 \text{ K}$$

18–9. Plot the vibrational contribution to the molar heat capacity of $Cl_2(g)$ from 250 K to 1000 K.

Use Equation 18.26 to write $\overline{C}_{V,vib}$ as a function of T:

$$\overline{C}_{V,vib} = R\left(\frac{\Theta_{vib}}{T}\right)^2 \frac{e^{-\Theta_{vib}/T}}{(1 - e^{-\Theta_{vib}/T})^2}$$

For Cl_2, we use $\Theta_{vib} = 805$ K (Table 18.2) in the above equation and plot $\overline{C}_{V,vib}$ versus T.

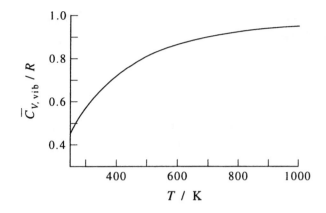

18–10. Plot the fraction of HCl(g) molecules in the first few vibrational states at 300 K and 1000 K.

Use Equation 18.28, substituting $\Theta_{vib} = 4227$ K (Table 18.2), to write f_v as a function of v, and plot. At 300 K $f_{v>0} = 7.6 \times 10^{-7}$ and at 1000 K $f_{v>0} = 1.46 \times 10^{-2}$.

$$f_v = (1 - e^{-\Theta_{vib}/T})e^{-v\Theta_{vib}/T}$$

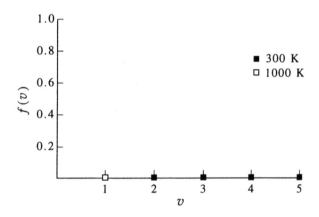

18–11. Calculate the fraction of molecules in the ground vibrational state and in all the excited states at 300 K for each of the molecules in Table 18.2.

The fraction of molecules in the ground vibrational state is given by (Equation 18.27)

$$f_0 = 1 - e^{-\Theta_{vib}/T}$$

and the fraction in all the excited states is $1 - f_0$, or $e^{-\Theta_{vib}/T}$. We can use the Θ_{vib} for the molecules given in Table 18.2 to find f_0 and $f_{v>0}$.

molecule	f_0	$f_{v>0}$
H_2	1.0000	1.0066×10^{-9}
D_2	1.0000	4.3555×10^{-7}
Cl_2	0.93167	6.8335×10^{-2}
Br_2	0.78633	0.21367
I_2	0.064180	0.35820
O_2	0.99946	5.4213×10^{-4}
N_2	0.99999	1.3051×10^{-5}
CO	0.99997	3.2207×10^{-5}
NO	0.99988	1.1584×10^{-4}
HCl	1.0000	7.5996×10^{-7}
HBr	1.0000	3.2942×10^{-6}
HI	0.99998	1.8706×10^{-5}
Na_2	0.053389	0.46611
K_2	0.35811	0.64189

18–12. Calculate the value of the characteristic rotational temperature Θ_{rot} for $H_2(g)$ and $D_2(g)$. (The bond lengths of H_2 and D_2 are 74.16 pm.) The atomic mass of deuterium is 2.014.

The reduced masses of hydrogen and deuterium, respectively, are

$$\mu(H_2) = \frac{(1.674 \times 10^{-27} \text{ kg})^2}{2(1.674 \times 10^{-27} \text{ kg})} = 8.370 \times 10^{-28} \text{ kg}$$

and

$$\mu(D_2) = \frac{(3.344 \times 10^{-27} \text{ kg})^2}{2(3.344 \times 10^{-27} \text{ kg})} = 1.672 \times 10^{-27} \text{ kg}$$

Now use the formula $\Theta_{rot} = \hbar^2/2\mu R^2 k_B$ (Equation 18.32) to find the value of Θ_{rot} for hydrogen and deuterium:

$$\Theta_{rot}(H_2) = \frac{(1.055 \times 10^{-34} \text{ J} \cdot \text{s})^2}{2(8.370 \times 10^{-28} \text{ kg})(74.16 \times 10^{-12} \text{ m})^2(1.38066 \times 10^{-23} \text{ J} \cdot \text{K}^{-1})}$$

$$= 87.56 \text{ K}$$

$$\Theta_{rot}(D_2) = \frac{(1.055 \times 10^{-34} \text{ J} \cdot \text{s})^2}{2(1.674 \times 10^{-27} \text{ kg})(74.16 \times 10^{-12} \text{ m})^2(1.38066 \times 10^{-23} \text{ J} \cdot \text{K}^{-1})}$$

$$= 43.78 \text{ K}$$

18–13. The average molar rotational energy of a diatomic molecule is RT. Show that typical values of J are given by $J(J+1) = T/\Theta_{rot}$. What are typical values of J for $N_2(g)$ at 300 K?

If $\langle E \rangle = RT$, then $\epsilon = k_B T$. From Equation 18.30a,

$$J(J+1) = \frac{2Ik_B T}{\hbar^2} = \frac{T}{\Theta_{rot}}$$

For N_2 at 300 K,

$$J(J+1) \approx 104$$

and $J \approx 9$ or 10.

18–14. There is a mathematical procedure to calculate the error in replacing a summation by an integral as we do for the translational and rotational partition functions. The formula is called the Euler-Maclaurin summation formula and goes as follows:

$$\sum_{n=a}^{b} f(n) = \int_a^b f(n)dn + \frac{1}{2}\{f(b) + f(a)\} - \frac{1}{12}\left\{ \frac{df}{dn}\Big|_{n=a} - \frac{df}{dn}\Big|_{n=b} \right\}$$

$$+ \frac{1}{720}\left\{ \frac{d^3f}{dn^3}\Big|_{n=a} - \frac{d^3f}{dn^3}\Big|_{n=b} \right\} + \cdots$$

Apply this formula to Equation 18.33 to obtain

$$q_{rot}(T) = \frac{T}{\Theta_{rot}}\left\{ 1 + \frac{1}{3}\left(\frac{\Theta_{rot}}{T}\right) + \frac{1}{15}\left(\frac{\Theta_{rot}}{T}\right)^2 + O\left[\left(\frac{\Theta_{rot}}{T}\right)^3\right]\right\}$$

Calculate the correction to replacing Equation 18.33 by an integral for $N_2(g)$ at 300 K; $H_2(g)$ at 300 K (being so light, H_2 is an extreme example).

$$q_{rot}(T) = \sum_{J=0}^{\infty}(2J+1)e^{-\Theta_{rot}J(J+1)/T}$$

$$= \int_0^{\infty} dJ(2J+1)e^{-\Theta_{rot}J(J+1)/T} + \frac{1}{2}(f(\infty) + f(0))$$

$$- \frac{1}{12}\left[\frac{df}{dJ}\Big|_{J=0} - \frac{df}{dJ}\Big|_{J=\infty}\right] + \frac{1}{720}\left[\frac{d^3f}{dJ^3}\Big|_{J=0} - \frac{d^3f}{dJ^3}\Big|_{J=\infty}\right] + \cdots$$

Let $u = J(J+1)$ and $du = (2J+1)dJ$. Then at $J = 0$, $u = 0$, and at $J = \infty$, $u = \infty$. Also, find the first and third derivatives of $f(J)$:

$$f(J) = (2J+1)e^{-\Theta_{rot}J(J+1)/T}$$

$$\frac{df}{dJ} = \frac{-\Theta_{rot}}{T}(2J+1)^2 e^{-\Theta_{rot}J(J+1)/T} + 2e^{-\Theta_{rot}J(J+1)/T}$$

$$\frac{d^2f}{dJ^2} = 4\left(\frac{-\Theta_{rot}}{T}\right)(2J+1)e^{-\Theta_{rot}J(J+1)/T} + \left(\frac{\Theta_{rot}}{T}\right)^2(2J+1)^3 e^{-\Theta_{rot}J(J+1)/T}$$

$$+ 2\left(\frac{-\Theta_{rot}}{T}\right)(2J+1)e^{-\Theta_{rot}J(J+1)/T}$$

$$\frac{d^3f}{dJ^3} = 8\left(\frac{-\Theta_{rot}}{T}\right)e^{-\Theta_{rot}J(J+1)/T} + 4\left(\frac{\Theta_{rot}}{T}\right)e^{-\Theta_{rot}J(J+1)/T} + O\left[\left(\frac{\Theta_{rot}}{T}\right)^2\right]$$

$$q_{rot} = \int_0^{\infty}(2J+1)e^{-\Theta_{rot}J(J+1)/T} + \frac{1}{2}(1) - \frac{1}{12}\left(\frac{-\Theta_{rot}}{T} + 2\right)$$

$$+\frac{1}{720}\left(-8\frac{\Theta_{\text{rot}}}{T}-4\frac{\Theta_{\text{rot}}}{T}\right)+O\left\{\left(\frac{\Theta_{\text{rot}}}{T}\right)^2\right\}$$

$$=\frac{T}{\Theta_{\text{rot}}}\left\{1+\frac{1}{3}\left(\frac{\Theta_{\text{rot}}}{T}\right)+\frac{1}{15}\left(\frac{\Theta_{\text{rot}}}{T}\right)^2+O\left[\left(\frac{\Theta_{\text{rot}}}{T}\right)^2\right]\right\}$$

For N_2 the correction factor to q_{rot} at 300 K is 0.32%; for H_2, the correction factor is 9.45%.

18–15. Apply the Euler-Maclaurin summation formula (Problem 18–14) to the one-dimensional version of Equation 18.4 to obtain

$$q_{\text{trans}}(a,\,T)=\left(\frac{2\pi mk_{\text{B}}T}{h^2}\right)^{1/2}a+\left[\frac{1}{2}+\frac{h^2}{48ma^2k_{\text{B}}T}\right]e^{-h^2/8ma^2k_{\text{B}}T}$$

Show that the correction amounts to about 10^{-8}% for $m=10^{-26}$ kg, $a=1$ dm, and $T=300$ K.

The one-dimensional version of Equation 18.4 is

$$q=\sum_{n=1}^{\infty}\exp\left(-\frac{\beta h^2n^2}{8ma^2}\right)=\sum_{n=1}^{\infty}e^{-bn^2}$$

where we let $b=\beta h^2/8ma^2$. The pertinent derivatives of q are

$$\frac{dq}{dn}=-2bne^{-bn^2}$$

$$\frac{d^2q}{dn^2}=-2be^{-bn^2}+4b^2n^2e^{-bn^2}$$

$$\frac{d^3q}{dn^3}=4b^2ne^{-bn^2}-8b^3n^3e^{-bn^2}$$

We can approximate

$$\int_1^{\infty}e^{-bn^2}\,dn\approx\int_0^{\infty}e^{-bn^2}\,dn$$

and use the Euler-Maclaurin summation formula from Problem 18–14 to find

$$q=\int_0^{\infty}e^{-bn^2}\,dn+\frac{1}{2}\left(e^{-b}-0\right)-\frac{1}{12}\left(-2be^{-b}\right)+\frac{1}{720}\left(4b^2e^{-b}-8b^3e^{-b}\right)$$

$$=\left(\frac{2\pi ma^2k_{\text{B}}T}{h^2}\right)^{1/2}+\left(\frac{1}{2}+\frac{1}{6}b+\frac{1}{180}b^2-\frac{1}{90}b^3\right)e^{-b}$$

$$=\left(\frac{2\pi mk_{\text{B}}T}{h^2}\right)^{1/2}a+\left[\frac{1}{2}+\frac{h^2}{48ma^2k_{\text{B}}T}+O(b^2)\right]e^{-h^2/8ma^2k_{\text{B}}T}$$

For $m=10^{-26}$ kg, $a=1$ dm, and $T=300$ K, simply replacing the sum by an integral gives a value of 2.43×10^9. The correction term is 0.5, which is 2×10^{-8}% of the value of the sum.

18–16. We were able to evaluate the vibrational partition function for a harmonic oscillator exactly by recognizing the summation as a geometric series. Apply the Euler-Maclaurin summation formula (Problem 18–14) to this case and show that

$$\sum_{v=0}^{\infty}e^{-\beta(v+\frac{1}{2})hv}=e^{-\Theta_{\text{vib}}/2T}\sum_{v=0}^{\infty}e^{-v\Theta_{\text{vib}}/T}$$

$$=e^{-\Theta_{\text{vib}}/2T}\left[\frac{T}{\Theta_{\text{vib}}}+\frac{1}{2}+\frac{\Theta_{\text{vib}}}{12T}+\cdots\right]$$

Show that the corrections to replacing the summation by an integration are very large for $O_2(g)$ at 300 K. Fortunately, we don't need to replace the summation by an integration in this case.

Recall that $\Theta_{vib} = h\nu/k_B$, so

$$\sum_{v=0}^{\infty} e^{-\beta(v+\frac{1}{2})h\nu} = e^{-\Theta_{vib}/2T} \sum_{v=0}^{\infty} e^{-v\Theta_{vib}/T}$$

Then

$$f(v) = e^{-v\Theta_{vib}/T} \qquad f'(v) = -\frac{\Theta_{vib}}{T} e^{-v\Theta_{vib}/T} \qquad f'''(v) = -\left(\frac{\Theta_{vib}}{T}\right)^3 e^{-v\Theta_{vib}/T}$$

and applying the Euler-Maclaurin summation formula yields

$$\sum_{v=0}^{\infty} e^{-v\Theta_{vib}/T} = \int_0^{\infty} e^{-v\Theta_{vib}/T} dv + \frac{1}{2}(1) - \frac{1}{12}\left(-\frac{\Theta_{vib}}{T}\right) + \frac{1}{720}\left(-\frac{\Theta_{vib}}{T}\right) + O(T^{-3})$$

$$= \frac{T}{\Theta_{vib}} + \frac{1}{2} + \frac{\Theta_{vib}}{12T} + O(T^{-3})$$

Then

$$\sum_{v=0}^{\infty} e^{-\beta(v+\frac{1}{2})h\nu} = e^{-\Theta_{vib}/2T}\left[\frac{T}{\Theta_{vib}} + \frac{1}{2} + \frac{\Theta_{vib}}{12T} + O(T^{-3})\right]$$

For O_2, $\Theta_{vib} = 2256$ K. Using Equation 18.23, we find that

$$q_{vib}(T) = \frac{e^{-2256\ \text{K}/600\ \text{K}}}{1 - e^{-2256\ \text{K}/600\ \text{K}}} = 0.0238$$

and using the correction, we have

$$q_{vib}(T) = e^{-2256\ \text{K}/600\ \text{K}}\left(\frac{300\ \text{K}}{2256\ \text{K}} + \frac{1}{2} + \frac{2256\ \text{K}}{3600\ \text{K}}\right) = 0.0293$$

which is about a 20% difference.

18–17. Plot the fraction of NO(g) molecules in the various rotational levels at 300 K and at 1000 K.

Use Equation 18.35, substituting $\Theta_{rot} = 2.39$ K (Table 18.2), to write f_J as a function of J, and plot.

$$f_J = (2J + 1)(\Theta_{rot}/T)e^{-\Theta_{rot}J(J+1)/T}$$

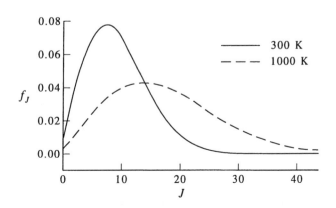

18–18. Show that the values of J at the maximum of a plot of f_J versus J (Equation 18.35) is given by

$$J_{max} \approx \left(\frac{T}{2\Theta_{rot}}\right)^{1/2} - \frac{1}{2}$$

Hint: Treat J as a continuous variable. Use this result to verify the values of J at the maxima in the plots in Problem 18–17.

$$f_J = (2J + 1)\frac{\Theta_{rot}}{T}e^{-\Theta_{rot}J(J+1)/T}$$

At the maximum of a plot of f_J versus J, the slope is zero, so

$$\frac{df}{dJ} = \frac{2\Theta_{rot}}{T}e^{-\Theta_{rot}J_{max}(J_{max}+1)/T} - (2J_{max} + 1)^2 \left(\frac{\Theta_{rot}}{T}\right)^2 e^{-\Theta_{rot}J_{max}(J_{max}+1)/T} = 0$$

We can solve this equation for J_{max}:

$$(2J_{max} + 1)^2 \left(\frac{\Theta_{rot}}{T}\right)^2 e^{-\Theta_{rot}J_{max}(J_{max}+1)/T} = \frac{2\Theta_{rot}}{T}e^{-\Theta_{rot}J_{max}(J_{max}+1)/T}$$

$$(2J_{max} + 1)^2 = \frac{2T}{\Theta_{rot}}$$

$$J_{max} = \left(\frac{T}{2\Theta_{rot}}\right)^{1/2} - \frac{1}{2}$$

For NO(g) at 300 K and 1000 K, $\Theta_{rot} = 2.39$ K, so the values of J_{max} given by the above equation are $J_{max} \approx 7$ and $J_{max} \approx 14$, respectively, in agreement with the plot in Problem 18–17.

18–19. The experimental heat capacity of $N_2(g)$ can be fit to the empirical formula

$$\overline{C}_V(T)/R = 2.283 + (6.291 \times 10^{-4}\ \text{K}^{-1})T - (5.0 \times 10^{-10}\ \text{K}^{-2})T^2$$

over the temperature range 300 K $< T <$ 1500 K. Plot $\overline{C}_V(T)/R$ versus T over this range using Equation 18.41, and compare your results with the experimental curve.

For N_2, $\Theta_{vib} = 3374$ K, so we plot the experimental equation given in the problem text and the theoretical equation

$$\frac{\overline{C}_V(T)}{R} = \frac{5}{2} + \left(\frac{\Theta_{vib}}{T}\right)^2 \frac{e^{-\Theta_{vib}T}}{(1 - e^{-\Theta_{vib}T})^2}$$

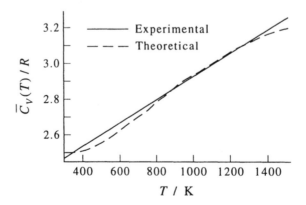

18–20. The experimental heat capacity of CO(g) can be fit to the empirical formula

$$\overline{C}_V(T)/R = 2.192 + (9.240 \times 10^{-4} \text{ K}^{-1})T - (1.41 \times 10^{-7} \text{ K}^{-2})T^2$$

over the temperature range 300 K < T < 1500 K. Plot $\overline{C}_V(T)/R$ versus T over this range using Equation 18.41, and compare your results with the experimental curve.

For CO, Θ_{vib} = 3103 K, so we plot the experimental equation given in the problem text and the theoretical equation

$$\frac{\overline{C}_V(T)}{R} = \frac{5}{2} + \left(\frac{\Theta_{vib}}{T}\right)^2 \frac{e^{-\Theta_{vib}T}}{(1 - e^{-\Theta_{vib}T})^2}$$

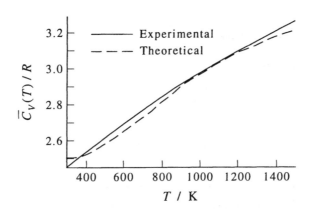

18–21. Calculate the contribution of each normal mode to the molar vibrational heat capacity of H$_2$O(g) at 600 K.

For H$_2$O, the values of Θ_{vib} for the three normal modes are 2290 K, 5160 K, and 5360 K. For $\Theta_{vib,j}$ = 5360 K,

$$\frac{\overline{C}_{V,j}}{R} = \left(\frac{5360}{600}\right)^2 \frac{e^{-5360/600}}{(1 - e^{-5360/600})^2} = 1.05 \times 10^{-2}$$

For $\Theta_{vib,j}$ = 5160 K,

$$\frac{\overline{C}_{V,j}}{R} = \left(\frac{5160}{600}\right)^2 \frac{e^{-5160/600}}{(1 - e^{-5160/600})^2} = 1.36 \times 10^{-2}$$

For $\Theta_{vib,j}$ = 2290 K,

$$\frac{\overline{C}_{V,j}}{R} = \left(\frac{2290}{600}\right)^2 \frac{e^{-2290/600}}{(1 - e^{-2290/600})^2} = 0.335$$

18–22. In analogy to the characteristic vibrational temperature, we can define a characteristic electronic temperature by

$$\Theta_{elec,j} = \frac{\varepsilon_{ej}}{k_B}$$

where ε_{ej} is the energy of the jth excited electronic state relative to the ground state. Show that if we define the ground state to be the zero of energy, then

$$q_{elec} = g_0 + g_1 e^{-\Theta_{elec,1}/T} + g_2 e^{-\Theta_{elec,2}/T} + \cdots$$

The first and second excited electronic states of O(g) lie 158.2 cm^{-1} and 226.5 cm^{-1} above the ground electronic state. Given $g_0 = 5$, $g_1 = 3$, and $g_2 = 1$, calculate the values of $\Theta_{elec,1}$, $\Theta_{elec,2}$, and q_{elec} (ignoring any higher states) for O(g) at 5000 K.

Substituting the values given in the problem into the definition of $\Theta_{elec,j}$ gives

$$\Theta_{elec,1} = \frac{158.2 \text{ cm}^{-1}}{0.69509 \text{ cm}^{-1}\cdot\text{K}^{-1}} = 227.6 \text{ K}$$

$$\Theta_{elec,2} = \frac{226.5 \text{ cm}^{-1}}{0.69509 \text{ cm}^{-1}\cdot\text{K}^{-1}} = 325.8 \text{ K}$$

We can write q_{elec} as (Equation 18.8)

$$q_{elec} = \sum g_{ej} e^{\varepsilon_{ej}/k_B T}$$
$$= g_0 + g_1 e^{-\Theta_{elec,1}/T} + g_2 e^{-\Theta_{elec,2}/T} + \cdots$$
$$q_{elec} = 5 + 3e^{-227.6/5000} + 1e^{-325.8/5000} = 8.803$$

18–23. Determine the symmetry numbers for H_2O, HOD, CH_4, SF_6, C_2H_2, and C_2H_4.

Symmetry numbers of selected molecules

Molecule	Symmetry Number
H_2O	2
HOD	1
CH_4	12
SF_6	24
C_2H_2	2
C_2H_4	4

18–24. The HCN(g) molecule is a linear molecule, and the following constants determined spectroscopically are $I = 18.816 \times 10^{-47}$ kg·m^2, $\tilde{\nu}_1 = 2096.7$ cm^{-1} (HC–N stretch), $\tilde{\nu}_2 = 713.46$ cm^{-1} (H–C–N bend, two-fold degeneracy), and $\tilde{\nu}_3 = 3311.47$ cm^{-1} (H–C stretch). Calculate the values of Θ_{rot} and Θ_{vib} and $\overline{C}_V$ at 3000 K.

We can use the definitions of Θ_{rot} and Θ_{vib} to write

$$\Theta_{rot} = \frac{\hbar^2}{2Ik_B} = 2.1405 \text{ K}$$

$$\Theta_{vib} = \frac{hc\tilde{\nu}}{k_B}$$

$$\Theta_{vib,1} = 3017 \text{ K} \quad \text{(HC-N stretch)}$$

$$\Theta_{vib,2} = 1026 \text{ K} \quad \text{(H-C-N bend)}$$

$$\Theta_{vib,3} = 4764 \text{ K} \quad \text{(H-C stretch)}$$

For linear polyatomic molecules, Equation 18.59 holds, and so

$$\frac{\overline{C}_V}{R} = \frac{5}{2} + \sum_{j=1}^{4} \overline{C}_{vib,j}$$

$$= \frac{5}{2} + \left(\frac{3017}{3000}\right)^2 \frac{e^{-3017/3000}}{(1 - e^{-3017/3000})^2} + 2\left(\frac{1026}{3000}\right)^2 \frac{e^{-1026/3000}}{(1 - e^{-1026/3000})^2}$$

$$+ \left(\frac{4764}{3000}\right)^2 \frac{e^{-4764/3000}}{(1 - e^{-4764/3000})^2}$$

$$= 2.5 + 0.92 + 1.98 + 0.81 = 6.21$$

18–25. The acetylene molecule is linear, the $C\equiv C$ bond length is 120.3 pm, and the C–H bond length is 106.0 pm. What is the symmetry number of acetylene? Determine the moment of inertia (Section 13–8) of acetylene and calculate the value of Θ_{rot}. The fundamental frequencies of the normal modes are $\tilde{\nu}_1 = 1975$ cm^{-1}, $\tilde{\nu}_2 = 3370$ cm^{-1}, $\tilde{\nu}_3 = 3277$ cm^{-1}, $\tilde{\nu}_4 = 729$ cm^{-1}, and $\tilde{\nu}_5 = 600$ cm^{-1}. The normal modes $\tilde{\nu}_4$ and $\tilde{\nu}_5$ are doubly degenerate. All the other modes are nondegenerate. Calculate $\Theta_{vib,j}$ and $\overline{C}_V$ at 300 K.

The symmetry number of acetylene is 2 (Problem 18.24). Choose the coordinate axis to bisect the center of the triple bond. Then

$$I = \sum_i m_i z_i^2 = 2\left(\frac{120.3 \times 10^{-12} \text{ m}}{2}\right)^2 (1.995 \times 10^{-26} \text{ kg})$$

$$+ 2\left(\frac{120.3 \times 10^{-12} \text{ m}}{2} + 106.0 \times 10^{-12} \text{ m}\right)^2 (1.67 \times 10^{-27} \text{ kg})$$

$$= 2.368 \times 10^{-46} \text{ kg} \cdot \text{m}^2$$

We can use the definitions of Θ_{rot} and Θ_{vib} to write

$$\Theta_{rot} = \frac{\hbar^2}{2Ik_B} = 1.701 \text{ K}$$

$$\Theta_{vib} = \frac{hc\tilde{\nu}}{k_B}$$

$$\Theta_{vib,2} = 2841 \text{ K}$$

$$\Theta_{vib,2} = 4849 \text{ K}$$

$$\Theta_{vib,3} = 4715 \text{ K}$$

$$\Theta_{vib,4} = 1049 \text{ K}$$

$$\Theta_{vib,5} = 863.3 \text{ K}$$

The vibrational molar heat capacities are given by (Equation 18.26)

$$\overline{C}_{V,vib} = R\left(\frac{\Theta_{vib}}{T}\right)^2 \frac{e^{-\Theta_{vib}/T}}{(1 - e^{-\Theta_{vib}/T})^2}$$

Because acetylene is a linear polyatomic molecule, we can use Equation 18.59 to find the molar heat capacity at 300 K:

$$\frac{\overline{C}_V}{R} = \frac{5}{2} + \sum_{j=1}^{7} \overline{C}_{vib,j}$$

$$= \frac{5}{2} + \overline{C}_{vib,1} + \overline{C}_{vib,2} + \overline{C}_{vib,3} + 2\overline{C}_{vib,4} + 2\overline{C}_{vib,5}$$

$$= \frac{5}{2} + 6.92 \times 10^{-3} + 2.50 \times 10^{-5} + 3.69 \times 10^{-5} + 2(0.394) + 2(0.523)$$

$$= 4.34$$

18–26. Plot the summand in Equation 18.53 versus J, and show that the most important values of J are large for $T \gg \Theta_{rot}$. We use this fact in going from Equation 18.53 to Equation 18.54.

The summand in Equation 18.53 is

$$(2J + 1)^2 e^{-\hbar^2 J(J+1)/2Ik_BT} = (2J + 1)^2 e^{-J(J+1)\Theta_{rot}/T}$$

Let $x = \Theta_{rot}/T$, so that for $\Theta_{rot} \ll T$ x is small, and then plot the summand versus J for different values of x.

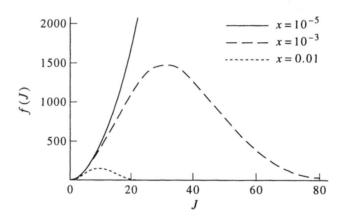

We can see that for $\Theta_{rot} \ll T$, the most important values of J (as far as contributions to the summand are concerned) are large, so Equation 18.54 holds quite well when this condition is met.

18–27. Use the Euler-Maclaurin summation formula (Problem 18–14) to show that

$$q_{rot}(T) = \frac{\pi^{1/2}}{\sigma} \left(\frac{T}{\Theta_{rot}} \right)^{3/2} + \frac{1}{6} + O\left(\frac{\Theta_{rot}}{T} \right)$$

for a spherical top molecule. Show that the correction to replacing Equation 18.53 by an integral is about 1% for CH_4 and 0.001% for CCl_4 at 300 K.

For a spherical top molecule,

$$q_{rot}(T) = \sum_{J=0}^{\infty} (2J + 1)^2 e^{-\hbar^2 J(J+1)/2Ik_BT} \qquad (18.53)$$

The pertinent derivatives are

$$f(J) = (2J+1)^2 e^{-\Theta_{rot} J(J+1)/T}$$

$$\frac{df}{dJ} = 4(2J+1)e^{-\Theta_{rot}(J^2+J)/T} - \frac{\Theta_{rot}}{T}(2J+1)^3 e^{-\Theta_{rot}(J^2+J)/T}$$

$$\frac{d^2 f}{dJ^2} = 8e^{-\Theta_{rot}(J^2+J)/T} - \frac{10\Theta_{rot}}{T}(2J+1)^2 e^{-\Theta_{rot}(J^2+J)/T}$$

$$+ \left(\frac{\Theta_{rot}}{T}\right)^2 (2J+1)^4 e^{-\Theta_{rot}(J^2+J)/T}$$

Applying the Euler-Maclaurin summation formula gives

$$q_{rot}(T) = \frac{1}{\sigma}\int_0^\infty e^{-\Theta_{rot}(J^2+J)/T} + \frac{1}{2}(1) - \frac{1}{12}\left[4 + O\left(\frac{\Theta_{rot}}{T}\right)\right] + O\left(\frac{\Theta_{rot}}{T}\right)$$

$$= \frac{\pi^{1/2}}{\sigma}\left(\frac{T}{\Theta_{rot}}\right)^{3/2} + \frac{1}{6} + O\left(\frac{\Theta_{rot}}{T}\right)$$

where we have included the symmetry number σ in the integral, as was done in the text. For CH_4 at 300 K the integral has a value of 37.07, and the correction term is about 1% of that; for CCl_4 at 300 K the integral has a value of about 32 500, and the correction term is about 0.001% of that.

18–28. The N–N and N–O bond lengths in the (linear) molecule N_2O are 109.8 pm and 121.8 pm, respectively. Calculate the center of mass and the moment of inertia of $^{14}N^{14}N^{16}O$. Compare your answer with the value obtained from Θ_{rot} in Table 18.4.

Choose the coordinate axis to bisect the central nitrogen atom. Then the moment of inertia (using the isotopic masses from the *CRC Handbook*) is

$$I = \sum_i m_i z_i^2 = (109.8 \times 10^{-12}\text{ m})^2(2.325 \times 10^{-26}\text{ kg})$$

$$+(121.8 \times 10^{-12}\text{ m})^2(2.656 \times 10^{-26}\text{ kg})$$

$$= 6.744 \times 10^{-46}\text{ kg·m}^2$$

and the center of mass of the molecule (relative to the central nitrogen atom) is

$$X = 0 + \frac{14.003}{44.001}(109.8 \times 10^{-12}\text{ m}) + \frac{15.995}{44.013}(121.8 \times 10^{-12}\text{ m})$$

$$= 7.922 \times 10^{-11}\text{ m}$$

The center of mass of the molecule is 79.22 pm away from the central nitrogen atom, along the N–O bond. The value of Θ_{rot} calculated from the above values is

$$\Theta_{rot} = \frac{\hbar^2}{2Ik_B} = 0.597\text{ K}$$

This value is within 1% of that in Table 18.4.

18–29. $NO_2(g)$ is a bent triatomic molecule. The following data determined from spectroscopic measurements are $\tilde{\nu}_1 = 1319.7$ cm^{-1}, $\tilde{\nu}_2 = 749.8$ cm^{-1}, $\tilde{\nu}_3 = 1617.75$ cm^{-1}, $\tilde{A}_0 = 8.0012$ cm^{-1}, $\tilde{B}_0 = 0.43304$ cm^{-1}, and $\tilde{C}_0 = 0.41040$ cm^{-1}. Determine the three characteristic vibrational temperatures and the characteristic rotational temperatures for each of the principle axes of $NO_2(g)$ at 1000 K. Calculate the value of $\overline{C}_V$ at 1000 K.

We can use Equation 18.49 to find $\Theta_{\text{vib},j}$:

$$\Theta_{\text{vib},j} = \frac{hc\nu_j}{k_B} = \frac{hc\tilde{\nu}_j}{k_B}$$

$$\Theta_{\text{vib},1} = \frac{(6.626 \times 10^{-34}\text{ J·s})(2.998 \times 10^{10}\text{ cm·s}^{-1})(1319.7\text{ cm}^{-1})}{1.381 \times 10^{-23}\text{ J·K}^{-1}} = 1899\text{ K}$$

$$\Theta_{\text{vib},2} = \frac{(6.626 \times 10^{-34}\text{ J·s})(2.998 \times 10^{10}\text{ cm·s}^{-1})(749.8\text{ cm}^{-1})}{1.381 \times 10^{-23}\text{ J·K}^{-1}} = 1079\text{ K}$$

$$\Theta_{\text{vib},3} = \frac{(6.626 \times 10^{-34}\text{ J·s})(2.998 \times 10^{10}\text{ cm·s}^{-1})(1617.75\text{ cm}^{-1})}{1.381 \times 10^{-23}\text{ J·K}^{-1}} = 2328\text{ K}$$

and Equation 18.32 for Θ_{rot}:

$$\Theta_{\text{rot},A} = \frac{(6.626 \times 10^{-34}\text{ J·s})(2.998 \times 10^{10}\text{ cm·s}^{-1})(8.0012\text{ cm}^{-1})}{1.381 \times 10^{-23}\text{ J·K}^{-1}} = 11.51\text{ K}$$

$$\Theta_{\text{rot},B} = \frac{(6.626 \times 10^{-34}\text{ J·s})(2.998 \times 10^{10}\text{ cm·s}^{-1})(0.43304\text{ cm}^{-1})}{1.381 \times 10^{-23}\text{ J·K}^{-1}} = 0.6230\text{ K}$$

$$\Theta_{\text{rot},C} = \frac{(6.626 \times 10^{-34}\text{ J·s})(2.998 \times 10^{10}\text{ cm·s}^{-1})(0.41040\text{ cm}^{-1})}{1.381 \times 10^{-23}\text{ J·K}^{-1}} = 0.5905\text{ K}$$

Finally, we use Equation 18.59 to determine the value of $\overline{C}_V$ at 1000 K.

$$\frac{\overline{C}_V}{R} = 2\left(\frac{3}{2}\right) + (1.899)^2 \frac{e^{-1.899}}{(1 - e^{-1.899})^2} + (1.079)^2 \frac{e^{-1.079}}{(1 - e^{-1.079})^2}$$

$$+ (2.328)^2 \frac{e^{-2.328}}{(1 - e^{-2.328})^2}$$

$$= 5.304$$

18–30. The experimental heat capacity of $NH_3(g)$ can be fit to the empirical formula

$$\overline{C}_V(T)/R = 2.115 + (3.919 \times 10^{-3}\text{ K}^{-1})T - (3.66 \times 10^{-7}\text{ K}^{-2})T^2$$

over the temperature range 300 K $< T <$ 1500 K. Plot $\overline{C}_V(T)/R$ versus T over this range using Equation 18.62 and the molecular parameters in Table 18.4, and compare your results with the experimental curve.

For NH_3, we plot the experimental equation given in the problem text and the theoretical equation

$$\frac{\overline{C}_V(T)}{R} = \frac{3}{2} + \frac{3}{2} + \sum_{j=1}^{6} \left(\frac{\Theta_{\text{vib},j}}{T}\right)^2 \frac{e^{-\Theta_{\text{vib},j}T}}{(1 - e^{-\Theta_{\text{vib},j}T})^2} \tag{18.62}$$

where $\Theta_{vib,1} = 4800$ K, $\Theta_{vib,2} = 1360$ K, $\Theta_{vib,3} = \Theta_{vib,4} = 4880$ K, and $\Theta_{vib,5} = \Theta_{vib,6} = 2330$ K (Table 18.4).

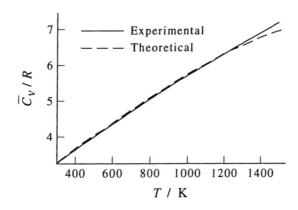

18–31. The experimental heat capacity of $SO_2(g)$ can be fit to the empirical formula

$$\overline{C}_V(T)/R = 6.8711 - \frac{1454.62 \text{ K}}{T} + \frac{160\,351 \text{ K}^2}{T^2}$$

over the temperature range 300 K $< T <$ 1500 K. Plot $\overline{C}_V(T)/R$ versus T over this range using Equation 18.62 and the molecular parameters in Table 18.4, and compare your results with the experimental curve.

For SO_2, we plot the experimental equation given in the problem text and the theoretical equation

$$\frac{\overline{C}_V(T)}{R} = \frac{3}{2} + \frac{3}{2} + \sum_{j=1}^{3} \left(\frac{\Theta_{vib,j}}{T}\right)^2 \frac{e^{-\Theta_{vib,j}T}}{(1 - e^{-\Theta_{vib,j}T})^2} \qquad (18.62)$$

where $\Theta_{vib,1} = 1660$ K, $\Theta_{vib,2} = 750$ K, and $\Theta_{vib,3} = 1960$ K (Table 18.4).

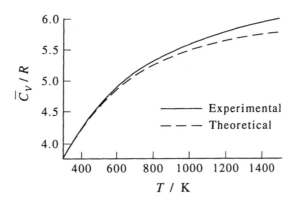

18–32. The experimental heat capacity of $CH_4(g)$ can be fit to the empirical formula

$$\overline{C}_V(T)/R = 1.099 + (7.27 \times 10^{-3} \text{ K}^{-1})T + (1.34 \times 10^{-7} \text{ K}^{-2})T^2$$
$$-(8.67 \times 10^{-10} \text{ K}^{-3})T^3$$

over the temperature range $300 \text{ K} < T < 1500 \text{ K}$. Plot $\overline{C}_V(T)/R$ versus T over this range using Equation 18.62 and the molecular parameters in Table 18.4, and compare your results with the experimental curve.

For CH_4, we plot the experimental equation given in the problem text and the theoretical equation

$$\frac{\overline{C}_V(T)}{R} = \frac{3}{2} + \frac{3}{2} + \sum_{j=1}^{9} \left(\frac{\Theta_{\text{vib},j}}{T}\right)^2 \frac{e^{-\Theta_{\text{vib},j}T}}{(1 - e^{-\Theta_{\text{vib},j}T})^2} \tag{18.62}$$

where $\Theta_{\text{vib},1} = 4170$ K, $\Theta_{\text{vib},2} = \Theta_{\text{vib},3} = 2180$ K, $\Theta_{\text{vib},4} = \Theta_{\text{vib},5} = \Theta_{\text{vib},6} = 4320$ K, and $\Theta_{\text{vib},7} = \Theta_{\text{vib},8} = \Theta_{\text{vib},9} = 1870$ K (Table 18.4).

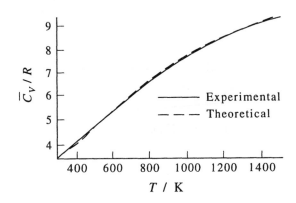

18–33. Show that the moment of inertia of a diatomic molecule is μR_e^2, where μ is the reduced mass, and R_e is the equilibrium bond length.

Let the point labelled in the figure be the center of mass of the molecule. r_1 and r_2 are the distances from the center of mass to masses 1 and 2 (with masses m_1 and m_2), respectively; R_0 is the bond length and M is the total mass of the molecule.

We can then write (by the definition of the center of mass)

$$r_1 = \frac{m_2 r_2}{m_1}$$

Now, because $R_0 = r_1 + r_2$ and $M = m_1 + m_2$,

$$R_0 = \left(\frac{m_2}{m_1} + 1\right) r_2 = \frac{M}{m_1} r_2$$

Finally, because $I = \sum_j m_j r_j^2$ and $\mu = m_1 m_2 / M$,

$$I = m_1 r_1^2 + m_2 r_2^2 = \left[m_1 \left(\frac{m_2}{m_1}\right)^2 + m_2\right] r_2^2$$

$$= \left(\frac{m_2^2}{m_1} + m_2\right) \frac{m_1^2 R_0^2}{M^2} = \left(\frac{m_2 M}{m_1}\right) \frac{m_1^2 R_0^2}{M^2}$$

$$= \frac{m_1 m_2}{M} R_0^2 = \mu R_0^2$$

18–34. Given that the values of Θ_{rot} and Θ_{vib} for H_2 are 85.3 K and 6332 K, respectively, calculate these quantities for HD and D_2. *Hint*: Use the Born-Oppenheimer approximation.

In the Born-Oppenheimer approximation, the potential curve of a diatomic molecule is independent of the isotopes of the constituent atoms. Then, in the formula $I = \mu R_0^2$, R_0 is the same for D_2, H_2, and DH. Therefore, in the harmonic oscillator-rigid rotator approximation,

$$\Theta_{vib} = \frac{h\nu}{k_B} \sim \mu^{-1/2} \quad \text{and} \quad \Theta_{rot} = \frac{\hbar^2}{2Ik_B} \sim \mu^{-1}$$

We can calculate the reduced masses of H_2, D_2, and DH in atomic masses:

$$\mu_{H_2} = \frac{1.008 \text{ amu}}{2} = 0.504 \text{ amu} \qquad \mu_{D_2} = \frac{2.014 \text{ amu}}{2} = 1.007 \text{ amu}$$

$$\mu_{DH} = \frac{(2.014 \text{ amu})(1.008 \text{ amu})}{3.022 \text{ amu}} = 0.672 \text{ amu}$$

Now we can use the relationships between μ and Θ_{vib} and μ and Θ_{rot} to find rotational and vibrational temperatures of D_2:

$$\frac{\Theta_{vib,D_2}}{\Theta_{vib,H_2}} = \left(\frac{\mu_{H_2}}{\mu_{D_2}}\right)^{1/2}$$

$$\Theta_{vib,D_2} = \left(\frac{0.504 \text{ amu}}{1.007 \text{ amu}}\right)^{1/2}(6332 \text{ K}) = 4480 \text{ K}$$

$$\frac{\Theta_{rot,D_2}}{\Theta_{rot,H_2}} = \frac{\mu_{H_2}}{\mu_{D_2}}$$

$$\Theta_{rot,D_2} = \frac{0.504 \text{ amu}}{1.007 \text{ amu}}(85.3 \text{ K}) = 42.7 \text{ K}$$

Similarly, for DH

$$\frac{\Theta_{vib,DH}}{\Theta_{vib,H_2}} = \left(\frac{\mu_{H_2}}{\mu_{DH}}\right)^{1/2}$$

$$\Theta_{vib,DH} = \left(\frac{0.504 \text{ amu}}{0.672 \text{ amu}}\right)^{1/2}(6332 \text{ K}) = 5484 \text{ K}$$

$$\frac{\Theta_{rot,DH}}{\Theta_{rot,H_2}} = \frac{\mu_{H_2}}{\mu_{DH}}$$

$$\Theta_{rot,DH} = \frac{0.504 \text{ amu}}{0.672 \text{ amu}}(85.3 \text{ K}) = 64.0 \text{ K}$$

18–35. Using the result for $q_{rot}(T)$ obtained in Problem 18–14, derive corrections to the expressions $\langle E_{rot} \rangle = RT$ and $C_{V,rot} = R$ given in Section 18–5. Express your result in terms of powers of Θ_{rot}/T.

From Problem 18–14, we write q_{rot} as

$$q_{rot}(T) = \frac{T}{\Theta_{rot}} \left\{ 1 + \frac{1}{3}\left(\frac{\Theta_{rot}}{T}\right) + \frac{1}{15}\left(\frac{\Theta_{rot}}{T}\right)^2 + O\left[\left(\frac{\Theta_{rot}}{T}\right)^3\right] \right\}$$

$$\ln q_{rot} = \ln\frac{T}{\Theta_{rot}} + \ln\left\{ 1 + \frac{1}{3}\frac{\Theta_{rot}}{T} + \frac{1}{15}\left(\frac{\Theta_{rot}}{T}\right)^2 + O\left[\left(\frac{\Theta_{rot}}{T}\right)^3\right] \right\}$$

We use the expansion

$$\ln(1 + x) = x - \frac{x^2}{2} + \frac{x^3}{3} - O(x^4) \tag{I.12}$$

to write $\ln q_{rot}$ as

$$\ln q_{rot} = \ln\frac{T}{\Theta_{rot}} + \left[\frac{1}{3}\frac{\Theta_{rot}}{T} + \frac{1}{15}\left(\frac{\Theta_{rot}}{T}\right)^2\right] - \frac{1}{2}\left(\frac{1}{3}\frac{\Theta_{rot}}{T}\right)^2 + O\left[\left(\frac{\Theta_{rot}}{T}\right)^3\right]$$

$$= \ln\frac{T}{\Theta_{rot}} + \frac{1}{3}\frac{\Theta_{rot}}{T} + \frac{1}{90}\left(\frac{\Theta_{rot}}{T}\right)^2 + O\left[\left(\frac{\Theta_{rot}}{T}\right)^3\right]$$

Now we use Equation 17.21 to write $\langle E \rangle$ as

$$\langle E \rangle = k_B T^2 \left(\frac{\partial \ln Q}{\partial T}\right)_{N,V} = N k_B T^2 \left(\frac{\partial \ln q}{\partial T}\right)_{N,V}$$

$$= RT^2 \left[\frac{1}{T} - \frac{1}{3}\frac{\Theta_{rot}}{T^2} - \frac{1}{45}\frac{\Theta_{rot}^2}{T^3} + O(T^{-4})\right]$$

$$= RT - \frac{R\Theta_{rot}}{3} - \frac{R\Theta_{rot}^2}{45T} + O\left[\left(\frac{\Theta_{rot}}{T}\right)^2\right]$$

Finally, use the definition of constant-volume heat capacity (Equation 17.25) to write

$$C_V = \left(\frac{\partial \langle E \rangle}{\partial T}\right)_{N,V}$$

$$= R + \frac{R}{45}\left(\frac{\Theta_{rot}}{T}\right)^2 + O\left[\left(\frac{\Theta_{rot}}{T}\right)^3\right]$$

18–36. Show that the thermodynamic quantities P and C_V are independent of the choice of a zero of energy.

Begin with Equation 17.14,

$$Q = \sum_j e^{-\beta E_j}$$

Now choose E_0 to be the zero of energy and define

$$Q_0 = \sum_j e^{-\beta(E_j - E_0)}$$

such that

$$Q = e^{-\beta E_0} Q_0$$

Now use Equation 17.32 to write

$$P = k_B T \left(\frac{\partial \ln Q}{\partial V}\right)_{N,\beta} = k_B T \left[\frac{\partial}{\partial V}(-\beta E_0 + \ln Q_0)\right]_{N,\beta} = k_B T \left(\frac{\partial \ln Q_0}{\partial V}\right)_{N,\beta}$$

and use Equation 17.21 to write

$$\langle E \rangle = k_B T^2 \left(\frac{\partial \ln Q}{\partial T}\right)_{N,V} = k_B T^2 \left[\frac{\partial}{\partial T}(-\beta E_0 + \ln Q_0)\right]_{N,V} = E_0 + k_B T^2 \left(\frac{\partial \ln Q_0}{\partial T}\right)_{N,V}$$

Because $\overline{C}_V = (\partial \langle E \rangle / \partial T)_{N,V}$ (Equation 17.25), we can write

$$\overline{C}_V = \left\{\frac{\partial}{\partial T}\left[E_0 + k_B T^2 \left(\frac{\partial \ln Q_0}{\partial T}\right)_{N,V}\right]\right\}_{N,V} = \left\{\frac{\partial}{\partial T}\left[k_B T^2 \left(\frac{\partial \ln Q_0}{\partial T}\right)_{N,V}\right]\right\}_{N,V}$$

Therefore, the values of P and $\overline{C}_V$ are independent of the choice of a zero of energy, as they must be.

18–37. Molecular nitrogen is heated in an electric arc. The spectroscopically determined relative populations of excited vibrational levels are listed below.

v	0	1	2	3	4	$\cdots$
$\dfrac{f_v}{f_0}$	1.000	0.200	0.040	0.008	0.002	$\cdots$

Is the nitrogen in thermodynamic equilibrium with respect to vibrational energy? What is the vibrational temperature of the gas? Is this value necessarily the same as the translational temperature? Why or why not?

At thermal equilibrium,

$$\frac{f_v}{f_0} = \frac{e^{-\beta h \nu (v+1/2)}}{q_{vib}}\left(\frac{e^{-\beta h \nu/2}}{q_{vib}}\right)^{-1} = e^{\beta h \nu v}$$

Thus, if nitrogen is in thermodynamic equilibrium with respect to vibrational energy, the graph of $\ln(f_v/f_0)$ vs. v will be a straight line with slope $-\beta h \nu$. The following figure shows the plot of $\ln(f_v/f_0)$ versus v.

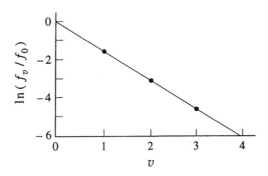

The slope of the line of best fit is -1.5648, which, using $\tilde{\nu}_{N_2} = 2330 \text{ cm}^{-1}$ (Table 5.1), corresponds to a vibrational temperature of 2140 K. The vibrational and translational temperatures need not be equal, because the time scale of the energy transfer between vibrational states and between translational states can be quite different (the energy transfer between vibrational states is usually orders of magnitude slower than that between translational states).

18–38. Consider a system of independent diatomic molecules constrained to move in a plane, that is, a two-dimensional ideal diatomic gas. How many degrees of freedom does a two-dimensional diatomic molecule have? Given that the energy eigenvalues of a two-dimensional rigid rotator are

$$\varepsilon_J = \frac{\hbar^2 J^2}{2I} \qquad J = 0, \ 1, \ 2, \ \ldots$$

(where I is the moment of inertia of the molecule) with a degeneracy $g_J = 2$ for all J except $J = 0$, derive an expression for the rotational partition function. The vibrational partition function is the same as for a three-dimensional diatomic gas. Write out

$$q(T) = q_{\text{trans}}(T) q_{\text{rot}}(T) q_{\text{vib}}(T)$$

and derive an expression for the average energy of this two-dimensional ideal diatomic gas.

A two-dimensional diatomic molecule has two translational degrees of freedom, one vibrational degree of freedom, and one rotational degree of freedom. We know that

$$\epsilon_J = \frac{\hbar^2 J^2}{2I} = \Theta_{\text{rot}} k_{\text{B}} J^2 \qquad J = 0, 1, 2, \cdots$$

So

$$q_{\text{rot}} = \frac{1}{\sigma} \int_0^\infty dJ\, g_J e^{-\epsilon_J / k_{\text{B}} T}$$

$$q_{\text{rot}} = 1 + 2 \int_0^\infty dJ\, e^{-J^2 \Theta_{\text{rot}}/T} \approx \left(\frac{\pi T}{\Theta_{\text{rot}}} \right)^{1/2}$$

We are told that q_{vib} is the same for a two-dimensional gas as it is for a three-dimensional gas, and we know q_{trans} for an ideal two-dimensional gas from Problem 18.3. We can now obtain an expression for the average energy of this gas:

$$q(T) = q_{\text{trans}}(T) q_{\text{rot}}(T) q_{\text{vib}}(T)$$

$$= \left(\frac{2a^2 \pi m k_{\text{B}} T}{h^2} \right) \left(\frac{\pi T}{\Theta_{\text{rot}}} \right)^{1/2} \left(\frac{e^{-\Theta_{\text{vib}}/2T}}{1 - e^{-\Theta_{\text{vib}}/T}} \right)$$

We now wish to find the temperature-dependent terms of $\ln q$:

$$\ln q = \ln T + \frac{1}{2} \ln T - \frac{\Theta_{\text{vib}}}{2T} - \ln(1 - e^{-\Theta_{\text{vib}}/2T}) + \text{terms not containing } T$$

Now, as in Example 18.5, we can take

$$\langle E \rangle = N k_{\text{B}} T^2 \left(\frac{\partial \ln q}{\partial T} \right)_V$$

$$= RT^2 \left(\frac{1}{T} + \frac{1}{2T} + \frac{\Theta_{\text{vib}}}{2T^2} + \frac{\Theta_{\text{vib}}}{T^2} \frac{e^{-\Theta_{\text{vib}}/T}}{1 - e^{-\Theta_{\text{vib}}/T}} \right)$$

$$= \frac{3RT}{2} + \frac{R\Theta_{\text{vib}}}{2} + R\Theta_{\text{vib}} \frac{e^{-\Theta_{\text{vib}}/T}}{1 - e^{-\Theta_{\text{vib}}/T}}$$

18–39. What molar constant-volume heat capacities would you expect under classical conditions for the following gases: (a) Ne, (b) O_2, (c) H_2O, (d) CO_2, and (e) $CHCl_3$?

Each of the gases has a contribution of $3R/2$ to its heat capacity from the translational partition function. In addition, there is a contribution of $R/2$ for each rotational degree of freedom and R for each vibrational degree of freedom. Therefore, the molar heat capacities are

a. $\frac{3}{2}R + 0R + 0R = \frac{3}{2}R$

b. $\frac{3}{2}R + \frac{2}{2}R + R = \frac{7}{2}R$

c. $\frac{3}{2}R + \frac{3}{2}R + 3R = 6R$

d. $\frac{3}{2}R + \frac{2}{2}R + 4R = \frac{13}{2}R$

e. $\frac{3}{2}R + \frac{3}{2}R + 9R = 12R$

18–40. In Chapter 13, we learned that the harmonic-oscillator model can be corrected to include anharmonicity. The energy of an anharmonic oscillator was given as (Equation 13.21)

$$\tilde{\varepsilon}_v = \left(v + \frac{1}{2}\right)\tilde{\nu}_e - \tilde{x}_e\tilde{\nu}_e\left(v + \frac{1}{2}\right)^2 + \cdots$$

where the frequency $\tilde{\nu}_e$ is expressed in cm^{-1}. Substitute this expression for $\tilde{\varepsilon}_v$ into the summation for the vibrational partition function to obtain

$$q_{vib}(T) = \sum_{v=0}^{\infty} e^{-\beta\tilde{\nu}_e(v+\frac{1}{2})} e^{\beta\tilde{x}_e\tilde{\nu}_e(v+\frac{1}{2})^2}$$

Now expand the second factor in the summand, keeping only the linear term in $\tilde{x}_e\tilde{\nu}_e$, to obtain

$$q_{vib}(T) = \frac{e^{-\Theta_{vib}/2T}}{1 - e^{-\Theta_{vib}/T}} + \beta\tilde{x}_e\tilde{\nu}_e e^{-\Theta_{vib}/2T} \sum_{v=0}^{\infty} \left(v + \frac{1}{2}\right)^2 e^{-\Theta_{vib}v/T} + \cdots$$

where $\Theta_{vib}/T = \beta\tilde{\nu}_e$. Given that (Problem I–15)

$$\sum_{v=0}^{\infty} vx^v = \frac{x}{(1-x)^2}$$

and

$$\sum_{v=0}^{\infty} v^2 x^v = \frac{x^2 + x}{(1-x)^3}$$

show that

$$q_{vib}(T) = q_{vib,ho}(T)\left[1 + \beta\tilde{x}_e\tilde{\nu}_e\left(\frac{1}{4} + 2q_{vib,ho}^2(T)\right) + \cdots\right]$$

where $q_{vib,ho}(T)$ is the harmonic-oscillator partition function. Estimate the magnitude of the correction for $Cl_2(g)$ at 300 K, for which $\Theta_{vib} = 805$ K and $\tilde{x}_e\tilde{\nu}_e = 2.675$ cm^{-1}.

Substituting the expression for ε_v given in the problem into the vibrational partition function summation gives the equation

$$q_{vib}(T) = \sum_{v=0}^{\infty} e^{-\beta\varepsilon_v} = \sum_{v=0}^{\infty} e^{-\beta\tilde{\nu}_e(v+\frac{1}{2})} \exp\left[\beta\tilde{x}_e\tilde{\nu}_e(v + \frac{1}{2})^2 + \cdots\right]$$

Expanding the second factor in the summand and keeping only the linear term in $\tilde{x}_e \tilde{\nu}_e$ gives

$$q_{vib}(T) = \sum_{v=0}^{\infty} e^{-\beta \tilde{\nu}_e (v+\frac{1}{2})} \left[1 + \beta \tilde{x}_e \tilde{\nu}_e (v + \frac{1}{2})^2 + \cdots \right]$$

$$= e^{-\Theta_{vib}/2T} \sum_{v=0}^{\infty} e^{-\Theta_{vib}/T} + \beta \tilde{x}_e \tilde{\nu}_e e^{-\Theta_{vib}/2T} \sum_{v=0}^{\infty} (v + \frac{1}{2})^2 e^{-\Theta_{vib} v/T}$$

We can use the geometric series (Equation I.3) to write this as

$$q_{vib}(T) = \frac{e^{-\Theta_{vib}/2T}}{1 - e^{-\Theta_{vib}/T}} + \beta \tilde{x}_e \tilde{\nu}_e e^{-\Theta_{vib}/2T} \sum_{v=0}^{\infty} (v + \frac{1}{2})^2 e^{-\Theta_{vib} v/T}$$

Using the sums given in the problem, we find that the sum in the equation for q_{vib} becomes

$$\sum_{v=0}^{\infty} (v + \frac{1}{2})^2 e^{-\Theta_{vib} v/T} = \frac{e^{-2\Theta_{vib}/T} + e^{-\Theta_{vib}/T}}{(1 - e^{-\Theta_{vib}/T})^3} + \frac{e^{-\Theta_{vib}/T}}{(1 - e^{-\Theta_{vib}/T})^2} + \frac{1}{4(1 - e^{-\Theta_{vib}/T})}$$

$$= \frac{2e^{-\Theta_{vib}/T}}{(1 - e^{-\Theta_{vib}/T})^3} + \frac{1}{4(1 - e^{-\Theta_{vib}/T})}$$

and q_{vib} is then

$$q_{vib} = \frac{e^{-\Theta_{vib}/2T}}{(1 - e^{-\Theta_{vib}/T})} \left\{ 1 + \beta \tilde{x}_e \tilde{\nu}_e \left[\frac{1}{4} + 2 \left(\frac{e^{-\Theta_{vib}/2T}}{1 - e^{-\Theta_{vib}/T}} \right)^2 \right] \right\}$$

$$= q_{vib,ho} \left[1 + \beta \tilde{x}_e \tilde{\nu}_e \left(\frac{1}{4} + 2q_{vib,ho}^2 \right) \right]$$

We can use the parameters given in the problem for Cl_2 at 300 K to find q_{vib} at this temperature.

$$\frac{q_{vib}}{q_{vib,ho}} = 1 + \frac{2.675 \text{ cm}^{-1}}{(0.695 \text{ cm}^{-1}\cdot\text{K}^{-1})(300 \text{ K})} \left[\frac{1}{4} + 2 \left(\frac{e^{-1.34}}{1 - e^{-2.68}} \right)^2 \right]$$

$$= 1 + (0.0128)(0.250 + 0.158) = 1 + 0.0052$$

The correction factor is 0.52% of q_{vib} for a harmonic oscillator.

18–41. Prove that

$$\int_0^{\infty} e^{-\alpha n^2} dn \approx \int_1^{\infty} e^{-\alpha n^2} dn$$

if α is very small. *Hint*: Prove that

$$\int_0^1 e^{-\alpha n^2} dn \ll \int_0^{\infty} e^{-\alpha n^2} dn$$

by expanding the exponential in the first integral.

Expanding the exponential in the first integral gives (Equation I.2)

$$e^{-\alpha n^2} = 1 + (-\alpha n^2) + O(\alpha^2)$$

Then the first integral becomes

$$\int_0^1 e^{-\alpha n^2} dn = \int_0^1 \left[1 - \alpha n^2 + O(\alpha^2) \right] dn = 1 - \frac{1}{3}\alpha + O(\alpha^2)$$

We know that

$$\int_0^\infty e^{-\alpha n^2} dn = \left(\frac{\pi}{4\alpha}\right)^{1/2}$$

As $\alpha \to 0$, $\int_0^\infty e^{-\alpha n^2} dn \to \infty$ and $\int_0^1 e^{-\alpha n^2} dn \to 1$. Thus, as $\alpha \to 0$,

$$\int_0^1 e^{-\alpha n^2} dn \ll \int_0^\infty e^{-\alpha n^2} dn$$

so

$$\int_0^\infty e^{-\alpha n^2} dn \approx \int_1^\infty e^{-\alpha n^2} dn$$

18–42. In this problem, we will derive an expression for the number of translational energy states with (translational) energy between ε and $\varepsilon + d\varepsilon$. This expression is essentially the degeneracy of the state whose energy is

$$\varepsilon_{n_x n_y n_z} = \frac{h^2}{8ma^2}(n_x^2 + n_y^2 + n_z^2) \qquad n_x, \ n_y, \ n_z = 1, \ 2, \ 3, \ \dots \tag{1}$$

The degeneracy is given by the number of ways the integer $M = 8ma^2\varepsilon/h^2$ can be written as the sum of the squares of three positive integers. In general, this is an erratic and discontinuous function of M (the number of ways will be zero for many values of M), but it becomes smooth for large M, and we can derive a simple expression for it. Consider a three-dimensional space spanned by n_x, n_y, and n_z. There is a one-to-one correspondence between energy states given by Equation 1 and the points in this n_x, n_y, n_z space with coordinates given by positive integers. Figure 18.8 shows a two-dimensional version of this space. Equation 1 is an equation for a sphere of radius $R = (8ma^2\varepsilon/h^2)^{1/2}$ in this space

$$n_x^2 + n_y^2 + n_z^2 = \frac{8ma^2\varepsilon}{h^2} = R^2$$

We want to calculate the number of lattice points that lie at some fixed distance from the origin in this space. In general, this is very difficult, but for large R we can proceed as follows. We treat R, or ε, as a continuous variable and ask for the number of lattice points between ε and $\varepsilon + \Delta\varepsilon$. To calculate this quantity, it is convenient to first calculate the number of lattice points consistent with an energy $\leq \varepsilon$. For large ε, an excellent approximation can be made by equating the number of lattice points consistent with an energy $\leq \varepsilon$ with the volume of one octant of a sphere of radius R.

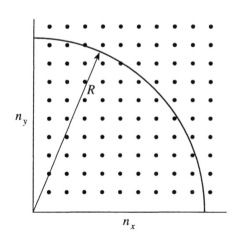

FIGURE 18.8
A two-dimensional version of the (n_x, n_y, n_z) space, the space with the quantum numbers n_x, n_y, and n_z as axes. Each point corresponds to an energy of a particle in a (two-dimensional) box.

We take only one octant because n_x, n_y, and n_z are restricted to be positive integers. If we denote the number of such states by $\Phi(\varepsilon)$, we can write

$$\Phi(\varepsilon) = \frac{1}{8}\left(\frac{4\pi R^3}{3}\right) = \frac{\pi}{6}\left(\frac{8ma^2\varepsilon}{h^2}\right)^{3/2}$$

The number of states with energy between ε and $\varepsilon + \Delta\varepsilon$ ($\Delta\varepsilon/\varepsilon \ll 1$) is

$$\omega(\varepsilon, \Delta\varepsilon) = \Phi(\varepsilon + \Delta\varepsilon) - \Phi(\varepsilon)$$

Show that

$$\omega(\varepsilon, \Delta\varepsilon) = \frac{\pi}{4}\left(\frac{8ma^2}{h^2}\right)^{3/2}\varepsilon^{1/2}\Delta\varepsilon + O[(\Delta\varepsilon)^2]$$

Show that if we take $\varepsilon = 3k_BT/2$, $T = 300$ K, $m = 10^{-25}$ kg, $a = 1$ dm, and $\Delta\varepsilon$ to be 0.010ε (in other words 1% of ε), then $\omega(\varepsilon, \Delta\varepsilon)$ is $O(10^{28})$. So, even for a system as simple as a single particle in a box, the degeneracy can be very large at room temperature.

We can express $\Phi(\varepsilon)$ and $\Phi(\varepsilon + \Delta\varepsilon)$, by definition, as

$$\Phi(\varepsilon) = \frac{\pi}{6}\left(\frac{8ma^2\varepsilon}{h^2}\right)^{3/2} \qquad \Phi(\varepsilon + \Delta\varepsilon) = \frac{\pi}{6}\left[\frac{8ma^2(\varepsilon + \Delta\varepsilon)}{h^2}\right]^{3/2}$$

Now we substitute these values into the expression given in the text for $\omega(\varepsilon, \Delta\varepsilon)$ and expand in $\Delta\varepsilon$:

$$\begin{aligned}
\omega(\varepsilon, \Delta\varepsilon) &= \Phi(\varepsilon + \Delta\varepsilon) - \Phi(\varepsilon) \\
&= \frac{\pi}{6}\left(\frac{8ma^2}{h^2}\right)^{3/2}\left[(\varepsilon + \Delta\varepsilon)^{3/2} - \varepsilon^{3/2}\right] \\
&= \frac{\pi}{6}\left(\frac{8ma^2}{h^2}\right)^{3/2}\left\{\varepsilon^{3/2}\left[\left(1 + \frac{\Delta\varepsilon}{\varepsilon}\right)^{3/2} - 1\right]\right\} \\
&= \frac{\pi}{6}\left(\frac{8ma^2}{h^2}\right)^{3/2}\varepsilon^{3/2}\left\{1 + \frac{3}{2}\frac{\Delta\varepsilon}{\varepsilon} + O\left[\left(\frac{\Delta\varepsilon}{\varepsilon}\right)^2\right] - 1\right\} \\
&= \frac{\pi}{4}\left(\frac{8ma^2}{h^2}\right)^{3/2}\varepsilon^{3/2}\frac{\Delta\varepsilon}{\varepsilon} + O\left[(\Delta\varepsilon)^2\right] \\
&= \frac{\pi}{4}\left(\frac{8ma^2}{h^2}\right)^{3/2}\varepsilon^{1/2}\Delta\varepsilon + O[(\Delta\varepsilon)^2]
\end{aligned}$$

Substituting the desired values in the above equations gives $\omega(\varepsilon, 0.01\varepsilon) = 9.5 \times 10^{27} = O(10^{28})$.

18–43. The translational partition function can be written as a single integral over the energy ε if we include the degeneracy

$$q_{\text{trans}}(V, T) = \int_0^\infty \omega(\varepsilon)e^{-\varepsilon/k_BT}d\varepsilon$$

where $\omega(\varepsilon)d\varepsilon$ is the number of states with energy between ε and $\varepsilon + d\varepsilon$. Using the result from the previous problem, show that $q_{\text{trans}}(V, T)$ is the same as that given by Equation 18.6.

As $\Delta\varepsilon \to 0$, $\omega(\varepsilon, \Delta\varepsilon) \to \omega(\varepsilon)d\varepsilon$, so we can use (from the above problem)

$$\omega(\varepsilon)d\varepsilon = \frac{\pi}{4}\left(\frac{8ma^2}{h^2}\right)^{3/2}\varepsilon^{1/2}d\varepsilon$$

Substituting this into the expression given for q_{trans} gives

$$q_{trans} = \int_0^\infty \frac{\pi}{4} \left(\frac{8ma^2}{h^2}\right)^{3/2} \varepsilon^{1/2} e^{-\varepsilon/k_B T} d\varepsilon$$

$$= \frac{\pi}{4} \left(\frac{8ma^2}{h^2}\right)^{3/2} \int_0^\infty \varepsilon^{1/2} e^{-\varepsilon/k_B T} d\varepsilon$$

$$= \frac{\pi}{4} \left(\frac{8ma^2}{h^2}\right)^{3/2} \int_0^\infty x e^{-x^2/k_B T} 2x dx$$

$$= \frac{\pi}{4} \left(\frac{8ma^2}{h^2}\right)^{3/2} \frac{k_B T}{2} (\pi k_B T)^{1/2}$$

$$= \left(\frac{2\pi m k_B T}{h^2}\right)^{3/2} a^3$$

which is the same as Equation 18.6.

CHAPTER 19

The First Law of Thermodynamics

PROBLEMS AND SOLUTIONS

19–1. Suppose that a 10-kg mass of iron at 20°C is dropped from a height of 100 meters. What is the kinetic energy of the mass just before it hits the ground? What is its speed? What would be the final temperature of the mass if all its kinetic energy at impact is transformed into internal energy? Take the molar heat capacity of iron to be $\overline{C}_P = 25.1 \; \text{J·mol}^{-1}·\text{K}^{-1}$ and the gravitational acceleration constant to be $9.80 \; \text{m·s}^{-2}$.

Just before the mass hits the ground, all of the potential energy that the mass originally had will be converted into kinetic energy. So

$$\text{PE} = mgh = (10 \; \text{kg})(9.80 \; \text{m·s}^{-2})(100 \; \text{m}) = 9.8 \; \text{kJ} = \text{KE}$$

Since kinetic energy can be expressed as $mv^2/2$, the speed of the mass just before hitting the ground is

$$v_f = \left(\frac{2\text{KE}}{m}\right)^{1/2} = \left[\frac{2(9.8 \; \text{kJ})}{10 \; \text{kg}}\right]^{1/2} = 44 \; \text{m·s}^{-1}$$

For a solid, the difference between $\overline{C}_V$ and $\overline{C}_P$ is small, so we can write $\Delta U = n\overline{C}_P \Delta T$ (Equation 19.39). Then

$$\Delta T = \frac{9.8 \; \text{kJ}}{\left(\dfrac{1 \times 10^4 \; \text{g}}{55.85 \; \text{g·mol}^{-1}}\right)(25.1 \; \text{J·mol}^{-1}·\text{K}^{-1})} = 2.2 \; \text{K}$$

The final temperature of the iron mass is then 22.2°C.

19–2. Consider an ideal gas that occupies 2.50 dm³ at a pressure of 3.00 bar. If the gas is compressed isothermally at a constant external pressure, P_{ext}, so that the final volume is 0.500 dm³, calculate the smallest value P_{ext} can have. Calculate the work involved using this value of P_{ext}.

Since the gas is ideal, we can write

$$P_2 = \frac{P_1 V_1}{V_2} = \frac{(3.00 \; \text{bar})(2.50 \; \text{dm}^3)}{0.500 \; \text{dm}^3} = 15.0 \; \text{bar}$$

The smallest possible value of P_{ext} is P_2. The work done in this case is (Equation 19.1)

$$w = -P_{\text{ext}} \Delta V = (-15.0 \; \text{bar})(-2.0 \; \text{dm}^3)\left(\frac{8.3145 \; \text{J·mol}^{-1}·\text{K}^{-1}}{0.083145 \; \text{bar·dm}^3·\text{mol}^{-1}·\text{K}^{-1}}\right) = 3000 \; \text{J}$$

19–3. A one-mole sample of $CO_2(g)$ occupies 2.00 dm^3 at a temperature of 300 K. If the gas is compressed isothermally at a constant external pressure, P_{ext}, so that the final volume is 0.750 dm^3, calculate the smallest value P_{ext} can have, assuming that $CO_2(g)$ satisfies the van der Waals equation of state under these conditions. Calculate the work involved using this value of P_{ext}.

The smallest value P_{ext} can have is P_2, where P_2 is the final pressure of the gas. We can use the van der Waals equation (Equation 16.5) and the constants given in Table 16.3 to find P_2:

$$
\begin{aligned}
P_2 &= \frac{RT_2}{\overline{V}_2 - b} - \frac{a}{\overline{V}_2^2} \\
&= \frac{(0.083145 \ dm^3 \cdot bar \cdot mol^{-1} \cdot K^{-1})(300 \ K)}{0.750 \ dm^3 \cdot mol^{-1} - 0.042816 \ dm^3 \cdot mol^{-1}} - \frac{3.6551 \ dm^6 \cdot bar \cdot mol^{-2}}{(0.750 \ dm^3 \cdot mol^{-1})^2} \\
&= 28.8 \ bar
\end{aligned}
$$

The work involved is (Equation 19.1)

$$
w = -P\Delta V = -(28.8 \times 10^5 \ Pa)(-1.25 \times 10^{-3} \ m^3) = 3.60 \ kJ
$$

19–4. Calculate the work involved when one mole of an ideal gas is compressed reversibly from 1.00 bar to 5.00 bar at a constant temperature of 300 K.

Using the ideal gas equation, we find that

$$
V_1 = \frac{nRT}{P_1} \qquad \text{and} \qquad V_2 = \frac{nRT}{P_2}
$$

We can therefore write $V_2/V_1 = P_1/P_2$. Now we substitute into Equation 19.2 to find

$$
\begin{aligned}
w &= -\int P_{ext} dV = -\int \frac{nRT}{V} dV \\
&= -nRT \ln\left(\frac{V_2}{V_1}\right) = -nRT \ln\left(\frac{P_1}{P_2}\right) \\
&= (-1 \ mol)(8.315 \ J \cdot mol^{-1} \cdot K^{-1})(300 \ K) \ln 0.2 = 4.01 \ kJ
\end{aligned}
$$

19–5. Calculate the work involved when one mole of an ideal gas is expanded reversibly from 20.0 dm^3 to 40.0 dm^3 at a constant temperature of 300 K.

We can integrate Equation 19.2 to find the work involved:

$$
\begin{aligned}
w &= -nRT \ln\left(\frac{V_2}{V_1}\right) \\
&= (-1 \ mol)(8.315 \ J \cdot mol^{-1} \cdot K^{-1})(300 \ K) \ln 2 = -1.73 \ kJ
\end{aligned}
$$

19–6. Calculate the minimum amount of work required to compress 5.00 moles of an ideal gas isothermally at 300 K from a volume of 100 dm^3 to 40.0 dm^3.

We note that the minimum amount of work required is the amount of work needed to reversibly compress the gas, so we can write Equation 19.2 as

$$w_{min} = w_{rev} = -nRT \ln\left(\frac{V_2}{V_1}\right)$$

$$= (-5.00 \text{ mol})(8.315 \text{ J}\cdot\text{mol}^{-1}\cdot\text{K}^{-1})(300 \text{ K}) \ln 0.400 = 11.4 \text{ kJ}$$

19-7. Consider an ideal gas that occupies 2.25 L at 1.33 bar. Calculate the work required to compress the gas isothermally to a volume of 1.50 L at a constant pressure of 2.00 bar followed by another isothermal compression to 0.800 L at a constant pressure of 3.75 bar (Figure 19.4). Compare the result with the work of compressing the gas isothermally and reversibly from 2.25 L to 0.800 L.

We can use Equation 19.2 to describe the work involved with the compressions under different circumstances.

a. Two-step process, each step at constant external pressure
  i. From (2.25 L, 1.33 bar) to (1.50 L, 2.00 bar),

$$w = -\int P_{ext}dV = (-2.00 \text{ bar})(1.50 \text{ L} - 2.25 \text{ L})(100 \text{ J}\cdot\text{bar}^{-1}\cdot\text{dm}^{-3}) = 150 \text{ J}$$

  ii. From (1.50 L, 2.00 bar) to (0.800 L, 2.50 bar),

$$w = -\int P_{ext}dV = (-3.75 \text{ bar})(0.800 \text{ L} - 1.50 \text{ L})(100 \text{ J}\cdot\text{bar}^{-1}\cdot\text{dm}^{-3}) = 263 \text{ J}$$

  The total work involved in the two-step process is +413 J.

b. Reversible process
  Because the gas is ideal, $PV = nRT$. We can then write

$$PV = (2.25 \text{ L})(1.33 \text{ bar})(100 \text{ J}\cdot\text{bar}^{-1}\cdot\text{dm}^{-3}) = 299.25 \text{ J} = nRT$$

and use Equation 19.2 to find w:

$$w = -\int PdV = -nRT \ln\left(\frac{V_2}{V_1}\right) = -(299.25 \text{ J}) \ln\left(\frac{0.800}{2.25}\right) = 309 \text{ J}$$

The total work involved in the reversible process is +309 J. Note that the work involved in the reversible process is less than the work involved at constant external pressure, as is expected.

19-8. Show that for an isothermal reversible expansion from a molar volume $\overline{V}_1$ to a final molar volume $\overline{V}_2$, the work is given by

$$w = -RT \ln\left(\frac{\overline{V}_2 - B}{\overline{V}_1 - B}\right) - \frac{A}{BT^{1/2}} \ln\left[\frac{(\overline{V}_2 + B)\overline{V}_1}{(\overline{V}_1 + B)\overline{V}_2}\right]$$

for the Redlich-Kwong equation.

For the Redlich-Kwong equation,

$$P = \frac{RT}{\overline{V} - B} - \frac{A}{T^{1/2}\overline{V}(\overline{V} + B)} \tag{16.7}$$

We can then use Equation 19.2 to find w.

$$
\begin{aligned}
w &= \int_{\overline{V}_1}^{\overline{V}_2} \left[-\frac{RT}{\overline{V} - B} + \frac{A}{T^{1/2}\overline{V}(\overline{V} + B)} \right] d\overline{V} \\
&= -RT \ln\left(\frac{\overline{V}_2 - B}{\overline{V}_1 - B} \right) - \frac{A}{T^{1/2}B} \left[\ln\left(\frac{\overline{V}_1}{\overline{V}_1 + B} \right) - \ln\left(\frac{\overline{V}_2}{\overline{V}_2 + B} \right) \right] \\
&= -RT \ln\left(\frac{\overline{V}_2 - B}{\overline{V}_1 - B} \right) - \frac{A}{BT^{1/2}} \ln\left[\frac{(\overline{V}_2 + B)\overline{V}_1}{(\overline{V}_1 + B)\overline{V}_2} \right]
\end{aligned}
$$

19–9. Use the result of Problem 19–8 to calculate the work involved in the isothermal reversible expansion of one mole of $CH_4(g)$ from a volume of $1.00 \text{ dm}^3 \cdot \text{mol}^{-1}$ to $5.00 \text{ dm}^3 \cdot \text{mol}^{-1}$ at 300 K. (See Table 16.4 for the values of A and B.)

From Table 16.4, $A = 32.205 \text{ dm}^6 \cdot \text{bar} \cdot \text{mol}^{-2} \cdot \text{K}^{1/2}$ and $B = 0.029850 \text{ dm}^3 \cdot \text{mol}^{-1}$. Then, using the equation for w from the previous problem,

$$
\begin{aligned}
w &= -RT \ln\left(\frac{\overline{V}_2 - B}{\overline{V}_1 - B} \right) - \frac{A}{BT^{1/2}} \ln\left[\frac{(\overline{V}_2 + B)\overline{V}_1}{(\overline{V}_1 + B)\overline{V}_2} \right] \\
&= -RT \ln 5.1231 - \frac{A}{BT^{1/2}} \ln 0.97681 \\
&= -(39.3 \text{ dm}^3 \cdot \text{bar} \cdot \text{mol}^{-1})(100 \text{ J} \cdot \text{bar}^{-1} \cdot \text{dm}^{-3}) = -3.93 \text{ kJ} \cdot \text{mol}^{-1}
\end{aligned}
$$

19–10. Repeat the calculation in Problem 19–9 for a van der Waals gas.

From Equation 16.5,

$$
P = \frac{RT}{\overline{V} - b} - \frac{a}{\overline{V}^2}
$$

Then (Equation 19.2)

$$
\begin{aligned}
w &= \int -P d\overline{V} = \int_{\overline{V}_1}^{\overline{V}_2} \left(-\frac{RT}{\overline{V} - b} + \frac{a}{\overline{V}^2} \right) d\overline{V} \\
&= -RT \ln \frac{\overline{V}_2 - b}{\overline{V}_1 - b} + \frac{a(\overline{V}_2 - \overline{V}_1)}{\overline{V}_2 \overline{V}_1}
\end{aligned}
$$

From Table 16.3, for methane $a = 2.3026 \text{ dm}^6 \cdot \text{bar} \cdot \text{mol}^{-2}$ and $b = 0.043067 \text{ dm}^3 \cdot \text{mol}^{-1}$. Then

$$
w = -39.18 \text{ dm}^3 \cdot \text{bar} \cdot \text{mol}^{-1} = -3.92 \text{ kJ} \cdot \text{mol}^{-1}
$$

19–11. Derive an expression for the reversible isothermal work of an expansion of a gas that obeys the Peng-Robinson equation of state.

The Peng-Robinson equation of state is

$$P = \frac{RT}{\overline{V} - \beta} - \frac{\alpha}{\overline{V}(\overline{V} + \beta) + \beta(\overline{V} - \beta)} \tag{16.7}$$

Substituting into Equation 19.2 gives

$$w_{\text{rev}} = -\int_1^2 d\overline{V} \left[\frac{RT}{\overline{V} - \beta} - \frac{\alpha}{\overline{V}(\overline{V} + \beta) + \beta(\overline{V} - \beta)} \right]$$

$$= -RT \ln\left(\frac{\overline{V}_2 - \beta}{\overline{V}_1 - \beta} \right) - \alpha \int_1^2 d\overline{V} \frac{1}{\overline{V}^2 + 2\overline{V}\beta - \beta^2}$$

$$= -RT \ln\left(\frac{\overline{V}_2 - \beta}{\overline{V}_1 - \beta} \right) - \frac{\alpha}{(8\beta^2)^{1/2}} \ln \left. \frac{2\overline{V} + 2\beta - (8\beta^2)^{1/2}}{2\overline{V} + 2\beta + (8\beta^2)^{1/2}} \right|_{\overline{V}_1}^{\overline{V}_2}$$

$$= -RT \ln\left(\frac{\overline{V}_2 - \beta}{\overline{V}_1 - \beta} \right) - \frac{\alpha}{(8\beta^2)^{1/2}} \ln \frac{(\overline{V}_2 - 0.4142\beta)(\overline{V}_1 + 2.414\beta)}{(\overline{V}_2 + 2.414\beta)(\overline{V}_1 - 0.4142\beta)}$$

19–12. One mole of a monatomic ideal gas initially at a pressure of 2.00 bar and a temperature of 273 K is taken to a final pressure of 4.00 bar by the reversible path defined by P/V = constant. Calculate the values of ΔU, ΔH, q, and w for this process. Take $\overline{C}_V$ to be equal to 12.5 $\text{J} \cdot \text{mol}^{-1} \cdot \text{K}^{-1}$.

Let $P/\overline{V} = C$. Then, since the gas is ideal, we can write

$$T_1 = \frac{P_1 V_1}{R} = \frac{P_1^2}{CR}$$

$$C = \frac{4.00 \text{ bar}^2}{(273 \text{ K})R}$$

Since $P/\overline{V}$ is constant throughout the process, we can also write

$$T_2 = \frac{P_2 V_2}{R} = \frac{P_2^2}{CR} = \frac{16 \text{ bar}^2}{CR} = \frac{(16.0 \text{ bar}^2)(273 \text{ K})R}{(4.00 \text{ bar}^2)R} = 1092 \text{ K}$$

Because the $\overline{C}_V$ we are given is temperature-independent, we can write (by the definition of molar heat capacity)

$$\Delta U = n \int_{T_1}^{T_2} \overline{C}_V dT$$

$$= (1 \text{ mol})(12.5 \text{ J} \cdot \text{mol}^{-1} \cdot \text{K}^{-1})(1092 \text{ K} - 273 \text{ K}) = 10.2 \text{ kJ}$$

Now we can use Equation 19.2 to calculate w, using the equality $P/\overline{V} = C$. Note that $\overline{V} = V$, since we are taking one mole of the gas.

$$w = -\int_{V_1}^{V_2} P dV = -\int_{V_1}^{V_2} CV dV = -\frac{C}{2}(V_2^2 - V_1^2)$$

$$= -\frac{C}{2}\left(\frac{P_2^2}{C^2} - \frac{P_1^2}{C^2} \right) = -\frac{16.0 \text{ bar}^2 - 4.00 \text{ bar}^2}{2C}$$

$$= -\frac{(12.0 \text{ bar}^2)(1 \text{ mol})(273 \text{ K})(0.083145 \text{ bar} \cdot \text{dm}^3 \cdot \text{mol}^{-1} \cdot \text{K}^{-1})}{2(4.00 \text{ bar}^2)}$$

$$= -3.40 \text{ kJ}$$

Finally, we can find q from Equation 19.10 and ΔH from Equation 19.35, letting $PV = nRT$.

$$q = \Delta U - w = 13.6 \text{ kJ}$$

$$\Delta H = \Delta U + nR\Delta T = 10.2 \text{ kJ} + (1 \text{ mol})(8.3145 \text{ J·mol}^{-1}\text{K}^{-1})(819 \text{ K})$$

$$= 17.0 \text{ kJ}$$

19–13. The isothermal compressibility of a substance is given by

$$\beta = -\frac{1}{V}\left(\frac{\partial V}{\partial P}\right)_T \tag{1}$$

For an ideal gas, $\beta = 1/P$, but for a liquid, β is fairly constant over a moderate pressure range. If β is constant, show that

$$\frac{V}{V_0} = e^{-\beta(P-P_0)} \tag{2}$$

where V_0 is the volume at a pressure P_0. Use this result to show that the reversible isothermal work of compressing a liquid from a volume V_0 (at a pressure P_0) to a volume V (at a pressure P) is given by

$$w = -P_0(V - V_0) + \beta^{-1}V_0\left(\frac{V}{V_0}\ln\frac{V}{V_0} - \frac{V}{V_0} + 1\right)$$

$$= -P_0V_0[e^{-\beta(P-P_0)} - 1] + \beta^{-1}V_0\{1 - [1 + \beta(P - P_0)]e^{-\beta(P-P_0)}\} \tag{3}$$

(You need to use the fact that $\int \ln x \, dx = x \ln x - x$.) The fact that liquids are incompressible is reflected by β being small, so that $\beta(P - P_0) \ll 1$ for moderate pressures. Show that

$$w = \beta P_0 V_0(P - P_0) + \frac{\beta V_0(P - P_0)^2}{2} + O(\beta^2)$$

$$= \frac{\beta V_0}{2}(P^2 - P_0^2) + O(\beta^2) \tag{4}$$

Calculate the work required to compress one mole of toluene reversibly and isothermally from 10 bar to 100 bar at 20°C. Take the value of β to be 8.95×10^{-5} bar^{-1} and the molar volume to be 0.106 L·mol^{-1} at 20°C.

We begin with Equation 1 and integrate both sides, letting β be constant with respect to pressure.

$$\int -\beta dP = \int V^{-1} dV$$

$$-\beta(P - P_0) = \ln\left(\frac{V}{V_0}\right)$$

$$\frac{V}{V_0} = e^{-\beta(P-P_0)}$$

$$P = -\beta^{-1}\ln\left(\frac{V}{V_0}\right) + P_0$$

Now we wish to find the reversible isothermal work of compressing a liquid from (P_0, V_0) to (P, V). We know that $\delta w = -P\,dV$ (Equation 19.2), so we use the expression we found above for P to write

$$\delta w = -\left[-\beta^{-1} \ln \left(\frac{V}{V_0} \right) + P_0 \right] dV$$

$$= -P_0\,dV + \beta^{-1} \ln \left(\frac{V}{V_0} \right) dV$$

Integrating both sides of this equation gives

$$w = -P_0(V - V_0) + \beta^{-1} \int \ln V\,dV - \beta^{-1} \int \ln V_0\,dV$$

$$= -P_0(V - V_0) + \beta^{-1} \left[V \ln V - V - (V_0 \ln V_0 - V_0) \right] - \beta^{-1}(V - V_0) \ln V_0$$

$$= -P_0(V - V_0) + \beta^{-1} \left(V \ln V - V - V_0 \ln V_0 + V_0 - V \ln V_0 + V_0 \ln V_0 \right)$$

$$= -P_0(V - V_0) + \beta^{-1} V_0 \left(\frac{V}{V_0} \ln \frac{V}{V_0} - \frac{V}{V_0} + 1 \right)$$

Substitution for V then yields the result

$$w = -P_0 V_0 \left[e^{-\beta(P - P_0)} - 1 \right] + \beta^{-1} V_0 \left\{ 1 - \left[1 + \beta(P - P_0) \right] e^{-\beta(P - P_0)} \right\}$$

which is Equation 3 in the text of the problem. Now let $x = -\beta(P - P_0)$. Because $-\beta(P - P_0) \ll 1$, $x \ll 1$. We can now write Equation 3 as

$$w = -P_0 V_0 (e^x - 1) + \beta^{-1} V_0 \left\{ 1 - \left[1 + \beta(P - P_0) \right] e^x \right\}$$

$$= -P_0 V_0 (e^x - 1) + \beta^{-1} V_0 - \beta^{-1} V_0 e^x - V_0 (P - P_0) e^x$$

Now, recall from MathChapter I that if x is small, we can write e^x as $1 + x + x^2/2 + O(x^3)$ (Equation I.2). Notice that to find w to $O(\beta^2)$, we must expand e^x to $O(x^3)$, since one of the above terms multiplies e^x by β^{-1}. Expanding the above equation gives

$$w = -P_0 V_0 \left[x + O(x^2) \right] + \beta^{-1} V_0 - \beta^{-1} V_0 \left[1 + x + \frac{x^2}{2} + O(x^3) \right] - V_0(P - P_0) \left[1 + x + O(x^2) \right]$$

$$= \beta P_0 V_0 (P_0 - P) + V_0 (P - P_0) - \frac{\beta V_0 (P - P_0)^2}{2} - V_0(P - P_0) + \beta V_0 (P - P_0)^2 + O(\beta^2)$$

$$= \beta P_0 V_0 (P - P_0) + \frac{\beta V_0 (P - P_0)^2}{2} + O(\beta^2)$$

$$= \frac{\beta V_0}{2} (P^2 - P_0^2) + O(\beta^2)$$

Now, for one mole of toluene [to $O(\beta^2)$], we use the parameters given in the problem to find

$$w = \frac{(8.95 \times 10^{-5}\ \text{bar}^{-1})(0.106\ \text{mol} \cdot \text{L}^{-1})^{-1}}{2} \left[(100\ \text{bar})^2 - (10\ \text{bar})^2 \right] = 418\ \text{J}$$

19–14. In the previous problem, you derived an expression for the reversible, isothermal work done when a liquid is compressed. Given that β is typically $O(10^{-4})$ bar^{-1}, show that $V/V_0 \approx 1$ for pressures up to about 100 bar. This result, of course, reflects the fact that liquids are not very compressible. We can exploit this result by substituting $dV = -\beta V\,dP$ from the defining equation of β into $w = -\int P\,dV$ and then treating V as a constant. Show that this approximation gives Equation 4 of Problem 19–13.

We are given that $\beta \sim O(10^{-4} \text{ bar}^{-1})$, and the largest pressure differential that can occur under the given conditions is on the order of $O(10^2 \text{ bar})$. Then, using Equation 2 of Problem 19.13, we find

$$\frac{V}{V_0} = e^{-\beta(P-P_0)} = e^{-O(10^{-2})} \approx 0.990$$

Therefore, $V/V_0 \approx 1$ for pressures ranging from 1 to 100 bar. Now $dV = -\beta V dP$, so

$$w = -\int P dV = \beta V_0 \int P dP = \frac{\beta V_0}{2}(P^2 - P_0^2)$$

as in Equation 4 of Problem 19.13.

19–15. Show that

$$\frac{T_2}{T_1} = \left(\frac{V_1}{V_2}\right)^{R/\overline{C}_V}$$

for a reversible adiabatic expansion of an ideal gas.

For an adiabatic expansion $\delta q = 0$, so $dU = \delta w$. By definition, $dU = n\overline{C}_V dT$, and for an ideal gas (Equation 19.2)

$$\delta w = -P dV = -nRT V^{-1} dV$$

We can then write

$$n\overline{C}_V dT = -nRT V^{-1} dV$$

$$\int \frac{\overline{C}_V}{T} dT = \int -\frac{R}{V} dV$$

$$\overline{C}_V \ln\left(\frac{T_2}{T_1}\right) = -R \ln\left(\frac{V_2}{V_1}\right)$$

$$\ln\left(\frac{T_2}{T_1}\right)^{\overline{C}_V} = \ln\left(\frac{V_2}{V_1}\right)^{-R}$$

Finally, exponentiating both sides gives

$$\frac{T_2}{T_1} = \left(\frac{V_2}{V_1}\right)^{-R/\overline{C}_V} = \left(\frac{V_1}{V_2}\right)^{R/\overline{C}_V}$$

19–16. Show that

$$\left(\frac{T_2}{T_1}\right)^{3/2} = \frac{\overline{V}_1 - b}{\overline{V}_2 - b}$$

for a reversible, adiabatic expansion of a monatomic gas that obeys the equation of state $P(\overline{V} - b) = RT$. Extend this result to the case of a diatomic gas.

For an adiabatic expansion $\delta q = 0$, so $dU = \delta w$. By definition, $dU = n\overline{C}_V dT$, and for this gas Equation 19.2 becomes

$$\delta w = -P dV = -n\frac{RT}{\overline{V} - b} d\overline{V}$$

Setting dU and δw equal to one another gives

$$n\overline{C}_V dT = -\frac{nRT}{\overline{V} - b}d\overline{V}$$

$$\int \frac{\overline{C}_V}{T}dT = \int -\frac{R}{\overline{V} - b}d\overline{V}$$

$$\overline{C}_V \ln\left(\frac{T_2}{T_1}\right) = -R \ln\left(\frac{\overline{V}_2 - b}{\overline{V}_1 - b}\right)$$

$$\left(\frac{T_2}{T_1}\right)^{\overline{C}_V/R} = \frac{\overline{V}_1 - b}{\overline{V}_2 - b}$$

For a monatomic gas, $\overline{C}_V = 3R/2$, and for a diatomic gas, $\overline{C}_V = 5R/2$. Thus

$$\left(\frac{T_2}{T_1}\right)^{3/2} = \frac{\overline{V}_1 - b}{\overline{V}_2 - b}$$

for a monatomic gas, and

$$\left(\frac{T_2}{T_1}\right)^{5/2} = \frac{\overline{V}_1 - b}{\overline{V}_2 - b}$$

for a diatomic gas.

19–17. Show that

$$\frac{T_2}{T_1} = \left(\frac{P_2}{P_1}\right)^{R/\overline{C}_P}$$

for a reversible adiabatic expansion of an ideal gas.

For an ideal gas, $\overline{C}_V + R = \overline{C}_P$ and

$$\frac{P_1 V_1}{P_2 V_2} = \frac{T_1}{T_2}$$

From Problem 19–15, we can write

$$\left(\frac{T_2}{T_1}\right)^{\overline{C}_V/R} = \left(\frac{V_1}{V_2}\right)$$

$$\frac{P_1}{P_2}\left(\frac{T_2}{T_1}\right)^{\overline{C}_V/R} = \frac{P_1 V_1}{P_2 V_2} = \frac{T_1}{T_2}$$

Then

$$\frac{P_1}{P_2} = \left(\frac{T_1}{T_2}\right)^{(\overline{C}_V + R)/R} = \left(\frac{T_1}{T_2}\right)^{\overline{C}_P/R}$$

$$\left(\frac{P_1}{P_2}\right)^{R/\overline{C}_P} = \frac{T_1}{T_2}$$

and, finally,

$$\frac{T_2}{T_1} = \left(\frac{P_2}{P_1}\right)^{R/\overline{C}_P}$$

19–18. Show that

$$P_1 V_1^{(\overline{C}_V + R)/\overline{C}_V} = P_2 V_2^{(\overline{C}_V + R)/\overline{C}_V}$$

for an adiabatic expansion of an ideal gas. Show that this formula reduces to Equation 19.23 for a monatomic gas.

For an ideal gas,

$$\frac{P_1 V_1}{P_2 V_2} = \frac{T_1}{T_2}$$

We can substitute this expression into the equation from Problem 19–15 to write

$$\frac{P_1 V_1}{P_2 V_2} = \left(\frac{V_1}{V_2}\right)^{R/\overline{C}_V}$$

Taking the reciprocal gives

$$\frac{P_1 V_1}{P_2 V_2} = \left(\frac{V_2}{V_1}\right)^{R/\overline{C}_V}$$

and rearranging yields

$$P_1 V_1^{\left(1 + R/\overline{C}_V\right)} = P_2 V_2^{\left(1 + R\overline{C}_V\right)}$$

For a monatomic ideal gas, $\overline{C}_V = \frac{3}{2}R$, so

$$P_1 V_1^{5/3} = P_2 V_2^{5/3} \tag{19.23}$$

19–19. Calculate the work involved when one mole of a monatomic ideal gas at 298 K expands reversibly and adiabatically from a pressure of 10.00 bar to a pressure of 5.00 bar.

Because this process is adiabatic, $\delta q = 0$. This means that

$$\delta w = dU = n\overline{C}_V dT$$

where $\overline{C}_V$ is temperature-independent (since the gas is ideal). We can use the equation from Problem 19–17 to write

$$T_2 = T_1 \left(\frac{P_2}{P_1}\right)^{R/\overline{C}_P}$$

For an ideal gas, $\overline{C}_P = 5R/2$, so

$$T_2 = (298 \text{ K}) \left(\frac{5.00 \text{ bar}}{10.00 \text{ bar}}\right)^{2/5} = 226 \text{ K}$$

Substituting into the expression for δw ($\overline{C}_V = 3R/2$) gives

$$w = n\overline{C}_V \int_{T_1}^{T_2} dT = \frac{3}{2}(8.314 \text{ J} \cdot \text{K}^{-1})(226 \text{ K} - 298 \text{ K}) = -900 \text{ J}$$

19–20. A quantity of $N_2(g)$ at 298 K is compressed reversibly and adiabatically from a volume of 20.0 dm^3 to 5.00 dm^3. Assuming ideal behavior, calculate the final temperature of the $N_2(g)$. Take $\overline{C}_V = 5R/2$.

Using $\overline{C}_V$ given in the problem, we find that (by definition)

$$dU = n\overline{C}_V dT = \frac{5}{2}nR\,dT$$

and Equation 19.2 gives δw as

$$\delta w = -P\,dV = -\frac{nRT}{V}dV$$

For a reversible adiabatic compression, $\delta q = 0$, and so $dU = dw$. Then

$$\frac{5}{2}nR\,dT = -\frac{nRT}{V}dV$$

$$\frac{5}{2}\frac{dT}{T} = -\frac{dV}{V}$$

$$\frac{5}{2}\ln\frac{T_2}{T_1} = -\ln\frac{V_2}{V_1}$$

$$\ln\frac{T_2}{T_1} = -\frac{2}{5}\ln\frac{5.00 \text{ dm}^3}{20.0 \text{ dm}^3}$$

$$T_2 = 519 \text{ K}$$

19–21. A quantity of $CH_4(g)$ at 298 K is compressed reversibly and adiabatically from 50.0 bar to 200 bar. Assuming ideal behavior, calculate the final temperature of the $CH_4(g)$. Take $\overline{C}_V = 3R$.

From Problem 19–17, we have

$$\frac{T_2}{T_1} = \left(\frac{P_2}{P_1}\right)^{R/\overline{C}_P}$$

Assuming ideal behavior, $\overline{C}_P = R + \overline{C}_V = 4R$. Then

$$T_2 = \left(\frac{200 \text{ bar}}{50.0 \text{ bar}}\right)^{1/4}(298 \text{ K}) = 421 \text{ K}$$

19–22. One mole of ethane at 25°C and one atm is heated to 1200°C at constant pressure. Assuming ideal behavior, calculate the values of w, q, ΔU, and ΔH given that the molar heat capacity of ethane is given by

$$\overline{C}_P/R = 0.06436 + (2.137 \times 10^{-2} \text{ K}^{-1})T$$

$$- (8.263 \times 10^{-6} \text{ K}^{-2})T^2 + (1.024 \times 10^{-9} \text{ K}^{-3})T^3$$

over the above temperature range. Repeat the calculation for a constant-volume process.

---

a. For a constant-pressure process, $q_p = \Delta H$ and $d\overline{H} = \overline{C}_p dT$. Then

$$\Delta\overline{H} = \int \overline{C}_p dT$$

$$= R\left[0.06436T + \frac{1}{2}(2.137 \times 10^{-2})T^2 - \frac{1}{3}(8.263 \times 10^{-6})T^3 \right.$$

$$\left. + \frac{1}{4}(1.024 \times 10^{-9})T^4\right]\Bigg|_{298\ K}^{1473\ K}$$

$$\Delta\overline{H} = 122.9\ \text{kJ·mol}^{-1}$$

We now use Equation 19.36, remembering that the gas is behaving ideally:

$$\Delta\overline{H} = \Delta\overline{U} + P\Delta\overline{V} = \Delta\overline{U} + R\Delta T$$

$$\Delta\overline{U} = 122.9\ \text{kJ·mol}^{-1} - (8.3145 \times 10^{-3}\ \text{kJ·mol}^{-1}\cdot\text{K}^{-1})(1473\ \text{K} - 298\ \text{K})$$

$$= 113.1\ \text{kJ·mol}^{-1}$$

Finally, we use the expression $\Delta U = q + w$ to write

$$w = \Delta\overline{U} - q = 113.1\ \text{kJ·mol}^{-1} - 122.9\ \text{kJ·mol}^{-1} = -9.8\ \text{kJ·mol}^{-1}$$

b. For a constant-volume process, $w = 0$, and so $\Delta\overline{U} = q$. $\Delta\overline{H}$ is the same as in the previous situation, so $\Delta\overline{H} = 122.9\ \text{kJ·mol}^{-1}$. We can use Equation 19.36, remembering that the gas behaves ideally, to write

$$\Delta\overline{H} = \Delta\overline{U} + \overline{V}\Delta P = \Delta\overline{U} + R\Delta T$$

$$\Delta U = 122.9\ \text{kJ·mol}^{-1} - (8.3145 \times 10^{-3}\ \text{kJ·mol}^{-1}\cdot\text{K}^{-1})(1473\ \text{K} - 298\ \text{K})$$

$$= 113.1\ \text{kJ·mol}^{-1}$$

Note that the value of $\Delta\overline{U}$ is the same as in part a, because U depends only on temperature for an ideal gas.

19–23. The value of $\Delta_r H°$ at 25°C and one bar is $+290.8$ kJ for the reaction

$$2\,\text{ZnO(s)} + 2\,\text{S(s)} \longrightarrow 2\,\text{ZnS(s)} + \text{O}_2(\text{g})$$

Assuming ideal behavior, calculate the value of $\Delta_r U°$ for this reaction.

Because both reactants are solid, $V_1 \approx 0$. The final volume will depend only on the amount of oxygen present; assuming it behaves ideally, we write

$$V_2 \approx \frac{nRT}{P} = \frac{(1\ \text{mol})(0.08314\ \text{dm}^3\cdot\text{bar·mol}^{-1}\cdot\text{K}^{-1})(298\ \text{K})}{1\ \text{bar}} = 24.78\ \text{dm}^3$$

Then $\Delta V \approx 24.78\ \text{dm}^3$ for the reaction. Using Equation 19.36, we write

$$\Delta_r U° = \Delta_r H° - P\Delta V$$

$$= 290.8\ \text{kJ} - (1\ \text{bar})(24.776\ \text{dm}^3)\left(\frac{1\ \text{kJ}}{10\ \text{dm}^3\cdot\text{bar}}\right)$$

$$= 288.3\ \text{kJ}$$

19–24. Liquid sodium is being considered as an engine coolant. How many grams of sodium are needed to absorb 1.0 MJ of heat if the temperature of the sodium is not to increase by more than 10°C. Take $\overline{C}_P = 30.8 \text{ J·K}^{-1}\text{·mol}^{-1}$ for Na(l) and 75.2 J·K^{-1}·mol^{-1} for H$_2$O(l).

We must have a coolant which can absorb 1.0×10^6 J without changing its temperature by more than 10 K. The smallest amount of sodium required will allow the temperature to change by exactly 10 K. We can consider this a constant-pressure process, because liquids are relatively incompressible. Then, substituting $\Delta T = 10$ K into Equation 19.40, we find

$$\Delta \overline{H} = \overline{C}_P \Delta T = 308 \text{ J·mol}^{-1}$$

We require one mole of sodium to absorb 308 J of heat. Therefore, to absorb 1.0 MJ of heat, we require

$$(1.0 \times 10^6 \text{ J}) \left(\frac{1 \text{ mol}}{308 \text{ J}} \right) \left(\frac{22.99 \text{ g}}{1 \text{ mol}} \right) = 74.6 \text{ kg}$$

74.6 kg of liquid sodium is needed.

19–25. A 25.0-g sample of copper at 363 K is placed in 100.0 g of water at 293 K. The copper and water quickly come to the same temperature by the process of heat transfer from copper to water. Calculate the final temperature of the water. The molar heat capacity of copper is 24.5 J·K^{-1}·mol^{-1} and that of water is 75.2 J·K^{-1}·mol^{-1}.

The heat lost by the copper is gained by the water. Since $\Delta H = n\overline{C}_P \Delta T$ (Equation 19.40), we can let x be the final temperature of the system and write the heat lost by the copper as

$$\left(\frac{25.0 \text{ g}}{63.546 \text{ g·mol}^{-1}} \right) (24.5 \text{ J·mol}^{-1}\text{·K}^{-1})(363 \text{ K} - x)$$

and the heat gained by the water as

$$\left(\frac{100.0 \text{ g}}{18.0152 \text{ g·mol}^{-1}} \right) (75.3 \text{ J·mol}^{-1}\text{·K}^{-1})(x - 293 \text{ K})$$

Equating these two expressions gives

$$3495 \text{ J} - (9.628 \text{ J·K}^{-1})x = (418.0 \text{ J·K}^{-1})x - 1.224 \times 10^5 \text{ J}$$
$$1.259 \times 10^5 \text{ K} = 427.6x$$
$$295 \text{ K} = x$$

The final temperature of the water is 295 K.

19–26. A 10.0-kg sample of liquid water is used to cool an engine. Calculate the heat removed (in joules) from the engine when the temperature of the water is raised from 293 K to 373 K. Take $\overline{C}_P = 75.2 \text{ J·K}^{-1}\text{·mol}^{-1}$ for H$_2$O(l).

We can use Equation 19.40, where $\Delta T = 373 \text{ K} - 293 \text{ K} = 80$ K. This gives

$$\Delta H = n\overline{C}_P \Delta T = \left(\frac{10.0 \times 10^3 \text{ g}}{18.0152 \text{ g·mol}^{-1}} \right) (75.2 \text{ J·mol}^{-1}\text{·K}^{-1})(80 \text{ K}) = 3340 \text{ kJ}$$

3340 kJ of heat is removed by the water.

19–27. In this problem, we will derive a general relation between C_p and C_V. Start with $U = U(P, T)$ and write

$$dU = \left(\frac{\partial U}{\partial P}\right)_T dP + \left(\frac{\partial U}{\partial T}\right)_P dT \tag{1}$$

We could also consider V and T to be the independent variables of U and write

$$dU = \left(\frac{\partial U}{\partial V}\right)_T dV + \left(\frac{\partial U}{\partial T}\right)_V dT \tag{2}$$

Now take $V = V(P, T)$ and substitute its expression for dV into Equation 2 to obtain

$$dU = \left(\frac{\partial U}{\partial V}\right)_T \left(\frac{\partial V}{\partial P}\right)_T dP + \left[\left(\frac{\partial U}{\partial V}\right)_T \left(\frac{\partial V}{\partial T}\right)_P + \left(\frac{\partial U}{\partial T}\right)_V\right] dT$$

Compare this result with Equation 1 to obtain

$$\left(\frac{\partial U}{\partial P}\right)_T = \left(\frac{\partial U}{\partial V}\right)_T \left(\frac{\partial V}{\partial P}\right)_T \tag{3}$$

and

$$\left(\frac{\partial U}{\partial T}\right)_P = \left(\frac{\partial U}{\partial V}\right)_T \left(\frac{\partial V}{\partial T}\right)_P + \left(\frac{\partial U}{\partial T}\right)_V \tag{4}$$

Last, substitute $U = H - PV$ into the left side of Equation (4) and use the definitions of C_p and C_V to obtain

$$C_p - C_V = \left[P + \left(\frac{\partial U}{\partial V}\right)_T\right] \left(\frac{\partial V}{\partial T}\right)_P$$

Show that $C_p - C_V = nR$ if $(\partial U/\partial V)_T = 0$, as it is for an ideal gas.

We can write the total derivatives of $V(P, T)$ and $U(V, T)$ as (MathChapter H)

$$dV = \left(\frac{\partial V}{\partial P}\right)_T dP + \left(\frac{\partial V}{\partial T}\right)_P dT \tag{a}$$

$$dU = \left(\frac{\partial U}{\partial V}\right)_T dV + \left(\frac{\partial U}{\partial T}\right)_V dT \tag{b}$$

Substituting dV from Equation a into Equation b gives

$$dU = \left(\frac{\partial U}{\partial V}\right)_T \left[\left(\frac{\partial V}{\partial P}\right)_T dP + \left(\frac{\partial V}{\partial T}\right)_P dT\right] + \left(\frac{\partial U}{\partial T}\right)_V dT$$

$$= \left(\frac{\partial U}{\partial V}\right)_T \left(\frac{\partial V}{\partial P}\right)_T dP + \left[\left(\frac{\partial U}{\partial V}\right)_T \left(\frac{\partial V}{\partial T}\right)_P + \left(\frac{\partial U}{\partial T}\right)_V\right] dT$$

We can also express U as a function of P and T, in which case the total derivative dU is

$$dU = \left(\frac{\partial U}{\partial P}\right)_T dP + \left(\frac{\partial U}{\partial T}\right)_P dT$$

Because the coefficients of dP and dT in both expressions for dU are equal, we can write

$$\left(\frac{\partial U}{\partial P}\right)_T = \left(\frac{\partial U}{\partial V}\right)_T \left(\frac{\partial V}{\partial P}\right)_T$$

and

$$\left(\frac{\partial U}{\partial T}\right)_P = \left(\frac{\partial U}{\partial V}\right)_T \left(\frac{\partial V}{\partial T}\right)_P + \left(\frac{\partial U}{\partial T}\right)_V \tag{c}$$

Substituting $H - PV$ for U into the left side of Equation c gives

$$\left(\frac{\partial [H - PV]}{\partial T}\right)_P = \left(\frac{\partial U}{\partial V}\right)_T \left(\frac{\partial V}{\partial T}\right)_P + \left(\frac{\partial U}{\partial T}\right)_V$$

$$\left(\frac{\partial H}{\partial T}\right)_P - P\left(\frac{\partial V}{\partial T}\right)_P - V\left(\frac{\partial P}{\partial T}\right)_P = \left(\frac{\partial U}{\partial V}\right)_T \left(\frac{\partial V}{\partial T}\right)_P + \left(\frac{\partial U}{\partial T}\right)_V$$

Using the definitions of C_P and C_V (Equations 19.39 and 19.40), this expression becomes

$$C_P - P\left(\frac{\partial V}{\partial T}\right)_P = \left(\frac{\partial U}{\partial V}\right)_T \left(\frac{\partial V}{\partial T}\right)_P + C_V$$

$$C_P - C_V = \left[P + \left(\frac{\partial U}{\partial V}\right)_T\right]\left(\frac{\partial V}{\partial T}\right)_P \tag{d}$$

If $(\partial U/\partial V) = 0$, then Equation d becomes

$$C_P - C_V = P\left(\frac{\partial V}{\partial T}\right)_P$$

Using the ideal gas equation to find $P(\partial V/\partial T)_P$, we find that

$$PV = nRT$$

$$P\left(\frac{\partial V}{\partial T}\right)_P + V\left(\frac{\partial P}{\partial T}\right)_P = nR\left(\frac{\partial T}{\partial T}\right)_P$$

$$P\left(\frac{\partial V}{\partial T}\right)_P = nR$$

$$C_P - C_V = nR$$

19–28. Following Problem 19–27, show that

$$C_P - C_V = \left[V - \left(\frac{\partial H}{\partial P}\right)_T\right]\left(\frac{\partial P}{\partial T}\right)_V$$

We can write the total derivatives of $P(V, T)$ and $H(P, T)$ as (MathChapter H)

$$dP = \left(\frac{\partial P}{\partial V}\right)_T dV + \left(\frac{\partial P}{\partial T}\right)_V dT \tag{a}$$

$$dH = \left(\frac{\partial H}{\partial P}\right)_T dP + \left(\frac{\partial H}{\partial T}\right)_P dT \tag{b}$$

Substituting dP from Equation a into Equation b gives

$$dH = \left(\frac{\partial H}{\partial P}\right)_T \left[\left(\frac{\partial P}{\partial V}\right)_T dV + \left(\frac{\partial P}{\partial T}\right)_V dT\right] + \left(\frac{\partial H}{\partial T}\right)_P dT$$

$$= \left(\frac{\partial H}{\partial P}\right)_T \left(\frac{\partial P}{\partial V}\right)_T dV + \left[\left(\frac{\partial H}{\partial P}\right)_T \left(\frac{\partial P}{\partial T}\right)_V + \left(\frac{\partial H}{\partial T}\right)_P\right] dT$$

We can also express H as a function of V and T, in which case the total derivative dH is

$$dH = \left(\frac{\partial H}{\partial V}\right)_T dV + \left(\frac{\partial H}{\partial T}\right)_V dT$$

Because the coefficients of dV and dT in both expressions for dH are equal, we can write

$$\left(\frac{\partial H}{\partial V}\right)_T = \left(\frac{\partial H}{\partial P}\right)_T \left(\frac{\partial P}{\partial V}\right)_T$$

and

$$\left(\frac{\partial H}{\partial T}\right)_V = \left(\frac{\partial H}{\partial P}\right)_T \left(\frac{\partial P}{\partial T}\right)_V + \left(\frac{\partial H}{\partial T}\right)_P \tag{c}$$

Substituting $U + PV$ for H into the left side of Equation c gives

$$\left(\frac{\partial [U + PV]}{\partial T}\right)_V = \left(\frac{\partial H}{\partial P}\right)_T \left(\frac{\partial P}{\partial T}\right)_V + \left(\frac{\partial H}{\partial T}\right)_P$$

$$\left(\frac{\partial U}{\partial T}\right)_V + P\left(\frac{\partial V}{\partial T}\right)_V + V\left(\frac{\partial P}{\partial T}\right)_V = \left(\frac{\partial H}{\partial P}\right)_T \left(\frac{\partial P}{\partial T}\right)_V + \left(\frac{\partial H}{\partial T}\right)_P$$

Using the definitions of C_P and C_V (Equations 19.39 and 19.40), this expression becomes

$$C_V + V\left(\frac{\partial P}{\partial T}\right)_V = \left(\frac{\partial H}{\partial P}\right)_T \left(\frac{\partial P}{\partial T}\right)_V + C_P$$

$$C_P - C_V = \left[V - \left(\frac{\partial H}{\partial P}\right)_T\right]\left(\frac{\partial P}{\partial T}\right)_V$$

which is the desired result.

19–29. Starting with $H = U + PV$, show that

$$\left(\frac{\partial U}{\partial T}\right)_P = C_P - P\left(\frac{\partial V}{\partial T}\right)_P$$

Interpret this result physically.

Take the partial derivative of both sides of this equation with respect to T, holding P constant, and substitute C_P for $(\partial H/\partial T)_P$.

$$H = U + PV$$

$$\left(\frac{\partial H}{\partial T}\right)_P = \left(\frac{\partial U}{\partial T}\right)_P + P\left(\frac{\partial V}{\partial T}\right)_P$$

$$C_P - P\left(\frac{\partial V}{\partial T}\right)_P = \left(\frac{\partial U}{\partial T}\right)_P$$

This expression tells us how the total energy of a constant-pressure system changes with respect to temperature. Recall that for a constant pressure process, $dH = \delta q$. Then $C_P = (\partial q/\partial T)$. Because $dU = \delta q + \delta w$, the work involved in the process must be $-P(\partial V/\partial T)_P$. The equation above is equivalent to the statement

$$\left(\frac{\partial U}{\partial T}\right)_P = \left(\frac{\partial [q + w]}{\partial T}\right)_P$$

19–30. Given that $(\partial U/\partial V)_T = 0$ for an ideal gas, prove that $(\partial H/\partial V)_T = 0$ for an ideal gas.

Begin with Equation 19.35 and use the ideal gas law to write

$$H = U + PV = U + nRT$$

Now take the partial derivative of both sides with respect to volume (note that for an ideal gas, U is dependent only upon temperature) to find

$$\left(\frac{\partial H}{\partial V}\right)_T = \left(\frac{\partial U}{\partial V}\right)_T + nR\left(\frac{\partial T}{\partial V}\right)_T = 0$$

19–31. Given that $(\partial U/\partial V)_T = 0$ for an ideal gas, prove that $(\partial C_V/\partial V)_T = 0$ for an ideal gas.

We define C_V as $(\partial U/\partial V)_T$ (Equation 19.39). Therefore,

$$\left(\frac{\partial C_V}{\partial V}\right)_T = \frac{\partial^2 U}{\partial V \partial T} = \frac{\partial}{\partial T}\left(\frac{\partial U}{\partial V}\right)_T$$

Since $(\partial U/\partial V)_T = 0$ for an ideal gas, $(\partial C_V/\partial V)_T = 0$.

19–32. Show that $C_P - C_V = nR$ if $(\partial H/\partial P)_T = 0$, as is true for an ideal gas.

From Problem 19.28,

$$C_P - C_V = \left[V - \left(\frac{\partial H}{\partial P}\right)_T\right]\left(\frac{\partial P}{\partial T}\right)_V = V\left(\frac{\partial P}{\partial T}\right)_V$$

where, as stated in the problem, $(\partial H/\partial P)_T = 0$. Substituting $P = nRTV^{-1}$ into the above expression gives

$$C_P - C_V = V\left(\frac{\partial[nRTV^{-1}]}{\partial T}\right)_V = nR$$

19–33. Differentiate $H = U + PV$ with respect to V at constant temperature to show that $(\partial H/\partial V)_T = 0$ for an ideal gas.

(Notice that this problem has you prove the same thing as Problem 19.30 without assuming that $(\partial U/\partial V)_T = 0$ for an ideal gas.) We can use the ideal gas equation to write Equation 19.35 as

$$H = U + PV = U + nRT$$

For an ideal gas, U is dependent only on temperature, and the product nRT is also dependent only on temperature. Therefore, H is a function only of temperature, and differentiating H at constant temperature will yield the result

$$\left(\frac{\partial H}{\partial V}\right)_T = \left(\frac{\partial U}{\partial V}\right)_T + nR\left(\frac{\partial T}{\partial V}\right)_T = 0$$

19–34. Given the following data for sodium, plot $\overline{H}(T) - \overline{H}(0)$ against T for sodium: melting point, 361 K; boiling point, 1156 K; $\Delta_{\text{fus}} H° = 2.60 \text{ kJ} \cdot \text{mol}^{-1}$; $\Delta_{\text{vap}} H° = 97.4 \text{ kJ} \cdot \text{mol}^{-1}$; $\overline{C}_P(\text{s}) = 28.2 \text{ J} \cdot \text{mol}^{-1} \cdot \text{K}^{-1}$; $\overline{C}_P(\text{l}) = 32.7 \text{ J} \cdot \text{mol}^{-1} \cdot \text{K}^{-1}$; $\overline{C}_P(\text{g}) = 20.8 \text{ J} \cdot \text{mol}^{-1} \cdot \text{K}^{-1}$.

We can use an extended form of Equation 19.46:

$$\overline{H}(T) - \overline{H}(0) = \int_0^{T_{\text{fus}}} \overline{C}_P(\text{s}) dT + \Delta_{\text{fus}} \overline{H} + \int_{T_{\text{fus}}}^{T_{\text{vap}}} \overline{C}_P(\text{l}) dT + \Delta_{\text{vap}} \overline{H} + \int_{T_{\text{vap}}}^{T} \overline{C}_P(\text{g}) dT$$

Notice the very large jump between the liquid and gaseous phases.

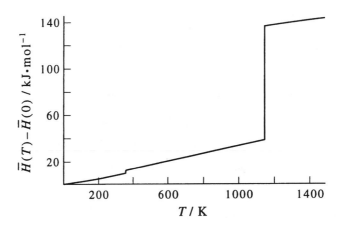

19–35. The $\Delta_r H°$ values for the following equations are

$$2 \text{ Fe(s)} + \tfrac{3}{2} O_2(\text{g}) \rightarrow Fe_2O_3(\text{s}) \quad \Delta_r H° = -206 \text{ kJ} \cdot \text{mol}^{-1}$$

$$3 \text{ Fe(s)} + 2 O_2(\text{g}) \rightarrow Fe_3O_4(\text{s}) \quad \Delta_r H° = -136 \text{ kJ} \cdot \text{mol}^{-1}$$

Use these data to calculate the value of $\Delta_r H$ for the reaction described by

$$4 \text{ Fe}_2O_3(\text{s}) + \text{Fe(s)} \longrightarrow 3 \text{ Fe}_3O_4(\text{s})$$

Set up the problem so that the summation of two reactions will give the desired reaction:

$$4[\text{Fe}_2O_3(\text{s}) \rightarrow 2 \text{ Fe(s)} + \tfrac{3}{2} O_2(\text{g})] \quad \Delta_r H = 4(206) \text{ kJ}$$

$$+ 3[3 \text{ Fe(s)} + 2 O_2(\text{g}) \rightarrow \text{Fe}_3O_4(\text{s})] \quad \Delta_r H = 3(-136) \text{ kJ}$$

$$\overline{4 \text{ Fe}_2O_3(\text{s}) + \text{Fe(s)} \longrightarrow 3 \text{ Fe}_3O_4(\text{s})} \quad \Delta_r H = 416 \text{ kJ}$$

19–36. Given the following data,

$$\tfrac{1}{2} H_2(\text{g}) + \tfrac{1}{2} F_2(\text{g}) \rightarrow \text{HF(g)} \quad \Delta_r H° = -273.3 \text{ kJ} \cdot \text{mol}^{-1}$$

$$H_2(\text{g}) + \tfrac{1}{2} O_2(\text{g}) \rightarrow H_2O(\text{l}) \quad \Delta_r H° = -285.8 \text{ kJ} \cdot \text{mol}^{-1}$$

calculate the value of $\Delta_r H$ for the reaction described by

$$2 F_2(\text{g}) + 2 H_2O(\text{l}) \longrightarrow 4 \text{ HF(g)} + O_2(\text{g})$$

Set up the problem so that the summation of two reactions will give the desired reaction:

$$4[\tfrac{1}{2}H_2(g) + \tfrac{1}{2}F_2(g) \rightarrow HF(g)] \qquad \Delta_r H = 4(-273.3) \text{ kJ}$$

$$+ \qquad 2[H_2O(l) \rightarrow H_2(g) + \tfrac{1}{2}O_2(g)] \quad \Delta_r H = 2(285.8) \text{ kJ}$$

$$2\,F_2(g) + 2\,H_2O(l) \longrightarrow 4\,HF(g) + O_2(g) \quad \Delta_r H = -521.6 \text{ kJ}$$

19–37. The standard molar heats of combustion of the isomers m-xylene and p-xylene are -4553.9 kJ·mol^{-1} and -4556.8 kJ·mol^{-1}, respectively. Use these data, together with Hess's Law, to calculate the value of $\Delta_r H°$ for the reaction described by

$$m\text{-xylene} \longrightarrow p\text{-xylene}$$

Because m-xylene and p-xylene are isomers, their combustion equations are stoichiometrically equivalent. We can therefore write

$$m\text{-xylene} \rightarrow \text{combustion products} \quad \Delta_r H = -4553.9 \text{ kJ}$$

$$+\,\text{combustion products} \rightarrow p\text{-xylene} \qquad \Delta_r H = +4556.8 \text{ kJ}$$

$$m\text{-xylene} \longrightarrow p\text{-xylene} \qquad\qquad \Delta_r H = +2.9 \text{ kJ}$$

19–38. Given that $\Delta_r H° = -2826.7$ kJ for the combustion of 1.00 mol of fructose at 298.15 K,

$$C_6H_{12}O_6(s) + 6\,O_2(g) \longrightarrow 6\,CO_2(g) + 6\,H_2O(l)$$

and the $\Delta_f H°$ data in Table 19.2, calculate the value of $\Delta_f H°$ for fructose at 298.15 K.

We are given $\Delta_r H°$ for the combustion of fructose in the statement of the problem. We use the values given in Table 19.2 for $CO_2(g)$, $H_2O(l)$, and $O_2(g)$:

$$\Delta_f H°[CO_2(g)] = -393.509 \text{ kJ·mol}^{-1} \qquad \Delta_f H°[H_2O(l)] = -285.83 \text{ kJ·mol}^{-1}$$

$$\Delta_f H°[O_2(g)] = 0$$

Now, by Hess's law, we write

$$\Delta_r H° = \sum \Delta_f H°[\text{products}] - \sum \Delta_f H°[\text{reactants}]$$

$$-2826.7 \text{ kJ·mol}^{-1} = 6(-393.509 \text{ kJ·mol}^{-1}) + 6(-285.83 \text{ kJ·mol}^{-1}) - \Delta_r H°[\text{fructose}]$$

$$\Delta_r H°[\text{fructose}] = 1249.3 \text{ kJ·mol}^{-1}$$

19–39. Use the $\Delta_f H°$ data in Table 19.2 to calculate the value of $\Delta_c H°$ for the combustion reactions described by the equations:

a. $CH_3OH(l) + \tfrac{3}{2}O_2(g) \longrightarrow CO_2(g) + 2\,H_2O(l)$

b. $N_2H_4(l) + O_2(g) \longrightarrow N_2(g) + 2\,H_2O(l)$

Compare the heat of combustion per gram of the fuels $CH_3OH(l)$ and $N_2H_4(l)$.

We will need the following values from Table 19.2:

$$\Delta_f H°[CO_2(g)] = -393.509 \text{ kJ·mol}^{-1} \qquad \Delta_f H°[H_2O(l)] = -285.83 \text{ kJ·mol}^{-1}$$
$$\Delta_f H°[N_2H_4(l)] = +50.6 \text{ kJ·mol}^{-1} \qquad \Delta_f H°[CH_3OH(l)] = -239.1 \text{ kJ·mol}^{-1}$$
$$\Delta_f H°[N_2(g)] = 0$$

a. Using Hess's law,

$$\Delta_r H° = \sum \Delta_f H°[\text{products}] - \sum \Delta_f H°[\text{reactants}]$$
$$= 2(-285.83 \text{ kJ}) + (-393.5 \text{ kJ}) - (-239.1 \text{ kJ})$$
$$= \left(\frac{-726.1 \text{ kJ}}{\text{mol methanol}}\right)\left(\frac{1 \text{ mol}}{32.042 \text{ g}}\right) = -22.7 \text{ kJ·g}^{-1}$$

b. Again, by Hess's law,

$$\Delta_r H° = \sum \Delta_f H°[\text{products}] - \sum \Delta_f H°[\text{reactants}]$$
$$= 2(-285.83 \text{ kJ}) - (+50.6 \text{ kJ})$$
$$= \left(\frac{-622.3 \text{ kJ}}{\text{mol N}_2\text{H}_4}\right)\left(\frac{1 \text{ mol}}{32.046 \text{ g}}\right) = -19.4 \text{ kJ·g}^{-1}$$

More energy per gram is produced by combusting methanol.

19–40. Using Table 19.2, calculate the heat required to vaporize 1.00 mol of $CCl_4(l)$ at 298 K.

$$CCl_4(l) \longrightarrow CCl_4(g)$$

We can subtract $\Delta_f H°[CCl_4(l)]$ from $\Delta_f H°[CCl_4(g)]$ to find the heat required to vaporize CCl_4:

$$\Delta_{vap} H° = -102.9 \text{ kJ} + 135.44 \text{ kJ} = 32.5 \text{ kJ}$$

19–41. Using the $\Delta_f H°$ data in Table 19.2, calculate the values of $\Delta_r H°$ for the following:

a. $C_2H_4(g) + H_2O(l) \longrightarrow C_2H_5OH(l)$
b. $CH_4(g) + 4\,Cl_2(g) \longrightarrow CCl_4(l) + 4\,HCl(g)$

In each case, state whether the reaction is endothermic or exothermic.

a. Using Hess's law,

$$\Delta_r H° = -277.69 \text{ kJ} - (-285.83 \text{ kJ} + 52.28 \text{ kJ}) = -44.14 \text{ kJ}$$

This reaction is exothermic.

b. Again, by Hess's law,

$$\Delta_r H° = 4(-92.31 \text{ kJ}) - 135.44 \text{ kJ} - (-74.81 \text{ kJ}) = -429.87 \text{ kJ}$$

This reaction is also exothermic.

19–42. Use the following data to calculate the value of $\Delta_{vap} H°$ of water at 298 K and compare your answer to the one you obtain from Table 19.2: $\Delta_{vap} H°$ at 373 K = 40.7 kJ·mol^{-1}; $\overline{C}_P(l) = 75.2$ J·mol^{-1}·K^{-1}; $\overline{C}_P(g) = 33.6$ J·mol^{-1}·K^{-1}.

We can create a figure similar to Figure 19.10 to illustrate this reaction:

$$H_2O(l) \xrightarrow{\Delta_{vap}H°,\,373\ K} H_2O(g)$$
$$\uparrow \Delta H_2 \qquad\qquad\qquad \downarrow \Delta H_3$$
$$H_2O(l) \xrightarrow{\Delta_{vap}H°,\,298\ K} H_2O(g)$$

Now we use Hess's Law to determine the enthalpy of vaporization.

$$\Delta_{vap}H°_{298\ K} = \Delta H_2 + \Delta H_3 + \Delta_{vap}H°_{373\ K}$$
$$= (75\ K)(75.2\ J\cdot mol^{-1}\cdot K^{-1}) + (-75\ K)(33.6\ J\cdot mol^{-1}\cdot K^{-1}) + 40.7\ kJ\cdot mol^{-1}$$
$$= 43.8\ kJ\cdot mol^{-1}$$

Using Table 19.2, we find

$$\Delta_{vap}H° = \Delta_f H°[H_2O(g)] - \Delta_f H°[H_2O(l)]$$
$$= -241.8\ kJ\cdot mol^{-1} + 285.83\ kJ\cdot mol^{-1} = 44.0\ kJ\cdot mol^{-1}$$

These values are fairly close. (Using values of $\overline{C}_P$ which include temperature-dependent terms may further improve the agreement.)

19–43. Use the following data and the data in Table 19.2 to calculate the standard reaction enthalpy of the water-gas reaction at 1273 K. Assume that the gases behave ideally under these conditions.

$$C(s) + H_2O(g) \longrightarrow CO(g) + H_2(g)$$

$$C_P°[CO(g)]/R = 3.231 + (8.379 \times 10^{-4}\ K^{-1})T - (9.86 \times 10^{-8}\ K^{-2})T^2$$

$$C_P°[H_2(g)]/R = 3.496 + (1.006 \times 10^{-4}\ K^{-1})T + (2.42 \times 10^{-7}\ K^{-2})T^2$$

$$C_P°[H_2O(g)]/R = 3.652 + (1.156 \times 10^{-3}\ K^{-1})T + (1.42 \times 10^{-7}\ K^{-2})T^2$$

$$C_P°[C(s)]/R = -0.6366 + (7.049 \times 10^{-3}\ K^{-1})T - (5.20 \times 10^{-6}\ K^{-2})T^2$$
$$+ (1.38 \times 10^{-9}\ K^{-3})T^3$$

We can create a figure similar to Figure 19.10 to illustrate this reaction.

$$C(s) + H_2O(g) \xrightarrow{\Delta_r H°,\,1273\ K} CO(g) + H_2(g)$$
$$\downarrow \Delta H_1 \qquad\qquad\qquad \uparrow \Delta H_2$$
$$C(s) + H_2O(g) \xrightarrow{\Delta_r H°,\,298\ K} CO(g) + H_2(g)$$

Now use Hess's Law to calculate the standard reaction enthalpy at 1273 K. To do the integrals, it is helpful to use a program like *Excel* or *Mathematica* (I used *Mathematica*), so that the tedium of adding and multiplying can be avoided.

$$\Delta_r H°_{1273} = \Delta_r H°_{298} + \Delta H_1 + \Delta H_2$$
$$= (-110.5\ kJ\cdot mol^{-1} + 241.8\ kJ\cdot mol^{-1})$$
$$+ R \int_{298}^{1273} \left\{ \overline{C}_P[CO(g)] + \overline{C}_P[H_2(g)] - \overline{C}_P[H_2O(g)] - \overline{C}_P[C(s)] \right\} dT$$
$$= 131.3\ kJ\cdot mol^{-1} + R\,[3725.01\ K + 3649.92\ K - 4542.43\ K - 2151.29\ K]$$
$$= 131.3\ kJ\cdot mol^{-1} + 5.664\ kJ\cdot mol^{-1} = 136.964\ kJ\cdot mol^{-1}$$

19–44. The standard molar enthalpy of formation of $CO_2(g)$ at 298 K is $-393.509 \text{ kJ·mol}^{-1}$. Use the following data to calculate the value of $\Delta_f H°$ at 1000 K. Assume the gases behave ideally under these conditions.

$$C_P°[CO_2(g)]/R = 2.593 + (7.661 \times 10^{-3} \text{ K}^{-1})T - (4.78 \times 10^{-6} \text{ K}^{-2})T^2$$
$$+ (1.16 \times 10^{-9} \text{ K}^{-3})T^3$$

$$C_P°[O_2(g)]/R = 3.094 + (1.561 \times 10^{-3} \text{ K}^{-1})T - (4.65 \times 10^{-7} \text{ K}^{-2})T^2$$

$$C_P°[C(s)]/R = -0.6366 + (7.049 \times 10^{-3} \text{ K}^{-1})T - (5.20 \times 10^{-6} \text{ K}^{-2})T^2$$
$$+ (1.38 \times 10^{-9} \text{ K}^{-3})T^3$$

We can create a figure similar to Figure 19.10 to illustrate this reaction.

$$C(s) + O_2(g) \xrightarrow{\Delta_r H°, 1000 \text{ K}} CO_2(g)$$
$$\downarrow \Delta H_1 \qquad\qquad \uparrow \Delta H_2$$
$$C(s) + O_2(g) \xrightarrow{\Delta_r H°, 298 \text{ K}} CO_2(g)$$

Now use Hess's Law to calculate the standard reaction enthalpy at 1000 K:

$$\Delta_r H°_{1000} = \Delta_r H°_{298} + \Delta H_1 + \Delta H_2$$
$$= -393.509 \text{ kJ·mol}^{-1} + R \int_{298}^{1000} \left\{ \overline{C}_P[CO_2(g)] - \overline{C}_P[O_2(g)] - \overline{C}_P[C(s)] \right\} dT$$
$$= -393.509 \text{ kJ·mol}^{-1} + R [4047.167 \text{ K} - 2732.278 \text{ K} - 1419.433 \text{ K}]$$
$$= -393.509 \text{ kJ·mol}^{-1} - 0.869 \text{ kJ·mol}^{-1} = -394.378 \text{ kJ·mol}^{-1}$$

19–45. The value of the standard molar reaction enthalpy for

$$CH_4(g) + 2 O_2(g) \longrightarrow CO_2(g) + 2 H_2O(g)$$

is -802.2 kJ at 298 K. Using the heat-capacity data in Problems 19–43 and 19–44 in addition to

$$C_P°[CH_4(g)]/R = 2.099 + (7.272 \times 10^{-3} \text{ K}^{-1})T + (1.34 \times 10^{-7} \text{ K}^{-2})T^2$$
$$- (8.66 \times 10^{-10} \text{ K}^{-3})T^3$$

to derive a general equation for the value of $\Delta_r H°$ at any temperature between 300 K and 1500 K. Plot $\Delta_r H°$ versus T. Assume that the gases behave ideally under these conditions.

We can create a figure similar to Figure 19.10 to illustrate this reaction.

$$CH_4(s) + 2O_2(g) \xrightarrow{\Delta_r H°} CO_2(g) + 2H_2O(g)$$
$$\downarrow \Delta H_1 \qquad\qquad \uparrow \Delta H_2$$
$$CH_4(s) + 2O_2(g) \xrightarrow{\Delta_r H°, 298 \text{ K}} CO_2(g) + 2H_2O(g)$$

Now use Hess's Law:

$$\Delta_r H^\circ = \Delta_r H^\circ_{298} + \Delta H_1 + \Delta H_2$$

$$= -802.2 \text{ kJ·mol}^{-1}$$

$$+ R \int_{298}^{T} \left\{ \overline{C}_P[CO_2(g)] + 2\overline{C}_P[H_2O(g)] - \overline{C}_P[CH_4(s)] - 2\overline{C}_P[O_2(g)] \right\} dT$$

$$= -802.2 \text{ kJ·mol}^{-1} + R \int_{298}^{T} \left[1.610 - (4.21 \times 10^{-4} \text{ K}^{-1})T \right] dT$$

$$+ R \int_{298}^{T} \left[-(3.70 \times 10^{-6} \text{ K}^{-2})T^2 + (2.03 \times 10^{-9} \text{ K}^{-3})T^3 \right] dT$$

$$= -805.8 \text{ kJ·mol}^{-1} + (1.339 \times 10^{-2} \text{ kJ·mol}^{-1}\cdot\text{K}^{-1})T$$

$$- (1.750 \times 10^{-6} \text{ kJ·mol}^{-1}\cdot\text{K}^{-2})T^2 - (1.025 \times 10^{-8} \text{ kJ·mol}^{-1}\cdot\text{K}^{-3})T^3$$

$$+ (4.211 \times 10^{-12} \text{ kJ·mol}^{-1}\cdot\text{K}^{-4})T^4$$

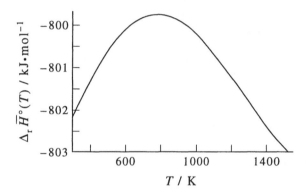

19–46. In all the calculations thus far, we have assumed the reaction takes place at constant temperature, so that any energy evolved as heat is absorbed by the surroundings. Suppose, however, that the reaction takes place under adiabatic conditions, so that all the energy released as heat stays within the system. In this case, the temperature of the system will increase, and the final temperature is called the *adiabatic flame temperature*. One relatively simple way to estimate this temperature is to suppose the reaction occurs at the initial temperature of the reactants and then determine to what temperature the products can be raised by the quantity $\Delta_r H^\circ$. Calculate the adiabatic flame temperature if one mole of $CH_4(g)$ is burned in two moles of $O_2(g)$ at an initial temperature of 298 K. Use the results of the previous problem.

We know from Problem 19–45 that 802.2 kJ·mol^{-1} of energy is produced when one mole of methane is burned in two moles of oxygen at 298 K. Now we determine how much the temperature of the products, one mole of CO_2 and two moles of H_2O, can be raised by this energy:

$$\Delta H(\text{products}) = R \int_{298}^{T} \left\{ \overline{C}_P[CO_2(g)] + 2\overline{C}_P[H_2O(g)] \right\} dT$$

$$802.2 \text{ kJ·mol}^{-1} = R \int_{298}^{T} \left\{ \overline{C}_P[CO_2(g)] + 2\overline{C}_P[H_2O(g)] \right\} dT$$

$$= R \int_{298}^{T} \left[9.897 + (9.97 \times 10^{-3} \text{ K}^{-1})T - (4.496 \times 10^{-6} \text{ K}^{-2})T^2 \right.$$

$$\left. + (1.160 \times 10^{-9} \text{ K}^{-3})T^3 \right] dT$$

$$= -27.89 \text{ kJ·mol}^{-1} + (8.23 \times 10^{-2} \text{ kJ·mol}^{-1}\cdot\text{K}^{-1})T + (4.15 \times 10^{-5} \text{ kJ·mol}^{-1}\cdot\text{K}^{-2})T^2$$

$$- (1.25 \times 10^{-8} \text{ kJ·mol}^{-1}\cdot\text{K}^{-3})T^3 + (2.41 \times 10^{-12} \text{ kJ·mol}^{-1}\cdot\text{K}^{-4})T^4$$

We can solve this polynomial using Simpson's rule or a numerical software package. Working in *Mathematica*, we find that the final temperature will be 4040 K.

19–47. Explain why the adiabatic flame temperature defined in the previous problem is also called the maximum flame temperature.

The adiabatic flame temperature is the temperature of the system if all the energy released as heat stays within the system. Since we are considering an isolated system, the adiabatic flame temperature is also the maximum temperature which the system can achieve.

19–48. How much energy as heat is required to raise the temperature of 2.00 moles of $O_2(g)$ from 298 K to 1273 K at 1.00 bar? Take

$$\overline{C}_P[O_2(g)]/R = 3.094 + (1.561 \times 10^{-3} \text{ K}^{-1})T - (4.65 \times 10^{-7} \text{ K}^{-2})T^2$$

We can use Equation 19.44:

$$\Delta H = \int_{T_1}^{T_2} n\overline{C}_p dT$$

$$= (2.00 \text{ mol})R \int_{298}^{1273} \left[3.094 + (1.561 \times 10^{-3} \text{ K}^{-1})T - (4.65 \times 10^{-7} \text{ K}^{-2})T^2\right] dT$$

$$= 64.795 \text{ kJ·mol}^{-1}$$

19–49. When one mole of an ideal gas is compressed adiabatically to one-half of its original volume, the temperature of the gas increases from 273 K to 433 K. Assuming that $\overline{C}_V$ is independent of temperature, calculate the value of $\overline{C}_V$ for this gas.

Equation 19.20 gives an expression for the reversible adiabatic expansion of an ideal gas:

$$\overline{C}_V dT = -\frac{RT}{V} dV$$

Integrating both sides and substituting the temperatures given, we find that

$$\int \frac{\overline{C}_V}{T} dT = -\int \frac{R}{V} dV$$

$$\overline{C}_V \ln \frac{T_2}{T_1} = -R \ln \frac{V_2}{V_1}$$

$$\overline{C}_V \ln \frac{433}{273} = -R \ln 2$$

$$\frac{\overline{C}_V}{R} = 1.50$$

19–50. Use the van der Waals equation to calculate the minimum work required to expand one mole of $CO_2(g)$ isothermally from a volume of 0.100 dm³ to a volume of 100 dm³ at 273 K. Compare your result with that which you calculate assuming ideal behavior.

In Problem 19–10, we found that the work done by a van der Waals gas was

$$w = -RT \ln \frac{\overline{V}_2 - b}{\overline{V}_1 - b} + \frac{a(\overline{V}_2 - \overline{V}_1)}{\overline{V}_2 \overline{V}_1}$$

Substituting $a = 3.6551 \text{ dm}^6 \cdot \text{bar} \cdot \text{mol}^{-2}$ and $b = 0.042816 \text{ dm}^3 \cdot \text{mol}^{-1}$ from Table 16.3 and using the parameters in the statement of the problem gives

$$w = -(0.083145 \text{ dm}^3 \cdot \text{bar} \cdot \text{mol}^{-1} \cdot \text{K}^{-1})(273 \text{ K}) \ln \frac{100 \text{ dm}^3 \cdot \text{mol}^{-1} - 0.042816 \text{ dm}^3 \cdot \text{mol}^{-1}}{0.100 \text{ dm}^3 \cdot \text{mol}^{-1} - 0.042816 \text{ dm}^3 \cdot \text{mol}^{-1}}$$

$$+3.6551 \text{ dm}^6 \cdot \text{bar} \cdot \text{mol}^{-2} \left[\frac{100 \text{ dm}^3 \cdot \text{mol}^{-1} - 0.100 \text{ dm}^3 \cdot \text{mol}^{-1}}{(100 \text{ dm}^3 \cdot \text{mol}^{-1})(0.100 \text{ dm}^3 \cdot \text{mol}^{-1})} \right]$$

$$= (-169.5 \text{ dm}^3 \cdot \text{bar} \cdot \text{mol}^{-1} + 36.5 \text{ dm}^3 \cdot \text{bar} \cdot \text{mol}^{-1})(0.1 \text{ kJ} \cdot \text{dm}^{-3} \cdot \text{bar}^{-1})(1 \text{ mol})$$

$$= -13.3 \text{ kJ}$$

For an ideal gas,

$$w = -\int P dV = -nRT \ln \left(\frac{V_2}{V_1} \right)$$

$$= (-156.80 \text{ dm}^3 \cdot \text{bar})(0.1 \text{ kJ} \cdot \text{bar}^{-1}) = -15.7 \text{ kJ}$$

The work needed to expand the van der Waals gas is greater than that needed for the ideal gas.

19–51. Show that the work involved in a reversible, adiabatic pressure change of one mole of an ideal gas is given by

$$w = \overline{C}_V T_1 \left[\left(\frac{P_2}{P_1} \right)^{R/\overline{C}_P} - 1 \right]$$

where T_1 is the initial temperature and P_1 and P_2 are the initial and final pressures, respectively.

For a reversible, adiabatic pressure change of an ideal gas, $\delta q = 0$, so $dU = dw$. Since $d\overline{U}$ is defined as $\overline{C}_V dT$,

$$dw = \overline{C}_V dT$$

for one mole of an ideal gas. Integrating, we find

$$w = \overline{C}_V T_2 - \overline{C}_V T_1 = \overline{C}_V (T_2 - T_1) = \overline{C}_V T_1 \left(\frac{T_2}{T_1} - 1 \right)$$

From Problem 19.17, we know that

$$\frac{T_2}{T_1} = \left(\frac{P_2}{P_1} \right)^{R/\overline{C}_P}$$

and so substituting gives

$$w = \overline{C}_V T_1 \left[\left(\frac{P_2}{P_1} \right)^{R/\overline{C}_P} - 1 \right]$$

19–52. In this problem, we will discuss a famous experiment called the *Joule-Thomson experiment*. In the first half of the 19th century, Joule tried to measure the temperature change when a gas is expanded into a vacuum. The experimental setup was not sensitive enough, however, and he found that there was no temperature change, within the limits of his error. Soon afterward, Joule and Thomson devised a much more sensitive method for measuring the temperature change upon expansion. In their experiments (see Figure 19.11), a constant applied pressure P_1 causes a quantity of gas to flow slowly from one chamber to another through a porous plug of silk or cotton. If a volume, V_1, of gas is pushed through the porous plug, the work done on the gas is $P_1 V_1$. The pressure on the other side of the plug is maintained at P_2, so if a volume V_2 enters the right-side chamber, then the net work is given by

$$w = P_1 V_1 - P_2 V_2$$

The apparatus is constructed so that the entire process is adiabatic, so $q = 0$. Use the First Law of Thermodynamics to show that

$$U_2 + P_2 V_2 = U_1 + P_1 V_1$$

or that $\Delta H = 0$ for a Joule-Thomson expansion. Starting with

$$dH = \left(\frac{\partial H}{\partial P}\right)_T dP + \left(\frac{\partial H}{\partial T}\right)_P dT$$

show that

$$\left(\frac{\partial T}{\partial P}\right)_H = -\frac{1}{C_P}\left(\frac{\partial H}{\partial P}\right)_T$$

Interpret physically the derivative on the left side of this equation. This quantity is called the *Joule-Thomson coefficient* and is denoted by μ_{JT}. In Problem 19–54 you will show that it equals zero for an ideal gas. Nonzero values of $(\partial T/\partial P)_H$ directly reflect intermolecular interactions. Most gases cool upon expansion [a positive value of $(\partial T/\partial P)_H$] and a Joule-Thomson expansion is used to liquefy gases.

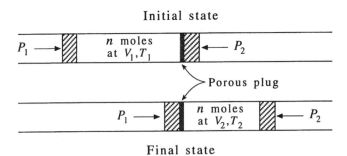

FIGURE 19.11
A schematic description of the Joule-Thomson experiment.

The net work is $w = P_1 V_1 - P_2 V_2$. Since $q = 0$, $U = w$, so

$$U_2 - U_1 = P_1 V_1 - P_2 V_2$$
$$U_2 + P_2 V_2 = U_1 + P_1 V_1$$

Since $\Delta H = U_1 + P_1V_1 - (U_2 + P_2V_2)$, $\Delta H = 0$. Now we write the total derivative of H as a function of P and T:

$$dH = \left(\frac{\partial H}{\partial P}\right)_T dP + \left(\frac{\partial H}{\partial T}\right)_P dT$$

Using the definition of C_P, we write this as

$$dH = \left(\frac{\partial H}{\partial P}\right)_T dP + C_P dT$$

$$-C_P dT = \left(\frac{\partial H}{\partial P}\right)_T dP - dH$$

$$dT = -\frac{1}{C_P}\left(\frac{\partial H}{\partial P}\right)_T dP + \frac{1}{C_P} dH$$

Keep H constant and divide through by dP to obtain

$$\left(\frac{\partial T}{\partial P}\right)_H = -\frac{1}{C_P}\left(\frac{\partial H}{\partial P}\right)_T + 0$$

The Joule-Thomson coefficient is a measure of the change of temperature of a gas with respect to the change in pressure in a Joule-Thomson expansion (or compression).

19–53. The Joule-Thomson coefficient (Problem 19–52) depends upon the temperature and pressure, but assuming an average constant value of $0.15 \text{ K} \cdot \text{bar}^{-1}$ for $N_2(g)$, calculate the drop in temperature if $N_2(g)$ undergoes a drop in pressure of 200 bar.

$$\left(0.15 \text{ K} \cdot \text{bar}^{-1}\right)\left(-200 \text{ bar}\right) = -30 \text{ K}$$

19–54. Show that the Joule-Thomson coefficient (Problem 19–52) can be written as

$$\mu_{JT} = \left(\frac{\partial T}{\partial P}\right)_H = -\frac{1}{C_P}\left[\left(\frac{\partial U}{\partial V}\right)_T\left(\frac{\partial V}{\partial P}\right)_T + \left(\frac{\partial(PV)}{\partial P}\right)_T\right]$$

Show that $(\partial T/\partial P)_H = 0$ for an ideal gas.

From Problem 19–52,

$$\mu_{JT} = -\frac{1}{C_P}\left(\frac{\partial H}{\partial P}\right)_T$$

Since $H = U + PV$,

$$\mu_{JT} = -\frac{1}{C_P}\left[\left(\frac{\partial U}{\partial P}\right)_T + \left(\frac{\partial(PV)}{\partial P}\right)_T\right]$$

$$= -\frac{1}{C_P}\left[\left(\frac{\partial U}{\partial V}\right)_T\left(\frac{\partial V}{\partial P}\right)_T + \left(\frac{\partial(PV)}{\partial P}\right)_T\right]$$

For an ideal gas, U and PV depend only on temperature, so $\mu_{JT} = 0$.

19–55. Use the rigid rotator-harmonic oscillator model and the data in Table 18.2 to plot $\overline{C}_P(T)$ for $CO(g)$ from 300 K to 1000 K. Compare your result with the expression given in Problem 19–43.

From Example 19–8, we know that for an ideal gas

$$\overline{C}_P = \overline{C}_V + R \tag{19.43}$$

And from Chapter 18, we know that for a linear polyatomic ideal gas

$$\frac{\overline{C}_V}{R} = \frac{5}{2} + \left(\frac{\Theta_{\text{vib}}}{T}\right)^2 \frac{e^{-\Theta_{\text{vib}}/T}}{(1 - e^{\Theta_{\text{vib}}/T})^2} \tag{18.41}$$

Therefore, since $\Theta_{\text{vib}}(CO) = 3103$ K, we wish to graph

$$\frac{\overline{C}_P}{R} = \frac{7}{2} + \left(\frac{3103 \text{ K}}{T}\right)^2 \frac{e^{-3103 \text{ K}/T}}{(1 - e^{-3103 \text{ K}/T})^2}$$

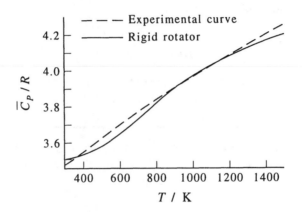

19–56. Use the rigid rotator-harmonic oscillator model and the data in Table 18.4 to plot $\overline{C}_P(T)$ for $CH_4(g)$ from 300 K to 1000 K. Compare your result with the expression given in Problem 19–45.

Again, for an ideal gas

$$\overline{C}_P = \overline{C}_V + R \tag{19.43}$$

And from Chapter 18, we know that for a nonlinear polyatomic ideal gas

$$\frac{\overline{C}_V}{R} = \frac{3}{2} + \frac{3}{2} + \sum_{j=1}^{3n-6} \left(\frac{\Theta_{\text{vib},j}}{T}\right)^2 \frac{e^{-\Theta_{\text{vib},j}/T}}{(1 - e^{-\Theta_{\text{vib},j}/T})^2} \tag{18.62}$$

Using the values given in the problem, we wish to graph

$$\frac{\overline{C}_P}{R} = 4 + \left(\frac{4170 \text{ K}}{T}\right)^2 \frac{e^{-4170 \text{ K}/T}}{(1 - e^{-4170 \text{ K}/T})^2} + 2\left(\frac{2180 \text{ K}}{T}\right)^2 \frac{e^{-2180 \text{ K}/T}}{(1 - e^{-2180 \text{ K}/T})^2}$$

$$+ 3\left(\frac{4320 \text{ K}}{T}\right)^2 \frac{e^{-4320 \text{ K}/T}}{(1 - e^{-4320 \text{ K}/T})^2} + 3\left(\frac{1870 \text{ K}}{T}\right)^2 \frac{e^{-1870 \text{ K}/T}}{(1 - e^{-1870 \text{ K}/T})^2}$$

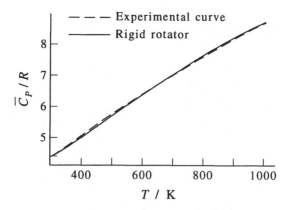

19–57. Why do you think the equations for the dependence of temperature on volume for a reversible adiabatic process (see Equation 19.22 and Example 19.6) depend upon whether the gas is a monatomic gas or a polyatomic gas?

For an adiabatic process, no energy is transferred as heat, so the change in internal energy is equal to the change in work. The internal energy of a monatomic gas (in the electronic ground state) is entirely in the translational degrees of freedom, which is directly related to the temperature of the gas. If the volume of the system increases, work is done by the system, and so the internal energy of the system must decrease. The only way for it to do so is by decreasing the amount of energy in the translational degrees of freedom, which decreases the observed temperature of the gas. The internal energy of a polyatomic gas (in the electronic ground state) is in the rotational, vibrational, and translational degrees of freedom. These vibrational, rotational, and translational energy levels are not necessarily in equilibrium (see Problem 18–37). If the volume of a polyatomic gas is increased, work is done by the system, as before, and the internal energy of the gas must again decrease. However, the gas can lose energy in the rotational or vibrational levels, which will not decrease the observed temperature of the gas. It can also lose energy in the translational levels, which will decrease the observed temperature of the gas, but the other available methods of decreasing the energy of the system will change the dependence of temperature on volume for a polyatomic gas from that observed for a monatomic gas.

The Multinominal Distribution and Stirling's Approximation

J–1. Use Equation J.3 to write the expansion of $(1 + x)^5$. Use Equation J.4 to do the same thing.

Using Equation J.3, we can write

$$(1 + x)^5 = \sum_{N_1=0}^{5} \frac{5!}{N_1!(5 - N_1)!} x^{5-N_1}$$

$$= 1 + 5x + 10x^2 + 10x^3 + 5x^4 + x^5$$

Equation J.4 gives the same result.

J–2. Use Equation J.6 to write out the expression for $(x + y + z)^2$. Compare your result to the one that you obtain by multiplying $(x + y + z)$ by $(x + y + z)$.

Using Equation J.6, we can write

$$(x + y + z)^2 = \sum_{N_1=0}^{2} \sum_{N_2=0}^{2} \sum_{N_3=0}^{2} {}^{*} \frac{N!}{N_1!N_2!N_3!} x^{N_1} y^{N_2} z^{N_3}$$

$$= \frac{2!}{2!0!0!} x^2 + \frac{2!}{1!0!1!} xz + \frac{2!}{1!1!0!} xy + \frac{2!}{0!1!1!} yz + \frac{2!}{0!2!0!} y^2 + \frac{2!}{0!0!2!} z^2$$

$$= 2yz + 2xz + 2xy + x^2 + y^2 + z^2$$

We obtain the same result when we multiply $(x + y + z)$ by $(x + y + z)$.

J–3. Use Equation J.6 to write out the expression for $(x + y + z)^4$. Compare your result to the one that you obtain by multiplying $(x + y + z)^2$ from Problem J–2 by itself.

Using Equation J.6, we can write

$$(x + y + z)^4 = \sum_{N_1=0}^{4} \sum_{N_2=0}^{4} \sum_{N_3=0}^{4} {}^{*} \frac{N!}{N_1!N_2!N_3!} x^{N_1} y^{N_2} z^{N_3}$$

$$= \frac{4!}{4!0!0!} x^4 + \frac{4!}{3!1!0!} x^3 y + \frac{4!}{3!0!1!} x^3 z + \frac{4!}{2!2!0!} x^2 y^2 + \frac{4!}{2!0!2!} x^2 z^2$$

$$+ \frac{4!}{2!1!1!} x^2 yz + \frac{4!}{1!3!0!} xy^3 + \frac{4!}{1!0!3!} xz^3 + \frac{4!}{1!2!1!} xy^2 z + \frac{4!}{1!1!2!} xyz^2$$

615

$$+ \frac{4!}{0!4!0!} y^4 + \frac{4!}{0!0!4!} z^4 + \frac{4!}{0!1!3!} yz^3 + \frac{4!}{0!3!1!} y^3 z + \frac{4!}{0!2!2!} y^2 z^2$$

$$= x^4 + 4x^3 y + 4x^3 z + 6x^2 y^2 + 6x^2 z^2 + 12x^2 yz + 4xy^3 + 4xz^3$$

$$+ 12xy^2 z + 12xyz^2 + y^4 + z^4 + 4yz^3 + 4y^3 z + 6y^2 z^2$$

We obtain the same result when we multiply $(x + y + z)^2$ by $(x + y + z)^2$.

J–4. How many permutations of the letters a, b, c are there?

$3! = 6$

J–5. The coefficients of the expansion of $(1 + x)^n$ can be arranged in the following form:

n								
0					1			
1				1		1		
2			1		2		1	
3		1		3		3		1
4	1		4		6		4		1

Do you see a pattern in going from one row to the next? The triangular arrangement here is called Pascal's triangle.

Each number in a row is the sum of the two numbers above it.

J–6. In how many ways can a committee of three be chosen from nine people?

$$\frac{9!}{3!6!} = 3 \times 7 \times 4 = 84$$

J–7. Calculate the relative error for $N = 50$ using the formula for Stirling's approximation given in Example J–3, and compare your result with that given in Table J.1 using Equation J.7. Take $\ln N!$ to be 148.47776 (*CRC Handbook of Chemistry and Physics*).

$$\ln N! = N \ln N - N + \ln(2\pi N)^{1/2}$$

$$= 195.6012 - 50 + 2.87495 = 148.4761$$

Compared to $\ln N!$, there is a 1.12×10^{-5} relative error. This can be compared with the 0.0194 relative error obtained when using Equation J.7 for $\ln N!$.

J–8. Prove that $x \ln x \rightarrow 0$ as $x \rightarrow 0$.

$$\lim_{x \to 0}(x \ln x) = \lim_{x \to 0}\left(\frac{\ln x}{x^{-1}}\right)$$

Since both numerator and denominator are undefined as $x \to 0$, we can apply l'Hôpital's rule:

$$\lim_{x \to 0}(x \ln x) = \lim_{x \to 0}\left(\frac{x^{-1}}{-x^{-2}}\right) = \lim_{x \to 0}\left(\frac{-x^2}{x}\right) = \lim_{x \to 0}(-x) = 0$$

J–9. Prove that the maximum value of $W(N, N_1) = N!/(N - N_1)!N_1!$ is given by $N_1 = N/2$. (*Hint*: Treat N_1 as a continuous variable.)

Let $N_1 = x$, with x being a continuous variable. Then

$$W = \frac{N!}{(N - x)!x!}$$

$$\ln W = \ln\left[\frac{N!}{(N - x)!x!}\right] = \ln N! - \ln(N - x)! - \ln x!$$

We are looking for the maximum value of W, so, at this value, $(dW/dx) = 0$. Also, we know that

$$\frac{1}{W}\frac{dW}{dx} = \frac{d(\ln W)}{dx}$$

At the maximum value of W, then, $d(\ln W)/dx = 0$. We can express $\ln W$ using Stirling's approximation (Equation J.7):

$$\ln W = \ln N! - \ln(N - x)! - \ln x!$$
$$= N \ln N - N - [(N - x) \ln(N - x) - (N - x)] - (x \ln x - x)$$

The derivative of $\ln W$ is then

$$\frac{d(\ln W)}{dx} = 0 + \frac{N - x}{N - x} + \ln(N - x) - 1 - \ln x - \frac{x}{x} + 1$$
$$= \ln(N - x) - \ln x$$

Setting $[d(\ln W)/dx] = 0$ gives

$$0 = \ln(N - x) - \ln x$$
$$\ln x = \ln(N - x)$$
$$2x = N$$
$$x = \frac{N}{2}$$

Recall that $N_1 = x$, and that $N_2 = N - N_1$. The maximum value of W then occurs when $N_1 = N/2$.

J–10. Prove that the maximum value of $W(N_1, N_2, \ldots, N_r)$ in Equation J.5 is given by
$N_1 = N_2 = \cdots = N_r = N/r$.

Given:

$$W = \frac{N!}{N_1!N_2! \cdots N_r!}$$

Take $N_r = N - N_1 - N_2 - \cdots - N_{r-1}$, because $\sum N_j = N$. Then we can use Stirling's approximation (Equation J.7) to write $\ln W$ as

$$\ln W = N \ln N - N - \sum_{j=1}^{r} N_j \ln N_j - N_j$$

$$= N \ln N - N - \sum_{j=1}^{r-1} N_j \ln N_j - N_j$$

$$- (N - N_1 - N_2 - \cdots - N_{r-1}) \ln(N - N_1 - N_2 - \cdots - N_{r-1}) + N - N_1 - N_2 - \cdots - N_{r-1}$$

Now take the partial derivative of $\ln W$ with respect to N_1. We find that

$$\frac{\partial \ln W}{\partial N_1} = -\ln N_1 + 1 - 1 + \ln(N - N_1 - N_2 - \cdots - N_{r-1}) + 1 - 1$$

As in Problem J–9, we set $d(\ln W)/dN_1 = 0$ to find

$$0 = -\ln N_1 - \ln(N - N_1 - N_2 - \cdots - N_{r-1})$$
$$\ln N_1 = \ln N_r$$
$$N_1 = N_r$$

We get a similar result for N_2 through N_{r-1}, so $N_1 = N_2 = \cdots = N_r$ and all are equal to N/r.

J–11. Prove that

$$\sum_{k=0}^{N} \frac{N!}{k!(N-k)!} = 2^N$$

We know that (Equation J.3)

$$(x + y)^N = \sum_{N_1=0}^{N} \frac{N!}{N_1!(N-N_1)!} x^{N_1} y^{N-N_1}$$

If $x = y = 1$, we can write this as

$$(1 + 1)^N = \sum_{N_1=0}^{N} \frac{N!}{N_1!(N-N_1)!} = \sum_{k=0}^{N} \frac{N!}{k!(N-k)!}$$

Thus

$$\sum_{k=0}^{N} \frac{N!}{k!(N-k)!} = (1 + 1)^N = 2^N$$

J–12. The quantity $n!$ as we have defined it is defined only for positive integer values of n. Consider now the function of x *defined* by

$$\Gamma(x) = \int_0^{\infty} t^{x-1} e^{-t} dt \qquad (1)$$

Integrate by parts (letting $u = t^{x-1}$ and $dv = e^{-t} dt$) to get

$$\Gamma(x) = (x - 1) \int_0^{\infty} t^{x-2} e^{-t} dt = (x - 1)\Gamma(x - 1) \qquad (2)$$

Now use Equation 2 to show that $\Gamma(x) = (x - 1)!$ if x is a positive integer. Although Equation 2 provides us with a general function that is equal to $(n - 1)!$ when x takes on integer values, it is defined just as well for non-integer values. For example, show that $\Gamma(3/2)$, which in a sense is $(\frac{1}{2})!$, is equal to $\pi^{1/2}/2$. Equation 1 can also be used to explain why $0! = 1$. Let $x = 1$ in Equation 1 to show that $\Gamma(1)$, which we can write as $0!$, is equal to 1. The function $\Gamma(x)$ defined by Equation 1 is called the *gamma function* and was introduced by Euler to generalize the idea of a factorial to general values of n. The gamma function arises in many problems in chemistry and physics.

We can integrate Γ by parts, letting $u = t^{x-1}$ and $v = -e^{-t}$, to find

$$\Gamma(x) = \int_0^\infty t^{x-1} e^{-t} dt$$

$$= -t^{x-1} e^{-t} \Big|_{t=0}^{t=\infty} - \int_0^\infty -e^{-t}(x-1)t^{x-2} dt$$

$$= (x-1) \int_0^\infty e^{-t} t^{x-2} dt$$

$$= (x-1)\,\Gamma(x-1)$$

Now if x is a positive integer, we can write

$$\Gamma(x) = (x-1)\,\Gamma(x-1) = (x-1)(x-2)\,\Gamma(x-2)$$

$$= \prod_{n=2}^{x-1} n\,\Gamma(1)$$

$$= \prod_{n=2}^{x-1} n \int_0^\infty e^{-t} dt = \prod_{n=2}^{x-1} n = (x-1)!$$

We can let $t = x^2$ and write $\Gamma(3/2)$ as

$$\Gamma(3/2) = \int_0^\infty t^{1/2} e^{-t} dt = \int_0^\infty 2x^2 e^{-x^2} dx$$

$$= 2\frac{1}{4}\sqrt{\pi} = \frac{\pi^{1/2}}{2}$$

Also, we can express $\Gamma(0)$ as

$$\Gamma(1) = \int_0^\infty t^0 e^{-t} dt = \int_0^\infty e^{-t} dt$$

$$= 1$$

CHAPTER 20

Entropy and the Second Law of Thermodynamics

20–1. Show that

$$\oint dY = 0$$

if Y is a state function.

If Y is a state function, dY must be an exact differential. This means that $\int_1^2 dY = Y_2 - Y_1$ and $\int_2^1 dY = Y_1 - Y_2$. Then

$$\oint dY = \int_1^2 dY + \int_2^1 dY = Y_2 - Y_1 + (Y_1 - Y_2) = 0$$

20–2. Let $z = z(x, y)$ and $dz = xydx + y^2dy$. Although dz is not an exact differential (why not?), what combination of dz and x and/or y is an exact differential?

The quantity dz is not an exact differential because the coefficient of the dx term is not independent of y. An exact differential would be dz/y, because the coefficient of dx is independent of y and the coefficient of dy is independent of x in

$$\frac{dz}{y} = xdx + ydy$$

20–3. Use the criterion developed in MathChapter H to prove that δq_{rev} in Equation 20.1 is not an exact differential (see also Problem H–11).

We can write δq_{rev} as

$$\delta q_{rev} = C_V(T)dT + \frac{nRT}{V}dV \tag{20.1}$$

The cross-derivatives of an exact differential are equal, so we will find the cross derivatives of δq_{rev} to determine its nature. These are the coefficient of dT differentiated with respect to V and the coefficient of dV differentiated with respect to T, or

$$\frac{\partial C_V(T)}{\partial V} = 0 \quad \text{and} \quad \frac{\partial}{\partial T}\left(\frac{nRT}{V}\right) = \frac{nR}{V}$$

Because these two quantities are not equal, δq_{rev} is an inexact differential.

20–4. Use the criterion developed in MathChapter H to prove that $\delta q_{rev}/T$ in Equation 20.1 is an exact differential.

We use Equation 20.2 to express $\delta q_{rev}/T$ as

$$\frac{\delta q_{rev}}{T} = \frac{C_V(T)}{T}dT + \frac{nR}{V}dV$$

The cross-derivatives of an exact differential are equal, so we will find the cross derivatives of $\delta q_{rev}/T$ to determine its nature. These are the coefficient of dT differentiated with respect to V and the coefficient of dV differentiated with respect to T, or

$$\frac{\partial}{\partial V}\left(\frac{C_V(T)}{T}\right) = 0 \qquad \text{and} \qquad \frac{\partial}{\partial T}\left(\frac{nR}{V}\right) = 0$$

Because these two quantities are equal, $\delta q_{rev}/T$ is an exact differential.

20–5. In this problem, we will prove that Equation 20.5 is valid for an arbitrary system. To do this, consider an isolated system made up of two equilibrium subsystems, A and B, which are in thermal contact with each other; in other words, they can exchange energy as heat between themselves. Let subsystem A be an ideal gas and let subsystem B be arbitrary. Suppose now that an infinitesimal reversible process occurs in A accompanied by an exchange of energy as heat $\delta q_{rev}(\text{ideal})$. Simultaneously, another infinitesimal reversible process takes place in B accompanied by an exchange of energy as heat $\delta q_{rev}(\text{arbitrary})$. Because the composite system is isolated, the First Law requires that

$$\delta q_{rev}(\text{ideal}) = -\delta q_{rev}(\text{arbitrary})$$

Now use Equation 20.4 to prove that

$$\oint \frac{\delta q_{rev}(\text{arbitrary})}{T} = 0$$

Therefore, we can say that the definition given by Equation 20.4 holds for any system.

We use the First Law as suggested in the problem and substitute Equation 20.1 for $\delta q_{rev}(\text{ideal})$ to write

$$\delta q_{rev}(\text{arbitrary}) = -\delta q_{rev}(\text{ideal})$$

$$= -C_V(T)dT - \frac{nRT}{V}dV$$

$$\frac{\delta q_{rev}(\text{arbitrary})}{T} = \frac{-C_V}{T}dT - \frac{nR}{V}dV$$

$$= d\left[\int \frac{-C_V(T)}{T}dT - nR\int \frac{dV}{V} + \text{constant}\right]$$

Then $\delta q_{rev}(\text{arbitrary})/T$ is the derivative of a state function. We know that the cyclic integral of a state function is equal to 0 (Problem 20–1). Therefore, we can write (as we did in Section 20–2 for ideal gases)

$$\oint \frac{\delta q_{rev}(\text{arbitrary})}{T} = 0$$

and Equation 20.4 holds for any system.

20–6. Calculate q_{rev} and ΔS for a reversible cooling of one mole of an ideal gas at a constant volume V_1 from P_1, V_1, T_1 to P_2, V_1, T_4 followed by a reversible expansion at constant pressure P_2 from P_2, V_1, T_4 to P_2, V_2, T_1 (the final state for all the processes shown in Figure 20.3). Compare your result for ΔS with those for paths A, B + C, and D + E in Figure 20.3.

Step 1. $P_1, V_1, T_1 \rightarrow P_2, V_1, T_4$
Because there is no change in the volume of the ideal gas, $\delta w = 0$, and we can write

$$dq_{rev,1} = dU = C_V(T)dT$$

$$q_{rev,1} = \int_{T_1}^{T_4} C_V(T)dT$$

$$\Delta S_1 = \int_{T_1}^{T_4} \frac{C_V(T)}{T}dT$$

Step 2. $P_2, V_1, T_4 \rightarrow P_2, V_2, T_1$
In this case, we write (by the First Law)

$$\delta q_{rev,2} = dU - \delta w = C_V(T)dT + PdV$$

$$q_{rev,2} = \int_{T_4}^{T_1} C_V(T)dT + \int_{V_1}^{V_2} P_2 dV$$

$$\Delta S_2 = \int_{T_4}^{T_1} \frac{C_V(T)}{T}dT + \int_{V_1}^{V_2} \frac{P_2}{T}dV$$

$$= \int_{T_4}^{T_1} \frac{C_V(T)}{T}dT + \int_{V_1}^{V_2} \frac{R}{V}dV$$

For the entire process, $P_1, V_1, T_1 \rightarrow P_2, V_2, T_1$, we have

$$q_{rev} = \int_{T_1}^{T_4} C_V(T)dT + \int_{T_4}^{T_1} C_V(T)dT + \int_{V_1}^{V_2} P_2 dV = P_2(V_2 - V_1)$$

$$\Delta S = \int_{T_1}^{T_4} \frac{C_V(T)}{T}dT + \int_{T_4}^{T_1} \frac{C_V(T)}{T}dT + \int_{V_1}^{V_2} \frac{R}{V}dV = R \ln \frac{V_2}{V_1}$$

The value of q_{rev} differs from those found for paths A, B + C, and D + E (Section 20–3), but the value of ΔS is the same (because entropy is a path-independent function).

20–7. Derive Equation 20.8 without referring to Chapter 19.

The temperature T_2 is the final temperature resulting from the reversible adiabatic expansion of one mole of an ideal gas. For a reversible expansion, $dw = -PdV = -nRTdV/V$, and for an adiabatic expansion $dU = dw$. Then, because $dU = C_V dT$,

$$\frac{C_V dT}{T} = \frac{-nRdV}{V}$$

$$\int_{T_1}^{T_2} \frac{C_V}{T}dT = -nR \int_{V_1}^{V_2} \frac{dV}{V}$$

$$\int_{T_1}^{T_2} \frac{C_V}{T}dT = -nR \ln \frac{V_2}{V_1}$$

which is Equation 20.8.

20–8. Calculate the value of ΔS if one mole of an ideal gas is expanded reversibly and isothermally from 10.0 dm^3 to 20.0 dm^3. Explain the sign of ΔS.

For an isothermal reaction of an ideal gas, $\delta w = -\delta q$, so $\delta q = PdV$. Then

$$\Delta S = \int \frac{\delta q_{rev}}{T} = \int \frac{P}{T} dV$$

Using T from the ideal gas equation gives

$$\Delta S = \int \frac{nR}{V} dV = nR \ln 2.00 = 5.76 \text{ J·K}^{-1}$$

The value of ΔS is positive because the gas is expanding.

20–9. Calculate the value of ΔS if one mole of an ideal gas is expanded reversibly and isothermally from 1.00 bar to 0.100 bar. Explain the sign of ΔS.

As in the previous problem, because the reaction is isothermal, $\delta q = PdV$. For an ideal gas,

$$dV = -\frac{nRT}{P^2} dP = -\frac{V}{P} dP$$

so we write ΔS as

$$\Delta S = \int \frac{P}{T} dV = \int -\frac{V}{T} dP = \int -\frac{nR}{P} dP = -nR \ln 0.1 = 19.1 \text{ J·K}^{-1}$$

The value of ΔS is positive because the gas expands.

20–10. Calculate the values of q_{rev} and ΔS along the path $D + E$ in Figure 20.3 for one mole of a gas whose equation of state is given in Example 20–2. Compare your result with that obtained in Example 20–2.

Path $D + E$ is the path described by $(P_1, V_1, T_1) \to (P_1, V_2, T_3) \to (P_2, V_2, T_1)$. For the first step,

$$\delta q_{rev} = dU - \delta w = C_V(T)dT - PdV$$

and for the second step (because the volume remains constant)

$$\delta q_{rev} = dU = C_V(T)dT$$

Then

$$\int \delta q_{rev,D+E} = \int_{T_1}^{T_3} C_V(T)dT - \int_{V_1}^{V_2} P_1 dV + \int_{T_3}^{T_1} C_V(T)dT$$

$$= -\int_{V_1}^{V_2} P_1 dV = -P_1(V_2 - V_1)$$

and

$$\Delta S_{rev,D+E} = -\int_{V_1}^{V_2} \frac{P_1}{T} dV$$

Substituting for T from the equation of state gives

$$\Delta S_{\text{rev,D+E}} = \int_{\overline{V}_1}^{\overline{V}_2} \frac{nP_1 R d\overline{V}}{P_1(\overline{V} - b)} = nR \ln\left(\frac{\overline{V}_2 - b}{\overline{V}_1 - b}\right)$$

The quantity q_{rev} differs from those for the two paths in Example 20–2, but ΔS for all three paths is the same.

20–11. Show that $\Delta S_{\text{D+E}}$ is equal to ΔS_A and $\Delta S_{\text{B+C}}$ for the equation of state given in Example 20–2.

From Example 20–3,

$$\Delta S_A = \Delta S_{\text{B+C}} = nR \ln \frac{\overline{V}_2 - b}{\overline{V}_1 - b}$$

and the equation of state used is

$$P = \frac{RT}{\overline{V} - b}$$

In Example 20–1, we calculated $\Delta S_{\text{D+E}}$ for an ideal gas. Without using the ideal gas equation of state, however, we found in Example 20–1 that

$$\Delta S_D = \int_{T_1}^{T_3} \frac{C_V(T)}{T} dT + P_1 \int_{V_1}^{V_2} \frac{dV}{T}$$

$$\Delta S_E = \int_{T_3}^{T_1} \frac{C_V(T)}{T} dT$$

and

$$\Delta S_{\text{D+E}} = P_1 \int_{V_1}^{V_2} \frac{dV}{T}$$

We can substitute T from the equation of state to write

$$\Delta S_{\text{D+E}} = P_1 \int_{\overline{V}_1}^{\overline{V}_2} \frac{nR d\overline{V}}{P_1(\overline{V} - b)} = nR \int_{\overline{V}_1}^{\overline{V}_2} \frac{d\overline{V}}{\overline{V} - b} = nR \ln \frac{\overline{V}_2 - b}{\overline{V}_1 - b}$$

Therefore $\Delta S_{\text{D+E}}$ is equal to ΔS_A and $\Delta S_{\text{B+C}}$.

20–12. Calculate the values of q_{rev} and ΔS along the path described in Problem 20–6 for one mole of a gas whose equation of state is given in Example 20–2. Compare your result with that obtained in Example 20–2.

For both steps, because the ideal gas equation was not used in calculating q_{rev}, q_{rev} is the same as it was in Problem 20–6:

$$q_{\text{rev}} = -P_2(V_2 - V_1)$$

Substituting the equation of state from Example 20–2 into the expression for ΔS from Problem 20–6 gives

$$\Delta S = \int_{\overline{V}_1}^{\overline{V}_2} \frac{nP_2}{T} d\overline{V} = \int_{\overline{V}_1}^{\overline{V}_2} \frac{nR}{\overline{V} - b} d\overline{V} = nR \ln\left(\frac{\overline{V}_2 - b}{\overline{V}_1 - b}\right)$$

This is (by no coincidence) the same value as that found for ΔS_A, ΔS_{B+C}, and ΔS_{D+E}.

20–13. Show that

$$\Delta S = C_P \ln \frac{T_2}{T_1}$$

for a constant-pressure process if C_P is independent of temperature. Calculate the change in entropy of 2.00 moles of $H_2O(l)$ ($\overline{C}_P = 75.2 \text{ J} \cdot \text{K}^{-1} \cdot \text{mol}^{-1}$) if it is heated from 10°C to 90°C.

Because ΔS is a state function, we can calculate it using a reversible process. For a constant-pressure reversible process (Equation 19.37), $\delta q_{rev} = dH = C_P dT$, and so

$$\Delta S = \frac{q_{rev}}{T} = \int_{T_1}^{T_2} \frac{C_P}{T} dT = n\overline{C}_P \ln \left(\frac{T_2}{T_1} \right)$$

For 2.00 mol of H_2O,

$$\Delta S = (2.00 \text{ mol})(75.2 \text{ J} \cdot \text{K}^{-1} \cdot \text{mol}^{-1}) \ln \frac{363}{283} = 37.4 \text{ J} \cdot \text{K}^{-1}$$

20–14. Show that

$$\Delta \overline{S} = \overline{C}_V \ln \frac{T_2}{T_1} + R \ln \frac{V_2}{V_1}$$

if one mole of an ideal gas is taken from T_1, V_1 to T_2, V_2, assuming that $\overline{C}_V$ is independent of temperature. Calculate the value of $\Delta \overline{S}$ if one mole of $N_2(g)$ is expanded from 20.0 dm^3 at 273 K to 300 dm^3 at 400 K. Take $\overline{C}_P = 29.4 \text{ J} \cdot \text{K}^{-1} \cdot \text{mol}^{-1}$.

For the path $(T_1, V_1) \rightarrow (T_2, V_2)$, $\delta w = -PdV$ and $\delta q = dU - \delta w = C_V dT + PdV$. We can then write $\Delta \overline{S}$ as

$$\Delta \overline{S} = \frac{1}{n} \left(\int \frac{C_V}{T} dT + \int \frac{P}{T} dV \right)$$

$$= \int \frac{\overline{C}_V}{T} dT + \int \frac{R}{V} dV$$

$$= \overline{C}_V \ln \frac{T_2}{T_1} + R \ln \frac{V_2}{V_1}$$

Because $\overline{C}_P - \overline{C}_V = R$ for an ideal gas, we can write this as

$$\Delta \overline{S} = (\overline{C}_P - R) \ln \frac{T_2}{T_1} + R \ln \frac{V_2}{V_1}$$

For N_2,

$$\Delta \overline{S} = (29.4 \text{ J} \cdot \text{mol}^{-1} \cdot \text{K}^{-1} - 8.314 \text{ J} \cdot \text{K}^{-1} \cdot \text{mol}^{-1}) \ln \frac{400}{273} + 8.314 \text{ J} \cdot \text{K}^{-1} \cdot \text{mol}^{-1} \ln \frac{300}{20.0}$$

$$= 30.6 \text{ J} \cdot \text{K}^{-1} \cdot \text{mol}^{-1}$$

20–15. In this problem, we will consider a two-compartment system like that in Figure 20.4, except that the two subsystems have the same temperature but different pressures and the wall that separates them is flexible rather than rigid. Show that in this case,

$$dS = \frac{dV_B}{T}(P_B - P_A)$$

Interpret this result with regard to the sign of dV_B when $P_B > P_A$ and when $P_B < P_A$.

We can use the First Law to write δq for each compartment as

$$\delta q_A = dU_A - \delta w_A = dU_A + P_A dV_A$$
$$\delta q_B = dU_B - \delta w_B = dU_B + P_B dV_B$$

We can write the total entropy of the system as

$$dS = dS_A + dS_B = \frac{\delta q_A}{T_A} + \frac{\delta q_B}{T_B}$$
$$= \frac{dU_A}{T_A} + \frac{P_A}{T_A}dV_A + \frac{dU_B}{T_B} + \frac{P_B}{T_B}dV_B$$

Because the two-compartment system is isolated, $dV_A = -dV_B$ and $dU_A = -dU_B$. Also, $T_A = T_B$. The quantity dS above then becomes

$$dS = (P_B - P_A)\frac{dV_B}{T}$$

When $P_B > P_A$, compartment B will expand, so, because dS must be greater than zero, dV_B is positive. Likewise, when $P_B < P_A$, compartment A will expand, so (again, because dS must be greater than zero) dV_B is negative.

20–16. In this problem, we will illustrate the condition $dS_{prod} \geq 0$ with a concrete example. Consider the two-component system shown in Figure 20.8. Each compartment is in equilibrium with a heat reservoir at different temperatures T_1 and T_2, and the two compartments are separated by a rigid heat-conducting wall. The total change of energy as heat of compartment 1 is

$$dq_1 = d_e q_1 + d_i q_1$$

where $d_e q_1$ is the energy as heat exchanged with the reservoir and $d_i q_1$ is the energy as heat exchanged with compartment 2. Similarly,

$$dq_2 = d_e q_2 + d_i q_2$$

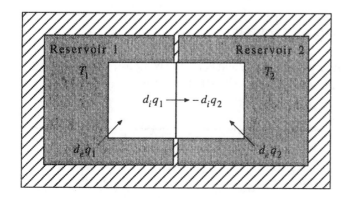

FIGURE 20.8
A two-compartment system with each compartment in contact with an (essentially infinite) heat reservoir, one at temperature T_1 and the other at temperature T_2. The two compartments are separated by a rigid heat-conducting wall.

Clearly,

$$d_i q_1 = -d_i q_2$$

Show that the entropy change for the two-compartment system is given by

$$dS = \frac{d_e q_1}{T_1} + \frac{d_e q_2}{T_2} + d_i q_1 \left(\frac{1}{T_1} - \frac{1}{T_2} \right)$$
$$= dS_{exchange} + dS_{prod}$$

where

$$dS_{exchange} = \frac{d_e q_1}{T_1} + \frac{d_e q_2}{T_2}$$

is the entropy *exchanged* with the reservoirs (surroundings) and

$$dS_{prod} = d_i q_1 \left(\frac{1}{T_1} - \frac{1}{T_2} \right)$$

is the entropy *produced* within the two-compartment system. Now show that the condition $dS_{prod} \geq 0$ implies that energy as heat flows spontaneously from a higher temperature to a lower temperature. The value of $dS_{exchange}$, however, has no restriction and can be positive, negative, or zero.

As stated in the text of the problem, we can write

$$d_i q_1 = -d_i q_2$$

The energy as heat exchanged between compartments 1 and 2 is involved in the entropy transferred between the two compartments. We can therefore express $dS_{exchange}$ as

$$dS_{exchange} = \frac{d_e q_1}{T_1} + \frac{d_e q_2}{T_2}$$

Similarly, the energy as heat exchanged between the compartments and the reservoirs is involved in the entropy produced within the two-compartment system, so we can write dS_{prod} as

$$dS_{prod} = \frac{d_i q_1}{T_1} + \frac{d_i q_2}{T_2} = d_i q_1 \left(\frac{1}{T_1} - \frac{1}{T_2} \right)$$

These are the only two means of changing the entropy of the system, so we can find dS_{tot} to be

$$dS_{tot} = dS_{exchange} + dS_{prod} = \frac{d_e q_1}{T_1} + \frac{d_e q_2}{T_2} + d_i q_1 \left(\frac{1}{T_1} - \frac{1}{T_2} \right)$$

Now take the condition $dS_{prod} \geq 0$. This is the same as saying that

$$d_i q_1 \left(\frac{1}{T_1} - \frac{1}{T_2} \right) \geq 0$$

Arbitrarily, let $T_1 > T_2$. Then $1/T_1 - 1/T_2 < 0$, so $d_i q_1 < 0$ and heat is flowing from compartment 1 to compartment 2. If $T_2 > T_1$, by the same reasoning, $d_i q_1 > 0$ and heat is flowing from compartment 2 to compartment 1.

20–17. Show that

$$\Delta S \geq \frac{q}{T}$$

for an isothermal process. What does this equation say about the sign of ΔS? Can ΔS decrease in a reversible isothermal process? Calculate the entropy change when one mole of an ideal gas is compressed reversibly and isothermally from a volume of 100 dm^3 to 50.0 dm^3 at 300 K.

We defined ΔS as (Equation 20.22)

$$\Delta S \geq \int \frac{\delta q}{T}$$

For an isothermal process T is constant, so we can write this expression as

$$\Delta S \geq \frac{1}{T} \int \delta q$$

and integrate over δq to write

$$\Delta S \geq \frac{q}{T}$$

We know that q can be positive or negative, while T is always positive; therefore, ΔS can be positive or negative for an isothermal process. For one mole of an ideal gas compressed reversibly and isothermally, $dU = 0$, so $\delta q_{rev} = -\delta w = P dV$. Then

$$\delta q_{rev} = P dV = \frac{nRT}{V} dV$$

$$q_{rev} = nRT \ln \frac{V_2}{V_1}$$

and the change in entropy is given by

$$\Delta S = nR \ln \frac{V_2}{V_1} = (1 \text{ mol})(8.3145 \text{ J} \cdot \text{K}^{-1} \cdot \text{mol}^{-1}) \ln 0.5 = -5.76 \text{ J} \cdot \text{K}^{-1}$$

The quantity ΔS is equal to q/T, rather than greater than it, because this is a reversible process.

20–18. Vaporization at the normal boiling point (T_{vap}) of a substance (the boiling point at one atm) can be regarded as a reversible process because if the temperature is decreased infinitesimally below T_{vap}, all the vapor will condense to liquid, whereas if it is increased infinitesimally above T_{vap}, all the liquid will vaporize. Calculate the entropy change when two moles of water vaporize at 100.0°C. The value of $\Delta_{vap}\overline{H}$ is 40.65 kJ·mol^{-1}. Comment on the sign of $\Delta_{vap}S$.

At constant pressure and temperature, $q_{rev} = n\Delta\overline{H}_{rev}$ (Equation 19.37). We know from the previous problem that, for a reversible isothermal process,

$$\Delta S = \frac{q}{T} = \frac{(2 \text{ mol})(40.65 \text{ kJ} \cdot \text{mol}^{-1})}{373.15 \text{ K}} = 217.9 \text{ J} \cdot \text{K}^{-1}$$

As the water becomes more disordered, changing from liquid to gas, the entropy increases.

20–19. Melting at the normal melting point (T_{fus}) of a substance (the melting point at one atm) can be regarded as a reversible process because if the temperature is changed infinitesimally from exactly T_{fus}, then the substance will either melt or freeze. Calculate the change in entropy when two moles of water melt at 0°C. The value of $\Delta_{fus}\overline{H}$ is 6.01 kJ·mol^{-1}. Compare your answer with the one you obtained in Problem 20–18. Why is $\Delta_{vap}S$ much larger than $\Delta_{fus}S$?

At constant pressure, $q = n\Delta\overline{H}$ (Equation 19.37). For a reversible isothermal process, we can express ΔS as (Problem 20.17)

$$\Delta S = \frac{q}{T} = \frac{(2 \text{ mol})(6.01 \text{ kJ·mol}^{-1})}{273.15 \text{ K}} = 44.0 \text{ J·K}^{-1}$$

The quantity $\Delta_{fus}S$ is much less than $\Delta_{vap}S$ because the difference in disorder between a solid and a liquid is much less than that between a liquid and a gas.

20–20. Consider a simple example of Equation 20.23 in which there are only two states, 1 and 2. Show that $W(a_1, a_2)$ is a maximum when $a_1 = a_2$. *Hint*: Consider $\ln W$, use Stirling's approximation, and treat a_1 and a_2 as continuous variables.

A simplified version of Equation 20.23, for two states only, is

$$W = \frac{(a_1 + a_2)!}{a_1! a_2!}$$

Let $a_1 = x$, with x being a continuous variable, and let $a_1 + a_2 = N$. Then we can express W in terms of N and x:

$$W = \frac{N!}{(N - x)! x!}$$
$$\ln W = \ln N! - \ln(N - x)! - \ln x!$$

Using Stirling's approximation for $\ln N!$ (MathChapter J), we can find the first derivative of $\ln W$ with respect to x:

$$\ln W = \ln N! - \ln(N - x)! - \ln x!$$
$$= N \ln N - N - [(N - x)\ln(N - x) - (N - x)] - (x \ln x - x)$$
$$\frac{d(\ln W)}{dx} = 0 + \frac{N - x}{N - x} + \ln(N - x) - 1 - \ln x - \frac{x}{x} + 1$$
$$= \ln(N - x) - \ln x$$

We are looking for the value of x that produces the maximum value of W, which is where $dW/dx = 0$. Because

$$\frac{1}{W}\frac{dW}{dx} = \frac{d(\ln W)}{dx}$$

the desired value of x will also give $[d(\ln W)/dx] = 0$. Setting this derivative equal to zero, we find

$$0 = \ln(N - x_{max}) - \ln x_{max}$$
$$\ln x_{max} = \ln(N - x_{max})$$
$$2x_{max} = N$$

And clearly $x_{max} = N/2$. Thus, the maximum value of W occurs when $a_1 = a_2 = N/2$.

20–21. Extend Problem 20–20 to the case of three states. Do you see how to generalize it to any number of states?

For the case of three states, Equation 20.23 becomes

$$W = \frac{(a_1 + a_2 + a_3)!}{a_1! a_2! a_3!}$$

Let $a_1 = x$ and $a_2 = y$, where x and y are continuous variables, and let $a_1 + a_2 + a_3 = A$. Then

$$W = \frac{A!}{(A - x - y)! x! y!}$$

$$\ln W = \ln A! - \ln(A - x - y)! - \ln x! - \ln y!$$

We can use Stirling's approximation for $\ln A!$ (MathChapter J) to write $\ln W$ and differentiate to find an expression for $d(\ln W)/dx$:

$$\ln W = A \ln A - A - [(A - x - y) \ln(A - x - y) - (A - x - y)] - (x \ln x - x) - (y \ln y - y)$$

$$\frac{d(\ln W)}{dx} = \ln(A - x - y) + 1 - 1 - \ln x - 1 + 1 = \ln(A - x - y) - \ln x$$

Similarly,

$$\frac{d(\ln W)}{dy} = \ln(A - x - y) - \ln y$$

As in the previous problem, the values of x and y which give the largest values of W occur where $d(\ln W)/dx = 0$ and $d(\ln W)/dy = 0$. Therefore,

$$0 = \ln(A - x_{max} - y) - \ln x_{max}$$

$$\ln x_{max} = \ln(A - x_{max} - y)$$

$$2x_{max} = A - y$$

Because we want the point at which both x and y are at their maxima, we substitute this value into the expression for $d(\ln W)/dy$ to find

$$0 = \ln(A - x_{max} - y_{max}) - \ln y_{max}$$

$$2y_{max} = \frac{A}{2} + \frac{y_{max}}{2}$$

$$y_{max} = \frac{A}{3}$$

Then (substituting back into the first equality) $x_{max} = A/3$. Thus, the maximum value of W occurs when $a_1 = a_2 = a_3 = A/3$. Problem J.10 generalizes this to any number of states.

20–22. Show that the system partition function can be written as a summation over levels by writing

$$Q(N, V, T) = \sum_E \Omega(N, V, E) e^{-E/k_B T}$$

Now consider the case of an isolated system, for which there is only one term in $Q(N, V, T)$. Now substitute this special case for Q into Equation 20.43 to derive the equation $S = k_B \ln \Omega$.

In Chapter 17, we defined the partition function $Q(N, V, T)$ as (Equation 17.14)

$$Q(N, V, T) = \sum_j e^{-E_j/k_B T}$$

Here a term representing an energy level of degeneracy Ω is written Ω times. We can write this, alternatively, as a sum over energy levels, where a term representing an energy level is written once and multiplied by its degeneracy Ω:

$$Q(N, V, T) = \sum_E \Omega(N, V, E)e^{-E/k_B T}$$

These two expressions are equivalent. For an isolated system, there will be only one term in $Q(N, V, T)$, so

$$Q = \Omega(N, V, E)e^{-E/k_B T}$$

$$\ln Q = \ln \Omega - \left(\frac{E}{k_B T}\right)$$

Applying Equation 20.43 allows us to write

$$S = k_B T \left(\frac{\partial \ln Q}{\partial T}\right)_{N,V} + k_B \ln Q$$

$$= k_B T \left(\frac{E}{k_B T^2}\right) + k_B \ln \Omega - \frac{E}{T}$$

$$= k_B \ln \Omega$$

which is Boltzmann's equation.

20-23. In this problem, we will show that $\Omega = c(N)f(E)V^N$ for an ideal gas (Example 20–3). In Problem 18–42 we showed that the number of translational energy states between ε and $\varepsilon + \Delta\varepsilon$ for a particle in a box can be calculated by considering a sphere in n_x, n_y, n_z space,

$$n_x^2 + n_y^2 + n_z^2 = \frac{8ma^2\varepsilon}{h^2} = R^2$$

Show that for an N-particle system, the analogous expression is

$$\sum_{j=1}^{N}(n_{xj}^2 + n_{yj}^2 + n_{zj}^2) = \frac{8ma^2 E}{h^2} = R^2$$

or, in more convenient notation

$$\sum_{j=1}^{3N} n_j^2 = \frac{8ma^2 E}{h^2} = R^2$$

Thus, instead of dealing with a three-dimensional sphere as we did in Problem 18–42, here we must deal with a $3N$-dimensional sphere. Whatever the formula for the volume of a $3N$-dimensional sphere is (it is known), we can at least say that it is proportional to R^{3N}. Show that this proportionality leads to the following expression for $\Phi(E)$, the number of states with energy $\leq E$,

$$\Phi(E) \propto \left(\frac{8ma^2 E}{h^2}\right)^{3N/2} = c(N)E^{3N/2}V^N$$

where $c(N)$ is a constant whose value depends upon N and $V = a^3$. Now, following the argument developed in Problem 18–42, show that the number of states between E and $E + \Delta E$ (which is essentially Ω) is given by

$$\Omega = c(N)f(E)V^N \Delta E$$

where $f(E) = E^{\frac{3N}{2}-1}$.

For an N-particle system, we wish to consider all the N particles in one $3N$-dimensional space, instead of the N particles in N individual three-dimensional spaces. The equation from Problem 18–42 then becomes

$$\sum_{j=1}^{3N} n_j^2 = \frac{8ma^2E}{h^2} = R^2$$

As in Problem 18–42, $\Phi(E) \propto$ the volume of the sphere, so

$$\Phi(E) \propto R^{3N} = \left(\frac{8ma^2E}{h^2}\right)^{3N/2}$$

Letting $c(N)$ be a proportionality constant allows us to write

$$\Phi(E) = c(N)E^{3N/2}V^N$$

Now, as in Problem 18–42, $\Omega = \Phi(E + \Delta E) - \Phi(E)$, so

$$\begin{aligned}
\Omega &= c(N)(E + \Delta E)^{3N/2}V^N - c(N)E^{3N/2}V^N \\
&= c(N)V^N\left[(E + \Delta E)^{3N/2} - E^{3N/2}\right] \\
&= c(N)V^N E^{3N/2}\left[\left(1 + \frac{\Delta E}{E}\right)^{3N/2} - 1\right] \\
&= c(N)V^N E^{3N/2}\left\{1 + \frac{3N}{2}\frac{\Delta E}{E} + O\left[\left(\frac{\Delta E}{E}\right)^2\right] - 1\right\} \\
&= \frac{3N}{2}c(N)V^N E^{3N/2}\frac{\Delta E}{E} + O\left[\left(\frac{\Delta E}{E}\right)^2\right] \\
&= c(N)V^N E^{\frac{3N}{2}-1}\Delta E = c(N)f(E)V^N\Delta E
\end{aligned}$$

where we have incorporated the factor of $3N/2$ into $c(N)$ and defined $f(E) = E^{\frac{3N}{2}-1}$.

20–24. Show that if a process involves only an isothermal transfer of energy as heat (*pure heat transfer*), then

$$dS_{\text{sys}} = \frac{dq}{T} \qquad \text{(pure heat transfer)}$$

The process involves only an isothermal transfer of energy as heat, so $\delta w = 0$ and $dU = \delta q$. Therefore,

$$dS_{\text{sys}} = \frac{\delta q}{T} = \frac{dU}{T} = \frac{dq}{T}$$

where we can write dq instead of δq because $\delta q = dU$, and U is a state function.

20–25. Calculate the change in entropy of the system and of the surroundings and the total change in entropy if one mole of an ideal gas is expanded isothermally and reversibly from a pressure of 10.0 bar to 2.00 bar at 300 K.

Because this is an isothermal reversible expansion, $\delta q = -\delta w = P dV$. We then use the ideal gas equation to write

$$\Delta S_{sys} = \int \frac{\delta q_{rev}}{T} = \int \frac{P}{T} dV = \int \frac{nR}{V} dV = nR \ln \frac{V_2}{V_1}$$

For an isothermal expansion of an ideal gas, $P_1 V_1 = P_2 V_2$. We can then write the change of entropy of the gas as

$$\Delta S_{sys} = (1 \text{ mol})(8.3145 \text{ J} \cdot \text{mol}^{-1} \cdot \text{K}^{-1}) \ln 5.00 = +13.4 \text{ J} \cdot \text{K}^{-1}$$

For a reversible expansion, $\Delta S_{tot} = 0$, so $\Delta S_{surr} = -\Delta S_{sys} = -13.4 \text{ J} \cdot \text{K}^{-1}$.

20–26. Redo Problem 20–25 for an expansion into a vacuum, with an initial pressure of 10.0 bar and a final pressure of 2.00 bar.

As in Problem 20–25, $\Delta S_{sys} = 13.4 \text{ J} \cdot \text{K}^{-1}$. However, because this is an irreversible expansion into a vacuum, $\Delta S_{surr} = 0$, so $\Delta S_{tot} = 13.4 \text{ J} \cdot \text{K}^{-1}$.

20–27. The molar heat capacity of 1-butene can be expressed as

$$\overline{C}_p(T)/R = 0.05641 + (0.04635 \text{ K}^{-1})T - (2.392 \times 10^{-5} \text{ K}^{-2})T^2 + (4.80 \times 10^{-9} \text{ K}^{-3})T^3$$

over the temperature range 300 K < T < 1500 K. Calculate the change in entropy when one mole of 1-butene is heated from 300 K to 1000 K at constant pressure.

At constant pressure, $\delta q = dH = n\overline{C}_p dT$. Then Equation 20.22 becomes (assuming a reversible process)

$$\Delta S = n \int_{300}^{1000} \frac{\overline{C}_p}{T} dT$$

$$= (1 \text{ mol}) R \int_{300}^{1000} \left[0.05641 T^{-1} + (0.04635 \text{ K}^{-1}) - (2.392 \times 10^{-5} \text{ K}^{-2})T \right.$$
$$\left. + (4.80 \times 10^{-9} \text{ K}^{-3})T^2 \right] dT$$

$$= 192.78 \text{ J} \cdot \text{K}^{-1}$$

20–28. Plot $\Delta_{mix}\overline{S}$ against y_1 for the mixing of two ideal gases. At what value of y_1 is $\Delta_{mix}\overline{S}$ a maximum? Can you give a physical interpretation of this result?

We can use Equation 20.30 for two gases:

$$\Delta_{mix}\overline{S} = -R\left(y_1 \ln y_1 - y_2 \ln y_2 \right)$$

Because $y_1 + y_2 = 1$, we can write this as

$$\Delta_{mix}\overline{S}/R = -y_1 \ln y_1 - (1 - y_1) \ln(1 - y_1)$$

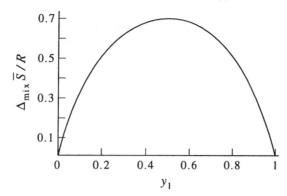

The quantity $\Delta_{mix}\overline{S}$ is a maximum when $y_1 = 0.5$. This means that the gases are most disordered when there are equal amounts of both present in a container.

20–29. Calculate the entropy of mixing if two moles of $N_2(g)$ are mixed with one mole $O_2(g)$ at the same temperature and pressure. Assume ideal behavior.

The mole fractions y are 2/3 for $N_2(g)$ and 1/3 for $O_2(g)$. Therefore,

$$\Delta_{mix}\overline{S} = -Ry_1 \ln y_1 - Ry_2 \ln y_2$$
$$= -\frac{2R}{3} \ln \frac{2}{3} - \frac{R}{3} \ln \frac{1}{3} = 5.29 \text{ J·K}^{-1}$$

20–30. Show that $\Delta_{mix}\overline{S} = R \ln 2$ if equal volumes of any two ideal gases under the same conditions are mixed.

Because there are equal volumes of ideal gases under the same conditions, $y_1 = y_2 = 0.5$ (Problem 20–28). Now Equation 20.30 gives

$$\Delta_{mix}\overline{S} = -Ry_1 \ln y_1 - Ry_2 \ln y_2$$
$$= -\frac{R}{2} \ln \frac{1}{2} - \frac{R}{2} \ln \frac{1}{2}$$
$$= -R \ln \frac{1}{2} = R \ln 2$$

20–31. Derive the equation $dU = TdS - PdV$. Show that

$$d\overline{S} = \overline{C}_V \frac{dT}{T} + R\frac{d\overline{V}}{\overline{V}}$$

for one mole of an ideal gas. Assuming that $\overline{C}_V$ is independent of temperature, show that

$$\Delta\overline{S} = \overline{C}_V \ln \frac{T_2}{T_1} + R \ln \frac{\overline{V}_2}{\overline{V}_1}$$

for the change from T_1, $\overline{V}_1$ to T_2, $\overline{V}_2$. Note that this equation is a combination of Equations 20.28 and 20.31.

From the definition of entropy, $\delta q_{rev} = TdS$, and by definition $\delta w_{rev} = -PdV$. The first law then gives $dU = \delta q_{rev} + \delta w_{rev} = TdS - PdV$. We divide through by T to obtain

$$dS = \frac{dU}{T} - \frac{P}{T}dV$$

Using the relation $dU = n\overline{C}_V dT$ and the ideal gas law, we find

$$dS = n\overline{C}_V\frac{dT}{T} - nR\frac{dV}{V}$$

or

$$d\overline{S} = \overline{C}_V\frac{dT}{T} - R\frac{d\overline{V}}{\overline{V}}$$

If $\overline{C}_V$ is temperature-independent, integrating gives

$$\int d\overline{S} = \overline{C}_V\int\frac{dT}{T} + R\int\frac{d\overline{V}}{\overline{V}}$$

$$\Delta\overline{S} = \overline{C}_V\ln\frac{T_2}{T_1} + R\ln\frac{\overline{V}_2}{\overline{V}_1}$$

20–32. Derive the equation $dH = TdS + VdP$. Show that

$$\Delta\overline{S} = \overline{C}_P\ln\frac{T_2}{T_1} - R\ln\frac{P_2}{P_1}$$

for the change of one mole of an ideal gas from T_1, P_1 to T_2, P_2, assuming that $\overline{C}_P$ is independent of temperature.

We derived the equation $dU = TdS - PdV$ in Problem 20–31. Now add $d(PV)$ to both sides of this equation to obtain

$$dU + d(PV) = dH = TdS + VdP$$

Now divide both sides by T to write

$$dS = \frac{dH}{T} - \frac{V}{T}dP$$

and use the relation $dH = n\overline{C}_P dT$ and the ideal gas law to obtain

$$dS = \frac{n\overline{C}_P}{T}dT - \frac{nR}{P}dP$$

or

$$d\overline{S} = \frac{\overline{C}_P}{T}dT - \frac{R}{P}dP$$

Assuming that $\overline{C}_P$ is temperature-independent, integrating gives

$$\Delta\overline{S} = \overline{C}_P\ln\frac{T_2}{T_1} - R\ln\frac{P_2}{P_1}$$

20–33. Calculate the change in entropy if one mole of $SO_2(g)$ at 300 K and 1.00 bar is heated to 1000 K and its pressure is decreased to 0.010 bar. Take the molar heat capacity of $SO_2(g)$ to be

$$\overline{C}_P(T)/R = 7.871 - \frac{1454.6\text{ K}}{T} + \frac{160351\text{ K}^2}{T^2}$$

We can use the result of the previous problem,

$$dS = \frac{n\overline{C}_P dT}{T} - \frac{nRdP}{P}$$

Then

$$\int dS = nR \int \frac{\overline{C}_P}{T} dT - nR \int \frac{dP}{P}$$

$$= (1.00\text{ mol})(8.314\text{ J·mol}^{-1}\text{·K}^{-1}) \left[\int_{300\text{ K}}^{1000\text{ K}} \left(\frac{7.871}{T} - \frac{1454.6\text{ K}}{T^2} + \frac{160351\text{ K}^2}{T^3} \right) dT - \ln\frac{0.010}{1.00} \right]$$

$$\Delta S = 95.6\text{ J·K}^{-1}$$

20–34. In the derivation of Equation 20.32, argue that $\Delta S_c > 0$ and $\Delta S_h < 0$. Now show that

$$\Delta S = \Delta S_c + \Delta S_h > 0$$

by showing that

$$\Delta S_c - |\Delta S_h| > 0$$

The two quantities $\Delta S_c > 0$ and $\Delta S_h < 0$ because the colder piece will become hotter and the hotter piece will become colder. Using the expressions for ΔS_c and ΔS_h in Section 20–6,

$$\Delta S_c = C_V \ln\frac{T_h + T_c}{2T_c} > 0 \qquad \text{and} \qquad \Delta S_h = C_V \ln\frac{T_h + T_c}{2T_h} < 0$$

Now, because $\Delta S_h < 0$, $|\Delta S_h| = -\Delta S_h$, and

$$|\Delta S_h| = C_V \ln\frac{2T_h}{T_h + T_c} > 0$$

The total change in entropy is given by

$$\Delta S = \Delta S_c + \Delta S_h = \Delta S_c - |\Delta S_h|$$

$$= C_V \ln\frac{(T_h + T_c)^2}{4T_h T_c} > 0$$

where we proved that $(T_h + T_c)^2 > 4T_h T_c$ in Section 20–6.

20–35. We can use the equation $S = k_B \ln W$ to derive Equation 20.28. First, argue that the probability that an ideal-gas molecule is found in a subvolume V_s of some larger volume V is V_s/V. Because the molecules of an ideal gas are independent, the probability that N ideal-gas molecules are found in V_s is $(V_s/V)^N$. Now show that the change in entropy when the volume of one mole of an ideal gas changes isothermally from V_1 to V_2 is

$$\Delta S = R \ln \frac{V_2}{V_1}$$

The probability that the molecule is in subvolume V_s is V_s/V, because the numerator represents the situations where the molecule is in V_s and the denominator represents all positions available to the molecule. Now we can write (using Boltzmann's equation)

$$S = k_B \ln W = k_B \ln \left(\frac{V_s}{V} \right)^N = R \ln \frac{V_s}{V}$$

Now take $V_s = V_1$ and an arbitrary V_2 and V. The change in entropy when an ideal gas goes from V_1 to V_2 isothermally is then

$$\Delta S = R \left(\ln \frac{V_2}{V} - \ln \frac{V_1}{V} \right) = R \ln \frac{V_2}{V_1}$$

20–36. The relation $n_j \propto e^{-\varepsilon_j/k_B T}$ can be derived by starting with $S = k_B \ln W$. Consider a gas with n_0 molecules in the ground state and n_j in the jth state. Now add an energy $\varepsilon_j - \varepsilon_0$ to this system so that a molecule is promoted from the ground state to the jth state. If the volume of the gas is kept constant, then no work is done, so $dU = dq$,

$$dS = \frac{dq}{T} = \frac{dU}{T} = \frac{\varepsilon_j - \varepsilon_0}{T}$$

Now, assuming that n_0 and n_j are large, show that

$$dS = k_B \ln \left\{ \frac{N!}{(n_0 - 1)! n_1! \cdots (n_j + 1)! \cdots} \right\} - k_B \ln \left\{ \frac{N!}{n_0! n_1! \cdots n_j! \cdots} \right\}$$

$$= k_B \ln \left\{ \frac{n_j!}{(n_j + 1)!} \frac{n_0!}{(n_0 - 1)!} \right\} = k \ln \frac{n_0}{n_j}$$

Equating the two expressions for dS, show that

$$\frac{n_j}{n_0} = e^{-(\varepsilon_j - \varepsilon_0)/k_B T}$$

From the problem,

$$dS = \frac{\varepsilon_j - \varepsilon_0}{T}$$

Recall that (Equation 20.24) $S = k_B \ln W$. For the initial state,

$$S_{\text{initial}} = k_B \ln W = k_B \ln \left[\frac{N!}{n_0! n_1! \cdots n_j! \cdots} \right]$$

and for the final state

$$S_{\text{final}} = k_B \ln W = k_B \ln \left[\frac{N!}{(n_0 - 1)! n_1! \cdots (n_j + 1)! \cdots} \right]$$

Then

$$dS = S_{final} - S_{initial}$$

$$= k_B \ln \left[\frac{N!}{(n_0 - 1)!n_1! \cdots (n_j + 1)! \cdots} \right] - k_B \ln \left[\frac{N!}{n_0!n_1! \cdots n_j! \cdots} \right]$$

$$= k_B \ln \left[\frac{n_j!}{(n_j + 1)!} \frac{n_0!}{(n_0 - 1)!} \right] = k_B \ln \frac{n_0}{n_j + 1}$$

$$= k_B \ln \frac{n_0}{n_j}$$

where the last equality holds because $n_j \gg 1$. Equating the two expressions for dS, we find that

$$k_B \ln \frac{n_0}{n_j} = \frac{\varepsilon_j - \varepsilon_0}{T}$$

$$\ln \frac{n_j}{n_0} = -\frac{\varepsilon_j - \varepsilon_0}{k_B T}$$

$$\frac{n_j}{n_0} = e^{-(\varepsilon_j - \varepsilon_0)/k_B T}$$

20–37. We can use Equation 20.24 to calculate the probability of observing fluctuations from the equilibrium state. Show that

$$\frac{W}{W_{eq}} = e^{-\Delta S/k_B}$$

where W represents the nonequilibrium state and ΔS is the entropy difference between the two states. We can interpret the ratio W/W_{eq} as the probability of observing the nonequilibrium state. Given that the entropy of one mole of oxygen is 205.0 J·K^{-1}·mol^{-1} at 25°C and one bar, calculate the probability of observing a decrease in entropy that is one millionth of a percent of this amount.

We can use Equation 20.24 to write

$$S_{eq} = k_B \ln W_{eq} \quad \text{and} \quad S = k_B \ln W$$

Then ΔS is

$$\Delta S = S - S_{eq} = k_B(\ln W - \ln W_{eq})$$

$$= k_B \ln \frac{W}{W_{eq}}$$

$$-\frac{\Delta S}{k_B} = \ln \frac{W}{W_{eq}}$$

$$e^{-\Delta S/k_B} = \frac{W}{W_{eq}}$$

The probability of observing a ΔS which is one millionth of one percent of 205 J·K^{-1}·mol^{-1} is

$$\exp\left[-\frac{(1 \text{ mol})(1.00 \times 10^{-8})(205.0 \text{ J·K}^{-1}\text{·mol}^{-1})}{1.381 \times 10^{-23} \text{ J·K}^{-1}} \right] = \exp[-1.485 \times 10^{17}] \approx 0$$

20–38. Consider one mole of an ideal gas confined to a volume V. Calculate the probability that all the N_A molecules of this ideal gas will be found to occupy one half of this volume, leaving the other half empty.

From Problem 20–35, we can write the probability as

$$\left(\frac{\frac{1}{2}V}{V}\right)^{N_A} = \left(\frac{1}{2}\right)^{N_A} \approx 0$$

20–39. Show that S_{system} given by Equation 20.40 is a maximum when all the p_j are equal. Remember that $\sum p_j = 1$, so that

$$\sum_j p_j \ln p_j = p_1 \ln p_1 + p_2 \ln p_2 + \cdots + p_{n-1} \ln p_{n-1}$$
$$+ (1 - p_1 - p_2 - \cdots - p_{n-1}) \ln(1 - p_1 - p_2 - \cdots - p_{n-1})$$

See also Problem J–10.

Begin with Equation 20.40,

$$S_{\text{system}} = -k_B \sum_j p_j \ln p_j$$

Substituting the expression given for $\sum_j p_j \ln p_j$, we find

$$S_{\text{system}} = -k_B \big[p_1 \ln p_1 + p_2 \ln p_2 + \cdots + p_{n-1} \ln p_{n-1}$$
$$+ (1 - p_1 - p_2 - \cdots - p_{n-1}) \ln(1 - p_1 - p_2 - \cdots - p_{n-1}) \big]$$

$$\frac{\partial S_{\text{system}}}{\partial p_j} = \ln p_j + 1 - \ln(1 - p_1 - \cdots - p_{j-1} - p_{j+1} - \cdots - p_{n-1}) - 1$$

$$0 = \ln p_j - \ln(1 - p_1 - \cdots - p_{j-1} - p_{j+1} - \cdots - p_{n-1})$$

$$p_j = 1 - p_1 - \cdots - p_{j-1} - p_{j+1} - \cdots - p_{n-1}$$

Because p_j can be any of p_1 to p_{n-1}, and the above equality holds for all p_j, all the p_j must be equal.

20–40. Use Equation 20.45 to calculate the molar entropy of krypton at 298.2 K and one bar, and compare your result with the experimental value of $164.1 \text{ J·K}^{-1}·\text{mol}^{-1}$.

This problem is like Example 20–6. We use Equation 20.45,

$$\bar{S} = \frac{5}{2}R + R \ln \left[\left(\frac{2\pi m k_B T}{h^2} \right)^{3/2} \frac{\bar{V}}{N_A} \right]$$

Assuming ideal behavior, at 298.2 K and one bar

$$\frac{N_A}{\bar{V}} = \frac{N_A P}{RT}$$

$$= \frac{(6.022 \times 10^{23} \text{ mol}^{-1})(1 \text{ bar})}{(0.08314 \text{ dm}^3 \cdot \text{bar·mol}^{-1} \cdot \text{K}^{-1})(298.2 \text{ K})}$$

$$= 2.429 \times 10^{22} \text{ dm}^{-3} = 2.429 \times 10^{25} \text{ m}^{-3}$$

and

$$\left(\frac{2\pi m k_B T}{h^2}\right)^{3/2} = \left[\frac{2\pi(0.08380 \text{ kg}\cdot\text{mol}^{-1})(1.3806 \times 10^{-23} \text{ J}\cdot\text{K}^{-1})(298.2 \text{ K})}{(6.022 \times 10^{23} \text{ mol}^{-1})(6.626 \times 10^{-34} \text{ J}\cdot\text{s})^2}\right]^{3/2}$$
$$= (8.199 \times 10^{21} \text{ m}^{-2})^{3/2} = 7.424 \times 10^{32} \text{ m}^{-3}$$

Then

$$\overline{S} = \frac{5}{2}R + 17.235R$$
$$= 164.1 \text{ J}\cdot\text{K}^{-1}\cdot\text{mol}^{-1}$$

This value is the same as the experimental value.

20–41. Use Equation 18.39 and the data in Table 18.2 to calculate the entropy of nitrogen at 298.2 K and one bar. Compare your result with the experimental value of 191.6 J·K^{-1}·mol^{-1}.

Recall from Chapter 18 that

$$Q = \frac{q^N}{N!}$$

Substituting into Equation 20.43 gives

$$S = Nk_B \ln q - k_B \ln N! + Nk_B T\left(\frac{\partial \ln q}{\partial T}\right)_V$$
$$= Nk_B \ln q - Nk_B \ln N + Nk_B + Nk_B T\left(\frac{\partial \ln q}{\partial T}\right)_V$$
$$= Nk_B + Nk_B \ln \frac{q}{N} + Nk_B T\left(\frac{\partial \ln q}{\partial T}\right)_V \tag{1}$$

For a diatomic ideal gas,

$$q = \left(\frac{2\pi M k_B T}{h^2}\right)^{3/2} V \frac{T}{\sigma\Theta_{\text{rot}}} \frac{e^{-\Theta_{\text{vib}}/2T}}{1 - e^{-\Theta_{\text{vib}}/T}} g_{e1} e^{D_e/k_B T} \tag{18.39}$$

Then

$$\ln \frac{q}{N} = \ln\left[\left(\frac{2\pi M k_B T}{h^2}\right)^{3/2} \frac{\overline{V}}{N_A}\right] + \ln \frac{T}{\sigma\Theta_{\text{rot}}} - \ln(1 - e^{-\Theta_{\text{vib}}/T})$$
$$- \frac{\Theta_{\text{vib}}}{2T} + \ln g_{e1} + \frac{D_e}{k_B T}$$

and

$$\left(\frac{\partial \ln q}{\partial T}\right) = \frac{3}{2T} + \frac{1}{T} + \frac{\Theta_{\text{vib}}}{2T^2} + \frac{(\Theta_{\text{vib}}/T^2)e^{-\Theta_{\text{vib}}/T}}{1 - e^{-\Theta_{\text{vib}}/T}} - \frac{D_e}{k_B T^2}$$

Substituting into Equation 1 above, we find that

$$\frac{\overline{S}}{R} = \frac{7}{2} + \ln\left[\left(\frac{2\pi M k_B T}{h^2}\right)^{3/2} \frac{\overline{V}}{N_A}\right] + \ln \frac{T}{\sigma\Theta_{\text{rot}}} - \ln(1 - e^{-\Theta_{\text{vib}}/T})$$
$$+ \frac{\Theta_{\text{vib}}/T}{e^{\Theta_{\text{vib}}/T} - 1} + \ln g_{e1}$$

For N_2, $\Theta_{vib} = 3374$ K, $\Theta_{rot} = 2.88$ K, $\sigma = 2$, and $g_{e1} = 1$. The various factors are as follows:

$$\left(\frac{2\pi M k_B T}{h^2}\right)^{3/2} = \left[\frac{2\pi(4.653 \times 10^{-26}\text{ kg})(1.381 \times 10^{-23}\text{ J·K}^{-1})(298.2\text{ K})}{(6.626 \times 10^{34}\text{ J·s})^2}\right]^{3/2}$$

$$= 1.435 \times 10^{32}\text{ m}^{-3}$$

$$\frac{\overline{V}}{N_A} = \frac{RT}{N_A P} = \frac{(0.08314\text{ dm}^3\text{·bar·mol}^{-1}\text{·K}^{-1})(298.2\text{ K})}{(6.022 \times 10^{23}\text{ mol}^{-1})(1\text{ bar})}$$

$$= 4.117 \times 10^{-23}\text{ dm}^3 = 4.117 \times 10^{-26}\text{ m}^3$$

$$\ln \frac{T}{\sigma \Theta_{rot}} = \ln \frac{298.2\text{ K}}{2(2.88\text{ K})} = 3.947$$

$$\ln(1 - e^{-\Theta_{vib}/T}) = -1.22 \times 10^{-5}$$

$$\frac{\Theta_{vib}/T}{e^{\Theta_{vib}/T} - 1} = \frac{11.31}{e^{11.31} - 1} = 1.380 \times 10^{-4}$$

The standard molar entropy is then

$$\frac{\overline{S}}{R} = 3.5 + 15.59 + 3.947 + 1.22 \times 10^{-5} + 1.380 \times 10^{-4}$$

$$= 23.04$$

This is 191.6 J·K^{-1}·mol^{-1}, which is also the experimental value.

20–42. Use Equation 18.57 and the data in Table 18.4 to calculate the entropy of $CO_2(g)$ at 298.2 K and one bar. Compare your result with the experimental value of 213.8 J·K^{-1}·mol^{-1}.

For a linear polyatomic ideal gas having three atoms,

$$q = \left(\frac{2\pi M k_B T}{h^2}\right)^{3/2} V \frac{T}{\sigma \Theta_{rot}} \left(\prod_{j=1}^{4} \frac{e^{-\Theta_{vib,j}/2T}}{1 - e^{-\Theta_{vib,j}/T}}\right) g_{e1} e^{D_e/k_B T} \tag{18.57}$$

Then

$$\ln \frac{q}{N} = \ln\left[\left(\frac{2\pi M k_B T}{h^2}\right)^{3/2} \frac{\overline{V}}{N_A}\right] + \ln \frac{T}{\sigma \Theta_{rot}} - \sum_{j=1}^{4} \ln(1 - e^{-\Theta_{vib,j}/T})$$

$$- \sum_{j=1}^{4} \frac{\Theta_{vib,j}}{2T} + \ln g_{e1} + \frac{D_e}{k_B T}$$

and

$$\left(\frac{\partial \ln q}{\partial T}\right) = \frac{3}{2T} + \frac{1}{T} + \sum_{j=1}^{4} \frac{\Theta_{vib,j}}{2T^2} + \sum_{j=1}^{4} \frac{(\Theta_{vib,j}/T^2)e^{-\Theta_{vib,j}/T}}{1 - e^{-\Theta_{vib,j}/T}} - \frac{D_e}{k_B T^2}$$

Substituting into Equation 1 from Problem 20–41, we find

$$\frac{\overline{S}}{R} = \frac{7}{2} + \ln\left[\left(\frac{2\pi M k_B T}{h^2}\right)^{3/2} \frac{\overline{V}}{N_A}\right] + \ln \frac{T}{\sigma \Theta_{rot}} - \sum_{j=1}^{4} \ln(1 - e^{-\Theta_{vib,j}/T})$$

$$+ \sum_{j=1}^{4} \left[\frac{(\Theta_{vib,j}/T)e^{-\Theta_{vib,j}/T}}{1 - e^{-\Theta_{vib,j}/T}}\right] + \ln g_{e1}$$

For CO_2, $\Theta_{vib,1} = 3360$ K, $\Theta_{vib,2} = \Theta_{vib,3} = 954$ K, $\Theta_{vib,4} = 1890$ K, $\Theta_{rot} = 0.561$ K, $\sigma = 2$, and $g_{e1} = 1$. The various factors are as follows:

$$\left(\frac{2\pi M k_B T}{h^2}\right)^{3/2} = \left[\frac{2\pi(7.308 \times 10^{-26} \text{ kg})(1.381 \times 10^{-23} \text{ J}\cdot\text{K}^{-1})(298.2 \text{ K})}{(6.626 \times 10^{34} \text{ J}\cdot\text{s})^2}\right]^{3/2}$$

$$= 2.825 \times 10^{32} \text{ m}^{-3}$$

$$\frac{\overline{V}}{N_A} = \frac{RT}{N_A P} = \frac{(0.08314 \text{ dm}^3\cdot\text{bar}\cdot\text{mol}^{-1}\cdot\text{K}^{-1})(298.2 \text{ K})}{(6.022 \times 10^{23} \text{ mol}^{-1})(1 \text{ bar})}$$

$$= 4.117 \times 10^{-23} \text{ dm}^3 = 4.117 \times 10^{-26} \text{ m}^3$$

$$\ln\frac{T}{\sigma\Theta_{rot}} = \ln\left[\frac{298.2 \text{ K}}{2(0.561 \text{ K})}\right] = 5.583$$

$$\sum_{j=1}^{4} \ln(1 - e^{-\Theta_{vib,j}/T}) = -0.08508$$

$$\sum_{j=1}^{4} \frac{\Theta_{vib,j}}{T}\left(\frac{e^{-\Theta_{vib,j}/T}}{1 - e^{-\Theta_{vib,j}/T}}\right) = 0.2835$$

The standard molar entropy is then

$$\frac{\overline{S}}{R} = 3.5 + 16.269 + 5.583 + 0.08508 + 0.2835$$

$$= 25.72$$

This is 213.8 $\text{J}\cdot\text{K}^{-1}\cdot\text{mol}^{-1}$, which is also the experimental value.

20–43. Use Equation 18.60 and the data in Table 18.4 to calculate the entropy of $NH_3(g)$ at 298.2 K and one bar. Compare your result with the experimental value of 192.8 $\text{J}\cdot\text{K}^{-1}\cdot\text{mol}^{-1}$.

For a nonlinear polyatomic ideal gas having four atoms,

$$q = \left(\frac{2\pi M k_B T}{h^2}\right)^{3/2} V \frac{\pi^{1/2}}{\sigma}\left(\frac{T}{\Theta_{rot,A}\Theta_{rot,B}\Theta_{rot,C}}\right)^{1/2}\left(\prod_{j=1}^{6}\frac{e^{-\Theta_{vib,j}/2T}}{1 - e^{-\Theta_{vib,j}/T}}\right)g_{e1}e^{D_e/k_B T} \quad (18.60)$$

Then

$$\ln\frac{q}{N} = \ln\left[\left(\frac{2\pi M k_B T}{h^2}\right)^{3/2}\frac{\overline{V}}{N_A}\right] + \ln\frac{\pi^{1/2}}{\sigma} + \frac{1}{2}\ln\left(\frac{T^3}{\Theta_{rot,A}\Theta_{rot,B}\Theta_{rot,C}}\right)$$

$$- \sum_{j=1}^{6}\ln(1 - e^{-\Theta_{vib,j}/T}) - \sum_{j=1}^{6}\frac{\Theta_{vib,j}}{2T} + \ln g_{e1} + \frac{D_e}{k_B T}$$

and

$$\left(\frac{\partial \ln q}{\partial T}\right) = \frac{3}{2T} + \frac{3}{2T} + \sum_{j=1}^{6}\frac{\Theta_{vib,j}}{2T^2} + \sum_{j=1}^{6}\frac{(\Theta_{vib,j}/T^2)e^{-\Theta_{vib,j}/T}}{1 - e^{-\Theta_{vib,j}/T}} - \frac{D_e}{k_B T^2}$$

Substituting into Equation 1 from Problem 20–41, we find

$$\frac{\overline{S}}{R} = 4 + \ln\left[\left(\frac{2\pi M k_B T}{h^2}\right)^{3/2}\frac{\overline{V}}{N_A}\right] + \ln\frac{\pi^{1/2}}{\sigma} + \frac{1}{2}\ln\frac{T^3}{\Theta_{rot,A}\Theta_{rot,B}\Theta_{rot,C}}$$

$$-\sum_{j=1}^{6} \ln(1 - e^{-\Theta_{vib,j}/T}) + \sum_{j=1}^{6}\left[\frac{(\Theta_{vib,j}/T)e^{-\Theta_{vib,j}/T}}{1 - e^{-\Theta_{vib,j}/T}}\right] + \ln g_{e1}$$

For NH_3, $\Theta_{vib,1} = 4800$ K, $\Theta_{vib,2} = 1360$ K, $\Theta_{vib,3} = \Theta_{vib,4} = 4880$ K, $\Theta_{vib,5} = \Theta_{vib,6} = 2330$ K, $\Theta_{rot,A} = \Theta_{rot,B} = 13.6$ K, $\Theta_{rot,C} = 8.92$ K, $\sigma = 3$, and $g_{e1} = 1$. The various factors are as follows:

$$\left(\frac{2\pi M k_B T}{h^2}\right)^{3/2} = \left[\frac{2\pi(2.828 \times 10^{-26} \text{ kg})(1.381 \times 10^{-23} \text{ J·K}^{-1})(298.2 \text{ K})}{(6.626 \times 10^{34} \text{ J·s})^2}\right]^{3/2}$$
$$= 6.801 \times 10^{31} \text{ m}^{-3}$$

$$\frac{\overline{V}}{N_A} = \frac{RT}{N_A P} = \frac{(0.08314 \text{ dm}^3 \cdot \text{bar·mol}^{-1} \cdot \text{K}^{-1})(298.2 \text{ K})}{(6.022 \times 10^{23} \text{ mol}^{-1})(1 \text{ bar})}$$
$$= 4.117 \times 10^{-23} \text{ dm}^3 = 4.117 \times 10^{-26} \text{ m}^3$$

$$\frac{1}{2}\ln\frac{T^3}{\Theta_{rot,A}\Theta_{rot,B}\Theta_{rot,C}} = 4.842$$

$$\sum_{j=1}^{6}\ln(1 - e^{-\Theta_{vib,j}/T}) = -0.01132$$

$$\sum_{j=1}^{6}\frac{\Theta_{vib,j}}{T}\left(\frac{e^{-\Theta_{vib,j}/T}}{1 - e^{-\Theta_{vib,j}/T}}\right) = 0.05451$$

The standard molar entropy is then

$$\frac{\overline{S}}{R} = 4 + 14.845 - 0.5262 + 4.842 + 0.01132 + 0.05451$$
$$= 23.23$$

This is 193.1 J·K^{-1}·mol^{-1}. The slight disagreement with the experimental value is due to our use of the rigid rotator-harmonic oscillator model.

20–44. Derive Equation 20.35.

The maximum efficiency is defined as

$$\text{maxmimum efficiency} = \frac{-w}{q_{rev,h}} = \frac{q_{rev,h} + q_{rev,c}}{q_{rev,h}} = 1 + \frac{q_{rev,c}}{q_{rev,h}}$$

Because the process is cyclic and reversible, $\Delta S_{engine} = 0$, and so (as in Equation 20.34)

$$q_{rev,c} = -q_{rev,h}\frac{T_c}{T_h}$$

The efficiency becomes

$$\text{maximum efficiency} = 1 - \frac{T_c}{T_h} = \frac{T_h - T_c}{T_h} \tag{20.35}$$

20–45. The boiling point of water at a pressure of 25 atm is 223°C. Compare the theoretical efficiencies of a steam engine operating between 20°C and the boiling point of water at 1 atm and at 25 atm.

At one atmosphere, using Equation 20.35 gives an efficiency of

$$\text{efficiency} = 1 - \frac{293}{373} = 21\%$$

At 25 atm, the same engine will give an efficiency of

$$\text{efficiency} = 1 - \frac{293}{496} = 41\%$$

CHAPTER 21

Entropy and the Third Law of Thermodynamics

21–1. Form the total derivative of H as a function of T and P and equate the result to dH in Equation 21.6 to derive Equations 21.7 and 21.8.

The total derivative of $S(T, P)$ is

$$dS = \left(\frac{\partial S}{\partial T}\right)_P dT + \left(\frac{\partial S}{\partial P}\right)_T dP$$

We can substitute this in Equation 21.6 to obtain

$$dH = TdS + VdP$$

$$dH = T\left(\frac{\partial S}{\partial T}\right)_P dT + T\left(\frac{\partial S}{\partial P}\right)_T dP + VdP$$

$$= T\left(\frac{\partial S}{\partial T}\right)_P dT + \left[V + T\left(\frac{\partial S}{\partial P}\right)_T\right]dP \tag{1}$$

We now write the total derivative of $H(T, P)$ as

$$dH = \left(\frac{\partial H}{\partial T}\right)_P dT + \left(\frac{\partial H}{\partial P}\right)_T dP$$

$$dH = C_P dT + \left(\frac{\partial H}{\partial P}\right)_T dP \tag{2}$$

Set the coefficients of dT in Equations 1 and 2 equal to each other to find Equation 21.7

$$\left(\frac{\partial S}{\partial T}\right)_P = \frac{C_P}{T}$$

and set the coefficients of dP equal to each other to obtain Equation 21.8:

$$\left[V + T\left(\frac{\partial S}{\partial P}\right)_T\right] = \left(\frac{\partial H}{\partial P}\right)_T$$

$$\left(\frac{\partial S}{\partial P}\right)_T = \frac{1}{T}\left[\left(\frac{\partial H}{\partial P}\right)_T - V\right] \tag{21.8}$$

21–2. The molar heat capacity of $H_2O(l)$ has an approximately constant value of $\overline{C}_P = 75.4\,\text{J·K}^{-1}\text{·mol}^{-1}$ from 0°C to 100°C. Calculate ΔS if two moles of $H_2O(l)$ are heated from 10°C to 90°C at constant pressure.

We use Equation 21.9 to write

$$\Delta S = \int_{T_1}^{T_2} \frac{n\overline{C}_P}{T} dT = \int_{283\ K}^{363\ K} \frac{(2\ \text{mol})(75.4\ \text{J·K}^{-1}\text{·mol}^{-1})}{T} dT$$

$$= (150.8\ \text{J·K}^{-1}) \ln \frac{363}{283} = 37.5\ \text{J·K}^{-1}$$

21–3. The molar heat capacity of butane can be expressed by

$$\overline{C}_P/R = 0.05641 + (0.04631\ \text{K}^{-1})T - (2.392 \times 10^{-5}\ \text{K}^{-2})T^2 + (4.807 \times 10^{-9}\ \text{K}^{-3})T^3$$

over the temperature range $300\ \text{K} \leq T \leq 1500\ \text{K}$. Calculate ΔS if one mole of butane is heated from 300 K to 1000 K at constant pressure.

We can use Equation 21.9 to write

$$\Delta S = \int_{T_1}^{T_2} \frac{n\overline{C}_P}{T} dT$$

$$= nR \int_{300\ K}^{1000\ K} \left[\frac{0.05641}{T} + 0.04631\ \text{K}^{-1} - (2.392 \times 10^{-5}\ \text{K}^{-2})T \right.$$

$$\left. + (4.807 \times 10^{-9}\ \text{K}^{-3})T^2 \right] dT$$

$$= (23.16R)(1\ \text{mol}) = 192.6\ \text{J·K}^{-1}$$

21–4 . The molar heat capacity of $C_2H_4(g)$ can be expressed by

$$\overline{C}_V(T)/R = 16.4105 - \frac{6085.929\ \text{K}}{T} + \frac{822\,826\ \text{K}^2}{T^2}$$

over the temperature range $300\ \text{K} < T < 1000\ \text{K}$. Calculate ΔS if one mole of ethene is heated from 300 K to 600 K at constant volume.

We can use Equation 21.5 to write

$$\Delta S = \int_{T_1}^{T_2} \frac{n\overline{C}_V}{T} dT$$

$$= nR \int_{300\ K}^{600\ K} \left[\frac{16.4105}{T} - \frac{6085.929\ \text{K}}{T^2} + \frac{822\,826\ \text{K}^2}{T^3} \right] dT$$

$$= (4.660R)(1\ \text{mol}) = 38.75\ \text{J·K}^{-1}$$

21–5. Use the data in Problem 21–4 to calculate ΔS if one mole of ethene is heated from 300 K to 600 K at constant pressure. Assume ethene behaves ideally.

For an ideal gas, $\overline{C}_P - \overline{C}_V = R$, so we can express $\overline{C}_P$ as

$$\overline{C}_P/R = 1 + 16.4105 - \frac{6085.929\ \text{K}}{T} + \frac{822\,826\ \text{K}^2}{T^2}$$

Then we can use Equation 21.9 to calculate ΔS:

$$\Delta S = \int_{T_1}^{T_2} \frac{n\overline{C}_P}{T} dT$$

$$= nR \int_{300 \text{ K}}^{600 \text{ K}} \left[\frac{17.4105}{T} - \frac{6085.929 \text{ K}}{T^2} + \frac{822\,826 \text{ K}^2}{T^3} \right] dT$$

$$= (5.353R)(1 \text{ mol}) = 44.51 \text{ J} \cdot \text{K}^{-1}$$

21–6. We can calculate the difference in the results of Problems 21–4 and 21–5 in the following way. First, show that because $\overline{C}_P - \overline{C}_V = R$ for an ideal gas,

$$\Delta \overline{S}_P = \Delta \overline{S}_V + R \ln \frac{T_2}{T_1}$$

Check to see numerically that your answers to Problems 21–4 and 21–5 differ by $R \ln 2 = 0.693R = 5.76 \text{ J} \cdot \text{K}^{-1} \cdot \text{mol}^{-1}$.

For an ideal gas, $\overline{C}_P - \overline{C}_V = R$. Equations 21.6 and 21.9 state that

$$\Delta \overline{S}_V = \int_{T_1}^{T_2} \frac{\overline{C}_V}{T} dT \qquad \text{and} \qquad \Delta \overline{S}_P = \int_{T_1}^{T_2} \frac{\overline{C}_P}{T} dT$$

Subtracting $\Delta \overline{S}_P$ from $\Delta \overline{S}_V$ gives

$$\Delta \overline{S}_P - \Delta \overline{S}_V = \int_{T_1}^{T_2} \frac{\overline{C}_P - \overline{C}_V}{T} dT = \int_{T_1}^{T_2} \frac{R}{T} dT = R \ln \frac{T_2}{T_1}$$

and so $\Delta \overline{S}_P$ can be written as

$$\Delta \overline{S}_P = \Delta \overline{S}_V + R \ln \frac{T_2}{T_1}$$

The answers to Problems 21–4 and 21–5 differ by $R \ln 2$, as required.

21–7. The results of Problems 21–4 and 21–5 must be connected in the following way. Show that the two processes can be represented by the diagram

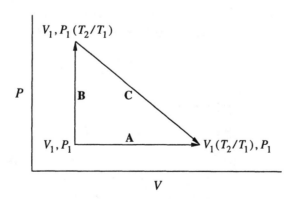

where paths A and B represent the processes in Problems 21–5 and 21–4, respectively.

Now, path A is equivalent to the sum of paths B and C. Show that ΔS_C is given by

$$\Delta S_C = nR \ln \frac{V_1 \left(\dfrac{T_2}{T_1} \right)}{V_1} = nR \ln \frac{P_1 \left(\dfrac{T_2}{T_1} \right)}{P_1} = nR \ln \frac{T_2}{T_1}$$

and that the result given in Problem 21–6 follows.

In Problem 21–4, ethane is heated at constant volume, so (assuming ideal behavior)

$$P_2 = P_1 \left(\frac{T_2}{T_1} \right)$$

Likewise, in Problem 21–5, the system is kept at constant pressure and so (assuming ideal behavior)

$$V_2 = V_1 \left(\frac{T_2}{T_1} \right)$$

These values correspond to those shown in the diagram. Now, path C represents an isothermal process. Because we are assuming ideal behavior, $dU = 0$, which means that $\delta q_{rev} = -\delta w = nRT/V \, dV$. Then we can write ΔS_C as (Equation 20.22)

$$\Delta S_C = \int \frac{\delta q_{rev}}{T} = nR \ln \frac{V_2}{V_1} = R \ln \frac{V_1(T_2/T_1)}{V_1} = R \ln \frac{T_2}{T_1}$$

Note that path A is equivalent to the sums of paths B and C, so $\Delta S_A = \Delta S_B + \Delta S_C$. Because $\Delta S_A = \Delta S_P$ and $\Delta S_B = \Delta S_V$, we can write

$$\Delta S_C = \Delta S_P - \Delta S_V$$

and the result given in Problem 21–6 follows.

21–8. Use Equations 20.23 and 20.24 to show that $S = 0$ at 0 K, where every system will be in its ground state.

We begin with Equations 20.23 and 20.24,

$$W = \frac{A}{a_1! a_2! \dots} \qquad \text{and} \qquad S = k_B \ln W$$

Let a_1 represent the ground state, so all other $a_j = 0$ when the system is in the ground state. Then $A = \sum_j a_j = a_1 + 0 = a_1$, and Equation 20.23 becomes

$$W = \frac{a_1!}{a_1!} = 1$$

Substitute this into Equation 20.24 for S to find

$$S = k_B \ln 1 = 0$$

21–9. Prove that $S = -k \sum p_j \ln p_j = 0$ when $p_1 = 1$ and all the other $p_j = 0$. In other words, prove that $x \ln x \to 0$ as $x \to 0$.

Let $p_1 = 1$ and all other $p_j = 0$. Then Equation 20.40 becomes

$$S = -k_B \sum p_j \ln p_j = 0 - k_B \sum x \ln x$$

where $x \to 0$. In Problem J–8, we proved that $x \ln x \to 0$ as $x \to 0$, so $S = 0 - 0 = 0$ under the conditions given.

21–10. It has been found experimentally that $\Delta_{vap}\overline{S} \approx 88 \ \mathrm{J \cdot K^{-1} \cdot mol^{-1}}$ for many nonassociated liquids. This rough rule of thumb is called *Trouton's rule*. Use the following data to test the validity of Trouton's rule.

Substance	$t_{fus}/°C$	$t_{vap}/°C$	$\Delta_{fus}\overline{H}/\mathrm{kJ \cdot mol^{-1}}$	$\Delta_{vap}\overline{H}/\mathrm{kJ \cdot mol^{-1}}$
Pentane	−129.7	36.06	8.42	25.79
Hexane	−95.3	68.73	13.08	28.85
Heptane	−90.6	98.5	14.16	31.77
Ethylene oxide	−111.7	10.6	5.17	25.52
Benzene	5.53	80.09	9.95	30.72
Diethyl ether	−116.3	34.5	7.27	26.52
Tetrachloromethane	−23	76.8	3.28	29.82
Mercury	−38.83	356.7	2.29	59.11
Bromine	−7.2	58.8	10.57	29.96

Use Equation 21.16,

$$\Delta_{vap}\overline{S} = \frac{\Delta_{vap}\overline{H}}{T_{vap}}$$

to construct a table of values of $\Delta_{vap}\overline{S}$.

Substance	$\Delta_{vap}\overline{S}/\mathrm{J \cdot mol^{-1} \cdot K^{-1}}$
Pentane	83.41
Hexane	84.39
Heptane	85.5
Ethylene oxide	89.9
Benzene	86.97
Diethyl ether	86.2
Tetrachloromethane	85.2
Mercury	93.85
Bromine	90.3

21–11. Use the data in Problem 21–10 to calculate the value of $\Delta_{fus}\overline{S}$ for each substance.

Use Equation 21.16,

$$\Delta_{fus}\overline{S} = \frac{\Delta_{fus}\overline{H}}{T_{fus}}$$

to construct a table of values of $\Delta_{fus}\overline{S}$.

Substance	$\Delta_{fus}\overline{S}/J\cdot mol^{-1}\cdot K^{-1}$
Pentane	58.7
Hexane	73.5
Heptane	77.6
Ethylene oxide	32.0
Benzene	35.7
Diethyl ether	46.3
Tetrachloromethane	13
Mercury	9.77
Bromine	40

21–12. Why is $\Delta_{vap}\overline{S} > \Delta_{fus}\overline{S}$?

$\Delta_{vap}\overline{S} \gg \Delta_{fus}\overline{S}$ because gases are essentially completely unordered; the molecules of a gas travel more or less randomly within the gas's container. Liquids, however, are much more cohesive and structured, and solids are very structured. The difference between the entropy of a liquid and that of a solid is less than the difference between the entropy of a liquid and that of a gas.

21–13. Show that if $C_P^s(T) \to T^\alpha$ as $T \to 0$, where α is a positive constant, then $S(T) \to 0$ as $T \to 0$.

We assume in the statement of the problem that

$$\lim_{T\to 0} C_P^s(T) = T^\alpha$$

where α is a positive constant. Then express S using Equation 21.10 and take the limit of S as $T \to 0$:

$$\lim_{T\to 0} S(T) = S(0\,K) + \lim_{T\to 0}\int_0^T \frac{C_P(T)}{T}dT = \lim_{T\to 0}\int_0^T \frac{T^\alpha}{T}dT = \lim_{T\to 0}\frac{T^\alpha}{T} = 0$$

as long as $S(0\,K) = 0$ and $\alpha > 0$ (as stipulated in the statement of the problem).

21–14. Use the following data to calculate the standard molar entropy of $N_2(g)$ at 298.15 K.

$$C_P^\circ[N_2(s_1)]/R = -0.03165 + (0.05460 \text{ K}^{-1})T + (3.520 \times 10^{-3} \text{ K}^{-2})T^2$$
$$- (2.064 \times 10^{-5} \text{ K}^{-3})T^3$$
$$10 \text{ K} \leq T \leq 35.61 \text{ K}$$

$$C_P^\circ[N_2(s_2)]/R = -0.1696 + (0.2379 \text{ K}^{-1})T - (4.214 \times 10^{-3} \text{ K}^{-2})T^2$$
$$+ (3.036 \times 10^{-5} \text{ K}^{-3})T^3$$
$$35.61 \text{ K} \leq T \leq 63.15 \text{ K}$$

$$C_P^\circ[N_2(l)]/R = -18.44 + (1.053 \text{ K}^{-1})T - (0.0148 \text{ K}^{-2})T^2$$
$$+ (7.064 \times 10^{-5} \text{ K}^{-3})T^3$$
$$63.15 \text{ K} \leq T \leq 77.36 \text{ K}$$

$C_P^\circ[N_2(g)]/R = 3.500$ from $77.36 \text{ K} \leq T \leq 1000 \text{ K}$, $C_p(T = 10.0 \text{ K}) = 6.15 \text{ J·K}^{-1}\text{·mol}^{-1}$, $T_{trs} = 35.61 \text{ K}$, $\Delta_{trs}\overline{H} = 0.2289 \text{ kJ·mol}^{-1}$, $T_{fus} = 63.15 \text{ K}$, $\Delta_{fus}\overline{H} = 0.71 \text{ kJ·mol}^{-1}$, $T_{vap} = 77.36 \text{ K}$, and $\Delta_{vap}\overline{H} = 5.57 \text{ kJ·mol}^{-1}$. The correction for nonideality (Problem 22–20) = 0.02 J·K^{-1}·mol^{-1}.

The easiest way to do this series of problems is to input the given data into a program like *Mathematica* and use Equation 21.17. For temperatures below the minimum value for which the formulae provided are valid, we can use the expression from Example 21–3, $\overline{S}(T) = \overline{C}_p(T)/3$. Here, we solve the formula

$$\overline{S}(T) = \frac{\overline{C}_P(10 \text{ K})}{3} + \int_{10}^{35.61} \frac{\overline{C}_P[N_2(s_1)]}{T}dT + \frac{\Delta_{trs}\overline{H}}{35.61 \text{ K}} + \int_{35.61}^{63.15} \frac{\overline{C}_P[N_2(s_2)]}{T}dT$$
$$+ \frac{\Delta_{fus}\overline{H}}{63.15 \text{ K}} + \int_{63.15}^{77.36} \frac{\overline{C}_P[N_2(l)]}{T}dT + \frac{\Delta_{vap}\overline{H}}{77.36 \text{ K}}$$
$$+ \int_{77.36}^{298.15} \frac{\overline{C}_P[N_2(g)]}{T}dT + \text{correction}$$

$$= 2.05 \text{ J·mol}^{-1}\text{·K}^{-1} + 25.86 \text{ J·mol}^{-1}\text{·K}^{-1} + 6.428 \text{ J·mol}^{-1}\text{·K}^{-1}$$
$$+ 23.41 \text{ J·mol}^{-1}\text{·K}^{-1} + 11.24 \text{ J·mol}^{-1}\text{·K}^{-1} + 11.78 \text{ J·mol}^{-1}\text{·K}^{-1}$$
$$+ 72.00 \text{ J·mol}^{-1}\text{·K}^{-1} + 39.26 \text{ J·mol}^{-1}\text{·K}^{-1} + 0.02 \text{ J·mol}^{-1}\text{·K}^{-1}$$

$$= 192.05 \text{ J·K}^{-1}\text{·mol}^{-1}$$

The literature value of the standard molar entropy is 191.6 J·K^{-1}·mol^{-1}. The slight discrepancy between these two values reflects the use of the ideal expression for $\overline{C}_P[N_2(g)]$. (Using the $\overline{C}_P[N_2(g)]$ that is given in the next problem, which is linear in T, gives a standard molar entropy of 191.04 J·mol^{-1}·K^{-1}.)

21–15. Use the data in Problem 21–14 and $\overline{C}_P[N_2(g)]/R = 3.307 + (6.29 \times 10^{-4} \text{ K}^{-1})T$ for $T \geq 77.36$ K to plot the standard molar entropy of nitrogen as a function of temperature from 0 K to 1000 K.

The function which describes the standard molar entropy of nitrogen must be defined differently for each phase and phase transition. Notice that the correction factor must be added to all functions to correct for nonideality.

From 0 K to 10 K,

$$\overline{S}(T) = \frac{\overline{C}_P(10 \text{ K})}{3} + \text{corr}$$

From 10 K to 35.61 K,

$$\overline{S}(T) = \frac{\overline{C}_P(10\ \text{K})}{3} + \int_{10}^{T} \frac{\overline{C}_P[\text{N}_2(\text{s}_1)]}{T} dT + \text{corr}$$

At 35.61 K,

$$\overline{S}(T) = \frac{\overline{C}_P(10\ \text{K})}{3} + \int_{10}^{35.61} \frac{\overline{C}_P[\text{N}_2(\text{s}_1)]}{T} dT + \frac{\Delta_{\text{trs}}\overline{H}}{35.61\ \text{K}} + \text{corr}$$

From 35.61 K to 63.15 K,

$$\overline{S}(T) = \frac{\overline{C}_P(10\ \text{K})}{3} + \int_{10}^{35.61} \frac{\overline{C}_P[\text{N}_2(\text{s}_1)]}{T} dT + \frac{\Delta_{\text{trs}}\overline{H}}{35.61\ \text{K}} + \int_{35.61}^{T} \frac{\overline{C}_P[\text{N}_2(\text{s}_2)]}{T} dT + \text{corr}$$

At 63.15 K,

$$\overline{S}(T) = \frac{\overline{C}_P(10\ \text{K})}{3} + \int_{10}^{35.61} \frac{\overline{C}_P[\text{N}_2(\text{s}_1)]}{T} dT + \frac{\Delta_{\text{trs}}\overline{H}}{35.61\ \text{K}} + \int_{35.61}^{63.15} \frac{\overline{C}_P[\text{N}_2(\text{s}_2)]}{T} dT$$
$$+ \frac{\Delta_{\text{fus}}\overline{H}}{63.15\ \text{K}} + \text{corr}$$

From 63.15 K to 77.36 K,

$$\overline{S}(T) = \frac{\overline{C}_P(10\ \text{K})}{3} + \int_{10}^{35.61} \frac{\overline{C}_P[\text{N}_2(\text{s}_1)]}{T} dT + \frac{\Delta_{\text{trs}}\overline{H}}{35.61\ \text{K}} + \int_{35.61}^{63.15} \frac{\overline{C}_P[\text{N}_2(\text{s}_2)]}{T} dT$$
$$+ \frac{\Delta_{\text{fus}}\overline{H}}{63.15\ \text{K}} + \int_{63.15}^{T} \frac{\overline{C}_P[\text{N}_2(\text{l})]}{T} dT + \text{corr}$$

At 77.36 K,

$$\overline{S}(T) = \frac{\overline{C}_P(10\ \text{K})}{3} + \int_{10}^{35.61} \frac{\overline{C}_P[\text{N}_2(\text{s}_1)]}{T} dT + \frac{\Delta_{\text{trs}}\overline{H}}{35.61\ \text{K}} + \int_{35.61}^{63.15} \frac{\overline{C}_P[\text{N}_2(\text{s}_2)]}{T} dT$$
$$+ \frac{\Delta_{\text{fus}}\overline{H}}{63.15\ \text{K}} + \int_{63.15}^{77.36} \frac{\overline{C}_P[\text{N}_2(\text{l})]}{T} dT + \frac{\Delta_{\text{vap}}\overline{H}}{77.36\ \text{K}} + \text{corr}$$

From 77.36 K to 1000 K,

$$\overline{S}(T) = \frac{\overline{C}_P(10\ \text{K})}{3} + \int_{10}^{35.61} \frac{\overline{C}_P[\text{N}_2(\text{s}_1)]}{T} dT + \frac{\Delta_{\text{trs}}\overline{H}}{35.61\ \text{K}} + \int_{35.61}^{63.15} \frac{\overline{C}_P[\text{N}_2(\text{s}_2)]}{T} dT$$
$$+ \frac{\Delta_{\text{fus}}\overline{H}}{63.15\ \text{K}} + \int_{63.15}^{77.36} \frac{\overline{C}_P[\text{N}_2(\text{l})]}{T} dT + \frac{\Delta_{\text{vap}}\overline{H}}{77.36\ \text{K}} + \int_{77.36}^{T} \frac{\overline{C}_P[\text{N}_2(\text{g})]}{T} dT + \text{corr}$$

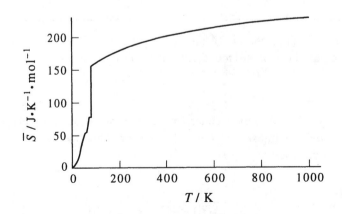

21–16. The molar heat capacities of solid, liquid, and gaseous chlorine can be expressed as

$$C_P^\circ[Cl_2(s)]/R = -1.545 + (0.1502 \text{ K}^{-1})T - (1.179 \times 10^{-3} \text{ K}^{-2})T^2$$
$$+ (3.441 \times 10^{-6} \text{ K}^{-3})T^3$$
$$15 \text{ K} \leq T \leq 172.12 \text{ K}$$

$$C_P^\circ[Cl_2(l)]/R = 7.689 + (5.582 \times 10^{-3} \text{ K}^{-1})T - (1.954 \times 10^{-5} \text{ K}^{-2})T^2$$
$$172.12 \text{ K} \leq T \leq 239.0 \text{ K}$$

$$C_P^\circ[Cl_2(g)]/R = 3.812 + (1.220 \times 10^{-3} \text{ K}^{-1})T - (4.856 \times 10^{-7} \text{ K}^{-2})T^2$$
$$239.0 \text{ K} \leq T \leq 1000 \text{ K}$$

Use the above molar heat capacities and $T_{fus} = 172.12$ K, $\Delta_{fus}\overline{H} = 6.406$ kJ·mol^{-1}, $T_{vap} = 239.0$ K, $\Delta_{vap}\overline{H} = 20.40$ kJ·mol^{-1}, and $\Theta_D = 116$ K. The correction for nonideality $= 0.502$ J·K^{-1}·mol^{-1} to calculate the standard molar entropy of chlorine at 298.15 K. Compare your result with the value given in Table 21.2.

$$\overline{S}(T) = \int_0^{15} \frac{12\pi^4}{5T} R \left(\frac{T}{\Theta_D}\right)^3 dT + \int_{15}^{172.12} \frac{\overline{C}_P[Cl_2(s)]}{T} dT + \frac{\Delta_{fus}\overline{H}}{172.12 \text{ K}}$$
$$+ \int_{172.12}^{239.0} \frac{\overline{C}_P[Cl_2(l)]}{T} dT + \frac{\Delta_{vap}\overline{H}}{239.0 \text{ K}} + \int_{239.0}^{298} \frac{\overline{C}_P[Cl_2(g)]}{T} dT + \text{correction}$$
$$= 1.401 \text{ J·K}^{-1}\cdot\text{mol}^{-1} + 69.37 \text{ J·K}^{-1}\cdot\text{mol}^{-1} + 37.22 \text{ J·K}^{-1}\cdot\text{mol}^{-1}$$
$$+21.86 \text{ J·K}^{-1}\cdot\text{mol}^{-1} + 85.36 \text{ J·K}^{-1}\cdot\text{mol}^{-1} + 7.54 \text{ J·K}^{-1}\cdot\text{mol}^{-1}$$
$$+0.502 \text{ J·K}^{-1}\cdot\text{mol}^{-1}$$
$$= 223.2 \text{ J·K}^{-1}\cdot\text{mol}^{-1}$$

The result is extremely close to the value of 223.1 J·K^{-1}·mol^{-1} found in Table 21.2.

21–17. Use the data in Problem 21–16 to plot the standard molar entropy of chlorine as a function of temperature from 0 K to 1000 K.

Do this in the same manner as Problem 21–15, using the appropriate values from Problem 21–16 and changing the limits of integration as required.

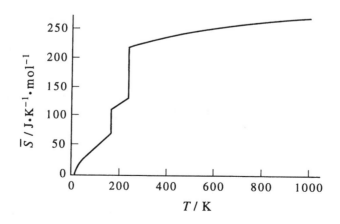

21–18. Use the following data to calculate the standard molar entropy of cyclopropane at 298.1 K.

$$C_P^\circ[C_3H_6(s)]/R = -1.921 + (0.1508\ K^{-1})T - (9.670 \times 10^{-4}\ K^{-2})T^2$$
$$+ (2.694 \times 10^{-6}\ K^{-3})T^3$$
$$15\ K \le T \le 145.5\ K$$

$$C_P^\circ[C_3H_6(l)]/R = 5.624 + (4.493 \times 10^{-2}\ K^{-1})T - (1.340 \times 10^{-4}\ K^{-2})T^2$$
$$145.5\ K \le T \le 240.3\ K$$

$$C_P^\circ[C_3H_6(g)]/R = -1.793 + (3.277 \times 10^{-2}\ K^{-1})T - (1.326 \times 10^{-5}\ K^{-2})T^2$$
$$240.3\ K \le T \le 1000\ K$$

$T_{fus} = 145.5$ K, $T_{vap} = 240.3$ K, $\Delta_{fus}\overline{H} = 5.44$ kJ·mol^{-1}, $\Delta_{vap}\overline{H} = 20.05$ kJ·mol^{-1}, and $\Theta_D = 130$ K. The correction for nonideality $= 0.54$ J·K^{-1}·mol^{-1}.

$$\overline{S}(T) = \int_0^{15} \frac{12\pi^4}{5T} R \left(\frac{T}{\Theta_D}\right)^3 dT + \int_{15}^{145.5} \frac{\overline{C}_P[C_3H_6(s)]}{T} dT + \frac{\Delta_{fus}\overline{H}}{145.5\ K}$$

$$+ \int_{145.5}^{240.3} \frac{\overline{C}_P[C_3H_6(l)]}{T} dT + \frac{\Delta_{vap}\overline{H}}{240.3\ K} + \int_{240.3}^{298.1} \frac{\overline{C}_P[C_3H_6(g)]}{T} dT$$

$$+ \text{correction}$$

$$= 0.995\ \text{J·K}^{-1}\text{·mol}^{-1} + 66.1\ \text{J·K}^{-1}\text{·mol}^{-1} + 37.4\ \text{J·K}^{-1}\text{·mol}^{-1}$$

$$+ 38.5\ \text{J·K}^{-1}\text{·mol}^{-1} + 83.4\ \text{J·K}^{-1}\text{·mol}^{-1} + 10.8\ \text{J·K}^{-1}\text{·mol}^{-1}$$

$$+ 0.54\ \text{J·K}^{-1}\text{·mol}^{-1}$$

$$= 237.8\ \text{J·K}^{-1}\text{·mol}^{-1}$$

This compares very well with the literature value of 237.5 J·K^{-1}·mol^{-1}.

21–19. Use the data in Problem 21–18 to plot the standard molar entropy of cyclopropane from 0 K to 1000 K.

Do this in the same manner as Problem 21–15, using the appropriate values from Problem 21–18 and changing the limits of integration as required.

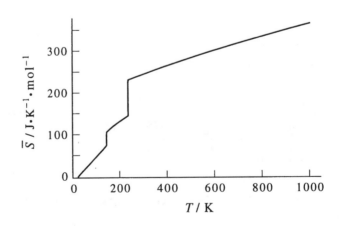

21–20. The constant-pressure molar heat capacity of N_2O as a function of temperature is tabulated below. Dinitrogen oxide melts at 182.26 K with $\Delta_{fus}\overline{H} = 6.54$ kJ·mol^{-1}, and boils at 184.67 K with $\Delta_{vap}\overline{H} = 16.53$ kJ·mol^{-1} at one bar. Assuming the heat capacity of solid dinitrogen oxide can be described by the Debye theory up to 15 K, calculate the molar entropy of $N_2O(g)$ at its boiling point.

T/K	$\overline{C}_P/\text{J·K}^{-1}\text{·mol}^{-1}$	T/K	$\overline{C}_P/\text{J·K}^{-1}\text{·mol}^{-1}$
15.17	2.90	120.29	45.10
19.95	6.19	130.44	47.32
25.81	10.89	141.07	48.91
33.38	16.98	154.71	52.17
42.61	23.13	164.82	54.02
52.02	28.56	174.90	56.99
57.35	30.75	180.75	58.83
68.05	34.18	182.26	Melting point
76.67	36.57	183.55	77.70
87.06	38.87	183.71	77.45
98.34	41.13	184.67	Boiling point
109.12	42.84		

We can do this problem in the same way we did Problems 21–14, 21–16, and 21–18. Because we are not given equations for the molar heat capacity, we can graph the heat capacity of the solid and liquid dinitrogen oxide, find a best-fit line, and use this to calculate the molar entropy of N_2O at the boiling point.

For solid dinitrogen oxide, a best-fit line gives the equation

$$\overline{C}_P[N_2O(s)]/\text{J·K}^{-1}\text{·mol}^{-1} = -13.153 + (1.1556\ \text{K}^{-1})T - (8.3372 \times 10^{-3}\ \text{K}^{-2})T^2 \\ + (2.3026 \times 10^{-5}\ \text{K}^{-3})T^3$$

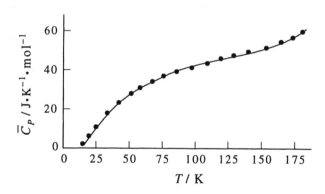

And for liquid dinitrogen oxide (with only two points), a line drawn between those two points has the equation

$$\overline{C}_P[N_2O(l)]/\text{J·K}^{-1}\text{·mol}^{-1} = 364.49 - (1.5625\ \text{K}^{-1})T$$

Note that, although we are given only two data points, the temperature varies by only 2° for dinitrogen oxide. From the Debye theory (Example 21–3) we can write the low temperature entropy as

$$\overline{S}(15\ \text{K}) = \frac{\overline{C}_P(15\ \text{K})}{3} \qquad 0 < T \le 15\ \text{K}$$

Now we can substitute into Equation 21.17, as before.

$$\overline{S}(T) = \frac{\overline{C}_P(15\text{ K})}{3} + \int_{15}^{182.26} \frac{\overline{C}_P[N_2O(s)]}{T}dT + \frac{\Delta_{fus}\overline{H}}{182.26\text{ K}}$$
$$+ \int_{182.26}^{184.67} \frac{\overline{C}_P[N_2O(l)]}{T}dT + \frac{\Delta_{vap}\overline{H}}{184.67\text{ K}}$$

$$= 0.967 \text{ J·K}^{-1}\text{·mol}^{-1} + 69.34 \text{ J·K}^{-1}\text{·mol}^{-1} + 35.9 \text{ J·K}^{-1}\text{·mol}^{-1}$$
$$+ 1.02 \text{ J·K}^{-1}\text{·mol}^{-1} + 89.5 \text{ J·K}^{-1}\text{·mol}^{-1}$$

$$= 196.7 \text{ J·K}^{-1}\text{·mol}^{-1}$$

21–21. Methylammonium chloride occurs as three crystalline forms, called β, γ, and α, between 0 K and 298.15 K. The constant-pressure molar heat capacity of methylammonium chloride as a function of temperature is tabulated below. The $\beta \to \gamma$ transition occurs at 220.4 K with $\Delta_{trs}\overline{H} = 1.779 \text{ kJ·mol}^{-1}$ and the $\gamma \to \alpha$ transition occurs at 264.5 K with $\Delta_{trs}\overline{H} = 2.818 \text{ kJ·mol}^{-1}$. Assuming the heat capacity of solid methylammonium chloride can be described by the Debye theory up to 12 K, calculate the molar entropy of methylammonium chloride at 298.15 K.

T/K	$\overline{C}_P/\text{J·K}^{-1}\text{·mol}^{-1}$	T/K	$\overline{C}_P/\text{J·K}^{-1}\text{·mol}^{-1}$
12	0.837	180	73.72
15	1.59	200	77.95
20	3.92	210	79.71
30	10.53	220.4	$\beta \to \gamma$ transition
40	18.28	222	82.01
50	25.92	230	82.84
60	32.76	240	84.27
70	38.95	260	87.03
80	44.35	264.5	$\gamma \to \alpha$ transition
90	49.08	270	88.16
100	53.18	280	89.20
120	59.50	290	90.16
140	64.81	295	90.63
160	69.45		

We can this problem in the same way as Problem 21–20, graphing the molar heat capacities of the β, α, and γ crystalline forms versus temperature:

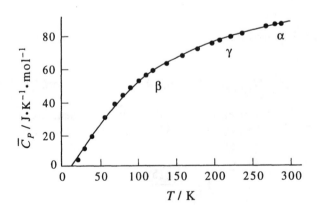

Fitting the curves to a polynomial, we find the following expressions for $\overline{C}_p$:

$$\overline{C}_p[\beta]/\text{J}\cdot\text{K}^{-1}\cdot\text{mol}^{-1} = -12.432 + (0.93892\ \text{K}^{-1})T - (3.4126 \times 10^{-3}\ \text{K}^{-2})T^2$$
$$+ (4.8562 \times 10^{-6}\ \text{K}^{-3})T^3$$
$$12\ \text{K} \leq T \leq 220.4\ \text{K}$$

$$\overline{C}_p[\gamma]/\text{J}\cdot\text{K}^{-1}\cdot\text{mol}^{-1} = 78.265 - (8.2955 \times 10^{-2}\ \text{K}^{-1})T + (4.4885 \times 10^{-4}\ \text{K}^{-2})T^2$$
$$220.4\ \text{K} \leq T \leq 264.5\ \text{K}$$

$$\overline{C}_p[\alpha]/\text{J}\cdot\text{K}^{-1}\cdot\text{mol}^{-1} = 35.757 + (0.28147\ \text{K}^{-1})T - (3.2362 \times 10^{-4}\ \text{K}^{-2})T^2$$
$$264.5\ \text{K} \leq T$$

From the Debye theory (Example 21–3),

$$\overline{S}(12\ \text{K}) = \frac{\overline{C}_p(12\ \text{K})}{3} \qquad 0 < T \leq 12\ \text{K}$$

Now we can write the molar entropy of methylammonium chloride as

$$\overline{S}(298.15\ \text{K}) = \frac{\overline{C}_p(12\ \text{K})}{3} + \int_{12}^{220.4} \frac{\overline{C}_p[\beta]}{T}dT + \frac{\Delta_{\beta\to\gamma}\overline{H}}{220.4\ \text{K}} + \int_{220.4}^{264.5} \frac{\overline{C}_p[\gamma]}{T}dT$$
$$+ \frac{\Delta_{\gamma\to\alpha}\overline{H}}{264.5\ \text{K}} + \int_{264.5}^{298.15} \frac{\overline{C}_p[\alpha]}{T}dT$$
$$= 0.279\ \text{J}\cdot\text{K}^{-1}\cdot\text{mol}^{-1} + 94.17\ \text{J}\cdot\text{K}^{-1}\cdot\text{mol}^{-1} + 8.07\ \text{J}\cdot\text{K}^{-1}\cdot\text{mol}^{-1}$$
$$+ 15.42\ \text{J}\cdot\text{K}^{-1}\cdot\text{mol}^{-1} + 10.65\ \text{J}\cdot\text{K}^{-1}\cdot\text{mol}^{-1} + 10.69\ \text{J}\cdot\text{K}^{-1}\cdot\text{mol}^{-1}$$
$$= 139.3\ \text{J}\cdot\text{K}^{-1}\cdot\text{mol}^{-1}$$

21–22. The constant-pressure molar heat capacity of chloroethane as a function of temperature is tabulated below. Chloroethane melts at 134.4 K with $\Delta_{\text{fus}}\overline{H} = 4.45\ \text{kJ}\cdot\text{mol}^{-1}$, and boils at 286.2 K with $\Delta_{\text{vap}}\overline{H} = 24.65\ \text{kJ}\cdot\text{mol}^{-1}$ at one bar. Furthermore, the heat capacity of solid chloroethane can be described by the Debye theory up to 15 K. Use these data to calculate the molar entropy of chloroethane at its boiling point.

T/K	$\overline{C}_p/\text{J}\cdot\text{K}^{-1}\cdot\text{mol}^{-1}$	T/K	$\overline{C}_p/\text{J}\cdot\text{K}^{-1}\cdot\text{mol}^{-1}$
15	5.65	130	84.60
20	11.42	134.4	90.83 (solid)
25	16.53		97.19 (liquid)
30	21.21	140	96.86
35	25.52	150	96.40
40	29.62	160	96.02
50	36.53	180	95.65
60	42.47	200	95.77
70	47.53	220	96.04
80	52.63	240	97.78
90	55.23	260	99.79
100	59.66	280	102.09
110	65.48	286.2	102.13
120	73.55		

Do this problem in the same way as Problem 21-20, graphing the molar heat capacities of solid and liquid chloroethane versus temperature:

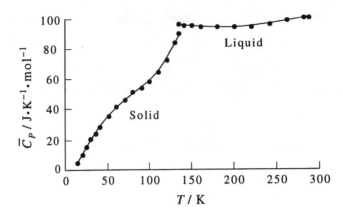

Fitting the curves to a polynomial, we find the following expressions for $\overline{C}_p$:

$$\overline{C}_p[\text{solid}]/\,\text{J·K}^{-1}\text{·mol}^{-1} = -19.195 + (1.863\ \text{K}^{-1})T - (1.8997 \times 10^{-2}\ \text{K}^{-2})T^2$$
$$+ (8.3132 \times 10^{-5}\ \text{K}^{-3})T^3$$
$$15\ \text{K} \leq T \leq 134.4\ \text{K}$$

$$\overline{C}_p[\text{liquid}]/\,\text{J·K}^{-1}\text{·mol}^{-1} = 118.15 - (0.24544\ \text{K}^{-1})T + (6.675 \times 10^{-4}\ \text{K}^{-2})T^2$$
$$134.4\ \text{K} \leq T \leq 298.15\ \text{K}$$

From the Debye theory (Example 21-3),

$$\overline{S}(15\ \text{K}) = \frac{\overline{C}_p(15\ \text{K})}{3} \qquad 0 < T \leq 15\ \text{K}$$

Now

$$\overline{S}(T) = \frac{\overline{C}_p(15\ \text{K})}{3} + \int_{15}^{134.4} \frac{\overline{C}_p[\text{solid}]}{T}dT + \frac{\Delta_{\text{fus}}\overline{H}}{134.4\ \text{K}} + \int_{134.4}^{286.2} \frac{\overline{C}_p[\text{liquid}]}{T}dT + \frac{\Delta_{\text{vap}}\overline{H}}{286.2\ \text{K}}$$

$$= 1.88\ \text{J·K}^{-1}\text{·mol}^{-1} + 78.1\ \text{J·K}^{-1}\text{·mol}^{-1} + 33.1\ \text{J·K}^{-1}\text{·mol}^{-1}$$
$$+73.4\ \text{J·K}^{-1}\text{·mol}^{-1} + 86.1\ \text{J·K}^{-1}\text{·mol}^{-1}$$

$$= 272.6\ \text{J·K}^{-1}\text{·mol}^{-1}$$

21-23. The constant-pressure molar heat capacity of nitromethane as a function of temperature is tabulated below. Nitromethane melts at 244.60 K with $\Delta_{\text{fus}}\overline{H} = 9.70\ \text{kJ·mol}^{-1}$, and boils at 374.34 K at one bar with $\Delta_{\text{vap}}\overline{H} = 38.27\ \text{kJ·mol}^{-1}$ at 298.15 K. Furthermore, the heat capacity of solid nitromethane can be described by the Debye theory up to 15 K. Use these data to calculate the molar entropy of nitromethane at 298.15 K and one bar. The vapor pressure of nitromethane is 36.66 torr at 298.15 K. (Be sure to take into account ΔS for the isothermal compression of nitromethane from its vapor pressure to one bar at 298.15 K).

T/K	$\overline{C}_P/\text{J·K}^{-1}\text{·mol}^{-1}$	T/K	$\overline{C}_P/\text{J·K}^{-1}\text{·mol}^{-1}$
15	3.72	200	71.46
20	8.66	220	75.23
30	19.20	240	78.99
40	28.87	244.60	melting point
60	40.84	250	104.43
80	47.99	260	104.64
100	52.80	270	104.93
120	56.74	280	105.31
140	60.46	290	105.69
160	64.06	300	106.06
180	67.74		

Do this problem in the same way as Problems 21–20, but include ΔS for the isothermal compression of nitromethane. Graph the molar heat capacities of solid and liquid chloroethane versus temperature:

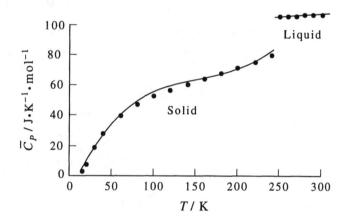

Fitting the curves to a polynomial, we find the following expressions for $\overline{C}_P$:

$$\overline{C}_P[\text{solid}]/\text{J·K}^{-1}\text{·mol}^{-1} = -11.177 + (1.1831\ \text{K}^{-1})T - (6.6826 \times 10^{-3}\ \text{K}^{-2})T^2$$
$$+ (1.3948 \times 10^{-5}\ \text{K}^{-3})T^3$$
$$15\ \text{K} \leq T \leq 244.60\ \text{K}$$

$$\overline{C}_P[\text{liquid}]/\text{J·K}^{-1}\text{·mol}^{-1} = 111.6 - (8.0557 \times 10^{-2}\ \text{K}^{-1})T + (2.0714 \times 10^{-4}\ \text{K}^{-2})T^2$$
$$244.60\ \text{K} \leq T \leq 300\ \text{K}$$

From the Debye theory (Example 21–3),

$$\overline{S}(T) = \frac{\overline{C}_P(15\ \text{K})}{3} \qquad 0 < T \leq 15\ \text{K}$$

Assuming that nitromethane behaves ideally, $dU = 0$ for the isothermal compression and so $\delta q = PdV$. Then we can express the change in entropy for the isothermal compression as

$$\Delta \overline{S} = \int Pd\overline{V} = \int_{P_1}^{P_2} \frac{R}{P} dP$$

We can now write the molar entropy of nitromethane at the given conditions as

$$\overline{S}(T) = \frac{\overline{C}_p(15\ \text{K})}{3} + \int_{15}^{244.60} \frac{\overline{C}_p[\text{solid}]}{T} dT + \frac{\Delta_{\text{fus}}\overline{H}}{244.60\ \text{K}} + \int_{244.60}^{374.34} \frac{\overline{C}_p[\text{liquid}]}{T} dT$$

$$+ \frac{\Delta_{\text{vap}}\overline{H}}{374.34\ \text{K}} - \int_{0.0489\ \text{bar}}^{1\ \text{bar}} \frac{R}{P} dP$$

$$= 1.24\ \text{J}\cdot\text{K}^{-1}\cdot\text{mol}^{-1} + 109.3\ \text{J}\cdot\text{K}^{-1}\cdot\text{mol}^{-1} + 39.66\ \text{J}\cdot\text{K}^{-1}\cdot\text{mol}^{-1}$$

$$+ 20.79\ \text{J}\cdot\text{K}^{-1}\cdot\text{mol}^{-1} + 128.4\ \text{J}\cdot\text{K}^{-1}\cdot\text{mol}^{-1} - 25.1\ \text{J}\cdot\text{K}^{-1}\cdot\text{mol}^{-1}$$

$$= 274.26\ \text{J}\cdot\text{K}^{-1}\cdot\text{mol}^{-1}$$

21–24. Use the following data to calculate the standard molar entropy of $CO(g)$ at its normal boiling point. Carbon monoxide undergoes a solid-solid phase transition at 61.6 K. Compare your result with the calculated value of $160.3\ \text{J}\cdot\text{K}^{-1}\cdot\text{mol}^{-1}$. Why is there a discrepancy between the calculated value and the experimental value?

$$\overline{C}_p[CO(s_1)]/R = -2.820 + (0.3317\ \text{K}^{-1})T - (6.408 \times 10^{-3}\ \text{K}^{-2})T^2$$
$$+ (6.002 \times 10^{-5}\ \text{K}^{-3})T^3$$
$$10\ \text{K} \leq T \leq 61.6\ \text{K}$$

$$\overline{C}_p[CO(s_2)]/R = 2.436 + (0.05694\ \text{K}^{-1})T$$
$$61.6\ \text{K} \leq T \leq 68.1\ \text{K}$$

$$\overline{C}_p[CO(l)]/R = 5.967 + (0.0330\ \text{K}^{-1})T - (2.088 \times 10^{-4}\ \text{K}^{-2})T^2$$
$$68.1\ \text{K} \leq T \leq 81.6\ \text{K}$$

and $T_{\text{trs}}(s_1 \rightarrow s_2) = 61.6\ \text{K}$, $T_{\text{fus}} = 68.1\ \text{K}$, $T_{\text{vap}} = 81.6\ \text{K}$, $\Delta_{\text{fus}}\overline{H} = 0.836\ \text{kJ}\cdot\text{mol}^{-1}$, $\Delta_{\text{trs}}\overline{H} = 0.633\ \text{kJ}\cdot\text{mol}^{-1}$, $\Delta_{\text{vap}}\overline{H} = 6.04\ \text{kJ}\cdot\text{mol}^{-1}$, $\Theta_D = 79.5\ \text{K}$, and the correction for nonideality $= 0.879\ \text{J}\cdot\text{K}^{-1}\cdot\text{mol}^{-1}$.

$$\overline{S}(T) = \int_0^{10} \frac{12\pi^4}{5T} R \left(\frac{T}{\Theta_D}\right)^3 dT + \int_{10}^{61.6} \frac{\overline{C}_p[CO(s_1)]}{T} dT + \frac{\Delta_{\text{trs}}\overline{H}}{61.6\ \text{K}}$$

$$+ \int_{61.6}^{68.1} \frac{\overline{C}_p[CO(s_2)]}{T} dT + \frac{\Delta_{\text{fus}}\overline{H}}{68.1\ \text{K}} + \int_{68.1}^{81.6} \frac{\overline{C}_p[CO(l)]}{T} dT + \frac{\Delta_{\text{vap}}\overline{H}}{81.6\ \text{K}} + \text{correction}$$

$$= 1.29\ \text{J}\cdot\text{K}^{-1}\cdot\text{mol}^{-1} + 40.0\ \text{J}\cdot\text{K}^{-1}\cdot\text{mol}^{-1} + 10.3\ \text{J}\cdot\text{K}^{-1}\cdot\text{mol}^{-1} + 5.11\ \text{J}\cdot\text{K}^{-1}\cdot\text{mol}^{-1}$$

$$+ 12.3\ \text{J}\cdot\text{K}^{-1}\cdot\text{mol}^{-1} + 10.9\ \text{J}\cdot\text{K}^{-1}\cdot\text{mol}^{-1} + 74.0\ \text{J}\cdot\text{K}^{-1}\cdot\text{mol}^{-1} + 0.879\ \text{J}\cdot\text{K}^{-1}\cdot\text{mol}^{-1}$$

$$= 154.7\ \text{J}\cdot\text{K}^{-1}\cdot\text{mol}^{-1}$$

We have found an experimental value for $\overline{S}$ of $154.7\ \text{J}\cdot\text{K}^{-1}\cdot\text{mol}^{-1}$. The difference between this and the calculated value is the residual entropy of the crystal, which is approximately $R\ln 2$, or $5.8\ \text{J}\cdot\text{K}^{-1}\cdot\text{mol}^{-1}$ (in agreement with the difference calculated here of $5.6\ \text{J}\cdot\text{K}^{-1}\cdot\text{mol}^{-1}$).

21–25. The molar heat capacities of solid and liquid water can be expressed by

$$\overline{C}_P[H_2O(s)]/R = -0.2985 + (2.896 \times 10^{-2} \text{ K}^{-1})T - (8.6714 \times 10^{-5} \text{ K}^{-2})T^2$$
$$+ (1.703 \times 10^{-7} \text{ K}^{-3})T^3$$
$$10 \text{ K} \leq T \leq 273.15 \text{ K}$$

$$\overline{C}_P[H_2O(l)]/R = 22.447 - (0.11639 \text{ K}^{-1})T + (3.3312 \times 10^{-4} \text{ K}^{-2})T^2$$
$$- (3.1314 \times 10^{-7} \text{ K}^{-3})T^3$$
$$273.15 \text{ K} \leq T \leq 298.15 \text{ K}$$

and $T_{fus} = 273.15$ K, $\Delta_{fus}\overline{H} = 6.007$ kJ·mol^{-1}, $\Delta_{vap}\overline{H}(T = 298.15$ K$) = 43.93$ kJ·mol^{-1}, $\Theta_D = 192$ K, the correction for nonideality = 0.32 J·K^{-1}·mol^{-1}, and the vapor pressure of H_2O at 298.15 K = 23.8 torr. Use these data to calculate the standard molar entropy of $H_2O(g)$ at 298.15 K. You need the vapor pressure of water at 298.15 K because that is the equilibrium pressure of $H_2O(g)$ when it is vaporized at 298.15 K. You must include the value of ΔS that results when you compress the $H_2O(g)$ from 23.8 torr to its standard value of one bar. Your answer should come out to be 185.6 J·K^{-1}·mol^{-1}, which does not agree exactly with the value in Table 21.2. There is a residual entropy associated with ice, which a detailed analysis of the structure of ice gives as $\Delta S_{residual} = R \ln(3/2) = 3.4$ J·K^{-1}·mol^{-1}, which is in good agreement with $\overline{S}_{calc} - \overline{S}_{exp}$.

We can do this problem in the same way as Problems 21–14, 21–16, and 21–18, taking into account ΔS for the isothermal compression of water. For an isothermal reaction, $\delta w = -\delta q$, so $\delta q = P dV$ and we assume that the gas is ideal. Then

$$\overline{S}(T) = \int_0^{10} \frac{12\pi^4}{5T} R \left(\frac{T}{\Theta_D}\right)^3 dT + \int_{10}^{273.15} \frac{\overline{C}_P[H_2O(s)]}{T} dT + \frac{\Delta_{fus}\overline{H}}{273.15 \text{ K}}$$
$$+ \int_{273.15}^{298.15} \frac{\overline{C}_P[H_2O(l)]}{T} dT + \frac{\Delta_{vap}\overline{H}}{298.15 \text{ K}}$$
$$- \int_{0.0317}^{1} \frac{R}{P} dP + \text{correction}$$

$$= 0.0915 \text{ J·K}^{-1}\text{·mol}^{-1} + 37.9 \text{ J·K}^{-1}\text{·mol}^{-1} + 22.0 \text{ J·K}^{-1}\text{·mol}^{-1}$$
$$+ 6.62 \text{ J·K}^{-1}\text{·mol}^{-1} + 147.3 \text{ J·K}^{-1}\text{·mol}^{-1} - 28.69 \text{ J·K}^{-1}\text{·mol}^{-1}$$
$$+ 0.32 \text{ J·K}^{-1}\text{·mol}^{-1}$$
$$= 185.6 \text{ J·K}^{-1}\text{·mol}^{-1}$$

Adding in the residual entropy gives a molar entropy of 189 J·K^{-1}·mol^{-1}.

21–26. Use the data in Problem 21–25 and the empirical expression

$$\overline{C}_P[H_2O(g)]/R = 3.652 + (1.156 \times 10^{-3} \text{ K}^{-1})T - (1.424 \times 10^{-7} \text{ K}^{-2})T^2$$
$$300 \text{ K} \leq T \leq 1000 \text{ K}$$

to plot the standard molar entropy of water from 0 K to 500 K.

Do this in the same manner as Problem 21–15, using the appropriate values from Problem 21–25 and changing the limits of integration as required.

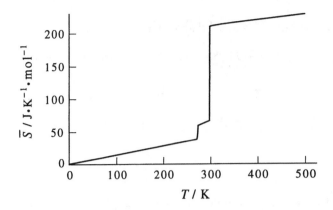

21–27. Show that

$$\overline{S} = R \ln \frac{qe}{N} + RT \left(\frac{\partial \ln q}{\partial T} \right)_V$$

We express the partition function of an ideal gas as (Equation 17.38)

$$Q(N, V, T) = \frac{q(V, T)^N}{N!}$$

and so

$$\ln Q(N, V, T) = N \ln q(V, T) - \ln N!$$

We substitute into Equation 21.19 to write

$$S = N k_B \ln q - k_B \ln N + N k_B T \left(\frac{\partial \ln q}{\partial T} \right)_V$$

We use Stirling's approximation and divide both sides of the equation by n to find

$$\overline{S} = N_A k_B \ln q - N_A k_B \ln N + k_B N_A + N_A k_B T \left(\frac{\partial \ln q}{\partial T} \right)_V$$

$$= R \ln \frac{q}{N} + R + RT \left(\frac{\partial \ln q}{\partial T} \right)_V$$

$$= R \ln \frac{qe}{N} + RT \left(\frac{\partial \ln q}{\partial T} \right)_V$$

21–28. Show that Equations 17.21 and 21.19 are consistent with Equations 21.2 and 21.3.

We begin with Equations 21.19 and 17.21,

$$S = k_B \ln Q + k_B T \left(\frac{\partial \ln Q}{\partial T} \right)_{N,V} \tag{21.19}$$

$$\langle E \rangle = k_B T^2 \left(\frac{\partial \ln Q}{\partial T} \right)_{N,V} \tag{17.21}$$

We can substitute $k_B T (\partial Q / \partial T)$ from Equation 17.21 into Equation 21.19 to find

$$S = k_B \ln Q + \frac{\langle E \rangle}{T} \tag{1}$$

$$\left(\frac{\partial S}{\partial T}\right)_V = k_B \left(\frac{\partial \ln Q}{\partial T}\right)_{N,V} + \frac{1}{T}\left(\frac{\partial \langle E \rangle}{\partial T}\right)_V - \frac{\langle E \rangle}{T^2}$$

Because $(\partial \langle E \rangle / \partial T)_V = C_V$, this becomes

$$\left(\frac{\partial S}{\partial T}\right)_V = k_B \left(\frac{\partial \ln Q}{\partial T}\right) + \frac{C_V}{T} - \frac{1}{T^2} k_B T^2 \left(\frac{\partial \ln Q}{\partial T}\right) = \frac{C_V}{T}$$

which is Equation 21.2. Now differentiate Equation 1 with respect to V:

$$S = k_B \ln Q + \frac{\langle E \rangle}{T}$$

$$\left(\frac{\partial S}{\partial V}\right)_T = k_B \left(\frac{\partial \ln Q}{\partial V}\right)_T + \frac{1}{T}\left(\frac{\partial \langle E \rangle}{\partial V}\right)_V$$

$$= \frac{1}{T}\left[k_B T \left(\frac{\partial \ln Q}{\partial V}\right)_T + \left(\frac{\partial U}{\partial V}\right)_T \right]$$

For an ideal gas, $k_B T = PV/N$, and $(\partial \ln Q/\partial V)_T = N/V$. Thus we have

$$\left(\frac{\partial S}{\partial V}\right)_T = \frac{1}{T}\left[P + \left(\frac{\partial U}{\partial V}\right)_T \right]$$

which is Equation 21.3.

21–29. Substitute Equation 21.23 into Equation 21.19 and derive the equation (Problem 20–31)

$$\Delta \overline{S} = \overline{C}_V \ln \frac{T_2}{T_1} + R \ln \frac{V_2}{V_1}$$

for one mole of a monatomic ideal gas.

We know that $Q = q^N/N!$ for an ideal gas (Equation 21.22), so substituting Equation 21.23 into this equation gives

$$Q = \frac{1}{N!}\left[\left(\frac{2\pi m k_B T}{h^2}\right)^{3/2} V \cdot g_{e1} \right]^N$$

Then

$$\ln Q = N \ln \left[\left(\frac{2\pi m k_B T}{h^2}\right)^{3/2} V \cdot g_{e1} \right] - N!$$

so, if only temperature and volume vary,

$$\ln Q_2 - \ln Q_1 = \frac{3}{2} N \ln \frac{T_2}{T_1} + \ln \frac{V_2}{V_1}$$

Also, we can find $(\partial \ln Q/\partial T)_V$:

$$\left(\frac{\partial \ln Q}{\partial T}\right)_{N,V} = \frac{3N}{2T}$$

Equation 21.19 states that

$$S = k_B \ln Q + k_B T \left(\frac{\partial \ln Q}{\partial T}\right)_{N,V}$$

Let $\Delta \overline{S} = \overline{S}_2 - \overline{S}_1$. Then

$$\Delta \overline{S} = \frac{1}{n}\left[k_B \ln Q_2 + k_B T_2 \left(\frac{\partial \ln Q_2}{\partial T}\right)_{N,V} - k_B \ln Q_1 - k_B T_1 \left(\frac{\partial \ln Q_1}{\partial T}\right)_{N,V}\right]$$

$$= \frac{3}{2} R \ln \frac{T_2}{T_1} + R \ln \frac{V_2}{V_1} + \frac{3R}{2T_2}T_2 - \frac{3R}{2T_1}T_1$$

$$= \frac{3}{2} R \ln \frac{T_2}{T_1} + R \ln \frac{V_2}{V_1}$$

where $3R/2$ is equal to $\overline{C}_V$ for a monatomic ideal gas (Equation 17.26).

21–30. Use Equation 21.24 and the data in Chapter 18 to calculate the standard molar entropy of $Cl_2(g)$ at 298.15 K. Compare your answer with the experimental value of 223.1 $J \cdot K^{-1} \cdot mol^{-1}$.

$$q = \left(\frac{2\pi M k_B T}{h^2}\right) V \frac{T}{\sigma \Theta_{rot}} \frac{e^{-\Theta_{vib}/2T}}{1 - e^{-\Theta_{vib}/T}} g_{e1} e^{D_e/k_B T} \tag{21.24}$$

Equation 21.28 (which is also for a diatomic ideal gas) can be written as

$$\frac{\overline{S}}{R} = \frac{7}{2} + \ln\left[\left(\frac{2\pi M k_B T}{h^2}\right)^{3/2} \frac{\overline{V}}{N_A}\right] + \ln \frac{T}{\sigma \Theta_{rot}} - \ln(1 - e^{-\Theta_{vib}/T}) + \frac{\Theta_{vib}/T}{e^{\Theta_{vib}/T} - 1} + \ln g_{e1}$$

(see also Problem 20–41).

For chlorine, $\Theta_{vib} = 805$ K, $\Theta_{rot} = 0.351$ K, $\sigma = 2$, and $g_{e1} = 1$. Then

$$\left(\frac{2\pi M k_B T}{h^2}\right)^{3/2} = \left[\frac{2\pi(1.1774 \times 10^{-25}\text{ kg})(1.381 \times 10^{-23}\text{ J} \cdot \text{K}^{-1})(298.15\text{ K})}{(6.626 \times 10^{34}\text{ J} \cdot \text{s})^2}\right]^{3/2} = 5.777 \times 10^{32}\text{ m}^{-3}$$

$$\frac{\overline{V}}{N_A} = \frac{RT}{N_A P} = \frac{(0.08314\text{ dm}^3 \cdot \text{bar} \cdot \text{mol}^{-1} \cdot \text{K}^{-1})(298.15\text{ K})}{(6.022 \times 10^{23}\text{ mol}^{-1})(1\text{ bar})} = 4.116 \times 10^{-26}\text{ m}^3$$

$$\ln\left(\frac{T}{\sigma \Theta_{rot}}\right) = \ln\left[\frac{298.15\text{ K}}{2(0.351\text{ K})}\right] = 6.051$$

$$\ln(1 - e^{-\Theta_{vib}/T}) = -0.06957$$

$$\frac{\Theta_{vib}/T}{e^{\Theta_{vib}/T} - 1} = 0.1945$$

The standard molar entropy is then

$$\frac{\overline{S}}{R} = 3.5 + 16.98 + 6.051 + 0.06954 + 0.1945 = 26.80$$

This is 222.8 $J \cdot K^{-1} \cdot mol^{-1}$, which is very close to the experimental value.

21–31. Use Equation 21.24 and the data in Chapter 18 to calculate the standard molar entropy of $CO(g)$ at its standard boiling point, 81.6 K. Compare your answer with the experimental value of 155.6 $J \cdot K^{-1} \cdot mol^{-1}$. Why is there a discrepancy of about 5 $J \cdot K^{-1} \cdot mol^{-1}$?

As in Problem 21–30,

$$\frac{\overline{S}}{R} = \frac{7}{2} + \ln\left[\left(\frac{2\pi M k_B T}{h^2}\right)^{3/2}\frac{\overline{V}}{N_A}\right] + \ln\frac{T}{\sigma\Theta_{rot}} - \ln(1 - e^{-\Theta_{vib}/T}) + \frac{\Theta_{vib}/T}{e^{\Theta_{vib}/T} - 1} + \ln g_{e1}$$

For carbon monoxide, $\Theta_{vib} = 3103$ K, $\Theta_{rot} = 2.77$ K, $\sigma = 1$, and $g_{e1} = 1$. Then

$$\left(\frac{2\pi M k_B T}{h^2}\right)^{3/2} = \left[\frac{2\pi(4.651 \times 10^{-26}\text{ kg})(1.381 \times 10^{-23}\text{ J·K}^{-1})(81.6\text{ K})}{(6.626 \times 10^{34}\text{ J·s})^2}\right]^{3/2} = 2.054 \times 10^{31}\text{ m}^{-3}$$

$$\frac{\overline{V}}{N_A} = \frac{RT}{N_A P} = \frac{(0.08314\text{ dm}^3\text{·bar·mol}^{-1}\text{·K}^{-1})(81.6\text{ K})}{(6.022 \times 10^{23}\text{ mol}^{-1})(1\text{ bar})} = 1.127 \times 10^{-26}\text{ m}^3$$

$$\ln\left(\frac{T}{\sigma\Theta_{rot}}\right) = \ln\left[\frac{81.6\text{ K}}{(2.77\text{ K})}\right] = 3.383$$

$$\ln(1 - e^{-\Theta_{vib}/T}) = -3.057 \times 10^{-17}$$

$$\frac{\Theta_{vib}/T}{e^{\Theta_{vib}/T} - 1} = 1.162 \times 10^{-15}$$

The standard molar entropy is then

$$\frac{\overline{S}}{R} = 3.5 + 12.352 + 3.383 + 3.057 \times 10^{-17} + 1.162 \times 10^{-15} = 19.23$$

This is $159.9\text{ J·K}^{-1}\text{·mol}^{-1}$, which is about $4\text{ J·K}^{-1}\text{·mol}^{-1}$ larger than the experimental value. The discrepancy is due to residual entropy.

21–32. Use Equation 21.26 and the data in Chapter 18 to calculate the standard molar entropy of $NH_3(g)$ at 298.15 K. Compare your answer with the experimental value of $192.8\text{ J·K}^{-1}\text{·mol}^{-1}$.

See Problem 20–43. The value calculated is $193.1\text{ J·K}^{-1}\text{·mol}^{-1}$.

21–33. Use Equation 21.24 and the data in Chapter 18 to calculate the standard molar entropy of $Br_2(g)$ at 298.15 K. Compare your answer with the experimental value of $245.5\text{ J·K}^{-1}\text{·mol}^{-1}$.

As in Problem 21–30,

$$\frac{\overline{S}}{R} = \frac{7}{2} + \ln\left[\left(\frac{2\pi M k_B T}{h^2}\right)^{3/2}\frac{\overline{V}}{N_A}\right] + \ln\frac{T}{\sigma\Theta_{rot}} - \ln(1 - e^{-\Theta_{vib}/T}) + \frac{\Theta_{vib}/T}{e^{\Theta_{vib}/T} - 1} + \ln g_{e1}$$

For bromine, $\Theta_{vib} = 463$ K, $\Theta_{rot} = 0.116$ K, $\sigma = 2$, and $g_{e1} = 1$. Then

$$\left(\frac{2\pi M k_B T}{h^2}\right)^{3/2} = \left[\frac{2\pi(2.654 \times 10^{-25}\text{ kg})(1.381 \times 10^{-23}\text{ J·K}^{-1})(298.15\text{ K})}{(6.626 \times 10^{34}\text{ J·s})^2}\right]^{3/2} = 1.955 \times 10^{33}\text{ m}^{-3}$$

$$\frac{\overline{V}}{N_A} = \frac{RT}{N_A P} = \frac{(0.08314\text{ dm}^3\text{·bar·mol}^{-1}\text{·K}^{-1})(298.15\text{ K})}{(6.022 \times 10^{23}\text{ mol}^{-1})(1\text{ bar})} = 4.116 \times 10^{-26}\text{ m}^3$$

$$\ln\left(\frac{T}{\sigma\Theta_{rot}}\right) = \ln\left[\frac{298.15\text{ K}}{2(0.116\text{ K})}\right] = 7.159$$

$$\ln(1 - e^{-\Theta_{vib}/T}) = -0.2378$$

$$\frac{\Theta_{vib}/T}{e^{\Theta_{vib}/T} - 1} = 0.417$$

The standard molar entropy is then

$$\frac{\overline{S}}{R} = 3.5 + 18.203 + 7.158 + 0.238 + 0.417 = 29.52$$

This is $245.4 \text{ J·K}^{-1}\text{·mol}^{-1}$, almost identical to the experimental value.

21–34. The vibrational and rotational constants for HF(g) within the harmonic oscillator-rigid rotator model are $\tilde{\nu}_0 = 3959 \text{ cm}^{-1}$ and $\tilde{B}_0 = 20.56 \text{ cm}^{-1}$. Calculate the standard molar entropy of HF(g) at 298.15 K. How does this value compare with that in Table 21.3?

We can use the equalities (given in Chapter 18)

$$\Theta_{vib} = \frac{hc\tilde{\nu}_0}{k_B} \qquad \Theta_{rot} = \frac{hc\tilde{B}_0}{k_B}$$

to find $\Theta_{vib} = 5696 \text{ K}$ and $\Theta_{rot} = 29.58 \text{ K}$, and then solve for entropy as we did in the previous problems.

As in Problem 21–30,

$$\frac{\overline{S}}{R} = \frac{7}{2} + \ln\left[\left(\frac{2\pi Mk_B T}{h^2}\right)^{3/2}\frac{\overline{V}}{N_A}\right] + \ln\frac{T}{\sigma\Theta_{rot}} - \ln(1 - e^{-\Theta_{vib}/T}) + \frac{\Theta_{vib}/T}{e^{\Theta_{vib}/T} - 1} + \ln g_{e1}$$

Substitute in to find the values of the various components of entropy. Note that $\sigma = 1$ and $g_{e1} = 1$.

$$\left(\frac{2\pi Mk_B T}{h^2}\right)^{3/2} = \left[\frac{2\pi(3.322 \times 10^{-26} \text{ kg})(1.381 \times 10^{-23} \text{ J·K}^{-1})(298.15 \text{ K})}{(6.626 \times 10^{34} \text{ J·s})^2}\right]^{3/2} = 8.658 \times 10^{31} \text{ m}^{-3}$$

$$\frac{\overline{V}}{N_A} = \frac{RT}{N_A P} = \frac{(0.08314 \text{ dm}^3\text{·bar·mol}^{-1}\text{·K}^{-1})(298.15 \text{ K})}{(6.022 \times 10^{23} \text{ mol}^{-1})(1 \text{ bar})} = 4.116 \times 10^{-26} \text{ m}^3$$

$$\ln\left(\frac{T}{\sigma\Theta_{rot}}\right) = 2.310$$

$$\ln(1 - e^{-\Theta_{vib}/T}) = -5.046 \times 10^{-9}$$

$$\frac{\Theta_{vib}/T}{e^{\Theta_{vib}/T} - 1} = 9.639 \times 10^{-8}$$

$$\frac{\overline{S}}{R} = 3.5 + 15.09 + 2.310 + 5.046 \times 10^{-9} + 9.639 \times 10^{-8} = 20.90$$

The standard molar entropy is $173.7 \text{ J·K}^{-1}\text{·mol}^{-1}$, which is very close to the value in Table 21.3.

21–35. Calculate the standard molar entropy of $H_2(g)$ and $D_2(g)$ at 298.15 K given that the bond length of both diatomic molecules is 74.16 pm and the vibrational temperatures of $H_2(g)$ and $D_2(g)$ are 6215 K and 4394 K, respectively. Calculate the standard molar entropy of HD(g) at 298.15 K ($R_e = 74.13$ pm and $\Theta_{vib} = 5496$ K).

We can use the relation $\Theta_{rot} = \hbar^2/2Ik_B$ (Equation 18.32) to find Θ_{rot} for HD, H_2, and D_2. Then we can solve for molar entropy as in Problem 21–30. For both H_2 and D_2, $\sigma = 2$ and $g_{e1} = 1$; for HD, $\sigma = 1$ and $g_{e1} = 1$. For H_2,

$$\Theta_{rot} = \frac{\hbar^2}{2\mu R_e^2 k_B} = \frac{\hbar^2}{2(8.368 \times 10^{-28} \text{ kg})(74.16 \times 10^{-12} \text{ m})^2 k_B} = 87.51 \text{ K}$$

and

$$\frac{\overline{S}}{R} = \frac{7}{2} + \ln\left[\left(\frac{2\pi M k_B T}{h^2}\right)^{3/2} \frac{\overline{V}}{N_A}\right] + \ln\frac{T}{\sigma \Theta_{rot}} - \ln(1 - e^{-\Theta_{vib}/T}) + \frac{\Theta_{vib}/T}{e^{\Theta_{vib}/T} - 1} + \ln g_{e1}$$

Substitute in to find the values of the various components of entropy.

$$\left(\frac{2\pi M k_B T}{h^2}\right)^{3/2} = \left[\frac{2\pi(3.348 \times 10^{-27} \text{ kg})(1.381 \times 10^{-23} \text{ J·K}^{-1})(298.15 \text{ K})}{(6.626 \times 10^{34} \text{ J·s})^2}\right]^{3/2} = 2.769 \times 10^{30} \text{ m}^{-3}$$

$$\frac{\overline{V}}{N_A} = \frac{RT}{N_A P} = \frac{(0.08314 \text{ dm}^3 \cdot \text{bar·mol}^{-1} \cdot \text{K}^{-1})(298.15 \text{ K})}{(6.022 \times 10^{23} \text{ mol}^{-1})(1 \text{ bar})} = 4.116 \times 10^{-26} \text{ m}^3$$

$$\ln\left(\frac{T}{\sigma \Theta_{rot}}\right) = 0.533$$

$$\ln(1 - e^{-\Theta_{vib}/T}) = -8.852 \times 10^{-10}$$

$$\frac{\Theta_{vib}/T}{e^{\Theta_{vib}/T} - 1} = 1.845 \times 10^{-8}$$

$$\frac{\overline{S}}{R} = 3.5 + 11.64 + 0.533 + 8.852 \times 10^{-10} + 1.845 \times 10^{-8} = 15.68$$

The standard molar entropy of $H_2(g)$ at 298.15 K is 130.3 J·K^{-1}·mol^{-1}.

For D_2,

$$\Theta_{rot} = \frac{\hbar^2}{2\mu R_e^2 k_B} = \frac{\hbar^2}{2(1.672 \times 10^{-27} \text{ kg})(74.16 \times 10^{-12} \text{ m})^2 k_B} = 43.79 \text{ K}$$

and

$$\frac{\overline{S}}{R} = \frac{7}{2} + \ln\left[\left(\frac{2\pi M k_B T}{h^2}\right)^{3/2} \frac{\overline{V}}{N_A}\right] + \ln\frac{T}{\sigma \Theta_{rot}} - \ln(1 - e^{-\Theta_{vib}/T}) + \frac{\Theta_{vib}/T}{e^{\Theta_{vib}/T} - 1} + \ln g_{e1}$$

Substitute in to find the values of the various components of entropy.

$$\left(\frac{2\pi M k_B T}{h^2}\right)^{3/2} = \left[\frac{2\pi(6.689 \times 10^{-27} \text{ kg})(1.381 \times 10^{-23} \text{ J·K}^{-1})(298.15 \text{ K})}{(6.626 \times 10^{34} \text{ J·s})^2}\right]^{3/2} = 7.822 \times 10^{30} \text{ m}^{-3}$$

$$\frac{\overline{V}}{N_A} = \frac{RT}{N_A P} = \frac{(0.08314 \text{ dm}^3 \cdot \text{bar·mol}^{-1} \cdot \text{K}^{-1})(298.15 \text{ K})}{(6.022 \times 10^{23} \text{ mol}^{-1})(1 \text{ bar})} = 4.116 \times 10^{-26} \text{ m}^3$$

$$\ln\left(\frac{T}{\sigma \Theta_{rot}}\right) = 1.225$$

$$\ln(1 - e^{-\Theta_{vib}/T}) = -3.977 \times 10^{-7}$$

$$\frac{\Theta_{vib}/T}{e^{\Theta_{vib}/T} - 1} = 5.861 \times 10^{-6}$$

$$\frac{\overline{S}}{R} = 3.5 + 12.682 + 1.226 + 3.977 \times 10^{-7} + 5.861 \times 10^{-6} = 17.41$$

The standard molar entropy of $D_2(g)$ at 298.15 K is 144.7 J·K^{-1}·mol^{-1}.

For HD,

$$\Theta_{\text{rot}} = \frac{\hbar^2}{2\mu R_e^2 k_B} = \frac{\hbar^2}{2(1.115 \times 10^{-27}\ \text{kg})(74.13 \times 10^{-12}\ \text{m})^2 k_B} = 65.71\ \text{K}$$

and

$$\frac{\overline{S}}{R} = \frac{7}{2} + \ln\left[\left(\frac{2\pi M k_B T}{h^2}\right)^{3/2} \frac{\overline{V}}{N_A}\right] + \ln\frac{T}{\sigma\Theta_{\text{rot}}} - \ln(1 - e^{-\Theta_{\text{vib}}/T}) + \frac{\Theta_{\text{vib}}/T}{e^{\Theta_{\text{vib}}/T} - 1} + \ln g_{e1}$$

Substitute in to find the values of the various components of entropy.

$$\left(\frac{2\pi M k_B T}{h^2}\right)^{3/2} = \left[\frac{2\pi(5.018 \times 10^{-27}\ \text{kg})(1.381 \times 10^{-23}\ \text{J}\cdot\text{K}^{-1})(298.15\ \text{K})}{(6.626 \times 10^{34}\ \text{J}\cdot\text{s})^2}\right]^{3/2} = 5.082 \times 10^{30}\ \text{m}^{-3}$$

$$\frac{\overline{V}}{N_A} = \frac{RT}{N_A P} = \frac{(0.08314\ \text{dm}^3\cdot\text{bar}\cdot\text{mol}^{-1}\cdot\text{K}^{-1})(298.15\ \text{K})}{(6.022 \times 10^{23}\ \text{mol}^{-1})(1\ \text{bar})} = 4.116 \times 10^{-26}\ \text{m}^3$$

$$\ln\left(\frac{T}{\sigma\Theta_{\text{rot}}}\right) = 1.512$$

$$\ln(1 - e^{-\Theta_{\text{vib}}/T}) = -9.871 \times 10^{-9}$$

$$\frac{\Theta_{\text{vib}}/T}{e^{\Theta_{\text{vib}}/T} - 1} = 1.820 \times 10^{-7}$$

$$\frac{\overline{S}}{R} = 3.5 + 12.251 + 1.512 + 9.871 \times 10^{-9} + 1.820 \times 10^{-7} = 17.26$$

The standard molar entropy of HD(g) at 298.15 K is 143.5 J·K^{-1}·mol^{-1}.

21–36. Calculate the standard molar entropy of HCN(g) at 1000 K given that $I = 1.8816 \times 10^{-46}$ kg·m^2, $\tilde{\nu}_1 = 2096.70$ cm^{-1}, $\tilde{\nu}_2 = 713.46$ cm^{-1}, and $\tilde{\nu}_3 = 3311.47$ cm^{-1}. Recall that HCN(g) is a linear triatomic molecule and therefore the bending mode, ν_2, is doubly degenerate.

In Problem 18–24, we found $\Theta_{\text{vib},j}$ and Θ_{rot} of HCN to be

$$\begin{aligned}
\Theta_{\text{vib},1} &= 3016\ \text{K} & \Theta_{\text{vib},4} &= 4764\ \text{K} \\
\Theta_{\text{vib},2,3} &= 1026\ \text{K} & \Theta_{\text{rot}} &= 2.135\ \text{K}
\end{aligned}$$

For a linear polyatomic ideal gas having three atoms,

$$q = \left(\frac{2\pi M k_B T}{h^2}\right)^{3/2} V \frac{T}{\sigma\Theta_{\text{rot}}} \left(\prod_{j=1}^{4} \frac{e^{-\Theta_{\text{vib},j}/2T}}{1 - e^{-\Theta_{\text{vib},j}/T}}\right) g_{e1} e^{D_e/k_B T} \tag{21.25}$$

Substituting into Equation 21.27, we find

$$\frac{\overline{S}}{R} = \frac{7}{2} + \ln\left[\left(\frac{2\pi M k_B T}{h^2}\right)^{3/2} \frac{\overline{V}}{N_A}\right] + \ln\frac{T}{\sigma\Theta_{\text{rot}}} - \sum_{j=1}^{4}\ln(1 - e^{-\Theta_{\text{vib},j}/T}) + \sum_{j=1}^{4}\left[\frac{(\Theta_{\text{vib},j}/T)e^{-\Theta_{\text{vib},j}/T}}{1 - e^{-\Theta_{\text{vib},j}/T}}\right] + \ln g_{e1}$$

Because HCN is asymmetrical, its symmetry number is unity. Then

$$\left(\frac{2\pi M k_B T}{h^2}\right)^{3/2} = \left[\frac{2\pi(4.488 \times 10^{-26}\ \text{kg})(1.381 \times 10^{-23}\ \text{J}\cdot\text{K}^{-1})(1000\ \text{K})}{(6.626 \times 10^{34}\ \text{J}\cdot\text{s})^2}\right]^{3/2} = 8.350 \times 10^{32}\ \text{m}^{-3}$$

$$\frac{\overline{V}}{N_A} = \frac{RT}{N_A P} = \frac{(0.08314\ \text{dm}^3\cdot\text{bar}\cdot\text{mol}^{-1}\cdot\text{K}^{-1})(1000\ \text{K})}{(6.022 \times 10^{23}\ \text{mol}^{-1})(1\ \text{bar})} = 1.381 \times 10^{-25}\ \text{m}^3$$

$$\ln\left(\frac{T}{\sigma\Theta_{rot}}\right) = \ln\left[\frac{1000\ K}{(2.135\ K)}\right] = 6.149$$

$$\sum_{j=1}^{4}\ln(1 - e^{-\Theta_{vib,j}/T}) = -0.9465$$

$$\sum_{j=1}^{4}\frac{\Theta_{vib,j}}{T}\left(\frac{e^{-\Theta_{vib,j}/T}}{1 - e^{-\Theta_{vib,j}/T}}\right) = 1.343$$

The standard molar entropy is then

$$\frac{\overline{S}}{R} = 3.5 + 18.563 + 6.149 + 0.9465 + 1.343 = 30.5$$

The standard molar entropy of HCN(g) at 1000 K is 253.6 $J \cdot K^{-1} \cdot mol^{-1}$. The experimentally observed value is 253.7 $J \cdot K^{-1} \cdot mol^{-1}$.

21–37. Given that $\tilde{v}_1 = 1321.3\ cm^{-1}$, $\tilde{v}_2 = 750.8\ cm^{-1}$, $\tilde{v}_3 = 1620.3\ cm^{-1}$, $\tilde{A}_0 = 7.9971\ cm^{-1}$, $\tilde{B}_0 = 0.4339\ cm^{-1}$, and $\tilde{C}_0 = 0.4103\ cm^{-1}$, calculate the standard molar entropy of $NO_2(g)$ at 298.15 K. (Note that $NO_2(g)$ is a bent triatomic molecule.) How does your value compare with that in Table 21.2?

In Problem 18–29, we found that

$$\Theta_{vib,1} = 1898.7\ K \qquad \Theta_{rot,A} = 11.512\ K$$
$$\Theta_{vib,2} = 1078.8\ K \qquad \Theta_{rot,B} = 0.62304\ K$$
$$\Theta_{vib,3} = 2327.6\ K \qquad \Theta_{rot,C} = 0.59047\ K$$

For a nonlinear polyatomic ideal gas having three atoms,

$$q = \left(\frac{2\pi M k_B T}{h^2}\right)^{3/2} V \frac{\pi^{1/2}}{\sigma}\left(\frac{T^3}{\Theta_{rot,A}\Theta_{rot,B}\Theta_{rot,C}}\right)^{1/2}\left(\prod_{j=1}^{3}\frac{e^{-\Theta_{vib,j}/2T}}{1 - e^{-\Theta_{vib,j}/T}}\right)g_{e1}e^{D_e/k_B T} \qquad (21.26)$$

Substituting into Equation 21.27, we find

$$\frac{\overline{S}}{R} = 4 + \ln\left[\left(\frac{2\pi M k_B T}{h^2}\right)^{3/2}\frac{\overline{V}}{N_A}\right] + \ln\left(\frac{\pi^{1/2}}{\sigma}\right) + \frac{1}{2}\ln\frac{T^3}{\Theta_{rot,A}\Theta_{rot,B}\Theta_{rot,C}}$$
$$- \sum_{j=1}^{3}\ln(1 - e^{-\Theta_{vib,j}/T}) + \sum_{j=1}^{3}\left[\frac{(\Theta_{vib,j}/T)e^{-\Theta_{vib,j}/T}}{1 - e^{-\Theta_{vib,j}/T}}\right] + \ln g_{e1}$$

From Table 18.4, $\sigma = 2$, and $g_{e1} = 1$. Then

$$\left(\frac{2\pi M k_B T}{h^2}\right)^{3/2} = \left[\frac{2\pi(7.639 \times 10^{-26}\ kg)(1.381 \times 10^{-23}\ J \cdot K^{-1})(298.15\ K)}{(6.626 \times 10^{34}\ J \cdot s)^2}\right]^{3/2} = 3.019 \times 10^{32}\ m^{-3}$$

$$\frac{\overline{V}}{N_A} = \frac{RT}{N_A P} = \frac{(0.08314\ dm^3 \cdot bar \cdot mol^{-1} \cdot K^{-1})(298.15\ K)}{(6.022 \times 10^{23}\ mol^{-1})(1\ bar)} = 4.116 \times 10^{-26}\ m^3$$

$$\frac{1}{2}\ln\left(\frac{T^3}{\Theta_{rot,A}\,\Theta_{rot,B}\,\Theta_{rot,C}}\right) = 7.825$$

$$\sum_{j=1}^{3}\ln(1 - e^{-\Theta_{vib,j}/T}) = -0.0293$$

$$\sum_{j=1}^{3}\frac{\Theta_{vib,j}}{T}\left(\frac{e^{-\Theta_{vib,j}/T}}{1 - e^{-\Theta_{vib,j}/T}}\right) = 0.114$$

The standard molar entropy is then

$$\frac{\overline{S}}{R} = 4 + 16.335 - 0.121 + 7.825 + 0.0293 + 0.114 = 28.18$$

This is 234.3 $J \cdot K^{-1} \cdot mol^{-1}$.

21-38. In Problem 21-48, you are asked to calculate the value of $\Delta_r S^\circ$ at 298.15 K using the data in Table 21.2 for the reaction described by

$$2\,CO(g) + O_2(g) \longrightarrow 2\,CO_2(g)$$

Use the data in Table 18.2 to calculate the standard molar entropy of each of the substances in this reaction [see Example 21-5 for the calculation of the standard molar entropy of $CO_2(g)$]. Then use these results to calculate the standard entropy change for the above reaction. How does your answer compare with what you obtained in Problem 21-48?

From Example 21-5, $S^\circ[CO_2(g)] = 213.8\ J \cdot K^{-1} \cdot mol^{-1}$. Because both CO and O_2 are diatomic molecules, we can write (as in Problem 21-30)

$$\frac{\overline{S}}{R} = \frac{7}{2} + \ln\left[\left(\frac{2\pi M k_B T}{h^2}\right)^{3/2}\frac{\overline{V}}{N_A}\right] + \ln\frac{T}{\sigma\Theta_{rot}} - \ln(1 - e^{-\Theta_{vib}/T}) + \frac{\Theta_{vib}/T}{e^{\Theta_{vib}/T} - 1} + \ln g_{e1}$$

Because CO is a heteronuclear diatomic molecule, $\sigma = 1$; because O_2 is homonuclear, $\sigma = 2$. For CO $\Theta_{vib} = 3103$ K and $\Theta_{rot} = 2.77$ K. Then

$$\left(\frac{2\pi M k_B T}{h^2}\right)^{3/2} = \left[\frac{2\pi(4.651 \times 10^{-26}\ kg)(1.381 \times 10^{-23}\ J \cdot K^{-1})(298.15\ K)}{(6.626 \times 10^{34}\ J \cdot s)^2}\right]^{3/2} = 1.434 \times 10^{32}\ m^{-3}$$

$$\frac{\overline{V}}{N_A} = \frac{RT}{N_A P} = \frac{(0.08314\ dm^3 \cdot bar \cdot mol^{-1} \cdot K^{-1})(298.15\ K)}{(6.022 \times 10^{23}\ mol^{-1})(1\ bar)} = 4.116 \times 10^{-26}\ m^3$$

$$\ln\left(\frac{T}{\sigma\Theta_{rot}}\right) = 4.679$$

$$\ln(1 - e^{-\Theta_{vib}/T}) = -3.02 \times 10^{-5}$$

$$\frac{\Theta_{vib}/T}{e^{\Theta_{vib}/T} - 1} = 3.14 \times 10^{-4}$$

$$\frac{\overline{S}}{R} = 3.5 + 15.591 + 4.679 + 3.02 \times 10^{-5} + 3.14 \times 10^{-4} = 23.77$$

The standard molar entropy of CO(g) at 298.15 K is 197.6 $J \cdot K^{-1} \cdot mol^{-1}$.

We follow the same procedure for O_2, with $\Theta_{vib} = 2256$ K and $\Theta_{rot} = 2.07$ K. Note that $g_{e1} = 3$ for O_2, so we cannot neglect the $\ln g_{e1}$ term!

$$\left(\frac{2\pi M k_B T}{h^2}\right)^{3/2} = \left[\frac{2\pi(5.313 \times 10^{-26} \text{ kg})(1.381 \times 10^{-23} \text{ J}\cdot\text{K}^{-1})(298.15 \text{ K})}{(6.626 \times 10^{34} \text{ J}\cdot\text{s})^2}\right]^{3/2} = 1.751 \times 10^{32} \text{ m}^{-3}$$

$$\frac{\overline{V}}{N_A} = \frac{RT}{N_A P} = \frac{(0.08314 \text{ dm}^3\cdot\text{bar}\cdot\text{mol}^{-1}\cdot\text{K}^{-1})(298.15 \text{ K})}{(6.022 \times 10^{23} \text{ mol}^{-1})(1 \text{ bar})} = 4.116 \times 10^{-26} \text{ m}^3$$

$$\ln\left(\frac{T}{\sigma\Theta_{rot}}\right) = 4.277$$

$$\ln(1 - e^{-\Theta_{vib}/T}) = -5.18 \times 10^{-4}$$

$$\frac{\Theta_{vib}/T}{e^{\Theta_{vib}/T} - 1} = 3.92 \times 10^{-3}$$

$$\ln g_{e1} = \ln 3 = 1.099$$

$$\frac{\overline{S}}{R} = 3.5 + 15.79 + 4.277 + 5.18 \times 10^{-4} + 3.92 \times 10^{-3} + 1.099 = 24.67$$

The standard molar entropy of $O_2(g)$ at 298.15 K is 205.1 J·K^{-1}·mol^{-1}.

We can calculate the entropy change for the above reaction easily using the method described in Section 21–9:

$$\Delta_r S° = 2S°[CO_2] - S°[O_2] - 2S°[CO]$$
$$= 2(213.8 \text{ J}\cdot\text{K}^{-1}\cdot\text{mol}^{-1}) - (205.1 \text{ J}\cdot\text{K}^{-1}\cdot\text{mol}^{-1}) - 2(197.6 \text{ J}\cdot\text{K}^{-1}\cdot\text{mol}^{-1})$$
$$= -172.7 \text{ J}\cdot\text{K}^{-1}\cdot\text{mol}^{-1}$$

This value is very close to that found in Problem 21–48.

21–39. Calculate the value of $\Delta_r S°$ for the reaction described by

$$H_2(g) + \tfrac{1}{2}O_2(g) \longrightarrow H_2O(g)$$

at 500 K using the data in Tables 18.2 and 18.4.

Because both H_2 and O_2 are diatomic molecules, we can write (as in the previous problem)

$$\frac{\overline{S}}{R} = \frac{7}{2} + \ln\left[\left(\frac{2\pi M k_B T}{h^2}\right)^{3/2}\frac{\overline{V}}{N_A}\right] + \ln\frac{T}{\sigma\Theta_{rot}} - \ln(1 - e^{-\Theta_{vib}/T}) + \frac{\Theta_{vib}/T}{e^{\Theta_{vib}/T} - 1} + \ln g_{e1}$$

Because both are homonuclear, $\sigma = 2$ for both H_2 and O_2. For H_2 $\Theta_{vib} = 6215$ K and $\Theta_{rot} = 85.3$ K. Then

$$\left(\frac{2\pi M k_B T}{h^2}\right)^{3/2} = \left[\frac{2\pi(3.347 \times 10^{-27} \text{ kg})(1.381 \times 10^{-23} \text{ J}\cdot\text{K}^{-1})(500 \text{ K})}{(6.626 \times 10^{34} \text{ J}\cdot\text{s})^2}\right]^{3/2} = 6.014 \times 10^{30} \text{ m}^{-3}$$

$$\frac{\overline{V}}{N_A} = \frac{RT}{N_A P} = \frac{(0.08314 \text{ dm}^3\cdot\text{bar}\cdot\text{mol}^{-1}\cdot\text{K}^{-1})(500 \text{ K})}{(6.022 \times 10^{23} \text{ mol}^{-1})(1 \text{ bar})} = 6.903 \times 10^{-26} \text{ m}^3$$

$$\ln\left(\frac{T}{\sigma\Theta_{rot}}\right) = 1.08$$

$$\ln(1 - e^{-\Theta_{vib}/T}) = -4.00 \times 10^{-6}$$

$$\frac{\Theta_{vib}/T}{e^{\Theta_{vib}/T} - 1} = 4.97 \times 10^{-5}$$

$$\frac{\overline{S}}{R} = 3.5 + 12.94 + 1.08 + 4.00 \times 10^{-6} + 4.97 \times 10^{-5} = 17.51$$

The standard molar entropy of H_2 at 500 K is 145.6 J·K^{-1}·mol^{-1}.

We can do the same for O_2 (with $\Theta_{vib} = 2256$ K and $\Theta_{rot} = 2.07$ K), keeping in mind that $g_{e1} = 3$:

$$\left(\frac{2\pi M k_B T}{h^2}\right)^{3/2} = \left[\frac{2\pi(5.313 \times 10^{-26}\text{ kg})(1.381 \times 10^{-23}\text{ J·K}^{-1})(500\text{ K})}{(6.626 \times 10^{34}\text{ J·s})^2}\right]^{3/2} = 3.803 \times 10^{32}\text{ m}^{-3}$$

$$\frac{\overline{V}}{N_A} = \frac{RT}{N_A P} = \frac{(0.08314\text{ dm}^3\text{·bar·mol}^{-1}\text{·K}^{-1})(500\text{ K})}{(6.022 \times 10^{23}\text{ mol}^{-1})(1\text{ bar})} = 6.903 \times 10^{-26}\text{ m}^3$$

$$\ln\left(\frac{T}{\sigma\Theta_{rot}}\right) = 4.79$$

$$\ln(1 - e^{-\Theta_{vib}/T}) = -0.0110$$

$$\frac{\Theta_{vib}/T}{e^{\Theta_{vib}/T} - 1} = 0.0501$$

$$\ln g_{e1} = \ln 3 = 1.099$$

$$\frac{\overline{S}}{R} = 3.5 + 17.08 + 4.79 + 0.0110 + 0.0501 + 1.099 = 26.54$$

The standard molar entropy of O_2 at 500 K is 220.6 J·K^{-1}·mol^{-1}.

Because H_2O is a bent polyatomic molecule, we treat it as we did NO_2 in Problem 21–37. From Table 18.4, $\sigma = 2$, $\Theta_{rot,A} = 40.1$ K, $\Theta_{rot,B} = 20.9$ K, $\Theta_{rot,C} = 13.4$ K, $\Theta_{vib,1} = 5360$ K, $\Theta_{vib,2} = 5160$ K, and $\Theta_{vib,3} = 2290$ K. Then

$$\left(\frac{2\pi M k_B T}{h^2}\right)^{3/2} = \left[\frac{2\pi(2.991 \times 10^{-26}\text{ kg})(1.381 \times 10^{-23}\text{ J·K}^{-1})(500\text{ K})}{(6.626 \times 10^{34}\text{ J·s})^2}\right]^{3/2} = 1.607 \times 10^{32}\text{ m}^{-3}$$

$$\frac{\overline{V}}{N_A} = \frac{RT}{N_A P} = \frac{(0.08314\text{ dm}^3\text{·bar·mol}^{-1}\text{·K}^{-1})(500\text{ K})}{(6.022 \times 10^{23}\text{ mol}^{-1})(1\text{ bar})} = 6.903 \times 10^{-26}\text{ m}^3$$

$$\frac{1}{2}\ln\left(\frac{T^3}{\Theta_{rot,A}\Theta_{rot,B}\Theta_{rot,C}}\right) = 4.66$$

$$\sum_{j=1}^{6}\ln(1 - e^{-\Theta_{vib,j}/T}) = -0.0104$$

$$\sum_{j=1}^{6}\frac{\Theta_{vib,j}}{T}\left(\frac{e^{-\Theta_{vib,j}/T}}{1 - e^{-\Theta_{vib,j}/T}}\right) = 0.048$$

The standard molar entropy is then

$$\frac{\overline{S}}{R} = 4 + \ln\left[\left(\frac{2\pi M k_B T}{h^2}\right)^{3/2}\frac{\overline{V}}{N_A}\right] + \ln\left(\frac{\pi^{1/2}}{\sigma}\right) + \frac{1}{2}\ln\frac{T^3}{\Theta_{rot,A}\Theta_{rot,B}\Theta_{rot,C}}$$

$$- \sum_{j=1}^{3}\ln(1 - e^{-\Theta_{vib,j}/T}) + \sum_{j=1}^{3}\left[\frac{(\Theta_{vib,j}/T)e^{-\Theta_{vib,j}/T}}{1 - e^{-\Theta_{vib,j}/T}}\right] + \ln g_{e1}$$

$$= 4 + 16.22 - 0.121 + 4.66 + 0.0104 + 0.048 = 24.82$$

which gives a value of $\overline{S} = 206.3$ J·K^{-1}·mol^{-1}.

Finally, we can calculate the value of $\Delta_r S°$ for the reaction above, as we did in the previous problem.

$$\Delta_r S° = S°[H_2O] - \tfrac{1}{2}S°[O_2] - S°[H_2]$$
$$= (206.3 \text{ J·K}^{-1}\text{·mol}^{-1}) - \tfrac{1}{2}(220.6 \text{ J·K}^{-1}\text{·mol}^{-1}) - (145.6 \text{ J·K}^{-1}\text{·mol}^{-1})$$
$$= -49.6 \text{ J·K}^{-1}\text{·mol}^{-1}$$

21–40. In each case below, predict which molecule of the pair has the greater molar entropy under the same conditions (assume gaseous species).

a. CO CO_2

b. $CH_3CH_2CH_3$

$$\begin{array}{c} H_2C\!\!-\!\!CH_2 \\ \diagdown\diagup \\ CH_2 \end{array}$$

c. $CH_3CH_2CH_2CH_2CH_3$

$$H_3C\!-\!\underset{\underset{CH_3}{|}}{\overset{\overset{CH_3}{|}}{C}}\!-\!CH_3$$

---

a. CO_2 (more atoms)
b. $CH_3CH_2CH_3$ (more flexibility)
c. $CH_3CH_2CH_2CH_2CH_3$ (more flexibility)

21–41. In each case below, predict which molecule of the pair has the greater molar entropy under the same conditions (assume gaseous species).

a. H_2O D_2O

b. CH_3CH_2OH

$$\begin{array}{c} H_2C\!\!-\!\!CH_2 \\ \diagdown\diagup \\ O \end{array}$$

c. $CH_3CH_2CH_2CH_2NH_2$

$$\begin{array}{c} \overset{H}{\underset{}{|}} \\ \overset{}{N} \\ H_2C\diagup\ \ \diagdown CH_2 \\ |\ \ \ \ \ \ \ \ | \\ H_2C\!-\!\!-\!\!-\!\!CH_2 \end{array}$$

---

a. D_2O (larger mass)
b. CH_3CH_2OH (more flexibility)
c. $CH_3CH_2CH_2CH_2NH_2$ (more flexibility)

21–42. Arrange the following reactions according to increasing values of $\Delta_r S°$ (do not consult any references).

a. $S(s) + O_2(g) \longrightarrow SO_2(g)$

b. $H_2(g) + O_2(g) \longrightarrow H_2O_2(l)$

c. $CO(g) + 3\,H_2(g) \longrightarrow CH_4(g) + H_2O(l)$

d. $C(s) + H_2O(g) \longrightarrow CO(g) + H_2(g)$

Recall that molar entropies of solids and liquids are much smaller than those of gases, so we can ignore the contribution of the solids and liquids to $\Delta_r S°$ when we order these reactions. Considering only the gaseous products and reactants, we can find Δn for each reaction to be

a. $\Delta n = 0$ **b.** $\Delta n = -2$ **c.** $\Delta n = -3$ **d.** $\Delta n = +1$

The correct ordering of the reactions is therefore **d > a > b > c.**

21–43. Arrange the following reactions according to increasing values of $\Delta_r S°$ (do not consult any references).

a. $2\,H_2(g) + O_2(g) \longrightarrow 2\,H_2O(l)$

b. $NH_3(g) + HCl(g) \longrightarrow NH_4Cl(s)$

c. $K(s) + O_2(g) \longrightarrow KO_2(s)$

d. $N_2(g) + 3\,H_2(g) \longrightarrow 2\,NH_3(g)$

Again, calculating Δn for each reaction for the gaseous products and reactants gives

a. $\Delta n = -3$ **b.** $\Delta n = -2$ **c.** $\Delta n = -1$ **d.** $\Delta n = -2$

The correct ordering of the reactions is therefore **c > b ≈ d > a.**

21–44. In Problem 21–40, you are asked to predict which molecule, $CO(g)$ or $CO_2(g)$, has the greater molar entropy. Use the data in Tables 18.2 and 18.4 to calculate the standard molar entropy of $CO(g)$ and $CO_2(g)$ at 298.15 K. Does this calculation confirm your intuition? Which degree of freedom makes the dominant contribution to the molar entropy of CO? Of CO_2?

In Problem 21–38 and Example 21–5, we used the data in Tables 18.2 and 18.4 to find that the standard molar entropy of $CO(g)$ is $197.6\ \mathrm{J\cdot K^{-1}\cdot mol^{-1}}$ and that of $CO_2(g)$ is $213.8\ \mathrm{J\cdot K^{-1}\cdot mol^{-1}}$. In both cases, the translational degrees of freedom make the dominant contribution to the molar entropy.

21–45. Table 21.2 gives $\overline{S}°[CH_3OH(l)] = 126.8\ \mathrm{J\cdot K^{-1}\cdot mol^{-1}}$ at 298.15 K. Given that $T_{vap} = 337.7$ K, $\Delta_{vap}\overline{H}(T_b) = 36.5\ \mathrm{kJ\cdot mol^{-1}}$, $\overline{C}_P[CH_3OH(l)] = 81.12\ \mathrm{J\cdot K^{-1}\cdot mol^{-1}}$, and $\overline{C}_P[CH_3OH(g)] = 43.8\ \mathrm{J\cdot K^{-1}\cdot mol^{-1}}$, calculate the value of $\overline{S}°[CH_3OH(g)]$ at 298.15 K and compare your answer with the experimental value of $239.8\ \mathrm{J\cdot K^{-1}\cdot mol^{-1}}$.

This is done in the same way as $\overline{S}°[Br_2(g)]$ was found in Section 21–7. First, we heat the methanol to its boiling point:

$$\Delta\overline{S}_1 = \overline{S}^l(337.7\ \mathrm{K}) - \overline{S}^l(298.15\ \mathrm{K}) = \overline{C}_P^l \ln\frac{T_2}{T_1}$$

$$= (81.12\ \mathrm{J\cdot K^{-1}\cdot mol^{-1}})\ln\frac{337.7}{298.15} = 10.10\ \mathrm{J\cdot K^{-1}\cdot mol^{-1}}$$

Then vaporize the methanol at its normal boiling point:

$$\Delta\overline{S}_2 = \overline{S}^g(337.7\ \mathrm{K}) - \overline{S}^l(337.7\ \mathrm{K}) = \frac{\Delta_{vap}\overline{H}}{T_{vap}}$$

$$= \frac{36\,500\ \mathrm{J\cdot mol^{-1}}}{337.7\ \mathrm{K}} = 108.1\ \mathrm{J\cdot K^{-1}\cdot mol^{-1}}$$

Finally, cool the gas back down to 298.15 K:

$$\Delta\overline{S}_3 = \overline{S}^g(298.15\ \mathrm{K}) - \overline{S}^g(337.7\ \mathrm{K}) = \overline{C}_P^g \ln\frac{T_2}{T_1}$$

$$= (43.8\ \mathrm{J\cdot K^{-1}\cdot mol^{-1}})\ln\frac{298.15}{337.7} = -5.456\ \mathrm{J\cdot K^{-1}\cdot mol^{-1}}$$

The sum of these three steps plus $\overline{S}^{\circ}_{298}[CH_3OH(l)] = 126.8 \text{ J} \cdot \text{K}^{-1} \cdot \text{mol}^{-1}$ will be the desired value:

$$\overline{S}^{\circ}_{298}[CH_3OH(g)] = \overline{S}^{\circ}_{298}[CH_3OH(l)] + \Delta\overline{S}_1 + \Delta\overline{S}_2 + \Delta\overline{S}_3$$
$$= 239.5 \text{ J} \cdot \text{K}^{-1} \cdot \text{mol}^{-1}$$

which is within 0.1% of the experimental value.

21–46. Given the following data, $T_{vap} = 373.15 \text{ K}$, $\Delta\overline{H}_{vap}(T_{vap}) = 40.65 \text{ kJ} \cdot \text{mol}^{-1}$, $\overline{C}_P[H_2O(l)] = 75.3 \text{ J} \cdot \text{K}^{-1} \cdot \text{mol}^{-1}$, and $\overline{C}_P[H_2O(g)] = 33.8 \text{ J} \cdot \text{K}^{-1} \cdot \text{mol}^{-1}$, show that the values of $\overline{S}^{\circ}[H_2O(l)]$ and $\overline{S}^{\circ}[H_2O(g)]$ in Table 21.2 are consistent.

This is done in the same way as Problem 21–45. First, we heat the water to its boiling point:

$$\Delta\overline{S}_1 = \overline{S}^l(373.15 \text{ K}) - \overline{S}^l(298.15 \text{ K}) = \overline{C}_P^l \ln\frac{T_2}{T_1}$$

$$= (75.3 \text{ J} \cdot \text{K}^{-1} \cdot \text{mol}^{-1}) \ln\frac{373.15}{298.15} = 16.90 \text{ J} \cdot \text{K}^{-1} \cdot \text{mol}^{-1}$$

Then vaporize the water at its normal boiling point:

$$\Delta\overline{S}_2 = \overline{S}^g(373.15 \text{ K}) - \overline{S}^l(373.15 \text{ K}) = \frac{\Delta_{vap}\overline{H}}{T_{vap}}$$

$$= \frac{40\,650 \text{ J} \cdot \text{mol}^{-1}}{373.15 \text{ K}} = 108.9 \text{ J} \cdot \text{K}^{-1} \cdot \text{mol}^{-1}$$

Finally, cool the gas back down to 298.15 K:

$$\Delta\overline{S}_3 = \overline{S}^g(298.15 \text{ K}) - \overline{S}^g(373.15 \text{ K}) = \overline{C}_P^g \ln\frac{T_2}{T_1}$$

$$= (33.8 \text{ J} \cdot \text{K}^{-1} \cdot \text{mol}^{-1}) \ln\frac{298.15}{373.15} = -7.584 \text{ J} \cdot \text{K}^{-1} \cdot \text{mol}^{-1}$$

The sum of these three steps plus $S^{\circ}_{298}[H_2O(l)] = 70.0 \text{ J} \cdot \text{K}^{-1} \cdot \text{mol}^{-1}$ will be the desired value:

$$\overline{S}^{\circ}_{298}[H_2O(g)] = \overline{S}^{\circ}_{298}[H_2O(l)] + \Delta\overline{S}_1 + \Delta\overline{S}_2 + \Delta\overline{S}_3$$
$$= 188.2 \text{ J} \cdot \text{K}^{-1} \cdot \text{mol}^{-1}$$

which is within 0.4% of the value in Table 21.2.

21–47. Use the data in Table 21.2 to calculate the value of $\Delta_r S^{\circ}$ for the following reactions at 25°C and one bar.

a. $C(s, \text{graphite}) + O_2(g) \longrightarrow CO_2(g)$

b. $CH_4(g) + 2\,O_2(g) \longrightarrow CO_2(g) + 2\,H_2O(l)$

c. $C_2H_2(g) + H_2(g) \longrightarrow C_2H_4(g)$

---

a. $\Delta_r S^{\circ} = S^{\circ}[\text{products}] - S^{\circ}[\text{reactants}]$
$$= 213.8 \text{ J} \cdot \text{K}^{-1} \cdot \text{mol}^{-1} - 205.2 \text{ J} \cdot \text{K}^{-1} \cdot \text{mol}^{-1} - 5.74 \text{ J} \cdot \text{K}^{-1} \cdot \text{mol}^{-1}$$
$$= 2.86 \text{ J} \cdot \text{K}^{-1} \cdot \text{mol}^{-1}$$

b. $\Delta_r S^\circ = S^\circ[\text{products}] - S^\circ[\text{reactants}]$

$= 2(70.0 \text{ J} \cdot \text{K}^{-1} \cdot \text{mol}^{-1}) + 213.8 \text{ J} \cdot \text{K}^{-1} \cdot \text{mol}^{-1} - 2(205.2 \text{ J} \cdot \text{K}^{-1} \cdot \text{mol}^{-1}) - 186.3 \text{ J} \cdot \text{K}^{-1} \cdot \text{mol}^{-1}$

$= -242.9 \text{ J} \cdot \text{K}^{-1} \cdot \text{mol}^{-1}$

c. $\Delta_r S^\circ = S^\circ[\text{products}] - S^\circ[\text{reactants}]$

$= 219.6 \text{ J} \cdot \text{K}^{-1} \cdot \text{mol}^{-1} - 130.7 \text{ J} \cdot \text{K}^{-1} \cdot \text{mol}^{-1} - 200.9 \text{ J} \cdot \text{K}^{-1} \cdot \text{mol}^{-1}$

$= -112.0 \text{ J} \cdot \text{K}^{-1} \cdot \text{mol}^{-1}$

21–48. Use the data in Table 21.2 to calculate the value of $\Delta_r S^\circ$ for the following reactions at 25°C and one bar.

a. $CO(g) + 2 H_2(g) \longrightarrow CH_3OH(l)$

b. $C(s, \text{graphite}) + H_2O(l) \longrightarrow CO(g) + H_2(g)$

c. $2 CO(g) + O_2(g) \longrightarrow 2 CO_2(g)$

---

a. $\Delta_r S^\circ = S^\circ[\text{products}] - S^\circ[\text{reactants}]$

$= 126.8 \text{ J} \cdot \text{K}^{-1} \cdot \text{mol}^{-1} - 197.7 \text{ J} \cdot \text{K}^{-1} \cdot \text{mol}^{-1} - 2(130.7 \text{ J} \cdot \text{K}^{-1} \cdot \text{mol}^{-1})$

$= -332.3 \text{ J} \cdot \text{K}^{-1} \cdot \text{mol}^{-1}$

b. $\Delta_r S^\circ = S^\circ[\text{products}] - S^\circ[\text{reactants}]$

$= 130.7 \text{ J} \cdot \text{K}^{-1} \cdot \text{mol}^{-1} + 197.7 \text{ J} \cdot \text{K}^{-1} \cdot \text{mol}^{-1} - 70.0 \text{ J} \cdot \text{K}^{-1} \cdot \text{mol}^{-1} - 5.74 \text{ J} \cdot \text{K}^{-1} \cdot \text{mol}^{-1}$

$= 252.66 \text{ J} \cdot \text{K}^{-1} \cdot \text{mol}^{-1}$

c. $\Delta_r S^\circ = S^\circ[\text{products}] - S^\circ[\text{reactants}]$

$= 2(213.8 \text{ J} \cdot \text{K}^{-1} \cdot \text{mol}^{-1}) - 205.2 \text{ J} \cdot \text{K}^{-1} \cdot \text{mol}^{-1} - 2(197.7 \text{ J} \cdot \text{K}^{-1} \cdot \text{mol}^{-1})$

$= -173.0 \text{ J} \cdot \text{K}^{-1} \cdot \text{mol}^{-1}$

CHAPTER 22

Helmholtz and Gibbs Energies

PROBLEMS AND SOLUTIONS

22–1. The molar enthalpy of vaporization of benzene at its normal boiling point (80.09°C) is 30.72 kJ·mol^{-1}. Assuming that $\Delta_{vap}\overline{H}$ and $\Delta_{vap}\overline{S}$ stay constant at their values at 80.09°C, calculate the value of $\Delta_{vap}\overline{G}$ at 75.0°C, 80.09°C, and 85.0°C. Interpret these results physically.

We can write (as in Section 22–2)

$$\Delta_{vap}\overline{G} = \Delta_{vap}\overline{H} - T\Delta_{vap}\overline{S}$$

At the boiling point of benzene, the liquid and vapor phases are in equilibrium, so $\Delta_{vap}\overline{G} = 0$. Thus, at 80.09°C,

$$0 = 30.72 \text{ kJ·mol}^{-1} - (353.24 \text{ K})\Delta_{vap}\overline{S}$$

$$\Delta_{vap}\overline{S} = 86.97 \text{ J·K}^{-1}\cdot\text{mol}^{-1}$$

Since $\Delta_{vap}\overline{H}$ and $\Delta_{vap}\overline{S}$ are assumed to stay constant at their boiling-point values, we know their numerical values and can substitute into our first equation:

$$\Delta_{vap}\overline{G}(75.0°C) = 30.72 \text{ kJ·mol}^{-1} - (348.15 \text{ K})(86.97 \text{ J·K}^{-1}\cdot\text{mol}^{-1}) = 441.4 \text{ J·mol}^{-1}$$

$$\Delta_{vap}\overline{G}(85.0°C) = 30.72 \text{ kJ·mol}^{-1} - (358.15 \text{ K})(86.97 \text{ J·K}^{-1}\cdot\text{mol}^{-1}) = -428.3 \text{ J·mol}^{-1}$$

From these values, we can see that at 75.0°C benzene will spontaneously condense, whereas at 85.0°C it will spontaneously evaporate (just as we would expect).

22–2. Redo Problem 22–1 without assuming that $\Delta_{vap}\overline{H}$ and $\Delta_{vap}\overline{S}$ do not vary with temperature. Take the molar heat capacities of liquid and gaseous benzene to be 136.3 J·K^{-1}·mol^{-1} and 82.4 J·K^{-1}·mol^{-1}, respectively. Compare your results with those you obtained in Problem 22–1. Are any of your physical interpretations different?

We wish to consider the temperature variation of $\Delta_{vap}\overline{G}$, so we must use Equation 22.31a,

$$\left(\frac{\partial \Delta_{vap}\overline{G}}{\partial T}\right)_P = -\Delta_{vap}\overline{S}(T)$$

where (as in Example 20–5)

$$\Delta_{vap}\overline{S}(T) = \Delta_{vap}\overline{S}(80.09°\ C) + \int_{353.24\ K}^{T} \frac{\Delta\overline{C}_P}{T}dT$$

$$= 86.97\ \text{J·K}^{-1}\text{·mol}^{-1} - (53.9\ \text{J·K}^{-1}\text{·mol}^{-1})\ln\frac{T}{353.24\ K}$$

$$= 403.2\ \text{J·K}^{-1}\text{·mol}^{-1} - (53.9\ \text{J·K}^{-1}\text{·mol}^{-1})\ln(T/K)$$

Substituting into Equation 22.31a, we can write

$$\Delta_{vap}\overline{G}(T) - \Delta_{vap}\overline{G}(353.24\ K) = -\int_{353.24\ K}^{T}\Big[403.2\ \text{J·K}^{-1}\text{·mol}^{-1}$$

$$-(53.9\ \text{J·K}^{-1}\text{·mol}^{-1})\ln(T/K)\Big]dT$$

$$= -(403.2\ \text{J·K}^{-1}\text{·mol}^{-1})(T - 353.24\ K)$$

$$+(53.9\ \text{J·K}^{-1}\text{·mol}^{-1})\left[T\ln(T/K) - T - 1719.3\ K\right]$$

Letting $T = 348.15\ K$ gives

$$\Delta_{vap}\overline{G}(T) = 2052\ \text{J·mol}^{-1} - 1608\ \text{J·mol}^{-1} = +444\ \text{J·mol}^{-1}$$

and letting $T = 358.15\ K$ gives

$$\Delta_{vap}\overline{G}(T) = -1980\ \text{J·mol}^{-1} + 1555\ \text{J·mol}^{-1} = -425\ \text{J·mol}^{-1}$$

Notice that taking the temperature variation of $\Delta_{vap}\overline{H}$ and $\Delta_{vap}\overline{S}$ into account made little difference over such a small temperature range.

22–3. Substitute $(\partial P/\partial T)_{\overline{V}}$ from the van der Waals equation into Equation 22.19 and integrate from $\overline{V}^{id}$ to $\overline{V}$ to obtain

$$\overline{S}(T,\overline{V}) - \overline{S}^{id}(T) = R\ln\frac{\overline{V} - b}{\overline{V}^{id} - b}$$

Now let $\overline{V}^{id} = RT/P^{id}$, $P^{id} = P° = 1$ bar, and $\overline{V}^{id} \gg b$ to obtain

$$\overline{S}(T,\overline{V}) - \overline{S}^{id}(T) = -R\ln\frac{RT/P°}{\overline{V} - b}$$

Given that $\overline{S}^{id} = 246.35\ \text{J·mol}^{-1}\text{·K}^{-1}$ for ethane at 400 K, show that

$$\overline{S}(\overline{V})/\text{J·mol}^{-1}\text{·K}^{-1} = 246.35 - 8.3145\ln\frac{33.258\ \text{L·mol}^{-1}}{\overline{V} - 0.065144\ \text{L·mol}^{-1}}$$

Calculate $\overline{S}$ as a function of $\rho = 1/\overline{V}$ for ethane at 400 K and compare your results with the experimental results shown in Figure 22.2. Show that

$$\overline{S}(\overline{V})/\text{J·mol}^{-1}\text{·K}^{-1} = 246.35 - 8.3145\ln\frac{33.258\ \text{L·mol}^{-1}}{\overline{V} - 0.045153\ \text{L·mol}^{-1}}$$

$$+ 13.68\ln\frac{\overline{V} + 0.045153\ \text{L·mol}^{-1}}{\overline{V}}$$

for the Redlich-Kwong equation for ethane at 400 K. Calculate $\overline{S}$ as a function of $\rho = 1/\overline{V}$ and compare your results with the experimental results shown in Figure 22.2.

From the van der Waals equation,

$$\left(\frac{\partial P}{\partial T}\right)_V = \frac{R}{\overline{V} - b}$$

We can substitute into Equation 22.19 to find an expression for $d\overline{S}$:

$$\left(\frac{\partial \overline{S}}{\partial \overline{V}}\right)_T = \left(\frac{\partial P}{\partial T}\right)_V$$

$$d\overline{S} = \frac{R}{\overline{V} - b}d\overline{V}$$

Integrating both sides of this equation gives

$$\overline{S}(T, \overline{V}) - \overline{S}^{\text{id}}(T) = R \ln \frac{\overline{V} - b}{\overline{V}^{\text{id}} - b}$$

Since $\overline{V}^{\text{id}}$ is quite large compared to b, we can neglect b in $\overline{V}^{\text{id}} - b$. Letting $\overline{V}^{\text{id}} = RT/P^{\text{id}}$ and $P^{\text{id}} = P^\circ = 1$ bar, we find

$$\overline{S}(T, \overline{V}) - \overline{S}^{\text{id}}(T) = -R \ln \frac{\overline{V}^{\text{id}}}{\overline{V} - b}$$

$$= -R \ln \frac{RT/P^\circ}{\overline{V} - b}$$

$$\overline{S}(T, \overline{V}) = \overline{S}^{\text{id}}(T) - R \ln \frac{RT}{\overline{V} - b}$$

where, in the last equality, the pressure units of R are in bars. For ethane, $a = 5.5818 \text{ dm}^6 \cdot \text{bar} \cdot \text{mol}^{-2}$ and $b = 0.065144 \text{ dm}^3 \cdot \text{mol}^{-1}$ (Table 16.3), so

$$\overline{S}(\overline{V}) = 246.35 \text{ J} \cdot \text{K}^{-1} \cdot \text{mol}^{-1} - (8.3145 \text{ J} \cdot \text{K}^{-1} \cdot \text{mol}^{-1}) \ln \frac{33.258 \text{ L} \cdot \text{mol}^{-1}}{\overline{V} - 0.065144 \text{ L} \cdot \text{mol}^{-1}}$$

To make graphing entropy vs. density easier, we can break up the logarithmic term and then graph, as shown:

$$\overline{S}/\text{J} \cdot \text{K}^{-1} \cdot \text{mol}^{-1} = 246.35 - (8.3145) \ln 33.258$$

$$+ (8.3145) \ln \left(\frac{\text{mol} \cdot \text{dm}^{-3}}{\rho} - 0.065144\right)$$

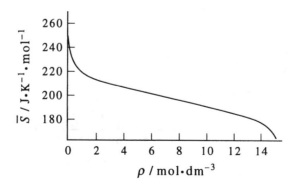

For a Redlich-Kwong gas,

$$P = \frac{RT}{\overline{V} - B} - \frac{A}{T^{1/2}\overline{V}(\overline{V} + B)}$$

and

$$\left(\frac{\partial P}{\partial T}\right)_V = \frac{R}{\overline{V} - B} + \frac{A}{2\overline{V}(\overline{V} + B)T^{3/2}}$$

Using Equation 22.19 and integrating (keeping temperature constant) gives

$$\left(\frac{\partial \overline{S}}{\partial \overline{V}}\right)_T = \frac{R}{\overline{V} - B} + \frac{A}{2\overline{V}(\overline{V} + B)T^{3/2}}$$

$$\int d\overline{S} = \int \left[\frac{R}{\overline{V} - B} + \frac{A}{2\overline{V}(\overline{V} + B)T^{3/2}}\right] d\overline{V}$$

$$\overline{S} - \overline{S}^{id} = R \ln \frac{\overline{V} - B}{\overline{V}^{id} - B} - \frac{A}{2T^{3/2}}\left[-\frac{1}{B}\ln\left(\frac{B + \overline{V}}{\overline{V}}\right) + \frac{1}{B}\ln\left(\frac{B + \overline{V}^{id}}{\overline{V}^{id}}\right)\right]$$

Since for an ideal gas B is negligible compared to $\overline{V}$,

$$\overline{S} - \overline{S}^{id} = R \ln \frac{\overline{V} - B}{\overline{V}^{id}} + \frac{A}{2BT^{3/2}} \ln \frac{\overline{V}}{\overline{V} + B}$$

Then, for ethane, since $A = 98.831$ dm^6·bar·mol^{-2}·K$^{1/2}$ and $B = 0.045153$ dm^3·mol^{-1} (Table 16.4),

$$\overline{S} = \overline{S}^{id} - R \ln \frac{\overline{V}^{id}}{\overline{V} - B} + \frac{A}{2BT^{3/2}} \ln \frac{\overline{V} + B}{\overline{V}}$$

$$\overline{S}/\text{J·K}^{-1}\text{·mol}^{-1} = 246.35 - 8.3145 \ln \frac{33.258 \text{ dm}^3\text{·mol}^{-1}}{\overline{V} - 0.045153 \text{ dm}^3\text{·mol}^{-1}}$$
$$+ 13.68 \ln \frac{\overline{V} + 0.045153 \text{ dm}^3\text{·mol}^{-1}}{\overline{V}}$$

To make graphing entropy vs. density easier, we can break up the logarithmic term and then graph, as shown. We have divided both numerator and denominator of the logarithmic terms in the previous expression by dm^3·mol^{-1}.

$$\overline{S}/\text{J·K}^{-1}\text{·mol}^{-1} = 246.35 - 8.3145\left[\ln 33.258 - \ln\left(\frac{1}{\rho} - 0.045153\right)\right]$$
$$+ 13.68\left[\ln\left(\frac{1}{\rho} + 0.045153\right) - \ln\frac{1}{\rho}\right]$$

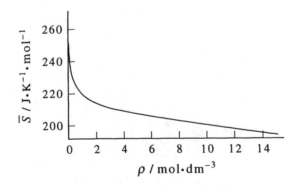

22–4. Use the van der Waals equation to derive

$$\overline{U}(T, \overline{V}) - \overline{U}^{id}(T) = -\frac{a}{\overline{V}}$$

Use this result along with the van der Waals equation to calculate the value of $\overline{U}$ as a function of $\overline{V}$ for ethane at 400 K, given that $\overline{U}^{id} = 14.55$ kJ·mol^{-1}. To do this, specify $\overline{V}$ (from 0.0700 L·mol^{-1} to 7.00 L·mol^{-1}, see Figure 22.2), calculate both $\overline{U}(\overline{V})$ and $P(\overline{V})$, and plot $\overline{U}(\overline{V})$ versus $P(\overline{V})$. Compare your result with the experimental data in Figure 22.3. Use the Redlich-Kwong equation to derive

$$\overline{U}(T, \overline{V}) - \overline{U}^{id}(T) = -\frac{3A}{2BT^{1/2}} \ln \frac{\overline{V} + B}{\overline{V}}$$

Repeat the above calculation for ethane at 400 K.

Begin with Equation 22.22,

$$\left(\frac{\partial U}{\partial V}\right)_T = -P + T\left(\frac{\partial P}{\partial T}\right)_V$$

In Problem 22–3, we found $(\partial P/\partial T)_V$ for the van der Waals and Redlich-Kwong equations. For the van der Waals equation, Equation 22.22 becomes

$$\left(\frac{\partial U}{\partial V}\right)_T = -\frac{RT}{\overline{V} - b} + \frac{a}{\overline{V}^2} + \frac{RT}{\overline{V} - b} = \frac{a}{\overline{V}^2}$$

$$\int_{\overline{U}^{id}}^{\overline{U}} dU = \int_{\overline{V}^{id}}^{\overline{V}} \frac{a}{\overline{V}^2} d\overline{V}$$

$$\overline{U} - \overline{U}^{id} = -\frac{a}{\overline{V}}$$

The van der Waals constants for ethane are listed in the previous problem. Substituting, we find that

$$\overline{U} = 14.55 \text{ kJ·mol}^{-1} - \frac{0.55818 \text{ kJ·dm}^3·\text{mol}^{-2}}{\overline{V}}$$

$$P = \frac{(0.083145 \text{ dm}^3·\text{bar·mol}^{-1}·\text{K}^{-1})(400 \text{ K})}{\overline{V} - 0.065144 \text{ dm}^3·\text{mol}^{-1}} - \frac{5.5818 \text{ dm}^6·\text{bar·mol}^{-1}}{\overline{V}^2}$$

We can use a parametric plot to plot $\overline{U}(\overline{V})$ vs. $P(\overline{V})$:

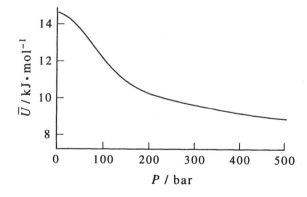

For the Redlich-Kwong equation, Equation 22.22 becomes

$$\left(\frac{\partial U}{\partial V}\right)_T = -\frac{RT}{\overline{V} - B} + \frac{A}{T^{1/2}\overline{V}(\overline{V} + B)} + \frac{RT}{\overline{V} - B} + \frac{A}{2T^{1/2}\overline{V}(\overline{V} + B)}$$

$$\int_{\overline{U}^{id}}^{\overline{U}} dU = \int_{\overline{V}^{id}}^{\overline{V}} \frac{3A}{2T^{1/2}\overline{V}(\overline{V}+B)} d\overline{V}$$

$$\overline{U} - \overline{U}^{id} = \frac{3A}{2T^{1/2}}\left(-\frac{1}{B}\ln\frac{\overline{V}+B}{\overline{V}} + \frac{1}{B}\ln\frac{\overline{V}^{id}+B}{\overline{V}^{id}}\right)$$

$$\overline{U} = \overline{U}^{id} - \frac{3A}{2BT^{1/2}}\ln\frac{\overline{V}+B}{\overline{V}}$$

The Redlich-Kwong constants for ethane are listed in the previous problem. Substituting, we find that

$$\overline{U} = 14.55 \text{ kJ}\cdot\text{mol}^{-1} - \frac{3(9.8831 \text{ kJ}\cdot\text{dm}^3\cdot\text{mol}^{-2}\cdot\text{K}^{1/2})}{2(0.045153 \text{ dm}^3\cdot\text{mol}^{-1})(400 \text{ K})^{1/2}}\ln\frac{\overline{V}+0.045153 \text{ dm}^3\cdot\text{mol}^{-1}}{\overline{V}}$$

$$P = \frac{(0.083145 \text{ dm}^3\cdot\text{bar}\cdot\text{mol}^{-1}\cdot\text{K}^{-1})(400 \text{ K})}{\overline{V} - 0.045153 \text{ dm}^3\cdot\text{mol}^{-1}} - \frac{98.831 \text{ dm}^6\cdot\text{bar}\cdot\text{mol}^{-2}\cdot\text{K}^{1/2}}{(400 \text{ K})^{1/2}\overline{V}(\overline{V}+0.045153 \text{ dm}^3\cdot\text{mol}^{-1})}$$

Again, use a parametric plot to plot $\overline{U}(\overline{V})$ vs. $P(\overline{V})$:

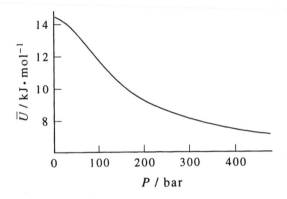

22–5. Show that $(\partial U/\partial V)_T = 0$ for a gas that obeys an equation of state of the form $Pf(V) = RT$. Give two examples of such equations of state that appear in the text.

We first take the partial derivative of P with respect to T, keeping V constant:

$$Pf(V) = RT$$

$$\left(\frac{\partial P}{\partial T}\right)_V = \frac{R}{f(V)}$$

From Equation 22.22, we can write

$$\left(\frac{\partial U}{\partial V}\right)_T = -P + T\left(\frac{\partial P}{\partial T}\right)_V = -P + \frac{RT}{f(V)} = -P + P = 0$$

Two such equations of state are the ideal gas equation and $P(\overline{V} - b) = RT$.

22–6. Show that

$$\left(\frac{\partial \overline{U}}{\partial \overline{V}}\right)_T = \frac{RT^2}{\overline{V}^2}\frac{dB_{2V}}{dT} + \frac{RT^2}{\overline{V}^3}\frac{dB_{3V}}{dT} + \cdots$$

Begin with Equation 22.22,

$$\left(\frac{\partial \overline{U}}{\partial \overline{V}}\right)_T = -P + T\left(\frac{\partial P}{\partial T}\right)_V$$

We can use the virial expansion in the volume to express Z as (Equation 16.22)

$$Z = 1 + B_{2V}\overline{V}^{-1} + B_{3V}\overline{V}^{-2} + O(\overline{V}^{-3})$$

$$P = RT\overline{V}^{-1} + B_{2V}RT\overline{V}^{-2} + B_{3V}RT\overline{V}^{-3} + O(\overline{V}^{-4})$$

$$\left(\frac{\partial P}{\partial T}\right)_V = \frac{R}{\overline{V}} + B_{2V}\frac{R}{\overline{V}^2} + \frac{dB_{2V}}{dT}\frac{RT}{\overline{V}^2} + B_{3V}\frac{R}{\overline{V}^3} + \frac{dB_{3V}}{dT}\frac{RT}{\overline{V}^3} + O(\overline{V}^{-4})$$

Substituting into Equation 22.22 gives

$$\left(\frac{\partial \overline{U}}{\partial \overline{V}}\right)_T = -\left(\frac{RT}{\overline{V}} + B_{2V}\frac{RT}{\overline{V}^2} + B_{3V}\frac{RT}{\overline{V}^3}\right) + \frac{RT}{\overline{V}} + O(\overline{V}^{-4})$$

$$+ B_{2V}\frac{RT}{\overline{V}^2} + \frac{dB_{2V}}{dT}\frac{RT^2}{\overline{V}^2} + B_{3V}\frac{RT}{\overline{V}^3} + \frac{dB_{3V}}{dT}\frac{RT^2}{\overline{V}^3} + O(\overline{V}^{-4})$$

$$= \frac{RT^2}{\overline{V}^2}\frac{dB_{2V}}{dT} + \frac{RT^2}{\overline{V}^3}\frac{dB_{3V}}{dT} + O(\overline{V}^{-4})$$

22–7. Use the result of the previous problem to show that

$$\Delta \overline{U} = -T\frac{dB_{2V}}{dT}(P_2 - P_1) + \cdots$$

Use Equation 16.41 for the square-well potential to show that

$$\Delta \overline{U} = -\frac{2\pi\sigma^3 N_A}{3}(\lambda^3 - 1)\frac{\varepsilon}{k_B T}e^{\varepsilon/k_B T}(P_2 - P_1) + \cdots$$

Given that $\sigma = 327.7$ pm, $\varepsilon/k_B = 95.2$ K, and $\lambda = 1.58$ for $N_2(g)$, calculate the value of $\Delta \overline{U}$ for a pressure increase from 1.00 bar to 10.0 bar at 300 K.

We integrate the equation we found in the previous problem (keeping T constant):

$$\left(\frac{\partial \overline{U}}{\partial \overline{V}}\right)_T = \frac{RT^2}{\overline{V}^2}\frac{dB_{2V}}{dT} + \frac{RT^2}{\overline{V}^3}\frac{dB_{3V}}{dT} + O(\overline{V}^{-4})$$

$$d\overline{U} = \frac{RT^2}{\overline{V}^2}\frac{dB_{2V}}{dT}d\overline{V} + O(\overline{V}^{-3})$$

$$\Delta \overline{U} = -\frac{RT^2}{\overline{V}}\frac{dB_{2V}}{dT}\bigg|_{\overline{V}_1}^{\overline{V}_2}$$

Substitute in for $\overline{V} = RT/P$ to get

$$\Delta \overline{U} = -T(P_2 - P_1)\frac{dB_{2V}}{dT} + \cdots$$

Using Equation 16.41,

$$B_{2V}(T) = \frac{2\pi\sigma^3 N_A}{3}\left[1 - (\lambda^3 - 1)\left(e^{\varepsilon/k_B T} - 1\right)\right]$$

$$\frac{dB_{2V}}{dT} = \frac{2\pi\sigma^3 N_A}{3}(\lambda^3 - 1)e^{\varepsilon/k_B T}\left(\frac{\varepsilon}{k_B T^2}\right)$$

Substitute into the equation for $\Delta\overline{U}$ to find

$$\Delta\overline{U} = -\frac{2\pi\sigma^3 N_A}{3}(\lambda^3 - 1)\frac{\varepsilon}{k_B T}e^{\varepsilon/k_B T}(P_2 - P_1) + \cdots$$

For $N_2(g)$ under the conditions specified,

$$\Delta\overline{U} = -\frac{2\pi\sigma^3 N_A}{3}(\lambda^3 - 1)\frac{\varepsilon}{k_B T}e^{\varepsilon/k_B T}(P_2 - P_1) + \cdots$$

$$= -\frac{2\pi(327.7 \times 10^{-12}\text{ m})^3(6.022 \times 10^{23}\text{ mol}^{-1})}{3}(1.58^3 - 1)\left(\frac{95.2}{300}\right)e^{95.2/300}(10.0 - 1.00)\text{ bar}$$

$$= -5.13 \times 10^{-4}\text{ bar}\cdot\text{m}^3\cdot\text{mol}^{-1}$$

$$= -51.3\text{ J}\cdot\text{mol}^{-1}$$

22–8. Determine $\overline{C}_P - \overline{C}_V$ for a gas that obeys the equation of state $P(\overline{V} - b) = RT$.

We can write, from the equation of state,

$$P(\overline{V} - b) = RT$$

$$\left(\frac{\partial P}{\partial T}\right)_V = \frac{R}{\overline{V} - b}$$

$$\left(\frac{\partial V}{\partial T}\right)_P = \frac{R}{P}$$

Now we substitute into Equation 22.23:

$$\overline{C}_P - \overline{C}_V = T\left(\frac{\partial P}{\partial T}\right)_V\left(\frac{\partial\overline{V}}{\partial T}\right)_P = \frac{RT}{\overline{V} - b}\left(\frac{R}{P}\right) = R$$

22–9. The coefficient of thermal expansion of water at 25°C is 2.572×10^{-4} K^{-1}, and its thermal compressibility is 4.525×10^{-5} bar^{-1}. Calculate the value of $C_P - C_V$ for one mole of water at 25°C. The density of water at 25°C is 0.99705 g$\cdot$mL^{-1}.

The molar volume of water is

$$\overline{V} = \left(\frac{1}{0.99705\text{ g}\cdot\text{mL}^{-1}}\right)\left(\frac{18.015\text{ g}}{1\text{ mol}}\right)\left(\frac{1\text{ dm}^3}{1000\text{ mL}}\right) = 0.018068\text{ dm}^3\cdot\text{mol}^{-1}$$

We can now substitute into Equation 22.27 to find $C_P - C_V$. For one mole,

$$C_P - C_V = \frac{\alpha^2 TV}{\kappa}$$

$$= \frac{(2.572 \times 10^{-4}\text{ K}^{-1})^2(298.15\text{ K})(0.018068\text{ dm}^3)}{4.525 \times 10^{-5}\text{ bar}^{-1}}$$

$$= 7.875 \times 10^{-3}\text{ dm}^3\cdot\text{bar}^{-1}\cdot\text{K}^{-1}$$

22–10. Use Equation 22.22 to show that

$$\left(\frac{\partial C_V}{\partial V}\right)_T = T\left(\frac{\partial^2 P}{\partial T^2}\right)_V$$

Show that $(\partial C_V/\partial V)_T = 0$ for an ideal gas and a van der Waals gas, and that

$$\left(\frac{\partial C_V}{\partial V}\right)_T = -\frac{3A}{4T^{3/2}\overline{V}(\overline{V}+B)}$$

for a Redlich-Kwong gas.

Recall that, by definition, $C_V = (\partial U/\partial T)_V$, so

$$\left(\frac{\partial C_V}{\partial V}\right)_T = \frac{\partial^2 U}{\partial V \partial T} = \frac{\partial^2 U}{\partial T \partial V} = \frac{\partial}{\partial T}\left(\frac{\partial U}{\partial V}\right)_T$$

Express $(\partial U/\partial V)$ using Equation 22.22 and write

$$\left(\frac{\partial U}{\partial V}\right)_T = -P + T\left(\frac{\partial P}{\partial T}\right)_V$$

$$\frac{\partial}{\partial T}\left(\frac{\partial U}{\partial V}\right)_T = -\left(\frac{\partial P}{\partial T}\right)_V + \left(\frac{\partial P}{\partial T}\right)_V + T\left(\frac{\partial^2 P}{\partial T^2}\right)_V$$

$$\left(\frac{\partial C_V}{\partial V}\right)_T = T\left(\frac{\partial^2 P}{\partial T^2}\right)_V$$

For an ideal gas and for a van der Waals gas,

$$P\overline{V} = RT \qquad\qquad P = \frac{RT}{\overline{V}-b} + \frac{a}{\overline{V}^2}$$

$$\left(\frac{\partial^2 P}{\partial T^2}\right)_V = 0 \qquad\qquad \left(\frac{\partial^2 P}{\partial T^2}\right)_V = 0$$

$$T\left(\frac{\partial^2 P}{\partial T^2}\right)_V = 0 = \left(\frac{\partial C_V}{\partial V}\right)_T \qquad T\left(\frac{\partial^2 P}{\partial T^2}\right)_V = 0 = \left(\frac{\partial C_V}{\partial V}\right)_T$$

For a Redlich-Kwong gas,

$$P = \frac{RT}{\overline{V}-B} - \frac{A}{T^{1/2}\overline{V}(\overline{V}+B)}$$

$$\left(\frac{\partial P}{\partial T}\right)_V = \frac{R}{\overline{V}-B} + \frac{A}{2\overline{V}(\overline{V}+B)T^{3/2}}$$

$$\left(\frac{\partial^2 P}{\partial T^2}\right)_V = -\frac{3}{4}\frac{A}{T^{5/2}\overline{V}(\overline{V}+B)}$$

$$T\left(\frac{\partial^2 P}{\partial T^2}\right)_V = -\frac{3A}{4T^{3/2}\overline{V}(\overline{V}+B)} = \left(\frac{\partial C_V}{\partial V}\right)_T$$

22–11. In this problem you will derive the equation (Equation 22.24)

$$C_P - C_V = -T\left(\frac{\partial V}{\partial T}\right)_P^2 \left(\frac{\partial P}{\partial V}\right)_T$$

To start, consider V to be a function of T and P and write out dV. Now divide through by dT at constant volume ($dV = 0$) and then substitute the expression for $(\partial P/\partial T)_V$ that you obtain into Equation 22.23 to get the above expression.

The total derivative of $V(T, P)$ is

$$dV(T, P) = \left(\frac{\partial V}{\partial T}\right)_P dT + \left(\frac{\partial V}{\partial P}\right)_T dP$$

Dividing through by dT at constant volume gives

$$0 = \left(\frac{\partial V}{\partial T}\right)_P + \left(\frac{\partial V}{\partial P}\right)_T \left(\frac{\partial P}{\partial T}\right)_V$$

$$\left(\frac{\partial P}{\partial T}\right)_V = -\left(\frac{\partial V}{\partial T}\right)_P \left(\frac{\partial P}{\partial V}\right)_T$$

Now substitute for $(\partial P/\partial T)_V$ into Equation 22.23:

$$C_P - C_V = T \left(\frac{\partial P}{\partial T}\right)_V \left(\frac{\partial V}{\partial T}\right)_P = -T \left(\frac{\partial V}{\partial T}\right)_P^2 \left(\frac{\partial P}{\partial V}\right)_T$$

which is Equation 22.24.

22–12. The quantity $(\partial U/\partial V)_T$ has units of pressure and is called the *internal pressure*, which is a measure of the intermolecular forces within the body of a substance. It is equal to zero for an ideal gas, is nonzero but relatively small for dense gases, and is relatively large for liquids, particularly those whose molecular interactions are strong. Use the following data to calculate the internal pressure of ethane as a function of pressure at 280 K. Compare your values with the values you obtain from the van der Waals equation and the Redlich-Kwong equation.

P/bar	(dP/dT)/bar·K^{-1}	$\overline{V}$/dm^3·mol^{-1}	P/bar	(dP/dT)/bar·K^{-1}	$\overline{V}$/dm^3·mol^{-1}
4.458	0.01740	5.000	307.14	6.9933	0.06410
47.343	4.1673	0.07526	437.40	7.9029	0.06173
98.790	4.9840	0.07143	545.33	8.5653	0.06024
157.45	5.6736	0.06849	672.92	9.2770	0.05882

Use Equation 22.22 to write

$$\left(\frac{\partial U}{\partial V}\right)_T = -P + T \left(\frac{\partial P}{\partial T}\right)_V$$

To find the experimental values of internal pressure, we can substitute the data given into the above equation. We expressed $(\partial U/\partial V)_T$ for the van der Waals equation in Problem 22–4 as

$$\left(\frac{\partial U}{\partial V}\right)_T = \frac{a}{\overline{V}^2}$$

and $(\partial U/\partial V)_T$ for the Redlich-Kwong equation as

$$\left(\frac{\partial U}{\partial V}\right)_T = \frac{3A}{2T^{1/2}\overline{V}(\overline{V} + B)}$$

We can use the molar volumes given in the statement of the problem and the constants from Tables 16.3 and 16.4 ($a = 5.5818$ dm^6·bar·mol^{-2}, $b = 0.065144$ dm^3·mol^{-1}, $A =$

98.831 $dm^6 \cdot bar \cdot mol^{-2} \cdot K^{1/2}$, $B = 0.045153$ $dm^3 \cdot mol^{-1}$) to create a table of values of internal pressures for each experimental pressure.

P/bar	Experimental	$(\partial U / \partial V)_T$/bar van der Waals	Redlich-Kwong
4.458	0.4140	0.2233	0.3512
47.343	1119.5	972.5	967.1
98.790	1296.7	1094	1064
157.45	1431.2	1190	1138
307.14	1651.0	1359	1265
437.40	1775.4	1465	1343
545.33	1853.0	1538	1395
672.92	1924.6	1613	1449

22–13. Show that

$$\left(\frac{\partial \overline{H}}{\partial P}\right)_T = -RT^2 \left(\frac{dB_{2P}}{dT} + \frac{dB_{3P}}{dT} P + \cdots\right)$$

$$= B_{2V}(T) - T\frac{dB_{2V}}{dT} + O(P)$$

Use Equation 16.41 for the square-well potential to obtain

$$\left(\frac{\partial \overline{H}}{\partial P}\right)_T = \frac{2\pi\sigma^3 N_A}{3}\left[\lambda^3 - (\lambda^3 - 1)\left(1 + \frac{\varepsilon}{k_B T}\right)e^{\varepsilon/k_B T}\right]$$

Given that $\sigma = 327.7$ pm, $\varepsilon/k_B = 95.2$ K, and $\lambda = 1.58$ for $N_2(g)$, calculate the value of $(\partial \overline{H}/\partial P)_T$ at 300 K. Evaluate $\Delta \overline{H} = \overline{H}(P = 10.0$ bar$) - \overline{H}(P = 1.0$ bar$)$. Compare your result with 8.724 $kJ \cdot mol^{-1}$, the value of $\overline{H}(T) - \overline{H}(0)$ for nitrogen at 300 K.

Use the virial expansion in the pressure (Equation 16.23):

$$Z = 1 + B_{2P}P + B_{3P}P^2 + O(P^3)$$

$$\overline{V} = \frac{RT}{P} + RTB_{2P} + PRTB_{3P} + O(P^2)$$

$$\left(\frac{\partial \overline{V}}{\partial T}\right)_P = \frac{R}{P} + RB_{2P} + RT\frac{dB_{2P}}{dT} + PRB_{3P} + PRT\frac{dB_{3P}}{dT} + O(P^2)$$

Substitute into Equation 22.34:

$$\left(\frac{\partial \overline{H}}{\partial P}\right)_T = \overline{V} - T\left(\frac{\partial \overline{V}}{\partial T}\right)_P$$

$$\left(\frac{\partial \overline{H}}{\partial P}\right)_T = \left[\frac{RT}{P} + RTB_{2P} + PRTB_{3P}\right] - \left[\frac{RT}{P} + RTB_{2P}\right.$$

$$\left. + RT^2\frac{dB_{2P}}{dT} + PRTB_{3P} + PRT^2\frac{dB_{3P}}{dT}\right] + O(P^2)$$

$$= -RT^2\left[\frac{dB_{2P}}{dT} + \frac{dB_{3P}}{dT} P + O(P^2)\right] \tag{1}$$

Since $B_{2V} = RTB_{2P}$ (Equation 16.24),

$$\frac{dB_{2V}}{dT} = RB_{2P} + RT\frac{dB_{2P}}{dT}$$

$$\frac{dB_{2V}}{dT}\frac{1}{RT} = \frac{B_{2V}}{RT^2} + \frac{dB_{2P}}{dT}$$

$$\frac{dB_{2P}}{dT} = \frac{1}{RT}\frac{dB_{2V}}{dT} - \frac{B_{2V}}{RT^2}$$

Then Equation 1 becomes

$$\left(\frac{\partial \overline{H}}{\partial P}\right)_T = -RT^2\left[\frac{dB_{2P}}{dT} + O(P)\right]$$

$$= B_{2V} - T\frac{dB_{2V}}{dT} + O(P) \tag{2}$$

Start with Equation 16.41 and find $-TdB_{2V}/dT$:

$$B_{2V}(T) = \frac{2\pi\sigma^3 N_A}{3}\left[1 - (\lambda^3 - 1)\left(e^{\varepsilon/k_B T} - 1\right)\right]$$

$$\frac{dB_{2V}}{dT} = \frac{2\pi\sigma^3 N_A}{3}(\lambda^3 - 1)e^{\varepsilon/k_B T}\left(\frac{\varepsilon}{k_B T^2}\right)$$

$$-T\frac{dB_{2V}}{dT} = -\frac{2\pi\sigma^3 N_A}{3}(\lambda^3 - 1)e^{\varepsilon/k_B T}\left(\frac{\varepsilon}{k_B T}\right)$$

Substituting this value into Equation 2 and ignoring terms of P or higher, we find

$$\left(\frac{\partial \overline{H}}{\partial P}\right)_T = \frac{2\pi\sigma^3 N_A}{3}\left[1 - (\lambda^3 - 1)\left(e^{\varepsilon/k_B T} - 1\right) - e^{\varepsilon/k_B T}\left(\frac{\varepsilon}{k_B T}\right)(\lambda^3 - 1)\right]$$

$$= \frac{2\pi\sigma^3 N_A}{3}\left[1 - \left(\lambda^3 e^{\varepsilon/k_B T} + 1 - \lambda^3 - e^{\varepsilon/k_B T}\right) - \lambda^3\frac{\varepsilon}{k_B T}e^{\varepsilon/k_B T} + \frac{\varepsilon}{k_B T}e^{\varepsilon/k_B T}\right]$$

$$= \frac{2\pi\sigma^3 N_A}{3}\left[\lambda^3 - e^{\varepsilon/k_B T}\left(\lambda^3 - 1 - \frac{\varepsilon}{k_B T} + \lambda^3\frac{\varepsilon}{k_B T}\right)\right]$$

$$= \frac{2\pi\sigma^3 N_A}{3}\left[\lambda^3 - (\lambda^3 - 1)\left(1 + \frac{\varepsilon}{k_B T}\right)e^{\varepsilon/k_B T}\right]$$

Using the parameters provided for nitrogen, this expression becomes

$$\left(\frac{\partial \overline{H}}{\partial P}\right)_T = \frac{2\pi(327.7 \times 10^{-12}\text{ m})^3(6.022 \times 10^{23}\text{ mol}^{-1})}{3}\left[1.58^3 - (1.58^3 - 1)\left(1 + \frac{95.2}{300}\right)e^{95.2/300}\right]$$

$$= -6.138 \times 10^{-5}\text{ m}^3\cdot\text{mol}^{-1}$$

Then

$$\Delta\overline{H} = (-6.138 \times 10^{-5}\text{ m}^3\cdot\text{mol}^{-1})\Delta P = -5.52 \times 10^{-4}\text{ m}^3\cdot\text{bar}\cdot\text{mol}^{-1} = -55.2\text{ J}\cdot\text{mol}^{-1}$$

22–14. Show that the enthalpy is a function of only the temperature for a gas that obeys the equation of state $P(\overline{V} - bT) = RT$.

$$\left(\frac{\partial \overline{H}}{\partial P}\right)_T = \overline{V} - T\left(\frac{\partial \overline{V}}{\partial T}\right)_P \qquad (22.34)$$

For a gas obeying the equation of state $P(\overline{V} - bT) = RT$,

$$\overline{V} = \frac{RT}{P} + bT$$

$$\left(\frac{\partial \overline{V}}{\partial T}\right)_P = \frac{R}{P} + b$$

Substituting into Equation 22.34, we find

$$\left(\frac{\partial \overline{H}}{\partial P}\right)_T = \frac{RT}{P} + bT - T\left(\frac{R}{P} + b\right) = 0$$

and so enthalpy does not depend on pressure for a gas with this equation of state.

22–15. Use your results for the van der Waals equation and the Redlich-Kwong equation in Problem 22–4 to calculate $\overline{H}(T, \overline{V})$ as a function of volume for ethane at 400 K. In each case, use the equation $\overline{H} = \overline{U} + P\overline{V}$. Compare your results with the experimental data shown in Figure 22.5.

$$\overline{H} = \overline{U}(T, \overline{V}) + P\overline{V}$$

From Problem 22–4, for a van der Waals gas

$$\overline{U} = 14.55 \text{ kJ} \cdot \text{mol}^{-1} - \frac{0.55818 \text{ kJ} \cdot \text{dm}^3 \cdot \text{mol}^{-2}}{\overline{V}}$$

$$P = \frac{(0.083145 \text{ dm}^3 \cdot \text{bar} \cdot \text{mol}^{-1} \cdot \text{K}^{-1})(400 \text{ K})}{\overline{V} - 0.065144 \text{ dm}^3 \cdot \text{mol}^{-1}} - \frac{5.5818 \text{ dm}^6 \cdot \text{bar} \cdot \text{mol}^{-1}}{\overline{V}^2}$$

and for a Redlich-Kwong gas

$$\overline{U} = 14.55 \text{ kJ} \cdot \text{mol}^{-1} - \frac{3(9.8831 \text{ kJ} \cdot \text{dm}^3 \cdot \text{mol}^{-2} \cdot \text{K}^{1/2})}{2(0.045153 \text{ dm}^3 \cdot \text{mol}^{-1})(400 \text{ K})^{1/2}} \ln \frac{\overline{V} + 0.045153 \text{ dm}^3 \cdot \text{mol}^{-1}}{\overline{V}}$$

$$P = \frac{(0.083145 \text{ dm}^3 \cdot \text{bar} \cdot \text{mol}^{-1} \cdot \text{K}^{-1})(400 \text{ K})}{\overline{V} - 0.045153 \text{ dm}^3 \cdot \text{mol}^{-1}} - \frac{98.831 \text{ dm}^6 \cdot \text{bar} \cdot \text{mol}^{-2} \cdot \text{K}^{1/2}}{(400 \text{ K})^{1/2} \overline{V}(\overline{V} + 0.045153 \text{ dm}^3 \cdot \text{mol}^{-1})}$$

Note that in using these values, we find $\overline{U}$ in terms of $\text{kJ} \cdot \text{mol}^{-1}$ and $P\overline{V}$ in terms of $\text{dm}^3 \cdot \text{bar} \cdot \text{mol}^{-1}$. Dividing $P\overline{V}$ by 10 will result in values of enthalpy given in $\text{kJ} \cdot \text{mol}^{-1}$. Using these values, we can produce plots of $\overline{H}$ vs. P.

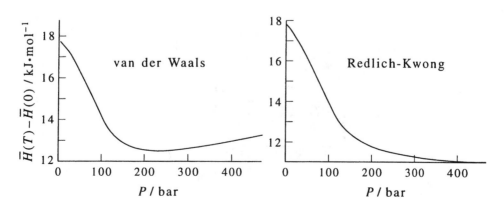

22–16. Use Equation 22.34 to show that

$$\left(\frac{\partial C_P}{\partial P}\right)_T = -T\left(\frac{\partial^2 V}{\partial T^2}\right)_P$$

Use a virial expansion in P to show that

$$\left(\frac{\partial \overline{C}_P}{\partial P}\right)_T = -T\frac{d^2 B_{2V}}{dT^2} + O(P)$$

Use the square-well second virial coefficient (Equation 16.41) and the parameters given in Problem 22–13 to calculate the value of $(\partial \overline{C}_P/\partial P)_T$ for $N_2(g)$ at 0°C. Now calculate $\overline{C}_P$ at 100 atm and 0°C, using $\overline{C}_P^{id} = 5R/2$.

We define C_P as $(\partial H/\partial T)_P$ (Equation 19.40). Starting with Equation 22.34,

$$\left(\frac{\partial H}{\partial P}\right)_T = V - T\left(\frac{\partial V}{\partial T}\right)_P$$

$$\frac{\partial}{\partial T}\left(\frac{\partial H}{\partial P}\right)_T = \left(\frac{\partial V}{\partial T}\right)_P - \left(\frac{\partial V}{\partial T}\right)_P - T\frac{\partial}{\partial T}\left(\frac{\partial V}{\partial T^2}\right)_P$$

$$\frac{\partial}{\partial P}\left(\frac{\partial H}{\partial T}\right)_P = -T\left(\frac{\partial^2 V}{\partial T^2}\right)_P$$

$$\left(\frac{\partial C_P}{\partial P}\right)_T = -T\left(\frac{\partial^2 V}{\partial T^2}\right)_P \tag{1}$$

Using a virial expansion in P, we find

$$\overline{V} = \frac{RT}{P} + RTB_{2P} + PRTB_{3P} + O(P^2)$$

$$\left(\frac{\partial \overline{V}}{\partial T}\right)_P = \frac{R}{P} + RB_{2P} + RT\frac{dB_{2P}}{dT} + O(P)$$

$$\left(\frac{\partial^2 \overline{V}}{\partial T^2}\right)_P = R\frac{dB_{2P}}{dT} + R\frac{dB_{2P}}{dT} + RT\frac{d^2 B_{2P}}{dT^2} + O(P)$$

$$= 2R\frac{dB_{2P}}{dT} + RT\frac{d^2 B_{2P}}{dT^2} + O(P)$$

Now, since $B_{2V} = RTB_{2P}$ (Equation 16.24),

$$B_{2P} = \frac{B_{2V}}{RT}$$

$$\frac{dB_{2P}}{dT} = \frac{1}{RT}\frac{dB_{2V}}{dT} - \frac{B_{2V}}{RT^2}$$

$$\frac{d^2 B_{2P}}{dT^2} = \frac{1}{RT}\frac{d^2 B_{2V}}{dT^2} - \frac{1}{RT^2}\frac{dB_{2V}}{dT} - \frac{1}{RT^2}\frac{dB_{2V}}{dT} + \frac{2B_{2V}}{RT^3}$$

$$= \frac{1}{RT}\frac{d^2 B_{2V}}{dT^2} - \frac{2}{RT^2}\frac{dB_{2V}}{dT} + \frac{2B_{2V}}{RT^3}$$

Now we solve Equation 1 for $(\partial \overline{C}_P / \partial P)_T$ in terms of B_{2V}:

$$\left(\frac{\partial \overline{C}_P}{\partial P} \right)_T = -T \left(\frac{\partial^2 \overline{V}}{\partial T^2} \right)_P$$

$$= -T \left[2R \frac{dB_{2P}}{dT} + RT \frac{d^2 B_{2P}}{dT^2} + O(P) \right]$$

$$= -T \left[2R \left(\frac{1}{RT} \frac{dB_{2V}}{dT} - \frac{B_{2V}}{RT^2} \right) + RT \left(\frac{1}{RT} \frac{d^2 B_{2V}}{dT^2} - \frac{2}{RT^2} \frac{dB_{2V}}{dT} + \frac{2B_{2V}}{RT^3} \right) \right] + O(P)$$

$$= -\frac{2d B_{2V}}{dT} + \frac{2B_{2V}}{T} - T \frac{d^2 B_{2V}}{dT^2} + \frac{2d B_{2V}}{dT} - \frac{2B_{2V}}{T} + O(P)$$

$$= -T \frac{d^2 B_{2V}}{dT^2} + O(P)$$

Using Equation 16.41,

$$B_{2V}(T) = \frac{2\pi \sigma^3 N_A}{3} \left[1 - (\lambda^3 - 1) \left(e^{\varepsilon/k_B T} - 1 \right) \right]$$

$$\frac{dB_{2V}}{dT} = \frac{2\pi \sigma^3 N_A}{3} (\lambda^3 - 1) e^{\varepsilon/k_B T} \left(\frac{\varepsilon}{k_B T^2} \right)$$

$$\frac{d^2 B_{2V}}{dT^2} = -\frac{2\pi \sigma^3 N_A}{3} (\lambda^3 - 1) e^{\varepsilon/k_B T} \frac{\varepsilon}{k_B T^3} \left(\frac{\varepsilon}{k_B T} + 2 \right)$$

$$\left(\frac{\partial \overline{C}_P}{\partial P} \right)_T = -T \frac{d^2 B_{2V}}{dT} = \frac{2\pi \sigma^3 N_A}{3} (\lambda^3 - 1) e^{\varepsilon/k_B T} \frac{\varepsilon}{k_B T^2} \left(\frac{\varepsilon}{k_B T} + 2 \right)$$

For nitrogen at 298.15 K,

$$\left(\frac{\partial \overline{C}_P}{\partial P} \right)_T = \frac{2\pi (327.7 \times 10^{-12} \text{ m})^3 (6.022 \times 10^{23} \text{ mol}^{-1})}{3} (1.58^3 - 1) e^{95.2/298.15} \frac{95.2}{(298.15)^2 \text{ K}} \left(\frac{95.2}{298.15} + 2 \right)$$

$$= 4.467 \times 10^{-7} \text{ m}^3 \cdot \text{mol}^{-1} \cdot \text{K}^{-1} = 4.47 \times 10^{-4} \text{ dm}^3 \cdot \text{mol}^{-1} \cdot \text{K}^{-1}$$

Finally,

$$\overline{C}_P - \overline{C}_P^{\text{id}} = (4.467 \times 10^{-4} \text{ dm}^3 \cdot \text{mol}^{-1} \cdot \text{K}^{-1})(P - P^{\text{id}})$$

$$\overline{C}_P = \frac{5R}{2} + (4.467 \times 10^{-4} \text{ dm}^3 \cdot \text{mol}^{-1} \cdot \text{K}^{-1})(99 \text{ atm})$$

$$= (2.5)(8.3145 \text{ J} \cdot \text{mol}^{-1} \cdot \text{K}^{-1}) + 4.42 \text{ J} \cdot \text{mol}^{-1} \cdot \text{K}^{-1}$$

$$= 25.21 \text{ J} \cdot \text{mol}^{-1} \cdot \text{K}^{-1}$$

22–17. Show that the molar enthalpy of a substance at pressure P relative to its value at one bar is given by

$$\overline{H}(T, P) = \overline{H}(T, P=1 \text{ bar}) + \int_1^P \left[\overline{V} - T \left(\frac{\partial \overline{V}}{\partial T} \right)_P \right] dP'$$

Calculate the value of $\overline{H}(T, P) - \overline{H}(T, P=1 \text{ bar})$ at 0°C and 100 bar for mercury given that the molar volume of mercury varies with temperature according to

$$\overline{V}(t) = (14.75 \text{ mL} \cdot \text{mol}^{-1})(1 + 0.182 \times 10^{-3} t + 2.95 \times 10^{-9} t^2 + 1.15 \times 10^{-10} t^3)$$

where t is the Celsius temperature. Assume that $\overline{V}(0)$ does not vary with pressure over this range and express your answer in units of kJ·mol^{-1}.

Begin with Equation 22.34:

$$\left(\frac{\partial \overline{H}}{\partial P}\right)_T = \overline{V} - T\left(\frac{\partial \overline{V}}{\partial T}\right)_P$$

$$d\overline{H} = \left[\overline{V} - T\left(\frac{\partial \overline{V}}{\partial T}\right)_P\right]dP$$

$$\overline{H}(T, P) - \overline{H}(T, 1 \text{ bar}) = \int_{1 \text{ bar}}^{P}\left[\overline{V} - T\left(\frac{\partial \overline{V}}{\partial T}\right)_P\right]dP'$$

where we have begun using P' as the quantity integrated over in order to distinguish it from P, the final pressure of the substance.

Using the values given for mercury,

$$\left(\frac{\partial \overline{V}}{\partial T}\right)_P = (14.75 \text{ mL·mol}^{-1})(0.182 \times 10^{-3} + 5.90 \times 10^{-9}t + 3.45 \times 10^{-10}t^2)$$

Then at 0° C and 100 bar,

$$\overline{H}(T, 100 \text{ bar}) - \overline{H}(T, 1 \text{ bar}) = (14.75 \text{ mL·mol}^{-1})\int_{1 \text{ bar}}^{100 \text{ bar}}\left[1 - T(0.182 \times 10^{-3})\right]dP'$$

$$= (14.75 \text{ mL·mol}^{-1})(99 \text{ bar})[1 - (298.15)(0.182 \times 10^{-3})]$$

$$= 1381 \text{ mL·bar·mol}^{-1} = 138.1 \text{ J·mol}^{-1}$$

22–18. Show that

$$dH = \left[V - T\left(\frac{\partial V}{\partial T}\right)_P\right]dP + C_P dT$$

What does this equation tell you about the natural variables of H?

Write the total derivative of $H(P, T)$:

$$dH = \left(\frac{\partial H}{\partial P}\right)_T dP + \left(\frac{\partial H}{\partial T}\right)_P dT$$

We can now use Equation 22–34 and the definition of C_P to write this as

$$dH = \left[V - T\left(\frac{\partial V}{\partial T}\right)_P\right]dP + C_P dT$$

Since the coefficients of dP and dT are not simple, this tells us that the natural variables of H are not P and T.

22–19. What are the natural variables of the entropy?

$$dS = PdV + \frac{1}{T}dU \tag{22.39}$$

Because the coefficients of dV and dU are simple thermodynamic quantities, we say that V and U are the natural variables of entropy.

22–20. Experimentally determined entropies are commonly adjusted for nonideality by using an equation of state called the (modified) Berthelot equation:

$$\frac{P\overline{V}}{RT} = 1 + \frac{9}{128}\frac{PT_c}{P_c T}\left(1 - 6\frac{T_c^2}{T^2}\right)$$

Show that this equation leads to the correction

$$S^\circ(\text{at one bar}) = \overline{S}(\text{at one bar}) + \frac{27}{32}\frac{RT_c^3}{P_c T^3}(1\text{ bar})$$

This result needs only the critical data for the substance. Use this equation along with the critical data in Table 16.5 to calculate the nonideality correction for $N_2(g)$ at 298.15 K. Compare your result with the value used in Table 21.1.

$$S^\circ(1\text{ bar}) - \overline{S}(1\text{ bar}) = \int_{P^{\text{id}}}^{1\text{ bar}}\left[\left(\frac{\partial\overline{V}}{\partial T}\right)_P - \frac{R}{P}\right]dP \qquad (22.54)$$

We find $(\partial\overline{V}/\partial T)$ from the modified Bethelot equation:

$$\frac{P\overline{V}}{RT} = 1 + \frac{9}{128}\frac{PT_c}{P_c T}\left(1 - 6\frac{T_c^2}{T^2}\right)$$

$$\overline{V} = \frac{RT}{P} + \frac{9R}{128}\frac{T_c}{P_c} - \frac{9\cdot 6}{128}\frac{RT_c^3}{P_c T^2}$$

$$\left(\frac{\partial\overline{V}}{\partial T}\right)_P = \frac{R}{P} + \frac{9\cdot 6\cdot 2}{128}\frac{RT_c^3}{P_c T^3}$$

Now substitute into Equation 22.54 to find $S^\circ(1\text{ bar}) - \overline{S}(1\text{ bar})$, neglecting P^{id} with respect to 1 bar:

$$S^\circ(1\text{ bar}) - \overline{S}(1\text{ bar}) = \int_{P^{\text{id}}}^{1\text{ bar}}\left[\frac{R}{P} + \frac{27}{32}\frac{RT_c^3}{P_c T^3} - \frac{R}{P}\right]dP$$

$$= \frac{27}{32}\frac{RT_c^3}{P_c T^3}(1\text{ bar})$$

For N_2 at 298.15 K, $T_c = 126.2$ K and $P_c = 34.00$ bar. Then the nonideality correction (the difference between the two values of S) is

$$S^\circ(1\text{ bar}) - \overline{S}(1\text{ bar}) = \frac{27}{32}\frac{(8.3145\text{ J}\cdot\text{mol}^{-1}\cdot\text{K}^{-1})(126.2\text{ K})^3}{(34.00\text{ bar})(298.15\text{ K})^3}(1\text{ bar})$$

$$= 0.0156\text{ J}\cdot\text{mol}^{-1}\cdot\text{K}^{-1}$$

This is essentially the value used in Table 21.1 ($0.02\text{ J}\cdot\text{K}^{-1}\cdot\text{mol}^{-1}$).

22–21. Use the result of Problem 22–20 along with the critical data in Table 16.5 to determine the nonideality correction for CO(g) at its normal boiling point, 81.6 K. Compare your result with the value used in Problem 21–24.

$$S^\circ(1\ \text{bar}) - \overline{S}(1\ \text{bar}) = \frac{27}{32}\frac{RT_c^3}{P_cT^3}(1\ \text{bar})$$

$$= \frac{27}{32}\frac{(8.3145\ \text{J·mol}^{-1}\text{·K}^{-1})(132.85\ \text{K})^3}{(34.935\ \text{bar})(81.6\ \text{K})^3}(1\ \text{bar})$$

$$= 0.867\ \text{J·mol}^{-1}\text{·K}^{-1}$$

This is comparable to the value used in Problem 21–24 (0.879 J·K^{-1}·mol^{-1}).

22–22. Use the result of Problem 22–20 along with the critical data in Table 16.5 to determine the nonideality correction for Cl$_2$(g) at its normal boiling point, 239 K. Compare your result with the value used in Problem 21–16.

$$S^\circ(1\ \text{bar}) - \overline{S}(1\ \text{bar}) = \frac{27}{32}\frac{RT_c^3}{P_cT^3}(1\ \text{bar})$$

$$= \frac{27}{32}\frac{(8.3145\ \text{J·mol}^{-1}\text{·K}^{-1})(416.9\ \text{K})^3}{(79.91\ \text{bar})(239\ \text{K})^3}(1\ \text{bar})$$

$$= 0.466\ \text{J·mol}^{-1}\text{·K}^{-1}$$

This is comparable to the value of 0.502 J·K^{-1}·mol^{-1} used in Problem 21–16.

22–23. Derive the equation

$$\left(\frac{\partial(A/T)}{\partial T}\right)_V = -\frac{U}{T^2}$$

which is a Gibbs-Helmholtz equation for A.

Begin with the definition of A (Equation 22.4):

$$A = U - TS$$

$$\frac{\partial}{\partial T}\left(\frac{A}{T}\right) = \frac{\partial}{\partial T}\left[\frac{U}{T} - S\right]$$

$$\left[\frac{\partial(A/T)}{\partial T}\right]_V = -\frac{U}{T^2} + \frac{1}{T}\left(\frac{\partial U}{\partial T}\right)_V - \left(\frac{\partial S}{\partial T}\right)_V$$

Now, by the definition of C_V,

$$\frac{C_V}{T} = \frac{1}{T}\left(\frac{\partial U}{\partial T}\right)_V$$

From Equation 21.2, we also know that

$$\frac{C_V}{T} = \left(\frac{\partial S}{\partial T}\right)_V$$

Therefore,

$$\left(\frac{\partial(A/T)}{\partial T}\right)_V = -\frac{U}{T^2} + \frac{C_V}{T} - \frac{C_V}{T} = -\frac{U}{T^2}$$

22–24. We can derive the Gibbs-Helmholtz equation directly from Equation 22.31a in the following way. Start with $(\partial G/\partial T)_P = -S$ and substitute for S from $G = H - TS$ to obtain

$$\frac{1}{T}\left(\frac{\partial G}{\partial T}\right)_P - \frac{G}{T^2} = -\frac{H}{T^2}$$

Now show that the left side is equal to $(\partial[G/T]/\partial T)_P$ to get the Gibbs-Helmholtz equation.

Begin with the equality in the statement of the problem and substitute for S as suggested (from Equation 22.13):

$$\left(\frac{\partial G}{\partial T}\right)_P = -S = \frac{G}{T} - \frac{H}{T}$$

$$\frac{1}{T}\left(\frac{\partial G}{\partial T}\right)_P - \frac{G}{T^2} = -\frac{H}{T^2} \tag{1}$$

Taking the partial derivative of G/T with respect to T gives

$$\frac{\partial}{\partial T}\left(\frac{G}{T}\right) = \frac{1}{T}\left(\frac{\partial G}{\partial T}\right)_P - \frac{G}{T^2}$$

so we can write Equation 1 as

$$\frac{\partial}{\partial T}\left(\frac{G}{T}\right) = -\frac{H}{T^2}$$

which is the Gibbs-Helmholtz equation.

22–25. Use the following data for benzene to plot $\overline{G}(T) - \overline{H}(0)$ versus T. [In this case we will ignore the (usually small) corrections due to nonideality of the gas phase.]

$$\overline{C}_P^s(T)/R = \frac{12\pi^4}{5}\left(\frac{T}{\Theta_D}\right)^3 \qquad \Theta_D = 130.5 \text{ K} \qquad 0\text{ K} < T < 13\text{ K}$$

$$\overline{C}_P^s(T)/R = -0.6077 + (0.1088\text{ K}^{-1})T - (5.345 \times 10^{-4}\text{ K}^{-2})T^2 + (1.275 \times 10^{-6}\text{ K}^{-3})T^3$$
$$13\text{ K} < T < 278.6\text{ K}$$

$$\overline{C}_P^l(T)/R = 12.713 + (1.974 \times 10^{-3}\text{ K}^{-1})T - (4.766 \times 10^{-5}\text{ K}^{-2})T^2$$
$$278.6\text{ K} < T < 353.2\text{ K}$$

$$\overline{C}_P^g(T)/R = -4.077 + (0.05676\text{ K}^{-1})T - (3.588 \times 10^{-5}\text{ K}^{-2})T^2 + (8.520 \times 10^{-9}\text{ K}^{-3})T^3$$
$$353.2\text{ K} < T < 1000\text{ K}$$

$$T_{\text{fus}} = 278.68\text{ K} \qquad \Delta_{\text{fus}}\overline{H} = 9.95\text{ kJ} \cdot \text{mol}^{-1}$$

$$T_{\text{vap}} = 353.24\text{ K} \qquad \Delta_{\text{vap}}\overline{H} = 30.72\text{ kJ} \cdot \text{mol}^{-1}$$

Use the Equation $\overline{G}(T) - \overline{H}(0) = \overline{H}(T) - \overline{H}(0) - T\overline{S}(T)$ (as in Section 22–7) and Equations 22.62 and 22.63 for $\overline{H}(T) - \overline{H}(0)$ and $\overline{S}(T)$ to plot. (See Problem 21.15 for an explanation of how to assign the values of entropy and enthalpy as functions of temperature.)

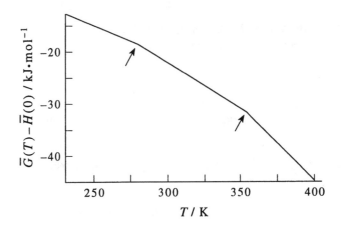

The discontinuities in the slope at the melting and boiling points are difficult to see and so are highlighted with arrows.

22–26. Use the following data for propene to plot $\overline{G}(T) - \overline{H}(0)$ versus T. [In this case we will ignore the (usually small) corrections due to nonideality of the gas phase.]

$$\overline{C}_P^s(T)/R = \frac{12\pi^4}{5}\left(\frac{T}{\Theta_D}\right)^3 \qquad \Theta_D = 100 \text{ K} \qquad 0 \text{ K} < T < 15 \text{ K}$$

$$\overline{C}_P^s(T)/R = -1.616 + (0.08677 \text{ K}^{-1})T - (9.791 \times 10^{-4} \text{ K}^{-2})T^2 + (2.611 \times 10^{-6} \text{ K}^{-3})T^3$$
$$15 \text{ K} < T < 87.90 \text{ K}$$

$$\overline{C}_P^l(T)/R = 15.935 - (0.08677 \text{ K}^{-1})T + (4.294 \times 10^{-4} \text{ K}^{-2})T^2 - (6.276 \times 10^{-7} \text{ K}^{-3})T^3$$
$$87.90 \text{ K} < T < 225.46 \text{ K}$$

$$\overline{C}_P^g(T)/R = 1.4970 + (2.266 \times 10^{-2} \text{ K}^{-1})T - (5.725 \times 10^{-6} \text{ K}^{-2})T^2$$
$$225.46 \text{ K} < T < 1000 \text{ K}$$

$$T_{\text{fus}} = 87.90 \text{ K} \qquad \Delta_{\text{fus}}\overline{H} = 3.00 \text{ kJ·mol}^{-1}$$

$$T_{\text{vap}} = 225.46 \text{ K} \qquad \Delta_{\text{vap}}\overline{H} = 18.42 \text{ kJ·mol}^{-1}$$

This is done in the same way as Problem 22–25.

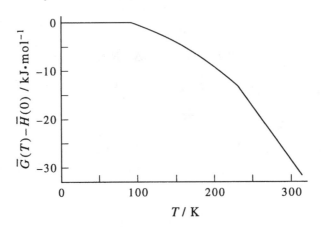

22–27. Use a virial expansion for Z to prove (a) that the integrand in Equation 22.74 is finite as $P \to 0$, and (b) that $(Z - 1)/P = 0$ for an ideal gas.

a. Use the virial expansion $Z = 1 + B_{2P}P + B_{3P}P^2 + O(P^3)$. Then

$$\frac{Z - 1}{P} = \frac{1 + B_{2P}P + B_{3P}P^2 + O(P^3) - 1}{P}$$
$$= B_{2P} + B_{3P}P + O(P^3)$$

As $P \to 0$, then, the integrand approaches B_{2P}, wich is finite.

b. For an ideal gas, $Z = P\overline{V}/RT$ and $P\overline{V} = RT$, so

$$\frac{Z - 1}{P} = \frac{1 - 1}{P} = 0$$

22–28. Derive a virial expansion in the pressure for $\ln \gamma$.

Begin with Equation 22.74 and expand Z as in Problem 22-27(a):

$$\ln \gamma = \int_0^P \frac{Z - 1}{P'} dP'$$
$$= \int_0^P \left(B_{2P} + B_{3P}P' + O(P'^2) \right) dP'$$
$$= B_{2P}P + \frac{B_{3P}P^2}{2} + O(P^3)$$

22–29. The compressibility factor for ethane at 600 K can be fit to the expression

$$Z = 1.0000 - 0.000612(P/\text{bar}) + 2.661 \times 10^{-6}(P/\text{bar})^2$$
$$- 1.390 \times 10^{-9}(P/\text{bar})^3 - 1.077 \times 10^{-13}(P/\text{bar})^4$$

for $0 \le P/\text{bar} \le 600$. Use this expression to determine the fugacity coefficient of ethane as a function of pressure at 600 K.

Substitute into Equation 22.74:

$$\ln \gamma = \int_0^P \frac{Z - 1}{P'} dP'$$
$$= \int_0^P \left[-6.12 \times 10^{-4} \text{ bar}^{-1} + (2.661 \times 10^{-6} \text{ bar}^{-2})P' \right.$$
$$\left. - (1.390 \times 10^{-9} \text{ bar}^{-3})P'^2 - (1.077 \times 10^{-13} \text{ bar}^{-4})P'^3 \right] dP'$$
$$= -(6.12 \times 10^{-4} \text{ bar}^{-1})P + \tfrac{1}{2}(2.661 \times 10^{-6} \text{ bar}^{-2})P^2$$
$$- \tfrac{1}{3}(1.390 \times 10^{-9} \text{ bar}^{-3})P^3 - \tfrac{1}{4}(1.077 \times 10^{-13} \text{ bar}^{-4})P^4$$
$$\gamma = \exp\left[-(6.12 \times 10^{-4} \text{ bar}^{-1})P + (1.3305 \times 10^{-6} \text{ bar}^{-2})P^2 \right.$$
$$\left. - (4.633 \times 10^{-10} \text{ bar}^{-3})P^3 - (2.693 \times 10^{-14} \text{ bar}^{-4})P^4 \right]$$

22–30. Use Figure 22.11 and the data in Table 16.5 to estimate the fugacity of ethane at 360 K and 1000 atm.

From Table 16.5, $T_c = 305.34$ K and $P_c = 48.077$ atm. Then at 360 K and 1000 atm, $T/T_c = 1.18$ and $P/P_c = 20.8$. Using Figure 22.11, it appears that $\gamma \approx 0.63$.

22–31. Use the following data for ethane at 360 K to plot the fugacity coefficient against pressure.

$\rho/\text{mol}\cdot\text{dm}^{-3}$	P/bar	$\rho/\text{mol}\cdot\text{dm}^{-3}$	P/bar	$\rho/\text{mol}\cdot\text{dm}^{-3}$	P/bar
1.20	31.031	6.00	97.767	10.80	197.643
2.40	53.940	7.20	112.115	12.00	266.858
3.60	71.099	8.40	130.149	13.00	381.344
4.80	84.892	9.60	156.078	14.40	566.335

Compare your result with the result you obtained in Problem 22–30.

By definition,

$$\frac{(Z-1)}{P} = \frac{1}{\rho RT} - \frac{1}{P} = y$$

Now we can plot $(Z-1)/P$ vs. P:

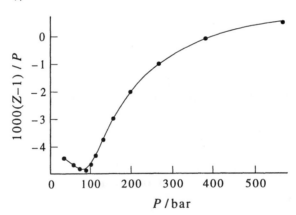

Numerical integration using the trapezoidal approximation allows us to graph f/P vs. P, in the same way that Figure 22.10 was created from Figure 22.9.

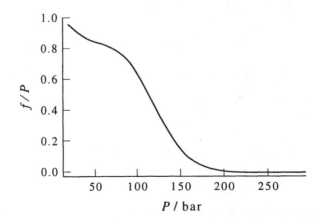

22–32. Use the following data for $N_2(g)$ at 0°C to plot the fugacity coefficient as a function of pressure.

P/atm	$Z = P\overline{V}/RT$	P/atm	$Z = P\overline{V}/RT$	P/atm	$Z = P\overline{V}/RT$
200	1.0390	1000	2.0700	1800	3.0861
400	1.2570	1200	2.3352	2000	3.3270
600	1.5260	1400	2.5942	2200	3.5640
800	1.8016	1600	2.8456	2400	3.8004

Again, plot $(Z - 1)/P$ vs. P:

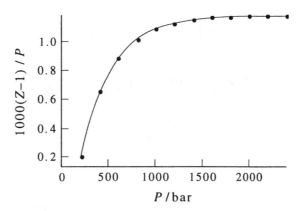

Then do numerical integration with the trapezoidal approximation to graph f/P vs. P:

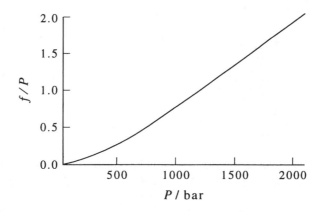

22–33. It might appear that we can't use Equation 22.72 to determine the fugacity of a van der Waals gas because the van der Waals equation is a cubic equation in $\overline{V}$, so we can't solve it analytically for $\overline{V}$ to carry out the integration in Equation 22.72. We can get around this problem, however, by integrating Equation 22.72 by parts. First show that

$$RT \ln \gamma = P\overline{V} - RT - \int_{\overline{V}^{id}}^{\overline{V}} P d\overline{V}' - RT \ln \frac{P}{P^{id}}$$

where $P^{id} \to 0$, $\overline{V}^{id} \to \infty$, and $P^{id}\overline{V}^{id} \to RT$. Substitute P from the van der Waals equation into the first term and the integral on the right side of the above equation and integrate to obtain

$$RT \ln \gamma = \frac{RT\overline{V}}{\overline{V} - b} - \frac{a}{\overline{V}} - RT - RT \ln \frac{\overline{V} - b}{\overline{V}^{id} - b} - \frac{a}{\overline{V}} - RT \ln \frac{P}{P^{id}}$$

Now use the fact that $\overline{V}^{id} \to \infty$ and that $P^{id}\overline{V}^{id} = RT$ to show that

$$\ln \gamma = -\ln\left[1 - \frac{a(\overline{V} - b)}{RT\overline{V}^2}\right] + \frac{b}{\overline{V} - b} - \frac{2a}{RT\overline{V}}$$

This equation gives the fugacity of a van der Waals gas as a function of $\overline{V}$. You can use the van der Waals equation itself to calculate P from $\overline{V}$, so the above equation, in conjunction with the van der Waals equation, gives $\ln \gamma$ as a function of pressure.

First we integrate Equation 22.72 by parts:

$$\ln \gamma = \int_{P^{id}}^{P} \left(\frac{\overline{V}}{RT} - \frac{1}{P'}\right) dP'$$

$$RT \ln \gamma = \int_{P^{id}}^{P} \overline{V} dP' - \int_{P^{id}}^{P} \frac{RT}{P'} dP'$$

$$= P'\overline{V}\Big|_{P^{id}}^{P} - \int_{\overline{V}^{id}}^{\overline{V}} P' d\overline{V} - RT \ln \frac{P}{P^{id}}$$

$$= P\overline{V} - (P\overline{V})^{id} - \int_{\overline{V}^{id}}^{\overline{V}} P' d\overline{V}' - RT \ln \frac{P}{P^{id}}$$

$$= P\overline{V} - RT - \int_{\overline{V}^{id}}^{\overline{V}} P' d\overline{V}' - RT \ln \frac{P}{P^{id}}$$

Substituting P from the van der Waals equation, we find that this equation becomes

$$RT \ln \gamma = \overline{V}\left[\frac{RT}{\overline{V} - b} - \frac{a}{\overline{V}^2}\right] - RT - \int_{\overline{V}^{id}}^{\overline{V}} \left[\frac{RT}{\overline{V}' - b} - \frac{a}{\overline{V}'^2}\right] d\overline{V}' - RT \ln \frac{P}{P^{id}}$$

$$= \frac{RT\overline{V}}{\overline{V} - b} - \frac{a}{\overline{V}} - RT - RT \ln \frac{\overline{V} - b}{\overline{V}^{id} - b} - \frac{a}{\overline{V}} + \frac{a}{\overline{V}^{id}} - RT \ln \frac{P}{P^{id}}$$

$$= \frac{RT\overline{V}}{\overline{V} - b} - \frac{a}{\overline{V}} - RT - RT \ln \frac{\overline{V} - b}{\overline{V}^{id}} - \frac{a}{\overline{V}} + \frac{a}{\overline{V}^{id}} - RT \ln \frac{P}{P^{id}}$$

Now, since $\overline{V}^{id} \longrightarrow \infty$, we can neglect b in the denominator of the logarithmic term and consider the $a/\overline{V}^{id}$ term negligible. Also, since $P^{id}\overline{V}^{id} = RT$, we write the above expression as

$$RT \ln \gamma = \frac{RT\overline{V}}{\overline{V} - b} - \frac{2a}{\overline{V}} - RT - RT\left(\ln \frac{\overline{V} - b}{\overline{V}^{id}} + \ln \frac{P}{P^{id}}\right)$$

$$\ln \gamma = \frac{\overline{V}}{\overline{V} - b} - \frac{2a}{\overline{V}RT} - 1 + \ln \overline{V}^{id} P^{id} - \ln P(\overline{V} - b)$$

$$= \frac{\overline{V} - (\overline{V} - b)}{\overline{V} - b} - \frac{2a}{RT\overline{V}} - \ln \frac{P(\overline{V} - b)}{RT}$$

$$= \frac{b}{\overline{V} - b} - \frac{2a}{RT\overline{V}} - \ln\left[\frac{(\overline{V} - b)}{RT}\left(\frac{RT}{\overline{V} - b} - \frac{a}{\overline{V}^2}\right)\right]$$

$$= \frac{b}{\overline{V} - b} - \frac{2a}{RT\overline{V}} - \ln\left[1 - \frac{a(\overline{V} - b)}{RT\overline{V}^2}\right]$$

22–34. Use the final equation in Problem 22–33 along with the van der Waals equation to plot $\ln \gamma$ against pressure for $CO(g)$ at 200 K. Compare your result with Figure 22.10.

From Table 16.3, $a = 1.4734$ dm^6·bar·mol^{-1}·K^{-1} and $b = 0.039523$ dm^3·mol^{-1} for CO. We are given $\ln \gamma$ as a function of $\overline{V}$ in Problem 22–33, and the van der Waals equation gives P as a function of $\overline{V}$. Therefore, we can choose values of $\overline{V}$ and calculate the corresponding values of $\ln \gamma$ and P. Then we can plot $\ln \gamma$ against P. The result is

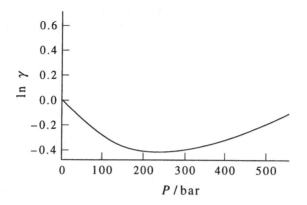

P / bar

22–35. Show that the expression for $\ln \gamma$ for the van der Waals equation (Problem 22–33) can be written in the reduced form

$$\ln \gamma = \frac{1}{3V_R - 1} - \frac{9}{4V_R T_R} - \ln\left[1 - \frac{3(3V_R - 1)}{8T_R V_R^2}\right]$$

Use this equation along with the van der Waals equation in reduced form (Equation 16.19) to plot γ against P_R for $T_R = 1.00$ and 2.00 and compare your results with Figure 22.11.

We can use Equations 16.12 to express a and b in terms of T_c and $\overline{V}_c$:

$$3\overline{V}_c = b + \frac{RT_c}{P_c} \qquad 3\overline{V}_c^2 = \frac{a}{P_c} \qquad \overline{V}_c^3 = \frac{ab}{P_c}$$

Combining the first and the second, and the second and the third, equations gives

$$3\overline{V}_c = b + \frac{\overline{V}_c^3 RT_c}{ab} \qquad 3\overline{V}_c^2 = \frac{a\overline{V}_c^3}{ab}$$

$$3\overline{V}_c = \frac{\overline{V}_c}{3} + \frac{3\overline{V}_c^2 RT_c}{a} \qquad b = \frac{\overline{V}_c}{3}$$

$$a = \frac{9}{8}\overline{V}_c RT_c$$

Now we can substitute into the expression for $\ln \gamma$ we found in the previous problem:

$$\begin{aligned}
\ln \gamma &= \frac{b}{\overline{V} - b} - \frac{2a}{RT\overline{V}} - \ln\left[1 - \frac{a(\overline{V} - b)}{RT\overline{V}^2}\right] \\[2mm]
&= \frac{\overline{V}_c}{3\overline{V} - \overline{V}_c} - \frac{2}{RT\overline{V}}\left(\frac{9}{8}\overline{V}_c RT_c\right) - \ln\left[1 - \frac{9}{8}\overline{V}_c RT_c \frac{(3\overline{V} - \overline{V}_c)}{3RT\overline{V}^2}\right] \\[2mm]
&= \frac{1}{3\overline{V}_R - 1} - \frac{9}{4T_R \overline{V}_R} - \ln\left[1 - \frac{3(3\overline{V}_R - 1)}{8T_R \overline{V}_R^2}\right]
\end{aligned}$$

Equation 16.19 gives

$$P_R = \frac{8}{3}T_R\left(\overline{V}_R - \frac{1}{3}\right)^{-1} - \frac{3}{\overline{V}_R^2}$$

We have now found expressions for both $\ln \gamma$ and P as functions of $\overline{V}_R$ and T_R. We can therefore choose values of $\overline{V}_R$ for a specific value of T_R and calculate the corresponding values of $\ln \gamma$ and P, and plot γ against P. The result is

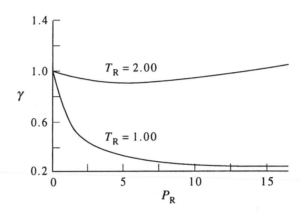

This looks very much like the experimental curves plotted in Figure 22.11.

22–36. Use the method outlined in Problem 22–33 to show that

$$\ln \gamma = \frac{B}{\overline{V} - B} - \frac{A}{RT^{3/2}(\overline{V} + B)} - \frac{A}{BRT^{3/2}}\ln\frac{\overline{V} + B}{\overline{V}} - \ln\left[1 - \frac{A(\overline{V} - B)}{RT^{3/2}\overline{V}(\overline{V} + B)}\right]$$

for the Redlich-Kwong equation. You need to use the standard integral

$$\int \frac{dx}{x(a + bx)} = -\frac{1}{a}\ln\frac{a + bx}{x}$$

For a Redlich-Kwong gas,

$$P = \frac{RT}{\overline{V} - B} - \frac{A}{T^{1/2}\overline{V}(\overline{V} + B)}$$

We can still use the first equation in Problem 22–43, since it was independent of the equation of state chosen. Also, realize that as $\overline{V}^{id} \to \infty$, B becomes negligible with respect to $\overline{V}^{id}$.

$$RT\ln\gamma = P\overline{V} - RT - \int_{\overline{V}^{id}}^{\overline{V}} P'\,d\overline{V} - RT\ln\frac{P}{P^{id}}$$

$$= \frac{RT\overline{V}}{\overline{V} - B} - \frac{A}{T^{1/2}(\overline{V} + B)} - RT - \int_{\overline{V}^{id}}^{\overline{V}}\left[\frac{RT}{\overline{V} - B} - \frac{A}{T^{1/2}\overline{V}(\overline{V} + B)}\right]d\overline{V} - RT\ln\frac{P}{P^{id}}$$

$$= RT\left(\frac{\overline{V}}{\overline{V} - B} - 1\right) - \frac{A}{T^{1/2}(\overline{V} + B)} - RT\ln\frac{\overline{V} - B}{\overline{V}^{id} - B}$$

$$- \frac{A}{T^{1/2}B}\left[\ln\frac{\overline{V} + B}{\overline{V}} - \ln\frac{\overline{V}^{id} + B}{\overline{V}^{id}}\right] - RT\ln\frac{P}{P^{id}}$$

$$\ln\gamma = \frac{B}{\overline{V} - B} - \frac{A}{RT^{3/2}(\overline{V} + B)} - \frac{A}{BRT^{3/2}}\ln\frac{\overline{V} + B}{\overline{V}} - \ln\frac{P(\overline{V} - B)}{P^{id}\overline{V}^{id}}$$

$$
= \frac{B}{\overline{V} - B} - \frac{A}{RT^{3/2}(\overline{V} + B)} - \frac{A}{BRT^{3/2}} \ln \frac{\overline{V} + B}{\overline{V}}
$$

$$
- \ln \left\{ \frac{(\overline{V} - B)}{RT} \left[\frac{RT}{\overline{V} - B} - \frac{A}{T^{1/2}\overline{V}(\overline{V} + B)} \right] \right\}
$$

$$
= \frac{B}{\overline{V} - B} - \frac{A}{RT^{3/2}(\overline{V} + B)} - \frac{A}{BRT^{3/2}} \ln \frac{\overline{V} + B}{\overline{V}} - \ln \left[1 - \frac{A(\overline{V} - B)}{RT^{3/2}\overline{V}(\overline{V} + B)} \right]
$$

22–37. Show that $\ln \gamma$ for the Redlich-Kwong equation (see Problem 22–36) can be written in the reduced form

$$
\ln \gamma = \frac{0.25992}{\overline{V}_R - 0.25992} - \frac{1.2824}{T_R^{3/2}(\overline{V}_R + 0.25992)}
$$

$$
- \frac{4.9340}{T_R^{3/2}} \ln \frac{\overline{V}_R + 0.25992}{\overline{V}_R} - \ln \left[1 - \frac{1.2824(\overline{V}_R - 0.25992)}{T_R^{3/2}\overline{V}_R(\overline{V}_R + 0.25992)} \right]
$$

From Problem 16–26, we can express A and B in terms of T_c and $\overline{V}_c$:

$$
3\overline{V}_c = \frac{RT_c}{P_c} \qquad B = 0.25992\overline{V}_c \qquad A = 0.42748\frac{R^2 T_c^{5/2}}{P_c}
$$

Then

$$
A = 0.42748\frac{R^2 T_c^{5/2} 3\overline{V}_c}{RT_c} = 1.2824 R\overline{V}_c T_c^{3/2}
$$

Now we can substitute into our expression for $\ln \gamma$ in the previous problem:

$$
\ln \gamma = \frac{B}{\overline{V} - B} - \frac{A}{RT^{3/2}(\overline{V} + B)} - \frac{A}{BRT^{3/2}} \ln \frac{\overline{V} + B}{\overline{V}} - \ln \left[1 - \frac{A(\overline{V} - B)}{RT^{3/2}\overline{V}(\overline{V} + B)} \right]
$$

$$
= \frac{0.25992\overline{V}_c}{\overline{V} - 0.25992\overline{V}_c} - \frac{1.2824 R\overline{V}_c T_c^{3/2}}{RT^{3/2}(\overline{V} + 0.25992\overline{V}_c)} - \frac{1.2824 R\overline{V}_c T_c^{3/2}}{RT^{3/2}(0.25992\overline{V}_c)} \ln \frac{\overline{V} + 0.25992\overline{V}_c}{\overline{V}}
$$

$$
- \ln \left[1 - \frac{1.2824 R\overline{V}_c T_c^{3/2}(\overline{V} - 0.25992\overline{V}_c)}{RT^{3/2}(\overline{V} + 0.25992\overline{V}_c)\overline{V}} \right]
$$

$$
= \frac{0.25992}{\overline{V}_R - 0.25992} - \frac{1.2824}{T_R^{3/2}(\overline{V}_R + 0.25992)}
$$

$$
- \frac{4.9340}{T_R^{3/2}} \ln \frac{\overline{V}_R + 0.25992}{\overline{V}_R} - \ln \left[1 - \frac{1.2824(\overline{V}_R - 0.25992)}{T_R^{3/2}\overline{V}_R(\overline{V}_R + 0.25992)} \right]
$$

22–38. Use the expression for $\ln \gamma$ in reduced form given in Problem 22–37 along with the Redlich-Kwong equation in reduced form (Example 16–7) to plot $\ln \gamma$ versus P_R for $T_R = 1.00$ and 2.00 and compare your results with those you obtained in Problem 22–35 for the van der Waals equation.

From Example 16–7, we have an expression for P_R as a function of T_R and $\overline{V}_R$:

$$
P_R = \frac{3T_R}{\overline{V}_R - 0.25992} - \frac{3.8473}{T_R^{1/2}\overline{V}_R(\overline{V}_R + 0.25992)}
$$

and from the previous problem we have an expression for $\ln \gamma$ as a function of T_R and $\overline{V}_R$. We can therefore choose values of $\overline{V}_R$ for a specific value of T_R and calculate the corresponding values of $\ln \gamma$ and P, and then plot γ against P. The result is

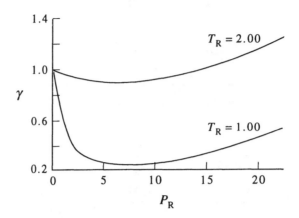

The upward curvature observed here is more marked than that in the plot we obtained from the van der Waals equation.

22–39. Compare $\ln \gamma$ for the van der Waals equation (Problem 22–33) with the values of $\ln \gamma$ for ethane at 600 K (Problem 22–29).

We can graph both the experimental and van der Waals $\ln \gamma$ vs. P, using $a = 5.5818 \text{ dm}^3 \cdot \text{bar} \cdot \text{mol}^{-2}$ and $b = 0.065144 \text{ dm}^3 \cdot \text{mol}^{-1}$. This is only a good fit at extremely low pressures.

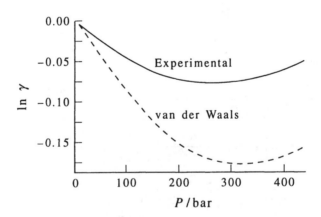

22–40. Compare $\ln \gamma$ for the Redlich-Kwong equation (Problem 22–36) with the values of $\ln \gamma$ for ethane at 600 K (Problem 22–29).

We can graph both the experimental and Redlich-Kwong $\ln \gamma$ vs. P, using $A = 98.831 \text{ dm}^3 \cdot \text{bar} \cdot \text{K}^{-1/2} \cdot \text{mol}^{-2}$ and $B = 0.045153 \text{ dm}^3 \cdot \text{mol}^{-1}$. The Redlich-Kwong equation provides a markedly better description of the behavior of $\ln \gamma$ than the van der Waals equation does.

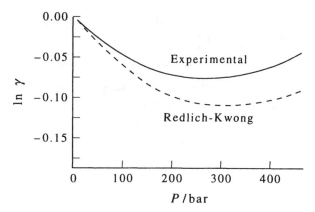

22–41. We can use the equation $(\partial S/\partial U)_V = 1/T$ to illustrate the consequence of the fact that entropy always increases during an irreversible adiabatic process. Consider a two-compartment system enclosed by rigid adiabatic walls, and let the two compartments be separated by a rigid heat-conducting wall. We assume that each compartment is at equilibrium but that they are not in equilibrium with each other. Because no work can be done by this two-compartment system (rigid walls) and no energy as heat can be exchanged with the surroundings (adiabatic walls),

$$U = U_1 + U_2 = \text{constant}$$

Show that

$$dS = \left(\frac{\partial S_1}{\partial U_1}\right) dU_1 + \left(\frac{\partial S_2}{\partial U_2}\right) dU_2$$

because the entropy of each compartment can change only as a result of a change in energy. Now show that

$$dS = dU_1 \left(\frac{1}{T_1} - \frac{1}{T_2}\right) \geq 0$$

Use this result to discuss the direction of the flow of energy as heat from one temperature to another.

We know that $U_1 + U_2$ is a constant, so $dU_1 = -dU_2$. Since the change of entropy of each compartment is dependent only on the energy change of each compartment (using the fact that we can express dS in terms of dU and dV, its natural variables), we can write

$$dS(U_1, U_2) = \left(\frac{\partial S_1}{\partial U_1}\right) dU_1 + \left(\frac{\partial S_2}{\partial U_2}\right) dU_2$$

Notice that this is a constant-volume process, so we use the expression $(\partial S/\partial U)_V = 1/T$ to write

$$\left(\frac{\partial S_1}{\partial U_1}\right)_V = \frac{1}{T_1} \qquad \text{and} \qquad \left(\frac{\partial S_2}{\partial U_2}\right)_V = \frac{1}{T_2}$$

Substituting into the expression for dS gives

$$dS = \frac{dU_1}{T_1} + \frac{dU_2}{T_2} = dU_1 \left(\frac{1}{T_1} - \frac{1}{T_2}\right) \geq 0$$

where the inequality holds because entropy always increases in an irreversible adiabatic process. If the energy flows from compartment 2 into compartment 1, dU_1 must be positive, and so $T_1 < T_2$ in order for the inequality to hold. Likewise, if the energy flows from compartment 2 into compartment 1, dU_1 is negative and so $T_1 > T_2$. The energy always flows from the higher temperature to the lower temperature.

22–42. Modify the argument in Problem 22–41 to the case in which the two compartments are separated by a nonrigid, insulating wall. Derive the result

$$dS = \left(\frac{P_1}{T_1} - \frac{P_2}{T_2}\right) dV_1$$

Use this result to discuss the direction of a volume change under an isothermal pressure difference.

Since the entire system is isolated, we know

$$U_1 + U_2 = \text{constant} \quad \text{and} \quad V_1 + V_2 = \text{constant}$$

This means that $dU_1 = -dU_2$ and $dV_1 = -dV_2$. Now entropy depends on the energy and the volume, so

$$dS = \left(\frac{\partial S_1}{\partial U_1}\right) dU_1 + \left(\frac{\partial S_1}{\partial V_1}\right) dV_1 + \left(\frac{\partial S_2}{\partial U_2}\right) dU_2 + \left(\frac{\partial S_2}{\partial V_2}\right) dV_1$$

From Equation 22.40, $(\partial S/\partial U)_V = 1/T$ and $(\partial S/\partial V)_U = P/T$, so

$$dS = \frac{dU_1}{T_1} - \frac{dU_1}{T_2} + \frac{P_1}{T_1} dV_1 - \frac{P_2}{T_2} dV_1$$

$$= dU\left(\frac{1}{T_1} - \frac{1}{T_2}\right) + dV_1\left(\frac{P_1}{T_1} - \frac{P_2}{T_2}\right)$$

For an isothermal process, this expression becomes

$$dS = \frac{dV_1}{T}(P_1 - P_2)$$

If the volume of compartment 1 increases, dV_1 is positive and so $P_1 > P_2$ in order for dS to be positive. If the volume of compartment 1 decreases, dV_1 is negative and $P_2 > P_1$. The higher pressure compartment will expand under an isothermal pressure difference.

22–43. In this problem, we will derive virial expansions for $\overline{U}, \overline{H}, \overline{S}, \overline{A}$, and $\overline{G}$. Substitute

$$Z = 1 + B_{2P}P + B_{3P}P^2 + \cdots$$

into Equation 22.65 and integrate from a small pressure, P^{id}, to P to obtain

$$\overline{G}(T, P) - \overline{G}(T, P^{\text{id}}) = RT \ln \frac{P}{P^{\text{id}}} + RT B_{2P}P + \frac{RT B_{3P}}{2} P^2 + \cdots$$

Now use Equation 22.64 (realize that $P = P^{\text{id}}$ in Equation 22.64) to get

$$\overline{G}(T, P) - G^\circ(T) = RT \ln P + RT B_{2P}P + \frac{RT B_{3P}}{2} P^2 + \cdots \tag{1}$$

at $P^\circ = 1$ bar. Now use Equation 22.31a to get

$$\overline{S}(T, P) - S^\circ(T) = -R \ln P - \frac{d(RT B_{2P})}{dT} P - \frac{1}{2}\frac{d(RT B_{3P})}{dT} P^2 + \cdots \tag{2}$$

at $P° = 1$ bar. Now use $\overline{G} = \overline{H} - T\overline{S}$ to get

$$\overline{H}(T, P) - H°(T) = -RT^2\frac{dB_{2P}}{dT}P - \frac{RT^2}{2}\frac{dB_{3P}}{dT}P^2 + \cdots \tag{3}$$

Now use the fact that $\overline{C}_P = (\partial\overline{H}/\partial T)_P$ to get

$$\overline{C}_P(T, P) - C_P°(T) = -RT\left[2\frac{dB_{2P}}{dT} + T\frac{d^2B_{2P}}{dT^2}\right]P - \frac{RT}{2}\left[2\frac{dB_{3P}}{dT} + T\frac{d^2B_{3P}}{dT^2}\right]P^2 + \cdots \tag{4}$$

We can obtain expansions for $\overline{U}$ and $\overline{A}$ by using the equation $\overline{H} = \overline{U} + P\overline{V} = \overline{U} + RTZ$ and $\overline{G} = \overline{A} + P\overline{V} = \overline{A} + RTZ$. Show that

$$\overline{U} - U° = -RT\left(B_{2P} + T\frac{dB_{2P}}{dT}\right)P - RT\left(B_{3P} + \frac{T}{2}\frac{dB_{3P}}{dT}\right)P^2 + \cdots \tag{5}$$

and

$$\overline{A} - A° = RT\ln P - \frac{RTB_{3P}}{2}P^2 + \cdots \tag{6}$$

at $P° = 1$ bar.

We can use the virial expansion in pressure to write

$$Z = 1 + B_{2P}P + B_{3P}P^2 + \cdots$$
$$\overline{V} = \frac{RT}{P} + B_{2P}RT + B_{3P}RTP + \cdots$$

Now substitute into Equation 22.65:

$$\left(\frac{\partial\overline{G}}{\partial P}\right)_T = \overline{V}$$

$$\int_{P^{id}}^P d\overline{G} = \int_{P^{id}}^P\left[\frac{RT}{P} + B_{2P}RT + B_{3P}RTP + O(P^2)\right]dP$$

$$\overline{G}(T, P) - \overline{G}(T, P^{id}) = RT\ln\frac{P}{P^{id}} + RTB_{2P}(P - P^{id}) + \frac{RTB_{3P}}{2}(P^2 - P^{id\,2}) + O(P^3)$$

Since P^{id} is very small, we can neglect it with respect to P in the last two terms and find

$$\overline{G}(T, P) - \overline{G}(T, P^{id}) = RT\ln\frac{P}{P^{id}} + RTB_{2P}P + \frac{RTB_{3P}}{2}P^2 + O(P^3)$$

Equation 22.64 states that

$$\overline{G}(T, P^{id}) = G°(T) + RT\ln\frac{P^{id}}{P°}$$

Substituting,

$$\overline{G}(T, P) - G°(T) = RT\ln\frac{P}{P°} + RTB_{2P}P + \frac{RTB_{3P}}{2}P^2 + O(P^3)$$

which is Equation 1 when $P° = 1$ bar. Now

$$\overline{S} = -\left(\frac{\partial \overline{G}}{\partial T}\right)_P$$

$$\overline{S}(T, P) - S°(T) = -\left(\frac{\partial [\overline{G}(T, P) - G°(T)]}{\partial T}\right)_P$$

$$= -R \ln P - \frac{d(RTB_{2P})}{dT} P - \frac{1}{2} \frac{d(TRB_{3P})}{dT} P^2 + O(P^3)$$

which is Equation 2. Since $\overline{G} = \overline{H} - T\overline{S}$, we can now write

$$\overline{G}(T, P) - G°(T) = \overline{H}(T, P) - H°(T) - T\left[\overline{S}(T, P) - S°(T)\right]$$

$$\overline{H}(T, P) - H°(T) = RT \ln P + RTB_{2P}P + \frac{RTB_{3P}}{2} P^2 + T[-R \ln P$$

$$- \frac{d(RTB_{2P})}{dT} P - \frac{1}{2} \frac{d(TRB_{3P})}{dT} P^2 + O(P^3)]$$

$$= -RT^2 \frac{dB_{2P}}{dT} P - \frac{RT^2}{2} \frac{dB_{3P}}{dT} P^2 + O(P^3)$$

This is Equation 3. Now we can take the partial derivative of enthalpy with respect to temperature to find $\overline{C}_P$:

$$\overline{C}_P = \left(\frac{\partial \overline{H}}{\partial T}\right)_P$$

$$\overline{C}_P(T, P) - C_P°(T) = \left(\frac{\partial [\overline{H}(T, P) - H°(T)]}{\partial T}\right)_P$$

$$= \left(-2RT \frac{dB_{2P}}{dT} - RT^2 \frac{d^2B_{2P}}{dT^2}\right) P - \left(RT \frac{dB_{3P}}{dT} + \frac{RT^2}{2} \frac{d^2B_{3P}}{dT^2}\right) P^2 + O(P^3)$$

$$= -RT \left(2\frac{dB_{2P}}{dT} + T\frac{d^2B_{2P}}{dT^2}\right) P - \frac{RT}{2} \left(2\frac{dB_{3P}}{dT} + T\frac{d^2B_{3P}}{dT^2}\right) P^2 + O(P^3)$$

This is Equation 4. Now use the fact that $\overline{U} = \overline{H} - P\overline{V}$:

$$\overline{U}(T, P) - U°(T) = \overline{H}(T, P) - H°(T) - P\left[\overline{V}(T, P) - V°(T)\right]$$

$$= \left[-RT^2 \frac{dB_{2P}}{dT} P - \frac{RT^2}{2} \frac{dB_{3P}}{dT} P^2\right] - P\left[\frac{ZRT}{P} - \frac{RT}{P}\right] + O(P^3)$$

$$= -RT^2 \frac{dB_{2P}}{dT} P - \frac{RT^2}{2} \frac{dB_{3P}}{dT} P^2 - ZRT - RT + O(P^3)$$

$$= -RT^2 \frac{dB_{2P}}{dT} P - \frac{RT^2}{2} \frac{dB_{3P}}{dT} P^2 - \left[1 + B_{2P}P + B_{3P}P^2\right] RT - RT + O(P^3)$$

$$= -RT \left(B_{2P} + T\frac{dB_{2P}}{dT}\right) P - RT \left(B_{3P} + \frac{T}{2}\frac{dB_{3P}}{dT}\right) P^2 + O(P^3)$$

In the above derivation, we realized that $PV° = RT$, or $Z = 1$ for these conditions. We can similarly use the fact that $\overline{G} = \overline{A} + RTZ$, and write

$$\overline{A} - A° = \overline{G} - G° - RT(Z - 1)$$

$$= \left[RT \ln P + RTB_{2P}P + \frac{RTB_{3P}}{2} P^2\right] - RT\left(1 + B_{2P}P + B_{3P}P^2 - 1\right) + O(P^3)$$

$$= RT \ln P + PRT B_{2P} + \frac{RT B_{3P}}{2} P^2 - PRT B_{2P} - RT B_{3P} P^2 + O(P^3)$$

$$= RT \ln P - \frac{RT B_{3P}}{2} P^2 + O(P^3)$$

22–44. In this problem, we will derive the equation

$$\overline{H}(T, P) - H°(T) = RT(Z - 1) + \int_{\overline{V}^{\mathrm{id}}}^{\overline{V}} \left[T \left(\frac{\partial P}{\partial T} \right)_V - P \right] d\overline{V}'$$

where $\overline{V}^{\mathrm{id}}$ is a very large (molar) volume, where the gas is sure to behave ideally. Start with $dH = TdS + VdP$ to derive

$$\left(\frac{\partial H}{\partial V} \right)_T = T \left(\frac{\partial S}{\partial V} \right)_T + V \left(\frac{\partial P}{\partial V} \right)_T$$

and use one of the Maxwell relations for $(\partial S/\partial V)_T$ to obtain

$$\left(\frac{\partial H}{\partial V} \right)_T = T \left(\frac{\partial P}{\partial T} \right)_V + V \left(\frac{\partial P}{\partial V} \right)_T$$

Now integrate by parts from an ideal-gas limit to an arbitrary limit to obtain the desired equation.

Start with Equation 22.49 and take the partial derivative of both sides with respect to V:

$$dH = TdS + VdP$$

$$\left(\frac{\partial H}{\partial V} \right)_T = T \left(\frac{\partial S}{\partial V} \right)_T + V \left(\frac{\partial P}{\partial V} \right)_T$$

Now use Equation 22.19 to write this as

$$\left(\frac{\partial H}{\partial V} \right)_T = T \left(\frac{\partial P}{\partial T} \right)_V + V \left(\frac{\partial P}{\partial V} \right)_T$$

We now integrate the above equation. Recall that $PV = nZRT$, and for an ideal gas $Z = 1$.

$$\int_{H^{\mathrm{id}}}^{H} dH = \int_{V^{\mathrm{id}}}^{V} \left[T \left(\frac{\partial P}{\partial T} \right)_V + V \left(\frac{\partial P}{\partial V} \right)_T \right] dV'$$

$$H - H° = \int_{V^{\mathrm{id}}}^{V} \left[T \left(\frac{\partial P}{\partial T} \right)_V - V \left(\frac{nZRT}{V^2} \right) + nRT \left(\frac{\partial Z}{\partial V} \right)_T \right] dV'$$

$$H - H° = \int_{V^{\mathrm{id}}}^{V} \left[T \left(\frac{\partial P}{\partial T} \right)_V - \frac{nZRT}{V} \right] dV' + \int_{1}^{Z} nRT dZ'$$

$$= nRT \int_{1}^{Z} dZ' + \int_{V^{\mathrm{id}}}^{V} \left[T \left(\frac{\partial P}{\partial T} \right)_V - P \right] dV'$$

$$= nRT(Z - 1) + \int_{V^{\mathrm{id}}}^{V} \left[T \left(\frac{\partial P}{\partial T} \right)_V - P \right] dV'$$

Dividing both sides of the above equation by n gives the desired equation.

22–45. Using the result of Problem 22–44, show that H is independent of volume for an ideal gas. What about a gas whose equation of state is $P(\overline{V} - b) = RT$? Does U depend upon volume for this equation of state? Account for any difference.

For an ideal gas,

$$\left(\frac{\partial P}{\partial T}\right)_V = \frac{nR}{V} \quad \text{and} \quad \left(\frac{\partial P}{\partial V}\right)_T = -\frac{nRT}{V^2}$$

Substituting into the equation from the previous problem,

$$\left(\frac{\partial H}{\partial V}\right)_T = T\left(\frac{\partial P}{\partial T}\right)_V + V\left(\frac{\partial P}{\partial V}\right)_T = \frac{nRT}{V} - \frac{nRTV}{V^2} = 0$$

For the second equation of state given in the problem,

$$\left(\frac{\partial P}{\partial T}\right)_{\overline{V}} = \frac{R}{\overline{V} - b} \quad \text{and} \quad \left(\frac{\partial P}{\partial \overline{V}}\right)_T = -\frac{RT}{(\overline{V} - b)^2}$$

$$\left(\frac{\partial \overline{H}}{\partial \overline{V}}\right)_T = T\left(\frac{\partial P}{\partial T}\right)_{\overline{V}} + \overline{V}\left(\frac{\partial P}{\partial \overline{V}}\right)_T$$

$$= \frac{RT}{\overline{V} - b} - \frac{\overline{V}RT}{(\overline{V} - b)^2} = \frac{RT}{\overline{V} - b}\left[1 - \frac{\overline{V}}{\overline{V} - b}\right]$$

Remember that $\overline{U} = \overline{H} - P\overline{V}$, so

$$\left(\frac{\partial \overline{U}}{\partial \overline{V}}\right)_T = \left(\frac{\partial \overline{H}}{\partial \overline{V}}\right)_T - P - \overline{V}\left(\frac{\partial P}{\partial \overline{V}}\right)_T$$

$$= \frac{RT}{\overline{V} - b}\left[1 - \frac{\overline{V}}{\overline{V} - b}\right] - \frac{RT}{\overline{V} - b} + \overline{V}\frac{RT}{(\overline{V} - b)^2} = 0$$

Therefore $\overline{U}$ does not depend on volume for a gas that obeys the equation of state $P(\overline{V} - b) = RT$.

22–46. Using the result of Problem 22–44, show that

$$\overline{H} - H° = \frac{RTb}{\overline{V} - b} - \frac{2a}{\overline{V}}$$

for the van der Waals equation.

For the van der Waals equation of state,

$$P = \frac{RT}{\overline{V} - b} - \frac{a}{\overline{V}^2} \quad \text{and} \quad \left(\frac{\partial P}{\partial T}\right)_{\overline{V}} = \frac{R}{\overline{V} - b}$$

Also,

$$Z = \frac{P\overline{V}}{RT} = \left[\frac{RT}{\overline{V} - b} - \frac{a}{\overline{V}^2}\right]\frac{\overline{V}}{RT}$$

Now we substitute these values into the equation from Problem 22–44:

$$\overline{H} - \overline{H}° = ZRT - RT + \int_{\overline{V}^{id}}^{\overline{V}}\left[T\left(\frac{\partial P}{\partial T}\right)_{\overline{V}} - P\right]d\overline{V}'$$

$$= \overline{V}\left[\frac{RT}{\overline{V} - b} - \frac{a}{\overline{V}^2}\right] - RT + \int_{\overline{V}^{id}}^{\overline{V}}\left[\frac{RT}{\overline{V} - b} - \frac{RT}{\overline{V} - b} + \frac{a}{\overline{V}^2}\right]d\overline{V}'$$

$$= RT\frac{\overline{V} - (\overline{V} - b)}{\overline{V} - b} - \frac{a}{\overline{V}} - \frac{a}{\overline{V}}\Big|_{\overline{V}^{id}}^{\overline{V}}$$

$$= \frac{RTb}{\overline{V} - b} - \frac{2a}{\overline{V}} + \frac{a}{\overline{V}^{id}}$$

$$= \frac{RTb}{\overline{V} - b} - \frac{2a}{\overline{V}}$$

because $\overline{V}^{id}$ is very large compared to $\overline{V}$.

22–47. Using the result of Problem 22–44, show that

$$\overline{H} - H° = \frac{RTB}{\overline{V} - B} - \frac{A}{T^{1/2}(\overline{V} + B)} - \frac{3A}{2BT^{1/2}} \ln \frac{\overline{V} + B}{\overline{V}}$$

for the Redlich-Kwong equation.

For the Redlich-Kwong equation of state,

$$P = \frac{RT}{\overline{V} - B} - \frac{A}{T^{1/2}\overline{V}(\overline{V} + B)} \qquad \left(\frac{\partial P}{\partial T}\right)_{\overline{V}} = \frac{R}{\overline{V} - B} + \frac{A}{2T^{3/2}\overline{V}(\overline{V} + B)}$$

Also, we write Z as

$$Z = \left[\frac{RT}{\overline{V} - B} - \frac{A}{T^{1/2}\overline{V}(\overline{V} + B)}\right]\frac{\overline{V}}{RT}$$

Now we substitute these values into the equation from Problem 22–44:

$$\overline{H} - \overline{H}° = ZRT - RT + \int_{\overline{V}^{id}}^{\overline{V}} \left[T\left(\frac{\partial P}{\partial T}\right)_{\overline{V}} - P\right]d\overline{V}'$$

$$= \overline{V}\left[\frac{RT}{\overline{V} - B} - \frac{A}{T^{1/2}\overline{V}(\overline{V} + B)}\right] - RT + \int_{\overline{V}^{id}}^{\overline{V}} \left[\frac{RT}{\overline{V}' - B} + \frac{A}{2T^{1/2}\overline{V}'(\overline{V}' + B)}\right.$$

$$\left. - \frac{RT}{\overline{V}' - B} + \frac{A}{T^{1/2}\overline{V}'(\overline{V}' + B)}\right]d\overline{V}'$$

$$= \frac{\overline{V}RT - RT(\overline{V} - B)}{\overline{V} - B} - \frac{A}{T^{1/2}(\overline{V} + B)} + \int_{\overline{V}^{id}}^{\overline{V}} \frac{3A}{2T^{1/2}\overline{V}'(\overline{V}' + B)}d\overline{V}'$$

$$= \frac{BRT}{\overline{V} - B} - \frac{A}{T^{1/2}(\overline{V} + B)} - \frac{3A}{2BT^{1/2}}\left(\ln \frac{\overline{V} + B}{\overline{V}} - \ln \frac{\overline{V}^{id} + B^{id}}{\overline{V}^{id}}\right)$$

$$= \frac{BRT}{\overline{V} - B} - \frac{A}{T^{1/2}(\overline{V} + B)} - \frac{3A}{2BT^{1/2}} \ln \frac{\overline{V} + B}{\overline{V}}$$

because $\overline{V}^{id}$ is very large compared to $\overline{V}$.

The following six problems involve the Joule-Thomson coefficient.

22–48. We introduced the Joule-Thomson effect and the Joule-Thomson coefficient in Problems 19–52 through 19–54. The Joule-Thomson coefficient is defined by

$$\mu_{JT} = \left(\frac{\partial T}{\partial P}\right)_H = -\frac{1}{C_P}\left(\frac{\partial H}{\partial P}\right)_T$$

and is a direct measure of the expected temperature change when a gas is expanded through a throttle. We can use one of the equations derived in this chapter to obtain a convenient working equation for μ_{JT}. Show that

$$\mu_{JT} = \frac{1}{C_P}\left[T\left(\frac{\partial V}{\partial T}\right)_P - V\right]$$

Use this result to show that $\mu_{JT} = 0$ for an ideal gas.

Start with

$$\left(\frac{\partial H}{\partial P}\right)_T = V - T\left(\frac{\partial V}{\partial T}\right)_P \qquad (22.34)$$

Substitute this into the expression for μ_{JT} to obtain

$$\mu_{JT} = \frac{1}{C_P}\left[T\left(\frac{\partial V}{\partial T}\right)_P - V\right]$$

For an ideal gas, $(\partial V/\partial T)_P = nR/P$, so

$$\mu_{JT} = \frac{1}{C_P}\left[\frac{nRT}{P} - V\right] = 0$$

since $PV = nRT$.

22–49. Use the virial equation of state of the form

$$\frac{P\overline{V}}{RT} = 1 + \frac{B_{2V}(T)}{RT}P + \cdots$$

to show that

$$\mu_{JT} = \frac{1}{C_P^{id}}\left[T\frac{dB_{2V}}{dT} - B_{2V}\right] + O(P)$$

It so happens that B_{2V} is negative and dB_{2V}/dT is positive for $T^* < 3.5$ (see Figure 16.15) so that μ_{JT} is positive for low temperatures. Therefore, the gas will cool upon expansion under these conditions. (See Problem 22–48.)

Use the virial equation of state to express V:

$$\frac{P\overline{V}}{RT} = 1 + \frac{B_{2V}(T)}{RT}P + O(P^2)$$

$$\overline{V} = \frac{RT}{P} + B_{2V}(T) + O(P)$$

$$\left(\frac{\partial \overline{V}}{\partial T}\right)_P = \frac{R}{P} + \frac{dB_{2V}}{dT} + O(P)$$

Substituting into the equation for μ_{JT} from Problem 22-48,

$$\mu_{JT} = \frac{1}{C_P}\left[T\left(\frac{\partial V}{\partial T}\right)_P - V\right]$$

$$= \frac{1}{C_P}\left[\frac{RT}{P} + T\frac{dB_{2V}}{dT} - \frac{RT}{P} - B_{2V} + O(P)\right]$$

$$= \frac{1}{C_P}\left[T\frac{dB_{2V}}{dT} - B_{2V}\right] + O(P)$$

22–50. Show that

$$\mu_{JT} = -\frac{b}{C_P}$$

for a gas that obeys the equation of state $P(\overline{V} - b) = RT$. (See Problem 22–48.)

For such a gas,

$$\left(\frac{\partial \overline{V}}{\partial T}\right)_P = \frac{R}{P} \qquad \text{and} \qquad \overline{V} = \frac{RT}{P} + b$$

Substituting into the equation for μ_{JT} from Problem 22-48,

$$\mu_{JT} = \frac{1}{C_P}\left[T\left(\frac{\partial V}{\partial T}\right)_P - V\right] = \frac{1}{C_P}\left[\frac{RT}{P} - \frac{RT}{P} - b\right] = -\frac{b}{C_P}$$

22–51. The second virial coefficient for a square-well potential is (Equation 16.41)

$$B_{2V}(T) = b_0[1 - (\lambda^3 - 1)(e^{\varepsilon/k_B T} - 1)]$$

Show that

$$\mu_{JT} = \frac{b_0}{C_P}\left[(\lambda^3 - 1)\left(1 + \frac{\varepsilon}{k_B T}\right)e^{\varepsilon/k_B T} - \lambda^3\right]$$

where $b_0 = 2\pi\sigma^3 N_A/3$. Given the following square-well parameters, calculate μ_{JT} at 0°C and compare your values with the given experimental values. Take $C_P = 5R/2$ for Ar and $7R/2$ for N_2 and CO_2.

Gas	$b_0/\text{mL}\cdot\text{mol}^{-1}$	λ	ε/k_B	$\mu_{JT}(\text{exptl})/\text{K}\cdot\text{atm}^{-1}$
Ar	39.87	1.85	69.4	0.43
N_2	45.29	1.87	53.7	0.26
CO_2	75.79	1.83	119	1.3

From Problem 22–49, we have

$$\mu_{JT} = \frac{1}{C_P}\left[T\frac{dB_{2V}}{dT} - B_{2V}\right] + O(P)$$

Now we find dB_{2V}/dT from Equation 16.41:

$$B_{2V} = b_0 \left[1 - (\lambda^3 - 1)(e^{\varepsilon/k_B T} - 1) \right]$$

$$\frac{dB_{2V}}{dT} = \frac{\varepsilon b_0}{k_B T^2}(\lambda^3 - 1)e^{\varepsilon/k_B T}$$

Substituting into the expression for μ_{JT}, we find

$$\mu_{JT} = \frac{1}{C_P}\left[T\frac{dB_{2V}}{dT} - B_{2V} \right] + O(P)$$

$$= \frac{1}{C_P}\left[\frac{b_0\varepsilon}{k_B T}(\lambda^3 - 1)e^{\varepsilon/k_B T} - b_0 + b_0(\lambda^3 - 1)(e^{\varepsilon/k_B T} - 1) \right]$$

$$= \frac{b_0}{C_P}\left\{ e^{\varepsilon/k_B T}\left[\frac{\varepsilon}{k_B T}(\lambda^3 - 1) \right] - 1 + \lambda^3 e^{\varepsilon/k_B T} - e^{\varepsilon/k_B T} - \lambda^3 + 1 \right\}$$

$$= \frac{b_0}{C_P}\left\{ e^{\varepsilon/k_B T}\left[\frac{\varepsilon}{k_B T}(\lambda^3 - 1) + (\lambda^3 - 1) \right] - \lambda^3 \right\}$$

$$= \frac{b_0}{C_P}\left[(\lambda^3 - 1)\left(1 + \frac{\varepsilon}{k_B T} \right)e^{\varepsilon/k_B T} - \lambda^3 \right]$$

We can now use the given values of λ, b_0, and ε/k_B to calculate μ_{JT}(theoretical) for Ar, N_2, and CO_2. We use $\overline{C}_P = 5R/2$ for argon and $\overline{C}_P = 7R/2$ for N_2 and CO_2.

Gas	Ar	N_2	CO_2
μ_{JT}(theor.)/K·atm^{-1}	0.44	0.24	1.39
μ_{JT}(exp.)/K·atm^{-1}	0.43	0.26	1.3
Percent Difference	3.4	7.3	6.6

22–52. The temperature at which the Joule-Thomson coefficient changes sign is called the *Joule-Thomson inversion temperature*, T_i. The low-pressure Joule-Thomson inversion temperature for the square-well potential is obtained by setting $\mu_{JT} = 0$ in Problem 22–51. This procedure leads to an equation for $k_B T/\varepsilon$ in terms of λ^3 that cannot be solved analytically. Solve the equation numerically to calculate T_i for the three gases given in the previous problem. The experimental values are 794 K, 621 K, and 1500 K for Ar, N_2, and CO_2, respectively.

$$0 = \frac{b_0}{C_P}\left[(\lambda^3 - 1)\left(1 + \frac{\varepsilon}{k_B T} \right)e^{\varepsilon/k_B T} - \lambda^3 \right]$$

$$\lambda^3 = (\lambda^3 - 1)\left(1 + \frac{\varepsilon}{k_B T} \right)e^{\varepsilon/k_B T}$$

We can use the experimental values as initial values and then use the Newton-Raphson method to find T_i. The inversion temperatures found are tabulated below.

Gas	Ar	N_2	CO_2
T_i(theor.)/K	791	634	1310
T_i(exp.)/K	794	621	1500
Percent Difference	0.378	2.09	12.7

22–53. Use the data in Problem 22–51 to estimate the temperature drop when each of the gases undergoes an expansion for 100 atm to one atm.

By definition, $\mu_{JT} = (\partial T/\partial P)_H$, so

$$\frac{1}{\mu_{JT}} = \left(\frac{\partial P}{\partial T}\right)_H$$

$$\int \mu_{JT} dP = \int dT$$

Let us assume that μ_{JT} does not change significantly over the pressure range. Then

$$\Delta T = \mu_{JT} \Delta P$$

Using the experimental values of μ_{JT}, we see that Ar(g) experiences a temperature drop of 42.6 K, $N_2(g)$ has a temperature drop of 25.7 K, and $CO_2(g)$ drops in temperature by 129 K.

22–54. When a rubber band is stretched, it exerts a restoring force, f, which is a function of its length L and its temperature T. The work involved is given by

$$w = \int f(L, T) dL \tag{1}$$

Why is there no negative sign in front of the integral, as there is in Equation 19.2 for P-V work? Given that the volume change upon stretching a rubber band is negligible, show that

$$dU = TdS + fdL \tag{2}$$

and that

$$\left(\frac{\partial U}{\partial L}\right)_T = T\left(\frac{\partial S}{\partial L}\right)_T + f \tag{3}$$

Using the definition $A = U - TS$, show that Equation 2 becomes

$$dA = -SdT + fdL \tag{4}$$

and derive the Maxwell relation

$$\left(\frac{\partial f}{\partial T}\right)_L = -\left(\frac{\partial S}{\partial L}\right)_T \tag{5}$$

Substitute Equation 5 into Equation 3 to obtain the analog of Equation 22.22

$$\left(\frac{\partial U}{\partial L}\right)_T = f - T\left(\frac{\partial f}{\partial T}\right)_L$$

For many elastic systems, the observed temperature-dependence of the force is linear. We define an *ideal rubber band* by

$$f = T\phi(L) \qquad \text{(ideal rubber band)} \tag{6}$$

Show that $(\partial U/\partial L)_T = 0$ for an ideal rubber band. Compare this result with $(\partial U/\partial V)_T = 0$ for an ideal gas.

Now let's consider what happens when we stretch a rubber band quickly (and, hence, adiabatically). In this case, $dU = dw = fdL$. Use the fact that U depends upon only the temperature for an ideal rubber band to show that

$$dU = \left(\frac{\partial U}{\partial T}\right)_L dT = fdL \tag{7}$$

The quantity $(\partial U/\partial T)_L$ is a heat capacity, so Equation 7 becomes

$$C_L dT = f dL \tag{8}$$

Argue now that if a rubber band is suddenly stretched, then its temperature will rise. Verify this result by holding a rubber band against your upper lip and stretching it quickly.

There is no negative sign in front of the integral because the force the rubber band exerts is a restoring force, which means that it is acting to contract the rubber band.

Since $dU = \delta q + \delta w$ and $\delta q = TdS$,

$$dU = TdS + fdL \tag{2}$$

Taking the partial derivative of both sides with respect to L at constant T gives

$$\left(\frac{\partial U}{\partial L}\right)_T = T\left(\frac{\partial S}{\partial L}\right)_T + f \tag{3}$$

Then, since $A = U - TS$, $U = A + TS$ and

$$dU = TdS + fdL$$
$$dA + TdS + SdT = TdS + fdL$$
$$dA = -SdT + fdL \tag{4}$$

We can also write dA as the total derivative of $A(T, L)$:

$$dA = \left(\frac{\partial A}{\partial T}\right)_L dT + \left(\frac{\partial A}{\partial L}\right)_T dL$$

Comparing the above equation and Equation 4, we see that

$$\left(\frac{\partial A}{\partial T}\right)_L = -S \quad \text{and} \quad \left(\frac{\partial A}{\partial L}\right)_T = f$$

and equating the second cross partial derivatives gives

$$-\left(\frac{\partial S}{\partial L}\right)_T = \left(\frac{\partial f}{\partial T}\right)_L \tag{5}$$

Substituting into Equation 3 gives

$$\left(\frac{\partial U}{\partial L}\right)_T = -T\left(\frac{\partial f}{\partial T}\right)_L + f$$

For an ideal rubber band,

$$f = T\phi(L) \quad \text{and so} \quad \left(\frac{\partial f}{\partial T}\right)_L = \phi_L$$

Then

$$\left(\frac{\partial U}{\partial L}\right)_T = -T\left(\frac{\partial f}{\partial T}\right)_L + f = -T\phi + T\phi = 0$$

Both this result and the result $(\partial U/\partial V)_T = 0$ essentially state that the energy of the system is independent of the length of the rubber band or the volume of the gas at constant temperature.

Now define $C_L = (\partial U/\partial T)_L$. For the ideal rubber band, U depends only on temperature, so we can write

$$U = \int \left(\frac{\partial U}{\partial T}\right)_L dT$$

$$dU = \left(\frac{\partial U}{\partial L}\right)_T dT = C_L dT$$

We know that $dU = fdL$ from the problem text, so we can now write

$$C_L dT = fdL$$

If we suddenly stretch a rubber band, we are applying force f over the distance we stretch the rubber band. Then

$$\int fdL = C_L \int dT$$

which is approximately

$$f\Delta L = C_L \Delta T$$

If ΔL is positive (we are stretching the rubber band), then ΔT must also be positive, and the rubber band heats up when we stretch it.

22–55. Derive an expression for ΔS for the reversible, isothermal expansion of one mole of a gas that obeys van der Waals equation. Use your result to calculate ΔS for the isothermal compression of ethane from $10.0 \text{ dm}^3 \cdot \text{mol}^{-1}$ to $1.00 \text{ dm}^3 \cdot \text{mol}^{-1}$ at 400 K. Compare your result to what you would get using the ideal-gas equation.

We can use the Maxwell relation (Equation 22.19)

$$\left(\frac{\partial \overline{S}}{\partial \overline{V}}\right)_T = \left(\frac{\partial P}{\partial T}\right)_{\overline{V}}$$

For the van der Waals equation,

$$P = \frac{RT}{\overline{V} - b} - \frac{a}{\overline{V}^2}$$

$$\left(\frac{\partial P}{\partial T}\right)_{\overline{V}} = \frac{R}{\overline{V} - b}$$

Substituting into the Maxwell equation above, we find that

$$\left(\frac{\partial \overline{S}}{\partial \overline{V}}\right)_T = \frac{R}{\overline{V} - b}$$

or

$$\Delta \overline{S} = R \ln \frac{\overline{V}_2 - b}{\overline{V}_1 - b}$$

For ethane, $b = 0.065144 \text{ dm}^3 \cdot \text{mol}^{-1}$, so

$$\Delta \overline{S} = (8.3145 \text{ J} \cdot \text{K}^{-1} \cdot \text{mol}^{-1}) \ln \frac{1.00 \text{ dm}^3 \cdot \text{mol}^{-1} - 0.065144 \text{ dm}^3 \cdot \text{mol}^{-1}}{10.0 \text{ dm}^3 \cdot \text{mol}^{-1} - 0.065144 \text{ dm}^3 \cdot \text{mol}^{-1}}$$

$$= -19.7 \text{ J} \cdot \text{K}^{-1} \cdot \text{mol}^{-1}$$

Using the ideal gas equation, we find

$$\left(\frac{\partial \overline{S}}{\partial \overline{V}}\right)_T = \left(\frac{\partial P}{\partial T}\right)_{\overline{V}} = \frac{R}{\overline{V}}$$

$$\Delta S = R \ln \frac{V_2}{V_1} = (8.3145 \text{ J}\cdot\text{K}^{-1}\cdot\text{mol}^{-1}) \ln \frac{1.00}{10.0} = -19.1 \text{ J}\cdot\text{K}^{-1}\cdot\text{mol}^{-1}$$

The van der Waals result is smaller than the value obtained with the ideal gas equation.

22–56. Derive an expression for ΔS for the reversible, isothermal expansion of one mole of a gas that obeys the Redlich-Kwong equation (Equation 16.7). Use your result to calculate ΔS for the isothermal compression of ethane from 10.0 dm³·mol⁻¹ to 1.00 dm³·mol⁻¹ at 400 K. Compare your result with the result you would get using the ideal-gas equation.

Because these are the same parameters as those used in the previous problem, the ideal gas equation of state gives a value of $-19.1 \text{ J}\cdot\text{K}^{-1}\cdot\text{mol}^{-1}$ for $\Delta \overline{S}$.

We can use the Maxwell relation (Equation 22.19)

$$\left(\frac{\partial \overline{S}}{\partial \overline{V}}\right)_T = \left(\frac{\partial P}{\partial T}\right)_{\overline{V}}$$

For the Redlich-Kwong equation,

$$P = \frac{RT}{\overline{V} - B} - \frac{A}{T^{1/2}\overline{V}(\overline{V} + B)}$$

$$\left(\frac{\partial P}{\partial T}\right)_{\overline{V}} = \frac{R}{\overline{V} - B} + \frac{A}{2T^{3/2}\overline{V}(\overline{V} + B)}$$

Substituting into the Maxwell equation (Equation 22.19), we find that

$$\left(\frac{\partial \overline{S}}{\partial \overline{V}}\right)_T = \left(\frac{\partial P}{\partial T}\right)_{\overline{V}} = \frac{R}{\overline{V} - B} + \frac{A}{2T^{3/2}\overline{V}(\overline{V} + B)}$$

or

$$\Delta \overline{S} = R \ln \frac{\overline{V}_2 - B}{\overline{V}_1 - B} - \frac{A}{2BT^{3/2}} \ln \frac{V_1(V_2 + B)}{V_2(V_1 + B)}$$

For ethane, $A = 98.831 \text{ dm}^6\cdot\text{bar}\cdot\text{mol}^{-2}\cdot\text{K}^{1/2}$ and $B = 0.045153 \text{ dm}^3\cdot\text{mol}^{-1}$, so

$$\Delta \overline{S} = (0.083145 \text{ dm}^3\cdot\text{bar}\cdot\text{mol}^{-1}\cdot\text{K}^{-1}) \ln \frac{1.00 \text{ dm}^3\cdot\text{mol}^{-1} - 0.045153 \text{ dm}^3\cdot\text{mol}^{-1}}{10.0 \text{ dm}^3\cdot\text{mol}^{-1} - 0.045153 \text{ dm}^3\cdot\text{mol}^{-1}}$$

$$- \frac{98.831 \text{ dm}^6\cdot\text{bar}\cdot\text{mol}^{-2}\cdot\text{K}^{1/2}}{2(0.045153 \text{ dm}^3\cdot\text{mol}^{-1})(400 \text{ K})^{3/2}} \ln \frac{(10.0 \text{ dm}^3\cdot\text{mol}^{-1})(1.045153 \text{ dm}^3\cdot\text{mol}^{-1})}{(1.00 \text{ dm}^3\cdot\text{mol}^{-1})(10.045153 \text{ dm}^3\cdot\text{mol}^{-1})}$$

$$= -0.200 \text{ dm}^3\cdot\text{bar} = -20.0 \text{ J}\cdot\text{K}^{-1}\cdot\text{mol}^{-1}$$

This is smaller than the value obtained with the ideal gas equation.

CHAPTER 23

Phase Equilibria

PROBLEMS AND SOLUTIONS

23–1. Sketch the phase diagram for oxygen using the following data: triple point, 54.3 K and 1.14 torr; critical point, 154.6 K and 37 828 torr; normal melting point, −218.4°C; and normal boiling point, −182.9°C. Does oxygen melt under an applied pressure as water does?

We can use the triple point and the normal melting point to construct the liquid-solid line and the triple point, normal boiling point, and critical point to construct the liquid-gas line. The liquid-gas line stops at the critical point. We produce the diagram

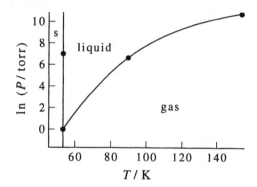

We can see that oxygen does not melt under an applied pressure, because its normal melting point temperature is higher than the triple point temperature.

23–2. Sketch the phase diagram for I_2 given the following data: triple point, 113°C and 0.12 atm; critical point, 512°C and 116 atm; normal melting point, 114°C; normal boiling point, 184°C; and density of liquid > density of solid.

We use the triple point and normal melting point to construct the liquid-solid line and the triple point, normal boiling point, and critical point to construct the liquid-gas line. Because the density of the liquid is greater than the density of the solid, the solid-liquid line has a positive slope.

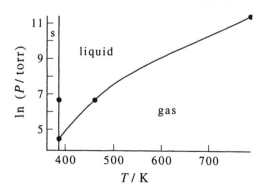

23–3. Figure 23.14 shows a density-temperature phase diagram for benzene. Using the following data for the triple point and the critical point, interpret this phase diagram. Why is the triple point indicated by a line in this type of phase diagram?

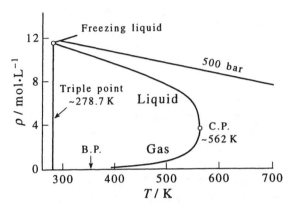

FIGURE 23.14
A density-temperature phase diagram of benzene.

	T/K	P/bar	$\rho/mol \cdot L^{-1}$ Vapor	Liquid
Triple point	278.680	0.04785	0.002074	11.4766
Critical point	561.75	48.7575	3.90	3.90
Normal freezing point	278.68	1.01325		
Normal boiling point	353.240	1.01325	0.035687	10.4075

The triple point is indicated by a line because it represents a temperature at which the solid, liquid, and gas phases all coexist at equilibrium. The line labelled triple point connects the densities of the liquid and vapor in equilibrium with each other. Notice that the liquid and gaseous densities become equal at the critical point. The line labelled 500 bar represents the density of benzene at 500 bar as a function of temperature. Below the information conveyed by the density-temperature phase diagram is represented in a pressure-temperature phase diagram.

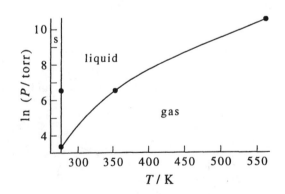

23–4. The vapor pressures of solid and liquid chlorine are given by

$$\ln(P^s/\text{torr}) = 24.320 - \frac{3777 \text{ K}}{T}$$

$$\ln(P^l/\text{torr}) = 17.892 - \frac{2669 \text{ K}}{T}$$

where T is the absolute temperature. Calculate the temperature and pressure at the triple point of chlorine.

This problem is done in the same way as Example 23–1. At the triple point, the two equations for the vapor pressure must be equivalent, since the solid and liquid coexist. Then

$$24.320 - \frac{3777 \text{ K}}{T_{tp}} = 17.892 - \frac{2669 \text{ K}}{T_{tp}}$$

$$(24.320 - 17.892)T_{tp} = -2669 \text{ K} + 3777 \text{ K}$$

$$T_{tp} = 172.4 \text{ K}$$

We can check this by substituting back into both expressions, and we find $\ln(P^s) = \ln(P^l) = 2.41$ torr and so $P_{tp} = 11.1$ torr.

23–5. The pressure along the melting curve from the triple-point temperature to an arbitrary temperature can be fit empirically by the Simon equation, which is

$$(P - P_{tp})/\text{bar} = a\left[\left(\frac{T}{T_{tp}}\right)^{\alpha} - 1\right]$$

where a and α are constants whose values depend upon the substance. Given that $P_{tp} = 0.04785$ bar, $T_{tp} = 278.68$ K, $a = 4237$, and $\alpha = 2.3$ for benzene, plot P against T and compare your result with that given in Figure 23.2.

Substituting into this expression, we find that we must plot

$$P/\text{bar} = 0.04785 + 4237\left[\left(\frac{T/\text{K}}{278.68 \text{ K}}\right)^{2.3} - 1\right]$$

$$P/\text{bar} = 0.04785 - 4237 + \frac{4237}{(278.68)^{2.3}}T^{2.3}$$

where T is on the y-axis and P is on the x-axis. The result looks very much like Figure 23.2.

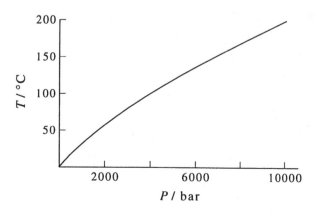

23–6. The slope of the melting curve of methane is given by

$$\frac{dP}{dT} = (0.08446 \text{ bar} \cdot \text{K}^{-1.85})T^{0.85}$$

from the triple point to arbitrary temperatures. Using the fact that the temperature and pressure at the triple point are 90.68 K and 0.1174 bar, calculate the melting pressure of methane at 300 K.

Integrating from the triple point to 300 K gives

$$\int_{0.1174 \text{ bar}}^{P_2} dP = \int_{90.68 \text{ K}}^{300 \text{ K}} 0.08446 \text{ bar} \cdot \text{K}^{-1.85}T^{0.85}dT$$

$$P_2 - 0.1174 \text{ bar} = \frac{0.08446 \text{ bar} \cdot \text{K}^{-1.85}}{1.85}\left[(300 \text{ K})^{1.85} - (90.68 \text{ K})^{1.85}\right]$$

$$P_2 = 1556 \text{ bar}$$

This is the melting pressure of methane at 300 K.

23–7. The vapor pressure of methanol along the entire liquid-vapor coexistence curve can be expressed very accurately by the empirical equation

$$\ln(P/\text{bar}) = -\frac{10.752849}{x} + 16.758207 - 3.603425x$$
$$+ 4.373232x^2 - 2.381377x^3 + 4.572199(1-x)^{1.70}$$

where $x = T/T_c$, and $T_c = 512.60$ K. Use this formula to show that the normal boiling point of methanol is 337.67 K.

At the normal boiling point, $P = 1$ atm $= 1.01325$ bar. If the normal boiling point of methanol is 337.67 K, then the equality below should hold when $x = 337.67/512.60$:

$$\ln(1.01325) \overset{?}{=} -\frac{10.752849}{x} + 16.758207 - 3.603425x$$
$$+ 4.373232x^2 - 2.381377x^3 + 4.572199(1-x)^{1.70}$$

$$0.013163 \overset{?}{=} -16.323364 + 16.758207 - 2.373719 + 1.897712 - 0.6807220 + 0.735141$$

$$0.013163 \approx 0.0132546$$

23–8. The standard boiling point of a liquid is the temperature at which the vapor pressure is exactly one bar. Use the empirical formula given in the previous problem to show that the standard boiling point of methanol is 337.33 K.

We do this in the same way as the previous problem, but substitute $x = 337.33/512.60$ into

$$\ln(1) \overset{?}{=} -\frac{10.752849}{x} + 16.758207 - 3.603425x$$
$$+ 4.373232x^2 - 2.381377x^3 + 4.572199(1-x)^{1.70}$$

$$0 \overset{?}{=} -16.339820 + 16.758207 - 2.371329 + 1.893892 - 0.678668 + 0.737572$$

$$0 \approx -0.000143$$

23–9. The vapor pressure of benzene along the liquid-vapor coexistence curve can be accurately expressed by the empirical expression

$$\ln(P/\text{bar}) = -\frac{10.655375}{x} + 23.941912 - 22.388714x$$
$$+ 20.2085593x^2 - 7.219556x^3 + 4.84728(1-x)^{1.70}$$

where $x = T/T_c$, and $T_c = 561.75$ K. Use this formula to show that the normal boiling point of benzene is 353.24 K. Use the above expression to calculate the standard boiling point of benzene.

This problem is essentially the same as Problem 23–7. We must substitute $x = 353.24/561.75$ into the equation

$$\ln(1.01325) \overset{?}{=} -\frac{10.655375}{x} + 23.941912 - 22.388714x$$
$$+ 20.2085593x^2 - 7.219556x^3 + 4.84728(1-x)^{1.70}$$

$$0.013163 \overset{?}{=} -16.945014 + 23.941912 - 14.078486 + 7.990776 - 1.795109 + 0.899064$$

$$0.013163 \approx 0.0131423$$

To calculate the standard boiling point of benzene, we must solve the polynomial equation for x when $P = 1$ bar, or when $\ln P/\text{bar} = 0$:

$$0 = -\frac{10.655375}{x} + 23.941912 - 22.388714x$$
$$+ 20.2085593x^2 - 7.219556x^3 + 4.84728(1-x)^{1.70}$$

Inputting this formula into a computational mathematics program such as *Mathematica* (or using the Newton-Raphson method) gives $x = 0.62806$, so the standard boiling point is $T = (561.75 \text{ K})x = 352.8$ K.

23–10. Plot the following data for the densities of liquid and gaseous ethane in equilibrium with each other as a function of temperature, and determine the critical temperature of ethane.

T/K	$\rho^l/\text{mol·dm}^{-3}$	$\rho^g/\text{mol·dm}^{-3}$	T/K	$\rho^l/\text{mol·dm}^{-3}$	$\rho^g/\text{mol·dm}^{-3}$
100.00	21.341	1.336×10^{-3}	283.15	12.458	2.067
140.00	19.857	0.03303	293.15	11.297	2.880
180.00	18.279	0.05413	298.15	10.499	3.502
220.00	16.499	0.2999	302.15	9.544	4.307
240.00	15.464	0.5799	304.15	8.737	5.030
260.00	14.261	1.051	304.65	8.387	5.328
270.00	13.549	1.401	305.15	7.830	5.866

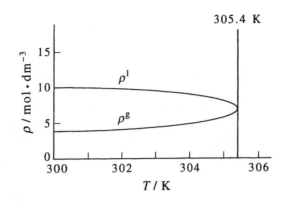

The critical temperature of ethane is about 305.4 K.

23–11. Use the data in the preceding problem to plot $(\rho^l + \rho^g)/2$ against $T_c - T$, with $T_c = 305.4$ K. The resulting straight line is an empirical law called the *law of rectilinear diameters*. If this curve is plotted on the same figure as in the preceding problem, the intersection of the two curves gives the critical density, ρ_c.

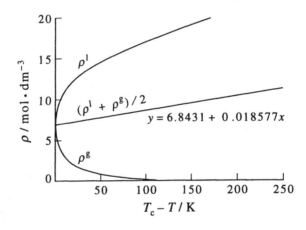

The critical density ρ_c is about 6.84 mol·dm^{-3}.

23–12. Use the data in Problem 23–10 to plot $(\rho^l - \rho^g)$ against $(T_c - T)^{1/3}$ with $T_c = 305.4$ K. What does this plot tell you?

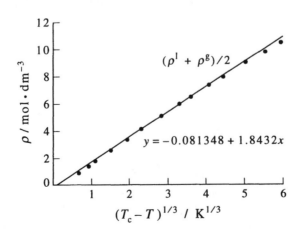

The linear nature of this plot tells us that $(\rho^l - \rho^g)$ varies as $(T - T_c)^{1/3}$ near the critical point.

23–13. The densities of the coexisting liquid and vapor phases of methanol from the triple point to the critical point are accurately given by the empirical expressions

$$\frac{\rho^l}{\rho_c} - 1 = 2.51709(1 - x)^{0.350} + 2.466694(1 - x)$$
$$- 3.066818(1 - x^2) + 1.325077(1 - x^3)$$

and

$$\ln \frac{\rho^g}{\rho_c} = -10.619689 \frac{1 - x}{x} - 2.556682(1 - x)^{0.350}$$
$$+ 3.881454(1 - x) + 4.795568(1 - x)^2$$

where $\rho_c = 8.40 \ \text{mol} \cdot \text{L}^{-1}$ and $x = T/T_c$, where $T_c = 512.60$ K. Use these expressions to plot ρ^l and ρ^g against temperature, as in Figure 23.7. Now plot $(\rho^l + \rho^g)/2$ against T. Show that this line intersects the ρ^l and ρ^g curves at $T = T_c$.

In this graph, the highest line represents ρ^l, the lowest line represents ρ^g, and the dashed line which comes between the two represents $(\rho^l + \rho^g)/2$. At $T = T_c$, $\rho^l = \rho^g$, and so $(\rho^l + \rho^g)/2 = \rho^l = \rho^g$. Therefore, the lines all meet at $T = T_c$.

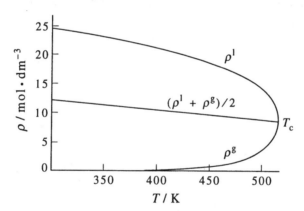

23–14. Use the expressions given in the previous problem to plot $(\rho^l - \rho^g)/2$ against $(T_c - T)^{1/3}$. Do you get a reasonably straight line? If not, determine the value of the exponent of $(T_c - T)$ that gives the best straight line.

We find a line which is reasonably straight (although the curvature shown here is marked, note the scale of the y-axis).

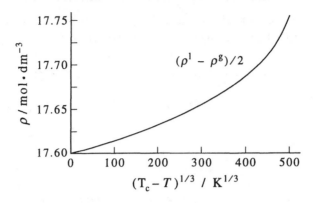

23–15. The molar enthalpy of vaporization of ethane can be expressed as

$$\Delta_{vap}\overline{H}(T)/\text{kJ·mol}^{-1} = \sum_{j=1}^{6} A_j x^j$$

where $A_1 = 12.857$, $A_2 = 5.409$, $A_3 = 33.835$, $A_4 = -97.520$, $A_5 = 100.849$, $A_6 = -37.933$, and $x = (T_c - T)^{1/3}/(T_c - T_{tp})^{1/3}$ where the critical temperature $T_c = 305.4$ K and the triple point temperature $T_{tp} = 90.35$ K. Plot $\Delta_{vap}\overline{H}(T)$ versus T and show that the curve is similar to that of Figure 23.8.

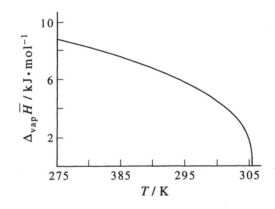

23–16. Fit the following data for argon to a cubic polynomial in T. Use your result to determine the critical temperature.

T/K	$\Delta_{vap}\overline{H}/J\cdot mol^{-1}$	T/K	$\Delta_{vap}\overline{H}/J\cdot mol^{-1}$
83.80	6573.8	122.0	4928.7
86.0	6508.4	126.0	4665.0
90.0	6381.8	130.0	4367.7
94.0	6245.2	134.0	4024.7
98.0	6097.7	138.0	3618.8
102.0	5938.8	142.0	3118.2
106.0	5767.6	146.0	2436.3
110.0	5583.0	148.0	1944.5
114.0	5383.5	149.0	1610.2
118.0	5166.5	150.0	1131.5

Fitting the data to a cubic polynomial in T gives the expression

$$\Delta_{vap}\overline{H}/J\cdot mol^{-1} = 39458.8 - (912.758\ K^{-1})T + (8.53681\ K^{-2})T^2 - (0.0276089\ K^{-3})T^3$$

Solving for T when $\Delta_{vap}\overline{H} = 0$ (at the critical temperature) gives a critical temperature of $T_c = 156.0\ K$. A better fit is to a fifth-order polynomial in T, which gives the expression

$$\Delta_{vap}\overline{H}/J\cdot mol^{-1} = 474232 - (21594.6\ K^{-1})T + (396.54\ K^{-2})T^2 - (3.61587\ K^{-3})T^3$$
$$+(1.63603 \times 10^{-2}\ K^{-4})T^4 - (2.94294 \times 10^{-5}\ K^{-5})T^5$$

We can solve this fifth-order equation for T when $\Delta_{vap}\overline{H} = 0$ by using a computational mathematics program or the Newton-Raphson method, which both give a critical temperature of $T_c = 153.2\ K$.

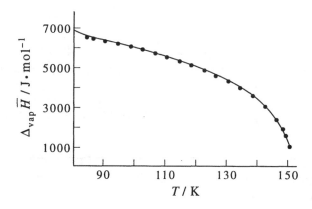

23–17. Use the following data for methanol at one atm to plot $\overline{G}$ versus T around the normal boiling point (337.668 K). What is the value of $\Delta_{vap}\overline{H}$?

T/K	$\overline{H}$/kJ·mol^{-1}	$\overline{S}$/J·mol^{-1}·K^{-1}
240	4.7183	112.259
280	7.7071	123.870
300	9.3082	129.375
320	10.9933	134.756
330	11.8671	137.412
337.668	12.5509	139.437
337.668	47.8100	243.856
350	48.5113	245.937
360	49.0631	247.492
380	50.1458	250.419
400	51.2257	253.189

We can use the formula $\overline{G} = \overline{H} - T\overline{S}$ to find $\overline{G}$ from this data and then plot $\overline{G}$ vs. T:

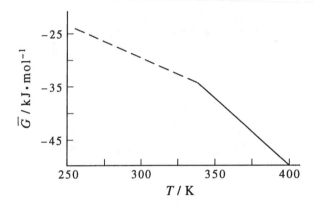

The line of best fit for the gaseous phase is $\overline{G}^{g}$/kJ·mol$^{-1} = 49.57 - 0249T$ and the line of best fit for the liquid phase is $\overline{G}^{l}$/kJ·mol$^{-1} = 8.1913 - 0.126T$. $\Delta_{vap}\overline{H}$ will simply be the change in enthalpy when going from a liquid to a gas:

$$\Delta_{vap}\overline{H}/kJ·mol^{-1} = 47.8100 - 12.5509 = 35.2591$$

23–18. In this problem, we will sketch $\overline{G}$ versus P for the solid, liquid, and gaseous phases for a generic ideal substance as in Figure 23.11. Let $\overline{V}^{s} = 0.600$, $\overline{V}^{l} = 0.850$, and $RT = 2.5$, in arbitrary units. Now show that

$$\overline{G}^{s} = 0.600(P - P_0) + \overline{G}_0^{s}$$
$$\overline{G}^{l} = 0.850(P - P_0) + \overline{G}_0^{l}$$

and

$$\overline{G}^{g} = 2.5\ln(P - P_0) + \overline{G}_0^{g}$$

where $P_0 = 1$ and $\overline{G}_0^{s}$, $\overline{G}_0^{l}$, and $\overline{G}_0^{g}$ are the respective zeros of energy. Show that if we (arbitrarily) choose the solid and liquid phases to be in equilibrium at $P = 2.00$ and the liquid and gaseous phases to be in equilibrium at $P = 1.00$, then we obtain

$$\overline{G}_0^{s} - \overline{G}_0^{l} = 0.250$$

and

$$\overline{G}_0^{\,l} = \overline{G}_0^{\,g}$$

from which we obtain

$$\overline{G}_0^{\,s} - \overline{G}_0^{\,g} = 0.250$$

Now we can express $\overline{G}^{\,s}$, $\overline{G}^{\,l}$, and $\overline{G}^{\,g}$ in terms of a common zero of energy, $\overline{G}_0^{\,g}$, which we must do to compare them with each other and to plot them on the same graph. Show that

$$\overline{G}^{\,s} - \overline{G}_0^{\,g} = 0.600(P - 1) + 0.250$$

$$\overline{G}^{\,l} - \overline{G}_0^{\,g} = 0.850(P - 1)$$

$$\overline{G}^{\,g} - \overline{G}_0^{\,g} = 2.5 \ln P$$

Plot these on the same graph from $P = 0.100$ to 3.00 and compare your result with Figure 23.11.

We know from Chapter 22 that $(\partial \overline{G} / \partial P)_T = \overline{V}$. This means that (for an ideal gas)

$$\overline{G} - \overline{G}_0 = \int_{P_0}^{P} \overline{V} dP = RT \ln \frac{P}{P_0}$$

For the solid and liquid phases, $\overline{V}$ is essentially constant with respect to pressure, and so $\overline{G} - \overline{G}_0 = \overline{V}(P - P_0)$. Therefore, we have

$$\overline{G}^{\,s} = 0.600(P - 1) + \overline{G}_0^{\,s}$$

$$\overline{G}^{\,l} = 0.850(P - 1) + \overline{G}_0^{\,l}$$

$$\overline{G}^{\,g} = RT \ln \frac{P}{P_0} + \overline{G}_0^{\,g} = 2.5 \ln P + \overline{G}_0^{\,g}$$

where the units are arbitrary. Now, at equilibrium, $\overline{G}^{\,1} = \overline{G}^{\,2}$. Since the solid and liquid are in equilibrium at $P = 2.00$ and the liquid and gas are in equilibrium at $P = 1.00$,

$$\overline{G}^{\,s}(P = 2.00) = \overline{G}^{\,l}(P = 2.00)$$

$$0.600 + \overline{G}_0^{\,s} = 0.850 + \overline{G}_0^{\,l}$$

$$\overline{G}_0^{\,s} - \overline{G}_0^{\,l} = 0.250$$

$$\overline{G}^{\,l}(P = 1.00) = \overline{G}^{\,g}(P = 1.00)$$

$$\overline{G}_0^{\,l} = \overline{G}_0^{\,g}$$

and so $\overline{G}_0^{\,s} - \overline{G}_0^{\,g} = 0.250$. Now substitute into the first equations we found:

$$\overline{G}^{\,s} = 0.600(P - 1) + \overline{G}_0^{\,s} = 0.600(P - 1) + 0.250 + \overline{G}_0^{\,g}$$

$$\overline{G}^{\,s} - \overline{G}_0^{\,g} = 0.600(P - 1) + 0.250$$

Also

$$\overline{G}^{\,l} - \overline{G}_0^{\,g} = 0.0850(P - 1)$$

and

$$\overline{G}^{\,g} - \overline{G}_0^{\,g} = 2.5 \ln P$$

A plot of the Gibbs energies of the gas, liquid, and solid using $\overline{G}_0^{\,g}$ as the zero of energy is shown:

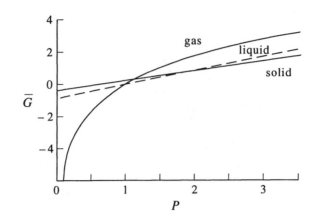

Since we see a gas-liquid-solid progression, we are looking at a temperature less than the triple point temperature (as explained in the caption of Figure 23.11).

23-19. In this problem, we will demonstrate that entropy always increases when there is a material flow from a region of higher concentration to one of lower concentration. (Compare with Problems 22-41 and 22-42.) Consider a two-compartment system enclosed by rigid, impermeable, adiabatic walls, and let the two compartments be separated by a rigid, insulating, but permeable wall. We assume that the two compartments are in equilibrium but that they are not in equilibrium with each other. Show that

$$U_1 = \text{constant}, \quad U_2 = \text{constant}, \quad V_1 = \text{constant}, \quad V_2 = \text{constant},$$

and

$$n_1 + n_2 = \text{constant}$$

for this system. Now show that

$$dS = \frac{dU}{T} + \frac{P}{T}dV - \frac{\mu}{T}dn$$

in general, and that

$$dS = \left(\frac{\partial S_1}{\partial n_1}\right)dn_1 + \left(\frac{\partial S_2}{\partial n_2}\right)dn_2$$

$$= dn_1\left(\frac{\mu_2}{T} - \frac{\mu_1}{T}\right) \geq 0$$

for this system. Use this result to discuss the direction of a (isothermal) material flow under a chemical potential difference.

The volume of each compartment cannot change, since the walls of the compartments are rigid. Thus $V_1 = $ constant and $V_2 = $ constant. Since the walls are adiabatic, $\delta q = 0$ for the gases in each compartment. For both components of the system $\delta w = 0$, since there is no change in volume, so $dU = 0$ for both compartments. Thus $U_1 = $ constant and $U_2 = $ constant. Finally, since the entire system is surrounded by impermeable walls, the total number of moles of gas in the system must remain constant, so $n_1 + n_2 = $ constant.

We have defined $\mu = (\partial G/\partial n)_{P,T}$ (Equation 23.3), so $\mu dn = dG$. Now recall (Equation 22.13) that

$$G = U - TS + PV$$

$$dG = dU - TdS + PdV$$

$$\frac{\mu dn}{T} = \frac{dU}{T} - dS + \frac{P}{T}dV$$

$$dS = \frac{dU}{T} + \frac{P}{T}dV - \frac{\mu}{T}dn$$

For this system, since $dU = dV = 0$,

$$dS_1 = -\frac{\mu_1}{T}dn_1 \qquad\qquad dS_2 = -\frac{\mu_2}{T}dn_2$$

Then

$$dS_{\text{system}} = dS_1 + dS_2$$

$$= \left(\frac{\partial S_1}{\partial n_1}\right) dn_1 + \left(\frac{\partial S_2}{\partial n_2}\right) dn_2$$

$$= -\frac{\mu_1}{T}dn_1 - \frac{\mu_2}{T}(-dn_1)$$

$$= dn_1 \left(\frac{\mu_2}{T} - \frac{\mu_1}{T}\right)$$

If molecules are flowing into compartment 1, then dn_1 is positive and $\mu_2 > \mu_1$ (since transfer occurs from the system with higher chemical potential to the system with lower chemical potential). Then both terms in the expression above are positive and $dS_{\text{system}} > 0$. If molecules are flowing into compartment 2, then dn_1 is negative and $\mu_2 < \mu_1$, making both terms negative and $dS_{\text{system}} > 0$. If dn_1 is 0 (no transfer occurs), then the two compartments are in equilibrium with respect to material flow, and $dS_{\text{system}} = 0$.

23–20. Determine the value of dT/dP for water at its normal boiling point of 373.15 K given that the molar enthalpy of vaporization is 40.65 kJ·mol^{-1}, and the densities of the liquid and vapor are 0.9584 g·L^{-1} and 0.6010 g·mL^{-1}, respectively. Estimate the boiling point of water at 2 atm.

First find $\overline{V}^{\text{g}} - \overline{V}^{\text{l}}$:

$$\overline{V}^{\text{g}} - \overline{V}^{\text{l}} = \left(\frac{1}{0.6010 \text{ g}\cdot\text{dm}^{-3}} - \frac{1}{958.4 \text{ g}\cdot\text{dm}^{-3}}\right)(18.015 \text{ g}\cdot\text{mol}^{-1})$$

$$= 29.96 \text{ dm}^3\cdot\text{mol}^{-1}$$

Now use Equation 23.10 to write

$$\frac{dT}{dP} = \frac{T\Delta_{\text{vap}}\overline{V}}{\Delta_{\text{vap}}\overline{H}}$$

$$= \frac{(373.15 \text{ K})(\overline{V}^{\text{g}} - \overline{V}^{\text{l}})}{40\,650 \text{ J}\cdot\text{mol}^{-1}}$$

$$= \left[\frac{(373.15 \text{ K})(29.96 \text{ dm}^3\cdot\text{mol}^{-1})}{40\,650 \text{ J}\cdot\text{mol}^{-1}}\right]\left(\frac{8.314 \text{ J}}{0.08206 \text{ dm}^3\cdot\text{atm}}\right)$$

$$= 27.9 \text{ K}\cdot\text{atm}^{-1}$$

To estimate the boiling point of water at 2 atm, we can find the change in temperature which accompanies a change in pressure of 1 atm (since we know the boiling point of water at 1 atm). That is $\Delta T = (27.9 \text{ K} \cdot \text{atm}^{-1})(1 \text{ atm}) = 27.9 \text{ K}$. Therefore, the boiling point of water at 2 atm is about 127.9°C.

23–21. The orthobaric densities of liquid and gaseous ethyl acetate are 0.826 g·mL^{-1} and 0.00319 g·mL^{-1}, respectively, at its normal boiling point (77.11°C). The rate of change of vapor pressure with temperature is 23.0 torr·K^{-1} at the normal boiling point. Estimate the molar enthalpy of vaporization of ethyl acetate at its normal boiling point.

First find $\overline{V}^{\text{g}} - \overline{V}^{\text{l}}$:

$$\overline{V}^{\text{g}} - \overline{V}^{\text{l}} = \left(\frac{1}{0.00319 \text{ g} \cdot \text{mL}^{-1}} - \frac{1}{0.826 \text{ g} \cdot \text{mL}^{-1}} \right) (88.102 \text{ g} \cdot \text{mol}^{-1})$$

$$\Delta_{\text{vap}} \overline{V} = 27\,510 \text{ cm}^3 \cdot \text{mol}^{-1} = 275.10 \text{ dm}^3 \cdot \text{mol}^{-1}$$

Now use Equation 23.10 to write

$$\Delta_{\text{vap}} \overline{H} = T \Delta_{\text{vap}} \overline{V} \left(\frac{dP}{dT} \right)$$

$$= (350.26 \text{ K})(27.51 \text{ dm}^3 \cdot \text{mol}^{-1})(23.0 \text{ torr} \cdot \text{K}^{-1}) \left(\frac{1 \text{ atm}}{760 \text{ torr}} \right) \left(\frac{8.314 \text{ J}}{0.08206 \text{ L} \cdot \text{atm}} \right)$$

$$= 29.5 \text{ kJ} \cdot \text{mol}^{-1}$$

23–22. The vapor pressure of mercury from 400°C to 1300°C can be expressed by

$$\ln(P/\text{torr}) = -\frac{7060.7 \text{ K}}{T} + 17.85$$

The density of the vapor at its normal boiling point is 3.82 g·L^{-1} and that of the liquid is 12.7 g·mL^{-1}. Estimate the molar enthalpy of vaporization of mercury at its normal boiling point.

If we express P using the above equation, we find that

$$\frac{dP}{dT} = P \left(\frac{7060.7 \text{ K}}{T^2} \right)$$

At the boiling point and one atmosphere of pressure,

$$\frac{dP}{dT} = (760 \text{ torr}) \left[\frac{7060.7 \text{ K}}{(629.88 \text{ K})^2} \right] = 13.52 \text{ torr} \cdot \text{K}^{-1}$$

We find $\Delta_{\text{vap}} \overline{V}$ by subtracting $\overline{V}^{\text{l}}$ from $\overline{V}^{\text{g}}$:

$$\overline{V}^{\text{g}} - \overline{V}^{\text{l}} = \left(\frac{1}{3.82 \text{ g} \cdot \text{dm}^{-3}} - \frac{1}{12700 \text{ g} \cdot \text{dm}^{-3}} \right) (200.59 \text{ g} \cdot \text{mol}^{-1})$$

$$\Delta_{\text{vap}} \overline{V} = 52.49 \text{ dm}^3 \cdot \text{mol}^{-1}$$

Now we use Equation 23.10 to estimate $\Delta_{vap}\overline{H}$.

$$\Delta_{vap}\overline{H} = T\Delta_{vap}\overline{V}\frac{dP}{dT}$$

$$= (629.88 \text{ K})(52.49 \text{ dm}^3\cdot\text{mol}^{-1})(13.52 \text{ torr}\cdot\text{K}^{-1})$$

$$= 447\,200 \text{ dm}^3\cdot\text{torr}\cdot\text{mol}^{-1}\left(\frac{1 \text{ atm}}{760 \text{ torr}}\right)\left(\frac{8.314 \text{ J}}{0.08206 \text{ dm}^3\cdot\text{atm}}\right)$$

$$= 59.62 \text{ kJ}\cdot\text{mol}^{-1}$$

23–23. The pressures at the solid-liquid coexistence boundary of propane are given by the empirical equation

$$P = -718 + 2.38565T^{1.283}$$

where P is in bars and T is in kelvins. Given that $T_{fus} = 85.46$ K and $\Delta_{fus}\overline{H} = 3.53$ kJ·mol^{-1}, calculate $\Delta_{fus}\overline{V}$ at 85.46 K.

At 85.46 K, the empirical equation gives

$$\frac{dP}{dT} = 3.06079(85.46)^{0.283} = 10.778 \text{ bar}\cdot\text{K}^{-1}$$

We substitute into Equation 23.10 to find

$$\Delta_{fus}\overline{V} = \frac{\Delta_{fus}\overline{H}}{T}\left(\frac{dP}{dT}\right)^{-1}$$

$$= \frac{35\,300 \text{ J}\cdot\text{mol}^{-1}}{85.46 \text{ K}}(10.778 \text{ bar}\cdot\text{K}^{-1})\left(\frac{10 \text{ bar}\cdot\text{cm}^3}{1 \text{ J}}\right)$$

$$= 383 \text{ cm}^3\cdot\text{mol}^{-1}$$

23–24. Use the vapor pressure data given in Problem 23–7 and the density data given in Problem 23–13 to calculate $\Delta_{vap}\overline{H}$ for methanol from the triple point (175.6 K) to the critical point (512.6 K). Plot your result.

We are given ρ in units of mol·dm^{-3} in Problem 23–13 and P in units of bars in Problem 23–7. We want to find $\Delta_{vap}\overline{H}$ using Equation 23.10:

$$\Delta_{vap}\overline{H} = T\Delta_{vap}\overline{V}\frac{dP}{dT}$$

Taking the derivative of the expression for P given in Problem 23–7 gives us

$$\frac{dP}{dT}/\text{bar}\cdot\text{K}^{-1} = P\left[-0.0070297 - 0.0151634(1 - 0.00195084T)^{0.7}\right.$$

$$\left. + \frac{5511.91}{T^2} + 3.32871 \times 10^{-5}T - 5.30412 \times 10^{-8}T^2\right]$$

Using $1/\rho^l$ for $\overline{V}^l$ and $1/\rho^g$ for $\overline{V}^g$ gives

$$\Delta_{vap}\overline{V}/\text{dm}^3\cdot\text{mol}^{-1} = \frac{1}{\rho^g} - \frac{1}{\rho^l}$$

or (where $\Delta_{vap}\overline{V}$ is in units of dm^3·mol^{-1}

$$\Delta_{vap}\overline{V} = 0.119048 \exp\left[2.55668(1 - 0.0019508T)^{0.35} - 3.88145(1 - 0.0019508T)\right.$$

$$\left. -4.79557(1 - 0.0019508T)^2 + \frac{5443.65(1 - 0.0019508)}{T}\right]$$

$$-0.119048\left[1 + 2.5171(1 - 0.0019508T)^{0.35} + 2.46669(1 - 0.0019508T)\right.$$

$$\left. -3.06682(1 - 0.0019508T)^2 + 1.32508(1 - 0.0019508T)^3\right]^{-1}$$

Substituting these expressions into Equation 23.10 gives $\Delta_{vap}\overline{H}$ in units of dm^3·bar·mol^{-1}. To convert this to kJ·mol^{-1} we must divide by 10.

Now graph

$$\Delta_{vap}\overline{H}/\text{kJ·mol}^{-1} = \frac{\Delta_{vap}\overline{V}}{10}\frac{dP}{dT}T$$

using the expressions found above for $\Delta_{vap}\overline{V}$ and dP/dT:

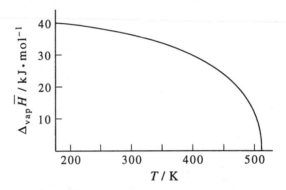

23–25. Use the result of the previous problem to plot $\Delta_{vap}\overline{S}$ of methanol from the triple point to the critical point.

Since at a transition point $\Delta_{trs}\overline{G} = 0$,

$$\frac{\Delta_{vap}\overline{H}}{T} = \Delta_{vap}\overline{S}$$

We can use the expression for $\Delta_{vap}\overline{H}$ given in Problem 23–24 (converting it to J·mol^{-1}, since these are the usual units of entropy) to graph $\Delta_{vap}\overline{S}$.

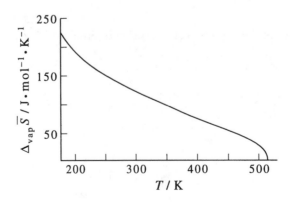

Notice that $\Delta_{vap}\overline{S} \to 0$ as $T \to T_c$.

23–26. Use the vapor pressure data for methanol given in Problem 23–7 to plot $\ln P$ against $1/T$. Using your calculations from Problem 23–24, over what temperature range do you think the Clausius-Clapeyron equation will be valid?

Use the formula for $\ln P$ given in Problem 23–7 to plot $\ln P$ vs. $1/T$.

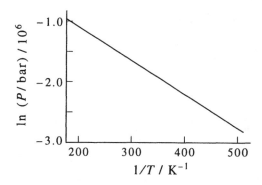

The slope of the line plotted should be constant for the Clausius-Clapeyron equation to be valid. It appears that the Clausius-Clapeyron equation is valid over the range plotted in Problem 23–31.

23–27. The molar enthalpy of vaporization of water is 40.65 kJ·mol⁻¹ at its normal boiling point. Use the Clausius-Clapeyron equation to calculate the vapor pressure of water at 110°C. The experimental value is 1075 torr.

Assuming $\Delta_{vap}\overline{H}$ remains constant with respect to temperature over this ten-degree temperature range, we can use Equation 23.13:

$$\ln \frac{P_2}{P_1} = \frac{\Delta_{vap}\overline{H}}{R}\left(\frac{T_2 - T_1}{T_1 T_2}\right)$$

$$\ln \frac{P_2}{1\text{ atm}} = \frac{40\,650\text{ J·mol}^{-1}}{8.3145\text{ J·mol}^{-1}\cdot\text{K}^{-1}}\left[\frac{10\text{ K}}{(373.15\text{ K})(383.15\text{ K})}\right]$$

$$\ln (P_2/\text{ atm}) = 0.342$$

$$P_2 = 1.408\text{ atm} = 1070\text{ torr}$$

23–28. The vapor pressure of benzaldehyde is 400 torr at 154°C and its normal boiling point is 179°C. Estimate its molar enthalpy of vaporization. The experimental value is 42.50 kJ·mol⁻¹.

Again, assuming $\Delta_{vap}\overline{H}$ does not vary over this temperature range, we can use Equation 23.13.

$$\ln \frac{P_2}{P_1} = \frac{\Delta_{vap}\overline{H}}{R}\left(\frac{T_2 - T_1}{T_1 T_2}\right)$$

$$\Delta_{vap}\overline{H} = \frac{RT_1 T_2}{T_2 - T_1}\ln\frac{P_2}{P_1}$$

$$= \frac{(8.3145\text{ J·mol}^{-1}\cdot\text{K}^{-1})(427.15\text{ K})(452.15\text{ K})}{25\text{ K}}\ln\frac{760}{400}$$

$$= 41.2\text{ kJ·mol}^{-1}$$

23–29. Use the following data to estimate the normal boiling point and the molar enthalpy of vaporization of lead.

T/K	1500	1600	1700	1800	1900
P/torr	19.72	48.48	107.2	217.7	408.2

Plot $\ln P$ vs. $1/T$:

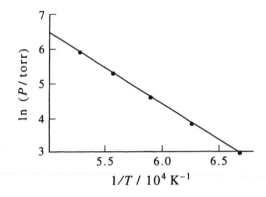

The equation for the line of best fit is $y = 17.3799 - 21597.6x$. At 1 atm (the normal boiling point pressure), $\ln 760 = y$ and so

$$17.3799 = 6.6333 + \frac{21597.6}{T}$$

$$T = 2010 \text{ K}$$

The normal boiling point is about 2010 K. Now recall that the slope of the plot we have created should be $-\Delta_{vap}\overline{H}/R$ (Equation 23.14). Then

$$\Delta_{vap}\overline{H} = (8.314 \text{ J·mol}^{-1}\text{·K}^{-1})(21597.6 \text{ K}) = 179.6 \text{ kJ·mol}^{-1}$$

23–30. The vapor pressure of solid iodine is given by

$$\ln(P/\text{atm}) = -\frac{8090.0 \text{ K}}{T} - 2.013 \ln(T/\text{K}) + 32.908$$

Use this equation to calculate the normal sublimation temperature and the molar enthalpy of sublimation of $I_2(s)$ at 25°C. The experimental value of $\Delta_{sub}\overline{H}$ is 62.23 kJ·mol^{-1}.

The sublimation temperature is found by setting $P = 1$ atm in the above equation:

$$0 = -\frac{8090.0 \text{ K}}{T_{sub}} - 2.013 \ln(T_{sub}/\text{K}) + 32.908$$

We can solve this equation for T using the Newton-Raphson method, and we find that $T_{sub} = 386.8$ K.

We can now use the equation provided and Equation 23.12 to find the molar enthalpy of sublimation:

$$\frac{\Delta_{sub}\overline{H}}{RT^2} = \frac{d \ln P}{dT} = \frac{8090.0 \text{ K}}{T^2} - \frac{2.013}{T}$$

$$\Delta_{sub}\overline{H} = R(8090.0 \text{ K} - 2.013T)$$

At 25°C,

$$\Delta_{sub}\overline{H} = (8.314 \text{ J}\cdot\text{mol}^{-1}\cdot\text{K}^{-1}) [8090.0 \text{ K} - 2.013(298.15 \text{ K})] = 62.27 \text{ kJ}\cdot\text{mol}^{-1}$$

23–31. Fit the following vapor pressure data of ice to an equation of the form

$$\ln P = -\frac{a}{T} + b \ln T + cT$$

where T is temperature in kelvins. Use your result to determine the molar enthalpy of sublimation of ice at 0°C.

$t/°C$	P/torr	$t/°C$	P/torr
−10.0	1.950	−4.8	3.065
− 9.6	2.021	−4.4	3.171
− 9.2	2.093	−4.0	3.280
− 8.8	2.168	−3.6	3.393
− 8.4	2.246	−3.2	3.509
− 8.0	2.326	−2.8	3.630
− 7.6	2.408	−2.4	3.753
− 7.2	2.493	−2.0	3.880
− 6.8	2.581	−1.6	4.012
− 6.4	2.672	−1.2	4.147
− 6.0	2.765	−0.8	4.287
− 5.6	2.862	−0.4	4.431
− 5.2	2.962	0.0	4.579

Fitting the data to an equation of this form gives

$$\ln P = -\frac{5686.7 \text{ K}}{T} + 4.4948 \ln T - (0.010527 \text{ K}^{-1}T)$$

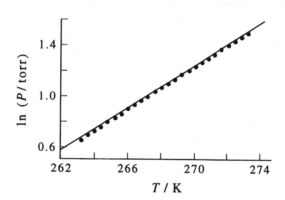

Using this equation and Equation 23.12, we find that

$$\frac{\Delta_{sub}\overline{H}}{RT^2} = \frac{d \ln P}{dT} = \frac{5686.7 \text{ K}}{T^2} + \frac{4.4948}{T} - 0.010527 \text{ K}^{-1}$$

$$\Delta_{sub}\overline{H} = R\left[5686.7 \text{ K} + 4.4948T - (0.010527 \text{ K}^{-1})T^2\right]$$

At 0°C,

$$\Delta_{sub}\overline{H} = (8.314 \text{ J}\cdot\text{mol}^{-1}\cdot\text{K}^{-1}) [5686.7 \text{ K} + 4.4948(273.15 \text{ K})$$
$$-(0.010527 \text{ K}^{-1})(273.15 \text{ K})^2]$$

$$= 50.96 \text{ kJ}\cdot\text{mol}^{-1}$$

23–32. The following table gives the vapor pressure data for liquid palladium as a function of temperature:

T/K	P/bar
1587	1.002×10^{-9}
1624	2.152×10^{-9}
1841	7.499×10^{-8}

Estimate the molar enthalpy of vaporization of palladium.

Plot $\ln P$ vs. $1/T$:

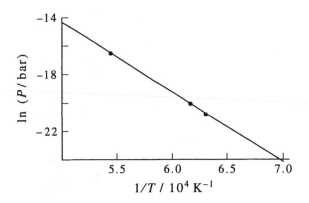

The line of best fit is $y = 10.4359 - 49407x$. Since the slope of this line is equal to $-\Delta_{vap}\overline{H}/R$,

$$\Delta_{vap}\overline{H} = (49407 \text{ K})(8.31451 \text{ J}\cdot\text{mol}^{-1}\cdot\text{K}^{-1}) = 410.8 \text{ kJ}\cdot\text{mol}^{-1}$$

23–33. The sublimation pressure of CO_2 at 138.85 K and 158.75 K is 1.33×10^{-3} bar and 2.66×10^{-2} bar, respectively. Estimate the molar enthalpy of sublimation of CO_2.

Substitute into Equation 23.13:

$$\ln \frac{P_2}{P_1} = \frac{\Delta_{sub}\overline{H}}{R}\left(\frac{T_2 - T_1}{T_2 T_1}\right)$$

$$\Delta_{sub}\overline{H} = (8.3145 \text{ J}\cdot\text{mol}^{-1}\cdot\text{K}^{-1}) \ln\left(\frac{2.66 \times 10^{-2}}{1.33 \times 10^{-3}}\right)\left[\frac{(138.85 \text{ K})(158.75 \text{ K})}{19.9 \text{ K}}\right]$$

$$= 27.6 \text{ kJ}\cdot\text{mol}^{-1}$$

23–34. The vapor pressures of solid and liquid hydrogen iodide can be expressed empirically as

$$\ln(P^s/\text{torr}) = -\frac{2906.2 \text{ K}}{T} + 19.020$$

and

$$\ln(P^l/\text{torr}) = -\frac{2595.7 \text{ K}}{T} + 17.572$$

Calculate the ratio of the slopes of the solid-gas curve and the liquid-gas curve at the triple point.

We can write the slopes of the solid-gas and liquid-gas curves as

$$\frac{dP^s}{dT} = P^s \left(\frac{2906.2 \text{ K}}{T^2}\right) \qquad \text{and} \qquad \frac{dP^l}{dT} = P^l \left(\frac{2595.7 \text{ K}}{T^2}\right)$$

where pressures are in units of torr. Since $P^s = P^l$ at the triple point, the ratio of the slopes at the triple point is

$$\frac{dP^s/dT}{dP^l/dT} = \frac{2906.2 \text{ K}}{2595.7 \text{ K}} = 1.120$$

23–35. Given that the normal melting point, the critical temperature, and the critical pressure of hydrogen iodide are 222 K, 424 K and 82.0 atm, respectively, use the data in the previous problem to sketch the phase diagram of hydrogen iodide.

The triple point is located where the solid and liquid vapor pressures are the same, so at the triple point

$$19.020 - \frac{2906.2 \text{ K}}{T_{tp}} = 17.572 - \frac{2595.7 \text{ K}}{T_{tp}}$$

$$1.448 T_{tp} = 310.5 \text{ K}$$

$$T_{tp} = 214.43 \text{ K}$$

Substituting to solve for P_{tp}, we find that $P_{tp} = 236.8$ torr. We also know that the normal melting point of HI is 222 K, so we can produce a phase diagram of hydrogen iodide by plotting the line between the solid and gas, the line between the vapor and gas, the critical point, the triple point, and the normal melting point. Note that the equation for the liquid-gas line is not completely accurate at high temperatures and pressures (it does not intersect the critical point).

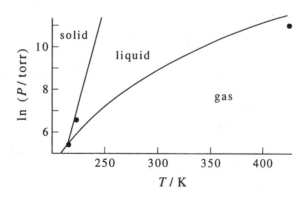

23–36. Consider the phase change

$$C(\text{graphite}) \rightleftharpoons C(\text{diamond})$$

Given that $\Delta_r G°/\text{J} \cdot \text{mol}^{-1} = 1895 + 3.363T$, calculate $\Delta_r H°$ and $\Delta_r S°$. Calculate the pressure at which diamond and graphite are in equilibrium with each other at 25°C. Take the density of diamond and graphite to be 3.51 g·cm^{-3} and 2.25 g·cm^{-3}, respectively. Assume that both diamond and graphite are incompressible.

To find the standard molar Gibbs entropy, we use the Maxwell relation (Equation 22.46)

$$\Delta_r \overline{S}^\circ = -\left(\frac{\partial \Delta_r \overline{G}^\circ}{\partial T}\right)_P = -3.363 \text{ J} \cdot \text{mol}^{-1}$$

Substituting into $\Delta_r \overline{G}^\circ = \Delta_r \overline{H}^\circ - T\Delta_r \overline{S}^\circ$ (Equation 22.13), we find that $\Delta_r \overline{H}^\circ = 1895 \text{ J} \cdot \text{mol}^{-1}$. Because both graphite and diamond are incompressible, we can write (as in Problem 23–18)

$$\overline{G}_{graph}^\circ = \overline{G}_{graph}^\circ + \overline{V}_{graph}(P - P_0)$$

and

$$\overline{G}_{diam}^\circ = \overline{G}_{diam}^\circ + \overline{V}_{diam}(P - P_0)$$

Combining these two equations gives

$$\Delta_r \overline{G} = \Delta_r \overline{G}^\circ + (\overline{V}_{diam} - \overline{V}_{graph})(P - P_0)$$

When graphite and diamond are in equilibrium, $\Delta_r \overline{G} = 0$. Substituting into the equation given in the problem, we see that at 25°C, $\Delta_r G^\circ = 2898 \text{ J} \cdot \text{mol}^{-1}$. Then

$$0 = \Delta_r \overline{G}^\circ + \left(\frac{1}{3510 \text{ g} \cdot \text{dm}^{-3}} - \frac{1}{2250 \text{ g} \cdot \text{dm}^{-3}}\right)(12.01 \text{ g} \cdot \text{mol}^{-1})(P - 1 \text{ bar})$$

$$= 2898 \text{ J} \cdot \text{mol}^{-1} - (1.916 \times 10^{-3} \text{ dm}^3 \cdot \text{mol}^{-1})(P - 1 \text{ bar})$$

$$P = \frac{1}{1.916 \times 10^{-3} \text{ dm}^3 \cdot \text{mol}^{-1}}(2898 \text{ J} \cdot \text{mol}^{-1})\left(\frac{0.08206 \text{ dm}^3 \cdot \text{bar}}{8.3145 \text{ J}}\right) + 1 \text{ bar}$$

$$= 15\,000 \text{ bar}$$

23–37. Use Equation 23.36 to calculate $\mu^\circ - E_0$ for Kr(g) at 298.15 K. The literature value is $-42.72 \text{ kJ} \cdot \text{mol}^{-1}$.

$$\mu^\circ - E_0 = -RT \ln\left[\left(\frac{q^0}{V}\right)\frac{k_B T}{P^\circ}\right] \tag{23.36}$$

We do this in the same way we found $\mu^\circ - E_0$ for Ar(g) in Section 23–5. First,

$$\frac{q^0(V, T)}{V} = \left(\frac{2\pi m k_B T}{h^2}\right)^{3/2}$$

$$= \left[\frac{(2\pi)(1.391 \times 10^{-25} \text{ kg} \cdot \text{mol}^{-1})(1.381 \times 10^{-23} \text{ J} \cdot \text{K}^{-1})(298.15 \text{ K})}{(6.626 \times 10^{-34} \text{ J} \cdot \text{s})^2}\right]^{3/2}$$

$$= 7.422 \times 10^{32} \text{ m}^{-3}$$

$$\frac{k_B T}{P^\circ} = \frac{(1.381 \times 10^{-23} \text{ J} \cdot \text{K}^{-1})(298.15 \text{ K})}{1.0 \times 10^5 \text{ Pa}} = 4.116 \times 10^{-26} \text{ m}^{-3}$$

Substituting into Equation 23.34 gives

$$\mu^\circ - E_0 = -RT \ln\left[\left(\frac{q^0}{V}\right)\frac{k_B T}{P^\circ}\right]$$

$$= -R(298.15 \text{ K}) \ln[(7.422 \times 10^{32} \text{ m}^{-3})(4.116 \times 10^{-26} \text{ m}^{-3})]$$

$$= -4.272 \times 10^4 \text{ J} \cdot \text{mol}^{-1}$$

23–38. Show that Equations 23.30 and 23.32 for $\mu(T, P)$ for a monatomic ideal gas are equivalent to using the relation $\overline{G} = \overline{H} - T\overline{S}$ with $\overline{H} = 5RT/2$ and S given by Equation 20.45.

Recall that μ for a pure substance is $\overline{G}$. Equation 20.45 is

$$\overline{S} = \frac{5}{2}R + R\ln\left[\left(\frac{2\pi mk_B T}{h^2}\right)^{3/2}\frac{\overline{V}}{N_A}\right]$$

Therefore,

$$\overline{G} = \overline{H} - T\overline{S} = \frac{5RT}{2} - TS$$

$$= \frac{5RT}{2} - \frac{5RT}{2} - RT\ln\left[\left(\frac{2\pi mk_B T}{h^2}\right)^{3/2}\frac{k_B T}{P}\right]$$

$$= -RT\ln\left[\left(\frac{2\pi mk_B T}{h^2}\right)^{3/2}\frac{k_B T}{P}\right] + RT\ln P$$

$$= -RT\ln\left[\left(\frac{q}{V}\right)k_B T\right] + RT\ln P$$

This is Equation 23.30. Equation 23.32 appears when we substitute $P^\circ = 1$ bar into this equation.

23–39. Use Equation 23.37 and the molecular parameters in Table 18.2 to calculate $\mu^\circ - E_0$ for $N_2(g)$ at 298.15 K. The literature value is -48.46 kJ·mol^{-1}.

$$\frac{q^0}{V} = \left(\frac{2\pi mk_B T}{h^2}\right)^{3/2}\frac{T}{\sigma\Theta_{rot}}\frac{1}{1 - e^{-\Theta_{vib}/T}}$$

$$= \left[\frac{2\pi(4.65\times10^{-26}\text{ kg·mol}^{-1})k_B(298.15\text{ K})}{h^2}\right]^{3/2}\frac{298.15\text{ K}}{2(2.88\text{ K})}\frac{1}{1 - e^{-3374/298.15}}$$

$$= 7.42\times10^{33}\text{ m}^{-3}$$

$$\frac{RT}{N_A P^\circ} = 4.116\times10^{-26}\text{ m}^3$$

where use the value of $RT/N_A P^\circ$ from Problem 23–36, since the fraction $RT/N_A P^\circ$ is independent of the substance. Now, substituting, we see that

$$\mu^\circ - E_0 = -RT\ln\left[\left(\frac{q^0}{V}\right)\frac{RT}{N_A P^\circ}\right]$$

$$= -R(298.15\text{ K})\ln[(7.42\times10^{33}\text{ m}^{-3})(4.116\times10^{-26}\text{ m}^3)]$$

$$= -48.43\text{ kJ·mol}^{-1}$$

23–40. Use Equation 23.37 and the molecular parameters in Table 18.2 to calculate $\mu^\circ - E_0$ for $CO(g)$ at 298.15 K. The literature value is -50.26 kJ·mol^{-1}.

23–40. Use Equation 23.37 and the molecular parameters in Table 18.2 to calculate $\mu° - E_0$ for CO(g) at 298.15 K. The literature value is -50.26 kJ·mol^{-1}.

$$\frac{q^0}{V} = \left(\frac{2\pi m k_B T}{h^2}\right)^{3/2} \frac{T}{\sigma \Theta_{\text{rot}}} \frac{1}{1 - e^{-\Theta_{\text{vib}}/T}}$$

$$= \left[\frac{2\pi (4.65 \times 10^{-26}\text{ kg·mol}^{-1}) k_B (298.15\text{ K})}{h^2}\right]^{3/2} \frac{298.15\text{ K}}{2(2.77\text{ K})} \frac{1}{1 - e^{-3103/298.15}}$$

$$= 1.54 \times 10^{34}\text{ m}^{-3}$$

$$\frac{RT}{N_A P°} = 4.116 \times 10^{-26}\text{ m}^3$$

Now, substituting, we see that

$$\mu° - E_0 = -RT \ln\left[\left(\frac{q^0}{V}\right) \frac{RT}{N_A P°}\right]$$

$$= -R(298.15\text{ K}) \ln[(1.54 \times 10^{34}\text{ m}^{-3})(4.116 \times 10^{-26}\text{ m}^3)]$$

$$= -50.25\text{ kJ·mol}^{-1}$$

23–41. Use Equation 18.60 [without the factor of $\exp(D_e/k_B T)$] and the molecular parameters in Table 18.4 to calculate $\mu° - E_0$ for CH$_4$(g) at 298.15 K. The literature value is -45.51 kJ·mol^{-1}.

$$\frac{q}{V} = \left(\frac{2\pi M k_B T}{h^2}\right)^{3/2} \frac{\pi^{1/2}}{\sigma} \left(\frac{T^3}{\Theta_{\text{rot,A}} \Theta_{\text{rot,B}} \Theta_{\text{rot,C}}}\right)^{1/2} \left[\prod_{j=1}^{9} \frac{e^{-\Theta_{\text{vib},j}/2T}}{(1 - e^{-\Theta_{\text{vib},j}/T})}\right] \qquad (18.60)$$

The ground-state energy must be considered for each vibrational state, so, in analogy to the derivation of q^0 in Section 23–5,

$$\frac{q^0}{V} = \left(\frac{2\pi M k_B T}{h^2}\right)^{3/2} \frac{\pi^{1/2}}{\sigma} \left(\frac{T^3}{\Theta_{\text{rot,A}} \Theta_{\text{rot,B}} \Theta_{\text{rot,C}}}\right)^{1/2} \left[\prod_{j=1}^{9} \frac{1}{(1 - e^{-\Theta_{\text{vib},j}/T})}\right]$$

$$= 2.30 \times 10^{33}\text{ m}^{-3}$$

$$\frac{RT}{N_A P°} = 4.116 \times 10^{-26}\text{ m}^3$$

$$\mu° - E_0 = -RT \ln\left[\left(\frac{q^0}{V}\right) \frac{RT}{N_A P°}\right]$$

$$= -R(298.15\text{ K}) \ln[(2.30 \times 10^{33}\text{ m}^{-3})(4.116 \times 10^{-26}\text{ m}^3)]$$

$$= -45.53\text{ kJ·mol}^{-1}$$

23–42. When we refer to the equilibrium vapor pressure of a liquid, we tacitly assume that some of the liquid has evaporated into a vacuum and that equilibrium is then achieved. Suppose, however, that we are able by some means to exert an additional pressure on the surface of the liquid. One way to do this is to introduce an insoluble, inert gas into the space above the liquid. In this problem, we will investigate how the equilibrium vapor pressure of a liquid depends upon the total pressure exerted on it.

Consider a liquid and a vapor in equilibrium with each other, so that $\mu^l = \mu^g$. Show that

$$\overline{V}^l dP^l = \overline{V}^g dP^g$$

because the two phases are at the same temperature. Assuming that the vapor may be treated as an ideal gas and that $\overline{V}^{\,l}$ does not vary appreciably with pressure, show that

$$\ln \frac{P^g(\text{at } P^l = P)}{P^g(\text{at } P^l = 0)} = \frac{\overline{V}^{\,l} P^l}{RT}$$

Use this equation to calculate the vapor pressure of water at a total pressure of 10.0 atm at 25°C. Take P^g (at $P^l = 0$) = 0.313 atm.

We start with the fact that $\mu^l = \mu^g$. Since μ can be written as a function of T and P, and since the temperature does not change, we can write

$$d\mu^l = \left(\frac{\partial \overline{G}}{\partial P}\right) dP^l = \overline{V}^{\,l} dP^l$$

Likewise, $d\mu^g = \overline{V}^{\,g} dP^g$, and so

$$\overline{V}^{\,l} dP^l = \overline{V}^{\,g} dP^g$$

follows naturally from the inital assumption. Now we assume that the vapor can be treated as an ideal gas and that $\overline{V}^{\,l}$ does not vary with respect to pressure, so

$$\overline{V}^{\,l} dP^l = \frac{RT}{P^g} dP^g$$

$$\frac{\overline{V}^{\,l}}{RT} dP^l = \frac{dP^g}{P^g}$$

$$\frac{\overline{V}^{\,l}}{RT} \int_0^P dP^l = \int_{P^g(P^l=0)}^{P^g(P^l=P)} \frac{dP^g}{P^g}$$

$$\frac{\overline{V}^{\,l} P}{RT} = \ln \frac{P^g(P^l = P)}{P^g(P^l = 0)}$$

The specific density of water is 1 g·cm^{-3}, so $\overline{V} = 0.018$ dm^3·mol^{-1}. For water at a total pressure of 10.0 atm at 298.15 K, since the vapor pressure of water expanding into a vacuum is 0.0313 atm at 298.15 K,

$$\ln \frac{P^g(P^l = P)}{0.0313 \text{ atm}} = \frac{\overline{V}^{\,l} P^l}{R(298.15 \text{ K})}$$

$$\ln \frac{P^g(P^l = P)}{0.0313 \text{ atm}} = \frac{(0.018 \text{ dm}^3 \cdot \text{mol}^{-1})(10.0 \text{ atm})}{(0.082058 \text{ dm}^3 \cdot \text{atm} \cdot \text{mol}^{-1} \cdot \text{K}^{-1})(298.15 \text{ K})}$$

$$\frac{P^g}{0.0313 \text{ atm}} = 1.007$$

$$P^g = 0.0315 \text{ atm}$$

The vapor pressure of water at a total pressure 10.0 atm at 298.15 K is 0.0315 atm, or a change of 2×10^{-4} atm.

23–43. Using the fact that the vapor pressure of a liquid does not vary appreciably with the total pressure, show that the final result of the previous problem can be written as

$$\frac{\Delta P^g}{P^g} = \frac{\overline{V}^{\,l} P^l}{RT}$$

Hint: Let $P^g(\text{at } P = P^l) = P^g(\text{at } P = 0) + \Delta P$ and use the fact that ΔP is small. Calculate ΔP for water at a total pressure of 10.0 atm at 25°C. Compare your answer with the one you obtained in the previous problem.

$$\ln \frac{P^g(P^l = P)}{P^g(P^l = 0)} = \frac{\overline{V}^l P^l}{RT}$$

$$\ln \frac{P^g(P^l = 0) + \Delta P^g}{P^g(P^l = 0)} = \frac{\overline{V}^l P^l}{RT}$$

Use the relation $\ln(1 + x) \approx x$ when x is small to obtain

$$\ln \left(1 + \frac{\Delta P^g}{P^g} \right) = \frac{\overline{V}^l P^l}{RT}$$

$$\ln \frac{\Delta P^g}{P^g} = \frac{\overline{V}^l P^l}{RT}$$

Again, the vapor pressure of water expanding into a vacuum is 0.0313 atm at 298.15 K and $\overline{V} = 0.018$ dm$^3 \cdot$mol^{-1}, so

$$\frac{\Delta P^g}{P^g} = \frac{\overline{V}^l P^l}{RT}$$

$$\Delta P^g = \frac{(0.0313 \text{ atm})(0.018 \text{ dm}^3 \cdot \text{mol}^{-1})(10.0 \text{ atm})}{(0.082058 \text{ dm}^3 \cdot \text{atm} \cdot \text{mol}^{-1} \cdot \text{K}^{-1})(298.15 \text{ K})}$$

$$= 2.30 \times 10^{-4} \text{ atm}$$

This would give a vapor pressure of 0.0315 atm, as in the previous problem.

23–44. In this problem, we will show that the vapor pressure of a droplet is not the same as the vapor pressure of a relatively large body of liquid. Consider a spherical droplet of liquid of radius r in equilibrium with a vapor at a pressure P, and a flat surface of the same liquid in equilibrium with a vapor at a pressure P_0. Show that the change in Gibbs energy for the isothermal transfer of dn moles of the liquid from the flat surface to the droplet is

$$dG = dn\, RT \ln \frac{P}{P_0}$$

This change in Gibbs energy is due to the change in surface energy of the droplet (the change in surface energy of the large, flat surface is negligible). Show that

$$dn\, RT \ln \frac{P}{P_0} = \gamma\, dA$$

where γ is the surface tension of the liquid and dA is the change in the surface area of a droplet. Assuming the droplet is spherical, show that

$$dn = \frac{4\pi r^2 dr}{\overline{V}^l}$$

$$dA = 8\pi r\, dr$$

and finally that

$$\ln \frac{P}{P_0} = \frac{2\gamma \overline{V}^l}{rRT} \tag{1}$$

Because the right side is positive, we see that the vapor pressure of a droplet is greater than that of a planar surface. What if $r \rightarrow \infty$?

For an isothermal process involving an ideal gas,

$$\Delta G = nRT \ln \frac{P_2}{P_1} \tag{22.58}$$

When a small amount of Gibbs energy goes from the flat surface (of vapor pressure P_0) to the droplet (with vapor pressure P), corresponding to adding dn moles to the droplet from the flat surface, the change in Gibbs energy dG is

$$G(\text{droplet}) - G(\text{surface}) = dn\,RT \ln \frac{P}{P_1} - dn\,RT \ln \frac{P_0}{P_1}$$

where P_1 is an arbitrary reference pressure. Therefore, we have

$$dG = dn\,RT \ln \frac{P}{P_0}$$

This change in surface energy is equal to γdA, so

$$dn\,RT \ln \frac{P}{P_0} = \gamma dA$$

If a spherical droplet contains n moles, then

$$n\overline{V}^{\,\text{l}} = \frac{4}{3}\pi r^3$$

$$dn = \frac{4\pi r^2 dr}{\overline{V}^{\,\text{l}}}$$

$$A = 4\pi r^2$$

$$dA = 8\pi r\,dr$$

Substituting these expressions back into $dn\,RT \ln(P/P_0) = \gamma dA$, we find that

$$\frac{4\pi r^2 dr}{\overline{V}^{\,\text{l}}} RT \ln \frac{P}{P_0} = \gamma 8\pi r\,dr$$

$$\ln \frac{P}{P_0} = \frac{2\gamma \overline{V}^{\,\text{l}}}{rRT}$$

If $r \rightarrow \infty$, then $\ln(P/P_0) \rightarrow 0$: the spherical droplet becomes more and more like the flat surface.

23–45. Use Equation 1 of Problem 23–44 to calculate the vapor pressure at 25°C of droplets of water of radius 1.0×10^{-5} cm. Take the surface tension of water to be 7.20×10^{-4} J·m^{-2}.

$$\ln \frac{P}{P_0} = \frac{2\gamma \overline{V}^{\,\text{l}}}{rRT}$$

$$\ln \frac{P}{0.0313 \text{ atm}} = \frac{2(7.20 \times 10^{-4} \text{ J·m}^{-2})(18.0 \times 10^{-6} \text{ m}^3 \cdot \text{mol}^{-1})}{(1.0 \times 10^{-7} \text{ m})(8.3145 \text{ J·mol}^{-1} \cdot \text{K}^{-1})(298.15 \text{ K})} = 1.046 \times 10^{-4}$$

Solving for P, we find that

$$P = (0.0313 \text{ atm})e^{1.046 \times 10^{-4}} = 0.0313 \text{ atm}$$

The vapor pressure of these droplets of water is not significantly different from the vapor pressure of water in a surface.

23–46. Figure 23.15 shows reduced pressure, P_R, plotted against reduced volume, $\overline{V}_R$, for the van der Waals equation at a reduced temperature, T_R, of 0.85. The so-called van der Waals loop apparent in the figure will occur for any reduced temperature less than unity and is a consequence of the simplified form of the van der Waals equation. It turns out that any analytic equation of state (one that can be written as a Maclaurin expansion in the reduced density, $1/\overline{V}_R$) will give loops for subcritical temperatures ($T_R < 1$). The correct behavior as the pressure is increased is given by the path abdfg in Figure 23.15. The horizontal region bdf, not given by the van der Waals equation, represents the condensation of the gas to a liquid at a fixed pressure. We can draw the horizontal line (called a *tie line*) at the correct position by recognizing that the chemical potentials of the liquid and the vapor must be equal at the points b and f. Using this requirement, Maxwell showed that the horizontal line representing condensation should be drawn such that the areas of the loops above and below the line must be equal. To prove *Maxwell's equal-area construction rule*, integrate $(\partial \mu / \partial P)_T = \overline{V}$ by parts along the path bcdef and use the fact that μ^l (the value of μ at point f) $= \mu^g$ (the value of μ at point b) to obtain

$$\mu^l - \mu^g = P_0(\overline{V}^l - \overline{V}^g) - \int_{bcdef} P d\overline{V}$$

$$= \int_{bcdef} (P_0 - P) d\overline{V}$$

where P_0 is the pressure corresponding to the tie line. Interpret this result.

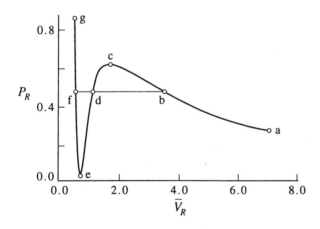

FIGURE 23.15
A plot of reduced pressure, P_R, versus reduced volume, $\overline{V}_R$, for the van der Waals equation at a reduced temperature, T_R, of 0.85.

Start with the equation

$$\int_{bcdef} d\mu = \int_{bcdef} \overline{V} dP$$

and integrate by parts to obtain

$$\mu^l - \mu^g = P_0(\overline{V}^l - \overline{V}^g) - \int_{bcdef} P d\overline{V}$$

Now combine the two terms on the right to get

$$\mu^l - \mu^g = \int_{bcdef} (P_0 - P) d\overline{V}$$

Now $\mu^{\text{l}} = \mu^{\text{g}}$, so

$$\int_{\text{bcdef}} (P_0 - P)d\overline{V} = 0$$

or the specified area in a plot of P against $\overline{V}$ is equal to zero.

23–47. The isothermal compressibility, κ_T, is defined by

$$\kappa_T = -\frac{1}{V}\left(\frac{\partial V}{\partial P}\right)_T$$

Because $(\partial P/\partial V)_T = 0$ at the critical point, κ_T diverges there. A question that has generated a great deal of experimental and theoretical research is the question of the manner in which κ_T diverges as T approaches T_c. Does it diverge as $\ln(T - T_c)$ or perhaps as $(T - T_c)^{-\gamma}$ where γ is some *critical exponent*? An early theory of the behavior of thermodynamic functions such as κ_T very near the critical point was proposed by van der Waals, who predicted that κ_T diverges as $(T - T_c)^{-1}$. To see how van der Waals arrived at this prediction, we consider the (double) Taylor expansion of the pressure $P(\overline{V}, T)$ about T_c and $\overline{V}_c$:

$$P(\overline{V}, T) = P(\overline{V}_c, T_c) + (T - T_c)\left(\frac{\partial P}{\partial T}\right)_c + \frac{1}{2}(T - T_c)^2\left(\frac{\partial^2 P}{\partial T^2}\right)_c$$

$$+ (T - T_c)(\overline{V} - \overline{V}_c)\left(\frac{\partial^2 P}{\partial V \partial T}\right)_c + \frac{1}{6}(\overline{V} - \overline{V}_c)^3\left(\frac{\partial^3 P}{\partial \overline{V}^3}\right)_c + \cdots$$

Why are there no terms in $(\overline{V} - \overline{V}_c)$ or $(\overline{V} - \overline{V}_c)^2$? Write this Taylor series as

$$P = P_c + a(T - T_c) + b(T - T_c)^2 + c(T - T_c)(\overline{V} - \overline{V}_c) + d(\overline{V} - \overline{V}_c)^3 + \cdots$$

Now show that

$$\left(\frac{\partial P}{\partial \overline{V}}\right)_T = c(T - T_c) + 3d(\overline{V} - \overline{V}_c)^2 + \cdots \qquad \left(\begin{array}{c}T \to T_c \\ V \to V_c\end{array}\right)$$

and that

$$\kappa_T = \frac{-1/\overline{V}}{c(T - T_c) + 3d(\overline{V} - \overline{V}_c)^2 + \cdots}$$

Now let $\overline{V} = \overline{V}_c$ to obtain

$$\kappa_T \propto \frac{1}{T - T_c} \qquad T \to (T_c)$$

Accurate experimental measurements of κ_T as $T \to T_c$ suggest that κ_T diverges a little more strongly than $(T - T_c)^{-1}$. In particular, it is found that $\kappa_T \to (T - T_c)^{-\gamma}$ where $\gamma = 1.24$. Thus, the theory of van der Waals, although qualitatively correct, is not quantitatively correct.

The Taylor expansion of $P(\overline{V}, T)$ about T_c and $\overline{V}_c$ is

$$P(\overline{V}, T) = P(\overline{V}_c, T_c) + (T - T_c)\left(\frac{\partial P}{\partial T}\right)_c + (\overline{V} - \overline{V}_c)\left(\frac{\partial P}{\partial \overline{V}}\right)_c$$

$$+ \frac{1}{2}(T - T_c)^2\left(\frac{\partial^2 P}{\partial T^2}\right)_c + (T - T_c)(\overline{V} - \overline{V}_c)\left(\frac{\partial^2 P}{\partial \overline{V} \partial T}\right)_c$$

$$+ \frac{1}{2}(\overline{V} - \overline{V}_c)^2\left(\frac{\partial^2 P}{\partial \overline{V}^2}\right)_c + \frac{1}{6}(\overline{V} - \overline{V}_c)^3\left(\frac{\partial^3 P}{\partial \overline{V}^3}\right)_c + \cdots$$

However, recall from Section 16–3 that, at the critical point, $(\partial P/\partial \overline{V})_c = (\partial^2 P/\partial \overline{V}^2)_c = 0$. Thus, the Taylor expansion becomes

$$P = P(\overline{V}_c, T_c) + (T - T_c)\left(\frac{\partial P}{\partial T}\right)_c + \frac{1}{2}(T - T_c)^2\left(\frac{\partial^2 P}{\partial T^2}\right)_c$$

$$+ (T - T_c)(\overline{V} - \overline{V}_c)\left(\frac{\partial^2 P}{\partial \overline{V}\partial T}\right)_c + \frac{1}{6}(\overline{V} - \overline{V}_c)^3\left(\frac{\partial^3 P}{\partial \overline{V}^3}\right)_c + \cdots$$

$$= P_c + a(T - T_c) + b(T - T_c)^2 + c(T - T_c)(\overline{V} - \overline{V}_c)$$

$$+ d(\overline{V} - \overline{V}_c)^3 + \cdots$$

$$\left(\frac{\partial P}{\partial \overline{V}}\right)_T = c(T - T_c) + 3d(\overline{V} - \overline{V}_c)^2 + \cdots$$

Note that in differentiating, we truncated our expansion and dropped terms of $O[(\overline{V} - \overline{V}_c)^4]$, $O[(T - T_c)^2]$, and third-order terms of $O[(\overline{V} - \overline{V}_c)^x(T - T_c)^y]$. Our partial derivative is thus accurate to $O[(\overline{V} - \overline{V}_c)^4]$ and $O[(T - T_c)^2]$. We can truncate these terms because when $\overline{V} \to \overline{V}_c$ and $T \to T_c$ the higher-order terms become negligible, so we find

$$\kappa_T = -\frac{1}{\overline{V}}\left(\frac{\partial P}{\partial \overline{V}}\right)_T^{-1}$$

$$= -\frac{1}{\overline{V}}\left[\frac{1}{c(T - T_c) + 3d(\overline{V} - \overline{V}_c)^2 + \cdots}\right] \qquad T \to T_c, \quad \overline{V} \to \overline{V}_c$$

Letting $\overline{V} = \overline{V}_c$, we find that

$$\kappa_T = -\frac{1}{\overline{V}_c c(T - T_c)}$$

$$\kappa_T \propto \frac{1}{(T - T_c)} \qquad T \to T_c$$

Again, this expression is only accurate to $O[(T - T_c)^2]$.

23–48. We can use the ideas of the previous problem to predict how the difference in the densities (ρ^l and ρ^g) of the coexisting liquid and vapor states (*orthobaric densities*) behave as $T \to T_c$. Substitute

$$P = P_c + a(T - T_c) + b(T - T_c)^2 + c(T - T_c)(\overline{V} - \overline{V}_c) + d(\overline{V} - \overline{V}_c)^3 + \cdots \qquad (1)$$

into the Maxwell equal-area construction (Problem 23–46) to get

$$P_0 = P_c + a(T - T_c) + b(T - T_c)^2 + \frac{c}{2}(T - T_c)(\overline{V}^l + \overline{V}^g - 2\overline{V}_c)$$

$$+ \frac{d}{4}[(\overline{V}^g - \overline{V}_c)^2 + (\overline{V}^l - \overline{V}_c)^2](\overline{V}^l + \overline{V}^g - 2\overline{V}_c) + \cdots \qquad (2)$$

For $P < P_c$, Equation 1 gives loops and so has three roots, $\overline{V}^l$, $\overline{V}_c$, and $\overline{V}^g$ for $P = P_0$. We can obtain a first approximation to these roots by assuming that $\overline{V}_c \approx \frac{1}{2}(\overline{V}^l + \overline{V}^g)$ in Equation 2 and writing

$$P_0 = P_c + a(T - T_c) + b(T - T_c)^2$$

To this approximation, the three roots to Equation 1 are obtained from

$$d(\overline{V} - \overline{V}_c)^3 + c(T - T_c)(\overline{V} - \overline{V}_c) = 0$$

Show that the three roots are

$$\overline{V}_1 = \overline{V}^{\,l} = \overline{V}_c - \left(\frac{c}{d}\right)^{1/2} (T_c - T)^{1/2}$$

$$\overline{V}_2 = \overline{V}_c$$

$$\overline{V}_3 = \overline{V}^{\,g} = \overline{V}_c + \left(\frac{c}{d}\right)^{1/2} (T_c - T)^{1/2}$$

Now show that

$$\overline{V}^{\,g} - \overline{V}^{\,l} = 2\left(\frac{c}{d}\right)^{1/2} (T_c - T)^{1/2} \qquad \left(\begin{array}{c} T < T_c \\ T \to T_c \end{array}\right)$$

and that this equation is equivalent to

$$\rho^l - \rho^g \longrightarrow A(T_c - T)^{1/2} \qquad \left(\begin{array}{c} T < T_c \\ T \to T_c \end{array}\right)$$

Thus, the van der Waals theory predicts that the critical exponent in this case is 1/2. It has been shown experimentally that

$$\rho^l - \rho^g \longrightarrow A(T_c - T)^{\beta}$$

where $\beta = 0.324$. Thus, as in the previous problem, although qualitatively correct, the van der Waals theory is not quantitatively correct.

We start with the result of Problem 23–46,

$$\mu^l - \mu^g = \int_{bcdef} (P_0 - P)d\overline{V}$$

$$0 = \int \left[P_0 - P_c - a(T - T_c) - b(T - T_c)^2 - c(T - T_c)(\overline{V} - \overline{V}_c) \right.$$
$$\left. - d(\overline{V} - \overline{V}_c)^3 - \cdots \right] d\overline{V}$$

$$= \left[P_0 - P_c - a(T - T_c) - b(T - T_c)^2 \right](\overline{V}^l - \overline{V}^g) - \frac{1}{2}c(T - T_c)\left[(\overline{V}^l - \overline{V}_c)^2 - (\overline{V}^g - \overline{V}_c)^2\right]$$

$$\quad - \frac{1}{4}d\left[(\overline{V}^l - \overline{V}_c)^4 - (\overline{V}^g - \overline{V}_c)^4\right]$$

$$P_0 = P_c + a(T - T_c) + b(T - T_c)^2$$

$$\quad + \frac{c}{2}\left[\frac{(\overline{V}^l)^2 - 2\overline{V}_c\overline{V}^l + \overline{V}_c^2 - (\overline{V}^g)^2 + 2\overline{V}^g\overline{V}_c - \overline{V}_c^2}{\overline{V}^l - \overline{V}^g}\right]$$

$$\quad + \frac{d}{4}\left[\frac{(\overline{V}^l - \overline{V}_c)^2 - (\overline{V}^g - \overline{V}_c)^2}{\overline{V}^l - \overline{V}^g}\right]\left[(\overline{V}^l - \overline{V}_c)^2 + (\overline{V}^g - \overline{V}_c)^2\right]$$

$$= P_c + a(T - T_c) + b(T - T_c)^2$$

$$\quad + \frac{c}{2}\left[\frac{(\overline{V}^l + \overline{V}^g)(\overline{V}^l - \overline{V}^g) - 2\overline{V}_c(\overline{V}^l - \overline{V}^g)}{\overline{V}^l - \overline{V}^g}\right]$$

$$\quad + \frac{d}{4}\left[\frac{(\overline{V}^l + \overline{V}^g)(\overline{V}^l - \overline{V}^g) - 2\overline{V}_c(\overline{V}^l - \overline{V}^g)}{\overline{V}^l - \overline{V}^g}\right]\left[(\overline{V}^l - \overline{V}_c)^2 + (\overline{V}^g - \overline{V}_c)^2\right]$$

$$= P_c + a(T - T_c) + b(T - T_c)^2 + \frac{c}{2}\left[\overline{V}^l + \overline{V}^g - 2\overline{V}_c\right]$$

$$\quad + \frac{d}{4}\left[\overline{V}^l + \overline{V}^g - 2\overline{V}_c\right]\left[(\overline{V}^l - \overline{V}_c)^2 + (\overline{V}^g - \overline{V}_c)^2\right]$$

Now assume that $\overline{V}_c \approx \frac{1}{2}(\overline{V}^g + \overline{V}^l)$ to get

$$P_0 = P_c + a(T - T_c) + b(T - T_c)^2$$

To this approximation, the three roots to Equation 1 are given by

$$d(\overline{V} - \overline{V}_c)^3 + c(T - T_c)(V - V_c) = 0$$
$$(\overline{V} - \overline{V}_c)[d(\overline{V} - \overline{V}_c)^2 + c(T - T_c)] = 0$$

This expression is accurate only to $O(T - T_c)$. We then find the three roots

$$\overline{V} - \overline{V}_c = 0$$
$$\overline{V}_2 = \overline{V}_c$$
$$\overline{V} - \overline{V}_c = \left[-\frac{c(T - T_c)}{d}\right]^{1/2}$$
$$\overline{V}_3 = \overline{V}_c + \left(\frac{c}{d}\right)^{1/2}(T_c - T)^{1/2}$$
$$\overline{V} - \overline{V}_c = -\left[-\frac{c(T - T_c)}{d}\right]^{1/2}$$
$$\overline{V}_1 = \overline{V}_c - \left(\frac{c}{d}\right)^{1/2}(T_c - T)^{1/2}$$

These values are only accurate to $O[(T - T_c)^{1/2}]$. We know that the largest root is the value of $\overline{V}^g$ and that the smallest root is the value of $\overline{V}^l$. Then, to the correct accuracy,

$$\overline{V}^g - \overline{V}^l = \overline{V}_c + \left(\frac{c}{d}\right)^{1/2}(T_c - T)^{1/2} - \left[\overline{V}_c - \left(\frac{c}{d}\right)^{1/2}(T_c - T)^{1/2}\right]$$
$$= 2\left(\frac{c}{d}\right)^{1/2}(T_c - T)^{1/2}$$

where $T \to T_c$, but $T < T_c$ (in order for the quantity $T_c - T$ to be real). Notice that the $(T - T_c)$ term in the product of the molar volumes drops out, since we have been truncating our expressions. The difference in densities is then

$$\rho^l - \rho^g = \frac{1}{\overline{V}^l} - \frac{1}{\overline{V}^g}$$
$$= \frac{\overline{V}^g - \overline{V}^l}{\overline{V}^l\overline{V}^g}$$
$$= 2\left(\frac{c}{d}\right)^{1/2}(T_c - T)^{1/2}\overline{V}_c^2$$
$$= A(T_c - T)^{1/2}$$

23–49. The following data give the temperature, the vapor pressure, and the density of the coexisting vapor phase of butane. Use the van der Waals equation and the Redlich-Kwong equation to calculate the vapor pressure and compare your result with the experimental values given below.

T/K	P/bar	$\rho^g/mol \cdot L^{-1}$
200	0.0195	0.00117
210	0.0405	0.00233
220	0.0781	0.00430
230	0.1410	0.00746
240	0.2408	0.01225
250	0.3915	0.01924
260	0.6099	0.02905
270	0.9155	0.04239
280	1.330	0.06008

For butane, from Tables 16.3 and 16.4, $a = 13.888$ dm$^6 \cdot$bar$\cdot$mol^{-2}, $b = 0.11641$ dm$^3 \cdot$mol^{-1}, $A = 290.16$ dm$^6 \cdot$bar$\cdot$mol$^{-2} \cdot$K$^{1/2}$, and $B = 0.080683$ dm$^3 \cdot$mol^{-1}. We can substitute the given density and temperature into the van der Waals and Redlich-Kwong equations and thus find the vapor pressure P that is given by each equation.

For the van der Waals approximation,

$$P = \frac{RT}{\overline{V} - b} - \frac{a}{\overline{V}^2} \tag{16.5}$$

and for the Redlich-Kwong approximation

$$P = \frac{RT}{\overline{V} - B} - \frac{A}{T^{1/2}\overline{V}(\overline{V} + B)} \tag{16.8}$$

T/K	P(van der Waals)/bar	$P($ Redlich-Kwong$)$/bar
200	0.0194	0.0194
210	0.0406	0.0406
220	0.0784	0.0783
230	0.1420	0.1417
240	0.2427	0.2419
250	0.3957	0.3938
260	0.6184	0.6143
270	0.9314	0.9233
280	1.3584	1.3432

23–50. The following data give the temperature, the vapor pressure, and the density of the coexisting vapor phase of benzene. Use the van der Waals equation and the Redlich-Kwong equation to calculate the vapor pressure and compare your result with the experimental values given below.

Use Equations 16.17 and 16.18 with $T_c = 561.75$ K and $P_c = 48.7575$ bar to calculate the van der Waals parameters and the Redlich-Kwong parameters.

T/K	P/bar	$\rho^g/\text{mol}\cdot\text{L}^{-1}$
290.0	0.0860	0.00359
300.0	0.1381	0.00558
310.0	0.2139	0.00839
320.0	0.3205	0.01223
330.0	0.4666	0.01734
340.0	0.6615	0.02399
350.0	0.9161	0.03248

We can do this in the same way as the previous problem after finding a, b, A, and B using Equations 16.17 and 16.18:

$$a = \frac{27(RT_c)^2}{64P_c} = \frac{27[(0.083145 \text{ dm}^3\cdot\text{bar}\cdot\text{mol}^{-1})(561.75)]^2}{64(48.7575 \text{ bar})} = 18.876 \text{ dm}^6\cdot\text{bar}\cdot\text{mol}^{-2}$$

$$b = \frac{RT_c}{8P_c} = \frac{(0.083145 \text{ dm}^3\cdot\text{bar}\cdot\text{mol}^{-1})(561.75)}{8(48.7575 \text{ bar})} = 0.11974 \text{ dm}^3\cdot\text{mol}^{-1}$$

$$A = 0.42748\frac{R^2 T_c^{5/2}}{P_c} = 0.42748\frac{(0.083145 \text{ dm}^3\cdot\text{bar}\cdot\text{mol}^{-1}\cdot\text{K}^{-1})^2(561.75 \text{ K})^{5/2}}{48.7575 \text{ bar}}$$

$$= 453.21 \text{ dm}^6\cdot\text{bar}\cdot\text{mol}^{-1}\cdot\text{K}^{1/2}$$

$$B = 0.086640\frac{RT_c}{P_c} = 0.086640\frac{(0.083145 \text{ dm}^3\cdot\text{bar}\cdot\text{mol}^{-1})(561.75)}{48.7575 \text{ bar}} = 0.082996 \text{ dm}^3\cdot\text{mol}^{-1}$$

Now substitute into the appropriate equations to find the vapor pressure P for each temperature and density:

T/K	$P(\text{van der Waals})/\text{bar}$	$P(\text{Redlich-Kwong})/\text{bar}$
290	0.00842	0.00757
300	0.13869	0.13843
310	0.21514	0.21459
320	0.32305	0.32194
330	0.47109	0.46897
340	0.66927	0.66541
350	0.92897	0.92225

CHAPTER 24

Solutions I
Liquid-Liquid Solutions

PROBLEMS AND SOLUTIONS

24–1. In the text, we went from Equation 24.5 to 24.6 using a physical argument involving varying the size of the system while keeping T and P fixed. We could also have used a mathematical process called *Euler's theorem*. Before we can learn about Euler's theorem, we must first define a *homogeneous function*. A function $f(z_1, z_2, \ldots, z_N)$ is said to be homogeneous if

$$f(\lambda z_1, \lambda z_2, \ldots, \lambda z_N) = \lambda f(z_1, z_2, \ldots, z_N)$$

Argue that extensive thermodynamic quantities are homogeneous functions of their extensive variables.

If we change extensive variables by a factor of λ, then we change an extensive function of these variables by a factor of λ.

24–2. Euler's theorem says that if $f(z_1, z_2, \ldots, z_N)$ is homogeneous, then

$$f(z_1, z_2, \ldots, z_N) = z_1 \frac{\partial f}{\partial \lambda_1} + z_2 \frac{\partial f}{\partial z_2} + \cdots + z_N \frac{\partial f}{\partial \lambda_N}$$

Prove Euler's theorem by differentiating the equation in Problem 24–1 with respect to λ and then setting $\lambda = 1$.

Apply Euler's theorem to $G = G(n_1, n_2, T, P)$ to derive Equation 24.6. (*Hint*: Because T and P are intensive variables, they are simply irrelevant variables in this case.)

Start with

$$\lambda f(z_1, z_2, \ldots, z_N) = f(\lambda z_1, \lambda z_2, \ldots, \lambda z_N)$$

differentiate with respect to λ to obtain

$$\begin{aligned}
f(z_1, z_2, \ldots, z_N) &= \frac{\partial f(\lambda z_1, \lambda z_2, \ldots, \lambda z_N)}{\partial \lambda z_1} \frac{\partial \lambda z_1}{\partial \lambda} + \cdots + \frac{\partial f(\lambda z_1, \lambda z_2, \ldots, \lambda z_N)}{\partial \lambda z_N} \frac{\partial \lambda z_N}{\partial \lambda} \\
&= z_1 \frac{\partial f(\lambda z_1, \lambda z_2, \ldots, \lambda z_N)}{\partial \lambda z_1} + \cdots + z_N \frac{\partial f(\lambda z_1, \lambda z_2, \ldots, \lambda z_N)}{\partial \lambda z_N}
\end{aligned}$$

Now set $\lambda = 1$

$$f(z_1, z_2, \ldots, z_N) = z_1 \frac{\partial f}{\partial z_1} + \cdots + z_N \frac{\partial f}{\partial z_N}$$

755

To apply this result to $G = G(n_1, n_2, T, P)$, we let $f = G$, $z_1 = n_1$, and $z_2 = n_2$ to write

$$G(n_1, n_2, T, P) = n_1 \left(\frac{\partial G}{\partial n_1} \right)_{T,P,n_2} + n_2 \left(\frac{\partial G}{\partial n_2} \right)_{T,P,n_1}$$

$$= n_1 \mu_1 + n_2 \mu_2$$

24–3. Use Euler's theorem (Problem 24–2) to prove that

$$Y(n_1, n_2, \ldots, T, P) = \sum n_j \overline{Y}_j$$

for any extensive quantity Y.

Simply let $Y = f$ and $n_j = z_j$ in Problem 24–2 to write

$$Y(n_1, n_2, \ldots, T, P) = n_1 \left(\frac{\partial Y}{\partial n_1} \right)_{T,P,n_{k \neq 1}} + n_2 \left(\frac{\partial Y}{\partial n_2} \right)_{T,P,n_{k \neq 2}} + \cdots$$

$$= n_1 \overline{Y}_1 + n_2 \overline{Y}_2 + \cdots$$

24–4. Apply Euler's theorem to $U = U(S, V, n)$. Do you recognize the resulting equation?

All three variables, S, V and n, are extensive. Using Euler's theorem (Problem 24–2) gives

$$U = S \left(\frac{\partial U}{\partial S} \right)_{V,n} + V \left(\frac{\partial U}{\partial V} \right)_{S,n} + n \left(\frac{\partial U}{\partial n} \right)_{S,V}$$

$$= S(T) + V(-P) + n(\mu)$$

$$= TS - PV + \mu n$$

or

$$G = \mu n = U - TS + PV = H - TS = A + PV$$

This is the defining equation for the Gibbs energy.

24–5. Apply Euler's theorem to $A = A(T, V, n)$. Do you recognize the resulting equation?

The extensive variables are V and n. Using Euler's theorem (Problem 24–2) gives

$$A = V \left(\frac{\partial A}{\partial V} \right)_{T,n} + n \left(\frac{\partial A}{\partial n} \right)_{T,V}$$

$$= V(-P) + n(\mu)$$

or

$$G = A + PV$$

This is the defining equation for the Gibbs energy.

24–6. Apply Euler's theorem to $V = V(T, P, n_1, n_2)$ to derive Equation 24.7.

Use the result of Problem 24–3 with $V = Y$.

24–7. The properties of many solutions are given as a function of the mass percent of the components. If we let the mass percent of component-2 be A_2, then derive a relation between A_2 and the mole fractions, x_1 and x_2.

$$A_2 = \frac{m_2}{m_1 + m_2} \times 100 = \frac{M_2 n_2}{M_1 n_1 + M_2 n_2} \times 100 \tag{1}$$

where the number of moles of component j is $n_j = m_j/M_j$ where M_j is its molar mass. Now divide numerator and denominator of Equation 1 by $n_1 + n_2$ to write

$$A_2 = \frac{M_2 x_2}{M_1 x_1 + M_2 x_2} \times 100$$

24–8. The *CRC Handbook of Chemistry and Physics* gives the densities of many aqueous solutions as a function of the mass percentage of solute. If we denote the density by ρ and the mass percentage of component-2 by A_2, the *Handbook* gives $\rho = \rho(A_2)$ (in $g \cdot mL^{-1}$). Show that the quantity $V = (n_1 M_1 + n_2 M_2)/\rho(A_2)$ is the volume of the solution containing n_1 moles of component 1 and n_2 moles of component-2. Now show that

$$\overline{V}_1 = \frac{M_1}{\rho(A_2)}\left[1 + \frac{A_2}{\rho(A_2)}\frac{d\rho(A_2)}{dA_2}\right]$$

and

$$\overline{V}_2 = \frac{M_2}{\rho(A_2)}\left[1 + \frac{(A_2 - 100)}{\rho(A_2)}\frac{d\rho(A_2)}{dA_2}\right]$$

Show that

$$V = n_1 \overline{V}_1 + n_2 \overline{V}_2$$

in agreement with Equation 24.7.

The mass of component j in the solution is $m_j = n_j M_j$, so the total mass is $n_1 M_1 + n_2 M_2$. Therefore, the volume is the mass divided by the density, or $V = (n_1 M_1 + n_2 M_2)/\rho(A_2)$. Now

$$\overline{V}_1 = \left(\frac{\partial V}{\partial n_1}\right)_{n_2} = \frac{M_1}{\rho(A_2)} - \frac{n_1 M_1 + n_2 M_2}{\rho^2(A_2)}\left[\frac{\partial \rho(A_2)}{\partial n_1}\right]_{n_2}$$

But, using Equation 1 of Problem 24–7,

$$\left[\frac{\partial \rho(A_2)}{\partial n_1}\right]_{n_2} = \left[\frac{d\rho(A_2)}{dA_2}\right]\left(\frac{\partial A_2}{\partial n_1}\right)_{n_2} = \left[\frac{d\rho(A_2)}{dA_2}\right]\left[-\frac{M_2 n_2 M_1}{(n_1 M_1 + n_2 M_2)^2} \times 100\right]$$

$$= \left[\frac{d\rho(A_2)}{dA_2}\right]\left(-\frac{A_2 M_1}{n_1 M_1 + n_2 M_2}\right)$$

Substitute this into $\overline{V}_1$ above to get

$$\overline{V}_1 = \frac{M_1}{\rho(A_2)} + \frac{A_2 M_1}{\rho^2(A_2)}\frac{d\rho(A_2)}{dA_2} = \frac{M_1}{\rho(A_2)}\left[1 + \frac{A_2}{\rho(A_2)}\frac{d\rho(A_2)}{dA_2}\right]$$

Similarly

$$\overline{V}_2 = \left(\frac{\partial V}{\partial n_2}\right)_{n_1} = \frac{M_2}{\rho(A_2)} - \frac{n_1 M_1 + n_2 M_2}{\rho^2(A_2)}\left[\frac{\partial\rho(A_2)}{\partial n_2}\right]_{n_1}$$

But

$$\left[\frac{\partial\rho(A_2)}{\partial n_2}\right]_{n_1} = \left[\frac{d\rho(A_2)}{dA_2}\right]\left(\frac{\partial A_2}{\partial n_2}\right)_{n_1}$$

$$= \left[\frac{d\rho(A_2)}{dA_2}\right]\left[\frac{100 M_2}{n_1 M_1 + n_2 M_2} - \frac{100 n_2 M_2^2}{(n_1 M_1 + n_2 M_2)^2}\right]$$

$$= \left[\frac{d\rho(A_2)}{dA_2}\right]\left[\frac{100 M_2 n_1 M_1}{(n_1 M_1 + n_2 M_2)^2}\right]$$

Substituting this into $\overline{V}_2$ gives

$$\overline{V}_2 = \frac{M_2}{\rho(A_2)}\left[1 - \frac{A_1}{\rho(A_2)}\frac{d\rho(A_2)}{dA_2}\right] = \frac{M_2}{\rho(A_2)}\left[1 + \frac{(A_2 - 100)}{\rho(A_2)}\frac{d\rho(A_2)}{dA_2}\right]$$

because $A_1 + A_2 = 100$. Finally,

$$n_1\overline{V}_1 + n_2\overline{V}_2 = \frac{n_1 M_1 + n_2 M_2}{\rho(A_2)} + \frac{(n_1 M_1 A_2 - n_2 M_2 A_1)}{\rho^2(A_2)}\frac{d\rho(A_2)}{dA_2}$$

But

$$n_1 M_1 A_2 - n_2 M_2 A_1 = \frac{n_1 M_1 M_2 n_2 - n_2 M_2 M_1 n_1}{n_1 M_1 + n_2 M_2} \times 100 = 0$$

so

$$n_1\overline{V}_1 + n_2\overline{V}_2 = \frac{n_1 M_1 + n_2 M_2}{\rho(A_2)} = V$$

24-9. The density (in $g\cdot mol^{-1}$) of a 1-propanol-water solution at 20°C as a function of A_2, the mass percentage of 1-propanol, can be expressed as

$$\rho(A_2) = \sum_{j=0}^{7}\alpha_j A_2^j$$

where

$\alpha_0 = 0.99823$	$\alpha_4 = 1.5312 \times 10^{-7}$
$\alpha_1 = -0.0020577$	$\alpha_5 = -2.0365 \times 10^{-9}$
$\alpha_2 = 1.0021 \times 10^{-4}$	$\alpha_6 = 1.3741 \times 10^{-11}$
$\alpha_3 = -5.9518 \times 10^{-6}$	$\alpha_7 = -3.7278 \times 10^{-14}$

Use this expression to plot $\overline{V}_{H_2O}$ and $\overline{V}_{\text{1-propanol}}$ versus A_2 and compare your values with those in Figure 24.1.

Substitute $\rho(A_2)$ into $\overline{V}_1 = \overline{V}_{1\text{-propanol}}$ in Problem 24–8 and into $\overline{V}_2 = \overline{V}_{H_2O}$ to obtain

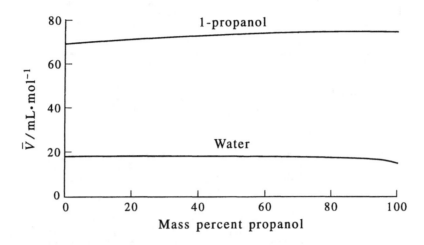

24–10. Given the density of a binary solution as a function of the mole fraction of component 2 $[\rho = \rho(x_2)]$, show that the volume of the solution containing n_1 moles of component 1 and n_2 moles of component 2 is given by $V = (n_1 M_1 + n_2 M_2)/\rho(x_2)$. Now show that

$$\overline{V}_1 = \frac{M_1}{\rho(x_2)}\left[1 + \left(\frac{x_2(M_2 - M_1) + M_1}{M_1}\right)\frac{x_2}{\rho(x_2)}\frac{d\rho(x_2)}{dx_2}\right]$$

and

$$\overline{V}_2 = \frac{M_2}{\rho(x_2)}\left[1 - \left(\frac{x_2(M_2 - M_1) + M_1}{M_2}\right)\frac{1 - x_2}{\rho(x_2)}\frac{d\rho(x_2)}{dx_2}\right]$$

Show that

$$V = n_1\overline{V}_1 + n_2\overline{V}_2$$

in agreement with Equation 24.7.

The total mass of the solution is $n_1 M_1 + n_2 M_2$, so its volume (mass/density) is $V = (n_1 M_1 + n_2 M_2)/\rho(x_2)$.

$$\overline{V}_1 = \left(\frac{\partial V}{\partial n_1}\right)_{n_2} = \frac{M_1}{\rho(x_2)} - \frac{n_1 M_1 + n_2 M_2}{\rho^2(x_2)}\left[\frac{\partial \rho(x_2)}{\partial n_1}\right]_{n_2}$$

But,

$$\left[\frac{\partial \rho(x_2)}{\partial n_1}\right]_{n_2} = \left[\frac{d\rho(x_2)}{dx_2}\right]\left(\frac{\partial x_2}{\partial n_1}\right)_{n_2} = \left[\frac{d\rho(x_2)}{dx_2}\right]\left[-\frac{n_2}{(n_1 + n_2)^2}\right]$$

$$= -\left[\frac{d\rho(x_2)}{dx_2}\right]\left(\frac{x_2}{n_1 + n_2}\right)$$

Substitute this result into $\overline{V}_1$ to get

$$\overline{V}_1 = \frac{M_1}{\rho(x_2)} + \frac{(n_1 M_1 + n_2 M_2)x_2}{(n_1 + n_2)\rho^2(x_2)}\frac{d\rho(x_2)}{dx_2} = \frac{M_1}{\rho(x_2)} + \frac{(x_1 M_1 + x_2 M_2)x_2}{\rho^2(x_2)}\frac{d\rho(x_2)}{dx_2}$$

$$= \frac{M_1}{\rho(x_2)}\left[1 + \frac{x_1 M_1 + x_2 M_2}{M_1}\frac{x_2}{\rho(x_2)}\frac{d\rho(x_2)}{dx_2}\right]$$

$$= \frac{M_1}{\rho(x_2)}\left[1 + \frac{M_1 + x_2(M_2 - M_1)}{M_1}\frac{x_2}{\rho(x_2)}\frac{d\rho(x_2)}{dx_2}\right]$$

Similarly

$$\overline{V}_2 = \left(\frac{\partial V}{\partial n_2}\right)_{n_1} = \frac{M_2}{\rho(x_2)} - \frac{n_1 M_1 + n_2 M_2}{\rho^2(x_2)}\left[\frac{\partial\rho(x_2)}{\partial n_2}\right]_{n_1}$$

But

$$\left[\frac{\partial\rho(x_2)}{\partial n_2}\right]_{n_1} = \frac{d\rho(x_2)}{dx_2}\left[\frac{\partial x_2}{\partial n_2}\right]_{n_1} = \frac{d\rho(x_2)}{dx_2}\left[\frac{1}{n_1 + n_2} - \frac{n_2}{(n_1 + n_2)^2}\right]$$

$$= \frac{d\rho(x_2)}{dx_2}\left[\frac{n_1}{(n_1 + n_2)^2}\right] = \frac{d\rho(x_2)}{dx_2}\frac{x_1}{n_1 + n_2}$$

Substitute this result into $\overline{V}_2$ to get

$$\overline{V}_2 = \frac{M_2}{\rho(A_2)} - \frac{(x_1 M_1 + x_2 M_2)x_1}{\rho^2(x_2)}\frac{d\rho(x_2)}{dx_2}$$

$$= \frac{M_2}{\rho(A_2)}\left\{1 - \left[\frac{M_1 + x_2(M_2 - M_1)}{M_2}\right]\frac{1 - x_2}{\rho(x_2)}\frac{d\rho(x_2)}{dx_2}\right\}$$

Finally,

$$n_1\overline{V}_1 + n_2\overline{V}_2 = \frac{n_1 M_1 + n_2 M_2}{\rho(x_2)} + \frac{M_1 + x_2(M_2 - M_1)}{\rho^2(A_2)}\frac{d\rho(A_2)}{dA_2}(n_1 x_2 - n_2 x_1)$$

But

$$n_1 x_2 - n_2 x_1 = \frac{n_1 n_2}{n_1 + n_2} - \frac{n_2 n_1}{n_1 + n_2} = 0$$

so

$$n_1\overline{V}_1 + n_2\overline{V}_2 = \frac{n_1 M_1 + n_2 M_2}{\rho(x_2)} = V$$

24–11. The density (in g·mol^{-1}) of a 1-propanol/water solution at 20°C as a function of x_2, the mole fraction of 1-propanol, can be expressed as

$$\rho(x_2) = \sum_{j=0}^{4}\alpha_j x_2^j$$

where

$$\alpha_0 = 0.99823 \qquad \alpha_3 = -0.17163$$

$$\alpha_1 = -0.48503 \qquad \alpha_4 = -0.01387$$

$$\alpha_2 = 0.47518$$

Use this expression to calculate the values of $\overline{V}_{H_2O}$ and $\overline{V}_{1\text{-propanol}}$ as a function of x_2 according to the equation in Problem 24–10.

Substitute $\rho(x_2)$ into $\overline{V}_1 = \overline{V}_{1\text{-propanol}}$ and $\overline{V}_2 = \overline{V}_{H_2O}$ in Problem 24–10 to obtain

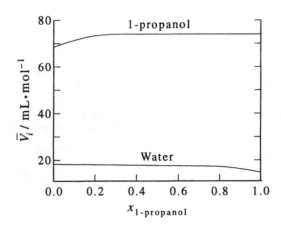

24–12. Use the data in the *CRC Handbook of Chemistry and Physics* to curve fit the density of a water/glycerol solution to a fifth-order polynomial in the mole fraction of glycerol, and then determine the partial molar volumes of water and glycerol as a function of mole fraction. Plot your result.

The curve fit of the density-mole fraction data gives (see the accompanying figure)

$$\rho(x_2) = 0.99849 + 1.1328x_2 - 2.7605x_2^2 + 4.1281x_2^3$$
$$- 3.2887x_2^4 + 1.0512x_2^5$$

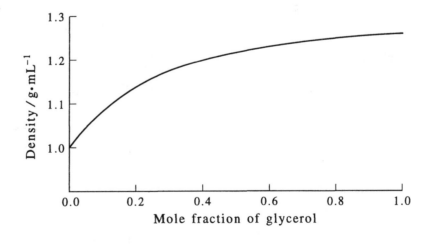

Substitute this result into the equations for $\overline{V}_1$ and $\overline{V}_2$ given in Problem 24–10 with $M_1 = 18.02$ and $M_2 = 92.09$ to get the following result:

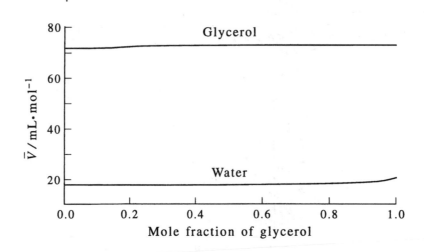

24–13. Just before Example 24–2, we showed that if one component of a binary solution obeys Raoult's law over the entire composition range, the other component does also. Now show that if $\mu_2 = \mu_2' + RT \ln x_2$ for $x_{2,\min} \leq x_2 \leq 1$, then $\mu_1 = \mu_1' + RT \ln x_1$ for $0 < x_1 < 1 - x_{2,\min}$). Notice that for the range over which μ_2 obeys the simple form given, μ_1 obeys a similarly simple form. If we let $x_{2,\min} = 0$, we obtain $\mu_1 = \mu_1^* + RT \ln x_1$ $(0 \leq x_1 \leq 1)$.

Start with the Gibbs-Duhem equation

$$x_1 d\mu_1 + x_2 d\mu_2 = 0$$

Solve for $d\mu_1$

$$d\mu_1 = -\frac{x_2}{x_1} d\mu_1 = -\frac{x_2}{x_1} \frac{RT}{x_1} dx_2 \qquad x_{2,\min} \leq x_2 \leq 1$$

$$= \frac{RT}{x_1} dx_1 \qquad 0 \leq x_1 \leq 1 - x_{2,\min}$$

Integrate to obtain

$$\mu_1 = \mu_1' + RT \ln x_1 \qquad 0 \leq x_1 \leq 1 - x_{2,\min}$$

24–14. Continue the calculations in Example 24–3 to obtain y_2 as a function of x_2 by varying x_2 from 0 to 1. Plot your result.

We use the equation

$$y_2 = \frac{P_2}{P_{total}} = \frac{x_2 P_2^*}{x_1 P_1^* + x_2 P_2^*} = \frac{x_2(45.2\ \text{torr})}{(1 - x_2)(20.9\ \text{torr}) + x_2(45.2\ \text{torr})}$$

A plot of y_2 against x_2 is

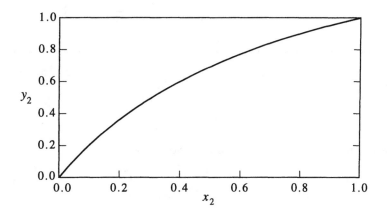

24–15. Use your results from Problem 24–14 to construct the pressure-composition diagram in Figure 24.4.

$$P_{total} = (1 - x_2)(20.9 \text{ torr}) + x_2(45.2 \text{ torr})$$

Solve the equation given in Problem 24–14 for x_2 in terms of y_2

$$x_2 = \frac{(20.9 \text{ torr})y_2}{45.2 \text{ torr} + (20.9 \text{ torr} - 45.2 \text{ torr})y_2}$$

Let x_2 vary from 0 to 1 in the first equation to calculate P_{total} as a function of x_2. Now let y_2 vary from 0 to 1 to calculate x_2 and then P_{total} to give P_{total} as a function of y_2. A plot of P_{total} against x_2 and y_2 is

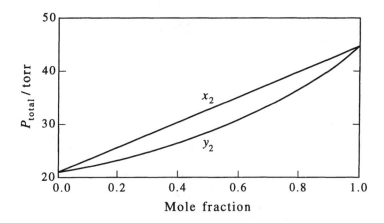

24–16. Calculate the relative amounts of liquid and vapor phases at an overall composition of 0.50 for one of the pair of values, $x_2 = 0.38$ and $y_2 = 0.57$, that you obtained in Problem 24–14.

We use Equation 24.19

$$\frac{n^{\mathrm{l}}}{n^{\mathrm{vap}}} = \frac{y_2 - x_a}{x_a - x_2} = \frac{0.57 - 0.50}{0.50 - 0.38} = 0.58$$

24–17. In this problem, we will derive analytic expressions for the pressure-composition curves in Figure 24.4. The liquid (upper) curve is just

$$P_{\text{total}} = x_1 P_1^* + x_2 P_2^* = (1 - x_2) P_1^* + x_2 P_2^* = P_1^* + x_2(P_2^* - P_1^*) \tag{1}$$

which is a straight line, as seen in Figure 24.4. Solve the equation

$$y_2 = \frac{x_2 P_2^*}{P_{\text{total}}} = \frac{x_2 P_2^*}{P_1^* + x_2(P_2^* - P_1^*)} \tag{2}$$

for x_2 in terms of y_2 and substitute into Equation (1) to obtain

$$P_{\text{total}} = \frac{P_1^* P_2^*}{P_2^* - y_2(P_2^* - P_1^*)}$$

Plot this result versus y_2 and show that it gives the vapor (lower) curve in Figure 24.4.

We solve Equation 2 for x_2 to obtain

$$x_2 = \frac{y_2 P_1^*}{P_2^* - y_2(P_2^* - P_1^*)}$$

Substitute this result into

$$P_{\text{total}} = P_1^* + x_2(P_2^* - P_1^*)$$

to get

$$P_{\text{total}} = \frac{P_1^* P_2^*}{P_2^* - y_2(P_2^* - P_1^*)}$$

The plots of P_{total} against x_2 and y_2 for $P_1^* = 20.9$ torr and $P_2^* = 45.2$ torr are

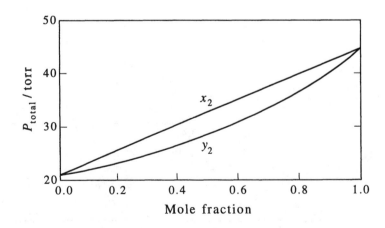

24–18. Prove that $y_2 > x_2$ if $P_2^* > P_1^*$ and that $y_2 < x_2$ if $P_2^* < P_1^*$. Interpret this result physically.

Start with

$$y_2 = \frac{x_2 P_2^*}{P_1^* + x_2(P_2^* - P_1^*)}$$

Divide both sides by x_2 and the numerator and denominator of the right side by P_1^* to obtain

$$\frac{y_2}{x_2} = \frac{P_2^*/P_1^*}{1 + x_2(P_2^*/P_1^* - 1)} = \frac{R}{1 + x_2(R - 1)}$$

where $R = P_2^*/P_1^*$. Now subtract 1 from both sides

$$\frac{y_2}{x_2} - 1 = \frac{R - 1 - x_2(R - 1)}{1 + x_2(R - 1)} = \frac{x_1(R - 1)}{1 + x_2(R - 1)}$$

If $R > 1$ ($P_2^* > P_1^*$), then the right side is always positive because $0 \le x_1 \le 1$ and $0 \le x_2 \le 1$ and so $y_2 > x_2$. If $R < 1$ ($P_2^* < P_1^*$), then the right side is always negative.

This result simply says that the mole fraction of a given component in the vapor phase will be greater than that of the other component if it is more volatile.

24–19. Tetrachloromethane and trichloroethylene form essentially an ideal solution at 40°C at all concentrations. Given that the vapor pressure of tetrachloromethane and trichloroethylene at 40°C are 214 torr and 138 torr, respectively, plot the pressure-composition diagram for this system (see Problem 24–17).

Plot

$$P_{total} = P_1^* + x_2(P_2^* - P_1^*) = 214 \text{ torr} - x_2(76 \text{ torr})$$

and

$$P_{total} = \frac{P_1^* P_2^*}{P_2^* - y_2(P_2^* - P_1^*)} = \frac{(214 \text{ torr})(138 \text{ torr})}{138 \text{ torr} + y_2(76 \text{ torr})}$$

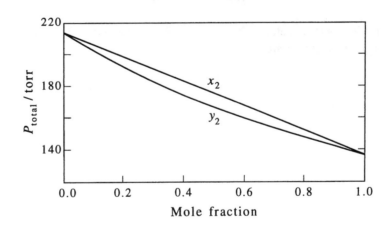

24–20. The vapor pressures of tetrachloromethane (1) and trichloroethylene (2) between 76.8°C and 87.2°C can be expressed empirically by the formulas

$$\ln(P_1^*/\text{torr}) = 15.8401 - \frac{2790.78}{t + 226.4}$$

and

$$\ln(P_2^*/\text{torr}) = 15.0124 - \frac{2345.4}{t + 192.7}$$

where t is the Celsius temperature. Assuming that tetrachloromethane and trichloroethylene form an ideal solution between 76.8°C and 87.2°C at all compositions, calculate the values of x_1 and y_1 at 82.0°C (at an ambient pressure of 760 torr).

Let 1 denote tetrachloromethane and 2 denote trichloroethylene.

$$\ln(P_1^*/\text{torr}) = 15.8401 - \frac{2790.78}{82.0 + 226.4} = 6.7919$$

or $P_1^* = 890$ torr. Similarly, $P_2^* = 648$ torr. Therefore, (see Example 24–5)

$$x_1 = \frac{P_2^* - 760 \text{ torr}}{P_2^* - P_1^*} = \frac{648 \text{ torr} - 760 \text{ torr}}{648 \text{ torr} - 890 \text{ torr}} = 0.463$$

$$y_1 = \frac{P_1}{760 \text{ torr}} = \frac{x_1 P_1^*}{760 \text{ torr}} = \frac{(0463)(890 \text{ torr})}{760 \text{ torr}} = 0.542$$

24–21. Use the data in Problem 24–20 to construct the entire temperature-composition diagram of a tetrachloromethane/trichlororethylene solution.

The vapor pressures of tetrachloromethane (1) and trichloroethylene (2) between 76.8°C and 87.2°C are given by

$$\ln(P_1^*/\text{torr}) = 15.8401 - \frac{2790.84}{t + 226.4}$$

$$\ln(P_2^*/\text{torr}) = 15.0124 - \frac{2345.4}{t + 192.7}$$

where t is the Celsius temperature. The mole fractions of tetrachloromethane (1) in the liquid and vapor phases at temperature t are given by

$$x_1 = \frac{P_2^* - 760 \text{ torr}}{P_2^* - P_1^*} \quad \text{and} \quad y_1 = \frac{x_1 P_1^*}{760 \text{ torr}}$$

Some data and the plot are given below.

$t/°C$	$P_1^*/torr$	$P_2^*/torr$	x_1	y_1
76.8	761.7	549.8	0.992	0.994
77.6	780.3	564.2	0.906	0.930
78.4	799.3	578.8	0.822	0.864
79.2	818.7	593.7	0.739	0.796
80.0	838.5	608.9	0.658	0.726
80.8	858.6	624.5	0.579	0.654
81.6	879.1	640.3	0.501	0.580
82.4	900.0	656.4	0.425	0.504
83.2	921.3	672.8	0.360	0.435
84.0	942.9	706.6	0.207	0.262
85.6	987.4	724.0	0.137	0.178
86.4	1010	741.7	0.068	0.091
87.2	1033	759.7	0.001	0.001

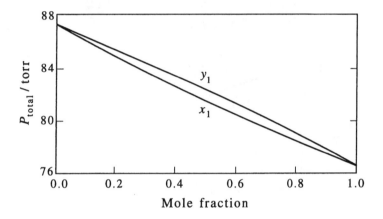

24–22. The vapor pressures of benzene and toluene between 80°C and 110°C as a function of the Kelvin temperature are given by the empirical formulas

$$\ln(P_{benz}^*/torr) = -\frac{3856.6\ K}{T} + 17.551$$

and

$$\ln(P_{tol}^*/torr) = -\frac{4514.6\ K}{T} + 18.397$$

Assuming that benzene and toluene form an ideal solution, use these formulas to construct a temperature-composition diagram of this system at an ambient pressure of 760 torr.

This problem is very similar to Problems 24–20 and 24–21. Some data and the plot are given below

T/K	$t/°C$	$P_1^*/torr$	$P_2^*/torr$	x_1	y_1
353.0	79.85	754.3	272.5	1.000	1.000
355.0	81.85	802.2	292.9	0.917	0.968
357.0	83.85	852.5	314.5	0.828	0.929
359.0	85.85	905.4	337.5	0.744	0.886
361.0	87.85	960.9	361.8	0.665	0.840
363.0	89.85	1019.2	387.6	0.590	0.791
365.0	91.85	1080.3	415.0	0.519	0.737
367.0	93.85	1144.3	443.9	0.451	0.680
369.0	95.85	1211.4	474.5	0.387	0.618
371.0	97.85	1281.6	506.9	0.327	0.551
373.0	99.85	1355.0	541.1	0.269	0.480
375.0	101.8	1431.9	577.1	0.214	0.403
377.0	103.8	1512.2	615.2	0.161	0.321
379.0	105.8	1596.0	655.3	0.111	0.234
381.0	107.8	1683.6	697.6	0.063	0.140
383.0	109.8	1775.0	742.1	0.017	0.040

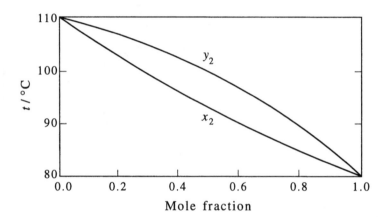

24–23. Construct the temperature-composition diagram for 1-propanol and 2-propanol in Figure 24.5 by varying t from 82.3°C (the boiling point of 2-propanol) to 97.2°C (the boiling point of 1-propanol), calculating (1) P_1^* and P_2^* at each temperature (see Example 24–5), (2) x_1 according to $x_1 = (P_2^* - 760)/(P_2^* - P_1^*)$, and (3) y_1 according to $y_1 = x_1 P_1^*/760$. Now plot t versus x_1 and y_1 on the same graph to obtain the temperature-composition diagram.

This problem is very similar to Problem 24–21. Some data and the plot are

$t/°C$	P_1^*/torr	P_2^*/torr	x_1	y_1
82.3	419.6	760.9	0.003	0.001
84.3	456.0	823.7	0.173	0.104
85.3	475.2	856.7	0.254	0.159
86.3	495.1	890.8	0.331	0.215
87.3	515.6	926.1	0.405	0.274
88.3	536.8	962.4	0.476	0.336
89.3	558.8	1000.0	0.544	0.400
90.3	581.5	1038.7	0.610	0.466
91.3	605.0	1078.7	0.673	0.536
92.3	629.2	1120.0	0.733	0.607
93.3	654.2	1162.5	0.792	0.682
94.3	680.1	1206.4	0.848	0.759
95.3	706.8	1251.7	0.902	0.839
96.3	734.3	1298.3	0.955	0.922
97.2	759.9	1341.5	1.000	1.000

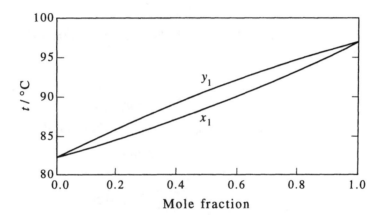

24–24. Prove that $\overline{V}_j = \overline{V}_j^*$ for an ideal solution, where $\overline{V}_j^*$ is the molar volume of pure component j.

The chemical potentials of compounds 1 and 2 of an ideal solution are given by

$$\mu_j = \mu_j^* + RT \ln x_j \qquad j = 1 \text{ and } 2$$

The Gibbs energy is given by

$$G = n_1\mu_1 + n_2\mu_2 = n_1\mu_1^* + n_2\mu_2^* + n_1 RT \ln x_1 + n_2 RT \ln x_2$$

The volume is given by

$$V = \left(\frac{\partial G}{\partial P}\right)_{T,n_1,n_2} = n_1\left(\frac{\partial \mu_1^*}{\partial P}\right)_T + n_2\left(\frac{\partial \mu_2^*}{\partial P}\right)_T$$
$$= n_1\overline{V}_1^* + n_2\overline{V}_2^*$$

By comparing this result to Equation 24.7, we see that $\overline{V}_j^* = \overline{V}_j$.

24–25. The volume of mixing of miscible liquids is defined as the volume of the solution minus the volume of the individual pure components. Show that

$$\Delta_{mix}\overline{V} = \sum x_i(\overline{V}_i - \overline{V}_i^*)$$

at constant P and T, where $\overline{V}_i^*$ is the molar volume of pure component i. Show that $\Delta_{mix}\overline{V} = 0$ for an ideal solution (see Problem 24–24).

Problem 24–24 shows that $\overline{V}_j^* = \overline{V}_j$ for an ideal solution, so $\Delta_{mix}\overline{V} = 0$.

24–26. Suppose the vapor pressures of the two components of a binary solution are given by

$$P_1 = x_1 P_1^* e^{x_2^2/2}$$

and

$$P_2 = x_2 P_2^* e^{x_1^2/2}$$

Given that $P_1^* = 75.0$ torr and $P_2^* = 160$ torr, calculate the total vapor pressure and the composition of the vapor phase at $x_1 = 0.40$.

$$\begin{aligned}
P_{total} &= P_1 + P_2 = x_1 P_1^* e^{x_2^2/2} + x_2 P_2^* e^{x_1^2/2} \\
&= (0.40)(75.0\text{ torr})e^{(0.60)^2/2} + (0.60)(160\text{ torr})e^{(0.40)^2/2} \\
&= 35.9\text{ torr} + 104\text{ torr} = 140\text{ torr}
\end{aligned}$$

$$y_1 = \frac{P_1}{P_{total}} = \frac{35.9\text{ torr}}{140\text{ torr}} = 0.26$$

24–27. Plot y_1 versus x_1 for the system described in the previous problem. Why does the curve lie below the straight line connecting the origin with the point $x_1 = 1$, $y_1 = 1$? Describe a system for which the curve would lie above the diagonal line.

We simply use

$$\begin{aligned}
y_1 &= \frac{P_1}{P_{total}} = \frac{x_1 P_1^* e^{x_2^2/2}}{x_1 P_1^* e^{x_2^2/2} + x_2 P_2^* e^{x_1^2/2}} \\
&= \frac{x_1(75.0\text{ torr})e^{(1-x_1)^2/2}}{x_1(75.0\text{ torr})e^{(1-x_1)^2/2} + (1-x_1)(160\text{ torr})e^{x_1^2/2}}
\end{aligned}$$

A plot of y_1 against x_1 is the curved line shown below.

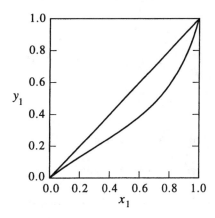

The straight line connecting the origin with the point $x_1 = y_1 = 1$ is given by $y_1 = x_1$. Therefore, component 1 is less volatile than component 2, so the vapor is richer in component 2; $y_1 < x_1$ because $P_1^* < P_2^*$.

24–28. Use the expressions for P_1 and P_2 given in Problem 24–26 to construct a pressure-composition diagram.

Start with

$$P_{total} = P_1 + P_2 = x_1 P_1^* e^{x_2^2/2} + x_2 P_2^* e^{x_1^2/2}$$
$$= x_1 (75.0 \text{ torr}) e^{(1-x_1)^2/2} + (1 - x_1)(160 \text{ torr}) e^{x_1^2/2}$$

and

$$y_1 = \frac{P_1}{P_{total}} = \frac{x_1 P_1^* e^{x_2^2/2}}{x_1 P_1^* e^{x_2^2/2} + x_2 P_2^* e^{x_1^2/2}}$$
$$= \frac{x_1 (75.0 \text{ torr}) e^{(1-x_1)^2/2}}{x_1 (75.0 \text{ torr}) e^{(1-x_1)^2/2} + (1 - x_1)(160 \text{ torr}) e^{x_1^2/2}}$$

Now calculate P_{total} and y_1 as a function of x_1 and then plot P_{total} against x_1 and y_1.

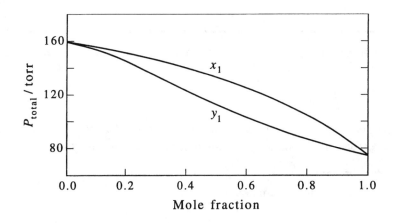

24–29. The vapor pressure (in torr) of the two components in a binary solution are given by

$$P_1 = 120x_1 e^{0.20x_2^2 + 0.10x_2^3}$$

and

$$P_2 = 140x_2 e^{0.35x_1^2 - 0.10x_1^3}$$

Determine the values of P_1^*, P_2^*, $k_{H,1}$, and $k_{H,2}$.

Use the fact that $P_j \rightarrow x_j P_j^*$ as $x_j \rightarrow 1$ to obtain

$$P_1 \longrightarrow 120x_1 \quad \text{as} \quad x_1 \longrightarrow 1 \quad \text{and} \quad P_2 \longrightarrow 140x_2 \quad \text{as} \quad x_2 \longrightarrow 1$$

or $P_1^* = 120$ torr and $P_2^* = 140$ torr. Now use the fact that $P_j \rightarrow k_{H,j} x_j$ as $x_j \rightarrow 0$ to obtain

$$P_1 \longrightarrow 120x_1 e^{0.30} = 162 \text{ torr} = k_{H,1}$$

and

$$P_2 \longrightarrow 140x_1 e^{0.25} = 180 \text{ torr} = k_{H,2}$$

24–30. Suppose the vapor pressure of the two components of a binary solution are given by

$$P_1 = x_1 P_1^* e^{\alpha x_2^2 + \beta x_2^3}$$

and

$$P_2 = x_2 P_2^* e^{(\alpha + 3\beta/2)x_1^2 - \beta x_1^3}$$

Show that $k_{H,1} = P_1^* e^{\alpha + \beta}$ and $k_{H,2} = P_2^* e^{\alpha + \beta/2}$.

Use the fact that $P_j \rightarrow k_{H,j} x_j$ as $x_j \rightarrow 0$ to obtain

$$P_1 \longrightarrow x_1 P_1^* e^{\alpha + \beta} \quad \text{and} \quad P_2 \longrightarrow x_2 P_2^* e^{\alpha + \beta/2}$$

or

$$k_{H,1} = P_1^* e^{\alpha + \beta} \quad \text{and} \quad k_{H,2} = P_2^* e^{\alpha + \beta/2}$$

24–31. The empirical expression for the vapor pressure that we used in Examples 24–6 and 24–7, for example,

$$P_1 = x_1 P_1^* e^{\alpha x_2^2 + \beta x_2^3 + \cdots}$$

is sometimes called the *Margules equation*. Use Equation 24.29 to prove that there can be no linear term in the exponential factor in P_1, for otherwise P_2 will not satisfy Henry's law as $x_2 \rightarrow 0$.

Assume that there is a linear term in the exponent of P_1.

$$P_1 = x_1 P_1^* e^{\alpha x_2}$$

Then

$$\frac{\partial \ln P_1}{\partial x_1} = \frac{1}{x_1} - \alpha$$

According to Equation 24.29,

$$\begin{aligned}
\frac{\partial \ln P_2}{\partial x_2} &= \frac{x_1}{x_2}\frac{\partial \ln P_1}{\partial x_1} \\
&= \frac{1}{x_2} - \alpha\frac{x_1}{x_2} = \frac{1}{x_2} - \alpha\frac{1-x_2}{x_2} \\
&= \frac{1-\alpha}{x_2} + \alpha
\end{aligned}$$

Integration with respect to x_2 gives

$$\ln P_2 = (1-\alpha)\ln x_2 + \alpha x_2 + \ln A$$

where $\ln A$ is an integration constant. Then

$$P_2 = A x_2^{1-\alpha}e^{\alpha x_2}$$

As $x_2 \to 0$, $P_2 \to A x_2^{1-\alpha}$. But according to Henry's law, $P_2 \to k_{H,2}x_2$ as $x_2 \to 0$, so α must equal zero.

24–32. In the text, we showed that the Henry's law behavior of component-2 as $x_2 \to 0$ is a direct consequence of the Raoult's law behavior of component 1 as $x_1 \to 1$. In this problem, we will prove the converse: the Raoult's law behavior of component 1 as $x_1 \to 1$ is a direct consequence of the Henry's law behavior of component-2 as $x_2 \to 0$. Show that the chemical potential of component-2 as $x_2 \to 0$ is

$$\mu_2(T, P) = \mu_2^\circ(T) + RT\ln k_{H,2} + RT\ln x_2 \quad x_2 \to 0$$

Differentiate μ_2 with respect to x_2 and substitute the result into the Gibbs-Duhem equation to obtain

$$d\mu_1 = RT\frac{dx_1}{x_1} \qquad x_2 \longrightarrow 0$$

Integrate this expression from $x_1 = 1$ to $x_1 \approx 1$ and use the fact that $\mu_1(x_1 = 1) = \mu_1^*$ to obtain

$$\mu_1(T, P) = \mu_1^*(T) + RT\ln x_1 \qquad x_1 \to 1$$

which is the Raoult's law expression for chemical potential. Show that this result follows directly from Equation 24.29.

Start with

$$\mu_2 = \mu_2^* + RT\ln\frac{P_2}{P_2^*} = \mu_2^* - RT\ln P_2^* + RT\ln P_2$$

As $x_2 \to 0$, $P_2 \to k_{H,2}x_2$, so

$$\begin{aligned}
\mu_2 &= \mu_2^* - RT\ln P_2^* + RT\ln k_{H,2} + RT\ln x_2 \quad (x_2 \longrightarrow 0) \\
&= \mu^\circ(T) + RT\ln k_{H,2} + RT\ln x_2 \quad (x_2 \longrightarrow 0)
\end{aligned}$$

Now

$$\frac{d\mu_2}{dx_2} = \frac{RT}{x_2}$$

and so according to the Gibbs-Duhem equation

$$d\mu_1 = -\frac{x_2}{x_1}d\mu_2 = -\frac{x_2}{x_1}\frac{RT}{x_2}dx_2 = RT\frac{dx_1}{x_1} \qquad (x_2 \longrightarrow 0)$$

Now integrate this expression from $x_1 = 1$ to $x_1 \approx 1$ (because the expression is valid only for $x_2 \approx 0$, or $x_1 \approx 1$) to obtain

$$\mu_1 = \mu_1^* + RT \ln x_1 \qquad (x_1 \longrightarrow 1)$$

24–33. In Example 24–7, we saw that if

$$P_1 = x_1 P_1^* e^{\alpha x_2^2 + \beta x_2^3}$$

then

$$P_2 = x_2 P_2^* e^{(\alpha + 3\beta/2)x_1^2 - \beta x_1^3} \qquad (x_2 \longrightarrow 0)$$

Show that this result follows directly from Equation 24.29.

Start with

$$\ln P_1 = \ln x_1 + \ln P_1^* + \alpha(1 - x_1)^2 + \beta(1 - x_1)^3$$

and differentiate with respect to x_1 to obtain

$$\frac{\partial \ln P_1}{\partial x_1} = \frac{1}{x_1} - 2\alpha(1 - x_1) - 3\beta(1 - x_1)^2$$

According to Equation 24.29

$$\frac{\partial \ln P_2}{\partial x_2} = \frac{x_1}{x_2}\frac{\partial \ln P_1}{\partial x_1} = \frac{1}{x_2} - 2\alpha x_1 - 3\beta x_1 x_2$$

$$= \frac{1}{x_2} - 2\alpha + 2\alpha x_2 - 3\beta x_2 + 3\beta x_2^2$$

Now integrate with respect to x_2 to obtain

$$\ln P_2 = \ln x_2 - 2\alpha x_2 + (2\alpha - 3\beta)\frac{x_2^2}{2} + \beta x_2^3 + A$$

where A is an integration constant. Substituting $x_2 = 1 - x_1$ in the last three terms gives

$$\ln P_2 = \ln x_2 - 2\alpha(1 - x_1) + \left(\frac{2\alpha - 3\beta}{2}\right)(1 - x_1)^2 + \beta(1 - x_1)^3 + A$$

$$= \ln x_2 - 2\alpha + 2\alpha x_1 + \alpha - 2\alpha x_1 + \alpha x_1^2 - \frac{3}{2}\beta + 3\beta x_1 - \frac{3}{2}\beta x_1^2$$

$$+ \beta - 3\beta x_1 + 3\beta x_1^2 - \beta x_1^3 + A$$

$$= \ln x_2 + \alpha x_1^2 + \frac{3}{2}\beta x_1^2 - \beta x_1^3 + \left(A - \alpha - \frac{\beta}{2}\right)$$

Rewrite this expression as

$$P_2 = x_2 B e^{\alpha x_1^2 + \frac{3}{2}\beta x_1^2 - \beta x_1^3}$$

where $B = A - \alpha - \beta/2$. Note that $B = P_2^*$ because $P_2 \to x_2 P_2^*$ as $x_2 \to 1$.

24–34. Suppose we express the vapor pressures of the components of a binary solution by

$$P_1 = x_1 P_1^* e^{\alpha x_2^2}$$

and

$$P_2 = x_2 P_2^* e^{\beta x_1^2}$$

Use the Gibbs-Duhem equation or Equation 24.29 to prove that α must equal β.

Start with

$$\ln P_1 = \ln x_1 + \ln P_1^* + \alpha x_2^2$$

Differentiate with respect to x_1 to obtain

$$\frac{\partial \ln P_1}{\partial x_1} = \frac{1}{x_1} - 2\alpha(1 - x_1)$$

Use Equation 24.29 to get

$$\frac{\partial \ln P_2}{\partial x_2} = \frac{x_1}{x_2}\frac{\partial \ln P_1}{\partial x_1} = \frac{1}{x_2} - 2\alpha x_1 = \frac{1}{x_2} - 2\alpha + 2\alpha x_2$$

Integrate with respect to x_2 to get

$$\begin{aligned}
\ln P_2 &= \ln x_2 - 2\alpha x_2 + \alpha x_2^2 + A \\
&= \ln x_2 - 2\alpha + 2\alpha x_1 + \alpha - 2\alpha x_1 + \alpha x_1^2 + A \\
&= \ln x_2 + \alpha x_1^2 + (A - \alpha)
\end{aligned}$$

where A is an integration constant. Therefore,

$$P_2 = B x_2 e^{\alpha x_1^2}$$

where $B = A - \alpha$. Clearly $B = P_2^*$ because $P_2 \to P_2^* x_2$ as $x_2 \to 1$.

24–35. Use Equation 24.29 to show that if one component of a binary solution obeys Raoult's law for all concentrations, then the other component also obeys Raoult's law for all concentrations.

According to Raoult's law

$$P_1 = x_1 P_1^*$$

Therefore,

$$\frac{\partial \ln P_1}{\partial x_1} = \frac{1}{x_1}$$

and Equation 24.29 gives

$$\frac{\partial \ln P_2}{\partial x_2} = \frac{1}{x_2}$$

which upon integration gives

$$\ln P_2 = \ln x_2 + A$$

$$P_2 = Ax_2 = P_2^* x_2$$

where A is an integration constant.

24–36. Use Equation 24.29 to show that if one component of a binary solution has positive deviations from Raoult's law, then the other component must also.

Equation 24.29 says that

$$x_1 \left(\frac{\partial \ln P_1}{\partial x_1} \right) = x_2 \left(\frac{\partial \ln P_2}{\partial x_2} \right)$$

or

$$\frac{\partial \ln P_1}{\partial \ln x_1} = \frac{\partial \ln P_2}{\partial \ln x_2}$$

For an ideal solution, $P_j^{id} = x_j P_j^*$ and so

$$\frac{\partial \ln P_1^{id}}{\partial \ln x_1} = \frac{\partial \ln P_2^{id}}{\partial \ln x_2} = 1$$

If $P_1 > P_1^{id} = x_1 P_1^*$ (positive deviation from ideality), then

$$\frac{\partial \ln P_1}{\partial \ln x_1} > 1$$

and so

$$\frac{\partial \ln P_2}{\partial \ln x_2} > 1$$

Conversely, if one component has negative deviations from ideality ($\partial \ln P_1 / \partial \ln x_1) < 1$, then the other must also.

24–37. If the vapor pressures of the two components in a binary solution are given by

$$P_1 = x_1 P_1^* e^{ux_2^2/RT} \quad \text{and} \quad P_2 = x_2 P_2^* e^{ux_1^2/RT}$$

show that

$$\Delta_{mix} \overline{G}/u = \Delta_{mix} G/(n_1 + n_2)u = \frac{RT}{u}(x_1 \ln x_1 + x_2 \ln x_2) + x_1 x_2$$

$$\Delta_{mix} \overline{S}/R = \Delta_{mix} S/(n_1 + n_2)R = -(x_1 \ln x_1 + x_2 \ln x_2)$$

and

$$\Delta_{\text{mix}}\overline{H}/u = \Delta_{\text{mix}}H/(n_1 + n_2)u = x_1 x_2$$

A solution that satisfies these equations is called a *regular solution*. A statistical thermodynamic model of binary solutions shows that u is proportional to $2\varepsilon_{12} - \varepsilon_{11} - \varepsilon_{22}$, where ε_{ij} is the interaction energy between molecules of components i and j. Note that $u = 0$ if $\varepsilon_{12} = (\varepsilon_{11} + \varepsilon_{22})/2$, which means that energetically, molecules of components 1 and 2 "like" the opposite molecules as well as their own.

Use the equations

$$G^{\text{sln}} = n_1 \mu_1 + n_2 \mu_2$$

and

$$\mu_j = \mu_j^* + RT \ln \frac{P_j}{P_j^*}$$

to write

$$G^{\text{sln}} = n_1 \mu_1^* + n_2 \mu_2^* + n_1 RT \ln(x_1 e^{ux_2^2/RT}) + n_2 RT \ln(x_2 e^{ux_1^2/RT})$$

But $n_1 \mu_1^* + n_2 \mu_2^*$ is the Gibbs energy of the two pure liquid components, so

$$\Delta_{\text{mix}} G = G^{\text{sln}} - n_1 \mu_1^* + n_2 \mu_2^* = n_1 RT \ln x_1 + n_2 RT \ln x_2 + u(n_1 x_2^2 + n_2 x_1^2)$$

Divide by the total number of moles, $n_1 + n_2$, to get

$$\Delta_{\text{mix}}\overline{G} = RT(x_1 \ln x_1 + x_2 \ln x_2) + ux_1 x_2(x_2 + x_1)$$

Now divide by u and use the fact that $x_1 + x_2 = 1$ to get

$$\Delta_{\text{mix}}\overline{G}/u = \frac{RT}{u}(x_1 \ln x_1 + x_2 \ln x_2) + x_1 x_2$$

Use the equation $\Delta_{\text{mix}}\overline{S} = -(\partial \Delta_{\text{mix}}\overline{G}/\partial T)$ to obtain

$$\Delta_{\text{mix}}\overline{S} = -R(x_1 \ln x_1 + x_2 \ln x_2)$$

Now use $\Delta_{\text{mix}}\overline{G} = \Delta_{\text{mix}}\overline{H} - T\Delta_{\text{mix}}\overline{S}$ to get

$$\Delta_{\text{mix}}\overline{H}/u = x_1 x_2$$

24–38. Prove that $\Delta_{\text{mix}}\overline{G}$, $\Delta_{\text{mix}}\overline{S}$, and $\Delta_{\text{mix}}\overline{H}$ in the previous problem are symmetric about the point $x_1 = x_2 = 1/2$.

Each expression is symmetric in x_1 and x_2. Therefore, they must be symmetric about $x_1 = x_2 = 1/2$.

24–39. Plot $P_1/P_1^* = x_1 e^{ux_2^2/RT}$ versus x_1 for $RT/u = 0.60, 0.50, 0.45, 0.40,$ and 0.35. Note that some of the curves have regions where the slope is negative. The following problem has you show that this behavior occurs when $RT/u < 0.50$. These regions are similar to the loops of the van der Waals equation or the Redlich-Kwong equation when $T < T_c$ (Figure 16.8), and in this case correspond to regions in which the two liquids are not miscible. The critical value $RT/u = 0.50$ corresponds to a solution critical temperature of $0.50u/R$.

See plot

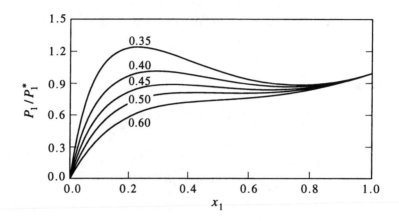

24-40. Differentiate $P_1 = x_1 P_1^* e^{u(1-x_1)^2/RT}$ with respect to x_1 to prove that P_1 has a maximum or a minimum at the points $x_1 = \frac{1}{2} \pm \frac{1}{2}(1 - \frac{2RT}{u})^{1/2}$. Show that $RT/u < 0.50$ for either a maximum or a minimum to occur. Do the positions of these extremes when $RT/u = 0.35$ correspond to the plot you obtained in the previous problem?

$$\frac{dP_1}{dx_1} = 0 = P_1^* e^{u(1-x_1)^2/RT} - \frac{2u}{RT} x_1(1 - x_1) P_1^* e^{u(1-x_1)^2/RT} = 0$$

Cancelling several factors and rearranging gives

$$x_1^2 - x_1 + \frac{RT}{2u} = 0$$

or

$$x_1 = \frac{1}{2} \pm \frac{1}{2}\left(1 - \frac{2RT}{u}\right)^{1/2}$$

The values of x_1 will not be real unless $2RT/u < 1$, or unless $RT/u < 0.50$. When $RT/u = 0.35$, $x_1 = 0.226$ (a maximum) and 0.774 (a minimum).

24-41. Plot $\Delta_{\text{mix}}\overline{G}/u$ in Problem 24–37 versus x_1 for $RT/u = 0.60, 0.50, 0.45, 0.40$, and 0.35. Note that some of the curves have regions where $\partial^2 \Delta_{\text{mix}}\overline{G}/\partial x_1^2 < 0$. These regions correspond to regions in which the two liquids are not miscible. Show that $RT/u = 0.50$ is a critical value, in the sense that unstable regions occur only when $RT/u < 0.50$. (See the previous problem.)

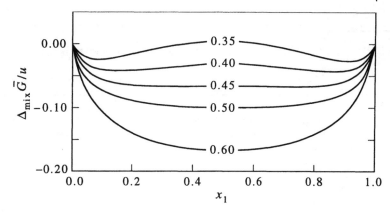

Differentiate $\Delta_{mix}\overline{G}$ with respect to x_1 to obtain

$$\frac{\partial \Delta_{mix}\overline{G}}{\partial x_1} = \frac{RT}{u}[\ln x_1 + 1 - \ln(1 - x_1) - 1] + 1 - 2x_1 = 0$$

or

$$\frac{RT}{u}\ln\frac{x_1}{1 - x_1} = 2x_1 - 1$$

Note that both sides of this equation equal zero when $x_1 = 1/2$. The unstable regions occur when $\partial^2 \Delta_{mix}\overline{G}/\partial x_1^2 < 0$:

$$\frac{\partial^2 \Delta_{mix}\overline{G}}{\partial x_1^2} = \frac{RT}{u}\left(\frac{1}{x_1} + \frac{1}{1 - x_1}\right) - 2$$

$$= \frac{RT}{u}\left[\frac{1}{x_1(1 - x_1)}\right] - 2$$

The unstable regions are centered at $x_1 = x_2 = 1/2$, so substituting $x_1 = x_2 = 1/2$ into $\partial^2 \Delta_{mix}\overline{G}/\partial x_1^2$ gives the inequality

$$\frac{4RT}{u} - 2 < 0$$

or

$$\frac{RT}{u} < \frac{1}{2}$$

24–42. Plot both $P_1/P_1^* = x_1 e^{ux_2^2/RT}$ and $P_2/P_2^* = x_2 e^{ux_1^2/RT}$ for $RT/u = 0.60$, 0.50, 0.45, 0.40, and 0.35. Prove that the loops occur for values of $RT/u < 0.50$.

See Problem 24–40 for proof that unstable regions occur only for $RT/u < 0.50$.

24–43. Plot both $P_1/P_1^* = x_1 e^{ux_2^2/RT}$ and $P_2/P_2^* = x_2 e^{ux_1^2/RT}$ against x_1 for $RT/u = 0.40$. The loops indicate regions in which the two liquids are not miscible, as explained in Problem 24–39. Draw a horizontal line connecting the left-side and the right-side intersections of the two curves. This line, which connects states in which the vapor pressure (or chemical potential) of each component is the same in the two solutions of different composition, corresponds to one of the horizontal lines in Figure 24.12. Now set $P_1/P_1^* = x_1 e^{ux_2^2/RT}$ equal to $P_2/P_2^* = x_2 e^{ux_1^2/RT}$ and solve for RT/u in terms of x_1. Plot RT/u against x_1 and obtain a coexistence curve like the one in Figure 24.13.

The plot of P_1/P_1^* and P_2/P_2^* against x_1 for $RT/u = 0.40$ is shown below.

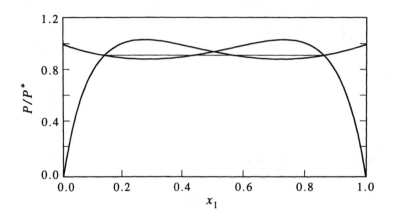

Write

$$x_1 e^{ux_2^2/RT} = x_2 e^{ux_1^2/RT}$$

as

$$e^{u(x_2^2-x_1^2)/RT} = \frac{x_2}{x_1}$$

and take logarithms to get

$$\frac{RT}{u} = \frac{x_2^2 - x_1^2}{\ln(x_2/x_1)} = \frac{1 - 2x_1}{\ln[(1 - x_1)/x_1]}$$

A plot of RT/u against x_1 follows.

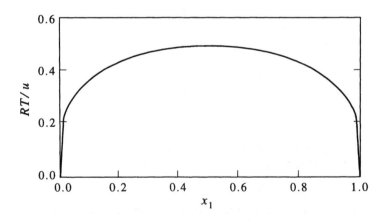

24-44. The molar enthalpies of mixing of solutions of tetrachloromethane (1) and cyclohexane (2) at 25°C are listed below.

x_1	$\Delta_{mix}\overline{H}/\text{J}\cdot\text{mol}^{-1}$
0.0657	37.8
0.2335	107.9
0.3495	134.9
0.4745	146.7
0.5955	141.6
0.7213	118.6
0.8529	73.6

Plot $\Delta_{mix}\overline{H}$ against $x_1 x_2$ according to Problem 24–37. Do tetrachloromethane and cyclohexane form a regular solution?

If tetrachloromethane and cyclohexane form a regular solution at 25°C, then a plot of $\Delta_{mix}\overline{H}/x_2$ against x_1 should be linear. The linearity of the following plot shows that they form a regular solution.

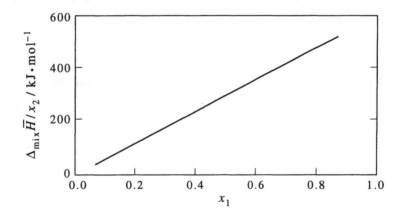

24-45. The molar enthalpies of mixing of solutions of tetrahydrofuran and trichloromethane at 25°C are listed below.

x_{THF}	$\Delta_{mix}\overline{H}/\text{J}\cdot\text{mol}^{-1}$
0.0568	−0.469
0.1802	−1.374
0.3301	−2.118
0.4508	−2.398
0.5702	−2.383
0.7432	−1.888
0.8231	−1.465
0.9162	−0.802

Do tetrahydrofuran and trichloromethane form a regular solution?

If tetrahydrofuran and trichloromethane form a regular solution at 25°C, then a plot of $\Delta_{mix}\overline{H}/x_2$ against x_1 should be linear. The nonlinearity of the following plot shows that they do not quite form a regular solution.

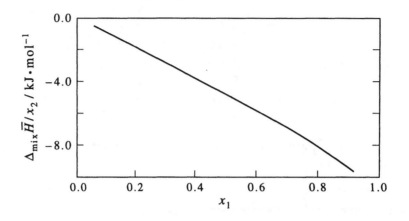

24–46. Derive the equation

$$x_1 d\ln\gamma_1 + x_2 d\ln\gamma_2 = 0$$

by starting with Equation 24.11. Use this equation to obtain the same result as in Example 24–8.

Equation 24.11 is

$$x_1 d\mu_1 + x_2 d\mu_2 = 0$$

Substitute $\mu_j = \mu_j^* + RT\ln\gamma_j x_j$ for μ_1 and μ_2 to obtain

$$x_1 RT\left(\frac{dx_1}{x_1} + d\ln\gamma_1\right) + x_2 RT\left(\frac{dx_2}{x_2} + d\ln\gamma_2\right) = 0$$

or

$$RT(dx_1 + dx_2) + RT x_1 d\ln\gamma_1 + RT x_2 d\ln\gamma_2 = 0$$

But $dx_1 + dx_2 = 0$ because $x_1 + x_2 = 1$, and so we have

$$x_1 d\ln\gamma_1 + x_2 d\ln\gamma_2 = 0$$

According to Example 24–8,

$$\gamma_1 = e^{\alpha x_2^2}$$

Therefore,

$$d\ln\gamma_2 = -\frac{x_1}{x_2}d\ln\gamma_1 = -\frac{x_1}{x_2}(2\alpha x_2 dx_2)$$
$$= -2\alpha(1 - x_2)dx_2$$

Integration from $x_2 = 1$ (where $\gamma_2 = 1$) to arbitrary x_2 gives

$$\ln \gamma_2 = -2\alpha \int_1^{x_2} (1 - x_2')dx_2'$$

$$= -2\alpha \left(x_2 - 1 - \frac{x_2^2 - 1}{2} \right) = \alpha(1 - 2x_2 + x_2^2).$$

$$= \alpha(1 - x_2)^2 = \alpha x_1^2$$

in agreement with Example 24–8.

24–47. The vapor pressure data for carbon disulfide in Table 24.1 can be curve fit by

$$P_1 = x_1(514.5 \text{ torr})e^{1.4967x_2^2 - 0.68175x_2^3}$$

Using the results of Example 24–7, show that the vapor pressure of dimethoxymethane is given by

$$P_2 = x_2(587.7 \text{ torr})e^{0.4741x_1^2 + 0.68175x_1^3}$$

Now plot P_2 versus x_2 and compare the result with the data in Table 24.1. Plot $\overline{G}^E$ against x_1. Is the plot symmetric about a vertical line at $x_1 = 1/2$? Do carbon disulfide and dimethoxymethane form a regular solution at 35.2°C?

According to Example 24–7, if

$$P_1 = x_2 P_1^* e^{\alpha x_2^2 + \beta x_2^3}$$

then

$$P_2 = x_2 P_2^* e^{(\alpha + 3\beta/2)x_1^2 - \beta x_1^3}$$

Therefore, since $\alpha = 1.4967$ and $\beta = -0.68175$,

$$P_2 = x_2 P_2^* e^{0.4741x_1^2 + 0.68175x_1^3}$$

A comparison of P_2 from Table 24.1 with that calculated from the above equation is shown below. The solid curve is the calculated curve and the dots represent the experimental data. The agreement is very good.

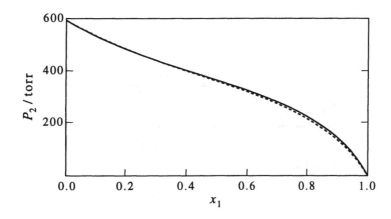

Now use Equation 24.51:

$$\overline{G}^E/RT = x_1 \ln \gamma_1 + x_2 \ln \gamma_2$$
$$= x_1(1.4967x_2^2 - 0.68175x_2^3 + 0.4741x_1^2 + 0.68175x_1^3)$$
$$= 0.8149x_1x_2(1 + 0.4183x_1)$$

The following plot shows that $\overline{G}^E$ is not symmetric about $x_1 = x_2 = 1/2$. Carbon disulfide and dimethoxymethane do not form a regular solution under the given conditions.

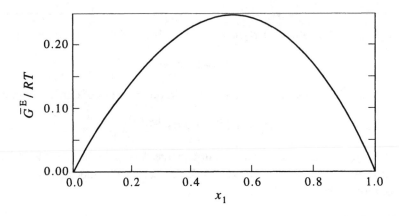

24–48. A mixture of trichloromethane and acetone with $x_{acet} = 0.713$ has a total vapor pressure of 220.5 torr at 28.2°C, and the mole fraction of acetone in the vapor is $y_{acet} = 0.818$. Given that the vapor pressure of pure trichloromethane at 28.2°C is 221.8 torr, calculate the activity and the activity coefficient (based upon a Raoult's law standard state) of trichloromethane in the mixture. Assume the vapor behaves ideally.

We have $x_{acet} = 0.713$, $y_{acet} = 0.818$, and $P_{total} = 220.5$ torr. Therefore,

$$P_{tri} = (1.000 - 0.818)(220.5 \text{ torr}) = 40.13 \text{ torr}$$

and

$$a_{tri}^{(R)} = \frac{P_{tri}}{P_{tri}^*} = \frac{40.13 \text{ torr}}{221.8 \text{ torr}} = 0.181$$

and

$$\gamma_{tri}^{(R)} = \frac{a_{tri}^{(R)}}{x_{tri}} = \frac{0.181}{1.000 - 0.713} = 0.631$$

24–49. Consider a binary solution for which the vapor pressure (in torr) of one of the components (say component 1) is given empirically by

$$P_1 = 78.8x_1 e^{0.65x_2^2 + 0.18x_2^3}$$

Calculate the activity and the activity coefficient of component 1 when $x_1 = 0.25$ based on a solvent and a solute standard state.

$$a_1^{(R)} = \frac{P_1}{P_1^*} = x_1 e^{0.65x_2^2 + 0.18x_2^3}$$

When $x_1 = 0.25$, $a_1^{(R)} = 0.25e^{0.4416} = 0.39$ and $\gamma_1^{(R)} = a_1^{(R)}/0.25 = 1.6$. The activity based upon a Henry's law standard state is given by

$$a_1^{(H)} = \frac{P_1}{k_{H,1}} = \frac{x_1 P_1^* e^{0.65x_2^2 + 0.18x_2^3}}{P_1^* e^{0.65 + 0.18}} = \frac{0.39}{2.29} = 0.17$$

and $\gamma_1^{(H)} = 0.17/0.25 = 0.68$

24–50. Some vapor pressure data for ethanol/water solutions at 25°C are listed below.

$x_{ethanol}$	$P_{ethanol}$/torr	P_{water}/torr
0.00	0.00	23.78
0.02	4.28	23.31
0.05	9.96	22.67
0.08	14.84	22.07
0.10	17.65	21.70
0.20	27.02	20.25
0.30	31.23	19.34
0.40	33.93	18.50
0.50	36.86	17.29
0.60	40.23	15.53
0.70	43.94	13.16
0.80	48.24	9.89
0.90	53.45	5.38
0.93	55.14	3.83
0.96	56.87	2.23
0.98	58.02	1.13
1.00	59.20	0.00

Plot these data to determine the Henry's law constant for ethanol in water and for water in ethanol at 25°C.

Henry's law constant of component j is given by the limiting slope of the vapor pressure of component j as $x_j \to 0$. The straight lines are shown in the following figure. The slopes of these lines give $k_{H,water} \approx 20 \text{ torr}/0.35 = 57 \text{ torr}$ and $k_{H,eth} \approx 25 \text{ torr}/0.10 = 250 \text{ torr}$

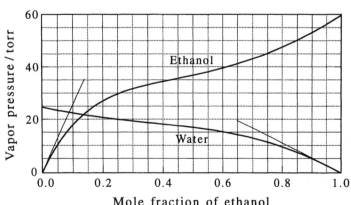

24–51. Using the data in Problem 24–50, plot the activity coefficients (based upon Raoult's law) of both ethanol and water against the mole fraction of ethanol.

The activity coefficients based upon Raoult's law are given by $\gamma_j = P_j/x_j P_j^*$, where all these quantities are given in the problem. The activities are shown in the following figure.

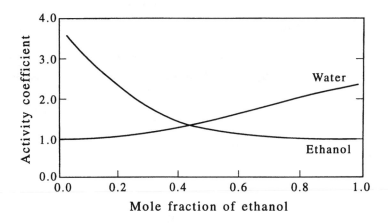

24–52. Using the data in Problem 24–50, plot $\overline{G}^{\mathrm{E}}/RT$ against $x_{\mathrm{H_2O}}$. Is a water/ethanol solution at 25°C a regular solution?

According to Equation 24.51,

$$\frac{\overline{G}^{\mathrm{E}}}{RT} = x_1 \ln \gamma_1 + x_2 \ln \gamma_2$$

The activity coefficients are calculated in Problem 24–51, and $\overline{G}^{\mathrm{E}}/RT$ is plotted against the mole fraction of water below. The plot is not symmetric about $x_1 = x_2 = 1/2$, and so water and ethanol do not form a regular solution under these conditions.

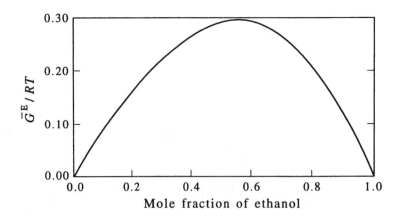

24–53. Some vapor pressure data for a 2-propanol/benzene solution at 25°C are

$x_{\text{2-propanol}}$	$P_{\text{2-propanol}}$/torr	P_{total}/torr
0.000	0.0	94.4
0.059	12.9	104.5
0.146	22.4	109.0
0.362	27.6	108.4
0.521	30.4	105.8
0.700	36.4	99.8
0.836	39.5	84.0
0.924	42.2	66.4
1.000	44.0	44.0

Plot the activities and the activity coefficients of 2-propanol and benzene relative to a Raoult's law standard state versus the mole fraction of 2-propanol.

The activity coefficients based upon Raoult's law are given by $\gamma_j = P_j / x_j P_j^*$, where all these quantities are given above. The activities and activity coefficients are shown in the following figures.

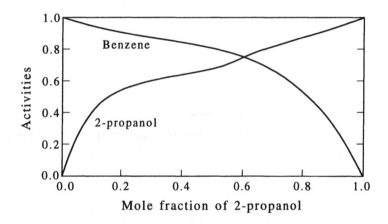

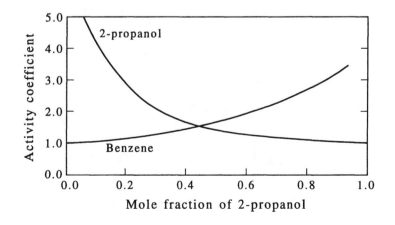

24–54. Using the data in Problem 24–53, plot $\overline{G}^E/RT$ versus $x_{\text{2-propanol}}$.

According to Equation 24.51,

$$\frac{\overline{G}^E}{RT} = x_1 \ln \gamma_1 + x_2 \ln \gamma_2$$

The activity coefficients are calculated in Problem 24–53, and $\overline{G}^E/RT$ is plotted against the mole fraction of 2-propanol below. The plot is not symmetric about $x_1 = x_2 = 1/2$, and so 2-propanol and benzene do not form a regular solution under these conditions.

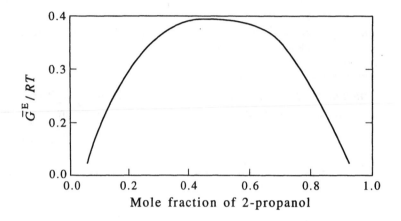

24–55. *Excess thermodynamic quantities* are defined relative to the values the quantities would have if the pure components formed an ideal solution at the same given temperature and pressure. For example, we saw that (Equation 24.51)

$$\frac{\overline{G}^E}{RT} = \frac{G^E}{(n_1 + n_2)RT} = x_1 \ln \gamma_1 + x_2 \ln \gamma_2$$

Show that

$$\frac{\overline{S}^E}{R} = \frac{\Delta S^E}{(n_1 + n_2)R} = -(x_1 \ln \gamma_1 + x_2 \ln \gamma_2)$$

$$- T\left(x_1 \frac{\partial \ln \gamma_1}{\partial T} + x_2 \frac{\partial \ln \gamma_2}{\partial T}\right)$$

Use the relation $S = -(\partial G/\partial T)_{P,n_1,n_2}$ to write

$$\overline{S}^E = -\left(\frac{\partial \overline{G}^E}{\partial T}\right)_{P,x_1} = -\frac{\partial}{\partial T}[RT(x_1 \ln \gamma_1 + x_2 \ln \gamma_2)]$$

$$= -R(x_1 \ln \gamma_1 + x_2 \ln \gamma_2) - RT\left(x_1 \frac{\partial \ln \gamma_1}{\partial T} + x_2 \frac{\partial \ln \gamma_2}{\partial T}\right)_P$$

24–56. Show that

$$\overline{G}^{E} = \frac{G^{E}}{n_1 + n_2} = ux_1x_2$$

$$\overline{S}^{E} = \frac{S^{E}}{n_1 + n_2} = 0$$

and

$$\overline{H}^{E} = \frac{H^{E}}{n_1 + n_2} = ux_1x_2$$

for a regular solution (see Problem 24–37).

A regular solution is defined in Problem 24–37. For example,

$$\Delta_{mix}\overline{G} = RT(x_1 \ln x_1 + x_2 \ln x_2) + ux_1x_2$$

But the first terms on the right side of this equation are $\Delta_{mix}\overline{G}^{id}$, and so

$$\overline{G}^{E} = \Delta_{mix}\overline{G} - \Delta_{mix}\overline{G}^{id} = ux_1x_2$$

Similarly,

$$\overline{S}^{E} = \Delta_{mix}\overline{S} - \Delta_{mix}\overline{S}^{id} = -R(x_1 \ln x_1 + x_2 \ln x_2) + R(x_1 \ln x_1 + x_2 \ln x_2)$$
$$= 0$$

and $\overline{G}^{E} = \overline{H}^{E} - T\overline{S}^{E}$ gives

$$\overline{H}^{E} = ux_1x_2$$

24–57. Example 24–7 expresses the vapor pressures of the two components of a binary solution as

$$P_1 = x_1 P_1^* e^{\alpha x_2^2 + \beta x_2^3}$$

and

$$P_2 = x_2 P_2^* e^{(\alpha + 3\beta/2)x_1^2 - \beta x_1^3}$$

Show that these expressions are equivalent to

$$\gamma_1 = e^{\alpha x_2^2 + \beta x_2^3} \quad \text{and} \quad \gamma_2 = e^{(\alpha + 3\beta/2)x_1^2 - \beta x_1^3}$$

Using these expressions for the activity coefficients, derive an expression for $\overline{G}^{E}$ in terms of α and β. Show that your expression reduces to that for $\overline{G}^{E}$ for a regular solution.

Start with Equation 24.51

$$\overline{G}^{E}/RT = x_1 \ln \gamma_1 + x_2 \ln \gamma_2$$

and $\gamma_j = P_j / x_j P_j^*$. For P_1 and P_2 given in the problem

$$\gamma_1 = \frac{P_1}{x_1 P_1^*} = e^{\alpha x_2^2 + \beta x_2^3} \quad \text{and} \quad \gamma_2 = \frac{P_2}{x_1 P_2^*} = e^{(\alpha + 3\beta/2)x_1^2 - \beta x_1^3}$$

and so

$$\overline{G}^E / RT = x_1(\alpha x_2^2 + \beta x_2^3) + x_2\left[\left(\alpha + \frac{3\beta}{2}\right)x_1^2 - \beta x_1^3\right]$$

$$= x_1 x_2\left(\alpha x_2 + \beta x_2^2 + \alpha x_1 + \frac{3\beta}{2}x_1 - \beta x_1^2\right)$$

$$= x_1 x_2\left[\alpha(x_1 + x_2) + \frac{3\beta}{2}x_1 + \beta(x_2^2 - x_1^2)\right]$$

$$= x_1 x_2\left[\alpha + \frac{3\beta}{2}x_1 + \beta(x_2 - x_1)(x_2 + x_1)\right]$$

$$= x_1 x_2\left[\alpha + \beta\left(1 - \frac{x_1}{2}\right)\right]$$

This expression reduces to that of a regular solution when $\beta = 0$.

24-58. Prove that the maxima or minima of $\Delta_{\text{mix}}\overline{G}$ defined in Problem 24–37 occur at $x_1 = x_2 = 1/2$ for any value of RT/u. Now prove that

$$\frac{\partial^2 \Delta_{\text{mix}}\overline{G}}{\partial x_1^2} \begin{cases} > 0 & \text{for } RT/u > 0.50 \\ = 0 & \text{for } RT/u = 0.50 \\ < 0 & \text{for } RT/u < 0.50 \end{cases}$$

at $x_1 = x_2 = 1/2$. Is this result consistent with the graphs you obtained in Problem 24–41?

Start with

$$\frac{\Delta_{\text{mix}}\overline{G}}{u} = \frac{RT}{u}(x_1 \ln x_1 + x_2 \ln x_2) + x_1 x_2$$

The maxima or minima are given by

$$\left(\frac{\partial \Delta_{\text{mix}}\overline{G}/u}{\partial x_1}\right) = \frac{RT}{u}[\ln x_1 + 1 - \ln(1 - x_2) - 1] + 1 - 2x_1 = 0$$

or by

$$\frac{RT}{u}\ln\frac{x_1}{1 - x_1} = 1 - 2x_1$$

Note that this equation is satisfied by $x_1 = x_2 = 1/2$ for any value of RT/u.

$$\left(\frac{\partial^2 \Delta_{\text{mix}}\overline{G}/u}{\partial x_1^2}\right) = \frac{RT}{u}\left(\frac{1}{x_1} + \frac{1}{1 - x_1}\right) - 2$$

$$= \frac{RT}{u}\left[\frac{1}{x_1(1 - x_1)}\right] - 2$$

$$= \frac{4RT}{u} - 2$$

at $x_1 = x_2 = 1/2$. This expression is greater than zero when $RT/u > 0.50$, less than zero when $RT/u < 0.50$, and equal to zero when $RT/w = 0.50$.

24–59. Use the data in Table 24.1 to plot Figures 24.15 through 24.17.

Use the relations $a_j^{(R)} = P_j/P_j^*$ and $\gamma_j^{(R)} = P_j/x_j P_j^*$. The results of the calculations are given below.

x_1	$P_{CS_2}/$torr	$P_{dimeth}/$torr	$a_{CS_2}^{(R)}$	$a_{dimeth}^{(R)}$	$\gamma_{CS_2}^{(R)}$	$\gamma_{dimeth}^{(R)}$	$\Delta G^E/kJ \cdot mol^{-1}$
0.0000	0.0	587.7	0.000	1.000	2.22	1.00	0.000
0.0489	54.5	558.3	0.106	0.950	2.17	1.00	0.037
0.1030	109.3	529.1	0.212	0.900	2.06	1.00	0.078
0.1640	159.5	500.4	0.310	0.851	1.89	1.02	0.120
0.2710	234.8	451.2	0.456	0.768	1.68	1.05	0.179
0.3470	277.6	412.7	0.540	0.706	1.55	1.08	0.204
0.4536	324.8	378.0	0.631	0.643	1.39	1.18	0.239
0.4946	340.2	360.8	0.661	0.614	1.34	1.21	0.242
0.5393	357.2	342.2	0.694	0.582	1.29	1.26	0.244
0.6071	381.9	313.3	0.742	0.533	1.22	1.36	0.242
0.6827	407.0	277.8	0.791	0.473	1.16	1.49	0.227
0.7377	424.3	250.1	0.825	0.426	1.12	1.62	0.209
0.7950	442.6	217.4	0.860	0.370	1.08	1.80	0.184
0.8445	458.1	184.9	0.890	0.315	1.05	2.02	0.154
0.9108	481.8	124.2	0.936	0.211	1.03	2.37	0.102
0.9554	501.0	65.1	0.974	0.111	1.02	2.48	0.059
1.0000	514.5	0.0	1.000	0.000	1.00	2.50	0.000

CHAPTER 25

Solutions II
Solid-Liquid Solutions

PROBLEMS AND SOLUTIONS

25–1. The density of a glycerol/water solution that is 40.0% glycerol by mass is 1.101 $g \cdot mL^{-1}$ at 20°C. Calculate the molality and the molarity of glycerol in the solution at 20°C. Calculate the molality at 0°C.

The mass of glycerol per millimeter of solution is

$$\text{g glycerol per mL} = (0.400)(1.101 \text{ g} \cdot \text{mL}^{-1}) = 0.4404 \text{ g} \cdot \text{mL}^{-1}$$

The number of moles of glycerol per liter of solution is

$$\text{molarity} = \frac{440.4 \text{ g} \cdot \text{L}^{-1}}{92.093 \text{ g} \cdot \text{mol}^{-1}} = 4.78 \text{ mol} \cdot \text{L}^{-1}$$

The number of grams of water per 0.4404 grams of glycerol is given by

$$1.101 \text{ g} - 0.4404 \text{ g} = 0.6606 \text{ g H}_2\text{O}$$

or 0.4404 g glycerol per 0.6606 g H_2O, or 0.6666 g glycerol per g H_2O. Therefore,

$$\text{molality} = \frac{666.6 \text{ g} \cdot \text{kg}^{-1}}{92.094 \text{ g} \cdot \text{mol}^{-1}} = 7.24 \text{ mol} \cdot \text{kg}^{-1}$$

25–2. Concentrated sulfuric acid is sold as a solution that is 98.0% sulfuric acid and 2.0% water by mass. Given that the density is 1.84 $g \cdot mL^{-1}$, calculate the molarity of concentrated sulfuric acid.

$$\text{g H}_2\text{SO}_4 \text{ per mL solution} = (0.980)(1.84 \text{ g} \cdot \text{mL}^{-1}) = 1.80 \text{ g} \cdot \text{mL}^{-1})$$

$$\text{molarity} = \frac{1800 \text{ g} \cdot \text{L}^{-1}}{98.08 \text{ g} \cdot \text{mol}^{-1}} = 18.4 \text{ mol} \cdot \text{L}^{-1}$$

25–3. Concentrated phosphoric acid is sold as a solution that is 85% phosphoric acid and 15% water by mass. Given that the molarity is 15 M, calculate the density of concentrated phosphoric acid.

A 15 molar solution implies that there are

$$(15 \text{ mol} \cdot \text{L}^{-1})(97.998 \text{ g} \cdot \text{mol}^{-1}) = 1470 \text{ g of phosphoric acid per liter of solution}$$

Therefore, the density of the solution is

$$\text{density} = \frac{1470 \text{ g} \cdot \text{L}^{-1}}{0.85} = 1700 \text{ g} \cdot \text{L}^{-1} = 1.7 \text{ g} \cdot \text{mL}^{-1}$$

25–4. Calculate the mole fraction of glucose in an aqueous solution that is $0.500 \text{ mol} \cdot \text{kg}^{-1}$ glucose.

There are 0.500 mol glucose per kg H_2O. so

$$x_2 = \frac{0.500 \text{ mol}}{0.500 \text{ mol} + \dfrac{1000 \text{ g } H_2O}{18.02 \text{ g} \cdot \text{mol}^{-1} \, H_2O}} = 0.00893$$

25–5. Show that the relation between molarity and molality for a solution with a single solute is

$$c = \frac{(1000 \text{ mL} \cdot \text{L}^{-1})\rho m}{1000 \text{ g} \cdot \text{kg}^{-1} + mM_2}$$

where c is the molarity, m is the molality, ρ is the density of the solution in $\text{g} \cdot \text{mL}^{-1}$, and M_2 is the molar mass $(\text{g} \cdot \text{mol}^{-1})$ of the solute.

Consider a solution of a certain molality, m, containing 1000 g of solvent. The total mass of the solution is $1000 \text{ g} \cdot \text{kg}^{-1} + mM_2$ and its volume (in mL) is $(1000 \text{ g} \cdot \text{kg}^{-1} + mM_2)/\rho$, where ρ is the density of the the solution in $\text{g} \cdot \text{mL}^{-1}$. The volume of the solution in liters is $(1000 \text{ g} \cdot \text{kg}^{-1} + mM_2)/\rho(1000 \text{ mL} \cdot \text{L}^{-1})$ liters. There are m moles of solute per $(1000 \text{ g} \cdot \text{kg}^{-1} + mM_2)/\rho(1000 \text{ mL} \cdot \text{L}^{-1})$ liters, so the molarity is

$$c = \frac{(1000 \text{ mL} \cdot \text{L}^{-1})\rho m}{1000 \text{ g} \cdot \text{kg}^{-1} + mM_2}$$

25–6. The *CRC Handbook of Chemistry and Physics* has tables of "concentrative properties of aqueous solutions" for many solutions. Some entries for $CsCl(s)$ are

$A/\%$	$\rho/\text{g} \cdot \text{mL}^{-1}$	$c/\text{mol} \cdot \text{L}^{-1}$
1.00	1.0058	0.060
5.00	1.0374	0.308
10.00	1.0798	0.641
20.00	1.1756	1.396
40.00	1.4226	3.380

where A is the mass percent of the solute, ρ is the density of the solution, and c is the molarity. Using these data, calculate the molality at each concentration.

Using the result of the previous problem,

$$m = \frac{(1000 \text{ g} \cdot \text{kg}^{-1})c}{(1000 \text{ mL} \cdot \text{L}^{-1})\rho - M_2 c}$$

We have then ($M_2 = 168.36 \text{ g} \cdot \text{mol}^{-1}$)

$c/\text{mol} \cdot \text{L}^{-1}$	$m/\text{mol} \cdot \text{kg}^{-1}$
0.060	0.060
0.308	0.313
0.641	0.660
1.396	1.484
3.380	3.960

25–7. Derive a relation between the mass percentage (A) of a solute in a solution and its molality (m). Calculate the molality of an aqueous sucrose solution that is 18% sucrose ($C_{12}H_{22}O_{11}$) by mass.

Mass percentage of solute, A_2, is given by

$$A_2 = \frac{\text{mass}_2}{\text{mass}_1 + \text{mass}_2} \times 100$$

If we take a solution containing 1000 g of solvent, then $\text{mass}_2 = mM_2$ and $\text{mass}_1 = 1000 \text{ g} \cdot \text{kg}^{-1}$, so

$$A_2 = \frac{mM_2}{1000 \text{ g} \cdot \text{kg}^{-1} + mM_2} \times 100$$

Solve for m to get

$$m = \frac{(1000 \text{ g} \cdot \text{kg}^{-1})A_2}{(100 - A_2)M_2}$$

For an aqueous sucrose solution that is 18% sucrose by mass,

$$m = \frac{(1000 \text{ g} \cdot \text{kg}^{-1})(18)}{(100 - 18)(342.3 \text{ g} \cdot \text{mol}^{-1})} = 0.73 \text{ mol} \cdot \text{kg}^{-1}$$

25–8. Derive a relation between the mole fraction of the solvent and the molality of a solution.

Start with

$$x_2 = \frac{n_2}{n_1 + n_2}$$

Now take a solution containing 1000 g of solvent, so that $n_2 = m$ and $n_1 = (1000 \text{ g})/M_1$, where M_1 is the molar mass of the solvent. Therefore,

$$x_2 = \frac{m}{\dfrac{1000 \text{ g} \cdot \text{kg}^{-1}}{M_1} + m} = \frac{mM_1}{1000 \text{ g} \cdot \text{kg}^{-1} + mM_1}$$

and

$$x_1 = 1 - x_2 = \frac{1000 \text{ g} \cdot \text{kg}^{-1}}{1000 \text{ g} \cdot \text{kg}^{-1} + mM_1}$$

25–9. The volume of an aqueous sodium chloride solution at 25°C can be expressed as

$$V/\text{mL} = 1001.70 + (17.298 \text{ kg} \cdot \text{mol}^{-1})m + (0.9777 \text{ kg}^2 \cdot \text{mol}^{-2})m^2$$

$$- (0.0569 \text{ kg}^3 \cdot \text{mol}^{-3})m^3$$

$$0 \leq m \leq 6 \text{ mol} \cdot \text{kg}^{-1}$$

where m is the molality. Calculate the molarity of a solution that is 3.00 molal in sodium chloride.

The volume of the solution at a 3.00 mol·kg^{-1} concentration is

$$V/\text{mL} = 1060.86$$

The mass of a 3.00 mol·kg^{-1} NaCl(aq) solution that contains 1000 g of solvent is

$$\text{mass} = 1000 \text{ g} \cdot \text{kg}^{-1} + (3.00 \text{ mol} \cdot \text{kg}^{-1})(58.444 \text{ g} \cdot \text{mol}^{-1})$$
$$= 1175.33 \text{ g}$$

The density of the solution is

$$\rho^{\text{sln}} = \frac{1175.33 \text{ g}}{1060.86 \text{ mL}} = 1.108 \text{ g} \cdot \text{mL}^{-1}$$

and so the molarity is (see Problem 25–5)

$$c = \frac{(1000 \text{ mL} \cdot \text{L}^{-1})\rho m}{1000 \text{ g} \cdot \text{kg}^{-1} + mM_2}$$

$$= \frac{(1000 \text{ mL} \cdot \text{L}^{-1})(1.108 \text{ g} \cdot \text{mL}^{-1})(3.00 \text{ mol} \cdot \text{kg}^{-1})}{1000 \text{ g} \cdot \text{kg}^{-1} + (3.00 \text{ mol} \cdot \text{kg}^{-1})(58.444 \text{ g} \cdot \text{mol}^{-1})}$$

$$= 2.83 \text{ mol} \cdot \text{L}^{-1}$$

25–10. If x_2^∞, m^∞, and c^∞ are the mole fraction, molality, and molarity, respectively, of a solute at infinite dilution, show that

$$x_2^\infty = \frac{m^\infty M_1}{1000 \text{ g} \cdot \text{kg}^{-1}} = \frac{c^\infty M_1}{(1000 \text{ mL} \cdot \text{L}^{-1})\rho_1}$$

where M_1 is the molar mass (g·mol^{-1}) and ρ_1 is the density (g·mL^{-1}) of the solvent. Note that mole fraction, molality, and molarity are all directly proportional to each other at low concentrations.

Start with $x_2 = n_2/(n_1 + n_2)$. At infinite dilution, $n_2 \to 0$, and so

$$x_2 = \frac{n_2}{n_1 + n_2} \longrightarrow \frac{n_2}{n_1} \quad \text{as} \quad n_2 \longrightarrow 0$$

Consider a solution containing 1000 g of solvent. In this case, $n_2 = m$ and $n_1 = (1000 \text{ g} \cdot \text{kg}^{-1})/M_1$, where M_1 is the molar mass of the solvent. Then

$$x_2^\infty = \frac{m^\infty}{\dfrac{1000 \text{ g} \cdot \text{kg}^{-1}}{M_1}} = \frac{m^\infty M_1}{1000 \text{ g} \cdot \text{kg}^{-1}}$$

According to Problem 25–5, $c^\infty \to (\text{mL} \cdot \text{L}^{-1})\rho m^\infty/(\text{g} \cdot \text{kg}^{-1})$, so

$$x_2^\infty = \frac{c^\infty}{\left(\dfrac{1000 \text{ mL} \cdot \text{L}^{-1}}{M_1}\right)\rho} = \frac{c^\infty M_1}{(1000 \text{ mL} \cdot \text{L}^{-1})\rho}$$

25–11. Consider two solutions whose solute activities are a_2' and a_2'', referred to the same standard state. Show that the difference in the chemical potentials of these two solutions is independent of the standard state and depends only upon the ratio a_2'/a_2''. Now choose one of these solutions to be at an arbitrary concentration and the other at a very dilute concentration (essentially infinitely dilute) and argue that

$$\frac{a_2'}{a_2''} = \frac{\gamma_{2x} x_2}{x_2^\infty} = \frac{\gamma_{2m} m}{m^\infty} = \frac{\gamma_{2c} c}{c^\infty}$$

Let

$$\mu_2' = (\mu_2^\circ)' + RT \ln a_2'$$
$$\mu_2'' = (\mu_2^\circ)'' + RT \ln a_2'' = (\mu_2^\circ)' + RT \ln a_2''$$

Therefore,

$$\Delta\mu = \mu_2' - \mu_2'' = RT \ln \frac{a_2'}{a_2''}$$

If the solution denoted by the double prime is very dilute, then $a_2'' = x_2^\infty$, m^∞, or c^∞. Therefore,

$$\frac{a_2'}{a_2''} = \frac{\gamma_{2x} x_2}{x_2^\infty} = \frac{\gamma_{2m} m}{m^\infty} = \frac{\gamma_{2c} c}{c^\infty}$$

25–12. Use Equations 25.4, 25.11, and the results of the previous two problems to show that

$$\gamma_{2x} = \gamma_{2m}\left(1 + \frac{mM_1}{1000 \text{ g} \cdot \text{kg}^{-1}}\right) = \gamma_{2c}\left(\frac{\rho}{\rho_1} + \frac{c[M_1 - M_2]}{\rho_1[1000 \text{ mL} \cdot \text{L}^{-1}]}\right)$$

where ρ is the density of the solution. Thus, we see that the three different activity coefficients are related to one another.

Using the result of the previous problem,

$$\gamma_{2x} = \frac{x_2^\infty}{m^\infty} \frac{m}{x_2} \gamma_{2m}$$

Using the result of Equation 25.4 and Problem 25–10, we have

$$\gamma_{2x} = \gamma_{2m}\left(\frac{M_1}{1000 \text{ g} \cdot \text{kg}^{-1}}\right)\left(\frac{1000 \text{ g} \cdot \text{kg}^{-1}}{M_1} + m\right) = \gamma_{2m}\left(1 + \frac{mM_1}{1000 \text{ g} \cdot \text{kg}^{-1}}\right)$$

Similarly, Problem 25–11 gives us

$$\gamma_{2x} = \gamma_{2c} \frac{x_2^\infty}{c^\infty} \frac{c}{x_2}$$

Using Equation 25.11 and the result of Problem 25–11, we have

$$\gamma_{2x} = \gamma_{2c} \left[\frac{M_1}{(1000 \text{ mL} \cdot \text{L}^{-1})\rho_1} \right] \left[\frac{(1000 \text{ mL} \cdot \text{L}^{-1})\rho + c(M_1 - M_2)}{M_1} \right]$$

$$= \gamma_{2c} \left[\frac{\rho}{\rho_1} + \frac{c(M_1 - M_2)}{(1000 \text{ mL} \cdot \text{L}^{-1})\rho_1} \right]$$

25–13. Use Equations 25.4, 25.11, and the results of Problem 25–12 to derive

$$\gamma_{2m} = \gamma_{2c} \left(\frac{\rho}{\rho_1} - \frac{cM_2}{\rho_1 [1000 \text{ mL} \cdot \text{L}^{-1}]} \right)$$

Given that the density of an aqueous citric acid ($M_2 = 192.12 \text{ g} \cdot \text{mol}^{-1}$) solution at 20°C is given by

$$\rho/\text{g} \cdot \text{mL}^{-1} = 0.99823 + (0.077102 \text{ L} \cdot \text{mol}^{-1})c$$

$$0 \leq c < 1.772 \text{ mol} \cdot \text{L}^{-1}$$

plot γ_{2m}/γ_{2c} versus c. Up to what concentration do γ_{2m} and γ_{2c} differ by 2%?

From Problem 25–11,

$$\gamma_{2m} = \gamma_{2c} \frac{m^\infty}{c^\infty} \frac{c}{m}$$

Using the results from Problems 25–5 and 25–10,

$$\gamma_{2m} = \gamma_{2c} \frac{1}{\rho_1} \left[\frac{(1000 \text{ mL} \cdot \text{L}^{-1})\rho - cM_2}{1000 \text{ mL} \cdot \text{L}^{-1}} \right] = \gamma_{2c} \left[\frac{\rho}{\rho_1} - \frac{cM_2}{\rho_1 (1000 \text{ mL} \cdot \text{L}^{-1})} \right]$$

The ratio γ_{2m}/γ_{2c} is plotted below.

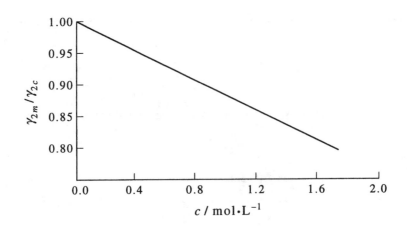

25–14. The *CRC Handbook of Chemistry and Physics* gives a table of mass percent of sucrose in an aqueous solution and its corresponding molarity at 25°C. Use these data to plot molality versus molarity for an aqueous sucrose solution.

Use the relation between mass percentage and molality that is derived in Problem 25–7 to calculate the molality at each mass percentage. Some representative values of A, c, and m and the plot of m against c are given below.

A	$c/\text{mol·L}^{-1}$	$m/\text{mol·kg}^{-1}$	A	$c/\text{mol·L}^{-1}$	$m/\text{mol·kg}^{-1}$
1.00	0.029	0.030	24.00	0.771	0.923
2.00	0.059	0.060	28.00	0.914	1.136
3.00	0.089	0.090	32.00	1.063	1.375
4.00	0.118	0.122	36.00	1.216	1.643
5.00	0.149	0.154	40.00	1.375	1.948
6.00	0.179	0.186	44.00	1.539	2.295
7.00	0.210	0.220	48.00	1.709	2.697
8.00	0.241	0.254	52.00	1.885	3.165
9.00	0.272	0.289	56.00	2.067	3.718
10.00	0.303	0.325	60.00	2.255	4.382
12.00	0.367	0.398	64.00	2.450	5.194
14.00	0.431	0.476	68.00	2.652	6.208
16.00	0.497	0.556	72.00	2.860	7.512
18.00	0.564	0.641	76.00	3.076	9.251
20.00	0.632	0.730	80.00	3.299	11.686

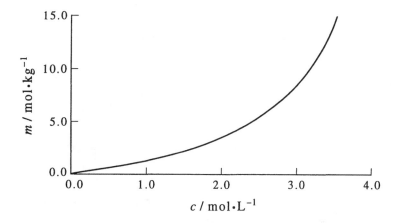

25–15. Using the data in Table 25.2, calculate the activity coefficient of water (on a mole fraction basis) at a sucrose concentration of 3.00 molal.

We use the equation $\gamma_1 = a_1/x_1$. The relation between molality and mole fraction is given by Equation 25.4:

$$x_1 = 1 - x_2 = \cfrac{1}{1 + \cfrac{mM_1}{1000 \text{ g·kg}^{-1}}} = \frac{1000 \text{ g·kg}^{-1}}{1000 \text{ g·kg}^{-1} + mM_1}$$

At $m = 3.00 \, \text{mol} \cdot \text{kg}^{-1}$ (with $M_1 = 18.02 \, \text{g} \cdot \text{mol}^{-1}$), we have $x_1 = 0.9487$. Therefore,

$$\gamma_1 = \frac{0.93276}{0.9487} = 0.983$$

25–16. Using the data in Table 25.2, plot the activity coefficient of water (on a mole fraction basis) against the mole fraction of water.

Calculate the mole fraction from the molality according to Problem 25–15, use the relation $\gamma_{1x} = a_1/x_1$, and plot the results to get

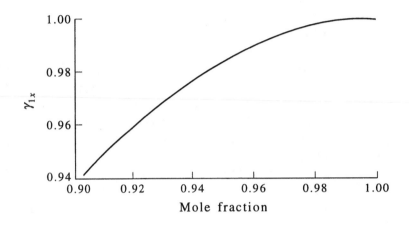

25–17. Using the data in Table 25.2, calculate ϕ at each value of m and reproduce Figure 25.2.

Use Equation 25.13,

$$\ln a_1 = -\frac{m\phi}{55.506 \, \text{mol} \cdot \text{kg}^{-1}}$$

25–18. Fit the data for the osmotic coefficient of sucrose in Table 25.2 to a fourth-degree polynomial and calculate γ_{2m} for a 1.00-molal solution. Compare your result with the one obtained in Example 25–5.

Suppressing the units of the coefficients, we get

$$\phi = 1 + 0.075329m + 0.016554m^2 - 0.0039647m^3 + 0.00024694m^4$$

Use Equation 25.15 to write

$$\ln \gamma_{2m} = \phi - 1 + \int_0^{1.00} \frac{\phi - 1}{m} dm = 0.08816 + 0.08235 = 0.1705$$

and so $\gamma_{2m} = 1.186$.

25–19. Using the data for sucrose given in Table 25.2, determine $\ln \gamma_{2m}$ at 3.00 molal by plotting $(\phi - 1)/m$ versus m and determining the area under the curve by numerical integration (Mathchapter G) rather than by curve fitting ϕ first. Compare your result with the value given in Table 25.2.

Using Kaleidagraph, we obtain

$$\int_0^{3.00} \frac{\phi - 1}{m} dm = 0.272$$

From Table 25.2, $\phi - 1 = 0.2879$, and so $\ln \gamma_{2m} = 0.560$, and $\gamma_{2m} = 1.75$.

25–20. Equation 25.18 can be used to determine the activity of the solvent at its freezing point. Assuming that ΔC_P^* is independent of temperature, show that

$$\Delta_{\text{fus}} \overline{H}(T) = \Delta_{\text{fus}} \overline{H}(T_{\text{fus}}^*) + \Delta \overline{C}_P^*(T - T_{\text{fus}}^*)$$

where $\Delta_{\text{fus}} \overline{H}(T_{\text{fus}}^*)$ is the molar enthalpy of fusion at the freezing point of the pure solvent (T_{fus}^*) and $\Delta \overline{C}_P^*$ is the difference in the molar heat capacities of liquid and solid solvent. Using Equation 25.18, show that

$$-\ln a_1 = \frac{\Delta_{\text{fus}} \overline{H}(T_{\text{fus}}^*)}{R(T_{\text{fus}}^*)^2} \theta + \frac{1}{R(T_{\text{fus}}^*)^2} \left(\frac{\Delta_{\text{fus}} \overline{H}(T_{\text{fus}}^*)}{T_{\text{fus}}^*} - \frac{\Delta \overline{C}_P^*}{2} \right) \theta^2 + \cdots$$

where $\theta = T_{\text{fus}}^* - T_{\text{fus}}$.

Use

$$\left(\frac{\partial \Delta_{\text{fus}} \overline{H}}{\partial T} \right)_P = \Delta \overline{C}_P^*$$

to derive

$$\Delta_{\text{fus}} \overline{H}(T) - \Delta_{\text{fus}} \overline{H}(T^*) = \Delta \overline{C}_P^*(T - T^*)$$

Using Equation 25.18

$$\begin{aligned}
\ln a_1 &= \int_{T_{\text{fus}}^*}^{T_{\text{fus}}} \frac{\Delta_{\text{fus}} \overline{H}(T)}{RT^2} dT \\
&= \frac{\Delta_{\text{fus}} \overline{H}(T_{\text{fus}}^*)}{R} \int_{T_{\text{fus}}^*}^{T_{\text{fus}}} \frac{dT}{T^2} + \frac{\Delta \overline{C}_P^*}{R} \int_{T_{\text{fus}}^*}^{T_{\text{fus}}} \frac{dT(T - T_{\text{fus}}^*)}{T^2} \\
&= \frac{\Delta_{\text{fus}} \overline{H}(T_{\text{fus}}^*)}{R} \left(-\frac{1}{T_{\text{fus}}} + \frac{1}{T_{\text{fus}}^*} \right) + \frac{\Delta \overline{C}_P^*}{R} \left(\ln \frac{T_{\text{fus}}}{T_{\text{fus}}^*} + \frac{T_{\text{fus}}^*}{T_{\text{fus}}} - \frac{T_{\text{fus}}^*}{T_{\text{fus}}^*} \right) \\
&= \frac{\Delta_{\text{fus}} \overline{H}(T_{\text{fus}}^*)}{RT_{\text{fus}}^*} \left(\frac{T_{\text{fus}} - T_{\text{fus}}^*}{T_{\text{fus}}} \right) + \frac{\Delta \overline{C}_P^*}{R} \left(\ln \frac{T_{\text{fus}}}{T_{\text{fus}}^*} + \frac{T_{\text{fus}}^* - T_{\text{fus}}}{T_{\text{fus}}} \right)
\end{aligned}$$

Now let $T_{\text{fus}} = T_{\text{fus}}^* - \theta$ and use $1/(1 - x) = 1 + x + x^2 + \cdots$ to get

$$\frac{1}{T_{\text{fus}}} = \frac{1}{T_{\text{fus}}^* - \theta} = \frac{1}{T_{\text{fus}}^*} \left[1 + \frac{\theta}{T_{\text{fus}}^*} + \frac{\theta^2}{(T_{\text{fus}}^*)^2} + \cdots \right]$$

and use $\ln(1 - x) = -x - x^2/2 + \cdots$ to get

$$\ln \frac{T_{fus}}{T_{fus}^*} = \ln \left(1 - \frac{\theta}{T_{fus}^*} \right) = -\frac{\theta}{T_{fus}^*} - \frac{1}{2} \frac{\theta^2}{(T_{fus}^*)^2} + \cdots$$

Finally, then

$$
\begin{aligned}
-\ln a_1 &= -\frac{\Delta_{fus}\overline{H}(T_{fus}^*)}{R(T_{fus}^*)^2} \left(-\theta - \frac{\theta^2}{T_{fus}^*} + \cdots \right) \\
&\quad - \frac{\Delta \overline{C}_P^*}{RT_{fus}^*} \left[-\theta - \frac{\theta^2}{2T_{fus}^*} + \cdots + \theta \left(1 + \frac{\theta}{T_{fus}^*} + \cdots \right) \right] \\
&= \frac{\Delta_{fus}\overline{H}(T_{fus}^*)}{R(T_{fus}^*)^2}\theta + \frac{1}{R(T_{fus}^*)^2} \left(\frac{\Delta_{fus}\overline{H}(T_{fus}^*)}{T_{fus}^*} - \frac{\Delta \overline{C}_P^*}{2} \right)\theta^2 + \cdots
\end{aligned}
$$

25–21. Take $\Delta_{fus}\overline{H}(T_{fus}^*) = 6.01 \text{ kJ·mol}^{-1}$, $\overline{C}_P^l = 75.2 \text{ J·K}^{-1}\text{·mol}^{-1}$, and $\overline{C}_P^s = 37.6 \text{ J·K}^{-1}\text{·mol}^{-1}$ to show that the equation for $-\ln a_1$ in the previous problem becomes

$$-\ln a_1 = (0.00969 \text{ K}^{-1})\theta + (5.2 \times 10^{-6} \text{ K}^{-2})\theta^2 + \cdots$$

for an aqueous solution. The freezing point depression of a 1.95-molal aqueous sucrose solution is 4.45°C. Calculate the value of a_1 at this concentration. Compare your result with the value in Table 25.2. The value you calculated in this problem is for 0°C, whereas the value in Table 25.2 is for 25°C, but the difference is fairly small because a_1 does not vary greatly with temperature (Problem 25–61).

Using the final equation in Problem 25–20, we have

$$
\begin{aligned}
-\ln a_1 &= \frac{(6.01 \text{ kJ·mol}^{-1})\theta}{(8.314 \text{ J·mol}^{-1}\text{·K}^{-1})(273.2 \text{ K})^2} \\
&\quad + \frac{1}{(8.314 \text{ J·mol}^{-1}\text{·K}^{-1})(273.2 \text{ K})^2} \left(\frac{6.01 \text{ kJ·mol}^{-1}}{273.2 \text{ K}} - 18.8 \text{ J·mol}^{-1}\text{·K}^{-1} \right) \\
&= (0.00969 \text{ K}^{-1})\theta + (5.2 \times 10^{-6} \text{ K}^{-2})\theta^2 + \cdots
\end{aligned}
$$

If $\theta = 4.45$ K, then

$$
\begin{aligned}
\ln a_1 &= -(0.00969 \text{ K}^{-1})(4.45 \text{ K}) - (5.2 \times 10^{-6} \text{ K}^{-2})(4.45 \text{ K})^2 \\
&= -0.0432
\end{aligned}
$$

and so $a_1 = 0.958$.

25–22. The freezing point of a 5.0-molal aqueous glycerol (1,2,3-propanetriol) solution is -10.6°C. Calculate the activity of water at 0°C in this solution. (See Problems 25–20 and 25–21.)

Use the equation derived in Problem 25–21

$$\ln a_1 = -(0.00969 \text{ K}^{-1})\theta - (5.2 \times 10^{-6} \text{ K}^{-2})\theta^2$$

with $\theta = 10.6$ K to get

$$\ln a_1 = -(0.00969\ \text{K}^{-1})(10.6\ \text{K}) - (5.2 \times 10^{-6}\ \text{K}^{-2})(10.6\ \text{K})^2$$
$$= -0.103$$

and so $a_1 = 0.902$.

25–23. Show that replacing T_{fus} by T_{fus}^* in the denominator of $(T_{\text{fus}} - T_{\text{fus}}^*)/T_{\text{fus}}^* T_{\text{fus}}$ (see Equation 25.20) gives $-\theta/(T_{\text{fus}}^*)^2 - \theta^2/(T_{\text{fus}}^*)^3 + \cdots$ where $\theta = T_{\text{fus}}^* - T_{\text{fus}}$.

$$\frac{T_{\text{fus}} - T_{\text{fus}}^*}{T_{\text{fus}} T_{\text{fus}}^*} = \frac{-\theta}{T_{\text{fus}}^*(T_{\text{fus}}^* - \theta)} = -\frac{\theta}{(T_{\text{fus}}^*)^2 \left(1 - \dfrac{\theta}{T_{\text{fus}}^*}\right)}$$

Now use the expansion $1/(1 - x) = x + x^2 + \cdots$ to write

$$\frac{T_{\text{fus}} - T_{\text{fus}}^*}{T_{\text{fus}} T_{\text{fus}}^*} = -\frac{\theta}{(T_{\text{fus}}^*)^2}\left[1 + \frac{\theta}{T_{\text{fus}}^*} + \frac{\theta^2}{(T_{\text{fus}}^*)^2} + \cdots\right]$$

$$= -\frac{\theta}{(T_{\text{fus}}^*)^2} - \frac{\theta^2}{(T_{\text{fus}}^*)^3} + \cdots$$

25–24. Calculate the value of the freezing point depression constant for nitrobenzene, whose freezing point is 5.7°C and whose molar enthalpy of fusion is $11.59\ \text{kJ} \cdot \text{mol}^{-1}$.

Using Equation 25.23, we write

$$K_f = \left(\frac{M_1}{1000\ \text{g} \cdot \text{kg}^{-1}}\right)\frac{R(T_{\text{fus}}^*)^2}{\Delta_{\text{fus}}\overline{H}}$$

$$= \left(\frac{123.11\ \text{g} \cdot \text{mol}^{-1}}{1000\ \text{g} \cdot \text{kg}^{-1}}\right)\left[\frac{(8.314\ \text{J} \cdot \text{mol}^{-1} \cdot \text{K}^{-1})(278.9\ \text{K})^2}{11.59 \times 10^3\ \text{J} \cdot \text{mol}^{-1}}\right]$$

$$= 6.87\ \text{K} \cdot \text{kg} \cdot \text{mol}^{-1}$$

25–25. Use an argument similar to the one we used to derive Equations 25.22 and 25.23 to derive Equations 25.24 and 25.25.

The condition for equilibrium at a temperature T is

$$\mu_1^{\text{g}}(T, P) = \mu_1^{\text{sln}}(T, P) = \mu_1^*(T, P) + RT \ln a_1 = \mu^l + RT \ln a_1$$

Solving for $\ln a_1$ gives

$$\ln a_1 = \frac{\mu_1^{\text{g}} - \mu_1^l}{RT}$$

Use the Gibbs-Helmholtz equation (Example 24–1) to get

$$\left(\frac{\partial \ln a_1}{\partial T}\right)_{P,x_1} = \frac{\overline{H}_1^l - \overline{H}_1^{\text{g}}}{RT^2} = -\frac{\Delta_{\text{vap}}\overline{H}}{RT^2}$$

This equation is similar to Equation 25.17 except for the negative sign, which occurs because boiling points of solutions are elevated whereas freezing points are lowered. The rest of the derivation follows Equations 25.18 through 25.23.

25–26. Calculate the boiling point elevation constant for cyclohexane given that $T_{vap} = 354$ K and that $\Delta_{vap}\overline{H} = 29.97$ kJ·mol^{-1}.

Using the analog of Equation 25.23, we have

$$K_b = \frac{(84.161 \text{ g·mol}^{-1})(8.314 \text{ J·mol}^{-1}\cdot\text{K}^{-1})(354 \text{ K})^2}{(1000 \text{ g·kg}^{-1})(29.97 \times 10^3 \text{ J·mol}^{-1})}$$

$$= 2.93 \text{ K·kg·mol}^{-1}$$

25–27. A solution containing 1.470 g of dichlorobenzene in 50.00 g of benzene boils at 80.60°C at a pressure of 1.00 bar. The boiling point of pure benzene is 80.09°C and the molar enthalpy of vaporization of pure benzene is 32.0 kJ·mol^{-1}. Determine the molecular mass of dichlorobenzene from these data.

The value of $\Delta_{vap}T$ is

$$\Delta_{vap}T = 80.60°C - 80.09°C = 0.51°C = 0.51 \text{ K}$$

Using the analog of Equation 25.23, we have

$$K_b = \frac{(78.108 \text{ g·mol}^{-1})(8.314 \text{ J·mol}^{-1}\cdot\text{K}^{-1})(353.2 \text{ K})^2}{(1000 \text{ g·kg}^{-1})(32.0 \times 10^3 \text{ J·mol}^{-1})}$$

$$= 2.53 \text{ K·kg·mol}^{-1}$$

The molality is given by

$$m = \frac{\Delta_{vap}T}{K_b} = \frac{0.51 \text{ K}}{2.53 \text{ K·kg}^{-1}\cdot\text{mol}^{-1}} = 0.20 \text{ mol·kg}^{-1}$$

Therefore,

$$1.470 \text{ g } C_6H_4Cl_2 \longleftrightarrow 50.0 \text{ g } C_6H_6$$

$$29.4 \text{ g } C_6H_4Cl_2 \longleftrightarrow 1000 \text{ g } C_6H_6 \longleftrightarrow 0.20 \text{ mol}$$

and so the molecular mass is 147.

25–28. Consider the following phase diagram for a typical pure substance. Label the region corresponding to each phase. Illustrate how this diagram changes for a dilute solution of a nonvolatile solute.

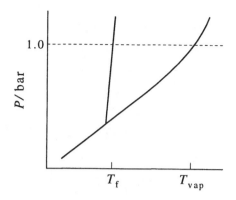

Now demonstrate that the boiling point increases and the freezing point decreases as a result of the dissolution of the solute.

Use the following figure for water, which is self-explanatory

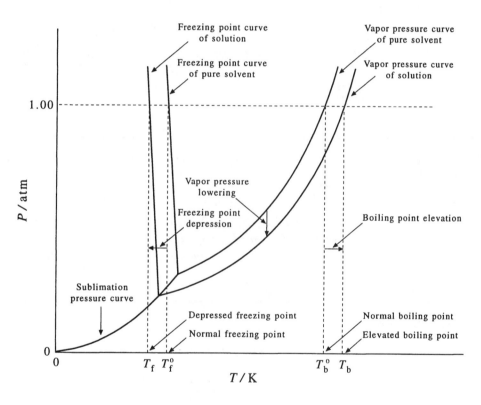

$T_f^o - T_f$ = freezing point depression $T_b - T_b^o$ = boiling point elevation

---

25–29. A solution containing 0.80 grams of a protein in 100 mL of a solution has an osmotic pressure of 2.06 torr at 25.0°C. What is the molecular mass of the protein?

We use Equation 25.31,

$$c = \frac{\Pi}{RT} = \frac{(2.06 \text{ torr})/(760 \text{ torr} \cdot \text{atm}^{-1})}{(0.08206 \text{ L} \cdot \text{atm} \cdot \text{mol}^{-1} \cdot \text{K}^{-1})(298.2 \text{ K})}$$

$$= 1.11 \times 10^{-4} \text{ mol} \cdot \text{L}^{-1} = 1.11 \times 10^{-5} \text{ mol}/100 \text{ mL}$$

Therefore, 1.11×10^{-5} mol corresponds to 0.80 g protein, and so the molecular mass of the protein is 72 000.

25–30. Show that the osmotic pressure of an aqueous solution can be written as

$$\Pi = \frac{RT}{V^*}\left(\frac{m}{55.506 \text{ mol·kg}^{-1}}\right)\phi$$

Simply substitute Equation 25.13 into Equation 25.30.

25–31. According to Table 25.2, the activity of the water in a 2.00 molal sucrose solution is 0.95807. What external pressure must be applied to the solution at 25.0°C to make the activity of the water in the solution the same as that in pure water at 25.0°C and 1 atm? Take the density of water to be 0.997 g·mL^{-1}.

Using Equation 25.30, we have

$$\Pi = -\frac{RT \ln a_1}{V_1^*} = -\frac{(0.08206 \text{ L·atm·mol}^{-1}\text{·K}^{-1})(298.2 \text{ K})(\ln 0.95807)}{0.01807 \text{ L·mol}^{-1}}$$

$$= 58.0 \text{ atm}$$

25–32. Show that $a_2 = a_\pm^2 = m^2\gamma_\pm^2$ for a 2–2 salt such as $CuSO_4$ and that $a_2 = a_\pm^4 = 27m^4\gamma_\pm^4$ for a 1–3 salt such as $LaCl_3$.

Equation 25.40 gives us $a_2 = a_\pm^\nu = m_\pm^\nu\gamma_\pm^\nu$. For a 2–2 salt, such as $MgSO_4$, $\nu_+ = 1$, $\nu_- = 1$, $m_+ = m$, and $m_- = m$, and so $a_2 = a_\pm^2 = m^2\gamma_\pm^2$, or $a_\pm = m\gamma_\pm$.

For a 1–3 salt, such as $LaCl_3$, $\nu_+ = 1$, $\nu_- = 3$, $m_+ = m$, and $m_- = 3m$, and so $a_2 = a_\pm^4 = [m^1(3m)^3]\gamma_\pm^4$, or $a_\pm = 27^{1/4}m\gamma_\pm$.

25–33. Verify the following table:

Type of salt	Example	I_m
1 − 1	KCl	m
1 − 2	$CaCl_2$	$3m$
2 − 1	K_2SO_4	$3m$
2 − 2	$MgSO_4$	$4m$
1 − 3	$LaCl_3$	$6m$
3 − 1	Na_3PO_4	$6m$

Show that the general result for I_m is $|z_+z_-|(\nu_1 + \nu_2)m/2$.

We use Equation 25.52 in terms of molality.

$$I_m = \frac{1}{2}\sum_{j=1}^s z_j^2 m_j = \frac{1}{2}(z_+^2 m_+ + z_-^2 m_-) \tag{1}$$

for a binary salt. Therefore, we have the following table.

type	I_m
1–1	$\frac{1}{2}(1m_+ + 1m_-) = \frac{1}{2}(2m) = m$
1–2	$\frac{1}{2}(4m_+ + 1m_-) = \frac{1}{2}(4m + 2m) = 3m$
2–1	$\frac{1}{2}(m_+ + 4m_-) = \frac{1}{2}(2m + 4m) = 3m$
2–2	$\frac{1}{2}(4m_+ + 4m_-) = \frac{1}{2}(8m) = 4m$
1–3	$\frac{1}{2}(9m_+ + 1m_-) = \frac{1}{2}(9m + 3m) = 6m$
3–1	$\frac{1}{2}(m_+ + 9m_-) = \frac{1}{2}(3m + 9m) = 6m$

To prove the general result, substitute $m_+ = v_+ m$ and $m_- = v_- m$ into Equation 1 to get

$$I_m = \frac{1}{2}(z_+^2 v_+ + z_-^2 v_-)m$$

Now use the electroneutrality condition $z_+ v_+ = |z_-|v_-$ to get

$$I_m = \frac{1}{2}(z_+|z_-|v_- + |z_-|^2 v_-)m$$

$$= \frac{z_+|z_-|}{2}\left(v_- + \frac{|z_-|v_-}{z_+}\right)m = \frac{|z_- z_+|}{2}(v_+ + v_-)m$$

25–34. Show that the inclusion of the factor v in Equation 25.41 allows $\phi \to 1$ as $m \to 0$ for solutions of electrolytes as well as nonelectrolytes. *Hint*: Realize that x_2 involves the total number of moles of solute particles (see Equation 25.44).

For a nonelectrolyte, $\ln a_1 \to \ln x_1 \to \ln(1 - x_2) \to -x_2$ as $x_2 \to 0$. According to Equation 25.21, $x_2 \to M_1 m/(1000 \text{ g}\cdot\text{kg}^{-1})$ as $m \to 0$, so ϕ defined by $\ln a_1 = -M_1 m\phi/(1000 \text{ g}\cdot\text{kg}^{-1})$ (Equation 25.13) becomes $\phi = (\ln a_1)/x_2 = 1$ as $x_2 \to 0$ or $m \to 0$. For an electrolyte, $x_2 \to vM_1 m/(1000 \text{ g}\cdot\text{kg}^{-1})$ as $x_2 \to 0$ or $m \to 0$. Therefore, ϕ defined by $\ln a_1 = -vM_1 m/(1000 \text{ g}\cdot\text{kg}^{-1})$ (Equation 25.41) becomes $\phi = (\ln a_1)/x_2 = 1$ as $x_2 \to 0$ or $m \to 0$.

25–35. Use Equation 25.41 and the Gibbs-Duhem equation to derive Equation 25.42.

Consider an aqueous solution consisting of 1000 g of water. The Gibbs-Duhem equation is

$$n_1 d \ln a_1 + n_2 \ln a_2 = 0$$

or

$$(55.506 \text{ mol}\cdot\text{kg}^{-1})d \ln a_1 + md \ln a_2 = 0$$

Use Equation 25.41 to obtain

$$-vd(m\phi) + md \ln a_2 = 0$$

Equation 25.37 gives $a_2 = a_\pm^v = m_\pm^v \gamma_\pm^v$, and so we have

$$-vd(m\phi) + mvd \ln m_\pm \gamma_\pm = 0$$

But generally $m_{\pm} = cm$, where c is a constant whose value depends upon the type of electrolyte (see Table 25.3), and so

$$\nu d(m\phi) = m\nu d \ln(cm\gamma_{\pm})$$

$$= m\nu d \ln(m\gamma_{\pm})$$

Thus

$$d(m\phi) = md \ln(m\gamma_{\pm})$$

or

$$md\phi + \phi dm = m(d \ln \gamma_{\pm} + d \ln m)$$

Division by m gives

$$d\phi + \phi \frac{dm}{m} = d \ln \gamma_{\pm} + \frac{dm}{m}$$

or

$$d \ln \gamma_{\pm} = d\phi + \frac{\phi - 1}{m} dm$$

Now integrate from $m = 0$ (where $\gamma_{\pm} = \phi = 1$) to m to obtain Equation 25.42.

25–36. The osmotic coefficient of $CaCl_2(aq)$ solutions can be expressed as

$$\phi = 1.0000 - (1.2083 \text{ kg}^{1/2} \cdot \text{mol}^{-1/2})m^{1/2} + (3.2215 \text{ kg} \cdot \text{mol}^{-1})m$$

$$- (3.6991 \text{ kg}^{3/2} \cdot \text{mol}^{-3/2})m^{3/2} + (2.3355 \text{ kg}^2 \cdot \text{mol}^{-2})m^2$$

$$- (0.67218 \text{ kg}^{5/2} \cdot \text{mol}^{-5/2})m^{5/2} + (0.069749 \text{ kg}^3 \cdot \text{mol}^{-3})m^3$$

$$0 \leq m \leq 5.00 \text{ mol} \cdot \text{kg}^{-1}$$

Use this expression to calculate and plot $\ln \gamma_{\pm}$ as a function of $m^{1/2}$.

Substitute the expression for ϕ given in the problem into Equation 25.42. The result is shown in the following figure.

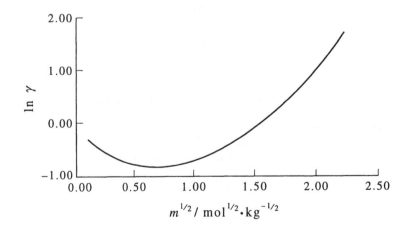

25-37. Use Equation 25.43 to calculate $\ln \gamma_{\pm}$ for NaCl(aq) at 25°C as a function of molality and plot it versus $m^{1/2}$. Compare your results with those in Table 25.4.

Substitute Equation 25.43 into Equation 25.42. The result is shown in the following figure. The calculated and experimental values are indistinguishable on the graph.

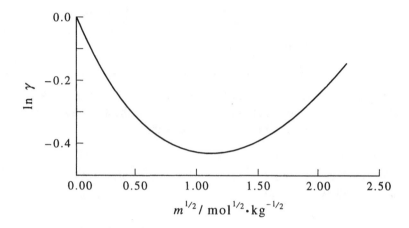

25-38. In Problem 25-19, you determined $\ln \gamma_{2m}$ for sucrose by calculating the area under the curve of $\phi - 1$ versus m. When dealing with solutions of electrolytes, it is better numerically to plot $(\phi - 1)/m^{1/2}$ versus $m^{1/2}$ because of the natural dependence of ϕ on $m^{1/2}$. Show that

$$\ln \gamma_{\pm} = \phi - 1 + 2 \int_0^{m^{1/2}} \frac{\phi - 1}{m^{1/2}} dm^{1/2}$$

Start with Equation 25.42, and let $x = m^{1/2}$ and $dx = dm/2m^{1/2} = dm/2x$ to obtain

$$\int_0^m \frac{\phi - 1}{m'} dm' = \int_0^x \frac{\phi - 1}{x'^2} 2x' dx'$$

$$= 2 \int_0^{m^{1/2}} \frac{\phi - 1}{m^{1/2}} dm^{1/2}$$

The full expression for $\ln \gamma_{\pm}$ is

$$\ln \gamma_{\pm} = \phi - 1 + 2 \int_0^{m^{1/2}} \frac{\phi - 1}{m^{1/2}} dm^{1/2}$$

25-39. Use the data in Table 25.4 to calculate $\ln \gamma_{\pm}$ for NaCl(aq) at 25°C by plotting $(\phi - 1)/m^{1/2}$ against $m^{1/2}$ and determine the area under the curve by numerical integration (Mathchapter G). Compare your values of $\ln \gamma_{\pm}$ with those you obtained in Problem 25-37, where you calculate $\ln \gamma_{\pm}$ from a curve-fit expression of ϕ as a polynomial in $m^{1/2}$.

The plot is essentially identical to the one obtained in Problem 25–37.

25–40. Don Juan Pond in the Wright Valley of Antarctica freezes at $-57°C$. The major solute in the pond is $CaCl_2$. Estimate the concentration of $CaCl_2$ in the pond water.

We say "estimate" because the concentration will be too large for Equation 25.45 to be quantitative. Nevertheless, we can "estimate" the molality to be

$$m \approx \frac{57 \text{ K}}{(3)(1.84 \text{ K·kg·mol}^{-1})} = 10 \text{ mol·kg}^{-1}$$

where the factor of 3 in the denominator results because $\nu = 3$ for $CaCl_2$.

25–41. A solution of mercury(II) chloride is a poor conductor of electricity. A 40.7-g sample of $HgCl_2$ is dissolved in 100.0 g of water, and the freezing point of the solution is found to be $-2.83°C$. Explain why $HgCl_2$ in solution is a poor conductor of electricity.

Because 40.7 g $HgCl_2$ corresponds to 0.150 mol $HgCl_2$, the molality of the solution is 1.50 mol·kg^{-1}. Using Equation 25.45, we find ν to be

$$\nu = \frac{\Delta T_{fus}}{K_f m} = \frac{2.83 \text{ K}}{(1.84 \text{ K·kg·mol}^{-1})(1.50 \text{ mol·kg}^{-1})} = 1.02$$

This result indicates that $HgCl_2$ is not dissociated under these conditions, and so is a poor conductor of electricity.

25–42. The freezing point of a 0.25-molal aqueous solution of Mayer's reagent, K_2HgI_4, is found to be $-1.41°C$. Suggest a possible dissociation reaction that takes place when K_2HgI_4 is dissolved in water.

Use Equation 25.45 to obtain $\nu = 3$. The equation for the dissociation reaction is

$$K_2KgI_4(aq) \longrightarrow 2 \text{ K}^+(aq) + HgI_4^{2-}(aq)$$

25–43. Given the following freezing-point depression data, determine the number of ions produced per formula unit when the indicated substance is dissolved in water to produce a 1.00-molal solution.

Formula	$\Delta T/\text{K}$
$PtCl_2 \cdot 4NH_3$	5.58
$PtCl_2 \cdot 3NH_3$	3.72
$PtCl_2 \cdot 2NH_3$	1.86
$KPtCl_3 \cdot NH_3$	3.72
K_2PtCl_4	5.58

Interpret your results.

Use Equation 25.45 to obtain

formula	ν	ions	
$PtCl_2 \cdot 4NH_3$	3	$Pt(NH_3)_4^{2+}$	$2\ Cl^-$
$PtCl_2 \cdot 3NH_3$	2	$Pt(NH_3)_3Cl^+$	Cl^-
$PtCl_2 \cdot 2NH_3$	1	$Pt(NH_3)_2Cl_2$	
$KPtCl_3 \cdot NH_3$	2	K^+	$Pt(NH_3)Cl_3^-$
K_2PtCl_4	3	$2\ K^+$	$PtCl_4^{2-}$

25–44. An aqueous solution of NaCl has an ionic strength of 0.315 mol·L^{-1}. At what concentration will an aqueous solution of K_2SO_4 have the same ionic strength?

The ionic strength, I_c, equals c for a 1–1 electrolyte and $3c$ for a 2–1 electrolyte. Therefore, a solution of K_2SO_4(aq) would have an ionic strength of 0.315 mol·L^{-1} when its molarity is 0.105 mol·L^{-1}.

25–45. Derive the "practical" formula for κ^2 given by Equation 25.53.

Start with

$$\kappa^2 = \frac{1}{\varepsilon_0 \varepsilon_r k_B T} \sum_{j=1}^{s} q_j^2 \frac{N_j}{V} = \frac{N_A e^2}{\varepsilon_0 \varepsilon_r k_B T} \sum_{j=1}^{s} z_j^2 \frac{n_j}{V}$$

Now

$$\frac{n_j}{V} = (1000\ \text{L·m}^{-3})c_j$$

because V, being in SI units, has units of m^3. Therefore,

$$\kappa^2 = \frac{2e^2 N_A (1000\ \text{L·m}^{-3})}{\varepsilon_0 \varepsilon_r k_B T}(I_c/\text{mol·L}^{-1})$$

25–46. Some authors define ionic strength in terms of molality rather than molarity, in which case

$$I_m = \frac{1}{2} \sum_{j=1}^{s} z_j^2 m_j$$

Show that this definition modifies Equation 25.53 for dilute solutions to be

$$\kappa^2 = \frac{2e^2 N_A (1000\ \text{L·m}^{-3})\rho}{\varepsilon_0 \varepsilon_r kT}(I_m/\text{mol·kg}^{-1})$$

where ρ is the density of the solvent (in g·mL^{-1}).

For dilute solutions, $c = \rho m$ (see Problem 25–5), and so $I_c = \rho I_m$. Therefore,

$$\kappa^2 = \frac{2e^2 N_A (1000\ \text{L·m}^{-3})\rho}{\varepsilon_0 \varepsilon_r kT}(I_m/\text{mol·kg}^{-1})$$

25–47. Show that

$$\ln \gamma_{\pm} = -1.171 |z_+ z_-| (I_m / \text{mol} \cdot \text{kg}^{-1})^{1/2}$$

for an aqueous solution at 25.0°C, where I_m is the ionic strength expressed in terms of molality. Take ε_r to be 78.54 and the density of water to be $0.99707 \, \text{g} \cdot \text{mL}^{-1}$.

We use the equation for κ^2 that is derived in Problem 25–46.

$$\kappa^2 = \frac{(2)(1.6022 \times 10^{-19} \, \text{C})^2 (6.0221 \times 10^{23} \, \text{mol}^{-1})(1000 \, \text{L} \cdot \text{m}^{-3})(0.99707 \, \text{g} \cdot \text{mL}^{-1})}{(8.8542 \times 10^{-12} \, \text{C}^2 \cdot \text{s}^2 \cdot \text{kg}^{-1} \cdot \text{m}^{-3})(78.54)(1.3806 \times 10^{-23} \, \text{J} \cdot \text{K}^{-1})(298.2 \, \text{K})}$$
$$\times (I_m / \text{mol} \cdot \text{kg}^{-1})$$
$$= (1.077 \times 10^{19} \, \text{g} \cdot \text{L} \cdot \text{mL}^{-1} \cdot \text{J}^{-1} \cdot \text{s}^{-2})(I_m / \text{mol} \cdot \text{kg}^{-1})$$
$$= (1.077 \times 10^{19} \, \text{g} \cdot \text{L} \cdot \text{mL}^{-1}) \frac{1 \, \text{kg}}{1000 \, \text{g}} \frac{1000 \, \text{mL}}{1 \, \text{L}} (\text{J}^{-1} \cdot \text{s}^{-2})(I_m / \text{mol} \cdot \text{kg}^{-1})$$
$$= (1.077 \times 10^{19} \, \text{m}^{-2})(I_m / \text{mol} \cdot \text{kg}^{-1})$$

The expression for $\ln \gamma_{\pm}$ is

$$\ln \gamma_{\pm} = -|z_+ z_-| \frac{e^2 \kappa}{8 \pi \varepsilon_0 \varepsilon_r kT}$$
$$= -|z_+ z_-| \frac{(1.6022 \times 10^{-19} \, \text{C})^2 (1.077 \times 10^{19} \, \text{m}^{-2})^{1/2} (I_m / \text{mol} \cdot \text{kg}^{-1})^{1/2}}{8 \pi (8.8542 \times 10^{-12} \, \text{C}^2 \cdot \text{s}^2 \cdot \text{kg}^{-1} \cdot \text{m}^{-3})(78.54)(1.3806 \times 10^{-23} \, \text{J} \cdot \text{K}^{-1})(298.2 \, \text{K})}$$
$$= -1.171 |z_+ z_-| (I_m / \text{mol} \cdot \text{kg}^{-1})^{1/2}$$

25–48. Use the Debye-Hückel theory to calculate $\ln \gamma_{\pm}$ for a 0.010-molar NaCl(aq) solution at 25.0°C. Take $\varepsilon_r = 78.54$ for $H_2O(l)$ at 25.0°C. The experimental value of $\ln \gamma_{\pm}$ is -0.103.

We can use Equation 25.56 directly.

$$\ln \gamma_{\pm} = -1.173(0.010)^{1/2} = -0.1173$$

and so $\gamma_{\pm} = 0.889$.

25–49. Derive the general equation

$$\phi = 1 + \frac{1}{m} \int_0^m m' d \ln \gamma_{\pm}$$

(*Hint:* See the derivation in Problem 25–35.) Use this result to show that

$$\phi = 1 + \frac{\ln \gamma_{\pm}}{3}$$

for the Debye-Hückel theory.

Start with (see Problem 25–35)

$$d(m\phi) = md \ln(m\gamma_{\pm}) = m(d \ln m + d \ln \gamma_{\pm})$$
$$= dm + md \ln \gamma_{\pm}$$

and integrate from $m = 0$ to arbitrary m to obtain

$$m\phi = m + \int_0^m m'd\ln\gamma_\pm$$

or

$$\phi = 1 + \frac{1}{m}\int_0^m m'd\ln\gamma_\pm \qquad (1)$$

Now use Equation 25.49 to write $\ln\gamma_\pm$ as

$$\ln\gamma_\pm = -|q_+q_-|\frac{(\kappa/m^{1/2})}{8\pi\varepsilon_0\varepsilon_r k_B T}m^{1/2}$$

where $\kappa/m^{1/2}$ is *independent of m*. Then

$$d\ln\gamma_\pm = -|q_+q_-|\frac{(\kappa/m^{1/2})}{8\pi\varepsilon_0\varepsilon_r k_B T}\frac{dm}{2m^{1/2}} = \frac{\ln\gamma_\pm}{2m}dm$$

and

$$md\ln\gamma_\pm = -\frac{1}{2}|q_+q_-|\frac{(\kappa/m^{1/2})}{8\pi\varepsilon_0\varepsilon_r k_B T}m^{1/2}dm = \frac{\ln\gamma_\pm}{2}dm$$

Substitute these results into Equation 1 to obtain

$$\begin{aligned}
\phi - 1 &= \frac{1}{m}\int_0^m m'd\ln\gamma_\pm \\
&= -\frac{1}{2}|q_+q_-|\frac{(\kappa/m^{1/2})}{8\pi\varepsilon_0\varepsilon_r k_B T}\frac{1}{m}\int_0^m m'^{1/2}dm' \\
&= \frac{\ln\gamma_\pm}{3}
\end{aligned}$$

25–50. In the Debye-Hückel theory, the ions are modeled as point ions and the solvent is modeled as a continuous medium (no structure) with a relative permittivity ε_r. Consider an ion of type i (i = a cation or an anion) situated at the origin of a spherical coordinate system. The presence of this ion at the origin will attract ions of opposite charge and repel ions of the same charge. Let $N_{ij}(r)$ be the number of ions of type j (j = a cation or an anion) situated at a distance r from the central ion of type i (a cation or anion). We can use a Boltzmann factor to say that

$$N_{ij}(r) = N_j e^{-w_{ij}(r)/k_B T}$$

where N_j/V is the bulk number density of j ions and $w_{ij}(r)$ is the interaction energy of an i ion with a j ion. This interaction energy will be electrostatic in origin, so let $w_{ij}(r) = q_j\psi_i(r)$, where q_j is the charge on the j ion and $\psi_i(r)$ is the electrostatic potential due to the central i ion.

A fundamental equation from physics that relates a spherically symmetric electrostatic potential $\psi_i(r)$ to a spherically symmetric charge density $\rho_i(r)$ is Poisson's equation

$$\frac{1}{r^2}\frac{d}{dr}\left[r^2\frac{d\psi_i(r)}{dr}\right] = -\frac{\rho_i(r)}{\varepsilon_0\varepsilon_r} \qquad (1)$$

where ε_r is the relative permittivity of the solvent. In our case, $\rho_i(r)$ is the charge density around the central ion (of type i). First, show that

$$\rho_i(r) = \frac{1}{V}\sum_j q_j N_{ij}(r) = \sum_j q_j C_j e^{-q_j\psi_i(r)/k_B T}$$

where C_j is the bulk number density of species j ($C_j = N_j/V$). Linearize the exponential term and use the condition of electroneutrality to show that

$$\rho_i(r) = -\psi_i(r) \sum_j \frac{q_j^2 C_j}{k_B T} \tag{2}$$

Now substitute $\rho_i(r)$ into Poisson's equation to get

$$\frac{1}{r^2} \frac{d}{dr} \left[r^2 \frac{d\psi_i(r)}{dr} \right] = \kappa^2 \psi_i(r) \tag{3}$$

where

$$\kappa^2 = \sum_j \frac{q_j^2 C_j}{\varepsilon_0 \varepsilon_r k_B T} = \sum_j \frac{q_j^2}{\varepsilon_0 \varepsilon_r k_B T} \left(\frac{N_j}{V} \right) \tag{4}$$

Show that Equation 3 can be written as

$$\frac{d^2}{dr^2} [r\psi_i(r)] = \kappa^2 [r\psi_i(r)]$$

Now show that the only solution for $\psi_i(r)$ that is finite for large values of r is

$$\psi_i(r) = \frac{Ae^{-\kappa r}}{r} \tag{5}$$

where A is a constant. Use the fact that if the concentration is very small, then $\psi_i(r)$ is just Coulomb's law and so $A = q_i/4\pi\varepsilon_0\varepsilon_r$ and

$$\psi_i(r) = \frac{q_i e^{-\kappa r}}{4\pi\varepsilon_0\varepsilon_r r} \tag{6}$$

Equation 6 is a central result of the Debye-Hückel theory. The factor of $e^{-\kappa r}$ modulates the resulting Coulombic potential, so Equation 6 is called a *screened Coulombic potential*.

The number of ions of type j situated at a distance r from the central ion of type i is given by

$$N_{ij}(r) = N_j e^{-q_j \psi_i(r)/k_B T}$$

The charge about a central ion of type i due to ions of type j is given by $q_j N_{ij}(r)$ and the net charge is given by $\sum_j q_j N_{ij}(r)$. The charge density at a distance r from the central ion of type i is given by

$$\rho_i(r) = \frac{1}{V} \sum_j q_j N_{ij}(r) = \sum_j q_j C_j e^{-q_j \psi_i(r)/k_B T}$$

Now expand the exponential using the expansion $e^{-x} = 1 - x + \cdots$ to obtain

$$\rho_i(r) = \sum_j q_j C_j - \frac{\psi_i(r)}{k_B T} \sum_j q_j^2 C_j + \cdots$$

$$= 0 \text{ (by electroneutrality)} - \frac{\psi_i(r)}{k_B T} \sum_j q_j^2 C_j^2 + \cdots$$

Now substitute $\rho_i(r)$ into Poisson's equation to get

$$\frac{1}{r^2} \frac{d}{dr} \left[r^2 \frac{d\psi_i(r)}{dr} \right] = \kappa^2 \psi_i(r) \tag{3}$$

where

$$\kappa^2 = \frac{1}{\varepsilon_0 \varepsilon_r k_B T} \sum_j q_j^2 C_j$$

Now

$$\frac{1}{r^2}\frac{d}{dr}\left[r^2 \frac{d\psi_i(r)}{dr}\right] = \frac{d^2\psi_i(r)}{dr^2} + \frac{2}{r}\frac{d\psi_i(r)}{dr}$$

and

$$\frac{d^2[r\psi_i(r)]}{dr^2} = r\frac{d^2\psi_i(r)}{dr^2} + 2\frac{d\psi_i(r)}{dr}$$

so

$$\frac{1}{r^2}\frac{d}{dr}\left[r^2 \frac{d\psi_i(r)}{dr}\right] = \frac{1}{r}\frac{d^2[r\psi_i(r)]}{dr^2}$$

Therefore, Equation 3 can be written as

$$\frac{d^2[r\psi_i(r)]}{dr^2} = \kappa^2[r\psi(r)]$$

This differential equation has the general solution

$$r\psi_i(r) = Be^{\kappa r} + Ae^{-\kappa r}$$

or

$$\psi_i(r) = \frac{B}{r}e^{\kappa r} + \frac{A}{r}e^{-\kappa r}$$

But B must be zero for $\psi_i(r)$ to be finite as $r \to \infty$. Therefore, we have simply

$$\psi_i(r) = \frac{A}{r}e^{-\kappa r}$$

If the concentration is very small, then $\kappa \to 0$ and $\psi_i(r) \to q_i/4\pi\varepsilon_0\varepsilon_r r$. Therefore,

$$\psi_i(r) \longrightarrow \frac{A}{r} = \frac{q_i}{4\pi\varepsilon_0\varepsilon_r r}$$

and we see that $A = q_i/4\pi\varepsilon_0\varepsilon_r$. Finally then, we have

$$\psi_i(r) = \frac{q_i e^{-\kappa r}}{4\pi\varepsilon_0\varepsilon_r r}$$

25–51. Use Equations 2 and 6 of the previous problem to show that the net charge in a spherical shell of radius r surrounding a central ion of type i is

$$p_i(r)dr = \rho_i(r)4\pi r^2 dr = -q_i\kappa^2 re^{-\kappa r}dr$$

as in Equation 25.54. Why is

$$\int_0^\infty p_i(r)dr = -q_i$$

Start with

$$p_i(r)dr = \rho_i(r)4\pi r^2 dr$$

Equations 2 and 4 of Problem 25–50 show that

$$-\frac{\rho_i(r)}{\varepsilon_0\varepsilon_r} = \kappa^2\psi_i(r)$$

so that

$$p_i(r)dr = -\varepsilon_0\varepsilon_r\kappa^2\psi_i(r)4\pi r^2 dr$$

Using Equation 6 of Problem 25–50, we have

$$p_i(r)dr = -q_i\kappa^2 re^{-\kappa r}dr$$

Therefore,

$$\int_0^\infty p_i(r)dr = -q_i\kappa^2\int_0^\infty re^{-\kappa r}dr = -q_i$$

which it must be because of electroneutrality.

25–52. Use the result of the previous problem to show that the most probable value of r is $1/\kappa$.

Problem 25–51 shows that $p_i(r) \approx re^{-\kappa r}$. Therefore, the most probable value of r is given by

$$\frac{dp_i}{dr} \approx e^{-\kappa r} - \kappa re^{-\kappa r} = 0$$

or $r_{mp} = 1/\kappa$.

25–53. Show that

$$r_{mp} = \frac{1}{\kappa} = \frac{304 \text{ pm}}{(c/\text{mol·L}^{-1})^{1/2}}$$

where c is the molarity of an aqueous solution of a 1–1 electrolyte at 25°C. Take $\varepsilon_r = 78.54$ for $H_2O(l)$ at 25°C.

Use Equation 25.53

$$\kappa^2 = \frac{2(1.602 \times 10^{-19} \text{ C})^2(6.022 \times 10^{23} \text{ mol}^{-1})(1000 \text{ L·m}^{-3})(I_c/\text{mol·L}^{-1})}{(8.8542 \times 10^{-12} \text{ C·s}^2\text{·kg}^{-1}\text{·m}^{-3})(78.54)(1.3806 \times 10^{-23} \text{ J·K}^{-1})(298.15 \text{ K})}$$

$$= (1.080 \times 10^{19} \text{ g·L·mL}^{-1}\text{·s}^{-2}\text{·J}^{-1})(I_c/\text{mol·L}^{-1})$$

$$= (1.080 \times 10^{19} \text{ g·L·mL}^{-1})\frac{1 \text{ kg}}{1000 \text{ g}} \cdot \frac{1000 \text{ mL}}{1 \text{ L}}(\text{s}^{-2}\text{·J}^{-1})(I_c/\text{mol·L}^{-1})$$

$$= (1.080 \times 10^{19} \text{ m}^{-2})(I_c/\text{mol·L}^{-1})$$

$$\kappa = (3.29 \times 10^9 \text{ m}^{-1})(I_c/\text{mol·L}^{-1})^{1/2}$$

For a 1–1 electrolyte, $I_c = c$, and so

$$\frac{1}{\kappa} = \frac{3.04 \times 10^{-10} \text{ m}}{(c/\text{mol·L}^{-1})^{1/2}} = \frac{304 \text{ pm}}{(c/\text{mol·L}^{-1})^{1/2}}$$

25–54. Show that

$$r_{mp} = \frac{1}{\kappa} = 430 \text{ pm}$$

for a 0.50-molar aqueous solution of a 1–1 electrolyte at 25°C. Take $\varepsilon_r = 78.54$ for $H_2O(l)$ at 25°C.

Use Equation 25.55 and the result of Problem 25–52:

$$r_{mp} = \frac{1}{\kappa} = \frac{304 \text{ pm}}{(c/\text{mol·L}^{-1})^{1/2}} = \frac{304 \text{ pm}}{(0.50)^{1/2}} = 430 \text{ pm}$$

25–55. How does the thickness of the ionic atmosphere compare for a 1–1 electrolyte and a 2–2 electrolyte?

Equation 25.50 shows that $\kappa_{2-2}^2 = 4\kappa_{1-1}^2$, or that $\kappa_{2-2} = 2\kappa_{1-1}$. Because $1/\kappa$ is a measure of the thickness of an ionic atmosphere, we see that the thickness of the ionic atmosphere of a 2–2 electrolyte is one half that of a 1–1 electrolyte.

25–56. In this problem, we will calculate the total electrostatic energy of an electrolyte solution in the Debye-Hückel theory. Use the equations in Problem 25–50 to show that the number of ions of type j in a spherical shell of radii r and $r + dr$ about a central ion of type i is

$$\left(\frac{N_{ij}(r)}{V}\right) 4\pi r^2 dr = C_j e^{-q_j \psi_i(r)/k_B T} 4\pi r^2 dr \approx C_j \left(1 - \frac{q_j \psi_i(r)}{k_B T}\right) 4\pi r^2 dr \qquad (1)$$

The total Coulombic interaction between the central ion of type i and the ions of type j in the spherical shell is $N_{ij}(r)u_{ij}(r)4\pi r^2 dr/V$ where $u_{ij}(r) = q_i q_j/4\pi\varepsilon_0\varepsilon_r r$. To determine the electrostatic interaction energy of all the ions in the solution with the central ion (of type i), U_i^{el}, sum $N_{ij}(r)u_{ij}(r)/V$ over all types of ions in a spherical shell and then integrate over all spherical shells to get

$$U_i^{el} = \int_0^\infty \left(\sum_j \frac{N_{ij}(r)u_{ij}(r)}{V}\right) 4\pi r^2 dr$$

$$= \sum_j \frac{C_j q_i q_j}{\varepsilon_0\varepsilon_r} \int_0^\infty \left(1 - \frac{q_j \psi_i(r)}{k_B T}\right) r dr$$

Use electroneutrality to show that

$$U_i^{el} = -q_i\kappa^2 \int_0^\infty \psi_i(r) r dr$$

Now, using Equation 6 of Problem 25–50, show that the interaction of all ions with the central ion (of type i) is given by

$$U_i^{el} = -\frac{q_i^2\kappa^2}{4\pi\varepsilon_0\varepsilon_r} \int_0^\infty e^{-\kappa r} dr = -\frac{q_i^2\kappa}{4\pi\varepsilon_0\varepsilon_r}$$

Now argue that the total electrostatic energy is

$$U^{\text{el}} = \frac{1}{2} \sum_i N_i U_i^{\text{el}} = -\frac{V k_{\text{B}} T \kappa^3}{8\pi}$$

Why is there a factor of 1/2 in this equation? Wouldn't you be overcounting the energy otherwise?

According to Problem 25–50, the number of ions of type j in a spherical shell of radii r and $r + dr$ about a central ion of type i is given by

$$\frac{N_{ij}(r) 4\pi r^2 dr}{V} = C_j e^{-q_j \psi_i(r)/k_{\text{B}} T} 4\pi r^2 dr$$

Linearize the exponential term to obtain

$$\frac{N_{ij}(r) 4\pi r^2 dr}{V} = C_j \left(1 - \frac{q_j \psi_i(r)}{k_{\text{B}} T} \right) 4\pi r^2 dr$$

The Coulombic interaction between the ions in the spherical shell and the central ion (of type i) is $u_{ij}(r) N_{ij}(r) 4\pi r^2 dr/V$, where $u_{ij}(r) = q_i q_j/4\pi\varepsilon_0\varepsilon_r r$. The interaction of all ions with the central ion is given by

$$U_i^{\text{el}} = \int_0^\infty \sum_j \frac{u_{ij}(r) N_{ij}(r) 4\pi r^2 dr}{V} = \sum_j \int_0^\infty \left(\frac{q_i q_j}{4\pi\varepsilon_0\varepsilon_r r} \right) C_j \left(1 - \frac{q_j \psi_i(r)}{k_{\text{B}} T} \right) 4\pi r^2 dr$$

$$= \frac{q_i}{4\pi\varepsilon_0\varepsilon_r} \sum_j \int_0^\infty q_j C_j 4\pi r\, dr - \frac{q_i}{4\pi\varepsilon_0\varepsilon_r k_{\text{B}} T} \sum_j q_j^2 C_j \int_0^\infty \psi_i(r) 4\pi r\, dr$$

$$= 0 \text{ (by electroneutrality)} - q_i \left(\sum_j \frac{q_j^2 C_j}{\varepsilon_0\varepsilon_r k_{\text{B}} T} \right) \int_0^\infty \psi_i(r) r\, dr$$

$$= -q_i \kappa^2 \int_0^\infty \psi_i(r) r\, dr$$

Using Equation 6 of Problem 25–50,

$$U_i^{\text{el}} = -\frac{q_i^2 \kappa^2}{4\pi\varepsilon_0\varepsilon_r} \int_0^\infty e^{-\kappa r} dr = -\frac{q_i^2 \kappa}{4\pi\varepsilon_0\varepsilon_r}$$

The total electrostatic energy is given by

$$U^{\text{el}} = \frac{1}{2} \sum_i N_i U_i^{\text{el}} = -\frac{V k_{\text{B}} T \kappa}{8\pi} \sum_i \frac{q_i^2 C_i}{\varepsilon_0\varepsilon_r k_{\text{B}} T} = -\frac{V k_{\text{B}} T \kappa^3}{8\pi}$$

The factor of 1/2 is needed in the second term in the above equation because in the summation over i, each ion occurs both as a central ion and as an ion in the spherical shell.

25–57. We derived an expression for U^{el} in the previous problem. Use the Gibbs-Helmholtz equation for A (Problem 22–23) to show that

$$A^{\text{el}} = -\frac{V k_{\text{B}} T \kappa^3}{12\pi}$$

Use the Gibbs-Helmholtz equation for A written in the form

$$\left(\frac{\partial \beta A^{\text{el}}}{\partial \beta} \right) = U^{\text{el}}$$

with (see Problem 25–56)

$$U^{el} = -\frac{V k_B T \kappa^3}{8\pi} = -\frac{V}{8\pi (k_B T)^{1/2}} \left(\sum_j \frac{q_j^2}{\varepsilon_0 \varepsilon_r} C_j \right)^{3/2}$$

$$= -\frac{V \beta^{1/2}}{8\pi} \left(\sum_j \frac{q_j^2 C_j}{\varepsilon_0 \varepsilon_r} \right)^{3/2}$$

Substitute this result into the Gibbs-Helmholtz equation and integrate from 0 to β to obtain

$$\beta A^{el} = -\frac{V \beta^{3/2}}{12\pi} \left(\sum_j \frac{q_j^2 C_j}{\varepsilon_0 \varepsilon_r} \right)^{3/2} = -\frac{V \kappa^3}{12\pi}$$

or

$$A^{el} = -\frac{V k_B T \kappa^3}{12\pi}$$

25–58. If we assume that the electrostatic interactions are the sole cause of the nonideality of an electrolyte solution, then we can say that

$$\mu_j^{el} = \left(\frac{\partial A^{el}}{\partial n_j} \right)_{T,V} = RT \ln \gamma_j^{el}$$

or that

$$\mu_j^{el} = \left(\frac{\partial A^{el}}{\partial N_j} \right)_{T,V} = k_B T \ln \gamma_j^{el}$$

Use the result you got for A^{el} in the previous problem to show that

$$k_B T \ln \gamma_j^{el} = -\frac{\kappa q_j^2}{8\pi \varepsilon_0 \varepsilon_r}$$

Use the formula

$$\gamma_\pm^\nu = \gamma_+^{\nu_+} \gamma_-^{\nu_-}$$

to show that

$$\ln \gamma_\pm = -\left(\frac{\nu_+ q_+^2 + \nu_- q_-^2}{\nu_+ + \nu_-} \right) \frac{\kappa}{8\pi \varepsilon_0 \varepsilon_r k_B T}$$

Use the electroneutrality condition $\nu_+ q_+ + \nu_- q_- = 0$ to rewrite $\ln \gamma_\pm$ as

$$\ln \gamma_\pm = -|q_+ q_-| \frac{\kappa}{8\pi \varepsilon_0 \varepsilon_r k_B T}$$

in agreement with Equation 25.49.

Using the final result from Problem 25–57,

$$
\begin{aligned}
\mu_j^{el} &= \left(\frac{\partial A^{el}}{\partial N_j}\right)_{\beta,V} \\
&= -\frac{V\beta^{1/2}}{12\pi}\frac{\partial}{\partial N_j}\left(\sum_j \frac{q_j^2 C_j}{\varepsilon_0\varepsilon_r V}\right)^{3/2} = -\frac{V\beta^{1/2}}{12\pi}\cdot\frac{3}{2}\left(\sum_j \frac{q_j^2 C_j}{\varepsilon_0\varepsilon_r}\right)^{1/2}\frac{q_j^2}{\varepsilon_0\varepsilon_r V} \\
&= -\frac{\kappa q_j^2}{8\pi\varepsilon_0\varepsilon_r} = k_B T \ln\gamma_j^{el}
\end{aligned}
$$

Now, take the logarithm of the equation $\gamma_\pm^\nu = \gamma_+^{\nu_+}\gamma_-^{\nu_-}$ and the previous result to obtain

$$
\ln\gamma_\pm = \frac{\nu_+ \ln\gamma_+ + \nu_- \ln\gamma_-}{\nu_+ + \nu_-} = -\left(\frac{\kappa}{8\pi\varepsilon_0\varepsilon_r k_B T}\right)\left(\frac{\nu_+ q_+^2 + \nu_- q_-^2}{\nu_+ + \nu_-}\right)
$$

But

$$
\begin{aligned}
\nu_+ q_+^2 + \nu_- q_-^2 &= q_+(\nu_+ q_+) + \nu_- q_-^2 = q_+(|\nu_- q_-|) + \nu_-|q_-|^2 \\
&= q_+|q_-|\left(\nu_- + \nu_-\frac{|q_-|}{q_+}\right) = |q_+ q_-|(\nu_- + \nu_+)
\end{aligned}
$$

where we have used the electroneutrality condition, $\nu_+ q_+ = \nu_-|q_-|$, and so finally

$$
\ln\gamma_\pm = -|q_+ q_-|\frac{\kappa}{8\pi\varepsilon_0\varepsilon_r k_B T}
$$

25–59. Derive Equation 25.56 from Equation 25.49.

See the solution to Problem 25–47, but do not include the factor $\rho = 0.99707 \text{ g}\cdot\text{mL}^{-1}$.

25–60. Show that Equation 25.59 reduces to Equation 25.49 for small concentrations.

We want to show that Equation 25.59 reduces to Equation 25.49 as $\rho \to 0$ or as $\kappa \to 0$. Let's consider $\ln\gamma_\pm^{el}$ first. Use the fact that $(1+x)^{1/2} = 1 + x/2 - x^2/8 + x^3/16 + \cdots$ to write

$$
\begin{aligned}
x(1+2x)^{1/2} - x - x^2 &= x\left[1 + x - \frac{(2x)^2}{8} + \frac{(2x)^3}{16} + O(x^4)\right] - x - x^2 \\
&= -\frac{x^3}{2}
\end{aligned}
$$

Using the fact that $x = \kappa d$, Equation 25.60 becomes

$$
\ln\gamma_\pm^{el} = -\frac{\kappa^3}{8\pi\rho}
$$

For a 1–1 electrolyte, $\kappa^2 = \rho/\varepsilon_0\varepsilon_r k_B T$, so we have

$$
\ln\gamma_\pm^{el} = -\frac{\kappa}{8\pi\varepsilon_0\varepsilon_r k_B T}
$$

in agreement with Equation 25.49 for a 1–1 electrolyte. The $\ln \gamma^{HS}$ contribution to Equation 25.59 is negligible when $\rho \to 0$ because $y = \pi \rho d^3/6$.

25–61. In this problem, we will investigate the temperature dependence of activities. Starting with the equation $\mu_1 = \mu_1^* + RT \ln a_1$, show that

$$\left(\frac{\partial \ln a_1}{\partial T}\right)_{P,x_1} = \frac{\overline{H}_1^* - \overline{H}_1}{RT^2}$$

where $\overline{H}_1^*$ is the molar enthalpy of the pure solvent (at one bar) and $\overline{H}_1$ is its partial molar enthalpy in the solution. The difference between $\overline{H}_1^*$ and $\overline{H}_1$ is small for dilute solutions, so a_1 is fairly independent of temperature.

Starting with $\mu_1 = \mu_1^* + RT \ln a_1$, differentiate μ_1/T with respect to T to obtain

$$\left(\frac{\partial \mu_1/T}{\partial T}\right)_{P,x_1} - \left(\frac{\partial \mu_1^*/T}{\partial T}\right)_P = R\left(\frac{\partial \ln a_1}{\partial T}\right)_{P,x_1}$$

Now use the equation (see Example 24–1)

$$\left(\frac{\partial \mu_j/T}{\partial T}\right)_{P,x_1} = -\frac{\overline{H}_j}{T^2}$$

to write

$$\left(\frac{\partial \ln a_1}{\partial T}\right)_{P,x_1} = \frac{\overline{H}_1^* - \overline{H}_1}{RT^2}$$

25–62. Henry's law says that the pressure of a gas in equilibrium with a non-electrolyte solution of the gas in a liquid is proportional to the molality of the gas in the solution for sufficiently dilute solutions. What form do you think Henry's law takes on for a gas such as $HCl(g)$ dissolved in water? Use the following data for $HCl(g)$ at 25°C to test your prediction.

$P_{HCl}/10^{-11}$ atm	$m_{HCl}/10^{-3}$ mol·kg^{-1}
0.147	1.81
0.238	2.32
0.443	3.19
0.663	3.93
0.851	4.47
1.080	5.06
1.622	6.25
1.929	6.84
2.083	7.12

A plot of pressure against molality is not a straight line, but a plot of pressure against molality squared is almost a straight line. This is due to the fact that $HCl(aq)$ dissociates into $H^+(aq)$ and $Cl^-(aq)$.

25–63. When the pressures in Problem 25–62 are plotted against molality squared, the result is almost a straight line. Curve fit the data to polynomials of the form

$$P = k_K m^2 (1 + c_1 m^{1/2} + c_2 m + c_3 m^{3/2} + \cdots)$$

of increasing degree and evaluate k_H.

If the data are fitted to $P = k_H m^2$, k_H turns out to be (in units of $\text{atm} \cdot \text{kg}^2 \cdot \text{mol}^{-2}$) 4.15×10^{-7}. The subsequent fits are (suppressing the units)

k_H	c_1	c_2	c_3	c_4
4.83×10^{-7}	-1.77			
4.92×10^{-7}	-2.24	3.48		
4.93×10^{-7}	-2.33	4.75	-6.21	
4.93×10^{-7}	-2.34	5.07	-9.45	10.3

Thus we see that $k_H = 4.93 \times 10^{-7} \text{ atm} \cdot \text{kg}^2 \cdot \text{mol}^{-2}$.

25–64. When the data in Problem 25–62 are plotted in the form of P/m^2 against $m^{1/2}$, the result is essentially a straight line with a negative slope. Why is this so? Use Debye-Hückel theory to calculate the slope of this line and compare your result with the final value of c_1 in Problem 25–63.

The activity of the HCl(aq) is given by $a_{HCl} = P/k_H$. Using the fact that $a_{HCl} = a_{\pm}^2 = m^2 \gamma_{\pm}^2$, we have

$$P = k_H m^2 \gamma_{\pm}^2$$

Note that as $m \to 0$, $\gamma_{\pm} \to 1$, and $P \to k_H m^2$, as expected. The Debye-Hückel expression for $\gamma_{\pm}$ in this case is

$$\ln \gamma_{\pm} = -1.171 m^{1/2}$$

Substitute this expression for $\gamma_{\pm}$ into $P = k_H m^2 \gamma_{\pm}^2$ and linearize the exponential according to $e^{-x} = 1 - x + \cdots$ to obtain

$$P = k_H m^2 (1 - 2.342 m^{1/2} + \cdots)$$

Thus, we predict that c_1 in Problem 25–63 is equal to -2.34, in excellent agreement.

CHAPTER 26

Chemical Equilibrium

PROBLEMS AND SOLUTIONS

26–1. Express the concentrations of each species in the following chemical equations in terms of the extent of reaction, ξ. The initial conditions are given under each equation.

a.

$$SO_2Cl_2(g) \; \rightleftharpoons \; SO_2(g) \; + \; Cl_2(g)$$

	SO_2Cl_2	SO_2	Cl_2
(1)	n_0	0	0
(2)	n_0	n_1	0

b.

$$2\,SO_3(g) \; \rightleftharpoons \; 2\,SO_2(g) \; + \; O_2(g)$$

(1)	n_0	0	0
(2)	n_0	0	n_1

c.

$$N_2(g) \; + \; 2\,O_2(g) \; \rightleftharpoons \; N_2O_4(g)$$

(1)	n_0	$2n_0$	0
(2)	n_0	n_0	0

We can use Equation 26.1 in all cases to express the concentrations of each species.

a.

$$SO_2Cl_2(g) \; \rightleftharpoons \; SO_2(g) \; + \; Cl_2(g)$$

(1)	$n_0 - \xi$	ξ	ξ
(2)	$n_0 - \xi$	$n_1 + \xi$	ξ

b.

$$2\,SO_3(g) \; \rightleftharpoons \; 2\,SO_2(g) \; + \; O_2(g)$$

(1)	$n_0 - 2\xi$	2ξ	ξ
(2)	$n_0 - 2\xi$	2ξ	$n_1 + \xi$

c.

$$N_2(g) \; + \; 2\,O_2(g) \; \rightleftharpoons \; N_2O_4(g)$$

(1)	$n_0 - \xi$	$2n_0 - 2\xi$	ξ
(2)	$n_0 - \xi$	$n_0 - 2\xi$	ξ

26–2. Write out the equilibrium-constant expression for the reaction that is described by the equation

$$2\,SO_2(g) + O_2(g) \rightleftharpoons 2\,SO_3(g)$$

Compare your result to what you get if the reaction is represented by

$$SO_2(g) + \frac{1}{2}O_2(g) \rightleftharpoons SO_3(g)$$

Using Equation 26.12, we write K_p for the first chemical equation as

$$K_P(T) = \frac{P_{SO_3}^2}{P_{O_2} P_{SO_2}^2}$$

For the second chemical equation, we again use Equation 26.12 to find

$$K_P'(T) = \frac{P_{SO_3}}{P_{O_2}^{1/2} P_{SO_2}}$$

which is the square root of K_P.

26–3. Consider the dissociation of $N_2O_4(g)$ into $NO_2(g)$ described by

$$N_2O_4(g) \rightleftharpoons 2\,NO_2(g)$$

Assuming that we start with n_0 moles of $N_2O_4(g)$ and no $NO_2(g)$, show that the extent of reaction, ξ_{eq}, at equilibrium is given by

$$\frac{\xi_{eq}}{n_0} = \left(\frac{K_P}{K_P + 4P}\right)^{1/2}$$

Plot ξ_{eq}/n_0 against P given that $K_P = 6.1$ at 100°C. Is your result in accord with Le Châtelier's principle?

At equilibrium, $n_{N_2O_4} = n_0 - \xi_{eq}$ and $n_{NO_2} = 2\xi_{eq}$. The partial pressures of the species are then

$$P_{N_2O_4} = \frac{n_0 - \xi_{eq}}{n_0 - \xi_{eq} + 2\xi_{eq}} P = \frac{n_0 - \xi_{eq}}{n_0 + \xi_{eq}} P$$

and

$$P_{NO_2} = \frac{2\xi_{eq}}{n_0 + \xi_{eq}} P = \frac{2\xi_{eq}}{n_0 + \xi_{eq}} P$$

Substituting into the expression for K_P, we find

$$K_P = \frac{P_{NO_2}^2}{P_{N_2O_4}} = \frac{4\xi_{eq}^2(n_0 + \xi_{eq})}{(n_0 + \xi_{eq})^2(n_0 - \xi_{eq})} P = \frac{4(\xi_{eq}/n_0)^2}{1 - (\xi_{eq}/n_0)^2} P$$

Solving this expression for ξ_{eq}/n_0, we find that

$$K_P - K_P(\xi_{eq}/n_0)^2 = 4P(\xi_{eq}/n_0)^2$$
$$(4P + K_P)(\xi_{eq}/n_0)^2 = K_P$$
$$\frac{\xi_{eq}}{n_0} = \left(\frac{K_P}{K_P + 4P}\right)^{1/2}$$

A plot of ξ_{eq}/n_0 against P is shown below.

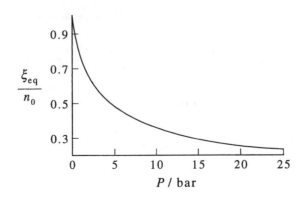

This is in accord with Le Châtelier's principle: as the pressure increases, the reaction occurs to a lesser extent and ξ_{eq} decreases.

26–4. In Problem 26–3 you plotted the extent of reaction at equilibrium against the total pressure for the dissociation of $N_2O_4(g)$ to $NO_2(g)$. You found that ξ_{eq} decreases as P increases, in accord with Le Châtelier's principle. Now let's introduce n_{inert} moles of an inert gas into the system. Assuming that we start with n_0 moles of $N_2O_4(g)$ and no $NO_2(g)$, derive an expression for ξ_{eq}/n_0 in terms of P and the ratio $r = n_{inert}/n_0$. As in Problem 26–3, let $K_P = 6.1$ and plot ξ_{eq}/n_0 versus P for $r = 0$ (Problem 26–3), $r = 0.50$, $r = 1.0$, and $r = 2.0$. Show that introducing an inert gas into the reaction mixture at constant pressure has the same effect as lowering the pressure. What is the effect of introducing an inert gas into a reaction system at constant volume?

At equilibrium, as before, $n_{N_2O_4} = n_0 - \xi_{eq}$ and $n_{NO_2} = 2\xi_{eq}$. However, the total number of moles present has changed to $n_0 + \xi_{eq} + n_{inert}$. The partial pressures of the species are then

$$P_{N_2O_4} = \frac{n_0 - \xi_{eq}}{n_0 + \xi_{eq} + n_{inert}} P \quad \text{and} \quad P_{NO_2} = \frac{2\xi_{eq}}{n_0 + \xi_{eq} + n_{inert}} P$$

Substituting into the expression for K_P, we find

$$K_P = \frac{P_{NO_2}^2}{P_{N_2O_4}} = \frac{4\xi_{eq}^2 P}{(n_0 - \xi_{eq})(n_0 + n_{inert} + \xi_{eq})} = \frac{4(\xi_{eq}/n_0)^2 P}{(1 - \xi_{eq}/n_0)(1 + r + \xi_{eq}/n_0)}$$

Solving for ξ_{eq}/n_0 gives

$$\frac{\xi_{eq}}{n_0} = -\frac{K_P r}{2(K_P + 4P)} \pm \frac{1}{2(K_P + 4P)}\left[K_P^2 r^2 + 4K_P(1 + r)(K_P + 4P)\right]^{1/2}$$

where we have let $r = n_{inert}/n_0$. When $r = 0$, as in Problem 26–3, this expression becomes

$$\frac{\xi_{eq}}{n_0} = \pm\frac{[4K_P(K_P + 4P)]^{1/2}}{2(K_P + 4P)} = \pm\left[\frac{K_P}{(K_P + 4P)}\right]^{1/2}$$

For ξ_{eq} to be positive, we take the positive root. Now we can plot this expression for various values of r:

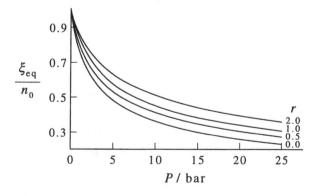

We see that introducing an inert gas into a constant-pressure reaction system increases the value of ξ_{eq} and so has the same effect as lowering the pressure. Introducing an inert gas into a constant-volume reaction system has no effect on the value of ξ_{eq}.

26–5. Re-do Problem 26–3 with n_0 moles of $N_2O_4(g)$ and n_1 moles of $NO_2(g)$ initially. Let $n_1/n_0 = 0.50$ and 2.0.

Now, at equilibrium, $n_{N_2O_4} = n_0 - \xi_{eq}$ and $n_{NO_2} = n_1 + 2\xi_{eq}$. The total number of moles of gas present will be $n_0 + n_1 + \xi_{eq}$. The partial pressures of the species are then, letting $s = n_1/n_0$,

$$P_{N_2O_4} = \frac{n_0 - \xi_{eq}}{n_0 + n_1 + \xi_{eq}} P = \frac{1 - \xi_{eq}/n_0}{1 + s + \xi_{eq}/n_0} P$$

and

$$P_{NO_2} = \frac{n_1 + 2\xi_{eq}}{n_0 + n_1 + \xi_{eq}} P = \frac{s + 2\xi_{eq}/n_0}{1 + s + \xi_{eq}/n_0} P$$

Substituting into the expression for K_P, we find

$$K_P = \frac{P_{NO_2}^2}{P_{N_2O_4}} = \frac{(s + 2\xi_{eq}/n_0)^2}{(1 + s + \xi_{eq}/n_0)(1 - \xi_{eq}/n_0)} P$$

and solving for ξ_{eq}/n_0 gives

$$\frac{\xi_{eq}}{n_0} = -\frac{s}{2} \pm \frac{1}{2}\left[s^2 - 4\left(\frac{Ps^2 - K_P s - K_P}{K_P + 4P} \right) \right]^{1/2}$$

When $s = 0$, as in Problem 26–3, this expression becomes

$$\frac{\xi_{eq}}{n_0} = \pm \frac{K_P^{1/2}}{2(K_P + 4P)^{1/2}} = \pm \left[\frac{K_P}{(K_P + 4P)} \right]^{1/2}$$

For ξ_{eq} to be positive, we take the positive root. Now we can plot this expression for various values of s:

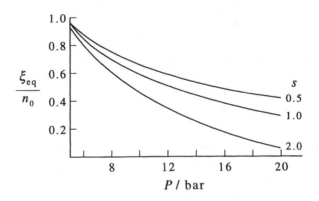

26–6. Consider the ammonia-synthesis reaction, which can be described by

$$N_2(g) + 3H_2(g) \rightleftharpoons 2NH_3(g)$$

Suppose initially there are n_0 moles of $N_2(g)$ and $3n_0$ moles of $H_2(g)$ and no $NH_3(g)$. Derive an expression for $K_P(T)$ in terms of the equilibrium value of the extent of reaction, ξ_{eq}, and the pressure, P. Use this expression to discuss how ξ_{eq}/n_0 varies with P and relate your conclusions to Le Châtelier's principle.

At equilibrium, there will be $n_0 - \xi_{eq}$ moles of $N_2(g)$, $3n_0 - 3\xi_{eq}$ moles of $H_2(g)$, and $2\xi_{eq}$ moles of $NH_3(g)$, yielding a total of $4n_0 - 2\xi_{eq}$ moles of gas. Then

$$P_{N_2} = \frac{n_0 - \xi_{eq}}{4n_0 - 2\xi_{eq}} P \qquad P_{H_2} = \frac{3n_0 - 3\xi_{eq}}{4n_0 - 2\xi_{eq}} P \qquad \text{and} \qquad P_{NH_3} = \frac{2\xi_{eq}}{4n_0 - 2\xi_{eq}} P$$

We then express K_P as

$$K_P = \frac{P_{NH_3}^2}{P_{H_2}^3 P_{N_2}}$$

$$= \frac{4\xi_{eq}^2 (4n_0 - 2\xi_{eq})^2}{(3n_0 - 3\xi_{eq})^3 (n_0 - \xi_{eq}) P^2}$$

$$= \frac{16(\xi_{eq}/n_0)^2 (2 - \xi_{eq}/n_0)^4}{27(2 - \xi_{eq}/n_0)^2 (1 - \xi_{eq}/n_0)^4 P^2}$$

$$= \frac{16(\xi_{eq}/n_0)^2 (2 - \xi_{eq}/n_0)^2}{27(1 - \xi_{eq}/n_0)^4 P^2}$$

The following plot of ξ_{eq}/n_0 against P shows that ξ_{eq}/n_0 increases as P increases, as Le Châtelier's principle would dictate.

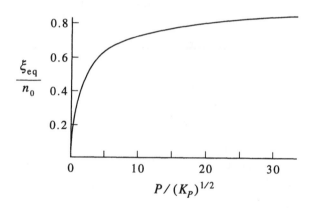

26–7. Nitrosyl chloride, NOCl, decomposes according to

$$2\,NOCl(g) \rightleftharpoons 2\,NO(g) + Cl_2(g)$$

Assuming that we start with n_0 moles of NOCl(g) and no NO(g) or $Cl_2(g)$, derive an expression for K_P in terms of the equilibrium value of the extent of reaction, ξ_{eq}, and the pressure, P. Given that $K_P = 2.00 \times 10^{-4}$, calculate ξ_{eq}/n_0 when $P = 0.080$ bar. What is the new value of ξ_{eq}/n_0 at equilibrium when $P = 0.160$ bar? Is this result in accord with Le Châtelier's principle?

At equilibrium, there will be $n_0 - 2\xi_{eq}$ moles of NOCl(g), $2\xi_{eq}$ moles of NO(g), and ξ_{eq} moles of $Cl_2(g)$, making a total of $n_0 + \xi_{eq}$ moles of gas present. Then

$$P_{NOCl} = \frac{n_0 - 2\xi_{eq}}{n_0 + \xi_{eq}} P \qquad P_{NO} = \frac{2\xi_{eq}}{n_0 + \xi_{eq}} P \qquad \text{and} \qquad P_{Cl_2} = \frac{\xi_{eq}}{n_0 + \xi_{eq}} P$$

We then write K_P as

$$K_P = \frac{P_{Cl_2} P_{NO}^2}{P_{NOCl}^2} = \frac{4\xi_{eq}^3 P}{(n_0 + \xi_{eq})(n_0 - 2\xi_{eq})^2} = \frac{4(\xi_{eq}/n_0)^3 P}{(1 + \xi_{eq}/n_0)(1 - 2\xi_{eq}/n_0)^2}$$

A plot of ξ_{eq}/n_0 against P looks like

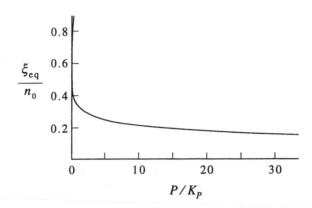

We see that ξ_{eq}/n_0 decreases as P increases, as Le Châtelier's principle would dictate. Letting $K_P = 2.00 \times 10^{-4}$, we find $\xi_{eq}/n_0 = 0.0783$ at $P = 0.080$ bar and $\xi_{eq}/n_0 = 0.0633$ at $P = 0.160$ bar, again in accord with Le Châtelier's principle.

26–8. The value of K_P at 1000°C for the decomposition of carbonyl dichloride (phosgene) according to

$$COCl_2(g) \rightleftharpoons CO(g) + Cl_2(g)$$

is 34.8 if the standard state is taken to be one bar. What would the value of K_P be if for some reason the standard state were taken to be 0.500 bar? What does this result say about the numerical values of equilibrium constants?

Use the definition of K_P to find the value of $K_P(0.500 \text{ bar})$ at the new standard state:

$$K_P(1 \text{ bar}) = \frac{(P_{CO}/1 \text{ bar})(P_{Cl_2}/1 \text{ bar})}{(P_{COCl_2}/1 \text{ bar})} = 34.8$$

$$K_P(0.500 \text{ bar}) = \frac{(P_{CO}/0.500 \text{ bar})(P_{Cl_2}/0.500 \text{ bar})}{(P_{COCl_2}/0.500 \text{ bar})}$$

$$= 0.500 K_P(1 \text{ bar}) = 17.4$$

The numerical values of equilibrium constants are dependent on the standard states chosen.

26–9. Most gas-phase equilibrium constants in the recent chemical literature were calculated assuming a standard state pressure of one atmosphere. Show that the corresponding equilibrium constant for a standard state pressure of one bar is given by

$$K_P(\text{bar}) = K_P(\text{atm})(1.01325)^{\Delta\nu}$$

where $\Delta\nu$ is the sum of the stoichiometric coefficients of the products minus that of the reactants.

Consider the reaction described by

$$\nu_A A(g) + \nu_B B(g) \rightleftharpoons \nu_Y Y(g) + \nu_Z Z(g)$$

(We can extend this case to include as many reactants and products as we desire.) Now write K_p:

$$K_p(\text{bar}) = \frac{(P_Z/1\ \text{bar})^{\nu_Z}(P_Y/1\ \text{bar})^{\nu_Y}}{(P_B/1\ \text{bar})^{\nu_B}(P_A/1\ \text{bar})^{\nu_A}}$$

$$= \frac{P_Z^{\nu_Z} P_Y^{\nu_Y}}{(1\ \text{bar})^{\nu_Z+\nu_Y}} \frac{(1\ \text{bar})^{\nu_A+\nu_B}}{P_B^{\nu_B} P_A^{\nu_A}}$$

$$= \frac{P_Z^{\nu_Z} P_Y^{\nu_Y}}{P_B^{\nu_B} P_A^{\nu_A}} \left(\frac{1}{1\ \text{bar}}\right)^{\Delta\nu} \left(\frac{1.01325\ \text{bar}}{1\ \text{atm}}\right)^{\Delta\nu}$$

$$= \frac{(P_Z/1\ \text{atm})^{\nu_Z}(P_Y/1\ \text{atm})^{\nu_Y}}{(P_B/1\ \text{atm})^{\nu_B}(P_A/1\ \text{atm})^{\nu_A}}(1.01325\ \text{bar})^{\Delta\nu}$$

$$= K_p(\text{atm})(1.01325\ \text{bar})^{\Delta\nu}$$

26–10. Using the data in Table 26.1, calculate $\Delta_r G°(T)$ and $K_p(T)$ at 25°C for

(a) $N_2O_4(g) \rightleftharpoons 2\,NO_2(g)$
(b) $H_2(g)+I_2(g) \rightleftharpoons 2\,HI(g)$
(c) $3\,H_2(g) + N_2(g) \rightleftharpoons 2\,NH_3(g)$

Use Equations 26.19 and 26.11 to find $\Delta_r G°$ and K_p.

a. $\Delta_r G° = 2(51.258\ \text{kJ·mol}^{-1}) - 97.787\ \text{kJ·mol}^{-1}$
$= 4.729\ \text{kJ·mol}^{-1}$
$K_p = e^{-\Delta_r G°/RT} = 0.148$

b. $\Delta_r G° = 2(1.560\ \text{kJ·mol}^{-1}) - 19.325\ \text{kJ·mol}^{-1}$
$= -16.205\ \text{kJ·mol}^{-1}$
$K_p = e^{-\Delta_r G°/RT} = 690$

c. $\Delta_r G° = 2(-16.637\ \text{kJ·mol}^{-1}) = -33.274\ \text{kJ·mol}^{-1}$
$K_p = e^{-\Delta_r G°/RT} = 6.80 \times 10^5$

26–11. Calculate the value of $K_c(T)$ based upon a one mol·L^{-1} standard state for each of the equations in Problem 26–10.

Use Equation 26.17, recalling that R must be in units of L·bar·K^{-1}·mol^{-1} because $c°$ is 1 mol·L^{-1} and $P°$ is 1 bar:

a. $K_c = K_p(RT)^{-\Delta\nu} = (0.148)[(298.15\ \text{K})(0.083145\ \text{L·bar·K}^{-1}\text{·mol}^{-1})]^{-1} = 5.97 \times 10^{-3}$
b. $K_c = K_p(RT)^{-\Delta\nu} = (690)[(298.15\ \text{K})(0.083145\ \text{L·bar·K}^{-1}\text{·mol}^{-1})]^0 = 690$
c. $K_c = K_p(RT)^{-\Delta\nu} = (6.80 \times 10^5)[(298.15\ \text{K})(0.083145\ \text{L·bar·K}^{-1}\text{·mol}^{-1})]^2 = 4.17 \times 10^8$

26–12. Derive a relation between K_p and K_c for the following:

a. $CO(g) + Cl_2(g) \rightleftharpoons COCl_2(g)$
b. $CO(g) + 3\,H_2(g) \rightleftharpoons CH_4(g) + H_2O(g)$
c. $2\,BrCl(g) \rightleftharpoons Br_2(g) + Cl_2(g)$

Again, use Equation 26.17.

a. $K_P = K_c \left(\frac{c°RT}{P°}\right)^{-1}$

b. $K_P = K_c \left(\frac{c°RT}{P°}\right)^{-2}$

c. $K_P = K_c \left(\frac{c°RT}{P°}\right)^{0} = K_c$

26–13. Consider the dissociation reaction of $I_2(g)$ described by

$$I_2(g) \rightleftharpoons 2\,I(g)$$

The total pressure and the partial pressure of $I_2(g)$ at 1400°C have been measured to be 36.0 torr and 28.1 torr, respectively. Use these data to calculate K_P (one bar standard state) and K_c (one $mol \cdot L^{-1}$ standard state) at 1400°C.

First we express P_{I_2} and P_I in bars:

$$P_{I_2} = 28.1 \text{ torr} \left(\frac{1.01325 \text{ bar}}{760 \text{ torr}}\right) = 0.0375 \text{ bar}$$

$$P_I = P_{tot} - P_{I_2} = 7.9 \text{ torr} \left(\frac{1.01325 \text{ bar}}{760 \text{ torr}}\right) = 0.0105 \text{ bar}$$

Now use the definitions of K_P and K_c to write

$$K_P = \frac{P_I^2}{P_{I_2}} = 2.94 \times 10^{-3}$$

$$K_c = K_P \left(\frac{P°}{c°RT}\right)$$

$$= (2.94 \times 10^{-3}) \left[\frac{1 \text{ bar}}{(1 \text{ mol} \cdot L^{-1})(1673 \text{ K})(0.083145 \text{ L} \cdot \text{bar} \cdot \text{mol}^{-1} \cdot \text{K}^{-1})}\right]$$

$$= 2.11 \times 10^{-5}$$

26–14. Show that

$$\frac{d \ln K_c}{dT} = \frac{\Delta_r U°}{RT^2}$$

for a reaction involving ideal gases.

We know that

$$K_P = K_c \left(\frac{c°RT}{P°}\right)^{\Delta\nu} \tag{26.17}$$

Now begin with Equation 26.29:

$$\frac{\Delta_r H°}{RT^2} = \frac{d \ln K_P}{dT} = \frac{d}{dT}\left[\ln K_c + \Delta\nu \ln \left(\frac{c°RT}{P°}\right)\right]$$

$$= \frac{d \ln K_c}{dT} + \frac{\Delta\nu}{T}$$

or

$$\frac{d \ln K_c}{dT} = \frac{\Delta_r H^\circ - \Delta \nu RT}{RT^2} = \frac{\Delta_r H^\circ - \Delta(PV)}{RT^2} = \frac{\Delta_r U^\circ}{RT^2}$$

because $U = H + PV$.

26–15. Consider the gas-phase reaction for the synthesis of methanol from $CO(g)$ and $H_2(g)$

$$CO(g) + 2\,H_2(g) \rightleftharpoons CH_3OH(g)$$

The value of the equilibrium constant K_P at 500 K is 6.23×10^{-3}. Initially equimolar amounts of $CO(g)$ and $H_2(g)$ are introduced into the reaction vessel. Determine the value of ξ_{eq}/n_0 at equilibrium at 500 K and 30 bar.

At equilibrium, the number of moles of $CO(g)$ will be $n_0 - \xi_{eq}$, the number of moles of $H_2(g)$ will be $n_0 - 2\xi_{eq}$, and the number of moles of $CH_3OH(g)$ will be ξ_{eq}. The total moles of gas present will therefore be $2n_0 - 2\xi_{eq}$. We can now find the partial pressures of each of the components of the mixture:

$$P_{CO} = \frac{n_0 - \xi_{eq}}{2(n_0 - \xi_{eq})} P = \frac{1}{2} P \qquad P_{H_2} = \frac{n_0 - 2\xi_{eq}}{2(n_0 - \xi_{eq})} P \qquad \text{and} \qquad P_{CH_3OH} = \frac{\xi_{eq}}{2(n_0 - \xi_{eq})} P$$

We then express K_P as

$$K_P = \frac{P_{CH_3OH}}{P_{H_2}^2 P_{CO}} = \frac{4\xi_{eq}(n_0 - \xi_{eq})}{P^2(n_0 - 2\xi_{eq})^2} = \frac{4x(1-x)}{P^2(1-2x)^2}$$

where $x = \xi_{eq}/n_0$. The value of K_P is 6.23×10^{-3}, so

$$\frac{4x(1-x)}{(1-2x)^2} = (30 \text{ bar})^2 (6.23 \times 10^{-3})$$

which we can solve numerically (using the Newton-Raphson method) to find

$$x = 0.305 \qquad \text{or} \qquad x = 0.695$$

Since $x < 0.50$ (otherwise the amount of H_2 present will be a negative quantity), $\xi_{eq}/n_0 = 0.31$.

26–16. Consider the two equations

(1) $\quad CO(g) + H_2O(g) \rightleftharpoons CO_2(g) + H_2(g) \qquad K_1$

(2) $\quad CH_4(g) + H_2O(g) \rightleftharpoons CO(g) + 3\,H_2(g) \qquad K_2$

Show that $K_3 = K_1 K_2$ for the sum of these two equations

(3) $\quad CH_4(g) + 2\,H_2O(g) \rightleftharpoons CO_2(g) + 4\,H_2(g) \qquad K_3$

How do you explain the fact that you would add the values of $\Delta_r G^\circ$ but multiply the equilibrium constants when adding Equations 1 and 2 to get Equation 3?

Use Equation 26.12 to express K_1, K_2, and K_3.

$$K_1 = \frac{P_{CO_2} P_{H_2}}{P_{H_2O} P_{CO}} \qquad\qquad K_2 = \frac{P_{H_2}^3 P_{CO}}{P_{H_2O} P_{CH_4}}$$

$$K_3 = \frac{P_{H_2}^4 P_{CO_2}}{P_{H_2O}^2 P_{CH_4}} = \left(\frac{P_{CO_2} P_{H_2}}{P_{H_2O} P_{CO}}\right)\left(\frac{P_{H_2}^3 P_{CO}}{P_{H_2O} P_{CH_4}}\right) = K_1 K_2$$

We multiply the equilibrium constants because of their logarithmic relationship with $\Delta_r G°$. Recall that (Equation 26.11) $\Delta_r G° = -RT \ln K_p$. Adding $\Delta_r G_1°$ and $\Delta_r G_2°$ would give

$$-RT \ln K_1 - RT \ln K_2 = -RT \ln K_1 K_2$$

26-17. Given:

$$
\begin{aligned}
2\,BrCl(g) &\rightleftharpoons Cl_2(g) + Br_2(g) &\qquad K_P &= 0.169 \\
2\,IBr(g) &\rightleftharpoons Br_2(g) + I_2(g) &\qquad K_P &= 0.0149
\end{aligned}
$$

Determine K_P for the reaction

$$BrCl(g) + \tfrac{1}{2} I_2(g) \rightleftharpoons IBr(g) + \tfrac{1}{2} Cl_2(g)$$

We number the equations in order of appearance. Equation 3 can be expressed by

$$\text{Equation } 3 = \tfrac{1}{2}\text{Equation } 1 - \tfrac{1}{2}\text{Equation } 2$$

This means that

$$\Delta_r G_3° = \tfrac{1}{2}\Delta_r G_1° - \tfrac{1}{2}\Delta_r G_2°$$

or

$$K_3 = \frac{K_1^{1/2}}{K_2^{1/2}} = \frac{(0.169)^{1/2}}{(0.0149)^{1/2}} = 3.37$$

26-18. Consider the reaction described by

$$Cl_2(g) + Br_2(g) \rightleftharpoons 2\,BrCl(g)$$

at 500 K and a total pressure of one bar. Suppose that we start with one mole each of $Cl_2(g)$ and $Br_2(g)$ and no $BrCl(g)$. Show that

$$G(\xi) = (1-\xi)G_{Cl_2}° + (1-\xi)G_{Br_2}° + 2\xi G_{BrCl}° + 2(1-\xi)RT \ln \frac{1-\xi}{2} + 2\xi RT \ln \xi$$

where ξ is the extent of reaction. Given that $G_{BrCl}° = -3.694$ kJ·mol^{-1} at 500 K, plot $G(\xi)$ versus ξ. Differentiate $G(\xi)$ with respect to ξ and show that the minimum value of $G(\xi)$ occurs at $\xi_{eq} = 0.549$. Also show that

$$\left(\frac{\partial G}{\partial \xi}\right)_{T,P} = \Delta_r G° + RT \ln \frac{P_{BrCl}^2}{P_{Cl_2} P_{Br_2}}$$

and that $K_P = 4\xi_{eq}^2/(1-\xi_{eq})^2 = 5.9$.

As the reaction progresses, the amount of $Cl_2(g)$ and $Br_2(g)$ can be expressed as $1 - \xi$ and the amount of $BrCl(g)$ will be 2ξ. We can then write the Gibbs energy of the reaction mixture as (Equation 26.20)

$$G(\xi) = (1-\xi)\overline{G}_{Cl_2} + (1-\xi)\overline{G}_{Br_2} + 2\xi \overline{G}_{BrCl}$$

Since the reaction is carried out at a total pressure of 1 bar, we can write

$$P_{Cl_2} = P_{Br_2} = \frac{1-\xi}{2} \qquad \text{and} \qquad P_{BrCl} = \frac{2\xi}{2} = \xi$$

We can use these expressions and Equation 22.59 to write $G(\xi)$ as

$$G(\xi) = (1-\xi)\left[G^\circ_{Cl_2} + RT \ln P_{Cl_2}\right] + (1-\xi)\left[G^\circ_{Br_2} + RT \ln P_{Br_2}\right]$$
$$+ 2\xi\left[G^\circ_{BrCl} + RT \ln P_{BrCl}\right]$$
$$= (1-\xi)G^\circ_{Cl_2} + (1-\xi)G^\circ_{Br_2} + 2\xi G^\circ_{BrCl}$$
$$+ 2(1-\xi)RT \ln \frac{1-\xi}{2} + 2\xi RT \ln \xi \tag{1}$$

Substituting the value given in the problem for G°_{BrCl} and zero for $G^\circ_{Cl_2}$ and $G^\circ_{Br_2}$ gives

$$G(\xi) = 2\xi(-3694 \text{ J·mol}^{-1}) + 2(1-\xi)(8.3145 \text{ J·mol}^{-1}·\text{K}^{-1})(500 \text{ K}) \ln \frac{1-\xi}{2}$$
$$+ 2\xi(8.3145 \text{ J·mol}^{-1}·\text{K}^{-1})(500 \text{ K}) \ln \xi$$
$$= (-7388 \text{ J·mol}^{-1})\xi + (8314.5 \text{ J·mol}^{-1})\left[(1-\xi)\ln\frac{1-\xi}{2} + \xi\ln\xi\right] \tag{2}$$

A plot of $G(\xi)$ against ξ is shown below.

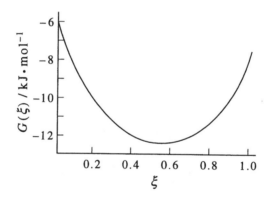

At equilibrium, $0 = (\partial G/\partial\xi)_{T,P}$. We use Equation 2 to express $G(\xi)$ and find that, at equilibrium,

$$0 = -7388 \text{ J·mol}^{-1} + (8314.5 \text{ J·mol}^{-1})\left[-\ln\frac{1-\xi_{eq}}{2} - 1 + \ln\xi_{eq} + 1\right]$$

$$0 = -7388 \text{ J·mol}^{-1} + (8314.5 \text{ J·mol}^{-1}) \ln\frac{2\xi_{eq}}{1-\xi_{eq}}$$

$$\frac{2\xi_{eq}}{1-\xi_{eq}} = 2.432$$

Solving for ξ_{eq} gives $\xi_{eq} = 0.549$.

Differentiating Equation 1 for $G(\xi)$ explicitly, we find that

$$\left(\frac{\partial G}{\partial \xi}\right)_{T,P} = -G^{\circ}_{Cl_2} - G^{\circ}_{Br_2} + 2G^{\circ}_{BrCl} - 2RT \ln \frac{1-\xi}{2}$$

$$+ 2(1-\xi)RT \left(\frac{2}{1-\xi}\right)\left(-\frac{1}{2}\right) + 2RT \ln \xi + \frac{2\xi RT}{\xi}$$

$$= \Delta_r G^{\circ} + RT \ln \frac{4}{(1-\xi)^2} - 2RT + RT \ln \xi^2 + 2RT$$

$$= \Delta_r G^{\circ} + RT \ln \left(\frac{2\xi}{1-\xi}\right)^2 = \Delta_r G^{\circ} + RT \ln \left(\frac{P^2_{BrCl}}{P_{Cl_2} P_{Br_2}}\right)$$

Note that

$$K_P = \frac{P^2_{BrCl}}{P_{Cl_2} P_{Br_2}} = \frac{4\xi^2_{eq}}{(1-\xi_{eq})^2}$$

so, at equilibrium,

$$K_P = \frac{4(0.549)^2}{(1-0.549)^2} = 5.9$$

26–19. Consider the reaction described by

$$2\,H_2O(g) \rightleftharpoons 2\,H_2(g) + O_2(g)$$

at 4000 K and a total pressure of one bar. Suppose that we start with two moles of $H_2O(g)$ and no $H_2(g)$ or $O_2(g)$. Show that

$$G(\xi) = 2(1-\xi)G^{\circ}_{H_2O} + 2\xi G^{\circ}_{H_2} + \xi G^{\circ}_{O_2} + 2(1-\xi)RT \ln \frac{2(1-\xi)}{2+\xi}$$

$$+ 2\xi RT \ln \frac{2\xi}{2+\xi} + \xi RT \ln \frac{\xi}{2+\xi}$$

where ξ is the extent of reaction. Given that $\Delta_f G^{\circ}[H_2O(g)] = -18.334\ \text{kJ·mol}^{-1}$ at 4000 K, plot $G(\xi)$ against ξ. Differentiate $G(\xi)$ with respect to ξ and show that the minimum value of $G(\xi)$ occurs at $\xi_{eq} = 0.553$. Also show that

$$\left(\frac{\partial G}{\partial \xi}\right)_{T,P} = \Delta_r G^{\circ} + RT \ln \frac{P^2_{H_2} P_{O_2}}{P^2_{H_2O}}$$

and that $K_P = \xi^3_{eq}/(2+\xi_{eq})(1-\xi_{eq})^2 = 0.333$ at one bar.

The amount of H_2O can be expressed by $2 - 2\xi$, the amount of H_2 as 2ξ, and the amount of O_2 as ξ. We can then write the Gibbs energy of the reaction mixture as (Equation 26.20)

$$G(\xi) = (2 - 2\xi)\overline{G}_{H_2O} + 2\xi \overline{G}_{H_2} + \xi \overline{G}_{O_2}$$

Since the reaction is carried out at a total pressure of one bar, we can write

$$P_{H_2O} = \frac{2(1-\xi)}{2+\xi} \qquad P_{H_2} = \frac{2\xi}{2+\xi} \qquad \text{and} \qquad P_{O_2} = \frac{\xi}{2+\xi}$$

We can use these expressions and Equation 8.59 to write $G(\xi)$ as

$$G(\xi) = 2(1-\xi)\left[G^\circ_{H_2O} + RT \ln P_{H_2O}\right] + 2\xi\left[G^\circ_{H_2} + RT \ln P_{H_2}\right]$$

$$+\xi\left[G^\circ_{O_2} + RT \ln P_{O_2}\right]$$

$$= 2(1-\xi)G^\circ_{H_2O} + 2\xi G^\circ_{H_2} + \xi G^\circ_{O_2} + 2(1-\xi)RT \ln \frac{2(1-\xi)}{2+\xi}$$

$$+2\xi RT \ln \frac{2\xi}{2+\xi} + \xi RT \ln \frac{\xi}{2+\xi}$$

Substituting the value given in the problem for $G^\circ_{H_2O}$ and zero for $G^\circ_{O_2}$ and $G^\circ_{H_2}$ gives

$$G(\xi) = 2(1-\xi)(-18\,334\,\text{J·mol}^{-1})$$

$$+2(1-\xi)(8.3145\,\text{J·mol}^{-1}\cdot\text{K}^{-1})(4000\,\text{K}) \ln \frac{2(1-\xi)}{2+\xi}$$

$$+2\xi(8.3145\,\text{J·mol}^{-1}\cdot\text{K}^{-1})(4000\,\text{K}) \ln \frac{2\xi}{2+\xi}$$

$$+\xi(8.3145\,\text{J·mol}^{-1}\cdot\text{K}^{-1})(4000\,\text{K}) \ln \frac{\xi}{2+\xi}$$

$$= (-36\,668\,\text{J·mol}^{-1})(1-\xi) + (66\,516\,\text{J·mol}^{-1})(1-\xi) \ln \frac{2(1-\xi)}{2+\xi}$$

$$+(66\,516\,\text{J·mol}^{-1})\xi \ln \frac{2\xi}{2+\xi} + (33\,258\,\text{J·mol}^{-1})\xi \ln \frac{\xi}{2+\xi}$$

A plot of $G(\xi)$ against ξ is shown below.

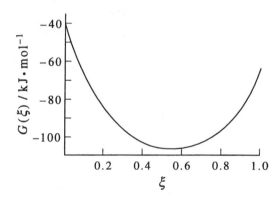

We now differentiate $G(\xi)$ with respect to ξ and find

$$\left(\frac{\partial G}{\partial \xi}\right)_{T,P} = (36\,668\,\text{J·mol}^{-1})$$

$$-(66\,516\,\text{J·mol}^{-1})\left\{\ln \frac{2(1-\xi)}{2+\xi} - \frac{(1-\xi)(2+\xi)}{2(1-\xi)}\left[-\frac{2}{2+\xi} - \frac{2(1-\xi)}{(2+\xi)^2}\right]\right\}$$

$$+(66\,516\,\text{J·mol}^{-1})\left\{\ln \frac{2\xi}{2+\xi} + \frac{\xi(2+\xi)}{2\xi}\left[\frac{2}{2+\xi} - \frac{2\xi}{(2+\xi)^2}\right]\right\}$$

$$+(33\,258\,\text{J·mol}^{-1})\left\{\ln \frac{\xi}{2+\xi} + (2+\xi)\left[\frac{1}{2+\xi} - \frac{\xi}{(2+\xi)^2}\right]\right\}$$

$$= (36\,668\,\text{J·mol}^{-1}) - (66\,516\,\text{J·mol}^{-1})\left[\ln \frac{2(1-\xi)}{2+\xi} + 1 + \frac{1-\xi}{2+\xi}\right]$$

$$+(66\,516\,\text{J·mol}^{-1})\left[\ln \frac{2\xi}{2+\xi} + 1 - \frac{\xi}{2+\xi}\right]$$

$$+(33\,258\,\mathrm{J\cdot mol^{-1}})\left[\ln\frac{\xi}{2+\xi}+1-\frac{\xi}{2+\xi}\right]$$

$$=36\,668\,\mathrm{J\cdot mol^{-1}}+(33\,258\,\mathrm{J\cdot mol^{-1}})\left[2\ln\frac{2+\xi}{2(1-\xi)}-\frac{2(1-\xi)}{2+\xi}+2\ln\frac{2\xi}{2+\xi}\right.$$

$$\left.-\frac{2\xi}{2+\xi}+\ln\frac{\xi}{2+\xi}+1-\frac{\xi}{2+\xi}\right]$$

$$=36\,668\,\mathrm{J\cdot mol^{-1}}+(33\,258\,\mathrm{J\cdot mol^{-1}})\ln\frac{\xi^3}{(1-\xi)^2(2+\xi)}\tag{1}$$

At equilibrium, $(\partial G/\partial\xi)_{T,P}=0$, so

$$\frac{\xi_{eq}^3}{(1-\xi_{eq})^2(2+\xi_{eq})}=\exp\left(\frac{-36\,668\,\mathrm{J\cdot mol^{-1}}}{33\,258\,\mathrm{J\cdot mol^{-1}}}\right)=0.332$$

Solving for ξ_{eq} gives $\xi_{eq}=0.553$.
Note that substituting for $\Delta_f G°[\mathrm{H_2O(g)}]$, R, T, $P_{\mathrm{H_2}}$, $P_{\mathrm{O_2}}$, and $P_{\mathrm{H_2O}}$ in Equation 1 gives

$$\left(\frac{\partial G}{\partial\xi}\right)_{T,P}=2\Delta_f G°[\mathrm{H_2O(g)}]+RT\ln\frac{P_{\mathrm{H_2}}^2\,P_{\mathrm{O_2}}}{P_{\mathrm{H_2O}}^2}=\Delta_r G°+RT\ln\frac{P_{\mathrm{H_2}}^2\,P_{\mathrm{O_2}}}{P_{\mathrm{H_2O}}}$$

Now

$$K_P=\frac{P_{\mathrm{H_2}}^2\,P_{\mathrm{O_2}}}{P_{\mathrm{H_2O}}^2}=\frac{\xi_{eq}^3}{(1-\xi_{eq})^2(2+\xi_{eq})}$$

so

$$K_P=\frac{(0.553)^3}{(1-0.553)^2(2+0.553)}=0.332$$

26–20. Consider the reaction described by

$$3\,\mathrm{H_2(g)}+\mathrm{N_2(g)}\rightleftharpoons 2\,\mathrm{NH_3(g)}$$

at 500 K and a total pressure of one bar. Suppose that we start with three moles of $\mathrm{H_2(g)}$, one mole of $\mathrm{N_2(g)}$, and no $\mathrm{NH_3(g)}$. Show that

$$G(\xi)=(3-3\xi)G_{\mathrm{H_2}}°+(1-\xi)G_{\mathrm{N_2}}°+2\xi\,G_{\mathrm{NH_3}}°$$

$$+(3-3\xi)RT\ln\frac{3-3\xi}{4-2\xi}+(1-\xi)RT\ln\frac{1-\xi}{4-2\xi}+2\xi\,RT\ln\frac{2\xi}{4-2\xi}$$

where ξ is the extent of reaction. Given that $G_{\mathrm{NH_3}}°=4.800\,\mathrm{kJ\cdot mol^{-1}}$ at 500 K (see Table 26.4), plot $G(\xi)$ versus ξ. Differentiate $G(\xi)$ with respect to ξ and show that the minimum value of $G(\xi)$ occurs at $\xi_{eq}=0.158$. Also show that

$$\left(\frac{\partial G}{\partial\xi}\right)_{T,P}=\Delta_r G°+RT\ln\frac{P_{\mathrm{NH_3}}^2}{P_{\mathrm{H_2}}^3\,P_{\mathrm{N_2}}}$$

and that $K_P=16\xi_{eq}^2(2-\xi_{eq})^2/27(1-\xi_{eq})^4=0.10$.

The amount of H_2 can be expressed by $3 - 3\xi$, the amount of N_2 as $1 - \xi$, and the amount of NH_3 as 2ξ. We can then write the Gibbs energy of the reaction mixture as

$$G(\xi) = (3 - 3\xi)\overline{G}_{H_2} + (1 - \xi)\overline{G}_{N_2} + 2\xi\overline{G}_{NH_3}$$

Since the reaction is carried out at a total pressure of one bar, we can write

$$P_{H_2} = \frac{3(1 - \xi)}{2(2 - \xi)} \qquad P_{N_2} = \frac{1 - \xi}{2(2 - \xi)} \qquad \text{and} \qquad P_{NH_3} = \frac{\xi}{(2 - \xi)}$$

We can use these expressions and Equation 8.59 to write $G(\xi)$ as (Equation 26.20)

$$\begin{aligned}
G(\xi) &= 3(1 - \xi)\left[G°_{H_2} + RT \ln P_{H_2}\right] + (1 - \xi)\left[G°_{N_2} + RT \ln P_{N_2}\right] \\
&\quad + 2\xi\left[G°_{NH_3} + RT \ln P_{NH_3}\right] \\
&= 3(1 - \xi)G°_{H_2} + (1 - \xi)G°_{N_2} + 2\xi G°_{NH_3} \\
&\quad + 3(1 - \xi)RT \ln \frac{3(1 - \xi)}{2(2 - \xi)} + (1 - \xi)RT \ln \frac{1 - \xi}{2(2 - \xi)} \\
&\quad + 2\xi RT \ln \frac{\xi}{2 - \xi}
\end{aligned}$$

Substituting the appropriate values of $G°$ gives

$$\begin{aligned}
G(\xi) &= \xi(9600 \text{ J·mol}^{-1}) + 3(1 - \xi)(4157.2 \text{ J·mol}^{-1}) \ln \frac{3(1 - \xi)}{2(2 - \xi)} \\
&\quad + (4157.2 \text{ J·mol}^{-1})\left[(1 - \xi) \ln \frac{1 - \xi}{2(2 - \xi)} + 2\xi \ln \frac{\xi}{2 - \xi}\right]
\end{aligned}$$

A plot of $G(\xi)$ against ξ is shown below.

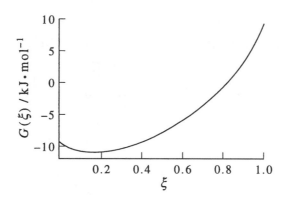

Now we differentiate $G(\xi)$ with respect to ξ:

$$\left(\frac{\partial G}{\partial \xi}\right)_{T,P} = 9600 \text{ J·mol}^{-1}$$

$$\begin{aligned}
&+ (4157.2 \text{ J·mol}^{-1})\left\{-3 \ln \frac{3(1 - \xi)}{2(2 - \xi)} + 2(2 - \xi)\left[-\frac{3}{2(2 - \xi)} + \frac{6(1 - \xi)}{4(2 - \xi)^2}\right] - \ln \frac{(1 - \xi)}{2(2 - \xi)}\right. \\
&\quad \left. + 2(2 - \xi)\left[-\frac{1}{2(2 - \xi)} + \frac{2(1 - \xi)}{4(2 - \xi)^2}\right] + 2 \ln \frac{\xi}{2 - \xi} + 2(2 - \xi)\left[\frac{1}{2 - \xi} + \frac{\xi}{(2 - \xi)^2}\right]\right\} \\
&= 9600 \text{ J·mol}^{-1} + (4157.2 \text{ J·mol}^{-1})\left[\ln \frac{16(2 - \xi)^2\xi^2}{27(1 - \xi)^4} - 3 + \frac{3 - 3\xi}{2 - \xi} - 1\right. \\
&\quad \left. + \frac{1 - \xi}{2 - \xi} + 2 + \frac{2\xi}{2 - \xi}\right]
\end{aligned}$$

$$= 9600 \text{ J·mol}^{-1} + (4157.2 \text{ J·mol}^{-1}) \ln \frac{16(2 - \xi)^2\xi^2}{27(1 - \xi)^4} \tag{1}$$

At equilibrium, $(\partial G/\partial \xi)_{T,P} = 0$, so

$$\frac{16(2 - \xi_{eq})^2 \xi_{eq}^2}{27(1 - \xi_{eq})^4} = \exp\left(\frac{-9600 \text{ J} \cdot \text{mol}^{-1}}{4157.2 \text{ J} \cdot \text{mol}^{-1}}\right) = 0.09934$$

Solving for ξ_{eq} gives $\xi_{eq} = 0.158$.
Note that substituting for $G^\circ_{NH_3}$, R, T, P_{NH_3}, P_{N_2}, and P_{N_2} in Equation 1 gives

$$\left(\frac{\partial G}{\partial \xi}\right)_{T,P} = 2G^\circ_{NH_3} + RT \ln \frac{P^2_{NH_3}}{P^3_{H_2} P_{N_2}} = \Delta_r G^\circ RT \ln \frac{P^2_{NH_3}}{P^3_{H_2} P_{N_2}}$$

Now

$$K_P = \frac{P^2_{NH_3}}{P^3_{H_2} P_{N_2}} = \frac{16\xi_{eq}^2 (2 - \xi_{eq})^2}{27(1 - \xi_{eq})^4}$$

so

$$K_P = \frac{16(0.158)^2 (2 - 0.158)^2}{27(1 - 0.158)^4} = 0.010$$

26–21. Suppose that we have a mixture of the gases $H_2(g)$, $CO_2(g)$, $CO(g)$, and $H_2O(g)$ at 1260 K, with $P_{H_2} = 0.55$ bar, $P_{CO_2} = 0.20$ bar, $P_{CO} = 1.25$ bar, and $P_{H_2O} = 0.10$ bar. Is the reaction described by the equation

$$H_2(g) + CO_2(g) \rightleftharpoons CO(g) + H_2O(g) \qquad K_P = 1.59$$

at equilibrium under these conditions? If not, in what direction will the reaction proceed to attain equilibrium?

Use Equation 26.25 to find Q_P and Equation 26.26 to find out which way the reaction will proceed.

$$Q_P = \frac{P_{H_2O} P_{CO}}{P_{H_2} P_{CO_2}} = \frac{(0.10)(1.25)}{(0.55)(0.20)} = 1.14$$

$$\Delta_r G = RT \ln \frac{Q_P}{K_P} = RT \ln \frac{1.14}{1.59}$$

$$\Delta_r G < 0$$

The reaction as written will proceed to the right.

26–22. Given that $K_P = 2.21 \times 10^4$ at 25°C for the equation

$$2 H_2(g) + CO(g) \rightleftharpoons CH_3OH(g)$$

predict the direction in which a reaction mixture for which $P_{CH_3OH} = 10.0$ bar, $P_{H_2} = 0.10$ bar, and $P_{CO} = 0.0050$ bar proceeds to attain equilibrium.

This is done in the same way as the previous problem.

$$Q_P = \frac{P_{CH_3OH}}{P^2_{H_2} P_{CO}} = \frac{10}{(0.10)^2(0.0050)} = 2.00 \times 10^5$$

$$\Delta_r G = RT \ln \frac{Q_P}{K_P} = RT \ln \frac{2.00 \times 10^5}{2.21 \times 10^4}$$

$$\Delta_r G > 0$$

Therefore, the reaction as written will proceed to the left.

26–23. The value of K_P for a gas-phase reaction doubles when the temperature is increased from 300 K to 400 K at a fixed pressure. What is the value of $\Delta_r H°$ for this reaction?

Use Equation 26.31, since we assume that $\Delta_r H°$ remains constant over this temperature range.

$$\ln \frac{K_P(T_2)}{K_P(T_1)} = -\frac{\Delta_r H°}{R} \left(\frac{1}{T_2} - \frac{1}{T_1} \right)$$

$$R \ln 2 = -\Delta_r H° \left(\frac{1}{400 \text{ K}} - \frac{1}{300 \text{ K}} \right)$$

$$\Delta_r H° = 6.91 \text{ kJ·mol}^{-1}$$

26–24. The value of $\Delta_r H°$ is 34.78 kJ·mol^{-1} at 1000 K for the reaction described by

$$H_2(g) + CO_2(g) \rightleftharpoons CO(g) + H_2O(g)$$

Given that the value of K_P is 0.236 at 800 K, estimate the value of K_P at 1200 K, assuming that $\Delta_r H°$ is independent of temperature.

Again, use Equation 26.31:

$$\ln \frac{K_P(T_2)}{K_P(T_1)} = -\frac{\Delta_r H°}{R} \left(\frac{1}{T_2} - \frac{1}{T_1} \right)$$

$$\ln \frac{K_P(1200 \text{ K})}{0.236} = -\frac{34\,780 \text{ J·mol}^{-1}}{8.3145 \text{ J·mol}^{-1} \cdot \text{K}^{-1}} \left(\frac{1}{1200 \text{ K}} - \frac{1}{800 \text{ K}} \right) = 1.743$$

$$K_P(1200 \text{ K}) = 1.35$$

26–25. The value of $\Delta_r H°$ is -12.93 kJ·mol^{-1} at 800 K for

$$H_2(g) + I_2(g) \rightleftharpoons 2 HI(g)$$

Assuming that $\Delta_r H°$ is independent of temperature, calculate K_P at 700 K given that $K_P = 29.1$ at 1000 K.

We do this as we did Problem 26–24, using Equation 26.31.

$$\ln \frac{K_P(T_2)}{K_P(T_1)} = -\frac{\Delta_r H°}{R} \left(\frac{1}{T_2} - \frac{1}{T_1} \right)$$

$$\ln \frac{29.1}{K_P(700 \text{ K})} = -\frac{12\,930 \text{ J·mol}^{-1}}{8.3145 \text{ J·mol}^{-1} \cdot \text{K}^{-1}} \left(\frac{1}{1000 \text{ K}} - \frac{1}{700 \text{ K}} \right) = 0.666$$

$$K_P(700 \text{ K}) = 14.9$$

26–26. The equilibrium constant for the reaction described by

$$2\,HBr(g) \rightleftharpoons H_2(g) + Br_2(g)$$

can be expressed by the empirical formula

$$\ln K = -6.375 + 0.6415 \ln(T/K) - \frac{11790\ K}{T}$$

Use this formula to determine $\Delta_r H^\circ$ as a function of temperature. Calculate $\Delta_r H^\circ$ at 25°C and compare your result to the one you obtain from Table 19.2.

Use Equation 26.29:

$$\frac{d \ln K}{dT} = \frac{\Delta_r H^\circ}{RT^2}$$

$$\frac{0.6415}{T} + \frac{11790\ K}{T^2} = \frac{\Delta_r H^\circ}{RT^2}$$

$$0.6145RT + (11790\ K)R = \Delta_r H^\circ$$

At 25°C, $\Delta_r H^\circ = 99.6\ \text{kJ·mol}^{-1}$. The value given in Table 19.2 for $\Delta_f H^\circ[HBr(g)]$ is $-36.3\ \text{kJ·mol}^{-1}$ and that given for $\Delta_f H^\circ[Br_2(g)]$ is $30.907\ \text{kJ·mol}^{-1}$. We can write $\Delta_r H^\circ$ (using these values) as

$$\Delta_r H^\circ = \Delta_f H^\circ[Br_2(g)] - 2(\Delta_f H^\circ[HBr(g)]) = 103.5\ \text{kJ·mol}^{-1}$$

in fairly good agreement with the value of $\Delta_r H^\circ$ found from the equilibrium constant.

26–27. Use the following data for the reaction described by

$$2\,HI(g) \rightleftharpoons H_2(g) + I_2(g)$$

to obtain $\Delta_r H^\circ$ at 400°C.

T/K	500	600	700	800
$K_P/10^{-2}$	0.78	1.24	1.76	2.31

We wish to express K_P in terms of $1/T$ and use Equation 26.29 to find $\Delta_r H^\circ$.

$1000K/T$	2	1.67	1.43	1.25
$\ln K_P$	−4.85	−4.39	−4.04	−3.77

A linear fit gives

$$\ln K_P = -1.9695 - \frac{1445.73\ K}{T}$$

$$\frac{d \ln K_P}{dT} = \frac{1445.73\ K}{T^2}$$

$$\Delta_r H^\circ = R(1445.73\ K) = 12.02\ \text{kJ·mol}^{-1}$$

and a fit of the form $a + b/T + c \ln T$ gives

$$\ln K_P = -2.33966 - \frac{1020.3\ K}{T} + 0.6833 \ln(T/K)$$

$$\frac{d \ln K_P}{dT} = \frac{1020.3\ K}{T^2} + \frac{0.6833}{T}$$

$$\Delta_r H^\circ = R(1020.3 + 0.6833T) = 12.31\ \text{kJ·mol}^{-1}$$

26–28. Consider the reaction described by

$$CO_2(g) + H_2(g) \rightleftharpoons CO(g) + H_2O(g)$$

The molar heat capacitites of $CO_2(g)$, $H_2(g)$, $CO(g)$, and $H_2O(g)$ can be expressed by

$$\overline{C}_P[CO_2(g)]/R = 3.127 + (5.231 \times 10^{-3} \text{ K}^{-1})T - (1.784 \times 10^{-6} \text{ K}^{-2})T^2$$

$$\overline{C}_P[H_2(g)]/R = 3.496 - (1.006 \times 10^{-4} \text{ K}^{-1})T + (2.419 \times 10^{-7} \text{ K}^{-2})T^2$$

$$\overline{C}_P[CO(g)]/R = 3.191 + (9.239 \times 10^{-4} \text{ K}^{-1})T - (1.41 \times 10^{-7} \text{ K}^{-2})T^2$$

$$\overline{C}_P[H_2O(g)]/R = 3.651 + (1.156 \times 10^{-3} \text{ K}^{-1})T + (1.424 \times 10^{-7} \text{ K}^{-2})T^2$$

over the temperature range 300 K to 1500 K. Given that

substance	$CO_2(g)$	$H_2(g)$	$CO(g)$	$H_2O(g)$
$\Delta_f H°/\text{kJ·mol}^{-1}$	-393.523	0	-110.516	-241.844

at 300 K and that $K_P = 0.695$ at 1000 K, derive a general expression for the variation of $K_P(T)$ with temperature in the form of Equation 26.34.

We first find the values of $\Delta_r C_P°$ and $\Delta_r H°$:

$$\Delta_r C_P° = C_P°[H_2O(g)] + C_P°[CO(g)] - C_P°[H_2(g)] - C_P°[CO_2(g)]$$

$$\Delta_r C_P°/R = 0.219 - (3.051 \times 10^{-3} \text{ K}^{-1})T + (1.544 \times 10^{-6} \text{ K}^{-2})T^2$$

$$\Delta_r H°(300 \text{ K}) = \Delta_f H°[H_2O(g)] + \Delta_f H°[CO(g)] - \Delta_f H°[H_2(g)] - \Delta_f H°[CO_2(g)]$$

$$= -241.844 \text{ kJ·mol}^{-1} - (-110.516 \text{ kJ·mol}^{-1}) - (-393.523 \text{ kJ·mol}^{-1})$$

$$= 262.195 \text{ kJ·mol}^{-1}$$

Now we use Equation 26.32 to find $\Delta_r H°(T)$:

$$\Delta_r H°(T) = \Delta_r H°(300 \text{ K}) + \int_{300}^{T} \Delta_r C_P°(T)dT$$

$$= 262.195 \text{ kJ·mol}^{-1} + R\int_{300}^{T} \Big[0.219 - (3.051 \times 10^{-3} \text{ K}^{-1})T$$

$$+(1.544 \times 10^{-6} \text{ K}^{-2})T^2\Big]dT$$

$$= 262.195 \text{ kJ·mol}^{-1} + R\Big[0.219(T - 300) - (1.525 \times 10^{-3} \text{ K}^{-1})(T^2 - 300^2)$$

$$+(5.145 \times 10^{-7} \text{ K}^{-2})(T^3 - 300^3)\Big]$$

$$= 262.195 \text{ kJ·mol}^{-1} + R\big[57.681 \text{ K} + 0.219T$$

$$-(1.525 \times 10^{-3} \text{ K}^{-1})T^2 + (5.145 \times 10^{-7} \text{ K}^{-2})T^3\big]$$

$$= 262.675 \text{ kJ·mol}^{-1} + (1.821 \times 10^{-3} \text{ kJ·mol}^{-1}\text{·K}^{-1})T$$

$$-(1.268 \times 10^{-5} \text{ kJ·mol}^{-1}\text{·K}^{-2})T^2 + (4.278 \times 10^{-9} \text{ kJ·mol}^{-1}\text{·K}^{-3})T^3$$

This equation is in the form $\alpha + \beta T + \gamma T^2 + \delta T^3$, as was expected (Equation 26.33). Substituting into Equation 26.34, we find that

$$\ln K_P(T) = -\frac{31592}{T} + 0.2190\ln(T/\text{K}) - (1.525 \times 10^{-3} \text{ K}^{-1})T$$

$$+(2.573 \times 10^{-7} \text{ K}^{-2})T^2 + A$$

At $T = 1000$ K, we know that $K_p = 0.695$, so

$$\ln 0.695 = -31.592 + 1.513 - 1.526 + 0.2573 + A$$
$$A = 30.984$$

and so

$$\ln K_p(T) = -\frac{31592}{T} + 0.2190 \ln(T/\text{K}) - (1.525 \times 10^{-3}\,\text{K}^{-1})T + (2.573 \times 10^{-7}\,\text{K}^{-2})T^2 + 30.984$$

26–29. The temperature dependence of the equilibrium constant K_p for the reaction described by

$$2\,\text{C}_3\text{H}_6(\text{g}) \rightleftharpoons \text{C}_2\text{H}_4(\text{g}) + \text{C}_4\text{H}_8(\text{g})$$

is given by the equation

$$\ln K_p(T) = -2.395 - \frac{2505\,\text{K}}{T} + \frac{3.477 \times 10^6\,\text{K}^2}{T^2} \qquad 300\,\text{K} < T < 600\,\text{K}$$

Calculate the values of $\Delta_r G°$, $\Delta_r H°$, and $\Delta_r S°$ for this reaction at 525 K.

Use Equation 26.11 to find $\Delta_r G°$, Equation 26.29 to find $\Delta_r H°$, and the relation $\Delta_r G° = \Delta_r H° - T\Delta_r S°$ to find $\Delta_r S°$:

$$\Delta_r G° = -RT \ln K_p = -R(525\,\text{K})\left[-2.395 - \frac{2505\,\text{K}}{525\,\text{K}} + \frac{3.477 \times 10^6\,\text{K}^2}{(525\,\text{K})^2}\right]$$

$$= -23.78\,\text{kJ}\cdot\text{mol}^{-1}$$

$$\Delta_r H° = RT^2\frac{d \ln K_p}{dT} = RT^2\left[\frac{2505\,\text{K}}{T^2} - \frac{6.954 \times 10^6\,\text{K}^2}{T^3}\right]$$

$$= -89.30\,\text{kJ}\cdot\text{mol}^{-1}$$

$$\Delta_r S° = \frac{\Delta_r H° - \Delta_r G°}{T} = -124.8\,\text{J}\cdot\text{mol}^{-1}\cdot\text{K}^{-1}$$

26–30. At 2000 K and one bar, water vapor is 0.53% dissociated. At 2100 K and one bar, it is 0.88% dissociated. Calculate the value of $\Delta_r H°$ for the dissociation of water at one bar, assuming that the enthalpy of reaction is constant over the range from 2000 K to 2100 K.

$$\text{H}_2\text{O}(\text{g}) \rightleftharpoons \text{H}_2(\text{g}) + \tfrac{1}{2}\text{O}_2(\text{g})$$

At 2000 K and one bar, there will be $0.9947n_0$ moles of $\text{H}_2\text{O}(\text{g})$, $0.0053n_0$ moles of $\text{H}_2(\text{g})$, and $0.00265n_0$ moles of $\text{O}_2(\text{g})$, for a total of $1.00265n_0$ moles. The partial pressures of the various gases are then

$$P_{\text{H}_2\text{O}} = \frac{0.9947}{1.00265}P = 0.9921P \qquad P_{\text{H}_2} = \frac{0.0053}{1.00265}P = 5.286 \times 10^{-3}P$$

and

$$P_{\text{O}_2} = \frac{0.00265}{1.00265}P = 2.643 \times 10^{-3}P$$

and $K_P(2000 \text{ K})$ at one bar is

$$K_P(2000 \text{ K}) = \frac{(2.643 \times 10^{-3})^{1/2}(5.286 \times 10^{-3})P^{1/2}}{0.9921} = 2.74 \times 10^{-4}$$

Likewise, at 2100 K and one bar, there will be $0.9912n_0$ moles of $H_2O(g)$, $0.0088n_0$ moles of $H_2(g)$, and $0.0044n_0$ moles of $O_2(g)$, for a total of $1.0044n_0$ moles. The partial pressures of the various gases are then

$$P_{H_2O} = \frac{0.9912}{1.0044}P = 0.9868P \qquad P_{H_2} = \frac{0.0088}{1.0044}P = 8.761 \times 10^{-3}P$$

and

$$P_{O_2} = \frac{0.0044}{1.0044}P = 4.381 \times 10^{-3}P$$

and $K_P(2000 \text{ K})$ at one bar is

$$K_P(2000 \text{ K}) = \frac{(4.381 \times 10^{-3})^{1/2}(8.761 \times 10^{-3})P^{1/2}}{0.9868} = 5.88 \times 10^{-4}$$

Now we can use Equation 26.31 to find $\Delta_r H°$:

$$\ln \frac{K_P(2100 \text{ K})}{K_P(2000 \text{ K})} = -\frac{\Delta_r H°}{R}\left(\frac{1}{2100 \text{ K}} - \frac{1}{2000 \text{ K}}\right)$$

$$\Delta_r H° = 266.5 \text{ kJ} \cdot \text{mol}^{-1}$$

26–31. The following table gives the standard molar Gibbs energy of formation of Cl(g) at three different temperatures.

T/K	1000	2000	3000
$\Delta_f G°/\text{kJ}\cdot\text{mol}^{-1}$	65.288	5.081	−56.297

Use these data to determine the value of K_P at each temperature for the reaction described by

$$\tfrac{1}{2}\text{Cl}_2(g) \rightleftharpoons \text{Cl}(g)$$

Assuming that $\Delta_r H°$ is temperature independent, determine the value of $\Delta_r H°$ from these data. Combine your results to determine $\Delta_r S°$ at each temperature. Interpret your results.

Use Equation 26.11 to find K_P at each temperature, then find $\ln K_P$ for use in determining $\Delta_r H°$.

T/K	$\Delta_f G°/\text{kJ}\cdot\text{mol}^{-1}$	K_P	$\ln K_P$
1000	65.288	3.889×10^{-4}	−7.852
2000	5.081	0.7367	−0.3056
3000	−56.297	9.554	2.257

The best-fit line to a plot of $\ln K_P$ vs. $1/T$ is $\ln K_P = 7.290 - 15148/T$. The slope of this line is $-\Delta_r H°/R$, so $\Delta_r H° = 125.9 \text{ kJ}\cdot\text{mol}^{-1}$. We can use the expression $\Delta_r G° = \Delta_r H° - T\Delta_r S°$ to find $\Delta_r S°$ at each temperature. These values are tabulated below.

T/K	1000	2000	3000
$\Delta_r S°/\text{J}\cdot\text{mol}^{-1}$	60.61	60.41	60.73

26–32. The following experimental data were determined for the reaction described by

$$SO_3(g) \rightleftharpoons SO_2(g) + \tfrac{1}{2}O_2(g)$$

T/K	800	825	900	953	1000
$\ln K_P$	-3.263	-3.007	-1.899	-1.173	-0.591

Calculate $\Delta_r G°$, $\Delta_r H°$, and $\Delta_r S°$ for this reaction at 900 K. State any assumptions that you make.

This is done in the same way as the previous problem. We assume that $\Delta_r H°$ does not vary significantly over the temperature range given. A best-fit line to to $\ln K_P$ in $1/T$ is $\ln K_P = 10.216 - (10851 \text{ K})/T$, which gives $\Delta_r H° = 90.2 \text{ kJ·mol}^{-1}$. Using Equation 26.11 and the relation $\Delta_r G° = \Delta_r H° - T\Delta_r S°$, we find that at 900 K $\Delta_r G° = 14.21 \text{ kJ·mol}^{-1}$ and $\Delta_r S° = 84.5 \text{ J·mol}^{-1}·\text{K}^{-1}$.

26–33. Show that

$$\mu = -RT \ln \frac{q(V,T)}{N}$$

if

$$Q(N, V, T) = \frac{[q(V,T)]^N}{N!}$$

Begin with Equation 23.27 and use Stirling's approximation for $N \ln N!$:

$$\mu = -RT \left(\frac{\partial \ln Q}{\partial N} \right)_{T,V}$$

$$= -RT \left[\frac{\partial}{\partial N} (N \ln q - N \ln N!) \right]_{T,V}$$

$$= -RT \left[\ln q - \frac{\partial}{\partial N} (-N \ln N + N) \right]$$

$$= -RT [\ln q - \ln N - 1 + 1] = -RT \ln \frac{q}{N}$$

26–34. Use Equation 26.40 to calculate $K(T)$ at 750 K for the reaction described by $H_2(g) + I_2(g) \rightleftharpoons 2\,HI(g)$. Use the molecular parameters given in Table 18.2. Compare your value to the one given in Table 26.2 and the experimental value shown in Figure 26.5.

We can use Equation 26.40 to calculate $K(T)$, substituting from Equation 18.39 for the partition functions of H_2, I_2, and HI.

$$K = \frac{q_{HI}^2}{q_{H_2} q_{I_2}}$$

$$= \left(\frac{m_{HI}^2}{m_{H_2} m_{I_2}} \right)^{3/2} \left[\frac{4\Theta_{rot}^{H_2}\Theta_{rot}^{I_2}}{(\Theta_{rot}^{HI})^2} \right] \frac{(1 - e^{-\Theta_{vib}^{H_2}/T})(1 - e^{-\Theta_{vib}^{I_2}/T})}{(1 - e^{-\Theta_{vib}^{HI}/T})^2} \exp \frac{2D_0^{HI} - D_0^{H_2} - D_0^{I_2}}{RT}$$

$$= \left[\frac{(127.9)^2}{(2.016)(253.8)}\right]^{3/2} \left[\frac{4(85.3 \text{ K})(0.0537 \text{ K})}{(9.25 \text{ K})^2}\right] \frac{(1 - e^{-6215/750})(1 - e^{-308/750})}{(1 - e^{-3266/750})^2}$$

$$\times \exp \frac{2(8.500 \text{ kJ} \cdot \text{mol}^{-1})}{(8.3145 \times 10^{-3} \text{ kJ} \cdot \text{mol}^{-1} \cdot \text{K}^{-1})(750 \text{ K})}$$

$$= 52.29$$

This is in good agreement with the values in the text.

26–35. Use the statistical thermodynamic formulas of Section 26–8 to calculate $K_P(T)$ at 900 K, 1000 K, 1100 K, and 1200 K for the association of Na(g) to form dimers, $Na_2(g)$ according to the equation

$$2 \text{ Na(g)} \rightleftharpoons Na_2(g)$$

Use your result at 1000 K to calculate the fraction of sodium atoms that form dimers at a total pressure of one bar. The experimental values of $K_P(T)$ are

T/K	900	1000	1100	1200
K_P	1.32	0.47	0.21	0.10

Plot $\ln K_P$ against $1/T$ to determine the value of $\Delta_r H°$.

We can calculate the partition function of Na using Equation 18.13 (for a monatomic ideal gas) and that of Na_2 using Equation 18.39 (for a diatomic ideal gas):

$$\frac{q_{\text{Na}}}{V} = \left[\frac{2\pi m k_B T}{h^2}\right]^{3/2} q_{\text{elec}} = \left[\frac{2\pi (0.022991 \text{ kg} \cdot \text{mol}^{-1}) RT}{N_A^2 h^2}\right]^{3/2} \quad (2)$$

$$= 2\left[7.543 \times 10^{18} \text{ m}^2 \cdot \text{K}^{-1} T\right]^{3/2}$$

$$\frac{q_{\text{Na}_2}}{V} = \left[\frac{2\pi M k_B T}{h^2}\right]^{3/2} \frac{T}{\sigma \Theta_{\text{rot}}} \frac{1}{1 - e^{-\Theta_{\text{vib}}/T}} e^{D_0/k_B T}$$

$$= \left[1.508 \times 10^{19} \text{ m}^2 \cdot \text{K}^{-1} T\right]^{3/2} \frac{T}{0.442 \text{ K}} \frac{e^{8707.7 \text{ K}/T}}{1 - e^{-229 \text{ K}/T}}$$

Using the procedure on page 1070 (where we calculate K_P for H_2O), we have

$$K_P(T) = \frac{(q_{\text{Na}_2}/V)}{(q_{\text{Na}}/V)^2} \left[\frac{(6.022 \times 10^{23} \text{ mol}^{-1})(10^5 \text{ Pa})}{(1 \text{ m}^{-3})(8.3145 \text{ J} \cdot \text{K}^{-1} \cdot \text{mol}^{-1})T}\right]$$

We can substitute into the above expressions to find K_P at 900 K, 1000 K, 1100 K, and 1200 K:

T/K	900	1000	1100	1200
K_P	1.47	0.52	0.22	0.11

Given that $K_P = 0.52$ at 1000 K, let us assume that we begin with n_0 moles of Na and no moles of the dimer. Then at equilibrium we will have $n_0 - 2\xi$ moles of Na and ξ moles of the dimer, so that for a total pressure of 1 bar,

$$P_{\text{Na}} = \frac{n_0 - 2\xi}{n_0 - \xi} P = \frac{1 - 2\xi/n_0}{1 - \xi/n_0} P$$

and

$$P_{Na_2} = \frac{\xi}{n_0 - \xi} P = \frac{\xi/n_0}{1 - \xi/n_0} P$$

Then

$$K_P = \frac{P_{Na_2}}{P_{Na}^2} = \frac{\xi/n_0(1 - \xi/n_0)}{(1 - 2\xi/n_0)^2 P}$$

$$0.52(1 - 2x)^2 = x - x^2$$

where $x = \xi/n_0$. Solving for x gives $x = 0.21$ or $x = 0.78$, but $x < 0.50$, so we find that 21% of the sodium atoms will form dimers at 1000 K.

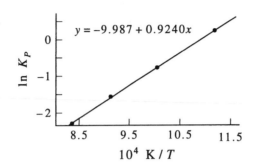

Plotting the experimental values, we find that $\ln K_P = -9.987 + (9240 \text{ K})/T$. Therefore, $\Delta_r H° = -R(9240 \text{ K}) = -76.8 \text{ kJ·mol}^{-1}$.

26–36. Using the data in Table 18.2, calculate K_P at 2000 K for the reaction described by the equation

$$CO_2(g) \rightleftharpoons CO(g) + \tfrac{1}{2} O_2(g)$$

The experimental value is 1.3×10^{-3}.

We can use Equation 18.39 to express the partition functions of O_2 and CO, and Equation 18.57 to express the partition function of CO_2. At 2000 K,

$$\frac{q_{CO}}{V} = \left[\frac{2\pi M k_B T}{h^2}\right]^{3/2} \frac{T}{\sigma \Theta_{rot}} \frac{1}{1 - e^{-\Theta_{vib}/T}} e^{D_0/k_B T}$$

$$= \left[\frac{2\pi(0.02801 \text{ kg·mol}^{-1})(8.3145 \text{ J·mol}^{-1}\cdot\text{K}^{-1})(2000 \text{ K})}{(6.022 \times 10^{23} \text{ mol}^{-1})^2(6.626 \times 10^{-34} \text{ J·s})^2}\right]^{3/2}$$

$$\times \frac{2000 \text{ K}}{2.77 \text{ K}} \frac{1}{1 - e^{-3103/2000}} e^{1072000/(8.3145)(2000)}$$

$$= 2.27 \times 10^{64} \text{ m}^{-3}$$

$$\frac{q_{O_2}}{V} = \left[\frac{2\pi M k_B T}{h^2}\right]^{3/2} \frac{T}{\sigma \Theta_{rot}} \frac{1}{1 - e^{-\Theta_{vib}/T}} 3e^{D_0/k_B T}$$

$$= \left[\frac{2\pi(0.03200 \text{ kg·mol}^{-1})(8.3145 \text{ J·mol}^{-1}\cdot\text{K}^{-1})(2000 \text{ K})}{(6.022 \times 10^{23} \text{ mol}^{-1})^2(6.626 \times 10^{-34} \text{ J·s})^2}\right]^{3/2}$$

$$\times \frac{2000 \text{ K}}{2(2.07 \text{ K})} \frac{1}{1 - e^{-2256/2000}} 3e^{494100/(8.3145)(2000)}$$

$$= 5.23 \times 10^{49} \text{ m}^{-3}$$

$$\frac{q_{CO_2}}{V} = \left[\frac{2\pi M k_B T}{h^2}\right]^{3/2} \frac{T}{\sigma \Theta_{rot}} \left[\prod_{j=1}^{4}(1 - e^{-\Theta_{vib,j}/T})^{-1}\right] e^{D_0/k_B T}$$

$$= \left[\frac{2\pi (0.04401 \text{ kg}\cdot\text{mol}^{-1})(8.3145 \text{ J}\cdot\text{mol}^{-1}\cdot\text{K}^{-1})(2000 \text{ K})}{(6.022 \times 10^{23} \text{ mol}^{-1})^2(6.626 \times 10^{-34} \text{ J}\cdot\text{s})^2}\right]^{3/2}$$

$$\times \frac{2000 \text{ K}}{2(0.561 \text{ K})}(1 - e^{-3360/2000})^{-1}(1 - e^{-954/2000})^{-2}(1 - e^{-1890/2000})^{-1}e^{1596\times10^3/(8.3145)(2000)}$$

$$= 5.88 \times 10^{79} \text{ m}^{-3}$$

We can now use Equation 26.39 to write (as we did in Section 26–8 for the reaction involving H_2O)

$$K_P(T) = \left[\frac{RT}{N_A(10^5 \text{ Pa})}\right]^{1/2} \frac{(q_{CO}/V)(q_{O_2}/V)^{1/2}}{(q_{CO_2}/V)}$$

$$= \left[\frac{(8.3145 \text{ J}\cdot\text{mol}^{-1}\cdot K^{-1})(2000 \text{ K})}{(6.022 \times 10^{23} \text{ mol}^{-1})(10^5 \text{ Pa})}\right]^{1/2} \frac{(2.27 \times 10^{64} \text{ m}^{-3})(5.23 \times 10^{49} \text{ m}^{-3})^{1/2}}{5.88 \times 10^{79} \text{ m}^{-3}}$$

$$= 1.46 \times 10^{-3}$$

26–37. Using the data in Tables 18.2 and 18.4, calculate the equilibrium constant for the water gas reaction

$$CO_2(g) + H_2(g) \rightleftharpoons CO(g) + H_2O(g)$$

at 900 K and 1200 K. The experimental values at these two temperatures are 0.43 and 1.37, respectively.

We have expressed the partition functions for CO_2 and CO in Problem 26–36, and those for H_2 and H_2O in Section 26–8. At a temperature T, these partition functions are

$$\frac{q_{CO_2}}{V} = \left[\frac{2\pi M k_B T}{h^2}\right]^{3/2} \frac{T}{\sigma \Theta_{rot}} \left[\prod_{j=1}^{4}(1 - e^{-\Theta_{vib,j}/T})^{-1}\right] e^{D_0/k_B T}$$

$$= \left[\frac{2\pi (0.04401 \text{ kg}\cdot\text{mol}^{-1})(8.3145 \text{ J}\cdot\text{mol}^{-1}\cdot\text{K}^{-1})T}{(6.022 \times 10^{23} \text{ mol}^{-1})^2(6.626 \times 10^{-34} \text{ J}\cdot\text{s})^2}\right]^{3/2} \frac{T}{2(0.561 \text{ K})}(1 - e^{-3360 \text{ K}/T})^{-1}$$

$$\times (1 - e^{-954 \text{ K}/T})^{-2}(1 - e^{-1890 \text{ K}/T})^{-1}e^{1596\times10^3 \text{ K}/8.3145T}$$

$$\frac{q_{H_2}}{V} = \left[\frac{2\pi M k_B T}{h^2}\right]^{3/2} \frac{T}{\sigma \Theta_{rot}} \frac{1}{1 - e^{-\Theta_{vib}/T}} e^{D_0/k_B T}$$

$$= \left[\frac{2\pi (2.016 \times 10^{-3} \text{ kg}\cdot\text{mol}^{-1})(8.3145 \text{ J}\cdot\text{mol}^{-1}\cdot\text{K}^{-1})T}{(6.022 \times 10^{23} \text{ mol}^{-1})^2(6.626 \times 10^{-34} \text{ J}\cdot\text{s})^2}\right]^{3/2} \frac{T}{2(85.3 \text{ K})}\frac{1}{1 - e^{-6215/T}} e^{431800 \text{ K}/8.3145T}$$

$$\frac{q_{CO}}{V} = \left[\frac{2\pi M k_B T}{h^2}\right]^{3/2} \frac{T}{\sigma \Theta_{rot}} \frac{1}{1 - e^{-\Theta_{vib}/T}} e^{D_0/k_B T}$$

$$= \left[\frac{2\pi (0.02801 \text{ kg}\cdot\text{mol}^{-1})(8.3145 \text{ J}\cdot\text{mol}^{-1}\cdot\text{K}^{-1})T}{(6.022 \times 10^{23} \text{ mol}^{-1})^2(6.626 \times 10^{-34} \text{ J}\cdot\text{s})^2}\right]^{3/2} \frac{T}{2.77 \text{ K}}\frac{1}{1 - e^{-3103 \text{ K}/T}} e^{1072000 \text{ K}/8.3145T}$$

$$\frac{q_{H_2O}}{V} = \left[\frac{2\pi M k_B T}{h^2}\right]^{3/2} \frac{\pi^{1/2}}{\sigma} \left(\frac{T^3}{\Theta_{rot,A}\Theta_{rot,B}\Theta_{rot,C}}\right)^{1/2} \left(\prod_{j=1}^{3}\frac{1}{1 - e^{-\Theta_{vib,j}/T}}\right) e^{D_0/k_B T}$$

$$= \left[\frac{2\pi (0.01801 \text{ kg}\cdot\text{mol}^{-1})(8.3145 \text{ J}\cdot\text{mol}^{-1}\cdot\text{K}^{-1})T}{(6.022 \times 10^{23} \text{ mol}^{-1})^2(6.626 \times 10^{-34} \text{ J}\cdot\text{s})^2}\right]^{3/2} \frac{\pi^{1/2}}{2} \left[\frac{T^3}{(40.1 \text{ K})(20.9 \text{ K})(13.4 \text{ K})}\right]^{1/2}$$

$$\times (1 - e^{-5360 \text{ K}/T})^{-1}(1 - e^{-5160 \text{ K}/T})^{-1}(1 - e^{-2290 \text{ K}/T})^{-1}e^{917600 \text{ K}/8.3145T}$$

Using Equation 26.39, we can write K_P in terms of the partition functions:

$$K_P = \frac{(q_{H_2O}/V)(q_{CO}/V)}{(q_{CO_2}/V)(q_{H_2}/V)}$$

Below are tabulated values for each partition function and K_P at 900 K and 1200 K.

	900 K	1200 K
$q_{CO_2}/V/m^{-3}$	1.38×10^{129}	3.22×10^{106}
$q_{H_2}/V/m^{-3}$	9.17×10^{56}	1.02×10^{51}
$q_{CO}/V/m^{-3}$	4.15×10^{97}	2.49×10^{82}
$q_{H_2O}/V/m^{-3}$	1.72×10^{88}	2.18×10^{75}
K_P	0.56	1.66

26–38. Using the data in Tables 18.2 and 18.4, calculate the equilibrium constant for the reaction

$$3\,H_2(g) + N_2(g) \rightleftharpoons 2\,NH_3(g)$$

at 700 K. The accepted value is 8.75×10^{-5} (see Table 26.4).

We have expressed the partition function of H_2 in Problem 26–37, and we can use Equation 18.39 to express the partition function of N_2 and Equation 18.60 to express that of NH_3. At 700 K, these partition functions are

$$\frac{q_{H_2}}{V} = \left[\frac{2\pi M k_B T}{h^2}\right]^{3/2} \frac{T}{\sigma \Theta_{rot}} \frac{1}{1 - e^{-\Theta_{vib}/T}} e^{D_0/k_B T}$$

$$= \left[\frac{2\pi(2.016 \times 10^{-3}\,kg \cdot mol^{-1})(8.3145\,J \cdot mol^{-1} \cdot K^{-1})(700\,K)}{(6.022 \times 10^{23}\,mol^{-1})^2(6.626 \times 10^{-34}\,J \cdot s)^2}\right]^{3/2}$$

$$\times \frac{700\,K}{2(85.3\,K)} \frac{1}{1 - e^{-6215/700}} e^{431800/(8.3145)(700)}$$

$$= 7.15 \times 10^{63}\,m^{-3}$$

$$\frac{q_{N_2}}{V} = \left[\frac{2\pi M k_B T}{h^2}\right]^{3/2} \frac{T}{\sigma \Theta_{rot}} \frac{1}{1 - e^{-\Theta_{vib}/T}} e^{D_0/k_B T}$$

$$= \left[\frac{2\pi(0.02802\,kg \cdot mol^{-1})(8.3145\,J \cdot mol^{-1} \cdot K^{-1})(700\,K)}{(6.022 \times 10^{23}\,mol^{-1})^2(6.626 \times 10^{-34}\,J \cdot s)^2}\right]^{3/2}$$

$$\times \frac{700\,K}{2(2.88\,K)} \frac{1}{1 - e^{-3374/700}} e^{941200/(8.3145)(700)}$$

$$= 1.08 \times 10^{105}\,m^{-3}$$

$$\frac{q_{NH_3}}{V} = \left[\frac{2\pi M k_B T}{h^2}\right]^{3/2} \frac{\pi^{1/2}}{\sigma} \left(\frac{T^3}{\Theta_{rot,A}\Theta_{rot,B}\Theta_{rot,C}}\right)^{1/2} \prod_1^6 (1 - e^{-\Theta_{vib,j}/T})^{-1} e^{D_0/k_B T}$$

$$= \left[\frac{2\pi(0.03104\,kg \cdot mol^{-1})(8.3145\,J \cdot mol^{-1} \cdot K^{-1})(700\,K)}{(6.022 \times 10^{23}\,mol^{-1})^2(6.626 \times 10^{-34}\,J \cdot s)^2}\right]^{3/2} \frac{\pi^{1/2}}{3} \left[\frac{(700\,K)^3}{(13.6\,K)^2(8.92\,K)}\right]^{1/2}$$

$$\times (1 - e^{-4800/700})^{-1}(1 - e^{-1360/700})^{-1}(1 - e^{-4880/700})^{-2}(1 - e^{-2330/700})^{-2} e^{1158000/(8.3145)(700)}$$

$$= 2.13 \times 10^{121}\,m^{-3}$$

Using Equations 26.39 and 26.17, we can express K_P as

$$K_P(T) = \left(\frac{k_B T}{10^5 \text{ Pa}}\right)^2 \frac{(q_{NH_3}/V)^2}{(q_{N_2}/V)(q_{H_2}/V)^3}$$

$$= \left[\frac{k_B(700 \text{ K})}{10^5 \text{ Pa}}\right]^2 \frac{(2.13 \times 10^{121} \text{ m}^{-3})^2}{(1.08 \times 10^{105} \text{ m}^{-3})(7.15 \times 10^{63} \text{ m}^{-3})^3}$$

$$= 1.23 \times 10^{-4} = 12.3 \times 10^{-5}$$

The discrepancy between the calculated value and the experimental value (about 40%) is due to the use of the rigid rotator-harmonic oscillator approximation.

26–39. Calculate the equilibrium constant K_P for the reaction

$$I_2(g) \rightleftharpoons 2 I(g)$$

using the data in Table 18.2 and the fact that the degeneracy of the ground electronic state of an iodine atom is 4 and that the degeneracy of the first excited electronic state is 2 and that its energy is 7580 cm^{-1}. The experimental values of K_P are

T/K	800	900	1000	1100	1200
K_P	3.05×10^{-5}	3.94×10^{-4}	3.08×10^{-3}	1.66×10^{-2}	6.79×10^{-2}

Plot $\ln K_p$ against $1/T$ to determine the value of $\Delta_r H°$. The experimental value is 153.8 kJ·mol^{-1}.

The degeneracy of the ground electronic state of an iodine atom is 4. The first excited state is 90.677 kJ·mol^{-1} above that and its degeneracy is 2, so (using Equations 18.13 and 18.39)

$$\frac{q_I}{V} = \left[\frac{2\pi m k_B T}{h^2}\right]^{3/2} \left(4 + 2e^{-\varepsilon_{e2}/k_B T}\right)$$

$$= \left[\frac{2\pi(0.1269 \text{ kg·mol}^{-1})(8.3145 \text{ J·mol}^{-1}·\text{K}^{-1})T}{(6.022 \times 10^{23} \text{ mol}^{-1})^2(6.626 \times 10^{-34} \text{ J·s})^2}\right]^{3/2} \left(4 + 2e^{-90677 \text{ K}/8.3145T}\right)$$

$$\frac{q_{I_2}}{V} = \left[\frac{2\pi M k_B T}{h^2}\right]^{3/2} \frac{T}{\sigma\Theta_{rot}} \frac{1}{1 - e^{-\Theta_{vib}/T}} e^{D_0/k_B T}$$

$$= \left[\frac{2\pi(0.2538 \text{ kg·mol}^{-1})(8.3145 \text{ J·mol}^{-1}·\text{K}^{-1})T}{(6.022 \times 10^{23} \text{ mol}^{-1})^2(6.626 \times 10^{-34} \text{ J·s})^2}\right]^{3/2} \frac{T}{2(0.0537 \text{ K})} \frac{1}{1 - e^{-308 \text{ K}/T}} e^{148800 \text{ K}/8.3145T}$$

Using Equations 26.39 and 26.17, we can write K_P as

$$K_P = \left(\frac{k_B T}{10^5 \text{ Pa}}\right) \frac{(q_I/V)^2}{(q_{I_2}/V)}$$

The calculated values of K_P using the partition functions above are

T/K	800	900	1000	1100	1200
K_P	3.14×10^{-5}	4.08×10^{-4}	3.19×10^{-3}	1.72×10^{-2}	7.07×10^{-2}

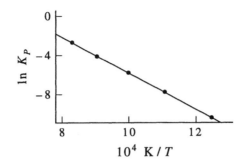

Plotting the calculated values, we find that $\ln K_P = 12.785 - (18498\ \text{K})/T$. Therefore, $\Delta_r H° = -R(-18498\ \text{K}) = 154.0\ \text{kJ·mol}^{-1}$.

26–40. Consider the reaction given by

$$H_2(g) + D_2(g) \rightleftharpoons 2\,HD(g)$$

Using the Born-Oppenheimer approximation and the molecular parameters in Table 18.2, show that

$$K(T) = 4.24e^{-77.7\ \text{K}/T}$$

Compare your predictions using this equation to the data in the JANAF tables.

We have an expression for K_P from Equation 26.39:

$$K_P = \frac{(q_{HD}/V)^2}{(q_{H_2}/V)(q_{D_2}/V)}$$

The relevant partition functions are (Equation 18.39)

$$\frac{q_{H_2}}{V} = \left[\frac{2\pi m_{H_2} k_B T}{h^2}\right]^{3/2} \frac{T}{2\Theta_{rot}^{H_2}} \frac{e^{-\Theta_{vib}^{H_2}/2T}}{1 - e^{-\Theta_{vib}^{H_2}/T}} e^{D_e^{H_2}/RT}$$

$$\frac{q_{D_2}}{V} = \left[\frac{2\pi m_{D_2} k_B T}{h^2}\right]^{3/2} \frac{T}{2\Theta_{rot}^{D_2}} \frac{e^{-\Theta_{vib}^{D_2}/2T}}{1 - e^{-\Theta_{vib}^{D_2}/T}} e^{D_e^{D_2}/RT}$$

$$\frac{q_{HD}}{V} = \left[\frac{2\pi m_{HD} k_B T}{h^2}\right]^{3/2} \frac{T}{\Theta_{rot}^{HD}} \frac{e^{-\Theta_{vib}^{HD}/2T}}{1 - e^{-\Theta_{vib}^{HD}/T}} e^{D_e^{HD}/RT}$$

Substituting into K_P gives

$$K_P = \left(\frac{2\pi m_{HD} k_B T}{h^2}\right)^{3/2} \left(\frac{2\pi m_{H_2} k_B T}{h^2}\right)^{-3/2} \left(\frac{2\pi m_{D_2} k_B T}{h^2}\right)^{-3/2} \left(\frac{T}{\Theta_{rot}^{HD}}\right)^2 \left(\frac{4\Theta_{rot}^{H_2}\Theta_{rot}^{D_2}}{T^2}\right)$$

$$\times \frac{e^{-2\Theta_{vib}^{HD}/2T}}{e^{-\Theta_{vib}^{H_2}/2T} e^{-\Theta_{vib}^{D_2}/2T}} (1 - e^{-\Theta_{vib}^{H_2}/T})(1 - e^{-\Theta_{vib}^{D_2}/T})(1 - e^{-\Theta_{vib}^{HD}/T})^{-2} e^{(2D_e^{HD} - D_e^{D_2} - D_e^{H_2})/RT}$$

$$= \left(\frac{m_{HD}^2}{m_{H_2} m_{D_2}}\right)^{3/2} \left[\frac{4\Theta_{rot}^{H_2}\Theta_{rot}^{D_2}}{(\Theta_{rot}^{HD})^2}\right] \left[\frac{(1 - e^{-\Theta_{vib}^{H_2}/T})(1 - e^{-\Theta_{vib}^{D_2}/T})}{(1 - e^{-\Theta_{vib}^{HD}/T})^2}\right] e^{-(2\Theta_{vib}^{HD} - \Theta_{vib}^{H_2} - \Theta_{vib}^{D_2})/2T}$$

$$\times e^{(2D_e^{HD} - D_e^{D_2} - D_e^{H_2})/RT} \tag{1}$$

Under the Born-Oppenheimer approximation, $D_e^{HD} = D_e^{D_2} = D_e^{H_2}$, so the last exponential term becomes 1. Also, k and R_e are the same for HD, H_2, and D_2. Then, since $\nu = (k/\mu)^{1/2}/2\pi$ and $I = \mu R_e^2$, we can write Θ_{vib} and Θ_{rot} as

$$\Theta_{vib} = \frac{h\nu}{k_B} = \frac{hk^{1/2}\pi}{2k_B}\mu^{-1/2} \propto \mu^{-1/2} \qquad \text{and} \qquad \Theta_{rot} = \frac{\hbar^2}{2Ik_B} = \frac{\hbar^2}{2R_e^2 k_B}\mu^{-1} \propto \mu^{-1}$$

Recall that $\mu_{AB} = (m_A m_B)/(m_A + m_B)$. Applying this formula, we find that $\mu_{H_2} = 0.5$ amu, $\mu_{HD} = 2/3$ amu, and $\mu_{D_2} = 1$ amu. We can now write

$$\frac{4\Theta_{rot}^{H_2}\Theta_{rot}^{D_2}}{(\Theta_{rot}^{HD})^2} = \frac{4\mu_{H_2}^{-1}\mu_{D_2}^{-1}}{\mu_{HD}^{-2}} = \frac{4(2/3\ \text{amu})^2}{(0.5\ \text{amu})(1\ \text{amu})} = \frac{32}{9}$$

We can also express Θ_{vib}^{HD} and $\Theta_{vib}^{D_2}$ in terms of $\Theta_{vib}^{H_2}$:

$$\frac{\Theta_{vib}^{HD}}{\Theta_{vib}^{H_2}} = \left(\frac{\mu_{H_2}}{\mu_{HD}}\right)^{1/2} = \left(\frac{1/2}{2/3}\right)^{1/2} = \frac{\sqrt{3}}{2}$$

$$\Theta_{vib}^{HD} = \frac{\sqrt{3}}{2}\Theta_{vib}^{H_2}$$

$$\frac{\Theta_{vib}^{D_2}}{\Theta_{vib}^{H_2}} = \left(\frac{\mu_{H_2}}{\mu_{D_2}}\right)^{1/2} = \left(\frac{1/2}{1}\right)^{1/2} = \frac{\sqrt{2}}{2}$$

$$\Theta_{vib}^{D_2} = \frac{\sqrt{2}}{2}\Theta_{vib}^{H_2}$$

Then

$$2\Theta_{vib}^{HD} - \Theta_{vib}^{H_2} - \Theta_{vib}^{D_2} = \left(3^{1/2} - 1 - 2^{-1/2}\right)\Theta_{vib}^{H_2} = 155.0\ \text{K}$$

where $\Theta_{vib}^{H_2} = 6332$ K. Substituting into Equation 1, we find

$$K = \left(\frac{9}{8}\right)^{3/2}\left(\frac{32}{9}\right)\left[\frac{(1 - e^{-\Theta_{vib}^{H_2}/T})(1 - e^{-\Theta_{vib}^{D_2}/T})}{(1 - e^{-\Theta_{vib}^{HD}/T})^2}\right]e^{-77.7\ \text{K}/T}$$

$$= 4.24 e^{-77.7\ \text{K}/T}$$

where we have neglected factors such as $1 - e^{-\Theta_{vib}/T}$, since they do not contribute significantly to K for $T < 1000$ K. The table below compares calculated values of K with values from the JANAF tables.

T/K	K_P(calc.)	K_P(JANAF)
200	2.88	2.90
400	3.49	3.48
600	3.73	3.72
800	3.85	3.84
1000	3.92	3.91

26–41. Using the harmonic oscillator-rigid rotator approximation, show that

$$K(T) = \left(\frac{m_{H_2}m_{Br_2}}{m_{HBr}^2}\right)^{3/2}\left(\frac{\sigma_{HBr}^2}{\sigma_{H_2}\sigma_{Br_2}}\right)\left(\frac{(\Theta_{rot}^{HBr})^2}{\Theta_{rot}^{H_2}\Theta_{rot}^{Br_2}}\right)\frac{(1 - e^{-\Theta_{vib}^{HBr}/T})^2}{(1 - e^{-\Theta_{vib}^{H_2}/T})(1 - e^{-\Theta_{vib}^{Br_2}/T})}e^{(D_0^{H_2}+D_0^{Br_2}-2D_0^{HBr})/RT}$$

for the reaction described by

$$2\,HBr(g) \rightleftharpoons H_2(g) + Br_2(g)$$

Using the values of Θ_{rot}, Θ_{vib}, and D_0 given in Table 18.2, calculate K at 500 K, 1000 K, 1500 K, and 2000 K. Plot $\ln K$ against $1/T$ and determine the value of $\Delta_r H°$.

We have an expression for K_P from Equation 26.39:

$$K_P = \frac{(q_{H_2}/V)(q_{Br_2}/V)}{(q_{HBr}/V)^2}$$

The relevant partition functions are

$$\frac{q_{H_2}}{V} = \left[\frac{2\pi m_{H_2} k_B T}{h^2}\right]^{3/2} \frac{T}{\sigma_{H_2}\Theta_{rot}^{H_2}} \frac{1}{1 - e^{-\Theta_{vib}^{H_2}/T}} e^{D_0^{H_2}/RT}$$

$$\frac{q_{Br_2}}{V} = \left[\frac{2\pi m_{Br_2} k_B T}{h^2}\right]^{3/2} \frac{T}{\sigma_{Br_2}\Theta_{rot}^{Br_2}} \frac{1}{1 - e^{-\Theta_{vib}^{Br_2}/T}} e^{D_0^{Br_2}/RT}$$

$$\frac{q_{HBr}}{V} = \left[\frac{2\pi m_{HBr} k_B T}{h^2}\right]^{3/2} \frac{T}{\sigma_{HBr}\Theta_{rot}^{HBr}} \frac{1}{1 - e^{-\Theta_{vib}^{HBr}/T}} e^{D_0^{HBr}/RT}$$

so we write K_P as

$$K(T) = \left(\frac{m_{H_2} m_{Br_2}}{m_{HBr}^2}\right)^{3/2} \left(\frac{\sigma_{HBr}^2}{\sigma_{H_2}\sigma_{Br_2}}\right) \left[\frac{(\Theta_{rot}^{HBr})^2}{\Theta_{rot}^{H_2}\Theta_{rot}^{Br_2}}\right] \frac{(1 - e^{\Theta_{vib}^{HBr}/T})^2}{(1 - e^{\Theta_{vib}^{H_2}/T})(1 - e^{\Theta_{vib}^{Br_2}/T})} e^{(D_0^{H_2} + D_0^{Br_2} - 2D_0^{HBr})/RT}$$

Using the values from Table 18.2, we find that

T/K	500	1000	1500	2000
K_P	8.96×10^{-12}	1.20×10^{-6}	6.63×10^{-5}	4.97×10^{-4}

We can use these values to create a graph of $\ln K_P$ vs. $1/T$ and curve-fit the points linearly to obtain the equation $\ln K_P = -1.70367 - (11876\,K)/T$.

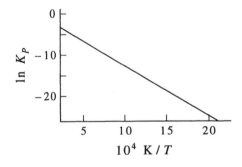

Therefore, $\Delta_r H° = -R(-11876\,K) = 98.8\,kJ\cdot mol^{-1}$ for the reaction (compared to an experimental value of $106.0\,kJ\cdot mol^{-1}$).

26–42. Use Equation 26.49b to calculate $H°(T) - H_0°$ for $NH_3(g)$ from 300 K to 6000 K and compare your values to those given in Table 26.4 by plotting them on the same graph.

$$H°(T) - H_0° = 4RT + \sum_j \frac{R\Theta_{vib,j}}{e^{-\Theta_{vib,j}/T} - 1} \qquad (26.49b)$$

We use Table 18.4 for the appropriate values of $\Theta_{vib,j}$ to produce the graph below. The data points are from the JANAF tables, and the line is the function represented by Equation 26.49b.

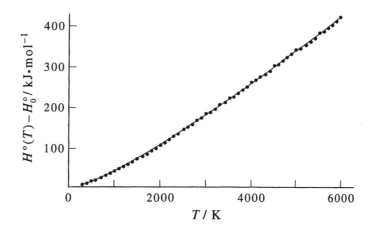

This is a very good fit to the JANAF data.

26–43. Use the JANAF tables to calculate K_p at 1000 K for the reaction described by

$$H_2(g) + I_2(g) \rightleftharpoons 2\,HI(g)$$

Compare your results to the value given in Table 26.2.

The JANAF tables give $\log K_f = 0.732$ for HI(g) at 1000 K. The equation given represents the formation of two moles of HI(g) from its consituent elements, and so $\log K = 2 \log K_f = 1.464$ and $\ln K = 3.37$. The value of $\ln K$ in Table 26.2 is 3.55.

26–44. Use the JANAF tables to plot $\ln K_p$ versus $1/T$ from 900 K to 1200 K for the reaction described by

$$2\,Na(g) \rightleftharpoons Na_2(g)$$

and compare your results to those obtained in Problem 26–35.

We can use Equation 26–11 to find K_p from the values given in the JANAF tables. From the JANAF tables,

T/K	900	1000	1100
$\Delta_f G°[Na_2(g)]/kJ \cdot mol^{-1}$	43.601	34.740	26.068
$\Delta_f G°[Na(g)]/kJ \cdot mol^{-1}$	43.601	34.740	26.068
$\Delta_r G°/kJ \cdot mol^{-1}$	−1.819	6.248	14.338
K_p(JANAF)	1.28	0.472	0.208

compared to $K_p(900\ K) = 1.47$, $K_p(1000\ K) = 0.52$, and $K_p(1100\ K) = 0.22$ from Problem 26–35.

26–45. In Problem 26–36 we calculated K_p for the decomposition of $CO_2(g)$ to $CO(g)$ and $O_2(g)$ at 2000 K. Use the JANAF tables to calculate K_p and compare your result to the one that you obtained in Problem 26–36.

$$CO_2(g) \rightleftharpoons CO(g) + \frac{1}{2} O_2(g)$$

From the JANAF tables,

	CO(g)	$O_2(g)$	$CO_2(g)$
$\Delta_f G°$/kJ·mol^{-1}	−286.034	0	−396.333

These values give a $\Delta_r G° = 110.299$ kJ·mol^{-1}, and (Equation 26.11)

$$K_p = e^{-\Delta_r G°/RT} = 1.32 \times 10^{-3}$$

compared to $K_p = 1.46 \times 10^{-3}$ from Problem 26–36.

26–46. You calculated K_p at 700 K for the ammonia synthesis reaction in Problem 26–38. Use the data in Table 26.4 to calculate K_p and compare your result to the one that you obtained in Problem 26–38.

$$3 H_2(g) + N_2(g) \rightleftharpoons 2 NH_3(g)$$

From the JANAF tables, we know that at 700 K $\Delta_f G°[NH_3(g)] = 27.190$ kJ·mol^{-1}. Therefore, for the reaction above, $\Delta_r G° = 2(27.190 \text{ kJ·mol}^{-1}) = 54.380$ kJ·mol^{-1}, and (Equation 26.11)

$$K_p = e^{-\Delta_r G°/RT} = 8.75 \times 10^{-5}$$

compared to $K_p = 12.3 \times 10^{-5}$ from Problem 26–38.

26–47. The JANAF tables give the following data for I(g) at one bar:

T/K	800	900	1000	1100	1200
$\Delta_f G°$/kJ·mol^{-1}	34.580	29.039	24.039	18.741	13.428

Calculate K_p for the reaction described by

$$I_2(g) \rightleftharpoons 2 I(g)$$

and compare your results to the values given in Problem 26–39.

The energy of the reaction above will be twice the energy of formation of iodine, or $\Delta_r G° = 2\Delta_f G°[I(g)]$. Then, using Equation 26.11, we can calculate K_p at each temperature above:

T/K	800	900	1000	1100	1200
K_p(JANAF)	3.05×10^{-5}	4.26×10^{-4}	3.08×10^{-3}	1.66×10^{-2}	6.78×10^{-2}
K_p(calc)	3.14×10^{-5}	4.08×10^{-4}	3.19×10^{-3}	1.72×10^{-2}	7.07×10^{-2}

where we calculated the values of K_p in Problem 26–39.

26–48. Use Equation 18.60 to calculate the value of $q^0(V, T)/V$ given in the text (page 1076) for $NH_3(g)$ at 500 K.

We can write Equation 18.60 in terms of D_0 as

$$q(V) = \left(\frac{2\pi M k_B T}{h^2}\right)^{3/2} V \frac{\pi^{1/2}}{\sigma} \left(\frac{T^3}{\Theta_{rot,A} \Theta_{rot,B} \Theta_{rot,C}}\right)^{1/2} \left[\prod_{j=1}^{6}(1 - e^{-\Theta_{vib,j}/T})^{-1}\right] e^{D_0/k_B T}$$

We can ignore the last exponential term when we look at q^0, since q^0 is the energy relative to the ground-state energy. For $NH_3(g)$, this becomes

$$\frac{q}{V} = \left[\frac{2\pi(0.01709 \text{ kg}\cdot\text{mol}^{-1})(8.3145 \text{ J}\cdot\text{mol}^{-1}\cdot\text{K}^{-1})(500 \text{ K})}{(6.022 \times 10^{23} \text{ mol}^{-1})^2(6.626 \times 10^{-34} \text{ J}\cdot\text{s})^2}\right]^{3/2} \frac{\pi^{1/2}}{3} \left[\frac{(500 \text{ K})^3}{(13.6 \text{ K})^2(8.92 \text{ K})}\right]^{1/2}$$

$$\times (1 - e^{-48/5})^{-1}(1 - e^{-136/50})^{-1}(1 - e^{-488/50})^{-2}(1 - e^{-233/50})^{-2}$$

$$= 2.59 \times 10^{34} \text{ m}^{-3}$$

26–49. The JANAF tables give the following data for $Ar(g)$ at 298.15 K and one bar:

$$-\frac{G^\circ - H^\circ(298.15 \text{ K})}{T} = 154.845 \text{ J}\cdot\text{mol}^{-1}\cdot\text{K}^{-1}$$

and

$$H^\circ(0 \text{ K}) - H^\circ(298.15 \text{ K}) = -6.197 \text{ kJ}\cdot\text{mol}^{-1}$$

Use these data to calculate $q^0(V, T)/V$ and compare your result to what you obtain using Equation 18.13.

Use Equation 26.52b to find the exponential term in Equation 26.52a:

$$-\frac{(G^\circ - H_0^\circ)}{T} = -\frac{[G^\circ - H^\circ(298.15 \text{ K})]}{T} + \frac{[H^\circ - H^\circ(298.15 \text{ K})]}{T}$$

$$= 154.845 \text{ J}\cdot\text{mol}^{-1}\cdot\text{K}^{-1} - \frac{6197 \text{ J}\cdot\text{mol}^{-1}}{298.15 \text{ K}}$$

$$= 134.06 \text{ J}\cdot\text{mol}^{-1}\cdot\text{K}^{-1}$$

Now substitute into Equation 26.52a:

$$\frac{q^0(V, T)}{V} = \frac{N_A P^\circ}{RT} e^{-(G^\circ - H_0^\circ)/RT}$$

$$= \frac{(6.022 \times 10^{23} \text{ mol}^{-1})(10^5 \text{ Pa})}{(8.3145 \text{ J}\cdot\text{mol}^{-1}\cdot\text{K}^{-1})(298.15 \text{ K})} e^{134.06/8.3145}$$

$$= 2.443 \times 10^{32} \text{ m}^{-3}$$

Using Equation 18.13 and looking at q^0 as we did in Section 23–5, we find that

$$\frac{q^0}{V} = \left(\frac{2\pi M k_B T}{h^2}\right)^{3/2}$$

$$= \left[\frac{2\pi(0.039948 \text{ kg}\cdot\text{mol}^{-1})(8.3145 \text{ J}\cdot\text{mol}^{-1}\cdot\text{K}^{-1})(298.15 \text{ K})}{(6.022 \times 10^{23} \text{ mol}^{-1})^2(6.626 \times 10^{-34} \text{ J}\cdot\text{s})^2}\right]^{3/2}$$

$$= 2.443 \times 10^{32} \text{ m}^{-3}$$

26–50. Use the JANAF tables to calculate $q^0(V, T)/V$ for $CO_2(g)$ at 500 K and one bar and compare your result to what you obtain using Equation 18.57 (with the ground state energy taken to be zero).

Using Equation 26.52b,

$$-\frac{(G° - H_0°)}{T} = 218.290 \text{ J·mol}^{-1}\text{·K}^{-1} + \frac{-9364 \text{ J·mol}^{-1}}{500 \text{ K}}$$
$$= 199.562 \text{ J·mol}^{-1}\text{·K}^{-1}$$

Now substitute into Equation 26.52a:

$$\frac{q^0(V, T)}{V} = \frac{N_A P°}{RT} e^{-(G° - H_0°)/RT}$$
$$= \frac{(6.022 \times 10^{23} \text{ mol}^{-1})(10^5 \text{ Pa})}{(8.3145 \text{ J·mol}^{-1}\text{·K}^{-1})(298.15 \text{ K})} e^{199.562/8.3145}$$
$$= 3.84 \times 10^{35} \text{ m}^{-3}$$

Using Equation 18.57 and looking at q^0 as we did in Section 23–5, we find that

$$\frac{q^0}{V} = \left(\frac{2\pi M k_B T}{h^2}\right)^{3/2} \left(\frac{T}{\sigma \Theta_{rot}}\right) \prod_{j=1}^{4}(1 - e^{-\Theta_{vib,j}/T})^{-1}$$
$$= \left[\frac{2\pi (0.04400 \text{ kg·mol}^{-1})(8.3145 \text{ J·mol}^{-1}\text{·K}^{-1})(500 \text{ K})}{(6.022 \times 10^{23} \text{ mol}^{-1})^2 (6.626 \times 10^{-34} \text{ J·s})^2}\right]^{3/2} \left[\frac{500 \text{ K}}{2(0.561 \text{ K})}\right]$$
$$\times (1 - e^{336/500})^{-1}(1 - e^{-954/500})^{-2}(1 - e^{-189/500})^{-1}$$
$$= 3.86 \times 10^{35} \text{ m}^{-3}$$

26–51. Use the JANAF tables to calculate $q^0(V, T)/V$ for $CH_4(g)$ at 1000 K and one bar and compare your result to what you obtain using Equation 18.60 (with the ground state energy taken to be zero).

Using Equation 26.52b,

$$-\frac{(G° - H_0°)}{T} = 209.370 \text{ J·mol}^{-1}\text{·K}^{-1} + \frac{-10024 \text{ J·mol}^{-1}}{1000 \text{ K}}$$
$$= 199.35 \text{ J·mol}^{-1}\text{·K}^{-1}$$

Now substitute into Equation 26.52a:

$$\frac{q^0(V, T)}{V} = \frac{N_A P°}{RT} e^{-(G° - H_0°)/RT}$$
$$= \frac{(6.022 \times 10^{23} \text{ mol}^{-1})(10^5 \text{ Pa})}{(8.3145 \text{ J·mol}^{-1}\text{·K}^{-1})(298.15 \text{ K})} e^{199.35/8.3145}$$
$$= 1.87 \times 10^{35} \text{ m}^{-3}$$

Using Equation 18.60 and looking at q^0 as we did in Section 23–5, we find that

$$\frac{q^0}{V} = \left(\frac{2\pi M k_B T}{h^2}\right)^{3/2} \frac{\pi^{1/2}}{\sigma} \left(\frac{T^3}{\Theta_{rot,A}\Theta_{rot,B}\Theta_{rot,C}}\right)^{1/2} \prod_{j=1}^{9}(1 - e^{-\Theta_{vib,j}/T})^{-1}$$

$$= \left[\frac{2\pi(0.01604\ kg\cdot mol^{-1})(8.3145\ J\cdot mol^{-1}\cdot K^{-1})(1000\ K)}{(6.022 \times 10^{23}\ mol^{-1})(6.626 \times 10^{-34}\ J\cdot s)^2}\right]^{3/2} \frac{\pi^{1/2}}{3}$$

$$\times \left[\frac{(1000\ K)^3}{(7.54\ K)^3}\right]^{1/2} (1 - e^{-417/100})^{-1}(1 - e^{-218/100})^{-2}(1 - e^{-432/100})^{-3}(1 - e^{-187/100})^{-3}$$

$$= 1.91 \times 10^{35}\ m^{-3}$$

26–52. Use the JANAF tables to calculate $q^0(V, T)/V$ for $H_2O(g)$ at 1500 K and one bar and compare your result to what you obtain using Equation 26.45. Why do you think there is some discrepancy?

Using Equation 26.52b,

$$-\frac{(G^\circ - H_0^\circ)}{T} = 218.520\ J\cdot mol^{-1}\cdot K^{-1} + \frac{-9904\ J\cdot mol^{-1}}{1500\ K}$$

$$= 211.9\ J\cdot mol^{-1}\cdot K^{-1}$$

Now substitute into Equation 26.52a:

$$\frac{q^0(V, T)}{V} = \frac{N_A P^\circ}{RT}e^{-(G^\circ - H_0^\circ)/RT}$$

$$= \frac{(6.022 \times 10^{23}\ mol^{-1})(10^5\ Pa)}{(8.3145\ J\cdot mol^{-1}\cdot K^{-1})(298.15\ K)}e^{211.9/8.3145}$$

$$= 5.66 \times 10^{35}\ m^{-3}$$

Using Equation 18.60 and looking at q^0 as we did in Section 23–5, we find that

$$\frac{q^0}{V} = \left(\frac{2\pi M k_B T}{h^2}\right)^{3/2} \frac{\pi^{1/2}}{\sigma} \left(\frac{T^3}{\Theta_{rot,A}\Theta_{rot,B}\Theta_{rot,C}}\right)^{1/2} \prod_{j=1}^{3}(1 - e^{-\Theta_{vib,j}/T})^{-1}$$

$$= \left[\frac{2\pi(0.018015\ kg\cdot mol^{-1})(8.3145\ J\cdot mol^{-1}\cdot K^{-1})(1500\ K)}{(6.022 \times 10^{23}\ mol^{-1})^2(6.626 \times 10^{-34}\ J\cdot s)^2}\right]^{3/2} \frac{\pi^{1/2}}{3}$$

$$\times \left[\frac{(1500\ K)^3}{(40.1\ K)(20.9\ K)(13.4\ K)}\right]^{1/2} (1 - e^{-536/150})^{-1}(1 - e^{-516/150})^{-1}(1 - e^{-229/150})^{-1}$$

$$= 5.51 \times 10^{35}\ m^{-3}$$

The small discrepancy between these two results is probably due to the use of the harmonic-oscillator approximation in obtaining Equation 18.60.

26–53. The JANAF tables give the following data:

	H(g)	Cl(g)	HCl(g)
$\Delta_f H^\circ(0\ K)/kJ\cdot mol^{-1}$	216.035	119.621	-92.127

Use these data to calculate D_0 for HCl(g) and compare your value to the one in Table 18.2.

$$\Delta_r H°/\text{kJ}\cdot\text{mol}^{-1}$$

$$\tfrac{1}{2}H_2(g) \rightleftharpoons H(g) \qquad 216.035 \qquad (1)$$

$$\tfrac{1}{2}Cl_2(g) \rightleftharpoons Cl(g) \qquad 119.621 \qquad (2)$$

$$\tfrac{1}{2}H_2(g) + \tfrac{1}{2}Cl_2(g) \rightleftharpoons HCl(g) \qquad -92.127 \qquad (3)$$

We can obtain the reaction HCl(g) $\rightleftharpoons$ H(g) + Cl(g) by subtracting Equation 3 from the sum of Equations 1 and 2, to find

$$D_0 = \Delta_r H° = (216.035 + 119.621 + 92.127)\ \text{kJ}\cdot\text{mol}^{-1} = 427.8\ \text{kJ}\cdot\text{mol}^{-1}$$

compared to a value of 427.8 kJ·mol^{-1} in Table 18.2.

26–54. The JANAF tables give the following data:

	C(g)	H(g)	CH$_4$(g)
$\Delta_f H°(0\ \text{K})/\text{kJ}\cdot\text{mol}^{-1}$	711.19	216.035	−66.911

Use these data to calculate D_0 for CH$_4$(g) and compare your value to the one in Table 18.4.

$$\Delta_r H°/\text{kJ}\cdot\text{mol}^{-1}$$

$$C(s) \rightleftharpoons C(g) \qquad 711.19 \qquad (1)$$

$$\tfrac{1}{2}H_2(g) \rightleftharpoons H(g) \qquad 216.035 \qquad (2)$$

$$C(s) + 2H_2(g) \rightleftharpoons CH_4(g) \qquad -66.911 \qquad (3)$$

We can obtain the reaction CH$_4$(g) $\rightleftharpoons$ 4H(g) + C(g) by subtracting Equation 3 from the sum of Equations 1 and four times Equation 2, to find

$$D_0 = \Delta_r H° = [66.911 + 711.19 + 4(216.035)]\ \text{kJ}\cdot\text{mol}^{-1} = 1642\ \text{kJ}\cdot\text{mol}^{-1}$$

compared to a value of 1642 kJ·mol^{-1} from Table 18.4.

26–55. Use the JANAF tables to calculate D_0 for CO$_2$(g) and compare your result to the one given in Table 18.4.

$$\Delta_r H°/\text{kJ}\cdot\text{mol}^{-1}$$

$$C(s) \rightleftharpoons C(g) \qquad 711.19 \qquad (1)$$

$$\tfrac{1}{2}O_2(g) \rightleftharpoons O(g) \qquad 246.790 \qquad (2)$$

$$C(s) + O_2(g) \rightleftharpoons CO_2(g) \qquad -393.115 \qquad (3)$$

We can obtain the reaction CO$_2$(g) $\rightleftharpoons$ 2O(g) + C(g) by subtracting Equation 3 from the sum of Equations 1 and two times Equation 2, to find

$$D_0 = \Delta_r H° = [393.115 + 711.19 + 2(246.790)]\ \text{kJ}\cdot\text{mol}^{-1} = 1598\ \text{kJ}\cdot\text{mol}^{-1}$$

compared to a value of 1596 kJ·mol^{-1} in Table 18.4.

26–56. A determination of K_γ (see Example 26–11) requires a knowledge of the fugacity of each gas in the equilibrium mixture. These data are not usually available, but a useful approximation is to take the fugacity coefficient of a gaseous constituent of a mixture to be equal to the value for the pure gas at the *total pressure of the mixture*. Using this approximation, we can use Figure 22.11 to determine γ for each gas and then calculate K_γ. In this problem we shall apply this approximation to the data in Table 26.5. First use Figure 22.11 to estimate that $\gamma_{H_2} = 1.05$, $\gamma_{N_2} = 1.05$, and that $\gamma_{NH_3} = 0.95$ at a total pressure of 100 bar and a temperature of 450°C. In this case $K_\gamma = 0.86$, in fairly good agreement with the value given in Example 26–11. Now calculate K_γ at 600 bar and compare your result with the value given in Example 26–11.

First, we must find the reduced temperatures and reduced pressures of each species at a pressure of 100 bar and a temperature of 450°C (we can use Table 16.5 for critical values):

$$P_R(N_2) = \frac{100\ \text{bar}}{34.0\ \text{bar}} = 2.94 \qquad T_R(N_2) = \frac{723\ \text{K}}{126.2\ \text{K}} = 5.73$$

$$P_R(H_2) = \frac{100\ \text{bar}}{12.838\ \text{bar}} = 7.79 \qquad T_R(H_2) = \frac{723\ \text{K}}{32.938\ \text{K}} = 22.0$$

$$P_R(NH_3) = \frac{100\ \text{bar}}{111.30\ \text{bar}} = 0.898 \qquad T_R(NH_3) = \frac{723\ \text{K}}{405.30\ \text{K}} = 1.78$$

Using Figure 22.11, it looks as if $\gamma_{H_2} = 1.05$, $\gamma_{N_2} = 1.05$, and $\gamma_{NH_3} = 0.95$. At 600 bar, $P_R(N_2) = 17.6$, $P_R(H_2) = 46.7$, and $P_R(NH_3) = 5.4$, so $\gamma_{H_2} = 1.3$, $\gamma_{N_2} = 1.3$, and $\gamma_{NH_3} = 0.9$. Then

$$K_\gamma = \frac{\gamma_{NH_3}}{\gamma_{H_2}^{3/2}\gamma_{N_2}^{3/2}} = 0.53$$

as compared to the value in Example 26–11 of 0.496. This is within the margin of error created by estimating the values of γ from Figure 22.11.

26–57. Recall from general chemistry that Le Châtelier's principle says that pressure has no effect on a gaseous equilibrium system such as

$$CO(g) + H_2O(g) \rightleftharpoons H_2(g) + CO_2(g)$$

in which the total number of moles of reactants is equal to the total number of moles of product in the chemical equation. The thermodynamic equilibrium constant in this case is

$$K_f = \frac{f_{CO_2} f_{H_2}}{f_{CO} f_{H_2O}} = \frac{\gamma_{CO_2}\gamma_{H_2}}{\gamma_{CO}\gamma_{H_2O}}\frac{P_{CO_2}P_{H_2}}{P_{CO}P_{H_2O}} = K_\gamma K_P$$

If the four gases behaved ideally, then pressure would have no effect on the position of equilibrium. However, because of deviations from ideal behavior, a shift in the equilibrium composition will occur when the pressure is changed. To see this, use the approximation introduced in Problem 26–56 to estimate K_γ at 900 K and 500 bar. Note that K_γ under these conditions is greater than K_γ at one bar, where $K_\gamma \approx 1$ (ideal behavior). Consequently, argue that an increase in pressure causes the equilibrium to shift to the left in this case.

At 900 K and 500 bar,

$$P_R(CO) = \frac{500 \text{ bar}}{34.935 \text{ bar}} = 14.3 \qquad T_R(CO) = \frac{900 \text{ K}}{132.85 \text{ K}} = 6.77$$

$$P_R(H_2O) = \frac{500 \text{ bar}}{220.55 \text{ bar}} = 2.27 \qquad T_R(H_2O) = \frac{900 \text{ K}}{647.126 \text{ K}} = 1.39$$

$$P_R(H_2) = \frac{500 \text{ bar}}{12.838 \text{ bar}} = 38.9 \qquad T_R(H_2) = \frac{900 \text{ K}}{32.938 \text{ K}} = 27.3$$

$$P_R(CO_2) = \frac{500 \text{ bar}}{73.843 \text{ bar}} = 6.77 \qquad T_R(CO_2) = \frac{900 \text{ K}}{304.14 \text{ K}} = 2.96$$

and so $\gamma_{CO} \approx 1.3$, $\gamma_{H_2O} \approx 0.8$, $\gamma_{H_2} \approx 1.15$, and $\gamma_{CO_2} \approx 1.1$. Then

$$K_\gamma = \frac{\gamma_{CO_2}\gamma_{H_2}}{\gamma_{H_2O}\gamma_{CO}} = 1.1$$

Since K_f must remain constant, K_P at 500 bar must be smaller than K_P at one bar and so the equilibrium will shift to the left.

26–58. Calculate the activity of $H_2O(l)$ as a function of pressure from one bar to 100 bar at 20.0°C. Take the density of $H_2O(l)$ to be 0.9982 g·mL^{-1} and assume that it is incompressible.

Use Equation 26.69,

$$
\begin{aligned}
\ln a &= \frac{\overline{V}}{RT}(P-1) \\
&= \left(\frac{1 \times 10^{-3} \text{ dm}^3}{0.9982 \text{ g}}\right)\left(\frac{18.015 \text{ g}}{1 \text{ mol}}\right)\left[\frac{1}{(0.083145 \text{ dm}^3 \cdot \text{bar} \cdot \text{mol}^{-1} \cdot \text{K}^{-1})(293.15 \text{ K})}\right](P-1) \\
&= (7.40 \times 10^{-4} \text{ bar}^{-1})(P-1)
\end{aligned}
$$

Below are some values of $\ln a$ and a at representative temperatures through the range 1 to 100 bar.

$P/$bar	$\ln a$	a
1	0	1.00
10	6.67×10^{-3}	1.01
50	3.63×10^{-2}	1.04
100	7.33×10^{-2}	1.08

26–59. Consider the dissociation of HgO(s,red) to Hg(g) and O_2(g) according to

$$\text{HgO(s, red)} \rightleftharpoons \text{Hg(g)} + \tfrac{1}{2}O_2(g)$$

If we start with only HgO(s,red), then assuming ideal behavior, show that

$$K_P = \frac{2}{3^{3/2}}P^{3/2}$$

where P is the total pressure. Given the following "dissociation pressure" of HgO(s,red) at various temperatures, plot $\ln K_P$ versus $1/T$.

$t/°C$	P/atm	$t/°C$	P/atm
360	0.1185	430	0.6550
370	0.1422	440	0.8450
380	0.1858	450	1.067
390	0.2370	460	1.339
400	0.3040	470	1.674
410	0.3990	480	2.081
420	0.5095		

An excellent curve fit to the plot of $\ln K_P$ against $1/T$ is given by

$$\ln K_P = -172.94 + \frac{4.0222 \times 10^5 \text{ K}}{T} - \frac{2.9839 \times 10^8 \text{ K}^2}{T^2} + \frac{7.0527 \times 10^{10} \text{ K}^3}{T^3}$$

$$630 \text{ K} < T < 750 \text{ K}$$

Use this expression to determine $\Delta_r H°$ as a function of temperature in the interval $630 \text{ K} < T < 750 \text{ K}$. Given that

$$C_P°[O_2(g)]/R = 4.8919 - \frac{829.931 \text{ K}}{T} - \frac{127962 \text{ K}^2}{T^2}$$

$$C_P°[Hg(g)]/R = 2.500$$

$$C_P°[HgO(s, red)]/R = 5.2995$$

in the interval $298 \text{ K} < T < 750 \text{ K}$, calculate $\Delta_r H°$, $\Delta_r S°$, and $\Delta_r G°$ at 298 K.

We can write K_P in terms of the partial pressures of mercury and oxygen (assuming an activity of unity for the solid):

$$K_P = P_{O_2}^{1/2} P_{Hg} = \left(\frac{1/2}{3/2}P\right)^{1/2}\left(\frac{P}{3/2}\right) = \frac{2}{3^{3/2}}P^{3/2}$$

Below is a plot of the experimental values of $\ln K_P$ against $1/T$.

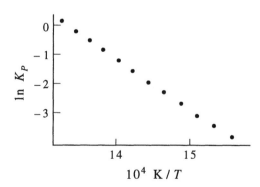

We can now use the equation given in the problem for $\ln K_P$ in Equation 26.29 to find $\Delta_r H°$:

$$\frac{d \ln K_P}{dT} = -\frac{4.0222 \times 10^5 \text{ K}}{T^2} + \frac{5.9678 \times 10^8 \text{ K}^2}{T^3} - \frac{2.1158 \times 10^{11} \text{ K}^3}{T^4}$$

$$\Delta_r H° = RT^2 \frac{d \ln K_P}{dT}$$

$$\Delta_r H° = R\left(-4.022 \times 10^5 \text{ K} + \frac{5.9678 \times 10^8 \text{ K}^2}{T} - \frac{2.1158 \times 10^9 \text{ K}^3}{T^2}\right)$$

Likewise, we can use Equation 26.11 to find $\Delta_r G°$:

$$\Delta_r G° = -RT \ln K_P$$

$$= R\left(172.94T - 4.0222 \times 10^5 \text{ K} + \frac{2.9839 \times 10^8 \text{ K}^2}{T} - \frac{7.0527 \times 10^{10} \text{ K}^3}{T^2}\right)$$

We can also find an empirical expression for $\Delta_r S°$ using the equation

$$\Delta_r S° = \frac{\Delta_r H° - \Delta_r G°}{T}$$

The expression for $\ln K_P$ used in the above equalities holds for temperatures ranging from 630 K to 750 K. To find the values of $\Delta_r H°$, $\Delta_r G°$, and $\Delta_r S°$ at 298 K, we can use the equation

$$\Delta_r H°(298 \text{ K}) = \Delta_r H°(700 \text{ K}) - \int_{298 \text{ K}}^{700 \text{ K}} \Delta C_P°(T)dT \tag{19.57}$$

and the similar equation

$$\Delta_r S°(298 \text{ K}) = \Delta_r S°(700 \text{ K}) - \int_{298 \text{ K}}^{700 \text{ K}} \frac{\Delta C_P°(T)}{T}dT$$

Substituting into the high-temperature expressions for $\Delta_r G°$ and $\Delta_r H°$ at 700 K, we find that $\Delta_r G°(700 \text{ K}) = 9.78 \text{ kJ·mol}^{-1}$, $\Delta_r H°(700 \text{ K}) = 154.0 \text{ kJ·mol}^{-1}$, and $\Delta_r S°(700 \text{ K}) = 206.1 \text{ J·mol}^{-1}·\text{K}^{-1}$. Then

$$\Delta_r S°(298 \text{ K}) = 206.1 \text{ J·mol}^{-1}·\text{K}^{-1}$$

$$- R\int_{298 \text{ K}}^{700 \text{ K}}\left[\frac{1}{2}\left(\frac{4.8919}{T} - \frac{829.931 \text{ K}}{T^2} - \frac{127962 \text{ K}^2}{T^3}\right) + \frac{2.500}{T} - \frac{5.2995}{T}\right]dT$$

$$= 206.1 \text{ J·mol}^{-1}·\text{K}^{-1}$$

$$- R\int_{298 \text{ K}}^{700 \text{ K}}\left(-\frac{0.35355}{T} - \frac{414.966 \text{ K}}{T^2} - \frac{63981 \text{ K}^2}{T^3}\right)dT$$

$$= 206.1 \text{ J·mol}^{-1}·\text{K}^{-1} + 11.6 \text{ J·mol}^{-1}·\text{K}^{-1} = 217.7 \text{ J·mol}^{-1}·\text{K}^{-1}$$

$$\Delta_r H°(298 \text{ K}) = 154.0 \text{ kJ·mol}^{-1}$$

$$- R\int_{298 \text{ K}}^{700 \text{ K}}\left[\frac{1}{2}\left(4.8919 - \frac{829.931 \text{ K}}{T} - \frac{127962 \text{ K}^2}{T^2}\right) + 2.500 - 5.2995\right]dT$$

$$= 154.0 \text{ kJ·mol}^{-1} - R\int_{298 \text{ K}}^{700 \text{ K}}\left(-0.35355 - \frac{414.966 \text{ K}}{T} - \frac{63981 \text{ K}^2}{T^2}\right)dT$$

$$= 154.0 \text{ kJ·mol}^{-1} + 5.15 \text{ kJ·mol}^{-1} = 159.2 \text{ kJ·mol}^{-1}$$

$$\Delta_r G°(298 \text{ K}) = \Delta_r H°(298 \text{ K}) - (298 \text{ K})\Delta_r S°(298 \text{ K})$$

$$= 159.2 \text{ kJ·mol}^{-1} - (298 \text{ K})(217.7 \text{ J·mol}^{-1}·\text{K}^{-1})$$

$$= 94.3 \text{ kJ·mol}^{-1}$$

26–60. Consider the dissociation of $Ag_2O(s)$ to $Ag(s)$ and $O_2(g)$ according to

$$Ag_2O(s) \rightleftharpoons 2\,Ag(s) + \frac{1}{2}\,O_2(g)$$

Given the following "dissociation pressure" data:

$t/°C$	173	178	183	188
P/torr	422	509	605	717

Express K_P in terms of P (in torr) and plot $\ln K_P$ versus $1/T$. An excelllent curve fit to these data is given by

$$\ln K_P = 0.9692 + \frac{5612.7\ \text{K}}{T} - \frac{2.0953 \times 10^6\ \text{K}^2}{T^2}$$

Use this expression to derive an equation for $\Delta_r H°$ from 445 K < T < 460 K. Now use the following heat capacity data:

$$C_P°[O_2(g)]/R = 3.27 + (5.03 \times 10^{-4}\ \text{K}^{-1})T$$

$$C_P°[Ag(s)]/R = 2.82 + (7.55 \times 10^{-4}\ \text{K}^{-1})T$$

$$C_P°[Ag_2O(s)]/R = 6.98 + (4.48 \times 10^{-3}\ \text{K}^{-1})T$$

to calculate $\Delta_r H°$, $\Delta_r S°$, and $\Delta_r G°$ at 298 K.

We can write K_P in terms of the partial pressure of oxygen (assuming an activity of unity for the solids):

$$K_P = P_{O_2}^{1/2} = P^{1/2}$$

Below is a plot of the experimental $\ln K_P$ versus $1/T$ for this reaction.

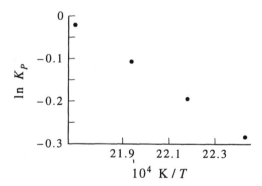

We can now use the equation given in the problem for $\ln K_P$ in Equation 26.29 to find $\Delta_r H°$ and in Equation 26.11 to find $\Delta_r G°$ (as in the previous problem):

$$\frac{d \ln K_P}{dT} = -\frac{5612.7\ \text{K}}{T^2} + \frac{4.1906 \times 10^6\ \text{K}^2}{T^3}$$

$$\Delta_r H° = RT^2 \frac{d \ln K_P}{dT}$$

$$\Delta_r H° = R\left(-5612.7\ \text{K} + \frac{4.1906 \times 10^6\ \text{K}^2}{T}\right)$$

$$\Delta_r G° = -RT \ln K_P$$

$$= R\left(-0.9692T - 5612.7\ \text{K} + \frac{2.0953 \times 10^6\ \text{K}^2}{T}\right)$$

$$\Delta_r S° = \frac{\Delta_r H° - \Delta_r G°}{T}$$

The expression for $\ln K_P$ used in the above equalities holds for temperatures ranging from 445 K to 460 K. To find the values of $\Delta_r H°$, $\Delta_r G°$, and $\Delta_r S°$ at 298 K, we can use the equation

$$\Delta_r H°(298 \text{ K}) = \Delta_r H°(450 \text{ K}) - \int_{298 \text{ K}}^{450 \text{ K}} \Delta C_P°(T) dT \qquad (19.57)$$

and the similar equation

$$\Delta_r S°(298 \text{ K}) = \Delta_r S°(450 \text{ K}) - \int_{298 \text{ K}}^{450 \text{ K}} \frac{\Delta C_P°(T)}{T} dT$$

Substituting into the high-temperature expressions for $\Delta_r S°$ and $\Delta_r H°$ at 450 K, we find that $\Delta_r S°(450 \text{ K}) = 94.09 \text{ J·mol}^{-1}\text{·K}^{-1}$ and $\Delta_r H°(450 \text{ K}) = 30.76 \text{ kJ·mol}^{-1}$. Then

$$\Delta_r S°(298 \text{ K}) = 94.09 \text{ J·mol}^{-1}\text{·K}^{-1} - R \int_{298 \text{ K}}^{450 \text{ K}} \left[\frac{1}{2}\left(\frac{3.27}{T} + 5.03 \times 10^{-4} \text{ K}^{-1} \right) \right.$$

$$\left. + 2\left(\frac{2.82}{T} + 7.55 \times 10^{-4} \text{ K}^{-1} \right) - \left(\frac{6.98}{T} + 4.48 \times 10^{-3} \text{ K}^{-1} \right) \right] dT$$

$$= 94.09 \text{ J·mol}^{-1}\text{·K}^{-1} - R \int_{298 \text{ K}}^{450 \text{ K}} \left(\frac{0.295}{T} + 1.31 \times 10^{-3} \text{ K}^{-1} \right) dT$$

$$= 94.09 \text{ J·mol}^{-1}\text{·K}^{-1} + 2.42 \text{ J·mol}^{-1}\text{·K}^{-1} = 96.51 \text{ J·mol}^{-1}\text{·K}^{-1}$$

$$\Delta_r H°(298 \text{ K}) = 30.76 \text{ kJ·mol}^{-1} - R \int_{298 \text{ K}}^{450 \text{ K}} \left\{ \frac{1}{2}\left[3.27 + (5.03 \times 10^{-4} \text{ K}^{-1})T \right] \right.$$

$$\left. 2\left[2.82 + (7.55 \times 10^{-4} \text{ K}^{-1})T \right] - \left[6.98 + (4.48 \times 10^{-3} \text{ K}^{-1})T \right] \right\} dT$$

$$= 30.76 \text{ kJ·mol}^{-1} - R \int_{298 \text{ K}}^{450 \text{ K}} \left[0.295 + (1.31 \times 10^{-3} \text{ K}^{-1})T \right] dT$$

$$= 30.76 \text{ kJ·mol}^{-1} + 0.912 \text{ kJ·mol}^{-1} = 31.67 \text{ kJ·mol}^{-1}$$

$$\Delta_r G°(298 \text{ K}) = \Delta_r H°(298 \text{ K}) - (298 \text{ K})\Delta_r S°(298 \text{ K})$$

$$= 31.67 \text{ kJ·mol}^{-1} - (298 \text{ K})(96.51 \text{ J·mol}^{-1}\text{·K}^{-1})$$

$$= 2.910 \text{ kJ·mol}^{-1}$$

26–61. Calcium carbonate occurs as two crystalline forms, calcite and aragonite. The value of $\Delta_r G°$ for the transition

$$CaCO_3(\text{calcite}) \rightleftharpoons CaCO_3(\text{aragonite})$$

is $+1.04 \text{ kJ·mol}^{-1}$ at 25°C. The density of calcite at 25°C is 2.710 g·cm^{-3} and that of aragonite is 2.930 g·cm^{-3}. At what pressure will these two forms of $CaCO_3$ be at equilbrium at 25°C.

The molar volume of aragonite is

$$\frac{1 \times 10^{-3} \text{ dm}^3}{2.930 \text{ g}} \left(\frac{100.09 \text{ g}}{1 \text{ mol}} \right) = 0.0342 \text{ dm}^3\text{·mol}^{-1}$$

and the molar volume of calcite is

$$\frac{1 \times 10^{-3} \text{ dm}^3}{2.710 \text{ g}} \left(\frac{100.09 \text{ g}}{1 \text{ mol}} \right) = 0.0369 \text{ dm}^3\text{·mol}^{-1}$$

We can use Equation 26.65 for $\Delta_r G°$ and Equation 26.69 to express the logarithmic terms. Note that when the forms are in equilibrium, the pressures will be equal.

$$\Delta_r G° = -RT \ln K_a = -RT \ln \frac{a_{\text{aragonite}}}{a_{\text{calcite}}}$$

$$1040 \text{ J·mol}^{-1} = -RT \left[\frac{(0.0342 \text{ dm}^3 \cdot \text{mol}^{-1})(P-1)}{RT} - \frac{(0.0369 \text{ dm}^3 \cdot \text{mol}^{-1})(P-1)}{RT} \right]$$

$$= (0.0027 \text{ dm}^3 \cdot \text{mol}^{-1})(P-1)$$

Solving this equation for P gives

$$P - 1 = \left(\frac{1040 \text{ J·mol}^{-1}}{0.0027 \text{ dm}^3 \cdot \text{mol}^{-1}} \right) \left(\frac{0.083145 \text{ dm}^3 \cdot \text{bar}}{8.3145 \text{ J}} \right)$$

$$P = 3800 \text{ bar}$$

26–62. The decomposition of ammonium carbamate, NH_2COONH_4 takes place according to

$$NH_2COONH_4(s) \rightleftharpoons 2 NH_3(g) + CO_2(g)$$

Show that if all the $NH_3(g)$ and $CO_2(g)$ result from the decomposition of ammonium carbamate, then $K_P = (4/27)P^3$, where P is the total pressure at equilibrium.

We can assume that the activity of the ammonium carbamate is unity, which means it makes no contribution to the equilibrium constant expression. We can write the number of moles of carbon dioxide present at equilibrium as ξ_{eq} and the number of moles of ammonia present at equilibrium as $2\xi_{\text{eq}}$, for a total of $3\xi_{\text{eq}}$ moles. (Since the ammonium carbamate is in solid phase and we have assumed that its activity is unity, it does not contribute to the total number of moles of gas present.) Then

$$P_{NH_3} = \frac{2\xi}{3\xi} P = \frac{2}{3} P \qquad \text{and} \qquad P_{CO_2} = \frac{\xi}{3\xi} P = \frac{1}{3} P$$

Then

$$K_P = P_{NH_3}^2 P_{CO_2} = \frac{4P^2}{9} \left(\frac{P}{3} \right) = \frac{4P^3}{27}$$

at equilibrium.

26–63. Calculate the solubility of LiF(aq) in water at $25°C$. Compare your result to the one you obtain by using concentrations instead of activities. Take $K_{sp} = 1.7 \times 10^{-3}$.

The equation for the dissolution of LiF(s) is

$$LiF(s) \rightleftharpoons Li^+(aq) + F^-(aq)$$

and the equilibrium-constant expression is

$$a_{Li^+} a_{F^-} = c_{Li^+} c_{F^-} \gamma_{\pm}^2 = K_{sp} = 1.7 \times 10^{-3}$$

or

$$c_{Li^+}c_{F^-} = \frac{K_{sp}}{\gamma_\pm^2}$$

Let the solubility of LiF(s) be s, then $c_{Li^+} = c_{F^-} = s$. Therefore, we have

$$s^2 = \frac{K_{sp}}{\gamma_\pm^2} \tag{1}$$

Set $\gamma_\pm = 1$ and solve Equation 1 for s to obtain $s = (1.7 \times 10^{-3})^{1/2} \text{mol·L}^{-1} = 0.0412 \text{ mol·L}^{-1}$. Now substitute this result into Equation 26.56 (with $I_c = s$) to calculate $\ln \gamma_\pm = -0.198$, or $\gamma_\pm = 0.820$. Substitute this value into Equation 1 to obtain $s = 0.0503 \text{ mol·L}^{-1}$. The next iteration gives $\gamma_\pm = 0.807$, and then $s = 0.0511 \text{ mol·L}^{-1}$. Once more gives $\gamma_\pm = 0.806$, so the final result is then $s = 0.0512 \text{ mol·L}^{-1}$. Thus, $s = 0.051 \text{ mol·L}^{-1}$ to two significant figures.

26–64. Calculate the solubility of CaF_2(aq) in a solution that is 0.0150 molar in $MgSO_4$(aq). Take $K_{sp} = 3.9 \times 10^{-11}$ for CaF_2(aq).

The equation for the dissolution of CaF_2(s) is

$$CaF_2(s) \rightleftharpoons Ca^{2+}(aq) + 2\,F^-(aq)$$

and the equilibrium-constant expression is

$$c_{Ca^{2+}}c_{F^-}^2\gamma_\pm^3 = K_{sp} = 3.9 \times 10^{-11}$$

Let the solubility of CaF_2(s) be s, then Ca^{2+}(aq) $= s$ and F^-(aq) $= 2s$. Therefore, we have

$$s(2s)^2 = 4s^3 = \frac{K_{sp}}{\gamma_\pm^3} \tag{1}$$

Set $\gamma_\pm = 1$ to obtain $s = 2.14 \times 10^{-4} \text{ mol·L}^{-1}$. This value of s gives

$$I_c = \frac{1}{2}\left[4s + 2s + (4)(0.0150 \text{ mol·L}^{-1}) + (4)(0.0150 \text{ mol·L}^{-1})\right]$$

$$= 0.000641 \text{ mol·L}^{-1} + 0.0600 \text{ mol·L}^{-1} = 0.0606 \text{ mol·L}^{-1}$$

Substituting this result into Equation 26.56 gives $\gamma_\pm = 0.629$. Substitute this value into Equation 1 to obtain $s = 3.40 \times 10^{-4} \text{ mol·L}^{-1}$. Now $I_c = 0.0610 \text{ mol·L}^{-1}$, and $\gamma_\pm = 0.628$. Now use this result in Equation 1 to get $s = 3.40 \times 10^{-4} \text{ mol·L}^{-1}$. Another iteration gives $\gamma_\pm = 0.628$ and $s = 3.40 \times 10^{-4} \text{ mol·L}^{-1}$. So, to two significant figures, $s = 3.4 \times 10^{-4} \text{ mol·L}^{-1}$.

26–65. Calculate the solubility of CaF_2(aq) in a solution that is 0.050 molar in NaF(aq). Compare your result to the one you obtain by using concentrations instead of activities. Take $K_{sp} = 3.9 \times 10^{-11}$ for CaF_2(aq).

The equation for the dissolution of CaF_2(s) is

$$CaF_2(s) \rightleftharpoons Ca^{2+}(aq) + 2\,F^-(aq)$$

and the equilibrium-constant expression is (Problem 26–64)

$$4s^3 = \frac{K_{sp}}{\gamma_{\pm}^3} \tag{1}$$

Set $\gamma_{\pm} = 1$ to obtain $s = 2.14 \times 10^{-4}$ mol·L^{-1}. This value of s gives

$$I_c = \frac{1}{2}\left[4s + 2s + (1)(0.050\,\text{mol·L}^{-1}) + (1)(0.050\,\text{mol·L}^{-1})\right] = 3s + 0.050\,\text{mol·L}^{-1}$$

Because s is much smaller than 0.050 mol·L^{-1}, we initially let $I_c = 0.050$ mol·L^{-1}. Substituting this value into Equation 26.56 gives $\gamma_{\pm} = 0.651$. Substitute this result into Equation 1 to obtain $s = 3.28 \times 10^{-4}$ mol·L^{-1}. Now $I_c = 0.0510$ mol·L^{-1}, $\gamma_{\pm} = 0.649$, and $s = 3.29 \times 10^{-4}$ mol·L^{-1}. Thus, $s = 3.3 \times 10^{-4}$ mol·L^{-1} to two significant figures.

CHAPTER 27

The Kinetic Theory of Gases

PROBLEMS AND SOLUTIONS

27–1. Calculate the average translational energy of one mole of ethane at 400 K, assuming ideal behavior. Compare your result to $\overline{U}^{id}$ for ethane at 400 K given in Figure 22.3.

From Section 18–1, for an ideal molecule

$$E_{trans} = \frac{3}{2}RT = 4.99 \text{ kJ}\cdot\text{mol}^{-1}$$

At $P = 0$ bar (ideal conditions), $\overline{U} \approx 14.6 \text{ kJ}\cdot\text{mol}^{-1}$ from Figure 22.3. Therefore, E_{trans} accounts for a third of the total energy.

27–2. Calculate the root-mean-square speed of a nitrogen molecule at 200 K, 300 K, 500 K, and 1000 K.

Using Equation 27.14, we find that

$$u_{rms} = \left(\frac{3RT}{M}\right)^{1/2} = \left[\frac{3(8.3145 \text{ J}\cdot\text{mol}^{-1}\cdot\text{K}^{-1})T}{0.02802 \text{ kg}\cdot\text{mol}^{-1}}\right]^{1/2}$$

Substituting for T, we find the values below.

T/K	$u_{rms}/\text{m}\cdot\text{s}^{-1}$
200	421.9
300	516.8
500	667.2
1000	943.5

27–3. If the temperature of a gas is doubled, by how much is the root-mean-square speed of the molecules increased?

Let the original temperature of the gas be T and the original root-mean-square speed be $u_{rms}(T)$. Then (Equation 27.14)

$$u_{rms}(2T) = \left[\frac{3R(2T)}{M}\right]^{1/2} = 2^{1/2}u_{rms}(T)$$

The root-mean-square speed is increased by a factor of $\sqrt{2}$.

27–4. The speed of sound in air at sea level at 20°C is about 770 mph. Compare this value with the root-mean-square speed of nitrogen and oxygen molecules at 20°C.

Using Equation 27.14, we write

$$u_{rms,N_2} = \left[\frac{3(8.31451 \text{ J}\cdot\text{mol}^{-1}\cdot\text{K}^{-1})(293 \text{ K})}{0.02802 \text{ kg}\cdot\text{mol}^{-1}}\right]^{1/2} = 511 \text{ m}\cdot\text{s}^{-1}$$

$$u_{rms,O_2} = \left[\frac{3(8.31451 \text{ J}\cdot\text{mol}^{-1}\cdot\text{K}^{-1})(293 \text{ K})}{0.03200 \text{ kg}\cdot\text{mol}^{-1}}\right]^{1/2} = 478 \text{ m}\cdot\text{s}^{-1}$$

The speed of sound is

$$770 \text{ mph} = 1239 \text{ km}\cdot\text{h}^{-1} = 344 \text{ m}\cdot\text{s}^{-1}$$

Nitrogen and oxygen molecules travel significantly faster (33% and 28% faster, respectively) than sound in air at 20°C.

27–5. Arrange the following gases in order of increasing root-mean-square speed at the same temperature: O_2, N_2, H_2O, CO_2, NO_2, $^{235}UF_6$, and $^{238}UF_6$.

The heavier the gas, the slower it will travel. (In Equation 27.14, the denominator of the root-mean-square speed increases with increasing mass.) Thus, the arrangement of gases requested is

$$^{238}UF_6 < {}^{235}UF_6 < NO_2 < CO_2 < O_2 < N_2 < H_2O$$

27–6. Consider a mixture of $H_2(g)$ and $I_2(g)$. Calculate the ratio of the root-mean-square speed of $H_2(g)$ and $I_2(g)$ molecules in the reaction mixture.

In the mixture, the two components have the same temperature, so (using Equation 27.14)

$$\frac{u_{rms,H_2}}{u_{rms,I_2}} = \left(\frac{3RT/M_{H_2}}{3RT/M_{I_2}}\right)^{1/2} = \left(\frac{M_{I_2}}{M_{H_2}}\right)^{1/2} = \left(\frac{253.8}{2.016}\right)^{1/2} = 11.2$$

27–7. The speed of sound in an ideal monatomic gas is given by

$$u_{sound} = \left(\frac{5RT}{3M}\right)^{1/2}$$

Derive an equation for the ratio u_{rms}/u_{sound}. Calculate the root-mean-square speed for an argon atom at 20°C and compare your answer to the speed of sound in argon.

Using Equation 27.14 and the definition of u_{sound} above, we find

$$\frac{u_{rms}}{u_{sound}} = \left(\frac{3RT/M}{5RT/3M}\right)^{1/2} = \left(\frac{9}{5}\right)^{1/2} = \frac{3}{\sqrt{5}}$$

The root-mean square speed of an argon atom at 20°C is

$$u_{rms} = \left(\frac{3RT}{M}\right)^{1/2} = 428 \text{ m} \cdot \text{s}^{-1}$$

and, using the relation between u_{rms} and u_{sound} given above, the speed of sound in argon is $319 \text{ m} \cdot \text{s}^{-1}$.

27–8. Calculate the speed of sound in argon at 25°C.

The speed of sound in argon at 25°C is (using the equivalence given in Problem 27–7)

$$u_{rms} = \left(\frac{5RT}{3M}\right)^{1/2} = 321 \text{ m} \cdot \text{s}^{-1}$$

27–9. The speed of sound in an ideal polyatomic gas is given by

$$u_{sound} = \left(\frac{\gamma RT}{M}\right)^{1/2}$$

where $\gamma = C_P/C_V$. Calculate the speed of sound in nitrogen at 25°C.

Recall from Chapter 17 that $\overline{C}_V = 5R/2$ for a diatomic ideal gas (neglecting the vibrational contribution), and that $\overline{C}_P - \overline{C}_V = R$ (Equation 19.43). Then

$$u_{sound} = \left(\frac{C_P RT}{C_V M}\right)^{1/2} = \left[\frac{7(8.3145 \text{ J} \cdot \text{mol}^{-1} \cdot \text{K}^{-1})(298.15 \text{ K})}{5(0.02802 \text{ kg} \cdot \text{mol}^{-1})}\right]^{1/2} = 352 \text{ m} \cdot \text{s}^{-1}$$

27–10. Use Equation 27.17 to prove that $\partial u/\partial u_x = u_x/u$.

Begin with Equation 27.17:

$$u^2 = u_x^2 + u_y^2 + u_z^2$$

$$\frac{\partial}{\partial u_x} u^2 = \frac{\partial}{\partial u_x}\left(u_x^2 + u_y^2 + u_z^2\right)$$

$$2u\frac{\partial u}{\partial u_x} = 2u_x$$

$$\frac{\partial u}{\partial u_x} = \frac{u_x}{u}$$

27–11. Give a physical argument why γ in Equation 27.24 must be a positive quantity.

Recall that $f(u_j)$ represents the probability distribution of the jth component of the velocity. As u_j increases, the probability of finding any molecule moving with speed u_j decreases, so that as $u_j \rightarrow \infty$, $f(u_j) \rightarrow 0$. If γ were negative $f(u_j)$ in Equation 27.24 would diverge as $u_j \rightarrow \infty$, so γ must be positive.

27–12. We can use Equation 27.33 to calculate the probability that the x-component of the velocity of a molecule lies within some range. For example, show that the probability that $-u_{x0} \le u_x \le u_{x0}$ is given by

$$\text{Prob}\{-u_{x0} \le u_x \le u_{x0}\} = \left(\frac{m}{2\pi k_B T}\right)^{1/2} \int_{-u_{x0}}^{u_{x0}} e^{-mu_x^2/2k_B T} du_x$$

$$= 2\left(\frac{m}{2\pi k_B T}\right)^{1/2} \int_0^{u_{x0}} e^{-mu_x^2/2k_B T} du_x$$

Now let $mu_x^2/2k_B T = w^2$ to get the cleaner-looking expression

$$\text{Prob}\{-u_{x0} \le u_x \le u_{x0}\} = \frac{2}{\pi^{1/2}} \int_0^{w_0} e^{-w^2} dw$$

where $w_0 = (m/2k_B T)^{1/2} u_{x0}$.

It so happens that the above integral cannot be evaluated in terms of any function that we have encountered up to now. It is customary to express the integral in terms of a new function called the *error function*, which is defined by

$$\text{erf}(z) = \frac{2}{\pi^{1/2}} \int_0^z e^{-x^2} dx \tag{1}$$

The error function can be evaluated as a function of z by evaluating its defining integral numerically. Some values of $\text{erf}(z)$ are

z	$\text{erf}(z)$	z	$\text{erf}(z)$
0.20	0.22270	1.20	0.91031
0.40	0.42839	1.40	0.95229
0.60	0.60386	1.60	0.97635
0.80	0.74210	1.80	0.98909
1.00	0.84270	2.00	0.99532

Now show that

$$\text{Prob}\{-u_{x0} \le u_x \le u_{x0}\} = \text{erf}(w_0)$$

Calculate the probability that $-(2k_B T/m)^{1/2} \le u_x \le (2k_B T/m)^{1/2}$.

$$f(u_x) = \left(\frac{m}{2\pi k_B T}\right)^{1/2} e^{-mu_x^2/2k_B T} \tag{27.33}$$

Since $f(u_x)$ is the probability that a molecule has velocity u_x,

$$\text{Prob}\{-u_{x0} \le u_x \le u_{x0}\} = \left(\frac{m}{2\pi k_B T}\right)^{1/2} \int_{-u_{x0}}^{u_{x0}} e^{-mu_x^2/2k_B T} du_x$$

$$= 2\left(\frac{m}{2\pi k_B T}\right)^{1/2} \int_0^{u_{x0}} e^{-mu_x^2/2k_B T} du_x$$

Now let $mu_x^2/2k_B T = w^2$. Then

$$\frac{mu_x du_x}{k_B T} = 2w dw$$

$$du_x = \frac{2k_B T}{m} \frac{w dw}{u_x} = \left(\frac{2k_B T}{m}\right)^{1/2} dw$$

Letting $w_0 = (m/2k_B T)^{1/2} u_{x0}$ and substituting into the probability expression above,

$$\text{Prob}\{-u_{x0} \leq u_x \leq u_{x0}\} = \frac{2}{\pi^{1/2}} \int_0^{w_0} e^{-w^2} dw$$

From the definition of an error function, it is easy to see that

$$\text{Prob}\{-u_{x0} \leq u_x \leq u_{x0}\} = \frac{2}{\pi^{1/2}} \int_0^{w_0} e^{-x^2} dx = \text{erf}(w_0)$$

To find the probability that $-(2k_B T/m)^{1/2} \leq u_x \leq (2k_B T/m)^{1/2}$, we first find w_0:

$$w_0 = \left(\frac{mu_{x0}^2}{2k_B T}\right)^{1/2} = \left[\frac{m(2k_B T)}{m(2k_B T)}\right]^{1/2} = 1$$

Since $\text{Prob}\{-u_{x0} \leq u_x \leq u_{x0}\} = \text{erf}(w_0)$, from the table of values of $\text{erf}(z)$ we find that

$$\text{Prob}\{-(2k_B T/m)^{1/2} \leq u_x \leq (2k_B T/m)^{1/2}\} = \text{erf}(1) = 0.84270$$

27–13. Use the result of Problem 27–12 to show that

$$\text{Prob}\{|u_x| \geq u_{x0}\} = 1 - \text{erf}(w_0)$$

$$\text{Prob}\{|u_x| \geq u_{x0}\} = 1 - \text{Prob}\{|u_x| \leq u_{x0}\}$$
$$= 1 - \text{Prob}\{-u_{x0} \leq u_x \leq u_{x0}\}$$

In Problem 27–12 we found that $\text{Prob}\{-u_{x0} \leq u_x \leq u_{x0}\} = \text{erf}(w_0)$, so

$$\text{Prob}\{|u_x| \geq u_{x0}\} = 1 - \text{erf}(w_0)$$

Notice that $\text{Prob}\{u_x \geq u_{x0}\} = \frac{1}{2}[1 - \text{erf}(w_0)]$ (see Problem 27–14).

27–14. Use the result of Problem 27–12 to calculate $\text{Prob}\{u_x \geq +(k_B T/m)^{1/2}\}$ and $\text{Prob}\{u_x \geq +(2k_B T/m)^{1/2}\}$.

We can write

$$\text{Prob}\{u_x \geq +(k_B T/m)^{1/2}\} = 1 - \text{Prob}\{u_x \leq +(k_B T/m)^{1/2}\}$$

Following the procedure used in Problem 27–12, this becomes

$$\text{Prob}\{u_x \geq +(k_B T/m)^{1/2}\} = 1 - \left(\frac{m}{2\pi k_B T}\right)^{1/2} \int_{-\infty}^{(k_B T/m)^{1/2}} e^{-mu_x^2/2k_B T} du_x$$

Letting $w = (m/2k_B T)^{1/2} u_x$ and $w_0 = (m/2k_B T)^{1/2} u_{x0}$, this becomes

$$\text{Prob}\{u_x \geq +(k_B T/m)^{1/2}\} = 1 - \left[\frac{1}{\pi^{1/2}} \int_{-\infty}^0 e^{-w^2} dw + \frac{1}{\pi^{1/2}} \int_0^{w_0 = 1/\sqrt{2}} e^{-w^2} dw\right]$$

$$= 1 - \left[\frac{1}{\pi^{1/2}}\left(\frac{\pi}{4}\right)^{1/2} + \frac{1}{2}\text{erf}\left(\frac{1}{\sqrt{2}}\right)\right]$$

$$= 0.159$$

Likewise,

$$\text{Prob}\{u_x \geq +(2k_BT/m)^{1/2}\} = 1 - \left[\frac{1}{\pi^{1/2}} \int_{-\infty}^{0} e^{-w^2} dw + \frac{1}{\pi^{1/2}} \int_{0}^{w_0=1} e^{-w^2} dw \right]$$

$$= 1 - \left[\frac{1}{\pi^{1/2}} \left(\frac{\pi}{4} \right)^{1/2} + \frac{1}{2} \text{erf}(1) \right]$$

$$= 0.0786$$

27–15. Use the result of Problem 27–12 to plot the probability that $-u_{x0} \leq u_x \leq u_{x0}$ against $u_{x0}/(2k_BT/m)^{1/2}$.

We found in Problem 27–12 that the above probability is given by $\text{erf}(w_0)$, where $w_0 = u_{x0}/(2k_BT/m)^{1/2}$. The required plot will be a plot of $\text{erf}(w_0)$ against w_0 and thus identical to the plot in the following problem.

27–16. Use Simpson's rule or any other numerical integration routine to verify the values of $\text{erf}(z)$ given in Problem 27–12. Plot $\text{erf}(z)$ against z.

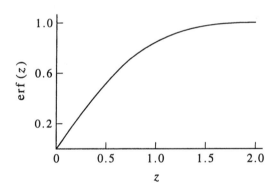

27–17. Derive an expression for the average value of the positive values of u_x.

$$f(u_x) = \left(\frac{m}{2\pi k_BT} \right)^{1/2} e^{-mu_x^2/2k_BT} \tag{27.33}$$

As in Equation 27.34, we can write the average value of the positive u_x as

$$\langle u_x + \rangle = \int_{0}^{\infty} u_x f(u_x) du_x = \left(\frac{m}{2\pi k_BT} \right)^{1/2} \int_{0}^{\infty} u_x e^{-mu_x^2/2k_BT} du_x$$

$$= \left(\frac{m}{2\pi k_BT} \right)^{1/2} \frac{2k_BT}{2m} = \left(\frac{k_BT}{2\pi m} \right)^{1/2}$$

27–18. This problem deals with the idea of the *escape velocity* of a particle from a body such as the Earth's surface. Recall from your course in physics that the potential energy of two masses, m_1 and m_2, separated by a distance r is given by

$$V(r) = -\frac{Gm_1m_2}{r}$$

(note the similarity with Coulomb's law) where $G = 6.67 \times 10^{-11}$ J·m·kg^{-1} is called the gravitional constant. Suppose a particle of mass m has a velocity u perpendicular to the Earth's surface. Show that the minimum velocity that the particle must have in order to escape the Earth's surface (its *escape velocity*) is given by

$$u = \left(\frac{2GM_{earth}}{R_{earth}}\right)^{1/2}$$

Given that $M_{earth} = 5.98 \times 10^{24}$ kg is the mass of the Earth and $R_{earth} = 6.36 \times 10^6$ m is its mean radius, calculate the escape velocity of a hydrogen molecule and a nitrogen molecule. What temperature would each of these molecules have to have so that their average speed exceeds their escape velocity?

The kinetic energy of the particle is equal and opposite to the potential energy between the two bodies when the particle has the minimum velocity required to escape the Earth's surface, so

$$\frac{1}{2}mv_{escape}^2 = \frac{GmM_{earth}}{R_{earth}}$$

$$v_{escape} = \left(\frac{2GM_{earth}}{R_{earth}}\right)^{1/2}$$

where we can consider the distance between the particle and the circumference of the earth negligible compared with the distance between the circumference of the earth and its center of mass. Since the escape velocity is independent of particle size, for both hydrogen and nitrogen

$$v_{escape} = \left[\frac{2(6.67 \times 10^{-11} \text{ J·m·kg}^{-1})(5.98 \times 10^{24} \text{ kg})}{6.36 \times 10^6 \text{ m}}\right]^{1/2} = 11\,200 \text{ m·s}^{-1}$$

From Equation 27.42,

$$T_{escape, H_2} = \frac{\pi M v_{escape}^2}{8R}$$

$$= \frac{\pi(0.002016 \text{ kg·mol}^{-1})(11\,200 \text{ m·s}^{-1})^2}{8(8.3145 \text{ J·K}^{-1}\cdot\text{mol}^{-1})} = 11.9 \times 10^3 \text{ K}$$

$$T_{escape, N_2} = \frac{\pi M v_{escape}^2}{8R}$$

$$= \frac{\pi(0.02802 \text{ kg·mol}^{-1})(11\,200 \text{ m·s}^{-1})^2}{8(8.3145 \text{ J·K}^{-1}\cdot\text{mol}^{-1})} = 16.6 \times 10^4 \text{ K}$$

27–19. Repeat the calculation in the previous problem for the moon's surface. Take the mass of the moon to be 7.35×10^{22} kg and its radius to be 1.74×10^6 m.

Now

$$v_{min} = \left(\frac{2GM_{moon}}{R_{moon}}\right)^{1/2} = 2370 \text{ m·s}^{-1}$$

Again, from Equation 27.42,

$$T_{escape,H_2} = \frac{\pi M v_{escape}^2}{8R}$$

$$= \frac{\pi (0.002016 \text{ kg·mol}^{-1})(2370 \text{ m·s}^{-1})^2}{8(8.3145 \text{ J·K}^{-1}\cdot\text{mol}^{-1})} = 537 \text{ K}$$

$$T_{escape,N_2} = \frac{\pi M v_{escape}^2}{8R}$$

$$= \frac{\pi (0.02802 \text{ kg·mol}^{-1})(2370 \text{ m·s}^{-1})^2}{8(8.3145 \text{ J·K}^{-1}\cdot\text{mol}^{-1})} = 7460 \text{ K}$$

27–20. Show that the variance of Equation 27.37 is given by $\sigma^2 = v_0^2 k_B T / mc^2$. Calcuate σ for the $3p\,^2P_{3/2}$ to $3s\,^2S_{1/2}$ transition in atomic sodium vapor (see Figure 8.4) at 500 K.

Equation 27.37 gives $I(\nu)$ as

$$I(\nu) = K \exp\left[\frac{-mc^2(\nu - \nu_0)^2}{2v_0^2 k_B T}\right] \tag{27.37}$$

Compare this equation to that of a general Gaussian curve (Example B–4), with variance $\sigma^2 = a^2$:

$$p(x)dx = c \exp\left[\frac{-(x - x_0)^2}{2a^2}\right]$$

Since we know that $I(\nu)$ has the shape of a Gaussian curve (by comparing the equations), the variance for $I(\nu)$ is given by

$$\sigma^2 = \frac{v_0^2 k_B T}{mc^2} = \frac{v_0^2 RT}{Mc^2}$$

The wavelength of the $3p\,^2P_{3/2}$ to $3s\,^2S_{1/2}$ transition is (from Figure 8.4) 5889.9×10^{-10} m. Since $\lambda\nu = c$, $\nu = 5.090 \times 10^{14} \text{ s}^{-1}$. Substituting, we find

$$\sigma^2 = \frac{(5.090 \times 10^{14} \text{ s}^{-1})^2(8.3145 \text{ J·mol}^{-1}\cdot\text{K}^{-1})(500 \text{ K})}{(0.02299 \text{ kg·mol}^{-1})(2.998 \times 10^8 \text{ m·s}^{-1})^2} = 5.21 \times 10^{17} \text{ s}^{-2}$$

or $\sigma = 7.22 \times 10^8 \text{ s}^{-1}$.

27–21. Show that the distribution of speeds for a two-dimensional gas is given by

$$F(u)du = \frac{m}{k_B T} u e^{-mu^2/2k_B T} du$$

(Recall that the area element in plane polar coordinates is $r\,dr\,d\theta$.)

The two-dimensional version of Equation 27.39 is

$$F(u)du = \left(\frac{m}{2\pi k_B T}\right) e^{-m(u_x^2 + u_y^2)/2k_B T} du_x du_y$$

Using a two-dimensional velocity space, we replace $u_x^2 + u_y^2$ by u^2 and $du_x du_y$ with $2\pi u du$ to obtain

$$F(u)du = \left(\frac{m}{k_B T}\right) u e^{-mu^2/2k_B T} du$$

27–22. Use the formula in the previous problem to derive formulas for $\langle u \rangle$ and $\langle u^2 \rangle$ for a two-dimensional gas. Compare your result for $\langle u^2 \rangle$ to $\langle u_x^2 \rangle + \langle u_y^2 \rangle$.

Because u is an intrinsically positive quantity, we can write our integrals over a positive range only. Then, using Equations B.12 and B.13 and Table 27.2 to evaluate the integrals,

$$\langle u \rangle = \int_0^\infty u F(u) du = \frac{m}{k_B T} \int_0^\infty u^2 e^{-mu^2/2k_B T} du$$

$$= \left(\frac{m}{k_B T}\right)\left(\frac{2k_B T}{4m}\right)\left(\frac{2\pi k_B T}{m}\right)^{1/2} = \left(\frac{\pi k_B T}{2m}\right)^{1/2} = \left(\frac{\pi RT}{2M}\right)^{1/2}$$

$$\langle u^2 \rangle = \int_0^\infty u^2 F(u) du = \frac{m}{k_B T} \int_0^\infty u^3 e^{-mu^2/2k_B T} du$$

$$= \left(\frac{m}{k_B T}\right)\frac{1}{2}\left(\frac{2k_B T}{m}\right)^2 = \frac{2k_B T}{m} = \frac{2RT}{M}$$

From Equations 27.5 and 27.6,

$$\langle u_x^2 \rangle = \langle u_y^2 \rangle = \frac{PV}{Nm} = \frac{RT}{Nm} = \frac{k_B T}{m}$$

So $\langle u^2 \rangle = \langle u_x^2 \rangle + \langle u_y^2 \rangle$.

27–23. Use the formula in Problem 27–21 to calculate the probability that $u \geq u_0$ for a two-dimensional gas.

The probability that $u \geq u_0$ for a two-dimensional gas is given by

$$\text{Prob}\{u \geq u_0\} = \int_{u_0}^\infty F(u) du = \frac{m}{k_B T} \int_{u_0}^\infty u e^{-mu^2/2k_B T} du$$

Let $x = (m/2k_B T)^{1/2} u$, so

$$\text{Prob}\{u \geq u_0\} = \frac{m}{k_B T} \int_{(m/2k_B T)^{1/2} u_0}^\infty \left(\frac{2k_B T}{m}\right)^{1/2} x e^{-x^2} \left(\frac{2k_B T}{m}\right)^{1/2} dx$$

$$= 2 \int_{(m/2k_B T)^{1/2} u_0}^\infty x e^{-x^2} dx = 2\left(-\frac{1}{2}\right) e^{-x^2} \Big|_{(m/2k_B T)^{1/2} u_0}^\infty$$

$$= e^{-mu_0^2/2k_B T}$$

27–24. Show that the probability that a molecule has a speed less than or equal to u_0 is given by

$$\text{Prob}\{u \leq u_0\} = \frac{4}{\pi^{1/2}} \int_0^{x_0} x^2 e^{-x^2} dx$$

where $x_0 = (m/2k_BT)^{1/2}u_0$. This integral cannot be expressed in closed form and must be integrated numerically. Use Simpson's rule or any other integration routine to evaluate $\text{Prob}\{u \le (2k_BT/m)^{1/2}\}$.

We use Equation 27.40 for $F(u)du$:

$$F(u)du = 4\pi \left(\frac{m}{2\pi k_B T} \right)^{3/2} u^2 e^{-mu^2/2k_B T} du$$

The probability that $u \le u_0$ is given by $\int_0^{u_0} F(u)du$, so

$$\text{Prob}\{u \le u_0\} = 4\pi \left(\frac{m}{2\pi k_B T} \right)^{3/2} \int_0^{u_0} u^2 e^{-mu^2/2k_B T} du$$

$$= 4\pi \left(\frac{m}{2\pi k_B T} \right)^{3/2} \int_0^{x=x_0} \left(\frac{2k_B T}{m} \right) x^2 e^{-x^2} \left(\frac{2k_B T}{m} \right)^{1/2} dx$$

$$= \frac{4}{\pi^{1/2}} \int_0^{x_0} x^2 e^{-x^2} dx$$

where we have let $x^2 = mu^2/2k_B T$ and $x_0 = (m/2k_B T)^{1/2}u_0$. To evaluate $\text{Prob}\{u \le (2k_B T/m)^{1/2}\}$, we let $x_0 = 1$, so

$$\text{Prob}\{u \le (2k_B T/m)^{1/2}\} = \frac{4}{\pi^{1/2}} \int_0^1 x^2 e^{-x^2} dx = 0.4276$$

27–25. Using Simpson's rule or any other integration routine, plot $\text{Prob}\{u \le u_0\}$ against $u_0/(m/2k_B T)^{1/2}$. (see Problem 27–24.)

We use the numerical integration package in *Mathematica* to plot $\text{Prob}\{u \le u_0\}$ against $u_0/(m/2k_B T)^{1/2}$.

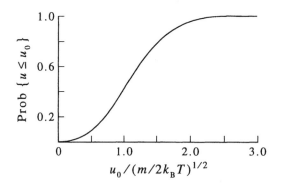

Note that $\text{Prob}\{u \le u_0\}$ goes to unity as $\left(\dfrac{2k_B T}{m} \right)^{1/2} u_0$ goes to infinity.

27–26. What is the most probable kinetic energy for a molecule in the gas phase?

The most probable kinetic energy for a molecule can be found by setting the derivative of $F(\varepsilon)$ equal to zero. Using Equation 27.44, we find

$$\frac{dF}{d\varepsilon} = \frac{2\pi}{(\pi k_B T)^{3/2}} \left[\frac{\varepsilon^{-1/2} e^{-\varepsilon/k_B T}}{2} - \frac{e^{-\varepsilon/k_B T} \varepsilon^{1/2}}{k_B T} \right]$$

$$0 = \frac{2\pi e^{-\varepsilon/k_B T}}{(\pi k_B T)^{3/2}} \left[\frac{1}{2\varepsilon^{1/2}} - \frac{\varepsilon^{1/2}}{k_B T} \right]$$

$$\varepsilon = \frac{k_B T}{2}$$

27–27. Derive an expression for $\sigma_\varepsilon^2 = \langle \varepsilon^2 \rangle - \langle \varepsilon \rangle^2$ from Equation 27.44. Now form the ratio $\sigma_\varepsilon / \langle \varepsilon \rangle$. What does this say about the fluctuations in ε?

$$F(\varepsilon)d\varepsilon = \frac{2\pi}{(\pi k_B T)^{3/2}} \varepsilon^{1/2} e^{-\varepsilon/k_B T} d\varepsilon \tag{27.44}$$

Using Equations B.12 and B.13,

$$\langle \varepsilon \rangle = \frac{2\pi}{(\pi k_B T)^{3/2}} \int_0^\infty \varepsilon^{3/2} e^{-\varepsilon/k_B T} d\varepsilon$$

$$= \frac{2\pi}{(\pi k_B T)^{3/2}} \left[\frac{3}{(2/k_B T)^2} \right] (\pi k_B T)^{1/2} = \frac{3}{2} k_B T$$

$$\langle \varepsilon^2 \rangle = \frac{2\pi}{(\pi k_B T)^{3/2}} \int_0^\infty \varepsilon^{5/2} e^{-\varepsilon/k_B T} d\varepsilon$$

$$= \frac{2}{(k_B T)^{3/2} \pi^{1/2}} \left[\frac{15}{(2/k_B T)^3} \right] (\pi k_B T)^{1/2} = \frac{15}{4} (k_B T)^2$$

$$\sigma_\varepsilon^2 = \langle \varepsilon^2 \rangle - \langle \varepsilon \rangle^2 = \frac{15 - 9}{4} (k_B T)^2 = \frac{3}{2} (k_B T)^2$$

$$\sigma_\varepsilon = \left(\frac{3}{2} \right)^{1/2} k_B T$$

$$\frac{\sigma_\varepsilon}{\langle \varepsilon \rangle} = \left(\frac{2}{3} \right)^{1/2}$$

This says that the fluctuations in ε are large compared to ε.

27–28. Compare the most probable speed of a molecule that collides with a small surface area with the most probable speed of a molecule in the bulk of the gas phase.

From Figure 27.6, we can see that the most probable speed of a molecule that collides with a small surface area is greater than the most probable speed of a molecule in the bulk of the gas. We have found (Equation 27.43) that u_{mp} for a molecule in the bulk of a gas is $(2k_B T/m)^{1/2}$. For a molecule colliding with a small surface area, we must find the u for which the probability of the molecule hitting the surface area is at a maximum, which is the u for which $d[uF(u)]/du = 0$.

$$uF(u) = u^3 e^{-mu^2/2k_\mathrm{B}T}$$

$$\frac{d[uF(u)]}{du} = \left[3u^2 - 2\frac{mu^4}{2k_\mathrm{B}T}\right]e^{-mu^2/2k_\mathrm{B}T}$$

$$\frac{2mu_{mp}^4}{2k_\mathrm{B}T} = 3u_{mp}^2$$

$$u_{mp} = \left(\frac{3k_\mathrm{B}T}{m}\right)^{1/2}$$

The ratio of the most probable speeds is

$$\frac{u_{mp}(\text{bulk})}{u_{mp}(\text{small area})} = \left(\frac{2}{3}\right)^{1/2}$$

27–29. Use Equation 27.48 to calculate the collision frequency per unit area for helium at 100 K and 10^{-6} torr.

This is much like Example 27–6. The number density is

$$\rho = \frac{N_\mathrm{A}P}{RT} = \frac{(6.022 \times 10^{23}\ \text{mol}^{-1})(10^{-6}\ \text{torr}/760\ \text{atm}\cdot\text{torr}^{-1})}{(0.082058\ \text{dm}^3\cdot\text{atm}\cdot\text{mol}^{-1}\cdot\text{K}^{-1})(100\ \text{K})}$$

$$= 9.656 \times 10^{13}\ \text{dm}^{-3} = 9.656 \times 10^{16}\ \text{m}^{-3}$$

and $\langle u \rangle$ is (Equation 27.42)

$$\langle u \rangle = \left(\frac{8RT}{\pi M}\right)^{1/2} = \left[\frac{8(8.314\ \text{J}\cdot\text{K}^{-1}\cdot\text{mol}^{-1})(100\ \text{K})}{\pi(4.0026 \times 10^{-3}\ \text{kg})}\right]^{1/2}$$

$$= 727\ \text{m}\cdot\text{s}^{-1}$$

Then, from Equation 27.48,

$$z_{\text{coll}} = \rho\frac{\langle u \rangle}{4} = 1.76 \times 10^{19}\ \text{m}^{-2}\cdot\text{s}^{-1}$$

27–30. Calculate the average speed of a molecule that strikes a small surface area. How does this value compare to the average speed of all the molecules?

The distribution of the speed of molecules that strike a small surface area goes as $u^3 e^{-mu^2/2k_\mathrm{B}T}$. This is not necessarily normalized, so to find the average speed of all the molecules we must divide the integral $\int u^4 e^{-mu^2/2k_\mathrm{B}T}\,du$ by the integral over all space of the speed of molecules that strike a small surface area:

$$\langle u \rangle = \frac{\int_0^\infty u^4 e^{-mu^2/2k_\mathrm{B}T}\,du}{\int_0^\infty u^3 e^{-mu^2/2k_\mathrm{B}T}\,du}$$

$$= \left[\frac{3}{8}\left(\frac{2k_\mathrm{B}T}{m}\right)^2\left(\frac{2\pi k_\mathrm{B}T}{m}\right)^{1/2}\right]\left[\frac{1}{2}\left(\frac{2k_\mathrm{B}T}{m}\right)^2\right]^{-1}$$

$$= \frac{3}{4}\left(\frac{2\pi k_\mathrm{B}T}{m}\right)^{1/2} = \left(\frac{9\pi k_\mathrm{B}T}{8m}\right)^{1/2}$$

This is larger than the average speed of the molecules in the bulk of the gas, given in Equation 27.42 as $(8k_BT/\pi m)^{1/2}$. The ratio is

$$\frac{\langle u \rangle_{\text{wall}}}{\langle u \rangle_{\text{bulk}}} = \left(\frac{9\pi^2}{64}\right)^{1/2} = \frac{3\pi}{8}$$

27–31. How long will it take for an initially clean surface to become 1.0% covered if it is bathed by an atmosphere of nitrogen at 77 K and one bar? Assume that every nitrogen molecule that strikes the surface sticks and that a nitrogen molecule covers an area of 1.1×10^5 pm^2.

As in Example 27–6,

$$\rho = \frac{P}{k_BT} = \frac{10^5 \text{ Pa}}{(1.3806 \times 10^{-23} \text{ J·K}^{-1})(77 \text{ K})} = 9.406 \times 10^{25} \text{ m}^{-3}$$

$$\langle u \rangle = \left(\frac{8RT}{\pi M}\right)^{1/2} = \left[\frac{8(8.314 \text{ J·mol}^{-1} \cdot \text{K}^{-1})(77 \text{ K})}{\pi(0.02802 \text{ kg·mol}^{-1})}\right]^{1/2} = 241.2 \text{ m·s}^{-1}$$

Then

$$z_{\text{coll}} = \rho \frac{\langle u \rangle}{4} = 5.67 \times 10^{27} \text{ m}^{-2} \cdot \text{s}^{-1}$$

This is the rate at which particles hit a surface of one square meter. For 1.0 % of the surface to be covered, 0.010 square meter must be covered. Since each molecule covers an area of 1.1×10^5 pm^2,

$$\frac{0.010 \text{ m}^2}{1.1 \times 10^{-19} \text{ m}^2} = 9.09 \times 10^{16} \text{ molecules}$$

must collide within the square meter.

$$\frac{9.09 \times 10^{16} \text{ molecules}}{5.67 \times 10^{27} \text{ molecule·s}^{-1}} = 1.60 \times 10^{-11} \text{ s}$$

or 1.60×10^{-11} s for 1.0% of each square meter of surface to be covered.

27–32. Calculate the number of methane molecules at 25°C and one torr that strike a 1.0 cm^2 surface in one millisecond.

First find z_{coll}:

$$\rho = \frac{P}{k_BT} = \frac{(1 \text{ torr}) (101325 \text{ Pa}/760 \text{ torr})}{(1.3806 \times 10^{-23} \text{ J·K}^{-1})(298.15 \text{ K})} = 3.24 \times 10^{22} \text{ m}^{-3}$$

$$\langle u \rangle = \left(\frac{8RT}{\pi M}\right)^{1/2} = \left[\frac{8(8.314 \text{ J·mol}^{-1} \cdot \text{K}^{-1})(298.15 \text{ K})}{\pi(0.01604 \text{ kg})}\right]^{1/2} = 627 \text{ m·s}^{-1}$$

$$z_{\text{coll}} = \rho \frac{\langle u \rangle}{4} = 5.08 \times 10^{24} \text{ m}^{-2} \cdot \text{s}^{-1}$$

The number of methane molecules striking a 1 cm^2 surface in 1 ms is given by

$$z_{\text{coll}}(1 \times 10^{-4} \text{ m}^{-2})(1.0 \times 10^{-3} \text{ s}^{-1}) = 5.08 \times 10^{17} \text{ molecules}$$

27–33. Consider the velocity selector shown in Figure 27.9. Let the distance between successive disks be h, the rotational frequency be ν (in units of Hz), and the angle between the slits of successive disks be θ (in degrees). Derive the following condition for a molecule traveling with speed u to pass through successive slits:

$$u = \frac{360\nu h}{\theta}$$

Typical values of h and θ are 2 cm and 2°, respectively, so $u = 3.6\nu$. By varying ν from 0 to about 500 Hz, you can select speeds from 0 to over 1500 m·s^{-1}.

Let $h = ut$, so that t is the time it takes for the molecule to travel between two disks. In order for the molecule to pass through both slits, the disks must have rotated by θ in time t. Since ν measures the number of revolutions per second of the disks, and there are 360° in one revolution, we can write $\theta = 360\nu t$, where θ is measured in degrees. Then

$$\frac{h}{u} = \frac{\theta}{360\nu} \quad \text{and so} \quad u = \frac{360\nu h}{\theta}$$

27–34. The figure below illustrates another method that has been used to determine the distribution of molecular speeds. A pulse of molecules collimated from a hot oven enter a rotating hollow drum. Let R be the radius of the drum, ν be its rotational frequency, and s be the distance through which the drum rotates during the time it takes for a molecule to travel from the entrance slit to the inner surface of the drum. Show that

$$s = \frac{4\pi R^2 \nu}{u}$$

where u is the speed of the molecule.

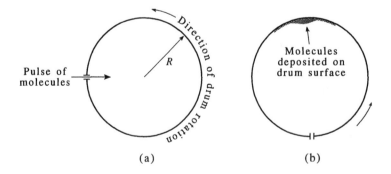

(a)   (b)

Use Equation 27.46 to show that the distribution of molecular speeds emerging from the oven is proportional to $u^3 e^{-mu^2/2k_B T}\,du$. Now show that the distribution of molecules striking the inner surface of the cylinder is given by

$$I(s)ds = \frac{A}{s^5}e^{-m(4\pi R^2 \nu)^2/2k_B Ts^2}\,ds$$

where A is simply a proportionality constant. Plot I versus s for various values of $4\pi R^2 \nu/(2k_B T/m)^{1/2}$, say 0.1, 1, and 3. Experimental data are quantitatively described by the above equation.

The molecule travels the length of the drum, $2R$, in time t, so $2R = ut$. The distance s that the drum rotates in time t is given by $R\theta$, where θ is the degree to which the drum has been rotated. Since ν measures the number of revolutions per second of the disks, and there are 2π radians in one revolution, we can write $\theta = 2\pi\nu t$, where θ is measured in degrees. Then

$$s = R\theta = 2R\pi\nu t = 2R\pi\nu\left(\frac{2R}{u}\right) = \frac{4\pi R^2 \nu}{u}$$

Equation 27.46 gives an expression for the number of collisions per time per unit area for molecules having speeds ranging from u to $u + du$. Therefore it is proportional to the distribution of molecular speeds in a pulse of molecules hitting a small unit area, and (defining this distribution as $I(u)du$)

$$I(u)du \propto u^3 e^{-mu^2/2k_B T}du$$

Using the expression found for s, we can write

$$ds = -\frac{4\pi R^2 \nu}{u^2}du$$

Now we can express $I(u)du$ in terms of s, as

$$I(s)ds \propto -\frac{u^5}{4\pi R^2 \nu}e^{-mu^2/2k_B T}ds$$

$$I(s)ds \propto -\left(\frac{4\pi R^2 \nu}{s}\right)^5 \frac{1}{4\pi R^2 \nu}e^{-m(4\pi R^2 \nu)^2/2k_B Ts^2}ds$$

$$I(s)ds = \frac{A}{s^5}e^{-m(4\pi R^2 \nu)^2/2k_B Ts^2}ds$$

where A is a proportionality constant.

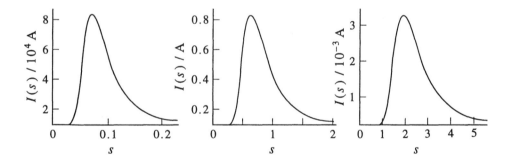

Note the different scales on the x and y-axes needed for different values of $4\pi R^2 \nu/(2k_B T/m)^{1/2}$.

27–35. Use Equation 27.49 to calculate the collision frequency of a single hydrogen molecule at 25°C and (a) one torr and (b) one bar.

Use Equation 27.49, substituting $\sigma = 0.230 \times 10^{-18}$ m^2 from Table 27.3:

$$z_A = \rho\sigma\left(\frac{8RT}{\pi M}\right)^{1/2}$$

$$= \rho(0.230 \times 10^{-18} \text{ m}^2)\left[\frac{8(8.3145 \text{ J·mol}^{-1}\text{·K}^{-1})(298.15 \text{ K})}{\pi(0.00202 \text{ kg·mol}^{-1})}\right]^{1/2} = \rho(4.070 \times 10^{-16} \text{ m}^3\text{·s}^{-1})$$

At 1 torr,

$$\rho = \frac{P}{k_B T} = \frac{(1 \text{ torr})}{(1.3806 \times 10^{-23} \text{ J·K}^{-1})(298.15 \text{ K})} \left(\frac{101325 \text{ Pa}}{760 \text{ torr}} \right) = 3.24 \times 10^{22} \text{ m}^{-3}$$

$$z_A = \rho(4.070 \times 10^{-16} \text{ m}^3 \cdot \text{s}^{-1}) = 1.32 \times 10^7 \text{ s}^{-1}$$

At 1 bar,

$$\rho = \frac{P}{k_B T} = \frac{(1 \text{ bar})}{(1.3806 \times 10^{-23} \text{ J·K}^{-1})(298.15 \text{ K})} \left(\frac{1 \times 10^5 \text{ Pa}}{1 \text{ bar}} \right) = 2.43 \times 10^{25} \text{ m}^{-3}$$

$$z_A = \rho(4.070 \times 10^{-16} \text{ m}^3 \cdot \text{s}^{-1}) = 9.89 \times 10^9 \text{ s}^{-1}$$

27–36. On the average, what is the time between collisions of a xenon atom at 300 K and (a) one torr and (b) one bar.

Use Equation 27.49, substituting $\sigma = 0.750 \times 10^{-18} \text{ m}^2$ from Table 27.3:

$$z_A = \rho \sigma \left(\frac{8RT}{\pi M} \right)^{1/2}$$

$$= \rho(0.750 \times 10^{-18} \text{ m}^2) \left[\frac{8(8.3145 \text{ J·mol}^{-1} \cdot \text{K}^{-1})(300 \text{ K})}{\pi(0.13130 \text{ kg·mol}^{-1})} \right]^{1/2}$$

$$= \rho(1.650 \times 10^{-16} \text{ m}^3 \cdot \text{s}^{-1})$$

At 1 torr,

$$\rho = \frac{P}{k_B T} = \frac{(1 \text{ torr})}{(1.3806 \times 10^{-23} \text{ J·K}^{-1})(300 \text{ K})} \left(\frac{101325 \text{ Pa}}{760 \text{ torr}} \right) = 3.22 \times 10^{22} \text{ m}^{-3}$$

$$z_A = \rho(1.650 \times 10^{-16} \text{ m}^3 \cdot \text{s}^{-1}) = 5.31 \times 10^6 \text{ s}^{-1}$$

$$z_A^{-1} = 1.88 \times 10^{-7} \text{ s}$$

(Recall that z_A^{-1} is the measure of the average time between collisions.) At 1 bar,

$$\rho = \frac{P}{k_B T} = \frac{(1 \text{ bar})}{(1.3806 \times 10^{-23} \text{ J·K}^{-1})(300 \text{ K})} \left(\frac{1 \times 10^5 \text{ Pa}}{1 \text{ bar}} \right) = 2.41 \times 10^{25} \text{ m}^{-3}$$

$$z_A = \rho(1.650 \times 10^{-16} \text{ m}^3 \cdot \text{s}^{-1}) = 3.98 \times 10^9 \text{ s}^{-1}$$

$$z_A^{-1} = 2.51 \times 10^{-10} \text{ s}$$

27–37. What is the probability that an oxygen molecule at 25°C and one bar will travel (a) 1.00×10^{-5} mm, (b) 1.00×10^{-3} mm, and (c) 1.00 mm without undergoing a collision?

Use Equation 27.55:

$$\int_0^d p(x)dx = \frac{1}{l} \int_0^d e^{-x/l}dx = 1 - e^{-d/l}$$

$$p(x) = 1 - e^{-d/l}$$

The probability of the oxygen molecule not colliding is $1 - p(x)$, since $p(x)$ is the probability of collision. We can use Equation 27.51 for l:

$$l = \frac{RT}{2^{1/2} N_A \sigma P}$$

$$= \frac{(8.3145 \text{ J·K}^{-1}\text{·mol}^{-1})(300 \text{ K})}{2^{1/2}(6.022 \times 10^{23} \text{ mol}^{-1})(0.410 \times 10^{-18} \text{ m}^2)(1.00 \times 10^5 \text{ Pa})}$$

$$= 7.143 \times 10^{-8} \text{ m} = 7.143 \times 10^{-5} \text{ mm}$$

a. $\text{Prob} = \exp\left(\dfrac{-1.00 \times 10^{-5}}{7.143 \times 10^{-5}}\right) = 0.869$

b. $\text{Prob} = \exp\left(\dfrac{-1.00 \times 10^{-3}}{7.143 \times 10^{-5}}\right) = 8.32 \times 10^{-7}$

c. $\text{Prob} = \exp\left(\dfrac{-1.00}{7.143 \times 10^{-5}}\right) \approx 0$

27–38. Repeat the calculation in the previous problem for a pressure of one torr.

Again, we can use Equation 27.51 for l:

$$l = \frac{RT}{2^{1/2} N_A \sigma P}$$

$$= \frac{(8.3145 \text{ J·K}^{-1}\text{·mol}^{-1})(300 \text{ K})}{2^{1/2}(6.022 \times 10^{23} \text{ mol}^{-1})(0.410 \times 10^{-18} \text{ m}^2)(1.00 \text{ torr})}\left(\frac{760 \text{ torr}}{101325 \text{ Pa}}\right)$$

$$= 5.36 \times 10^{-5} \text{ m} = 5.36 \times 10^{-2} \text{ mm}$$

a. $\text{Prob} = \exp\left(\dfrac{-1.00 \times 10^{-5}}{5.36 \times 10^{-2}}\right) \approx 1.00$

b. $\text{Prob} = \exp\left(\dfrac{-1.00 \times 10^{-3}}{5.36 \times 10^{-2}}\right) = 0.982$

c. $\text{Prob} = \exp\left(\dfrac{-1.00}{5.36 \times 10^{-2}}\right) = 7.84 \times 10^{-9}$

27–39. At an altitude of 150 km, the pressure is about 2×10^{-6} torr and the temperature is about 500 K. Assuming for simplicity that the air consists entirely of nitrogen, calculate the mean free path under these conditions. What is the average collision frequency?

Using Equation 27.51,

$$l = \frac{RT}{2^{1/2} N_A \sigma P}$$

$$= \frac{(8.3145 \text{ J·mol}^{-1}\text{·K}^{-1})(500 \text{ K})}{2^{1/2}(6.022 \times 10^{23} \text{ mol}^{-1})(0.450 \times 10^{-18} \text{ m}^2)(2 \times 10^{-6} \text{ torr})}\left(\frac{760 \text{ torr}}{101325 \text{ Pa}}\right)$$

$$= 40.7 \text{ m}$$

Then

$$z_A = \frac{\langle u \rangle}{l} = \left(\frac{8RT}{\pi M}\right)^{1/2}\frac{1}{l} = \frac{615 \text{ m·s}^{-1}}{40.7 \text{ m}} = 15.1 \text{ s}^{-1}$$

27–40. The following table gives the pressure and temperature of the Earth's upper atmosphere as a function of altitude:

altitude/km	P/mbar	T/K
20.0	56	220
40.0	3.2	260
60.0	0.28	260
80.0	0.013	180

Assuming for simplicity that air consists entirely of nitrogen, calculate the mean free path at each of these conditions.

Again, use Equation 27.51 to express l in terms of T and P:

$$l = \frac{R}{2^{1/2} N_A \sigma} \left(\frac{T}{P}\right)$$

$$= \frac{8.3145 \text{ J·mol}^{-1}\text{·K}^{-1}}{2^{1/2}(6.022 \times 10^{23} \text{ mol}^{-1})(0.450 \times 10^{-18} \text{ m}^2)} \left(\frac{1 \text{ Pa}}{100 \text{ mbar}}\right)\left(\frac{T}{P}\right)$$

$$= (2.1699 \times 10^{-7} \text{ K·mbar}^{-1})\frac{T}{P}$$

We can substitute the values given in the table to calculate the mean free path for each altitude:

altitude/km	l/m
20.0	8.52×10^{-7}
40.0	1.76×10^{-5}
60.0	2.01×10^{-4}
80.0	3.00×10^{-3}

27–41. Interstellar space has an average temperature of about 10 K and an average density of hydrogen atoms of about one hydrogen atom per cubic meter. Compute the mean free path of a hydrogen atom in interstellar space. Take the diameter of a hydrogen atom to be 100 pm.

We defined $\sigma = \pi d^2$ in Section 27–6, so $\sigma = \pi(100 \times 10^{-12} \text{ m})^2 = 3.142 \times 10^{-20} \text{ m}^2$. The density of hydrogen atoms is 1 m^{-3}, so, from the equations directly preceding Equation 27.51,

$$l = \frac{1}{2^{1/2}\sigma\rho} = \frac{1}{2^{1/2}(3.14 \times 10^{-20} \text{ m}^2)(1 \text{ m}^{-3})} = 2 \times 10^{19} \text{ m}$$

27–42. Calculate the pressures at which the mean free path of a hydrogen molecule will be 100 μm, 1.00 mm, and 1.00 m at 20°C.

Using Equation 27.51 and σ from Table 27.3, we find

$$P = \frac{RT}{2^{1/2} N_A \sigma l} = \frac{(8.314 \text{ J·mol}^{-1}\text{·K}^{-1})(293.15 \text{ K})}{2^{1/2}(6.022 \times 10^{23} \text{ mol}^{-1})(0.230 \times 10^{-18} \text{ m}^2)} l^{-1}$$

$$P/\text{Pa} = \frac{0.0124}{l/\text{m}}$$

l/m	P/Pa	P/bar
1.00×10^{-4}	124	1.24×10^{-3}
1.00×10^{-3}	12.4	1.24×10^{-4}
1.00	0.0124	1.24×10^{-7}

27–43. Derive an expression for the distance, d, at which a fraction f of the molecules will have been scattered from a beam consisting initially of n_0 molecules. Plot d against f.

The probability of one of the n_0 molecules colliding between 0 and d is (Equation 27.55)

$$\text{Prob} = f = \int_0^d p(x)dx = \int_0^d \frac{1}{l}e^{-x/l}dx = 1 - e^{-d/l}$$

Solving for d in terms of f gives

$$d = -l\ln(1 - f)$$

Below is a plot of d versus f.

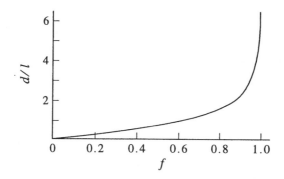

27–44. Calculate the frequency of nitrogen–oxygen collisions per dm^3 in air at the conditions given in Problem 27–40. Assume in this case that 80% of the molecules are nitrogen molecules.

We use Equations 27.58 to find $\sigma_{\text{N}_2,\text{O}_2}$, ρ_{N_2}, and ρ_{O_2} in terms of P and T:

$$\sigma_{\text{N}_2,\text{O}_2} = \pi\left(\frac{d_{\text{N}_2} + d_{\text{O}_2}}{2}\right)^2$$

$$= \pi\left(\frac{380 \times 10^{-12}\text{ m} + 360 \times 10^{-12}\text{ m}}{2}\right)^2 = 4.30 \times 10^{-19}\text{ m}^2$$

$$\langle u_r\rangle = \left(\frac{8k_\text{B}T}{\pi\mu}\right)^{1/2} = \left[\frac{8(8.3145\text{ J·K}^{-1}\cdot\text{mol}^{-1})(0.06002\text{ kg})T}{\pi(0.02802\text{ kg})(0.03200\text{ kg})}\right]^{1/2}$$

$$= 37.65T^{1/2}$$

$$\rho_{\text{N}_2} = \frac{N_\text{A}P_{\text{N}_2}}{RT} = \frac{(6.022 \times 10^{23}\text{ mol}^{-1})(0.80P)}{(8.3145\text{ J·mol}^{-1}\cdot\text{K}^{-1})T} = 5.79 \times 10^{22}\left(\frac{P}{T}\right)$$

$$\rho_{\text{O}_2} = \frac{N_\text{A}P_{\text{N}_2}}{RT} = \frac{(6.022 \times 10^{23}\text{ mol}^{-1})(0.20P)}{(8.3145\text{ J·mol}^{-1}\cdot\text{K}^{-1})T} = 1.44 \times 10^{22}\left(\frac{P}{T}\right)$$

where P is measured in Pa and T is measured in K. Substituting into Equation 27.57, we find

$$Z_{N_2,O_2} = \sigma_{N_2,O_2}\langle u_r\rangle \rho_{N_2}\rho_{O_2} = (4.30 \times 10^{-19}\ \text{m}^2)(37.65T^{1/2})(5.79 \times 10^{22})(1.44 \times 10^{22})\left(\frac{P}{T}\right)^2$$

altitude/km	$P/100$ Pa	T/K	$Z_{N_2,O_2}/\text{s}^{-1}\cdot\text{m}^{-3}$
20.0	56	220	1.31×10^{32}
40.0	3.2	260	3.32×10^{29}
60.0	0.28	260	2.54×10^{27}
80.0	0.013	180	9.51×10^{24}

27–45. Use Equation 27.58 to show that

$$\langle u_r\rangle = (\langle u_A\rangle^2 + \langle u_B\rangle^2)^{1/2}$$

Begin with Equation 27.58 and write μ as $m_A m_B/(m_A + m_B)$:

$$\langle u_r\rangle = \left(\frac{8k_B T}{\pi\mu}\right)^{1/2} = \left[\frac{8k_B T}{\pi}\left(\frac{m_A + m_B}{m_A m_B}\right)\right]^{1/2} = \left[\frac{8\pi k_B T}{\pi m_B} + \frac{8\pi k_B T}{\pi m_A}\right]^{1/2} = \left[\langle u_B\rangle^2 + \langle u_A\rangle^2\right]^{1/2}$$

27–46. Modify the derivation of Equation 27.49 to consider the collision frequency of a molecule of type A with B molecules in a mixture of A and B. Derive Equation 27.57 directly from your answer.

The molecule of type A will sweep out a cylinder of diameter $2d_A$. Since the type B molecules have a radius d_B, the effective target radius is given by $(d_A + d_B)/2$. The number density of the B molecules is given by ρ_B and we replace m by μ, so the average speed is given by $\langle u_r\rangle$. Equation 27.49 then becomes

$$z_A = \rho_B\langle u_r\rangle\pi\left(\frac{d_A + d_B}{2}\right)^2 = \rho_B\langle u_r\rangle\sigma_{AB}$$

Since $Z_{AB} = \rho_A z_A$, we find

$$Z_{AB} = \sigma_{AB}\langle u_r\rangle\rho_A\rho_B \tag{27.57}$$

27–47. Consider a mixture of methane and nitrogen in a 10.0 dm³ container at 300 K with partial pressures $P_{CH_4} = 65.0$ mbar and $P_{N_2} = 30.0$ mbar. Use the equation that you derived in the previous problem to calculate the collision frequency of a methane molecule with nitrogen molecules. Also calculate the frequency of methane–nitrogen collisions per dm³.

First find the necessary values of ρ, σ, and $\langle u_r\rangle$:

$$\rho_{CH_4} = \frac{P_{CH_4}}{k_B T} = \frac{6500\ \text{Pa}}{(1.3806 \times 10^{-23}\ \text{J}\cdot\text{K}^{-1})(300\ \text{K})} = 1.57 \times 10^{24}\ \text{m}^{-3}$$

$$\rho_{N_2} = \frac{P_{CH_4}}{k_B T} = \frac{3000\ \text{Pa}}{(1.3806 \times 10^{-23}\ \text{J}\cdot\text{K}^{-1})(300\ \text{K})} = 7.24 \times 10^{23}\ \text{m}^{-3}$$

$$\sigma_{CH_4, N_2} = \pi \left(\frac{410 \times 10^{-12} \text{ m} + 380 \times 10^{-12} \text{ m}}{2} \right)^2 = 4.90 \times 10^{-19} \text{ m}^2$$

$$\langle u_r \rangle = \left[\frac{8RT(m_1 + m_2)}{\pi m_1 m_2} \right]^{1/2} = \left[\frac{8R(300 \text{ K})(0.044062 \text{ kg})}{\pi(0.02802 \text{ kg})(0.016042 \text{ kg})} \right]^{1/2} = 789 \text{ m} \cdot \text{s}^{-1}$$

Now, using the equations from the previous problem,

$$z_{CH_4} = \rho_{N_2} \sigma_{CH_4, N_2} \langle u_r \rangle = 2.80 \times 10^8 \text{ s}^{-1}$$

$$Z_{CH_4, N_2} = \rho_{CH_4} z_{CH_4} = 4.40 \times 10^{32} \text{ m}^{-3} \cdot \text{s}^{-1}$$

27–48. Calculate the average relative kinetic energy with which the molecules in a gas collide.

We know that the normalized average kinetic energy is given by

$$\langle \varepsilon_K \rangle = \frac{\int \varepsilon f(\varepsilon) d\varepsilon}{\int f(\varepsilon) d\varepsilon}$$

where $f(\varepsilon) d\varepsilon$ is the distribution of kinetic energy. We can substitute $\varepsilon_r e^{-\varepsilon_r / k_B T}$ for $f(\varepsilon)$ to write

$$\langle \varepsilon_K \rangle = \frac{\int \varepsilon_r^2 e^{-\varepsilon_r / k_B T} d\varepsilon_r}{\int \varepsilon_r e^{-\varepsilon_r / k_B T} d\varepsilon_r} = \frac{2(k_B T)^3}{(k_B T)^2} = 2k_B T$$

The following four problems deal with molecular effusion.

27–49. Equation 27.48 gives us the frequency of collisions that the molecules of a gas make with a surface area of the walls of the container. Suppose now we make a very small hole in the wall. If the mean free path of the gas is much larger than the width of the hole, any molecule that strikes the hole will leave the container without undergoing any collisions along the way. In this case, the molecules leave the container individually, independently of the others. The rate of flow through the hole will be small enough that the remaining gas is unaffected, and remains essentially in equilibrium. This process is called *molecular effusion*. Equation 27.48 can be applied to calculate the rate of molecular effusion. Show that Equation 27.48 can be expressed as

$$\text{effusion flux} = \frac{P}{(2\pi m k_B T)^{1/2}} = \frac{N_A P}{(2\pi M R T)^{1/2}} \tag{1}$$

where P is the pressure of the gas. Calculate the number of nitrogen molecules that effuse per second through a round hole of 0.010 mm diameter if the gas is at 25°C and one bar.

We begin with Equation 27.48:

$$z_{coll} = \frac{1}{4} \rho \langle u \rangle = \frac{1}{4} \left(\frac{P}{k_B T} \right) \left(\frac{8 k_B T}{\pi m} \right)^{1/2}$$

$$= \frac{P}{(2\pi m k_B T)^{1/2}} = \frac{N_A P}{(2\pi M R T)^{1/2}}$$

$$= \frac{(6.022 \times 10^{23} \text{ mol}^{-1})^{1/2}(10^5 \text{ Pa})}{[2\pi(0.02802 \text{ kg} \cdot \text{mol}^{-1})(1.38 \times 10^{-23} \text{ J} \cdot \text{K}^{-1})(298 \text{ K})]^{1/2}}$$

$$= 2.88 \times 10^{27} \text{ m}^{-2} \cdot \text{s}^{-1}$$

This will be the effusion flux of the gas. For a round hole of 0.010 mm diameter,

$$\text{molecules per second} = z_{\text{coll}}[\pi(0.005 \times 10^{-3} \text{ m})^2] = 2.26 \times 10^{17} \text{ s}^{-1}$$

27–50. Equation 1 of the previous problem can be used to determine vapor pressures of substances with very low vapor pressures. This was done by Irving Langmuir to measure the vapor pressure of tungsten at various temperatures in his investigation of tungsten filaments in light bulbs and vacuum tubes. (Langmuir, who was awarded the Nobel Prize in chemistry in 1932, worked for General Electric.) He estimated the rate of effusion by weighing the tungsten filament at the beginning and the end of each experimental run. Langmuir did these experiments around 1913, but his data appear in the *CRC Handbook of Chemistry and Physics* to this day. Use the following data to determine the vapor pressure of tungsten at each temperature and then determine the molar enthalpy of vaporization of tungsten.

T/K	effusion flux/$\text{g} \cdot \text{m}^{-2} \cdot \text{s}^{-1}$
1200	3.21×10^{-23}
1600	1.25×10^{-14}
2000	1.76×10^{-9}
2400	4.26×10^{-6}
2800	1.10×10^{-3}
3200	6.38×10^{-3}

Note that the flux is given in units of $\text{g} \cdot \text{m}^{-2} \cdot \text{s}^{-1}$, so we divide the flux by 1000 to have units of $\text{kg} \cdot \text{m}^{-2} \cdot \text{s}^{-1}$. Using Equation 1 of Problem 27–49 gives

$$P = (2\pi m k_{\text{B}} T)^{1/2} \times \frac{\text{effusion flux}}{1000 \text{ g} \cdot \text{kg}^{-1}}$$
$$= A T^{1/2} \times \text{effusion flux}$$

Because we are going to plot $\ln P$ against $1/T$, we can write the above equation as

$$\ln P = \ln A + \ln T^{1/2} + \ln(\text{effusion flux})$$

and, because A is a constant, we do not need to evaluate it. Thus we form the table of values

T/K	$T^{1/2} \times$ effusion flux
1200	1.11×10^{-21}
1600	5.00×10^{-13}
2000	7.87×10^{-8}
2400	2.08×10^{-4}
2800	5.89×10^{-2}
3200	0.361

Since $P \propto T^{1/2} \times$ effusion flux, the slope of the best-fit line to $\ln P$ versus $1/T$ will be equal to the slope of the best-fit line to $\ln(T^{1/2} \times$ effusion flux) versus $1/T$. Recall (Chapter 23) that

$$\ln P = -\frac{\Delta_{\text{vap}} H}{RT} + \text{constant}$$

Therefore, the slope of the best-fit line to either plot described above will be equal to $-\Delta_{\text{vap}} H/R$.

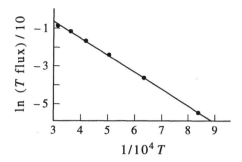

The slope of this plot is -92830, and so $\Delta_{vap}H = 92830R = 772 \text{ kJ} \cdot \text{mol}^{-1}$.

27–51. The vapor pressure of mercury can be determined by the effusion technique described in the previous problem. Given that 0.126 mg of mercury passed through a small hole of area 1.65 mm² in 2.25 hours at 0°C, calculate the vapor pressure of mercury in torr.

$$\text{effusion} = \left(\frac{0.126 \times 10^{-3} \text{ g}}{200.6 \text{ g} \cdot \text{mol}^{-1}}\right)(6.022 \times 10^{23} \text{ molecule} \cdot \text{mol}^{-1})\left(\frac{1}{1.65 \times 10^{-6} \text{ m}^2}\right)$$

$$\times \left(\frac{1}{2.25 \text{ hr}}\right)\left(\frac{1 \text{ hr}}{3600 \text{ s}}\right)$$

$$= 2.83 \times 10^{19} \text{ molecule} \cdot \text{m}^{-2} \cdot \text{s}^{-1}$$

$$P = \text{effusion} \times (2\pi m k_B T)^{1/2}$$

$$= (2.83 \times 10^{19} \text{ molecule} \cdot \text{m}^{-2} \cdot \text{s}^{-1})\left[2\pi\left(\frac{0.2006 \text{ kg} \cdot \text{mol}^{-1}}{6.022 \times 10^{23} \text{ molecule} \cdot \text{mol}^{-1}}\right)\right.$$

$$\left.\times (1.38 \times 10^{-23} \text{ J} \cdot \text{K}^{-1})(273.15 \text{ K})\right]^{1/2}$$

$$= 2.51 \times 10^{-3} \text{ Pa}\left(\frac{760 \text{ torr}}{101325 \text{ Pa}}\right) = 1.89 \times 10^{-5} \text{ torr}$$

27–52. We can use Equation 1 of Problem 27–49 to derive an expression for the pressure as a function of time for an ideal gas that is effusing from its container. First show that

$$\text{rate of effusion} = -\frac{dN}{dt} = \frac{PA}{(2\pi m k_B T)^{1/2}}$$

where N is the number of molecules effusing and A is the area of the hole. At constant T and V,

$$-\frac{dN}{dt} = \frac{d}{dt}\left(\frac{PV}{k_B T}\right) = \frac{V}{k_B T}\frac{dP}{dt}$$

Now show that

$$P(t) = P(0)e^{-\alpha t}$$

where $\alpha = (k_B T/2\pi m)^{1/2} A/V$. Note that the pressure of the gas decreases exponentially with time.

The rate of effusion is given by dN/dt, since this represents how many molecules effuse per unit time. From Problem 27–49, we know the rate of effusion per unit area (effusion flux), so multiplying by the area of the hole from which effusion occurs will give the rate of effusion and

$$\text{rate of effusion} = -\frac{dN}{dt} = \frac{PA}{(2\pi m k_B T)^{1/2}}$$

For an ideal gas, $nN_A = PV/k_BT$ and so $N = PV/k_BT$. If T and V are held constant, we find

$$\frac{dN}{dt} = \frac{d}{dt}\left(\frac{PV}{k_BT}\right) = \frac{V}{k_BT}\frac{dP}{dt}$$

Use the first equation for rate of effusion to write

$$-\frac{V}{k_BT}\frac{dP}{dt} = \frac{PA}{(2\pi mk_BT)^{1/2}}$$

$$-\frac{dP}{dt} = \left(\frac{k_BT}{2\pi m}\right)^{1/2}\frac{A}{V}P = \alpha P$$

The solution to this equation is (as in the derivation of Equation 27.53)

$$P(t) = P(0)e^{-\alpha t}$$

27–53. How would you interpret the velocity distribution

$$h(v_x, v_y, v_z) = \left(\frac{m}{2\pi k_BT}\right)^{3/2}\exp\left[-\frac{m}{2k_BT}\left\{(v_x - a)^2 + (v_y - b)^2 + (v_z - c)^2\right\}\right]$$

This is the velocity distribution relative to an overall translational motion of the system with velocity $a\mathbf{i} + b\mathbf{j} + c\mathbf{k}$.

CHAPTER 28

Chemical Kinetics I: Rate Laws

PROBLEMS AND SOLUTIONS

28–1. For each of the following chemical reactions, calculate the equilibrium extent of reaction at 298.15 K and one bar. (See Section 26–4.)

a. $H_2(g) + Cl_2(g) \rightleftharpoons 2\,HCl(g)$ $\qquad \Delta_r G^\circ = -190.54 \text{ kJ} \cdot \text{mol}^{-1}$

Initial amounts: one mole of $H_2(g)$ and $Cl_2(g)$ and no $HCl(g)$.

b. $N_2(g) + O_2(g) \rightleftharpoons 2\,NO(g)$ $\qquad \Delta_r G^\circ = 173.22 \text{ kJ} \cdot \text{mol}^{-1}$

Initial amounts: one mole of $N_2(g)$ and $O_2(g)$ and no $NO(g)$.

---

a. At equilibrium, there will be $(1 - \xi_{eq})$ moles of $H_2(g)$ and $Cl_2(g)$, and $2\xi_{eq}$ moles of $HCl(g)$. We can write the partial pressures of each species at equilibrium as

$$P_{H_2} = P_{Cl_2} = \frac{(1 - \xi_{eq})}{2}P \qquad \text{and} \qquad P_{HCl} = \frac{2\xi_{eq}}{2}P = \xi_{eq}P$$

where P is the total pressure. We can then write K_P, by definition, as

$$K_P = \frac{P_{HCl}^2}{P_{H_2}P_{Cl_2}} = \frac{(2\xi_{eq})^2}{(1 - \xi_{eq})^2}$$

Equating $\ln K_P$ to $-\Delta_r G^\circ / RT$ (Equation 26.11) allows us to write

$$\ln K_P = \frac{190.54 \text{ kJ} \cdot \text{mol}^{-1}}{(8.3145 \times 10^{-3} \text{ kJ} \cdot \text{mol}^{-1} \cdot \text{K}^{-1})(298.15 \text{ K})} = 76.862$$

$$K_P = 2.40 \times 10^{33}$$

$$\frac{(2\xi_{eq})^2}{(1 - \xi_{eq})^2} = 2.40 \times 10^{33}$$

Solving this equation for ξ_{eq} gives a value of $\xi_{eq} = 1$ mol.

b. At equilibrium, there will be $(1 - \xi_{eq})$ moles of $N_2(g)$ and $O_2(g)$, and $2\xi_{eq}$ moles of $NO(g)$. We can write the partial pressures of each species at equilibrium as

$$P_{N_2} = P_{O_2} = \frac{(1 - \xi_{eq})}{2}P \qquad \text{and} \qquad P_{NO} = \frac{2\xi_{eq}}{2}P = \xi_{eq}P$$

where P is the total pressure. We can then write K_P, by definition, as

$$K_P = \frac{P_{NO}^2}{P_{N_2} P_{O_2}} = \frac{(2\xi_{eq})^2}{(1 - \xi_{eq})^2}$$

Equating $\ln K_P$ to $-\Delta_r G^\circ / RT$ (Equation 26.11) allows us to write

$$\ln K_P = \frac{-173.22 \text{ kJ} \cdot \text{mol}^{-1}}{(8.3145 \times 10^{-3} \text{ kJ} \cdot \text{mol}^{-1} \cdot \text{K}^{-1})(298.15 \text{ K})} = -69.87$$

$$K_P = 4.50 \times 10^{-31}$$

$$\frac{(2\xi_{eq})^2}{(1 - \xi_{eq})^2} = 4.50 \times 10^{-31}$$

Solving this equation for ξ_{eq} gives a value of $\xi_{eq} = 3.3 \times 10^{-16}$ mol.

28–2. Dinitrogen oxide, N_2O, decomposes according to the equation

$$2 N_2O(g) \longrightarrow 2 N_2(g) + O_2(g)$$

Under certain conditions at 900 K, the rate of reaction is 6.16×10^{-6} mol·dm^{-3}·s^{-1}. Calculate the values of $d[N_2O]/dt$, $d[N_2]/dt$, and $d[O_2]/dt$.

Using Equation 28–7, we see that for this reaction,

$$6.16 \times 10^{-6} \text{ mol} \cdot \text{dm}^{-3} \cdot \text{s}^{-1} = \frac{1}{2} \frac{d[N_2]}{dt} = -\frac{1}{2} \frac{d[N_2O]}{dt} = \frac{d[O_2]}{dt}$$

We find that $d[N_2O]/dt = -1.23 \times 10^{-5}$ mol·dm^{-3}·s^{-1}, $d[N_2]/dt = 1.23 \times 10^{-5}$ mol·dm^{-3}·s^{-1}, and $d[O_2]/dt = 6.16 \times 10^{-6}$ mol·dm^{-3}·s^{-1}.

28–3. Suppose the reaction in Problem 28–2 is carried out in a 2.67 dm^3 container. Calculate the value of $d\xi/dt$ corresponding to the rate of reaction of 6.16×10^{-6} mol·dm^{-3}·s^{-1}.

From Problem 28–2, we know that $d[O_2]/dt = 6.16 \times 10^{-6}$ mol·dm^{-3}·s^{-1}. Using Equation 28.5 gives

$$\frac{1}{V} \frac{d\xi}{dt} = \frac{d[O_2]}{dt}$$

$$\frac{d\xi}{dt} = (6.16 \times 10^{-6} \text{ mol} \cdot \text{dm}^{-3} \cdot \text{s}^{-1})(2.67 \text{ dm}^3) = 1.64 \times 10^{-5} \text{ mol} \cdot \text{s}^{-1}$$

28–4. The oxidation of hydrogen peroxide by permanganate occurs according to the equation

$$2 \text{KMnO}_4(aq) + 3 \text{H}_2\text{SO}_4(aq) + 5 \text{H}_2\text{O}_2(aq) \longrightarrow 2 \text{MnSO}_4(aq)$$

$$+ 8 \text{H}_2\text{O}(l) + 5 \text{O}_2(g) + \text{K}_2\text{SO}_4(aq)$$

Define v, the rate of reaction, in terms of each of the reactants and products.

Using Equation 28.6, we write

$$v(t) = -\frac{1}{2}\frac{d[KMnO_4]}{dt} = -\frac{1}{3}\frac{d[H_2SO_4]}{dt} = -\frac{1}{5}\frac{d[H_2O_2]}{dt} = \frac{1}{2}\frac{d[MnSO_4]}{dt}$$
$$= \frac{1}{8}\frac{d[H_2O]}{dt} = \frac{1}{5}\frac{d[O_2]}{dt} = \frac{d[K_2SO_4]}{dt}$$

28–5. The second-order rate constant for the reaction

$$O(g) + O_3(g) \longrightarrow 2\,O_2(g)$$

is 1.26×10^{-15} $cm^3 \cdot molecule^{-1} \cdot s^{-1}$. Determine the value of the rate constant in units of $dm^3 \cdot mol^{-1} \cdot s^{-1}$.

$$\left(\frac{1.26 \times 10^{-15}\ cm^3}{molecule \cdot s}\right)\left(\frac{6.022 \times 10^{23}\ molecule}{mol}\right)\left(\frac{1\ dm}{10\ cm}\right)^3 = 7.59 \times 10^5\ dm^3 \cdot mol^{-1} \cdot s^{-1}$$

28–6. The definition of the rate of reaction in terms of molar concentration (Equation 28.5) assumes that the volume remains constant during the course of the reaction. Derive an expression for the rate of reaction in terms of the molar concentration of a reactant A for the case in which the volume changes during the course of the reaction.

From the definition of molar concentration, we know that $n_A = [A]V$, and so

$$dn_A = V d[A] + [A]dV \tag{1}$$

The rate of reaction is defined as

$$v = -\frac{1}{v_A V}\frac{dn_A}{dt}$$

and substituting dn_A from Equation 1 gives

$$v = -\frac{1}{v_A}\frac{d[A]}{dt} - \frac{[A]}{v_A V}\frac{dV}{dt}$$

28–7. Derive the integrated rate law for a reaction that is zero order in reactant concentration.

The rate law for a zero-order reaction is $v(t) = k$, or $-d[A]/dt = k$. Integrating gives the integrated rate law

$$\int -d[A] = \int kdt$$
$$[A] - [A]_0 = -kt$$

28–8. Determine the rate law for the reaction described by

$$NO(g) + H_2(g) \longrightarrow products$$

from the initial rate data tabulated below.

$P_0(H_2)$/torr	$P_0(NO)$/torr	v_0/torr·s^{-1}
400	159	34
400	300	125
289	400	160
205	400	110
147	400	79

Calculate the rate constant for this reaction.

We do this problem in the same way as Example 28–2. The rate law has the form

$$v = k[NO]^{m_{NO}}[H_2]^{m_{H_2}}$$

To determine m_{NO}, we use Equation 28.19 and the first two entries in the data table to find

$$m_{NO} = \frac{\ln(34/125)}{\ln(159/300)} = 2.05 \approx 2$$

Likewise, using the third and fourth entries in the data table, we find that

$$m_{H_2} = \frac{\ln(160/110)}{\ln(289/205)} = 1.09 \approx 1$$

Assuming the orders are integer valued gives the rate law

$$v = k[NO]^2[H_2]$$

We can substitute each set of data given in the table into the rate law to calculate the average rate constant. Below is a table of the rate constant found for each v_0 in the table.

v_0/torr·s^{-1}	34	125	160	110	79
k/torr^{-2}·s^{-1}	3.36×10^{-6}	3.47×10^{-6}	3.46×10^{-6}	3.35×10^{-6}	3.36×10^{-6}

The average value of k is 3.40×10^{-6} torr^{-2}·s^{-1}.

28–9. Sulfuryl chloride decomposes according to the equation

$$SO_2Cl_2(g) \longrightarrow SO_2(g) + Cl_2(g)$$

Determine the order of the reaction with respect to $SO_2Cl_2(g)$ from the following initial-rate data collected at 298.15 K.

$[SO_2Cl_2]_0$/mol·dm^{-3}	0.10	0.37	0.76	1.22
v_0/mol·dm^{-3}·s^{-1}	2.24×10^{-6}	8.29×10^{-6}	1.71×10^{-5}	2.75×10^{-5}

Calculate the rate constant for this reaction at 298.15 K.

We do this problem in the same way as Example 28–2. The rate law has the form

$$v = k[SO_2Cl_2]^{m_{SO_2Cl_2}}$$

To determine $m_{SO_2Cl_2}$, we use Equation 28.19 and the second and third entries in the data table to find

$$m_{SO_2Cl_2} = \frac{\ln(8.29 \times 10^{-6}/1.71 \times 10^{-5})}{\ln(0.37/0.76)} = 1.01 \approx 1$$

Assuming the order is integer valued gives the rate law

$$v = k[SO_2Cl_2]$$

We can substitute the data given in the table into the rate law to calculate the average rate constant. Below is a table of the rate constant found for each tabulated v_0.

$v_0/\text{mol}\cdot\text{dm}^{-3}\cdot\text{s}^{-1}$	2.24×10^{-6}	8.29×10^{-6}	1.71×10^{-5}	2.75×10^{-5}
k/s^{-1}	2.24×10^{-5}	2.24×10^{-5}	2.25×10^{-5}	2.25×10^{-5}

The average value of k is $2.25 \times 10^{-5}\ \text{s}^{-1}$.

28–10. Consider the reaction described by

$$Cr(H_2O)_6^{3+}(aq) + SCN^-(aq) \longrightarrow Cr(H_2O)_5(SCN)^{2+}(aq) + H_2O(l)$$

for which the following initial rate data were obtained at 298.15 K.

$[Cr(H_2O)_6^{3+}]_0/\text{mol}\cdot\text{dm}^{-3}$	$[SCN^-]_0/\text{mol}\cdot\text{dm}^{-3}$	$v_0/\text{mol}\cdot\text{dm}^{-3}\cdot\text{s}^{-1}$
1.21×10^{-4}	1.05×10^{-5}	2.11×10^{-11}
1.46×10^{-4}	2.28×10^{-5}	5.53×10^{-11}
1.66×10^{-4}	1.02×10^{-5}	2.82×10^{-11}
1.83×10^{-4}	3.11×10^{-5}	9.44×10^{-11}

Determine the rate law for the reaction and the rate constant at 298.15 K. Assume the orders are integers.

The rate law has the form

$$v = k[Cr(H_2O)_6^{3+}]^{m_{Cr(H_2O)_6^{3+}}}[SCN^-]^{m_{SCN^-}}$$

The first and third entries in the data table are for an essentially constant concentration of $[SCN^-]$, so, proceeding as in Example 28–2,

$$m_{Cr(H_2O)_6^{3+}} = \frac{\ln(2.82 \times 10^{-11}/2.11 \times 10^{-11})}{\ln(1.66 \times 10^{-4}/1.21 \times 10^{-4})} = 0.917 \approx 1$$

Using this result and the first two entries of the data table, we have the rate equations

$$2.11 \times 10^{-11}\ \text{mol}\cdot\text{dm}^{-3}\cdot\text{s}^{-1} = k(1.21 \times 10^{-4}\ \text{mol}\cdot\text{dm}^{-3})(1.05 \times 10^{-5}\ \text{mol}\cdot\text{dm}^{-3})^{m_{SCN^-}}$$

$$5.53 \times 10^{-11}\ \text{mol}\cdot\text{dm}^{-3}\cdot\text{s}^{-1} = k(1.46 \times 10^{-4}\ \text{mol}\cdot\text{dm}^{-3})(2.28 \times 10^{-5}\ \text{mol}\cdot\text{dm}^{-3})^{m_{SCN^-}}$$

Taking the ratio of these two equations gives

$$0.460 = (0.460)^{m_{SCN^-}}$$

from which we find $m_{SCN^-} = 1$. The rate law is first order in each reagent and second order overall:

$$v = k[Cr(H_2O)_6^{3+}][SCN^-]$$

We can substitute the data given in the table into the rate law to calculate the average rate constant. The four sets of data give rate constants of

$v_0/\text{mol}\cdot\text{dm}^{-3}\cdot\text{s}^{-1}$	2.11×10^{-11}	5.53×10^{-11}	2.82×10^{-11}	9.44×10^{-11}
$k/\text{dm}^3\cdot\text{mol}^{-1}\cdot\text{s}^{-1}$	1.66×10^{-2}	1.66×10^{-2}	1.67×10^{-2}	1.66×10^{-2}

and the average value of k is $1.66 \times 10^{-2} \text{ dm}^3\cdot\text{mol}^{-1}\cdot\text{s}^{-1}$.

28–11. Consider the base-catalyzed reaction

$$OCl^-(aq) + I^-(aq) \longrightarrow OI^-(aq) + Cl^-(aq)$$

Use the following initial-rate data to determine the rate law and the corresponding rate constant for the reaction.

$[OCl^-]/\text{mol}\cdot\text{dm}^{-3}$	$[I^-]/\text{mol}\cdot\text{dm}^{-3}$	$[OH^-]/\text{mol}\cdot\text{dm}^{-3}$	$v_0/\text{mol}\cdot\text{dm}^{-3}\cdot\text{s}^{-1}$
1.62×10^{-3}	1.62×10^{-3}	0.52	3.06×10^{-4}
1.62×10^{-3}	2.88×10^{-3}	0.52	5.44×10^{-4}
2.71×10^{-3}	1.62×10^{-3}	0.84	3.16×10^{-4}
1.62×10^{-3}	2.88×10^{-3}	0.91	3.11×10^{-4}

Because the reaction is base-catalyzed, the rate will depend on the basicity of the solution, or on the concentration of OH^- present. Therefore, the rate law has the form

$$v = k[OCl^-]^{m_{OCl^-}} [I^-]^{m_{I^-}} [OH^-]^{m_{OH^-}}$$

The first two entries in the data table hold $[OCl^-]$ and $[OH^-]$ constant, so we write (as in Example 28–2)

$$m_{I^-} = \frac{\ln(3.06 \times 10^{-4}/5.44 \times 10^{-4})}{\ln(1.62 \times 10^{-3}/2.88 \times 10^{-3})} = 1$$

The second and fourth entries in the data table hold $[OCl^-]$ and $[I^-]$ constant, so we write

$$m_{OH^-} = \frac{\ln(5.44 \times 10^{-4}/3.11 \times 10^{-4})}{\ln(0.52/0.91)} = -0.9991 \approx -1$$

Using these results and the first and third entries of the data table, we have the rate equations

$$3.06 \times 10^{-4} \text{ mol}\cdot\text{dm}^{-3}\cdot\text{s}^{-1} = k\frac{(1.62 \times 10^{-3} \text{ mol}\cdot\text{dm}^{-3})^{m_{OCl^-}} (1.62 \times 10^{-3} \text{ mol}\cdot\text{dm}^{-3})}{0.52 \text{ mol}\cdot\text{dm}^{-3}}$$

$$3.16 \times 10^{-4} \text{ mol}\cdot\text{dm}^{-3}\cdot\text{s}^{-1} = k\frac{(2.71 \times 10^{-3} \text{ mol}\cdot\text{dm}^{-3})^{m_{OCl^-}} (1.62 \times 10^{-3} \text{ mol}\cdot\text{dm}^{-3})}{0.84 \text{ mol}\cdot\text{dm}^{-3}}$$

Taking the ratio of these two equations gives

$$0.599 = (0.598)^{m_{OCl^-}}$$

from which we find $m_{OCl^-} = 1$. The rate law is therefore

$$v = k\frac{[OCl^-][I^-]}{[OH^-]}$$

We can substitute the data given in the table into the rate law to calculate the average rate constant. The four sets of data give rate constants of

$v_0/\text{mol} \cdot \text{dm}^{-3} \cdot \text{s}^{-1}$	3.06×10^{-4}	5.44×10^{-4}	3.16×10^{-4}	3.11×10^{-4}
k/s^{-1}	60.6	60.6	60.5	60.7

and the average value of k is 60.6 s^{-1}.

28–12. The reaction

$$SO_2Cl_2(g) \longrightarrow SO_2(g) + Cl_2(g)$$

is first order and has a rate constant of $2.24 \times 10^{-5} \text{ s}^{-1}$ at 320°C. Calculate the half-life of the reaction. What fraction of a sample of $SO_2Cl_2(g)$ remains after being heated for 5.00 hours at 320°C? How long will a sample of $SO_2Cl_2(g)$ need to be maintained at 320°C to decompose 92.0% of the initial amount present?

The integrated rate law for a first order reaction is

$$\ln \frac{[SO_2Cl_2]}{[SO_2Cl_2]_0} = -kt \tag{28.22}$$

At $t = t_{1/2}$, $[SO_2Cl_2] = \frac{1}{2}[SO_2Cl_2]_0$, and substituting into Equation 28.22 gives

$$\ln \tfrac{1}{2} = -(2.24 \times 10^{-5} \text{ s}^{-1})t_{1/2}$$
$$3.09 \times 10^4 \text{ s} = t_{1/2}$$

After being heated for 5.00 hours at 320°C, the amount of SO_2Cl_2 present can be found by solving the equation

$$\ln \frac{[SO_2Cl_2]}{[SO_2Cl_2]_0} = -(2.24 \times 10^{-5} \text{ s}^{-1})(5.00 \text{ hr}) \left(3600 \text{ s} \cdot \text{hr}^{-1}\right)$$
$$[SO_2Cl_2] = 0.668[SO_2Cl_2]_0$$

In other words, 68.8% of the sample will remain. The time it takes to decompose 92.0% of SO_2Cl_2 can be found by solving

$$\ln \frac{[SO_2Cl_2]}{[SO_2Cl_2]_0} = -(2.24 \times 10^{-5} \text{ s}^{-1})t$$
$$\ln(1 - 0.920) = -(2.24 \times 10^{-5} \text{ s}^{-1})t$$
$$1.13 \times 10^5 \text{ s} = t$$

It takes 31.3 hours to decompose 92.0% of the intial amount of SO_2Cl_2 present.

28–13. The half-life for the following gas-phase decomposition reaction

$$\begin{array}{l} H_2C \!\!-\!\! CHCH_2CH_2CH_3 \\ \quad | \qquad | \qquad\qquad\qquad \longrightarrow \quad H_2C \!\!=\!\! CHCH_2CH_2CH_3 \ + \ H_2C \!\!=\!\! CH_2 \\ H_2C \!\!-\!\! CH_2 \end{array}$$

is found to be independent of the initial concentration of the reactant. Determine the rate law and integrated rate law for this reaction.

This is a first-order reaction, because the half-life is independent of the initial reactant concentration. The rate law and integrated rate law for the reaction are $v = k[A]$ and $[A] = [A]_0 e^{-kt}$, respectively, where A represents the reactant.

28–14. Hydrogen peroxide, H_2O_2, decomposes in water by a first-order kinetic process. A 0.156-mol·dm^{-3} solution of H_2O_2 in water has an initial rate of 1.14×10^{-5} mol·dm^{-3}·s^{-1}. Calculate the rate constant for the decomposition reaction and the half-life of the decomposition reaction.

We can find the rate constant by substituting into the first-order rate law (Equation 28.21):

$$v_0 = k[H_2O_2]_0$$
$$k = \frac{1.14 \times 10^{-5}\ \text{mol·dm}^{-3}\text{·s}^{-1}}{0.156\ \text{mol·dm}^{-3}} = 7.31 \times 10^{-5}\ \text{s}^{-1}$$

For a first-order reaction, we can use Equation 28.25 to determine the half-life of the reaction:

$$t_{1/2} = \frac{\ln 2}{k} = 9.48 \times 10^3\ \text{s} = 2.63\ \text{hr}$$

28–15. A first-order reaction is 24.0% complete in 19.7 minutes. How long will the reaction take to be 85.5% complete? Calculate the rate constant for the reaction.

To find the rate constant of the reaction, we can write (from Equation 28.22)

$$\ln \frac{[A]}{[A]_0} = -k(19.7\ \text{min})$$
$$\ln(1 - 0.24) = -k(19.7\ \text{min})$$
$$1.39 \times 10^{-2}\ \text{min}^{-1} = k$$

To find the time it takes for the reaction to be 85.5% complete, we solve the equation

$$\ln \frac{[A]}{[A]_0} = -(1.39 \times 10^{-2}\ \text{min}^{-1})t$$
$$\ln(1 - 0.855) = -(1.39 \times 10^{-2}\ \text{min}^{-1})t$$
$$139\ \text{min} = t$$

28–16. The nucleophilic substitution reaction

$$PhSO_2SO_2Ph(sln) + N_2H_4(sln) \longrightarrow PhSO_2NHNH_2(sln) + PhSO_2H(sln)$$

was studied in cyclohexane solution at 300 K. The rate law was found to be first order in $PhSO_2SO_2Ph$. For an initial concentration of $[PhSO_2SO_2Ph]_0 = 3.15 \times 10^{-5}$ mol·dm^{-3}, the following rate data were observed. Determine the rate law and the rate constant for this reaction.

$[N_2H_4]_0/10^{-2}$ mol·dm^{-3}	0.5	1.0	2.4	5.6
v/mol·dm^{-3}·s^{-1}	0.085	0.17	0.41	0.95

We know that the rate law is first order in $PhSO_2SO_2Ph$, so we can write the rate law as

$$v = k[PhSO_2SO_2Ph][N_2H_2]^{m_{N_2H_2}}$$

To determine $m_{N_2H_2}$, we use Equation 28.19 and the first two sets of data in the table to find

$$m_{N_2H_2} = \frac{\ln(0.085/0.17)}{\ln(0.5/1.0)} = 1$$

This gives the rate law

$$v = k[PhSO_2SO_2Ph][N_2H_2]$$

We can substitute each set of data given in the table into the rate law to calculate the average rate constant:

$v/mol \cdot dm^{-3} \cdot s^{-1}$	0.085	0.17	0.41	0.95
$k/dm^3 \cdot s^{-1} \cdot mol^{-1}$	5.4×10^5	5.4×10^5	5.4×10^5	5.4×10^5

The average value of k is $5.4 \times 10^5 \ dm^3 \cdot s^{-1} \cdot mol^{-1}$.

28–17. Show that if A reacts to form either B or C according to

$$A \xrightarrow{k_1} B \quad \text{or} \quad A \xrightarrow{k_2} C$$

then

$$[A] = [A]_0 e^{-(k_1+k_2)t}$$

Now show that $t_{1/2}$, the half-life of A, is given by

$$t_{1/2} = \frac{0.693}{k_1 + k_2}$$

Show that $[B]/[C] = k_1/k_2$ for all times t. For the set of initial conditions $[A] = [A]_0$, $[B]_0 = [C]_0 = 0$, and $k_2 = 4k_1$, plot [A], [B], and [C] as a function of time on the same graph.

If A simultaneously reacts to form B and C, we can write

$$\frac{d[A]}{dt} = -k_1[A] - k_2[A]$$

$$\frac{d[A]}{[A]} = -(k_1 + k_2)dt$$

$$\ln \frac{[A]}{[A]_0} = -(k_1 + k_2)t$$

$$[A] = [A]_0 e^{-(k_1+k_2)t}$$

Now, at the half life of the reaction ($t = t_{1/2}$), we have

$$\ln \tfrac{1}{2} = -(k_1 + k_2)t_{1/2}$$

$$\frac{0.693}{k_1 + k_2} = t_{1/2}$$

We can write

$$\frac{d[\text{B}]}{dt} = k_1[\text{A}] = k_1[\text{A}]_0 e^{-(k_1+k_2)t}$$

and

$$\frac{d[\text{C}]}{dt} = k_2[\text{A}] = k_2[\text{A}]_0 e^{-(k_1+k_2)t}$$

Because $[\text{B}]_0 = [\text{C}]_0 = 0$, we integrate the above expressions to get

$$[\text{B}] = \frac{k_1[\text{A}]_0}{k_1 + k_2}\left[1 - e^{-(k_1+k_2)t}\right]$$

$$[\text{C}] = \frac{k_2[\text{A}]_0}{k_1 + k_2}\left[1 - e^{-(k_1+k_2)t}\right]$$

from which we see that $[\text{B}]/[\text{C}] = k_1/k_2$ for all times t. The plot below shows [A], [B], and [C] as functions of time.

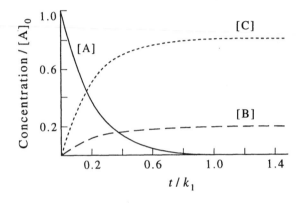

The following six problems deal with the decay of radioactive isotopes, which is a first-order process. Therefore, if $N(t)$ is the number of a radioactive isotope at time t, then $N(t) = N(0)e^{-kt}$. In dealing with radioactive decay, the half-life, $t_{1/2} = 0.693/k$, is almost exclusively used to describe the rate of decay (the kinetics of decay).

28–18. You order a sample of Na_3PO_4 containing the radioisotope phosphorus–32 ($t_{1/2} = 14.3$ days). If the shipment is delayed in transit for two weeks, how much of the original activity will remain when you receive the sample?

The rate constant for this reaction is

$$k = \frac{0.693}{t_{1/2}} = \frac{0.693}{14.3 \text{ days}} = 4.85 \times 10^{-2} \text{ days}^{-1}$$

Then

$$N(t) = N(0)e^{-kt}$$

$$\frac{N(t)}{N(0)} = e^{-(0.485 \times 10^{-2} \text{ days}^{-1})(14 \text{ days})} = 0.507$$

50.7% of the original activity will remain when you receive the sample.

28–19. Copper–64 ($t_{1/2} = 12.8$ h) is used in brain scans for tumors and in studies of Wilson's disease (a genetic disorder characterized by the inability to metabolize copper). Calculate the number of days required for an administered dose of copper–64 to drop to 0.10% of the initial value injected. Assume no loss of copper–64 except by radioactive decay.

To find the time required, we solve the equation

$$\frac{N(t)}{N(0)} = 0.0010 = e^{-0.693t/12.8 \text{ h}}$$

to find $t = 128$ hour $= 5.32$ days.

28–20. Sulfur–38 can be incorporated into proteins to follow certain aspects of protein metabolism. If a protein sample initially has an activity of $10\,000$ disintegrations·min^{-1}, calculate the activity 6.00 h later. The half-life of sulfur–38 is 2.84 h. *Hint*: Use the fact that the rate of decay is proportional to $N(t)$ for a first-order process.

Because the rate of decay is proportional to $N(t)$, we can write

$$\frac{N(6.00 \text{ h})}{N(0 \text{ h})} = \frac{\text{rate of decay } (6.00 \text{ h})}{\text{rate of decay } (0 \text{ h})} = e^{-(0.693)(6.00 \text{ h})/2.84 \text{ h}}$$

$$\text{decay rate } (6.00 \text{ h}) = (10\,000 \text{ disintintegrations·min}^{-1})(0.231)$$

$$= 2310 \text{ disintegrations·min}^{-1}$$

28–21. The radioisotope phosphorus–32 can be incorporated into nucleic acids to follow certain aspects of their metabolism. If a nucleic acid sample initially has an activity of $40\,000$ disintegrations ·· min^{-1}, calculate the activity 220 h later. The half-life of phosphorus–32 is 14.28 d. *Hint*: Use the fact that the rate of decay is proportional to $N(t)$ for a first-order process.

As in Problem 28–20, we can write

$$\frac{\text{rate of decay } (220 \text{ h})}{\text{rate of decay } (0 \text{ h})} = e^{-(0.693)(220 \text{ h})/(342.72 \text{ h})}$$

$$\text{decay rate } (220 \text{ h}) = (40\,000 \text{ disintegrations·min}^{-1})(0.64)$$

$$= 26\,000 \text{ disintegrations·min}^{-1}$$

28–22. Uranium–238 decays to lead–206 with a half-life of 4.51×10^9 y. A sample of ocean sediment is found to contain 1.50 mg of uranium–238 and 0.460 mg of lead-206. Estimate the age of the sediment assuming that lead–206 is formed only by the decay of uranium and that lead–206 does not itself decay.

The amount of U–238 which must have decayed to form 0.460 mg of Pb–206 is

$$(0.460\text{mg Pb}) \left(\frac{1 \text{ mol Pb}}{206 \text{ g Pb}} \right) \left(\frac{1 \text{ mol U}}{1 \text{ mol Pb}} \right) \left(\frac{238 \text{ g U}}{1 \text{ mol U}} \right) = 0.531 \text{ mg U}$$

At $t = 0$, therefore, there are $1.50 \text{ mg} + 0.531 \text{ mg} = 2.03 \text{ mg}$ of uranium in the sediment. Then

$$\ln \frac{1.50 \text{ mg}}{2.03 \text{ mg}} = -\frac{\ln 2}{4.51 \times 10^9 \text{ y}} t$$

$$1.97 \times 10^9 \text{ y} = t$$

28–23. Potassium-argon dating is used in geology and archeology to date sedimentary rocks. Potassium–40 decays by two different paths

$$^{40}_{19}\text{K} \longrightarrow {}^{40}_{20}\text{Ca} + {}^{0}_{-1}e \quad (89.3\%)$$

$$^{40}_{19}\text{K} \longrightarrow {}^{40}_{18}\text{Ar} + {}^{0}_{1}e \quad (10.7\%)$$

The overall half-life for the decay of potassium–40 is 1.3×10^9 y. Estimate the age of sedimentary rocks with an argon–40 to potassium–40 ratio of 0.0102. (See Problem 28–17.)

From Problem 28–17, we have the expressions

$$[\text{K}] = [\text{K}]_0 e^{-(k_1 + k_2)t} \quad \text{and} \quad t_{1/2} = \frac{\ln 2}{k_1 + k_2}$$

We can substitute the overall half-life of potassium–40 into the second equation to find $k_1 + k_2$:

$$k_1 + k_2 = \frac{\ln 2}{t_{1/2}} = 5.33 \times 10^{-10} \text{ y}^{-1}$$

For every mole of potassium which decays, 0.107 moles of argon are formed. Assuming that the only source of argon is potassium decay, we can find the ratio of potassium present at time t to potassium orginally present:

$$\left(\frac{0.0102 \text{ mol Ar at } t}{1 \text{ mol K at } t} \right) \left(\frac{1 \text{ mol K decayed}}{0.107 \text{ mol Ar at } t} \right) = \frac{0.0953 \text{ mol K decayed}}{1 \text{ mol K at } t}$$

There were 1.0953 mol of potassium originally present for every mole present at time t. Now substitute this value into Equation 1:

$$[\text{K}] = [\text{K}]_0 e^{-(5.33 \times 10^{-10} \text{ y}^{-1})t}$$

$$\ln \frac{1}{1.0953} = -(5.33 \times 10^{-10} \text{ y}^{-1})t$$

$$1.71 \times 10^8 \text{ y} = t$$

28–24. In this problem, we will derive Equation 28.32 from the rate law (Equation 28.31)

$$-\frac{d[\text{A}]}{dt} = k[\text{A}][\text{B}] \tag{1}$$

Use the reaction stoichiometry of Equation 28.30 to show that $[\text{B}] = [\text{B}]_0 - [\text{A}]_0 + [\text{A}]$. Use this result to show that Equation 1 can be written as

$$-\frac{d[\text{A}]}{dt} = k[\text{A}]\{[\text{B}]_0 - [\text{A}]_0 + [\text{A}]\}$$

Now separate the variables and then integrate the resulting equation subject to its initial conditions to obtain the desired result, Equation 28.32:

$$kt = \frac{1}{[A]_0 - [B]_0} \ln \frac{[A][B]_0}{[B][A]_0}$$

The amount of [B] present is equal to the amount of B reacted subtracted from $[B]_0$. Because one mole of A reacts with one mole of B, we can express the amount of B reacted as $[A]_0 - [A]$, so $[B] = [B]_0 - ([A]_0 - [A]) = [B]_0 - [A]_0 + [A]$. Substituting into Equation 1 gives

$$-\frac{d[A]}{dt} = k[A]\{[B]_0 - [A]_0 + [A]\}$$

$$-\frac{d[A]}{[A]\{[B]_0 - [A]_0 + [A]\}} = k\,dt$$

$$-\frac{1}{[B]_0 - [A]_0} \ln \frac{[A]}{[B]_0 - [A]_0 + [A]} \bigg|_{[A]_0}^{[A]} = kt$$

$$\frac{1}{[A]_0 - [B]_0} \ln \frac{[A][B]_0}{[A]_0[B]} = kt$$

28–25. Equation 28.32 is indeterminate if $[A]_0 = [B]_0$. Use L'Hopital's rule to show that Equation 28.32 reduces to Equation 28.33 when $[A]_0 = [B]_0$. (*Hint*: Let $[A] = [B] + x$ and $[A]_0 = [B]_0 + x$.).

We begin with Equation 28.32,

$$kt = \frac{1}{[A]_0 - [B]_0} \ln \frac{[A][B]_0}{[B][A]_0}$$

Let $[A] = [B] + x$ and $[A]_0 = [B]_0 + x$. Then we can write Equation 28.32 as

$$kt = \frac{1}{[B]_0 + x - [B]_0} \ln \frac{([B] + x)[B]_0}{[B]([B]_0 + x)} = \frac{1}{x} \ln \frac{([B] + x)[B]_0}{[B]([B]_0 + x)}$$

Now, as $x \to 0$,

$$\lim_{x \to 0} kt = \lim_{x \to 0} \frac{1}{x} \ln \frac{([B] + x)[B]_0}{[B]([B]_0 + x)}$$

$$= \lim_{x \to 0} \frac{\ln([B] + x) - \ln[B] + \ln[B]_0 - \ln([B]_0 + x)}{x}$$

This fraction is indeterminate, so we use L'Hopital's rule to write

$$\lim_{x \to 0} kt = \lim_{x \to 0} \left(\frac{1}{[B] + x} - \frac{1}{[B]_0 + x} \right)$$

$$= \frac{1}{[B]} - \frac{1}{[B]_0}$$

28–26. Uranyl nitrate decomposes according to

$$UO_2(NO_3)_2(aq) \longrightarrow UO_3(s) + 2\,NO_2(g) + \frac{1}{2}O_2(g)$$

The rate law for this reaction is first order in the concentration of uranyl nitrate. The following data were recorded for the reaction at 25.0°C.

t/min	0	20.0	60.0	180.0	360.0
$[\text{UO}_2(\text{NO}_3)_2]/\text{mol·dm}^{-3}$	0.01413	0.01096	0.00758	0.00302	0.00055

Calculate the rate constant for this reaction at 25.0°C.

Because the reaction is first-order, a plot of $\ln[\text{UO}_2(\text{NO}_3)_2]$ versus time should give a straight line of slope $-k$.

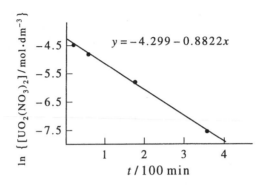

The slope of the best-fit line is $-8.82 \times 10^{-3}\ \text{min}^{-1}$, so $k = 8.82 \times 10^{-3}\ \text{min}^{-1}$.

28–27. The data for the decomposition of uranyl nitrate (Problem 28–26) at 350°C are tabulated below

t/min	0	6.0	10.0	17.0	30.0	60.0
$[\text{UO}_2(\text{NO}_3)_2]/\text{mol·dm}^{-3}$	0.03802	0.02951	0.02089	0.01259	0.00631	0.00191

Calculate the rate constant for this reaction at 350°C.

Again, because the reaction is first-order, a plot of $\ln[\text{UO}_2(\text{NO}_3)_2]$ versus time should give a straight line of slope $-k$.

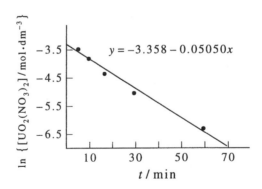

The slope of the best fit line is $-0.0505\ \text{min}^{-1}$, so $k = 5.05 \times 10^{-2}\ \text{min}^{-1}$.

28–28. The following data are obtained for the reaction

$$2\,\text{N}_2\text{O}(g) \longrightarrow 2\,\text{N}_2(g) + \text{O}_2(g)$$

at 900 K.

t/s	0	3146	6494	13 933
$[N_2O]/\text{mol}\cdot\text{dm}^{-3}$	0.521	0.416	0.343	0.246

The rate law for this reaction is second order in N_2O concentration. Calculate the rate constant for this decomposition reaction.

Because the reaction is second-order in N_2O, a plot of $1/[N_2O]$ versus time should give a straight line of slope k.

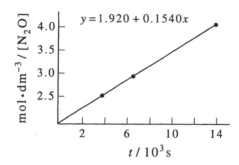

The slope of the best-fit line is 1.54×10^{-4} $\text{mol}\cdot\text{dm}^{-3}\cdot\text{s}^{-1}$, so this is the value of k for the decomposition reaction.

28–29. Consider a chemical reaction

$$A \longrightarrow \text{products}$$

that obeys the rate law

$$-\frac{d[A]}{dt} = k[A]^n$$

where n, the reaction order, can be any number except $n = 1$. Separate the concentration and time variables and then integrate the resulting expression assuming the concentration of A is $[A]_0$ at time $t = 0$ and is $[A]$ at time t to show that

$$kt = \frac{1}{n-1}\left(\frac{1}{[A]^{n-1}} - \frac{1}{[A]_0^{n-1}}\right) \qquad n \neq 1 \qquad (1)$$

Use Equation 1 to show that the half-life of a reaction of order n is

$$kt_{1/2} = \frac{1}{n-1}\frac{2^{n-1}-1}{[A]_0^{n-1}} \qquad n \neq 1 \qquad (2)$$

Show that this result reduces to Equation 28.29 when $n = 2$.

Separating the variables in the rate law and integrating gives

$$-\frac{d[A]}{[A]^n} = kdt$$

$$\frac{1}{n-1}\frac{1}{[A]^{n-1}}\bigg|_{[A]_0}^{[A]} = kt$$

$$\frac{1}{n-1}\left(\frac{1}{[A]^{n-1}} - \frac{1}{[A]_0^{n-1}}\right) = kt \qquad (1)$$

where $n \neq 1$ (if $n = 1$, the integration is incorrect). At $t = t_{1/2}$, $[A] = [A]_0/2$, so Equation 1 becomes

$$kt_{1/2} = \frac{1}{n-1}\left[\frac{1}{(\frac{1}{2}[A])^{n-1}} - \frac{1}{[A]_0^{n-1}}\right]$$

$$= \frac{1}{n-1}\frac{2^{n-1}-1}{[A]_0^{n-1}} \tag{2}$$

where, again, $n \neq 1$. When $n = 2$, this result becomes

$$kt_{1/2} = \frac{1}{[A]_0} \tag{28.29}$$

28–30. Show that Equation 1 of Problem 28–29 can be written in the form

$$\frac{\left(\dfrac{[A]_0}{[A]}\right)^x - 1}{x} = k[A]_0^x t$$

where $x = n - 1$. Now use L'Hopital's rule to show that

$$\ln \frac{[A]}{[A]_0} = -kt$$

for $n = 1$. (Remember that $da^x/dx = a^x \ln a$.)

Starting with Equation 1 of Problem 28–29, we can write

$$kt = \frac{1}{n-1}\left(\frac{1}{[A]^{n-1}} - \frac{1}{[A]_0^{n-1}}\right)$$

$$(n-1)kt = \frac{[A]_0^{n-1} - [A]^{n-1}}{[A]^{n-1}[A]_0^{n-1}}$$

$$(n-1)k[A]_0^{n-1}t = \frac{[A]_0^{n-1}}{[A]^{n-1}} - 1$$

If we let $x = n - 1$, then we have the desired result. We can now use L'Hopital's rule to find the value of kt as $x \to 0$ ($n \to 1$):

$$\lim_{x \to 0} kt = \lim_{x \to 0} \frac{\left(\dfrac{1}{[A]}\right)^x - \left(\dfrac{1}{[A]_0}\right)^x}{x}$$

$$\lim_{x \to 0} kt = \lim_{x \to 0}\left[\left(\frac{1}{[A]}\right)^x \ln \frac{1}{[A]} - \left(\frac{1}{[A]_0}\right)^x \ln \frac{1}{[A]_0}\right] = \ln \frac{[A]_0}{[A]}$$

$$-kt = \ln \frac{[A]}{[A]_0}$$

as $n \to 1$.

28–31. The following data were obtained for the reaction

$$N_2O(g) \longrightarrow N_2(g) + \frac{1}{2}O_2(g)$$

$[N_2O]_0/mol \cdot dm^{-3}$	1.674×10^{-3}	4.458×10^{-3}	9.300×10^{-3}	1.155×10^{-2}
$t_{1/2}/s$	1200	470	230	190

Assume the rate law for this reaction is

$$-\frac{d[N_2O]}{dt} = k[N_2O]^n$$

and use Equation 2 of Problem 28–29 to determine the reaction order of N_2O by plotting $\ln t_{1/2}$ against $\ln[A]_0$. Calculate the rate constant for this decomposition reaction.

We can take the natural logarithm of both sides of Equation 2 of Problem 28–29 to write the equation as

$$\ln k + \ln t_{1/2} = (1 - n)\ln[A]_0 + \ln\left[\frac{1}{n-1}\left(2^{n-1} - 1\right)\right]$$

The slope of a plot of $\ln t_{1/2}$ versus $\ln[A]_0$ will be $(1 - n)$. The data provided are plotted in this form in the following graph.

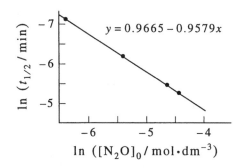

The slope of the plot of $\ln t_{1/2}$ versus $\ln[N_2O]_0$ is -1, so $n = 2$ and the reaction is second-order in $[N_2O]$. We use Equation 28.29 and the data provided to find the value of the rate constant:

$[N_2O]_0/mol \cdot dm^{-3}$	1.674×10^{-3}	4.458×10^{-3}	9.300×10^{-3}	1.155×10^{-2}
$k/dm^3 \cdot mol^{-1} \cdot s^{-1}$	0.50	0.48	0.47	0.46

The average value of the rate constant is $k = 0.47\ dm^3 \cdot mol^{-1} \cdot s^{-1}$.

28–32. We will derive Equation 28.39 from Equation 28.38 in this problem. Rearrange Equation 28.38 to become

$$\frac{d[A]}{(k_1 + k_{-1})[A] - k_{-1}[A]_0} = -dt$$

and integrate to obtain

$$\ln\{(k_1 + k_{-1})[A] - k_{-1}[A]_0\} = -(k_1 + k_{-1})t + \text{constant}$$

or

$$(k_1 + k_{-1})[A] - k_{-1}[A]_0 = ce^{-(k_1+k_{-1})t}$$

where c is a constant. Show that $c = k_1[A]_0$ and that

$$(k_1 + k_{-1})[A] - k_{-1}[A]_0 = k_1[A]_0 e^{-(k_1+k_{-1})t} \tag{1}$$

Now let $t \to \infty$ and show that

$$[A]_0 = \frac{(k_1 + k_{-1})[A]_{eq}}{k_{-1}}$$

and

$$[A]_0 - [A]_{eq} = \frac{k_1[A]_{eq}}{k_{-1}} = \frac{k_1[A]_0}{k_1 + k_{-1}}$$

Use these results in Equation 1 to obtain Equation 28.39.

We begin by separating the variables in Equation 28.38:

$$-\frac{d[A]}{dt} = (k_1 + k_{-1})[A] - k_{-1}[A]_0$$

$$\frac{d[A]}{(k_1 + k_{-1})[A] - k_{-1}[A]_0} = -dt$$

We then integrate to find

$$\int \frac{d[A]}{(k_1 + k_{-1})[A] - k_{-1}[A]_0} = -\int dt$$

$$\frac{\ln\{(k_1 + k_{-1})[A] - k_{-1}[A]_0\}}{k_1 + k_{-1}} = -t + \text{constant}$$

$$\ln\{(k_1 + k_{-1})[A] - k_{-1}[A]_0\} = -(k_1 + k_{-1})t + \text{constant}$$

$$(k_1 + k_{-1})[A] - k_{-1}[A]_0 = ce^{-(k_1+k_{-1})t} \qquad (a)$$

where c is a constant. At $t = 0$, $[A] = [A]_0$, so Equation a becomes

$$(k_1 + k_{-1} - k_{-1})[A]_0 = c$$

$$k_1[A]_0 = c$$

Substituting this result back into Equation a gives Equation 1:

$$(k_1 + k_{-1})[A] - k_{-1}[A]_0 = k_1[A]_0 e^{-(k_1+k_{-1})t} \qquad (1)$$

Now let $t \to \infty$, in which case $[A] \to [A]_{eq}$. The exponential term in Equation 1 then goes to zero, and so

$$(k_1 + k_{-1})[A]_{eq} - k_{-1}[A]_0 = 0$$

$$[A]_0 = \frac{(k_1 + k_{-1})[A]_{eq}}{k_{-1}}$$

Subtracting $[A]_{eq}$ from both sides gives

$$[A]_0 - [A]_{eq} = \frac{(k_1 + k_{-1} - k_{-1})[A]_{eq}}{k_{-1}} = \frac{k_1[A]_{eq}}{k_{-1}} = \frac{k_1[A]_0}{k_1 + k_{-1}}$$

Substituting into Equation 1 gives

$$(k_1 + k_{-1})[A] - k_{-1}[A]_0 = k_1[A]_0 e^{-(k_1+k_{-1})t}$$

$$[A] = \frac{k_{-1}[A]_0}{k_1 + k_{-1}} + \frac{k_1[A]_0}{k_1 + k_{-1}} e^{-(k_1+k_{-1})t}$$

$$[A] = [A]_{eq} + ([A]_0 - [A]_{eq})e^{-(k_1+k_{-1})t}$$

which is Equation 28.39.

28–33. Consider the general chemical reaction

$$A + B \underset{k_{-1}}{\overset{k_1}{\rightleftharpoons}} P$$

If we assume that both the forward and reverse reactions are first order in their respective reactants, the rate law is given by (Equation 28.52)

$$\frac{d[P]}{dt} = k_1[A][B] - k_{-1}[P] \tag{1}$$

Now consider the response of this chemical reaction to a temperature jump. Let $[A] = [A]_{2,eq} + \Delta[A]$, $[B] = [B]_{2,eq} + \Delta[B]$, and $[P] = [P]_{2,eq} + \Delta[P]$, where the subscript "2,eq" refers to the new equilibrium state. Now use the fact that $\Delta[A] = \Delta[B] = -\Delta[P]$ to show that Equation 1 becomes

$$\frac{d\Delta[P]}{dt} = k_1[A]_{2,eq}[B]_{2,eq} - k_{-1}[P]_{2,eq}$$
$$- \{k_1([A]_{2,eq} + [B]_{2,eq}) + k_{-1}\}\Delta[P] + O(\Delta[P]^2)$$

Show that the first terms on the right side of this equation cancel and that Equations 28.53 and 28.54 result.

After the temperature jump to the new equilibrium state, we can write Equation 1 in terms of the new equilibrium state and the changes in concentration:

$$\frac{d([P]_{2,eq} + \Delta[P])}{dt} = k_1([A]_{2,eq} + \Delta[A])([B]_{2,eq} + \Delta[B]) - k_{-1}([P]_{2,eq} + \Delta[P])$$

Because $[P]_{2,eq}$ does not vary with time and $\Delta[A] = \Delta[B] = -\Delta[P]$, we can simplify the above expression to

$$\frac{d\Delta[P]}{dt} = k_1([A]_{2,eq} - \Delta[P])([B]_{2,eq} - \Delta[P]) - k_{-1}([P]_{2,eq} + \Delta[P])$$

$$\frac{d\Delta[P]}{dt} = k_1[A]_{2,eq}[B]_{2,eq} - k_{-1}[P]_{2,eq} - \{k_1([A]_{2,eq} + [B]_{2,eq}) + k_{-1}\}\Delta[P] + O\{(\Delta[P])^2\}$$

At equilibrium, $d[P]/dt = 0$, so $k_{-1}[P]_{2,eq} = k_1[A]_{2,eq}[B]_{2,eq}$ and the first terms on the right hand side of the equation cancel. Then (disregarding second-order and higher terms in $\Delta[P]$)

$$\frac{d\Delta[P]}{dt} = -\{k_1([A]_{2,eq} + [B]_{2,eq}) + k_{-1}\}\Delta[P]_{2,eq}$$

$$\frac{d\Delta[P]}{\Delta[P]} = -dt\{k_1([A]_{2,eq} + [B]_{2,eq}) + k_{-1}\}$$

$$\ln\frac{\Delta[P]}{\Delta[P]_0} = -t\{k_1([A]_{2,eq} + [B]_{2,eq}) + k_{-1}\}$$

$$\Delta[P] = \Delta[P]_0 e^{-t/\tau}$$

where

$$\tau = \frac{1}{k_1([A]_{2,eq} + [B]_{2,eq}) + k_{-1}}$$

28–34. The equilibrium constant for the reaction

$$H^+(aq) + OH^-(aq) \underset{k_{-1}}{\overset{k_1}{\rightleftharpoons}} H_2O(l)$$

at 25°C is $K_c = [H_2O]/[H^+][OH^-] = 5.49 \times 10^{15}$ mol$^{-1} \cdot$dm^3. The time-dependent conductivity of the solution following a temperature jump to a final temperature of 25°C shows a relaxation time of $\tau = 3.7 \times 10^{-5}$ s. Determine the values of the rate constants k_1 and k_{-1}. At 25°C, the density of water is $\rho = 0.997$ g$\cdot$cm^{-3}.

The concentration of water at 25°C is

$$\frac{0.997 \text{ g} \cdot \text{cm}^{-3}}{18.015 \text{ g} \cdot \text{mol}^{-1}} \left(\frac{10 \text{ cm}}{1 \text{ dm}}\right)^3 = 55.3 \text{ mol} \cdot \text{dm}^{-3}$$

Let x be the number of moles of water that dissociate per liter. Then, at equilibrium, $[H_2O] = 55.3$ mol$\cdot$dm$^{-3} - x$, and $[H^+] = [OH^-] = x$. We can use the K_c given in the problem and these values to write

$$K_c = \frac{[H_2O]}{[H^+][OH^-]} = \frac{55.3 \text{ mol} \cdot \text{dm}^{-3} - x}{x^2}$$

$$5.49 \times 10^{15} \text{ mol}^{-1} \cdot \text{dm}^3 = \frac{55.3 \text{ mol} \cdot \text{dm}^{-3} - x}{x^2}$$

$$x = 1.00 \times 10^{-7} \text{ dm}^{-3} \cdot \text{mol}$$

where we have taken the positive root as x. Then $[OH^-] = [H^+] = 1.00 \times 10^{-7}$ mol$\cdot$dm^{-3}. We also note that $K_c = k_1/k_{-1}$, so

$$k_1 = 5.49 \times 10^{15} \text{ mol}^{-1} \cdot \text{dm}^3 k_{-1} \tag{1}$$

We can use Equation 28.54 to write

$$k_{-1} + k_1([H^+] + [OH^-]) = \frac{1}{\tau}$$

Substituting for $[H^+]$, $[OH^-]$, and k_1, we find

$$k_{-1} = \frac{1}{(3.7 \times 10^{-5} \text{ s})[1 + (5.49 \times 10^{15} \text{ mol}^{-1} \cdot \text{dm}^3)(2.00 \times 10^{-7} \text{ mol} \cdot \text{dm}^{-3})]}$$

$$= 2.5 \times 10^{-5} \text{ s}^{-1}$$

and (substituting back into Equation 1) $k_1 = 1.4 \times 10^{11}$ mol$^{-1} \cdot$dm$^3 \cdot$s^{-1}.

28–35. The equilibrium constant for the reaction

$$D^+(aq) + OD^-(aq) \underset{k_{-1}}{\overset{k_1}{\rightleftharpoons}} D_2O(l)$$

at 25°C is $K_c = 4.08 \times 10^{16}$ mol$^{-1} \cdot$dm^3. The rate constant k_{-1} is independently found to be 2.52×10^{-6} s^{-1}. What do you predict for the observed relaxation time for a temperature-jump experiment to a final temperature of 25°C? The density of D_2O is $\rho = 1.104$ g$\cdot$cm^{-3} at 25°C.

The concentration of D_2O at 25°C is

$$\frac{1.104 \text{ g} \cdot \text{cm}^{-3}}{20.027 \text{ g} \cdot \text{mol}^{-1}} \left(\frac{10 \text{ cm}}{1 \text{ dm}}\right)^3 = 55.13 \text{ mol} \cdot \text{dm}^{-3}$$

Let x be the number of moles of D_2O that dissociate per liter. Then, at equilibrium, $[D_2O] = 55.13$ mol·dm$^{-3} - x$, and $[D^+] = [OD^-] = x$. We can use the K_c given in the problem and these values to write

$$K_c = \frac{[H_2O]}{[H^+][OH^-]} = \frac{55.13 \text{ mol·dm}^{-3} - x}{x^2}$$

$$4.08 \times 10^{16} \text{ mol}^{-1} \cdot \text{dm}^3 = \frac{55.13 \text{ mol·dm}^{-3} - x}{x^2}$$

$$x = 3.68 \times 10^{-8} \text{ dm}^{-3} \cdot \text{mol}$$

where we have taken the positive root as x. Then $[OD^-] = [D^+] = 3.68 \times 10^{-8}$ mol·dm^{-3}. We can also write K_c as k_1/k_{-1}, so (using the given value of k_{-1})

$$k_1 = k_{-1}K_c = (2.52 \times 10^{-6} \text{ s}^{-1})(4.08 \times 10^{16} \text{ mol}^{-1} \cdot \text{dm}^3) = 1.03 \times 10^{11} \text{ mol}^{-1} \cdot \text{dm}^3 \cdot \text{s}^{-1}$$

We can now use Equation 28.54 to write

$$\tau = \frac{1}{k_{-1} + k_1([D^+] + [OD^-])}$$

$$= \frac{1}{2.52 \times 10^{-6} \text{ s}^{-1} + (1.03 \times 10^{11} \text{ mol}^{-1} \cdot \text{dm}^3 \cdot \text{s}^{-1})(7.36 \times 10^{-8} \text{ mol·dm}^{-3})}$$

$$= 1.32 \times 10^{-4} \text{ s}$$

28–36. Consider the chemical reaction described by

$$2\,A(aq) \underset{k_{-1}}{\overset{k_1}{\rightleftharpoons}} D(aq)$$

If we assume the forward reaction is second order and the reverse reaction is first order, the rate law is given by

$$\frac{d[D]}{dt} = k_1[A]^2 - k_{-1}[D] \tag{1}$$

Now consider the response of this chemical reaction to a temperature jump. Let $[A] = [A]_{2,eq} + \Delta[A]$ and $[D] = [D]_{2,eq} + \Delta[D]$, where the subscript "2,eq" refers to the new equilibrium state. Now use the fact that $\Delta[A] = -2\Delta[D]$ to show that Equation 1 becomes

$$\frac{d\Delta[D]}{dt} = -(4k_1[A]_{2,eq} + k_{-1})\Delta[D] + O(\Delta[D]^2)$$

Show that if we ignore the $O(\Delta[D]^2)$ term, then

$$\Delta[D] = \Delta[D]_0 e^{-t/\tau}$$

where $\tau = 1/(4k_1[A]_{2,eq} + k_{-1})$.

We let $[A] = [A]_{2,eq} + \Delta[A]$ and $[D] = [D]_{2,eq} + \Delta[D]$ in Equation 1:

$$\frac{d[D]}{dt} = k_1[A]^2 - k_{-1}[D]$$

$$\frac{d([D]_{2,eq} + \Delta[D])}{dt} = k_1([A]_{2,eq} + \Delta[A])^2 - k_{-1}([D]_{2,eq} + \Delta[D])$$

Now use the relation $\Delta[A] = -2\Delta[D]$ to write

$$\frac{d\Delta[D]}{dt} = k_1([A]_{2,eq} - 2\Delta[D])^2 - k_{-1}([D]_{2,eq} + \Delta[D])$$

$$= k_1[A]_{2,eq}^2 - 4k_1[A]_{2,eq}\Delta[D] - k_{-1}[D]_{2,eq} - k_{-1}\Delta[D] + O\{(\Delta[D])^2\}$$

At equilibrium, $d[D]/dt = 0$, so (from Equation 1) $k_{-1}[D]_{2,eq} = k_1[A]_{2,eq}^2$. Then

$$\frac{d\Delta[D]}{dt} = -(4k_1[A]_{2,eq} + k_{-1})\Delta[D] + O\{(\Delta[D])^2\}$$

Ignoring second-order and higher terms in $\Delta[D]$ and integrating gives

$$\frac{d\Delta[D]}{dt} = -(4k_1[A]_{2,eq} + k_{-1})\Delta[D]$$

$$\frac{d\Delta[D]}{\Delta[D]} = -dt(4k_1[A]_{2,eq} + k_{-1})$$

$$\ln\frac{\Delta[D]}{\Delta[D]_0} = t(4k_1[A]_{2,eq} + k_{-1})$$

$$\Delta[D] = \Delta[D]_0 e^{-t/\tau}$$

where $\tau = 1/(4k_1[A]_{2,eq} + k_{-1})$.

28–37. In Problem 28–36, you showed that the relaxation time for the dimerization reaction

$$2\,A(aq) \underset{k_{-1}}{\overset{k_1}{\rightleftharpoons}} D(aq)$$

is given by $\tau = 1/(4k_1[A]_{2,eq} + k_{-1})$. Show that this equation can be rewritten as

$$\frac{1}{\tau^2} = k_{-1}^2 + 8k_1 k_{-1}[S]_0$$

where $[S]_0 = 2[D] + [A] = 2[D]_{2,eq} + [A]_{2,eq}$.

We write the equilibrium constant in terms of the rate constants and the concentrations of D and A:

$$K = \frac{[D]_{2,eq}}{[A]_{2,eq}^2} = \frac{k_1}{k_{-1}}$$

$$[D]_{2,eq} = \frac{[A]_{2,eq}^2 k_1}{k_{-1}} \tag{1}$$

Because $[S]_0$ is a constant, $[S]_0 = [A]_{2,eq} + 2[D]_{2,eq}$. Substituting for $[D]_{2,eq}$ from Equation 1 gives

$$[S]_0 = [A]_{2,eq} + \frac{2[A]_{2,eq}^2 k_1}{k_{-1}}$$

$$k_{-1}[S]_0 = 2k_1[A]_{2,eq}^2 + k_{-1}[A]_{2,eq} \tag{2}$$

From Problem 28–36, we can write $1/\tau^2$ as

$$\frac{1}{\tau^2} = (4k_1[A]_{2,eq} + k_{-1})^2$$

$$= 16k_1^2[A]_{2,eq}^2 + 8k_1 k_{-1}[A]_{2,eq} + k_{-1}^2$$

$$= 8k_1\left(2k_1[A]_{2,eq}^2 + k_{-1}[A]_{2,eq}\right) + k_{-1}^2$$

Substituting from Equation 2, we obtain

$$\frac{1}{\tau^2} = 8k_1 k_{-1}[S]_0 + k_{-1}^2$$

28-38. The first step in the assembly of the protein yeast phosphoglycerate mutase is a reversible dimerization of a polypeptide,

$$2\,A(aq) \underset{k_{-1}}{\overset{k_1}{\rightleftharpoons}} D(aq)$$

where A is the polypeptide and D is the dimer. Suppose that a 1.43×10^{-5} mol·dm^{-3} solution of A is prepared and allowed to come to equilibrium at 280 K. Once equilibrium is achieved, the temperature of the solution is jumped to 293 K. The rate constants k_1 and k_{-1} for the dimerization reaction at 293 K are 6.25×10^3 dm^3·mol^{-1}·s^{-1} and 6.00×10^{-3} s^{-1}, respectively. Calculate the value of the relaxation time observed in the experiment. (*Hint*: See Problem 28-37.)

From Problem 28-37, we have

$$\frac{1}{\tau^2} = 8k_1 k_{-1}[S]_0 + k_{-1}^2$$

We are given $[A]_0 = 1.43 \times 10^{-5}$ mol·dm^{-3}. Because $[D]_0 = 0$, $[A]_0 = [A]_{2,eq} + 2[D]_{2,eq}$, so $[S]_0 = 1.43 \times 10^{-5}$ mol·dm^{-3}. Then

$$\frac{1}{\tau^2} = 8(6.25 \times 10^3 \text{ dm}^3 \cdot \text{mol}^{-1} \cdot \text{s}^{-1})(6.00 \times 10^{-3} \text{ s}^{-1})(1.43 \times 10^{-5} \text{ mol} \cdot \text{dm}^{-3}) + (6.00 \times 10^{-3} \text{ s}^{-1})^2$$

$$\frac{1}{\tau^2} = 4.33 \times 10^{-3} \text{ s}^{-2}$$

$$\tau = 15.2 \text{ s}$$

28-39. Does the Arrhenius A factor always have the same units as the reaction rate constant?

Yes. The Arrhenius equation is

$$k = A e^{-E_a/RT} \tag{28.57}$$

Because the exponential term is unitless, k and A have the same units.

28-40. Use the results of Problems 28-26 and 28-27 to calculate the values of E_a and A for the decomposition of uranyl nitrate.

We can use the expression from Example 28-8,

$$E_a = R\left(\frac{T_1 T_2}{T_1 - T_2}\right) \ln \frac{k(T_1)}{k(T_2)}$$

In Problems 28–26 and 28–27, we found that $k = 8.82 \times 10^{-3}$ min^{-1} at 298 K and that $k = 0.0505$ min^{-1} at 623 K for the decomposition of uranyl nitrate. Substituting into the above equation, we find

$$E_a = (8.315 \text{ J·K}^{-1}\text{·mol}^{-1}) \left[\frac{(298 \text{ K})(623 \text{ K})}{298 \text{ K} - 623 \text{ K}} \right] \ln \frac{8.82 \times 10^{-3} \text{ min}^{-1}}{0.05050 \text{ min}^{-1}} = 8290 \text{ J·mol}^{-1}$$

The value of A is given by Equation 28.57. Using the values at 298 K, we find that

$$A = ke^{E_a/RT} = (8.82 \times 10^{-3} \text{ min}^{-1}) \exp \left[\frac{8290 \text{ J·mol}^{-1}}{(8.315 \text{ J·mol}^{-1}\text{·K}^{-1})(298 \text{ K})} \right] = 0.250 \text{ min}^{-1}$$

28–41. The experimental rate constants for the reaction described by

$$\text{OH(g)} + \text{ClCH}_2\text{CH}_2\text{Cl(g)} \longrightarrow \text{H}_2\text{O(g)} + \text{ClCHCH}_2\text{Cl(g)}$$

at various temperatures are tabulated below.

T/K	292	296	321	333	343	363
$k/10^8$ dm^3·mol^{-1}·s^{-1}	1.24	1.32	1.81	2.08	2.29	2.75

Determine the values of the Arrhenius parameters A and E_a for this reaction.

We can write the Arrhenius equation as

$$\ln k = \ln A - \frac{E_a}{R} \frac{1}{T} \tag{28.56}$$

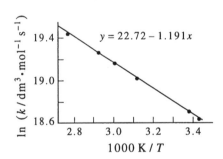

The best-fit line to a plot of $\ln k$ versus $1/T$ will have an intercept of $\ln A$ and a slope of $-E_a/R$. Here, the best fit of the experimental data gives $A = 7.37 \times 10^9$ dm^3·mol^{-1}·s^{-1} and $E_a = 9.90$ kJ·mol^{-1}.

28–42. The Arrhenius parameters for the reaction described by

$$\text{HO}_2\text{(g)} + \text{OH(g)} \longrightarrow \text{H}_2\text{O(g)} + \text{O}_2\text{(g)}$$

are $A = 5.01 \times 10^{10}$ dm^3·mol^{-1}·s^{-1} and $E_a = 4.18$ kJ·mol^{-1}. Determine the value of the rate constant for this reaction at 298 K.

We can use Equation 28.57:

$$k = Ae^{-E_a/RT}$$

$$= (5.01 \times 10^{10} \text{ dm}^3 \cdot \text{mol}^{-1} \cdot \text{s}^{-1}) \exp\left[-\frac{4180 \text{ J} \cdot \text{mol}^{-1}}{(8.315 \text{ J} \cdot \text{K}^{-1} \cdot \text{mol}^{-1})(298 \text{ K})}\right]$$

$$= 9.27 \times 10^9 \text{ dm}^3 \cdot \text{mol}^{-1} \cdot \text{s}^{-1}$$

28–43. At what temperature will the reaction described in Problem 28–42 have a rate constant that is twice that at 298 K?

Again, use Equation 28.57:

$$T = \frac{-E_a}{R} \frac{1}{\ln(k/A)}$$

$$= \frac{-4180 \text{ J} \cdot \text{mol}^{-1}}{8.3145 \text{ J} \cdot \text{mol}^{-1} \cdot \text{K}^{-1}} \frac{1}{\ln[2(9.27 \times 10^9)/5.01 \times 10^{10}]}$$

$$= 506 \text{ K}$$

28–44. The rate constants for the reaction

$$CHCl_2(g) + Cl_2(g) \longrightarrow CHCl_3(g) + Cl(g)$$

at different temperatures are tabulated below

T/K	357	400	458	524	533	615
$k/10^7 \text{ dm}^3 \cdot \text{mol}^{-1} \cdot \text{s}^{-1}$	1.72	2.53	3.82	5.20	5.61	7.65

Calculate the values of the Arrhenius parameters A and E_a for this reaction.

We can write the Arrhenius equation as

$$\ln k = \ln A - \frac{E_a}{RT} \tag{28.56}$$

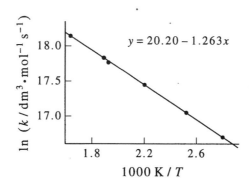

The best-fit line to a plot of $\ln k$ versus $1/T$ will have an intercept of $\ln A$ and a slope of $-E_a/R$. Here, the best fit of the experimental data gives $A = 5.93 \times 10^8 \text{ dm}^3 \cdot \text{mol}^{-1} \cdot \text{s}^{-1}$ and $E_a = 10.5 \text{ kJ} \cdot \text{mol}^{-1}$.

28–45. The rate constant for the chemical reaction

$$2\,N_2O_5(g) \longrightarrow 4\,NO_2(g) + O_2(g)$$

doubles from 22.50°C to 27.47°C. Determine the activation energy of the reaction. Assume the pre-exponential factor is independent of temperature.

We can use the expression for E_a derived in Example 28–8:

$$
\begin{aligned}
E_a &= R\left(\frac{T_1 T_2}{T_1 - T_2}\right)\ln\frac{k(T_1)}{k(T_2)}\\[4pt]
&= (8.3145\ \text{J}\cdot\text{mol}^{-1}\cdot\text{K}^{-1})\left[\frac{(295.65\ \text{K})(300.62\ \text{K})}{(295.65 - 300.62)\ \text{K}}\right]\ln\frac{1}{2}\\[4pt]
&= 103.1\ \text{kJ}\cdot\text{mol}^{-1}
\end{aligned}
$$

28–46. Show that if A reacts to form either B or C according to

$$A \xrightarrow{k_1} B \quad \text{or} \quad A \xrightarrow{k_2} C$$

then E_a, the observed activation energy for the disappearance of A, is given by

$$E_a = \frac{k_1 E_1 + k_2 E_2}{k_1 + k_2}$$

where E_1 is the activation energy for the first reaction and E_2 is the activation energy for the second reaction.

The Arrhenius equations for each reaction are (Equation 28.57)

$$k_1 = A_1 e^{-E_1/RT} \quad \text{and} \quad k_2 = A_2 e^{-E_2/RT}$$

The observed rate constant for the disappearance of A is going to be the sum of k_1 and k_2. Therefore, the Arrhenius equation for the net effect of both reactions can be written as (Equation 28.55)

$$\frac{d\ln(k_1 + k_2)}{dT} = \frac{E_a}{RT^2}$$

Substituting for k_1 and k_2 and taking the derivative gives

$$
\begin{aligned}
\frac{1}{k_1 + k_2}\frac{d}{dT}\left(A_1 e^{-E_1/RT} + A_2 e^{-E_2/RT}\right) &= \frac{E_a}{RT^2}\\[4pt]
\frac{1}{k_1 + k_2}\left(\frac{A_1 E_1}{RT^2}e^{-E_1/RT} + \frac{A_2 E_2}{RT^2}e^{-E_2/RT}\right) &= \frac{E_a}{RT^2}\\[4pt]
\frac{1}{k_1 + k_2}\left(A_1 E_1 e^{-E_1/RT} + A_2 E_2 e^{-E_2/RT}\right) &= E_a\\[4pt]
\frac{E_1 k_1 + E_2 k_2}{k_1 + k_2} &= E_a
\end{aligned}
$$

28–47. Cyclohexane interconverts between a "chair" and a "boat" structure. The activation parameters for the reaction from the chair to the boat form of the molecule are $\Delta^{\ddagger}H^{\circ} = 31.38\ \text{kJ}\cdot\text{mol}^{-1}$ and

$\Delta^{\ddagger}S^{\circ} = 16.74 \text{ J} \cdot \text{K}^{-1}$. Calculate the standard Gibbs energy of activation and the rate constant for this reaction at 325 K.

We can use Equation 28.73 to find $\Delta^{\ddagger}G^{\circ}$:

$$\Delta^{\ddagger}G^{\circ} = \Delta^{\ddagger}H^{\circ} - T\Delta^{\ddagger}S^{\circ}$$
$$= 31.38 \text{ kJ} \cdot \text{mol}^{-1} - (325 \text{ K})(16.74 \text{ J} \cdot \text{mol}^{-1})$$
$$= 25.94 \text{ kJ} \cdot \text{mol}^{-1}$$

The rate constant is given by Equation 28.72 ($c^{\circ} = 1$):

$$k = \frac{k_{\text{B}}T}{h}e^{-\Delta^{\ddagger}G^{\circ}/RT}$$
$$= (6.77 \times 10^{12} \text{ s}^{-1}) \exp\left[\frac{-25\,940 \text{ J} \cdot \text{mol}^{-1}}{(8.315 \text{ J} \cdot \text{K}^{-1} \cdot \text{mol}^{-1})(325 \text{ K})}\right]$$
$$= 4.59 \times 10^{8} \text{ s}^{-1}$$

28–48. The gas-phase rearrangement reaction

$$\text{vinyl allyl ether} \longrightarrow \text{allyl acetone}$$

has a rate constant of $6.015 \times 10^{-5} \text{ s}^{-1}$ at 420 K and a rate constant of $2.971 \times 10^{-3} \text{ s}^{-1}$ at 470 K. Calculate the values of the Arrhenius parameters A and E_a. Calculate the values of $\Delta^{\ddagger}H^{\circ}$ and $\Delta^{\ddagger}S^{\circ}$ at 420 K. (Assume ideal-gas behavior.)

We can use the expression derived in Example 28–8 to find E_a:

$$E_a = R\left(\frac{T_1 T_2}{T_1 - T_2}\right)\ln\frac{k(T_1)}{k(T_2)}$$
$$= (8.315 \text{ J} \cdot \text{K}^{-1} \cdot \text{mol}^{-1})\left[\frac{(420 \text{ K})(470 \text{ K})}{420 \text{ K} - 470 \text{ K}}\right]\ln\frac{6.015 \times 10^{-5} \text{ s}^{-1}}{2.971 \times 10^{-3} \text{ s}^{-1}}$$
$$= 128.0 \text{ kJ} \cdot \text{mol}^{-1}$$

Using this value in the Arrhenius equation with the rate constant at 420 K gives

$$6.015 \times 10^{-5} \text{ s}^{-1} = A \exp\left[-\frac{128\,000 \text{ J} \cdot \text{mol}^{-1}}{(8.315 \text{ J} \cdot \text{K}^{-1} \cdot \text{mol}^{-1})(420 \text{ K})}\right] \qquad A = 5.01 \times 10^{11} \text{ s}^{-1}$$

We can write (Equation 28.55)

$$\frac{d\ln k}{dt} = \frac{E_a}{RT^2}$$

and, for an ideal-gas system, we also have (Equation 28.76)

$$\frac{d\ln k}{dt} = \frac{1}{T} + \frac{\Delta^{\ddagger}U^{\circ}}{RT^2}$$

For a unimolecular reaction, $\Delta^{\ddagger}H^{\circ} = \Delta^{\ddagger}U^{\circ}$, and so equating these two expressions gives

$$\frac{E_a}{RT^2} = \frac{1}{T} + \frac{\Delta^{\ddagger}H^{\circ}}{RT^2}$$
$$E_a = RT + \Delta^{\ddagger}H^{\circ}$$

We can solve this equation for $\Delta^{\ddagger}H^{\circ}$ at 420 K:

$$\Delta^{\ddagger}H^{\circ} = 128.0\,\text{kJ}\cdot\text{mol}^{-1} - (8.315 \times 10^{-3}\,\text{kJ}\cdot\text{mol}^{-1})(420\,\text{K}) = 124.5\,\text{kJ}\cdot\text{mol}^{-1}$$

Because $E_a = \Delta^{\ddagger}H^{\circ} + RT$ for a reaction of type $A \rightarrow P$, Equation 28.80 becomes

$$A = \frac{ek_BT}{h}e^{\Delta^{\ddagger}S^{\circ}/R}$$

Solving for $\Delta^{\ddagger}S^{\circ}$ gives

$$\Delta^{\ddagger}S^{\circ} = R\ln\left[\frac{Ah}{ek_BT}\right]$$

$$= (8.315\,\text{J}\cdot\text{K}^{-1}\cdot\text{mol}^{-1})\ln\left[\frac{(4.98 \times 10^{11}\,\text{s}^{-1})(6.626 \times 10^{-34}\,\text{J}\cdot\text{s})}{e(1.381 \times 10^{-23}\,\text{J}\cdot\text{K}^{-1})(420\,\text{K})}\right]$$

$$= -32.1\,\text{J}\cdot\text{mol}^{-1}\cdot\text{K}^{-1}$$

28–49. The kinetics of a chemical reaction can be followed by a variety of experimental techniques, including optical spectroscopy, NMR spectroscopy, conductivity, resistivity, pressure changes, and volume changes. When using these techniques, we do not measure the concentration itself but we know that the observed signal is proportional to the concentration; the exact proportionality constant depends on the experimental technique and the species present in the chemical system. Consider the general reaction given by

$$\nu_A A + \nu_B B \longrightarrow \nu_Y Y + \nu_Z Z$$

where we assume that A is the limiting reagent so that $[A] \rightarrow 0$ as $t \rightarrow \infty$. Let p_i be the proportionality constant for the contribution of species i to S, the measured signal from the instrument. Explain why at any time t during the reaction, S is given by

$$S(t) = p_A[A] + p_B[B] + p_Y[Y] + p_Z[Z] \tag{1}$$

Show that the initial and final readings from the instrument are given by

$$S(0) = p_A[A]_0 + p_B[B]_0 + p_Y[Y]_0 + p_Z[Z]_0 \tag{2}$$

and

$$S(\infty) = p_B\left([B]_0 - \frac{\nu_B}{\nu_A}[A]_0\right) + p_Y\left([Y]_0 + \frac{\nu_Y}{\nu_A}[A]_0\right) + p_Z\left([Z]_0 + \frac{\nu_Z}{\nu_A}[A]_0\right) \tag{3}$$

Combine Equations 1 through 3 to show that

$$[A] = [A]_0\frac{S(t) - S(\infty)}{S(0) - S(\infty)}$$

(*Hint*: Be sure to express [B], [Y], and [Z] in terms of their initial values, [A] and $[A]_0$.)

Equation 1 holds because the total measured signal is the sum of the signals from each species. Because $p_i[i]$ is the amount of signal that species i contributes, we have Equation 1. Initially, therefore, we have

$$S(0) = p_A[A]_0 + p_B[B]_0 + p_Y[Y]_0 + p_Z[Z]_0 \tag{2}$$

As $t \to \infty$ (at the time of the final reading), $[A] \to 0$. From the stoichiometry of the reaction, we know that at $t = \infty$, $\nu_B [A]_0 / \nu_A$ moles of B have reacted, $\nu_Y [A]_0 / \nu_A$ moles of Y have been produced, and $\nu_Z [A]_0 / \nu_A$ moles of Z have been produced. We then have

$$S(\infty) = p_B \left([B]_0 - \frac{\nu_B}{\nu_A}[A]_0 \right) + p_Y \left([Y]_0 + \frac{\nu_Y}{\nu_A}[A]_0 \right) + p_Z \left([Z]_0 + \frac{\nu_Z}{\nu_A}[A]_0 \right) \qquad (3)$$

At time t, when the concentration of A is $[A]$,

$$[B] = [B]_\infty + \frac{\nu_B}{\nu_A}[A]$$

$$[Y] = [Y]_\infty - \frac{\nu_Y}{\nu_A}[A]$$

$$[Z] = [Z]_\infty - \frac{\nu_Z}{\nu_A}[A]$$

We can combine these equations with Equations 1 and 3 to find

$$S(t) = p_A[A] + p_B \left([B]_0 - \frac{\nu_B}{\nu_A}[A]_0 + \frac{\nu_B}{\nu_A}[A] \right)$$

$$+ p_Y \left([Y]_0 + \frac{\nu_Y}{\nu_A}[A]_0 - \frac{\nu_Y}{\nu_A}[A] \right) + p_Z \left([Z]_0 + \frac{\nu_Z}{\nu_A}[A]_0 - \frac{\nu_Z}{\nu_A}[A] \right)$$

Then

$$S(t) - S(\infty) = p_A[A] + \frac{\nu_B}{\nu_A}p_B[A] - \frac{\nu_Y}{\nu_A}p_Y[A] - \frac{\nu_Z}{\nu_A}p_Z[A]$$

$$S(0) - S(\infty) = p_A[A]_0 + \frac{\nu_B}{\nu_A}p_B[A]_0 - \frac{\nu_Y}{\nu_A}p_Y[A]_0 - \frac{\nu_Z}{\nu_A}p_Z[A]_0$$

and

$$\frac{S(t) - S(\infty)}{S(0) - S(\infty)} = \frac{[A]}{[A]_0}$$

28–50. Use the result of Problem 28–49 to show that for the first-order rate law, $v = k[A]$, the time-dependent signal is given by

$$S(t) = S(\infty) + [S(0) - S(\infty)]e^{-kt}$$

For the first-order rate law

$$[A] = [A]_0 e^{-kt} \qquad (28.23)$$

We can use the final result of Problem 28–49 to write this as

$$\frac{S(t) - S(\infty)}{S(0) - S(\infty)} = \frac{[A]}{[A]_0} = e^{-kt}$$

or

$$S(t) = S(\infty) + [S(0) - S(\infty)]e^{-kt}$$

28–51. Use the result of Problem 28–49 to show that for the second-order rate law, $v = k[A]^2$, the time-dependent signal is given by

$$S(t) = S(\infty) + \frac{S(0) - S(\infty)}{1 + [A]_0 kt}$$

For the second order rate law

$$\frac{1}{[A]} = \frac{1}{[A]_0} + kt \tag{28.33}$$

We can write this equation in terms of $[A]/[A]_0$:

$$[A] = \frac{[A]_0}{1 + [A]_0 kt}$$

$$\frac{[A]}{[A]_0} = \frac{1}{1 + [A]_0 kt}$$

and then use the final result of Problem 28–49 to write

$$\frac{S(t) - S(\infty)}{S(0) - S(\infty)} = \frac{[A]}{[A]_0} = \frac{1}{1 + [A]_0 kt}$$

or

$$S(t) = S(\infty) + \frac{S(0) - S(\infty)}{1 + [A]_0 kt}$$

28–52. Because there is a substantial increase in the volume of the solution as the reaction proceeds, the decomposition of diacetone alcohol can be followed by a dilatometer, a device that measures the volume of a sample as a function of time. The instrument readings at various times are tabulated below.

Time/s	0	24.4	35.0	48.0	64.8	75.8	133.4	∞
S/arbitrary units	8.0	20.0	24.0	28.0	32.0	34.0	40.0	43.3

Use the expressions derived in Problems 28–50 and 28–51 to determine if the decomposition reaction is a first- or second-order process.

If the decomposition reaction is a first-order process, then

$$\ln \frac{S(t) - S(\infty)}{S(0) - S(\infty)} = -kt$$

and a plot of $\ln[S(t) - S(\infty)]/[S(0) - S(\infty)]$ versus t will be linear. If the decomposition reaction is a second-order process, then

$$1 + [A]_0 kt = \left[\frac{S(t) - S(\infty)}{S(0) - S(\infty)} \right]^{-1}$$

and a plot of $[S(0) - S(\infty)]/[S(t) - S(\infty)]$ versus t will be linear.

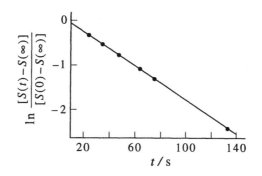

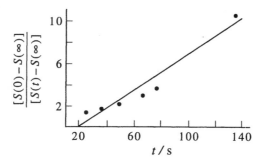

This is a first-order process.

28–54. In Problem 28–49, we assumed that A reacted completely so that $[A] \to 0$ as $t \to \infty$. Show that if the reaction does not go to completion but establishes an equilibrium instead, then

$$[A] = [A]_{2,eq} + \{[A]_0 - [A]_{2,eq}\} \frac{S(t) - S(\infty)}{S(0) - S(\infty)}$$

where $[A]_{2,eq}$ is the equilibrium concentration of A.

Because $S(t)$ and $S(0)$ are the same as in Problem 28–49, we can still use Equations 1 and 2 from that problem. If the reaction establishes an equilibrium, Equation 3 becomes

$$S(\infty) = p_A[A]_{2,eq} + p_B\left([B]_0 - \frac{\nu_B}{\nu_A}[A]_0 + \frac{\nu_B}{\nu_A}[A]_{2,eq}\right) + p_Y\left([Y]_0 + \frac{\nu_Y}{\nu_A}[A]_0 - \frac{\nu_Y}{\nu_A}[A]_{2,eq}\right)$$

$$+ p_Z\left([Z]_0 + \frac{\nu_Z}{\nu_A}[A]_0 - \frac{\nu_Z}{\nu_A}[A]_{2,eq}\right)$$

Then

$$S(t) - S(\infty) = p_A\left([A] - [A]_{2,eq}\right) + p_B\left(\frac{\nu_B}{\nu_A}[A] - \frac{\nu_B}{\nu_A}[A]_{2,eq}\right)$$

$$- p_Y\left(\frac{\nu_Y}{\nu_A}[A] - \frac{\nu_Y}{\nu_A}[A]_{2,eq}\right) - p_Z\left(\frac{\nu_Z}{\nu_A}[A] - \frac{\nu_Z}{\nu_A}[A]_{2,eq}\right)$$

$$= \left(p_A + \frac{\nu_B}{\nu_A}p_B - \frac{\nu_Y}{\nu_A}p_Y - \frac{\nu_Z}{\nu_A}p_Z\right)\left([A] - [A]_{eq}\right)$$

and

$$S(0) - S(\infty) = p_A\left([A]_0 - [A]_{2,eq}\right) + p_B\left(\frac{\nu_B}{\nu_A}[A]_0 - \frac{\nu_B}{\nu_A}[A]_{2,eq}\right)$$

$$- p_Y\left(\frac{\nu_Y}{\nu_A}[A]_0 - \frac{\nu_Y}{\nu_A}[A]_{2,eq}\right) - p_Z\left(\frac{\nu_Z}{\nu_A}[A]_0 - \frac{\nu_Z}{\nu_A}[A]_{2,eq}\right)$$

$$= \left(p_A + \frac{\nu_B}{\nu_A}p_B - \frac{\nu_Y}{\nu_A}p_Y - \frac{\nu_Z}{\nu_A}p_Z\right)\left([A]_0 - [A]_{eq}\right)$$

Dividing the first equation by the second gives

$$\frac{S(t) - S(\infty)}{S(0) - S(\infty)} = \frac{[A] - [A]_{2,eq}}{[A]_0 - [A]_{2,eq}}$$

or

$$[A] = [A]_{2,eq} + ([A]_0 - [A]_{2,eq})\frac{S(t) - S(\infty)}{S(0) - S(\infty)}$$

CHAPTER 29

Chemical Kinetics II: Reaction Mechanisms

PROBLEMS AND SOLUTIONS

29–1. Give the units of the rate constant for a unimolecular, bimolecular, and termolecular reaction.

Since v is in units of $mol \cdot dm^{-3} \cdot s^{-1}$, the units of k are

$$k_{unimolecular}: (mol \cdot dm^{-3} \cdot s^{-1})(dm^3 \cdot mol^{-1}) = s^{-1}$$
$$k_{bimolecular}: (mol \cdot dm^{-3} \cdot s^{-1})(dm^3 \cdot mol^{-1})^2 = dm^3 \cdot mol^{-1} \cdot s^{-1}$$
$$k_{termolecular}: (mol \cdot dm^{-3} \cdot s^{-1})(dm^3 \cdot mol^{-1})^3 = dm^6 \cdot mol^{-2} \cdot s^{-1}$$

29–2. Determine the rate law for the following reaction

$$F(g) + D_2(g) \xrightarrow{k} FD(g) + D(g)$$

Give the units of k. Determine the molecularity of this reaction.

Because this is an elementary reaction, the rate law is

$$v = k[F][D_2]$$

This is a bimolecular reaction, so the units of k are $dm^3 \cdot mol^{-1} \cdot s^{-1}$.

29–3. Determine the rate law for the reaction

$$I(g) + I(g) + M(g) \xrightarrow{k} I_2(g) + M(g)$$

where M is any molecule present in the reaction container. Give the units of k. Determine the molecularity of this reaction. Is this reaction identical to

$$I(g) + I(g) \xrightarrow{k} I_2(g)$$

Explain.

Because this is an elementary reaction, the rate law is

$$v = k[M][I]^2$$

This is a termolecular reaction, so the units of k are $dm^6 \cdot mol^{-2} \cdot s^{-1}$. There is a difference between the molecularity of the two reactions (the second one is bimolecular), so they are not identical.

29–4. For $T < 500$ K, the reaction

$$NO_2(g) + CO(g) \xrightarrow{k_{obs}} CO_2(g) + NO(g)$$

has the rate law

$$\frac{d[CO_2]}{dt} = k_{obs}[NO_2]^2$$

Show that the following mechanism is consistent with the observed rate law

$$NO_2(g) + NO_2(g) \xrightarrow{k_1} NO_3(g) + NO(g) \quad \text{(rate determining)}$$

$$NO_3(g) + CO(g) \xrightarrow{k_2} CO_2(g) + NO_2(g)$$

Express k_{obs} in terms of k_1 and k_2.

If the first step of the mechanism is the rate-determining step, then the rate law is

$$\frac{d[CO_2]}{dt} = k_1[NO_2]^2$$

and $k_{obs} = k_1$. The rate constant k_2 does not affect k_{obs}.

29–5. Solve Equation 29.21 to obtain $[A] = [A]_0 e^{-k_1 t}$, and substitute this result into Equation 29.22 to obtain

$$\frac{d[I]}{dt} + k_2[I] = k_1[A]_0 e^{-k_1 t}$$

This equation is of the form (see the *CRC Handbook of Standard Mathematical Tables*, for example)

$$\frac{dy(x)}{dx} + p(x)y(x) = q(x)$$

a linear, first-order differential equation whose general solution is

$$y(x)e^{h(x)} = \int q(x)e^{h(x)}dx + c$$

where $h(x) = \int p(x)dx$ and c is a constant. Show that this solution leads to Equation 29.25.

The solution to Equation 29.21 (letting the concentration of $[A]$ at $t = 0$ be $[A]_0$) is $[A] = [A]_0 e^{-k_1 t}$. Substituting into Equation 29.22 gives

$$\frac{d[I]}{dt} = k_1[A] - k_2[I]$$

$$= k_1[A]_0 e^{-k_1 t} - k_2[I]$$

$$\frac{d[I]}{dt} + k_2[I] = k_1[A]_0 e^{-k_1 t}$$

The solution to this differential equation is

$$[I]e^{k_2 t} = k_1[A]_0 \int e^{(k_2 - k_1)t}\, dt + c$$

$$[I] = \frac{k_1[A]_0}{k_2 - k_1} e^{-k_1 t} + c e^{-k_2 t}$$

At $t = 0$, $[I] = 0$, so

$$c = -\frac{k_1[A]_0}{k_2 - k_1}$$

and we obtain

$$[I] = \frac{k_1[A]_0}{k_2 - k_1}(e^{-k_1 t} - e^{-k_2 t}) \tag{29.25}$$

29–6. Verify that Equation 29.32 is obtained if Equation 29.30 is substituted into Equation 29.23 and the resulting expression is integrated.

Start with Equations 29.23 and 29.30:

$$\frac{d[P]}{dt} = k_2[I] \qquad \text{and} \qquad [I] = \frac{k_1}{k_2}[A]_0 e^{-k_1 t}$$

Substituting Equation 29.30 into Equation 29.23 gives

$$\frac{d[P]}{dt} = k_1[A]_0 e^{-k_1 t}$$

and integrating gives

$$[P] = -[A]_0 e^{-k_1 t} + c$$

At $t = 0$, $[P] = 0$, and so $c = [A]_0$. Thus

$$[P] = -[A]_0 e^{-k_1 t} + [A]_0 = [A]_0(1 - e^{-k_1 t})$$

which is Equation 29.32.

29–7. Consider the reaction mechanism

$$A \xrightarrow{k_1} I \xrightarrow{k_2} P$$

where $[A] = [A]_0$ and $[I]_0 = [P]_0 = 0$ at time $t = 0$. Use the exact solution to this kinetic scheme (Equations 29.24 through 29.26) to plot the time dependence of $[A]/[A]_0$, $[I]/[A]_0$, and $[P]/[A]_0$ versus $\log k_1 t$ for the case $k_2 = 2k_1$. On the same graph, plot the time dependence of $[A]/[A]_0$, $[I]/[A]_0$, and $[P]/[A]_0$ using the expressions for $[A]$, $[I]$, and $[P]$ obtained assuming the steady-state approximation for $[I]$. Based on your results, can you use the steady-state approximation to model the kinetics of this reaction mechanism when $k_2 = 2k_1$?

Letting $x = k_1 t$ and $k_2 = 2k_1$, Equations 29.24 through 29.26 become

$$\frac{[A]}{[A]_0} = e^{-x}$$

$$\frac{[I]}{[A]_0} = \frac{k_1}{2k_1 - k_1}(e^{-x} - e^{-2x}) = e^{-x} - e^{-2x}$$

$$\frac{[P]}{[A]_0} = 1 - \frac{[A]}{[A]_0} - \frac{[I]}{[I]_0} = 1 - 2e^{-x} - e^{-2x}$$

Likewise, letting $x = k_1 t$ and $k_2 = 2k_1$, the steady-state approximations for [A], [I], and [P] (Equations 29.24, 29.30, and 29.32) become

$$\frac{[A]}{[A]_0} = e^{-x} \qquad \frac{[I]}{[A]_0} = \frac{k_1}{2k_1}e^{-x} = \frac{e^{-x}}{2} \qquad \frac{[P]}{[A]_0} = 1 - e^{-x}$$

In the plot, the dashed lines represent the steady state approximations and the solid lines represent the exact values. Both give the same values for $[A]/[A]_0$, but the steady-state approximations of $[I]/[A]_0$ and $[P]/[A]_0$ differ significantly from the exact solutions. We can conclude that the steady-state approximation is not a good model for the kinetics of a reaction mechanism for which $k_2 = 2k_1$.

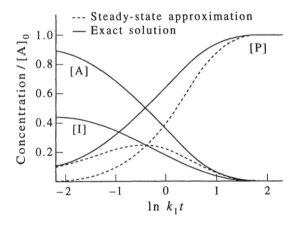

29-8. Consider the mechanism for the decomposition of ozone presented in Example 29–5. Explain why either (a) $v_{-1} \gg v_2$ and $v_{-1} \gg v_1$ or (b) $v_2 \gg v_{-1}$ and $v_2 \gg v_1$ must be true for the steady-state approximation to apply. The rate law for the decomposition reaction is found to be

$$\frac{d[O_3]}{dt} = -k_{obs}[O_3][M]$$

Is this rate law consistent with the conditions given by either (a) or (b) or both?

The mechanism presented in Example 29–5 is

$$M(g) + O_3(g) \underset{k_{-1}}{\overset{k_1}{\rightleftharpoons}} O_2(g) + O(g) + M(g)$$

$$O(g) + O_3(g) \overset{k_2}{\Longrightarrow} 2O_2(g)$$

For the steady state approximation to apply, the concentration of O must be negligible. This means that the rates of reaction of [O] must be greater than the rates of formation of [O]. We examine each of the proposed conditions to determine whether this is true.

a. If $v_{-1} \gg v_2$ and $v_{-1} \gg v_1$, then very little O(g) can accumulate. When O(g) is generated, it reacts quickly back to reactants, and less often to the products.

b. If $v_2 \gg v_{-1}$ and $v_2 \gg v_1$, then when O(g) is formed it quickly goes to product, and (once again) very little is accumulated.

For the steady state in O, we have (Example 29–5)

$$\frac{d[O_3]}{dt} = -\frac{2k_1 k_2 [O_3]^2 [M]}{k_{-1}[O_2][M] + k_2[O_3]}$$

Because all these reactions are elementary, we can write

$$v_2 = k_2[O][O_3] \qquad v_1 = k_1[M][O_3] \quad \text{and} \quad v_{-1} = k_{-1}[O_2][O][M]$$

Under the conditions given by (a),

$$k_{-1}[O_2][O][M] \gg k_2[O][O_3]$$

$$k_{-1}[O_2][M] \gg k_2[O_3]$$

Then

$$\frac{d[O_3]}{dt} = -\frac{2k_1 k_2 [O_3]^2 [M]}{k_{-1}[O_2][M] + k_2[O_3]} \approx -\frac{2k_1 k_2 [O_3]^2 [M]}{k_{-1}[O_2][M]} = \frac{2k_1 k_2}{k_{-1}} \frac{[O_3]^2}{[O_2]}$$

which is not consistent with the observed rate law. Under the conditions given by (b),

$$k_2[O][O_3] \gg k_{-1}[O_2][O][M]$$

$$k_2[O_3] \gg k_{-1}[O_2][M]$$

Then

$$\frac{d[O_3]}{dt} = -\frac{2k_1 k_2 [O_3]^2 [M]}{k_{-1}[O_2][M] + k_2[O_3]} \approx -\frac{2k_1 k_2 [O_3]^2 [M]}{k_2[O_3]} = 2k_1[O_3][M]$$

which is consistent with the observed rate law.

29–9. Consider the reaction mechanism

$$A + B \underset{k_{-1}}{\overset{k_1}{\rightleftharpoons}} C \tag{1}$$

$$C \overset{k_2}{\Longrightarrow} P \tag{2}$$

Write the expression for $d[P]/dt$, the rate of product formation. Assume equilibrium is established in the first reaction before any appreciable amount of product is formed, and thereby show that

$$\frac{d[P]}{dt} = k_2 K_c[A][B]$$

where K_c is the equilibrium constant for step (1) of the reaction mechanism. This assumption is called the *fast-equilibrium approximation*.

Product is formed only in step (2) of the reaction mechanism, so the rate of product formation can be written as

$$\frac{d[P]}{dt} = k_2[C]$$

If equilibrium is established in the first reaction before any appreciable amount of product is formed, then

$$K_c = \frac{k_1}{k_{-1}} = \frac{[C]}{[A][B]}$$

Solving for [C] and substituting gives

$$\frac{d[P]}{dt} = k_2 K_c[A][B]$$

29–10. The rate law for the reaction of *para*-hydrogen to *ortho*-hydrogen

$$para\text{-}H_2(g) \xrightarrow{k_{obs}} ortho\text{-}H_2(g)$$

is

$$\frac{d[ortho\text{-}H_2]}{dt} = k_{obs}[para\text{-}H_2]^{3/2}$$

Show that the following mechanism is consistent with this rate law.

$$para\text{-}H_2(g) \underset{k_{-1}}{\overset{k_1}{\rightleftharpoons}} 2\,H(g) \qquad \text{(fast equilibrium)} \tag{1}$$

$$H(g) + para\text{-}H_2(g) \xrightarrow{k_2} ortho\text{-}H_2(g) + H(g) \tag{2}$$

Express k_{obs} in terms of the rate constants for the individual steps of the reaction mechanism.

The rate law is

$$\frac{d[ortho\text{-}H_2]}{dt} = k_2[para\text{-}H_2][H]$$

Because the first step achieves a fast equilibrium, we can write

$$K_c = \frac{[H]^2}{[para\text{-}H_2]}$$

where K_c is the equilibrium constant for step (1) of the reaction mechanism. Solving this equation for [H] gives

$$[H] = K_c^{1/2}[para\text{-}H_2]^{1/2}$$

Substituting into the observed rate law,

$$\frac{d[ortho\text{-}H_2]}{dt} = k_2 K_c^{1/2}[para\text{-}H_2]^{3/2}$$

and therefore

$$k_{obs} = k_2 K_c^{1/2} = k_2 \left(\frac{k_1}{k_{-1}}\right)^{1/2}$$

29–11. Consider the decomposition reaction of $N_2O_5(g)$

$$2\,N_2O_5(g) \xrightarrow{k_{obs}} 4\,NO_2(g) + O_2(g)$$

A proposed mechanism for this reaction is

$$N_2O_5(g) \underset{k_{-1}}{\overset{k_1}{\rightleftharpoons}} NO_2(g) + NO_3(g)$$

$$NO_2(g) + NO_3(g) \xrightarrow{k_2} NO(g) + NO_2(g) + O_2(g)$$

$$NO_3(g) + NO(g) \xrightarrow{k_3} 2\,NO_2(g)$$

Assume that the steady-state approximation applies to both the $NO(g)$ and $NO_3(g)$ reaction intermediates to show that this mechanism is consistent with the experimentally observed rate law

$$\frac{d[O_2]}{dt} = k_{obs}[N_2O_5]$$

Express k_{obs} in terms of the rate constants for the individual steps of the reaction mechanism.

Under the steady state approximation for NO, we can write

$$\frac{d[NO]}{dt} = k_2[NO_2][NO_3] - k_3[NO][NO_3] = 0$$

Solving this equation for [NO] gives

$$[NO] = \frac{k_2[NO_2]}{k_3} \tag{1}$$

Using the steady state approximation for NO_3, we can also write

$$\frac{d[NO_3]}{dt} = k_1[N_2O_5] - k_{-1}[NO_2][NO_3] - k_2[NO_2][NO_3] - k_3[NO_3][NO] = 0$$

Solving this equation for $[NO_3]$ gives

$$[NO_3](k_{-1}[NO_2] + k_2[NO_2] + k_3[NO]) = k_1[N_2O_5]$$

or

$$[NO_3] = \frac{k_1[N_2O_5]}{k_{-1}[NO_2] + k_2[NO_2] + k_3[NO]}$$

Substituting for [NO] from Equation 1, we find

$$[NO_3] = \frac{k_1[N_2O_5]}{k_{-1}[NO_2] + k_2[NO_2] + k_2[NO_2]} \tag{2}$$

Now, from step (2) of the reaction mechanism, we can express $d[O_2]/dt$ as

$$\frac{d[O_2]}{dt} = k_2[NO_2][NO_3]$$

Substituting for $[NO_3]$ from Equation 2, we find

$$\frac{d[O_2]}{dt} = k_2[NO_2]\frac{k_1[N_2O_5]}{k_2[NO_2] + k_2[NO_2] + k_{-1}[NO_2]}$$

$$= \frac{k_1 k_2}{2k_2 + k_{-1}}[N_2O_5]$$

and $k_{obs} = k_1 k_2/(2k_2 + k_{-1})$.

29–12. The rate law for the reaction between CO(g) and Cl_2(g) to form phosgene (Cl_2CO)

$$Cl_2(g) + CO(g) \xrightarrow{k_{obs}} Cl_2CO(g)$$

is

$$\frac{d[Cl_2CO]}{dt} = k_{obs}[Cl_2]^{3/2}[CO]$$

Show that the following mechanism is consistent with this rate law.

$$Cl_2(g) + M(g) \underset{k_{-1}}{\overset{k_1}{\rightleftharpoons}} 2\,Cl(g) + M(g) \qquad \text{(fast equilibrium)}$$

$$Cl(g) + CO(g) + M(g) \underset{k_{-2}}{\overset{k_2}{\rightleftharpoons}} ClCO(g) + M(g) \qquad \text{(fast equilibrium)}$$

$$ClCO(g) + Cl_2(g) \xrightarrow{k_3} Cl_2CO(g) + Cl(g) \qquad \text{(slow)}$$

where M is any gas molecule present in the reaction container. Express k_{obs} in terms of the rate constants for the individual steps of the reaction mechanism.

For the first two steps, the fast equilibrium allows us to write

$$K_{c,1} = \frac{[M][Cl]^2}{[Cl_2][M]} = \frac{[Cl]^2}{[Cl_2]}$$

and

$$K_{c,2} = \frac{[ClCO][M]}{[Cl][CO][M]} = \frac{[ClCO]}{[Cl][CO]}$$

From these equations, we can write

$$[Cl] = K_{c,1}^{1/2}[Cl_2]^{1/2} \tag{1}$$

and

$$[ClCO] = K_{c,2}[Cl][CO] \tag{2}$$

Using the third step of the reaction mechanism to write $d[Cl_2CO]/dt$ and Equations 1 and 2, we have

$$\begin{aligned}
\frac{d[Cl_2CO]}{dt} &= k_3[ClCO][Cl_2] \\
&= k_3 K_{c,2}[Cl][CO][Cl_2] \\
&= k_3 K_{c,2} K_{c,1}^{1/2}[Cl_2]^{3/2}[CO]
\end{aligned}$$

Using the principle of detailed balance (Equation 29.6), we find that

$$k_{obs} = \frac{k_3 k_2 k_1^{1/2}}{k_{-2} k_{-1}^{1/2}}$$

29–13. Nitramide (O_2NNH_2) decomposes in water according to the chemical equation

$$O_2NNH_2(aq) \xrightarrow{k_{obs}} N_2O(g) + H_2O(l)$$

The experimentally determined rate law for this reaction is

$$\frac{d[N_2O]}{dt} = k_{obs} \frac{[O_2NNH_2]}{[H^+]}$$

A proposed mechanism for this reaction is

$$O_2NNH_2(aq) \underset{k_{-1}}{\overset{k_1}{\rightleftharpoons}} O_2NNH^-(aq) + H^+(aq) \qquad \text{(fast equilibrium)}$$

$$O_2NNH^-(aq) \xrightarrow{k_2} N_2O(g) + OH^-(aq) \qquad \text{(slow)}$$

$$H^+(aq) + OH^-(aq) \xrightarrow{k_3} H_2O(l) \qquad \text{(fast)}$$

Is this mechanism consistent with the observed rate law? If so, what is the relationship between k_{obs} and the rate constants for the individual steps of the mechanism?

From step (1) of the reaction mechanism, the fast equilibrium allows us to write

$$K_c = \frac{[O_2NNH^-][H^+]}{[O_2NNH_2]}$$

and so the rate equation for the second step of the reaction mechanism becomes

$$\frac{d[N_2O]}{dt} = k_2[O_2NNH^-] = k_2 K_c \frac{[O_2NNH_2]}{[H^+]}$$

This is consistent with the observed rate law, and

$$k_{obs} = k_2 K_c = \frac{k_2 k_1}{k_{-1}}$$

29–14. What would you predict for the rate law for the reaction mechanism in Problem 29–13 if, instead of a fast equilibrium followed by a slow step, you assumed that the concentration of $O_2NNH^-(aq)$ was such that the steady-state approximation could be applied to this reaction intermediate?

The rate equation for the second step of the reaction mechanism is still

$$\frac{d[N_2O]}{dt} = k_2[O_2NNH^-]$$

Applying the steady-state approximation to find $[O_2NNH^-]$, we find that

$$\frac{d[O_2NNH^-]}{dt} = 0 = k_1[O_2NNH_2] - k_{-1}[O_2NNH^-][H^+] - k_2[O_2NNH^-]$$

This gives

$$[O_2NNH^-](k_{-1}[H^+] + k_2) = k_1[O_2NNH_2]$$

$$[O_2NNH^-] = \frac{k_1[O_2NNH_2]}{k_{-1}[H^+] + k_2}$$

Then

$$\frac{d[\mathrm{N_2O}]}{dt} = \frac{k_2 k_1 [\mathrm{O_2NNH_2}]}{k_{-1}[\mathrm{H^+}] + k_2}$$

which differs from the experimentally observed rate law. Note that if $k_2 \ll k_{-1}[\mathrm{H^+}]$, the first step essentially achieves a fast equilibrium and we find the rate law given in the previous problem.

29–15. The rate law for the hydrolysis of ethyl acetate by aqueous sodium hydroxide at 298 K

$$\mathrm{CH_3COOCH_2CH_3(aq)} + \mathrm{OH^-(aq)} \xrightarrow{k_{obs}} \mathrm{CH_3CO_2^-(aq)} + \mathrm{CH_3CH_2OH(aq)}$$

is

$$\frac{d[\mathrm{CH_3CH_2OH}]}{dt} = k_{obs}[\mathrm{OH^-}][\mathrm{CH_3COOCH_2CH_3}]$$

Despite the form of this rate law, this reaction is not an elementary reaction but is believed to occur by the following mechanism

$$\mathrm{CH_3COOCH_2CH_3(aq)} + \mathrm{OH^-(aq)} \underset{k_{-1}}{\overset{k_1}{\rightleftharpoons}} \mathrm{CH_3CO^-(OH)OCH_2CH_3(aq)}$$

$$\mathrm{CH_3CO^-(OH)OCH_2CH_3(aq)} \xrightarrow{k_2} \mathrm{CH_3CO_2H(aq)} + \mathrm{CH_3CH_2O^-(aq)}$$

$$\mathrm{CH_3CO_2H(aq)} + \mathrm{CH_3CH_2O^-(aq)} \xrightarrow{k_3} \mathrm{CH_3CO_2^-(aq)} + \mathrm{CH_3CH_2OH(aq)}$$

Under what conditions does this mechanism give the observed rate law? For those conditions, express k_{obs} in terms of the rate constants for the individual steps of the reaction mechanism.

From step (3) of the reaction mechanism, we can write

$$\frac{d[\mathrm{CH_3CH_2OH}]}{dt} = k_3[\mathrm{CH_3CO_2H}][\mathrm{CH_3CH_2O^-}] \tag{1}$$

If we assume that equilibrium is quickly reached in step (1) of the reaction mechanism, we find that

$$K_{c,1} = \frac{k_1}{k_{-1}} = \frac{[\mathrm{CH_3CO^-(OH)OCH_2CH_3}]}{[\mathrm{OH^-}][\mathrm{CH_3COOCH_2CH_3}]}$$

$$[\mathrm{CH_3CO^-(OH)OCH_2CH_3}] = \frac{k_1}{k_{-1}}[\mathrm{OH^-}][\mathrm{CH_3COOCH_2CH_3}] \tag{2}$$

We now use the steady-state approximation for $\mathrm{CH_3CO_2H}$ to write

$$\frac{d[\mathrm{CH_3CO_2H}]}{dt} = k_2[\mathrm{CH_3CO^-(OH)OCH_2CH_3}] - k_3[\mathrm{CH_3CO_2H}][\mathrm{CH_3CH_2O^-}] = 0$$

Then

$$k_3[\mathrm{CH_3CO_2H}][\mathrm{CH_3CH_2O^-}] = k_2[\mathrm{CH_3CO^-(OH)OCH_2CH_3}] \tag{3}$$

and the rate law becomes

$$\frac{d[\mathrm{CH_3CH_2OH}]}{dt} = k_2[\mathrm{CH_3CO^-(OH)OCH_2CH_3}]$$

$$= \frac{k_2 k_1}{k_{-1}}[\mathrm{OH^-}][\mathrm{CH_3CHOOCH_2CH_3}]$$

where we have substituted Equations 2 and 3 into Equation 1. This will give the observed rate law, with $k_{obs} = k_2 k_1 / k_{-1}$.

29–16. The decomposition of perbenzoic acid in water

$$2\,C_6H_5CO_3H(aq) \rightleftharpoons 2\,C_6H_5CO_2H(aq) + O_2(g)$$

is proposed to occur by the following mechanism

$$C_6H_5CO_3H(aq) \underset{k_{-1}}{\overset{k_1}{\rightleftharpoons}} C_6H_5CO_3^-(aq) + H^+(aq)$$

$$C_6H_5CO_3H(aq) + C_6H_5CO_3^-(aq) \overset{k_2}{\longrightarrow} C_6H_5CO_2H(aq)$$
$$+ C_6H_5CO_2^-(aq) + O_2(g)$$

$$C_6H_5CO_2^-(aq) + H^+(aq) \overset{k_3}{\longrightarrow} C_6H_5CO_2H(aq)$$

Derive an expression for the rate of formation of O_2 in terms of the reactant concentration and $[H^+]$.

From step (2) of the reaction mechanism, we write

$$\frac{d[O_2]}{dt} = k_2[C_6H_5CO_3H][C_6H_5CO_3^-]$$

Assuming that the first step reaches equilibrium rapidly, we have

$$K_c = \frac{k_1}{k_{-1}} = \frac{[H^+][C_6H_5CO_3^-]}{[C_6H_5CO_3H]}$$

$$[C_6H_5CO_3^-] = \frac{k_1}{k_{-1}} \frac{[C_6H_5CO_3H]}{[H^+]}$$

Substituting into the above rate law gives

$$\frac{d[O_2]}{dt} = \frac{k_2 k_1}{k_{-1}} \frac{[C_6H_5CO_3H]^2}{[H^+]}$$

29–17. The rate law for the reaction described by

$$2\,H_2(g) + 2\,NO(g) \overset{k_{obs}}{\longrightarrow} N_2(g) + 2\,H_2O(g)$$

is

$$\frac{d[N_2]}{dt} = k_{obs}[H_2][NO]^2$$

Below is a proposed mechanism for this reaction

$$H_2(g) + NO(g) + NO(g) \overset{k_1}{\Longrightarrow} N_2O + H_2O(g)$$

$$H_2(g) + N_2O(g) \overset{k_2}{\Longrightarrow} N_2(g) + H_2O(g)$$

Under what conditions does this mechanism give the observed rate law? Express k_{obs} in terms of the rate constants for the individual steps of the mechanism.

From step (2) of the reaction mechanism,

$$\frac{d[N_2]}{dt} = k_2[H_2][N_2O] \tag{1}$$

We assume we can use the steady-state approximation for N_2O. Then

$$\frac{d[N_2O]}{dt} = k_1[H_2][NO]^2 - k_2[H_2][N_2O] = 0$$

or

$$[N_2O] = \frac{k_1}{k_2}[NO]^2$$

Substituting into Equation 1, we find that

$$\frac{d[N_2]}{dt} = k_1[H_2][NO]^2$$

If this mechanism is followed, then $k_{obs} = k_1$.

29–18. A second proposed mechanism for the reaction discussed in Problem 27-17 is

$$NO(g) + NO(g) \underset{k_{-1}}{\overset{k_1}{\rightleftharpoons}} N_2O_2(g)$$

$$H_2(g) + N_2O_2(g) \overset{k_2}{\Longrightarrow} N_2O(g) + H_2O(g)$$

$$H_2(g) + N_2O(g) \overset{k_3}{\Longrightarrow} N_2(g) + H_2O(g)$$

Under what conditions does this mechanism give the observed rate law? Express k_{obs} in terms of the rate constants for the individual steps of the mechanism. Do you favor this mechanism or that given in Problem 27-17? Explain your reasoning.

From step (3) of the reaction mechanism, we can write (as in the previous problem)

$$\frac{d[N_2]}{dt} = k_3[H_2][N_2O] \tag{1}$$

If we use the steady-state approximation for N_2O,

$$\frac{d[N_2O]}{dt} = 0 = k_2[H_2][N_2O_2] - k_3[H_2][N_2O]$$

$$[N_2O] = \frac{k_2}{k_3}[N_2O_2]$$

and Equation 1 becomes

$$\frac{d[N_2]}{dt} = k_2[H_2][N_2O_2] \tag{2}$$

Now, if we assume that the first step establishes a fast equilibrium, we can write

$$K_{c,1} = \frac{k_1}{k_{-1}} = \frac{[N_2O_2]}{[NO]^2}$$

$$\frac{k_1}{k_{-1}}[NO]^2 = [N_2O_2]$$

Substituting into Equation 2 then gives

$$\frac{d[N_2]}{dt} = \frac{k_2 k_1}{k_{-1}}[H_2][NO]^2$$

which would give an observed rate constant of $k_2 k_1 / k_{-1}$. It seems more probable that this mechanism is followed, because each step of this mechanism is a bimolecular reaction. The mechanism of Problem 29–17 requires a termolecular reaction to occur.

29–19. An alternative mechanism for the chemical reaction

$$Cl_2(g) + CO(g) \xrightarrow{k_{obs}} Cl_2CO(g)$$

(see Problem 29–12) is

$$Cl_2(g) + M(g) \underset{k_{-1}}{\overset{k_1}{\rightleftharpoons}} 2\,Cl(g) + M(g) \qquad \text{(fast equilibrium)}$$

$$Cl(g) + Cl_2(g) \underset{k_{-2}}{\overset{k_2}{\rightleftharpoons}} Cl_3(g) \qquad \text{(fast equilibrium)}$$

$$Cl_3(g) + CO(g) \xrightarrow{k_3} Cl_2CO(g) + Cl(g)$$

where M is any molecule present in the reaction chamber. Show that this mechanism also gives the observed rate law. How would you go about determining whether this mechanism or the one given in Problem 29–12 is correct?

The observed rate law is (from Problem 29–12)

$$\frac{d[Cl_2CO]}{dt} = k_{obs}[Cl_2]^{3/2}[CO]$$

From step (3) of the reaction mechanism, we write

$$\frac{d[Cl_2CO]}{dt} = k_3[Cl_3][CO] \tag{1}$$

Because steps (2) and (1) establish fast equilibria, we write

$$K_{c,2} = \frac{k_2}{k_{-2}} = \frac{[Cl_3]}{[Cl][Cl_2]}$$

which gives

$$[Cl_3] = \frac{k_2}{k_{-2}}[Cl][Cl_2] \tag{2}$$

Also,

$$K_{c,1} = \frac{k_1}{k_{-1}} = \frac{[Cl]^2[M]}{[Cl_2][M]}$$

and so

$$[Cl] = \left(\frac{k_1}{k_{-1}}\right)^{1/2}[Cl_2]^{1/2}$$

Substituting this expression for [Cl] into Equation 2 gives

$$[Cl_3] = \frac{k_2}{k_{-2}}\left(\frac{k_1}{k_{-1}}\right)^{1/2}[Cl_2]^{3/2}$$

and substituting this expression into Equation 1 gives

$$\frac{d[Cl_2CO]}{dt} = \frac{k_3k_2k_1^{1/2}}{k_{-2}k_{-1}^{1/2}}[Cl_2]^{3/2}[CO]$$

which corresponds to the observed rate law. To determine whether this mechanism occurs or that in Problem 29–12, we might check to see whether Cl_3 is produced during the reaction.

29–20. The Lindemann reaction mechanism for the isomerization reaction

$$CH_3NC(g) \longrightarrow CH_3CN(g)$$

is

$$CH_3NC(g) + M(g) \underset{k_{-1}}{\overset{k_1}{\rightleftharpoons}} CH_3NC^*(g) + M(g)$$

$$CH_3NC^*(g) \overset{k_2}{\Longrightarrow} CH_3CN(g)$$

Under what conditions does the steady-state approximation apply to CH_3NC^*?

The steady-state approximation will apply if $d[CH_3NC^*]/dt \approx 0$. This will be true if $v_{-1} \gg v_1$, when $k_{-1}[CH_3NC^*] \gg k_1[CH_3NC]$, or if $v_2 \gg v_1$, when $k_2[CH_3NC^*] \gg k_1[CH_3NC][M]$. In both cases, the CH_3NC^* is slowly formed and quickly reacted, and the concentration of CH_3NC^* at any given time is very small.

29–21. In Section 29–6 we examined the unimolecular reaction

$$CH_3NC(g) \Longrightarrow CH_3CN(g)$$

Consider this reaction carried out in the presence of a helium buffer gas. The collision of a CH_3NC molecule with either another CH_3NC molecule or a helium atom can energize the molecule, thereby leading to reaction. If the energizing reactions involving a CH_3NC molecule and a He atom occur with different rates, the reaction mechanism would be given by

$$CH_3NC(g) + CH_3NC(g) \underset{k_{-1}}{\overset{k_1}{\rightleftharpoons}} CH_3NC^*(g) + CH_3NC(g)$$

$$CH_3NC(g) + He(g) \underset{k_{-2}}{\overset{k_2}{\rightleftharpoons}} CH_3NC^*(g) + He(g)$$

$$CH_3NC^*(g) \overset{k_3}{\Longrightarrow} CH_3CN$$

Apply the steady-state approximation to the intermediate species, $CH_3NC^*(g)$, to show that

$$\frac{d[CH_3CN]}{dt} = \frac{k_3(k_1[CH_3NC]^2 + k_2[CH_3NC][He])}{k_{-1}[CH_3NC] + k_{-2}[He] + k_3}$$

Show that this equation is equivalent to Equation 29.55 when [He] = 0.

Applying the steady-state approximation gives

$$\frac{d[CH_3NC^*]}{dt} = 0 = k_1[CH_3NC]^2 - k_{-1}[CH_3NC^*][CH_3NC] + k_2[CH_3NC][He]$$
$$-k_{-2}[CH_3NC^*][He] - k_3[CH_3NC^*]$$

Then

$$(k_{-2}[He] + k_3 + k_{-1}[CH_3NC])[CH_3NC^*] = k_1[CH_3NC]^2 + k_2[CH_3NC][He]$$

and

$$[CH_3NC^*] = \frac{k_1[CH_3NC]^2 + k_2[CH_3NC][He]}{k_{-2}[He] + k_3 + k_{-1}[CH_3NC]}$$

From step (3), we have

$$\frac{d[CH_3CN]}{dt} = k_3[CH_3NC^*]$$
$$= \frac{k_3(k_1[CH_3NC]^2 + k_2[CH_3NC][He])}{k_{-2}[He] + k_3 + k_{-1}[CH_3NC]}$$

If $[He] = 0$, then

$$\frac{d[CH_3CN]}{dt} = \frac{k_3 k_1[CH_3NC]^2}{k_3 + k_{-1}[CH_3NC]} = k_{obs}[CH_3NC]$$

which is Equation 29.55.

29–22. Consider the reaction and mechanism given in Problem 29–10. The activation energy for the dissociation of $H_2(g)$ [step (1)] is given by D_0, the dissociation energy. If the activation energy of step (2) of the mechanism is E_2, show that $E_{a,obs}$, the experimentally determined activation energy, is given by

$$E_{a,obs} = E_2 + \frac{D_0}{2}$$

Also show that A_{obs}, the experimentally determined Arrhenius pre-exponential factor, is given by

$$A_{obs} = A_2 \left(\frac{A_1}{A_{-1}}\right)^{1/2}$$

where A_i is the Arrhenius pre-exponential factor corresponding to the rate constant k_i.

From Problem 29–10, $k_{obs} = k_2 k_1^{1/2}/k_{-1}^{1/2}$. If each step of the reaction mechanism shows Arrhenius behavior, we can write (as in Example 29–7)

$$k_1 = A_1 e^{-D_0/RT} \qquad k_{-1} = A_{-1} \qquad k_2 = A_2 e^{-E_2/RT}$$

(Note that the activation energy for the formation of H_2 is 0.) Substituting into an Arrhenius equation for the reaction rate constant gives

$$k_{obs} = \frac{k_2 k_1^{1/2}}{k_{-1}^{1/2}} = A_{obs} e^{-E_{a,obs}/RT}$$

$$\frac{A_2 A_1^{1/2}}{A_{-1}^{1/2}} e^{-E_2/RT} e^{-D_0/2RT} = A_{obs} e^{-E_{a,obs}/RT}$$

$$A_2 \left(\frac{A_1}{A_{-1}}\right)^{1/2} e^{-(E_2+D_0/2)/RT} = A_{\text{obs}} e^{-E_{\text{a,obs}}/RT}$$

so

$$E_{\text{a,obs}} = E_2 + \frac{D_0}{2} \quad \text{and} \quad A_{\text{obs}} = A_2 \left(\frac{A_1}{A_{-1}}\right)^{1/2}$$

29–23. The thermal decomposition of ethylene oxide occurs by the mechanism

$$H_2COCH_2(g) \overset{k_1}{\Longrightarrow} H_2COCH(g) + H(g)$$

$$H_2COCH(g) \overset{k_2}{\Longrightarrow} CH_3(g) + CO(g)$$

$$CH_3(g) + H_2COCH_2(g) \overset{k_3}{\Longrightarrow} H_2COCH(g) + CH_4(g)$$

$$CH_3(g) + H_2COCH(g) \overset{k_4}{\Longrightarrow} \text{products}$$

Which of these reaction(s) are the initiation, propagation, and termination step(s) of the reaction mechanism? Show that if the intermediates CH_3 and H_2COCH are treated by the steady-state approximation, the rate law, $d[\text{products}]/dt$, is first order in ethylene oxide concentration.

Initiation step: (1)
Propagation steps: (2), (3)
Termination step: (4)
From the termination step we have

$$\frac{d[\text{products}]}{dt} = k_4[CH_3][H_2COCH]$$

Treating CH_3 and H_2COCH by the steady-state approximation gives

$$\frac{d[CH_3]}{dt} = 0 = k_2[H_2COCH] - k_3[CH_3][H_2COCH_2] - k_4[CH_3][H_2COCH]$$

and

$$\frac{d[H_2COCH]}{dt} = 0 = -k_2[H_2COCH] + k_3[CH_3][H_2COCH_2] - k_4[CH_3][H_2COOH] + k_1[H_2COCH_2]$$

Adding these two expressions together gives

$$0 = -2k_4[CH_3][H_2COCH] + k_1[H_2COCH_2]$$

$$[CH_3][H_2COCH] = \frac{k_1}{2k_4}[H_2COCH_2]$$

so

$$\frac{d[\text{products}]}{dt} = \frac{k_1}{2}[H_2COCH_2]$$

which is first-order in $[H_2COCH_2]$.

The next six problems examine the kinetics of the thermal decomposition of acetaldehyde.

29–24. A proposed mechanism for the thermal decomposition of acetaldehyde

$$CH_3CHO(g) \overset{k_{\text{obs}}}{\longrightarrow} CH_4(g) + CO(g)$$

is

$$CH_3CHO(g) \overset{k_1}{\Longrightarrow} CH_3(g) + CHO(g) \tag{1}$$

$$CH_3(g) + CH_3CHO(g) \overset{k_2}{\Longrightarrow} CH_4(g) + CH_3CO(g) \tag{2}$$

$$CH_3CO(g) \overset{k_3}{\Longrightarrow} CH_3(g) + CO(g) \tag{3}$$

$$2\,CH_3(g) \overset{k_4}{\Longrightarrow} C_2H_6 \tag{4}$$

Is this reaction a chain reaction? If so, identify the initiation, propagation, inhibition, and termination step(s). Determine the rate laws for $CH_4(g)$, $CH_3(g)$, and $CH_3CO(g)$. Show that if you assume the steady-state approximation for the intermediate species, $CH_3(g)$ and $CH_3CO(g)$, the rate law for methane formation is given by

$$\frac{d[CH_4]}{dt} = \left(\frac{k_1}{k_4}\right)^{1/2} k_2[CH_3CHO]^{3/2}$$

This is a chain reaction.

    Initiation step: (1)
    Propagation steps: (2), (3)
    Termination step: (4)

The rate laws for $CH_4(g)$, $CH_3(g)$, and $CH_3CO(g)$ are

$$\frac{d[CH_4]}{dt} = k_2[CH_3][CH_3CHO]$$

$$\frac{d[CH_3]}{dt} = k_1[CH_3CHO] - k_2[CH_3][CH_3CHO] + k_3[CH_3CO] - k_4[CH_3]^2$$

$$\frac{d[CH_3CO]}{dt} = k_2[CH_3][CH_3CHO] - k_3[CH_3CO]$$

Assuming the steady-state approximation for the appropriate intermediates, we find

$$\frac{d[CH_3CO]}{dt} = 0 = k_2[CH_3][CH_3CHO] - k_3[CH_3CO]$$

$$[CH_3CO] = \frac{k_2}{k_3}[CH_3][CH_3CHO] \tag{1}$$

and (using Equation 1 to express $[CH_3CO]$)

$$\frac{d[CH_3]}{dt} = k_1[CH_3CHO] - k_2[CH_3][CH_3CHO] + k_3[CH_3CO] - k_4[CH_3]^2$$

$$0 = k_1[CH_3CHO] - k_2[CH_3][CH_3CHO] + k_2[CH_3][CH_3CHO] - k_4[CH_3]^2$$

$$[CH_3] = \left(\frac{k_1}{k_4}\right)^{1/2} [CH_3CHO]^{1/2}$$

Substituting this expression into the rate law for methane gives

$$\frac{d[CH_4]}{dt} = k_2 \left(\frac{k_1}{k_4}\right)^{1/2} [CH_3CHO]^{3/2}$$

29–25. Suppose that we replace the termination step (Equation 4) of the mechanism in Problem 29–24 with the termination reaction

$$2\,CH_3CO(g) \overset{k_4}{\Longrightarrow} CH_3COCOCH_3$$

Determine the rate laws for $CO(g)$, $CH_3(g)$, and $CH_3CO(g)$. Once again, assume that the steady-state approximation can be applied to the intermediates $CH_3(g)$ and $CH_3CO(g)$, and show that in this case the rate of formation of CO is given by

$$\frac{d[CO]}{dt} = \left(\frac{k_1}{k_4}\right)^{1/2} k_3[CH_3CHO]^{1/2}$$

The rate laws are

$$\frac{d[CO]}{dt} = k_3[CH_3CO]$$

$$\frac{d[CH_3]}{dt} = k_1[CH_3CHO] - k_2[CH_3][CH_3CHO] + k_3[CH_3CO]$$

$$\frac{d[CH_3CO]}{dt} = k_2[CH_3][CH_3CHO] - k_3[CH_3CO] - k_4[CH_3CHO]^2$$

Using the steady-state approximation gives

$$
\begin{array}{llll}
0 = & k_1[CH_3CHO] & -k_2[CH_3][CH_3CHO] & +k_3[CH_3CO] \\
+\,0 = & & +k_2[CH_3][CH_3CHO] & -k_3[CH_3CO] & -k_4[CH_3CHO]^2 \\
\hline
0 = & k_1[CH_3CHO] & & & -k_4[CH_3CHO]^2
\end{array}
$$

We can then write

$$k_4[CH_3CHO]^2 = k_1[CH_3CHO]$$

$$[CH_3CHO] = \left(\frac{k_1}{k_4}\right)^{1/2} [CH_3CHO]^{1/2}$$

Finally, substituting into the rate law for CO gives

$$\frac{d[CO]}{dt} = \left(\frac{k_1}{k_4}\right)^{1/2} k_3[CH_3CHO]^{1/2}$$

29–26. The chain length γ of a chain reaction is defined as the rate of the overall reaction divided by the rate of the initiation step. Give a physical interpretation of the chain length. Show that γ for the reaction mechanism and rate law given in Problem 29–25 is

$$\gamma = k_3 \left(\frac{1}{k_1 k_4}\right)^{1/2} [CH_3CHO]^{-1/2}$$

The chain length γ is the average number of times the propagation steps are repeated before a termination step occurs. For a decomposition reaction, it is the number of molecules decomposed by a single carrier molecule. (A carrier molecule is an intermediate responsible for chain propagation.) For the reaction mechanism given in Problem 29–25,

$$\gamma = \frac{\left(\dfrac{k_1}{k_4}\right)^{1/2} k_3[CH_3CHO]^{1/2}}{k_1[CH_3CHO]} = \left(\frac{1}{k_1 k_4}\right)^{1/2} k_3[CH_3CHO]^{-1/2}$$

29–27. Show that the chain length γ (see Problem 29–26) for the reaction mechanism and the rate law given in Problem 29–24 is

$$\gamma = k_2 \left(\frac{1}{k_1 k_4} \right)^{1/2} [CH_3CHO]^{1/2}$$

For the reaction mechanism given in Problem 29–24,

$$\gamma = \frac{\left(\dfrac{k_1}{k_4} \right)^{1/2} k_2[CH_3CHO]^{3/2}}{k_1[CH_3CHO]} = \left(\frac{1}{k_1 k_4} \right)^{1/2} k_2[CH_3CHO]^{1/2}$$

29–28. Consider the mechanism for the thermal decomposition of acetaldehyde given in Problem 29–24. Show that E_{obs}, the measured Arrhenius activation energy for the overall reaction, is given by

$$E_{obs} = E_2 + \tfrac{1}{2}(E_1 - E_4)$$

where E_i is the activation energy of the ith step of the reaction mechanism. How is A_{obs}, the measured Arrhenius pre-exponential factor for the overall reaction, related to the Arrhenius pre-exponential factors for the individual steps of the reaction mechanism?

The rate law is

$$\frac{d[CH_4]}{dt} = \left(\frac{k_1}{k_4} \right)^{1/2} k_2[CH_3CHO]^{3/2}$$

so

$$k_{obs} = \left(\frac{k_1}{k_4} \right)^{1/2} k_2$$

Using the Arrhenius equation, we can write

$$k_1 = A_1 e^{-E_1/RT} \qquad k_4 = A_4 e^{-E_4/RT} \qquad k_2 = A_2 e^{-E_2/RT}$$

Then

$$A_{obs} e^{-E_{obs}/RT} = k_{obs} = \left(\frac{k_1}{k_4} \right)^{1/2} k_2 = A_2 \left(\frac{A_1}{A_4} \right)^{1/2} e^{-(E_1-E_4)/2RT - E_2/RT}$$

From this, we find

$$A_{obs} = A_2 \left(\frac{A_1}{A_4} \right)^{1/2} \qquad \text{and} \qquad E_{obs} = E_2 + \frac{1}{2}(E_1 - E_4)$$

29–29. Consider the mechanism for the thermal decomposition of acetaldehyde given in Problem 29–25. Show that E_{obs}, the measured Arrhenius activation energy for the overall reaction, is given by

$$E_{obs} = E_3 + \tfrac{1}{2}(E_1 - E_4)$$

where E_i is the activation energy of the ith step of the reaction mechanism. How is A_{obs}, the measured Arrhenius pre-exponential factor for the overall reaction, related to the Arrhenius pre-exponential factors for the individual steps of the reaction mechanism?

The rate law is

$$\frac{d[CO]}{dt} = \left(\frac{k_1}{k_4}\right)^{1/2} k_3 [CH_3CHO]^{1/2}$$

so

$$k_{obs} = \left(\frac{k_1}{k_4}\right)^{1/2} k_3$$

This is identical to k_{obs} in the previous problem, except that k_2 has been replaced by k_3. Thus

$$A_{obs} = A_3 \left(\frac{A_1}{A_4}\right)^{1/2} \quad \text{and} \quad E_{obs} = E_3 + \frac{1}{2}(E_1 - E_4)$$

29–30. Consider the reaction between $H_2(g)$ and $Br_2(g)$ discussed in Section 29–7. Justify why we ignored the $H_2(g)$ dissociation reaction in favor of the $Br_2(g)$ dissociation reaction as being the initiating step of the reaction mechanism.

The bond strength of the H_2 bond is much greater than that of the Br_2 bond. From Table 18.2, we can see that D_0 for H_2 is 432 $kJ \cdot mol^{-1}$, while D_0 for Br_2 is only 190 $kJ \cdot mol^{-1}$. It is therefore more likely that the Br_2 bond will dissociate than that the H_2 bond will dissociate.

29–31. In Section 29-7, we considered the chain reaction between $H_2(g)$ and $Br_2(g)$. Consider the related chain reaction between $H_2(g)$ and $Cl_2(g)$.

$$Cl_2(g) + H_2(g) \longrightarrow 2 HCl(g)$$

The mechanism for this reaction is

$$Cl_2(g) + M(g) \overset{k_1}{\Longrightarrow} 2 Cl(g) + M(g) \tag{1}$$

$$Cl(g) + H_2(g) \overset{k_2}{\Longrightarrow} HCl(g) + H(g) \tag{2}$$

$$H(g) + Cl_2(g) \overset{k_3}{\Longrightarrow} HCl(g) + Cl(g) \tag{3}$$

$$2 Cl(g) + M(g) \overset{k_4}{\Longrightarrow} Cl_2(g) + M(g) \tag{4}$$

Label the initiation, propagation, and termination step(s). Use the following bond dissociation data to explain why it is reasonable not to include the analogous inhibition steps in this mechanism that are included in the mechanism for the chain reaction involving $Br_2(g)$.

Molecule	$D_0/kJ \cdot mol^{-1}$
H_2	432
HBr	363
HCl	428
Br_2	190
Cl_2	239

Initiation step: (1)
Propagation steps: (2), (3)
Termination step: (4)
The inhibition steps for the chain reaction between $H_2(g)$ and $Br_2(g)$ are (Section 29–7)

$$HBr(g) + H(g) \xrightarrow{k_{-2}} Br(g) + H_2(g) \tag{4a}$$

$$HBr(g) + Br(g) \xrightarrow{k_3} H(g) + Br_2(g) \tag{5a}$$

These reaction steps have molar enthalpies of

$$\Delta_{r,4a}\overline{H} = -D_0(H_2) + D_0(HBr) = -69 \text{ kJ·mol}^{-1}$$
$$\Delta_{r,5a}\overline{H} = -D_0(Br_2) + D_0(HBr) = 173 \text{ kJ·mol}^{-1}$$

We neglected the contribution of step (5a) because this reaction is so much more endothermic than Reaction 4a. The corresponding inhibition steps for the chain reaction between $H_2(g)$ and $Cl_2(g)$ would be

$$HCl(g) + H(g) \xrightarrow{k_{-2}} Cl(g) + H_2(g) \tag{i}$$

$$HCl(g) + Br(g) \xrightarrow{k_3} H(g) + Cl_2(g) \tag{ii}$$

with corresponding molar enthalpies

$$\Delta_{r,i}\overline{H} = -D_0(H_2) + D_0(HCl) = -4 \text{ kJ·mol}^{-1}$$
$$\Delta_{r,ii}\overline{H} = -D_0(Cl_2) + D_0(HCl) = 189 \text{ kJ·mol}^{-1}$$

Again, the contribution of the second step shown here is highly endothermic, and so we disregard this reaction. We also see that the first step shown is only slightly exothermic (in contrast with the corresponding step in the bromine reaction), so we can also neglect this step when describing the chain reaction involving Cl_2.

29–32. Derive the rate law for $v = (1/2)(d[HCl]/dt)$ for the mechanism of the

$$Cl_2(g) + H_2(g) \longrightarrow 2\,HCl(g)$$

reaction given in Problem 29–31.

From steps (2) and (3), we write

$$\frac{d[HCl]}{dt} = k_2[Cl][H_2] + k_3[H][Cl_2] \tag{1}$$

Applying the steady-state approximation to Cl and H gives

$$\frac{d[Cl]}{dt} = 0 = 2k_1[Cl_2][M] - 2k_4[Cl]^2[M] - k_2[Cl][H_2] + k_3[H][Cl_2]$$

$$+ \frac{d[H]}{dt} = 0 = + k_2[Cl][H_2] - k_3[H][Cl_2]$$

$$\rule{9cm}{0.4pt}$$

$$0 = 2k_1[Cl_2][M] - 2k_4[Cl]^2[M]$$

so

$$[Cl] = \left(\frac{k_1}{k_4}\right)^{1/2}[Cl_2]^{1/2}$$

Substituting this back into the steady-state approximation for H gives

$$[H] = \frac{k_2}{k_3} \frac{[Cl][H_2]}{[Cl_2]} = \frac{k_2}{k_3} \left(\frac{k_1}{k_4}\right)^{1/2} \frac{[Cl_2]^{1/2}[H_2]}{[Cl_2]} = \frac{k_2}{k_3} \left(\frac{k_1}{k_4}\right)^{1/2} \frac{[H_2]}{[Cl_2]^{1/2}}$$

Then Equation 1 becomes

$$\frac{d[HCl]}{dt} = k_2[Cl][H_2] + k_3[H][Cl_2]$$

$$= k_2 \left(\frac{k_1}{k_4}\right)^{1/2} [Cl_2]^{1/2}[H_2] + k_2 \left(\frac{k_1}{k_4}\right)^{1/2} [H_2][Cl_2]^{1/2}$$

$$= 2k_2 \left(\frac{k_1}{k_4}\right)^{1/2} [H_2][Cl_2]^{1/2}$$

or

$$\frac{1}{2} \frac{d[HCl]}{dt} = k_2 \left(\frac{k_1}{k_4}\right)^{1/2} [H_2][Cl_2]^{1/2}$$

29–33. It is possible to initiate chain reactions using photochemical reactions. For example, in place of the thermal initiation reaction for the $Br_2(g) + H_2(g)$ chain reaction

$$Br_2(g) + M \overset{k_1}{\Longrightarrow} 2\,Br(g) + M$$

we could have the photochemical initiation reaction

$$Br_2(g) + h\nu \Longrightarrow 2\,Br(g)$$

If we assume that all the incident light is absorbed by the Br_2 molecules and that the quantum yield for photodissociation is 1.00, then how does the photochemical rate of dissociation of Br_2 depend on I_{abs}, the number of photons per unit time per unit volume? How does $d[Br]/dt$, the rate of formation of Br, depend on I_{abs}? If you assume that the chain reaction is initiated only by the photochemical generation of Br, then how does $d[HBr]/dt$ depend on I_{abs}?

The quantum yield is the number of Br_2 molecules that react for each photon absorbed. The quantity $I_{abs}t$ is the concentration of photons available, so

$$\frac{d[Br_2]}{dt} = \frac{d(I_{abs}\Phi t)}{dt} = I_{abs}$$

The rate of formation of Br in this step would be twice the rate of dissociation of Br_2, so

$$\frac{d[Br]}{dt} = 2I_{abs}$$

The expressions for $d[HBr]/dt$ and $d[H]/dt$ that were found in Section 29–7 (Equations 29.61 and 29.62) still apply, but now applying the steady-state approximation to Br and H gives

$$\frac{d[Br]}{dt} = 0 = 2I_{abs} - k_2[Br][H_2] + k_{-2}[HBr][H] - k_{-1}[Br]^2 + k_3[H][Br_2]$$

$$+ \frac{d[H]}{dt} = 0 = \qquad k_2[Br][H_2] - k_{-2}[HBr][H] \qquad\qquad - k_3[H][Br_2]$$

$$\overline{\qquad\qquad\qquad\qquad\qquad\qquad\qquad\qquad\qquad\qquad\qquad\qquad}$$

$$0 = 2I_{abs} \qquad\qquad\qquad\qquad\qquad\qquad\qquad\qquad - k_{-1}[Br]^2$$

so

$$[Br] = \left(\frac{2I_{abs}}{k_{-1}}\right)^{1/2}$$

Substituting this back into the steady-state approximation for H gives

$$[H] = \frac{k_2[Br][H_2]}{k_{-2}[HBr] + k_3[Br_2]} = \left(\frac{2I_{abs}}{k_{-1}}\right)^{1/2} \frac{k_2[H_2]}{k_{-2}[HBr] + k_3[Br_2]}$$

Then Equation 29.61 becomes

$$\begin{aligned}
\frac{d[HBr]}{dt} &= k_2[Br][H_2] - k_{-2}[HBr][H] + k_3[H][Br_2] \\
&= k_2\left(\frac{2I_{abs}}{k_{-1}}\right)^{1/2}[H_2] - (k_{-2}[HBr] - k_3[Br_2])\left(\frac{2I_{abs}}{k_{-1}}\right)^{1/2}\frac{k_2[H_2]}{k_{-2}[HBr] + k_3[Br_2]} \\
&= k_2\left(\frac{2I_{abs}}{k_{-1}}\right)^{1/2}[H_2]\left[1 + \frac{k_3[Br_2] - k_{-2}[HBr]}{k_{-2}[HBr] + k_3[Br_2]}\right] \\
&= k_2\left(\frac{2I_{abs}}{k_{-1}}\right)^{1/2}\left[\frac{2[H_2]}{(k_{-2}/k_3)[HBr]/[Br_2] + 1}\right]
\end{aligned}$$

29–34. In Section 29–9, we derived the Michaelis-Menton rate law for enzyme catalysis. The derivation presented there is limited to the case in which only the rate of the initial reaction is measured so that $[S] = [S]_0$ and $[P] = 0$. We will now determine the Michaelis-Menton rate law by a different approach. Recall that the Michaelis-Menton mechanism is

$$E + S \underset{k_{-1}}{\overset{k_1}{\rightleftharpoons}} ES$$

$$ES \overset{k_2}{\Longrightarrow} E + P$$

The rate law for this reaction is $v = k_2[ES]$. Write the rate expression for [ES]. Show that if you apply the steady-state approximation to this intermediate, then

$$[ES] = \frac{[E][S]}{K_m} \tag{1}$$

where K_m is the Michaelis constant. Now show that

$$[E]_0 = [E] + \frac{[E][S]}{K_m} \tag{2}$$

(*Hint*: The enzyme is not consumed.) Solve Equation 2 for [E] and substitute the result into Equation 1 and thereby show that

$$v = \frac{k_2[E]_0[S]}{K_m + [S]} \tag{3}$$

If the rate is measured during a time period when only a small amount of substrate is consumed, then $[S] = [S]_0$ and Equation 3 reduces to the Michaelis-Menton rate law given by Equation 29.78.

Applying the steady-state approximation to the rate law for ES gives

$$\frac{d[ES]}{dt} = 0 = k_1[E][S] - k_{-1}[ES] - k_2[ES]$$

$$[ES] = \frac{k_1[E][S]}{k_{-1} + k_2} = \frac{[E][S]}{K_m}$$

Now using Equation 29.74 gives

$$[E]_0 = [ES] + [E] = \frac{[E][S]}{K_m} + [E]$$

$$= [E]\left(1 + \frac{[S]}{K_m}\right)$$

$$[E] = [E]_0 \frac{K_m}{K_m + [S]}$$

Substituting into Equation 1 gives

$$[ES] = \frac{[E]_0[S]}{K_m + [S]}$$

$$v = k_2[ES] = \frac{k_2[E]_0[S]}{K_m + [S]}$$

If $[S] = [S]_0$, this reduces to Equation 29.78.

29–35. The ability of enzymes to catalyze reactions can be hindered by *inhibitor molecules*. One of the mechanisms by which an inhibitor molecule works is by competing with the substrate molecule for binding to the active site of the enzyme. We can include this inhibition reaction in a modified Michaelis-Menton mechanism for enzyme catalysis.

$$E + S \underset{k_{-1}}{\overset{k_1}{\rightleftharpoons}} ES \tag{1}$$

$$E + I \underset{k_{-2}}{\overset{k_2}{\rightleftharpoons}} EI \tag{2}$$

$$ES \overset{k_3}{\Longrightarrow} E + P \tag{3}$$

In Equation 2, I is the inhibitor molecule and EI is the enzyme-inhibitor complex. We will consider the case where reaction (2) is always in equilibrium. Determine the rate laws for [S], [ES], [EI], and [P]. Show that if the steady-state assumption is applied to ES, then

$$[ES] = \frac{[E][S]}{K_m}$$

where K_m is the Michaelis constant, $K_m = (k_{-1} + k_3)/k_1$. Now show that material balance for the enzyme gives

$$[E]_0 = [E] + \frac{[E][S]}{K_m} + [E][I]K_I$$

where $K_I = [EI]/[E][I]$ is the equilibrium constant for step (2) of the above reaction mechanism. Use this result to show that the initial reaction rate is given by

$$v = \frac{d[P]}{dt} = \frac{k_3[E]_0[S]}{K_m + [S] + K_m K_I[I]} \approx \frac{k_3[E]_0[S]_0}{K_m' + [S]_0} \tag{4}$$

where $K_m' = K_m(1 + K_I[I])$. Note that the second expression in Equation 4 has the same functional form as the Michaelis-Menton equation. Does Equation 4 reduce to the expected result when $[I] \to 0$?

$$\frac{d[S]}{dt} = -k_1[E][S] + k_{-1}[ES]$$

$$\frac{d[ES]}{dt} = k_1[E][S] - k_{-1}[ES] - k_3[ES]$$

$$\frac{d[EI]}{dt} = k_2[E][I] - k_{-2}[EI]$$

$$\frac{d[P]}{dt} = k_3[ES]$$

Applying the steady-state assumption to ES gives

$$0 = k_1[E][S] - (k_{-1} + k_3)[ES]$$

$$[ES] = \frac{k_1}{k_{-1} + k_3}[E][S] = \frac{[E][S]}{K_m}$$

Summing the concentrations of all components containing E gives

$$[E]_0 = [E] + [ES] + [EI] = [E] + \frac{[E][S]}{K_m} + \frac{[EI]}{[E][I]}[E][I]$$

$$= [E] + \frac{[E][S]}{K_m} + [E][I]K_I$$

$$= [E]\left(\frac{K_m + [S] + [I]K_m K_I}{K_m}\right)$$

and so

$$[E] = \frac{K_m[E]_0}{K_m + [S] + [I]K_m K_I}$$

The rate of formation of product (the reaction rate) is given by $v = k_3[ES]$, so

$$v = \frac{d[P]}{dt} = k_3[ES] = k_3\frac{[E][S]}{K_m}$$

$$= k_3\frac{[E]_0[S]}{K_m + [S] + K_m K_I[I]}$$

$$\approx \frac{k_3[E]_0[S]_0}{K_m(1 + K_I[I]) + [S]_0} \approx \frac{k_3[E]_0[S]_0}{K_{m'} + [S]_0}$$

If $[I] \to 0$, then $K_{m'} \to K_m$ and so

$$v \to \frac{k_3[E]_0[S]_0}{K_m + [S]_0}$$

as we would expect.

29–36. Antibiotic-resistant bacteria have an enzyme, penicillinase, that catalyzes the decomposition of the antibiotic. The molecular mass of penicillinase is $30\,000$ g·mol^{-1}. The turnover number of the enzyme at 28°C is 2000 s^{-1}. If 6.4 μg of penicillinase catalyzes the destruction of 3.11 mg of amoxicillin, an antibiotic with a molecular mass of 364 g·mol^{-1}, in 20 seconds at 28°C, how many active sites does the enzyme have?

$$\frac{6.4 \times 10^{-6}\ \text{g}}{30\,000\ \text{g·mol}^{-1}} = 2.13 \times 10^{-10}\ \text{mol penicillinase}$$

$$\frac{3.11 \times 10^{-3}\ \text{g}}{364\ \text{g·mol}^{-1}} = 8.54 \times 10^{-6}\ \text{mol amoxicillin}$$

This is the amount of amoxicillin catalyzed in twenty seconds. Since the catalyst is not destroyed in the reaction, the same amount of penicillinase will destroy 2.56×10^{-5} mol of amoxicillin in one minute. Now

$$\text{turnover number} \times \text{\# active sites} = \frac{2.56 \times 10^{-5}\ \text{mol·min}^{-1}}{2.13 \times 10^{-10}\ \text{mol}}$$

$$2000\ \text{s}^{-1} \times \text{\# active sites} = 1.20 \times 10^5\ \text{min}^{-1}$$

$$\text{\# active sites} = \frac{2.00 \times 10^3\ \text{s}^{-1}}{2000\ \text{s}^{-1}} = 1$$

29–37. Show that the inverse of Equation 29.78 is

$$\frac{1}{v} = \frac{1}{v_{max}} + \frac{K_m}{v_{max}} \frac{1}{[S]_0} \tag{1}$$

This equation is called the *Lineweaver-Burk equation*. In Example 29–9, we examined the reaction for the hydration of CO_2 that is catalyzed by the enzyme carbonic anhydrase. For a total enzyme concentration of 2.32×10^{-9} mol·dm^{-3}, the following data were obtained.

v/mol·dm^{-3}·s^{-1}	$[CO_2]_0/10^{-3}$ mol·dm^{-3}
2.78×10^{-5}	1.25
5.00×10^{-5}	2.50
8.33×10^{-5}	5.00
1.66×10^{-4}	20.00

Plot these data according to Equation 1, and determine the values of K_m, the Michaelis constant, and k_2, the rate constant for product formation from the enzyme-substrate complex from the slope and intercept of the best-fit line to the plotted data.

Recall that $v_{max} = k_2[E]_0$, so we find

$$v = \frac{k_2[S]_0[E]_0}{K_m + [S]_0}$$

$$\frac{1}{v} = \frac{K_m}{k_2[S]_0[E]_0} + \frac{1}{k_2[E]_0}$$

$$\frac{1}{v} = \frac{K_m}{v_{max}[S]_0} + \frac{1}{v_{max}}$$

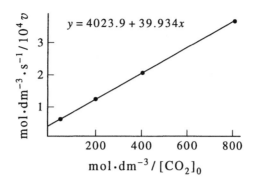

We can now solve for v_{max} and K_m:

$$\frac{1}{v_{max}} = 4020 \text{ dm}^3 \cdot \text{s} \cdot \text{mol}^{-1}$$

$$v_{max} = 2.49 \times 10^{-4} \text{ mol} \cdot \text{dm}^{-3} \cdot \text{s}^{-1}$$

$$\frac{K_m}{v_{max}} = 39.9 \text{ s}$$

$$K_m = 9.94 \times 10^{-3} \text{ mol} \cdot \text{dm}^{-3}$$

Because we are given that $[E]_0 = 2.32 \times 10^{-9} \text{ mol} \cdot \text{dm}^{-3}$, we find

$$k_2 = \frac{v_{max}}{[E]_0} = \frac{2.49 \times 10^{-4} \text{ mol} \cdot \text{dm}^{-3} \cdot \text{s}^{-1}}{2.32 \times 10^{-9} \text{ mol} \cdot \text{dm}^{-3}} = 1.07 \times 10^5 \text{ s}^{-1}$$

29–38. Carbonic anhydrase catalyzes the reaction

$$H_2O(l) + CO_2(g) \rightleftharpoons H_2CO_3(aq)$$

Data for the reverse dehydration reaction using a total enzyme concentration of $2.32 \times 10^{-9} \text{ mol} \cdot \text{dm}^{-3}$ are given below

$v/\text{mol} \cdot \text{dm}^{-3} \cdot \text{s}^{-1}$	$[H_2CO_3]_0/10^{-3} \text{ mol} \cdot \text{dm}^{-3}$
1.05×10^{-5}	2.00
2.22×10^{-5}	5.00
3.45×10^{-5}	10.00
4.17×10^{-5}	15.00

Use the approach discussed in Problem 29–37 to determine the values of K_m, the Michaelis constant, and k_2, the rate of product formation from the enzyme substrate complex.

We use the method developed in Problem 29–37, and plot $1/v$ versus $1/[H_2CO_3]_0$:

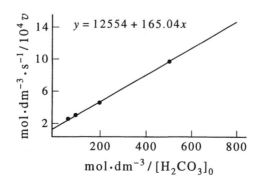

As before,

$$\frac{1}{v_{max}} = 12\,600 \text{ dm}^3 \cdot \text{s} \cdot \text{mol}^{-1}$$

$$v_{max} = 7.94 \times 10^{-5} \text{ mol} \cdot \text{dm}^{-3} \cdot \text{s}^{-1}$$

$$\frac{K_m}{v_{max}} = 165 \text{ s}$$

$$K_m = 1.31 \times 10^{-2} \text{ mol} \cdot \text{dm}^{-3}$$

Because we are given $[E]_0 = 2.32 \times 10^{-9}$ mol·dm^{-3}, we find

$$k_2 = \frac{v_2}{[E]_0} = 3.43 \times 10^4 \text{ s}^{-1}$$

29–39. Show that the Michaelis-Menton mechanism for enzyme catalysis gives $v = (1/2)v_{max}$ when $[S]_0 = K_m$.

The Michaelis-Menton mechanism gives the rate expression

$$v = \frac{k_2[S]_0[E]_0}{K_m + [S]_0} \tag{29.78}$$

If $[S]_0 = K_m$, then this becomes

$$v = \frac{k_2 K_m [E]_0}{2K_m} = \frac{k_2[E]_0}{2} = \frac{v_{max}}{2}$$

29–40. The protein catalase catalyzes the reaction

$$2\,H_2O_2(aq) \longrightarrow 2\,H_2O(l) + O_2(g)$$

and has a Michaelis constant of $K_m = 25 \times 10^{-3}$ mol·dm^{-3} and a turnover number of 4.0×10^7 s^{-1}. Calculate the initial rate of this reaction if the total enzyme concentration is 0.016×10^{-6} mol·dm^{-3} and the initial substrate concentration is 4.32×10^{-6} mol·dm^{-3}. Calculate v_{max} for this enzyme. Catalase has a single active site.

The Michaelis-Menton equation is

$$v = \frac{k_2[S]_0[E]_0}{K_m + [S]_0} = \frac{v_{max}[S]_0}{K_m + [S]_0} \tag{29.78}$$

The turnover number for a single active site catalyst is equal to $v_{max}/[E]_0$, so

$$v_{max} = (4.0 \times 10^7 \text{ s}^{-1})(0.016 \times 10^{-6} \text{ mol} \cdot \text{dm}^{-3})$$

$$= 0.64 \text{ mol} \cdot \text{dm}^{-3} \cdot \text{s}^{-1}$$

and

$$v = \frac{(0.64 \text{ mol} \cdot \text{dm}^{-3} \cdot \text{s}^{-1})(4.32 \times 10^{-6} \text{ mol} \cdot \text{dm}^{-3})}{(25 \times 10^{-3} \text{ mol} \cdot \text{dm}^{-3}) + (4.32 \times 10^{-6} \text{ mol} \cdot \text{dm}^{-3})} = 1.11 \times 10^{-4} \text{ mol} \cdot \text{dm}^{-3} \cdot \text{s}^{-1}$$

29–41. The presence of 4.8×10^{-6} $mol \cdot dm^{-3}$ of a competitive inhibitor decreases the initial rate calculated in Problem 29–40 by a factor of 3.6. Calculate K_I, the equilibrium constant for the binding reaction between the enzyme and the inhibitor. (*Hint:* See Problem 29–35.)

The inital rate is now given by (from Problem 29–35)

$$v = \frac{k_3[E]_0[S]_0}{K_m(1 + K_I[I]) + [S]} = \frac{v_{max}[S]_0}{K_m(1 + K_I[I]) + [S]}$$

(recall that, in this case, $k_3[E]_0 = v_{max}$). The maximum rate is the same for this problem as the previous problem, and v can be found by dividing the initial rate found in Problem 29–40 by 3.6, so

$$3.08 \times 10^{-5} = \frac{(0.64)(4.32 \times 10^{-6})}{(25 \times 10^{-3})[1 + K_I(4.8 \times 10^{-6} \, mol \cdot dm^{-3})] + 4.32 \times 10^{-6}}$$

$$1 + K_I(4.8 \times 10^{-6} \, mol \cdot dm^{-3}) = \frac{(0.64)(4.32 \times 10^{-6}) - (4.32 \times 10^{-6})(3.08 \times 10^{-5})}{(3.08 \times 10^{-5})(25 \times 10^{-3})}$$

$$K_I = 5.4 \times 10^5 \, dm^3 \cdot mol^{-1}$$

29–42. The turnover number for acetylcholinesterase, an enzyme with a single active site that metabolizes acetylcholine, is 1.4×10^4 s^{-1}. How many grams of acetylcholine can 2.16×10^{-6} g of acetylcholinesterase metabolize in one hour? (Take the molecular mass of the enzyme to be 4.2×10^4 $g \cdot mol^{-1}$; acetylcholine has the molecular formula $C_7NO_2H_{16}^+$.)

$$\frac{2.16 \times 10^{-6} \, g}{4.2 \times 10^4 \, g \cdot mol^{-1}} = 5.1 \times 10^{-11} \, mol \text{ acetocholinesterase}$$

The enzyme has a single active site, so

$$v_{max} = (\text{turnover number})(\text{mol acetocholinesterase})$$

$$= (1.4 \times 10^4 \, s^{-1})(5.1 \times 10^{-11} \, mol)\left(\frac{3600 \, hr^{-1}}{1 \, s^{-1}}\right) = 2.6 \times 10^{-3} \, mol \cdot hr^{-1}$$

The molecular mass of acetylcholine is 146.21 $g \cdot mol^{-1}$, so 0.38 g of acetylcholine are metabolized in one hour.

29–43. Consider the following mechanism for the recombination of bromine atoms to form molecular bromine

$$2 \, Br(g) \underset{k_{-1}}{\overset{k_1}{\rightleftharpoons}} Br_2^*(g)$$

$$Br_2^*(g) + M(g) \overset{k_2}{\Longrightarrow} Br_2(g) + M(g)$$

The first step results in formation of an energized bromine molecule. This excess energy is then removed by a collision with a molecule M in the sample. Show that if the steady-state approximation is applied to $Br_2^*(g)$, then

$$\frac{d[Br]}{dt} = -\frac{2k_1k_2[Br]^2[M]}{k_{-1} + k_2[M]}$$

Determine the limiting expression for $d[Br]/dt$ when $v_2 \gg v_{-1}$. Determine the limiting expression for $d[Br]/dt$ when $v_2 \ll v_{-1}$.

We can write the rate laws for Br and Br_2^* as

$$\frac{d[Br]}{dt} = -2k_1[Br]^2 + 2k_{-1}[Br_2^*]$$

$$\frac{d[Br_2^*]}{dt} = k_1[Br]^2 - k_{-1}[Br_2^*] - k_2[Br_2^*][M]$$

Using the steady-state approximation for Br_2^* gives

$$[Br_2^*] = \frac{k_1[Br]^2}{k_{-1} + k_2[M]}$$

Then

$$\frac{d[Br]}{dt} = -2k_1[Br]^2 + \frac{2k_1 k_{-1}[Br]^2}{k_{-1} + k_2[M]}$$

$$= \frac{-2k_1 k_{-1}[Br]^2 - 2k_1[Br]^2 k_2[M] + 2k_{-1}k_1[Br]^2}{k_{-1} + k_2[M]}$$

$$= \frac{-2k_1 k_2[Br]^2[M]}{k_{-1} + k_2[M]}$$

If $v_2 \gg v_{-1}$, then $k_2[Br_2^*][M] \gg k_{-1}[Br_2^*]$ and so $k_2[M] \gg k_{-1}$. The rate equation then becomes

$$\frac{d[Br]}{dt} = -2k_1[Br]^2$$

If $v_2 \ll v_{-1}$, then $k_2[M] \ll k_{-1}$. The rate equation then becomes

$$\frac{d[Br]}{dt} = -\frac{2k_1 k_2}{k_{-1}}[Br]^2[M]$$

29–44. A mechanism for the recombination of bromine atoms to form molecular bromine is given in Problem 29–43. When this reaction occurs in the presence of a large excess of buffer gas, a negative activation energy is measured. Because M(g), the buffer gas molecule, is responsible for the deactivation of $Br_2^*(g)$ but is not consumed itself by the reaction, we can consider it to be a catalyst. Below are the measured rate constants for this reaction in the presence of the same concentration of excess Ne(g) and CCl_4(g) buffer gases at several temperatures. Which gas is the better catalyst for this reaction?

	Ne	CCl_4
T/K	k_{obs}/mol$^{-2} \cdot$dm$^6 \cdot$s^{-1}	k_{obs}/mol$^{-2} \cdot$dm$^6 \cdot$s^{-1}
367	1.07×10^9	1.01×10^{10}
349	1.15×10^9	1.21×10^{10}
322	1.31×10^9	1.64×10^{10}
297	1.50×10^9	2.28×10^{10}

Why do you think there is a difference in the "catalytic" behavior of these two buffer gases?

CCl_4 is the better catalyst for this reaction. There is a difference in the catalytic behavior of the gases because CCl_4 has vibrational and rotational energy levels, which make it more effective in absorbing energy from the excited bromine molecule.

29–45. The standard Gibbs energy change of reaction for

$$2H_2(g) + O_2(g) \longrightarrow 2H_2O(g)$$

is -457.4 kJ at 298 K. At room temperature, however, this reaction does not occur and mixtures of gaseous hydrogen and oxygen are stable. Explain why this is so. Is such a mixture indefinitely stable?

Although the reaction is spontaneous, it has a very high energy of activation, and so it is very unlikely that the reactants will have enough energy to create water vapor. This mixture will be indefinitely stable as long as no external catalysts are present to lower the activation energy and facilitate the reaction. (It will eventually react even in the absence of catalysts, however, as the reaction is spontaneous).

29–46. The HF(g) chemical laser is based on the reaction

$$H_2(g) + F_2(g) \longrightarrow 2HF(g)$$

The mechanism for this reaction involves the elementary steps

		$\Delta_r H°/\text{kJ·mol}^{-1}$ at 298K
(1)	$F_2(g) + M(g) \underset{k_{-1}}{\overset{k_1}{\rightleftharpoons}} 2F(g) + M(g)$	$+159$
(2)	$F(g) + H_2(g) \underset{k_{-2}}{\overset{k_2}{\rightleftharpoons}} HF(g) + H(g)$	-134
(3)	$H(g) + F_2(g) \overset{k_3}{\Longrightarrow} HF(g) + F(g)$	-411

Comment on why the reaction $H_2(g) + M(g) \to 2H(g) + M(g)$ is not included in the mechanism of the HF(g) laser even though it produces a reactant that could participate in step (3) of the reaction mechanism. Derive the rate law for $d[\text{HF}]/dt$ for the above mechanism assuming that the steady-state approximation can be applied to both intermediate species, F(g) and H(g).

The reaction $H_2(g) + M(g) \to 2H(g) + M(g)$ is not included in the mechanism of the laser because a very large energy would be needed to break the H_2 bond (≈ 432 kJ·mol^{-1}). Now we can write the rate laws for HF, F, and H:

$$\frac{d[\text{HF}]}{dt} = k_2[\text{F}][\text{H}_2] - k_{-2}[\text{HF}][\text{H}] + k_3[\text{H}][\text{F}_2]$$

$$\frac{d[\text{F}]}{dt} = 2k_1[\text{F}_2][\text{M}] - 2k_{-1}[\text{F}]^2[\text{M}] - k_2[\text{F}][\text{H}_2] + k_{-2}[\text{HF}][\text{H}] + k_3[\text{H}][\text{F}_2]$$

$$\frac{d[\text{H}]}{dt} = k_2[\text{F}][\text{H}_2] - k_{-2}[\text{HF}][\text{H}] - k_3[\text{H}][\text{F}_2]$$

Applying the steady-state approximation gives

$$0 = 2k_1[F_2][M] - 2k_{-1}[F]^2[M] - k_2[F][H_2] + k_{-2}[HF][H] + k_3[H][F_2]$$

$$+ 0 = \qquad\qquad\qquad + k_2[F][H_2] - k_{-2}[HF][H] - k_3[H][F_2]$$

$$\overline{0 = 2k_1[F_2][M] - 2k_{-1}[F]^2[M]}$$

so

$$[F] = \frac{k_1^{1/2}[F_2]^{1/2}}{k_{-1}^{1/2}}$$

Substituting into the steady-state approximation for H gives

$$0 = k_2 \frac{k_1^{1/2}[F_2]^{1/2}[H_2]}{k_{-1}^{1/2}} - k_{-2}[HF][H] - k_3[H][F_2]$$

$$[H] = \frac{k_2 \left(\dfrac{k_1}{k_{-1}}\right)^{1/2}[F_2]^{1/2}[H_2]}{k_{-2}[HF] + k_3[F_2]}$$

Finally, substitute into the rate equation for HF to get

$$\frac{d[HF]}{dt} = k_2[F][H_2] - k_{-2}[HF][H] + k_3[H][F_2]$$

$$= \left(\frac{k_1}{k_{-1}}\right)^{1/2} k_2[F_2]^{1/2}[H_2] + (k_3[F_2] - k_{-2}[HF]) \frac{k_2 \left(\dfrac{k_1}{k_{-1}}\right)^{1/2}[F_2]^{1/2}[H_2]}{k_{-2}[HF] + k_3[F_2]}$$

$$= \left(\frac{k_1}{k_{-1}}\right)^{1/2} k_2[F_2]^{1/2}[H_2] \left(1 + \frac{k_3[F_2] - k_{-2}[HF]}{k_{-2}[HF] + k_3[F_2]}\right)$$

$$= \left(\frac{k_1}{k_{-1}}\right)^{1/2} k_2[F_2]^{1/2}[H_2] \frac{2k_3[F_2]}{k_{-2}[HF] + k_3[F_2]}$$

$$\frac{1}{2}\frac{d[HF]}{dt} = \left(\frac{k_1}{k_{-1}}\right)^{1/2} k_2[F_2]^{1/2}[H_2] \frac{1}{1 + (k_{-2}/k_3)[HF][F_2]^{-1}}$$

29–47. A mechanism for ozone creation and destruction in the stratosphere is

$$O_2(g) + h\nu \xrightarrow{j_1} O(g) + O(g)$$

$$O(g) + O_2(g) + M(g) \xrightarrow{k_2} O_3(g) + M(g)$$

$$O_3(g) + h\nu \xrightarrow{j_3} O_2(g) + O(g)$$

$$O(g) + O_3(g) \xrightarrow{k_4} O_2(g) + O_2(g)$$

where we have used the symbol j to indicate that the rate constant is for a photochemical reaction. Determine the rate expressions for $d[O]/dt$ and $d[O_3]/dt$. Assume that both intermediate species, $O(g)$ and $O_3(g)$, can be treated by the steady-state approximation and thereby show that

$$[O] = \frac{2j_1[O_2] + j_3[O_3]}{k_2[O_2][M] + k_4[O_3]} \tag{1}$$

and

$$[O_3] = \frac{k_2[O][O_2][M]}{j_3 + k_4[O]} \tag{2}$$

Now substitute Equation 1 into Equation 2 and solve the resulting quadratic formula for $[O_3]$ to obtain

$$[O_3] = [O_2]\frac{j_1}{2j_3}\left\{\left(1 + 4\frac{j_3}{j_1}\frac{k_2}{k_4}[M]\right)^{1/2} - 1\right\}$$

Typical values for these parameters at an altitude of 30 km are $j_1 = 2.51 \times 10^{-12}$ s^{-1}, $j_3 = 3.16 \times 10^{-4}$ s^{-1}, $k_2 = 1.99 \times 10^{-33}$ $cm^6 \cdot molecule^{-2} \cdot s^{-1}$, $k_4 = 1.26 \times 10^{-15}$ $cm^3 \cdot molecule^{-1} \cdot s^{-1}$, $[O_2] = 3.16 \times 10^{17}$ $molecule \cdot cm^{-3}$, and $[M] = 3.98 \times 10^{17}$ $molecule \cdot cm^{-3}$. Find $[O_3]$ and $[O]$ at an altitude of 30 km using Equations 1 and 2. Was the use of the steady-state assumption justified?

The rate expressions for the intermediate species O and O_3 are

$$\frac{d[O]}{dt} = 2j_1[O_2] + j_3[O_3] - k_2[O][O_2][M] - k_4[O][O_3]$$

$$\frac{d[O_3]}{dt} = k_2[O][O_2][M] - j_3[O_3] - k_4[O][O_3]$$

Applying the steady-state approximation gives (Equation 1)

$$k_4[O][O_3] + k_2[O][O_2][M] = 2j_1[O_2] + j_3[O_3]$$

$$[O] = \frac{2j_1[O_2] + j_3[O_3]}{k_4[O_3] + k_2[O_2][M]}$$

and (Equation 2)

$$j_3[O_3] + k_4[O][O_3] = k_2[O][O_2][M]$$

$$[O_3] = \frac{k_2[O][O_2][M]}{j_3 + k_4[O]}$$

Substituting Equation 1 into Equation 2 gives

$$[O_3]\left(j_3 + k_4[O]\right) = k_2[O][O_2][M]$$

$$j_3[O_3] + k_4[O_3]\frac{2j_1[O_2] + j_3[O_3]}{k_4[O_3] + k_2[O_2][M]} = k_2[O_2][M]\frac{2j_1[O_2] + j_3[O_3]}{k_4[O_3] + k_2[O_2][M]}$$

$$j_3k_2[O_3][O_2][M] + k_4j_3[O_3]^2 + 2j_1k_4[O_3][O_2] + k_4j_3[O_3]^2 = 2k_2j_1[O_2]^2[M]$$

$$+ j_3k_2[O_3][O_2][M]$$

This equation can be written as

$$0 = 2k_4j_3[O_3]^2 + 2j_1k_4[O_3][O_2] - 2k_2j_1[O_2]^2[M]$$

$$0 = [O_3]^2 + \frac{j_1}{j_3}[O_2][O_3] - \frac{k_2j_1}{k_4j_3}[O_2]^2[M]$$

$$[O_3] = -\frac{j_1}{2j_3}[O_2] \pm \frac{1}{2}\sqrt{\left(\frac{j_1}{j_3}\right)^2[O_2]^2 + 4\frac{k_2}{k_4}\frac{j_1}{j_3}[O_2]^2[M]}$$

$$= -\frac{j_1}{2j_3}[O_2] \pm \sqrt{\left(\frac{j_1}{2j_3}\right)^2[O_2]^2 + \frac{k_2}{k_4}\frac{j_1}{j_3}[O_2]^2[M]}$$

$$= -\frac{j_1}{2j_3}[O_2] \pm \frac{j_1}{2j_3}[O_2]\sqrt{1 + \frac{4k_2}{k_4}\frac{j_3}{j_1}[M]}$$

We must have a positive value of $[O_3]$, so we have

$$[O_3] = -\frac{j_1}{2j_3}[O_2] + \frac{j_1}{2j_3}[O_2]\sqrt{1 + \frac{4k_2}{k_4}\frac{j_3}{j_1}[M]}$$

$$= [O_2]\frac{j_1}{2j_3}\left[\left(1 + 4\frac{j_3}{j_1}\frac{k_2}{k_4}[M]\right)^{1/2} - 1\right]$$

For the given parameters,

$$[O_3] = (3.16 \times 10^{17} \text{ molecule} \cdot \text{cm}^{-3})\frac{2.51 \times 10^{-12} \text{ s}^{-1}}{2(3.16 \times 10^{-4} \text{ s}^{-1})}$$

$$\left\{\left[1 + 4\frac{(3.16 \times 10^{-4} \text{ s}^{-1})(1.99 \times 10^{-33} \text{ cm}^6 \cdot \text{molecule}^{-2} \cdot \text{s}^{-1})}{(2.51 \times 10^{-12} \text{ s}^{-1})(1.26 \times 10^{15} \text{ cm}^3 \cdot \text{molecule}^{-1} \cdot \text{s}^{-1})}\right.\right.$$

$$\left.\left.\times (3.98 \times 10^{17} \text{ molecule} \cdot \text{cm}^{-3})\right]^{1/2} - 1\right\}$$

$$= (1.25 \times 10^9 \text{ molecule} \cdot \text{cm}^{-3})\left(1.78 \times 10^4\right)$$

$$= 2.23 \times 10^{13} \text{ molecule} \cdot \text{cm}^{-3}$$

This value is about 10^4 less than the value of $[M]$. Using this value for $[O_3]$ in Equation 1 gives (we drop the units for convenience)

$$[O] = \frac{2(2.51 \times 10^{-12})(3.16 \times 10^{17}) + (3.16 \times 10^{-4})(2.23 \times 10^{13})}{(1.99 \times 10^{-33})(3.16 \times 10^{17})(3.98 \times 10^{17}) + (1.26 \times 10^{-15})(2.23 \times 10^{13})}$$

$$= 2.82 \times 10^7 \text{ molecules} \cdot \text{cm}^{-3}$$

so the use of the steady-state approximation is justified.

In the next four problems we shall examine the explosive reaction

$$2 H_2(g) + O_2(g) \rightleftharpoons 2 H_2O(g)$$

29–48. A simplified mechanism for this reaction is

$$\text{electric spark} + H_2(g) \Longrightarrow 2 H(g) \tag{1}$$

$$H(g) + O_2(g) \xrightarrow{k_1} OH(g) + O(g) \tag{2}$$

$$O(g) + H_2(g) \xrightarrow{k_2} OH(g) + H(g) \tag{3}$$

$$H_2(g) + OH(g) \xrightarrow{k_3} H_2O(g) + H(g) \tag{4}$$

$$H(g) + O_2(g) + M(g) \xrightarrow{k_4} HO_2(g) + M(g) \tag{5}$$

A reaction that produces more molecules that can participate in chain-propagation steps than it consumes is called a branching chain reaction. Label the branching chain reaction(s), initiation reaction(s), propagation reaction(s), and termination reaction(s) for this mechanism. Use the following bond dissociation energies to evaluate the energy change for steps (2) and (3).

Molecule	$D_0/\text{kJ} \cdot \text{mol}^{-1}$
H_2	432
O_2	493
OH	424

Branching chain: (2), (3)
Initiation: (1)
Propagation: (4)
Termination: (5)

Using the bond dissocation energies, we find that

$$\Delta_2 \text{Energy} = -D_0(\text{OH}) + D_0(\text{O}_2) = 69 \text{ kJ} \cdot \text{mol}^{-1}$$

$$\Delta_3 \text{Energy} = -D_0(\text{OH}) + D_0(\text{H}_2) = 8 \text{ kJ} \cdot \text{mol}^{-1}$$

29–49. Using the mechanism given in Problem 29–48, determine the rate expression for [H] when the initiation step involves an electric spark that gives rise to a rate I_0 of the hydrogen atom production. Determine the rate expresions for [OH] and [O]. Assume that $[\text{O}] \approx [\text{OH}] \ll [\text{H}]$, so now we can apply the steady-state approximation to the intermediate species, O(g) and OH(g). Show that this use of the steady-state approximation gives

$$[\text{O}] = \frac{k_1[\text{H}][\text{O}_2]}{k_2[\text{H}_2]} \quad \text{and} \quad [\text{OH}] = \frac{2k_1[\text{H}][\text{O}_2]}{k_3[\text{H}_2]}$$

Use these results and your rate expression for [H] to show that

$$\frac{d[\text{H}]}{dt} = I_0 + (2k_1[\text{O}_2] - k_4[\text{O}_2][\text{M}])[\text{H}]$$

We must add the rate I_0 to the rate expression for [H] we would find without the electric spark, so

$$\frac{d[\text{H}]}{dt} = I_0 - k_1[\text{H}][\text{O}_2] + k_2[\text{O}][\text{H}_2] + k_3[\text{OH}][\text{H}_2] - k_4[\text{H}][\text{O}_2][\text{M}]$$

$$\frac{d[\text{OH}]}{dt} = k_1[\text{H}][\text{O}_2] + k_2[\text{O}][\text{H}_2] - k_3[\text{OH}][\text{H}_2]$$

$$\frac{d[\text{O}]}{dt} = k_1[\text{H}][\text{O}_2] - k_2[\text{O}][\text{H}_2]$$

Using the steady-state approximation for O and OH gives

$$0 = k_1[\text{H}][\text{O}_2] - k_2[\text{O}][\text{H}_2]$$

$$[\text{O}] = \frac{k_1[\text{H}][\text{O}_2]}{k_2[\text{H}_2]}$$

$$0 = k_1[\text{H}][\text{O}_2] + k_2[\text{O}][\text{H}_2] - k_3[\text{OH}][\text{H}_2]$$

$$[\text{OH}] = \frac{k_1[\text{H}][\text{O}_2] + k_2[\text{O}][\text{H}_2]}{k_3[\text{H}_2]}$$

$$= \frac{2k_1[\text{H}][\text{O}_2]}{k_3[\text{H}_2]}$$

Substituting into the rate expression for [H] gives

$$\frac{d[\text{H}]}{dt} = I_0 - k_1[\text{H}][\text{O}_2] + k_2[\text{O}][\text{H}_2] + k_3[\text{OH}][\text{H}_2] - k_4[\text{H}][\text{O}_2][\text{M}]$$

$$= I_0 - k_1[\text{H}][\text{O}_2] + \frac{k_1 k_2[\text{H}][\text{H}_2][\text{O}_2]}{k_2[\text{H}_2]} + \frac{2k_1 k_3[\text{H}_2][\text{H}][\text{O}_2]}{k_3[\text{H}_2]} - k_4[\text{H}][\text{O}_2][\text{M}]$$

$$= I_0 - k_1[\text{H}][\text{O}_2] + k_1[\text{H}][\text{O}_2] + 2k_1[\text{H}][\text{O}_2] - k_4[\text{H}][\text{O}_2][\text{M}]$$

$$= I_0(2k_1[\text{O}_2] - k_4[\text{O}_2][\text{M}])[\text{H}]$$

29–50. Consider the result of Problem 29–49. The rate of hydrogen atom production has a functional dependence of

$$\frac{d[\text{H}]}{dt} = I_0 + (\alpha - \beta)[\text{H}] \tag{1}$$

Which step(s) of the chemical reaction are responsible for the magnitudes of α and β? We can envision two solutions to this rate law, one for $\alpha > \beta$ and one for $\alpha < \beta$. For $\alpha < \beta$ show that the solution to Equation 1 becomes

$$[\text{H}] = \frac{I_0}{\beta - \alpha}(1 - e^{-(\beta - \alpha)t})$$

Plot [H] as a function of time. Determine the slope of the plot at short times. Determine the final steady-state value of [H].

Write the rate expression for [H] as

$$\frac{d[\text{H}]}{dt} + (\beta - \alpha)[\text{H}] = I_0$$

This is of the form discussed in Problem 29–5, so

$$[\text{H}] = e^{-\int (\beta - \alpha)dt}\left[\int e^{\int (\beta - \alpha)dt} I_0 dt\right]$$

$$= e^{-(\beta - \alpha)t}\left[\int e^{(\beta - \alpha)t} I_0 dt\right]$$

$$= e^{-(\beta - \alpha)t}\left[\frac{e^{(\beta - \alpha)t}}{\beta - \alpha}I_0 - \frac{I_0}{\beta - \alpha}\right]$$

$$= \frac{I_0}{\beta - \alpha} - \frac{I_0}{\beta - \alpha}e^{-(\beta - \alpha)t} = \frac{I_0}{\beta - \alpha}\left[1 - e^{-(\beta - \alpha)t}\right]$$

We plot $[\text{H}](\beta - \alpha)$ versus t below.

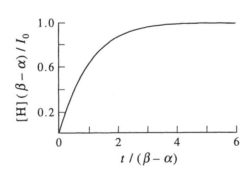

At short times the slope of the plot is $I_0 t$ and the final steady-state value of [H] is $I_0/(\beta - \alpha)$.

29–51. We now consider the solution to the equation (Problem 29–50)

$$\frac{d[\text{H}]}{dt} = I_0 + (\alpha - \beta)[\text{H}]$$

when $\alpha > \beta$. Show that the solution to this differential equation is given by

$$[\text{H}] = \frac{I_0}{\alpha - \beta}(e^{(\alpha - \beta)t} - 1)$$

Plot [H] as a function of time. Describe the differences observed between this plot and that obtained in Problem 29–50. Which case do you think is characteristic of a chemical explosion?

The solution is the same as in the previous problem, since the relative magnitudes of α and β were not considered when solving the differential equation. However, since $\alpha > \beta$, to make their difference positive we can express the solution above in terms of $\alpha - \beta$, instead of $\beta - \alpha$. Then

$$[H] = \frac{I_0}{\beta - \alpha}\left[1 - e^{-(\beta-\alpha)t}\right]$$

$$= \frac{-I_0}{\alpha - \beta}\left[1 - e^{(\alpha-\beta)t}\right]$$

$$= \frac{I_0}{\alpha - \beta}\left[e^{(\alpha-\beta)t} - 1\right]$$

We plot $[H](\alpha - \beta)$ versus t:

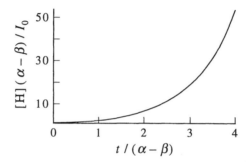

In this case the concentration of H does not converge as $t \to \infty$, and so this case is characteristic of a chemical explosion.

CHAPTER 30

Gas-Phase Reaction Dynamics

PROBLEMS AND SOLUTIONS

30–1. Calculate the hard-sphere collision theory rate constant for the reaction

$$NO(g) + Cl_2(g) \Longrightarrow NOCl(g) + Cl(g)$$

at 300 K. The collision diameters of NO and Cl_2 are 370 pm and 540 pm, respectively. The Arrhenius parameters for the reaction are $A = 3.981 \times 10^9 \, dm^3 \cdot mol^{-1} \cdot s^{-1}$ and $E_a = 84.9 \, kJ \cdot mol^{-1}$. Calculate the ratio of the hard-sphere collision theory rate constant to the experimental rate constant at 300 K.

We can use the first of Equations 27.58 to determine σ_{AB}:

$$\sigma_{AB} = \pi d_{AB}^2 = \pi \left(\frac{370 \, pm + 540 \, pm}{2} \right)^2 = 6.50 \times 10^{-19} \, m^2$$

and the second of Equations 27.58 to determine $\langle u_r \rangle$:

$$\langle u_r \rangle = \left(\frac{8 k_B T}{\pi \mu} \right)^{1/2}$$

$$= \left\{ \frac{8(1.381 \times 10^{-23} \, J \cdot K^{-1})(300 \, K)}{\pi \left[\frac{(70.906 \, amu)(30.006 \, amu)}{(100.912 \, amu)} \right] (1.661 \times 10^{-27} \, kg \cdot amu^{-1})} \right\}^{1/2}$$

$$= 549 \, m \cdot s^{-1}$$

Now substitute into Equation 30.5 to find the rate constant, k, for the hard-sphere collision theory.

$$k_{theor} = (1000 \, dm^3 \cdot m^{-3}) N_A \sigma_{AB} \langle u_r \rangle$$

$$= (1000 \, dm^3 \cdot m^{-3})(6.022 \times 10^{23} \, mol^{-1})(6.50 \times 10^{-19} \, m^2)(549 \, m \cdot s^{-1})$$

$$= 2.15 \times 10^{11} \, dm^3 \cdot mol^{-1} \cdot s^{-1}$$

We use Equation 28.57 to find the experimental value of k:

$$k_{exp} = A e^{-E_a/RT}$$

$$= (3.981 \times 10^9 \, dm^3 \cdot mol^{-1} \cdot s^{-1}) \exp \left[-\frac{84900 \, J \cdot mol^{-1}}{(8.315 \, J \cdot K^{-1} \cdot mol^{-1})(300 \, K)} \right]$$

$$= 6.58 \times 10^{-6} \, dm^3 \cdot mol^{-1} \cdot s^{-1}$$

The ratio of the theoretical rate constant to the experimental rate constant is 3.27×10^{16}.

30–2. Compare a plot of $\sigma_r(E_r)/\pi d_{AB}^2$ given by Equation 30.14 to the data shown in Figure 30.2.

$$\sigma_r(E_r) = \begin{cases} 0 & E_r < E_0 \\ \pi d_{AB}^2 \left(1 - \dfrac{E_0}{E_r}\right) & E_r \geq E_0 \end{cases} \qquad (30.14)$$

Plotting Equation 30.14 produces the following figure:

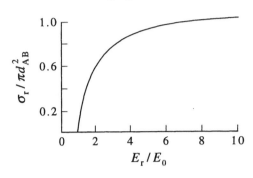

We see that the function $\sigma_r(E_r)/\pi d_{AB}^2$ is a fairly good approximation to the behavior of the data shown in Figure 30.2.

30–3. Show that Equation 30.15, the rate constant for the line-of-centers model, is obtained by substituting Equation 30.14, the reaction cross section for the line-of-centers model, into Equation 30.11 and then integrating the resulting expression.

Substituting $\sigma_r(E_r)$ from Equation 30.14 into Equation 30.11 gives

$$k = \left(\frac{2}{k_B T}\right)^{3/2} \left(\frac{1}{\mu\pi}\right)^{1/2} \int_{E_0}^{\infty} dE_r E_r e^{-E_r/k_B T} \pi d_{AB}^2 \left(1 - \frac{E_0}{E_r}\right)$$

$$= \left(\frac{2}{k_B T}\right)^{3/2} \left(\frac{1}{\mu\pi}\right)^{1/2} \int_{E_0}^{\infty} dE_r E_r e^{-E_r/k_B T} \pi d_{AB}^2$$

$$\quad - \left(\frac{2}{k_B T}\right)^{3/2} \left(\frac{1}{\mu\pi}\right)^{1/2} \int_{E_0}^{\infty} dE_r E_0 e^{-E_r/k_B T} \pi d_{AB}^2$$

The first integral here is the same as that in the derivation of Equation 30.13 (Section 30–1), so

$$k = \langle u_r \rangle \sigma_{AB} e^{-E_0/k_B T} \left(1 + \frac{E_0}{k_B T}\right) - \left(\frac{2}{k_B T}\right)^{3/2} \left(\frac{1}{\mu\pi}\right)^{1/2} \int_{E_0}^{\infty} dE_r E_0 e^{-E_r/k_B T} \pi d_{AB}^2$$

$$= \langle u_r \rangle \sigma_{AB} e^{-E_0/k_B T} \left(1 + \frac{E_0}{k_B T}\right) - \left(\frac{2}{k_B T}\right)^{3/2} \left(\frac{1}{\mu\pi}\right)^{1/2} \pi d_{AB}^2 e^{-E_0/k_B T} E_0 k_B T$$

$$= \langle u_r \rangle \sigma_{AB} e^{-E_0/k_B T} \left(1 + \frac{E_0}{k_B T}\right) - \langle u_r \rangle \sigma_{AB} e^{-E_0/k_B T} \frac{E_0}{k_B T}$$

$$= \langle u_r \rangle \sigma_{AB} e^{-E_0/k_B T}$$

30–4. The Arrhenius parameters for the reaction

$$NO(g) + O_3(g) \Longrightarrow NO_2(g) + O_2(g)$$

are $A = 7.94 \times 10^9$ $dm^3 \cdot mol^{-1} \cdot s^{-1}$ and $E_a = 10.5$ $kJ \cdot mol^{-1}$. Assuming the line-of-centers model, calculate the values of E_0, the threshold energy, and σ_{AB}, the hard-sphere reaction cross section, for this reaction at 1000 K.

In Example 30–3, we found that, for the line-of-centers model,

$$E_a = \tfrac{1}{2}k_B T + E_0 \qquad \text{and} \qquad A = \langle u_r \rangle \sigma_{AB} e^{1/2} \tag{1}$$

We can find E_0 using the first of these equations:

$$
\begin{aligned}
E_0 &= E_a - \tfrac{1}{2}k_B N_A T = E_a - \tfrac{1}{2}RT \\
&= 10\,500\ \mathrm{J \cdot mol^{-1}} - \tfrac{1}{2}(8.3145\ \mathrm{J \cdot mol^{-1} \cdot K^{-1}})(1000\ \mathrm{K}) \\
&= 6.34\ \mathrm{kJ \cdot mol^{-1}}
\end{aligned}
$$

We now use the second of Equations 27.58 to determine $\langle u_r \rangle$:

$$\langle u_r \rangle = \left\{ \frac{8(1.381 \times 10^{-23}\ \mathrm{J \cdot K^{-1}})(1000\ \mathrm{K})}{\pi \left[\dfrac{(30.006\ \mathrm{amu})(47.998\ \mathrm{amu})}{(78.004\ \mathrm{amu})} \right](1.661 \times 10^{-27}\ \mathrm{kg \cdot amu^{-1}})} \right\}^{1/2} = 1070\ \mathrm{m \cdot s^{-1}}$$

Finally, we can solve the second of Equations 1 for σ_{AB}.

$$\sigma_{AB} = \frac{A}{\langle u_r \rangle e^{1/2}} = \frac{(7.94 \times 10^9\ \mathrm{dm^3 \cdot mol^{-1} \cdot s^{-1}})}{e^{1/2}(1070\ \mathrm{m \cdot s^{-1}})(6.022 \times 10^{23}\ \mathrm{mol^{-1}})} \left(\frac{1\ \mathrm{m}}{10\ \mathrm{dm}} \right)^3 = 7.47 \times 10^{-21}\ \mathrm{m^2}$$

30–5. Consider the following bimolecular reaction at 3000 K:

$$CO(g) + O_2(g) \Longrightarrow CO_2(g) + O(g)$$

The experimentally determined Arrhenius pre-exponential factor is $A = 3.5 \times 10^9$ $dm^3 \cdot mol^{-1} \cdot s^{-1}$, and the activation energy is $E_a = 213.4$ $kJ \cdot mol^{-1}$. The hard-sphere collision diameter of O_2 is 360 pm and that for CO is 370 pm. Calculate the value of the hard sphere line-of-centers model rate constant at 3000 K and compare it with the experimental rate constant. Also compare the calculated and experimental A values.

First, use Equation 28.57 to find the experimental value of k:

$$
\begin{aligned}
k_{exp} &= A e^{-E_a/RT} \\
&= (3.5 \times 10^9\ \mathrm{dm^3 \cdot mol^{-1} \cdot s^{-1}}) \exp\left[-\frac{213\,400\ \mathrm{J \cdot mol^{-1}}}{(8.315\ \mathrm{J \cdot mol^{-1} \cdot K^{-1}})(3000\ \mathrm{K})} \right] \\
&= 6.7 \times 10^5\ \mathrm{dm^3 \cdot mol^{-1} \cdot s^{-1}}
\end{aligned}
$$

We can use the first of Equations 27.58 to find σ_{AB}:

$$\sigma_{AB} = \pi \left[\frac{(360 \times 10^{-12}\ \mathrm{m} + 370 \times 10^{-12}\ \mathrm{m})}{2} \right]^2 = 4.19 \times 10^{-19}\ \mathrm{m^2}$$

and the second of Equations 27.58 to find $\langle u_r \rangle$:

$$\langle u_r \rangle = \left(\frac{8k_B T}{\pi \mu} \right)^{1/2} = \left\{ \frac{8(1.381 \times 10^{-23} \text{ J} \cdot \text{K}^{-1})(3000 \text{ K})}{\pi \left[\dfrac{(28.010 \text{ amu})(31.999 \text{ amu})}{(60.009 \text{ amu})} \right] (1.661 \times 10^{-27} \text{ kg} \cdot \text{amu}^{-1})} \right\}^{1/2} = 2060 \text{ m} \cdot \text{s}^{-1}$$

We can also use the expression derived in Example 30–3 to find E_0:

$$\begin{aligned} E_0 &= E_a - \tfrac{1}{2}k_B N_A T = E_a - \tfrac{1}{2}RT \\ &= 213400 \text{ J} \cdot \text{mol}^{-1} - \tfrac{1}{2}(8.3145 \text{ J} \cdot \text{mol}^{-1} \cdot \text{K}^{-1})(3000 \text{ K}) \\ &= 200.9 \text{ kJ} \cdot \text{mol}^{-1} \end{aligned}$$

Now substitute into Equation 30.15:

$$\begin{aligned} k_{\text{theor}} &= \langle u_r \rangle \sigma_{AB} e^{-E_0/k_B T} \\ &= (2060 \text{ m} \cdot \text{s}^{-1})(4.19 \times 10^{-19} \text{ m}^2) \exp\left[-\frac{200\,900 \text{ J} \cdot \text{mol}^{-1}}{(8.314 \text{ J} \cdot \text{mol}^{-1} \cdot \text{K}^{-1})(3000 \text{ K})} \right] \\ &= 2.74 \times 10^{-19} \text{ m}^3 \cdot \text{molecule}^{-1} \cdot \text{s}^{-1} = 1.65 \times 10^8 \text{ dm}^3 \cdot \text{mol}^{-1} \cdot \text{s}^{-1} \end{aligned}$$

The ratio of the theoretical rate constant to the experimental rate constant is 250. The theoretical value of A is (Example 30–3)

$$A = \langle u_r \rangle \sigma_{AB}(1000 N_A) e^{1/2} = 8.57 \times 10^{11} \text{ dm}^3 \cdot \text{mol}^{-1} \cdot \text{s}^{-1}$$

which is 250 times greater than the experimental A.

30–6. The threshold energy, E_0, for the reaction

$$H_2^+(g) + He(g) \Longrightarrow HeH^+(g) + H(g)$$

is 70.0 kJ·mol⁻¹. Determine the lowest vibrational level of $H_2^+(g)$ such that the internal vibrational energy of the reactants exceeds E_0. The spectroscopic constants for H_2^+ are $\tilde{\nu}_e = 2321.7 \text{ cm}^{-1}$ and $\tilde{\nu}_e \tilde{x}_e = 66.2 \text{ cm}^{-1}$.

Recall from Chapter 13 that the vibrational energy of a molecule can be expressed by

$$G(v) = \tilde{\nu}_e(v + \tfrac{1}{2}) - \tilde{x}_e \tilde{\nu}_e(v + \tfrac{1}{2})^2 \tag{13.21}$$

We want to determine the lowest vibrational quantum number such that the vibrational energy is greater than 70.0 kJ·mol⁻¹. Converting this value to cm⁻¹ gives

$$(70.0 \text{ kJ} \cdot \text{mol}^{-1}) \left(\frac{83.60 \text{ cm}^{-1}}{\text{kJ} \cdot \text{mol}^{-1}} \right) = 5850 \text{ cm}^{-1}$$

Now

$$5850 \text{ cm}^{-1} < (2321.7 \text{ cm}^{-1})(v + \tfrac{1}{2}) - (66.2 \text{ cm}^{-1})(v + \tfrac{1}{2})^2$$

Solving this quadratic equation for v gives $v > 2.2$, so the lowest value of v such that the internal vibrational energy of the reactants is greater than E_0 is 3.

30–7. Calculate the total kinetic energy of an F(g) atom moving at a speed of 2500 m·s⁻¹ toward a head-on collision with a stationary $D_2(g)$ molecule. (Assume the reactants are hard spheres.)

Use Equation 30.18:

$$KE = \tfrac{1}{2}m_A u_A^2 + \tfrac{1}{2}m_B u_B^2$$
$$= \tfrac{1}{2}(18.998 \text{ amu})(1.661 \times 10^{-27} \text{ kg·amu}^{-1})(2500 \text{ m·s}^{-1})^2 + 0$$
$$= 9.86 \times 10^{-20} \text{ J}$$

30–8. A F(g) atom and a D_2(g) molecule are moving toward a head-on collision with one another. The F(g) atom has a speed of 1540 m·s^{-1}. Calculate the speed of the D_2(g) molecule so that the total kinetic energy is the same as that in Problem 30–7. (Assume the reactants are hard spheres.)

Use Equation 30.18 again:

$$KE = \tfrac{1}{2}m_A u_A^2 + \tfrac{1}{2}m_B u_B^2$$
$$9.86 \times 10^{-20} \text{ J} = \tfrac{1}{2}(18.998 \text{ amu})(1.661 \times 10^{-27} \text{ kg·amu}^{-1})(1540 \text{ m·s}^{-1})^2$$
$$+ \tfrac{1}{2}(4.028 \text{ amu})(1.661 \times 10^{-27} \text{ kg·amu}^{-1})u_B^2$$
$$6.12 \times 10^{-20} \text{ J} = \tfrac{1}{2}(4.028 \text{ amu})(1.661 \times 10^{-27} \text{ kg·amu}^{-1})u_B^2$$
$$u_B = 4280 \text{ m·s}^{-1}$$

30–9. In Problem 30–7, you calculated the total kinetic energy for a F(g) atom moving at a speed of 2500 m·s^{-1} toward a head-on collision with a stationary $D_2(v = 0)$ molecule. Determine the ratio of the total kinetic energy to the zero-point vibrational energy of the D_2(g) molecule given that $\tilde{v}_{D_2} = 2990 \text{ cm}^{-1}$.

The total kinetic energy from Problem 30–7 is 9.86×10^{-20} J. Recall from Chapter 13 that $G(v) = (v + \tfrac{1}{2})\tilde{v}$, so the zero-point vibrational energy is

$$G(0) = \tfrac{1}{2}(3118.4 \text{ cm}^{-1})\left(\frac{\text{kJ·mol}^{-1}}{83.60 \text{ cm}^{-1}}\right)\left(\frac{1000}{N_A}\right) = 3.10 \times 10^{-20} \text{ J}$$

The ratio of the total kinetic energy to the zero-point vibrational energy is 3.2.

30–10. Consider the head-on collision between a F(g) atom and a stationary D_2(g) molecule. Estimate the minimum speed of the F(g) atom so that its kinetic energy exceeds the bond dissociation energy of D_2(g). (The value of D_0 for D_2 is 435.6 kJ·mol^{-1}.)

The dissociation energy of a D_2(g) molecule is

$$\frac{435.6 \text{ kJ·mol}^{-1}}{N_A} = 7.23 \times 10^{-19} \text{ J}$$

This is the minimum energy needed by the fluorine atom. We use Equation 30.18 again:

$$KE = \tfrac{1}{2}m_A u_A^2 + \tfrac{1}{2}m_B u_B^2$$
$$7.23 \times 10^{-19} \text{ J} \leq \tfrac{1}{2}(18.998 \text{ amu})(1.661 \times 10^{-27} \text{ kg·amu}^{-1})u_F^2 + 0$$
$$u_F \geq 6770 \text{ m·s}^{-1}$$

The minimum speed is 6770 m·s^{-1}.

30–11. Following Example 30–4, show that the equations

$$\mathbf{u}_{cm} = \frac{m_C}{M}\mathbf{u}_C + \frac{m_D}{M}\mathbf{u}_D$$

and

$$\mathbf{u}_r = \mathbf{u}_C - \mathbf{u}_D$$

lead to

$$\mathbf{u}_C = \mathbf{u}_{cm} + \frac{m_D}{M}\mathbf{u}_r \tag{1}$$

and

$$\mathbf{u}_D = \mathbf{u}_{cm} - \frac{m_C}{M}\mathbf{u}_r \tag{2}$$

Multiply $\mathbf{u}_{cm}$ by M/m_D and add the resulting expression to that for $\mathbf{u}_r$ to obtain

$$\frac{M}{m_D}\mathbf{u}_{cm} + \mathbf{u}_r = \frac{m_C + m_D}{m_D}\mathbf{u}_C$$

Solving this equation for $\mathbf{u}_C$ gives

$$\mathbf{u}_C = \frac{M}{m_C + m_D}\mathbf{u}_{cm} + \frac{m_D}{m_C + m_D}\mathbf{u}_r$$

or (because $m_C + m_D = M$)

$$\mathbf{u}_C = \mathbf{u}_{cm} + \frac{m_D}{M}\mathbf{u}_r \tag{1}$$

Now multiply $\mathbf{u}_{cm}$ by M/m_C and subtract the resulting expression from that for $\mathbf{u}_r$ to obtain

$$\frac{M}{m_C}\mathbf{u}_{cm} - \mathbf{u}_r = \frac{m_D + m_C}{m_C}\mathbf{u}_D$$

Solving this equation for $\mathbf{u}_D$ gives

$$\mathbf{u}_D = \frac{M}{m_C + m_D}\mathbf{u}_{cm} - \frac{m_C}{m_C + m_D}\mathbf{u}_r$$

which is

$$\mathbf{u}_D = \mathbf{u}_{cm} - \frac{m_C}{M}\mathbf{u}_r \tag{2}$$

30–12. Derive Equation 30.22.

Begin with the equation

$$KE_{prod} = \tfrac{1}{2}m_C u_C^2 + \tfrac{1}{2}m_D u_D^2 \tag{1}$$

Recall that

$$\mathbf{u}_r' = \mathbf{u}_C - \mathbf{u}_D$$

and

$$\mathbf{u}_{cm} = \frac{m_C \mathbf{u}_C + m_D \mathbf{u}_D}{M} \tag{30.21}$$

Multiply $\mathbf{u}_{cm}$ by M/m_D and add the resulting expression to the equation for $\mathbf{u}_r'$ to obtain

$$\mathbf{u}_C = \mathbf{u}_{cm} + \frac{m_D}{M}\mathbf{u}_r' \tag{2}$$

Next, multiply $\mathbf{u}_{cm}$ by M/m_C and subtract the resulting equation from that for $\mathbf{u}_r'$ to obtain

$$\mathbf{u}_D = \mathbf{u}_{cm} - \frac{m_C}{M}\mathbf{u}_r' \tag{3}$$

Finally, substitute Equations 2 and 3 into Equation 1 to find

$$\begin{aligned}
KE_{prod} &= \frac{m_C}{2}\left(\mathbf{u}_{cm} + \frac{m_D}{M}\mathbf{u}_r'\right)^2 + \frac{m_D}{2}\left(\mathbf{u}_{cm} - \frac{m_C}{M}\mathbf{u}_r'\right)^2 \\
&= \frac{1}{2}(m_C + m_D)u_{cm}^2 + \frac{1}{2}\left(\frac{m_C m_D^2}{M^2} + \frac{m_D m_C^2}{M^2}\right)u_r'^2 \\
&= \frac{1}{2}Mu_{cm}^2 + \frac{1}{2}\left(\frac{m_C m_D}{M}\right)\left(\frac{m_C + m_D}{M}\right)u_r'^2 \\
&= \tfrac{1}{2}Mu_{cm}^2 + \tfrac{1}{2}\mu' u_r'^2
\end{aligned}$$

where $\mu' = m_C m_D / M$. This is Equation 30.22.

30–13. The speed of sound, u_s, in a fluid is given by

$$u_s^2 = \frac{\gamma \overline{V}}{M \kappa_T} \tag{1}$$

where $\gamma = C_P/C_V$, M is the molar mass, and $\kappa_T = -(1/V)(\partial V/\partial P)_T$ is the isothermal compressibility of the fluid. Assuming ideal behavior, calculate the speed of sound in $N_2(g)$ at 25°C. Take $\overline{C}_P = 7R/2$. The measured value is 348 m·s⁻¹.

For an ideal gas, $\overline{C}_P - \overline{C}_V = R$ (Equation 19.43), so $\overline{C}_V = 5R/2$. Then $\gamma = \overline{C}_P/\overline{C}_V = 1.4$. For an ideal gas $\kappa_T = 1/P$ (Problem H–1), so we substitute into Equation 1 to find

$$u_s^2 = \frac{\gamma}{M}P\overline{V} = \frac{\gamma RT}{M}$$

or

$$u_s = \left(\frac{\gamma RT}{M}\right)^{1/2}$$

We substitute into this expression to find u_s for nitrogen:

$$u_s = \left[\frac{(1.4)(8.3145\ \text{J·mol}^{-1}\text{·K}^{-1})(298\ \text{K})}{0.02802\ \text{kg·mol}^{-1}}\right]^{1/2} = 352\ \text{m·s}^{-1}$$

30–14. The speed of sound, u_s, in a fluid is given by Equation 1 of Problem 30–13. In addition, $\overline{C}_P$ and $\overline{C}_V$ are related by (Equation 22.27)

$$\overline{C}_P - \overline{C}_V = \frac{\alpha^2 T \overline{V}}{\kappa_T}$$

where $\alpha = (1/V)(\partial V/\partial T)_P$ is the coefficient of thermal expansion. Given that $\overline{C}_P = 135.6 \text{ J·K}^{-1}\text{·mol}^{-1}$, $\kappa_T = 9.44 \times 10^{-10} \text{ Pa}^{-1}$, $\alpha = 1.237 \times 10^{-3} \text{ K}^{-1}$, and the density $\rho = 0.8765 \text{ g·mL}^{-1}$ for benzene at one atm and 20°C, calculate the speed of sound in benzene. The measured value is 1320 m·s^{-1}.

First, we find the molar volume of benzene:

$$\overline{V} = \left(\frac{78.114 \text{ g}}{1 \text{ mol}}\right)\left(\frac{1 \text{ mL}}{0.8765 \text{ g}}\right)\left(\frac{1 \text{ m}^3}{1 \times 10^6 \text{ mL}}\right) = 8.912 \times 10^{-5} \text{ m}^3\text{·mol}^{-1}$$

Now use the equation given in the problem to find C_V:

$$\overline{C}_V = \overline{C}_P - \frac{\alpha^2 T \overline{V}}{\kappa_T}$$

$$= 135.6 \text{ J·K}^{-1}\text{·mol}^{-1} - \frac{(1.237 \times 10^{-3} \text{ K}^{-1})^2(293 \text{ K})(8.912 \times 10^{-5} \text{ m}^3\text{·mol}^{-1})}{9.44 \times 10^{-10} \text{ Pa}^{-1}}$$

$$= 135.6 \text{ J·K}^{-1}\text{·mol}^{-1} - 42.35 \text{ J·K}^{-1}\text{·mol}^{-1}$$

$$= 93.25 \text{ J·K}^{-1}\text{·mol}^{-1}$$

Then

$$\gamma = \frac{\overline{C}_P}{\overline{C}_V} = \frac{135.6 \text{ J·K}^{-1}\text{·mol}^{-1}}{93.25 \text{ J·K}^{-1}\text{·mol}^{-1}} = 1.454$$

Finally, we substitute into the equation for u_s given in the previous problem.

$$u_s = \left(\frac{\gamma \overline{V}}{M\kappa_T}\right)^{1/2} = \left[\frac{(1.454)(8.912 \times 10^{-5} \text{ m}^3\text{·mol}^{-1})}{(0.078114 \text{ kg})(9.44 \times 10^{-10} \text{ Pa}^{-1})}\right]^{1/2} = 1330 \text{ m·s}^{-1}$$

30–15. The peak speed of the molecules in a supersonic molecular beam of a carrier gas is well approximated by

$$u_{\text{peak}} = \left(\frac{2RT}{M}\right)^{1/2}\left(\frac{\gamma}{\gamma - 1}\right)^{1/2}$$

where T is the temperature of the source chamber of the gas mixture, M is the molar mass of the carrier gas, and γ is the ratio of the heat capacities, $\gamma = C_P/C_V$, of the carrier gas. Determine the peak velocity for a benzene molecule in a supersonic neon beam in which the source chamber of the gas is maintained at 300 K. Repeat the calculation for a helium beam under the same conditions. Assume that He(g) and Ne(g) can be treated as ideal gases.

For an ideal monatomic gas,

$$C_V = \tfrac{3}{2}Rn \qquad \text{and} \qquad C_P = C_V + nR = \tfrac{5}{2}Rn$$

Then

$$\gamma = \frac{C_P}{C_V} = \frac{5}{3}$$

For neon,

$$u_{peak} = \left(\frac{2RT}{M}\right)^{1/2}\left(\frac{\gamma}{\gamma-1}\right)^{1/2} = \left[\frac{2(8.3145\,\text{J}\cdot\text{K}^{-1}\cdot\text{mol}^{-1})(300\,\text{K})}{0.02018\,\text{kg}\cdot\text{mol}^{-1}}\right]^{1/2}\left(\frac{5}{2}\right)^{1/2} = 786\,\text{m}\cdot\text{s}^{-1}$$

For helium,

$$u_{peak} = \left(\frac{2RT}{M}\right)^{1/2}\left(\frac{\gamma}{\gamma-1}\right)^{1/2} = \left[\frac{2(8.3145\,\text{J}\cdot\text{K}^{-1}\cdot\text{mol}^{-1})(300\,\text{K})}{0.004003\,\text{kg}\cdot\text{mol}^{-1}}\right]^{1/2}\left(\frac{5}{2}\right)^{1/2} = 1770\,\text{m}\cdot\text{s}^{-1}$$

30–16. Estimate the temperature required so that the average speed of a benzene molecule in a gas cell is the same as that for a benzene molecule in a helium supersonic molecular beam generated under the conditions stated in Problem 30–15.

In Problem 30–15, we found that the $u_{peak} = 1770\,\text{m}\cdot\text{s}^{-1}$ for the heium beam. Setting $u_{peak} = \langle u_r\rangle$, we use Equation 28.57 to write

$$\langle u_r\rangle = \left(\frac{8k_BT}{\pi m}\right)^{1/2} = \left(\frac{8RT}{\pi M}\right)^{1/2}$$

$$1770\,\text{m}\cdot\text{s}^{-1} = \left[\frac{8(8.3145\,\text{J}\cdot\text{mol}^{-1}\cdot\text{K}^{-1})T}{\pi(0.078114\,\text{kg}\cdot\text{mol}^{-1})}\right]^{1/2}$$

$$T = 11\,500\,\text{K}$$

30–17. Show that for the general reaction

$$A(g) + BC(g) \Longrightarrow AB(g) + C(g)$$

Equation 30.28 can be written as

$$E_{tot} = \tfrac{1}{2}\mu u_r^2 + F(J) + G(v) + T_e$$
$$= \tfrac{1}{2}\mu u_r'^2 + F'(J) + G'(v) + T_e'$$

within the harmonic oscillator-rigid rotator approximation where T_e, $G(v)$, and $F(J)$ are the electronic, vibrational, and rotational terms of the diatomic reactant, BC(g), and T_e', $G'(v)$, and $F'(J)$ are the corresponding terms for the diatomic product, AB(g).

The total energy is

$$E_{tot} = E_{trans} + E_{rot} + E_{vib} + E_{elec}$$

Using the relationships in Chapter 13 for the harmonic oscillator-rigid rotator model, we can rewrite this sum as

$$E_{tot} = \tfrac{1}{2}\mu u_r^2 + F(R) + G(v) + T_e$$

Also,

$$E_{tot} = E_{trans}' + E_{rot}' + E_{vib}' + E_{elec}'$$

We can use the relationships in Chapter 13 to write this as

$$E_{\text{tot}} = \tfrac{1}{2}\mu u_r'^2 + F'(R) + G'(v) + T_e'$$

30–18. Consider the reaction

$$\text{Cl(g)} + \text{H}_2(v = 0) \Longrightarrow \text{HCl}(v) + \text{H(g)}$$

where $D_e(\text{H}_2) - D_e(\text{HCl}) = 12.4 \text{ kJ}\cdot\text{mol}^{-1}$. Assume there is no activation barrier to the reaction. Model the reactants as hard spheres (no vibrational motion) and calculate the minimum value of the relative speed required for reaction to occur. If we model $\text{H}_2(g)$ and $\text{HCl}(g)$ as hard-sphere harmonic oscillators with $\tilde{\nu}_{\text{H}_2} = 4159 \text{ cm}^{-1}$ and $\tilde{\nu}_{\text{HCl}} = 2886 \text{ cm}^{-1}$, calculate the minimum value of the relative speed required for reaction to occur.

For the minimum relative speed required for the reaction to occur, u_r' must equal zero. Therefore (Equation 30.24)

$$E_{\text{prod,int}} - E_{\text{react,int}} = \tfrac{1}{2}\mu u_r^2$$

Treating the reactants and products as hard spheres gives $E_{\text{prod,int}} - E_{\text{react,int}} = D_e(\text{H}_2) - D_e(\text{HCl})$, so solving the above equation for u_r gives

$$
\begin{aligned}
u_r &= \left\{ \frac{2[D_e(\text{H}_2) - D_e(\text{HCl})]}{\mu} \right\}^{1/2} \\[2mm]
&= \left\{ \frac{2(12\,400 \text{ J}\cdot\text{mol}^{-1})}{\left[\dfrac{(35.453 \text{ g}\cdot\text{mol}^{-1})(2.016 \text{ g}\cdot\text{mol}^{-1})}{(37.469 \text{ g}\cdot\text{mol}^{-1})} \right](10^{-3} \text{ kg}\cdot\text{g}^{-1})} \right\}^{1/2} \\[2mm]
u_r &= 3610 \text{ m}\cdot\text{s}^{-1}
\end{aligned}
$$

Using the hard-sphere model, the minimum relative speed required is $3610 \text{ m}\cdot\text{s}^{-1}$. Now let the molecules be hard sphere harmonic oscillators. The energy of the $v = 0$ vibrational state is given by $(1/2)\tilde{\nu}$, so for H_2

$$E_{\text{vib}} = \tfrac{1}{2}(4159 \text{ cm}^{-1})\left(\frac{\text{kJ}\cdot\text{mol}^{-1}}{83.60 \text{ cm}^{-1}} \right) = 24.87 \text{ kJ}\cdot\text{mol}^{-1}$$

and for HCl

$$E_{\text{vib}}' = \tfrac{1}{2}(2886 \text{ cm}^{-1})\left(\frac{\text{kJ}\cdot\text{mol}^{-1}}{83.60 \text{ cm}^{-1}} \right) = 17.26 \text{ kJ}\cdot\text{mol}^{-1}$$

Let $E_{\text{trans}}' = 0$ (as we did above), and write (assuming the products are in their ground electronic and rotational states)

$$
\begin{aligned}
E_{\text{trans}} + E_{\text{vib}} - D_e(\text{H}_2) &= E_{\text{vib}}' - D_e(\text{HCl}) \\
E_{\text{trans}} &= 17.26 \text{ kJ}\cdot\text{mol}^{-1} - 24.87 \text{ kJ}\cdot\text{mol}^{-1} + 12.4 \text{ kJ}\cdot\text{mol}^{-1} \\
&= 4.79 \text{ kJ}\cdot\text{mol}^{-1}
\end{aligned}
$$

In order for the reaction to proceed, the translational energy of the reactants must be at least $4.79 \text{ kJ}\cdot\text{mol}^{-1}$. Therefore

$$\tfrac{1}{2}\mu u_r^2 = 4.79 \text{ kJ}\cdot\text{mol}^{-1}$$

and

$$u_r = \left[\frac{2(4790 \text{ J} \cdot \text{mol}^{-1})}{1.907 \times 10^{-3} \text{ kg} \cdot \text{mol}^{-1}} \right]^{1/2} = 2240 \text{ m} \cdot \text{s}^{-1}$$

30–19. The reaction $H(g) + F_2(v = 0) \Rightarrow HF(g) + F(g)$ produces vibrationally excited HF molecules. Determine the minimum value of the relative kinetic energy such that HF(g) molecules in the $v = 12$ vibrational state are produced. The following are the vibrational spectroscopic constants for HF and F_2: $\tilde{\nu}_e(HF) = 4138.32 \text{ cm}^{-1}$, $\tilde{\nu}_e(F_2) = 916.64 \text{ cm}^{-1}$, $\tilde{\nu}_e \tilde{x}_e(HF) = 89.88 \text{ cm}^{-1}$, $\tilde{\nu}_e \tilde{x}_e(F_2) = 11.24 \text{ cm}^{-1}$, $D_0(HF) = 566.2 \text{ kJ} \cdot \text{mol}^{-1}$, and $D_0(F_2) = 154.6 \text{ kJ} \cdot \text{mol}^{-1}$.

The zero-point energy of F_2 is given by (Equation 13.21)

$$E_{vib} = \tfrac{1}{2}(916.64 \text{ cm}^{-1}) - \tfrac{1}{4}(11.24 \text{ cm}^{-1}) = 455.5 \text{ cm}^{-1} = 5.45 \text{ kJ} \cdot \text{mol}^{-1}$$

To determine the relative kinetic energy needed to produce HF molecules in the $v = 12$ vibrational energy level, we first calculate the difference in energy between the $v = 0$ and $v = 12$ energy levels of HF (Equation 13.21):

$$\Delta E = \left[(4138.32 \text{ cm}^{-1})(12 + \tfrac{1}{2}) - (89.88 \text{ cm}^{-1})(12 + \tfrac{1}{2})^2 \right]$$
$$- \left[(4138.32 \text{ cm}^{-1})(\tfrac{1}{2}) - (89.88 \text{ cm}^{-1})(\tfrac{1}{2})^2 \right]$$
$$= 37\,685 \text{ cm}^{-1} - 2047 \text{ cm}^{-1} = 35\,639 \text{ cm}^{-1} = 426.30 \text{ kJ} \cdot \text{mol}^{-1}$$

When the product molecules have zero relative velocity (at the minimum possible value of the relative kinetic energy), the kinetic energy of the products is zero, and we can write (Equation 30.24)

$$E_{prod,int} - E_{react,int} = KE_{react} - KE_{prod} = KE_{react}$$

The internal energy of the reactants is given by $-D_0(F_2)$ (the zero-point vibrational energy) and that of the products is given by the sum of $-D_0(HF)$ and ΔE. We can, therefore, write

$$KE_{react} = D_0(F_2) - D_0(HF) + \Delta E$$
$$= 154.6 \text{ kJ} \cdot \text{mol}^{-1} - 566.2 \text{ kJ} \cdot \text{mol}^{-1} + 426.30 \text{ kJ} \cdot \text{mol}^{-1}$$
$$= +15.7 \text{ kJ} \cdot \text{mol}^{-1}$$

30–20. Consider the energetics of the reaction

$$F(g) + H_2(v = 0) \Longrightarrow HF(v) + H(g)$$

where the relative translational energy of the reactants is $7.62 \text{ kJ} \cdot \text{mol}^{-1}$, and $D_e(H_2) - D_e(HF) = -140 \text{ kJ} \cdot \text{mol}^{-1}$. Determine the range of possible vibrational states of the product HF(g) molecule. Assume the vibrational motion of both $H_2(g)$ and HF(g) is harmonic with $\tilde{\nu}_{H_2} = 4159 \text{ cm}^{-1}$ and $\tilde{\nu}_{HF} = 3959 \text{ cm}^{-1}$.

This problem is similar to Example 30–6. Assuming that the reactants and products are in the ground electronic and rotational states and using Equation 30.28 gives

$$E_{trans} + E_{vib} - D_e(H_2) = E'_{trans} + E'_{vib} - D_e(HF)$$

Because $E_{vib} = \frac{1}{2}h\nu_{H_2} = 24.88\,\text{kJ}\cdot\text{mol}^{-1}$, we can write the above equation as

$$E'_{trans} = 7.62\,\text{kJ}\cdot\text{mol}^{-1} + 24.88\,\text{kJ}\cdot\text{mol}^{-1} + 140\,\text{kJ}\cdot\text{mol}^{-1} - E'_{vib}$$
$$= 170\,\text{kJ}\cdot\text{mol}^{-1} - E'_{vib}$$

Translational energy is an intrinsically positive quantity, so for the reaction to occur, $E'_{vib} < 170\,\text{kJ}\cdot\text{mol}^{-1}$. Using the harmonic oscillator approximation, we obtain

$$E'_{vib} = \left(v + \tfrac{1}{2}\right)h\nu_{HF} < 170\,\text{kJ}\cdot\text{mol}^{-1}$$
$$\left(v + \tfrac{1}{2}\right) < \frac{170\,\text{kJ}\cdot\text{mol}^{-1}}{47.35\,\text{kJ}\cdot\text{mol}^{-1}}$$
$$v \le 3.6$$

The possible vibrational states of the product are therefore $v = 0, 1, 2$, and 3.

30–21. In Example 30–5 we calculated the speeds of the products relative to the center of mass, $|\mathbf{u}_{DF} - \mathbf{u}_{cm}|$ and $|\mathbf{u}_D - \mathbf{u}_{cm}|$, for the reaction $F(g) + D_2(g) \Longrightarrow DF(g) + D(g)$ assuming that the reactants and products could be treated as hard spheres. Now calculate these quantities taking into account the zero-point vibrational energies of $D_2(g)$ and $DF(g)$. Assume that the vibrational motion of $D_2(g)$ and $DF(g)$ is harmonic with $\tilde{\nu}_{D_2} = 2990\,\text{cm}^{-1}$ and $\tilde{\nu}_{DF} = 2907\,\text{cm}^{-1}$, respectively. How different are your results from the hard-sphere calculations presented in Example 30–5?

As in Example 30–5, $\mu = 5.52 \times 10^{-27}\,\text{kg}$, $u_r = 2.14 \times 10^3\,\text{m}\cdot\text{s}^{-1}$, $KE_{react} = 7.62\,\text{kJ}\cdot\text{mol}^{-1}$, and $\mu' = 3.05 \times 10^{-27}\,\text{kg}$. Using Equation 30.28, we have

$$E'_{trans} = E_{trans} + E_{vib} - E'_{vib} - \left[D_e(D_2) - D_e(DF)\right]$$
$$= 7.62\,\text{kJ}\cdot\text{mol}^{-1} + E_{vib} - E'_{vib} + 140\,\text{kJ}\cdot\text{mol}^{-1} \qquad (1)$$

The quantities E_{vib} and E'_{vib} are given by

$$E_{vib} = \tfrac{1}{2}\tilde{\nu}_{D_2} = \frac{2990\,\text{cm}^{-1}}{2}\left(\frac{\text{kJ}\cdot\text{mol}^{-1}}{83.60\,\text{cm}^{-1}}\right) = 17.88\,\text{kJ}\cdot\text{mol}^{-1}$$

$$E'_{vib} = \tfrac{1}{2}\tilde{\nu}_{DF} = \frac{2907\,\text{cm}^{-1}}{2}\left(\frac{\text{kJ}\cdot\text{mol}^{-1}}{83.60\,\text{cm}^{-1}}\right) = 17.39\,\text{kJ}\cdot\text{mol}^{-1}$$

Then Equation 1 becomes

$$E'_{trans} = 7.62\,\text{kJ}\cdot\text{mol}^{-1} + 17.88\,\text{kJ}\cdot\text{mol}^{-1} - 17.39\,\text{kJ}\cdot\text{mol}^{-1} + 140\,\text{kJ}\cdot\text{mol}^{-1}$$
$$= 148\,\text{kJ}\cdot\text{mol}^{-1} = 2.46 \times 10^{-19}\,\text{J}\cdot\text{molecule}^{-1}$$

Because $E'_{trans} = \frac{1}{2}\mu'u_r'^2$, we now have

$$\tfrac{1}{2}\mu'u_r'^2 = 2.46 \times 10^{-19}\,\text{J}$$
$$u_r' = \left[\frac{(2.46 \times 10^{-19}\,\text{J})(2)}{3.05 \times 10^{-27}\,\text{kg}}\right]^{1/2}$$
$$= 12\,700\,\text{m}\cdot\text{s}^{-1}$$

This is the same value we found for u_r' in Example 30–5. For the case considered here, the zero-point vibrational energy does not affect the results for $|\mathbf{u}_{DF} - \mathbf{u}_{cm}|$ and $|\mathbf{u}_D - \mathbf{u}_{cm}|$, which are $1.11 \times 10^3\,\text{m}\cdot\text{s}^{-1}$ and $1.16 \times 10^4\,\text{m}\cdot\text{s}^{-1}$, respectively.

The following four problems consider the reaction

$$Cl(g) + HBr(v = 0) \Longrightarrow HCl(v) + Br(g)$$

where the relative translational energy of the reactants is 9.21 kJ·mol^{-1}, the difference $D_e(HBr) - D_e(HCl) = -67.2$ kJ·mol^{-1}, and the activation energy for this reaction is ≈ 6 kJ·mol^{-1}.

30–22. Determine the range of possible vibrational states of the product molecule, HCl(g). The spectroscopic constants for HBr(g) and HCl(g) are

	$\tilde{\nu}_e/cm^{-1}$	$\tilde{\nu}_e\tilde{x}_e/cm^{-1}$
HBr	2648.98	45.22
HCl	2990.95	52.82

Draw a diagram for this reaction that is similar to that shown in Figure 30.8 for the F(g) + D$_2$(g) reaction.

This problem is similar to Example 30–6. Using Equation 30.28 and assuming that the molecules are in their ground rotational and electronic states, we write

$$E'_{trans} = E_{trans} + E_{vib} - E'_{vib} - \left[D_e(HBr) - D_e(HCl) \right]$$

We can use Equation 13.21 to find the ground vibrational state of HBr:

$$\begin{aligned} E_{vib} &= \tilde{\nu}_e(v + \tfrac{1}{2}) - \tilde{x}_e\tilde{\nu}_e(v + \tfrac{1}{2})^2 \\ &= (2648.98 \text{ cm}^{-1})(\tfrac{1}{2}) - (45.22 \text{ cm}^{-1})(\tfrac{1}{2})^2 \\ &= 1313.2 \text{ cm}^{-1} = 15.71 \text{ kJ·mol}^{-1} \end{aligned}$$

Then

$$\begin{aligned} E'_{trans} &= 9.21 \text{ kJ·mol}^{-1} + 15.71 \text{ kJ·mol}^{-1} - E'_{vib} + 67.2 \text{ kJ·mol}^{-1} \\ &= 92.12 \text{ kJ·mol}^{-1} - E'_{vib} \end{aligned}$$

For the reaction to occur, $E'_{trans} > 0$, so $E'_{vib} < 92.12$ kJ·mol^{-1}:

$$E'_{vib} = \tilde{\nu}_e(v + \tfrac{1}{2}) - \tilde{x}_e\tilde{\nu}_e(v + \tfrac{1}{2})^2 < 92.12 \text{ kJ·mol}^{-1}$$
$$(2990.95 \text{ cm}^{-1})(v + \tfrac{1}{2}) - (52.82 \text{ cm}^{-1})(v + \tfrac{1}{2})^2 < 92.12 \text{ kJ·mol}^{-1}$$

Because v must be an integer, we find that $v \leq 2$. Therefore, $v = 0$, $v = 1$, and $v = 2$ are the possible vibrational states of the product.

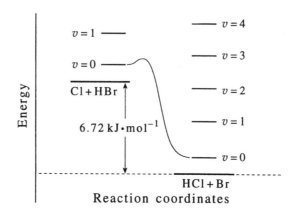

30–23. Calculate the value of $|\mathbf{u}_{HCl} - \mathbf{u}_{cm}|$, the speed of the HCl(g) molecule relative to the center of mass, for each of the possible vibrational states of HCl(g) in Problem 30–22.

This problem is similar to Example 30–7. In Problem 30–22, we showed that

$$E'_{trans} + E'_{vib} = 92.12 \text{ kJ·mol}^{-1}$$

Now substitute for E'_{trans} and E'_{vib} to find u'_r in terms of v:

$$E'_{trans} + E'_{vib} = \tfrac{1}{2}\mu' u'^2_r + \tilde{v}_e(v + \tfrac{1}{2}) - \tilde{v}_e \tilde{x}_e(v + \tfrac{1}{2})^2$$

or

$$92.12 \text{ kJ·mol}^{-1} = \tfrac{1}{2}(2.504 \times 10^{-2} \text{ kg·mol}^{-1})u'^2_r + (35.772 \text{ kJ·mol}^{-1})(v + \tfrac{1}{2})$$
$$- (0.6317 \text{ kJ·mol}^{-1})(v + \tfrac{1}{2})^2 \qquad (1)$$

where the reduced mass of the products is given by

$$\mu' = \left[\frac{(36.461 \text{ g·mol}^{-1})(79.904 \text{ g·mol}^{-1})}{(116.365 \text{ g·mol}^{-1})}\right](1 \times 10^{-3} \text{ kg·g}^{-1}) = 2.504 \times 10^{-2} \text{ kg·mol}^{-1}$$

Solving Equation 1 for u'_r gives

$$u'_r = \left\{\left(\frac{2}{2.504 \times 10^{-2}}\right)\left[92\,120 - 35\,772(v + \tfrac{1}{2}) - (631.7)(v + \tfrac{1}{2})^2\right]\right\}^{1/2} \text{ m·s}^{-1} \qquad (2)$$

From Example 30–5, the speed of the HCl(g) molecule relative to the center of mass is given by

$$|\mathbf{u}_{HCl} - \mathbf{u}_{cm}| = \frac{m_{Br}}{M}u'_r = \frac{79.904 \text{ amu}}{116.365 \text{ amu}}u'_r \qquad (3)$$

We can calculate u'_r for $v = 0, 1,$ and 2 using Equation 2, and then use Equation 3 to obtain $|\mathbf{u}_{HCl} - \mathbf{u}_{cm}|$. The results are tabulated below.

v	$u'_r/10^3 \text{m·s}^{-1}$	$|\mathbf{u}_{HCl} - \mathbf{u}_{cm}|/10^3 \text{m·s}^{-1}$
0	2.438	1.674
1	1.785	1.225
2	0.7275	0.500

30–24. Determine the speeds for a HCl(g) molecule relative to the center of mass $|\mathbf{u}_{HCl} - \mathbf{u}_{cm}|$ in the $v = 0, J = 0$ and $v = 0, J = 1$ states. The rotational constants for HCl(g) are $\tilde{B}_e = 10.59 \text{ cm}^{-1}$ and $\tilde{\alpha}_e = 0.307 \text{ cm}^{-1}$.

Recall from Chapter 13 that the rotational energy is $\tilde{B}_v J(J + 1)$ (Equation 13.8), where $\tilde{B}_v = \tilde{B}_e - \tilde{\alpha}_e(v + \tfrac{1}{2})$ (Equation 13.17). In this case, Equation 30.28 becomes

$$E'_{trans} + E'_{vib} + E'_{rot} = E_{trans} + E_{vib} - \left[D_e(HBr) - D_e(HCl)\right]$$
$$= 9.21 \text{ kJ·mol}^{-1} + 15.71 \text{ kJ·mol}^{-1} + 67.2 \text{ kJ·mol}^{-1}$$
$$= 92.1 \text{ kJ·mol}^{-1}$$

The reduced mass of the products is, as in Problem 30–23, $2.504 \times 10^{-2} \text{ kg} \cdot \text{mol}^{-1}$. We now express the energy of the products as

$$E'_{\text{trans}} + E'_{\text{vib}} + E'_{\text{rot}} = \tfrac{1}{2}\mu' u_r'^2 + \tilde{v}'_e(v + \tfrac{1}{2}) - \tilde{v}'_e\tilde{x}'_e(v + \tfrac{1}{2})^2 + \left[\tilde{B}'_e - \tilde{\alpha}'_e(v + \tfrac{1}{2})\right] J(J + 1)$$

For $v = 0$, this becomes

$$92.12 \text{ kJ} \cdot \text{mol}^{-1} = \tfrac{1}{2}(2.504 \times 10^{-2} \text{ kg} \cdot \text{mol}^{-1})u_r'^2 + (35.772 \text{ kJ} \cdot \text{mol}^{-1})(\tfrac{1}{2})$$
$$-(0.6317 \text{ kJ} \cdot \text{mol}^{-1})(\tfrac{1}{2})^2 + (0.1249 \text{ kJ} \cdot \text{mol}^{-1})J(J + 1)$$

or

$$u_r^2 = 5.942 \times 10^6 \text{ m}^2 \cdot \text{s}^{-2} - 9973 J(J + 1) \text{ m}^2 \cdot \text{s}^{-2}$$

As in Problem 30–23, the speed of the HCl(g) molecule relative to the center of mass is given by

$$|\mathbf{u}_{\text{HCl}} - \mathbf{u}_{\text{cm}}| = \frac{79.904 \text{ amu}}{116.365 \text{ amu}} u_r'$$

For $v = 0$, $J = 0$, $|\mathbf{u}_{\text{HCl}} - \mathbf{u}_{\text{cm}}| = 1674 \text{ m} \cdot \text{s}^{-1}$; for $v = 0$, $J = 1$, $|\mathbf{u}_{\text{HCl}} - \mathbf{u}_{\text{cm}}| = 1671 \text{ m} \cdot \text{s}^{-1}$.

30–25. Using the data in Problem 30–24, determine the value of J_{min}, the minimum value of J, such that the kinetic energy of a HCl($v = 0$, $J = J_{\text{min}}$) molecule is greater than the kinetic energy of an HCl($v = 1$, $J = 0$) molecule. [Note that if this reaction produces HCl($v = 0$, $J \geq J_{\text{min}}$), then these molecules have relative speeds characteristic of an HCl($v = 1$) molecule, affecting the analysis of the product velocity contour plots.]

From Problem 30–24, we have

$$92.12 \text{ kJ} \cdot \text{mol}^{-1} = E'_{\text{trans}} + (35.772 \text{ kJ} \cdot \text{mol}^{-1})(\tfrac{1}{2} + v)$$
$$-(0.6317 \text{ kJ} \cdot \text{mol}^{-1})(\tfrac{1}{2} + v)^2$$
$$+(0.1249 \text{ kJ} \cdot \text{mol}^{-1})J(J + 1)$$

For $v = 1$, $J = 0$, E'_{trans} (the kinetic energy of the HCl molecule) is $39.87 \text{ kJ} \cdot \text{mol}^{-1}$. For a molecule in the $v = 0$, $J = J_{\text{min}}$ state to have a greater kinetic energy,

$$39.87 \text{ kJ} \cdot \text{mol}^{-1} < 92.12 \text{ kJ} \cdot \text{mol}^{-1} - (35.772 \text{ kJ} \cdot \text{mol}^{-1})(\tfrac{1}{2})$$
$$+(0.6317 \text{ kJ} \cdot \text{mol}^{-1})(\tfrac{1}{2})^2$$
$$-(0.1249 \text{ kJ} \cdot \text{mol}^{-1})J_{\text{min}}(J_{\text{min}} + 1)$$

Because J must be an integer, $J \geq 17$.

30–26. Using the data given in Table 13.2, estimate the minimum value of the relative speed of the reactants so that the following reactions occur:

$$\text{HCl}(v = 0) + \text{Br}(g) \Longrightarrow \text{HBr}(v = 0) + \text{Cl}(g)$$

and

$$\text{HCl}(v = 1) + \text{Br}(g) \Longrightarrow \text{HBr}(v = 0) + \text{Cl}(g)$$

We assume that $E_{rot} = E'_{rot} = 0$. From Problem 30–22, the ground vibrational state of HBr has an energy of 15.71 kJ·mol^{-1}. Assuming that the reactants and products are in their ground electronic states, Equation 30.28 becomes

$$E'_{trans} + E'_{vib} = E_{trans} + E_{vib} + \left[D_e(\text{HBr}) - D_e(\text{HCl}) \right]$$

Because we want to find the minimum value of the relative speed of the reactants, we let $E'_{trans} = 0$. Then the above equation becomes

$$15.71 \text{ kJ·mol}^{-1} = E_{trans} + (35.772 \text{ kJ·mol}^{-1})(v + \tfrac{1}{2})$$
$$-(0.6317 \text{ kJ·mol}^{-1})(v + \tfrac{1}{2})^2 - 67.2 \text{ kJ·mol}^{-1}$$

so that

$$E_{trans}(v = 0) = 65.2 \text{ kJ·mol}^{-1}$$
$$E_{trans}(v = 1) = 30.7 \text{ kJ·mol}^{-1}$$

The reduced mass of the products is

$$\mu' = 2.465 \times 10^{-2} \text{ kg·mol}^{-1}$$

So

$$u_r(v = 0) = \left[\frac{2E_{trans}(v = 0)}{\mu} \right]^{1/2}$$
$$= \left[\frac{2(65\,200 \text{ J·mol}^{-1})}{2.465 \times 10^{-2} \text{ kg·mol}^{-1}} \right]^{1/2} = 2300 \text{ m·s}^{-1}$$
$$u_r(v = 1) = \left[\frac{2E_{trans}(v = 0)}{\mu} \right]^{1/2}$$
$$= \left[\frac{2(30\,700 \text{ J·mol}^{-1})}{2.465 \times 10^{-2} \text{ kg·mol}^{-1}} \right]^{1/2} = 1580 \text{ m·s}^{-1}$$

30–27. Do the values of the radii of the dashed circles in Figure 30.11 increase, decrease, or remain the same as the relative translational energy of the reactants is increased from 7.62 kJ·mol^{-1}? Determine the percentage change, if any, in the radius of the dashed circle for $v = 0$ if the relative translational energy is doubled from 7.62 kJ·mol^{-1} to 15.24 kJ·mol^{-1}.

The circles correspond to the maximum relative speeds a product molecule can have. Because speed and translational energy are directly related, and because the translational energy of the product is directly related to the translational energy of the reactants (Equation 30.28), the radii of the dashed circles will increase as the translational energy of the reactants is increased. If the relative translational energy were doubled, then u_r would increase by $\sqrt{2}$. Because the radius of the circle is proportional to u_r, the radius would also increase by $\sqrt{2}$.

30–28. Figure 30.11 presents the contour map for the product molecule DF(g) for the reaction between F(g) and $D_2(v = 0)$. The dashed lines correspond to the expected speeds for DF(g) molecules in those vibrational states when J, the rotational quantum number, equals 0. The regions between two circles then correspond to molecules that are rotationally excited. Determine the minimum value for J such that a DF($v = 2$) molecule has a relative speed expected for a DF($v = 3$) molecule. The spectroscopic constants for DF(g) are given in Example 30–8. Does your result suggest that there could be a problem encountered in the analysis of the scattering data for this reaction?

From Example 30–7, we know that

$$E'_{trans} + E'_{vib} + E'_{rot} = E_{trans} + E_{vib} - \left[D_e(D_2) - D_e(DF) \right]$$
$$= 166 \text{ kJ} \cdot \text{mol}^{-1} = 13\,877.6 \text{ cm}^{-1}$$

The largest value of E_{trans} expected for a DF($v=3$) molecule would correspond to $J = 0$, giving

$$E'_{trans} = 13\,877.6 \text{ cm}^{-1} - \tilde{\nu}_e(\tfrac{7}{2}) + \tilde{\nu}_e\tilde{x}_e(\tfrac{7}{2})^2 = 3943.7 \text{ cm}^{-1}$$

Now consider a DF($v = 2$) molecule in the rotational state J. We want to determine the smallest value of J such that $E'_{trans} = 3943.7$ cm^{-1}. Because energy is conserved, this requires that

$$E'_{rot} = 13\,877.6 \text{ cm}^{-1} - 3943.7 \text{ cm}^{-1} - \tilde{\nu}_e(\tfrac{5}{2}) + \tilde{\nu}_e\tilde{x}_e(\tfrac{5}{2})^2$$

or

$$E'_{rot} = 2723.96 \text{ cm}^{-1}$$

Using Equation 13.17,

$$E'_{rot} = \left[\tilde{B}_e - \tilde{\alpha}_e \left(v + \tfrac{1}{2} \right) \right] J(J + 1)$$

so

$$2\,723.96 \text{ cm}^{-1} = \left[11.007 \text{ cm}^{-1} - (0.293 \text{ cm}^{-1})(\tfrac{5}{2}) \right] J(J + 1)$$

and (since J is an integer) $J = 16$. This is a rather large value of J, so this effect can generally be ignored in the analysis of the scattering data.

30–29. For the reaction $Cl(g) + H_2(g) \Rightarrow HCl(g) + H(g)$, $D_e(H_2) - D_e(HCl) = 12.4$ kJ·mol^{-1}. Assuming that the relative kinetic energy is 8.52 kJ· mol^{-1} and that the $H_2(g)$ reactant is prepared in a $v = 3$, $J = 0$ state, what are the possible vibrational states of HCl(g)? The vibrational spectroscopic constants for $H_2(g)$ and HCl(g) are $\tilde{\nu}_e(H_2) = 4401.21$ cm^{-1}, $\tilde{\nu}_e(HCl) = 2990.95$ cm^{-1}, $\tilde{\nu}_e\tilde{x}_e(H_2) = 121.34$ cm^{-1}, and $\tilde{\nu}_e\tilde{x}_e(HCl) = 52.82$ cm^{-1}.

If the H_2 is prepared in the $v = 3$ state, then

$$E_{vib} = \tilde{\nu}_e(v + \tfrac{1}{2}) - \tilde{\nu}_e\tilde{x}_e(v + \tfrac{1}{2})^2 = (4401.21 \text{ cm}^{-1})(\tfrac{7}{2}) - (121.34 \text{ cm}^{-1})(\tfrac{7}{2})^2 = 13\,917.82 \text{ cm}^{-1}$$

Using Equation 30.28, we find that (taking $J = 0$)

$$E'_{trans} + E'_{vib} = E_{trans} + E_{vib} - \left[D_e(H_2) - D_e(HCl) \right]$$
$$E'_{trans} = 703.91 \text{ cm}^{-1} + 13\,917.82 \text{ cm}^{-1} - 1036.64 \text{ cm}^{-1} - E'_{vib}$$
$$E'_{trans} = 13\,585.09 \text{ cm}^{-1} - E'_{vib}$$

The value of E'_{trans} must be positive, and so $E'_{vib} < 13\,585.09$ cm^{-1}, giving

$$13\,585.90 \text{ cm}^{-1} > (2990.95 \text{ cm}^{-1})(v + \tfrac{1}{2}) - (52.82 \text{ cm}^{-1})(v + \tfrac{1}{2})^2$$
$$4.5 > v$$

The possible vibrational states of the product HCl(g) molecule are 0, 1, 2, 3, and 4.

30–30. Suppose that the reaction given in Problem 30–29 produces HCl(g) in $v = v_{max}$, the highest possible vibrational state under the given conditions. Determine the largest possible value of J for the HCl($v = v_{max}$, J) molecule. The rotational constants of HCl(g) are $\tilde{B}_e = 10.59 \text{ cm}^{-1}$ and $\tilde{\alpha}_e = 0.307 \text{ cm}^{-1}$.

From Problem 30–29, we have

$$E'_{trans} = 13\,585.09 \text{ cm}^{-1} - E'_{vib} - E'_{rot} \tag{1}$$

The largest value of v for the HCl(g) product molecule is $v = 4$, so

$$E'_{vib} = (2990.95 \text{ cm}^{-1})(4 + \tfrac{1}{2}) - (52.82 \text{ cm}^{-1})(4 + \tfrac{1}{2})^2 = 12\,389.67 \text{ cm}^{-1}$$

Recall (Equation 13.17) that

$$\tilde{B}_v = \tilde{B}_e - \tilde{\alpha}_e(v + \tfrac{1}{2}) = 10.59 \text{ cm}^{-1} - (0.307 \text{ cm}^{-1})(4.5) = 9.21 \text{ cm}^{-1}$$

and (Equation 13.18)

$$E'_{rot} = \tilde{B}J(J + 1) = (9.21 \text{ cm}^{-1})J(J + 1)$$

Substituting into Equation 1, we find

$$E'_{trans} = 1195.42 \text{ cm}^{-1} - (9.21 \text{ cm}^{-1})J(J + 1)$$

For the reaction to occur, $E'_{trans} > 0$, so

$$1195.42 \text{ cm}^{-1} > (9.21 \text{ cm}^{-1})J(J + 1)$$

$$10.9 > J$$

The largest value of J is 10.

30–31. Consider the product velocity distribution for the reaction between K(g) and $I_2(v = 0)$ at a relative translational energy of $15.13 \text{ kJ} \cdot \text{mol}^{-1}$ shown in Figure 30.13. Assume that the vibrational motion of I_2(g) and KI(g) is harmonic with $\tilde{v}_{I_2} = 213 \text{ cm}^{-1}$ and $\tilde{v}_{KI} = 185 \text{ cm}^{-1}$. Given that $D_e(I_2) - D_e(KI) = -171 \text{ kJ} \cdot \text{mol}^{-1}$, determine the maximum vibrational quantum number for the product KI(g). Now determine the speed of a KI($v = 0$) molecule relative to the center of mass. Repeat the calculation for the KI($v = 1$) molecule. Do the data in the contour map support a conclusion that KI(g) is produced in a distribution of vibrational levels?

For $I_2(v = 0)$,

$$E_{vib} = \tilde{v}_{I_2}(v + \tfrac{1}{2})$$
$$= (213 \text{ cm}^{-1})(\tfrac{1}{2}) = 107 \text{ cm}^{-1}$$

Again using Equation 30.28, we find that

$$E'_{trans} + E'_{vib} = E_{trans} + E_{vib} - \left[D_e(I_2) - D_e(KI)\right]$$

$$E'_{trans} = (15.13 \text{ kJ} \cdot \text{mol}^{-1})\left(\frac{83.60 \text{ cm}^{-1}}{\text{kJ} \cdot \text{mol}^{-1}}\right) + 106.65 \text{ cm}^{-1} + (171 \text{ kJ} \cdot \text{mol}^{-1})\left(\frac{83.60 \text{ cm}^{-1}}{\text{kJ} \cdot \text{mol}^{-1}}\right) - E'_{vib}$$

$$E'_{trans} = 15\,667.12 \text{ cm}^{-1} - E'_{vib}$$

As long as $E'_{vib} < 15\,667$ cm^{-1}, the reaction can occur. Then (using the constants given)

$$15\,667 \text{ cm}^{-1} > \tilde{\nu}_{KI}(v + \tfrac{1}{2}) = (185.4 \text{ cm}^{-1})(v + \tfrac{1}{2})$$
$$84.7 > v$$

The maximum vibrational quantum number of the product molecule is 84. For a KI($v = 0$) product molecule, we can write

$$E'_{trans} = 15\,667 \text{ cm}^{-1} - (\tfrac{1}{2})(185.4 \text{ cm}^{-1}) = 15\,575 \text{ cm}^{-1} = 3.0937 \times 10^{-19} \text{ J}$$

The reduced mass of the products is $\mu' = 1.19 \times 10^{-25}$ kg, so

$$\frac{\mu' u_r'^2}{2} = 3.093 \times 10^{-19} \text{ J}$$
$$u_r' = 2280 \text{ m·s}^{-1}$$

The relative speed of a KI($v = 0$) molecule relative to the center of mass (as in Example 30–5) is

$$|\mathbf{u}_{KI} - \mathbf{u}_{cm}| = \frac{m_I}{M}u' = \frac{126.904 \text{ amu}}{292.9073 \text{ amu}}(2280 \text{ m·s}^{-1}) = 988 \text{ m·s}^{-1}$$

For a KI($v = 1$) molecule,

$$E'_{trans} = 15\,667.12 \text{ cm}^{-1} - (1.5)(185.4 \text{ cm}^{-1})$$
$$= 15\,389.02 \text{ cm}^{-1} = 3.057 \times 10^{-19} \text{ J}$$
$$\frac{\mu' u_r'^2}{2} = 3.057 \times 10^{-19} \text{ J}$$
$$u_r' = 2267 \text{ m·s}^{-1}$$

and

$$|\mathbf{u}_{KI} - \mathbf{u}_{cm}| = \frac{m_I}{M}u_r' = \frac{166.0028 \text{ amu}}{292.907 \text{ amu}}(2267 \text{ m·s}^{-1}) = 982 \text{ m·s}^{-1}$$

These calculations and the contour plot in Figure 30.13 support the conclusion that KI(g) is produced in a distribution of vibrational levels.

30–32. Below is a plot of the LiCl(g) product velocity distribution for the reaction

$$\text{Li(g)} + \text{HCl}(v = 0) \Longrightarrow \text{LiCl}(v) + \text{H(g)}$$

recorded at a relative translational energy of 38.49 kJ·mol^{-1}. Is this reaction an example of a rebound reaction, a stripping reaction, or a reaction in which a long-lived complex (relative to the rotational period of the complex) is formed between the reactants before any product molecules are produced? Explain your reasoning.

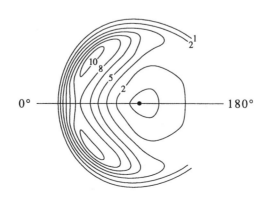

Because the scattering is localized in the forward direction, this reaction is an example of a stripping reaction.

30–33. Shown below are the velocity contour plots of the $N_2D^+(g)$ product recorded at two different relative translational energies for the reaction

$$N_2^+(g) + D_2(v = 0) \Longrightarrow N_2D^+(v) + D(g)$$

The scale between the two plots indicating $1000 \text{ m} \cdot \text{s}^{-1}$ applies to both plots.

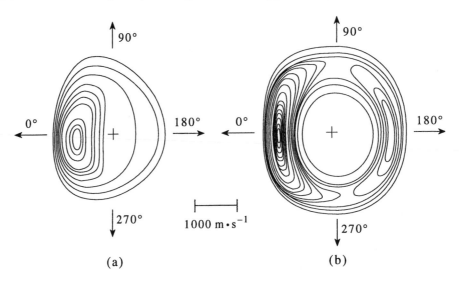

(a)   (b)

The value of $D_e(N_2D^+) - D_e(N_2^+) - D_e(D_2)$ is equal to $96 \text{ kJ} \cdot \text{mol}^{-1}$. The relative translational energy of the reactants is $301.02 \text{ kJ} \cdot \text{mol}^{-1}$ and $781.49 \text{ kJ} \cdot \text{mol}^{-1}$ for the left and right contour plots, respectively. Propose an explanation for why $N_2D^+(g)$ product molecules observed with low relative velocities are present in (a) but absent in (b).

There is more total energy available to the products at a collision energy of $781.49 \text{ kJ} \cdot \text{mol}^{-1}$ than at a collision energy of $301.02 \text{ kJ} \cdot \text{mol}^{-1}$. Therefore, low relative velocity products for the $781.49 \text{ kJ} \cdot \text{mol}^{-1}$ have a greater internal energy than the low relative velocity products for the $301.02 \text{ kJ} \cdot \text{mol}^{-1}$ case. Under the experimental conditions given, product molecules with a low relative translational energy do not appear on the right contour plot because such molecules would have internal energies greater than the dissocation energy of N_2D^+, and therefore fragment.

30–34. The reaction between $Ca(g)$ and $F_2(g)$ generates an electronically excited product according to the equation

$$Ca(^1S_0) + F_2(g) \Longrightarrow CaF^*(B^2\Sigma^+) + F(g)$$

The radii of $Ca(^1S_0)$ and $F_2(g)$ are 100 pm and 370 pm, respectively. Determine the hard-sphere collision cross section. The cross section for this reaction is $> 10^6 \text{ pm}^2$. Propose a mechanism for this reaction.

Use Equation 27.58:

$$\sigma_{AB} = \pi \left[\frac{(200 \times 10^{-12} \text{ m} + 740 \times 10^{-12} \text{ m})}{2} \right]^2 = 6.94 \times 10^{-19} \text{ m}^2 = 6.94 \times 10^5 \text{ pm}^2$$

The hard sphere collision cross section is smaller than the measured cross section. The following harpoon mechanism is consistent with the above comparison.

$$Ca(^1S_0) + F_2(g) \Longrightarrow Ca^+ + F_2^-(g) \Longrightarrow CaF^*(B^2\Sigma^+) + F(g)$$

30–35. Consider the reaction described in Problem 30–34. The product $CaF^*(B^2\Sigma^+)$ relaxes to its electronic ground state by fluorescence. Explain how you could determine the vibrational states of the product from a measurement of the fluorescence spectrum.

Given the spectroscopic parameters T_e, $\tilde{\nu}'_e$, and $\tilde{\nu}'_e \tilde{x}'_e$ for the $B^2\Sigma^+$ excited state of CaF(g), and the values of $\tilde{\nu}''_e$ and $\tilde{\nu}''_e \tilde{x}''_e$, an observed fluorescence line at $\tilde{\nu}_{obs}$ would be given by

$$\tilde{\nu}_{obs} = T_e - E_{vib,lower} + E_{vib,upper}$$
$$= T_e - \left[\tilde{\nu}''_e (v'' + \tfrac{1}{2}) - \tilde{\nu}''_e \tilde{x}''_e (v'' + \tfrac{1}{2})^2 \right] + \left[\tilde{\nu}'_e (v' + \tfrac{1}{2}) - \tilde{\nu}'_e \tilde{x}'_e (v' + \tfrac{1}{2})^2 \right] \tag{1}$$

The vibrational states of the product can be determined by fitting the experimental data to Equation 1, realizing that v'' and v' must be integers.

30–36. For the reaction

$$Ca(^1S_0) + F_2(g) \Longrightarrow CaF^*(B^2\Sigma^+) + F(g)$$

the peak of the fluorescence spectrum corresponds to emission from the $v' = 10$ level of the $B^2\Sigma^+$ state to the $v'' = 10$ level of the ground electronic state of CaF^*. Calculate the wavelength of this emission line. The spectroscopic constants for the $B^2\Sigma^+$ state are $T_e = 18\,844.5$ cm^{-1}, $\tilde{\nu}'_e = 566.1$ cm^{-1}, and $\tilde{\nu}'_e \tilde{x}'_e = 2.80$ cm^{-1} and those for the ground state are $\tilde{\nu}''_e = 581.1$ cm^{-1} and $\tilde{\nu}''_e \tilde{x}''_e = 2.74$ cm^{-1}. In what part of the electromagnetic spectrum will you observe this emission?

Using Equation 1 from Problem 30–35, we find that

$$\tilde{\nu}_{obs} = T_e - \left[\tilde{\nu}''_e (v'' + \tfrac{1}{2}) - \tilde{\nu}''_e \tilde{x}''_e (v'' + \tfrac{1}{2})^2 \right] + \left[\tilde{\nu}'_e (v' + \tfrac{1}{2}) - \tilde{\nu}'_e \tilde{x}'_e (v' + \tfrac{1}{2})^2 \right]$$
$$= 18\,844.5 \text{ cm}^{-1} - (581.1 \text{ cm}^{-1})(\tfrac{21}{2})$$
$$+ (2.74 \text{ cm}^{-1})(\tfrac{21}{2})^2 + (566.1 \text{ cm}^{-1})(\tfrac{21}{2}) - (2.80 \text{ cm}^{-1})(\tfrac{21}{2})^2$$
$$= 18\,680.4 \text{ cm}^{-1} \approx 535 \text{ nm}$$

The emission is in the visible region.

30–37. Describe the potential-energy surfaces for the reactions

$$I(g) + H_2(v = 0) \Longrightarrow HI(v) + H(g)$$

and

$$I(g) + CH_4(v = 0) \Longrightarrow HI(v) + CH_3(g)$$

The potential energy surface for the first reaction will be similar to that described in the chapter for $F(g) + D_2(g)$. The potential energy surface is four-dimensional. The potential energy surface for the second reaction depends on the relative orientation of five atoms, and so is $(3N - 6)$-dimensional, or nine-dimensional. As a result, calculating the reaction coordinates for the second reaction is much more involved.

30–38. The following plot depicts the potential-energy surface for the isomerization reaction

$$OClO(g) \Longrightarrow ClOO(g)$$

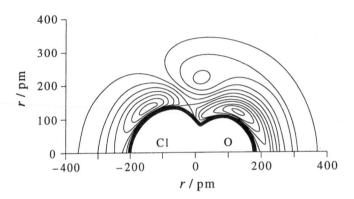

The contour map is a plot of the potential energy as a function of the location of an oxygen atom around a diatomic ClO of fixed bond length. The energy spacing betwen contour lines is 38.6 kJ·mol^{-1}. Label the location of the oxygen atom in the reactant (OClO) and product (ClOO) molecules. Draw the minimum energy path for the isomerization reaction. Which isomer is more stable? Estimate the range for the height of the activation barrier for this isomerization reaction from the potential-energy surface. Is the energy barrier to isomerization less than, greater than, or equal to the barrier for dissociation into $O(g) + ClO(g)$?

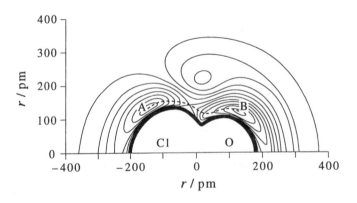

Point A is the location of the oxygen atom in OClO, and point B is the location of the oxygen atom in ClOO. ClOO is the more stable isomer, because there are more contour lines between the oxygen atom and the transition state for the isomerization reaction from in this configuration. There are six contour lines to cross to reach the transition state from the minimum energy configuration of OClO, and eight to cross to reach the transition state from the minimum energy conformation of ClOO. The height of the activation barrier is greater than the energy of six contour lines and less than the energy of eight contour lines, and so is somewhere between 309 kJ·mol^{-1} and 232 kJ·mol^{-1}. The energy barrier to isomerization is less than the energy barrier to dissociation, because there are six contour lines to cross to reach the transition state and at least seven to dissociate into $O(g)$ and $ClO(g)$. The dashed line depicts the minimum energy path for the isomerization reaction.

30–39. The *opacity function* $P(b)$ is defined as the fraction of collisions with impact parameter b that lead to reaction. The reaction cross section is related to the opacity function by

$$\sigma_r = \int_0^\infty 2\pi b P(b)\,db$$

Justify this expression. Assume that the opacity function is given by

$$P(b) = \begin{cases} 1 & b \le d_{AB} \\ 0 & b > d_{AB} \end{cases}$$

Show that this opacity function gives the hard-sphere collision theory model for σ_r.

The integrand is the product of the hard sphere collision cross section $2\pi b$ and the fraction of collisions with impact parameter b that lead to a reaction. If we integrate over all values of b, then we take into account all possible values of the hard sphere cross section weighted by their probability of leading to reaction. This will give the average reaction cross section (see MathChapter B). Substituting the $P(b)$ given, when $b \le d_{AB}$

$$\sigma_r = \int_0^{d_{AB}} 2\pi b P(b)\,db = \pi b^2 \Big|_0^{d_{AB}} = \pi d_{AB}^2$$

and when $b > d_{AB}$ $\sigma_r = 0$. These results are the hard-sphere collision theory model for σ_r.

30–40. The opacity function is defined in Problem 30–39. Determine an expression for b_{max} in terms of d_{AB}, E_0, and E_r so that an opacity function given by

$$P(b) = \begin{cases} 1 & b \le b_{max} \\ 0 & b > b_{max} \end{cases}$$

yields the reaction cross section $\sigma_r(E_r)$ for the line-of-centers model (Equation 30.14).

Using the relationship between σ_r and $P(b)$ given in Problem 30–39, we find for the opacity function stated in this problem, $\sigma_r = \pi b_{max}^2$ for $b \le b_{max}$ and $\sigma_r = 0$ for $b > b_{max}$. Comparing to Equation 30.14, we see that

$$\pi b_{max}^2 = \pi d_{AB}^2 \left(1 - \frac{E_0}{E_r}\right)$$

and so

$$b_{max} = d_{AB} \left(1 - \frac{E_0}{E_r}\right)^{1/2}$$

30–41. For the reaction $H(g) + H_2(g)$, the opacity function (defined in Problem 30–39) is

$$P(b) = \begin{cases} A \cos \dfrac{\pi b}{2b_{max}} & b \le b_{max} \\ 0 & b > b_{max} \end{cases}$$

where A is a constant. Derive an expression for the reaction cross section in terms of b_{max}.

The quantity $\sigma_r = 0$ for $b > b_{max}$. For $b \leq b_{max}$,

$$\sigma_r = \int_0^{b_{max}} 2\pi b \left(A \cos \frac{\pi b}{2b_{max}} \right) db$$

Let $x = b/b_{max}$, so that

$$\sigma_r = 2\pi A b_{max}^2 \int_0^1 x \cos \frac{\pi x}{2} dx$$

$$= 2\pi A b_{max}^2 \left[\frac{4}{\pi^2} \cos \frac{\pi x}{2} + \frac{2x}{\pi} \sin \frac{\pi x}{2} \right]_0^1$$

$$= 4A b_{max}^2 \left(1 - \frac{2}{\pi} \right)$$

30–42. Explain how the $F(g) + D_2(g)$ reaction can be exploited to make a chemical laser. (*Hint:* See Table 30.2 and Section 15–4.)

Table 30.2 shows that the product $DF(g)$ is formed with a population inversion between the state $v = 3$ and the states $v = 0$, 1, and 2. There is also a population inversion between the state $v = 2$ and the states $v = 0$ and 1, as well as between the states $v = 1$ and $v = 0$. Therefore, there are many possible transitions that could be used to achieve lasing.

30–43. A quantum-mechanical calculation of the potential-energy surface for the collinear hydrogen atom exchange reaction described by $H_A(g) + H_B H_C(g) \Longrightarrow H_A H_B(g) + H_C(g)$ gives a reaction barrier that lies $58.75 \text{ kJ} \cdot \text{mol}^{-1}$ above the bottom of the potential well of the reactants. Calculate the minimum relative speed for the collision between $H(g)$ and $H_2(v = 0)$ so that the hydrogen-atom exchange reaction occurs. Assume that the vibrational motion of $H_2(g)$ is harmonic.

To overcome a reaction barrier of $58.75 \text{ kJ} \cdot \text{mol}^{-1}$ the sum of the translational and vibrational energies of the reactants must exceed this value, or

$$E_{trans} + E_{vib}(H_2, v = 0) > 58.75 \text{ kJ} \cdot \text{mol}^{-1}$$

Using $\tilde{\nu}_{H_2} = 4401 \text{ cm}^{-1}$ from Table 5.1, we have

$$E_{trans} > (58.75 \text{ kJ} \cdot \text{mol}^{-1}) \left(\frac{83.60 \text{ cm}^{-1}}{\text{kJ} \cdot \text{mol}^{-1}} \right) - E_{vib}^{H_2}$$

$$> 4912 \text{ cm}^{-1} - (4401 \text{ cm}^{-1})(\tfrac{1}{2}) = 2712 \text{ cm}^{-1} = 32.43 \text{ kJ} \cdot \text{mol}^{-1}$$

Now, since $E_{trans} = \mu u^2/2$, we have

$$32.43 \text{ kJ} \cdot \text{mol}^{-1} < \frac{1}{2} \left[\frac{(1.008 \text{ amu})(2.016 \text{ amu})}{(3.024 \text{ amu})} \right] (1.661 \times 10^{-27} \text{ kg} \cdot \text{amu}^{-1}) u_r^2$$

or $u_r = 9820 \text{ m} \cdot \text{s}^{-1}$.

30–44. Below is a drawing of the contour plot of the potential-energy surface of the collinear H(g) + H$_2$(g) reaction in the vicinity of the transition state. We take r_{12} and r_{23} to be the bond length of H$_2$ reactant and product, respectively. Label the location of the transition state. Draw a dashed line that indicates the lowest energy path for the reaction. Draw a two-dimensional representation of the reaction path in which you plot $V(r_{12}, r_{23})$ as a function of $r_{12} - r_{23}$.

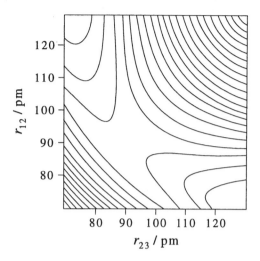

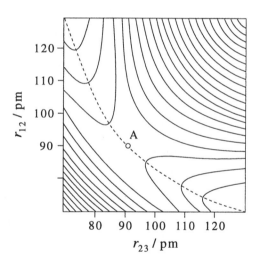

Point A is the location of the transition state.

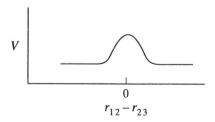

30–45. Repeat the calculation in Problem 30–43 for the reaction

$$H(g) + D_2(v = 0) \Longrightarrow HD(v = 0) + D(g)$$

Assume that the vibrational motion of $D_2(g)$ is harmonic.

The calculation is identical to that of Problem 30–43, except that $\tilde{\nu}_{D_2} = 2990 \text{ cm}^{-1}$ (Table 5.1) and $\mu = 1.339 \times 10^{-27}$ kg. Following the same procedures, we obtain $u_{min} = 10\,070 \text{ m·s}^{-1}$.

CHAPTER 31

Solids and Surface Chemistry

PROBLEMS AND SOLUTIONS

31–1. Polonium is the only metal that exists as a simple cubic lattice. Given that the length of a side of the unit cell of polonium is 334.7 pm at 25°C, calculate the density of polonium.

Polonium exists as a cubic lattice, so there is only one atom per unit cell.

$$\text{mass unit cell} = \frac{(1)(209 \text{ g·mol}^{-1})}{6.022 \times 10^{23} \text{ mol}^{-1}} = 3.47 \times 10^{-22} \text{ g}$$

$$V = (334.7 \times 10^{-10} \text{ cm})^3 = 3.749 \times 10^{-23} \text{ cm}^3$$

$$\rho = \frac{m}{V} = \frac{3.47 \times 10^{-22} \text{ g}}{3.749 \times 10^{-23} \text{ cm}^3} = 9.26 \text{ g·cm}^{-3}$$

31–2. Consider the packing of hard spheres of radius R in a primitive cubic lattice, a face-centered cubic lattice, and a body-centered cubic lattice. Show that a, the length of the unit cell, and f, the fraction of the volume of the unit cell occupied by the spheres, are given as listed.

Unit cell	a	f
Primitive cubic	$2R$	$\pi/6$
Face-centered cubic	$4R/\sqrt{2}$	$\pi\sqrt{2}/6$
Body-centered cubic	$4R/\sqrt{3}$	$\pi\sqrt{3}/8$

a. Primitive cubic

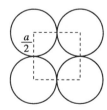

In a primitive cubic unit cell, the atoms touch along the edge of the cube, and there are two atomic radii per edge. Therefore $R + R = 2R = a$. The volume of the unit cell is given by $a^3 = (2R)^3 = 8R^3$. There is one atom per unit cell, so

$$f = \frac{\frac{4}{3}\pi R^3}{8R^3} = \frac{\pi}{6}$$

b. Face-centered cubic

This type of cell was treated in Example 31–3. There we found that $(4R)^2 = a^2 + a^2$, so

$$a = \left(\frac{16R^2}{2}\right)^{1/2} = \frac{4R}{\sqrt{2}}$$

and (because there are four atoms per unit cell)

$$f = \frac{4\left(\frac{4}{3}\pi R^3\right)}{\frac{32}{\sqrt{2}}R^3} = \frac{\sqrt{2}\pi}{6}$$

c. Body-centered cubic

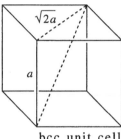

bcc unit cell

In a body-centered cubic unit cell, the atoms touch along the main diagonal, and so there are four atomic radii along the diagonal that contains the center of the cube. Thus

$$a^2 + (\sqrt{2}a)^2 = (4R)^2$$

$$3a^2 = 16R^2$$

$$a = \frac{4}{\sqrt{3}}R$$

and, because there are two atoms per unit cell,

$$f = 2\left(\frac{\frac{4}{3}\pi R^3}{\frac{64}{3\sqrt{3}}R^3}\right) = \frac{\sqrt{3}\pi}{8}$$

31–3. Tantalum forms a body-centered cubic unit cell with $a = 330.2$ pm. Calculate the crystallographic radius of a tantalum atom.

A body-centered cubic unit cell, as represented in Figure 31.4, has two atoms per unit cell, and the effective radius of an atom is given by one-fourth of the main diagonal of the cell. The quantity m, the length of the main diagonal of the tantalum atom, is

$$m = \left[(a)^2 + (\sqrt{2}a)^2\right]^{1/2} = 3^{1/2}a$$

(This is illustrated in Problem 31–2, part c.) Dividing by four, we find the crystallographic radius of a tantalum atom to be

$$r = \frac{3^{1/2}a}{4} = 143.0 \text{ pm}$$

31–4. Nickel forms a face-centered cubic unit cell with $a = 351.8$ pm. Calculate the crystallographic radius of a nickel atom.

As in Example 31–2, the crystallographic radius of an atom in a face-centered cubic unit cell is one-fourth of the length of the diagonal of a face. Then

$$r = \frac{(2)^{1/2}a}{4} = \frac{(2)^{1/2}(351.8 \text{ pm})}{4} = 124.4 \text{ pm}$$

31–5. Copper, which crystallizes as a face-centered cubic lattice, has a crystallographic radius of 127.8 pm. Calculate the density of copper.

The mass of one unit cell (containing four copper atoms) is

$$m = \frac{4(63.55 \text{ g} \cdot \text{mol}^{-1})}{6.022 \times 10^{23} \text{ mol}^{-1}} = 4.221 \times 10^{-22} \text{ g}$$

The length of the unit cell is given by (Problem 31–4)

$$a = \frac{4r}{2^{1/2}} = \frac{4(127.8 \text{ pm})}{2^{1/2}} = 361.5 \text{ pm}$$

Then the volume of one unit cell is

$$V = a^3 = 4.723 \times 10^{-23} \text{ cm}^3$$

and the density of copper is 8.937 g·cm^{-3}.

31–6. Europium, which crystallizes as a body-centered cubic lattice, has a density of 5.243 g·cm^{-3} at 20°C. Calculate the crystallographic radius of a europium atom at 20°C.

The molar mass of europium is 152.0 g·mol^{-1} and a body-centered cubic unit cell contains 2 atoms, so

$$V = \frac{m}{\rho} = \frac{2(152.0 \text{ g} \cdot \text{mol}^{-1})}{(6.022 \times 10^{23} \text{ mol}^{-1})(5.243 \text{ g} \cdot \text{cm}^{-3})}$$
$$= 9.628 \times 10^{-23} \text{ cm}^3$$
$$a = V^{1/3} = 4.583 \times 10^{-8} \text{ cm} = 458.3 \text{ pm}$$

From Problem 31–2, we know that $a = 4R/\sqrt{3}$, so $R = 198.5$ pm.

31–7. Potassium crystallizes as a body-centered cubic lattice, and the length of a unit cell is 533.3 pm. Given that the density of potassium is 0.8560 g·cm^{-3}, calculate the Avogadro constant.

The cubic volume of a unit cell is $V = a^3$, or (for potassium)

$$V = (533.3 \text{ pm})^3 = 1.517 \times 10^8 \text{ pm}^3 = 1.517 \times 10^{-22} \text{ cm}^3$$

Recall that there are two potassium atoms per unit cell. The molar mass of potassium is 39.10 g·mol^{-1}, so we can write

$$V = \frac{m}{\rho} = \frac{2(39.10 \text{ g·mol}^{-1})}{N_A(0.8560 \text{ g·cm}^{-3})}$$

$$N_A = \frac{2(39.10 \text{ g·mol}^{-1})}{(0.8560 \text{ g·cm}^{-3})(1.517 \times 10^{-22} \text{ cm}^3)} = 6.023 \times 10^{23} \text{ mol}^{-1}$$

31–8. Cerium crystallizes as a face-centered cubic lattice, and the length of a unit cell is 516.0 pm. Given that the density of cerium is 6.773 g·cm^{-3}, calculate the Avogadro constant.

The volume of a cubic unit cell is $V = a^3$, or (for cerium)

$$V = (516.0 \text{ pm})^3 = 1.374 \times 10^8 \text{ pm}^3 = 1.374 \times 10^{-22} \text{ cm}^3$$

There are four atoms per unit cell for a face-centered cubic unit cell. The molar mass of cerium is 140.1 g·mol^{-1}, so we can write

$$V = \frac{m}{\rho} = \frac{4(140.1 \text{ g·mol}^{-1})}{N_A(6.773 \text{ g·cm}^{-3})}$$

$$N_A = \frac{4(140.1 \text{ g·mol}^{-1})}{(6.773 \text{ g·cm}^{-3})(1.374 \times 10^{-22} \text{ cm}^3)} = 6.022 \times 10^{23} \text{ mol}^{-1}$$

31–9. Given that the density of KBr is 2.75 g·cm^{-3} and that the length of an edge of a cubic unit cell is 654 pm, determine how many formula units of KBr there are in a unit cell. Does the unit cell have a NaCl or a CsCl structure? (See Figure 31.18.)

Potassium bromide has a molar mass of 119.0 g·mol^{-1}. Then

$$m = \rho V = (2.75 \text{ g·cm}^{-3})(654 \times 10^{-10} \text{ cm})^3 = 7.692 \times 10^{-22} \text{ g}$$

If we let the number of KBr units per unit cell be x, then

$$\frac{x(119.0 \text{ g·mol}^{-1})}{6.022 \times 10^{23} \text{ mol}^{-1}} = 7.692 \times 10^{-22} \text{ g}$$

$$x = 3.9$$

We conclude that there are four formula units of KBr in a unit cell, which corresponds to an NaCl type of unit cell structure.

31–10. Crystalline potassium fluoride has the NaCl type of structure shown in Figure 31.18a. Given that the density of KF(s) is 2.481 g·cm^{-3} at 20°C, calculate the unit cell length and the nearest-neighbor distance in KF(s). (The nearest-neighbor distance is the shortest distance between the centers of any two adjacent ions in the lattice.)

There are four formula units of KF per unit cell, and the molar mass of KF is 58.10 g·mol^{-1}. Then

$$\rho = \frac{m}{V} = \frac{4(58.10 \text{ g·mol}^{-1})}{(6.022 \times 10^{23} \text{ mol}^{-1})a^3}$$

so

$$a^3 = \frac{3.859 \times 10^{-22} \text{ g}}{2.481 \text{ g·cm}^{-3}} = 1.555 \times 10^{-22} \text{ cm}^3$$

$$a = 5.378 \times 10^{-8} \text{ cm} = 537.8 \text{ pm}$$

The nearest-neighbor distance is $\frac{1}{2}a$ (see Figure 31.18a), or 268.9 pm.

31-11. The crystalline structure of sodium chloride can be described by two interpenetrating face-centered cubic structures (see Figure 31.18a) with four formula units per unit cell. Given that the length of a unit cell is 564.1 pm at 20°C, calculate the density of NaCl(s). The literature value is 2.163 g·cm^{-3}.

The molar mass of NaCl is 58.44 g·mol^{-1}. There are four formula units per unit cell, so

$$\rho = \frac{m}{V} = \frac{4(58.44 \text{ g·mol}^{-1})}{(6.022 \times 10^{23} \text{ mol}^{-1})(564.1 \times 10^{-10} \text{ cm})^3} = 2.162 \text{ g·cm}^{-3}$$

31-12. Determine the Miller indices of each set of lines shown in the figure below.

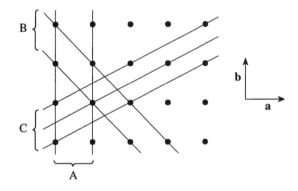

(a) 10   (b) 11   (c) $1\bar{2}$

31-13. Determine the Miller indices of each set of lines shown in the figure below.

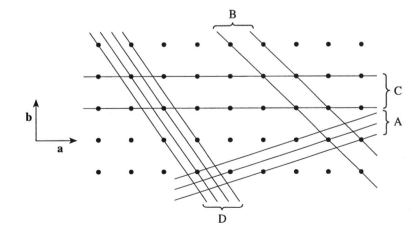

(a) $1\bar{3}$ (b) 11   (c) 01   (d) 32

31–14. Sketch the following planes in a two-dimensional square lattice: (a) 01, (b) 21, (c) $1\bar{1}$, (d) 32.

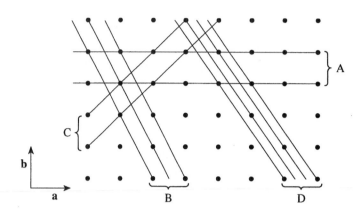

31–15. What is the relation between the 11 planes and the $1\bar{1}$ planes of a two-dimensional square lattice?

The 11 planes are perpendicular to the $1\bar{1}$ planes.

31–16. What is the relation between the $1\bar{1}$ planes and the $\bar{1}1$ planes of a two-dimensional square lattice?

The $1\bar{1}$ planes and $\bar{1}1$ planes are indistinguishable.

31–17. In this problem, we will derive a two-dimensional version of Equation 31.2. Using the figure below, show that

$$\tan\alpha = \frac{b/k}{a/h} \quad \text{and} \quad \sin\alpha = \frac{d}{a/h}$$

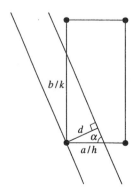

Now show that

$$\sin^2 \alpha = \frac{\tan^2 \alpha}{1 + \tan^2 \alpha}$$

and that

$$\frac{1}{d^2} = \frac{h^2}{a^2} + \frac{k^2}{b^2}$$

Equation 31.2 is the extension of this result to three dimensions.

From the figure, using the right triangle with hypotenuse a/h we find that

$$\sin \alpha = \frac{d}{a/h}$$

Likewise, using the right triangle with sides a/h and b/k we find that

$$\tan \alpha = \frac{b/k}{a/h}$$

In general, letting r be the hypotenuse of a right triangle and x and y be the sides adjacent and opposite angle α, respectively, $\sin \alpha = y/r$ and $\tan \alpha = y/x$. Then

$$\sin^2 \alpha = \frac{y^2}{r^2} = \frac{y^2}{(x^2 + y^2)} = \frac{(y/x)^2}{1 + (y/x)^2} = \frac{\tan^2 \alpha}{1 + \tan^2 \alpha}$$

Then

$$\sin^2 \alpha = \frac{\tan^2 \alpha}{1 + \tan^2 \alpha}$$

$$\frac{d^2}{(a/h)^2} = \frac{(b/k)^2}{(a/h)^2 + (b/k)^2}$$

$$\frac{(a/h)^2}{d^2} = \frac{(a/h)^2 + (b/k)^2}{(b/k)^2}$$

$$\frac{1}{d^2} = \frac{k^2}{b^2} + \frac{h^2}{a^2}$$

31–18. Determine the Miller indices of the four planes shown in the figure below.

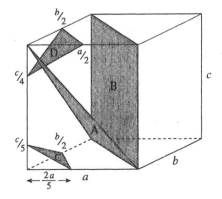

(a) 111   (b) 110   (c) 54 10   (d) $22\bar{4}$

31–19. Sketch the following planes in a three-dimensional cubic lattice: (a) 011, (b) $1\bar{1}0$, (c) 211, (d) 222.

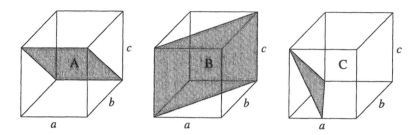

The (222) planes are depicted in Example 31–5.

31–20. Determine the Miller indices of the plane that intersects the crystal axes at (a) $(a, 2b, 3c)$, (b) $(a, b, -c)$, and (c) $(2a, b, c)$.

(a) 632   (b) $11\bar{1}$ (c) 122

31–21. Calculate the separation between the (a) 100 planes, (b) 111 planes, and (c) $12\bar{1}$ planes in a cubic lattice whose unit cell length is 529.8 pm.

Use Equation 31.3 to calculate the separation between planes. For the 100 planes,

$$\frac{1}{d^2} = \frac{h^2 + k^2 + l^2}{a^2} = \frac{1}{(529.8 \text{ pm})^2}$$
$$d = 529.8 \text{ pm}$$

For the 111 planes,

$$\frac{1}{d^2} = \frac{h^2 + k^2 + l^2}{a^2} = \frac{3}{(529.8 \text{ pm})^2}$$
$$d = 305.9 \text{ pm}$$

For the $12\bar{1}$ planes,

$$\frac{1}{d^2} = \frac{h^2 + k^2 + l^2}{a^2} = \frac{6}{(529.8 \text{ pm})^2}$$
$$d = 216.3 \text{ pm}$$

31–22. The distance between the 211 planes in barium is 204.9 pm. Given that barium forms a body-centered cubic lattice, calculate the density of barium.

We use Equation 31.3 to find the length of a unit cell.

$$\frac{1}{d^2} = \frac{h^2 + k^2 + l^2}{a^2} = \frac{6}{a^2}$$
$$a = [6(204.9 \text{ pm})^2]^{1/2} = 501.9 \text{ pm}$$

A body-centered cubic lattice contains two atoms per unit cell, so

$$\rho = \frac{m}{V} = \frac{2(137.3 \text{ g}\cdot\text{mol}^{-1})}{(6.022 \times 10^{23} \text{ mol}^{-1})(501.9 \times 10^{-10} \text{ cm})^3}$$
$$= 3.607 \text{ g}\cdot\text{cm}^{-3}$$

31–23. Gold crystallizes as a face-centered cubic crystal. Calculate the surface number density of gold atoms in the 100 planes. Take the length of the unit cell (Figure 31.3) to be 407.9 pm.

The (100) planes run along the sides of the unit cell perpendicular to the **a** axis. Therefore, in the area of one such face, there are

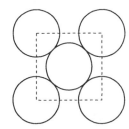

$$4\left(\frac{1}{4}\right) + 1 = 2 \text{ atoms per face}$$

Then

$$\text{surface number density} = \frac{2}{(407.9 \text{ pm})^2} = 1.20 \times 10^{15} \text{ cm}^{-2}$$

31–24. Chromium crystallizes as a body-centered cubic structure with a density of 7.20 g·cm^{-3} at 20°C. Calculate the length of a unit cell and the distance between successive 110, 200, and 111 planes.

First, find the length of a unit cell of chromium:

$$V = a^3 = \frac{m}{\rho} = \frac{2(51.996 \text{ g}\cdot\text{mol}^{-1})}{(6.022 \times 10^{23} \text{ mol}^{-1})(7.20 \text{ g}\cdot\text{cm}^{-3})}$$
$$a = 288.4 \text{ pm}$$

Then use Equation 31.3 for each set of planes:

$$\frac{1}{d_{110}^2} = \frac{1+1}{(288.4 \text{ pm})^2}$$
$$d_{110} = 203.9 \text{ pm}$$
$$\frac{1}{d_{200}^2} = \frac{4}{(288.4 \text{ pm})^2}$$
$$d_{200} = 144.2 \text{ pm}$$
$$\frac{1}{d_{111}^2} = \frac{1+1+1}{(288.4 \text{ pm})^2}$$
$$d_{111} = 166.5 \text{ pm}$$

31–25. A single crystal of NaCl is oriented such that the incident X-rays are perpendicular to the **a** axis of the crystal. The distance between the spots corresponding to diffraction from the origin and 100 planes is 14.8 mm, and the detector is located 52.0 mm from the crystal. Calculate the value of a, the length of the unit cell along the **a** axis. Take the wavelength of the X-radiation to be $\lambda = 154.433$ pm.

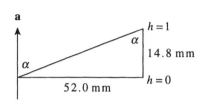

Note that according to the diagram, $\tan \alpha = 52.0/14.8$, so $\alpha = 74.11°$. Now, from Equation 31.9,

$$a = \frac{h\lambda}{\cos \alpha} = \frac{154.433 \text{ pm}}{\cos 74.11°} = 564.1 \text{ pm}$$

31–26. Silver crystallizes as a face-centered cubic structure with a unit cell length of 408.6 pm. The single crystal of silver is oriented such that the incident X-rays are perpendicular to the **c** axis of the crystal. The detector is located 29.5 mm from the crystal. What is the distance between the diffraction spots from the 001 and 002 planes on the face of the detector for (a) the $\lambda = 154.433$-pm line of copper, and (b) the $\lambda = 70.926$-pm line of a molybdenum X-ray source? Which X-ray source gives you the better spatial resolution between the diffraction spots?

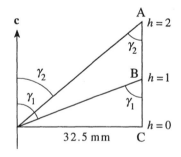

Recall that $a \cos \gamma = l\lambda$ (Equation 31.11), so

$$\gamma_1 = \cos^{-1}\left(\frac{\lambda}{408.6 \text{ pm}}\right) \qquad\qquad \gamma_2 = \cos^{-1}\left(\frac{2\lambda}{408.6 \text{ pm}}\right)$$

For part (a), $\gamma_1 = 67.79°$ and $\gamma_2 = 40.89°$, and for part (b), $\gamma_1 = 80.00°$ and $\gamma_2 = 69.68°$. Now let line segment $\overline{BC}$ be x_1 and line segment $\overline{AC}$ be x_2. Then

$$\tan \gamma_1 = \frac{29.5 \text{ mm}}{x_1} \qquad\qquad \tan \gamma_2 = \frac{29.5 \text{ mm}}{x_2}$$

Because the distance between the 001 and 002 diffraction spots is $x_2 - x_1$, we can write

$$d = x_2 - x_1 = \frac{29.5 \text{ mm}}{\tan \gamma_2} - \frac{29.5 \text{ mm}}{\tan \gamma_1} = 29.5 \text{ mm} \left[(\tan \gamma_2)^{-1} - (\tan \gamma_1)^{-1}\right]$$

Substituting, we find that $d = 22.02$ mm for $\lambda = 154.433$ pm and that $d = 5.721$ mm for $\lambda = 70.926$ pm. The copper X-ray source gives the better spatial resolution between the diffraction spots.

31–27. The X-ray diffraction angles for the first-order diffraction spot from the 111 planes of a cubic crystal with $a = 380.5$ pm are observed to be $\alpha = 18.79°$, $\beta = 0°$, and $\gamma = 0°$. How is the crystal oriented? Take the wavelength of the X-radiation to be $\lambda = 154.433$ pm.

We can use Equations 31.6 through 31.8 to find the values of α_0, β_0, and γ_0. Because we are examining the first-order diffraction spot, $n = 1$, and because the crystal is cubic, $a = b = c = 380.5$ pm. Then

$$a(\cos\alpha - \cos\alpha_0) = h\lambda$$

$$\cos\alpha_0 = \cos\alpha - \frac{h\lambda}{a}$$

$$= \cos 18.79° - \frac{154.433}{380.5}$$

$$\alpha_0 = 57.26°$$

$$\cos\beta_0 = \cos\beta - \frac{k\lambda}{b}$$

$$= \cos 0° - \frac{154.433}{380.5}$$

$$\beta_0 = 53.55°$$

$$\cos\gamma_0 = \cos\gamma - \frac{l\lambda}{c}$$

$$= \cos 0° - \frac{154.433}{380.5}$$

$$\gamma_0 = 53.55°$$

31–28. The unit cell of topaz is orthorhombic with $a = 839$ pm, $b = 879$ pm, and $c = 465$ pm. Calculate the values of the Bragg X-ray diffraction angles from the 110, 101, 111, and 222 planes. Take the wavelength of the X-radiation to be $\lambda = 154.433$ pm.

Use the Bragg equation for the first-order diffraction angle:

$$\lambda = 2\left(\frac{d}{n}\right)\sin\theta = 2d\sin\theta \tag{31.12}$$

We now find d by substituting the values of hkl and a, b and c into the expression

$$\frac{1}{d^2} = \frac{h^2}{a^2} + \frac{k^2}{b^2} + \frac{l^2}{c^2} \tag{31.2}$$

The values for d, $\sin\theta$, and θ for each set of planes are tabulated below.

	d/pm	$\sin\theta$	θ
110	606.9	0.1272	7.309°
101	406.7	0.1899	10.94°
111	369.1	0.2092	12.08°
222	184.6	0.4183	24.73°

31–29. In this problem, we will derive the Bragg equation, Equation 31.12. William and Lawrence Bragg (father and son) assumed that X-rays are scattered by successive planes of atoms within a crystal (see the following figure).

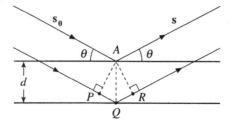

Each set of planes reflects the X-rays specularly; that is, the angle of incidence is equal to the angle of reflection, as shown in the figure. The X-radiation reflected from the lower plane in the figure travels a distance PQR longer than the X-radiation reflected by the upper layer. Show that $PQR = 2d \sin \theta$, and argue that $2d \sin \theta$ must be an integral number of wavelengths for constructive interference and hence a diffraction pattern to be observed.

The congruent right triangles $\triangle APQ$ and $\triangle ARQ$ provide us with the equations

$$\sin \theta = \frac{PQ}{d} \quad \text{and} \quad \sin \theta = \frac{QR}{d}$$

Because $PQ + PR = PQR$, we see that $PQR = 2d \sin \theta$. For constructive interference to occur, the waves (which are in phase when they enter the crystal) must once again be in phase when they emerge from the crystal, so PQR (and therefore $2d \sin \theta$) must be an integral number of wavelengths.

31–30. The observed Bragg diffraction angle of the second-order reflection from the 222 planes of a potassium crystal is $\theta = 27.43°$ when X-radiation of wavelength $\lambda = 70.926$ pm is used. Given that potassium exists as a body-centered cubic lattice, determine the length of the unit cell and the density of the crystal.

We can use the Bragg equation for a cubic unit cell (given by Equation 31.13):

$$\sin^2 \theta = \frac{n^2 \lambda^2 (h^2 + k^2 + l^2)}{4a^2}$$

$$a = \left[\frac{12(70.926 \text{ pm})^2}{\sin^2 27.43°} \right] = 533.4 \text{ pm}$$

The molar mass of potassium is 39.10 g·mol^{-1}, and the body-centered cubic lattice has two atoms per unit cell. The density of the crystal is thus

$$\frac{2(39.098 \text{ g·mol}^{-1})}{(6.022 \times 10^{23} \text{ mol}^{-1})(533.4 \times 10^{-10} \text{ cm})^3} = 0.8558 \text{ g·cm}^{-3}$$

31–31. The crystalline structure of $CuSO_4(s)$ is orthorhombic with unit cell dimensions of $a = 488.2$ pm, $b = 665.7$ pm, and $c = 831.6$ pm. Calculate the value of θ, the first-order Bragg diffraction angle, from the 100 planes, the 110 planes, and the 111 planes if $CuSO_4(s)$ is irradiated with X-rays with $\lambda = 154.433$ pm.

Using the Bragg equation (Equation 31.12), we have

$$\frac{\lambda}{2d} = \sin \theta$$

for the first-order diffraction angle. Because

$$\frac{1}{d^2} = \frac{h^2}{a^2} + \frac{k^2}{b^2} + \frac{l^2}{c^2} \tag{31.2}$$

we can write

$$\frac{1}{d_{100}^2} = \frac{1}{(488.2 \text{ pm})^2}$$
$$d_{100} = 488.2 \text{ pm}$$

Substituting this value of d_{100} into the Bragg equation and solving for θ gives

$$\theta_{100} = \sin^{-1} \frac{\lambda}{2d_{100}} = \sin^{-1} \frac{154.433 \text{ pm}}{2(488.2 \text{ pm})} = 9.100°$$

Likewise, we can find d_{110} and θ_{110}

$$\frac{1}{d_{110}^2} = \frac{1}{(488.2 \text{ pm})^2} + \frac{1}{(665.7 \text{ pm})^2}$$
$$d_{110} = 393.7 \text{ pm}$$
$$\theta_{110} = \sin^{-1} \frac{\lambda}{2d_{110}} = 11.31°$$

and d_{111} and θ_{111}:

$$\frac{1}{d_{111}^2} = \frac{1}{(488.2 \text{ pm})^2} + \frac{1}{(665.7 \text{ pm})^2} + \frac{1}{(831.6 \text{ pm})^2}$$
$$d_{111} = 355.8 \text{ pm}$$
$$\theta_{111} = \sin^{-1} \frac{\lambda}{2d_{111}} = 12.53°$$

31–32. One experimental method of collecting X-ray diffraction data, (called the *powder method*) involves irradiating a crystalline powder rather than a single crystal. The various sets of reflecting planes in a powder will be essentially randomly oriented so that there will always be planes oriented such that they will reflect the monochromatic X-radiation. The crystallites whose particular *hkl* planes are oriented at the Bragg diffraction angle, θ, to the incident beam will reflect the beam constructively. In this problem, we will illustrate the procedure that can be used for indexing the planes that give rise to observed reflections and consequently leads to the determination of the type of unit cell. This method is limited to cubic, tetragonal, and orthorhombic crystals (all unit cell angles are 90°). We will illustrate the method for a cubic unit cell.

First, show that the Bragg equation can be written as

$$\sin^2 \theta = \frac{\lambda^2}{4a^2}(h^2 + k^2 + l^2)$$

for a cubic unit cell. Then we tabulate the diffraction-angle data in order of increasing values of $\sin^2 \theta$. We then search for the smallest sets of h, k, and l that are in the same ratios as the values of $\sin^2 \theta$. We then compare these values of h, k, and l with the allowed values given in Problem 31–38 to determine the type of unit cell.

Lead is known to crystallize in one of the cubic structures. Suppose that a powder sample of lead gives Bragg reflections at the following angles: 15.66°, 18.17°, 26.13°, 31.11°, 32.71°, and 38.59°, using X-radiation with $\lambda = 154.433$ pm. Now form a table of increasing values of $\sin^2 \theta$, divide by the smallest value, convert the resulting values to integer values by multiplying by a common integer factor, and then determine the possible values of h, k, and l. For example, the first two entries in such a table are listed below.

$\sin^2 \theta$	Division by 0.0729	Conversion to integer value	Possible value of hkl
0.0729	1	3	111
0.0972	1.33	4	200

Complete this table, determine the type of cubic unit cell for lead, and determine its length.

The Bragg equation is (Equation 31.12)

$$\lambda = \frac{2d}{n} \sin \theta$$

We use this equation to find an expression for $\sin^2 \theta$:

$$\sin \theta = \frac{n\lambda}{2d}$$

$$\sin^2 \theta = \frac{n^2 \lambda^2}{4} \left(\frac{1}{d^2} \right)$$

$$= \frac{n^2 \lambda^2}{4} \left(\frac{h^2 + k^2 + l^2}{a^2} \right)$$

$$= \frac{\lambda^2}{4a^2} (h^2 + k^2 + l^2)$$

assuming first-order reflection angles. Continuing the table in the problem gives

$\sin^2 \theta$	Division by 0.0729	Conversion to integer value	Possible value of hkl
0.0729	1	3	111
0.0972	1.33	4	200
0.1940	2.66	8	220
0.2670	3.66	11	311
0.2920	4	12	222
0.3890	5.34	16	400

Using the table in Problem 31–38, we see that this corresponds to a face-centered cubic unit cell. Substituting the first entry (for example) in the above table into the Bragg equation, we find that

$$a^2 = \frac{\lambda^2}{4 \sin^2 \theta} (h^2 + k^2 + l^2) = \frac{3(154.433 \text{ pm})^2}{4 \sin^2 15.66°}$$

$$a = 495.5 \text{ pm}$$

31–33. The X-ray powder diffraction patterns of NaCl(s) and KCl(s), both of which have the structures given in Figure 31.18a, are shown below.

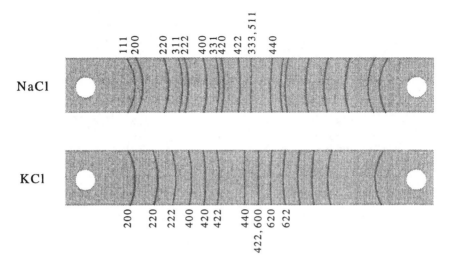

Given that NaCl and KCl have the same crystal structure, explain the differences between the two sets of data. Realize that the value of f_{K^+} is almost equal to f_{Cl^-} because K^+ and Cl^- are isoelectronic.

Problem 31–41 shows that the structure factor for an NaCl type unit cell is

$$
\begin{aligned}
F(hkl) &= 4(f_+ + f_-) & h, k, l \text{ all even} \\
&= 4(f_+ - f_-) & h, k, l \text{ all odd} \\
&= 0 & \text{otherwise}
\end{aligned}
$$

For NaCl, $f_+ \neq f_-$ and so all possible reflections can be observed. For KCl, however, $f_+ = f_-$ and so the hkl odd reflections have an essentially zero structure factor. This accounts for the observed differences between the two powder diffraction patterns. Notice that reflections associated with the all-odd Miller indices are missing in the KCl data.

31–34. Iridium crystals have a cubic unit cell. The first six observed Bragg diffraction angles from a powered sample using X-rays with $\lambda = 165.8$ pm are 21.96°, 25.59°, 37.65°, 45.74°, 48.42°, and 59.74°. Use the method outlined in Problem 31–32 to determine the type of cubic unit cell and its length.

Using the method of Problem 31–32, we find

$\sin^2 \theta$	Division by smallest	Conversion to integer value	Possible value of hkl
0.1398	1	3	111
0.1866	1.33	4	200
0.3731	2.67	8	220
0.5129	3.67	11	311
0.5595	4	12	222
0.7461	5.33	16	400

Comparing these values to the values in the table in Problem 31–38, we see that these data correspond to a face-centered cubic unit cell. Substituting the third entry (for example) in the above table into the Bragg equation, we find that

$$
a^2 = \frac{\lambda^2}{4 \sin^2 \theta} (h^2 + k^2 + l^2) = \frac{8(165.8 \text{ pm})^2}{4 \sin^2 37.65°}
$$
$$
a = 383.9 \text{ pm}
$$

31–35. The density of tantallum at 20°C is 16.69 g·cm⁻³, and its unit cell is cubic. Given that the first five observed Bragg diffraction angles are $\theta = 19.31°, 27.88°, 34.95°, 41.41°$, and 47.69°, find the type of unit cell and its length. Take the wavelength of the X-radiation to be $\lambda = 154.433$ pm.

$\sin^2\theta$	0.1093	0.2187	0.3282	0.4375	0.5469
Division by smallest	1	2	3	4	5

Polonium is the only metal that crystallizes as a primitive cubic lattice, so the smallest possible integer that we can have in the table is 2. Therefore, we write the values in the above table as

$\sin^2\theta$	Conversion to integer value	Possible value of hkl
0.1093	2	110
0.2187	4	200
0.3282	6	211
0.4375	8	220
0.5469	10	310

Comparing these values to the values in the table in Problem 31–38, we see that these data correspond to a body-centered cubic unit cell. Substituting the first entry (for example) in the above table into the Bragg equation, we find that

$$a^2 = \frac{\lambda^2}{4\sin^2\theta}(h^2+k^2+l^2) = \frac{2(154.433 \text{ pm})^2}{4\sin^2 19.31°}$$
$$a = 330.2 \text{ pm}$$

31–36. The density of silver at 20°C is 10.50 g·cm⁻³, and its unit cell is cubic. Given that the first five observed Bragg diffraction angles are $\theta = 19.10°, 22.17°, 32.33°, 38.82°$, and 40.88°, find the type of unit cell and its length. Take the wavelength of the X-radiation to be $\lambda = 154.433$ pm.

$\sin^2\theta$	Division by smallest	Conversion to integer value	Possible value of hkl
0.1071	1	3	111
0.1424	1.33	4	200
0.2860	2.67	8	220
0.3931	3.67	11	311
0.4283	4	12	222

Comparing these values to the values in the table in Problem 31–38, we see that these data correspond to a face-centered cubic unit cell. Substituting the fourth entry (for example) in the above table into the Bragg equation, we find that

$$a^2 = \frac{\lambda^2}{4\sin^2\theta}(h^2+k^2+l^2) = \frac{11(154.433 \text{ pm})^2}{4\sin^2 38.82°}$$
$$a = 408.6 \text{ pm}$$

31–37. Derive an expression for the structure factor of a primitive cubic unit cell and a face-centered cubic unit cell. Show that there will be observed reflections for a primitive unit cell for all integer values of h, k, and l and reflections for a face-centered-cubic unit cell only if h, k, and l are either all even or all odd.

The coordinates of the lattice points in a primitive unit cell are $(0, 0, 0)$, $(1, 0, 0)$, $(0, 1, 0)$, $(0, 0, 1)$, $(1, 1, 0)$, $(1, 0, 1)$, $(0, 1, 1)$, and $(1, 1, 1)$. (This was determined in the beginning of Section 31–2.) All of the atoms in a primitive unit cell are shared among eight unit cells, so the scattering efficiency of each lattice point should be multiplied by 1/8. The unit of distance is a, so using Equation 31.24 gives (as in Example 31–9)

$$F(hkl) = \frac{1}{8} f \left[e^{2\pi i(0)} + e^{2\pi i(h)} + e^{2\pi i(k)} + e^{2\pi i(l)} + e^{2\pi i(h+k)} + e^{2\pi i(k+l)} \right.$$
$$\left. + e^{2\pi i(h+l)} + e^{2\pi i(h+k+l)} \right]$$
$$= \frac{1}{8} f \left[1^0 + 1^h + 1^k + 1^l + 1^{h+k} + 1^{k+l} + 1^{h+l} + 1^{h+k+l} \right]$$
$$= \frac{1}{8}(8f) = f$$

All integer values will give a nonzero $F(hkl)$. For a face-centered-cubic unit cell, the coordinates of the lattice points are $(0, 0, 0)$, $(1, 0, 0)$, $(0, 1, 0)$, $(0, 0, 1)$, $(1, 1, 0)$, $(1, 0, 1)$, $(0, 1, 1)$, $(1, 1, 1)$, $(\frac{1}{2}, \frac{1}{2}, 1)$, $(1, \frac{1}{2}, \frac{1}{2})$, $(\frac{1}{2}, \frac{1}{2}, 0)$, $(0, \frac{1}{2}, \frac{1}{2})$, $(\frac{1}{2}, 0, \frac{1}{2})$, and $(\frac{1}{2}, 1, \frac{1}{2})$. The corner atoms are shared among eight unit cells and the face-centered atoms are shared among two unit cells, so we must multiply the scattering efficiency of the corner lattice points by 1/8 and the scattering efficiency of the face-centered lattice points by 1/2. The unit of distance is a. Using Equation 31.24 gives

$$F(hkl) = \frac{1}{8} f(8) + \frac{1}{2} f \left[e^{2\pi i(h/2+k/2)} + e^{2\pi i(k/2+l/2)} + e^{2\pi i(h/2+l/2)} + e^{2\pi i(h/2+k/2+l)} \right.$$
$$\left. + e^{2\pi i(h+k/2+l/2)} + e^{2\pi i(h/2+k+l/2)} \right]$$
$$= f + \frac{1}{2} f \left[(-1)^{h+k} + (-1)^{k+l} + (-1)^{h+l} \right.$$
$$\left. + (-1)^{h+k+2l} + (-1)^{2h+k+l} + (-1)^{h+2k+l} \right]$$
$$= f + f \left[(-1)^{h+k} + (-1)^{k+l} + (-1)^{h+l} \right]$$

If h, k, and l are all odd, then $F(hkl) = 4f$. If one of h, k, and l is even or one of h, k, and l is odd, then $F(hkl) = 0$. If h, k, and l, are all even, then the second term is nonzero and $F(hkl) = 4f$.

31–38. Use the results of the previous problem and Example 31–9 to verify the entries in the following table.

Miller indices (hkl)	Cubic lattice type for which a reflection is observed		
100	pc		
110	pc		bcc
111	pc	fcc	
200	pc	fcc	bcc
210	pc		
211	pc		bcc
220	pc	fcc	bcc
300	pc		
221	pc		
310	pc		bcc
311	pc	fcc	
222	pc	fcc	bcc
320	pc		
321	pc		bcc
400	pc	fcc	bcc

Examination of the table reveals that it tabulates specific cases of the general rules described more succinctly in the previous problem and Example 31–9.

31–39. The X-ray diffraction pattern of a cubic crystalline substance shows data that correspond to reflections from the 110, 200, 220, 310, 222, and 400 planes. What type of cubic unit cell does the substance have? (*Hint*: See the table in Problem 31–38.)

Examination of the table shows that the substance has a bcc lattice type.

31–40. Chromium is either a face-centered cubic or a body-centered cubic crystalline solid. Given that it has the following observed successive values of d: 203.8 pm, 144.2 pm, 117.7 pm, 102.0 pm, 91.20 pm, and 83.25 pm, determine the type of cubic unit cell, the length of the unit cell, and the density. (*Hint*: See the table in Problem 31–38.)

We know that

$$\frac{1}{d^2} = \frac{h^2 + k^2 + l^2}{a^2} \tag{31.3}$$

so $d^{-2} \propto (h^2 + k^2 + l^2)$.

$10^5 \text{ pm}^2/d^2$	2.41	4.81	7.22	9.57	12.0	14.4
Division by smallest	1	2	3	4	5	6

Polonium is the only metal that crystallizes as a primitive cubic lattice, so the smallest possible integer that we can have in the table is 2. Therefore, we write the values in the above table as

$10^5 \text{ pm}^2/d^2$	Conversion to integer value	Possible value of hkl
2.41	2	110
4.81	4	200
7.22	6	211
9.57	8	220
12.0	10	310
14.4	12	222

Comparing these values to the values in the table in Problem 31–38, we see that these data correspond to a body-centered cubic unit cell. Substituting the first entry (for example) in the above table into the Bragg equation, we find that

$$a = \left[2(203.8 \text{ pm})^2\right]^{1/2} = 288.2 \text{ pm}$$

and the density of the cell is

$$\rho = \frac{m}{a^3} = \frac{2(52.00 \text{ g})}{N_A(288.2 \times 10^{-10} \text{ cm})^3} = 7.215 \text{ g·cm}^{-3}$$

31–41. In this problem, we will derive the structure factor for a sodium chloride-type unit cell. First, show that the coordinates of the cations at the eight corners are $(0,0,0)$, $(1,0,0)$, $(0,1,0)$, $(0,0,1)$, $(1,1,0)$, $(1,0,1)$, $(0,1,1,)$, and $(1,1,1)$ and those at the six faces are $(\frac{1}{2},\frac{1}{2},0)$, $(\frac{1}{2},0,\frac{1}{2})$, $(0,\frac{1}{2},\frac{1}{2})$, $(\frac{1}{2},\frac{1}{2},1)$, $(\frac{1}{2},1,\frac{1}{2})$, and $(1,\frac{1}{2},\frac{1}{2})$. Similarly, show that the coordinates of the anions along the 12 edges are $(\frac{1}{2},0,0)$, $(0,\frac{1}{2},0)$, $(0,0,\frac{1}{2})$, $(\frac{1}{2},1,0)$, $(1,\frac{1}{2},0)$, $(0,\frac{1}{2},1)$, $(\frac{1}{2},0,1)$, $(1,0,\frac{1}{2})$, $(0,1,\frac{1}{2})$, $(\frac{1}{2},1,1)$, $(1,\frac{1}{2},1)$, and $(1,1,\frac{1}{2})$ and those of the anion at the center of the unit cell are $(\frac{1}{2},\frac{1}{2},\frac{1}{2})$. Now show that

$$F(hkl) = \frac{f_+}{8}[1 + e^{2\pi ih} + e^{2\pi ik} + e^{2\pi il} + e^{2\pi i(h+k)} + e^{2\pi i(h+l)} + e^{2\pi i(k+l)} + e^{2\pi i(h+k+l)}]$$

$$+ \frac{f_+}{2}[e^{\pi i(h+k)} + e^{\pi i(h+l)} + e^{\pi i(k+l)} + e^{\pi i(h+k+2l)} + e^{\pi i(h+2k+l)} + e^{\pi i(2h+k+l)}]$$

$$+ \frac{f_-}{4}[e^{\pi ih} + e^{\pi ik} + e^{\pi il} + e^{\pi i(h+2k)} + e^{\pi i(2h+k)} + e^{\pi i(k+2l)} + e^{\pi i(h+2l)}$$

$$+ e^{\pi i(2h+l)} + e^{\pi i(2k+l)} + e^{\pi i(h+2k+2l)} + e^{\pi i(2h+k+2l)} + e^{\pi i(2h+2k+l)}]$$

$$+ f_- e^{\pi i(h+k+l)}$$

$$= f_+[1 + (-1)^{h+k} + (-1)^{h+l} + (-1)^{k+l}]$$

$$+ f_-[(-1)^h + (-1)^k + (-1)^l + (-1)^{h+k+l}]$$

Finally, show that

$$F(hkl) = 4(f_+ + f_-)$$

if h, k, and l are all even; that

$$F(hkl) = 4(f_+ - f_-)$$

if h, k, and l are all odd, and that $F(hkl) = 0$ otherwise.

Examining Figure 31.18 shows that the coordinates of the cations and anions are as listed above. Substituting into Equation 31.24 gives

$$F(hkl) = \frac{f_+}{8}[1 + e^{2\pi ih} + e^{2\pi ik} + e^{2\pi il} + e^{2\pi i(h+k)} + e^{2\pi i(h+l)} + e^{2\pi i(k+l)} + e^{2\pi i(h+k+l)}]$$

$$+ \frac{f_+}{2}[e^{\pi i(h+k)} + e^{\pi i(h+l)} + e^{\pi i(k+l)} + e^{\pi i(h+k+2l)} + e^{\pi i(h+2k+l)} + e^{\pi i(2h+k+l)}]$$

$$+ \frac{f_-}{4}[e^{\pi ih} + e^{\pi ik} + e^{\pi il} + e^{\pi i(h+2k)} + e^{\pi i(2h+k)} + e^{\pi i(k+2l)} + e^{\pi i(h+2l)}$$

$$+ e^{\pi i(2h+l)} + e^{\pi i(2k+l)} + e^{\pi i(h+2k+2l)} + e^{\pi i(2h+k+2l)} + e^{\pi i(2h+2k+l)}]$$

$$+ f_- e^{\pi i(h+k+l)}$$

$$= f_+[1 + (-1)^{h+k} + (-1)^{h+l} + (-1)^{k+l}] + f_-[(-1)^h + (-1)^k + (-1)^l + (-1)^{h+k+l}]$$

If h, k, and l are all even, then this expression becomes

$$F(hkl) = f_+(4) + f_-(4) = 4(f_+ + f_-)$$

If h, k, and l are all odd, then

$$F(hkl) = f_+(4) + f_-(-4) = 4(f_+ - f_-)$$

If one of h, k, or l is even, then

$$F(hkl) = f_+(1 - 1 + 1 - 1) + f_-(1 - 1 - 1 + 1) = 0$$

and if one of h, k, or l is odd, then

$$F(hkl) = f_+(1 + 1 - 1 - 1) + f_-(1 + 1 - 1 - 1) = 0$$

31–42. Show that

$$
\begin{aligned}
F(hkl) &= f_+ + f_- & &\text{if } h, k, \text{ and } l \text{ are all even} \\
& & &\text{or just one of them is even} \\
&= f_+ - f_- & &\text{if all are odd} \\
& & &\text{or just one is odd}
\end{aligned}
$$

for the CsCl(s) crystal structure shown in Figure 31.18b. Cesium bromide and cesium iodide have the same crystal structure as cesium chloride. Compare the expected diffraction patterns of cesium chloride and cesium iodide. Recall that Cs^+ and I^- are isoelectronic.

The coordinates of the anion are $(\frac{1}{2}, \frac{1}{2}, \frac{1}{2})$, and the coordinates of the cations are $(0, 0, 0)$, $(0, 0, 1)$, $(1, 0, 0)$, $(0, 1, 0)$, $(0, 1, 1)$, $(1, 1, 0)$, $(1, 0, 1)$, and $(1, 1, 1)$. Using Equation 31.24 as in the previous problem, we find that

$$
\begin{aligned}
F(hkl) &= f_- e^{\pi i(h+k+l)} + \frac{1}{8} f_+ [e^{2\pi i} + e^{2\pi ih} + e^{2\pi ik} + e^{2\pi il} + e^{2\pi i(h+k)} \\
&\qquad\qquad\qquad\qquad + e^{2\pi i(k+l)} + e^{2\pi i(h+l)} + e^{2\pi i(h+k+l)}] \\
&= f_-(-1)^{h+k+l} + \frac{1}{8} f_+ [8] \\
&= f_+ + f_- & &\text{if } (h + k + l) \text{ is even} \\
&= f_+ - f_- & &\text{if } (h + k + l) \text{ is odd}
\end{aligned}
$$

Because $(h + k + l)$ is even if h, k, and l are all even, or if only one of them is even, and $(h + k + l)$ is odd if they are all odd or if only one is odd, the equalities above are the same as those given in the problem text. Because Cs^+ and I^- are isoelectronic, their scattering factors are identical, and so the structure factor corresponding to the planes where $(h + k + l)$ is odd will be zero. However, Cs^+ and Cl^- are not isoelectronic, so the lattice planes such that $(h + k + l)$ is odd give rise to diffraction spots.

31–43. In this problem, we will prove that a crystal lattice can have only one-, two-, three-, four-, and six-fold axes of symmetry. Consider the following figure, where P_1, P_2, and P_3 are three lattice points, each separated by the lattice vector $\mathbf{a}$.

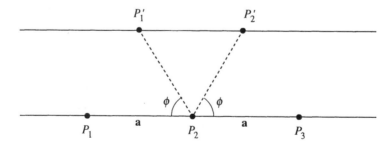

If the lattice has n-fold symmetry, then both a clockwise and a counter-clockwise rotation by $\phi = 360°/n$ about the point P_2 will lead to the points P_1' and P_2', which must be lattice points (because of the fact that the lattice has an n-fold axis of symmetry). Show that the vector distance $P_1'P_2'$ must satisfy the relation

$$2\mathbf{a}\cos\phi = N\mathbf{a}$$

where N is a positive or negative integer. Now show that the only values of ϕ that satsify the above relation are $360°$ ($n = 1$), $180°$ ($n = 2$), $120°$ ($n = 3$), $90°$ ($n = 4$), and $60°$ ($n = 6$), corresponding to $N = 2, -2, -1, 0,$ and 1, respectively.

If P_1' and P_2' are lattice points, then they must be separated by an integer number of lattice vectors $\mathbf{a}$, or $P_1'P_2' = N\mathbf{a}$, where N is an integer. Because the P_1' and P_2' are obtained by rotating P_1 into P_1' and P_2 into P_2', the distances $P_1P_2 = P_2'P_2 = \mathbf{a}$. Therefore,

$$\cos\phi = \frac{N\mathbf{a}/2}{\mathbf{a}} = \frac{N}{2}$$

Because N is an integer, and the expression $N/2$ cannot be greater than 1 or less than -1, we must have $-2 \le N \le 2$. We tabulate the possible values of N, ϕ, and n below.

N	$\cos\phi$	ϕ	n
2	1	360°	1
1	0.5	60°	6
0	0	90°	4
−1	−0.5	120°	3
−2	−1	180°	2

31–44. The von Laue equations are often expressed in vector notation. The following figure illustrates the X-ray scattering from two lattice points P_1 and P_2.

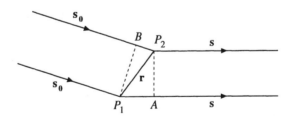

Let $\mathbf{s}_0$ be a unit vector in the direction of the incident radiation and $\mathbf{s}$ be a unit vector in the direction of the scattered X-radiation. Show that the difference in the path lengths of the waves scattered from P_1 and P_2 is given by

$$\delta = P_1A - P_2B = \mathbf{r}\cdot\mathbf{s} - \mathbf{r}\cdot\mathbf{s}_0 = \mathbf{r}\cdot\mathbf{S}$$

where $\mathbf{S} = \mathbf{s} - \mathbf{s}_0$. Because P_1 and P_2 are lattice points, $\mathbf{r}$ must be expressible as $m\mathbf{a} + n\mathbf{b} + p\mathbf{c}$, where m, n, and p are integers, and $\mathbf{a}$, $\mathbf{b}$, and $\mathbf{c}$ are the unit cell axes. Show that the fact that δ must be an integral multiple of the wavelength λ leads to the equations

$$\mathbf{a}\cdot\mathbf{S} = h\lambda$$

$$\mathbf{b}\cdot\mathbf{S} = k\lambda$$

$$\mathbf{c}\cdot\mathbf{S} = l\lambda$$

where h, k, and l are integers. These equations are the von Laue equations in vector notation.

The difference in the distance travelled by the two waves is $P_1 A - P_2 B$. Because $\mathbf{s}$ and $\mathbf{s}_0$ are unit vectors, we can write the distance $P_1 A$ as $\mathbf{r} \cdot \mathbf{s}$ and the distance $P_2 B$ as $\mathbf{r} \cdot \mathbf{s}_0$ (see the definition of dot product, MathChapter C) and so, letting δ be the difference in the distance travelled by the two waves and $\mathbf{S} = \mathbf{s} - \mathbf{s}_0$, we obtain

$$\delta = \mathbf{r} \cdot \mathbf{S}$$

The quantity δ must be an integral multiple of λ in order to have constructive interference. We can therefore write

$$N\lambda = \delta = \mathbf{r} \cdot \mathbf{S}$$
$$N\lambda = (m\mathbf{a} + n\mathbf{b} + p\mathbf{c}) \cdot \mathbf{S}$$
$$N\lambda = m(\mathbf{a} \cdot \mathbf{S}) + n(\mathbf{b} \cdot \mathbf{S}) + p(\mathbf{c} \cdot \mathbf{S})$$

This last relation must be true for all sets of integers N, m, n, and p. Therefore, each term on the right side of the equation must equal an integral multiple of λ. We then write

$$\mathbf{a} \cdot \mathbf{S} = h\lambda$$
$$\mathbf{b} \cdot \mathbf{S} = k\lambda$$
$$\mathbf{c} \cdot \mathbf{S} = l\lambda$$

where h, k, and l are integers.

31–45. We can derive the Bragg equation from the von Laue equations derived in the previous problem. First show that $\mathbf{S} = \mathbf{s} - \mathbf{s}_0$ bisects the angle between $\mathbf{s}_0$ and $\mathbf{s}$ and is normal to the plane from which the X-radiation would be specularly reflected (the angle of incidence equals the angle of reflection). Now show that the distance from the origin of the $\mathbf{a}$, $\mathbf{b}$, and $\mathbf{c}$ axes to the hkl plane is given by

$$d = \frac{\mathbf{a}}{h} \cdot \frac{\mathbf{S}}{|\mathbf{S}|} = \frac{\mathbf{b}}{k} \cdot \frac{\mathbf{S}}{|\mathbf{S}|} = \frac{\mathbf{c}}{l} \cdot \frac{\mathbf{S}}{|\mathbf{S}|} = \frac{\lambda}{|\mathbf{S}|}$$

Last, show that $|\mathbf{S}| = [(\mathbf{s} - \mathbf{s}_0) \cdot (\mathbf{s} - \mathbf{s}_0)]^{1/2} = [2 - 2\cos 2\theta]^{1/2} = 2\sin\theta$, which leads to the Bragg equation $d = \lambda/2\sin\theta$.

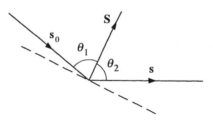

We know from the previous problem that $\mathbf{S} = \mathbf{s} - \mathbf{s}_0$, and that both $\mathbf{s}$ and $\mathbf{s}_0$ are unit vectors of length 1. Using the definition of dot product, we can write

$$\cos\theta_2 = \frac{\mathbf{S} \cdot \mathbf{s}}{|\mathbf{S}|} = \frac{s^2 - \mathbf{S} \cdot \mathbf{s}_0}{|\mathbf{S}|} = \frac{1 - \mathbf{S} \cdot \mathbf{s}_0}{|\mathbf{S}|}$$

and (because the angle between $\mathbf{S}$ and $\mathbf{s}_0$ is $\pi - \theta_1$)

$$\cos(\pi - \theta_1) = -\cos\theta_1 = \frac{\mathbf{S} \cdot \mathbf{s}_0}{|\mathbf{S}|} = \frac{\mathbf{S} \cdot \mathbf{s}_0 - s_0^2}{|\mathbf{S}|} = \frac{\mathbf{S} \cdot \mathbf{s}_0 - 1}{|\mathbf{S}|}$$

We see that $-\cos\theta_1 = -\cos\theta_2$, and, because both θ_1 and θ_2 are acute angles, $\theta_1 = \theta_2$. The plane from which radiation would be specularly reflected is represented in the figure by a dashed line; as is shown in the figure, because $2\theta + 2\phi = 180°$, $\theta + \phi = 90°$ and $\mathbf{S}$ is normal to the plane of specular reflection.

$$\lambda = \frac{\mathbf{a}\cdot\mathbf{S}}{h} = \frac{\mathbf{b}\cdot\mathbf{S}}{k} = \frac{\mathbf{c}\cdot\mathbf{S}}{l}$$

In the previous problem, we found that

$$\mathbf{a}\cdot\mathbf{S} = |\mathbf{a}||\mathbf{S}|\cos\alpha = h\lambda$$

$$\mathbf{b}\cdot\mathbf{S} = |\mathbf{b}||\mathbf{S}|\cos\beta = k\lambda$$

$$\mathbf{c}\cdot\mathbf{S} = |\mathbf{c}|\mathbf{S}|\cos\gamma = l\lambda$$

where α, β, and γ are illustrated below.

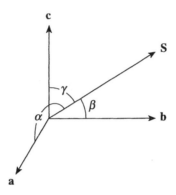

Note that

$$\cos\alpha = \frac{\mathbf{S}\cdot\mathbf{i}}{|\mathbf{S}|} = \frac{S_x}{|\mathbf{S}|} \qquad \cos\beta = \frac{\mathbf{S}\cdot\mathbf{j}}{|\mathbf{S}|} = \frac{S_y}{|\mathbf{S}|} \qquad \cos\gamma = \frac{\mathbf{S}\cdot\mathbf{k}}{|\mathbf{S}|} = \frac{S_z}{|\mathbf{S}|}$$

where S_x, S_y, and S_z are the components of $\mathbf{S}$ in the x, y, and z-directions. Therefore,

$$\cos\alpha^2 + \cos\beta^2 + \cos\gamma^2 = \frac{S_x^2 + S_y^2 + S_z^2}{|\mathbf{S}|^2} = 1 \tag{1}$$

We now substitute the von Laue equations into Equation 31.2 and use Equation 1 to write

$$\frac{1}{d^2} = \frac{h^2}{a^2} + \frac{k^2}{b^2} + \frac{l^2}{c^2}$$

$$= \frac{|S|^2}{\lambda^2}[\cos^2\alpha + \cos^2\beta + \cos^2\gamma]$$

$$\frac{1}{d} = \frac{|S|}{\lambda}$$

We can now use the vector properties delineated in MathChapter C to find $|\mathbf{S}|$, again recalling that $\mathbf{s}$ and $\mathbf{s}_0$ are unit vectors.

$$|\mathbf{S}| = [(\mathbf{s} - \mathbf{s}_0)\cdot(\mathbf{s} - \mathbf{s}_0)]^{1/2}$$

$$= [1^2 + 1^2 - 2\mathbf{s}\cdot\mathbf{s}_0]^{1/2} = [2 - 2\cos 2\theta]^{1/2}$$

$$= [2 - 2\cos^2\theta + 2\sin^2\theta]^{1/2} = [4\sin^2\theta]^{1/2} = 2\sin\theta$$

So $1/d = 2\sin\theta/\lambda$, or $d = \lambda/2\sin\theta$.

31–46. The enthalpy of adsorption for H_2 adsorbed on a surface of copper is $-54.4 \text{ kJ} \cdot \text{mol}^{-1}$. The activation energy for going from the physisorbed state to the chemisorbed state is $31.3 \text{ kJ} \cdot \text{mol}^{-1}$, and the curve crossing between these two potentials occurs at $V(z) = 21 \text{ kJ} \cdot \text{mol}^{-1}$. Draw a schematic representation similar to that in Figure 31.20 for the case of H_2 interacting with copper.

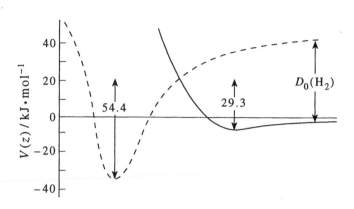

31–47. In Section 27–4, we showed that the collision frequency per unit area is (Equation 27.48)

$$z_{\text{coll}} = \frac{\rho \langle u \rangle}{4} \qquad (1)$$

Use Equation 1 and the ideal-gas law to show that J_N, the number of molecules striking a surface of unit area (1 m^2) in one second, is

$$J_N = \frac{P N_A}{(2\pi M R T)^{1/2}}$$

where M is the molar mass of the molecule, P is the pressure of the gas, and T is the temperature. How many nitrogen molecules will strike a 1.00-cm^2 surface in 1.00 s at 298.1 K and a gas pressure of 1.05×10^{-6} Pa?

The average speed is given by

$$\langle u \rangle = \left(\frac{8RT}{\pi M} \right)^{1/2} \qquad (27.42)$$

Using the ideal gas law, we find that $\rho = N_A/V = P N_A/RT$, so

$$z_{\text{coll}} = \frac{\rho \langle u \rangle}{4} = \frac{\rho}{4} \left(\frac{8RT}{\pi M} \right)^{1/2} = \frac{N_A P}{4RT} \left(\frac{8RT}{\pi M} \right)^{1/2} = \frac{N_A P}{(2\pi M R T)^{1/2}} = J_N$$

Under the conditions given,

$$
\begin{aligned}
J_N &= \frac{N_A P}{(2\pi M R T)^{1/2}} \\
&= \frac{(6.022 \times 10^{23} \text{ mol}^{-1})(1.05 \times 10^{-6} \text{ Pa})}{[2\pi (0.02802 \text{ kg} \cdot \text{mol}^{-1})(8.315 \text{ J} \cdot \text{K}^{-1} \cdot \text{mol}^{-1})(298.15 \text{ K})]^{1/2}} \\
&= \frac{6.323 \times 10^{17} \text{ kg} \cdot \text{m}^{-1} \cdot \text{s}^{-2}}{20.89 \text{ kg} \cdot \text{m} \cdot \text{s}^{-1}} \\
&= 3.03 \times 10^{16} \text{ s}^{-1} \cdot \text{m}^{-2} = 3.03 \times 10^{12} \text{ cm}^{-2} \cdot \text{s}^{-1}
\end{aligned}
$$

so 3.03×10^{12} molecules will collide with the 1.00-cm^2 surface in one second.

31–48. One *langmuir* corresponds to an exposure of a surface to a gas at a pressure of 1.00×10^{-6} torr for 1 second at 298.15 K. Define one langmuir in units of pascals instead of torr. How many nitrogen molecules will strike a surface of area 1.00 cm² when exposed to 1.00 langmuir? (See Problem 31–47.)

$$1.00 \times 10^{-6} \text{ torr} \left(\frac{1.01325 \times 10^5 \text{ Pa}}{760 \text{ torr}} \right) = 1.33 \times 10^{-4} \text{ Pa}$$

A langmuir thus corresponds to an exposure of a surface to a gas at a pressure of 1.33×10^{-4} Pa for one second at 298.15 K. Using these conditions, we can substitute into our expression for J_N from the previous problem:

$$
\begin{aligned}
J_N &= \frac{N_A P}{(2\pi M R T)^{1/2}} \\
&= \frac{(6.022 \times 10^{23} \text{ mol}^{-1})(1.33 \times 10^{-4} \text{ Pa})}{[2\pi(0.02802 \text{ kg} \cdot \text{mol}^{-1})(8.315 \text{ J} \cdot \text{K}^{-1} \cdot \text{mol}^{-1})(298.15 \text{ K})]^{1/2}} \\
&= \frac{8.029 \times 10^{19} \text{ kg} \cdot \text{m}^{-1} \cdot \text{s}^{-2}}{20.89 \text{ kg} \cdot \text{m} \cdot \text{s}^{-1}} \\
&= 3.83 \times 10^{18} \text{ s}^{-1} \cdot \text{m}^{-2} = 3.83 \times 10^{14} \text{ cm}^{-2} \cdot \text{s}^{-1}
\end{aligned}
$$

A total of 3.83×10^{14} nitrogen molecules will strike the 1.00-cm² surface in one second.

31–49. If the density of surface sites is 2.40×10^{14} cm^{-2} and every molecule that strikes the surface adsorbs to one of these sites, determine the fraction of a monolayer created by the exposure of a 1.00-cm² surface to 1.00×10^{-4} langmuir of $N_2(g)$ at 298.15 K.

In the previous problem, we found that the exposure of a 1.00-cm² surface to one langmuir resulted in 3.83×10^{14} molecules colliding with the surface. Then exposure of the same surface to 1.00×10^{-4} L would result in 3.83×10^{10} collisions. Assuming that all these molecules are adsorbed,

$$\frac{3.83 \times 10^{10}}{2.40 \times 10^{14}} = 0.016\%$$

of the surface will be covered. This is the fraction of monolayer created.

31–50. For conducting surface experiments it is important to maintain a clean surface. Suppose that a 1.50-cm² surface is placed inside a high-vacuum chamber at 298.15 K and the pressure inside the chamber is 1.00×10^{-12} torr. If the density of the surface sites is 1.30×10^{16} cm^{-2} and we assume that the only gas in the chamber is H_2O and that each of the H_2O molecules that strike the surface adsorb to one surface site, how long will it be until 1.00% of the surface sites are occupied by water?

The pressure inside the chamber in Pa is

$$1 \times 10^{-12} \text{ torr} \left(\frac{1.01325 \times 10^5 \text{ Pa}}{760 \text{ torr}} \right) = 1.33 \times 10^{-10} \text{ Pa}$$

The rate of collision per unit of surface area is given by J_N from Problem 31–47 is

$$J_N = \frac{N_A P}{(2\pi MRT)^{1/2}}$$

$$= \frac{(6.022 \times 10^{23} \text{ mol}^{-1})(1.33 \times 10^{-10} \text{ Pa})}{[2\pi(0.01801 \text{ kg}\cdot\text{mol}^{-1})(8.315 \text{ J}\cdot\text{K}^{-1}\cdot\text{mol}^{-1})(298.15 \text{ K})]^{1/2}}$$

$$= \frac{8.029 \times 10^{13} \text{ kg}\cdot\text{m}^{-1}\cdot\text{s}^{-2}}{16.79 \text{ kg}\cdot\text{m}\cdot\text{s}^{-1}}$$

$$= 4.79 \times 10^{12} \text{ s}^{-1}\cdot\text{m}^{-2} = 4.79 \times 10^{8} \text{ cm}^{-2}\cdot\text{s}^{-1}$$

Now if 1.00% of the surface sites are occupied, the density of occupied surface sites will be 1.30×10^{14} cm^{-2}. Then

$$(4.79 \times 10^{8} \text{ cm}^{-2}\cdot\text{s}^{-1})t = 1.30 \times 10^{14} \text{ cm}^{-2}$$

$$t = 2.71 \times 10^{5} \text{ s}$$

31–51. Use the results of Example 31–12 to determine the rate of desorption of CO from palladium at 300 K and 500 K.

Because $k_d = 1/\tau$, at 300 K $k_d = 3.8 \times 10^{-14}$ s^{-1} and at 500 K $k_d = 5.6 \times 10^{-4}$ s^{-1}.

31–52. The following data were obtained for the adsorption of $N_2(g)$ to a piece of solid graphite at 197 K. The tabulated volumes are the volumes that the adsorbed gas would occupy at 0.00°C and one bar

P/bar	3.54	10.13	16.92	26.04	29.94
$V/10^{-4}$ m^3	328	456	497	527	536

Determine the values of V_m and b using the Langmuir adsorption isotherm. The total mass of the carbon solid is 1325 g. Determine the fraction of the carbon atoms that are accessible as binding sites if you assume that each surface atom can adsorb one N_2 molecule.

Use the equation from Example 31–10 to fit the above data:

$$\frac{1}{V} = \frac{1}{PbV_m} + \frac{1}{V_m}$$

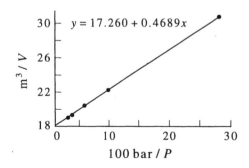

Comparing the line of best fit to the equation above, we see that

$$\frac{1}{V_m} = 17.260 \text{ m}^{-3}$$

and

$$V_m = 5.79 \times 10^{-2} \text{ m}^3$$

Likewise,

$$\frac{1}{bV_m} = 46.89 \text{ bar} \cdot \text{m}^{-3}$$

and

$$b = \frac{1}{46.89 V_m} = 0.368 \text{ bar}$$

At $0.00°$ C and one bar, one mole of gas occupies about 22.4×10^{-3} m^3. The number of moles of gas in V_m is then

$$\frac{5.59 \times 10^{-2} \text{ m}^3}{22.4 \times 10^{-3} \text{ m}^3 \cdot \text{mol}^{-1}} = 2.58 \text{ mol}$$

If each surface atom adsorbs one N$_2$ molecule, then 2.58 mol of carbon atoms on the surface are accessible to N$_2$(g). The total number of moles of carbon is 1325 g/12.01 g$\cdot$mol^{-1} = 110 mol, and so the fraction of carbon atoms accessible is 0.023, or 2.3%.

31–53. The first-order surface reaction

$$A(g) \Longrightarrow A(ads) \Longrightarrow B(g)$$

has a rate of 1.8×10^{-4} mol$\cdot$dm$^{-3} \cdot$s^{-1}. The surface has a dimension of 1.00 cm by 3.50 cm. Calculate the rate of reaction if the dimensions of the two sides of the surface were each doubled. [Assume that A(g) is in excess.]

Because A(g) is in excess, the second step is the rate-determining step and

$$v = k_d \theta \sigma_0$$

If the dimensions of each side of the surface are doubled, the surface area will increase by a factor of 4. The quantity σ_0 will therefore increase by a factor of 4, and the the reaction rate on the larger reaction surface will be $v = 7.2 \times 10^{-4}$ mol$\cdot$dm$^{-3} \cdot$s^{-1}.

31–54. Consider the reaction scheme

$$A(g) + S \overset{k_1}{\Longrightarrow} A\text{–}S \overset{k_2}{\Longrightarrow} P(g)$$

for which the rate law is

$$v = k_2 \theta_A$$

where θ_A is the fraction of surface sites occupied by A molecules. Use the Langmuir adsorption isotherm (Equation 31.35) to obtain an expression for the reaction rate in terms of K_c and [A]. Under what conditions will the reaction be first order in the concentration of A?

Begin with the Langmuir adsorption isotherm (Equation 31.35) and solve for θ_A:

$$\frac{1}{\theta_A} = 1 + \frac{1}{K_c[A]}$$

$$\theta_A = \frac{K_c[A]}{K_c[A] + 1}$$

Substituting this result into the given rate law gives

$$v = \frac{K_c k_2 [A]}{K_c[A] + 1}$$

The reaction will be first order in [A] if $K_c[A] \ll 1$, in which case the expression above reduces to $v = k_2 K_c[A]$.

31–55. Consider a surface-catalyzed bimolecular reaction between molecules A and B that has a rate law of the form

$$v = k_3 \theta_A \theta_B$$

where θ_A is the fraction of surface sites occupied by reactant A and θ_B is the fraction of surface sites occupied by reactant B. A mechanism consistent with this reaction is as follows:

$$A(g) + S(s) \underset{k_d^A}{\overset{k_a^A}{\rightleftharpoons}} A\text{–}S(s) \qquad \text{(fast equilibrium)} \qquad (1)$$

$$B(g) + S(s) \underset{k_d^B}{\overset{k_a^B}{\rightleftharpoons}} B\text{–}S(s) \qquad \text{(fast equilibrium)} \qquad (2)$$

$$A\text{–}S(s) + B\text{–}S(s) \overset{k_3}{\Longrightarrow} \text{products}$$

Take K_A and K_B to be the equilibrium constants for Equations 1 and 2, respectively. Derive expressions for θ_A and θ_B in terms of [A], [B], K_A, and K_B. Use your results to show that the rate law can be written as

$$v = \frac{k_3 K_A K_B [A][B]}{(1 + K_A[A] + K_B[B])^2}$$

At any given time, the fraction of available surface sites is $1 - \theta_A - \theta_B$.

From Equation 1, we have the equations (analogous to Equations 31.33 and 31.34)

$$v_d^A = k_d^A \theta_A \sigma_0$$

$$v_a^A = k_a^A [A](1 - \theta_A - \theta_B)\sigma_0$$

Likewise, for reactant B,

$$v_d^B = k_d^B \theta_B \sigma_0$$

$$v_a^B = k_a^B [B](1 - \theta_A - \theta_B)\sigma_0$$

At equilibrium,

$$k_d^A \theta_A = k_a^A [A](1 - \theta_A - \theta_B)$$

$$\frac{\theta_A}{1 - \theta_A - \theta_B} = K_A[A] \qquad (3)$$

and

$$k_d^B \theta_B = k_a^B [B](1 - \theta_A - \theta_B)$$

$$\frac{\theta_B}{1 - \theta_A - \theta_B} = K_B [B] \qquad (4)$$

We can solve Equations 3 and 4 simultaneously to obtain

$$\theta_A = \frac{K_A[A]}{1 + K_A[A] + K_B[B]}$$

$$\theta_B = \frac{K_B[B]}{1 + K_A[A] + K_B[B]}$$

Substituting these expressions for θ_A and θ_B into the rate law gives

$$v = \frac{k_3 K_A K_B [A][B]}{(1 + K_A[A] + K_B[B])^2}$$

31–56. Reconsider the surface-catalyzed bimolecular reaction in Problem 31–55. If A(g) and B(g) do not compete for surface sites, but instead each molecule uniquely binds to a different type of surface site, show that the rate law is given by

$$v = \frac{k_3 K_A K_B [A][B]}{(1 + K_A[A])(1 + K_B[B])}$$

If A and B do not compete for surface sites, the fraction of surface sites available for A is $1 - \theta_A$ and the fraction available for B is $1 - \theta_B$. Then, at equilibrium,

$$k_a^A[A](1 - \theta_A) = k_d^A \theta_A$$
$$k_a^B[B](1 - \theta_B) = k_d^B \theta_B$$

Solving for θ_A and θ_B gives

$$\theta_A = \frac{K_A[A]}{1 + K_A[A]}$$

$$\theta_B = \frac{K_B[B]}{1 + K_B[B]}$$

and substituting these expressions into the rate law gives

$$v = k_3 \theta_A \theta_B = \frac{k_3 K_A K_B [A][B]}{(1 + K_A[A])(1 + K_B[B])}$$

31–57. In this problem we derive Equation 31.45, the rate law for the oxidation reaction $2\,CO(g) + O_2(g) \longrightarrow 2\,CO_2(g)$ assuming that the reaction occurs by the Langmuir-Hinshelwood mechanism. The overall rate law for this mechanism is

$$v = k_3 \theta_{CO} \theta_{O_2}$$

Show that

$$\theta_{O_2} = \frac{(K_{O_2}[O_2])^{1/2}}{1 + (K_{O_2}[O_2])^{1/2} + K_{CO}[CO]}$$

and

$$\theta_{CO} = \frac{K_{CO}[CO]}{1 + (K_{O_2}[O_2])^{1/2} + K_{CO}[CO]}$$

Use these expressions and the relationship $b = K_c/k_B T$ to obtain the rate law given by Equation 31.45. (Assume ideal-gas behavior.)

For CO, as in Problem 31–55,

$$k_a^{CO}[CO](1 - \theta_{CO} - \theta_{O_2}) = k_d^{CO}\theta_{CO}$$

Because the second reaction of the mechanism produces 2 O atoms,

$$k_a^{O_2}[O_2](1 - \theta_{CO} - \theta_{O_2})^2\sigma_0^2 = k_d\theta_{O_2}^2\sigma_0^2$$

or

$$k_a^{O_2}[O_2](1 - \theta_{CO} - \theta_{O_2})^2 = k_d\theta_{O_2}^2$$

Taking the square root of both sides of this equation gives

$$\left(k_a^{O_2}[O_2]\right)^{1/2}(1 - \theta_{CO} - \theta_{O_2}) = \left(k_d^{O_2}\right)^{1/2}\theta_{O_2}$$

We can substitute K_{CO} and K_{O_2} for the adsorption and desorption constants in the above expression and find

$$\left(K_{O_2}[O_2]\right)^{1/2} = \frac{\theta_{O_2}}{1 - \theta_{CO} - \theta_{O_2}}$$

and

$$K_{CO}[CO] = \frac{\theta_{CO}}{1 - \theta_{CO} - \theta_{O_2}}$$

Solving these two equations simultaneously gives

$$\theta_{O_2} = \frac{\left(K_{O_2}[O_2]\right)^{1/2}}{1 + \left(K_{O_2}[O_2]\right)^{1/2} + K_{CO}[CO]}$$

$$\theta_{CO} = \frac{K_{CO}[CO]}{1 + \left(K_{O_2}[O_2]\right)^{1/2} + K_{CO}[CO]}$$

Substituting these results into the overall rate law, we find that

$$v = k_3\theta_{CO}\theta_{O_2} = \frac{k_3 K_{CO}[CO]K_{O_2}^{1/2}[O_2]^{1/2}}{(1 + K_{O_2}^{1/2}[O_2]^{1/2} + K_{CO}[CO])^2}.$$

Assuming ideal behavior, $[O_2] = P_{O_2}/k_B T$ and $[CO] = P_{CO}/k_B T$, and from the text $b_{CO} = K_{CO}/k_B T$ and $b_{O_2} = K_{O_2}/k_B T$. Substituting these expressions into the rate law above gives

$$v = \frac{k_3 b_{O_2}^{1/2} P_{O_2}^{1/2} b_{CO} P_{CO}}{(1 + b_{O_2}^{1/2} P_{O_2}^{1/2} + b_{CO} P_{CO})^2} \tag{31.45}$$

31–58. In this problem we derive Equation 31.46, the rate law for the oxidation reaction $2\,CO(g) + O_2(g) \longrightarrow 2\,CO_2(g)$ assuming that the reaction occurs by the Eley-Rideal mechanism. The overall rate law for this mechanism is

$$v = k_3 \theta_{O_2}[CO]$$

Assuming that both $CO(g)$ and $O_2(g)$ compete for adsorption sites, show that

$$v = \frac{k_3 K_{O_2}^{1/2}[O_2]^{1/2}[CO]}{1 + K_{O_2}^{1/2}[O_2]^{1/2} + K_{CO}[CO]}$$

Use the relationship between K_c and b and the ideal-gas law to show that this equation is equivalent to Equation 31.46.

The step of the mechanism for the absorption of O_2 is the same as that discussed in the previous problem for O_2, so θ_{O_2} is

$$\theta_{O_2} = \frac{\left(K_{O_2}[O_2]\right)^{1/2}}{1 + \left(K_{O_2}[O_2]\right)^{1/2} + K_{CO}[CO]}$$

Substituting into the overall rate law,

$$v = k_3 \theta_{O_2}[CO] = \frac{k_3 K_{O_2}^{1/2}[O_2]^{1/2}[CO]}{1 + K_{O_2}^{1/2}[O_2]^{1/2} + K_{CO}[CO]}$$

Assuming ideal behavior, $[O_2] = P_{O_2}/k_B T$ and $[CO] = P_{CO}/k_B T$, and from the text $b_{CO} = K_{CO}/k_B T$ and $b_{O_2} = K_{O_2}/k_B T$. Substituting these expressions into the above rate law gives

$$v = \frac{k_3 b_{O_2}^{1/2} P_{O_2}^{1/2} b_{CO} P_{CO}}{1 + b_{O_2}^{1/2} P_{O_2}^{1/2} + b_{CO} P_{CO}} \tag{31.46}$$

31–59. The hydrogenation of ethene on copper obeys the rate law

$$v = \frac{k[H_2]^{1/2}[C_2H_4]}{(1 + K[C_2H_4])^2}$$

where k and K are constants. Mechanistic studies show that the reaction occurs by the Langmuir-Hinshelwood mechanism. How are k and K related to the rate constants for the individual steps of the reaction mechanism? What can you conclude about the relative adsorption of $H_2(g)$ and $C_2H_4(g)$ to the copper surface from the form of the observed rate law?

The rate law for the Langmuir-Hinshelwood mechanism is (Problem 31–57)

$$v = k_3 \theta_{C_2H_4} \theta_{H_2} = \frac{k_3 K_{C_2H_4}[C_2H_4]K_{H_2}^{1/2}[H_2]^{1/2}}{(1 + K_{H_2}^{1/2}[H_2]^{1/2} + K_{C_2H_4}[C_2H_4])^2}$$

The $K_{H_2}^{1/2}[H_2]^{1/2}$ term does not appear in the denominator of the rate law provided in the problem, so $1 + K_{C_2H_4}[C_2H_4] \gg K_{H_2}^{1/2}[H_2]^{1/2}$. This means that C_2H_4 must adsorb much more extensively than

H_2. Comparing the given rate law and the above expression for v, we find that $k = k_3 K_{H_2}^{1/2} K_{C_2H_4}$ and $K = K_{C_2H_4}$.

31–60. The iron-catalyzed exchange reaction

$$NH_3(g) + D_2(g) \longrightarrow NH_2D(g) + HD(g)$$

obeys the rate law

$$v = \frac{k[D_2]^{1/2}[NH_3]}{(1 + K[NH_3])^2}$$

Is this rate law consistent with either the Eley-Rideal or Langmuir-Hinshelwood mechanisms? How are k and K related to the rate constants of the individual steps of the mechanism you chose? What does the rate law tell you about the relative adsorption of $D_2(g)$ and $NH_3(g)$ to the iron surface?

This rate law is consistent with the Langmuir-Hinshelwood mechanism (Problem 31–57). The total rate law for the mechanism would be

$$v = \frac{k_3 K_{NH_3}[NH_3] K_{D_2}^{1/2}[D_2]^{1/2}}{(1 + K_{D_2}^{1/2}[D_2]^{1/2} + K_{NH_3}[NH_3])^2}$$

If $K_{D_2}^{1/2}[D_2]^{1/2} \ll 1 + K_{NH_3}[NH_3]$, this reduces to the rate law given in the problem, in which case $K = K_{NH_3}$ and $k = k_3 K_{NH_3} K_{D_2}^{1/2}$. The above inequality tells us that NH_3 adsorbs to the surface much more extensively than D_2.

31–61. Consider the surface-catalyzed exchange reaction

$$H_2(g) + D_2(g) \longrightarrow 2\,HD(g)$$

Experimental studies show that this reaction occurs by the Langmuir-Hinshelwood mechanism by which both $H_2(g)$ and $D_2(g)$ first dissociatively chemisorb to the surface. The rate-determining step is the reaction between the adsorbed H and D atoms. Derive an expression for the rate law for this reaction in terms of the gas-phase pressures of $H_2(g)$ and $D_2(g)$. (Assume ideal-gas behavior.)

Because both reactions produce two atoms, we can write (as in Problem 31–57)

$$k_a^{D_2}[D_2](1 - \theta_{H_2} - \theta_{D_2})^2 = k_d \theta_{D_2}^2$$

Taking the square root of both sides gives

$$\left(k_a^{D_2}[D_2]\right)^{1/2}(1 - \theta_{H_2} - \theta_{D_2}) = \left(k_d^{D_2}\right)^{1/2}\theta_{D_2}$$

or

$$K_{D_2}^{1/2}[D_2]^{1/2} = \frac{\theta_{D_2}}{1 - \theta_{H_2} - \theta_{D_2}}$$

with an equivalent expression in K_{H_2}. Solving for θ_{D_2} and θ_{H_2} gives

$$\theta_{D_2} = \frac{K_{D_2}^{1/2}[D_2]^{1/2}}{1 + K_{D_2}^{1/2}[O_2]^{1/2} + K_{H_2}^{1/2}[H_2]^{1/2}}$$

$$\theta_{H_2} = \frac{K_{H_2}^{1/2}[H_2]^{1/2}}{1 + K_{D_2}^{1/2}[O_2]^{1/2} + K_{H_2}^{1/2}[H_2]^{1/2}}$$

Finally,

$$v = k_3 \theta_{D_2} \theta_{H_2} = \frac{k_3 K_{D_2}^{1/2} K_{H_2}^{1/2}[D_2]^{1/2}[H_2]^{1/2}}{(1 + K_{D_2}^{1/2}[O_2]^{1/2} + K_{H_2}^{1/2}[H_2]^{1/2})^2} = \frac{(k_3 b_{D_2} b_{H_2} P_{D_2} P_{H_2})^{1/2}}{(1 + b_{D_2}^{1/2} P_{D_2}^{1/2} + b_{H_2}^{1/2} P_{H_2}^{1/2})^2}$$

31–62. LEED spectroscopy records the intensities and locations of electrons that are diffracted from a surface. For an electron to diffract, its de Broglie wavelength must be less than twice the distance between the atomic planes in the solid (see Section 31–9). Show that the de Broglie wavelength of an electron accelerated through a potential difference of ϕ volts is given by

$$\lambda/pm = \left(\frac{1.504 \times 10^6 \text{ V}}{\phi}\right)^{1/2}$$

The de Broglie equation is $\lambda = h/p$ (Equation 1.12). Because $E = p^2/2m$, and the energy of an electron accelerated through a potential difference of ϕ volts is $e\phi$,

$$\lambda = \frac{h}{(2mE)^{1/2}} = \left(\frac{h^2}{2me\phi}\right)^{1/2}$$

$$= \left[\frac{(6.626 \times 10^{-34} \text{ J} \cdot \text{s})^2}{2(9.109 \times 10^{-31} \text{ kg})(1.602 \times 10^{-19} \text{ C})\phi}\right]^{1/2}$$

$$\lambda = \left(\frac{1.504 \times 10^{-18} \text{ V} \cdot \text{m}^2}{\phi}\right)^{1/2}$$

$$\lambda/pm = \left[\frac{(1.504 \times 10^{-18} \text{ V} \cdot \text{m}^2)(10^{12} \text{ pm} \cdot \text{m}^{-1})^2}{\phi}\right]^{1/2} = \left(\frac{1.504 \times 10^6 \text{ V}}{\phi}\right)^{1/2}$$

31–63. The distance between the 100 planes of a nickel substrate, whose surface is a 100 plane, is 351.8 pm. Calculate the minimum accelerating potential so that electrons can diffract from the crystal. Calculate the kinetic energy of these electrons. (*Hint*: See Problem 31–62.)

From Problem 31–62, we know that the de Broglie wavelength must be less than twice the distance between the atomic planes. Therefore, the minimum accelerating potential so that electrons can diffract from the crystal is

$$2(351.8) = \left(\frac{1.504 \times 10^6 \text{ V}}{\phi_{min}}\right)^{1/2}$$

$$\phi_{min} = \frac{1.504 \times 10^6 \text{ V}}{(703.6)^2} = 3.04 \text{ V}$$

and the kinetic energy of the diffracted electrons is

$$E = e\phi = 4.87 \times 10^{-19} \text{ J} = 293 \text{ kJ} \cdot \text{mol}^{-1}$$

31–64. The distance between the 111 surface of silver and the second layer of atoms is 235 pm, the same as in the bulk. If electrons with a kinetic energy of 8.77 eV strike the surface, will an electron diffraction pattern be observed? (*Hint*: See Problem 31–62.)

An electron diffraction pattern will be observed if the de Broglie wavelength is less than twice the interplanar distance, or (in this case) less than 470 pm. The electrons with kinetic energy 8.77 eV have been accelerated through a potential difference of 8.77 V. We can find λ using the equations developed in Problem 31–62:

$$\lambda/pm = \left(\frac{1.504 \times 10^6 \text{ V}}{\phi}\right)^{1/2} = \left[\frac{1.504 \times 10^6 \text{ V}}{8.77 \text{ V}}\right]^{1/2} = 414 \text{ pm}$$

An electron diffraction pattern is observed.

31–65. Figure 31.28 shows the relative rates of ammonia synthesis for five different surfaces of iron. Iron crystallizes as a body-centered cubic structure. Draw a schematic representation of the atomic arrangement of the 100, 110, and 111 surfaces. (*Hint*: See Figure 31.9.) Determine the center-to-center distance between nearest neighbor atoms on the surface in units of a, the dimension of the unit cell.

Below is a schematic representation of the appropriate surfaces. The distances marked are calculated by observing how the surface corresponds to the body-centered cubic structure.

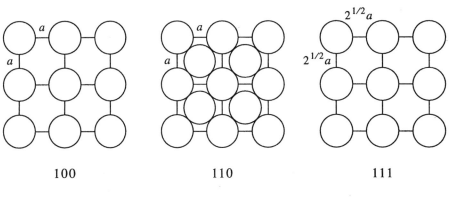

surface	nearest-neighbor distance
100	a
110	$\dfrac{\sqrt{2}a}{2}$
111	$\sqrt{2}a$

31–66. The *Freundlich adsorption isotherm* is given by

$$V = kP^a$$

where k and a are constants. Can the data in Problem 31–52 be described by the Freundlich adsorption isotherm? Determine the best-fit values of k and a to the data.

We fit the data in Problem 31–52 to a Freundlich adsorption isotherm below.

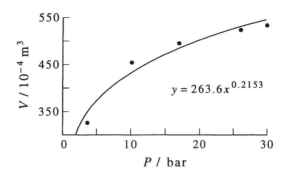

The data can be reasonably described by the Freundlich adsorption isotherm. From the best fit to the data, we find that $a = 0.22$ and that $k = 0.026 \text{ m}^{-3}$.

31–67. Show that if $\theta \ll 1$, the Langmuir adsorption isotherm reduces to the Freundlich adsorption isotherm (Problem 31–66) with $k = bV_m$ and $a = 1$.

The Langmuir adsorption isotherm is (Example 31–10)

$$\frac{1}{V} = \frac{1}{PbV_m} + \frac{1}{V_m}$$

If $\theta \ll 1$, then $1/V \gg 1$ and $1/bPV_m \gg 1$. Then

$$\frac{1}{V} = \frac{1}{bPV_m}$$

$$V = bPV_m$$

which is the Freundlich adsorption isotherm when $k = bV_m$ and $a = 1$.

31–68. Multilayer physisorption is often described by the *BET adsorption isotherm*

$$\frac{P}{V(P^* - P)} = \frac{1}{cV_m} + \frac{(c-1)P}{V_m c P^*}$$

where P^* is the vapor pressure of the adsorbate at the temperature of the experiment, V_m is the volume corresponding to a monolayer of coverage on the surface, V is the total volume adsorbed at pressure P, and c is a constant. Rewrite the equation for the BET adsorption isotherm in the form

$$\frac{V}{V_m} = f(P/P^*)$$

Plot V/V_m versus P/P^* for $c = 0.1, \ 1.0, \ 10,$ and 100. Discuss the shapes of the curves.

Let $x = P/P^*$. Then the BET isotherm can be expressed as

$$\frac{x}{V(1-x)} = \frac{1}{cV_m} + \frac{(c-1)x}{cV_m}$$

$$\frac{x}{V(1-x)} = \frac{1}{cV_m}[1 + (c-1)x]$$

$$\frac{x}{1-x} = \frac{V}{cV_m}[1 + (c-1)x]$$

$$\frac{cx}{1-x} = \frac{V}{V_m}[1 + (c-1)x]$$

Solving for V/V_m gives

$$\frac{V}{V_m} = \frac{cx}{(1-x)[1+(c-1)x]}$$

Below we plot V/V_m versus P/P^* for different values of c.

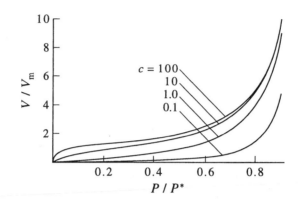

When $c \ll 1$, adsorption in the first layer is strongly favored relative to higher layers, so the first layer is almost completely filled before higher layers begin to fill. This accounts for the shape of the $c = 0.1$ curve. The behavior of all curves as $P \to P^*$ reflects the fact that liquid is condensing on the surface of the solid.

31–69. The energy of adsorption, E_{ads}, can be measured by the technique of *temperature programmed desorption* (TPD). In a TPD experiment, the temperature of a surface with bound adsorbate is changed according to the equation

$$T = T_0 + \alpha t \tag{1}$$

where T_0 is the initial temperature, α is a constant that determines the rate at which the temperature is changed, and t is the time. A mass spectrometer is used to measure the concentration of molecules that desorb from the surface. The analysis of TPD data depends on the kinetic model for desorption. Consider a first-order desorption process

$$M\text{–}S(s) \xrightarrow{k_d} M(g) + S(s)$$

Write an expression for the rate law for desorption. Use Equation 1, Equation 31.37, and your rate law to show that your rate law can be written as

$$\frac{d[M\text{–}S]}{dT} = -\frac{[M\text{–}S]}{\alpha}(\tau_0^{-1}e^{-E_{ads}/RT}) \tag{2}$$

With increasing temperature, $d[M\text{–}S]/dt$ initially increases, then reaches a maximum, after which it decreases. Let $T = T_{max}$ be the temperature corresponding to the maximum rate of desorption. Use Equation 2 to show that at T_{max}

$$\frac{E_{ads}}{RT_{max}^2} = \frac{\tau_0^{-1}}{\alpha}e^{-E_{ads}/RT_{max}} \tag{3}$$

(*Hint*: Remember that [M–S] is a function of T.)

The rate law for desorption is given by

$$\frac{d[M\text{–}S]}{dt} = -k_d[M\text{–}S]$$

Substituting for k_d from Equation 31.37, we find

$$\frac{d[M-S]}{dt} = -\tau_0^{-1} e^{-E_{ads}/RT}[M-S]$$

From Equation 1, we can write

$$t = \frac{T - T_0}{\alpha}$$

and so

$$\frac{dt}{dT} = \frac{1}{\alpha}$$

Now we can find $d[M-S]/dT$ by using the above expressions for $d[M-S]/dt$ and dT/dt:

$$\frac{d[M-S]}{dT} = \frac{d[M-S]}{dt}\frac{dt}{dT} = -\frac{[M-S]}{\alpha}\tau_0^{-1} e^{-E_{ads}/RT} \qquad (a)$$

At T_{max},

$$\frac{d}{dT}\left[\frac{d[M-S]}{dt}\right] = 0$$

Then

$$-\frac{1}{\tau_0}\frac{d}{dT}\left\{e^{-E_{ads}/RT}[M-S]\right\} = 0$$

which gives

$$e^{-E_{ads}/RT_{max}}\left\{\frac{E_{ads}}{RT_{max}^2}[M-S] + \frac{d[M-S]}{dt}\right\} = 0$$

Using Equation a for $d[M-S]/dT$ in the above expression, we find that

$$\frac{E_{ads}}{RT_{max}^2} = \frac{\tau_0^{-1}}{\alpha}e^{-E_{ads}/RT_{max}}$$

which is Equation 3.

31–70. Show that Equation 3 of Problem 31–69 can be written as

$$2\ln T_{max} - \ln\alpha = \frac{E_{ads}}{RT_{max}} + \ln\frac{E_{ads}}{R\tau_0^{-1}}$$

What are the slope and intercept of a plot of $(2\ln T_{max} - \ln\alpha)$ versus $1/T_{max}$? The maximum desorption rates of CO from the 111 surface of palladium as a function of the rate of heating of the palladium surface are given below. Determine the values of E_{ads} and τ_0^{-1} from these data. Use the results to determine k_d, the desorption rate constant, at 600 K.

$\alpha/K \cdot s^{-1}$	T_{max}/K
26.0	500
20.1	496
16.5	493
11.0	487

Begin with Equation 3 of Problem 31–69, take the natural logarithm of both sides, and manipulate to obtain the equation in the text of the problem:

$$\frac{E_{ads}}{T_{max}^2} = \frac{R\tau_0^{-1}}{\alpha} e^{-E_{ads}/RT_{max}}$$

$$\ln E_{ads} - \ln T_{max}^2 = \ln R\tau_0^{-1} - \ln \alpha - \frac{E_{ads}}{RT_{max}}$$

$$-2\ln T_{max} + \ln \alpha = \ln R\tau_0^{-1} - \ln E_{ads} - \frac{E_{ads}}{RT_{max}}$$

$$2\ln T_{max} - \ln \alpha = \frac{E_{ads}}{RT_{max}} + \ln \frac{E_{ads}}{R\tau_0^{-1}}$$

The slope of a plot of $(2\ln T_{max} - \ln \alpha)$ versus $1/T_{max}$ is E_{ads}/R and the intercept is $\ln(E_{ads}/R\tau_0^{-1})$.

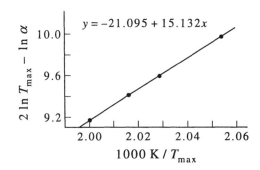

From the slope of the line of best fit, we find that

$$1.513 \times 10^4 \text{ K} = \frac{E_{ads}}{R}$$

$$E_{ads} = 125 \text{ kJ} \cdot \text{mol}^{-1}$$

and from the y-intercept of the line best fit,

$$-21.095 = \ln \frac{E_{ads}}{R\tau_0^{-1}}$$

$$\tau_0^{-1} = \frac{E_{ads}}{Re^{-21.1}} = 2.15 \times 10^{13} \text{ s}^{-1}$$

Therefore, we find (from Equation 31.37)

$$k_d = \tau_0^{-1} e^{-E_{ads}/RT} = 280 \text{ s}^{-1}$$

31–71. At a heating rate of 10 K·s^{-1}, the maximum rate of desorption of CO from a Pd(s) surface occurs at 625 K. Calculate the value of E_{ads}, assuming that the desorption is a first-order process and that $\tau_0 = 1.40 \times 10^{-12}$ s. (See Problems 31–69 and 31–70).

Substitute into the equation found in Problem 31–70.

$$2\ln T_{max} - \ln \alpha = \frac{E_{ads}}{RT_{max}} + \ln \frac{E_{ads}}{R\tau_0^{-1}}$$

$$2\ln(625) - \ln 10 = +\frac{E_{ads}}{R(625 \text{ K})} + \ln \frac{E_{ads}}{R(1.40 \times 10^{-12} \text{ s})^{-1}}$$

Using Newton's method to solve this equation gives $E_{ads} = 146 \text{ kJ} \cdot \text{mol}^{-1}$.

Some Mathematical Formulas

$\sin(x \pm y) = \sin x \cos y \pm \cos x \sin y$

$\cos(x \pm y) = \cos x \cos y \mp \sin x \sin y$

$\sin x \sin y = \frac{1}{2}\cos(x - y) - \frac{1}{2}\cos(x + y)$

$\cos x \cos y = \frac{1}{2}\cos(x - y) + \frac{1}{2}\cos(x + y)$

$\sin x \cos y = \frac{1}{2}\sin(x + y) + \frac{1}{2}\sin(x - y)$

$e^{\pm ix} = \cos x \pm i \sin x$

$$\cos x = \frac{e^{ix} + e^{-ix}}{2} \qquad\qquad \sin x = \frac{e^{ix} - e^{-ix}}{2i}$$

$$\cosh x = \frac{e^{x} + e^{-x}}{2} \qquad\qquad \sinh x = \frac{e^{x} - e^{-x}}{2}$$

$$f(x) = f(a) + f'(a)(x - a) + \frac{1}{2!}f''(a)(x - a)^2 + \frac{1}{3!}f'''(a)(x - a)^3 + \cdots$$

$$e^x = 1 + x + \frac{x^2}{2!} + \frac{x^3}{3!} + \frac{x^4}{4!} + \cdots$$

$$\cos x = 1 - \frac{x^2}{2!} + \frac{x^4}{4!} - \frac{x^6}{6!} + \cdots$$

$$\sin x = x - \frac{x^3}{3!} + \frac{x^5}{5!} - \frac{x^7}{7!} + \cdots$$

$$\ln(1 + x) = x - \frac{x^2}{2} + \frac{x^3}{3} - \frac{x^4}{4} + \cdots \qquad -1 < x \le 1$$

$$\frac{1}{1 - x} = 1 + x + x^2 + x^3 + x^4 + \cdots \qquad x^2 < 1$$

$$(1 \pm x)^n = 1 \pm nx + \frac{n(n - 1)}{2!}x^2 \pm \frac{n(n - 1)(n - 2)}{3!}x^3 + \cdots \qquad x^2 < 1$$

$$\int_0^\infty x^n e^{-ax}\,dx = \frac{n!}{a^{n+1}} \qquad (n \text{ positive integer})$$

$$\int_0^\infty e^{-ax^2}\,dx = \left(\frac{\pi}{4a}\right)^{1/2}$$

$$\int_0^\infty x^{2n} e^{-ax^2}\,dx = \frac{1 \cdot 3 \cdot 5 \cdots (2n - 1)}{2^{n+1}a^n}\left(\frac{\pi}{a}\right)^{1/2} \qquad (n \text{ positive integer})$$

$$\int_0^\infty x^{2n+1} e^{-ax^2}\,dx = \frac{n!}{2a^{n+1}} \qquad (n \text{ positive integer})$$

$$\int_0^\infty \sin\frac{n\pi x}{a}\sin\frac{m\pi x}{a} = \int_0^a \cos\frac{n\pi x}{a}\cos\frac{m\pi x}{a} = \frac{a}{2}\delta_{nm}$$

$$\int_0^a \cos\frac{n\pi x}{a}\sin\frac{m\pi x}{a} = 0 \qquad (m \text{ and } n \text{ integers})$$